Edible Medicinal and Non-Medicinal Plants

T.K. Lim

Edible Medicinal and Non-Medicinal Plants

Volume 10, Modified Stems, Roots, Bulbs

Springer

ISBN 978-94-017-7275-4 ISBN 978-94-017-7276-1 (eBook)
DOI 10.1007/978-94-017-7276-1

Library of Congress Control Number: 2016930034

Springer Dordrecht Heidelberg New York London

Printed on acid-free paper

Springer Science+Business Media B.V. Dordrecht is part of Springer Science+Business Media (www.springer.com)

Acknowledgements

Special thanks for the use of digital images are accorded to: Olivia Lopez for ahipa (*Pachyrhizus ahipa*), and GF Chung for purple nut sedge (*Cyperus rotundus*).

Disclaimer

The author and publisher of this work have checked with sources believed to be reliable in their efforts to confirm the accuracy and completeness of the information presented herein and that the information is in accordance with the standard practices accepted at the time of publication. However, neither the author nor publishers warrant that the information is in every aspect accurate and complete, and they are not responsible for errors or omissions or for consequences from the application of the information in this work. This book is a work of reference and is not intended to supply nutritive or medical advice to any individual. The information contained in the notes on edibility, uses, nutritive values, medicinal attributes and medicinal uses and suchlike included here is recorded information and does not constitute recommendations. No responsibility will be taken for readers' own actions.

Contents

Introduction 1

Amaranthaceae

Beta vulgaris 26

Cannaceae

Canna indica 69

Cibotiaceae

Cibotium barometz 82

Convolvulaceae

Ipomoea batatas 92

Cyperaceae

Cyperus papyrus 172

Cyperus rotundus 178

Eleocharis dulcis 209

Dioscoreaceae

Dioscorea alata 218

Dioscorea bulbifera 235

Dioscorea esculenta 253

Dioscorea hamiltonii 261

Dioscorea hispida 266

Dioscorea oppositifolia 275

Dioscorea rotundata 291

Tacca leontopetaloides 301

Euphorbiaceae

Manihot esculenta 308

Fabaceae

Glycyrrhiza glabra 354

Pachyrhizus ahipa 458

Pachyrhizus erosus 465

Pueraria montana var. *lobata* 482

Medical Glossary 541

Scientific Glossary 617

Common Name Index 641

Scientific Name Index 649

Introduction

This book, volume 10, is a continuation of a multi-compendium, *Edible Medicinal and Non-Medicinal Plants.* It covers plants with edible modified storage subterranean stems (corms, rhizomes, stem tubers) and unmodified subterranean stem stolons, above-ground swollen stems and hypocotyls, storage roots (tap root, lateral roots, root tubers) and bulbs, which are eaten as conventional or functional food as vegetables and spices, and as herbal teas, and may provide a source of food additives or nutraceuticals. A list of such edible plant species from families Acanthaceae to Zygophyllaceae are presented in a tabular form and 32 such edible species from the families Alismataceae, Amaryllidaceae, Apiaceae, Araceae, Araliaceae, Asparagaceae, Asteraceae, Basellaceae, Brassicaceae and Campanulaceae have been covered in detail in the preceding volume. Volume 10 covers in detail 19 edible species in the families Amaranthaceae, Cannaceae, Cibotiaceae, Convolvulaceae, Cyperaceae, Dioscoreaceae, Euphorbiaceae and, Fabaceae. Other species in these families with similar edible plant parts are listed in Table 1. Many plants with edible plant parts, known for their edible fruits or flowers, have been covered in earlier volumes and for those that are better known for other non-reproductive plant parts will be covered in later volumes.

As in the preceding nine volumes, topics covered include: taxonomy (botanical name and synonyms); common English and vernacular names; origin and distribution; agro-ecological requirements; edible plant part and uses; plant botany; nutritive and medicinal/pharmacological properties with up-to-date research findings, traditional medicinal uses; other non-edible uses; and selected/cited references for further reading.

A corm or bulbotuber is defined as a short, vertical, swollen underground plant stem that serves as a storage organ used by some plants to survive unfavourable adverse periods. It bears membranous or scaly leaves and buds. Some examples of plants with edible corms are found in *Amorphophallus* spp., *Colocasia esculenta* (taro), *Eleocharis dulcis* (Chinese water chestnut), and *Sagittaria* spp. (arrowhead or wapato) and *Xanthosoma* spp. (cocoyam or tannia). Corms often give rise to many small secondary corms or cormlet called cormels at the end of very short stolons.

Rhizome is a modified subterranean stem of a plant that is usually found underground, producing roots and shoots. It is used by the plant as storage organ and whole rhizome or pieces of the rhizome serve as vegetative propagules to give rise to new plants. Examples of plants with edible rhizomes include gingers (*Zingiber* spp.), turmeric (*Cucurma longa*), greater galangal (*Alpinia galanga*), lesser galangal (*Alpinia officinarum*), sand ginger or kencur (*Kaempferia galanga*), lotus root (*Nelumbo nucifera*), *Typha* spp., finger root (*Boesenbergia rotunda*) and arrowroot (*Maranta arundinacea*).

T.K. Lim, *Edible Medicinal and Non-Medicinal Plants: Volume 10, Modified Stems, Roots, Bulbs,*
DOI 10.1007/978-94-017-7276-1_1

A stem tuber is a modified plant storage organ that is formed from thickened rhizome or stolon. The tops or sides of the tuber produce shoots that grow into typical stems and leaves and the undersides produce roots. The stem tuber has all the parts of a normal stem, including nodes (eyes) and internodes. A stem tuber may start off as an enlargement of the hypocotyls of the seedling and may include the epicotyl or upper section of the root as in the case of maca (*Lepidium meyenii*). More commonly, as in *Plectranthus esculenta* in the Lamiaceae family, numerous tubers are formed on short stolons that arise from the base of the stem or, as in potatoes, tubers are formed as enlarged stolons thickened and enlarged into storage organs. In some *Cyperus* species, e.g. tigernut or chufa (*C. esculentus*), the stolons end with the growth of tubers that can give rise to new plants. Other striking examples of plants with stem tubers include hog potato or groundnut (*Apios americana*), Jerusalem artichoke or sunchoke (*Helianthus tuberosus*), earthnut pea (*Lathyrus tuberosus*), oca or New Zealand yam (*Oxalis tuberosa*), Chinese artichoke or crosne (*Stachys affinis*), mashua or añu (*Tropaeolum tuberosum*) and ulluco (*Ullucus tuberosus*). In Botany, a stolon is a horizontal modified stem arising from the base of a plant that produces new plants from buds at its tip or nodes and forms adventitious roots at the nodes; it can be creeping above the ground surface or underground. An example of a plant with edible stolon is *Imperata cylindrica*. However, some botanists used the term stolons for stem branches that arise from the base of the stem that creeps above the ground and those that creep horizontally underground as rhizomes. An example of a plant with swollen, aboveground storage stem is the kohlrabi.

Tap root is the true main root of the plant and in some species the tap root is modified and fleshy, rich in stored nutrients; they may or may not be fused with the hypocotyl or basal stem tissues and maybe napiform, globose, conical, fusiform or cylindrical in shape. Notable examples of plants with edible tap roots are: *Abelmoschus* spp., beet (*Beta vulgaris*), rutabaga, turnip, *Bunium persicum*, burdock, carrot, radish and daikon, celeriac, jicama and ahipa (*Pachyrhizus* spp.), parsnips, parsley, skirret (*Sium sisarum*), bush potato (*Vigna lanceolata*), salsify (*Tragopogon porrifolius*), black salsify (*Scorzonera hispanica*), tongkat Ali (*Eurycoma longifolia*) and many others. Examples of plants with edible root tubers or tuberous roots with enlarged root and lateral roots that function as storage organs, lacking nodes, internodes and adventitious buds include, notably, pignut or earthnut (*Conopodium majus*), sweet potato (*Ipomoea batatas*), desert yam (*Ipomoea costata*), cassava or yuca or manioc (*Manihot esculenta*), yams (*Dioscorea* spp.), mauka or chago (*Mirabilis expansa*), breadroot, tipsin, or prairie turnip (*Psoralea esculenta*), and yacón (*Smallanthus sonchifolius*).

Bulb is a much reduced underground stem bearing at its apex a growing or floral primordium surrounded by thick, fleshy modified scale leaves or leaf bases that serve as food storage organs during dormancy and enable the plant to survive through adverse periods. The fleshy leaves are arranged in a concentric manner. Bulbs can be tunicate, i.e. with membranous papery covering (scale leaves) or tunic that protects the inner fleshy scale leaves from drying and mechanical injury. Examples of tunicate bulbs are the Alliums, onions, leeks, hyacinth and tulips. In imbricate or non-tunicate bulbs, the fleshy scale leaves are not in concentric rings but are loosely arranged or spreading, overlapping one another at the margin. Such a bulb is not a compact body and not usually covered by a common tunic. Examples are the garlic (*Allium sativum*) and some *Lilium* lilies.

Table 1 Plants with edible modified storage subterranean stems (corms, rhizomes, stem tubers) and unmodified subterranean stem stolons, above-ground swollen stems and hypocotyls and storage roots (tap root, lateral roots, root tubers) in the families: Amaranthaceae, Cannaceae, Cibotiaceae, Convolvulaceae, Cyperaceae, Dioscoreaceae, Euphorbiaceae and Fabaceae

Family	Scientific Name	Common/Vernacular Names	Edible Part Use	Reference
Amaranthaceae	*Beta vulgaris* Cicla Group	Swiss Chard, Spinach Beet, Foliage Beet Seakale Beet	Some cultivars of Swiss chard have edible root	Facciola (1990)
Amaranthaceae	*Beta vulgaris* cv Crassa Group	Beet, Beet Root, Sugar Beet, Mangel Wurzel	Globose root boiled or cooked as vegetables. Fermented beetroot juice is commercially available. Sugar beet is a source of sugar, syrup and molasses	Larkcom (1984), Facciola (1990)
Amaranthaceae	*Beta vulgaris* cv Group Garden Beet	Beet Root, Garden Beet, Field Beet	As above	Oyen (1994)
Amaranthaceae	*Beta vulgaris* cv. Group Spinach Beet	Foliage Beet, Leaf Beet	As above	Oyen (1994)
Amaranthaceae	*Beta vulgaris* L.	Beet, Beetroot, Garden Beet	Globose root boiled or cooked as vegetables	Oyen (1994)
Amaranthaceae	*Beta vulgaris L.* var. saacharifera = *Beta vulgaris* L.	Sugar Beet	Swollen, fleshy globose root processed for sugar	Codex (2014)
Amaranthaceae	*Beta vulgaris* L. var.*conditiva* = *Beta vulgaris* L.	Beetroot	As above	Codex (2014)
Amaranthaceae	*Beta vulgaris* var *esculenta* = *Beta vulgaris* L.	Beet Root, Garden Bee, Field Beet	Peeled and cooked before eating, can be roasted, added to soups and pickled. Pickled beet roots used in salads as side dish or as a condiment; slices used in hamburgers	Facciola (1990), van Wyk (2006), Santich et al. (2008), Phillips and Rix (1993)
Amaranthaceae	*Beta vulgaris* var. *rapa* Dumont	Garden Beet, Beet Root	As above	Hu (2005)
Amaranthaceae	*Beta vulgaris* var. *vulgaris* = *Beta vulgaris* L.	Beet, Garden Beet	Swollen, fleshy globose root processed for sugar	Phillips and Rix (1993)
Cannaceae	*Canna achiras* Gill. = *Canna indica* L	Indian Shot, Canna	Rhizome eaten; source of Indian arrowroot called *rous le mois* in Chile and Argentina	Hedrick (1972), Facciola (1990)
Cannaceae	*Canna bidentata* Bertol. = *Canna indica* L.	Indian Shot, Balisier	In West Africa, starchy rhizome eaten	Irvine (1952), Uphof (1968)

(continued)

Table 1 (continued)

Family	Scientific Name	Common/Vernacular Names	Edible Part Use	Reference
Cannaceae	*Canna edulis* Ker Gawl.	Edible Canna, Queensland Arrowroot; Achira; Gruya, Par Baul Fauk, Nung Gum (Assamese)	Starch obtained from rhizome used to make translucent noodles in Vietnam. Rhizome eaten in Assam	Burkill (1966), Facciola (1990), Flores et al. (2003), Santich et al. (2008), Medhi and Borthakur (2012), Codex (2014)
Cannaceae	*Canna indica* L.	Indian Shot	As above	Ong and Siemonsma (1996)
Cannaceae	*Canna lutea* Mill. = *Canna indica* (L	K 'Uuwaap (Teenek)	In Yucatan, rhizome eaten as a famine food by the Huastec Maya	Alcorn (1984)
Cannaceae	*Canna orientalis* Rosc. = *Canna indica* L.	Indian Shot	Rhizome cooked and eaten as food	Burkill (1966)
Cibotiaceae	*Cibotium barometz* (L.) J.Sm.	Woolly Fern, Golden Chicken Fern, Golden Moss	Roots are eaten	Cui (1998), Dai et al. (2003), Cao et al. (2007), Yun et al. (2009a, b), Liu et al. (2012)
Convolvulaceae	*Calystegia japonica* (Thunb.) Choisy = *Calystegia pubescens* Lindl.	California Rose	In China, leafy shoots and roots eaten. Roots reported to be purgative	Read (1946)
Convolvulaceae	*Calystegia sepium* (L.) R. Br.	Large Bindweed, Hedge Bindweed	In China, root washed and steamed, or sun-dried, then broken into fragments. Eaten with rice or ground into a meal and steamed in the form of cakes.	Read (1946)
Convolvulaceae	*Convolvulus chinensis* Ker.-Gawl. = *Convolvulus arvensis* L.	Chinese Bindweed; Fu-Fu-Miao, Tian Xuan Hua (Chinese)	Rhizomes eaten in a gruel	Hu (2005)
Convolvulaceae	*Convolvulus erubescens* Sims	Australian Bindweed	Fibrous and not especially tasty roots	Low (1989, 1991)
Convolvulaceae	*Ipomoea graminea* R.Br.	Bush Potato	Large tubers roasted and eaten	Cribb and Cribb (1987)
Convolvulaceae	*Ipomoea racemosa* Poir. = *Turbina racemosa* (Poir.) D.F. Austin	Soh Lah	Tuber eaten in Meghalaya	Sawian et al. (2007)
Convolvulaceae	*Ipomoea aquatica* Forsk	Water Spinach, Kangkong, River Spinach, Water Morning Glory	Roots are occasionally cooked and eaten	Facciola (1990)

Convolvulaceae	*Ipomoea batatas* (Linn.) Lam	Sweet Potato, Phan Karo (Meghalaya) Ruidok (Assamese)	Tuber eaten in Meghalaya; tuber eaten in Karbi Assam; sweet potatoes always eaten cooked – boiled, baked, roasted or fried. Work well in stews, soups, and braised dishes.	Burkill (1966), Facciola (1990), Phillips and Rix (1993), Takagi et al. (1996), Hu (2005), van Wyk (2006), Sawian et al. (2007), Walter and Lebot (2007), Kar and Borthakur (2008), Santich et al. (2008), Codex (2014)
Convolvulaceae	*Ipomoea cairica* (L.) Sweet	Morning Glory, Mile-A-Minute Vine, Messina Creeper, Cairo Morning Glory, Coast Morning Glory; Wu Zhao Jin Long (Chinese)	Roots used to extract starch in Yunnan	Hu (2005)
Convolvulaceae	*Ipomoea calobra* F.Muell.	Bush Potato, Goolabura	Large tubers roasted and eaten	Cribb and Cribb (1987), Low (1991)
Convolvulaceae	*Ipomoea costata* F.Muell. ex Benth.	Bush Potato, Desert Yam	As above	Low (1991)
Convolvulaceae	*Ipomoea digitata* L. = *Ipomoea cheirophylla* O'Donell	Spanish Woodbine	Oblong tubers eaten like sweet potatoes	Hedrick (1972), Tanaka (1976), Facciola (1990)
Convolvulaceae	*Ipomoea eriocarpa* R.Br.	B Tiny Morning Glory; Buta (Hindi); Mulli Balli (Kannada)	Large tubers roasted and eaten	Cribb and Cribb (1987)
Convolvulaceae	*Ipomoea gracilis* R.Br.	Almor-Ira	Large tubers roasted and eaten	Cribb and Cribb (1987)
Convolvulaceae	*Ipomoea mauritiana* Jacq.	Giant Potato; Qi Zhao Long (Chinese)	Roots used to extract starch	Hu (2005)
Convolvulaceae	*Ipomoea pandurata* (L.) G. Meyer	Wild Potato Vine, Man-Of-The-Earth	Huge, tuberous root weighing some times 20 lb	Saunders (1920)
Convolvulaceae	*Ipomoea pes-caprae* subsp. *brasiliensis*	Goat's Foot Convolvulus, Beach Morning Glory	Fleshy tap root eaten after baking and pounding	Cribb and Cribb (1987)
Convolvulaceae	*Ipomoea polpha* R.W. Johnson	Bush Potato, Wier Vine	Large tubers roasted and eaten	Low (1991)
Convolvulaceae	*Ipomoea polymorpha* Roem. & Schult.	Silky Cow Vine	Large tubers roasted and eaten	Cribb and Cribb (1987)
Convolvulaceae	*Ipomoea staphylina* Roem. & Schult.	Lesser Glory; Hai Nan Shu (Chinese)	Fleshy tubers eaten in Hainan island	Hu (2005)
Convolvulaceae	*Ipomoea velutina* R.Br.	Velvety Morning Glory	Large tubers roasted and eaten	Cribb and Cribb (1987)

(continued)

Table 1 (continued)

Family	Scientific Name	Common/Vernacular Names	Edible Part Use	Reference
Convolvulaceae	*Ipomoea violacea* L.	Beach Moonflower, Sea Moonflower	Large tubers roasted and eaten	Cribb and Cribb (1987)
Convolvulaceae	*Merremia hungaiensis* (Lingels. & Borza) RC Fang	Huang Hua Tu Gua, Shan Tu Gua (Chinese)	Enlarged root eaten in Yunnan	Hu (2005)
Cyperaceae	*Bolboschoenus caldwellii* (V.J.) Soják	Sea Club Rush	Grape-sized, sweet, fibrous tubers eaten	Low (1989, 1991)
Cyperaceae	*Bolboschoenus fluviatilis* (Torr.) Soják	Marsh Club Rush	As above	Low (1991)
Cyperaceae	*Bolboschoenus maritimus* (L.) Palla	Sea Club Rush	Tubers eaten after treatment	Cribb and Cribb (1987)
Cyperaceae	*Carex* spp.	Ware Sedge	Sweet, enlarged underground stem eaten	Schofield (2003)
Cyperaceae	*Cyperus bifax* C.B.Clarke = *Cyperus rotundus* L.	Downs Nutgrass	Tubers produced on rhizomes are dried, coat removed, shaken with hot ashes, eaten raw or rubbed to a powder and eaten as porridge	Cribb and Cribb (1987), Harden (1993)
Cyperaceae	*Cyperus bulbosus* Vahl.	Nalgoo (Australia)	Tubers pleasantly starchy. In India (Bombay Presidency), bulbs dried and pulverized, then mixed with *jowar*, *bajra* (millet) or wheat flour to make bread	Gammie (1902), Paton and Dunlop (1904), Irvine (1957), Burkill (1966), Gupta and Kanodia (1968), Saxena (1979), Low (1991), Jansen and Aguilar (1996)
Cyperaceae	*Cyperus esculentus* L.	Chufa, Tiger Nut, Yellow Nutgrass	Chufa's hard tubers are sweet and tasty. In Zimbabwe, tubers eaten raw or cooked	Saunders (1920), Burkill (1966), Cribb and Cribb (1987), Zinyama et al. (1990), Jansen and Aguilar (1996), Codex (2014)
Cyperaceae	*Cyperus esculentus* L. *var. sativus* Boeckeler = *Cyperus esculentus* L.	Yellow Nut Grass, Chufa; You Sha Cao (Chinese)	Tubers eaten in China	Hu (2005)
Cyperaceae	*Cyperus jeminicus* Rottb.	NF	In India, tubers ground into flour.	Watt (1908)
Cyperaceae	*Cyperus papyrus* L.	Egyptian Reed, Paper Reed	Starchy rhizomes and culms are edible	Mahr (2011)

Cyperaceae	*Cyperus rotundus* L.	Nutgrass, Purple Nutgrass; Mothee, Motha (Rajasthan)	In France, root recommended as famine food. Can be eaten raw or cooked. Can be dried and reduced to a flour. In India (Jaisalmer district, Rajasthan), fibre and cuticle of root removed, root dried, ground and made into bread and sometimes mixed with other flour. In Western Rajasthan, tubers roasted; also boiled, outer skin peeled off, and the starchy rhizome eaten with spices	Parmentier 1781 (cited by Freedman (2009)), Saunders (1920), Burkill (1966), Saxena (1979), Cribb and Cribb (1987), Harden (1993), Jansen and Aguilar (1996)
Cyperaceae	*Eleocharis dulcis* (Burm.f.) Trinius ex Henschel	Water Chestnut, Ground Chestnut	Corms eaten raw or fresh or from canned material. Used in salad or as snack, pickled and as a condiment; also baked. Goes well in soups, stir fries, dumplings or as garnish for vegetable dishes. Starch obtained from tubers for domestic use, mixed with sugar to prepare a refreshing morning drink as well as for pastry	Cribb and Cribb (1987), Low (1989), Facciola (1990), Phillips and Rix (1993), Paisooksantivatana (1966), Hu (2005), van Wyk (2006), Santich et al. (2008), Codex (2014)
Cyperaceae	*Eleocharis kuroguwai* Ohwi	Kuro Gawai (Japanese)	Corm edible	Codex (2014)
Cyperaceae	*Eriophorum gracile* Koch	Mousenuts, Alaska Cotton, Swamp Cotton	Rootstock used raw or cooked	Schofield (2003)
Cyperaceae	*Fimbristylis kysoor* (Roxb.) Dalz. & Gibs.	NF	Root eaten in India	Watt (1908)
Cyperaceae	*Fimbristylis subbispicata* Nees = *Fimbristylis tristachya* var. *subbispicata* (Nees) T. Koyama	Sedge, Pond Onion	Shoots and roots eaten in China	Read (1946)
Cyperaceae	*Lepironia articulata* (Retz.) Domin	Grey Sedge	Underground stem eaten Rhizome edible	Cribb and Cribb (1987), Low (1989)

(continued)

Table 1 (continued)

Family	Scientific Name	Common/Vernacular Names	Edible Part Use	Reference
Cyperaceae	*Mariscus sieberianus* Nees ex C.B.Clarke = *Cyperus cyperoides* (L.) Kuntze	Tall Sedge	In China, roots and seeds made into flour	Read (1946)
Cyperaceae	*Schoenoplectus litoralis* (Schrad.) Palla	Daly River Club Rush	Edible roots	Cribb and Cribb (1987)
Cyperaceae	*Scirpus americanus* Pers. = *Schoenoplectus americanus* (Pers.) Volkart.	Bulrush	Rootstock eaten raw, boiled, baked or roasted, also pounded into flour	Schofield (2003)
Cyperaceae	*Scirpus californicus* (C.A.Mey.) Steud. = *Schoenoplectus californicus* (C.A.Mey.) Soják.	California Bulrush, Totora	Rhizomes peeled, baked and eaten	Facciola (1990)
Cyperaceae	*Scirpus grossus* L. f. = *Actinoscirpus grossus* (L.f.) Goetgh. & D.A.Simpson.	Giant Bulrush, Greater Club Rush	India and Kumaon region, Western Himalayas roots burnt, then ground into flour from which bread is prepared	Paton and Dunlop (1904), Bhargava (1960)
Cyperaceae	*Scirpus lacustris* L. = *Schoenoplectus lacustris* (L.) Palla.	Tule, Great Bulrush	In China: shoots and roots eaten. Rootstock can be processed into syrup or flour	Saunders (1920), Read (1946), Uphof (1968), Harrington (1974), Gibbons and Tucker (1979), Facciola (1990)
Cyperaceae	*Scirpus microcarpus* J.Presl & C.Presl	Bulrush	Rootstock eaten raw, boiled, baked or roasted	Schofield (2003)
Cyperaceae	*Scirpus paludosus* A.Nelson = *Bolboschoenus maritimus* subsp. *paludosus* (A.Nelson) T. Koyama.	Alkali Bulrush, Nutgrass	Rhizome eaten raw or made into flour for bread	Uphof (1968), Fernald et al. (1985), Facciola (1990)
Cyperaceae	*Scirpus subterminalis* Torr. = *Schoenoplectus subterminalis* (Torr.) Soják.	Bulrush	Rootstock eaten raw, boiled, baked or roasted	Schofield (2003)
Cyperaceae	*Scirpus tuberosus* Roxb. (illeg.) = *Eleocharis dulcis* (Burm.f.) Trin. ex Hensch.	Chechur (Assamese); Khitro (Bodo)	Root tubers are sweet and eaten fresh especially in Upper Assam, also can be cooked as vegetable with potato and chicken	Patiri and Borah (2007)

Cyperaceae	*Scirpus validus* Vahl = *Schoenoplectus tabernaemontani* (C.C. Gmel.) Palla	Bulrush, Tule, River Club Rush	White underground shoots after boiling are of excellent flavour; roots eaten cooked or preserved in rice bran	Tanaka (1976), Cribb and Cribb (1987), Fernald et al. (1985), Facciola (1990), Schofield (2003)
Dioscoreaceae	*Dioscorea daunea* Prain & Burkill	Suna (Thai Sakai)	Tuber used as substitute *Dioscorea* food species during famine	Maneenoon et al. (2008)
Dioscoreaceae	*Dioscorea membranacea* Pierre	Chatong (Thai Sakai)	Tuber used as substitute *Dioscorea* food species during famine	Maneenoon et al. (2008)
Dioscoreaceae	*Dioscorea stemonoides* Prain & Burkill	Kungkwad (Thai Sakai)	Tuber main source of carbohydrate for the Sakai tribe in Peninsular Thailand	Maneenoon et al. (2008)
Dioscoreaceae	*Dioscorea transversa* R.Br.	Long Yam	Tuber eaten raw or roasted, also aerial bulbils	Cribb and Cribb (1987), Wightman and Andrews (1989), Low (1989), Harden (1993)
Dioscoreaceae	*Dioscorea aculeata* Balb. ex Kunth = *Dioscorea cayennensis* Lam.	Fancy Yam, Potato Yam, Lesser Yam, Lesser Asiatic Yam, Igname; Man-Alu (Kumaon Region, Western Himalayas)	Roots cut, boiled prior to eating	Bhargava (1960), Ochse and van den Brink (1980)
Dioscoreaceae	*Dioscorea alata* L.	Purple Yam, Greater Yam, Winged Yam, Water Yam, White Yam; Kath Alu (Assamese); Yams Kalung. (Tamil); Niluva Pendalum (Telugu)	Tubers are eaten cooked as vegetable. In China, tubers are used in soups. In Oceania, tuber cut into pieces and baked or roasted whole or boiled in marmite or grated and used for *lap-lap*	Burkill (1966), Ochse and van den Brink (1980), Low (1991), Onwueme and Ganga (1996), Onwueme (1996a), Hu (2005), Patiri and Borah (2007), Walter and Lebot (2007), Codex (2014)
Dioscoreaceae	*Dioscorea anguina* Roxb. = *Dioscorea pubera* Blume	Kakalu (Bengali); Savida Dumpa (Telugu)	Tuber eaten in India	Watt (1908)
Dioscoreaceae	*Dioscorea belophylla* (Prain) Voigt ex Haines	Spear-Leaved Yam	In India (Garhwal Himalayas), tuber eaten after repeated boiling, washing and baking	Gupta (1962)

(continued)

Table 1 (continued)

Family	Scientific Name	Common/Vernacular Names	Edible Part Use	Reference
Dioscoreaceae	*Dioscorea bulbifera* L.	Aerial Yam, Air Potato, Air Yam, Bitter Yam, Cheeky Yam, Potato Yam, Wild Yam; Gosh Alu (Assamese); Ho (Hawaiian); Genthi (Kumaon Region, Western Himalayas); Kapuang (Thai Sakai)	Root tubers are eaten cooked as vegetable during winter India (Kumaon region, Western Himalayas): axillary tubers cut into pieces, steeped in water, and boiled prior to eating. Hawaii: aerial bulbils eaten. Yams baked, boiled or fried slices of tuber or pureed tuber may be added to soups, stew, soufflés, fritter and various sweet dishes	Patiri and Borah (2007), Cribb and Cribb (1987), Wightman and Andrews (1989), Low (1989), Facciola (1990), Bhargava (1960), Hu (2005), Onwueme (1996a), Walter and Lebot (2007)
Dioscoreaceae	*Dioscorea calcicola* Prain & Burkill	Bayae (Thai Sakai)	Tuber main source of carbohydrate for the Sakai tribe in Peninsular Thailand	Maneenoon et al. (2008)
Dioscoreaceae	*Dioscorea cayenensis* Lam.	Yellow Guinea Yam, Yellow Yam, White Yan	Yams used to make fufu, may also be used in same way as potatoes or sweet potatoes	Burkill (1966), Facciola (1990), van Wyk (2006), Codex (2014)
Dioscoreaceae	*Dioscorea cumingii* Prain & Burkill	Lima-Lima (Tagalog, Kasi (Igorot), Pari (Bagobo)	Tuber used as food in Luzon (Philippines)	Groen et al. (1996)
Dioscoreaceae	*Dioscorea deltoidea* Wall. ex Griseb.	Nepal Yam; Gun (Kumaon Region, Western Himalayas)	Tubers cut into pieces, steeped in water, and boiled and/or baked prior to eating	Bhargava (1960), Gupta (1962)
Dioscoreaceae	*Dioscorea divaricata* Blanco	Pakit, Kiroi (Tagalog), Sulian (Iloko), Baklaiakang (Bisaya)	Tuber used as food in the Philippines, baked, boiled or fried	Groen et al. (1996)
Dioscoreaceae	*Dioscorea dumetorum* Kunth (Pax)	Cluster Yam; Ma-Nyeny, E-Dyeny (Bedik, Gold Coast); Rogon Biri (Hausa, Nigeria)	In Gold Coast, tuber used as a famine food. The tuber is boiled, peeled, sliced, pounded and steeped in running (preferably salt) water	Irvine (1952), Burkill (1966), Ferry et al. (1974), Mortimore (1989)

Dioscoreaceae	*Dioscorea esculenta* (Lour.) Burkill	Lesser Asiatic Yam, Sweet Yam, Potato Yam; Ruipheng Selu (Assamese)	Tuber eaten in Karbi Assam. Yams baked, boiled or fried slices of tuber or pureed tuber may be added to soups, stew, soufflés, fritters and various sweet dishes	Burkill (1966), Ochse and van den Brink (1980), Facciola (1990), Onwueme (1996a, b), Hu (2005), van Wyk (2006), Walter and Lebot (2007), Kar and Borthakur (2008), Codex (2014)
Dioscoreaceae	*Dioscorea esculenta* (Lour.) Burkill var. *fasciculata* (Roxb.) Prain & Burkill = *Dioscorea esculenta* (Lour.) Burkill	Moa Alu (Assamese)	Tubers are used as vegetable	Patiri and Borah (2007)
Dioscoreaceae	*Dioscorea filiformis* Blume	Wauh (Malaysia); Aroi Huwi Curuk (Sumatra); Dudung (Java); Balun (Thai Sakai)	Tubers boiled and eaten in Malaysia. Tuber main source of carbohydrate for the Sakai tribe in Peninsular Thailand	Groen et al. (1996), Maneenoon et al. (2008)
Dioscoreaceae	*Dioscorea gibbiflora* Hook.f. = *Dioscorea filiformis* Blume	Wild Yam	Tuber eaten	Burkill (1966)
Dioscoreaceae	*Dioscorea glabra* Roxb.	Mandong (Thai); Luntak (Thai Sakai)	Tubers (glutinous and starchy) used as food in Peninsular Malaysia and Andaman Islands. Tuber main source of carhohydrate for the Sakai tribe in Peninsular Thailand	Groen et al. (1996), Maneenoon et al. (2008)
Dioscoreaceae	*Dioscorea hastifolia* Nees	Warrine	Tuber eaten	Low (1989)
Dioscoreaceae	*Dioscorea hemsleyi* Prain & Burkill	Glutinous; Nian-Shan-Yao (Chinese)	Tuber eaten	Hu (2005)
Dioscoreaceae	*Dioscorea hirtiflora* Benth.	Mng'oko (Tanzania)	Sierra Leone; Nigeria (northern): eaten as a famine food	Irvine (1952)
Dioscoreaceae	*Dioscorea hispida* Dennst.	Intoxicating Yam, Asiatic Bitter Yam; Gadog, Gadong, Gadong Lilin, Gadong Mabuk, Gadung (Malay)	Tuber eaten	Burkill (1966), Onwueme (1996c), Ochse and van den Brink (1980), Codex (2014)
Dioscoreaceae	*Dioscorea japonica* Thunb.	Glutinous Yam, Chinese Yam, Japanese Yam, Taiwanese Yam, Yama-No-Imo	Tuber eaten in China	Read (1946), Facciola (1990), Burkill (1966), Hu (2005), Codex (2014)

(continued)

Table 1 (continued)

Family	Scientific Name	Common/Vernacular Names	Edible Part Use	Reference
Dioscoreaceae	*Dioscorea laurifolia* Wall. ex Hook.f.	Clangporn (Thai Sakai)	Tubers eaten in Peninsular Malaysia	Burkill (1966), Groen et al. (1996)
Dioscoreaceae	*Dioscorea luzonensis* Schauer	Pakit, Mayabang (Tagalog) Kamengeg (Iloko)	Tubers used for food	Groen et al. (1996)
Dioscoreaceae	*Dioscorea macrostachya* Benth. = *Dioscorea mexicana* Scheidw.	Panil Book	Tuber cooked and eaten in Mexico	Alcorn (1984), Kunkel (1984), Facciola (1990)
Dioscoreaceae	*Dioscorea macroura* Harms = *Dioscorea sansibarensis* Pax	Zanzibar Yam	Tropical Africa: eaten as a famine food	Irvine (1952), Uphof (1968)
Dioscoreaceae	*Dioscorea minutiflora* Engl.	Aha Bayere (Twi, Gold Coast); Magoraza, Hazara (Hausa, Nigeria)	Gold Coast: eaten as a famine food. Nigeria (Kano State, northern): tuber eaten	Irvine (1952), Mortimore (1989)
Dioscoreaceae	*Dioscorea nummularia* Lam.	Prickly Yam	Tubers eaten	Low (1991), Onwueme (1996a), Walter and Lebot (2007)
Dioscoreaceae	*Dioscorea opposita* Thunb = *Dioscorea oppositifolia* L.	Chinese Yam, Korean Yam, Japanese Mountain Yam, Nagaimo, Yamaimo	Tuber eaten	Burkill (1966), Facciola (1990), Hu (2005), van Wyk (2006), Codex (2014)
Dioscoreaceae	*Dioscorea oppositifolia* L.	As Above	India (Deccan): tuber eaten	Watt (1908), Facciola (1990)
Dioscoreaceae	*Dioscorea orbiculata* Hook.f.	Ubi Garam (Indonesia), Takob, Ubi Garam (Malaysia) Man Tayong (Thai), Takob (Thai Sakai)	Tubers eaten in Peninsular Malaysia. Tuber main source of carbohydrate for the Sakai tribe in Peninsular Thailand	Burkill (1966), Groen et al. (1996), Maneenoon et al. (2008)
Dioscoreaceae	*Dioscorea owenii* Prain & Burkill	NF	Tubers eaten	Burkill (1966)
Dioscoreaceae	*Dioscorea pentaphylla* L.	Five-Leaf Yam; Pi'A (Hawaiian); Ser (Thai Sakai); Pachpotia Alu, Ruipheng (Assamese); Chai, Chavi, Alshi, Shahada, Kala Kand, Jaglia Che Kand, Kadu Kand (Bombay Presidency); Taigun, Takuli (Kumaon Region, Western Himalayas); Kanta-Alu (Western Rajasthan	In India, tubers cut into pieces, steeped in water, and boiled or baked prior to eating, In Hawaii, tuber steamed and eaten warm. In Oceania, tubers boiled baked or used in *lap-lap*. In Peninsular Thailand, tubers main source of carbohydrate for the Sakai	Gammie (1902), Watt (1908), Handy (1940), Bhargava (1960), Gupta (1962), Gupta and Neal (1965), Burkill (1966), Gupta and Kanodia (1968), Low (1989), Onwueme (1996a), Patiri and Borah (2007), Walter and Lebot (2007), Ochse and van den Brink (1980), Maneenoon et al. (2008), Kar and Borthakur (2008)
Dioscoreaceae	*Dioscorea persimilis* Prain & Burkill = *Dioscorea hamiltonii* Hook f.	Khoai Mai, Ciu Mai (Vietnamese)	Tuber eaten, boiled in soups	Tanaka and Nguyen (2007)

Dioscoreaceae	*Dioscorea piscatorum* Prain & Burkill	Fish Poison Yam, Tuba Gunjo (Indonesia), Tuba Ubi (Malaysia); Kiyak (Thai Sakai)	Tubers eaten boiled, baked or roasted. Tuber used as substitute *Dioscorea* food species during famine in Peninsular Thailand	Burkill (1966), Groen et al. (1996), Maneenoon et al. (2008)
Dioscoreaceae	*Dioscorea polyclados* Hook.f.	Kedut (Sumatra), Kedut (Malaysia)	Tubers eaten after several boilings or baked in Peninsular Malaysia	Burkill (1966), Groen et al. (1996)
Dioscoreaceae	*Dioscorea polystachya* Turz.	Chinese Yam	Tubers eaten	Codex (2014)
Dioscoreaceae	*Dioscorea prainiana* R. Kunth	Ubi Kelonak, Kelunoh, Kelana (Malaysia)	Tubers eaten in Peninsular Malaysia	Burkill (1966)
Dioscoreaceae	*Dioscorea prazeri* Prain & Burkill	Sehod (Thai Sakai)	Tubers used as substitute *Dioscorea* food species during famine in Peninsular Thailand	Maneenoon et al. (2008)
Dioscoreaceae	*Dioscorea preussii* Pax.	Preuss' Dioscorea	In tropical, Central Africa, tuber eaten in times of famine	Irvine (1952)
Dioscoreaceae	*Dioscorea pubera* Blume	Rui-Chilong	Bubils and tubers eaten in Karbi, Assam	Groen et al. (1996), Kar and Borthakur (2008)
Dioscoreaceae	*Dioscorea pyrifolia* Kunth	Ubi Babi, Badak, Akar Kemnyan Paya, Huwi Upas (Sundanese), Ilus (Javanese); Hngo (Tahi Sakai)	Tubers eaten after several boilings or baked in Peninsular Malaysia. Tuber main source of carbohydrate for the Sakai tribe in Peninsular Thailand	Burkill (1966), Groen et al. (1996), Maneenoon et al. (2008)
Dioscoreaceae	*Dioscorea quinata* Walter = *Dioscorea villosa* L.	Magiya, Munia (Kumaon Region, Western Himalayas India)	Tubers cut into pieces, steeped in water and boiled, prior to eating	Bhargava (1960)
Dioscoreaceae	*Dioscorea rotundata* Poir = *Dioscorea cayennensis* subsp. *rotundata* (Poir.) J. Miege	Eight Month Yam, Round White Yam, White Yam, White Guinea Yam	Yams used to make fufu, may also be used in same way as potatoes or sweet potatoes	Akinwande et al. (2007), Codex (2014)
Dioscoreaceae	*Dioscorea sagitata* Poir.	Five-Leaved Yam; Tarur (Kumaon Region, Western Himalayas, India)	Axillary tubers cut into pieces, steeped in water, and boiled prior to eating	Bhargava (1960)

(continued)

Table 1 (continued)

Family	Scientific Name	Common/Vernacular Names	Edible Part Use	Reference
Dioscoreaceae	*Dioscorea sativa* L. = *Dioscorea villosa* L.	Kath Alu (Assamese); Kadia Kand, Mano Kand, Vaj Kand, Kadawa Kand, Kedvo Kand (Bombay Presidency)	Tubers are eaten cooked as vegetable by boiling or toasting. After neutralizing toxic substances, the tuber may be mixed with *konda,* or some other flour, and then eaten	Gammie (1902), Patiri and Borah (2007)
Dioscoreaceae	*Dioscorea schimperiana* Hochst. ex Kunth.	Yagniat (Kipsigis, Kenya)	Root tubers eaten	Kabuye (1986)
Dioscoreaceae	*Dioscorea trifida* L.f.	Indian Yam, Cush-Cush, And Yampee	Tubers eaten boiled, baked	Facciola (1990), Walter and Lebot (2007), Codex (2014)
Dioscoreaceae	*Dioscorea triphylla* L. = *Dioscorea pentaphylla* L.	See *D. Pentaphylla*	Tubers eaten in India (Deccan)	Watt (1908)
Dioscoreaceae	*Dioscorea tuberosa* Vell. = *Dioscorea cinnamomifolia* Hook.	NF	India (Garhwal Himalayas): tuber eaten after repeated boiling, washing and baking	Gupta (1962)
Dioscoreaceae	*Dioscorea wallichii* Hook.f.	Rui Nihang (Assamese); Yarex (Thai Sakai)	Tuber main source of carbohydrate for the Sakai tribe in Peninsular Thailand	Maneenoon et al. (2008)
Dioscoreaceae	*Tacca involucrata* Schumach. & Thonn. = *Tacca leontopetaloides* (L.) O Kuntze	As Below	In Nigeria, the Munshi first boil the tuber to remove the toxic element. A coarse flour, called *amara*, is prepared from it.	Irvine (1952)
Dioscoreaceae	*Tacca leontopetaloides* (L.) O Kuntze	Polynesian Arrowroot, East Indian Arrowroot, Salep	Root tuber grated, pounded, soaked and baked	Cribb and Cribb (1987), Wightman and Andrews (1989), Low (1989, 1991), Facciola (1990), Jukema and Paisooksantivatana (1996), Hu (2005), Codex (2014)
Dioscoreaceae	*Tacca pinnatifida* J.R. Forst & G. Forst. = *Tacca leontopetaloides* (L.) O Kuntze	As Above	Tubers usually eaten after preparation ; bitter when raw	Burkill (1966)

Euphorbiaceae	*Manihot esculenta* Crantz	Cassava, Manioc, Tapioca, Yuca	Tuberous roots eaten boiled, fried, baked, roasted and also processed into flour, farina, sweetmeats, bread, syrup, pasties, fufu, chips pastries and cakes. Flour also used for thickening soups and sauces. Root sap boiled to make the condiment cassareep or fermented into chicha and other alcoholic beverages	Burkill (1966), Uphof (1968), Hedrick (1972), Tanaka (1976), Ochse and van den Brink (1980), Facciola (1990), Veltkamp and De Bruijn (1996), Hu (2005)
Euphorbiaceae	*Manihot glaziovii* Müll.Arg = *Manihot carthaginensis* subsp. *glaziovii* (Müll.Arg.) Allem	Manicoba, Mandioca Brava, Cerea Rubber	Tubers eaten sometimes, also a source of starch	Tanaka (1976), Facciola (1990)
Euphorbiaceae	*Manihot utilissima* Pohl.= *Manihot esculenta* Crantz	Cassava, Bitter Cassava; Ingwese, Aloti (Gabon); Bafra (Arabic, Sudan)	As for *M. esculenta*	Ochse and van den Brink (1980), Abdelmuti (1991), Burkill (1966), Codex (2014)
Fabaceae	*Acacia bidwillii* Benth.	Corkwood Wattle	Young roots cooked as food by the aborigines	Cribb and Cribb (1982, 1987)
Fabaceae	*Acacia crassicarpa* Benth.	Northern Wattle, Thick-Podded Salwood, Brown Salwood, Papua New Guinea Red Wattle, Red Wattle	As above	Cribb and Cribb (1982, 1987)
Fabaceae	*Acacia holosericea* G. Don	Silver Leaf Wattle	As above	Cribb and Cribb (1987)
Fabaceae	*Apios fortunei* Maxim.	Potato Bean, Groundnut	Thick tubers eaten as emergency food in China	Read (1946), Hu (2005)
Fabaceae	*Apios tuberosa* Moench. = *Apios americana* Medik.	Groundnut, Wild Bean, Potato Bean, American Potato Bean, Indian Potato	Starchy tuber eaten	Saunders (1920)
Fabaceae	*Argyrolobium marginatum* Bolus	Izi Ntondo (Zulu)	In Zululand (Ubombo district), roots eaten cooked or uncooked	Hely-Hutchinson (1898)

(continued)

Table 1 (continued)

Family	Scientific Name	Common/Vernacular Names	Edible Part Use	Reference
Fabaceae	*Astragalus fraxinifolius* DC	Astragal Yasenelistnyi (Russian)	Starch of root recommended as a famine food for extending bread flour, after removal of bitter element	Parmentier (1781) (cited by Freedman (2009))
Fabaceae	*Astragalus membranaceus* (Fisch.) Bunge = *Astragalus propinquus* Schischkin	Astragalus; Huangchoy (Chinese)	Dried root slices used in combination with Codonopsis (Dan shen) goji berries for a tonic soup with spare ribs	Hu (2005)
Fabaceae	*Atylosia reticulata* (Dryand.) Taubert ex Ewart & Davies = *Cantharospermum reticulatum* (Dryand.) Taubert ex Ewart & Davies.	NF	Roots roasted	Cribb and Cribb (1987)
Fabaceae	*Bauhinia hupehana* Craib = *Bauhinia glauca* subsp. *hupehana* (Craib) T.Chen	Shen Zi Ye, Hubei Yang Ti Jia (Chinese)	Root and stem stewed with pig kidneys, intestines or cooked with pork as special health food in Huber and Sichuan	Hu (2005)
Fabaceae	*Butea frondosa* Roxb. = *Butea monosperma* (Lam.) Taub.	Dangs (Bombay)	In India, roots toasted and eaten	Gammie (1902), Watt (1908)
Fabaceae	*Butea monosperma* (Lam.) Taub.	Dhak, Pala (Rajasthan, Western India)	Succulent young roots roasted or boiled and eaten with salt	Gupta and Kanodia (1968), Shankarnarayan and Saxena (1987)
Fabaceae	*Dalea candida* (Michx.) Willd.	White Prairie Clover	Root eaten raw or chewed, eaten as delicacy by children	Yanovsky (1936), Tanaka (1976), Facciola (1990)
Fabaceae	*Dalea gattingeri* (A. Heller) Barneby	Purpletassels	Root eaten raw or chewed	Yanovsky (1936), Uphof (1968), Tanaka (1976)
Fabaceae	*Dalea purpurea* Vent	Purple Prairie Clover	Root eaten raw or chewed	Facciola (1990)
Fabaceae	*Dolichos biflorus* L.= *Vigna unguiculata* (L.) Walp.	Horse Gram	In Australia (North Queensland), rootstock roasted and eaten	Watt (1908), Irvine (1957)
Fabaceae	*Eriosema chinense* Vogel	Chinese Eriosoma; Katil (Indonesia)	Tubers eaten cooked, enlarged fleshy tuberous roots used in tonifying broth with pork	Cribb and Cribb (1987), Groen et al. (1996), Hu (2005)
Fabaceae	*Erythrina vespertilio* Benth.	Bat's Wing Coral Tree	Roots eaten raw by aborigines	Cribb and Cribb (1987)

Fabaceae	*Flemingia procumbens* Roxb.	Sohplong (India)	Starch rib tubers eaten raw	Groen et al. (1996)
Fabaceae	*Flemingia vestita* Baker = *Flemingia procumbens* Roxb.	Soh Phlang	Tuber eaten in Meghalaya	Sawian et al. (2007)
Fabaceae	*Galactia tenuiflora* (Willd.) Wight & Arn.	Florida Hammock Milkpea	Root eaten after treatment	Cribb and Cribb (1987)
Fabaceae	*Glycyrrhiza glabra* L.	Licorice, Liquorice	Pieces of rhizomes used as flavourant or sweetener, can be chewed as sweet snack used in confectionery, sweets, drinks, dark beers (stouts) and liqueurs	van Wyk (2006)
Fabaceae	*Glycyrrhiza lepidota* Pursh	American Licorice	Roots chewed, added to foods for flavouring, or dried and brewed in tea	Uphof (1968), Fernald et al. (1985), Facciola (1990)
Fabaceae	*Hardenbergia retusa* Benth. = *Vandasina retusa* (Benth.) Rauschert	Sarsaparilla Vine	Roots roasted	Cribb and Cribb (1987)
Fabaceae	*Hedysarum boreale* Nutt.	Licorice-Root, Sweet-Root	Young sweet root have a licorice flavour and are eaten raw, boiled, baked or added in soups	Uphof (1968), Gibbons and Tucker (1979), Fernald et al. (1985), Facciola (1990)
Fabaceae	*Hedysarum mackenzii* Richardson = *Hedysarum boreale* subsp. *mackenzii* (Richardson) S.L. Welsh	Licorice-Root, Sweet Broom	Young sweet root have a licorice flavour and are eaten	Uphof (1968), Hedrick (1972), Fernald et al. (1985), Facciola (1990)
Fabaceae	*Hedysarum occidentale* Greene	Licorice Root, Sweet Vetch	As above	Facciola (1990)
Fabaceae	*Labichea buettneriana* F. Muell.	NF	Roots roasted	Cribb and Cribb (1987)
Fabaceae	*Lablab purpureus* (L.) Sweet	Bonavista Bean, Hyacinth Bean, Dolichos Bean, Seim Bean, Lablab Bean	Large, starchy root edible	Hedrick (1972), Facciola (1990)
Fabaceae	*Lathyrus tuberosus* L.	Earthnut Pea, Tuberous Pea	France: root tuber recommended as a famine food cooked, or dried and reduced to flour for use in baking bread	Parmentier (1781) (cited by Freedman (2009)), Hedrick (1972), Fernald et al. (1985), Facciola (1990), Codex (2014)

(continued)

Table 1 (continued)

Family	Scientific Name	Common/Vernacular Names	Edible Part Use	Reference
Fabaceae	*Lotus siloquosus* L.	NF	France: farinaceous root recommended as a famine food	Parmentier (1781) (cited by Freedman (2009))
Fabaceae	*Lupinus arcticus* S. Watson	Arctic Lupine	Roots used as survival food after roasting	Schofield (2003)
Fabaceae	*Lupinus nootkatensis* Sims	Nootka Lupine	As above	Schofield (2003)
Fabaceae	*Macrotyloma uniflorum* (Lam.) Verd.	Madras Gram, Horse Gram	Fleshy root roasted and eaten by Aborigines in Australia	Cribb and Cribb (1987), Facciola (1990)
Fabaceae	*Melilotus albus* Medik.	Sweet Clover	Roots prized as food by some native groups in North America	Schofield (2003)
Fabaceae	*Melilotus officinalis* (L.) Pall.	Yellow Sweet Clover	As above	Facciola (1990), Schofield (2003)
Fabaceae	*Millettia speciosa* Champ.ex Benth.	Showy Millettia; Shan Lian Ou (Chinese)	Root fresh or dried boiled with pork for soup that strengthens bones; used in southern China and Hong Kong	Hu (2005)
Fabaceae	*Moghania philippinensis* (Merr. & Rolfe) Li = *Flemingia prostrata* Roxb.	Southern Astragalus; Qian Jin Ba (Chinese)	Sliced roots cooked with pigfeet in water and cooking wine, a special southern Chinese cuisine	Hu (2005)
Fabaceae	*Mucuna glabra* (Reinecke) Wilmot-Dear	Tupe	Brazil (northeast): flour is made from both the seeds and roots. The flour or starch thus obtained is made into a variety of Brazilian foods including *farofa*, in which the meal is sautéed and mixed with bits of meat, crisp fat, chopped egg, etc.; *boijus*, which are small, sweet cakes; and *angus*, which are dumplings, the flour being merely boiled in water	De Castro (1952)

Fabaceae	*Orobus tuberosus* L. = *Lathyrus linifolius* (Reichard) Bassler.	Tuberous Bitter Vetch	In France, boiled root eaten as a famine food	Parmentier (1781) (cited by Freedman (2009))
Fabaceae	*Pachyrhizus ahipa* (Wedd.) Parodi	Ahipa, Yam bean	The roots are sweet and crispy; when eaten raw it can be peeled like banana or eaten as snacks or in green and fruit salads or prepared as juice	Popenoe et al. (1989), Sørensen et al. (1997), Hermann and Heller (1997), Codex (2014)
Fabaceae	*Pachyrhizus angulatus* Rich. ex DC. = *Pachyrhizus erosus* (L.) Urb.	As Below	In India, root eaten	Watt (1908)
Fabaceae	*Pachyrhizus erosus* (L.) Urban	Yam bean, Jicama, Sengkuang, Bangkwaun	Sweetish, subglobose tuberous root eaten fresh raw as a snack, or cooked, stir fried, stewed and in other dishes	Burkill (1966), Facciola (1990), Sørensen (1996), Sørensen and van Hoof (1966), Santich et al. (2008), van Wyk (2006), Codex (2014)
Fabaceae	*Pachyrhizus tuberosus* (Lam.) Spreng.	Amazonian Yam Bean, Jicama, Jactupe	As above	Hedrick (1972), Sørensen (1996)
Fabaceae	*Pediomelum cuspidatum* (Pursh) Rydb. = *Psoralea cuspidata* Pursh	Indian Turnip, Largebract Indian Breadroot	As for *P. esculenta*	Yanovsky (1936), Harrington (1974)
Fabaceae	*Pediomelum esculentum* (Pursh) Rydb. = *Psoralea esculenta* Pursh	Prairie Turnip, Indian Breadroot, Tipsin, Scurfpea; Timpsula	Tuber edible, eaten raw or in stews, grind to flour for soups and bread	Yanovsky (1936), Kaldy et al. (1980)
Fabaceae	*Pediomelum hypogaeum* var. *hypogaeum* (Nutt.) Rydb. = *Psoralea hypogaea* Torr. & A. Gray	Little Indian Breadroot, Subterranean Indian Breadroot	As above	Yanovsky (1936), Harrington (1974)
Fabaceae	*Pediomelum tenuiflorum* (pursh) A.N. Egan	Slimflower Scurfpea	As above	Yanovsky (1936)
Fabaceae	*Pediomelum subacaule* (Torr. & A. Gray) Rydb.	Whiterim Scurfpea	As above	Yanovsky (1936)
Fabaceae	*Petalostemon candidum* (Willd.) Michx. = *Dalea candida* Willd.	White Prairie Clover	Roots eaten or chewed for the sweet flavour	Yanovsky (1936), Uphof (1968), Facciola (1990)
Fabaceae	*Phaseolus adenanthus* G. Mey. = *Vigna adenantha* (G. Mey.) Marechal & al.	Adzuki Bean, Moth Bean, Wild Pea Adzuki Bean, Moth Bean, Wild Pea	In India, root eaten cooked	Watt (1908), Hedrick (1972), Facciola (1990)

(continued)

Table 1 (continued)

Family	Scientific Name	Common/Vernacular Names	Edible Part Use	Reference
Fabaceae	*Phaseolus coccineus* L.	Runner Bean, Scarlet Runner Bean, Dutch Runner Bean	Starchy tuberous root eaten	Popenoe et al. (1989), Facciola (1990)
Phaseolus coccineus L. Fabaceae	*Phaseolus rostratus* Wall. = *Vigna adenantha* (G. Mey.) Marechal & al.	Karalsona (Tamil); Karalasana, Karu Alachandra (Telugu)	In India (Madras Presidency), tuberous roots eaten cooked	Shortt (1887–88) (cited by Freedman (2009))
Fabaceae	*Psophocarpus palustris* Desv.	African Winged Bean	Tuberous root eaten in some parts of Africa	Dalziel (1955), Uphof (1968), Facciola (1990)
Fabaceae	*Psophocarpus tetragonolobus* (L.) DC.	Winged Bean Root, Asparagus Bean Root, Goa Bean Root	Tuberous root eaten	Hu (2005) Facciola (1990), Khan (1994), Lim (2012), Codex (2014)
Fabaceae	*Psoralea badocana* (Blanco) Benth. = *Cullen badocanum* (Blanco) Verdc.	NF	Roots eaten roasted	Cribb and Cribb (1987)
Fabaceae	*Psoralea argophylla* Pursh	Silver Leaf Scurf Pea, Silver Leaf Indian Breadroot	Root edible raw or cooked in stews, ground into flour for thick soups and bread	Yanovsky (1936), Tanaka (1976)
Fabaceae	*Psoralea canescens* Michx.	Buckroot	As above	Yanovsky (1936), Hedrick (1972), Tanaka (1976);
Fabaceae	*Psoralea cuspidata* Pursh	Indian Breadroot, Largebract Indian Breadroot	As above	Tanaka (1976), Fernald et al. (1985)
Fabaceae	*Psoralea esculenta* Pursh.	Bread-Root, Prairie Turnip, Prairie Potato	Fresh tubers may be eaten raw with a dressing of oil, vinegar and salt, or they may be boiled or roasted	Saunders (1920), Yanovsky (1936), Uphof (1968), Hedrick (1972), Kaldy et al. (1980), Facciola (1990), Groen et al. (1996)
Fabaceae	*Psoralea hypogaea* Torr. & A. Gray	Small Indian Breadroot	As above	Saunders (1920)
Fabaceae	*Psoralea tenuiflora* Pursh.	Slender Scurfy Pea, Slimflower Scurfpea	Root edible raw or cooked in stews, ground into flour for thick soups and bread	Yanovsky (1936), Tanaka (1976)

Fabaceae	*Pueraria lobata* (Willd.) Ohwi = *Pueraria montana* var. *lobata* (Willd.) Sanjappa & Pradeep.	Chik, Kudzu, Pueraria	Tubers edible, starch from tubers used for sauces, porridges, jelly puddings, confectionary and beverages in China, Japan and Papua New Guinea. Elsewhere used as famine food. Roots cooked with Chinese dates, sliced yams for a soup	Facciola (1990), Phillips and Rix (1993), Groen et al. (1996), Hu (2005), Walter and Lebot (2007), Codex (2014)
Fabaceae	*Pueraria lobata* (Willd.) Ohwi var. *lobata* = *Pueraria phaseoloides* var. *phaseoloides* (Roxb.) Benth.	Kudzu, Japanese Arrowroot	As above	Groen et al. (1996)
Fabaceae	*Pueraria lobata* (Willd.) Ohwi var. *thomsonii* (Benth.) van der Maesen = *Pueraria montana* var. *chinensis* (Ohwi) Sanjappa & Pradeep.	Kudzu, Thomson Kudzu	Tubers edible	Groen et al. (1996)
Fabaceae	*Pueraria lobata* (Willd.) Ohwi var. *montana* (Lour.) van der Maesen = *Pueraria montana* var. *montana* (Lour.) Merr.	Kudzu, Taiwan Kudzu	Tubers edible	Groen et al. (1996)
Fabaceae	*Pueraria edulis* Pamp.	Edible Kudzu; Ge Gen Fen (Chinese)	Starch from enlarged root extracted for noodles	Hu (2005)
Fabaceae	*Pueraria hirsuta* Kurz. = *Pueraria stricta* Kurz.	Cudzú-Tropical (Portuguese, Brazil)	China: root is steamed and eaten. Japan: processed into flour during the period of scarcity immediately following World War II	Read (1946), Uphof (1968)
Fabaceae	*Pueraria montana* (Lour.) Merr.	Ge Ma Mu (Chinese)	Starch obtained from roots	Hu (2005)
Fabaceae	*Pueraria phaseoloides* (Roxb.) Benth.	Pani Alu (Assamese)	Tuber is fleshy and tasty, it is often eaten raw	Facciola (1990), Patiri and Borah (2007)
Fabaceae	*Pueraria thomsonii* Benth. = *Pueraria montana* var. *chinensis* (Ohwi) Sanjappa & Pradeep	Sweet Kudzu; Pani Alu (Assamese); Gang E Teng (Chinese)	Roots sliced used in soup with pork chops in China. In Assam, tuberous roots are eaten cooked	Hu (2005), Patiri and Borah (2007)

(continued)

Table 1 (continued)

Family	Scientific Name	Common/Vernacular Names	Edible Part Use	Reference
Fabaceae	*Pueraria thunbergiana* (Siebold & Zucc.) Benth. = *Pueraria montana* var. *lobata* (Willd.) Sanjappa & Pradeep	Kudzu	Fleshy tuber eaten	Read (1946)
Fabaceae	*Pueraria tuberosa* (Willd.) DC.	Indian Kudzu, Nepalese Kudzu; Urahi Alu, Pani Alu (Assamese); Bilai-Kand, Biralu, Birali Panwa, Sural (Kumaon Region, Western Himalayas)	Tuberous roots eaten	Bhargava (1960), Gupta (1962), Patiri and Borah (2007)
Fabaceae	*Rhynchosia comosa* Baker = *Pseudeminia comosa* (Baker) Verdc.	L 'Indo (Sandawe)	In Central Tanzania, roots chewed for juices	Newman (1975)
Fabaceae	*Sphenostylis stenocarpa* (A. Rich.) Harms	Africa Yam Bean	Tuber eaten raw or cooked like potato	Kay (1973), Popenoe et al. (1989), Facciola (1990)
Fabaceae	*Trifolium repens* L.	White Clover	Roots prized by some native groups in North America	Schofield (2003)
Fabaceae	*Tylosema esculentum* (Burch.) A. Schreib.	Marama Bean, Gemsbok Bean	Sweet tuber baked, boiled or roasted	Fox et al. (1982), Popenoe et al. (1989), Facciola (1990)
Fabaceae	*Vigna lanceolata* Benth.	Pencil Yam Maloga Bean	The tuberous roots were used after roasting, and is reportedly one of the best vegetables available to the natives	Cribb and Cribb (1987), Harden (1991)
Fabaceae	*Vigna luteola* (Jacq.) Benth.	Dalrymple Vigna, Hairypod Cowpea, Deer Pea; Kuanga	Roots edible	Fox et al. (1982), Facciola (1990)
Fabaceae	*Vigna marina* (Burm.) Merr.	Sea Bean, Notched Cowpea	Roots edible	Cribb and Cribb (1987)
Fabaceae	*Vigna radiata* (L.) R. Wilczek	Beach Pea; Notched Cowpea	As above	Cribb and Cribb (1987)
Fabaceae	*Vigna vexillata* (L.) A. Rich.	Wild Cowpea; Chaoli, Halgia (Bombay)	In India (Bombay Presidency), tubers eaten. In West Africa, the rootstock is eaten. In Australia (North Queensland), roots roasted and eaten	Gammie (1902), Irvine (1952), Irvine (1957), Cribb and Cribb (1987), Facciola (1990)
Fabaceae	*Apios americana* Medik	American Potato Bean	Tuberous roots eaten raw or cooked	Facciola (1990), Phillips and Rix (1993), Codex (2014)

NF Not found

Selected References

Abdelmuti OMS (1991) Biochemical and nutritional evaluation of famine foods of the Sudan. Doctoral dissertation in Biochemistry and Nutrition. Faculty of Agriculture. University of Khartoum, Khartoum, Sudan

Akinwande BA, Asiedu R, Adeyemi IA, Maziya-Dixon B (2007) Influence of time of harvest on the yield and sensory attributes of white yam (*Dioscorea rotundata*) in southwest Nigeria. J Food Agric Environ 5(2):179–184

Alcorn JB (1984) Huastec Mayan ethnobotany. University of Texas Press, Austin

Bhargava KS (1960) Unusual and supplementary food plants of Kumaon. Bombay Nat Hist Soc J 56:26–31

Burkill IH (1966) A dictionary of the economic products of the Malay Peninsula. Revised reprint, 2 vol. Ministry of Agriculture and Co-operatives, Kuala Lumpur. Vol. 1 (A-H) pp. 1–1240, Vol. 2 (I-Z). pp. 1241–2444

Cao QM, Zhong HY, Li ZH, Sun CB (2007) Exploitation and utilization study of fern starch in China. Food Res Dev 28(12):168–171

Codex (2014) Draft revision of the classification of foods and animal feeds: selected vegetable commodity groups. CX/PR 14/46/7. ftp://ftp.fao.org/codex/meetings/ccpr/ccpr46/pr46_07e.pdf

Cribb AB, Cribb JW (1987) Wild food in Australia, 2nd edn. Fontana Collins, Sydney, 240 pp

Cribb AB, Cribb JW (1982) Useful wild plants in Australia. Fontana Collins, Sydney, 269 pp

Cui GY (1998) Edible fern resources and development in China. Chin Cuisine Res 6(1):21–28

Dai XL, Li XG, Wu SF (2003) List of Chinese edible pteridophytes. Quart Forest By-Prod Spec Chin 4:5–6

Dalziel JM (1955) The useful plants of west Tropical Africa (Reprint of 1937 ed.) Crown Agents for Overseas Governments and Administrations, London, 612 pp

De Castro J (1952) The geography of hunger. Little, Brown, Boston, 337 pp

Facciola S (1990) Cornucopia. A source book of edible plants. Kampong Publ., Vista, 677 pp

Fernald ML, Kinsey AC, Rollins RC (1985) Edible wild plants of Eastern North America, 2nd edn. Harper & Row, New York

Ferry MP, Gessain M, Gessain R (1974) Ethno-botanique Tenda. Centre de Recherches Anthropologiques de Musée de l' Homme, Paris, 356 pp

Flores HE, Walker TS, Guimaraes RL, Bais HP, Vivanco JM (2003) Andean root and tuber crops: underground rainbows. HortSci 38(2):161–167

Freedman RL (2009) Famine foods database. Purdue University. http://www.hort.purdue.edu/newcrop/FamineFoods/ff_home.html

Fox FW, Young MMN, Hallowes D (1982) Food from the veld: edible wild plants of southern Africa botanically identified and described. Delta Books, Johannesburg, 399pp

Gammie GA (1902) A note on plants used for food during famines and seasons of scarcity in the Bombay Presidency. India Bot Surv Rec 2(2):171–196

Gibbons E, Tucker G (1979) Euell Gibbons' handbook of edible wild plants. The Doning Co, Virginia Beach

Groen LE, Siemonsma JS, Jansen PCM (1996) Minor species yielding non-seed carbohydrates. In: Flach M, Rumawas F (eds) Plant resources of south east Asia no. 9. Plants yielding non-seed carbohydrates. Prosea Foundation, Bogor, pp 165–186

Gupta RK (1962) Some unusual and interesting food plants of the Garhwal Himalayas. J Agric d'Agric Trop Bot Appl 9(11–12):532–535

Gupta RK, Kanodia KC (1968) Plants used during scarcity and famine periods in the dry regions of India. J Agric Trop Bot Appl 15:265–285

Handy ESC (1940) The Hawaiian planter. Volume 1. His plants, methods, and areas of cultivation, Bulletin 161. Bernice Pauahi Bishop Museum, Honolulu

Harden GJ (ed) (1991) Flora of New South Wales, vol 2. New South Wales University Press, Sydney, 574 pp

Harden GJ (ed) (1993) Flora of New South Wales, vol 4. New South Wales University Press, Sydney, 775 pp

Harrington HD (1974) Edible native plants of the rocky mountains. University New Mexico Press, Albuquerque, 400pp

Hedrick UP (1972) Sturtevant's edible plants of the world. Dover Publications, New York, 686 pp

Hely-Hutchinson WF (1898) Famine plants in Zululand. Great Britain. Royal Botanic Gardens at Kew. Bull Misc Inf 135:52–54

Hermann M, Heller J (eds) (1997) Andean roots and tubers: ahipa, arracacha, maca and yacon. Promoting the conservation and use of underutilized and neglected crops. Institute of Plant Genetics and Crop Plant Research/International Plant Genetic Resources Institute (IPGRI), Gatersleben/Rome, 256 pp

Hu SY (2005) Food plants of China. The Chinese University Press, Hong Kong, 844 pp

Irvine FR (1952) Supplementary and emergency food plants of West Africa. Econ Bot 6(1):23–40

Irvine FR (1957) Wild and emergency foods of Australian and Tasmanian aborigines. Oceania 23(2): 113–142

Jansen PCM, Aguilar NO (1996) *Cyperus* L. In: Flach M, Rumawas F (eds) Plant resources of south east Asia no. 9. Plants yielding non-seed carbohydrates. Prosea Foundation, Bogor, pp 78–82

Jukema J, Paisooksantivatana Y (1996) *Tacca leontopetaloides* (L.) O. Kuntze. In: Flach M, Rumawas F (eds) Plant resources of south east Asia no. 9. Plants yielding non-seed carbohydrates. Prosea Foundation, Bogor, pp 156–159

Kabuye CHS (1986) Edible roots from wild plants in arid and semi-arid Kenya. J Arid Environ 11:65–73

Kaldy MS, Johnston A, Wilson DB (1980) Nutritive value of Indian bread-root, squaw-root, and Jerusalem artichoke. Econ Bot 34(4):352–357

Kar A, Borthakur SK (2008) Wild vegetables of Karbi – Anglong district, Assam. Nat Prod Rad 7(5):448–460

Kay DE (1973) Root crops. Tropical Products Institute, London (2), 245 pp

Khan TN (1994) *Psophocarpus tetragonolobus* (L.) DC. In: Siemonsma JS, Piluek K (eds) Plant resources of south-east Asia no 8. Vegetables. Prosea Foundation, Bogor, pp 229–233

Kunkel G (1984) Plants for human consumption. An annotated checklist of the edible phanerogams and ferns. Koeltz Scientific Books, Koenigstein

Larkcom J (1984) The salad Garden. The Viking Press, New York

Lim TK (2012) Edible medicinal and Non-medicinal plants. Volume 2, Fruits. Springer, Dordrecht

Liu YJ, Wujisguleng Long CL (2012) Food uses of ferns in China: a review. Acta Soc Bot Pol 81(4):263–270

Low T (1989) Bush tucker – Australia's wild food harvest. Angus & Robertson, Sydney, 233pp

Low T (1991) Wild food plants of Australia. Angus & Robertson, North Ryde, 240 pp

Mahr S (2011) Papyrus, *Cyperus papyrus*. http://wimastergardener.org/?q=papyrus

Maneenoon K, Sirirugsa P, Sridith K (2008) Ethnobotany of *Dioscorea* L. (Dioscoreaceae), a major food plant of the Sakai tribe at Banthad Range, Peninsular Thailand. Ethnobot Res Appl 6:385–394

Medhi P, Borthakur SK (2012) Phytoresources from North Cachar Hills of Assam – III: edible plants sold at Haflong market. Indian J Nat Prod Res 3(1):84–109

Mortimore M (1989) Adapting to drought. Farmers, famines and desertification in West Africa. Cambridge University Press, Cambridge, 324 pp

Newman JL (1975) Dimensions of Sandawe diet. Ecol Food Nutr 4(1):33–39

Ochse JJ, van den Brink RCB (1980) Vegetables of the Dutch Indies, 3rd edn. Ascher & Co., Amsterdam, 1016 pp

Ong HC, Siemonsma JS (1996) *Canna indica* L. In: Flach M, Rumawas F (eds) Plant resources of south east Asia no. 9. Plants yielding non-seed carbohydrates. Prosea Foundation, Bogor, pp 63–66

Onwueme IC (1996a) *Dioscorea* L. In: Flach M, Rumawas F (eds) Plant resources of south east Asia no. 9. Plants yielding non-seed carbohydrates. Prosea Foundation, Bogor, pp 85–90

Onwueme IC (1996b) *Dioscorea esculenta* (Lour.) Burkill. In: Flach M, Rumawas F (eds) Plant resources of south east Asia no. 9. Plants yielding non-seed carbohydrates. Prosea Foundation, Bogor, pp 93–95

Onwueme IC (1996c) *Dioscorea hispida* Dennst. In: Flach M, Rumawas F (eds) Plant resources of south east Asia no. 9. Plants yielding non-seed carbohydrates. Prosea Foundation, Bogor, pp 95–97

Onwueme IC, Ganga ZN (1996) *Dioscorea alata* L. In: Flach M, Rumawas F (eds) Plant resources of south east Asia no. 9. Plants yielding non-seed carbohydrates. Prosea Foundation, Bogor, pp 90–93

Oyen LPA (1994) *Beta vulgaris* L. In: Siemonsma JS, Piluek K (eds) Plant resources of southeast Asia No. 8. Vegetables. Prosea Foundation, Bogor, pp 97–101

Paisooksantivatana Y (1966) *Eleocharis dulcis* (Burm.f.) Trinius ex Henschel. In: Flach M, Rumawas F (eds) Plant resources of south east Asia no. 9. Plants yielding non-seed carbohydrates. Prosea Foundation, Bogor, pp 97–100

Parmentier AA (1781) Recherches sur les végétaux nourrisans qui, dans les temps de disette, peuvent remplacer les alimens ordinaires. Avec de nouvelles observations su la culture des pommes de terre. Imprimerie royale, Paris, 599pp

Paton DN, Dunlop JC (1904) The nutritive values of some uncultivated foods used by the Bhils during recent famines. Agric Ledger 6:37–73

Patiri B, Borah A (2007) Wild edible plants of Assam. Geetakhi Printers and Publishers, Guwahati

Phillips R, Rix M (1993) Vegetables. Pan Books, London, 270pp

Popenoe H, King SR, Léon J, Kalinkowski LS, Vietmeyer ND, Dafforn M (1989) Lost crops of the Incas. Little known plants of the Andes with promise for worldwide cultivation. National Academy Press, Washington, pp 37–46

Read BE (ed) (1946) Famine foods listed in the Chiu huang pen ts'ao [of Ting Wang Chou]: giving their identity, nutritional values and notes on their preparation. Henry Lester Institute of Medical Research, Shanghai, 93pp

Santich B, Bryant G, Bacon, Clifton C, Connor D, Felder A, Foster S, Graue J, Loyer J, morrachian M, Raab C, Sandall P Santich S, Stybe K, Stybe R (13) (2008) Edible: the illustrated guide to the world's food plants. Cameron House, South Australia, 360 pp

Saxena SK (1979) Plant foods of Western Rajasthan. Man Environ 3:35–43

Saunders CF (1920) Useful wild plants of the United States and Canada. Robert M. Mcbride & Co, New York

Sawian JT, Jeeva S, Lyndem FG, Mishra BP, Laloo RC (2007) Wild edible plants of Meghalaya, North-east India. Nat Prod Rad 6(5):410–462

Schofield J (2003) Discovering wild plants: Alaska, Western Canada, The Northwest, 1st edn. Alaska Northwest Books, Bothell, 353 pp

Shankarnarayan KA, Saxena SK (1987) Lie supporting arid zone plants in famine periods. In: Poorada R, Kapoor P, Arora R (eds) Plants in arid lands. Bhagmed, National Bureau of Plant Genetic Resources, New Delhi

Shortt J (1887–1888) List of wild plants and vegetables used as food by people in famine times. Indian Forester 3:232–238

Sørensen M (1996) Yam Bean: *Pachyrhizus* DC. – Promoting the conservation and use of underutilized and neglected crops. 2. Bioversity Internationa, 141pp

Sørensen M, van Hoof WCH (1966) *Pachyrhizus erosus* (L.) Urban. In: Flach M, Rumawas F (eds) Plant resources of south east Asia no. 9. Plants yielding non-seed carbohydrates. Prosea Foundation, Bogor, pp 137–141

Sørensen M, Gruneberg WJ, Ørting B (1997) Ahipa (*Pachyrhizus ahipa* (Wedd.) Parodi). In: Herman M, Heller J (eds) Andean roots and tubers: ahipa, arracacha, maca and yacon. Promoting the conservation and use of underutilized and neglected crops. International Plant Genetic Resources Institute, Rome, pp 13–73

Takagi H, Kuo CG, Sakamoto S (1996) *Ipomoea batatas* (L.) Lamk. In: Flach M, Rumawas F (eds) Plant resources of south east Asia no. 9. Plants yielding non-seed carbohydrates. Prosea Foundation, Bogor, pp 102–107

Tanaka T (1976) Tanaka's cyclopaedia of edible plants of the world. Keigaku Publishing, Tokyo, 924 pp

Tanaka Y, Nguyen VK (2007) Edible wild plants of Vietnam: the bountiful garden. Orchid Press, Bangkok, 175 pp

The Plant List (2014) http://www.theplantlist.org

Uphof JCT (1968) Dictionary of economic plants. Verlag von Cremer, Lehrer

Van Wyk BE (2006) Food plants of the world an illustrated guide. Timber Press, Portland, 480 pp

Veltkamp HJ, De Bruijn GH (1996) *Manihot esculenta* Crantz. In: Flach M, Rumawas F (eds) Plant resources of south east Asia no. 9. Plants yielding non-seed carbohydrates. Prosea Foundation, Bogor, pp 107–113

Walter A, Lebot V (2007) Gardens of Oceania, ACIAR monograph no. 122. Australian Centre for International Agricultural Research, Canberra, 326 pp

Watt G (1908) The commercial products of India: being an abridgement of "the dictionary of the economic products of India". J. Murray, London, 1189 pp

Wightman G, Andrews M (1989) Plants of the northern territory vine forests. Conservation Commission of the Northern Territory, Darwin, 163 pp

Yanovsky E (1936) Food plants of the North American Indians, United States Department of Agriculture. Miscellaneous publication. no. 237. U.S. Department of Agriculture, Washington, DC

Yun XL, Zhao NW, Pan LT, Zhao JH (2009a) Resource distribution and exploitation of edible ferns in Guizhou (I). J Anhui Agric Sci 37(32):15911–15912

Yun XL, Zhao NW, Pan LT, Zhao JH (2009b) Resource distribution and exploitation of edible ferns in Guizhou (II). J Anhui Agric Sci 37(33):16317–16319

Zinyama LM, Matiza T, Campbell DJ (1990) The use of wild foods during periods of food shortage in rural Zimbabwe. Ecol Food Nutr 24(4):231–255

Beta vulgaris

Scientific Name

***Beta vulgaris* L.**

Synonyms

Beta alba DC., *Beta altissima* Steud., *Beta atriplicifolia* Rouy, *Beta bengalensis* Roxb., *Beta brasiliensis* Voss (inval.), *Beta carnulosa* Gren., *Beta cicla* (L.) L., *Beta cicla* (L.) Pers., *Beta cicla* var. *argentea* Krassochkin & Burenin, *Beta cicla* var. *viridis* Krassochkin & Burenin, *Beta crispa* Tratt., *Beta decumbens* Moench, *Beta esculenta* Salisb., *Beta foliosa* Ehrenb. ex Steud. (inval.), *Beta hortensis* Mill., *Beta hybrida* Andrz., *Beta incarnata* Steud., *Beta lutea* Steud., *Beta marina* Crantz, *Beta maritima* L., *Beta maritima* var. *atriplicifolia* Krassochkin, *Beta maritima* subsp. *atriplicifolia* (Rouy) Burenin, *Beta maritima* subsp. *danica* Krassochkin, *Beta maritima* var. *erecta* Krassochkin, *Beta maritima* var. *glabra* Delile, *Beta maritima* subsp. *marcosii* (O.Bolòs & Vigo) Juan & M.B.Crespo, *Beta maritima* subsp. *orientalis* (Roth) Burenin, *Beta maritima* var. *pilosa* Delile, *Beta maritima* var. *prostrata* Krassochkin, *Beta noeana* Bunge ex Boiss., *Beta orientalis* Roth, *Beta orientalis* L., *Beta purpurea* Steud., *Beta rapa* Dumort., *Beta rapacea* Hegetschw., *Beta rosea* Steud., *Beta sativa* Bernh., *Beta stricta* K.Koch, *Beta sulcata* Gasp., *Beta triflora* Salisb., *Beta vulgaris* var. *altissima* Döll, *Beta vulgaris* var. *annua* Asch. & Graebn., *Beta vulgaris* subsp. *asiatica* Krassochkin ex Burenin, *Beta vulgaris* var. *asiatica* Burenin, *Beta vulgaris* var. *atriplicifolia* (Rouy) Krassochkin, *Beta vulgaris* var. *aurantia* Burenin, *Beta vulgaris* subsp. *cicla* (L.) Schübl. & G. Martens, *Beta vulgaris* var. *cicla* L., *Beta vulgaris* subsp. *cicla* (L.) W.D.J. Koch, *Beta vulgaris* var. *coniciformis* Burenin, *Beta vulgaris* var. *crassa* Alef., *Beta vulgaris* var. *debeauxii* Clary, *Beta vulgaris* var. *foliosa* (Asch. & Schweinf.) Aellen, *Beta vulgaris* subsp. *foliosa* Asch. & Schweinf., *Beta vulgaris* var. *glabra* (Delile) Aellen, *Beta vulgaris* var. *grisea* Aellen, *Beta vulgaris* subsp. *lomatogonoides* Aellen, *Beta vulgaris* var. *lutea* DC., *Beta vulgaris* var. *marcosii* O.Bolòs & Vigo, *Beta vulgaris* var. *maritima* (L.) Moq., *Beta vulgaris* subsp. *maritima* (L.) Arcang., *Beta vulgaris* subsp. *maritima* (L.) Thell., *Beta vulgaris* var. *mediasiatica* Burenin, *Beta vulgaris* var. *orientalis* (Roth) Moq., *Beta vulgaris* subsp. *orientalis* (Roth) Aellen, *Beta vulgaris* var. *ovaliformis* Burenin, *Beta vulgaris* var. *perennis* L., *Beta vulgaris* var. *pilosa* (Delile) Moq., *Beta vulgaris* subsp. *provulgaris* Ford-Lloyd & J.T. Williams, *Beta vulgaris* var. *rapacea* W.D.J.Koch, *Beta vulgaris* var. *rosea* Moq., *Beta vulgaris* var. *rubidus* Burenin, *Beta vulgaris* var. *rubra* L., *Beta vulgaris* var. *rubrifolia* Krassochkin ex Burenin, *Beta vulgaris* var. *saccharifera* Alef., *Beta vulgaris* var. *virescens* Burenin, *Beta vulgaris* var. *viridifolia*

T.K. Lim, *Edible Medicinal and Non-Medicinal Plants: Volume 10, Modified Stems, Roots, Bulbs*,
DOI 10.1007/978-94-017-7276-1_2

Krassochkin ex Burenin, *Beta vulgaris* var. *vulgaris, Beta vulgaris* subsp. *vulgaris*

Family

Amaranthaceae, also placed in Chenopodiaceae

Common/English Names

Beet, Beetroot, Chard, Chioggia Beet, Field Beet, Fodder Beet, Fodder Sugar Beet, Foliage Beet, Forage Beet, Garden Beetroot, Golden Beet, Indian Spinach, Italian Beetroot, Leaf Beet, Mangel, Mangel-Wurzel, Mangold, Ordinary Beet, Ordinary Garden Beet, Perpetual Beet, Roman Kale, Rainbow Chard, Red Beet, Red Beetroot, Red-Fleshed Beetroot, Root Beet, Savoy Beet, Sea Beet, Seakale Beet, Semi-Sweet Sugarbeet, Sicilian Broad-Rib Beet, Silver Beet Spinach Beet, Spinach Chard, Silver Beet, Sugarbeet, Sugar Beet, Swiss Chard, Target Beetroot, White Beet, White Beetroot, White-Fleshed Beetroot, Yellow Beet, Yellow Beetroot, Yellow-Fleshed Beetroot

Vernacular Names

Arabic: Bangar, Bangar Albi, Salq, Shamandar

Bosnian: Cikla

Catalan: Remolatxa

Chinese: Bai Gen Tian Cai, Guan Shang Ye Tian Cai, Hong Gen Tian Cai, Hong Tou Cai, Hou Pic Cai, Huang Gen Tian Cai, Jun Da Cai, Kuan Bing Ye Tian Cai, Kuan Jing Ye Tian Cai, Si Niu Tian Cai, Si Yong Tian Cai, Tang Tian Cai, Tang Luo Bo, Tang Yong Tian Cai, Tian Cai, An Cai, Ye Yong Tian Cai, Xia Jing Ye Tian Cai, Zhu Na Cai

Croatian: Blitva, Cikla, Kravlja Repa, Obicna Blitva, Primorska Blitva, Šećerna Repa

Czech: Cukrová Řepa, Flepa Bengálsky, Řepa Cukrová, Řepa Obecná Červená

Danish: Bladbede, Foderbede, Foderroe, Fodersukkerroer, Guldbede, Hvidbede, Rødbede, Runkelroe, Soelvbede, Sukkerroe

Dutch: Biet, Gele Biet, Kroot, Kroten, Rode Biet, Snijbiet, Strandbiet, Suhkrupeet, Suikerbiet, Voederbiet, Zoete Voederbiet

Estonian: Harilik, Lehtpeet, Punapeet, Söödapeet, Suhkrupeet

Euskara: Beterraba

Finnish: Lehtimangoldi, Lehtijuurikas, Punajuurikas, Rantajuurikas, Rehujuurikas, Rehusokerijuurikas, Sokerijuurikas

French: Arde, Bette À Carde, Bette À Carde Du Chili, Bette À Cardes Multicolour, Bette À Côtes, Bette À Couper, Betterave, Betterave Blanche Potagère, Betterave Champêtre, Betterave Demi-Sucrière, Betterave Fourragère, Betterave Jaune, Betterave Jaune Potagère, Betterave Maritime, Betterave Potagère, Betterave Potagère À Chair Jaune, Betterave Rose d'Italie, Betterave Rouge, Betterave Rouge Potagère, Betterave À Sucre, Betterave De Distillerie, Betterave Sucrière, Disette, Feuille De Bette, Feuille De Blette, Poirée, Poirée À Carde, Poirée À Couper, Poirée À Carde Du Chili, Poirée Bette, Poirée Ordinaire

Gaelic (Scottish): Biotais

Galician: Remolacha

German: Weißkohl, Blatt-Mangold, Burgunderrübe, Cardonen-Bete, Futterrübe, Futterzuckerrübe, Gehaltsrübe, Gelbe Bete, Gelbe Rübe, Halbzuckerrübe, Krautstiel, Mangel Wurzel, Mangold, Rippenmangold, Römische Bete, Römischer Kohl, Rote Beete, Rote Bete, Rote Rübe, Rote Rüben, Rübe, Rübenmangold, Runkeln, Runkelrübe, Schnittmangold, See-Mangold, Schweizer Mangold, Stengelmangold, Stielmangold, Weiße Schlesische Zuckerrübe, Wild-Bete, Wilde Rübe, Wilde Runkelrübe, Zuckerhaltige Rübe, Zuckerrübe

Haitian Creole: Bètrav

Hebrew: Selek Alim, Selek Adom, Selek Mispo, Selek Sukar

India: Chukandar, Chukkndan, Mīṭhē Cuqandara, Palak, Palangsag, Palanki (Hindi), (Malayalam), Palakya (Sanskrit), Carkkaraivaḷḷikkiḻaṅku (Tamil)

Indonesia: Bit, Bit Gula

Iraq: Salk

Italian: Barba, Barbabietola, Barbabietola Bianca, Barbabietola Da Foraggio, Barbabietola Da Insalata, Barbabietole Da Orto, Barbabietola Da Zucchero, Barbabietola Di Chioggia, Barbabietola Rossa, Barbabietola Semizuccherina, Barbabietola Zuccherina, Bieta A Foglia, Bieta Da Coste, Bietola A Radice Rossa, Bietola Bianca, Bietola Comune, Bietola Da Coste, Bietola Da Orto, Bietola Da Taglio, Bietola Del Brasile, Bietola Rossa, Bietola Rossa E Gialla

Japanese: Aka Kabu, Biito, Biitsu, Hama Fudansou, Kaensai, Satou Daikon, Shiryouyou Biito, Shokuyou Biito, Teeburu Biito, Tensai

Kashubian: Rąkla

Korean: Geundae, Kuntae

Kurdish: Silk

Ladino: Kuchundúrya

Latvian: Parastā Biete

Ligurian: Giæa

Luxembourgish: Rout Rommel

Malaysia: Akar Bit, Bit

Nepalese: Bangaalii Paaluugo, Cukandar, Guliyo Muulaa

Norwegian: Bladbete, Sukkerbete

Philippines: Remolatsa, Remolatsa Pul

Polish: Burak Cukrowy, Burak Liściowy, Burak Pastewny, Burak Zwyczajny, Ćwikła Cukrowa

Portuguese: Acelga, Acelga Brava, Beterraba, Beterraba-Açucareira, Beterraba-Forrageira, Beterraba Forraginosa, Beterraba-Sacarina, Beterraba-Vermelha, Beterraba De Salada, Beterraba Hortícola, Beterraba De Mesa, Beterraba Vermelha, Patarrábia, Terraba

Quechuan: Rimulacha

Romanian: Sfeclă Roşie

Russian: Mangol'd, Primorskaia, Sakharnaia Svëkla, Svëkla Kormováia, Svëkla Krasnaia, Svëkla Listovaia, Svëkla Listovaja, Svëkla Obyknovennaia, Svekla Obyknovennaia Korneplodnaia, Svëkla Polusakharnaia, Svëkla Sakharnaia, Svëkla Shpinatanaia, Svëkla Stolóvaia, Svëkla Zheltaia Salatnaia

Slovakian: Cvikľa, Repa Cvikľa, Repa Obyčajná, Repa Obyčajná Cvikľa

Slovenian: Krmna Pesa, Mangold, Navadna Pesa, Primorska Pesa, Rdezhe Pesa, Sladkorna Pesa

Sorbian: Wšědna Rěpa

Spanish: Acelga, Acelga Cardo, Betarraga Azucarera, Betarraga Forrajera, Remolacha, Remolacha Amarilla, Remolacha Azucarera, Remolacha Blanca, Remolacha Colorada, Remolacha De Azucar, Remolacha De Mesa, Remolacha Forrajera, Remolacha Roja, Remolacha Semiazucarera

Swedish: Beta, Foderbeta, Fodersockerbeta Mangold, Rödbeta, Sockerbeta, Strandbeta;

Thai: Bīt Daeng, Bītrūt, Chaa Rót, Hua Bee Tor, Náam Bit, Phakkat Farang (Bangkok), Pàk-Gàat-Daeng, Phak Kat Daeng (Central Thailand)

Turkish: Pancar

Vietnamese: Chi Củ Cải Ngọt

Walon: Betrâle

Origin/Distribution

Sugar beets and other *B. vulgaris* cultivars, such as beetroot and chard, share a common wild ancestor, the sea beet (*Beta vulgaris* subsp. *maritima*). All cultivated beets fall into the subspecies *Beta vulgaris* subsp. *vulgaris*. The centre of origin of beet (*Beta*) is believed to be the Middle East, near the Tigris and Euphrates Rivers. It is thought that wild beets spread west into the Mediterranean and north along the Atlantic sea coast. Beets and their relatives are grown throughout the world for human and animal food.

Agroecology

Beets are a cool climate vegetable and are able to withstand mild frost but cannot tolerate hot, dry conditions. Optimum temperatures are from 15 to 19 °C. Temperatures below10 °C for 2 weeks cause a physiological shift from vegetative (rosette) to reproductive (flower bolting) growth. Cooler weather will promote the development of deep-red pigmentation. They tolerate annual precipitation of 230–320 cm. Irrigation is useful from planting to harvest to sustain proper hypocotyl growth and development.

They grow well in a variety of soils, growing best in a deep, friable well-drained, light-textured,

Italian: Barba, Barbabietola, Barbabietola Bianca, Barbabietola Da Foraggio, Barbabietola Da Insalata, Barbabietole Da Orto, Barbabietola Da Zucchero, Barbabietola Di Chioggia, Barbabietola Rossa, Barbabietola Semizuccherina, Barbabietola Zuccherina, Bieta A Foglia, Bieta Da Coste, Bietola A Radice Rossa, Bietola Bianca, Bietola Comune, Bietola Da Coste, Bietola Da Orto, Bietola Da Taglio, Bietola Del Brasile, Bietola Rossa, Bietola Rossa E Gialla
Japanese: Aka Kabu, Biito, Biitsu, Hama Fudansou, Kaensai, Satou Daikon, Shiryouyou Biito, Shokuyou Biito, Teeburu Biito, Tensai
Kashubian: Rąkla
Korean: Geundae, Kuntae
Kurdish: Silk
Ladino: Kuchundúrya
Latvian: Parastā Biete
Ligurian: Giæa
Luxembourgish: Rout Rommel
Malaysia: Akar Bit, Bit
Nepalese: Bangaalii Paaluugo, Cukandar, Guliyo Muulaa
Norwegian: Bladbete, Sukkerbete
Philippines: Remolatsa, Remolatsa Pul
Polish: Burak Cukrowy, Burak Liściowy, Burak Pastewny, Burak Zwyczajny, Ćwikła Cukrowa
Portuguese: Acelga, Acelga Brava, Beterraba, Beterraba-Açucareira, Beterraba-Forrageira, Beterraba Forraginosa, Beterraba-Sacarina, Beterraba-Vermelha, Beterraba De Salada, Beterraba Hortícola, Beterraba De Mesa, Beterraba Vermelha, Patarrábia, Terraba
Quechuan: Rimulacha
Romanian: Sfeclă Roşie
Russian: Mangol'd, Primorskaia, Sakharnaia Svëkla, Svëkla Kormováia, Svëkla Krasnaia, Svëkla Listovaia, Svëkla Listovaja, Svëkla Obyknovennaia, Svekla Obyknovennaia Korneplodnaia, Svëkla Polusakharnaia, Svëkla Sakharnaia, Svëkla Shpinatanaia, Svëkla Stolóvaia, Svëkla Zheltaia Salatnaia
Slovakian: Cvikľa, Repa Cvikľa, Repa Obyčajná, Repa Obyčajná Cvikľa
Slovenian: Krmna Pesa, Mangold, Navadna Pesa, Primorska Pesa, Rdezhe Pesa, Sladkorna Pesa
Sorbian: Wšědna Rěpa
Spanish: Acelga, Acelga Cardo, Betarraga Azucarera, Betarraga Forrajera, Remolacha, Remolacha Amarilla, Remolacha Azucarera, Remolacha Blanca, Remolacha Colorada, Remolacha De Azucar, Remolacha De Mesa, Remolacha Forrajera, Remolacha Roja, Remolacha Semiazucarera
Swedish: Beta, Foderbeta, Fodersockerbeta Mangold, Rödbeta, Sockerbeta, Strandbeta;
Thai: Bīt Daeng, Bītrūt, Chaa Rót, Hua Bee Tor, Náam Bit, Phakkat Farang (Bangkok), Pàk-Gàat-Daeng, Phak Kat Daeng (Central Thailand)
Turkish: Pancar
Vietnamese: Chi Củ Cải Ngọt
Walon: Betråle

Origin/Distribution

Sugar beets and other *B. vulgaris* cultivars, such as beetroot and chard, share a common wild ancestor, the sea beet (*Beta vulgaris* subsp. *maritima*). All cultivated beets fall into the subspecies *Beta vulgaris* subsp. *vulgaris*. The centre of origin of beet (*Beta*) is believed to be the Middle East, near the Tigris and Euphrates Rivers. It is thought that wild beets spread west into the Mediterranean and north along the Atlantic sea coast. Beets and their relatives are grown throughout the world for human and animal food.

Agroecology

Beets are a cool climate vegetable and are able to withstand mild frost but cannot tolerate hot, dry conditions. Optimum temperatures are from 15 to 19 °C. Temperatures below10 °C for 2 weeks cause a physiological shift from vegetative (rosette) to reproductive (flower bolting) growth. Cooler weather will promote the development of deep-red pigmentation. They tolerate annual precipitation of 230–320 cm. Irrigation is useful from planting to harvest to sustain proper hypocotyl growth and development.

They grow well in a variety of soils, growing best in a deep, friable well-drained, light-textured,

sandy-loam soil rich in organic matter, but poorly on clay. Optimum pH is 6.0–6.8, but neutral and alkaline soils are tolerated in some areas. Potassium and phosphorous fertilisation is needed in large portions to ensure hypocotyl growth and development.

Edible Plant Parts and Uses

Swiss chard and Spinach beet leaves are eaten as pot herb like spinach or in salads. Young leaves of the garden beet are sometimes used similarly. The midribs of Swiss chard are eaten boiled, while the whole leaf blades are eaten as spinach beet. Swiss chard leaves are utilized in 'tammia' (Egyptian food), pan bread and pizza (Bakry et al. 2014). In Africa, the whole leaf blades are usually prepared with the midribs as one dish. Older leaves and stems are stir-fried.

Beetroot can be peeled, steamed and then eaten warm with butter as a delicacy; cooked, pickled and then eaten cold as a condiment; or peeled, shredded raw and then eaten as a salad. It is also common in Australia and New Zealand for pickled beetroot to be consumed in a burger. In Eastern Europe, beet soup, such as cold 'borscht', is a popular dish. Canning beets have a globe shape, with the cultivar 'Ruby Queen', being the most popular; it has excellent quality and is very sweet. The globe red cultivars, especially the 'Detroit Dark Red', are the most popular in production for table beets. Yellow-coloured garden beets are grown on a very small scale for home consumption.

Sugar beet, cultivated *Beta vulgaris*, is a plant whose root contains a high concentration of sucrose and is grown commercially for sugar production. By-products of sugar beet processing include pulp and molasses. Most of the molasses produced is processed further to remove the remaining sucrose. The pulp and most of the remaining molasses are mixed together, dried and sold as livestock feed.

Betanins, obtained from beetroots, are used industrially as red food colourants, e.g. to improve the colour of tomato paste, sauces, desserts, jams and jellies, ice cream, sweets and breakfast cereals. Red beet pigments, in the form of industrially produced concentrate, are used as a soft drink colorant (Havlíková et al. 1985). Beet concentrate in the concentration used (2.5 g per 100 g of drink) did not influence taste or flavour of drink.

Botany

An annual or biennial, robust, succulent herb (Plates 1, 6–9); main root long, stout, tapered, side roots forming a dense, extensive root system; hypocotyl and upper part of the main root conspicuously swollen; fleshy, tuber-like, being globular, napiform, fusiform, cylindrical, or branched and not-tuber-like, dark red, white, or yellow (Plates 2–5). Stem erect, distally branched, ribbed, striate, 60–120 (–200 cm). Leaves arising from the crown of the hypocotyl. Basal leaves long petiolate; petiole stout, abaxially convex, adaxially flattened or slightly concave, red, white, green, yellow coloured; lamina oblong, 20–30 × 10–15 cm, adaxially crisped, subglossy, subglabrous, green, dark green or red, abaxially with strongly raised veins, base cuneate, truncate,

Plate 1 Beetroot plant habit

Plate 2 Globose beetroot

Plate 3 Fusiform beetroot

Plate 4 Napiform beetroot

Plate 5 Orange-yellow beetroot

Plate 6 Rainbow chard

or slightly cordate, margin entire or undulate, apex obtuse (Plates 1, 6–9). Cauline leaves alternate, smaller than basal ones; lamina ovate, rhombic or lanceolate-oblong, base gradually narrowed into petiole, apex attenuate. Flowers sessile, bisexual, usually 2–3(–5) together in cymes, small, greenish, subtended by minute bracts; perianth 5-partite, united at base; segments linear or narrowly oblong, becoming leathery and incurved in fruit, stamens 5; ovary 1-celled, surrounded by a disk, with 2–3 stigmas. Fruit, a nut enclosed within the swollen corky perianth-bases, 3–7 mm in diameter, 1–6 fruits adhering in glomerulus. Seed lenticular to kidney-shaped, red-brown, 1.5–3 mm in diameter.

Nutritive/Medicinal Properties

The following proximate nutrient composition (per 100 g edible portion) of raw beet was reported by USDA-ARS (2014): water 87.58 g, energy 43 kcal (180 kJ), protein 1.61 g, total lipid 0.17 g, ash 1.08 g, carbohydrate 9.56 g, total

Plate 7 White-stemmed chard

Plate 8 Yellow-stemmed chard

Plate 9 Red-stemmed chard

dietary fibre 2.8 g, total sugars 6.76 g, Ca 16 mg, Fe 0.80 mg, Mg 23 mg, P 40 mg, K 325 mg, Na 78 mg, Zn 0.35 mg, Cu 0.075 mg, Mn 0.329 mg, Se 0.7 μg, vitamin C 4.9 mg, thiamine 0.031 mg, riboflavin 0.040 mg, niacin 0.334 mg, pantothenic acid 0.155 mg, vitamin B-6 0.067 mg, total folate 109 μg, total choline 6 mg, betaine 128.7 mg, vitamin A 33 IU, vitamin A 2 μg RAE, vitamin E (α-tocopherol) 0.04 mg, vitamin K (phylloquinone) 0.2 μg, total saturated fatty acids 0.027 g, 16:0 0.026 g, 18:0 0.001 g, total monounsaturated fatty acids 0.032 g, 18:1 undifferentiated 0.032 g, total polyunsaturated fatty acids 0.060 g, 18:2 undifferentiated 0.055 g, 18:3 undifferentiated 0.005 g, phytosterols 25 mg, tryptophan 0.019 g, threonine 0.047 g, isoleucine 0.048 g, leucine 0.068 g, lysine 0.058 g, methionine 0.018 g, cystine 0.019 g, phenylalanine 0.046 g, tyrosine 0.038 g, valine 0.056 g, arginine 0.042 g, histidine 0.021 g, alanine 0.060 g, aspartic acid 0.116 g, glutamic acid 0.428 g, glycine 0.031 g, proline 0.042 g, serine 0.059 g and β-carotene 20 μg. Raw beet also contained the flavones, luteolin 0.4 mg and the flavonol quercetin 0.1 mg/100 g (Hertog et al. 1992; Lugasi and Hovari 2000).

The following proximate nutrient composition (per 100 g edible portion) of raw Swiss chard was reported by USDA-ARS (2014): water 92.66 g, energy 19 kcal (79 kJ), protein 1.80 g, total lipid 0.20 g, ash 1.60 g, carbohydrate 3.74 g, total dietary fibre 1.6 g, total sugars 1.10 g, Ca 51 mg,

Fe 1.80 mg, Mg 81 mg, P 46 mg, K 379 mg, Na 213 mg, Zn 0.36 mg, Cu 0.179 mg, Mn 0.366 mg, Se 0.9 μg, vitamin C 30 mg, thiamine 0.040 mg, riboflavin 0.090 mg, niacin 0.4 mg, pantothenic acid 0.172 mg, vitamin B-6 0.099 mg, total folate 14 μg, total choline 18 mg, betaine 0.3 mg, vitamin A 6116 IU, vitamin A 306 μg RAE, *β*-carotene 3647 μg, *α*-carotene 45 μg, lutein + zeaxanthin 11000 μg, vitamin E (*α*-tocopherol) 1.89 mg, vitamin K (phylloquinone) 830 μg, total saturated fatty acids 0.030 g, 16:0 0.030 g, total monounsaturated fatty acids 0.040 g, 18:1 undifferentiated 0.040 g, total polyunsaturated fatty acids 0.070 g, 18:2 undifferentiated 0.063 g, 18:3 undifferentiated 0.007 g, tryptophan 0.017 g, threonine 0.083 g, isoleucine 0.147 g, leucine 0.130 g, lysine 0.099 g, methionine 0.019 g, cystine 0.019 g, phenylalanine 0.110 g, valine 0.110 g, arginine 0.117 g, histidine 0.036 g. Swiss chard contained (mg/100 g) the flavan-3-ol, (+)-catechin 1.5 mg; the flavonols, kaempferol 5.8 mg, myricetin 3.1 mg and quercetin 2.2 mg (Pyo et al. 2004).

The comparative nutrient profile of air-dried (AD – 3 samples), freeze-dried (FD 2 – samples), spray-dried (SD – 2 samples) red beetroot concentrates and betalain-enriched red beetroot extract (FC– 1 sample) was determined respectively as follows (Nemzer et al. 2011): energy value 333–348; 334–354; 364–366; 102 cal/100 g, energy value from fat 2.34–51.5; 3.42–4.93; <1.0; <1.0 cal/100 g; moisture 5.3–8.4; 2.3–4.5; 3.84–4.12; 5.84 mg/100 g; total carbohydrate 75.7–76.6; 75.4–81.74; 88.6–89; 23.6 g/100 g; total dietary fibre 18–20.8; 17.2–21.8; <1.0; 4.55 g/100 g; fructose 0.9–1.3; 0.8–1.2;1.7–1.9; <0.1 g/100 g; glucose 1.1–1.3; 0.8–1.2 g, 1.7–19; <0.1 g/100 g; sucrose 51.5–55.9; 50.7–60.5; 26.2–33.2; 0.3 g/100 g; lactose <0.1; <0.1; <0.1; <0.1 g/100 g; maltose <0.1; <0.1; 0.7–0.8; <0.1 g/100 g; galactose <0.1; <0.1; <0.1; <0.1 g/100 g; total sugars 53.5–58.1; 53.5–62.5; 30.8–37.6; 0.3 g/100 g; protein as total nitrogen (N x 6.25, Kjeldahl method) 8.13–10.3; 8.63–12.4; 3.81–4.13; 64.7 g/100 g; protein as nitrogen adjusted for NPN 1.85–2.80; 2.34–4.28; 0.09–0.38; 1.58 g/100 g; protein as amino acids (total amino acids) 5.33–6.88; 6.05–6.05; 2.41–2.44; 1.99 g /100 g; total nitrogen 1.3–1.65; 1.38–1.99; 0.61–0.66; 10.4 %; non-protein nitrogen 1.00–1.21; 1.00–1.31; 0.6; 10.1 %; saturated fatty acids 0.073–1.26; 0.09–0.129; <0.007–0.019; <0.007 g/100 g; monounsaturated fatty acids 0.047–0.290; 0.061–0.063; <0.007–0.022; <0.007 g/100 g; polysaturated fatty acids 0.067–0.401; 0.115–0.333; <0.007–0.021; <0.007 g/100 g; *trans*-fatty acids 0.02–0.139; <0.007–0.097; <0.007; <0.007 g/100 g; total fatty acids 0.248–1.53; 0.365–0.523; <0.007–0.065; <0.007 g/100 g; total vitamin A <100; <100; <100; <100 IU/100 g; vitamin A from carotenes <35; <35; <35; <35 IU/100 g; *β*-carotene <0.02; 0.02; 0.02; 0.02 mg/100 g and ash 5.32–6.08; 5.82–6.14; 2.17–2.79; 1.18 g/100 g. The mineral profile of AD,SD, FD and FC (mg/100 g) was reported respectively as follows: aluminium 0.47–3.88; 1.37–5.47;0.59–2.07; 3.71 mg/100 g; barium 1.27–2.71; 1.72–3.09; <0.10; 0.72 mg/100 g; boron 0.74–0.97; 0.31–0.69; <0.02; 0.21 mg/100 g; calcium 97.9–224; 98.5–106; 39.7–61.0; 74.8 mg/100 g; copper 0.39–0.82; 0.60–0.86; <0.05–0.10; 6.11 mg/100 g; iron 2.47–6.95; 3.43–9.20; 1.30–2.05; 28.9 mg/100 g; magnesium 135–195; 130–204; 54.9–75.4; 152.6 mg/100 g; manganese 2.13–4.76; 2.93–4.64; 1.19–1.22; 13.41 mg/100 g; potassium 103–2120; 1790–2090; 1010–1090; 124 mg/100 g; sodium 192–1320; 196–315; 125–158; 16 mg/100 g; zinc 1.58–2.17; 2.18–2.89; 0.28–0.35; 11.1 mg/100 g. The amino acid profile of AD,SD, FD and FC (mg/100 g) was reported respectively as follows: aspartic acid 0.502–0.643; 0.557–0.603; 0.248–0.282; 0.132 mg; threonine 0.186–0.230; 0.200–0.238; 0.037–0.049; 0.049 mg; serine 0.261–0.349; 2.41–2.78; 0.103–1.123; 0.057 mg; glutamic acid 2.00–2.84; 2.41–2.78; 1.39–1.60; 0.741; proline 0.141–0.191; 0.167–0.191; 0.016–0.027; <0.01 mg; glycine 0.179–0.247; 0.197–0.247; 0.034–0.048; 0.297 mg; alanine 0.452–0.519; 0.407–0.432; 0.119–0.128; 0.059 mg; valine 0.237–0.290; 0.267–0.317; 0.051–0.071; 0.057 mg; isoleucine 0.211–0.246; 0.213–0.268; 0.065–0.076; 0.026 mg; leucine 0.228–0.287; 0.260–0.360; 0.046–0.072; 0.040 mg; tyrosine 0.166–0.200; 0.196–0.197; <0.001–0.015; <0.01 mg; phenylalanine 0.114–0.157; 0.132–0.186; <0.01–0.020; 0.045 mg; lysine 0.242–

0.309; 0.257–0.328; <0.01–0.57; 0.062 mg; histidine 0.104–0.136; 0.126–0.140; 0.016–0.023; 0.046 mg; arginine 0.152–0.193; 0.172–0.260; 0.032–0.038; 0.386 mg; cystine 0.034–0.041; 0.035–0.051; <0.01; 0.195 mg; methionine 0.058–0.084; 0.052–0.099; 0.017–0.019; 0.068 mg and tryptophan 0.056–0.069; 0.068–0.077; <0.01–0.023; 0.644 mg.

Among vegetable plants, red beet had been found to contain a relatively high level of the B vitamin folic acid (Wang and Goldman 1997). Free folic acid content (FFAC) in shoot tissue was significantly greater than root tissue for both inbreds W384 and W357. FFAC accumulation in shoot tissue increased sharply from 60 to 80 DAP (days after planting) but decreased sharply from 80 to 100 DAP.

In Swiss chard leaves and stalks, mean content of vitamin C was 307 and 72 mg/kg of fresh matter (FM), potassium content 4198 and 4848 mg, sodium content 2101 and 966 mg, calcium content 481 and 310 mg and magnesium content was 361 and 113 mg/kg FM, respectively (Pokluda and Kuben 2002). Proximate nutrient composition of Swiss chard leaves comprised: moisture 91 %, protein 2.3 %, ether extract 0.4 %, crude fibre 0.9 %, carbohydrates 3.7 %, ash 1.7 %, minerals mg/100 g: Na 164 mg, K 505 mg, P 32 mg, Ca 85 mg, Fe 3.05 and Zn 2 mg (Bakry et al. 2014).

Besides amino acids: glycine, alanine, leucine, isoleucine, valine, glutamic acid, aspartic acid, threonine and traces of tyrosine and phenylalanine, the following organic acids were found in sugar beet diffusion juices: citric, oxalic, lactic, succinic, malic, mucic, glycolic, glutaric and pyrrolidone carboxylic acids (Stark et al. 1950). γ-Amino-butyric acid was isolated from beetroot (Westall 1950). Organic acids in sugar beets diffusion juice appeared in the following order: citric > oxalic > malic > glycolic (Owens et al. 1953). Pyrrolidone carboxylic and lactic acids also appeared in diffusion juices, the former arising from glutamine and the latter from lactic fermentation. Organic acids and inorganic anions found in beet sugar included sulfates, nitrates, chloride, acetic, lactic and formic acids (Magne et al. 1998). Root exudates of sugar beet were found to contain citramalic acid and salicylic acid (Khorassani et al. 2011). Both metabolites solubilized soil P and their exudation by roots was stimulated by P deficiency.

Two D1-D2-Cyt b559 complex forms called RCIIa and RCIIb with different pigment stoichiometry were isolated from *B. vulgaris* and characterized (Montoya et al. 1991; Montoya et al. 1993). The isolated D1-D2-Cyt b559 complex presented a pigment stoichiometry of six chlorophyll a, two β-carotene and one cytochrome b559 per two pheophytin a. Electronic absorption spectra of the RCIIb at 277 K showed significant differences compared to RCIIa, i.e. a strong decrease in the absorbance due to carotenoid and chlorophyll for the same amount of pheophytin. Calculation of the relative area under the gaussians together with pigment stoichiometry data suggested that the 680, 672 and 669–670 nm components contained, respectively, two chlorophylls, two pheophytins and four chlorophylls for the RCIIa, and two chlorophylls, two pheophytins and two chlorophylls for the RCIIb.

Proteins and Peptides

Studies by Leigh et al. (1979) found that in beetroot, most of the sucrose and much of the acid invertase were in the vacuoles. An inverse relationship between sucrose content and acid invertase activity was found. Both enzymatic betacyanine and betaxanthin decolourizing activity were found in subcellular tissue extracted from the beetroot (Shih and Wiley 1982). The optimum pH for betaxanthin degradation was pH 3.4, similar to that for betacyanin. Glucoamylase and α-amylase were found in callus and suspension cultures of sugar beets as well as in mature roots (Masuda et al. 1988). Both mature roots and callus contained α-amylase and glucoamylase in the soluble fraction. Dedifferentiation of leaves of young plants to callus was accompanied by induction of glucoamylase and repression of some α-amylases and the debranching enzyme. A 170-kDa polypeptide was identified associated with the vacuolar Na+/H+ antiport tonoplast vesicles isolated from sugar beet cell suspensions (Barkla and Blumwald 1991). Polyclonal antibodies against the 170 kDa polypeptide almost completely inhibited the anti-

port activity. A cytochrome oxidase II gene was identified and characterized in mitochondria of the sugar beet (Mann et al. 1991). A family of plasma membrane (PM) intrinsic proteins was purified and characterized from PM vesicles derived from storage tissue of *Beta vulgaris* (Qi et al. 1995). This PM intrinsic protein-enriched fraction also contained high levels of UDP-glucose:(1,3)-β-glucan (callose) synthase activity. Sequence analysis of tryptic fragments derived from polypeptides of 31 and 27 kD revealed significant homologies to plant major intrinsic proteins (MIPs) identified from cloned sequences. These MIPs included clone 7a from pea and RD28 from Arabidopsis, both of water-stress proteins; a tomato-ripening-associated membrane protein; and PIP 2b, a PM-bound water channel protein from Arabidopsis. MIPs, therefore, represented abundantly occurring components of PMs derived from beet storage tissue. A gene (Ch1) encoding a novel type of chitinase was isolated from *Beta vulgaris* (Berglund et al. 1995). The Ch1 protein was found to consist of an N-terminal hydrophobic prepeptide of 25 amino acids followed by a hevein-like domain of 22 amino acid residues, an unusually long proline-rich domain of 131 amino acid residues with 90 prolines, and finally a catalytic domain of 261 amino acid residues. Two novel antifungal proteins, AX1 and AX2, were isolated from sugar beet leaves infected with *Cercospora beticola*. AX1 (MW = 5078 D) and AX2 (MW = 5193 D) were N-terminally sequenced and identified as monomeric, basic proteins consisting of 46 amino acid residues, of which 8 are cysteines (Kragh et al. 1995). Based on primary sequence homology (24–46 % identity), they are related to the superfamily of gamma-thionins. High concentrations of AX proteins were detected extracellularly in cell walls and in globular bodies around necrotic lesions in sugar beet leaves infected with *C. beticola*, suggesting that AX proteins were involved in antifungal defence. Furthermore, AX proteins or serologically related proteins were detected in xylem, stomata, and stomatal cells as well as in sugar beet styles. Two novel, identical antifungal proteins, IWF1 and IWF2, were isolated from the intercellular washing fluid (IWF) of sugar beet leaves (Nielsen et al. 1996). The proteins were basic, monomeric proteins of 91 amino acid residues, 89 of which were identical. Based on primary sequence homology, including the presence of 8 conserved cysteine residues, IWF1 and IWF2 were related to the family of plant non-specific lipid transfer proteins (nsLTPs). Another potent antifungal peptide, designated IWF4, was identified from sugar beet leaves (Nielsen et al. 1997). The 30-amino-acid residue sequence of IWF4 was rich in cysteines and glycines and had a highly basic isoelectric point. IWF4 showed homology to the chitin-binding (hevein) domain of chitin-binding proteins, e.g. class I and IV chitinases. Notably, it bound chitin more strongly than the chitin-binding chitinases. IWF4 mRNA was expressed in the aerial parts of the plant only, with a constitutive expression in young and mature leaves and in young flowers. Another antifungal protein, designated IWF6, was isolated from sugar beet washing (Kristensen et al. 2000). The protein, IWF6, comprised 37 amino acids with 6 cysteines. The cDNA clone isolated encoded a precursor protein with an N-terminal putative signal sequence of 45 amino acids, followed by the mature protein of 37 amino acids. IWF6 showed less than 26 % identity to any previously described protein. β-glucosidases were detected in the leaves and roots of common beet, *Beta vulgaris*; both had been demonstrated to catalyze hydrolysis of native betacyanins (Zakharova and Petrova 2000). Catalase, a major primary antioxidant defence component that primarily catalyzes the decomposition of hydrogen peroxide to water was purified from chard (*Beta vulgaris* var. *cicla*) (Dinçer and Aydemir 2001). The molecular weight of the purified catalase and its subunit were determined to be 235 000 and 58 500 Da, indicating chard catalase to be a tetramer. Polyphenol oxidase (PPO) was extracted from beetroot, in both soluble and membrane fractions (Gandía-Herrero et al. 2004). The genetically transformed hairy root cultures of red beet were shown to produce very high levels of peroxidase (Rudrappa and Neelwarne 2005).

Studies by Shin et al. (2003) found that invertase played a crucial role in the control of metabolic fluxes, downstream sucrose partitioning and ultimately hairy root growth of red beet

cultures. Sugar beet, a non-fructan plant, was found to contain fructan 6-exohydrolases (6-FEHs) (levanases) (Van den Ende et al. 2003). Sugar beet 6-FEH was more related to cell wall invertases than to vacuolar invertases and had a low isoelectric point (pI), clearly different from typical high pI cell wall invertases. Several acid phosphatases were found in beetroot plasma membranes (Armienta-Aldana and González De La Vara 2004). BE27 and BE29 two forms of beetin, a virus-inducible type 1 ribosome-inactivating protein, were isolated from *B. vulgaris* leaves (Iglesias et al. 2008). Beetin was found to be formed only in adult plants but not in germ or young plants. Two full-length cDNAs encoding flavonoid-specific glucosyltransferases, UGT73A4 and UGT71F1, were isolated from a cDNA library of *B. vulgaris* cell suspension cultures (Isayenkova et al. 2006). Both enzymes exhibited a broad substrate specificity, but a distinct regioselectivity, glucosylating a variety of flavonols, flavones, flavanones and coumarins. UGT73A4 showed a preference for the 4′- and 7-OH position in the flavonoids, whereas UGT71F1 preferentially glucosylated the 3- or the 7-OH position. Glucosylation of betanidin, the aglycone of the major betacyanin, betanin, in *B. vulgaris* was also observed to a low extent by both enzymes. Several *O*-glycosylated vitexin derivatives isolated from leaves of young *B. vulgaris* plants and rutin obtained from *B. vulgaris* tissue culture were discussed as potential endogenous products of UGT73A4 and UGT71F1.

Ceballos-Laita et al. (2015) found that proteins identified in the non-redundant proteome of sugar beet leaf apoplastic fluid could be broadly classified into those associated with protein metabolism: proteasome subunit alpha type, proteasome subunit alpha type-5, proteasome subunit alpha type-6-like isoform X1, proteasome subunit beta type-6, proteasome subunit alpha type-7, cystein proteinase RD19a/like, cysteine protease, aspartic protease, serine carboxypeptidase, serine carboxypeptidase-like 20-like, unknown protein with peptidase domain, subtilisin-like protease-like, chaperonin 20, peptidyl-prolyl *cis-trans* isomerize, rubisco subunit binding protein alpha subunit, heat shock 70- protein, predicted heat shock cognate 70 kDa protein 2-like, elongation factor Tu; carbon metabolism: enolase, triosephosphate isomerase, fructose-bisphosphate aldolase, 2,3bisphosphoglycerate-independent phosphoglycerate mutase, phosphoglucomutase, glyceraldehydes-3-phosphate dehydrogenase, ribose-5-phosphate isomerize, ribulose-phosphate 3-epimerase, transketolase, phosphoglycerate kinase, sedoheptulose-1,7-bisphosphatase, phosphoribulokinase, carboxylase/oxygenase, 23 KDa OEC protein, malate dehydrogenase (cytoplasmic), carbonic anhydrase, oxaloacetase (*Dianthus caryophyllus*); stress and defence: osmotin-like protein, thaumatin-like protein, abscisic acid stress ripening-related protein, protein IN2-1-homolog B-like, uncharacterized protein with Bet-v-l-allergen domain, ascorbate peroxidase, monohydroascorbate reductase, peroxidase superfamily protein, peroxidase, peroxiredoxin, type II peroxiredoxin, Cu/Zn superoxide dismutase, lactoylglutathione lyase, lactoylglutathione lyase isoform X2, predicted isoflavone reductase homolog; polysaccharide metabolism: 3-glucanase family protein, acidic endochitinase SP2, acidic endochitinase SE2, chitinase, beta-xylosidase/alpha-L-arabinofuranosidase, UDP-glucuronic acid decarboxylase 1, beta-fructofuranosidase, uncharacterized protein with hydrolase domain, unknown protein with hydrolase domain; amino acid metabolism: serine hydroxymethyltransferase, serine transhydroxymethyltransferase, aminomethyltransferase, glutamine synthetase, aspartate aminotransferase; lipid metabolism: 3-hydroxybutyryl-CoA dehydratase, uncharacterized protein with lipase domain, uncharacterized protein with lipase domain; others: ferredoxin-NADP reductase, alcohol dehydrogenase, flavoprotein WrbA-like, acylpyruvase FAHD1, cytosolic ATP sulfurylase, nucleoside diphosphate kinase 2, thiamine thiazole synthase, uncharacterized germin protein, soluble inorganic pyrophosphatase 1; and unknown function: putative protein (*Hordeum vulgare*), uncharacterized protein (*Vitis vinifera*), uncharacterized protein (*Jatropha curcas*), uncharacterized protein (*Jatropha curcas)*, jasmonate-induced protein (*Atriplex canescens*), CSP41A protein, hypothetical protein CICLE-v10029208mg (*Citrus clementin*a) and

no blast result. Functional classification of the non-redundant proteins indicated that stress and defense, protein metabolism, cell wall and carbon metabolism accounted for approximately 75 % of the identified proteome. The identification of three chitinase isoforms among proteins increasing in relative abundance with Fe-deficiency suggested that one of the few effects of Fe-deficiency in the leaf apoplast proteome included cell wall modifications.

PIP (plasma membrane intrinsic proteins) aquaporins are important water transporters; they could interact to form functional heterooligomeric assemblies specially combining PIP2 monomers with PIP1 monomers (Jozefkowicz et al. 2013). They found that of *B. vulgaris* PIP aquaporins, BvPIP2;2 was able to interact with BvPIP1;1 but BvPIP2;1 showed no functional interaction. The lack of functional interaction between BvPIP2;1 and BvPIP1;1 was further corroborated by dose-response curves of water permeability due to aquaporin activity exposed to different acidic conditions. They also found that loop A was relevant for PIP1-PIP2 functional interaction since mutations in this loop modified the behaviour of *Bv*PIP2;1-*Bv*PIP1;1 co-expression.

Protein-associated polysaccharides extracted from sugar beet pulp were characterized (Fishman et al. 2013). For pectin, recovery ranged from 8 to 14 %, degree of methy-esterification 66–73 %, crude protein 1.3–1.7 %, M(w) 262–318 kDa, η(w) 0.22–0.23 dL/g, Rg(z) 36–39 nm and Rh(z) 41–42 nm. For alkaline soluble polysaccharides (ASP I), recovery ranged from 4.0 to 6.5 %, crude protein 1.2 to 4.8 %, weight average molar mass (M(w)) 66 to 68 kDa, weight average intrinsic viscosity (η(w)) 0.27 to 0.30 dL/g, z-average radius of gyration (Rg(z)) 25 to 29 nm and z-average hydrated radius (Rh(z)) 10 to 11 nm. ASP II recovery ranged from 2.0 to 8.6 %, crude protein 1.2 to 4.8 %, M(w) 299 to 339 kDa, η(w) 0.22 to 0.33 dL/g, Rg(z) 33 to 34 nm and Rh(z) 30 to 34 nm. Recovery of the residue mainly cellulose, ranged from 20.3 to 22.3 %. The cellulose in this fraction was converted to carboxymethyl cellulose (CMC). The CMC fraction contained 0.33–0.43 crude protein and had an M(w) ranging from 127 to 263 kDa, η(w) 3.6 to 8.0 dL/g, Rg(z) 35 to 45 nm and Rh(z) 27 to 40 n.

Pectins

Pectins sequentially extracted from sugar beet pulp with water (WSP), oxalate (OXP), hot acid (HP) and cold dilute alkali (OHP) had fairly low molecular weights, a high degree of acetylation and relatively high contents of neutral sugars, but there were clear differences between the four fractions (Rombouts and Thibault 1986a,b,c). The main neutral sugars in each pectin were arabinose and galactose, and rhamnose, fucose, xylose, mannose and glucose were also present. Polyphenols (1–2 %) and possibly proteins (3–6 %) were associated with the purified pectins. In addition, feruloyl groups (up to 0.6 %) were linked mainly to the acid-soluble and alkali-soluble pectins. After enzymatic degradation the pectins contained various amounts of degradation-resistant (hairy) fragments in which the molar ratios of neutral sugar residues to galacturonic acid residues were 4.8, 4.6, 3.8 and 3.7 for WSP, OXP, HP and OHP, respectively. The molar ratios of rhamnose residues to galacturonic acid residues in these fragments were 0.15, 0.20, 0.38 and 0.35, respectively. The pectins also contained sequences of galacturonic acid residues with relatively little neutral sugar residues attached (smooth fragments). Methyl ester and acetyl groups were distributed fairly regularly along the smooth fragments. Feruloyl groups located in the hairy fragments appeared to be linked to the neutral sugar side chains. Other phenolic compounds, associated with the purified pectins, appear not to be covalently linked.

In the pectin fraction from red beet, galactose, arabinose and glucose dominated; xylose, rhamnose and mannose were also present but in smaller amounts (Dongowski 1996). A higher galacturonic acid (GalA) content was found in the soluble fractions with the exception of the alkali extract. Pectins from red beet were moderately or lowly esterified and partially acetylated. The composition of the alcohol-insoluble substance (AIS) and of the residue after pectin extraction (RE) comprised 14.6 and 9.5 % pectin; 10.6 and 17.6 % protein; 58.7 and 64.9 % total polysaccharides respectively. In the AIS, 23.3 % soluble and 54.7 % insoluble dietary fibre were estimated, whereas in RE 15.3 and

54.7 % were found (enzymatic method). The following dietary fibre fractions were determined by the detergent method for both preparations: 39.0 and 52.7 % neutral detergent fibre (NDF); 6.3 and. 4.5 % NDF filtrate; 23.6 and 41.8 % acid detergent fibre (ADF); 1.2 and 1.8 % lignin. The water binding capacity decreased from 19.85 g water (AIS) to 11. 53 g water (RE). The cell walls of sugar beet and beetroot were similar in yield and neutral carbohydrate composition; the cell-wall galacturonic acid content of beetroot was 50 % higher as compared with sugar beet (Waldron et al. 1997). They were also rich in ferulic acid (FA) and its derivatives (6–7 mg/g CWM). In sugar beet, over 20 % of the FA was in dimer form. In beetroot, however, the value was only 10 %. The main FA dimers were 8-*O*-4′DiFA and 8,5′DiFA (benzofuran form). The results indicated that the degree of thermal stability of cell–cell adhesion and, therefore, texture in *B. vulgaris* tissues was related to the degree of FA-cross-linking between pectic polysaccharides.

Pectins were recovered from fresh sugar beet by acidic extraction of AIR (alcohol-insoluble residues) (Levigne et al. 2002). The amount of acid-extracted sugar beet pectins varied from 23 to 354 mg/g of initial material. The galacturonic acid content varied from 295 to 538 mg/g extract, arabinose 3 to 477 mg/g, rhamnose 11 to 54 mg/g, galactose 26 to 85 mg/g, degree of methylation (DM) 34 to- 94, and degree of acetylation (DA) 6 to 43. *Cis*- and *trans*-ferulic acid monomer contents varied from 2.6 to 16.7 mg/g, dehydroferulic acids 0.1 to 1.8 mg/g. Feruloyl dimers resulting from 8-5′, 5-5′ and 8-O-4′ radical couplings were detected. The 8–5′, and 8-O-4′ radical couplings remained the main dimers. Intrinsic viscosity varied from 172 to 493 mg/g; molar mass varied from 70.00 to 355.000 g/mol, the highest molar masses were observed at pH 2 and at 75 °C. Polydispersity index varied from 1.7 to 4.7. Mark-Houwink coefficient varied from 0.28 to 1.19. Side chains of sugar beet pectins, mainly composed of arabinose (Ara) and galactose (Gal) residues, were esterified by ferulic acid units (Levigne et al. 2004). Enzymatic hydrolysis of beet cell walls yielded several feruloylated oligosaccharides. Two new oligomers arabinotriose and an arabinotetraose esterified by two ferulic acid residues were obtained. It was shown that feruloyl groups were linked to *O*-5 of Ara residues, in addition to the known *O*-2 position. The two neighbouring Ara units may be esterified by two ferulic acid units. Homogalacturonan-derived partly methylated and/or acetylated oligogalacturonates were recovered after enzymatic hydrolysis (endo-polygalacturonase + pectin methyl esterase + side-chain degrading enzymes) of sugar beet pectin followed by anion-exchange and size exclusion chromatography (Ralet et al. 2005). Around 90 % of the GalA and 75 % of the acetyl groups present in the initial sugar beet pectin were recovered as homogalacturonan-derived oligogalacturonates, the remaining GalA and acetyl belonging to rhamnogalacturonic regions. 2-*O*-acetyl- and 3-*O*-acetyl-GalA were detected in roughly similar amounts but 2,3-di-*O*-acetylation was absent. Methyl-esterified GalA residues occurred mainly upstream 2-*O*-acetyl GalA. Oligogalacturonates containing GalA residues that were once methyl- and acetyl-esterified were recovered in very limited amounts.

Fishman et al. (2008) demonstrated that microwave-assisted extraction of sugar beet pulp under moderate pressure and at relatively low temperature could extract under acid conditions high molar mass, moderate-viscosity pectin in minutes rather than hours as required by conventional heating. Trypsin digestion of commercial sugar beet pectin afforded extension protein and corresponded to previously reported extensin peptides found in sugar beet cell suspension cultures (Nuñez et al. 2009). Pulsed electric field-assisted modification of pectin from sugar beet pulp increased degree of esterification (DE) ranging from 49 to 84 (Ma and Wang 2013). Thermogravimetric investigation of modified pectin indicated a higher thermal stability than the untreated one. The optimal temperature, time and pH value of sugar beet pulp pectin (SBPP) extraction were 93.7 °C, 3 h and 1.21, respectively (Lv et al. 2013). The yield, yield stress and tint value of the SBPP extracted at the optimal condition were 24.45 %, above 0.1 Pa and -6.0, respectively. Ma et al. (2013) demonstrated that

the yields of pectins were directly correlated with the decrease of pH and reaction time, and the optimum yield of 17.2 % was obtained at pH 1.5 and 2 h. Furthermore, the acid type (citric acid, malic acid and lactic acid) also affected the physicochemical characteristics of pectin, especially on the esterification degree (42–71 °C), galacturonic acid content (60.2–77.8 %), emulsion activity (35.2–40.1 %) and emulsion stability (62.1–79.4 %), and a relatively single pectin mainly consisting of homogalacturonan could be obtained under a suitable reaction condition, which was an excellent crude material for the production of emulsion activity.

Galactose was found to be the most abundant neutral sugar, followed by arabinose and rhamnose in sugar beet pectin (SBP) (Combo et al. 2013). Other neutral sugars, such as xylose, mannose and glucose were present, but in low amounts. Galacturonic acid constituted 59.5 % of SBP, degree of methylation and degree of acetylation was 44 % and 35.8 %, respectively, and molecular weight was 63 kDa. During enzymatic hydrolysis of sugar beet pectin by combining endopolygalacturonase and pectin methylesterase, three pectic oligosaccharides (POS) D-galacturonic acid monohydrate, digalacturonic and trigalacturonic were obtained. Also, during hydrolysis higher degree of polymerisation (DP 6-9) were hydrolyzed into smallers DP.

Pectin-enriched products extracted from red beet wastes were found to have potential thickening properties for usage in food formulation, as well as with some healthy physiological effects (Fissore et al. 2011). Red beet tissue contained ferulic acid, which cross-links pectin macromolecules through arabinose residues to anchor them into the cell wall.

Sugar Beet Pulp

Sugar beet pulp is a by-product of sugar production and consists mainly of cellulose, hemicellulose and pectin (Hutnan et al. 2000). Sugar beet pulp had been reported to be especially rich in pectins (15–30 %), containing high amounts of galacturonic acid (21.1 %), arabinose (20.9 %) and glucose (21.1 %). Other components include rhamnose 2.4 %, fucose 0.2 %, xylose 1.7 %, mannose 1.1 %, galactose 5.1 %, methanol 1.8 %, acetic acid 3.9 %, ferulic acid 0.8 %, diferulic acid 0.8 %, protein 11.4 % and ash 3.6 % (Micard et al. 1996). Beet pectins like other pectins, possess a backbone of α-(1→4)-linked galacturonic acid residues forming long, smooth regions which may be interrupted by 'hairy' region consisting of galacturonic acid and (1→2)-linked rhamnose (Micard et al. 1996). Some rhamnose residues carry side chains consisting mainly of (1→5)-linked arabinans with branches attached to position 3. Other structural features may include (1→4)-linked β-galactans of low polymerisation degree and highly branched (1→3,6)-linked galactans. Sugar-beet pulp was found to contain dehydrodimers of ferulic acid, derived from 8-5′, 5-5′, 8-8′ and 8-*O*-4′ coupling, the 8-5′ form being preponderant; no 4-*O*-5′ dimer was detected (Micard et al. 1997). Total dehydrodimers represented 0.14 % (w/w) of the pulp. The lignocellulosic fraction of dried sugar beet pulp was reported to contain 22–30 % cellulose, 24–32 % hemicelluloses (especially arabinan), 24–32 % pectin substances and 3–4 % lignin (Spagnuolo et al. 1997). Enzymatic hydrolysis by cellulase, hemicellulase and pectinase was as effective in degrading the beet pulp as the acid hydrolysis. Total hydrolysis of beet pulp afforded monosaccharides D-glucose from cellulose, L-arabinose from hemicellulose and D-galacturonic acid from pectin. Pectinase appeared to be the most important enzyme, since by hydrolysing the pectic surface of the lignocellulosic substrate it favoured the degradation of cellulose and hemicellulose by the respective enzymes. Enzymatic depolymerisation of the solid residue in sugar beet pulp (SBP) resulted in 30.6 % hemicellulose, 25.95 % cellulose, 12.1 % lignin, 0.83 % pectin and 4.86 % protein (Olmos and Hansen 2012). The total carbohydrates and reduced sugars in the hydrolysate were 143.5 and 5.24 g/L, respectively. The extracted pectin infrared spectrum revealed that it corresponded to low methoxyl pectin. It was noted that enzymatic treatment (with commercial pectinases) of SBP can be used in the recovery of valuable by-

products, such as pectin oligosaccharides and solid fractions rich in hemicellulose. The pectic extract from enzymatic hydrolysis of sugar beet pulp showed a degree of polymerisation (DP) profile of 55.8 % with DP $\geqslant 7$; 4.9 % with DP6; 5.8 % between DP2 and DP6; 4.7 % with DP2; and 28.8 % with DP1 (Concha et al. 2013).

Weiland (1993) reported that sugar beet pulp could be used for biogas production by anaerobic treatment and contained 150–180 g/kg total solid, 180–220 g/L total COD (chemical oxygen demand), 90–95 % total volatile solids, 40–41 % total carbon, 1–1.2 % total Kjeldahl nitrogen, 5–40 kg/kg C/N 3 and pH 3.9–4.0.

Betalains

Two betaxanthins (vulgaxanthin-I and vulgaxanthin-II) were isolated from beetroot (*Beta vulgaris*) (Piattelli et al. 1965). The yellow pigment, vulgaxanthin 1, was isolated and purified from yellow beets (*Beta vulgaris var. lutea*) (Singer and von Elbe 1980). Vulgaxanthin I was more stable at pH 5.0 than 3.0 or 7.0, and less stable in purified form than in beet juice and was quite heat labile in an oxygen-free system. An orange-coloured water-soluble pigment identified as 5-*O*-*β*-D-glucopyranosylneobetanidin (neobetanin) was isolated from red beet (*B. vulgaris*) roots (Alard et al. 1985). Cyclodopa glucoside (=(2*S*)-5-(*β*-D-glocopyranosyloxy)-6-hydroxyindoline-2-carboxylic acid) was found to occur in red beet (*Beta vulgaris* var. *rubra*) (Wyler et al. 1984). Cyclodopa glucoside occurred in considerable amount in red beet juice varying from 0.07 to 1.9 mmol/kg or from 12 to 46 % relative to the content of betanin. The occurrence of free cyclodopa glucoside supported its role as intermediate in betanin biosynthesis. Betanin and feruloylbetanin (lamparanthin II) were identified in *B. vulgaris* cell cultures (Bokern et al. 1991).

Betaxanthins from hairy root cultures of *B. vulgaris* var. *lutea* showed two predominating substances, tyrosine-betaxanthin (portulaxanthin II) and glutamine-betaxanthin (vulgaxanthin I) ((Hempel and Böhm 1997). The presence of histidine betaxanthin (muscaaurin VII), proline-betaxanthin (indicaxanthin), dopa-betaxanthin (dopaxanthin), glutamic acid-betaxanthin (vulgaxanthin II), asparagine-betaxanthin (vulgaxanthin III) and leucine-betaxanthin (vulgaxanthin IV) were also found.

Administration of nine L-amino acids to individual hairy root strains led to an increase in concentration or the appearance of one betaxanthin in each case. Also, the d-isomers of amino acids were incorporated into corresponding betaxanthins not only by hairy root cultures but also by seedlings of *B. vulgaris* var. *lutea*. The betacyanin content increased with increasing cell growth during the log phase of the cell suspension culture of table beet (*Beta vulgaris*) (Akita et al. 2000). Reducing the total nitrogen concentration (30 mM) and modifying the ratio of ammonium to nitrate (1:14) resulted in an increased betacyanin content. Supplementation of Fe^{2+} to the Linsmaier-Skoog (LS) medium also promoted betacyanin production. The maximal betacyanin yield was achieved with a 2 mM Fe^{2+} concentration. Combining these conditions, we established a revised LS medium to improve betacyanin productivity (250 mg/l for a 14-day culture). The revised Linsmaier-Skoog medium enabled the maximum betacyanin yield of 550 mg/l to be obtained from a 14-day suspension culture of table beet (Akita et al. 2002). This medium promoted the betacyanin production in three types of cell lines differing in the betacyanin productivity. A protein fraction with peroxidase activity against guaiacol from *B. vulgaris* roots oxidized both betanidin and betanin (betanidin 5-*O*-*β*-d-glucoside), the former being the more efficient substrate for the enzyme (Martinez-Parra and Munoz 2001). The protein fraction contained three strongly basic peroxidase isoenzymes. Betanidin quinone was formed as the only product in the course of enzymatic betanidin oxidation, whereas betalamic acid and several oxidized *cyclo*-DOPA 5-*O*-*β*-d-glucoside polymers were generated during the oxidation of betanin.

Stintzing et al. (2002) found the presence of seven further betaxanthins, namely serine-betaxanthin, *γ*-aminobutyric acid-betaxanthin, dopamine-betaxanthin (miraxanthin V), valine-betaxanthin, isoleucine-betaxanthin,

phenylalanine-betaxanthin, and tryptophan-betaxanthin in yellow beetroot cultivar 'BejoZaden' but could not confirmed the earlier reported occurrence of asparagine-betaxanthin (vulgaxanthin III), histidine-betaxanthin (muscaaurin VII), glutamic acid-betaxanthin (vulgaxanthin II), dopa-betaxanthin (dopaxanthin), and tyrosine-betaxanthin (portulacaxanthin II). Twenty-five betaxanthins were found in yellow Swiss chard petioles: histidine betaxanthin (muscaaurin VII), asparagine-betaxanthin (vulgaxanthin III), serine-betaxanthin, histamine-betaxanthin, glutamine betaxanthin (vulgaxanthin I), aspartic acid-betaxanthin (miraxanthin II), glycine-betaxanthin (portulacaxanthin III), ethanolamine-betaxanthin, lysine-betaxanthin, threonine-betaxanthin, glutamic acid-betaxanthin (vulgaxanthin II), alanine-betaxanthin, γ-aminobutyric-betaxanthin, proline-betaxanthin (indicaxanthin), dopa-betaxanthin (dopaxanthin), tyrosine-betaxanthin (portulacaxanthin II), dopamine-betaxanthin (miraxanthin V), methionine-betaxanthin, valine-betaxanthin, tyramine-betaxanthin, 3-methoxytyramine betaxanthin, isoleucine-betaxanthin, leucine-betaxanthin (vulgaxanthin IV), phenylalanine-betaxanthin and tryptophan-betaxanthin (Kugler et al. 2007). All except histamine betaxanthin were found in red beetroot and yellow beetroot. Red beetroots displayed a much higher betaxanthin content of 720.9 mg/kg fresh weight as compared to their yellow counterparts again demonstrating the underestimated importance of betaxanthins in red beetroots. Strikingly, an identical set of betaxanthins was found in both yellow and red beetroots, thus underlining their close phylogenetic relationship. In both beet varieties, glutamine-betaxanthin (vulgaxanthin I) was found to be the distinctly dominating betaxanthin amounting to 275.0 and 135.9 mg/kg fresh weight, followed by dopamine-betaxanthin (miraxanthin V) with 77.6 and 22.1 mg/kg fresh weight in red and yellow beetroot, respectively. Earlier, Kugler et al. (2004) found 19 betaxanthins in yellow Swiss chard petioles from the cultivar 'Bright Lights': histidine betaxanthin (muscaaurin VII), asparagine-betaxanthin (vulgaxanthin III), serine-betaxanthin, histamine-betaxanthin, glutamine betaxanthin (vulgaxanthin I), glycine-betaxanthin (portulacaxanthin III), glutamic acid-betaxanthin (vulgaxanthin II), alanine-betaxanthin, γ-aminobutyric-betaxanthin, proline-betaxanthin (indicaxanthin), tyrosine-betaxanthin (portulacaxanthin II), dopamine-betaxanthin (miraxanthin V), valine-betaxanthin, tyramine-betxanthin, 3-methoxytyramine betaxanthin, isoleucin-betaxanthin, leucine-betaxanthin (vulgaxanthin IV), phenylalanine-betaxanthin and tryptophan-betaxanthin. Glutamine-betaxanthin (vulgaxanthin I) and dopamine-betaxanthin (miraxanthin V) clearly dominated over all other betaxanthins amounting to 41.6 and 33.5 mg/kg fresh weight, respectively. The betacyanin pattern of Swiss chard purple petioles was composed of betanin, isobetanin, betanidin, and isobetanidin. Phyllocactin was present in only trace amounts, further acylated structures such as betanidin – monoferuloyl-5-*O*-β-diglucoside and lampranthin II, accompanied by their corresponding C_{15}-epimers were identified (Kugler et al. 2007).

Beta vulgaris hairy root culture grown in submerged culture yielded 42 mg betalains (16 mg betacyanins and 26 betaxanthins) and 1.5 g dry biomass in 40 ml medium (Pavlov et al. 2003). Betalains found in hairy root cultures of *B. vulgaris* subsp *vulgaris* (Garden beet group) beatinin, batalamic acid, isobetanin, miraxanthin V, phyllocactin, 2-descarboxy-betanin, unknown betacyanin,2-descarboxy-betanidin,6′*O*-malonyl-2-descarboxy-betanin and dopamine-derived betacyanin (Kobayashi et al. 2001). The betalains found in beetroot (*B. vulgaris*) peel extract were vulgaxanthin I, vulgaxanthin II, indicaxanthin, betanin, prebetanin, isobetanin and neobetanin (Kujala et al. 2001b). Also cyclodopa glucoside, N-formylcyclodopa glucoside, glucoside of dihydroxyindole-carboxylic acid, betalamic acid, L-tryptophan, *p*-coumaric acid, ferulic acid and traces of unidentified flavonoids were detected. The following betalains were indentified in four red beetroot cultivars: betaxanthins: vulgaxanthinI, vulgaxanthin II; and betacyanins: betanin, isobetain (Kujala et al. 2002).

Red beetroot dried extract was found to contain the betacyanins and their derivatives and betaxanthin pigments (Nemzer et al. 2011). Betacyanins

and their derivatives detected included prebetanin, 2′-*O*-glucosyl-betanin; betanin; isoprebetanin; 17-decarboxy-betanin; 2′-*O*-glucosyl-isobetanin; isobetanin;, 17-decarboxy-isobetanin; 15-decarboxy-betanin; 2-decarboxy-betanin; 2-decarboxy-isobetanin; betanidin; 17-decarboxy-betanin; 2,17-bidecarboxy-betanidin/isobetanidin; 17-decarboxy-neobetanin; isobetanidin; 17-decarboxy-isoneobetanin; 2,15,17-tridecarboxy-2,3-de hydro-neobetanin; neobetanin; 15-decarboxy-betanidin; 2-decarboxy-betanidin/isobetanidin; 2,17-bidecarboxy-2,3-dehydro-neobetanin; 2,17-bidecarboxy-neobetanin; 2-decarboxy-neobetanin; 6′-*O*-feruloyl-2′-*O*-glucosyl-betanin; 6′-*O*-feruloyl-2′-*O*-glucosyl-isobetanin; 2-decarboxy-2,3-dehydro-neobetanin; 6′-*O*-feruloyl-betanin/isobetanin; and betaxanthins: glutamine-betaxanthin (vulgaxanthin I); glutamic acid-betaxanthin (vulgaxanthin II); γ-aminobutyric acid- betaxanthin; proline-isobetaxanthin (isoindicaxanthin); proline-betaxanthin (indicaxanthin); dopamine-betaxanthin (miraxanthin V); tyrosine isobetaxanthin (isoportulacaxanthin II); tyrosine-betaxanthin (portulacaxanthin II); valine-isobetaxanthin; valine-betaxanthin; isoleucine-isobetaxanthin; isoleucine-betaxanthin; leucine-isobetaxanthin (isovulgaxanthin IV), leucine-betaxanthin (vulgaxanthin IV), phenylalanine-betaxanthin and tryptophan-betaxanthin. The principal betanin/isobetanin accounted for 41 % of total betalain content. Four predominant betalains, two betacyanins (betanin and isobetanin) and two betaxanthins (vulgaxanthin I and miraxanthin V) were isolated and quantified in nine beetroot cultivars produced in the greenhouse or field (Lee et al. 2014). Betanin and vulgaxanthin I were the major compounds in red and yellow beetroot extracts, respectively, and they comprised >90 % of the betalain content in the tested cultivars. The total betalain content of beetroots produced from the field was between 650 and 800 μg/g fresh weight, approximately 25 % higher than those from the greenhouse. The betaine content of the beetroot grown in the field was between 3.0 and 4.8 mg/g fresh weight, approximately 20 % higher than in plants from the greenhouse. There was great variation among the cultivars with respect to their contents of betalains and betaine.

Two mixtures of decarboxylated and dehydrogenated betacyanins from processed red beetroot (*B. vulgaris*) juice were fractionated by high-performance counter-current chromatography (HPCCC) producing a range of isolated components (Spórna-Kucab et al. 2013). Mixture 1 contained mainly betacyanins, 14,15-dehydro-betanin (neobetanin) and their decarboxylated derivatives, while mixture 2 consisted of decarboxy- and dehydro-betacyanins. 2-decarboxy-betanin/-isobetanin, 2,17-bidecarboxy-betanin/-isobetanin and neobetanin were purified from mixture 1 and 17-decarboxy-neobetanin, 2,15,17-tridecarboxy-2,3-dehydro-neobetanin, 2-decarboxy-neobetanin and 2,15,17-tridecarboxy-neobetanin from mixture 2. A mixture of betalains: betanin/isobetanin, decarboxybetanins (17-decarboxy-betanin/-isobetanin, 2-decarboxy-betanin/-isobetanin, 2,17-bidecarboxy-betanin/-isobetanin pairs) and neobetanin from processed red beetroot (*B. vulgaris*) juice was separated in food-grade, gradient solvent systems using HPCCC (Spórna-Kucab et al. 2015).

The degradation of both red beet pigments (betanine and vulgaxanthin–I) followed first-order reaction kinetics (Saguy 1979). The energies of activation were 19.2 and 16.3 Kcal/mol for betanine and vulgaxanthin-I, respectively, and were independent of pH. Maximum thermostability was observed at pH 5.8 for both pigments. In beet juice under atmospheric conditions vulgaxanthin–I was more sensitive than betanine. Betacyanins were less stable at pH 3 than at pH 5 and were highly light sensitive (Sapers and Hornstein 1979). Differences in betacyanin stability at 25 °C between 48 beet cultivars were significant at the 0.10 level, but not at 0.05. Betaxanthins were similarly affected by pH and light; betaxanthins were much less stable than betacyanins at 25 °C. Betanine was most stable at an intermediate pH range (4.0–5.0) (Huang and Elbe 1987). The influence of pH on two betanine degradation products showed that betalamic acid was most stable as the pH of the reaction increased, while cyclodopa-5–*O*-glucoside was more stable as the pH decreased. The thermal degradation of red beet major pigment betanin to betalamic acid and the degradation of the latter to

cleavage produced was investigated by Saguy et al. (1978). The degradation of both pigments followed a first-order reaction. Betalamic acid was thermally more stable than betanin. Mixtures of mono-, bi- and tridecarboxylated betacyanins together with their corresponding neobetacyanins were obtained from *Beta vulgaris* root juice as heating degradation products of betacyanins (Wybraniec 2005). Two monodecarboxy-betacyanin pairs of diastereomers were detected after the decarboxylation in ethanolic and aqueous solutions. Generation of 17-decarboxy-betacyanins and 2-decarboxy-betacyanins was suggested to be attributed to betacyanin thermal degradation products. Other main products of decarboxylation were 2,17-bidecarboxybetanin, its isoform, and 14,15-dehydrogenated (neobetacyanin) derivatives of all the decarboxylated betacyanins. Studies showed that addition of sucrose to the medium of transformed sugar beet cells augmented betanin yield more strongly than calcium and yeast extract used as abiotic and biotic elicitors (Križnik and Pavoković 2010). Yeast extract could be used for reverse sequestration of betanin where the cells can be used over an extended period.

Phenolic Compounds

Two phytoalexins were isolated from sugar beet leaves infected with *Cercospora beticola* and their structures were shown to be 2′,5-dimethoxy-6,7-methylenedioxyflavanone and 2′-hydroxy-5-methoxy-6,7-methylenedioxyisoflavone (Geigert et al. 1973). Leaves of sugar beet (*Beta vulgaris*) infected with *Cercospora beticola* yielded two compounds, the flavanone betagarin (5,2′-dimethoxy-6,7-methylenedioxyflavanone) and the isoflavone betavulgarin (2′-hydroxy-5-methoxy-6,7-methylenedioxyisoflavone) (Johnson et al. 1976; Martin 1977; Martin 1989). Two phenolic amides *N*-(4-hydroxy-3-methoxyphenethyl) ferulamide (*N*-trans-feroylhomovanillylamine) and *N*-trans-feruloyltyramine were isolated from the seed balls (Chiji et al. 1984). 5,2′-Dihydroxy-6,7-methylenedioxyisoflavone was identified from the seed balls of sugar beet (Chiji et al. 1986). A neolignan, 6-oxo-2-(4-hydroxy-3,5-dimethoxyphenyl)-3,7-dioxabicyclo-[3.3.0]-octane, and indole-3-carboxylic acid were also isolated. New flavonoid compounds, 3,5-dihydroxy-6,7-methylenedioxyflavanone, 2′,5-dihydroxy-6,7-methylenedioxyisoflavone, and 5-hydroxy-6,7-methylenedioxyflavone were isolated from sugar beet roots infected with *Rhizoctonia solani* (Takahashi et al. 1987). Sugar beet roots produced new *β*-xyloside and *β*-glucoside of 2′-hydroxy-6,7-methylenedioxy-5- methoxyisoflavone (betavulgarin) in response to infection by *Rhizoctonia solani* (Elliger and Halloin 1994). Also formed were 6,7-methylenedioxy-5-methoxyflavone, 6,7-methylenedioxy-5-methoxydihydroflavonol and 3′-methoxy-4′,5,7-trihydroxy-dihydroflavonol, not reported previously in *Beta vulgaris*. Two major flavins that accumulated in roots of iron-deficient sugar beet were identified as riboflavin 3′-sulfate and riboflavin 5′-sulfate, accounting for 82 and 15 % of the total flavin concentration in deficient roots, respectively (Susín et al. 1993).

Waldron et al. (1997) found ferulic acid, ferulic acid dehydrodimers (8-*O*-4′DiFA and 8,5′DiFA (benzofuran form)), *p*-hydroxybenzoic acid, vanillic acid and *trans*-*p*-coumaric acid in cell walls of beetroot. Bokern et al. (1991) isolated five ferulic acid conjugates: 1-*O*-(*E*)-feruloyl-*β*-glucopyranoside; *β*-D-fructofuranosyl-*α*-(6-*O*-[*E*]-feruloylglucopyranoside) ; 6-*O*-(*E*)-feruloyl-*β*-glucopyranoside; 1-*O*-(*E*)-feruloyl-3-*α*-glucuronosylglycerol and *N*-(*E*)-feruloylaspartic acid from cell cultures of beetroot, when freeze-dried cells were extracted with 50 % aqueous methanol. Ferulic acid was reported in *Beta vulgaris* leaves (Harborne et al. 1999). Winter and Herrmann (1986) reported several phenolic acids from different parts of beetroot extracted with 80 % aqueous methanol (3′-caffeoylquinic acid, 3′-*p*-coumaroylquinic acid, 3′-feruloylquinic acid, 1-*O*-*p*-coumaroyl-*β*--D-glucose, 1-*O*-feruloyl-*β*-D-glucose, 1-*O*-sinapoyl-*β*-D-glucose). Also dihydrocaffeic acid had been detected in beetroot (Harborne et al. 1999).

Five flavonoids from Swiss chard leaves (*Beta vulgaris* subsp. *cycla*) cultivar 'green' were isolated and identified as vitexin 2″-xyloside, 6″-malonyl-2″-xylosyl vitexin, kaempferol 3-gentiobioside, isorhamnetin 3-gentiobioside,

and isorhamnetin 3-vicianoside (Gil et al. 1998). The flavonoid content of fresh leaves (green cultivar) was in the range 2.4–3.0 mg/g fresh weight (f.w.). Cultivar 'yellow' contained only flavone *C*-glycosides (2.1–2.3 mg/g f.w.), while the flavonols were not detected. Their vitamin C content was between 0.4 and 0.5 mg/g f.w. and, after domestic processing, an 80 % loss was observed. Modified atmosphere packaging (MAP) (7 % O2 and 10 % CO2) had no effect on total flavonoid content after 8 days of storage, although it increased flavonoid extraction during cooking in boiling water. In contrast, vitamin C content decreased, especially in MAP-stored Swiss chard, to reach levels below 50 % of the initial content after 8 days of cold storage. The total phenolic contents in various red beetroot parts were found to decrease in the order peel, crown, flesh (Kujala et al. 2000). Significant differences in the contents of total phenolics, main betacyanins (betanin and isobetanin), and the main known ferulic acid ester (*β*-D-fructofuranosyl-*α*-D-(6-*O*-(*E*)-feruloylglucopyranoside) were found when the effect of cold storage (5 °C, 0–196 days) on the constituents of the peel from intact roots was examined. In addition to the betacyanins of red beetroot peel found, tentative identifications of betanidin and feruloylamaranthin were made.

The following phenolics were reported in four beetroot cultivars: 5,5′,6,6′-tetrahydroxy-3,3′-biindolyl; feruloylglucose; *β*-D-fructofuranosyl-*α*-D-(6-*O*-(*E*)-feruloylglucopyranoside), two phenolic amides: *N-trans*-Feruloyltyramine; *N-trans*-Feruloylhomovanillylamine; four flavonoids: betagarin, betavulgarin, cochliophilin A, dihydroisorhamnetin (Kujala et al. 2002). Four phenolic compounds were isolated from *Beta vulgaris* var. *cicla* and identified as *N-cis*-feruloyl 3-*O*-methyldopamine; *N-cis*-feruloyl tyramine; *N-trans*-feruloyl 3-*O*-methyldopamine; *N-trans*-feruloyl tyramine (Kim et al. 2003). Two flavone glycosides vitexin 7-*O*-*β*-D-glucopyranoside and vitexin 2″-*O*-*β*-D-glucopyranoside were isolated as new constituents from the aerial parts of *Beta vulgaris* var. *cycla* (Kim et al. 2004).

Thirteen phenolics were detected in Swiss chard leaves, the major phenolic acid and flavonoid were syringic acid and kaempferol (Pyo et al. 2004). The phenolic fraction of Swiss chard (*B. vulgaris* subsp. *cycla*) extract contained vitexin-2-*O*-rhamnoside, its demethylated form 2″-*O*-xylosylvitexin, isorhamnetin 3-gentiobioside, and rutin (Ninfali et al. 2007). 2,4,5-trihydroxybenzaldehyde, 2,5-dihydroxybenzaldehyde, vanillic acid, xylosylvitexin, glucopyranosyl-glucopyrasyl-rhamnetin and glucopyranosyl-xylosyl-rhamnetin were purified and identified from ethyl acetate fraction of *B. vulgaris* seed extract (Gennari et al. 2011). Swiss chard leaves were found to have the following phenolic compounds: p-benzoic acid, chlorogenic acid, ferulic acid, catechol, syringic acid, caffeic acid, gallic acid, vanillic acid, protocatechuic acid, *p*-coumaric acid, catechin, myricetin, quercetin and kaempferol (Bakry et al. 2014). *Beta vulgaris* var. *cicla* was reported to contain a significant amount of phenolics, catechin hydrate, epicatechin, ferulic, protocatechuic, vanillic, *p*-coumaric, *p*-hydroxybenzoic, caffeic and syringic acids, flavonoids and proline (Nidhal et al. 2014). The ethyl acetate fraction of *Beta vulgaris* subsp. *perennis* afforded the isolation of quercetin, 4′-hydroxy-5-methoxy-6,7-methylenedioxy flavanone, quercetrin and rutin, phenolic acids: syringic, ferulic acids and the monoterpene dehydro-vomifoliol (Abdel-Monem et al. 2014).

Saponins

Eleven triterpene saponins were found in red beet, *Beta vulgaris* (Mroczek et al. 2012). All the saponins present in red beetroots were 3-*O*-glucuronides of oleanolic acid or hederagenin with different constituents of glucose or xylose. The structures of the saponins were elucidated as saponin 1 aglycone oleanolic acid with sugar chain of glucuronic acid (R1) and H (R2); saponin 2 aglycone oleanolic acid with sugar chain of xylose and glucuronic acid (R1) and H (R2); saponin 3 aglycone oleanolic acid with sugar chain of glucose and glucuronic acid (R1) and H (R2); saponin 4 aglycone hederagenin with sugar chain of glucose and glucuronic acid (R1) and H

(R2); saponin 5 aglycone oleanolic acid with sugar chain of xylose, glucose, glucuronic acid (R1) and H (R2); saponin 6 aglycone oleanolic acid with sugar chain of xylose, glucuronic acid (R1) and glucose(R2); saponin 7 unknown; saponin 8 aglycone oleanolic acid with sugar chain of glucose, glucuronic acid (R1) and glucose (R2); saponin 9 aglycone hederagenin with sugar chain of glucose and glucuronic acid (R1) and glucose (R2); saponin 10 aglycone oleanolic acid with sugar chain of glucose, glucuronic acid, xylose (R1) and glucose (R2); saponin 11 aglycone oleanolic acid with sugar chain of glucose, glucuronic acid, glucose (R1) and glucose (R2).

Extraction of sugar beet (*Beta vulgaris*) molasses afforded the known 3-*O*-6-D-glucuronopyranosyl-3*β*-hydroxy-olean-12-en-28-oic acid (1) and a novel compound which 3-*O*-(*β*-D-glucopyranosyl-(1 → 2) (*β*-D-xylopyranosyl-(1 → 3))*β*-D-glucuronopyranosyl-3*β*-hydroxyolean-12-en-28-oic acid (Ridout et al. 1994). These compounds were not present in the root of the fresh plant but were found in the leaves along with another new saponin: 3-*O*-[*β*-D-glucopyranosyl-(1 → 2)-(*β*-D-xylopyranosyl-(1 → 3))*β*-D-glucuronopyranosyl] -28-*O*-*β*-D-glucopyranosyl-3*β*-hydroxyolean-12-en-28-oic acid. Two new oleanolic acid saponins were isolated from the leaves and roots of *Beta vulgaris* (Massiot et al. 1994). Both contained the unusual feature of a 3,4-seco-glycopyranosyl moiety.

From *B. vulgaris* (sugar beet) roots the following glucuronide saponins named betavulgarosides I, II, III, IV, V, VI, VII, VIII (Yoshikawa et al. 1995, 1996; 1997; 1998; Murakami et al. 1999) and from the leaves betavulgarosides I, II, III, IV, V, IX and X (Yoshikawa et al. 1995, 1998) were isolated and their structures elucidated. Four acidic acetal-type substituent analogues were synthesized from L- and D-arabinose (Yoshikawa et al. 1997). *α*-L-arabinopyranosyl moiety of momordin I was converted to a dioxolane-type substituent of betavulgaroside II or to an acetal-type substituent of betavulgaroside IV (Murakami et al. 1999). Additionally, the 2′-diastereoisomer of betavulgaroside IV was synthesized from momordin I, and four acetal-type substituent analogues were also synthesized from L- and D-arabinose. Saponins were solated by means of alkaline extraction from table beetroot (Atamanova et al. 2005). Preliminary extraction stage eliminated the problem of removal of lipophilic components. With large beetroot pre-extraction, yield of saponins was 36 % and purity 72 %; without pre-extraction, yield was 42 % and purity 48 %. With small beetroots without pre-extraction, yield was 52 % and purity 76 %.

Subrin

The major classes of monomers from the depolymerisation of root suberin from red beet comprised 20 % soluble, 13 % phenolic and 3 % aliphatic fractions (Kolattukudy et al. 1975). The aliphatic monomers of suberin in red beet comprised 18 % fatty acids, 16 % dicarboxylic acids, 6 % fatty alcohol, 29 % *ω*-hydroxy acid and 3 % polar fraction. Among the fatty acids, very long chain acids ($>C_{20}$) were the dominant components. In the alcohol fraction C_{18}, C_{20}, C_{22} and C_{24}, saturated primary alcohols were the major components. C_{16} and C_{18} dicarboxylic acids were the major dicarboxylic acids of the suberin and in all cases octadec-9-ene-1, 18-dioic acid was the major component. Compounds which would be derived from 18-hydroxyoctadec-9-enoic acid and octadec-9-ene-1, 18-dioic acid by epoxidation followed by hydration of the epoxide, were also detected in most of the suberin samples. The amount of suberin-associated wax extracted from the periderm of red beet was 7.2 mg/kg of chloroform extractable material, equivalent to 34 % weight of chloroform extract or 4 µg/cm2 was surface area (Espelie et al. 1980). The composition of red beet wax comprised 20 % hydrocarbon, 1.2 % wax ester, 34 % fatty alcohol, 8 % fatty acids and 37 % unknown component. Chain length distribution of the fatty alcohols, fatty acids, *ω*-hydroxy acids and dicarboxylic acids in the polar fraction of the chloroform extract of red beet storage organ were 2.91 %, 1.93 %, 0.59 % and 0.91 %, respectively.

Miscellaneous Phytochemicals

A compound of unusual structure was isolated from red beetroot peel extract and identified as 5,5′,6,6′-tetrahydroxy-3,3′-biindolyl (Kujala et al. 2001a). Two norisoprenoids (+)-dehydrovomifoliol and 3-hydroxy-5α,6α-epoxy-β-ionone were isolated from the aerial parts of *Beta vulgaris* var. *cicla* (Kim et al. 2004).

A complexed poly(3-hydroxyalkanoate)-component derived from a complex with calcium polyphosphate was isolated from sugar beet (*Beta vulgaris*) and determined to be a homopolymer composed of 3-hydroxybutyrate with number-average molecular weight (Mn = 9,124 Da) and polydispersity index (PDI = 1.01) (Suzuki et al. 2005). Sugar beet leaves contained virus-inducible type 1 (single chain) ribosome-inactivating proteins (RIP) named beetins (Iglesias et al. 2005). Beetins comprised of a single polypeptide chain with a varying degree of glycosylation and strongly inhibited in-vitro protein synthesis in rabbit reticulocyte lysates (IC_{50} = 1.15 ng/ml). The external application of purified beetin to sugar beet leaves prevented infection by artichoke mottled crinkle virus (AMCV), which supported the hypothesis that beetins could be involved in plant systemic acquired resistance subjected to induction by phytopathogens.

Volatiles and Aroma Compounds

The characteristic earthy aroma of beets was attributed to its volatile component, geosmin (*trans*-l, l0-dimethyl-*trans*-(9)-decalol) (Murray et al. 1975; Acree et al. 1976). Beet peels were found to contain the highest concentration of geosmin (Tyler et al. 1978a). They also detected geosmin in beet juice (Tyler et al. 1978b). A sensory panel estimated threshold of 5.8 parts geosmin in 109 parts beet juice was 35 times higher than the odour threshold of geosmin in water (Tyler et al. 1979). An increase in geosmin content was perceived as an increase in beet-like aroma up to a 5.8 ng/g concentration of geosmin. Juice made from cooked and peeled beets showed a 56–60 % reduction in geosmin content in comparison with juice prepared from raw beets (with geosmin concentration below 5.8 ng/g). It was found that processing may reduce the characteristic aroma of beets. Of the primary volatile chemicals responsible for the characteristic malodour of beet sugar, only geosmin, 2,5-dimethylpyrazine, furfural, butyric acid, and isovaleric acid were deemed to be likely contributors to the characteristic odour defect (Marsili et al. 1994).

Freidig and Goldman (2014) reported that in the production of geosmin (2β,6α-dimethylbicyclo[4.4.0]decan-1β-ol) in table beet a significant interaction between cultivar and environment was found, but generalisations could be made for high- or low-producing cultivars, demonstrating that geosmin levels were cultivar-specific. 'Bull's Blood', 'Chioggia', and sugar beet exhibited the highest geosmin levels. Cultivars grown in the field had the smallest range of geosmin production, from 4.84 to 20.82 μg geosmin/kg root tissue. Earlier, Lu et al. (2003a) also reported geosmin concentration in beet to be cultivar-dependent; the geosmin concentrations in four beet cultivars ranged from 9.69 to 26.7 μg/kg. Lu et al. (2003b) confirmed the presence of geosmin in beet seedlings. The relative recovery of geosmin from beet seedling extracts was 72 %. The concentration of geosmin in beet seedlings remained constant for up to 5 months but increased at 6 months.

Seventeen volatile constituents of cooked beets (*Beta vulgaris*) were identified by Parliment et al. (1977). Major findings were that 4-methylpyridine and pyridine (both constituting about 60 %), dimethyl sulfide, isovaleraldehyde, and furfural appeared in great abundance. Geosmin and 2-methoxy-3-sec-butylpyrazine were found in lower amounts; however, with the knowledge that these two compounds exhibit low odour thresholds, it was hypothesized that these contribute heavily to the overall flavour impact of the beet. Beet peels were found to have the highest concentration of geosmin (Tyler et al. 1978a). Geosmin was extracted from beet juice with Freon 113 (1,1,2-trichloro-1,2,2-trifluoroethane), concentrated, and purified by adsorption on Florisil (Tyler et al. 1978b).

Approximately 10 compounds were found to be correlated with the defined sensory attributes

of liquid beet sugar (Pihlsgård et al. 1999). Among these compounds associated with the sensory attributes, dimethyl disulfide, 2,6-dimethylpyrazine, 4-methoxyphenol, and 2, 5-dimethylfuran. Butanone, heptane-2-one, 2,6-dimethylpyrazine, 2-ethyl-5-methylpyrazine, 3-ethyl-2,5-dimethylpyrazine, 2,5-dimethylfuran, dimethyldisulfide, and *p*-methoxyphenol were positively correlated to the analyzed sensory attributes for aroma and flavour in liquid beet sugar production (Pihlsgård 2000). It was shown that compounds giving sour and earthy aromas together with fruity aromas predominated in the early parts of the process and that caramel-like and burnt aromas increased in the process. Aldehydes, ketones, and alkylpyrazines were the groups of compounds that increased the most in the liquid sugar manufacturing process. In the final product, the amounts of alkylpyrazines, ketones, aldehydes, cyclotene and dimethyldisulfide were still high. Some compounds, such as small ketones, even exhibited higher concentrations in the final product than in refinery samples taken further upstream. It was, furthermore, shown that butanal, 2-hexanol and *p*-methoxyphenol were significantly less released from the liquid sugar matrix to the headspace when macromolecules of refinery origin were included in the matrix. The behaviour of different volatile compounds varied greatly throughout the production process of liquid beet sugar, with some compounds such as geosmin (*trans*-1,10-dimethyl-*trans*-9-decalol), dimethyl disulfide, and propionic and hexanoic acid present at the beginning of the process but disappearing rapidly after further processing (Pihlsgård et al. 2000). Other compounds, such as indole, dihydrobenzofuran and 2-phenylethanol, were not detected at the start of the process but were formed later on and removed in the final product. In the final product, three pyrazines remained at fairly low concentrations, together with 3-methylcyclopentadione, ethylhexanol, and methyl pyrrole ketone. In general, earthy and sour aromas were often present in the raw sugar beet juice sample, whereas caramel aromas were mainly present in the samples taken further downstream in the process ((Pihlsgård et al. 2001). For fruity, floral and solvent-like aromas, different parallel trends were noted. Some aromas were present only at the beginning of the process, whereas others developed towards the end of the process.

Volatile fatty acids found in white beet sugar and associated off-odours were: acetic acid (sour, vinegar), butanoic acid (butyric acid) (sour, rancid, cheesy), isovaleric acid (3-methyl butanoic acid) (rancid, cheesy, sweaty), pentanoic acid (valeric acid) (rancid, sweaty), hexanoic acid (n-caproic acid) (rancid, sweaty, cheesy, fatty), heptanoic acid (rancid, tallow, disagreeable), octanoic acid (caprylic acid) (slightly unpleasant, rancid, fruity-acid, oily) and nonanoic acid (pelargonic acid) (fatty, cheesy, waxy) (Moore et al. 2003).

Antioxidant Activity

Betalains (betacyanins and betaxanthins), the main pigments of red beet (*Beta vulgaris*) roots, showed an anti-radical effect when measured by the destruction of the 2,2′-azino-bis(3-ethylbenzthiazoline-6-sulphonic acid (ABTS) free radical generated by the horseradish peroxidase/hydrogen peroxide-mediated oxidation of ABTS (Escribano et al. 1998). The anti-radical activity of betacyanins was greater than that of the betaxanthins and increased with the pH of the reaction medium.

The ethanol extract, derived from hairy roots of *Beta vulgaris* cv. Detroit Dark Red culture grown for 15 days under submerged conditions, showed a high anti-radical activity (83 % of inhibition of the stable DPPH) (Pavlov et al. 2002). The hairy root culture of cultivar Detroit Dark Red was the best producer of betalains (13.27 mg/g dry weight total pigment production). Aqueous and ethanolic tissue extracts of the regular red, white, orange and high-pigment red beet phenotypes were most capable of inhibiting metmyoglobin/H(2)O(2)-mediated oxidation of 2-2′-azino-bis(3-ethylbenzthiazoline-6-sulfonic acid) (ABTS) and 2,2′-azobis-(2-amidinopropane) dihydrochloride (AAPH)-mediated bleaching of β-carotene (Wettasinghe et al. 2002). These same extracts were also most efficient at reducing ABTS radical cation and inducing quinone reductase in

murine hepatoma (Hepa 1c1c7) cells in-vitro. Significant differences in the antioxidant activity were found between leaves and stems of hydroponically grown Swiss chard (*Beta vulgaris* subspecies *cycla*) (Pyo et al. 2004). Phenolic content and composition of the leaves and the stems also were found to be different. A positive linear correlation ($R^2=0.943$) was demonstrated between radical scavenging activity and total phenolic content of each extract.

Jiratanan and Liu (2004) found that the antioxidant activity of beets processed under typical commercial processing conditions remained constant despite an 8 % loss of vitamin C, a 60 % loss of colour, and 30 % loss of dietary folate. There was a slight but significant 5 % increase in phenolic content of processed beets. At in-vitro conditions of gastrointestinal tract, betalains were found to be relatively stable, as their radical scavenging activity decreased from 75 % inhibition of 2,2-diphenyl-1-picrylhydrazyl (DPPH*) to about 38 % (Pavlov et al. 2005). It was established that pH below 3 and concentrations of the bile salts up to 4 % had no great influence on betalains stability. Crude aqueous and ethanolic extracts of root tissue of red (Rd) and high-pigment (HP) beet (*Beta vulgaris*) strains exhibited antioxidant and phase II enzyme-inducing activities (Lee et al. 2005). Active fractions were found to contain vulgaxanthins I and II, and isobetanin, but other components remained unidentified. No statistically significant effect of diet on phase II enzyme induction in various organs was obtained when two of the isolated active fractions were incorporated into rodent diets at 10–150 ppm over a 2-month period, and wide ranges of tissue enzyme levels among individual rodent animal were observed. This lack of effect and diversity in response to diet may be related to the wide range in absorptive capacity of and/or insufficient level or enrichment of the active agents or to difficulties in assessing such activity in-vivo. Subsequent to the animal studies, betanin was isolated in pure form, and confirmed to be quinone reductase (QR) inducers in the bioassay Studies by Sacan and Yanardag (2010) indicated that chard (*Beta vulgaris*) may provide a natural source of antioxidant and anti-acetylcholinesterase activities and proline content.

Georgiev et al. (2010) found that betalain extracts obtained from hairy root cultures of the red beetroot *B. vulgaris* cv. Detroit Dark Red had higher antioxidant activity than extracts obtained from mature beetroots: sixfold higher 2,2-dyphenyl-1-picrylhydrazyl radical scavenging ability (90.7 % inhibition, $EC_{50}=0.11$ mg, vs. 14.2 % inhibition, $EC_{50}=0.70$ mg) and 3.28-fold higher oxygen radical absorbance capacity (4,100 μM TE/g dry extract, vs. 1,250 μM TE/g dry extract). The high antioxidant activity of the hairy root extracts was associated with increased concentrations (more than 20-fold) of total phenolic concomitant compounds, which may have synergistic effects with betalains. The presence of 4-hydroxybenzoic acid, caffeic acid, catechin hydrate, and epicatechin were detected in both types of extract, but at different concentrations. Rutin was only present at high concentration (1.096 mg/g) dry extract) in betalain extracts from the hairy root cultures, whereas chlorogenic acid was only detected at measurable concentrations in extracts from intact beetroot plants.

B. vulgaris ethyl acetate extract exhibited relatively good DPPH antioxidant activity with SC_{50} (scavenging capacity) 8.5 μg/ml, compared with that of vitamin C (SC_{50} 1.24 μg/ml) (Abdel-Monem et al. 2014). *Beta vulgaris* root ethanol extract exhibited in-vitro DPPH radical scavenging activity from 4.34 to 18.55 % at concentration range of 200–1000 μg/ml (Chakole et al. 2011). The DPPH-free radical scavenging activity of ethanol, acetone and water extracts of beetroot pomace extracts, expressed as EC_{50}, ranged from 0.133 mg/ml to 0.275 mg/ml (Čanadanović-Brunet et al. 2011). Beetroot pomace was found to contain total phenolic contents of 316.30–564.50 mg GAE/g of dry extract, flavonoids 316.30–564.50 mg RE/g of dry extract, betacyanins 18.78–24.18 mg/g of dry extract, and betaxanthins 11.19–22.90 mg/g of dry extract. Significant correlation was observed between all phytochemical components and scavenging activity; 0.5 mg/ml of ethanol extract completely eliminated hydroxyl radical, which had been gen-

erated in Fenton system, while the same concentration of this extract scavenged 75 % of superoxide anion radicals. The reducing power of all the beetroot pomace extracts increased with increasing concentrations. Betanin, a beetroot betacyanin, inhibited the peroxynitrite (0.5 mM)-dependent nitration of tyrosine (0.1 mM) (Sakihama et al. 2012). Additionally, the IC_{50} value of betanin (19.2 μM) was lower than that of ascorbate (79.6 μM). The presence of betanin (0.05–1.0 mM) also inhibited peroxynitrite (0.5 mM)-dependent DNA strand cleavage in a concentration-dependent manner. The results suggested that betalains could protect cells from nitrosative stress in addition to protecting them from oxidative stresses. Beetroot pomace extract (BPE) exhibited antioxidant activity; the anti-radical activity on DPPH radical was (EC_{50}=0.0797 mg/ml), hydroxyl radical (EC_{50}=0.0655 mg/ml) and superoxide anion radical (EC_{50}=1.0625 mg/ml) (Vulić et al. 2013).

Alternative farming system, namely conventional (CON), integrated (INT), organic (ORG), biodynamic (BD), and control farming systems was found to influence the antioxidant activity, sugars, organic acids and total phenolic content of red beet (Bavec et al. 2010). Significant differences were measured for malic acid, total phenolic content (TPC) and total antioxidant activity, where malic acid content ranged from 2.39 g/kg FW (control) to 1.63 g/kg FW (CON, ORG, and INT). The highest TPC was measured in BD and control samples (0.677 and 0.672 mg GAE(gallic acid equivalent)/g, respectively), and the lowest in CON samples (0.511 mg GAE/ g). Antioxidant activity was positively correlated with TPC (R^2=0.6187) and ranged from 0.823 μM TE trolox equivalent)/g FW to 1.270 μM TE/g FW in CON and BD samples, respectively, whereas total sugar content ranged from 21.03 g/kg FW (CON) to 31.58 g/kg FW (BD).

Drying beetroot at high temperatures (100+90 °C/5.6 h; 90 °C/6 h) increased the antioxidant potential of the processed products while mild drying conditions decreased it (80 °C/6 h; 100+70 °C/6 h) or had no effect on it (70 °C/7 h; 100+80 °C/6 h) (Raupp et al. 2011). For the canned products, the antioxidant potential did not differ according to the pH (4.2 to 3.8) for any of the four acids tested. Some processing methods influenced the antioxidant potential of the processed products, and this was also dependent on changes in the total phenolics content.

Beetroot (*Beta vulgaris*) leaves showed significant levels of protein and lipids in all developmental stages, and all proximate composition nutrients decreased during maturation stages; the highest content was observed at 60 days (Biondo et al. 2014). The Fe content decreased during the developmental stages (from 342.75 to 246.30 mg/kg), while the content of K increased (from 13,367.64 to 20,784.90 mg/kg). With regard to fatty acid composition, linolenic acid was present in the greatest quantity, and it increase up to 2.58 mg/g (*in natura*) and 40.11 mg/g (dehydrated) at 100 days of development. The n-6/n-3 ratios were low in all stages. The TPC and antioxidant activity by DPPH$^{\bullet}$ changed during the developmental stages. The TPC was highest in the 100-day dehydrated leaves (15.27 mg GAE/g FW), and the 50 % inhibition of DPPH$^{\bullet}$ (IC_{50} 89.52 μg/mL) was better in the 60-day *in natura* leaves. The antioxidant activity and the chemical constituents, mainly the ω-3fatty acid, increased during the stages of development.

Hypoglycaemic Activity

The anti-diabetic activity of red beet had been reported to be attributed to its fibre, flavonoids, betalains (betacyanin, betaxanthin), pectin and other cell phytochemicals (Murthy and Manchali 2012). The dietary constituents and the bioactive molecules were reported to prevent diabetes mainly by increases in insulin production/sensitivity, increases in energy metabolism, alerting the activity of digestive/metabolic enzymes, activation of detoxification (phase II) enzymes, and decreasing LDL, cholesterol, and triglyceride levels. The antioxidant abilities of bioactive molecules in red beet were also reported to play a key role.

Betavulgarosides II, III, IV and the prosapogenol from *B. vulgaris* roots were found to

exhibit hypoglycaemic activity in an oral glucose tolerance test in rats (Yoshikawa et al. 1995, 1996, 1997). Betavulgarosides III and IV exhibited inhibitory activity on glucose absorption (Yoshikawa et al. 1997). Lipid peroxidation was increased and glutathione levels were decreased in both aorta and heart tissue of streptozotocin-induced diabetic rats, but treatment with chard (*Beta vulgaris* var. *cicla*) extract reversed the effects of diabetes on blood glucose and tissue lipid peroxidation and glutathione levels (Sener et al. 2002). They also reported that serum urea and creatinine levels significantly increased in the streptozotocin-induced diabetic groups, but the chard extracts significantly reduced serum urea and creatinine levels (Yanardağ et al. 2002). It was concluded that chard extract may reduce serum urea and creatinine levels and confer a protective effect on the kidney of diabetic rats. Administration of chard (*Beta vulgaris* var. *cicla*) extract inhibited diabetic increases in non-enzymatic glycosylation of skin proteins and blood glucose level, effects except the increase in lipid peroxidation (Tunali et al. 1998). The data indicated that the use of chard may be effective in preventing or at least retarding the development of some diabetic complications. In streptozotocin-induced diabetic animals, administration of chard extract, caused an increase in the number of B cells of Langerhans islets and in the secretory granules, together with many hypertrophic Golgi apparatus and granules of low densities (Bolkent et al. 2000). In contrast, in the diabetic animals a decrease in the number of B cells of Langerhans islets and in the secretory materials, a swollen granular endoplasmic reticulum cisternae and widened intercellular areas in some of B cells were observed. Chard extract, while having no effect on blood glucose and body weight in the normal group, reduced the blood glucose value in streptozotocin-induced hyperglycaemic animals. But, in a diabetic group given chard, the body weight significantly increased in comparison to the diabetic group; maximum reduction in blood glucose levels was observed on the 42nd day. According to the morphological and biochemical results obtained, it was concluded that the extract of this plant when administered by gavage may reduce blood glucose levels by regeneration of pancreatic B cells. Administration of chard extract to diabetic rats decreased blood glucose levels, liver lipid peroxidation (LPO), non-enzymatic glycosylation (NEG) levels, serum alanine, aspartate transaminase, alkaline phosphatase activities, total lipid level, sialic and uric acid levels, and increased blood glutathione, body weight, and liver glutathione (GSH) levels (Ozsoy-Sacan et al. 2004). The results suggested that chard extract had a protective effect on the liver in diabetes mellitus.

The aqueous and ethyl acetate extracts of *B. vulgaris* and glibenclamide reduced blood glucose levels of the diabetic rats significantly as compared to the untreated diabetic group (Abdel-Monem et al. 2014). The ethyl acetate extract (400 mg/kg) had nearly the same potency as glibenclamide (5 mg/kg), but the aqueous extract (400 mg/kg) had higher potency than glibenclamide. Administration of chard aqueous extract to streptozotocin-induced diabetic rats elicited a significant reduction in the fasting blood glucose and increased glycogen levels in liver (Gezginci-Oktayoglu et al. 2014). Additionally, activity of adenosine deaminase, an important enzyme for modulating the bioactivity of insulin, was decreased by chard treatment. Immunostaining analysis showed increased nuclear translocation of Akt2 and synthesis of GLUT2 in the hepatocytes of chard or/and insulin-treated hyperglycaemic rats. The oxidative stress was decreased and antioxidant defence was increased by chard extract or/and insulin treatment to hyperglycaemic rats as evidenced by the decreased malondialdehyde formation, the activities of catalase, superoxide dismutase, myeloperoxidase and increased glutathione levels. Treatment of aqueous fraction of *B. vulgaris* extract ameliorated hyperglycaemia in diabetic mice due to enhanced glucose-stimulated insulin secretion mediated by acetylcholine and glucagon-like peptide-1 (GLP-1) (Ul Kabir et al. 2015). Elevated glucose uptake via increased membrane bound GLUT4 transporters in the skeletal muscles and subsequent glycogen synthesis may also play a part in the anti-hyperglycaemic activity of *B. vulgaris*.

Anti-cancer Activity

The in-vitro inhibitory effect of *Beta vulgaris* (beet root) extract on Epstein-Barr virus early antigen (EBV-EA) induction using Raji cells revealed a high order of activity compared to capsanthin, cranberry, red onion skin and short and long red bell peppers (Kapadia et al. 1996). A crude extract of Indian spinach (*Beta vulgaris* var. *bengalensis*) leaves administered as a dietary supplement was observed to reduce significantly the cytotoxic effects of a known carcinogen, lead subacetate (20, 30 mg/kg b.w.) in mice in-vivo (Nandi et al. 1997). An in-vivo anti-tumour promoting activity evaluation against the mice skin and lung cancer bioassays also revealed a significant tumour inhibitory effect. Red beetroot extract exhibited synergistic anti-proliferative activity with doxorubicin against human pancreatic (PaCa), breast (MCF-7) and prostate (PC-3) tumour cells in-vitro (Kapadia et al. 2013). The synergistic cytotoxicity was best when the beetroot/doxorubicin ratio of 1:5 was used in PaCa cells at IC_{50}, IC_{75} and IC_{90} dose levels and in MCF-7 cells at IC_{90} dose level. The ethanolic extract of *Beta vulgaris* subsp. *perennis* exhibited a mild cytotoxic activity (IC_{50} 60.26 μg/ml compared with doxorubicin IC_{50} 21.4 μg/ml) against Hep-G2 cells (Abdel-Monem et al. 2014).

The phenolic fraction of Swiss chard (*B. vulgaris* subsp. *cycla*) extract contained vitexin-2″-*O*-rhamnoside, its demethylated form 2″-*O*-xylosylvitexin, isorhamnetin 3-gentiobioside, and rutin (Ninfali et al. 2007). Lechner et al. (2010) found that drinking water with red beetroot food colour (78 μg/ml) antagonized oesophageal carcinogenesis in *N*-nitrosomethylbenzylamine (NMBA)-treated rats. The number of NMBA-induced oesophageal papillomas was reduced by 45 % in animals that received the food colour compared to controls. The treatment also resulted in reduced rates of cell proliferation in both precancerous oesophageal lesions and in papillomas of NMBA-treated rats. Based on the fact that red beetroot colour contained betanins, with strong antioxidant activity, it was postulated that these effects were mediated through inhibition of oxygen radical-induced signal transduction.

A fraction of ethyl-acetate extract of *B. vulgaris* seed labelled as P4, exhibited marked anti-proliferative activity of human colon cancer (RKO) cells (Gennari et al. 2011). Of the active constituents, xylosylvitexin exhibited the strongest anti-proliferative activity on RKO cells, together with an enhancement of the apoptosis, an increase of cells in the G_1 phase and a reduction of cells in the S phase; on the contrary, the proliferation of normal human fibroblasts was significantly stimulated. In a study of 41 irradiated patients with head and neck cancer, supplementation of *Beta vulgaris* did not worsen survival times, intensification of acute radiation reactions was reduced, and the level markers of oxidative stress/DNA damage were also not affected (Roszkowski 2012). The results suggested that supplementation of *Beta vulgaris* in irradiated patients was safe for assisting therapy.

Swiss chard (*Beta vulgaris cicla*, BVc) and beetroot (*Beta vulgaris rubra*, BVr) extracts possessed anti-hypertensive and hypoglycaemic activity as well as excellent antioxidant activity (Ninfali and Angelino 2013). BVc contained apigenin flavonoids, namely vitexin, vitexin-2-*O*-rhamnoside and vitexin-2-*O*-xyloside, which showed anti-proliferative activity on cancer cell lines. BVr contained secondary metabolites, called betalains, used as natural dyes in food industry, and showed anti-cancer activity. *B. vulgaris* methanolic root extracts exhibited in-vitro anti-proliferative activity against MCF7 breast cancer cell line (Tripathy and Pradhan 2013). The maximum tumour cell growth inhibition was observed with 1000 μg/ml. Betanin exerted a 49 % inhibition of HepG2 cell proliferation at 200 μg/mL, and betaine exerted a 25 % inhibition at 800 μg/mL, implying a higher cytotoxicity of betanin compared with betaine (Lee et al. 2014).

Taxane-administered patients ($N=18$, 68 ± 8 years old) with metastatic prostate cancer, treated with table beet (*Beta vulgaris* ssp. *esculenta* var. *rubra*) at a dose of 10 g, twice daily for a month, decreased the high levels of Zn- and free protoporphyrin and improved the transmethylating ability (Nyirády et al. 2010). It appeared that moderate and permanent consumption of table beet product affected the life expectancy of patients favourably; however, due to the increas-

ing values of EGF (epidermal growth factor), medical control was necessary for patients with prostate cancer treated by chemotherapy. Both doxorubicin and the beetroot extract exhibited a dose-dependent cytotoxic effect in the human prostate (PC-3) and breast (MCF-7) cancer cell lines tested (Kapadia et al. 2011). Although the cytotoxicity of the beetroot extract was significantly lower when compared to doxorubicin, it continued to decrease the growth rate of the PC-3 cells (3.7 % in 3 days vs. 12.5 % in 7 days) when tested at the concentration of 29 μg/ml. In contrast, doxorubicin, at the same concentration level, completely inhibited the growth of the PC-3 cells in 3 days. Similarly, comparative studies in the normal human skin FC and liver HC cell lines showed that the beetroot extract had significantly lower cytotoxic effect than doxorubicin (8.6 % vs. 100 %, respectively, at 29 μg/ml concentration of each, 3-day test period). The results suggested that betanin, the major betacyanin constituent, may play an important role in the cytotoxicity exhibited by the red beetroot extract. Beetroot pomace extract (BPE) exhibited cytotoxic properties against Ehrlich carcinoma (EAC) cells in-vivo due to induction of oxidative stress (Vulić et al. 2013). The largest decreases in EAC cell numbers were observed in the pretreated male (approximately 53 %) and female (approximately 47 %) mice, and also the EAC cell viability was decreased after administration of BPE. The activities of the antioxidant enzymes, xanthine oxidase (XOD) and peroxidase (Px) were significantly different between the untreated EAC control group and all other groups that were treated with BPE. The XOD and Px activities were very low in untreated malignant cells, but increased significantly after administration of BPE.

At a concentration range of 12.5–25 mg/mL the pectic extract from sugar beet pulp killed 80.6 % of MCF-7 breast cancer cells, exhibiting a higher anti-proliferative activity than 4-hydroxytamoxifen (4-OHT), a classical anticancer drug, which killed 56.5 % of the cells (Concha et al. 2013). Studies found that organically produced and conventionally produced beetroots and fermented beetroot juices had different chemical properties and different impacts on cancer cells (Kazimierczak et al. 2014). It was found that anti-cancer activity was stronger in the case of organically produced fermented juices when compared with conventionally produced ones. Injectable and biodegradable sugar beet pectin/gelatin hydrogels were found useful for biomedical applications (Takei et al. 2013). The gels containing doxorubicin, an anti-cancer drug, successfully suppressed the growth of a solid tumour created by subcutaneous injection of mouse melanoma B16F1 cells into nude mice.

Anti-inflammatory Activity

Potent anti-bradykinin activity was found in the beetroot (Nagase et al. 1975). About 4–7 μg of bradykinin (an inflammatory mediator) was antagonized by the crude beet juice equivalent to 1 g of the root in the assay of contraction of the guinea-pig ileum. The inhibitory effects of this substance were observed on the smooth muscle contraction of guinea-pig, the vasodilatation in dog and the increase in the capillary permeability in rat, all of which were induced by bradykinin. The carrageenan-induced oedema in the rat's paw and some of the above biological responses by histamine and 5-hydroxytryptamine were also suppressed distinctly or slightly by the beet bradykinin antagonist. The anti-bradykinin substance from beetroot was suggested to be norepinephrine or its related catecholamine, judging from the results obtained regarding its chemical and other properties (Ikekita et al. 1983).

The phenolic amides *N-cis*-feruloyl 3-*O*-methyldopamine; *N-cis*-feruloyl tyramine; *N-trans*-feruloyl 3-*O*-methyldopamine; *N-trans*-feruloyl tyramine isolated from *Beta vulgaris* var. *cicla,* exhibited modest inhibitory activity on LPS-activated nitric oxide production dose-dependently in RAW 264.7 cells (Kim et al. 2003). Aqueous leaf extract of *B. vulgaris* at a dose level of 1000 mg/kg exhibited significant anti-inflammatory activity which was comparable to that of the standard indomethacin (10 mg/kg) in both carrageenan-induced rat paw oedema method for acute inflammation and cotton pellet

granuloma method for chronic inflammation (Jain et al. 2011). *Beta vulgaris* root ethanol extract exhibited anti-inflammatory activity in the carrageenan-induced rat paw oedema method (Chakole et al. 2011). The aqueous extract of *B. vulgaris* significantly decreased rat hind paw oedema thickness compared to control group, while the ethanolic extract had no anti-inflammatory effect (Abdel-Monem et al. 2014).

Intraperitoneal (i.p.) administration of beetroot betalain (30–300 mg/kg) reduced carrageenan (100 μg/paw)-induced paw oedema and neutrophil migration to the paw skin tissue (Martinez et al. 2015). Betalain (100 mg/kg) treatment by subcutaneous or per oral routes also inhibited the carrageenan-induced paw oedema. Importantly, the post-treatment with betalain (100 mg/kg, i.p.) significantly inhibited carrageenan-induced and complete Freund's adjuvant (10 μl/paw)-induced paw oedema. Betalain (100 mg/kg) also reduced carrageenan (500 μg/cavity)-induced recruitment of total leukocytes, including mononuclear cells and neutrophils, as well as increasing vascular permeability in the peritoneal cavity. Additionally, betalain significantly reduced carrageenan-induced superoxide anion, tumour necrosis factor-alpha (TNF-α) and interleukin (IL)-1β levels in the peritoneal fluid, as well as augmenting IL-10 levels. The results indicated that this compound exerted prominent anti-inflammatory effect on carrageenan-induced paw oedema and peritonitis by reducing the production of superoxide anion and the cytokines TNF-α and IL-1β, in addition to increasing IL-10 levels, thus suggesting that betalain possessed therapeutic potential that could be utilized in the treatment of inflammation-associated diseases.

Hypotensive Activity

Webb et al. (2008) reported that in healthy volunteers, approximately 3 h after ingestion of a dietary nitrate load (beetroot juice 500 mL), blood pressure (BP) was substantially reduced (Δmax −10.4/8 mmHg); an effect that correlated with peak increases in plasma nitrite concentration. The dietary nitrate load also prevented endothelial dysfunction induced by an acute ischaemic insult in the human forearm and significantly attenuated ex-vivo platelet aggregation in response to collagen and ADP. Interruption of the enterosalivary conversion of nitrate to nitrite (facilitated by bacterial anaerobes situated on the surface of the tongue) prevented the rise in plasma nitrite, blocked the decrease in BP, and abolished the inhibitory effects on platelet aggregation, confirming that these vasoprotective effects were attributable to the activity of nitrite converted from the ingested nitrate. Two separate randomly controlled, single-blind, crossover, postprandial studies in normotensive volunteers showed that beetroot juice (BJ) consumption significantly, and in a near dose-dependent manner, lowered systolic BP (SBP) and diastolic BP (DBP) over a period of 24 h, compared with water control (Hobbs et al. 2012). Further, bread products enriched with 100 g red or white beetroot lowered SBP and DBP over a period of 24 h (red-beetroot-enriched bread), with no statistical differences between the varieties. Total urinary NO(x) significantly increased following the consumption of 100, 250 and 500 g BJ and after red-beetroot-enriched bread ingestion, but did not reach significance for white-beetroot-enriched bread compared with the no-beetroot condition. In a double-blind, randomized, placebo-controlled, crossover study of 15 men and 15 women, consumption of beetroot juice demonstrated a significant reduction in systolic blood pressure in men at 6 h after consumption (Coles and Clifton 2013). The results suggested that consumption of beetroot juice will lower blood pressure in men when consumed as part of a normal diet in free-living healthy adults. The systematic review and meta-analysis of randomized clinical trials ($n = 16$) examining the effects of inorganic nitrate and beetroot supplementation on blood pressure conducted by Siervo et al. (2013) found that inorganic nitrate and beetroot juice supplementation was associated with a significant reduction in systolic blood pressure (BP). Inorganic nitrate and beetroot juice consumption were associated with greater changes in systolic BP than diastolic BP.

Immunomodulatory Activity

B. vulgaris methanolic root extracts exhibited in-vivo immunomodulatory activity (Tripathy and Pradhan 2013). The immunomodulatory effect of the extract on specific antibody production at 1000 μg/ml was comparable to the effect of Levamisole (2 mg/kg) as a positive control at both primary (10.9) and secondary (11.6) delayed-type hypersensitivity response.

Antihypercholesterolemic Activity

Diet supplementation of high-cholesterol-fed hypercholesterolemic chemically induced colon cancer rats, with 5 % and 15 % cellulose or with 15 % fibre isolated from red beet for 14 weeks caused a reduction of serum cholesterol and triacylglycerol levels (by 30 and 40 %, respectively) and a significant increase in the fraction of cholesterol carried in HDL (Bobek et al. 2000). This diet induced also a significant decrease (almost by 30 %) of cholesterol content in aorta. Higher cellulose content in the diet and even more so the administration of red beet fibre caused a significant reduction of conjugated dienes content in plasma, erythrocytes and in liver. Also observed were increases in the activities of superoxide dismutase and catalase in erythrocytes and in colon and activities of glutathione peroxidase and glutathione-S-transferase in liver. The presence of both higher cellulose content and red beet fibre in the diet significantly reduced the incidence of precancerous lesion-aberrant crypt foci in the colon. The diet containing red beet fibre did not affect significantly the incidence of colon tumours, although the number of animals bearing tumours was reduced by 30 %. Studies by Khalili and Mahdavi (2004) found that the elevated triglyceride and cholesterol levels in streptozotocin-induced diabetic rats were significantly decreased by treatment with *B. vulgaris* extract.

Lee et al. (2009) found that red beet leaf (RBL) supplementation improved antioxidant status in C57BL/6 J mice fed high-fat high-cholesterol diet. In RBL mice, lipid peroxidation determined as 2-thiobarbituric acid-reactive substances (TBARS value) was significantly reduced in the plasma and selected organs (liver, heart and kidney). Levels of antioxidants (glutathione and *β*-carotene) and the activities of antioxidant enzyme (glutathione peroxidase) in plasma and liver were considerably increased, suggesting that antioxidant defences were improved by RBL diet. Comet parameters such as tail DNA (%), tail extent moment, olive tail moment and tail length were significantly reduced by 25.1 %, 49.4 %, 35.4 %, and 23.7 %, respectively, in plasma lymphocyte DNA of RBL mice compared with control mice, and indicated the increased resistance of lymphocyte DNA to oxidative damage. In addition, the RBL diet controlled body weight together with a significant reduction of fat pad (retroperitoneal, epididymal, inguinal fat and total fat).

Administration of lyophilized aqueous extract of *Beta vulgaris* (beetroot) at doses of 250 and 500 mg/kg body weight for 70 consecutive days to cholesterol-rich diet-induced hypercholesterolemic rats, showed a significant decrease in total cholesterol and triglycerides and significant increase in high-density lipoprotein-cholesterol (HDL-C) (Al-Dosari et al. 2011). Beetroot treatment significantly decreased malondialdehyde (MDA)A level and significantly replenished the reduced NP-SH (non-protein-sulfhydryl) content in both liver and heart tissue. The acute toxicity test of BVE showed no mortality or morbidity in rats. The results indicated *B. vulgaris* root extract had a significant antihypercholesterolemic and antioxidant potential and/or free radical scavenging properties in hypercholesterolemic rats, possibly exerted by the phytoconstituents present in the beetroot.

Hepatoprotective Activity

Ethanolic extract of *Beta vulgaris* roots administered orally at doses of 1000, 2000 and 4000 mg/kg exhibited significant dose-dependent hepatoprotective activity against carbon tetrachloride (CCl_4)-induced hepatotoxicity in rats (Agarwal et al. 2006). Two flavone glycosides, vitexin 7-*O*-β-D-glucopyranoside and vitexin 2″-*O*-β-

D-glucopyranoside isolated from the aerial parts of *Beta vulgaris* var. *cicla*, demonstrated hepatoprotective activity with values of 65.8 and 56.1 %, respectively, in primary cultured rat hepatocytes with CCl4-induced cell toxicity, compared to controls (Kim et al. 2004). This was comparable to that of silibinin (69.8 %) which was used as a positive control. Dietary administration of table beet protected the duodenum during hepatic ischaemia-reperfusion of the rat liver (Váli et al. 2006). Beneficial effect of table beet treatment was found in changes of total scavenging capacity of the duodenum during ischaemia-reperfusion injury. H-donating ability and reducing power of the gut decreased in the table-beet-fed group during ischaemia-reperfusion compared to normal group with ischaemia-reperfusion. Other antioxidant parameters of the plasma increased in rats fed table beet diet, and change in reducing power was significant. As a result of table beet feeding, global parameters (H-donating ability, reducing power, free SH group concentration) and enzymatic antioxidants (glutathione peroxidase and superoxide dismutase) of the rat liver were found to increase significantly, which indicated that the treatment had a positive effect on its redox state (Váli et al. 2007). The increase found in zinc and copper content may protect the hepatocytes against oxidative stress because these elements are required for the function of superoxide dismutase enzymes. In the table-beet-fed group, the concentration of short-chain fatty acids decreased, whereas that of long-chain fatty acids increased. The changes in metal element and fatty acid concentrations confirmed that these elements had an essential function in cellular pathways. Treatment of ethanolic extract of *Beta vulgaris* (EEBV) root protected against CCl_4-induced hepatic damage in rats (Pal et al. 2010). Hepatoprotective activity of EEBV was studied by estimating serum enzyme levels of total protein and bilirubin. Treatment with EEBV elicited significant reduction of CCl4-induced elevated serum levels of aspartate aminotransferase (AST), alanine aminotransferase (ALT), alkaline phosphatase (ALP) activities and bilirubin with parallel significant increase in total protein, indicating the extract could preserve the normal functional status of the liver. Histology of the liver sections of the animals treated with the extracts showed the presence of normal hepatic cords, absence of necrosis and fatty infiltration, which further evidenced the hepatoprotective activity.

Beta vulgaris leaf n-butanol fraction exhibited hepatoprotective activity against ethanol-mediated hepatotoxicity in rat hepatocyte cultures and in-vivo evaluation, the fraction at doses of 50,100 and 200 mg/kg showed marked protective action against ethanol-induced hepatic toxicity in rats as evident by restoration of biochemical changes induced by ethanol (Jain and Singhai 2012).

Hematopoietic Activity

Oral administration of *B. vulgaris* root methanol extract to rats for 16 days exerted significant increases in packed cell volume (PCV), haemoglobin concentration, red blood cell counts (RBCs), and total lymphocyte count (Indhumathi and Kannikaparameswari 2012). The other parameters such as mean corpuscular volume (MCV), mean corpuscular haemoglobin (MCH), mean corpuscular haemoglobin concentration (MCMH), also showed dose-dependent increases.

Adaptogenic Activity

Experiments on Wistar rats provided a preliminary evaluation of the adaptogen activity of saponins extracted from sugar beet in comparison to analogous properties of dry *Eleutherococcus* extract, a classical adaptogen preparation (Brezhneva et al. 2001). In some cases, the two substances produced similar effects, but were somewhat less pronounced for sugar beet saponins. Atamanova et al. (2005) also found adaptogenic activity of saponins from table beetroots. The adaptogen activity was evaluated by the method of external adverse factor of Brezhneva et al. (2001), with reference to a dry residue of the *Eleutherococcus* extract. They demonstrated that introduction of table beetroot saponins into a medium in concentrations between $1–10^{-6}$ and $1–10^{-8}$ g/ml increased by 30 % the resistance of

test cells of *Paramecium caudatum* infusoria to an external adverse factor (15 % aqueous NaCl).

Ergogenic Activity

In a double-blind crossover study, consumption of nitrate-rich, whole beetroot improved endurance running performance in healthy adults (Murphy et al. 2012). Since nitrate ingestion from other sources may have detrimental health effects, it would be prudent for individuals seeking performance benefits to obtain nitrates from whole vegetables, such as beetroot. In a balanced crossover design involving ten healthy men, ingestion of beetroot juice (BR) caused a dose-dependent increase in plasma nitrite and reduced the steady-state oxygen (O_2) uptake during moderate-intensity exercise but there was no additional improvement in exercise tolerance after ingesting BR containing 16.8 mmol compared with 8.4 mmol nitrate (Wylie et al. 2013). In a randomized, double-blind, crossover design of nine healthy, physically active subjects dietary supplementation with nitrate-rich beetroot juice accelerated oxygen uptake (VO2 kinetics) and enhanced exercise tolerance during severe-intensity exercise when initiated from an elevated metabolic rate (Breese et al. 2013). Studies by Pinna et al. (2014) found that beetroot juice supplementation for 6 days positively affected performance of swimmers as it reduced the aerobic energy cost and increased the workload at anaerobic threshold.

Nephroprotective Activity

Treatment of rodents with beet root ethanolic extract (BVEE) ameliorated gentamicin-induced nephrotoxicity associated oxidative stress, inflammation, and apoptosis (El Gamal et al 2014). BVEE (250 and 500 mg/kg) treatment along with gentamicin restored/increased the renal endogenous antioxidant status. Gentamicin-induced increased renal inflammatory cytokines (TNF-α and IL-6), nuclear protein expression of NF-κB (p65), NF-κB-DNA binding activity, myeloperoxidase (MPO) activity, and nitric oxide level were significantly down regulated upon BVEE treatment. In addition, BVEE treatment significantly reduced the amount of cleaved caspase 3 and Bax, protein expression and increased the Bcl-2 protein expression.

Antimicrobial Activity

Repeated intranasal administration of an aqueous *Beta vulgaris* extract prior to intranasal instillation inoculation of influenza virus A/PR/8/34 (H1N1) to mice conferred a partial protection against the experimental influenza infection (Prahoveanu et al. 1986). There was a significant decrease in the hemagglutination titres recorded in mouse lung homogenates, a decrease in mortality rate and an increase in the mean survival time as compared with the untreated, virus-inoculated controls.

Two novel antifungal proteins, AX1 and AX2, isolated from sugar beet leaves, strongly inhibited growth of *Cercospora beticola* and other filamentous fungi, but have little or no effect against bacteria (Kragh et al. 1995). Two novel, identical antifungal proteins, IWF1 and IWF2, isolated from sugar beet leaves exhibited strong in-vitro antifungal activity against *Cercospora beticola*, the causal agent of leaf spot disease in sugar beet (Nielsen et al. 1996). Another antifungal protein, designated IWF6, isolated from sugar beet was able to inhibit the growth of the pathogen *C. beticola* in-vitro, by 75 % after 120 h of growth at a concentration of 20 μg/m (Kristensen et al. 2000).

In antibacterial tests, *Staphylococcus aureus* and *Bacillus cereus* showed higher susceptibility to beetroot pomace extracts than *Escherichia coli* and *Pseudomonas aeruginosa* (Čanadanović-Brunet et al. 2011). Beetroot pomace extract (BPE) exhibited antimicrobial activity; Gram(-) bacteria (*Salmonella typhimurium, Citrobacter freundii*) and Gram(+) bacteria (*Staphylococcus aureus, S. sciuri, Bacillus cereus*) showed high susceptibility, while yeasts and moulds were resistant (Vulić et al. 2013).

Beetin 27 (BE27), a ribosome-inactivating protein (RIP) from sugar beet (*Beta vulgaris* L.) leaves, was reported to be an antiviral protein induced by virus and signalling compounds such as hydrogen peroxide and salicylic acid (Iglesias et al. 2015). Its role as a defence protein had been attributed to its RNA polynucleotide:adenosine

glycosidase activity. BE27 could have a defensive role against pathogens as it displayed rRNA N-glycosidase activity against yeast and *Agrobacterium tumefaciens* ribosomes, DNA polynucleotide:adenosine glycosidase activity against herring sperm DNA, and magnesium-dependent endonuclease activity against the supercoiled plasmid PUC19 (nicking activity). Furthermore, BE27 possessed superoxide dismutase activity, thus being able to produce the signal compound hydrogen peroxide. BE27 was also toxic to COLO 320 cells, inducing apoptosis in these cells by either activating the caspase pathways and/or inhibiting protein synthesis.

Skin Disease Ameliorating Activity

From aqueous and methanolic *B. vulgaris* leaf extracts solutions and ointments were prepared with different concentrations and their clinical effects for the treatment of acne and psoriasis were evaluated using 360 patients (Nidhal et al. 2014). The clinical study showed that *B. vulgaris* extract gave significant healing effect for the treatment of acne within 2 weeks using aqueous solutions prepared from extracts of fresh and dried leaves with more predominant effect for the fresh leaves and the solution dosage form was better than ointment in the treatment of acne. Similar results were obtained for psoriasis but the ointment dosage form was much more effective.

Photoprotective Activity

Gel formulation prepared from beetroot (*B. vulgaris*) extract (containing 66.5 mg/g total phenolic content and 29.05 mg/g flavonoids) exhibited photoprotective activity as evaluated using the model system based on UV-induced discoloration of paprika gel (Kapur et al. 2012). The SPF (sun protection factor) was determined as 1.34. The absorbance value of the 150 μg/ml dilution of the gel formulation had higher absorbance value in comparison to 200 μg/ml dilution of commercially available sunscreens. The methanol beetroot extract also exhibited photoprotective activity. The results indicated beetroot to have potential as a sunscreen agent.

Antiparasitic Activity

Pigs fed insulin plus sugar beet fibre (SBF) diet (Diet 4) and on SBF diet (Diet 3) had 86 % and 70 % adult worm *Oesophagostomum dentatum* reductions compared with the controls, respectively (Petkevicius et al. 2003). The worm recoveries from the pigs on the insulin supplemented diet (Diet 2) were reduced by 97 % compared to the controls (Diet 1). The results from this study indicated that highly degradable and rapidly fermentable carbohydrates such as dietary insulin have a profound deworming effect on *O. dentatum* infection.

Estrogenic Activity

Elghamry et al. (1971) detected significant estrogenic potency of sugar beets. The crude methanolic extract of the plant leaves yielded 185.63 mg% of a pure compound which was confirmed to be *β*-sitosterol. The minimum estrogenic dose of the isolated substance reached 2.5 μg daily as tested subcutaneously in mice. The relative potency of this estrogenic principle in terms of estradiol was 1:8400.

Pharmacokinetic Studies

Frank et al. (2005) conducted a study on the urinary pharmacokinetics of betalains on six nonsmoking female volunteers administered a single oral dose of red beetroot juice. The amount of intact betalains (betanin and isobetanin) recovered in urine was 10,013 μg corresponding to 0.28 % of the administered dose. Maximum excretion rates were observed after a median tmax, R of 3.0 h (range 2.5–8.0 h) amounting to 91.7 μg/h. The terminal elimination rate constant (lambda-z) and the corresponding half-life were 0.097/h and 7.43 h, respectively. Using the lambda-z estimates, the expected total betalain amount excreted in urine was 1228 μg. Based on the results obtained, it was assumed that either

the bioavailability of the betalains was low or that renal clearance was a minor route of systemic elimination for these compounds. After red beet juice (RBJ) ingestion, the betacyanins betanin and isobetanin were excreted in the volunteers' urine (0.28 % of the administered dose) (Netzel et al. 2005). Compared to the ingestion of water, RBJ consumption resulted in a significantly increased urinary excretion of total phenolics (51.1 % of the administered phenolics) and other antioxidant compounds (43.9 % of the administered compounds) within 24 h.

Adverse Issues

Blázovics et al. (2007) found that extreme consumption of *Beta vulgaris* var. *rubra* could cause metal ion (Cu, Fe, Mg, Mn, Zn and P) accumulation in the rat's liver. They suggested that extreme consumption of table beetroot could cause several disturbances not only in cases of healthy patients but in patients suffering with metal accumulating diseases, e.g. porphyria cutanea tarda, haemochromatosis or Wilson disease; although moderate consumption may be beneficial in iron-deficiency anaemia and inflammatory bowel diseases.

Allergy to sugar beet pollen had been reported by Peck and Moffat (1959) and as an occupational disease by Ursing (1968). During production of sugar beet (*Beta vulgaris*) seeds in greenhouses, workers frequently develop allergic symptoms (Luoto et al. 2008). 17 kDa and 14 kDa protein homologues were identified in sugar beet pollen showing sequence similarity with *Chenopodium* allergens, Che a 1 and Che a 2. Sequence data were obtained by mass spectrometry (70 and 25 %, respectively for Beta v 1 and Beta v 2), and could be used for cloning and recombinant expression of the allergens.

Traditional Medicinal Uses

According to Bruntz and Jalaoux (1918) beet leaves are official in the French, Spanish and Mexican Pharmacopoeias. Although little used in modern herbalism, beet has a long history of folk use, especially in the treatment of cancerous tumours (Duke 1983; Chevallier 1996). Beet juice, seed decoction or other parts of the plant have been used in the treatment of tumours, leukaemia and other forms of cancer such as cancer of the breast, oesophagus, glands, head, intestines, leg, lip, lung, prostate, rectum, spleen, stomach, and uterus. In bygone days, beet juice was recommended as a remedy for anaemia and yellow jaundice, and, put into the nostrils to purge the head, clear ringing ears, and alleviate toothache, used with vinegar to rid the scalp of dandruff as scurf, and was recommended to prevent falling hair (Duke 1983). Beet juice has been applied to ulcers; a decoction is used as a purgative by those who suffer from haemorrhoids in South Africa. Leaves and roots were used as an emmenagogue. The plant was found effective in the treatment of feline ascariasis. The root of white-rooted forms contain betaine which promotes the regeneration of liver cells and the metabolism of fat cells, and that of red-rooted forms contains betanin – which is partly responsible for red beet's immune-enhancing effect (Chevallier 1996). The root can be used as part of the diet, or the juice can be extracted and used as a health-promoting tonic drink. The root is carminative, haemostatic, stomachic and a tonic for women (Duke and Ayensu 1985). Chard (*Beta vulgaris* var. *cicla*) is used as a hypoglycaemic agent by diabetic patients in Turkey (Bolkent et al. 2000). Nadkarni (2000) stated beet, especially red beet, to be an active emmenagogue, acting as a resolvent on the vitiated secretions of stomach and bowel, while the white beet is laxative and diuretic. He added that externally a decoction with a little vinegar added heals the itch, cleanses scurf and dandruff from the head and is excellent for all kinds of ulcerous and running sores. Kirtikar and Basu (1975) record that the seeds have cooling and diaphoretic properties. The fresh leaves are applied to burns and bruises.

Other Uses

Food for Livestock

Fodder beet originated in the Middle East and was being used as cattle feed. Beet pulp is a

by-product from the processing of sugar beet, which is used as fodder for horses and other livestock (Wikipedia 2015). It is supplied either as dried flakes or as compressed pellets, but when fed to horses it is usually soaked in water first. Beet pulp is commonly used in beef cattle diets as a supplement or roughage replacement in finishing diets.

Bioethanol Production

Sugar beet juice concentrate was found to be suitable for ethanol production yielding, depending on the yeast strain, ca. 85–87 g/L ethanol with ca. 82 % practical yield and more than 95 % of sugars consumption after 72 h of fermentation (Kawa-Rygielska et al. 2013). Nutrients enrichment further increased ethanol yield. The best results were obtained for media supplemented with urea + $Mg_3(PO_4)_2$ yielding 91.16–92.06 g/L ethanol with practical yield ranging 84.78–85.62 % and full sugars consumption. Sugar beet pulp is an abundant industrial waste material that holds a great potential for bioethanol production owing to its high content of cellulose, hemicelluloses and pectin (Rezić et al. 2013). Integrated enzymatic hydrolyzation and *Saccharomyces cerevisiae* fermentation of sugar beet pulp in a single-tank bioreactor afforded maximum ethanol yield of 0.1 g ethanol/g of dry weight (0.25 g ethanol/g total sugar content); the efficiency of ethanol production was 49 %, and the productivity of the bioprocess was 0.29 g/l/h. Volatile fatty acids were produced during the acidification of beet sugar vinasse in fixed bed reactors (Gil-Pena et al. 1986). The daily output of acids ranged between 63 and 75 Kg/m^2, and the joint acid production amounted to 50–60 % butyric acid, 29–35 % acetic acid and 8–9 % propionic acid.

Biogas Production

Continuous anaerobic degradation of sugar beet pulp afforded a biogas yield of more than 85 % of the theoretical value for total carbon conversion and resulting in a corresponding COD reduction (Stoppok and Buchholz 1985). Hutnan et al. (2000) found sugar beet pulp to be a suitable material for anaerobic methanogenic biodegradation for biogas production besides being utilized for cattle feeding. Beet pulp was treated by a two-step anaerobic process; the first step consisted of hydrolysis and acidification. The second step was methanogenesis. Methane yield was over 0.360 m^3/kg dried pulp and excess sludge production was 0.094 g per gram COD added. Brooks et al. (2008) reported that the first large-scale biogas plant was established at a Hungarian sugar beet factory in 2007. Digesting approximately 50 % of the sugar beet pulp (800 tonne/day, 22 %TS), the biogas produced could substitute about 40 % of the natural gas required for the thermal energy supply within the sugar processing.

Plant Germination/Growth Inhibition

The following germination inhibiting compounds were isolated from *B. vulgaris*: *cis*-4-cyclohexene-1, 2-dicarboximide from sugar beet fruits on lettuce seeds (Mitchell and Tolbert 1968); *p*-hydroxybenzoic acid, vanillic acid *p*-coumaric acid and ferulic acid from sugar beet clusters (Massart 1957; Battle and Whittington 1969; Inoue and Yamamoto 1975a); salicylic acid, syringic acid (Inoue and Yamamoto 1975a); gallic acid from sugar beet fruits (Sebeson and Snyder 1969); potassium nitrate, oxalic acid from sugar beet seed balls (Inoue and Yamamoto 1975b); vanillin, *p*-hydroxybenzaldehyde, syringaldehyde, protocatechualdehyde, vanillic acid, *p*-hydroxybenzoic acid syringic acid from red beet seed balls, all inhibiting lettuce seeds (Chiji et al. 1980) and phenolic amides *N*-(4-hydroxy-3-methoxyphenethyl) ferulamide (*N-trans*-feroylhomovanillylamine) and *N-trans*-feruloyltyramine (Chiji et al. 1984). Water extract of sugar beet exhibited allelopathic effect on purslane, *Portulaca oleracea* (Dadkhah 2013). The extract did not inhibit germination but strongly reduced seed vigour and seedling growth (shoot length, root length and leaf area) of purslane.

Pest Control

Results of studies by Smigocki et al. (2013) suggested that the *Beta vulgaris* serine proteinase inhibitor gene (BvSTI) may prove useful for effective control of several different lepidopteran insect pests in genetically modified tobacco and other plants.

Biomedical Uses

Injectable and biodegradable sugar beet pectin/gelatin hydrogels were found useful for biomedical applications (Takei et al. 2013). Studies indicated *Beta vulgaris* pulp powder to be a good pharmaceutical adjuvant, specifically as disintegrating agent (Bablu et al. 2010).

Comments

Letschert (1993) typified *Beta vulgaris* in the same way as already proposed by Hammer (1986) so that for leafy beets (convar. *cicla)* and tuberous sugar, fodder and garden beets (convar. *vulgaris*) as two morphologically distinct entities names of common use can be accepted. They subdivided *B. vulgaris* into three subspecies: subsp. *vulgaris*, containing all cultivated materials; subsp. *maritima*, a large and variable group of plant types; and subsp. *adanensis* (Letschert et al. 1994). Further subdivision is considered to be of little or no use. A new classification for the cultivated and weed forms of beet was proposed by Lange et al. (1999). The proposal distinguished four beet cultivar groups: Leaf Beet Group, Garden Beet Group, Fodder Beet Group and Sugar Beet Group.

Subspecies *vulgaris* is valued as a source of sugar, a domestic vegetable, and food for livestock. Beets will grow in alkaline soils and tolerate arid conditions, making them an especially valuable crop in marginal lands. Aristotle referred to the use of red greens from beets ca. 350 b.c.

The classification of both the wild and cultivated forms of *Beta vulgaris* is confusing (Oyen 2004). The cultivated forms of *Beta vulgaris* are all classified in subsp. *vulgaris* (Germeier and Frese 2001; Frese et al. 2004) The classification of the cultivated forms as accepted by the World Beta Network and the International Data Base for *Beta* recognises four denomination classes within *Beta vulgaris* subsp. *vulgaris*:

- Garden Beet Group (also named Conditiva Group).
- Leaf Beet Group (in other cultivar-group classifications divided into Swiss chard or Flavescens Group, and Spinach beet or Cicla Group, but many intermediate types exist).
- Fodder Beet Group (also named Crassa Group).
- Sugar Beet Group (also named Altissima Group).

For the vegetable types, some cultivars recommended for the tropics are:

- Garden Beet Group: 'Crimson Globe' and 'Detroit Dark Red';
- Leaf Beet Group (Swiss chard): 'Fordhook Giant', 'Lucullus' and 'Verte à carde blanche de Nice';
- Leaf Beet Group (Spinach beet): 'All Green' and 'Palak'

Selected References

Abdel-Monem A, Fawkeya A, Rasha H (2014) Biological activity of the Egyptian medicinal plants: Part 3 Antioxidant, cytotoxicity, anti-diabetic activities and constituents of *Beta vulgaris* subsp. *perennis*. Indian Streams Res J 4(1):P1

Acree TE, Lee CY, Butts RM, Barnard J (1976) Geosmin, the earthy component of table beet odor. J Agric Food Chem 24(2):430–431

Agarwal M, Srivastava VK, Saxena KK, Kumar A (2006) Hepatoprotective activity of *Beta vulgaris* against CCl4-induced hepatic injury in rats. Fitoterapia 77(2):91–93

Akita T, Hina Y, Nishi T (2000) Production of betacyanins by a cell suspension culture of table beet (*Beta vulgaris* L.). Biosci Biotechnol Biochem 64(9):1807–1812

Akita T, Hina Y, Nishi T (2002) New medium composition for high betacyanin production by a cell suspension culture of table beet (*Beta vulgaris* L.). Biosci Biotechnol Biochem 66(4):902–905

Alard D, Wray V, Grotjahn L, Reznik H, Strack D (1985) Neobetanin: isolation and identification from *Beta vulgaris*. Phytochemistry 24(10):2383

Al-Dosari M, Alqasoumi S, Ahmed M, Al-Yahya M, Ansari MN, Rafatullah S (2011) Effect of *Beta vulgaris* L. on cholesterol rich diet-induced hypercholesterolemia in rats. Farmacia 59:669–678

Armienta-Aldana E, González De La Vara LE (2004) Acid phosphatases from beet root (*Beta vulgaris*) plasma membranes. Physiol Plant 121(2):223–230

Atamanova SA, Brezhneva TA, Slivkin AI, Nikolaevskii1 VA, Selemenev VF, Mironenko NV (2005) Isolation of saponins from table beetroot and primary evaluation of their pharmacological activity. Pharm Chem J 39(12):650–665

Bablu R, Rishabha M, Mayank B, Kumar SP (2010) Preparation and characterization of *Beta vulgaris* pulp powder as a pharmaceutical excipient. Pharmacogn J 2(11):395–399

Bakry AA, Youssed ME, El Gindy AA (2014) Utilization of Swiss chard as functional food. Nat Sci 12(12):104–111

Barkla BJ, Blumwald E (1991) Identification of a 170-kDa protein associated with the vacuolar Na+/H+ antiport of *Beta vulgaris*. Proc Natl Acad Sci U S A 88(24):11177–11181

Battle JP, Whittington WJ (1969) The relation between inhibitory substances and variability in time to germination of sugar beet clusters. J Agric Sci 73(3):337–346

Bavec M, Turinek M, Grobelnik-Mlakar S, Slatnar A, Bavec F (2010) Influence of industrial and alternative farming systems on contents of sugars, organic acids, total phenolic content, and the antioxidant activity of red beet (*Beta vulgaris* L. ssp. *vulgaris* Rote Kugel). J Agric Food Chem 58(22):11825–11831

Berglund L, Brunstedt J, Nielsen KK, Chen Z, Mikkelsen JD, Marcker KA (1995) A proline-rich chitinase from *Beta vulgaris*. Plant Mol Biol 27(1):211–216

Biondo PBF, Boeing JS, Barizão EO, de Souza NE, Matsushita M, de Oliveira CC, Boroski M, Visentainer JV (2014) Evaluation of beetroot (*Beta vulgaris* L.) leaves during its developmental stages: a chemical composition study. Food Sci Technol 34(1):94–101

Blázovics A, Sárdi E, Szentmihályi K, Váli L, Takács-Hájos M, Stefanovits-Bányai E (2007) Extreme consumption of *Beta vulgaris* var. *rubra* can cause metal ion accumulation in the liver. Acta Biol Hung 58(3):281–286

Bobek P, Galbavý S, Mariássyová M (2000) The effect of red beet (*Beta vulgaris* var. *rubra*) fiber on alimentary hypercholesterolemia and chemically induced colon carcinogenesis in rats. Nahrung 44(3):184–187

Bokern M, Heuer S, Wray V, Witte L, Macek T, Vanek T, Strack D (1991) Ferulic acid conjugates and betacyanins from cell cultures of *Beta vulgaris*. Phytochemistry 30:3261–3265

Bolkent S, Yanardağ R, Tabakoğlu-Oğuz A, Ozsoy-Saçan O (2000) Effects of chard (*Beta vulgaris* L. var. *cicla*) extract on pancreatic B cells in streptozotocin-diabetic rats: a morphological and biochemical study. J Ethnopharmacol 73(1-2):251–259

Breese BC, McNarry MA, Marwood S, Blackwell JR, Bailey SJ, Jones AM (2013) Beetroot juice supplementation speeds O2 uptake kinetics and improves exercise tolerance during severe-intensity exercise initiated from an elevated metabolic rate. Am J Physiol Regul Integr Comp Physiol 305(12):R1441–R1450

Brezhneva TA, Nikolaevskii VA, Selemenev VF, Slivkin AI, Muad AA, Khind T, SafonovaEF (2001) Isolation of saponins from sugar beet roots and preliminary characterization of their adaptogen properties. Pharm Chem J 35(3):159–161

Brooks L, Parravicini V, Svardal K, Kroiss H, Prendl L (2008) Biogas from sugar beet press pulp as substitute of fossil fuel in sugar beet factories. Water Sci Technol 58(7):1497–1504

Bruntz L, Jalaoux M (1918) Plantes officinales et plantes à drogues médicamenteuses, nomenclature méthodique dressée d'après toutes les éditions des codes officiels des médicaments des divers États du monde, précédée de considérations générales sur les pharmacopées. Vigot frères, 268pp

Čanadanović-Brunet JM, Savatović SS, Ćetković GS, Vulić JJ, Djilas SM, Markov SL, Cvetković DD (2011) Antioxidant and antimicrobial activities of beet root pomace extracts. Czech J Food Sci 29(6):575–585

Ceballos-Laita L, Gutierrez-Carbonell E, Lattanzio G, Vázquez S, Contreras-Moreira B, Abadía A, Abadía J, López-Millán AF (2015) Protein profile of *Beta vulgaris* leaf apoplastic fluid and changes induced by Fe deficiency and Fe resupply. Front Plant Sci 6:145

Chakole R, Zade S, Charde M (2011) Antioxidant and anti-inflammatory activity of ethanolic extract of *Beta vulgaris* Linn. roots. Int J Biomed Adv Res 2(4):124–130

Chevallier A (1996) The encyclopedia of medicinal plants. Dorling Kindersley, London, 336 pp

Chiji H, Tanaka S, Izawa M (1980) Phenolic germination inhibitors in the seed balls of red beet (*Beta vulgaris* L. var. *rubra*. Agric Biol Chem 44(1):205–207

Chiji H, Giga T, Izawa M, Kiriyama S (1984) Two phenolic amides in the seed balls of sugar beet (*Beta vulgaris* L. var. *saccharifera* Alefeld). Agric Biol Chem 48(6):1653–1654

Chiji H, Arakawa Y, Ueda S, Kuroda M, Izawa M (1986) 5,2′-dihydroxy-6,7-methylenedioxyisoflavone from seed balls of sugar beet. Phytochemistry 25(1):281–282

Coles LT, Clifton PM (2013) Effect of beetroot juice on lowering blood pressure in free-living, disease-free adults: a randomized, placebo-controlled trial. Nutr J 11:106

Combo AM, Aguedo M, Quiévy N, Danthine S, Goffin D, Jacquet N, Blecker C, Devaux J, Paquot M (2013) Characterization of sugar beet pectic-derived oligosaccharides obtained by enzymatic hydrolysis. Int J Biol Macromol 52:148–156

Concha J, Weinstein C, Zúñiga ME (2013) Production of pectic extracts from sugar beet pulp with antiproliferative activity on a breast cancer cell line. Front Chem Sci Eng 7(4):482–489

Dadkhah A (2013) Allelopathic effect of sugar beet (*Beta vulgaris*) and eucalyptus (*Eucalyptus camaldulensis*) on seed germination and growth of *Portulaca oleracea*. Russ Agric Sci 39(2):117–123

Dinçer A, Aydemir T (2001) Purification and characterization of catalase from chard (*Beta vulgaris* var. *cicla*). J Enzym Inhib 16(2):165–175

Dongowski G (1996) Untersuchungen an Ballaststoff-und Pektinfraktionen von Beta vulgaris var. conditiva. Z Lebensm Unters Forsch 202(4):285–293. (In German). [Dietary fiber and pectin fractions of *Beta vulgaris* var. conditiva.]

Duke JA (1983) Handbook of energy crops. http://www.hort.purdue.edu/newcrop/duke_energy/dukeindex.html

Duke JA, Ayensu ES (1985) Medicinal plants of china, vols 1&2. Reference Publications, Inc., Algonac, 705 pp

El Gamal AA, Al Said MS, Raish M, Al-Sohaibani M, Al-Massarani SM, Ahmad A, Hefnawy M, Al-Yahya M, Basoudan OA, Rafatullah S (2014) Beetroot (*Beta vulgaris* L.) extract ameliorates gentamicin-induced nephrotoxicity associated oxidative stress, inflammation, and apoptosis in rodent model. Mediat Inflamm 2014:983952

Elghamry MI, Grunert E, Aehnelt E (1971) An active principle responsible for estrogenicity in the leaves of *Beta vulgaris*. Isolation, identification and estimation. Planta Med 19(3):208–214

Elliger CA, Halloin JM (1994) Phenolics induced in *Beta vulgaris* by *Rhizoctonia solani* infection. Phytochemistry 37(3):691–693

Escribano J, Pedreño MA, García-Carmona F, Muñoz R (1998) Characterization of the antiradical activity of betalains from *Beta vulgaris* L. roots. Phytochem Anal 9(3):124–127

Espelie KE, Sadek NZ, Kolattukudy PE (1980) Composition of suberin-associated waxes from the subterranean storage organs of seven plants: Parsnip, carrot, rutabaga, turnip, red beet, sweet potato and potato. Planta 148(5):468–476

Fishman ML, Chau HK, Cooke PH, Hotchkiss AT Jr (2008) Global structure of microwave-assisted flash-extracted sugar beet pectin. J Agric Food Chem 56(4):1471–1478

Fishman ML, Chau HK, Qi PX, Hotchkiss AT Jr, Yadav MP (2013) Physico-chemical characterization of protein-associated polysaccharides extracted from sugar beet pulp. Carbohydr Polym 92(2):2257–2266

Fissore EN, Ponce NM, Matkovic L, Stortz CA, Rojas AM, Gerschenson LN (2011) Isolation of pectin-enriched products from red beet (*Beta vulgaris* L. var. *conditiva*) wastes: composition and functional properties. Food Sci Technol Int 17(6):517–527

Frank T, Stintzing FC, Carle R, Bitsch I, Quaas D, Strass G, Bitsch R, Netzel M (2005) Urinary pharmacokinetics of betalains following consumption of red beet juice in healthy humans. Pharmacol Res 52(4):290–297

Freidig AK, Goldman IL (2014) Geosmin (2β,6α-dimethylbicyclo[4.4.0]decan-1β-ol) production associated with *Beta vulgaris* ssp. *vulgaris* is cultivar specific. J Agric Food Chem 62(9):2031–2036

Frese L, Germeir C, Lipman E, Maggioni L (Compliers) (2004) Report of a Working Group on *Beta* and World *Beta* Network. Second Joint meeting 23–26 October 2002, Bologna, Italy. International Plant Genetic Resources Institute, Rome, Italy

Gandía-Herrero F, García-Carmona F, Escribano J (2004) Purification and characterization of a latent polyphenol oxidase from beet root (*Beta vulgaris* L.). J Agric Food Chem 52(3):609–615

Geigert J, Stermitz FR, Johnson G, Maag DD, Johnson DK (1973) Two phytoalexins from sugarbeet (*Beta vulgaris*) leaves. Tetrahedron 29(18):2703–2706

Gennari L, Felletti M, Blasa M, Angelino D, Celeghini C, Corallini A, Ninfali P (2011) Total extract of *Beta vulgaris* var. *cicla* seeds versus its purified phenolic components: antioxidant activities and antiproliferative effects against colon cancer cells. Phytochem Anal 22:272–279

Georgiev VG, Weber J, Kneschke EM, Denev PN, Bley T, Pavlov AI (2010) Antioxidant activity and phenolic content of betalain extracts from intact plants and hairy root cultures of the red beetroot *Beta vulgaris* cv. Detroit dark red. Plant Foods Hum Nutr 65(2):105–111

Germeier CU, Frese L (2001) The International Database of *Beta* – an expert system for *Beta* genetic resources. In: Proc 64th Congress, Institut International de Recherches Betteravieres, Bruges, Belgium, 26–27 June 2001. pp. 123–132

Gezginci-Oktayoglu S, Sacan O, Bolkent S, Ipci Y, Kabasakal L, Sener G, Yanardag R (2014) Chard (*Beta vulgaris* L. var. *cicla*) extract ameliorates hyperglycemia by increasing GLUT2 through Akt2 and antioxidant defense in the liver of rats. Acta Histochem 116(1):32–39

Gil MI, Ferreres F, Tomás-Barberán FA (1998) Effect of modified atmosphere packaging on the flavonoids and vitamin C content of minimally processed Swiss chard (*Beta vulgaris* subspecies *cycla*). J Agric Food Chem 46(5):2007–2012

Gil-Pena ML, Sardinero E, Garcia-Serrano P, Schnabel I, Garrido J (1986) Continuous production of volatile fatty acids by acidogenesis of sugar beet vinasse. Environ Technol 7:479–486

Hammer K (1986) *Beta* L. In: Schultze-Motel J (ed) Rudolf Mansfelds Verzeichnis landwirtschaftlicher und gärtnerischer Kulturpflanzen (ohne Zierpflanzen). Akademie-Verlag, Berlin, pp 145–151

Harborne JB, Baxter H, Moss GP (eds) (1999) Phytochemical dictionary. TJ International Ltd, Padstow

Havlíková L, Míková K, Kyzlink V (1985) Red beet pigments as soft drink colorants. Nahrung 29(8):723–730

Hempel J, Böhm H (1997) Betaxanthin pattern of hairy roots from *Beta vulgaris* var. *lutea* and its alteration by feeding of amino acids. Phytochemistry 44:847–852

Hertog MGL, Hollman PCH, Katan MB (1992) Content of potentially anticarcinogenic flavonoids of 28 vege-

tables and fruits commonly consumed in The Netherlands. J Agric Food Chem 40:2379–2383
Hobbs DA, Kaffa N, George TW, Methven L, Lovegrove JA (2012) Blood pressure-lowering effects of beetroot juice and novel beetroot-enriched bread products in normotensive male subjects. Br J Nutr 108(11):2066–2074
Huang AS, Elbe JV (1987) Effect of pH on the degradation and regeneration of betanine. J Food Sci 52(6):1689–1693
Hutnan M, Drtil M, Mrafkova L (2000) Anaerobic biodegradation of sugar beet pulp. Biodegradation 11(4):203–211
Iglesias R, Pérez Y, de Torre C, Ferreras JM, Antolín P, Jiménez P, Rojo MA, Méndez E, Girbés T (2005) Molecular characterization and systemic induction of single-chain ribosome-inactivating proteins (RIPs) in sugar beet (*Beta vulgaris*) leaves. J Exp Bot 56(416):1675–1684
Iglesias R, Pérez Y, Citores L, Ferreras JM, Méndez E, Girbés T (2008) Elicitor-dependent expression of the ribosome-inactivating protein beetin is developmentally regulated. J Exp Bot 59(6):1215–1223
Iglesias R, Citores L, Di Maro A, Ferreras JM (2015) Biological activities of the antiviral protein BE27 from sugar beet (*Beta vulgaris* L.). Planta 241(2):421–433
Ikekita M, Moriya H, Moriwaki C, Rurikawa T (1983) Some properties of anti-bradykinin substance from beet (*Beta vulgaris* L. var. *rapa* Dumort. f. *rubra* DC.) roots. Yakugaku Zasshi 99(6):607–611 (In Japanese)
Indhumathi T, Kannikaparameswari K (2012) Hematopoietic study of the methanolic root extract of *Beta vulgaris* on albino rats-an in vivo study. Int J Pharm Biol Sci 3(4):1005–1015
Inoue K, Yamamoto R (1975a) Growth inhibitor in sugar beet species sphere: for phenolic substance. Nihon Sakumotsu Gakkai-shibu Kenkyu Kogai 72:41–44
Inoue K, Yamamoto R (1975b) The growth inhibitors in sugar beet seed balls: ii. isolation of potassium nitrate as the germination inhibitor and hypocotyl stimulating substance. Jpn J Crop Sci 44(4):465–470 (In Japanese)
Isayenkova J, Wray V, Nimtz M, Strack D, Vogt T (2006) Cloning and functional characterisation of two regioselective flavonoid glucosyltransferases from *Beta vulgaris*. Phytochemistry 67(15):1598–1612
Jain NK, Singhai AK (2012) Protective role of *Beta vulgaris* L. leaves extract and fractions on ethanol-mediated hepatic toxicity. Acta Pol Pharm 69(5):945–950
Jain S, Garg VK, Sharma PK (2011) Anti-inflammatory activity of aqueous extract of *Beta vulgaris* L. J Basic Clin Pharm 2(2):83–86
Jiratanan T, Liu RH (2004) Antioxidant activity of processed table beets (*Beta vulgaris* var, *conditiva*) and green beans (*Phaseolus vulgaris* L.). J Agric Food Chem 52(9):2659–2670
Johnson G, Maag DD, Johnson DK, Thomas RD (1976) The possible role of phytoalexins in the resistance of sugarbeet (*Beta vulgaris*) to *Cercospora beticola*. Physiol Plant Pathol 8:225–230
Jozefkowicz C, Rosi P, Sigaut L, Soto G, Pietrasanta LI, Amodeo G, Alleva K (2013) Loop A is critical for the functional interaction of two *Beta vulgaris* PIP aquaporins. PLoS One 8(3), e57993
Kapadia GJ, Tokuda H, Konoshima T, Nishino H (1996) Chemoprevention of lung and skin cancer by *Beta vulgaris* (beet) root extract. Cancer Lett 100(1-2):211–214
Kapadia GJ, Azuine MA, Rao GS, Arai T, Iida A, Tokuda H (2011) Cytotoxic effect of the red beetroot (*Beta vulgaris* L.) extract compared to doxorubicin (adriamycin) in the human prostate (PC-3) and breast (MCF-7) cancer cell lines. Anticancer Agents Med Chem 11(3):280–284
Kapadia GJ, Rao GS, Ramachandran C, Iida A, Suzuki N, Tokuda H (2013) Synergistic cytotoxicity of red beetroot (*Beta vulgaris* L.) extract with doxorubicin in human pancreatic, breast and prostate cancer cell lines. J Complement Integr Med 10(1):113–122
Kapur A, Sati S, Ranjan A, Gupta P (2012) Screening methanolic extracts of *Beta vulgaris* roots for photoprotective activity. Int J Pharm Pharm Sci 4(4):124–127
Kawa-Rygielska J, Pietrzak W, Regiec P, Stencel P (2013) Utilization of concentrate after membrane filtration of sugar beet thin juice for ethanol production. Bioresour Technol 133:134–141
Kazimierczak R, Hallmann E, Lipowski J, Drela N, Kowalik A, Püssa T, Matt D, Luik A, Gozdowski D, Rembiałkowska E (2014) Beetroot (*Beta vulgaris* L.) and naturally fermented beetroot juices from organic and conventional production: metabolomics, antioxidant levels and anticancer activity. J Sci Food Agric 94(13):2618–2629
Khalili M, Mahdavi MRV (2004) Effect of *Beta vulgaris* extract on triglyceride and cholesterol in diabetic male rats. Iran J Pharm Res 3(Supp 2):55
Khorassani R, Hettwer U, Ratzinger A, Steingrobe B, Karlovsky P, Claassen N (2011) Citramalic acid and salicylic acid in sugar beet root exudates solubilize soil phosphorus. BMC Plant Biol 11:121
Kobayashi N, Schmidt J, Wray V, Schliemann W (2001) Formation and occurrence of dopamine-derived betacyanins. Phytochemistry 56(5):429–436
Kim YL, Han MS, Lee JS, Kim JW, Kim YC (2003) Inhibitory phenolic amides on lipopolysaccharide-induced nitric oxide production in RAW 264.7 cells from *Beta vulgaris* var. *cicla* seeds. Phytother Res 17(8):983–985
Kim IK, Chin YW, Lim SW, Kim YC, Kim JW (2004) Norisoprenoids and hepatoprotective flavone glycosides from the aerial parts of *Beta vulgaris* var. *cicla*. Arch Pharm Res 27(6):600–603
Kirtikar KR, Basu BD (1975) Indian medicinal plants, vol 3, 2nd edn. Interntional Book Distributors, Dehra Dun
Kolattukudy PE, Kronman K, Poulose AJ (1975) Determination of structure and composition of suberin from the roots of carrot, parsnip, rutabaga, turnip, red beet, and sweet potato by combined gas-liquid chro-

matography and mass spectrometry. Plant Physiol 55(3):567–573

Kragh KM, Nielsen JE, Nielsen KK, Dreboldt S, Mikkelsen JD (1995) Characterization and localization of new antifungal cysteine-rich proteins from *Beta vulgaris*. Mol Plant Microbe Interact 8(3):424–434

Kristensen AK, Brunstedt J, Nielsen JE, Kreiberg JD, Mikkelsen JD, Roepstorff P, Nielsen KK (2000) Partial characterization and localization of a novel type of antifungal protein (IWF6) isolated from sugar beet leaves. Plant Sci 159(1):29–38

Křižnik B, Pavoković D (2010) Enhancement of betanin yield in transformed cells of sugar beet (*Beta vulgaris* L.). Acta Bot Croat 69(2):173–182

Krumbiegel A (1998) Morphology and anatomy in annual taxa of *Beta vulgaris* s.l. (Chenopodiaceae). Nord J Bot 18(2):159–167

Kugler F, Stintzing FC, Carle R (2004) Identification of betalains from petioles of differently colored Swiss chard (*Beta vulgaris* L. ssp. *cicla* [L.] Alef. Cv. Bright Lights) by high-performance liquid chromatography-electrospray ionization mass spectrometry. J Agric Food Chem 52(10):2975–2981

Kugler F, Graneis S, Stintzing FC, Carle R (2007) Studies on betaxanthin profiles of vegetables and fruits from the Chenopodiaceae and Cactaceae. Z Naturforsch C 62(5-6):311–318

Kujala TS, Loponen JM, Klika KD, Pihlaja K (2000) Phenolics and betacyanins in red beetroot (*Beta vulgaris*) root: distribution and effect of cold storage on the content of total phenolics and three individual compounds. J Agric Food Chem 48(11):5338–5342

Kujala TS, Klika KD, Ovcharenko V, Loponen JM, Vienola MS, Pihlaja K (2001a) 5,5′,6,6′-Tetrahydroxy-3,3′-biindolyl from beetroot (*Beta vulgaris*) peel extract. Z Naturforsch C 56(9-10):714–718

Kujala TS, Loponen JM, Pihlaja K (2001b) Betalains and phenolics in red beetroot (*Beta vulgaris*) peel extracts: extraction and characterisation. Z Naturforsch C 56(5-6):343–348

Kujala TS, Vienola MS, Klika KD, Loponen JM, Pihlaja K (2002) Betalain and phenolic compositions of four beetroot (Beta vulgaris) cultivars. Eur Food Res Technol 214(6):505–510

Lange W, Brandenburg WA, de Bock TSM (1999) Taxonomy and cultonomy of beet (*Beta vulgaris* L.). Bot J Linn Soc 130(1):81–96

Lechner JF, Wang LS, Rocha CM, Larue B, Henry C, McIntyre CM, Riedl KM, Schwartz SJ, Stoner GD (2010) Drinking water with red beetroot food color antagonizes esophageal carcinogenesis in N-nitrosomethylbenzylamine-treated rats. J Med Food 13(3):733–739

Lee CH, Wettasinghe M, Bolling BW, Ji LL, Parkin KL (2005) Betalains, phase II enzyme-inducing components from red beetroot (*Beta vulgaris* L.) extracts. Nutr Cancer 53(1):91–103

Lee JH, Son CW, Kim MY, Kim MH, Kim HR, Kwak ES, Kim S, Kim MR (2009) Red beet (*Beta vulgaris* L.) leaf supplementation improves antioxidant status in C57BL/6J mice fed high fat high cholesterol diet. Nutr Res Pract 3(2):114–121

Lee EJ, An D, Nguyen CT, Patil BS, Kim J, Yoo KS (2014) Betalain and betaine composition of greenhouse- or field-produced beetroot (*Beta vulgaris* L.) and inhibition of HepG2 cell proliferation. J Agric Food Chem 62(6):1324–1331

Leigh RA, Rees T, Fuller WA, Banfield J (1979) The location of acid in ivertase activity and sucrose in the vacuoles of storage roots of beetroot (Beta vulgaris). Biochem J 178(3):539–547

Letschert JP (1993) Beta section beta: biogeographical patterns of variation, and taxonomy. Wageningen Agric Univ Pap 93:1

Letschert JPW, Lange W, Frese L, Van Der Berg D (1994) Taxonomy of the section Beta. J Sugar Beet Res 31:69–85

Levigne S, Ralet MC, Thibault JF (2002) Characterisation of pectins extracted from fresh sugar beet under different conditions using an experimental design. Carbohydr Polym 49(2):145–153

Levigne SV, Ralet MC, Quéméner BC, Pollet BN, Lapierre C, Thibault JF (2004) Isolation from sugar beet cell walls of arabinan oligosaccharides esterified by two ferulic acid monomers. Plant Physiol 134(3):1173–1180

Lu G, Fellman JK, Edwards CG, Mattinson DS, Navazio J (2003a) Quantitative determination of geosmin in red beets (*Beta vulgaris* L.) using headspace solid-phase microextraction. J Agric Food Chem 51(4):1021–1025

Lu G, Edwards CG, Fellman JK, Mattinson DS, Navazio J (2003b) Biosynthetic origin of geosmin in red beets (*Beta vulgaris* L.). J Agric Food Chem 51(4):1026–1029

Lugasi A, Hovari J (2000) Flavonoid aglycons in foods of plant origin I. Vegetables. Acta Aliment 29:345–352

Luoto S, Lambert W, Blomqvist A, Emanuelsson C (2008) The identification of allergen proteins in sugar beet (*Beta vulgaris*) pollen causing occupational allergy in greenhouses. Clin Mol Allerg 6:7

Lv C, Wang Y, Wang LJ, Li D, Adhikari B (2013) Optimization of production yield and functional properties of pectin extracted from sugar beet pulp. Carbohydr Polym 95(1):233–240

Ma S, Wang ZH (2013) Pulsed electric field-assisted modification of pectin from sugar beet pulp. Carbohydr Polym 92(2):1700–1704

Ma S, Yu SJ, Zheng XL, Wang XX, Bao QD, Guo XM (2013) Extraction, characterization and spontaneous emulsifying properties of pectin from sugar beet pulp. Carbohydr Polym 98(1):750–753

Magne V, Mathlouthi M, Robilland B (1998) Determination of some organic acids and inorganic anions in beet sugar by ionic HPLC. Food Chem 61(4):449–453

Mann V, Ekstein I, Nissen H, Hiser C, McIntosh L, Hirschberg J (1991) The cytochrome oxidase II gene in mitochondria of the sugar-beet *Beta vulgaris* L. Plant Mol Biol 17(3):559–566

Marsili RT, Miller N, Kilmer GJ, Simmons RE (1994) Identification and quantitation of the primary chemi-

cals responsible for the characteristic malodor of beet sugar by purge and trap GC-MS-OD techniques. J Chromatogr Sci 32(5):165–171

Martin SS (1977) Accumulation of the flavonoids betagarin and betavulgarin in *Beta vulgaris* infected by the fungus *Cercospora beticola*. Physiol Plant Pathol 11:297–303

Martin SS (1989) Analysis of constitutive and induced phenolics of *Beta vulgaris* by high performance liquid chromatography. J Sugar Beet Res 26(2):33–39

Martinez RM, Longhi-Balbinot DT, Zarpelon AC, Staurengo-Ferrari L, Baracat MM, Georgetti SR, Sassonia RC, Verri WA Jr, Casagrande R (2015) Anti-inflammatory activity of betalain-rich dye of *Beta vulgaris*: effect on edema, leukocyte recruitment, superoxide anion and cytokine production. Arch Pharm Res 38(4):494–504

Martinez-Parra J, Munoz R (2001) Characterization of betacyanin oxidation catalyzed by a peroxidase from *Beta vulgaris* L. roots. J Agric Food Chem 49(8):4064–4068

Massart L (1957) Inhibitors of germination of sugar-beet roots and other dry fruit and seeds. Biokhimiya 22:417–420

Massiot G, Dijoux MG, Lavaud C, Le Men-Olivier L, Connolly JD, Sheeley DM (1994) Seco-glycosides of oleanolic acid from *Beta vulgaris*. Phytochemistry 37(6):1667–1670

Masuda H, Murata M, Takahashi T, Sugawara S (1988) Purification and properties of glucoamylase from sugar beet cells in suspension culture. Plant Physiol 88(1):172–177

Micard V, Renard CMGC, Thibault J-F (1996) Enzymatic saccharification of sugarbeet pulp. Enzym Microbial Technol 19:162–170

Micard V, Grabber JH, Ralph J, Renard CMGC, Thibault J-F (1997) Dehydrodiferulic acids from sugar-beet pulp. Phytochemistry 44(7):1365–1368

Mitchell ED, Tolbert NE (1968) Isolation from sugar beet fruit and characterization of *cis*-4-cyclohexene-1, 2-dicarboximide as a germination inhibitor. Biochemistry 7(3):1019–1025

Montoya G, Yruela I, Picorel R (1991) Pigment stoichiometry of a newly isolated D1-D2-Cyt b559 complex from the higher plant *Beta vulgaris* L. FEBS Lett 283(2):255–258

Montoya G, Cases R, Yruela I, Picorel R (1993) Spectroscopic characterization of two forms of the D1-D2-cytochrome b559 complex from sugar beet. Photochem Photobiol 58(5):724–729

Moore SJ, Godshall MA, Grimm CC (2003) Comparison of two methods of volatile analysis for determining the causes of off-odors in white beet sugars SPME and headspace. Int Sugar J 105(1253):224–229

Mroczek A, Kapusta I, Janda B, Janiszowska W (2012) Triterpene saponin content in the roots of red beet (*Beta vulgaris* L.) cultivars. J Agric Food Chem 60(50):12397–12402

Murakami T, Matsuda H, Inadzuki M, Hirano K, Yoshikawa M (1999) Medicinal Foodstuffs. XVI. (1) Sugar beet. (3): Absolute stereostructures of betavulgarosides II and IV, hypoglycemic saponins having a unique substituent, from the roots of *Beta vulgaris* L. Chem Pharm Bull 47(12):1717–1724

Murphy M, Eliot K, Heuertz RM, Weiss E (2012) Whole beetroot consumption acutely improves running performance. J Acad Nutr Diet 112(4):548–552

Murray KE, Bannister PA, Buttery RG (1975) Geosmin: an important volatile constituent of beetroot (*Beta vulgaris*). Chem Ind (Lond) 22:973–974

Murthy KNC, Manchali S (2012) Anti-diabetic potentials of red beet pigments and other constituents. In: Neelwarne B (ed) Red beet biotechnology: food and pharmaceutical applications. Springer, New York, pp 155–174

Nadkarni KM (2000) Indian materia medica Vol I & II. Popular Prakashan, Bombay

Nagase H, Hojima Y, Moriwaki C, Moriya H (1975) Anti-bradykinin activity found in beet (*Beta vulgaris* L. var *rapa* Dumort. f. *rubra* DC.). Chem Pharm Bull 23(5):971–979

Nandi P, Talukder G, Sharma A (1997) Dietary supplementation with leaf extract of *Beta vulgaris* L. var. *benghalensis* Hort. in modifying cytotoxicity of lead subacetate in mouse *in vivo*. Phytother Res 11:273–276

Nemzer B, Pietrzkowski Z, Spórna A, Stalica P, Thresher W, Michałowski T, Wybraniec S (2011) Betalainic and nutritional profiles of pigment-enriched red beet root (*Beta vulgaris* L.) dried extracts. Food Chem 127(1):42–53

Netzel M, Stintzing FC, Quaas D, Strass G, Carle R, Bitsch R, Bitsch I, Frank T (2005) Renal excretion of antioxidative constituents from red beet in humans. Food Res Int 38(8):1051–1058

Nidhal KM, Abdul-Jalil TZ, Alhamdany AT, Janabi HA (2014) Phytochemical study of the Iraqi *Beta vulgaris* leaves and its clinical applications for the treatment of different dermatological diseases. World J Pharm Pharm Sci 3(8):5–19

Nielsen KK, Nielsen JE, Madrid SM, Mikkelsen JD (1996) New antifungal proteins from sugar beet (*Beta vulgaris* L.) showing homology to non-specific lipid transfer proteins. Plant Mol Biol 31(3):539–552

Nielsen KK, Nielsen JE, Madrid SM, Mikkelsen JD (1997) Characterization of a new antifungal chitin-binding peptide from sugar beet leaves. Plant Physiol 113(1):83–91

Ninfali P, Angelino D (2013) Nutritional and functional potential of *Beta vulgaris* cicla and rubra. Fitoterapia 89:188–199

Ninfali P, Bacchiocca M, Antonelli A, Biagiotti E, Di Gioacchino AM, Piccoli G, Stocchi V, Brandi G (2007) Characterization and biological activity of the main flavonoids from Swiss Chard (*Beta vulgaris* subspecies *cycla*). Phytomedicine 14(2-3):216–221

Nuñez A, Fishman ML, Fortis LL, Cooke PH, Hotchkiss AT Jr (2009) Identification of extensin protein associated with sugar beet pectin. J Agric Food Chem 57(22):10951–10958

Nyirády P, Sárdi E, Beko G, Szucs M, Horváth A, Székely E, Szentmihályi K, Romics I, Blázovics A (2010) Effects of bioactive molecules of *Beta vulgaris* L. ssp. *esculenta* var. *rubra* on metastatic prostate cancer. Orv Hetil 151(37):1495–1503 (In Hungarian)

Olmos JC, Hansen MZ (2012) Enzymatic depolymerization of sugar beet pulp: Production and characterization of pectin and pectic-oligosaccharides as a potential source for functional carbohydrates. Chem Eng J 192:29–36

Owens HS, Goodban AE, Stark JB (1953) Fractionation of organic acids in sugar beets by ion exchange resins. Anal Chem 25(10):1507–1511

Oyen LPA (1994) *Beta vulgaris* L. In: Siemonsma JS, Piluek K (eds). Plant resources of South-east Asia No 8 Vegetables. PROSEA Foundation, Bogor, pp 97–101

Oyen LPA (2004) *Beta vulgaris* L. [Internet] Record from PROTA4U. Grubben GJH, Denton OA (eds) PROTA (Plant Resources of Tropical Africa / Ressources végétales de l'Afrique tropicale), Wageningen. http://www.prota4u.org/search.asp

Ozsoy-Sacan O, Karabulut-Bulan O, Bolkent S, Yanardag R, Ozgey Y (2004) Effects of chard (*Beta vulgaris* L. var *cicla*) on the liver of the diabetic rats: a morphological and biochemical study. Biosci Biotechnol Biochem 68(8):1640–1648

Pal R, Girhepunje K, Gevariya HB, Thirumoorthy N (2010) Hepatoprotective activity of *Beta vulgaris* against CCl4 induced acute hepatotoxicity in rats. Arch Appl Sci Res 2(1):14–18

Parliment TH, Kolor MG, Maing IY (1977) Identification of the major volatile components of cooked beets. J Food Sci 42(6):1592–1593

Pavlov A, Kovatcheva P, Georgiev V, Koleva I, Ilieva M (2002) Biosynthesis and radical scavenging activity of betalains during the cultivation of red beet (*Beta vulgaris*) hairy root cultures. Z Naturforsch C 57(7-8):640–644

Pavlov A, Georgiev V, Kovatcheva P (2003) Relationship between type and age of the inoculum cultures and betalains biosynthesis by *Beta vulgaris* hairy root culture. Biotechnol Lett 25(4):307–309

Pavlov A, Kovatcheva P, Tuneva D, Ilieva M, Bley T (2005) Radical scavenging activity and stability of betalains from *Beta vulgaris* hairy root culture in simulated conditions of human gastrointestinal tract. Plant Foods Hum Nutr 60(2):43–47

Peck GA, Moffat DA (1959) Allergy to the pollen of the common sugar beet (*Beta vulgaris*). J Allerg 30(2):140–150

Petkevicius S, Bach Knudsen KE, Murrell KD, Wachmann H (2003) The effect of inulin and sugar beet fibre on *Oesophagostomum dentatum* infection in pigs. Parasitology 127(Pt 1):61–68

Piattelli M, Minale L, Prota G (1965) Pigments of Centrospermae. III. Betaxanthins from *Beta vulgaris* L. Phytochemistry 4(1):121–125

Pihlsgård P (2000) Aroma and flavor in liquid beet sugar processing. Chalmers Tekniska Högsk, Gothenburg, 44 pp

Pihlsgård P, Larsson M, Leufvén A, Lingnert H (1999) Chemical and sensory properties of liquid beet sugar. J Agric Food Chem 47(10):4346–4352

Pihlsgård P, Larsson M, Leufvén A, Lingnert H (2000) Volatile compounds in the production of liquid beet sugar. J Agric Food Chem 48(10):4844–4850

Pihlsgård P, Leufvén A, Lingnert H (2001) Aroma compounds in the production of liquid beet sugar. J Agric Food Chem 49(8):3875–3880

Pinna M, Roberto S, Milia R, Marongiu E, Olla S, Loi A, Migliaccio GM, Padulo J, Orlandi C, Tocco F, Concu A, Crisafulli A (2014) Effect of beetroot juice supplementation on aerobic response during swimming. Nutrients 6(2):605–615

Pokluda R, Kuben J (2002) Comparison of selected Swiss chard (*Beta vulgaris* ssp. *cicla* L.) varieties. Hortic Sci 29:114–118

Porcher MH et al. (1995–2020) Searchable world wide web multilingual multiscript plant name database. Published by The University of Melbourne. Australia. http://www.plantnames.unimelb.edu.au/Sorting/Frontpage.html

Prahoveanu E, Eşanu V, Anton G, Frunzulică S (1986) Prophylactic effect of a *Beta vulgaris* extract on experimental influenza infection in mice. Virologie 37(2):121–123

Pyo YH, Lee TC, Logendra L, Rosen RT (2004) Antioxidant activity and phenolic compounds of Swiss chard (*Beta vulgaris* subspecies *cycla*) extracts. Food Chem 85:19–26

Qi X, Tai CY, Wasserman BP (1995) Plasma membrane intrinsic proteins of *Beta vulgaris* L. Plant Physiol 108(1):387–392

Ralet MC, Cabrera JC, Bonnin E, Quéméner B, Hellìn P, Thibault JF (2005) Mapping sugar beet pectin acetylation pattern. Phytochemistry 66(15):1832–1843

Raupp DDS, Rodrigues E, Rockenbach II, Carbonar A, Campos PFD, Borsato AV, Fett R (2011) Effect of processing on antioxidant potential and total phenolics content in beet (*Beta vulgaris* L.). Food Sci Technol 31(3):688–693

Rezić T, Oros D, Marković I, Kracher D, Ludwig R, Santek B (2013) Integrated hydrolyzation and fermentation of sugar beet pulp to bioethanol. J Microbiol Biotechnol 23(9):1244–1252

Ridout CL, Price KR, Parkin G, Dijoux MG, Lavaud C (1994) Saponins from sugar beet and the floc problem. J Agric Food Chem 42(2):279–282

Rombouts FM, Thibault JF (1986a) Enzymic and chemical degradation and the fine structure of pectins from sugar-beet pulp. Carbohydr Res 154(1):189–203

Rombouts FM, Thibault JF (1986b) Feruloylated pectic substances from sugar-beet pulp. Carbohydr Res 154(1):177–187

Rombouts FM, Thibault JF (1986c) Sugar beet pectins: chemical structure and gelation through oxidative coupling. In: Fishman ML, Jen JJ (eds) Chemistry and function of pectins. Am. Chem. Soc. Chapter 5, pp 49–60

Roszkowski K (2012) Influence of *Beta vulgaris* supplementation on the level of oxidative and damage/oxidative stress in head and neck cancer patients undergoing radiotherapy. J Nucl Med Radiat Ther 3:4

Rudrappa T, Neelwarne B (2005) Peroxidase production from hairy root cultures of red beet (*Beta vulgaris*). Electr J Biotechnol 8:2

Sacan O, Yanardag R (2010) Antioxidant and antiacetylcholinesterase activities of chard (*Beta vulgaris* L. var. *cicla*). Food Chem Toxicol 48(5):1275–1280

Saguy I (1979) Thermostability of red beet pigments (betanine and vulgaxanthin–I): influence of pH and temperature. J Food Sci 44(5):1554–1555

Saguy I, Kopelman IJ, Mizrahi S (1978) Thermal kinetic degradation of betanin and betalamic acid. J Agric Food Chem 26(2):360–362

Sakihama Y, Maeda M, Hashimoto M, Tahara S, Hashidoko Y (2012) Beetroot betalain inhibits peroxynitrite-mediated tyrosine nitration and DNA strand cleavage. Free Radic Res 46(1):93–99

Sapers GM, Hornstein JS (1979) Varietal differences in colorant properties and stability of red beet pigments. J Food Sci 44(4):1245–1248

Sebeson JM, Snyder FW (1969) Gallic acid in sugarbeet fruits. J Am Soc Sugar Beet Technol 15:617–624

Sener G, Saçan O, Yanardağ R, Ayanoğlu-Dülger G (2002) Effects of chard (*Beta vulgaris* L. var. *cicla*) extract on oxidative injury in the aorta and heart of streptozotocin-diabetic rats. J Med Food 5(1):37–42

Shih CC, Wiley RC (1982) Betacyanine and betaxanthine decolorizing enzymes in the beet (*Beta vulgaris* L.) root. J Food Sci 47(1):164–166

Shin KS, Chakrabarty D, Ko JY, Han SS, Paek KY (2003) Sucrose utilization and mineral nutrient uptake during hairy root growth of red beet (*Beta vulgaris* L.) in liquid culture. Plant Growth Regul 39(2):187–193

Siervo M, Lara J, Ogbonmwan I, Mathers JC (2013) Inorganic nitrate and beetroot juice supplementation reduces blood pressure in adults: a systematic review and meta-analysis. J Nutr 143(6):818–826

Singer JW, von Elbe JH (1980) Degradation rates of vulgaxanthine I. J Food Sci 45:489–491

Smigocki AC, Ivic-Haymes S, Li H, Savić J (2013) Pest protection conferred by a *Beta vulgaris* serine proteinase inhibitor gene. PLoS One 8(2), e57303

Spagnuolo M, Crecchio C, Pizzigallo MDR, Ruggiero P (1997) Synergistic effects of cellulotic and pectinolytic enzymes in degrading sugar beet pulp. Biores Technol 60:215–222

Spórna-Kucab A, Ignatova S, Garrard I, Wybraniec S (2013) Versatile solvent systems for the separation of betalains from processed *Beta vulgaris* L. juice using counter-current chromatography. J Chromatogr B 941:54–61

Spórna-Kucab A, Garrard I, Ignatova S, Wybraniec S (2015) New solvent systems for gradient counter-current chromatography in separation of betanin and its derivatives from processed *Beta vulgaris* L. juice. J Chromatogr A 1380:29–37

Stark JB, Goodban AE, Owens HS (1950) Organic acids in sugar beet diffusion juices. In: Proc Sixth Gen Meeting, Am Soc Sugar Beet Technologists, pp 578–583

Stintzing FC, Schieber A, Carle R (2002) Identification of betalains from yellow beet (*Beta vulgaris* L.) and cactus pear [*Opuntia ficus-indica* (L.) Mill.] by high-performance liquid chromatography-electrospray ionization mass spectrometry. J Agric Food Chem 50(8):2302–2307

Stoppok E, Buchholz K (1985) Continuous anaerobic conversion of sugar beet pulp to biogas. Biotechnol Lett 7(2):119–124

Susín S, Abián J, Sánchez-Baeza F, Peleato ML, Abadía A, Gelpí E, Abadía J (1993) Riboflavin 3′- and 5′-sulfate, two novel flavins accumulating in the roots of iron-deficient sugar beet (*Beta vulgaris*). J Biol Chem 268(28):20958–20965

Suzuki Y, Esumi Y, Koshino H, Doi Y (2005) Isolation and structure determination of complexed poly(3-hydroxyalkanoate) from beet (*Beta vulgaris* L.). Macromol Biosci 5(9):853–862

Takahashi H, Sasaki T, Ito M (1987) New flavonoids isolated from infected sugarbeet roots. Bull Chem Soc Jpn 60(6):2261–2262

Takei T, Sugihara K, Yoshida M, Kawakami K (2013) Injectable and biodegradable sugar beet pectin/gelatin hydrogels for biomedical applications. J Biomater Sci Polym Ed 24(11):1333–1342

The Plant List (2013) Version 1.1. Published on the Internet; http://www.theplantlist.org/. Accessed 1 Jan

Tripathy G, Pradhan D (2013) Evaluation of in-vitro antiproliferative activity and in-vivo immunomodulatory activity of *Beta vulgaris*. Asian J Pharm Clin Res 6(1):127–130

Tunali T, Yarat A, Yanardağ R, Ozçelík F, Ozsoy O, Ergenekon G, Emeklí N (1998) The effect of chard (*Beta vulgaris* L. var. *cicla*) on the skin of streptozotocin induced diabetic rats. Pharmazie 53(9):638–640

Tyler LD, Acree TE, Becker R, Nelson RR, Butts RM (1978a) Effect of maturity, cultivar, field history, and the operations of peeling and coring on the geosmin content of *Beta vulgaris*. Agric Food Chem 26(6):1466–1469

Tyler LD, Acree TE, Nelson RR, Butts RM (1978b) Determination of geosmin in beet juice by gas chromatography. J Agric Food Chem 26(3):774–775

Tyler LD, Acree TE, Smith NL (1979) Sensory evaluation of geosmin in juice made from cooked beets. J Food Sci 44(1):79–81

U.S. Department of Agriculture, Agricultural Research Service (USDA-ARS) (2014) USDA national nutrient database for standard reference, Release 27. Nutrient Data Laboratory Home Page, http://www.ars.usda.gov/ba

Ul Kabir A, Samad MB, Ahmed A, Jahan MR, Akhter F, Tasnim J, Hasan SM, Sayfe SS, Hannan JM (2015) Aqueous fraction of *Beta vulgaris* ameliorates hyperglycemia in diabetic mice due to enhanced glucose stimulated insulin secretion, mediated by acetylcho-

line and GLP-1, and elevated glucose uptake via increased membrane bound GLUT4 transporters. PLoS One 10(2), e0116546

Ursing BO (1968) Sugar beet pollen allergy as an occupational disease. Allergy 23(5):396–399

USDA, ARS, National Genetic Resources Program (2014) Germplasm resources information network – *(GRIN)* [Online Database]. National Germplasm Resources Laboratory, Beltsville, Maryland. http://www.ars-grin.gov. Accessed 9 Apr 2015

Váli L, Fébel H, Stefanovits-Bányai É, Sárdi É, Lugasi A, Szentmihályi K, Blázovics A (2006) Duodenum protecting effects of table beet (*Beta vulgaris* L. ssp. *esculenta* var. *rubra*) during hepatic ischaemia-reperfusion. Acta Aliment 35(4):445–453

Váli L, Stefanovits-Bányai E, Szentmihályi K, Fébel H, Sárdi E, Lugasi A, Kocsis I, Blázovics A (2007) Liver-protecting effects of table beet (*Beta vulgaris* var. *rubra*) during ischemia-reperfusion. Nutrition 23(2):172–178

Van den Ende W, De Coninck B, Clerens S, Vergauwen R, Van Laere A (2003) Unexpected presence of fructan 6-exohydrolases (6-FEHs) in non-fructan plants: characterization, cloning, mass mapping and functional analysis of a novel "cell-wall invertase-like" specific 6-FEH from sugar beet (*Beta vulgaris* L.). Plant J 36(5):697–710

Vulić JJ, Cebović TN, Canadanović VM, Cetković GS, Djilas SM, Canadanović-Brunet JM, Velićanski AS, Cvetković DD, Tumbas VT (2013) Antiradical, antimicrobial and cytotoxic activities of commercial beetroot pomace. Food Funct 4(5):713–721

Waldron KW, Ng A, Parker ML, Parr AJ (1997) Ferulic acid dehydrodimers in the cell walls of *Beta vulgaris* and their possible role in texture. J Sci Food Agric 74:221–228

Wang M, Goldman IL (1997) Accumulation and distribution of free folic acid content in red beet (*Beta vulgaris* L.). Plant Foods Hum Nutr 50(1):1–8

Webb AJ, Patel N, Loukogeorgakis S, Okorie M, Aboud Z, Misra S, Rashid R, Miall P, Deanfield J, Benjamin N, MacAllister R, Hobbs AJ, Ahluwalia A (2008) Acute blood pressure lowering, vasop®rotective, and antiplatelet properties of dietary nitrate via bioconversion to nitrite. Hypertension 51:784–790

Weiland P (1993) One-and two-step anaerobic digestion of solid agroindustrial residues. Water SciTechnol 27:145–151

Westall RG (1950) Isolation of gamma-amino-butyric acid from beetroot (*Beta vulgaris*). Nature 165(4201):717

Wettasinghe M, Bolling B, Plhak L, Xiao H, Parkin K (2002) Phase II enzyme-inducing and antioxidant activities of beetroot (*Beta vulgaris* L.) extracts from phenotypes of different pigmentation. J Agric Food Chem 50(23):6704–6707

Wikipedia (2015) Beet pulp. http://en.wikipedia.org/wiki/Beet_pulp

Winter M, Herrmann K (1986) Esters and glucosides of hydroxycinnamic acids in vegetables. J Agric Food Chem 34:616–620

Wybraniec S (2005) Formation of decarboxylated betacyanins in heated purified betacyanin fractions from red beet root (*Beta vulgaris* L.) monitored by LC-MS/MS. J Agric Food Chem 53(9):3483–3487

Wyler H, Meuer U, Bauer J, Stravs-Mombelli L (1984) Cyclodopa glucoside (=(2S)-5-(β-D-glocopyranosyloxy)-6-hydroxyindoline-2-carboxylic acid) and its occurrence in red beet (beta vulgaris var. rubra L.). Helv Chim Acta 67(5):1348–1355

Wylie LJ, Kelly J, Bailey SJ, Blackwell JR, Skiba PF, Winyard PG, Jeukendrup AE, Vanhatalo A, Jones AM (2013) Beetroot juice and exercise: pharmacodynamic and dose-response relationships. J Appl Physiol (1985) 115(3):325–336

Yanardağ R, Bolkent S, Ozsoy-Saçan O, Karabulut-Bulan O (2002) The effects of chard (*Beta vulgaris* L. var. *cicla*) extract on the kidney tissue, serum urea and creatinine levels of diabetic rats. Phytother Res 16(8):758–761

Yoshikawa M, Murakami T, Kadoya M, Matsuda H, Yamahara J, Muraoka O, Murakami N (1995) Betavulgarosides I, II, III, IV, and V, hypoglycemic glucuronide saponins from the roots and leaves of *Beta vulgari*s L. (sugar beet). Heterocycles 41(8):621–1626

Yoshikawa M, Murakami T, Kadoya M, Matsuda H, Muraoka O, Yamahara J, Murakami N (1996) Medicinal Foodstuffs. III. Sugar beet. (1): hypoglycemic oleanolic acid oligoglycosides, betavulgaroside I, II, III and IV from the root of *Beta vulgaris* L. (Chenopodiaceae). Chem Pharm Bull 44(6):1212–1217

Yoshikawa M, Murakami T, Inaduki M, Hirano K, Yamahara J, Matsuda H (1997) Absolute stereostuctures of betavulgarosides III and IV inhibitors of glucose absorption from the roots of Beta vulgaris L. (sugar beet). Chem Pharm Bull 45(3):561–563

Yoshikawa M, Murakami T, Kadoya M, Yamahara J, Matsuda H (1998) Medicinal foodstuffs. XV. Sugar beet. (2): structures of betavulgarosides V, VI, VII, VIII, IX and X from the root and leaves of sugar beet (*Beta vulgaris* L., Chenopodiaceae). Chem Pharm Bull 46(11):1758–1763

Zakharova NS, Petrova TA (2000) beta-Glucosidase of leaves and roots of the common beet *Beta vulgaris*. Prikl Biokhim Mikrobiol 36(4):458–461 (In Russian)

Zhu GL, Mosyakin SL, Clemants SE (2003) Chenopodiaceae Ventenat. In: Wu ZY, Raven PH, Hong DY (eds) Flora of China. Vol. 5 (Ulmaceae through Basellaceae). Science Press, Beijing; Missouri Botanical Garden Press, St. Louis

Canna indica

Scientific Name

Canna indica **L.**

Synonyms

Canna achiras Gillies ex D. Don, *Canna achiras* Gill., *Canna altensteinii* Bouché, *Canna amabilis* T.Koyama & Nob.Tanaka, *Canna ascendens* Ciciar.,*Canna aurantiaca* Roscoe, *Canna aureovittata* Lodd., *Canna barbadica* Bouché (inval.), *Canna bidentata* Bertol., *Canna bifida* Roem. & Schult., *Canna brasiliensis* Roscoe ex Spreng., *Canna carnea* Roscoe, *Canna cearensis* Huber, *Canna chinensis* Willd. (illeg.), *Canna cinnabarina* Bouché, *Canna coccinea* Mill., *Canna coccinea* Roscoe, *Canna coccinea* var. *bicolor* Kraenzl., *Canna coccinea* var. *concolor* Regel, *Canna coccinea* f. *flaviflora* Chodat & Hassl., *Canna coccinea* var. *floribunda* (Bouché) Regel, *Canna coccinea* var. *limbata* Regel, *Canna commutata* Bouché, *Canna compacta* Roscoe, *Canna concinna* Bouche, *Canna crocea* Lag. ex Rchb., *Canna crocea* Roem. & Schult., *Canna densiflora* Bouché, *Canna densifolia* Bouché, *Canna discolor* Lindl., *Canna discolor* var. *rubripunctata* Nob.Tanaka, *Canna discolor* var. *viridifolia* Nob.Tanaka, *Canna edulis* Ker Gawl., *Canna ehrenbergii* Bouché, *Canna elegans* Raf. (illeg.), *Canna ellipticifolia* Stokes (illeg.), *Canna ellipticifolia* var. *coccinea* (Mill.) Stokes, *Canna ellipticifolia* var. *lutea* (Mill.) Stokes, *Canna ellipticifolia* var. *patens* (Aiton) Stokes, *Canna ellipticifolia* var. *rubra* Stokes, *Canna esculenta* Loudon (inval.), *Canna exigua* Bouché, *Canna eximia* Bouché ex Horan., *Canna flavescens* Link, *Canna floribunda* Bouché, *Canna formosa* Bouché, *Canna fulgida* Bouché, *Canna heliconiifolia* Bouché, *Canna heliconiifolia* var. *xalapensis* (Bouché) Kraenzl., *Canna humilis* Bouché, *Canna indica* Curtis, *Canna indica* var. *coccinea* (Mill.) Aiton, *Canna indica* var. *coccinea* Willd., *Canna indica* var. *edwardsii* Regel, *Canna indica* var. *flava* (Roscoe) Baker, *Canna indica* var. *karsteniana* Regel, *Canna indica* var. *limbata* (Regel) Petersen, *Canna indica* var. *lutea* (Mill.) Aiton, *Canna indica* var. *maculata* Hook., *Canna indica* var. *nepalensis* (Bouché) Baker, *Canna indica* subsp. *orientalis* Baker, *Canna indica* var. *patens* Aiton, *Canna indica* var. *rubra* Aiton, *Canna indica* f. *rubroaurantiaca* Makino, *Canna indica* var. *sanctaerosae* (Kraenzl.) Nob.Tanaka, *Canna indica* var. *variegata* Regel, *Canna indica* var. *warszewiczii* Nob.Tanaka, *Canna juncea* Retz., *Canna laeta* Bouché, *Canna lagunensis* Lindl., *Canna lambertii* Lindl. ex Ker Gawl., *Canna lanuginosa* Roscoe, *Canna leptochila* Bouché, *Canna limbata* Roscoe (illeg.), *Canna lutea* Mill., *Canna lutea* Larrañaga (illeg.), *Canna lutea* Baker, *Canna lutea* var. *aurantiaca* (Roscoe) Regel, *Canna lutea* var. *maculata* (Hook.) Regel, *Canna lutea* var. *pallida* (Roscoe) Regel, *Canna macro-*

T.K. Lim, *Edible Medicinal and Non-Medicinal Plants: Volume 10, Modified Stems, Roots, Bulbs*,
DOI 10.1007/978-94-017-7276-1_3

phylla Horan., *Canna maculata* (Hook.) Link, *Canna maxima* Lodd. ex Roscoe (inval.), *Canna montana* Blume, *Canna moritziana* Bouché, *Canna nepalensis* Bouché, *Canna nepalensis* D. Dietr., *Canna occidentalis* Ker Gawl., *Canna orientalis* Bouché (illeg.), *Canna orientalis* Roscoe (illeg.), *Canna orientalis* var. *flava* Roscoe, *Canna pallida* Roscoe, *Canna pallida* var. *maculata* (Hook.) Roscoe, *Canna patens* (Aiton) Roscoe, *Canna patens* Hook., *Canna patens* var. *limbata* (Regel) Baker, *Canna pentaphylla* D. Dietr. (Spelling variant), *Canna platyphylla* Nees & Mart., *Canna plurituberosa* T.Koyama & Nob.Tanaka, *Canna poeppigii* Bouché, *Canna polyclada* Wawra, *Canna polymorpha* Bouché, *Canna portoricensis* Bouché, *Canna pruinosa* Hoffmanns., *Canna pulchra* Bouché ex Horan., *Canna pulchra* Hassk., *Canna recurvata* Bouché, *Canna roscoeana* Bouché (illeg.), *Canna rotundifolia* André, *Canna rubra* Willd. (illeg.), *Canna rubricaulis* Link, *Canna sanctae-rosae* Kraenzl., *Canna sanguinea* Bouché (inval.), *Canna sanguinea* Warsz. ex Otto & A.Dietr., *Canna saturate-rubra* Bouché ex K.Koch, *Canna schubertii* Horan., *Canna seleriana* Kraenzl., *Canna sellowii* Bouché, *Canna speciosa* Roscoe ex Sims (illeg.), *Canna speciosa* Hegetschw (illeg.), *Canna spectabilis* Bouché, *Canna sulphurea* Bouché (inval.), *Canna surinamensis* Bouché, *Canna tenuiflora* Bouché ex A. Dietr., *Canna texensis* Regel, *Canna textoria* Noronha (inval.), *Canna thyrsiflora* Hegetschw. (illeg.), *Canna tinei* Tod. (inval.), *Canna variabilis* Willd. (illeg.), *Canna variegata* Besser, *Canna variegata* Bouché (illeg.), *Canna variegatifolia* Ciciar., *Canna ventricosa* Bouché, *Canna warszewiczii* A.Dietr. (illeg.), *Canna warszewiczii* var. *flameus* Ram.Goyena, *Canna xalapensis* Bouché, *Cannacorus indicus* (L.) Medik., *Cannacorus ovatus* Moench (illeg.), *Distemon brasiliensis* (Roscoe ex Spreng.) Bouché, *Distemon grandis* Horan. (illeg.), *Xyphostylis lutea* (Mill.) Raf.

Family

Cannaceae

Common/English Names

Achira, African Arrowfoot, African Arrowroot, African Turmeric, Calenda, Canna, Canna Lily, edible Canna, English Shot, Edible Canna, Edible Arrowroot, English Shot (South America), Indian Canna Arrowroot, Indian Shot, Indian-Shot, Purple Arrowroot, Queensland Arrowroot, Sierra Leone Arrowroot, Wild Tapioca.

Vernacular Names

Afrikaans: Indiese Kanna;
Argentina: Achira, Achera, Achira Colorada;
American Samoa: Fa'I Masoa, Fagamanu;
Arabic: Kanna;
Brazil: Albará, Araruta Bastarda, Bananeirinha-Da-Índia, Bananeirinha-De-Flor, Bandua De Uribe, Beri, Bery, Birú Manso, Caeté-Dos-Jardns, Cana-Da-Índia Merú, Erva-de-conteira, Imbiry (Portuguese);
Bolivia: Achira, Achera;
Burmese: Adalut, Butsarana;
Caribbean Islands: Balisier À Chapelets, Balisier Jaune, Toloman, Tous-Les-Mois (French);
Chamorro: Mongos Halum-Tano;
Chinese: Shi Yong Mei Ren Jiao, Jiao Yu, Chiao Yu, Ba Jiao Yu, Pa Chiao Yu;
Colombia: Capacho, Sugú, Chisqua, Adura, Chumbima, Bandera De Uriba;
Cook Islands: Tiare Papa'A, Nūāenga, Nuāēnga, Pia Renga;
Costa Rica: Tikas, Punyapong, Kaska, Piriquitoya;
Chuukese: Apeellap, Oruuru, Yapellap, Yoruuru;
Czech: Dosna Indická;
Danish: Almindelig Kanna, Kanna Spiselig Kanna;
Dominican Republic: Tolumán;
Dutch: Eetbaar Bloemriet, Indisch Bloemriet, Ganjong;
Estonian: India Kanna;
Ecuador: Luano;
Fiji: Gasau Ni Ga, Ngasau Ni Nga;

French: Balisier Comestible, Balisier Rouge, Balisier Des Indes, Canna D'inde, Canna, Canna Floriféré, Tous-Les-Mois, Safran Marron, Sagou;
French Polynesia: Pia-Raroto'A;
French Reunion: Conflor;
Futuna: Fagafaga;
German: Achira, Blumenrohr, Indisches, Indisches Blumenrohr, Kapocho, Westindisches Blumenrohr;
Guam: Mongos Halum-Tano;
Guyana: Sakasiri;
Hawaii: Ali'Ipoe, Li'Ipoe, Poloka, Poloke;
India: Sarbajaya (Bengali), Sabbajaya, Sarvajjya, Sudarson, Keli (Hindi), Hudingana, Kaelahoo, Kalahu, Sugandharaju, Chare Gundina Gida, Hoo Dingana, Kyaanaa Gida, Kela Hoo, Sarva Jaya, Canna Gida, Kelahu, Hoodingala, Kaabaale, Kaalahoo, Kaela Hoo, Kare Gulaganji, Kyana Gida (Kannada), Kattuvala, Katu-Bala, Katuvara, Katoobala, Kattuvazha, Katubala (Malayalam), Devakeli, Daevakeli, Kardai, Kardali, Dev-Keli, Ran-Keli (Marathi), Kat Champa (Oriya), Devakili, Kamakshee, Kamakshi, Krishnatamara, Sarvajaya, Shilarambha, Silarumba, Vanakadali (Sanskrit), Kalvalai, Kalvalaimani, Kundimani, Puvalai, Siramalai, Kal Valai, Kantamani, Kalvazhai, Kandamani-Cheddi, Kandamanu, Poovalai, Cilaivalai, Kallankari, Kallankaricceti, Kalvalaicceti, Kanalvalai 1, Kattuvalai, Lakiya, Lakiyavalai, Patacitacceti, Patacitam, Tiranapakitam, Tiranapakitavalai, Nilavalai (Tamil), Guruginja, Krishnatamara, Mettatamara, Guriginzda (Telugu);
Indonesia: Ganjong, Lembong Nyidra, Senitra, Buah Tasbeh, Sebeh, Sigi-Sigi, Kembang Gedang, Puspa Nidra, Puspa Midra, Puspa Nyidra (Java), Ganjol, Ganjong, Sebe, Sebeh, Tasbeh (Sundanese), Banjur, Ghnajong, Manjong (Madurese), Laos Jambe, Laos Mekah, Ubi Pikul (Sumatra);
Italian: Canna Comestible, Canna Dolce, Canna Edule, Canna D'india;
Japanese: Akabana Dan Doku, Choukuyou Kana, Kana Indika;
Khmer: Chek Tehs;
Kiribati: Riti;
Laos: Kwayz Ke, Son, Kwayz Phutta Son;
Malaysia: Daun Tasbeh, Gangjong, Pisang Sebaik, Ubi Gereda, Kenyong;
Mexico: Papatla;
Nepalese: Gaane Sarvadaa, Gane Sarvada;
Nigeria: Bakalele, Bakare Kare (Hausa);
Panama: Café Cimarron;
Philippines: Balunsaying, Tapuranga Kolintasan, Tikas-Tikas (Bisaya), Bangali (Bikol), Lasa (Ivatan), Kiuingam (Ifugao), Kakuentasa, Kuentas-Kuentasan Saging-Saging, Tikas, Tikas-Tikas, Tikis-Tikis, Tukas-Tukas, Zembu (Tagalog);
Pohnpei: Gwangwa, Gwangwaama, Luiuenwai;
Polish: Paciorecznik Indyjski;
Portuguese: Cana Comestível, Cana Da Índia, Birú Manso, Merú, Albará, Albará, Araruta Bastarda, Araruta De Porco, Bananeirinha Da India, Bananeirinha Da India De Flor, Bananeirinha-Da-Índia, Bananeirinha-De-Flor, Beri;
Puerto Rico: Gruya;
Puluwat: Apeellap, Oruuru;
Quechua: Achira;
Russian: Kanna Indiiskaia;
South America: Toolima (Creole), Tasca, Tikas-Tikas, Calenda;
Samoa: Fa'I Masoa Fagamanu;
São Tomé e Príncipe: Salaconta;
Spanish: Achir, Achera, Achira, Achira roja, Caña Comestible, Caña De Las Indias, Cañacoro, Chupa Flor, Frutilla, Hierba del Rosario, Platanillo de Cuba, Platanillo, Yuquilla;
Tahitian: Pia-Raroto'A, Re'A Pua'Aniho;
Thai: Sakhu Chin, Puttharksa, Putthason;
Tonga: Te Misimisi;
Venezuela: Capacho;
Vietnamese: Chuoi Hoa, Dong Rieng, Khoai Dao;
West Indies: Tous-Les-Mois, Toloman (French), Maraca, Imocona, Platanillo, Cañacoros Maraca, Imocona, Platanillo, Cañacoros (Spanish).

Origin/Distribution

The plant is native to central, south America (Peru, Chile) and the Caribbean (Antilles). It has widely naturalised in other warm tropical and subtropical countries and is now widely cultivated pan-tropically. It is commonly cultivated as food plant in South and Central America, Libya, tropical Africa, India, Vietnam, the Philippines, Cambodia and Papua New Guinea.

Agroecology

Canna indica grows in various climatic conditions from sea level to 2900 m altitude at the equator. The plant is apparently day-length neutral, and it appears to grow under a broad range of light environments. Areas with a well-distributed mean annual rainfall of 1000–2000 mm is preferred but it tolerates rainfall from 250 to 4000 mm. It thrives in full sun but will tolerate partial shade and is sensitive to drought. It tolerates light frost down to 0 °C and will grow normally at low temperatures above 10 °C and high temperature from 30 to 32 °C. It thrives on many soil types including heavy soils and weathered, acidic, tropical latosols, but prefers deep, fertile, well-drained and well-aerated sandy loams rich in humus. The rhizomes develop poorly in compacted clays. It has a wide pH tolerance from 4.0 to 8.0. In its native range it occurs in the high mountains, the warm Andean valley where temperatures of 20–25 °C are normal and in the warmer plateau with temperature up to 32 °C. It also grows well in Asia and in the Pacific where mean annual temperature ranges from 25 to 35 °C. It occurs in thickets, in wet areas along riversides, riparian zones, water courses, in wastelands, disturbed areas and fallow fields and in rubber plantations as a weed.

Edible Plant Parts and Uses

Edible rhizome can be eaten raw; it is usually eaten cooked like potatoes, cassava or taro as food and as a commercial source of starch (Tanaka 1998, 2004). Once baked, the rhizome becomes translucent, mucilaginous and sweet. In Thailand, boiled canna rhizomes are valued by the locals for their starch content. In Vietnam, a small proportion of canna starch is utilised for unspecific household uses or processed into minor products such as candies, cakes and rice papers (Hermann et al. 1999). However, canna starch is commonly processed into noodles. Canna noodles belong to the class of cellophane or glass noodles with a glossy, transparent appearance. They are rectangular and are 0.8-mm thick. Bland and slippery, they add texture to a variety of Vietnamese dishes, especially to soups and stir-fried dishes. They are regarded as a luxury item and reserved for celebrations. Being less wasteful and less costly to extract, canna starch has totally replaced mung bean starch as the raw material for glass noodle production. In Vietnam, canna noodles made from canna rhizome are popular, usually in soup with chicken pieces (Tanaka 1998, 2004). In China, a village in Guangxi Province, the residents brew an alcoholic beverage called 'Ping fa lou' from the rhizome starch (Tanaka 2004). In Indonesia, cooked young rhizomes are a delicacy and are frequently sold by street hawkers in cities in Java. A traditional Andean feast is baked achira, roast guinea pig, yacon, and quinoa beer. In Hawaii, canna rhizome was formerly used in making 'haupia', a traditional dessert (of coconut milk, starch, sugar and gelatin) served particularly at 'luaus' and local gatherings. In Colombia, the flour is used to make salted crackers in homes and in factories for commercial distribution. It is also mixed with cheese (*colaciones*). The starch is easily digestible and is particularly used as food for children, invalids and convalescing patients. The flour obtained is also used for making cakes, jellies and veal broth. The bakery products prepared from canna starch are much lighter, spongier and crispier than those from wheat products. Young shoots are eaten as green vegetable and young seeds used as an addition to tortillas. In Cambodia, ripe fruits are eaten fresh or as sweets. Red canna flowers serve as a potential source of natural antioxidant and a natural food colorant. Rhizomes

and leaves are used as food material in Myanmar (Indrayan et al. 2011).

Botany

An upright perennial rhizomatous herb with succulent lateral-spreading subterranean rhizomes with fibrous roots (Plates 1 and 2). Stems 1–2 m high, sturdy, glabrous, green. Leaves large, attractive, 20–60 cm long by 10–35 cm wide, ovate–oblong, alternate, thick, green often with reddish tinge along the thin entire margin, with arching parallel veins, acute apex and cuneate base which clasp the stem like a sheath (Plate 3). Flowers in a lax, terminal raceme that arises from leaf sheath, often in pairs, red, yellow or red and yellow (Plate 4). Flowers with three types of

Plate 1 Rhizome with fibrous roots

Plate 2 Cooked rhizome

Plate 3 Large ovate, oblong leaves

pseudo-petals (modified staminodes and petaloid filaments). The three outer staminodes are relatively broad (3.5–6 cm long and 0.5–1.5 cm wide) while the inner staminode (i.e. labellum) is usually somewhat narrower and recurved. The single petaloid filament is narrower again (3–4 mm wide) and has an anther about 10 mm long about half way up one of its sides. The petals are reduced to three, narrow, bract-like structure below the pseudo-petals, 4–6.5 cm long, 0.4–0.7 cm wide, fused together into perianth tube at the base. Sepals 3 narrow, 9–17 mm × 2–5 mm enclosed by floral bract and bracteole. Capsule broadly ovoid to subglobose, 2 × 1.5 cm, green, verrucose (covered with short projections), subtended by persistent sepals, turning brown at maturity (Plate 5). Seeds globose to ovoid, shiny, and black 5–8 mm × 4–7 mm.

Nutritive/Medicinal Properties

Flower Phytochemicals

Fresh red *C. indica* flowers were found to contain 121 mg/kg total carotenes, total xanthophylls 189 mg/kg, total carotenoids 310 mg/kg (Tinoi et al. 2006). The carotenoids were:

Plate 4 Plant in flower

Plate 5 Verrucose, green fruit

β-carotene 15.9 mg/kg, *β*-cryptoxanthin 5.60 mg/kg, lutein 64.5 mg/kg, violaxanthin 25.8 mg/kg, zeaxanthin 15.2 mg/kg and unidentified 183 mg/kg. Four anthocyanin pigments cyanidin-3-*O*-(6″-*O*-*α*-rhamnopyranosyl)-*β*-glucopyranoside; cyanidin-3-*O*-(6″-*O*-*α*-rhamnopyranosyl)-*β*-galactopyranoside; cyanidin-3-*O*-*β*-glucopyranoside; and cyanidin-*O*-*β*-galactopyranoside along with quercetin and lycopene were isolated from *C. indica* red flowers (Srivastava and Vankar 2010a). Methylated anthocyanin glycosides were isolated from red *Canna indica* flower and identified as malvidin 3-*O*-(6-*O*-acetyl-*β*-D-glucopyranoside)-5-*O*-*β*-D-glucopyranoside, malvidin3,5-*O*-β-D-diglucopyranoside,cyanidin-3-*O*-(6″-O-*α*-rhamnopyranosyl-*β*-glucopyranoside, cyanidin-3-*O*-(6″-*O*-*α*-rhamnopyranosyl)-*β*-galactopyranoside, cyanidin-3-*O*-*β*-glucopyranoside and cyanidin-*O*-*β*-galactopyranoside (Srivastava and Vankar 2010b).

Leaf Nutrients/Phytochemicals

The approximate nutrient composition of the fresh leaves used as fodder was reported as: water 90 %, protein 1 %, fat 0.2 % carbohydrates 7 %, ash 1.4 % and digestibility about 20 % (Ong and Siemonsma 1996). Earlier, Arellano et al. (1993) reported the chemical composition and in-vitro digestibility of leaf and stem of *C. indica* for animal feeding as: in-vitro dry matter digestibility 55.4 %, lignin 15.5 %, cell wall 54.4 %, Ca 600 mg/100 g, Ca 421 mg/100 g, tannic acid 518 mg/100 g, trypsin inhibitor 4228 UIT/g; alkaloids were detected but saponins and cyanogenic glucosides were not detected.

Carbon-labelled studies showed that infiltration of canna leaves with C^{14}-labelled fructose resulted in a decrease of C^{14}-fructose and a simultaneous increase of C^{14} activity in sucrose (Putman and Hassid 1954). The infiltrated radioactive sugars also produced labelled aspartic and *γ*-aminobutyric acids, alanine, glutamine, citric, malic, glyceric, succinic and lactic acids.

Rhizome Nutrients/Phytochemicals

The nutritive composition of the rhizome per 100 g edible portion was reported to contain: water 75 g, protein 1 g, fat 0.1 g, carbohydrates 22.6 g, Ca 21 mg, P 70 mg, Fe 20 mg, vitamin B1 0.1 mg and vitamin C 10 mg (Ong and Siemonsma 1996). The carbohydrates comprised >90 % starch and about 10 % sugars (glucose and sucrose). The starch is highly soluble and easily digestible.

Lectin isolated from canna rhizome was found to immobilise mannose (Winter et al. 1995). The lectin was weakly inhibited by glycosides and agglutinated rabbit erythrocytes. A steroid containing β-unsaturated, 5-member lactone ring designated cannagenin was isolated from the ether extract of the root (Motawe 1995). The petroleum ether extract of the rhizome afforded the following hydrocarbons: 5, 8- henicosdiene (3.27 %), 7-henicosane (3.70 %), 3, 15-dihydroxy-2-octadecene (45.12 %), 6-hydroxy eicosane (5.18 %), tricosane (2.40 %)

and tetracosane (1.89 %) (Nirmal et al. 2008). Two pure compounds, stigmasterol and 6 β-hydroxystigmasta-4, 22-diene-3-one were isolated from the rhizome (Chainakul et al. 2001).

Forty-three compounds representing 95.32 % of the oil were isolated from canna rhizome essential oil (Indrayan et al. 2011). The sesquiterpene hydrocarbons and their derivatives formed the major part (52.56 %). Prominent among them were γ-eudesmol (9.79 %), δ-cadinol (6.33 %), γ-selinene (5.23 %) and luciferin (5.05 %), caryophyllene oxide (4.96 %), α-caryophyllene (4.78 %) and *trans*-nerolidol (3.23 %). The major monoterpenes and derivatives were 1-terpinen-4-ol (4.60 %), α-fenchyl acetate (3.26 %), 1,8-cineole (3.17 %) and β-pinene (3.13 %). Four diterpenes in fair quantity (7.14 %) were unidentified. All were oxygenated diterpenes, the major being manool and geranyl linalool (2.75 % each). Geranyl linalool was also found. Palmitic acid was found in fair amount (8.53 %). Other minor compounds included: hexanal 0.05 %, furfural 0.04 %, heptanol 0.06 %, α-pinene 0.73 %, camphene 0.48 %, γ-terpinene 0.71 %, α-terpinolene 0.22 %, *cis*-sabinene hydrate 0.19 %, β-linalool 0.63 %, fenchol 0.46 %, *trans*-pinocarveol 0.40 %, bornyl acetate 0.52 %, 2,3-pinanediol 0.08 %, isobornyl acetate 2.56 %, unidentified 0.60 %, unidentified 1.30 %, geosmin 0.49 %, β-caryophyllene 2.00 %, selina-3,7(11)-diene 1.72 %, carotol 2.72 %, 9-cedranone 2.43 %, α-acorenol 2.49 %, myristic acid 1.83 %, dibutyl phthalate 0.85 %, unidentified 0.65 %, methyl linoleate 0.63 %, unidentified 0.49 %, dodecenyl succinic anhydride 0.98, oleic acid 0.33 %, stearic acid 0.38 %, unidentified 1.29 %, geranyl geraniol acetate 0.50 %, unidentified 0.35 % and 4,8,13-duvatriene-1,3-diol 1.14 %.

Compared to potato starch, edible canna starch was found to have larger granular size (50.8 × 34.5 μm), higher resistance to enzymatic digestion, gelatinized at higher temperature and its paste retrograded more extensively than that of potato starch (Inatsu et al. 1983). It had higher amylose content (27.1 %) but its amylose (DP ≅ 3, 100) was considerably smaller than potato starch amylose (5, 500), whereas no significant difference was observed between unit chain profiles of amylopectins of both starches. Edible canna starch contained almost the same amount of P (52 mg%) but contained higher amount of Ca (25 mg%) and could be deemed a Ca-starch. When this Ca was substituted for K, the resulting K-starch showed pasting properties of high maximum viscosity (760 B.U.) and large breakdown (580 B.U.), quite high transmittance and high swelling power. In contrast, the original Ca-starch showed depressed gelatinisation, i.e. it showed low (450 B.U.) but stable viscosity, low transmittance and low swelling power. In essence, edible canna K-starch was close to potato starch but its Ca-starch had pasting properties similar to those of cereal starches.

C. edulis starch granules were oval-shaped with granular sizes between 35 and 101 μm, amylose content of 23.8 %, amylopectin with average chain length of 21.9 and β-amylolysis limit of 67.6 %, B-type X-ray diffraction pattern, gelatinisation enthalpy of 15.7 J/g and peak temperature of the endothermic DSC (differential scanning calorimetry)-transition was 61.2 °C (Santacruz et al. 2002). *C. edulis* showed the highest swelling power and particle rigidity (the elastic modulus, G'), forming stronger gels than *Oxalis tuberosa* and *Arracacia xanthorrhiza* (Santacruz et al. 2003). *C. edulis* gel kept at 4 °C showed a high increase in G' during the first day of storage. A decrease in pH from 6.5 to 4.0 produced a loss of structure in the starch gel, as was shown by the reduction of G'. Storage at freezing temperature (−20 °C) produced higher changes in G' than refrigeration. The edible canna starch (processed from the rhizomes) was shown to have a larger granule size, slightly higher amylose content, longer amylopectin unit-chains, a lower gelatinisation temperature, and higher viscosity and retrogradation in comparison with maize (Tanaka et al. 2006). Achira flours showed high dietary fibre content (16.5–32.2 % db) and high concentration of starch in the case of the smaller particle size fraction (Andrade-Mahecha et al. 2012). Significant differences in protein and starch content, swelling power, solubility, and thermal properties were observed between the Brazilian and the Colombian starch. Results showed the starch and flour produced from achira

rhizomes to have great technological potential for use as functional ingredient in the food industry.

Antioxidant Activity

Red canna flowers were found to serve as a potential source of natural antioxidant exhibiting good scavenging ability of DPPH radicals and a natural food colorant with good stability of colour and tinctorial strength (Vankar and Srivastava 2008). At 100 μg/ml, the methanolic extract of aerial plant parts showed maximum inhibition of 76.70 % in the DPPH radical scavenging assay, 74.36 % in the hydroxyl radical scavenging assay, 61.37 % in the hydrogen peroxide assay and 62.84 % in the nitric oxide assay (Joshi et al. 2009a). The results indicated the aerial parts of *Canna indica* to be effective in scavenging free radicals and to have the potential to be a powerful antioxidant.

Six solvent fractions of aqueous extract of *C. edulis* rhizome showed 2,2-diphenyl-1-picryl hydrazyl (DPPH) free radical inhibition above 75 % (Mishra et al. 2012). Diethyl ether: ethyl acetate (1:3) fraction showed the maximum inhibition per cent and contained the highest amount of total flavonol and total proanthocyanidins. Maximum NO scavenging activity and hydroxyl radical inhibition activity were observed in bioactive diethyl ether: ethyl acetate (1:1) fraction. Inhibition of lipid peroxides was maximum in the ethyl acetate fraction.

Antinociceptive Activity

The benzene and methanol extracts of various plant parts, namely the leaves, flowers, rhizomes and seeds showed significant central and peripheral analgesic activity in hot plate method and acetic-acid-induced writhing test, respectively, at the dose of 50 mg/kg intraperitoneally (Nirmal et al. 2007). Methanolic leaf extract showed highest increase in reaction time in hot plate method while benzene leaf extract showed more inhibitory effect on writhing induced by acetic acid.

Anti-inflammatory Activity

Studies by Chen et al. (2013) demonstrated that *Canna indica* ethanolic extract suppressed lipopolysaccharide (LPS)-induced inflammatory mediator production in RAW 264.7 macrophages and also inhibited high glucose-induced inflammatory mediator expression by the regulation of MAPK pathway.

Anthelmintic Activity

The methanolic rhizome extract took less time to cause paralysis of the earthworm, *Pheretima posthuma* (Nirmal et al. 2007).

Antimicrobial Activity

Canna rhizome essential oil showed good antibacterial activity against *Staphylococcus aureus* but mild activity against *Bacillus subtilis* and no activity against *Eschericia coli* (Indrayan et al. 2011).

Antiviral Activity

Canna indica rhizome was found to have anti-HIV activity (Woradulayapinij et al. 2005). Recent studies in Thailand showed that aqueous Canna rhizome extract exhibited HIV-1 RT (reverse transcriptase) inhibitory activity higher than 90 % at a 200 μg/ml concentration. The water extract (IC_{50} 22.56 μg/ml) yielded two proteins, Cip31 (31 kDa) and Cip14 (14 kDa) with IC_{50} of 17.41 and 19.25 μg/ml and isoelectric point (pI) of 3.5 and 6.35, respectively. Both proteins showed significant HIV-1 RT inhibition. A novel 10 kDa protein with anti-HIV-1 reverse transcriptase (RT) inhibitory activity was isolated from *Canna indica* leaves (Thepouyporn et al. 2012). The leaf protein was identified to be a putative plastocyanin.

Anti-cancer Activity

Crude rhizome hexane and ethanol extract exhibited cytotoxicity against P388 leukaemia cells with ED_{50} 64.50 and 133.50 μg/m L, respectively (Chainakul et al. 2001). Two pure compounds, stigmasterol and β-hydroxystigmasta-4, 22-diene-3-one, and other two toxic minor components were isolated. They showed cytotoxicity against P388 leukaemia cells with ED_{50} values of 55.50, 37.50, 19.00 and 21.50 μg/m L, respectively.

Hepatoprotective Activity

The methanol extract of the aerial parts exhibited hepatoprotective effect against carbon-tetrachloride-induced hepatotoxicity in rats (Joshi et al. 2009b). The extract at doses (100 and 200 mg/kg) restored the levels of all serum parameters like aspartate aminotransferase, alanine aminotransferase, total bilirubin which were elevated in CCl4 administrated rats. Lipid peroxidation, reduced glutathione, catalase levels were observed normal in extract treated rats.

Molluscidal Activity

Successive ether and chloroform extracts of different plant parts and fatty acids from *Canna indica* demonstrated molluscicidal activity on snails *Biomphalaria alexandrina* (Motawe 1995). Root extracts were the most active with LC_{50} values of 110, 160 and 198 ppm for the ether, chloroform and fatty acid extracts, respectively. An ether extract of the dry plant contained chlorophyll and an oily material identified as a steroid containing *β*-unsaturated, 5-member lactone ring designated cannagenin. Cannagenin had a highly synergistic effect with chlorophyll on the mortality of snails. Canna rhizome was found to have molluscidal activity (Tripathi et al. 2004). The lethal concentration LC_{50} for the snail, *Lymnaea acuminata* was 6.54 mg/l. Active compounds in the rhizomes significantly inhibited the activity of acetylcholinesterase, acid/alkaline phosphatase, $Na^{+}K^{+}$ATPase and lactic dehydrogenase in the nervous tissue of *Lymnaea acuminata*. The molluscidal activity was found to be time- and dose-dependent (Tripathi and Singh 2000).

Larvicidal Activity

Leaf extracts of *C. indica* was found to be mosquitocidal to the mosquito, *Culex quinquefasciatus* Say, the major lymphatic filariasis vector (Rahuman et al. 2009). The highest larval mortality was found in the leaf acetone, chloroform, methanol and petroleum ether extracts (LC_{50} = 29.62, 59.18, 40.77 and 44.38 ppm; LC_{90} = 148.55, 267.87, 165.00 and 171.91 ppm) against second instar larvae, respectively, and (LC_{50} = 121.88, 118.25, 69.76 and 56.31 ppm; LC_{90} = 624.35, 573.93, 304.27 and 248.24 ppm) against fourth instar larvae, respectively.

Antimalarial Activity

The dichloromethane fraction of the rhizome extract exhibited antimalarial activity in-vitro against *Plasmodium falciparum* with IC_{50} of 25 μg/ml (do Céu de Madureira et al. 2002).

Traditional Medicinal Uses

Canna indica has been widely used as a folklore medicine in tropical and subtropical areas with beneficial effects in numerous diseases, including infection, rheumatism and hepatitis (Chen et al. 2013).

In traditional medicine, the plant is considered to be demulcent, diaphoretic, antipyretic and diuretic (Duke and Ayensu 1985). It is reported to relieve gastrointestinal disorders. The plant has been used in the treatment of women's complaints. A decoction of the root with fermented rice has been used in the treatment of gonorrhoea and amenorrhoea. The dried flowers have been used in a decoction to stop external wound bleeding. The root is given as a demulcent and stimu-

lant and used as a diaphoretic and diuretic in fevers and dropsy (Kirtikar and Basu 2001). In Kampuchea, leaf bases are medicinally used for furuncles, syphilis and a mush was made of the rhizomes and drunk for yaws as it has depurative action (Burkill 1966). In Java, the seeds were pounded into a paste for poulticing headache, and extract of the rhizome used as therapy for diarrhoea. In India, the rhizomes are diuretic, diaphoretic, stimulant and demulcent; a decoction of the rhizomes has been used in fevers, dropsy and dyspepsia. The seeds are cordial and vulnerary. In São Tomé e Príncipe, the roots have been used for treatment of malaria and fever (do Céu de Madureira et al. 2002). In Okeigbo, Ondo state, southwest Nigeria, leaves are used to treat malaria (Odugbemi et al. 2007) In the Philippines, the rhizome in decoction was used as a diuretic, and, when macerated in water, is said to relieve nosebleed (Stuart 2012). In Costa Rica, infusion of leaves used as diuretic; rhizomes used as emollient and in Bangladesh, paste of plant used for tonsillitis (Stuart 2012).

Other Uses

Cannas are one of the most popular garden/landscape ornamental plants. They are used widely in landscape settings, gardens, border plantings in parks and in patios as the flowers come in various colours and blends of mixed rainbow colours and for its beautiful, lustrous attractive foliage and as living fence and windbreak. The plant is also grown for magic purposes and as fetish plant. In Toi Long village in Vietnam, the inflorescences are sold as household decorations for the New Year (Tanaka 2004). In a village in Guangxi Province, China, leftover rhizomes, leaves and stems are used to a make a feed for livestock, often swines. Farmers near Sao Paulo, Brazil also grow the crop as pig feed.

The seeds are frequently used for necklaces in Africa and for making artefacts such as strings of holy rosary beads called *tasbih* in Malaysia and Indonesia (Burkill 1966). The dried seeds are also used as the mobile elements of the *kayamb*, a musical instrument from French Réunion, as well as the *hosho*, a gourd rattle from Zimbabwe, where the seeds are known as *hota* seeds. In goldmine regions of Bolivia, the seeds have been used for weighing small quantities of gold. The waste product of rhizome after the extraction of starch is used as a soil improver. Fibre obtained from the leaves is used for making paper. Rhizomes, foliage and stem are also used for animal fodder for cattle and pigs (Popenoe et al. 1989). The rhizomes are also used as a source of dye. In Ecuador, Africa, Malaysia and India the leaves are used for wrapping of food and also as plates in India.

Cannas are employed to extract many undesirable pollutants in a wetland environment as it has a high tolerance to contaminants. Aquatic plants including *C. indica* are widely used for phytoremediation (Zang et al. 2010). Cheng et al. (2007)) found the production and release of phosphatase to be the key mechanism for *C. indica* to degrade triazophos (O, O-diethyl-O-(1-phenyl-1, 2, 4-triazole-3-base) sulfur phosphate (TAP), indicating its potential of phytoremediation of TAP, from contaminated water in conjunction with the development of constructed wetland. Bose et al. (2008) found *C. indica* to be well-adapted in industrial sludge amendments and also suitable for phytoremediation of most of the studied metals: Fe>Cr>Mn>Zn>Ni>Cu>Cd>Pb.

Using an air-cathode microbial fuel cell (MFC) inoculated with rumen microorganisms, electricity could be directly produced with a maximum power density of 0.405 W/m^3 from *Canna indica*, a lignocellulosic aquatic plant rich in cellulose, hemicellulose and lignin, without pretreatment (Zang et al. 2010). The study suggested a promising potential to utilise lignocellulosic biomass for energy generation.

Comments

Molecular cytogenetical studies by Matoba et al. (2011) supported the hybrid (triploid) origin of edible canna *C. discolor* between *C. indica* var. *indica* and *C. plurituberosa*.

Selected References

Andrade-Mahecha MM, Tapia-Blácido DR, Menegalli FC (2012) Physical–chemical, thermal, and functional properties of achira (*Canna indica* L.) flour and starch from different geographical origin. Starch 64:348–358

Arellano ML, Carranco JM, Pérez-Gil RF, Hernández PE, Partida IH, Ripoll SH (1993) Chemical composition of 6 unconventional plants from Oaxaca State, Mexico, as potential resources for animal feed. Arch Latinoam Nutr 43(3):264–268 (In Spanish)

Bose S, Jain A, Rai V, Ramanathan AL (2008) Chemical fractionation and translocation of heavy metals in *Canna indica* L. grown on industrial waste amended soil. J Hazard Mater 160(1):187–193

Burkill IH (1966) A dictionary of the economic products of the Malay Peninsula. Revised reprint. 2 volumes. Ministry of Agriculture and Co-operatives, Kuala Lumpur, Malaysia. Vol. 1 (A—H) pp. 1—1240, Vol. 2 (I—Z), pp 1241–2444

Chainakul S, Soukharath V, Tambunchong C, Picha P (2001) Study on cytotoxicity of the hexane crude extracted from the rhizome of *Canna indica* Linn. on cancer cells. Srinakharinwirot Univ Sci J 17(1):25–34 (In Thai)

Chen HJ, Chen CN, Sung ML, Wu YC, Ko PL, Tso TK (2013) *Canna indica* L. attenuates high-glucose- and lipopolysaccharide-induced inflammatory mediators in monocyte/macrophage. J Ethnopharmacol 148(1):317–321

Cheng S, Xiao J, Xiao H, Zhang L, Wu Z (2007) Technical note phytoremediation of triazophos by *Canna indica* Linn. in a hydroponic system. Int J Phytorem 9(6):453–463

do Céu de Madureira M, Paula Martins A, Gomes M, Paiva J, Proença da Cunha A, do Rosário V (2002) Antimalarial activity of medicinal plants used in traditional medicine in S. Tomé and Príncipe islands. J Ethnopharmacol 81(1):23–29

Duke JA, Ayensu ES (1985) Medicinal plants of China, vols 1 & 2. Reference Publications, Inc. Algonac, 705 p

Foundation for Revitalisation of Local Health Traditions (2008) FRLHT database. http://envis.frlht.org

French BR (1986) Food plants of Papua New Guinea – a compendium. Australia and Pacific Science Foundation, Ashgrove, 408 p

Gade DW (1966) Achira, the edible canna, its cultivation and use in the Peruvian Andes. Econ Bot 20:407–415

Hermann M, Quynh NK, Peters D (1999) Reappraisal of edible canna as a high value starch crop in Vietnam. Impact on a changing world. CIP program report 1997–98. CIP, Lima, Peru, pp 415–424

Hu SY (2005) Food plants of China. The Chinese University Press, Hong Kong, 844pp

Inatsu O, Maeda I, Jimi N, Takahashi K, Taniguchi H, Kawabata M, Nakamura M (1983) Edible canna starch. I. Some properties of edible canna starch produced in Taiwan. J Jpn Soc Starch Sci 30(1):38–47

Indrayan AK, Bhojak NK, Kumar N, Shatru A, Gaur A (2011) Chemical composition and antimicrobial activity of the essential oil from the rhizome of *Canna indica* Linn. Indian J Chem 50B:1136–1139

Joshi YM, Kadam VJ, Kaldhone PR (2009a) In-vitro antioxidant activity of methanolic extract of aerial parts of *Canna indica* L. J Pharm Res 2(11):712–1715

Joshi YM, Kadam VJ, Patil YV, Kaldhone PR (2009b) Investigation of hepatoprotective activity of aerial parts of *Canna indica* L. on carbon tetrachloride treated rats. J Pharm Res 2(12):1879–1882

Kirtikar KR, Basu BD (2001) Indian medicinal plants, vol 4, 2nd edn. Oriental Enterprises, Uttranchal, p 2450

Matoba H, Tanaka N, Uchiyama H, Koyama T (2011) The origin of southeastern Asian triploid edible canna (*Canna discolor* Lindl.) revealed by molecular cytogenetical study. Econ Bot 65(3):308–314

Mishra T, Das AP, Sen A (2012) Phytochemical screening and in-vitro antioxidant profiling of solvent fractions of *Canna edulis* Ker Gawler. Free Rad Antioxid 2(1):13–20

Motawe HM (1995) Cannagenin: a new molluscicidal agent from *Canna indica* L. J Herbs Spice Med Plants 2(4):3–10

Nirmal SA, Kolhe NM, Pal SC, Mandal SC (2008) Nonpolar compounds from *Canna indica* rhizomes. Facta Univ Ser Phys Chem Technol 6(1):141–144

Nirmal SA, Shelke SM, Gagare PB, Jadhav PR, Dethe PM (2007) Antinociceptive and anthelmintic activity of *Canna indica*. Nat Prod Res 21(12):1042–1047

Ochse JJ, Bakhuizen van den Brink RC (1980) Vegetables of the Dutch Indies, 3rd edn. Ascher & Co., Amsterdam, 1016 pp

Odugbemi TO, Akinsulire OR, Aibinu I, Fabeku PO (2007) Medicinal plants useful for malarial therapy in Okeigbo, Ondo state, southwest Nigeria. Afr J Tradit Complement Altern Med 4(2):191–198

Ong HC, Siemonsma JS (1996) *Canna indica* L. In: Flach M, Rumawas F (eds) Plants yielding non-seed carbohydrates, Plant resources of South-East Asia No. 9. Prosea Foundation, Bogor, pp 63–66

Pacific Island Ecosystems at Risk (PIER) (1999) *Canna indica* L. Cannaceae. http://www.hear.org/pier/species/canna_indica.htm

Popenoe H, King SR, Léon J, Kalinkowski LS, Vietmeyer ND, Dafforn M (1989) Lost crops of the Incas. Little known plants of the Andes with promise for worldwide cultivation. National Academy Press, Washington

Putman EW, Hassid WZ (1954) Sugar transformation in leaves of *Canna indica*. I. Synthesis and inversion of sucrose. J Biol Chem 207(2):885–902

Rahuman AA, Bagavan A, Kamaraj C, Saravanan E, Zahir AA, Elango G (2009) Efficacy of larvicidal botanical extracts against *Culex quinquefasciatus* Say (Diptera: Culicidae). Parasitol Res 104(6):1365–1372

Santacruz S, Koch K, Svensson E, Ruales J, Eliasson AC (2002) Three underutilised sources of starch from the Andean region in Ecuador Part I. Physico-chemical characterisation. Carbohydr Polym 49(1):63–70

Santacruz S, Ruales J, Eliasson AC (2003) Three under-utilised sources of starch from the Andean region in Ecuador. Part II. Rheological characterisation. Carbohydr Polym 51(1):85–92

Srivastava J, Vankar PS (2010a) *Canna indica* flower: new source of anthocyanins. Plant Physiol Biochem 48(12):1015–1019

Srivastava J, Vankar PS (2010b) Methylated anthocyanidin glycosides from flowers of *Canna indica*. Carbohydr Res 345(14):2023–2029

Stuart GU (2012) Philippine alternative medicine. Manual of some Philippine medicinal plants. http://www.stuartxchange.org/OtherHerbals.html

Tanaka N (1998) Economic botanical notes on edible Canna (Cannaceae) in South Vietnam. J Jpn Bot 73:319–324

Tanaka N (2004) The utilization of edible *Canna* plants in southeastern Asia and southern China. Econ Bot 58(1):112–114

Tanaka N, Inouchi N, Koyama T (2006) Edible canna and its starch: an under-exploited starch-producing plant resource. Foods Food Ingred J Jpn 211(4):319–325

The Plant List (2014) *Canna indica* L. http://www.theplantlist.org/

Thepouyporn A, Yoosook C, Chuakul W, Thirapanmethee K, Napaswad C, Wiwat C (2012) Purification and characterization of anti-HIV-1 protein from *Canna indica* L. leaves. Southeast Asian J Trop Med Public Health 43(5):1153–1160

Tinoi J, Rakariyatham N, Deming RL (2006) Determination of major carotenoid constituents in petal extract of eight selected flowering in the north of Thailand. Chiang Mai J Sci 33(2):327–334

Tripathi SM, Singh DK (2000) Molluscicidal activity of *Punica granatum* bark and *Canna indica* root. Braz J Med Biol Res 33(11):1351–1355

Tripathi SM, Singh VK, Singh S, Singh DK (2004) Enzyme inhibition by the molluscicidal agent *Punica granatum* Linn. bark and *Canna indica* Linn. root. Phytother Res 18(7):501–506

Tropicos Org. Missouri botanical garden. Jan 2013 http://www.tropicos.org

Vankar PS, Srivastava J (2008) Comparative study of total phenol, flavonoid contents and antioxidant activity in *Canna indica* and *Hibiscus rosa sinensis*: Prospective natural food dyes. Int J Food Eng 4(3): Article 4

Winter HC, Kishida E, Goldstein IJ (1995) Novel lectin and mannose-binding protein from rhizomes of *Canna* (Cannaceae). Curr Top Plant Physiol 15:310–311

Woradulayapinij W, Soonthornchareonnon N, Wiwat C (2005) In vitro HIV type 1 reverse transcriptase inhibitory activities of Thai medicinal plants and *Canna indica* L. rhizomes. J Ethnopharmacol 101(1–3):84–89

Wu D, Kress WJ (2000) Cannaceae. In: Wu ZY, Raven PH (eds) Flora of China. Vol. 24 (Flagellariaceae through Marantaceae). Science Press/Missouri Botanical Garden Press, Beijing/St. Louis

Zang GL, Sheng GP, Tong ZH, Liu XW, Teng SX, Li WW, Yu HQ (2010) Direct electricity recovery from *Canna indica* by an air-cathode microbial fuel cell inoculated with rumen microorganisms. Environ Sci Technol 44(7):2715–2720

Cibotium barometz

Scientific Name

***Cibotium barometz* (L.) J. Sm.**

Synonyms

Aspidium barometz (L.) Willd., *Dicksonia barometz* (L.) Link, *Nephrodium barometz* (L.) Sweet, *Polypodium barometz* L.

Family

Cibotiaceae

Common/English Names

Chain Fern, Chain Fern Rhizome, Cibot Rhizome, Cibota, Cibotium, Golden Chicken Fern (Plate 2), Golden Hair Dog Fern, Golden Moss, Golden Lamb, Lamb of Tartary, Scythian Lamb, Tartarian Lamb, Vegetable lamb of Tartary, Wooly Fern.

Vernacular Names

Chinese: Guo Ji, Jin Mao Guo Ji, Huang Gou Tou;
Dutch: Goudharige Hondsvaren;
French: Agneau De Scythie, Cibotie, Pitchawar;
Hawaiian: Pulu-Pulu;
Indonesian: Paku Simpai;
Italian: Felci Arboree;
Japanese: Hitsuji Shida, Taka-Warbi, Shishiba (Okinawa);
Malaysia: Paku Kidang, Penawar Jambi, Bulu Pusi, Bulu Empusi;
Philippines: Salagisog (Bikol), Tinampa (Igorot), Borabor, Borabor Ta Paku, Sabong To Borabor (Iloko);
Thailand: La Ong Faifaa (Central), Kut Phi Pa (Northern), Kut Suea, Pho Si (Pattani), Khon Kai Noi (Loei), Taet Ling (Trat), Ninla Pho Si (Songkhla, Yala), Wankai Noi (General), Hatsa Daeng (Nakhon Ratchasima);
Vietnamese: Cau Tich, Kim Mao (Vietnamese Chinese), Cut Bang (Tay), Co Cut Pa (Thai Ethnic), Nhai Cu Vang (Dao Ethnic), Dang Pam (K'ho Ethnic), Long Cu Ly, Cu Lan, Long Khi

Origin/Distribution

Wild distribution of the species had been reported in southern parts of China, northeast India, western Malay Peninsula, Indonesia (from Java to Sumatra), Myanmar, Thailand, Vietnam, Laos, Cambodia (Qin and Dong 2003; Zhang et al. 2008; Nguyen et al. 2009a; Zhang

T.K. Lim, *Edible Medicinal and Non-Medicinal Plants: Volume 10, Modified Stems, Roots, Bulbs*,
DOI 10.1007/978-94-017-7276-1_4

Plate 1 *Cibotium barometz* plant habit

and Nishida 2013), Taiwan (Van Steenis and Holttum 1959) and Philippines (Nguyen et al. 2009a).

Agroecology

This fern species has a relatively widespread distribution, occurring in the valley, edges of the forest, along stream-banks in the lowlands, wet mountainous ravines from 100 to 1500 m elevation in tropical and sub-tropical zones in China, Indochina and southeast Asia (WHO 1990; NIMM 1999; Qin and Dong 2003). The plant is hygrophilous and shade enduring when young. It is well adapted to warm and humid conditions. Optimum average temperature varies between 20 and 23 °C, the rainfall ranges from 1800 to 2600 mm every year. It thrives on red-brown ferralitic and acid soils but will tolerate marginally alkaline soils.

Edible Plant Parts and Uses

The rhizome starch is used for making cakes and liquor in China (Cui 1998; Dai et al. 2003; Cao et al. 2007; Yun et al. 2009a, b; Liu et al. 2012).

Botany

A large tree fern reaching height of 1–3 m high (Plate 1) with massive, prostrate to erect caudex (trunk), the young tops and base covered with dense, stiff, golden-brown, long hairs (Plate 4). Fronds in a tuft at the apex of the trunk, 1–2 m long, bipinnately compound, ovate to elliptical in outline, up to 2 × 1 m, under side glaucous, upper side darker green (Plate 3), with stipes thick, up to 1 m long or more, triangular in transverse section at base, densely bearing caducous adpressed hairs, stipe and rachis green, turning purplish beneath with age; base of stipe with a mass of long (1–1.5 cm) hairs, upper part of stipe and rachis covered with small, appressed flaccid hairs becoming glabrescent; pinnae many, alternating, pinnate pinnatifid, in outline oblong to lanceolate, apex acuminate; pinnules numerous, often with a few pairs of tertiary leaflets at the base, deeply pinnatifid throughout, very shortly stalked or sub-sessile at distal parts of pinnae, pinnule-segments slight falcate, apiculate, margins crenulate to serrulate-serrate. Sori 1–5 pairs on pinnule-segments; indusia bivalvate, outer indusia round, inner ones more or less oblong; outer valve of indusium usually large; paraphyses dark

Plate 2 *Cibotium barometz* plant label

reddish brown, long and numerous. Spores pale yellowish, with equatorial ridge.

Nutritive/Medicinal Properties

Rhizome Phytochemicals

Pterosin R (Murakami et al. 1980), pterosin Z (Zhao et al. 2011) and ptaquiloside (Potter and Baird 2000) were found in *C. barometz*. Yang et al. (2010) reported *C. barometz* to contain volatile oils, pterosins, aromatic compounds, water-soluble phenolic compounds, flavonoids, amino acids, inorganic elements, volatile oils, pterosins, aromatic compounds, water-soluble phenolic compounds, flavonoids, amino acids, inorganic elements. Xu et al. (2012a) reported *C. barometz* processed products (drying, cutting, roasting, boiled, etc.) to contain phenolic compounds, volatile oil, sterols, saccharides, glucosides, amino acids, mineral elements and phospholipids.

Pterosin R (Murakami et al. 1980), pterosin Z (Zhao et al. 2011) and ptaquiloside (Potter and Baird 2000) were found in *C. barometz*. Twelve constituents, mainly organic acids, were isolated from *C. barometz* rhizome volatile oil with palmitic acid and linoleic acid as the major acids (Jia et al. 1996). The contents of protocatechuic acid and caffeic acid in all the samples of *C. barometz* rhizomes collected from various areas were above 0.020 and 0.029 %, respectively, and the highest content was found in the processed products (Yuan et al. 2000). Twenty-five compounds were identified in the essential oil from *C. barometz* rhizome; the main constituents were oleic acid, linoleic acid, palmitic acid, pentadecanoic acid, 7,10,13-hexadecatrienoic acid methyl ester, linolenic acid methyl ester with relative content over 10 %, respectively (Xu et al. 2000). The content of phosphatidylcholine was 0.198 % in *C. barometz* (Xu et al. 2001). Zhang and Wang (2001) found the following compounds in the rhizomes, onitin, protocatechuic acid, *β*-sitosterol, daucosterol and 2-furancarboxaldehyde-50-hydroxylmethyl. Nine compounds were isolated and identified from the rhizomes: palmitic acid, *β*-sitosterol, 1-mono-palmitin, daucosterol, caffeic acid, protocatechuic acid, protocatechuic aldehyde, *n*-butyl-*β*-D-fructopyranoside and D-glucose (Cheng et al. 2003). Twelve compounds palmitic acid, palmitic acid methyl ester, linoleic acid, oleic acid, stearic acid ethyl ester, 4′-hydroxyacetanilide, sucrose, C_{27} saturated fatty acid, protocatechualdehyde, *β*-sitosterol, vanillin and 3,4,5,7-tetrahydroxyflavone were isolated from *C. barometz* rhizome (Xu et al. 2004). *C. barometz* rhizomes were reported to contain flavonoids, kaempferol and onychin (Xu et al. 2004; Hu and Yu 2006).

Ye et al. (2006) reported *C. barometz* to have flavonoids, and to be rich in the trace elements such as Fe, Ca, Zn, Mg, Ni, Mn, Cu. Wu et al. (2007) found the following compounds in the

Plate 3 Close-up of frond

Plate 4 Golden-brown hairy caudex top with frond removed

rhizomes: β-sitosterol, daucosterol, onitin, alternariol, (3*R*)-des-*O*-methyl lasiodiplodin, protocatechuic acid. Ryu and Lee (2008) found protocatechuic acid, caffeic acid and shinbarometin (2-*O*-(9Z,12Z-octadecadienoyl)-3-*O*-[α-D-galactopyranosyl-(1$''$ $\rightarrow$ 6$'$)-*O*-β-D-galactopyranosyl]glycerol) in the rhizomes. Eight compounds including two new furan derivatives, cibotiumbarosides A and B, corchoionoside C and a new glycoglycerolipid, cibotiglycerol were isolated from a methanol extract of *Cibotium barometz* rhizomes (Nguyen et al. 2009b). Three unusual sesquiterpenes having 1-indanone nucleus (1, 3 and 4) and an unusual orthoester spiropyranosyl derivative of protocatechuic acid (2) were isolated from *Cibotium barometz* rhizomes (Wu and Yang 2009).

Xie et al. (2011) reported 5-hydroxymethyl furfural, protocatechuic acid and protocatechuic aldehyde in raw material and processed products of *C. barometz*.

The content of total phenolic acid in *C. barometz* rhizome from different areas in China varied from 3.72–6.16 %, and it was 3.09–5.09 %

in its processed products (Ju et al. 2012). Ten compounds were purified from 70 % alcohol extract of *C. barometz* and their structure were identified as 1-*O*-caffeyl-D-glucopyranose; 6-*O*-caffeyl-D-glucopyranose; 3-*O*-caffeyl-D-glucopyranose; 3-hydroxymethyl-2(5*H*)-furanone; *β*-miroside; cibotiumbaroside A; protocatechuic acid; glucose; mannose; Corchoionoside C and kojic acid (Xu et al. 2012b). Xu et al. (2013b) isolated two hydrolyzable tannins, 4-*O*-caffeoyl-*α*-D-glucopyranose and 4-*O*-caffeoyl-*β*-D-glucopyranose, from the rhizomes. The content of tannins in the rhizomes were found to decrease after different processing (Jia et al. 2001). Six compounds isolated from *C. barometz* were identified as 6-*O*-protocatechuoyl-D-glucopyranose; 3-*O*-caffeoyl-D-glucopyranose; 1-*O*-caffeoyl-*β*-D-glucopyranose; caffeic acid; protocatechuic acid and *p*-hydroxybenzoic acid (Xu et al. 2013c). Compound 1 was a novel phenolic glucoside named cibotiumbaroside D.

Xu et al. (2012a, b) reported the following pharmacological effects of *C. barometz* and its processed products: analgesic, haemostatic, anti-inflammatory, anti-bone loss, antioxidative, anti-platelet, hepatoprotective, antihyperlipidemic and central nervous system (CNS). Earlier, Yang et al. (2010) reported the following pharmacological effects of *C. barometz*: antiosteoporotic, anti-inflammatory, haemostatic, analgesic and anti-rheumatic. Maillard reaction was found to be involved in the processing of C. *barometz* rhizome, which may be contributed to the variation in chemicals and activity between raw and processed C. *barometz* (Xu and Jia 2011).

Antioxidant Activity

The chloroform and *n*-butanol fractions of the ethanol rhizome extract exhibited significant 1,1-diphenyl-2-picrylhydrazyl (DPPH) radical scavenging activity (Ryu and Lee 2008). Baked processed products of *C. barometz* exhibited stronger antioxidative activity than the raw material (Xu et al. 2011). The methanol extract of C barometz rhizome exhibited effective antioxidant activity in a dose-dependent manner with IC_{50} values of 44.2, 19.84, 137.66, 22.94, 289.73, 53.52 μg/mL for DPPH•, ABTS•$^+$, super anion •O_2^-, hydroxyl •OH scavenging assays, Fe^{3+} reducing power, Cu^2+ reducing power assays, respectively (Mai et al. 2012). Its total phenolic content was 50.88 mg CAE (caffeic acid equivalent)/g and the caffeic acid content (the major contributing compound) was 1.82 mg/g. Three rhizome compounds 1-*O*-caffeyl-D-glucopyranose, 3-*O*-caffeyl-D-glucopyranose compound 3 and cibotiumbaroside A showed significant DPPH antioxidant activity, and the scavenging activity of 1-*O*-caffeyl-D-glucopyranose was similar to that of vitamin C (Xu et al. 2012b).

Lai et al. (2009) reported *C. barometz* leaf extract to have antioxidative (1,1-diphenyl-2-picrylhydrazyl (DPPH), ferric ion reducing power (FRP), *β*-carotene bleaching (BCB) and ferrous ion chelating (FIC)) and antibacterial activity in-vitro. The IC_{50} value for DPPH radical scavenging assay was 0.25 mg/mL and 1542 mg ascorbic acid/100 g and ferric reducing power was 804 mg GAe/100 g and its total phenolic content was 1589 mg GAE/100 g. The order of the fern extracts in these antioxidative activities was similar to that for TPC, i.e., *Blechnum orientale* ≥ *Drynaria linearis* > *C. barometz* > *Asplenium aureum* > *Asplenium nidus*. The antioxidative activity in terms of its ability to inhibit lipid peroxidation as measured by beta carotene bleaching assay showed that at the highest concentration, the BCB antioxidative activity decreased in the order D. linearis (99 %) > *B. orientale* (73 %) = *C. barometz* (69 %) > *A. aureum* (51 %) = *A. nidus* (47 %). All ferns showed very low ferrous ion chelating activity (<22 %), except for *A. aureum* (58 % at a concentration of 6.7 mg/mL).

Antiosteoporotic Activity

Cibotium barometz was found to have high inducing ability on alkaline phosphatase activity in human foetal osteoblast (Lee et al. 2003). Several compounds isolated from the rhizome including cibotiumbaroside B and cibotiglycerol showed inhibition of the bone-destroying osteoclast for-

mation with no effect on bone marrow-derived macrophage (BMM) cell viability (Nguyen et al 2009b). Yu et al. (2011) found the *n*-butanol extract of *C. barometz* processed products had the most significant effect on cell proliferation of osteoblasts in the rat and in-vitro culture. *Cibotium barometz* rhizome extract prevented total bone mineral density decrease in the femur induced by ovariectomy in female rats, which was accompanied by a significant decrease in skeletal remodelling, as was evidenced by the decreased levels of the bone turnover markers, such as osteocalcin (OC), alkaline phosphatase (ALP), deoxypyridinoline (DPD) and urinary Ca and P excretions (Zhao et al. 2011). The treatment could also enhance the bone strength and prevent the deterioration of trabecular microarchitecture. The results indicated that *C. barometz* extract might be a potential alternative medicine for the prevention and treatment of post-menopausal osteoporosis.

Various *n*-butanol extract fractions from different processed products of *C. barometz* showed a significant proliferative effect on osteoblasts in the order of the wined > the heated > the salted > the sand-heated and wined system > the alcohol-processed > the steamed > the crude material (Xu et al. 2013a). The q test showed no significant difference among sand-heated, alcohol-processed and steamed *C. barometz*; no significant difference between heated and salted *C. barometz*. Various control substances also showed a certain proliferative effect on osteoblasts in the order of the mixed control > protocatechuic aldehyde > protocatechuic acid > kojic acid. The q test showed no significant difference between protocatechuic aldehyde and protocatechuic acid. All of *n*-butanol extract fractions from different processed products of *C. barometz* showed a significant effect on osteoblast proliferation, of which wined *C. barometz* showed the best effect. All of phenolic compounds such as protocatechuic aldehyde, protocatechuic acid and kojic acid showed a significant proliferative effect on osteoblasts. Raw rhizome slices of *C. barometz* steamed with rice wine and its index constituents like protocatechuic acid and protocatechuic aldehyde were found to promote proliferation and differentiation of primary rat osteoblasts cultured in-vitro (Xu et al. 2014b). *Cibotium barometz* was reported to be active in preventing post-menopausal osteoporosis in ovariectomized rat (Rufus et al. 2013). Among the different processed rhizomes of *C. barometz* namely raw, sand-baked, wined, steamed and salted, sand-baked and wined processed rhizomes were better than the steamed, salted and raw rhizomes in inhibiting retinoic acid-induced osteoporosis in male rats evidenced by their effects on s-(TRAP) tartrate-resistant acid phosphatase and total scores of OPG (osteoprotegerin), Ca, P, interleukins IL-6, IL-1 and TNF-alpha (Xu et al. 2014a).

Antiviral Activity

Six herbal extracts, including two from *C. barometz* rhizome (designated as CBE and CBM), were found to be potent inhibitors of severe acute respiratory syndrome associated coronavirus (SARS-CoV) at concentrations between 25 and 200 μg/mL (Wen et al. 2011). Among the extracts, CBM also showed significant inhibition of SARS-CoV 3CL protease activity with IC_{50} value of 39 μg/mL.

Antibacterial Activity

The ranking of the antibacterial activity based on the number of test bacteria inhibited of fern leaf extracts was *Asplenium nidus* (4) > *Drynaria linearis* (3) = *Blechnum orientale* (3) > *C. barometz* (2) (Lai et al. 2009). *C. barometz* leaf extract was inhibitory in-vitro *to Staphylococcus aureus* and *Bacillus cereus*.

Anticancer Activity

Many traditional applications or phytotherapeutic concepts propose to inhibit the proliferation of prostate cancer cells. Fractions from *Cibotium barometz* exhibited hormonal influences on LNCaP and PC-3 prostate cancer cells (Bobach

et al. 2014). The differential behaviour of the two prostate cancer cell lines allowed the discrimination between potential androgenic or antiandrogenic activities and effects on the oestrogen or glucocorticoid receptor.

Hepatoprotective Activity

Onychin exhibited antioxidation and protective effect from liver damage induced by lipid peroxide in mice (Yang et al. 2002). It significantly decreased the level of lipid peroxide malondialdehyde in the liver homogenate.

Antityrosinase Activity

C. barometz leaf extract showed moderate antityrosinase activity (35 % inhibition), equivalent to 102 mg of quercetin/g and 11 mg of kojic acid/g.

Pharmacokinetic Studies

Three unusual sesquiterpenes having 1-indanone nucleus (1, 3 and 4) and an unusual orthoester spiropyranosyl derivative of protocatechuic acid (2) were isolated from *Cibotium barometz* rhizomes (Wu and Yang 2009). Compound 1 was well-absorbed, and 2 and 3 were poorly absorbed compounds in the human intestine. Using human Caco-2 cell monolayer model, the permeation rates of 1, 3 and 4 increased linearly as a function of time up to 180 min and with the concentration within the test range of 25–200 μM.

Traditional Medicinal Uses

Cibotium barometz has been traditionally used as anti-inflammatory and anodyne (Wu and Yang 2009). Its rhizomes and roots were reported to be collected for medicinal uses, including use as blood coagulant and treatment of ulcers, rheumatism, typhoid and coughs (Puri 1970; May 1978; Nguyen et al. 2009a, b; Wu et al. 2007). This fern has been reported to be used in traditional medicine to treat fainting, wounds, ulcers, cough, rheumatism and used as kidney and liver tonic (May 1978; Piggott 1988). According to Nguyen et al. (2009a, b)), Zhang et al. (2008), the rhizome of *C. barometz* is believed to replenish the liver and the kidney, strengthen the tendons, muscles and bones and relieve rheumatic conditions. It is widely used to cure rheumatism, limb-ache, lumbago, neuralgia and pollakiuria in aged humans, leucorrhoea, sciatica, micturition, enuresis and body-ache in pregnant women. The golden hair covering the rhizome is used for styptic for poulticing the wounds and cuts in the limbs to stop bleeding in Peninsular Malaysia and China (Burkill 1966) (Plate 2). In Vietnam, rhizomes are used to treat rheumatism, lumbago, neuralgia, sciatica, enuresis and body aches in pregnant women (WHO 1990; NIMM 1999).

According to Stuart (2014), the rhizome is considered as anodyne, anti-inflammatory, antirheumatic, tonic, styptic. In Chinese medicine, the rhizome is used to tonify *yang*; used as antirheumatic, for strengthening the bones and muscles, and to replenish the liver, kidneys and the male generative organs. It is recommended as an 'old man's remedy'. The roots are also used for the treatment of lumbar pain, numbness, hemiplegia, leucorrhoea, spermatorrhoea, tumours and bleeding in women. In Philippines, the rhizomes are used as topical for wounds and ulcers, and as haemostatic poultice for wounds, osteodynia, leucorrhoea, dysuria and polyuria. *C. barometz* is one of 30 components in a Chinese herb pill used in regimen of therapy – herb therapy, foot massage, leg traction and exercise – for femoral head necrosis (Stuart 2014).

Other Uses

C. barometz is reported to be used in southeast Asia for medicinal purposes and as food and fibre (Lemmens et al. 1989). The hairs that cover the rhizome were reported to be used for stuffing cushions (Chandra 1970; Van Steenis and Holttum 1959) or as packing material (May 1978). The fibres of the stem have been used by the Annamites for weaving into hates, etc. in

Indochina (Burkill 1966). Oldfield (1995) noted that tree ferns were used in the horticultural market as pot plants as well as for landscaping, and to act as substrate material for orchids. In general, all *Cibotium* species have also ornamental value and, e.g. crowns with croziers are cut for table decoration. In China, a diluted solution of plant parts is used to control aphids and spider mites.

Comments

Dried rhizome of this fern is in demand as medicine. It is known to be in trade and to be collected from the wild in China and Vietnam. Nguyen et al. (2009a) reported that the species was listed as Threatened in the 1996 Red Data Book of Vietnam.

Selected References

Beddome RH (1883) Handbook to the ferns of British India, Ceylon and the Malay Peninsula. Thacker. Spink and Co., London, p 500

Bobach C, Schurwanz J, Franke K, Denkert A, Sung TV, Kuster R, Mutiso PC, Seliger B, Wessjohann LA (2014) Multiple readout assay for hormonal (androgenic and antiandrogenic) and cytotoxic activity of plant and fungal extracts based on differential prostate cancer cell line behavior. J Ethnopharmacol 155(1):721–730

Burkill IH (1966) A dictionary of the economic products of the Malay Peninsula. Revised reprint of 1st ed, 1935, 2 vols. Ministry of Agriculture and Co-operatives, Kuala Lumpur, Malaysia. Vol. 1 (A-H) pp 1–1240, Vol. 2 (I-Z), pp 1241–2444

Cao QM, Zhong HY, Li ZH, Sun CB (2007) Exploitation and utilization study of fern starch in China. Food Res Dev 28(12):168–171

Chandra S (1970) Vascular organization of the rhizome of *Cibotium barometz*. Am Fern J 60(2):68–72

Cheng QH, Yang ZL, Hu YM (2003) Studies on the chemical constituents of *Cibotium barometz*. Prog Pharm Sci 14(5):298–299 (In Chinese)

Cui GY (1998) Edible fern resources and development in China. Chin Cuisine Res 6(1):21–28

Dai XL, Li XG, Wu SF (2003) List of Chinese edible pteridophytes. Quart Forest By Prod Spec Chin 4:5–6

Hoshizaki BJ (1964) Cultivated tree ferns of the genus. Cibotium (Dicksoniaceae) Baileya 12(4):137–146

Hu YW, Yu JL (2006) Advances in study on chemical constituents and pharmacological effects of rhizoma cibotii. Lishizhen Med Mat Med Res 17(2):275–276

Jia TZ, Li J, Xie SQ, Zhang JP (1996) Comparative research on the constituents of the volatile oil in rhizome of *Cibotium barometz* (L.) J. Sm and processed product. Zhongguo Zhong Yao Za Zhi 21(4):216–217 (In Chinese)

Jia TZ, Hu FY, Li J, Xie SQ, Han JL (2001) Content determination of tannins in different processed samples of *Citotium barometz*. Liaoning J Tradit Chin Med 28(5):305–306 (In Chinese)

Jones DL (1987) Encyclopaedia of ferns, an introduction to ferns, their structure, biology, economic importance, cultivation and propagation. Lothian Publishing Company PTY Ltd, Melbourne, p 433

Ju CG, Xu G, Song YJ, Zhao WL, Jia TZ (2012) Comparison of the content of total phenolic acid in *Cibotium barometz* and its processed products from different areas. China J Exp Trad Med Formulae 18(10):24–26 (In Chinese)

Lai HY, Lim YY, Tan SP (2009) Antioxidative, tyrosinase inhibiting and antibacterial activities of leaf extracts from medicinal ferns. Biosci Biotechnol Biochem 73(6):1362–1366

Lee MK, Choi HI, You HK, Shin HS (2003) Effects of several herbal medicines on alkaline phosphatase activity in human fetal osteoblasts. J Korean Acad Periodontol 33(1):49–60 (In Korean)

Lemmens RHMJ, Jansen PCM, Siemonsma JS, Stavast FM (1989) Plant resources of South-East Asia. Basic list of species and commodity grouping. PROSEA Project, Wageningen

Liu Y, Wujisguleng W, Long C (2012) Food uses of ferns in China: a review. Acta Soc Bot Pol 81(4):263–270

Mai WQ, Chen DF, Li XC (2012) Antioxidant activity of rhizome cibotii in vitro. Adv Pharm Bull 2(1):107–114

May LW (1978) The economic uses and associated folklore of ferns and fern allies. Bot Rev 44(4):491–528

Murakami T, Satake T, Ninomiya K, Iida H, Yamauchi K, Tanaka N, Saiki Y, Chen CM (1980) Pterosin-derivate aus der familie pteridaceae. Phytochemistry 19(8):1743–1746

National Institute of Materia Medica (NIMM) (1999) Selected medicinal plants, volume 1. Science and technology. Publishing House, Hanoi

Nguyen T (2006) The Red List of medicinal plants of Viet Nam – 2006. J Med Mat 11(3):97–105, National Institute of Medicinal materials, Hanoi. (In Vietnamese)

Nguyen T, Le TS, Ngo DP, Nguyen QN, Pham TH, Nguyen TH (2009a) Non-detriment finding for *Cibotium barometz* in Viet Nam. SC58 Doc. 21.1. Annex 2

Nguyen XC, Chau VM, Phan VK, Hoang TH, Ninh KB, Nguyen XN, Nguyen HT, Jung JW, Kim HJ, Kim SY, Kim JA, Kim YH (2009b) Inhibitors of osteoclast formation from rhizomes of *Cibotium barometz*. J Nat Prod 72(9):1673–1677

Ogle BM, Ho TT, Hoang ND, Nguyen NXD (2003) Food, feed or medicine: the multiple functions of edible wild plants in Vietnam. Econ Bot 57(1):103–117

Oldfield S (1995) Significant trade in CITES Appendix II plants. Tree ferns. Prepared for the CITES Secretariat

by the World Conservation Monitoring Centre, Cambridge, UK

Piggott AG (1988) Ferns of Malaysia in colour. Tropical Press SDN. BHD, Kuala Lumpur, p 458

Potter DM, Baird M (2000) Carcinogenic effects of ptaquiloside in bracken fern and related compounds. Brit J Cancer 83(7):924–920

Puri HS (1970) Indian Pteridophytes used in folk remedies. Am Fern J 60(4):137–143

Qin HN, Dong SY (2003) Proposed revision of resolution conference. 9.24 (CoP12 Com. I.3). Criteria for listing on Appendix I and Appendix II. Test of the applicability of the criteria. *Cibotium barometz*. http://www.cites.org/common/prog/criteria/flora/C_barometz_CN.pdf

Rufus P, Mohamed N, Shuid AN (2013) Beneficial effects of traditional Chinese medicine on the treatment of osteoporosis on ovariectomised rat models. Curr Drug Targets 14(14):1689–1693

Ryu M, Lee IS (2008) Antioxidant constituents from the rhizomes of *Cibotium barometz*. Planta Med 74:PA221

Stuart GU (2014) Philippine alternative medicine. Manual of some Philippine medicinal plants. http://www.stuartxchange.org/OtherHerbals.html

Van Steenis CGGJ, Holttum RE (eds) (1959) Flora Malesiana, vol 1. M. Nijhoff/W. Junk, Boston, pp 164–166

Wen CC, Shyur LF, Jan JT, Liang PH, Kuo CJ, Arulselvan P, Wu JB, Kuo SC, Yang NS (2011) Traditional Chinese medicine herbal extracts of *Cibotium barometz, Gentiana scabra, Dioscorea batatas, Cassia tora,* and *Taxillus chinensis* inhibit SARS-CoV replication. J Tradit Complement Med 1(1):41–50

WHO (1990) Medicinal plants in Viet Nam. Regional Publications, Western Pacific Series No 3. World Health Organization, Regional Office for the Western Pacific, Manila, p 95

Wu Q, Yang XW (2009) The constituents of *Cibotium barometz* and their permeability in the human Caco-2 monolayer cell model. J Ethnopharmacol 125(3):417–422

Wu Q, Yang XW, Yang SH, Zou L, Yan J (2007) Chemical constituents of *Cibotium barometz*. Nat Prod Res Dev 19:240–243 (In Chinese)

Xie Y, Luo R, He ZJ (2011) The study of processing technology of *Cibotium barometz*. Med Healthc Apparatus 18(7):1096–1098 (In Chinese)

Xu N, Jia T (2011) Chemical reaction and products in processed *Cibotium barometz*. Zhongguo Zhong Yao Za Zhi 36(15):2066–2070 (In Chinese)

Xu ZY, Chen ZD, Chen ZL, Hou LB, Zhang K (2000) Studies on the chemical constituents of *Cibotium barometz* (II). Pharm J Chin People Liberat Army 16:65–68 (In Chinese)

Xu ZY, Chen ZL, Chen ZD, Ge FH, Zhang K (2001) Determination of phospholipid in *Cibotium barometz* by supercritical fluid extraction and RP-HPLC. Zhong Yao Cai 24(3):174–175 (In Chinese)

Xu ZY, Yan Y, Chen ZD, Chen ZL, Zhang K (2004) Studies on the chemical constituents of *Cibotium barometz* (III). Pharm J Chin People Liberat Army 20(5):337–339 (In Chinese)

Xu N, Cao Y, Zhou L, Jia TZ (2011) Study on antioxidation of raw and processed *Cibotium barometz*. Chin Arch Tradit Chin Med 11:2424–2426 (In Chinese)

Xu G, Ju CG, Yu HT, Zhang Q, Jia TZ (2012a) Advanced on processing *Cibotium barometz*. China J Exp Tradit Med Formulae 18(5):238–242 (In Chinese)

Xu N, Zhang Q, Cao Y, Jia TZ, Zhong X (2012b) Chemical constituents from *Cibotium barometz* and their scavenging activities for DPPH radical. China Exp Tradit Med Formulae 18(24):162–166

Xu G, Pei QY, Ju CG, Zhang F, Jia TZ (2013a) Detection on effect of different processed *Cibotium barometz* on osteoblasts by CCK-8. Zhongguo Zhong Yao Za Zhi 38(24):4319–4323 (In Chinese)

Xu N, Cao Y, Shi YN, Zhong X, Jia TZ (2013b) New hydrolysable tannin from *Cibotium barometz*. China J Chin Mat Med 38(5):698–702 (In Chinese)

Xu N, Shi YN, Xie X, Zhang Q, Cao Y, Jia TZ (2013c) A novel phenolic glycoside from *Cibotium barometz*. Chin Trad Herbal Drug 44(4):383–386 (In Chinese)

Xu G, Sun N, Zhao MJ, Ju CG, Jia TZ (2014a) Study on decoction's effect of different processed rhizomes of *Cibotium barometz* on retinoic acid induced male rats osteoporosis. Zhongguo Zhong Yao Za Zhi 39(6):1011–1015 (In Chinese)

Xu G, Zhao MJ, Sun N, Ju CG, Jia TZ (2014b) Effect of the RW-Cb and its active ingredient like P-acid and P-aldehyde on primary rat osteoblasts. J Ethnopharmacol 151(1):237–241

Yang FX, Zhu BY, Zheng X, Hu JF, Ge XL, Huang HL, Liao DF (2002) Protective effect of onychin on lipid peroxidation in experimental liver damage of mice. J Nanhua Univ (Med Ed) 30:4–6 (In Chinese)

Yang HJ, Wu Q, Yang SH (2010) Advances in research of chemical constituents and pharmacological activities of *Cibotium barometz*. Chin J Exp Tradit Med Formulae 15:23–234 (In Chinese)

Ye XQ, Yang XF, Qiu PH, Huang KX (2006) Determination of flavonoids and trace elements in *Cibotium barometz*. Chin J Spectr Lab 1:39–43 (In Chinese)

Yu HT, Ju CG, Zhang Q, Jia Z (2011) Effect of different extracts from *Cibotium barometz* on culturing osteoblasts. China J Exp Tradit Med Formulae 17(24):36–39, In Chinese

Yuan Z, Yu JT, Su SW (2000) Determination of the contents of protocatechuic acid and caffeic acid in Rhizoma Cibotii and Rhizoma Pteris by TLCs. J Shenyang Pharm Univ 26(5):388–340 (In Chinese)

Yun XL, Zhao NW, Pan LT, Zhao JH (2009a) Resource distribution and exploitation of edible ferns in Guizhou (I). J Anhui Agric Sci 37(32):15911–15912

Yun XL, Zhao NW, Pan LT, Zhao JH (2009b) Resource distribution and exploitation of edible ferns in Guizhou (II). J Anhui Agric Sci 37(33):16317–16319

Zhang XC, Nishida H (2013) Cibotiaceae. In: Wu ZY, Raven PH, Hong DY (eds) Flora of China. Vol. 2–3 (Lycopodiaceae through Polypodiaceae). Science

Press/Missouri Botanical Garden Press, Beijing/St. Louis

Zhang CL, Wang ZX (2001) Studies on the constituents of *Cibotium barometz* (L.) J.Sm. rhizome. Chin J Med Chem 11(5):279–280 (In Chinese)

Zhang XC, Jia JS, Zhang GM (2008) Non-detriment finding for *Cibotium barometz* in China. In: International NDF workshop case studies. WG 2 – perennials Case Study 1. China, pp 1–8

Zhao X, Wu ZX, Zhang Y, Yan YB, He Q, Cao PC, Lei W (2011) Anti-osteoporosis activity of *Cibotium barometz* extract on ovariectomy-induced bone loss in rats. J Ethnopharmacol 137(3):1083–1088

Zhou JJ, Xie GR, Yan XJ (2011) Encyclopedia of traditional Chinese medicines – molecular structures, pharmacological activities, natural sources and applications: Vol. 4: isolated compounds N-S. Springer, Berlin/Heidelberg

Ipomoea batatas

Scientific Name

Ipomoea batatas **(L.) Lamk**

Synonyms

Batatas edulis (Thunb.) Choisy, *Batatas edulis* (Thunberg ex Murray) Choisy; *Convolvulus batatas* L.; *Convolvulus chrysorrhizus* Soland. ex G. Forster, *Convolvulus candicans* Solander ex Sims; *Convolvulus edulis* Thunb., *Convolvulus edulis* Thunberg ex Murray, *Ipomoea batatas* var. *edulis* (Thunb.) Kuntze, *Ipomoea batatas* var. *edulis* (Thunberg ex Murray) Makino, *Ipomoea batatas* var. *lobata* Gagnepain & Courchet, *Ipomoea chrysorrhiza* (Soland. ex G. Forster) Peter, *Ipomoea edulis* (Thunberg ex Murray) Makino; *Ipomoea fastigiata* Sweet.

Family

Convolvulaceae

Common/English Names

Sweet Potato

Vernacular Names

Afrikaans: Patat;
Albanian: Patate E Ëmbël;
Angola: €Kapa (Umubumbu), Batata-Doce, Batata-Da-Ilha (Portuguese);
Arabic: Ba Ta Tah Helua;
Argentina: Yety;
Aymara: Pua, Tipali, Tuctuca;
Benin: Kukundu (Yoruba);
Brazil: Batata Doce, Batata Da Ilha, Batata Da Terra (Portuguese);
Bolivia: Apichu;
Burkina Faso: Tingo (Baoulé), Téguibo, Déguibo, Bassi (Guéré);
Burmese: Myonk-Ni;
Cameroon: Makeu, Ton'ho-La (Bamileke);
Chamorro: Kamote;
Chinese: Bai Shu, Di Gua, Fan Shu, Gan Shu;
Comoros: Maniyambatsi (Great Comoros);
Chuukese: Kómwu, Kómwuti, Kómwuutiy, Omwuti, Sachúmayimwo, Yomwuutiy;
Cuba: Boniato;
Cook Islands: Ku'Ara, Kūara, Kūmala, Kūmara, Kumara (Maori);
Croatian: Indijski Krumpir, Slatki Krumpir;
Danish: Batat, Sød Kartoffel;
Democratic Republic of Congo: Matembele Bangi (Lingala), Mabelenge, Mibabanga

T.K. Lim, *Edible Medicinal and Non-Medicinal Plants: Volume 10, Modified Stems, Roots, Bulbs*,
DOI 10.1007/978-94-017-7276-1_5

(Ngombe), Matembele (Swahili), Tshilunga (Tshiluba);
Dutch: Bataat, Zoete Aardappel;
Estonian: Bataat, Okariin;
Ethiopia: Siquar-Dinich (Omotic);
Fiji: Kawai Ni Vavalangi, Kumala, Kumara, Ndambithi, Wa Uvi;
French: Batate, Patate Douce;
Gabon: Émongo (Baduma), Amongo (Bakèlè), Gémongo (Baléngi), Mongo (Balumbu, Bavové), Imongo (Banzabi), Lungu (Bapunu), Lifita (Bavili), Mbongo (Benga), Imongo (Béséki, Ngowé), Égwèta (Ivéa, Mitsogo), Mongu (Bavarama, Bavungu, Eshira), Lémongho, Futa (Mindumu), Mongo Y'onigi (Mpongwè, Nkomi, Namôngha (Fang),Orungu), Ofogola (Galoa);
German: Batate, Süsskartoffel;
Greek: Glikopatata;
Guatemala: Chin, Sis, Is;
Haiti: Patate Dôce;
Hawaiian: 'Uala, 'Wala;
Honduras: Mabi;
Hungarian: Édes Burgonya;
I-Kiribati: Te Kumara;
India: Lalalu (Bengali), Batata, Bataka, Kannagi, Sakariu (Gujarati), Mitha Alu, Shakar Kanda, Sakarkand (Hindi), Genasu (Kanada), Chakkare Kilangu (Malayalam), Batata, Bataka, Ratala, Ratalu (Marathi), Kanh (Oriya), Carkkaraivalli, Ciignikkilangu, Sakkaravelleikilangu, Vattaalagnkilangu, Vellikkilangu (Tamil), Chelagadda (Telugu);
Indonesia: Ubi Jalar, Ubi Keladek, Ketel Rambat (Javanese), Huwi Boled (Sundanese);
Italian: Patata Americana, Patata Dolce, Batata;
Ivory Coast: Tingo (Baoulé), Gonenbi (Ebrie), Téguibo, Déguibo, Bassi (Guéré);
Japanese: Imo, Kara Imo, Kan Sho, Satsuma-Imo, Ryuukyuu Imo;
Kenya: Ngwaci (Kikiyu), Mîrîyo (Central Province),
Khmer: Dâmlô:Ng Chviĕ;
Korean: Goguma;
Kwara'Ae: Butete;
Laotian: Man Kè:W;
Lithuanian: Saldžiųjų Bulvių;
Malaysia: Ubi Keladi, Ubi Keladek, Keledek;
Marquesan: Kūma'a;
Nigeria: Ipotato (Ndokwa Delta State), Odunkun (Ogun State);
Niuean: Kumara, Simala;
Madagascar: Vomanga (Agnalazaha Forest, Southeastern Madagascar);
Marshallese: Piteto;
Mexico: Camotli;
Norwegian: Søtpoteter;
Palauan: Emutii;
Panama: Kuwas:
Papua New Guinea: Kaukau, Kaema;
Paraguay: Yety;
Peru: Apichu, Kumara;
Philippines: Tigsi (Bisaya), Tugi (Bontok), Lapni (Ifugao), Kamote (Tagalog), Panggi-Bagun (Sulu);
Polish: Słodki Ziemniak;
Popular Republic of Congo: Osokoro (Kôyô);
Portuguese: Camote, Papas, Batata Doce;
Pukapukan: Kūmala;
Rarotongan: Kūmara;
Russian: Batat;
Samoan: 'Umala;
Senegal: Juifata (Diola);
Serbian: Slatki Krompir;
Slovak: Sladký Zemiak, Ocarina;
South Africa: Ubhatata (Northern Maputaland, Kwazulu-Natal Province);
Spanish: Batata, Boniato, Papas, Camote, Papa Dulce, Moniato, Cumala Huasca, Cumal Huasca, Cumara, Curiti, Jarissi Jabo;
Swahili: Anago, Dankoli, Dantiu, Kiazi, Kinkio, Veeazee, Viazi Vitamu;
Swedish: Sötpotatis;
Tahitian: Umara, 'Umara;
Thai: Mam-Thet;
Togo: Patate Douce (French);
Tongan: Kumala, Kumara;
Tuamotuan: Kūmara;
Turkish: Sarmaşık Patatesi, Tatlı Patates, Yer Elması;
Uganda: Mboli (Bulamogo Country), Pot-Ecok (Luganda), Akarandura, Enkoora (Rutooro), Lumonde (Sango Bay, Southern Uganda);
Ulithian: Komoti;
Uruguay: Boniato;
Vanuatu: Kūmala;

Venezuela: Chaco;
Vietnamese: Khoai Lang, Khoai Mon;
Welsh: Tatws Melys;
Woleaisan: Gamwuutiy;
Yapese: Gamuti, Kamote, Kamuut

Origin/Distribution

Available evidence shows that sweet potato originated from the neotropics, either Central or South American lowlands with subsequent dispersal to North America, Europe, Africa, Asia and the Pacific islands (Zhang et al. 2000; Gichuki et al. 2003). Results of AFLP molecular studies supported the hypothesis that Central America was the primary centre of diversity and most likely the centre of origin of sweet potato and Peru-Ecuador should be regarded as a secondary centre of sweet potato diversity (Zhang et al. 2000). This was confirmed by subsequent polymorphic RAPD molecular studies that found the primary centre of diversity probably had two distinct genepools (Gichuki et al. 2003). It was proposed that the dispersal of the sweet potato from its origin may have mainly involved varieties from Central America/Caribbean as opposed to varieties from South America. There was an indication that new genepools may be evolving in Africa and Asia due to hybridization and adaptation to the local environments. New Guinea is considered to be the most important secondary centre of genetic diversity for sweet potato, particularly the highlands region (Yen 1974). Combining nuclear and chloroplast data, Roullier et al. (2013) showed that New Guinea sweet potato landraces were principally derived from the northern neotropical genepool (Caribbean and Central America), but that some South American clones may also have been introduced, either early by Polynesians themselves or (more likely) later by whalers and missionaries and through twentieth-century movements across the Pacific. It is cultivated worldwide and widely naturalized in the tropics and sub-tropics.

Srisuwan et al. (2006), using cytogenetic approaches, found that *Ipomoea trifida* might be the progenitor of *I. batatas*, and *I. tabascana*, inter-specific hybrid between these two species.

Agroecology

Sweet potato is adaptable to a wide range of climatic conditions from the warm humid tropics to mild sub-temperate zones from near sea level to 2000 m elevation. It has tolerance to low temperature in the higher altitudes, but is frost-sensitive. It grows best on sandy loams that are well-drained and fertile and is intolerant of water logging and is usually grown in on mounds or ridges. It can be grown in clayey soils, but tuberous yield can be affected. It can be grown in semi-arid conditions. Soils below a pH of 5 are at a risk of aluminium toxicity and below pH 4.5 (other than organic soils). Aluminium will severely reduce the growth and development of sweet potato.

Edible Plant Parts and Uses

The tuberous roots and leaves and young shoots are eaten. The starchy tuberous roots are by far the most important product. In some tropical areas, sweet potato is a staple food-crop.

Sweet potato leaves are a common side dish in Chinese cuisines, often boiled with garlic and vegetable oil and dashed with salt before serving. In Taiwan, they are commonly found at *biàndāng* restaurants, as well as dishes featuring the sweet potato root. In Malaysia, the leaves are fried with garlic and *sambal belacan*. The young leaves and vine tips of sweet potato leaves are also widely consumed as a vegetable in west African countries (Guinea, Sierra Leone and Liberia, for example). In PNG and the south Pacific islands, the young leaves and shoots are sometimes eaten as greens. Leaves eaten cooked in water, coconut cream or stir fried with chillies, garlic and dried shrimps, onion. One common recipe using shoot tips is 'sweet potato tip soup' made up of sweet potato shoot tips, butter/cooking oil, chopped onion, flour, milk . The leaves are a good addition to soups and are reported to be an excellent food for babies, pregnant women, and breast-feeding

mothers. One recipe is called Baby's delight – sweet potato leaves, small pieces of pumpkin, fish, coconut cream.

Tubers can be eaten raw and taste somewhat like a sweet carrot. Tubers are usually eaten cooked, boiled or baked or roasted. They are used in various food dishes, cakes, buns, soups, casseroles, pies, fries and chips.

Sweet potatoes may be baked in an earth oven or they may be boiled or steamed. Steamed/boiled chunks, for a simple and healthy snack, of sweet potato may be boiled in water or cooked in the microwave. Baked sweet potato serves as an alternative to potato and can be eaten as it is or with brown sugar and butter. They may be eaten as they are or mashed with milk or coconut cream. In the Dominican Republic, they are served for breakfast. Sweet potato butter can be cooked into a gourmet spread. In America, caramelized sweet potatoes are prepared with brown sugar, marshmallows, maple syrup, molasses or other sweet ingredients, and served as a side dish, a traditional dish especially on Thanksgiving. Sweet potato pie is another favourite American dish. Sweet potato fries are another common preparation, and are made by julienning and deep frying sweet potatoes. Sweet potato chips can be sliced, fried and eaten. Sweet potato is a staple food for people in north-eastern Uganda where sun-dried slices called 'Amukeke' or sun-dried crushed tubers called 'Inginyo' are the important modes of preparation. Amukeke is mainly for breakfast, eating it with peanut sauce and a cup of tea. Inginyo will be mixed with cassava flour and tamarind, this food is called 'atapa'. People eat 'atapa' with smoked fish cooked in peanut sauce or with dried cowpea leaves cooked in peanut sauce.

Cooked sweet potato can be made into a variety of dishes in soups, stews, curries and stir fry. Mashed with a little coconut cream, fish and green vegetables; it makes a good baby food. One popular dish in Fiji is called 'Meal in coconut shell' which comprises clean shelled and halved coconut, small sweet potato, coconut milk, green leaves, tomato and spring onions. Sweet potato buns are made with sweet potato cooked and mashed, milk, self-rising flour and lemon juice.

In Japan, sweet potato is used as a vegetable in tempura, in *Yaki-imo* (roasted sweet potato) a delicacy in winter, and in *Daigaku-imo* a baked sweet potato dessert. In *Imo-gohan,* slices or small chunks of sweet potato are cooked in rice. Sweet potato is also served in *nimono* (Japanese foods such as fish, meat and vegetables that are simmered in a seasoned broth) or *nitsuke* (a sea food dish) boiled and flavoured with typically soy sauce, *Mirin* (sweetened Japanese rice wine used for cooking) and *Dashi* (Japanese soup stock). Sweet potato is also used in sweet potato paste ball called *Imo-kinton* or a Japanese confectionary called *wagashi*.

Significant amounts of sweet potato are being processed into industrial starch, alcohol, noodles (Plate 7) and other products, especially in China and Korea. In China, sweet potato is also processed into a myriad of products for local consumption and export such as sweet potato powder, starch, flour, frozen cubes, sweet potato puree, dried and dehydrated slices, chips, candied dried slices and noodles. In the Shandong and Sichuan provinces of China, transparent cellophane noodles are made from starch extracted from sweet potato, an important crop in China's small farming systems. In Korea, sweet potato starch is commonly employed to make the popular Korean sweet potato vermicelli, a type of cellophane noodles called *dang myun, dangmyun, tang myun, tangmyu*. The *dang myun* noodles are glassy and transparent and have a very interesting chewy texture and are very long, slippery and almost elastic. The most common Korean dish using *dang myun* is *japchae*, a beef stir-fry using sesame and soy. Another favourite dish using *dang myun* is '*Cold dang myun noodle salad*'. Sweet potato is also fermented and made into alcohol and spirits, e.g., *Shōchū* – a Japanese spirit made from fermentation of rice and sweet potato; *Soju* – Korean alcoholic beverage, often mistaken as rice wine, but actually almost always made in combination with other ingredients such as wheat, barley or sweet potatoes. Sweet potato bread and flour (100 %) are also marketed as hypoallergenic for people who cannot tolerate grain breads and flours.

Purple-fleshed sweet potato is rich in anthocyanins and is used as a natural food colourant with nutritious and health benefits; the deep purple paste and flour made from cv. 'Ayamurasaki' are used for the preparation of noodles, bread, jams, sweet potato chips, confectionery, juices and alcoholic beverages (Yamakawa et al. 1998; Suda et al. 2003). These foods and beverages using purple-fleshed are being sold in shops at stations, airports and tourist resorts in the Kyushu-Okinawa area in Japan. Studies found that extruded ready-to-eat breakfast cereals (RTEBCs) made from 100 % of sweet potato flour (SPF), and 75 %/25 % SPF/whole-wheat bran (WWB) and extrusion cooking were well-liked and acceptable to sixth graders attending an elementary school in Auburn, Alabama, but the 100 % WWB was unacceptable (Dansby and Bovell-Benjamin 2003).

Tubers of Turkish sweet potato cv. Hatay Kirmizi can be consumed not only in steamed, boiled and fried forms but also can be processed into food products, such as muffins, cookies, biscuits, breakfast foods with longer shelf-life, and improved characteristics (Tokusoglu and Yildirim 2012). Also, sweet potato Hatay Kirmizi can be processed into flour and used as a thickener, antioxidant enhancer and colour source in industrial powder soups, gravy, extruder snacks and some bakery products. Sweet potato tubers were processed into non-alcoholic beverages flavoured with citrus lime and ginger in Ghana (Wireko-Manu et al. 2010).

Vacuum frying (1.33 kPa), with the aid of a de-oiling mechanism, was used to produce low-fat sweet potato chips (Ravli et al. 2013). The final oil content of the vacuum-fried chips was 60 % lower than those found in traditionally fried sweet potato chip, which indicated that the de-oiling mechanism was crucial in vacuum-frying processing. Studies by Johnson et al. (2010) found that high fructose syrup, a highly valued sweetener for the food and beverage industries, could be produced from sweet potato flours and their blends with cereal flours.

Botany

A sprawling, low, herbaceous annual (Plate 1), with fusiform or elongated subterranean tubers; tuber skin colour ranging from between red, purple, brownish, yellowish-brown and white and tuber flesh colour white yellow, orange and

Plate 1 Sweet potato planting

Plate 2 (**a–d**) Variously coloured sweet potato tubers

purple (Plates 2, 3). Stems prostrate or ascending, green or purplish, glabrous or pilose, much branched, rooting at nodes and with milky sap. Leaves alternate, ovate-orbicular, entire or palmately 3-7-lobed or -parted, subcordate or cordate, 4–15 cm long, 3–11 cm wide, on 3–15 cm long petioles (Plates 4, 5, and 6). Inflorescences axillary; peduncle 3–18 cm long, 1–several-flowered; pedicels 3–12 mm long; bracts lanceolate and deciduous. Sepals sub-equal, the inner somewhat longer, oblong to elliptic-oblong, 7–12 mm long by 3–5 mm wide, acute and mucronate, subcoriaceous. Corolla violet or lilac (Plate 5), white above, campanulate, 3–4.7 cm long. Stamens 5, adnate to the perianth, free and alternating with the corolla lobes. Anthers dehiscing via longitudinal slits. Ovary syncarpous, superior with two ovules per cell and one simple style. Fruit a novoid or depressed globose, non-fleshy capsule. Seeds glabrous.

Nutritive/Medicinal Properties

Tuber Nutrients/Phytochemicals

Raw sweet potato tuber (per 100 g edible portion proximate) was reported to have the following nutrient values: water 77.28 g, energy 86 kcal (359 kJ), protein 1.57 g, total lipid 0.05 g, ash 0.99 g, carbohydrate 20.12 g, total dietary fibre 3.0 g, total sugars 4.180 g, minerals – Ca 30 mg, Fe 0.61 mg, Mg 25 mg, P 47 mg, K 337 mg, Na 55 mg, Zn 0.30 mg, Cu 0.151 mg, Mn 0.258 mg,

Plate 3 White-purple-fleshed tuber

Plate 4 Bronze-coloured young foliage

Se 0.6 mcg; vitamin C 2.4 mg, thiamin 0.078 mg, riboflavin 0.061 mg, niacin 0.557 mg, pantothenic acid 0.800 mg, vitamin B-6 0.209 mg, total folate 11 μg, total choline 12.3 mg, vitamin A 14187 IU, vitamin A RAE 709 μg, β-carotene 8509 μg, α-carotene 7 μg, vitamin E (α-tocopherol) 0.26 mg, β-tocopherol 0.01 mg, vitamin K (phylloquinone) 1.8 μg; total saturated fatty acids 0.018 g, 16:0 (palmitic acid) 0.018 g, 18:0 (stearic acid) 0.001 g, total monounsaturated fatty acids 0.001 g, 18:1 undifferentiated (oleic acid) 0.001 g, total polyunsaturated fatty acids 0.014 g, 18:2 undifferentiated (linoleic acid) 0.013 g, 18:3 undifferentiated (linolenic acid) 0.001 g, phytosterols 12 mg, amino acids – tryptophan 0.031 g, threonine 0.083 g, isoleucine

Plate 5 Old leaves and flower

Plate 6 Purplish-stalked sweet potato leaves sold in markets

0.055 g, leucine 0.092 g, lysine 0.066 g, methionine 0.029 g, cystine 0.022 g, phenylalanine 0.089 g, tyrosine 0.034 g, valine 0.0.086 g, arginine 0.055 g, histidine 0.031 g, alanine 0.077 g, aspartic acid 0.382 g, glutamic acid 0.155 g, glycine 0.063, proline 0.052 g and serine 0.088 g (USDA–ARS 2014). Sweet potato storage roots of 10 sweet potato varieties were found to contain the following flavonoid (%DW) quercetin 0.68 %, myricetin 0.09 %, luteolin 0.01 %, apigenin 0 %, kaempferol 0.01 % and total flavonoids 0.85 % (Ojong et al. 2008).

The proximate nutrient composition of fresh, boiled and baked sweet potatoes were respectively as follows: dry matter 31.07–33.76 g, 33–37.65 g, 36.55–40.65 g; ash 2.13–2.54 g, 2.19–2.60 g, 2.31–2.62 g; crude fibre 2.33–2.65 g, 2.45–2.76 g, 2.11–2.64 g; protein 4.29–5.08 g, 4.36–5.03 g, 3.54–4.56 g; starch 63.90–64.89 g, 49.22–57.43 g, 55.80–60.22 g; ascorbic acid 14.07–20.18, 24.77–37.15, 19.43–27.88 mg/100 g dw; glucose 2.73–4.68 mg, 1.34–3.94 mg, 1.72–4.90 mg/g dw; fructose 1.43–4 mg, 1.42–3.75 mg, 1.24–3.38 mg/g dw; sucrose 56.94–59.97 mg, 48.99–61.50 mg, 55.52–64.36 mg/gdw; maltose not detected, 48.13–122.81 mg, 48.52–56.27 mg/g dw; β-carotene content 5.63–15.63 mg; 3.28–12.64 mg, 1.15–10.07 mg/g dw; retinol activity equivalent (RAE) values 169.5–439.6 µg, 102.3–353.1 µg, 31.65–260.5 µg/100 g wet basis, respectively (Dincer

Plate 7 Sweet potato noodles

et al. 2011). Both the boiling and baking treatments resulted in a significant reduction in the β-carotene content and therefore in the RAE values of the sweet potato samples. Generally, the treated samples had a higher dry matter content than the fresh samples. The protein content of all samples decreased significantly during baking, while it did not change during boiling. The boiling process led to significant degradation in starch. DPPH antiradical activity of sweet potatoes was enhanced by heat treatment via the formation of phenolic compounds. Total phenolic content of the samples increased with both baking and boiling treatments.

Total sugars in stored, raw staple sweet potato types range from 2.9 to 3.2 % on a fresh weight (FW) basis (Picha 1985). The major sugars in raw roots: sucrose, fructose and glucose and their contents range from 1.3 to 2.5 %, 0.4 to 0.7 % and 0.4 to 1.0 %, respectively. Total sugars in stored, raw dessert sweet potato types range from 4.6 to 5.5 % on a FW basis (Picha 1985). Corresponding sucrose, fructose and glucose contents range from 2.8 to 4.1 %, 0.3 to 1.2 % and 0.2 to 1.5 %, respectively. Maltose was the major sugar and sucrose the secondary sugar in all cultivars at harvest (Picha 1986). Maltose decreased during curing and over long-term storage. Sucrose, glucose and fructose concentrations increased during curing and through at least 4 weeks of storage in the orange-flesh cultivars. Sucrose concentration was always greater than either monosaccharide. Dessert sweet potato types generally have a cream coloured to orange flesh and dry weight content ranging from 17.7 to 26.3 % with starch contents ranging from about 13.0 to 22.0 % (Picha 1987). Dry matter (DM) content and free sugar composition (g/100 g FW) of raw tubers of nine sweet potato genotypes were: DM 18–30.6 %, total sugars 3.37–6.80 g, fructose 0.23–1.48 g, glucose 0.33–1.87 g and sucrose 1.36–4.14 g (Lewthwaite et al. 1997). In cooked tuber of the same lines, total sugars was 6.69–10.24 g, fructose 0.22–1.39 g, glucose 0.34–1.72 g, sucrose 1.31–4.15 g and maltose 2.44–5.51 g. A strong linear relationship exists between fructose, glucose and sucrose content in raw and baked roots All the clones produced considerable amounts of maltose during cooking, which was significantly related to % dry weight. Sucrose was the major sugar during all stages of development from 7 to 19 weeks after transplanting, representing at least 68 % of total sugars across all sweet potato cultivars and dates (La Bronte et al. 2000). Fructose and glucose content profiles varied among and within cultivars during development. Cultivars ranking the highest in total sugars had either more monosaccharides to compensate for lower sucrose content or more sucrose to compensate for a lower monosaccharide content. Cultivars with high dry weight and alcohol insoluble (solids (starch)) could be

selected early during storage root development. Sugar and non-volatile acid contents (g/100 g dry matter) in raw and baked roots of six sweet potato cultivars were reported by Wang and Kays (2003) respectively as: sucrose 5.61–21.34 g; 5.67–26.25 g; glucose 0.31–6.31 g; 0.65–6.31 g; fructose 0.25–5.50 g; 0.44–6.75 g; galactose 0.03–0.13 g; 0–0.18 g; inositol 0.09–0.22 g; 0.06–0.31 g; maltose 0–0.03 g; 5.18–29.08 g; malic acid 0.39–1.59 g; 0.40–2.01 g; citric acid 0.07–0.41 g; 0.05–0.49 g and quinic acid 0.24–0.62 g; 0.4–0.72 g.

Fresh sweet potato and sweet potato flour were reported to have the following nutrient composition (w/w db), respectively: moisture content 73.1, 7.7 %; total sugars in glucose equivalent 75.0, 77 %, glucose 2.4, 2.1 %; fructose 2.6, 1.6 %, sucrose 8.0, 15.8 %, fibre 10, 3 %; proteins 3.5, 6.6 %, lipids 0.4, 1.8 % and ash 4.1, 2.7 % (Lareo et al. 2013). The mean concentration (mg/kg wet weight) of mineral nutrients and toxic metal determined in white-, red- and orange-fleshed sweet potato varieties were respectively reported as: Na 564, 324, 559.6 mg; K 4510, 4094, 4551 mg; Ca 536.3, 789.1, 321.7 mg; Mg 706.2, 562.2, 422.2 mg; Cu 1.264, 1.452, 1.156 mg; Fe 7.195, 6.637, 4.817 mg; Mn 2.713, 2.58, 1.742 mg; Zn 2.499, 2.523, 1.742 mg; Cr 0.034, 0.023, 0.022 mg; Ni 0.056, 0.045, 0.030 mg; Cd 0.001, 0.004, 0.001 mg and PB 0.002, 0.005, 0.001 mg (Luis et al. 2014). Commercial sweet potato fibre was found to be mainly composed of glucose (88.4 %), but small amounts of other sugars were also detected (Salvador et al. 2000). Sweet potato cell wall materials had the highest amount of pectin and galacturonic acid.

The average yield and dietary fibre (DF) content of DF products from 10 sweet potato varieties were 9.97 and 75.19 %, respectively (Mei et al. 2010). Average contents of cellulose, lignin, pectin and hemicellulose were 31.19, 16.85, 15.65 and 11.38 g/100 g of dry matter in DF products, respectively. The relative monosaccharide contents of DF were in the order glucose > uronic acid > galactose > arabinose > xylose > rhamnose > mannose. Swelling capacity, water-holding capacity, oil-holding capacity and glucose absorption capacity determinations of the DF of sweet potato varieties had respective ranges of 8.11–12.56 mL/g, 3.54–6.15 g/g, 1.43–2.48 g/g and 0.54–1.27 mmol/g.

Major components of non-protein-nitrogen fraction of 'Jewel' sweet potato at 107 storage days were asparagine 61 %, aspartic acid 11 %, glutamic acid 4 %, serine 4 % and threonine 3 % (Purcell and Walter 1980). Levels of most amino acids changed with time. The dry matter, protein and starch contents of the sweet potatoes were significantly changed by baking and boiling while the ash and crude fibre contents did not differ as significantly (Dincer et al. 2011). The β-carotene contents of baked and boiled sweet potatoes were lower than those of fresh sweet potatoes; however, the total phenolic and ascorbic acid contents of the baked and boiled sweet potatoes were higher than those of the fresh samples. Generally, the antiradical activity of the sweet potatoes increased with the treatments. Sucrose, glucose and fructose were quantified as free sugars in all fresh sweet potatoes; however, maltose was determined in the treated samples.

Studies by Roxas et al. (1985) found the absence of oligosaccharides in the storage roots of sweet potatoes. They found only monosaccharides and sucrose in the samples. The sum of glucose, fructose and sucrose accounted for 85–96 % and 17–54 % of total soluble sugars identified in extracted fractions of raw and cooked sweet potato tubers (Truong et al. 1986). Verbascose was found in trace amounts, stachyose was not detected. Starch content of raw and cooked samples ranged from 33–73 to 32–61 %. Starch degradation products maltose, maltotriose were present in cooked samples coeluted with cellobiose and raffinose, respectively. The tubers contained 0.23–0.4 % cellobiose and negligible raffinose. The concentrations of indigestible oligosaccharides were too low to account for the flatulence that accompanied sweet potato as a staple food. The sugar composition of fresh and baked sweet potatoes (% DW basis) were determined respectively as: total sugars (4.50–8.41 %, 15.11–19.14 %), fructose (0.24–1.06 %, 0.26–0.86 %), sucrose (2.52–7.77 %, 1.53–5.02 %), glucose (0.39–2.02 %, 0.31–1.37 %), maltose

(0–0.39 %, 8.81–13.97 %) (Lai et al. 2013). Maltose increased dramatically after baking. The starch granules of fresh sweet potato were oval-shaped and generally <20 μm, but after the baking treatment starch granules completely gelatinized.

Purple sweet potato wine was found to have the following proximate compositions (per 100 mL): total soluble sugar (TSS), 2.25° Brix; starch, 0.15 g; total sugar, 1.35 g; total acidity, 1.34 g tartaric acid; phenol, 0.36 g (caffeic acid equivalent); anthocyanin, 55.09 mg; tannin, 0.64 mg; lactic acid, 1.14 mg; ethanol, 9.33 %(v/v) and pH, 3.6 (Ray et al. 2012).

Lipids

Changes in fatty acid composition of Georgia Red and Centennial varieties of sweet potatoes were observed during storage and appeared to be mere pronounced at low storage temperatures (10 and 4.5 °C) (Boggess et al. 1967). The most consistent changes found were an increase in tetracosanoic acid and a decrease in short-chain saturated acids. The Centennial variety contained higher levels of total lipids, which were generally reflected in higher levels of the three fractions, (1) non-phospholipids, (2) cephalin and (3) lecithin. The increase in total lipids and the individual lipid fractions with storage is indicative of two processes that may have occurred in the stored roots. The lipids may have become more extractable as the respiring potato underwent compositional changes, or lipids were being synthesized from non-lipid components.

Lipid composition (% total lipid weight) of cured Centennial sweet potatoes consisted of: neutral lipids 42.1 – triglycerides 26.9 %, steryl esters 6.1 %, diglycerides 3.8 %, hydrocarbons 2.8 %, sterols (free) 2.5 %; glycolipids 30.8 % – monogalactosyl diglycerides 13.6 %, digalactosyl diglyceride 6.3 %, cerebroside 4.7 %, esterified steryl glucoside 3.5 %, unknown 2.1 %, steryl glucoside 0.6 %; phospholipids 27.1 % – phosphatidyl ethanolamine 7.8 %, phosphatidyl chlorine 7.0 %, phosphatidyl inositol 5.1 %, unknown 3.0 %, cardiolipid 1.6 %, phosphatidyl glyceride 1.2 %, phosphatidyl serine 1.1 % and phosphatidic acid 0.4 % (Walter et al. 1971). The predominant fatty acids were stearic, palmitic, oleic, linoleic and linolenic. The amount of lipophilic extractives in sweet potato root ranged from 0.87 to 1.32 % dry weight (Cordeiro et al. 2013). Fatty acids and sterols were the major families of compounds identified. The most abundant saturated and unsaturated fatty acids were hexadecanoic acid (182–428 mg/kg) and octadeca-9,12-dienoic acid (133–554 mg/kg), respectively. β-Sitosterol was the principal phytosterol, representing 55.2–77.6 % of this family, followed by campesterol. Long-chain aliphatic alcohols and α-tocopherol were also detected, but in smaller amounts.

Sugawara and Miyazawa (1999) reported sweet potato roots to contain (mg/100 g) 67 mg total glycolipids comprising 9.7 mg monogalactosyldiacylglycerols (MGDG), 22.6 mg digalactosyldiacylglycerols (DGDG), 15 mg acylated steryl glucoside (ASG), 5.6 mg steryl glucoside (SG) and 14.1 mg ceramide monohexoside (CMH).

Starch

The physicochemical and functional properties of starch of 14 sweet potato lines (10 white), (2 yellow), (1 dark purple), (1 dark orange) were determined as: organic matter 85.45–99.48 %, ash 2.81–6.45 %, protein 3.96–5.13 %, fibre 2.11–3.96 %, starch 15.2–28.1 %, sugars 1.51–3.9 %, dry matter 26.8–33.3 %, extractable starch 16.8–21.1 %, starch swelling volume 32.5–50 mg/g, solubility 7.15–13.65 %, amylose 0.311–0.337, peak viscosity 2924–3902 cP; breakdown 1080–2541 cP, final viscosity 2098–4530 cP, setback 800–1683.5 cP and pasting temp 65.9–77.50 °C (Moorthy et al. 2010).

The proximate compositions and physicochemical properties of 21 Caribbean sweet potatoes were reported by Aina et al. (2012) as: moisture (8.0–12.4 %), protein (0.0–0.2 %), ash (0.1–0.5 %), and reducing (0.3–2.3 %) and non-reducing sugar (0.1–0.2 %) contents of starches were significantly different among the cultivars. Amylose content varied significantly between 12.8 and 21.3 %. Swelling power and solubility ranged between 7.8–31.1 % and 1.5–9.6 %, respectively. Pasting properties such as peak vis-

cosity measured in Rapid Visco Units (143.2–288.8 RVU), breakdown viscosity (29.4–162.6 RVU) and setback viscosity (15.0–78.8 RVU), pasting temperature (73.5–87.7 °C) and time to pasting temperature (3.6–4.5 min) varied significantly among the cultivars. Breakdown viscosity was poorly correlated with final viscosity attained ($R^2 = -0.0507$); however, pasting temperature was correlated ($R^2 = 0.479$) with setback viscosity. The variability observed in the physicochemical properties of the starches was related to specific requirements for use in the production of noodles, pasta and inclusion in bread and weaning food formulations. The physico-chemical properties of sweet potato starch of Nigerian varieties were reported by Nwokocha et al. (2014) as follows: 12.03 % moisture, 0.11 % ash, 0.12 % fat, 0.05 % nitrogen, 0.07 % phosphorus, 27.7 % amylose; particle characteristics: particle number 97, maximum diameter 27 mm, minimum diameter 2.0 mm, mean diameter 9.15 mm, length/diameter 1.17, roundness 0.77; gelatinization properties: onset temperature 72.8 °C, peak temperature 74.9 °C, completion temperature 76.6 °C, gelatinization range 4.2 °C, endothermic enthalpy 11.85 J/g; pasting properties: pasting temperature 75 °C, temperature at peak viscosity 89 °C, peak viscosity during heating (PV) 265BU, viscosity at 95 °C 250 BU, viscosity after 30 min holding at 95 °C (HPV) 180 BU, viscosity on cooling to 50 °C (CPV) 350 BU, stability ratio (HPV/PV) 0.7`, setback ratio (CPV/HPV) 1.94. Irish potato had a paste clarity of 4.9 and syneresis of 10.75 % based on 1 % and syneresis on 5 % aqueous starch pastes. Irish potato had larger starch granules, higher phosphorus and lower amylose contents than sweet potato starch. It also exhibited a lower gelatinization temperature, higher swelling power and amylose leaching compared to sweet potato starch. Sweet potato starch exhibited a higher pasting temperature, higher paste stability, and setback ratio and greater stability to shear thinning than Irish potato starch. The rheological properties indicated non-Newtonian behaviour for the two starch pastes. The storage and loss moduli of the two starch pastes were frequency-dependent with values higher for sweet potato at all points within the angular frequency range employed. Irish potato starch paste exhibited higher paste clarity and lower syneresis than sweet potato starch paste. Irish potato has superior properties for application as thickener while sweet potato is better in withstanding severe processing conditions. The pasting properties showed wide ranges of variation among twenty sweet potato cultivars and lines, and the amylose content ranged between 13.3 and 17.2 % (Katayama et al. 1999). Analysis of variance showed that the varietal differences were significant at 0.1 % level for the pasting temperature, setback, amylose content and starch content. The differences among lines and years were significant at 1 % level for the pasting temperature, peak viscosity and breakdown. The estimated heritability values of the pasting temperature, peak viscosity, setback, amylose content and starch content were 0.80, 0.49, 0.77, 0.88 and 0.85, respectively. The amylose content showed significant positive correlations in both years with the pasting temperature, the peak viscosity temperature and the setback. The starch content did not show any significant correlation with the pasting properties and the amylose content. High amylopectin content of sweet potato starch was associated with a high gelatinization temperature and correspondingly less susceptibility to α-amylase attack (Zhang and Oates 1999). The hydrolysis pattern was correlated with the degree of hydrolysis. Extensive surface erosion was shown to indicate a high degree of hydrolysis, whereas less surface erosion indicated less degradation.

Sweet potato SP1-W-YR variety showed the highest extracted starch with about 17.52 %, followed by SP2-P-P cultivar with 15.54 % (Thao and Noomhorm 2011). The starches from all sweet potato varieties were high in apparent amylose content ranging from 28.06 to 34.52 %. The protein content in starches from all sweet potato cultivars ranged from 0.15 to 0.23 % db, ash content was 0.110–0.282 % and lipid content in SP2-P-P and SP3-P-Y starches, which was 0.084 and 0.061 %, respectively. The protein and lipid played important roles in retention of amylose in starch noodles during cooking, resulting in minimizing cooking losses. For all sweet potato vari-

eties, the starch granule shapes were heterogeneous and no noticeable difference, including small or largely polygonal and circle-shaped particles. The mean length of SPS granules ranged from 14 to 17 μm. The pasting temperature for sweet potato starches ranged from 80.1 °C (SP4-OP-O starch) to 82.3 °C (SP1-W-YR starch). The peak viscosity of all sweet potato starches ranged from 403.06 RVU for SP1-W-YR starch to 473.63 for SP4-OP-O starch. Among sweet potato starches, SP1-W-YR starch was the most stable to temperature and shear treatment, followed by SP4-OP-O, SP2-P-P and SP3-P-Y starches in descending order. In another study, moisture, protein, ash, lipid and phosphorus content of the starches from 11 sweet potato cultivars varied from 3.86 to 6.52 %, 0.28 to 0.75 %, 0.10 to 0.47 % and 0.00 to 0.02 %, respectively (Abegunde et al. 2013). Amylose content varied between 13.33 and 26.83 %. The starches differed in their mean granule sizes, particle size distribution and susceptibility to pancreatin hydrolysis. Swelling power and solubility ranged from 13.46 to 26.13 g/g and 8.56 to 18.77 %, respectively. Higher retrogradation tendency was observed in pastes of starches of high amylose content. Gelatinization temperature and enthalpy ranged from 55.54 to 69.11 °C and 6.40 to 11.89 J/g, respectively. Pasting properties including peak viscosity (134–255 BU), breakdown viscosity (91–162 BU), setback viscosity (26–112 BU), peak time (5.97–7.03 min) and pasting temperature (67.20–73.00 °C) varied significantly among the sweet potato starches. Phosphorus content of the starches had substantial effect on their swelling power showing positive correlations. There was significant positive correlation between swelling power and solubility of the starches. Thermal and pasting parameters also showed significant correlations. Zhang et al. (2002) reported that most sweet potato genotypes in their study showed a slight decrease in starch content during 0–180 days of storage, but in the genotype Hi-dry, it decreased substantially. Alpha-amylase activity increased during the first 2 months of storage, followed by a decrease with continued storage to a level similar to that at harvest. The decline in starch content was correlated with α-amylase activity in the first 60 days storage. Trypsin inhibitor activity (TIA) in the fresh roots varied among genotypes from 3.90 to 21.83 U/mg. Storage had little influence on TIA level. There was considerable genotypic variation in digestibility, with up to 27 % reduction in digestibility after 120 days in storage. Glucose and sucrose concentration increased early in storage and then remained fairly constant. Storage reduced flour pasting viscosities, with up to nearly a 30 % decline in peak viscosity. Earlier, Takahata et al. (1994) reported that although β-amylase activity decreased with an increase in temperature in all sweet potato lines, it had greater heat stability in the high maltose line than in the other lines. Starch gelatinization of high maltose lines occurred at a lower temperature than did that of other lines. The flour and raw starches isolated from red and white sweet potato cultivars had high amylose content (32–34 %), similar gelatinization characteristics with onset temperature of 67 °C and enthalpy of 10.5–11.0 J/g and exhibited a Ca-type X-ray diffraction pattern (Osundahunsi et al. 2003). Both starches had well-correlated and high solubilization and swelling temperatures, starting at 80 °C. Pasting properties of the white cultivar exhibit lower tendency for retrogradation. Water and oil absorption capacities were low for both red and white flours. When parboiled, both cultivars showed improved water absorption capacity and decreased least gelation concentration. It was concluded that the white cultivar should be preferred when low retrogradation tendency was required.

Unit chain length distributions of amylopectins and their φ,β-limit dextrins (reflecting amylopectin internal part) from 11 Chinese sweet potato genotypes were characterized and found to be highly correlated to the thermal and pasting properties of granular starches (Zhu et al. 2011). The weight-based unit chain length profiles of whole amylopectin and their internal parts both had three distinguishable major groups with approximate range of DP 6–36, 37–68 and >69 for amylopectins and DP 3–25, 26–55, and >55 for φ,β-limit dextrins. Among different genotypes, two different patterns of B_{fp} (fingerprint

B-chains, DP 3–7) were observed for φ,β-limit dextrins, whereas A_{fp} (fingerprint A-chains, DP 6–8) for whole amylopectins were consistent. Reconstruction of amylopectins from their φ,β-limit dextrins revealed that B-chains with internal DP > 20 possessed an external chain length corresponding to the average value DP 12.8. A pomegranate concept was proposed for sweet potato starch granule (Lian et al. 2012). Similar to the structure of pomegranate, out-layer of the starch granule was deemed equivalent to skin of pomegranate, the double-helix blocklets represented the garnet of pomegranate, the amylopectin clusters with one reducing end at hilum, equivalent to primary body of pomegranate, constitute the basic structure of the starch granule, in the special parts of the clusters, lots of blocklets formed and increased, similar to the formation of garnet of pomegranate.

Carotenoids

Major carotenes and epoxide derivatives identified in Centennial sweet potatoes included *β*-carotene 86.35 %, phytoene 2.55 %, phytofluene 1.95 %, ζ carotene 1.77 %, *α*-carotene 0.9 %, *α*-carotene 5′,6′-epoxide 1.05 %, luteochrome 0.21 %, mutatochrome 0.84 %, *γ*-carotene 0.77 %, *cis*-*γ*-carotene 0.13 %, aurochrome 0.05 %, *β*-carotene 5,6,5′,6′-diepoxide 0.05 % and eight unknowns (Purcell and Walter 1968). Total carotene content determined was 0.13 mg/g fw or 0.45 mg/g dw. Monohydroxycarotenoids and polyhydroxycarotenoids isolated from Centennial sweet potato were cryptoxanthin 5,6,5′6′-diepoxide 0.06 %, hydroxyl-*α*-carotene 5′,8′-epoxide 0.01 %, cryptoxanthin 5,6,5′,8′-diepoxide 0.25 %, cryptoxanthin 5,8-epoxide 1.10 %, hydroxyl–ζ-carotene 0.06 %, lutein 5,6-epoxide 0.17 %, *cis*-violaxanthin 0.09 %, violaxanthin 0.06 %, 10 unknowns, 2 hydrocarbon mixtures and a monohydroxy mixture.

Seven carotenoids: *β*-carotene; *β*-carotene-5,6,5′6′-diepoxide; *β*-carotene-5,6-epoxide; luteochrome; *α*-zeacarotene; *β*-zeacarotene and aurochrome were identified in 10 Brazilian raw and cooked sweet potatoes (Almeida-Muradian and Penteado 1993). *β*-carotene was the main carotenoid of the following cultivars: Centennial, Heart Gold, Anapolis, Acadian, Morada Inta, Vineland Bush and clone CNPH. Luteochrome was the principal carotenoid of Monalisa, IAC-2-71 and SRT-252 cultivars. For raw roots, the vitamin A value varied from 1 retinol equivalents/100 g for IAC-2-71 cultivar to 3703 retinol equivalents/100 g for Acadian. For cooked roots, Acadian cultivar presented the highest provitamin A activity, with 4021 retinol equivalents/100 g. A series of carotenoids with a 5,6-dihydro-5,6-dihydroxy-*β*-end group, named ipomoeaxanthins A (1), B (2), C1 (3) and C2(4) were isolated from the flesh of yellow sweet potato 'Benimasari' (Maoka et al. 2007). Their structures were determined to be (5R,6S,3′R)-5,6-dihydro-*β*,*β*-carotene-5,6,3′-triol (1), (5*R*,6*S*,5′*R*,6′*S*)-5,6,5′,6′-tetrahydro-*β*,*β*-carotene-5,6,5′6′-tetrol (2),(5*R*,6*S*,5′*R*,8′*R*)-5′,8′-epoxy-5,6,5′,8′-tetrahydro-*β*,*β*-carotene-5,6-diol (3) and (5*R*,6*S*,5′*R*,8′*S*)-5′,8′-epoxy-5,6,5′,8′-tetrahydro-*β*,*β*-carotene-5,6-diol (4).

All-*trans*-*β*-carotene was the major provitamin A carotenoid in the roots and the mean content of seven improved orange-fleshed sweet potato (OFSP) cultivars ranged from 108 to 315 mg/g dry matter (Bengtsson et al. 2008). The retention of all-*trans*-*β*-carotene was 78 % when OFSP were boiled in water for 20 min. When OFSP were steamed for 30 min the retention was 77 %, whereas deep-frying OFSP roots for 10 min resulted in retention levels of 78 %. Drying slices of OFSP roots at 57 °C in a forced-air oven for 10 h reduced the all-*trans*-*β*-carotene content by 12 %. Solar drying and open-air sun drying OFSP slices to a moisture content of 10 % resulted in all-*trans*-*β*-carotene losses of 9 and 16 %, respectively. The *cis*-isomer 13-*cis*-*β*-carotene was found in noticeable amounts in all processed samples, but not in any raw samples. The formation of 13-*cis*-*β*-carotene correlated with the original amount of all-*trans*-*β*-carotene found in the raw OFSP root. The main carotenoid identified in Nigerian sweet potato varieties was pro-vitamin A carotenoid, (*β*-carotene) in its trans-cis isomers, namely: all-*trans*, 9-*cis*, 13-*cis* and 15-*cis* *β*-carotene isomers (Ukom et al.

2011). *Trans-β*-carotene had the highest concentration in all four varieties followed by 9-*cis-β*-carotene and 13-*cis-β*-carotene, respectively. Nitrogen fertilizer significantly increased *trans-cis* isomers of *β*-carotene with incremental nitrogen fertilizer application up to 80 kg N/ha.

From sweet potato variety CYY95-26, 278.1 mg/kg of *β*-carotene was extracted (Lien et al. 2012). The major carotene of CYY95-26 was all-*trans-β*-carotene. CYY95-26 also contained small amount of 9-*cis-β*-carotene and 13-*cis-β*-carotene which were carotene isomers. Four Brazilian sweet potato cultivars presented high levels of carotenoids in raw roots, predominantly all-*trans-β*-carotene (79.1–128.5 mg/100 g DW) (Donado-Pestana et al. 2012). The other carotenoids in raw roots were 13-*cis-β*-carotene (8.8–9.6 mg), 9-*cis-β*-carotene (4.9–6.1 mg), 5,6-epoxy-*β*-carotene (7.0–11.3 mg), lutein (0.1–0.4 mg) and zeaxanthin (0.1–0.2 mg). The total phenolic compounds varied among cultivars raw roots (1.30–1.93 mg GAE/g DW), flour (0.96–1.56 mg/g) and heat treatments (1.05–2.05 mg/g DW). In most cases, the heat treatments resulted in a significant decrease in the carotenoids and phenolic compounds contents as well as antioxidant capacity (DPPH and ABTS assays). Processing of flour presented the greatest losses of major carotenoids and phenolics. The carotenoid profile of sweet potato flour was all-*trans-β*-carotene (45.4–79.7 mg/100 g DW), 13-*cis-β*-carotene (2.7–4.7 mg), 9-*cis-β*-carotene (1.2–2.1 mg), 5,6-epoxy-*β*-carotene (3.8–6.5 mg), lutein (0.1–0.3 mg) and zeaxanthin (0.1–0.2 mg). Orange-fleshed cv. Benihayto contained 18.7 mg/100 g fresh weight of *β*-carotene comparable to US cultivars (Takahata et al. 1993). The average retinol equivalent of representative cultivars (Resisto, Benihayto, Santo Amaro, Caromex and Red Jeweel) was 2.8 which equaled the maximum value of carrot cultivars. No carotenoids were found in the yellowish-white cultivars. Orange-fleshed sweet potato was found to have 85 μg/g all-*trans-β*-carotene and 4 μg/g *α*-carotene and no lycopene (Pacheco et al. 2014).

Studies found that the extent of retention of carotenoid in sweet potato roots varied with the method of processing (Vimala et al. 2011). The highest retention was observed in oven drying (total carotenoids 90–91 % and *β*-carotene 89–96 %) followed by boiling (total carotenoids 85–90 % and *β*-carotene 84–90 %) and frying (total carotenoids 77–85 % and *β*-carotene 72–86 %). The lowest retention of total carotenoids (63–73 %) and *β*-carotene (63–73 %) was recorded in the sun drying method. Studies by Bechoff et al. (2010) found that losses of carotenoids (about 70 %) of orange-fleshed sweet potatoes during 4 months storage were considered to be more of a nutritional constraint to the utilization of dried sweet potato than losses occurring during drying (15 % or less). Two orange-fleshed varieties (Resisto and W-119) contained significantly more *β*-carotene, chlorogenic acid and vitamin C than the two cream-fleshed varieties (Bosbok and Ndou) (Rautenbach et al. 2010). Thermal processing decreased the carotenoid and vitamin C content of all the varieties, but increased the chlorogenic acid content and antioxidant capacity (ORAC, FRAP and ABTS). Drought stress appeared to increase the *β*-carotene, vitamin C and chlorogenic acid contents as well as the antioxidant capacity of some of the sweet potato varieties, especially W-119.

Polysaccharides and Glycosides

Soluble pectins of sweet potatoes increased during curing and protopectin decreased correspondingly (Heinze and Appleman 1943). At storage temperature, the protopectin increased again while pectin decreased. The polysaccharide PSPP (purified sweet potato polysaccharide), isolated and purified from sweet potato root, was found to be a (1→6)-*α*-D-glucan with a molecular weight of 53.2 kDa (Zhao et al. 2005). Four polysaccharide components named as PPSP, PPSPII, PPSPIII and PPSPIV were purified from purple sweet potato (Jiang et al. 2011). PSPI was mainly composed of glucose and galactose, PPSP II was composed of glucose and had a typical absorption peak of *β*-D-glucose chitosan pyranose, PPSP III was a glycoprotein showing a protein absorption peak.

Batatins I (1) and II (2), two ester-type dimers of acylated pentasaccharides, were isolated from

the hexane-soluble extract of sweet potato (Escalante-Sánchez and Pereda-Miranda 2007). Both polymers 1 and 2 represented dimers of compound batatinoside I, a new polyacylated macroyclic pentasaccharide also isolated from the plant. The hexane-soluble extract from sweet potato roots yielded five new lipophilic oligosaccharides of jalapinolic acid, batatinosides II-VI (1–5), as well as the known pescapreins I (6) and VII (7) and murucoidin I (8), which are part of the purgative resin glycoside mixture (Escalante-Sánchez et al. 2008). Compounds 1 and 2 were tetraglycosidic lactones of operculinic acid C. The pentasaccharide structures for compounds 3 and 4 were confirmed to be macrolactones of simonic acid B, and that characterized for 5 was derived from operculinic acid A. All compounds contained an esterifying residue that was composed of a long-chain fatty acid, *n*-decanoic acid (capric) or *n*-dodecanoic acid (lauric). In compound 3, an additional short-chain fatty acid, (2*S*)-methylbutyric acid, was also identified. Batatins III–VI (1–4), glycolipid ester-type dimers, were isolated from sweet potato tuberous roots (Rosas-Ramírez et al. 2011). These ester-type dimers consisted of two units of the hetero-tetrasaccharide operculinic acid C. Each unit was esterified by a different amount and type of acid residues: (2*S*)-methylbutanoic, cinnamic, decanoic (capric) and dodecanoic (lauric) acids. Purification of the chloroform-soluble resin glycosides from yellow-skinned sweet potato roots yielded six oligosaccharides, batatin VII (1) and batatinosides VII–IX (2–4), together with the known resin glycosides pescaprein I and batatinoside IV (Rosas-Ramírez and Pereda-Miranda 2013). Operculinic acid A was identified as pentasaccharide glycosidic core structure for compounds 2 and 4, and simonic acid B for 3. Batatin VII (1) represented a dimer of the known batatinoside IV, consisting of two units of simonic acid B. Four factors were found to significantly affect the pectin yield from sweet potato in the following order: solution pH > extraction time > extraction temperature > liquid/solid ratio (Zhang and Mu 2011). The selected optimal extraction conditions were liquid/solid ratio 20:1, extraction time 3.3 h, extraction temperature 66 °C and solution pH 7.9. These conditions yielded about 10.24 % of pectin vs. 10.27 % for the predicted value. The degree of esterification and gel strength of extracted pectin with disodium phosphate solution in the optimized condition were 11.2 % and 115.6 g, respectively.

Linalyl-*β*-glucoside (LBG), *α*-terpinyl-*β*-glucoside (TBG), neryl-*β*-glucoside (NBG) and geranyl-*β*-glucoside (GBG) in sweet potatoes were identified as trimethylsilyl derivatives (Ohta et al. 1991). Amounts of these monoterpene alcohol-*β*-glucosides were from 36.9 μg/kg sweet potato of LBG to 189 μg/kg sweet potato of TBG. In contrast, 75.8 μg/1 distillation residue of TBG and traces of other monoterpene alcohol-*β*-glucosides occurred in the distillation residue. *β*-glucosidase (*β*-D-glucoside glucohydrolase) in *shiro-koji* were purified using *p*-nitrophenyl-*β*-glucoside (PNPG), and three active fractions, one major fraction and two very minor fractions, were found. *hiro-koji* *β*-glucosidases were active on tested *β*-glucosides except for LBG and TBG. Five new ether-soluble resin glycosides (jalapins), simonins I–V, were isolated from sweet potato roots (Noda et al. 1992). Simonin I was characterized as an example of resin glycoside with aromatic acid (*trans-cinnamic* acid) as a component organic acid.

Two new resin glycosides, batatosides I and II, and five known compounds, friedelin, scopoletin, octadecyl caffeate, *β*-sitosterol and daucosterol, were isolated sweet potato roots (Yin and Kong 2008). Three new pentasaccharide resin glycosides, batatosides III–V (1–3), were isolated from sweet potato (Yin et al. 2008a). Batasin III was elucidated as (*S*)-jalapinolic acid 11-*O*-*α*-L-rhamnopyranosyl-(1→3)-*O*-[3-*O*-*trans*-4-*O*-(*S*)-2-methylbutyryl-*α*-L-rhamnopyranosyl-(1→4)]-*O*-[2-*O*-(*S*)-2-methylbutyryl]-*α*-L-rhamnopyranosyl-(1→4)-*O*-*α*-L-rhamnopyranosyl-(1→2)-*O*-*β*-D-fucopyranoside, intramolecular 1,2″-ester. Batatoside IV was elucidated as (*S*)-jalapinolic acid 11-*O*-*α*-L-rhamnopyranosyl-(1→3)-*O*-*α*-L-rhamnopyranosyl (1→4)]-*O*-[2-*O*-(*S*)-2-methylbutyryl]-*α*-L-rhamnopyranosyl-(1→4)-*O*-*α*-L-rhamnopyranosyl-(1→2)-*O*-*β*-D-fucopyranoside, intramolecular 1,3″-ester. Batatoside V was characterized as (*S*)-jalapinolic acid 11-*O*-*α*-L-rhamnopyranosyl-

(1→3)-*O*-[2-*O*-*trans*-cinnamoyl-4-*O*-*n*-decanoyl-α-L-rhamnopyranosyl-(1→4)]-*O*-[2-*O*-*n*-dodecanoyl]-α-L-rhamnopyranosyl-(1→4)-*O*-α-L-rhamnopyranosyl-(1→2)-*O*-β-D-fucopyranoside, intramolecular 1,3″-ester. Seven new resin glycosides, batatosides A–G, along with two known compounds, batatinoside I and simonin IV were isolated from the tubers (Yin et al. 2008b). Four compounds were isolated from sweet potato and identified as citrusin C, caffeic acid, 3,4-di-*O*-caffeoylquinic acid and 1,2,3,4-tetrahydro-β-carboline-3-carboylic acid (Yin et al. 2008c). Nine new lipo-oligosaccharides, batatosides H–P, were isolated from sweet potato tubers (Yin et al. 2009). They were characterized as tetra- or pentasaccharides that formed a macrolactone with the aglycone, (11*S*)-hydroxyhexadecanoic acid (jalapinolic acid). Six compounds batatinoside I, citrusin C, octadecyl caffeate, β-amyrin acetate, caffeic acid and scopoletin were isolated from the chloroform extract of sweet potato tubers (Yin et al. 2012).

Two known ether-soluble resin glycosides simon IV and operculin VII were isolated from sweet potato roots with four new glycosides which were tetra- or pentasaccharide monomers in which the sugar moieties were partially acylated by organic acids and combine with the aglycone, jalapinolic acid, to form a macrocyclic ester (Noda and Horiuchi 2008). Compound I was characterized as (11*S*)-[[*O*-4-*O*-*n*-dodecanoyl-α-L-rhamnopyranosyl-(1→4)-*O*-2-*O*-*n*-decanoyl-α-L-rhamnopyranosyl-(1→4)-*O*-α-L-rhamnopyranosyl-(1→2)-β-D-fucopyranosyl]oxy]-jalapinolic acid, intramolecular 1,2-ester (1) and compound 2 elucidated as (11*S*)-[[*O*-α-L-rhamnopy-ranosyl-(1→3)-*O*-[(2-*O*-*trans*-cinnamoyl)-(4-*O*-isobutanoyl) a-L-rhamnopyranosyl-(1→4)]-*O*-2-*O*-(2*S*)-methylbuthanoyl-α-L-rhamnopyranosyl-(1→4)-*O*-α-L-rhamnopyranosyl-(1→2)-β-D-fucopyranosyl]oxy]-jalapinolic acid, intramolecular 1,2-ester. Compounds 3 and 4 consisted of the same glycosidic acid, simonic acid B and differed only in the organic acid residue; compound 3 had isobutyric, *trans*-cinnamic and *n*-decanoic acids, while compound 4 had methylbutyric, *trans*-cinnamic and *n*-decanoic acids.

Proteinaceous Compounds

Ipomoein, a globulin, was isolated from sweet potato by enzymic action and found to have properties similar to those of albumin (Jones and Gersdorff 1931). Sucrose synthetase with molecular weight of 540,000 was isolated from sweet potato roots (Murata 1971). The enzymes were postulated to be involved in the breakdown of sucrose in sweet potato root tissues instead of the sucrose-synthesizing reaction. A marked rise in the phenylalanine ammonia-lyase activity and the polyphenol synthesis was observed in sliced roots of a sweet potato (Minamikawa and Uritani 1965). The enzyme activity was found to be localized in the root tissue adjacent to the sliced surface. The results suggested the important role of phenylalanine ammonia-lyase in the polyphenol synthesis and de novo synthesis of the enzyme protein molecule in the sliced tissues. Phenylalanine ammonia-lyase (PAL) was not present in fresh sweet potato tissue, but appeared in response to cut injury, reaching a maximum, and then decreasing, in parallel with PAL activity (Tanaka and Uritani 1976, 1977).

Under all curing conditions, NPN increased and protein N decreased indicating protein hydrolysis, which was higher at higher curing temperatures. Fairly consistent increases in basic N, amide N, residual N occurred in all lots, but a consistent increase in amino-N occurred only in those lots cured at high temperatures of 35 and 40 °C. The nitrogen distribution during storage at 10–12 °C remained fairly stable during storage. The amide N, basic N and residual N increased slightly in most lots towards the end of storage period. A temperature of 30 °C was found optimal for curing; higher temperature was unsuitable because of internal break down during curing. Nearly half of the nitrogen contained in sweet potato may be recovered as a concentrate containing > 80 % protein, which were found to be limiting in sulphur-containing amino acids but having excess of lysine (Purcell et al. 1978).

Highly polymerized free polysomes were formed in sweet potato root tissue in response to wounding (Oba et al. 1982). The degree of polymerization also increased to greater than 15-mers. Polysomes were organized from pre-existing free ribosomes in fresh tissues immediately after wounding; synthesis of new ribosomes may occur after 6 h when the rate of polysome formation decreased.

Ipomeamarone 15-hydroxylase was found to be a cytochrome P-450-dependent, mixed-function oxygenase (Fujita et al. 1982). Its activity was found in a microsomal fraction from cut-injured and *Ceratocystis fimbriata*-infected sweet potato root tissues, but was not found in fresh tissue of sweet potato roots. The major soluble protein of sweet potato roots had an apparent molecular weight of 25 KDa and accounted for 60–70 % of the total soluble protein extracted from fresh tissue (Li and Oba 1985). This protein was identified as antigenic component A which was degraded into peptides of lower molecular weights (9500–20,000) after storage, cutting or fungal infection of the roots. Another major protein, β-amylase, was also identified. Cutting, infection or storage of root tissue also resulted in the production of new isozymes of peroxidase, acid phosphatase and esterase. Acid phosphatase of sweet potato root tissue was found to consist of five components (Asahi et al. 1967). All components hydrolyzed various phosphate compounds including phosphomonoester- and pyrophosphate-linkages. Their optimum pH values were in the range of pH 5 to 6. In sweet potato root tissue, cinnamic acid 4-hydroxylase activity increased markedly in response to cut injury, and reached a maximum after 1 day of incubation (Tanaka et al. 1974). The patterns of development and successive decline were similar to those for phenylalanine ammonia-lyase activity. The enzyme may be involved in the cytochrome P-450 mediated electron transport system. Sweet potato storage roots were found to contain high amounts of extractable amylolytic enzymes (Hagenimana et al. 1994). These storage roots also have a very high starch content. Three major amylolytic enzymes were identified in sweet potato storage roots: α-amylase, β-amylase and starch phosphorylase. β-Amylase was abundant throughout the root at all times, and its high levels did not directly affect starch degradation rates. Starch phosphorylase protein level remained constant, while its extractable activity increased. Starch content decreased during sweet potato seed root germination.

Two forms of trypsin inhibitors with molecular weights of 31 and 21 kDa were found in sweet potato roots and they were different from those found in the leaves (Wang and Yeh 1996). The level of trypsin inhibitory activity was closely related with pest resistance. Trypsin inhibitors (TIs) (73, 38 and 22 kDa) were purified from storage roots, sprouted roots and sprouts of sweet potato variety Tainong 57 (Hou and Lin 1997b). Polyamines cadaverine, spermidine and spermine were found in all TI hydrolysates with different amounts in storage roots, sprouted roots and sprouts. TIs purified from the sprouts had higher polyamine titers, which were expressed as nmol/μg protein, than those from sprouted roots or storage roots. Trypsin inhibitors from sweet potato were found to have dehydroascorbate reductase and monodehydroascorbate reductase activities (Hou and Lin 1997b). Sweet potato trypsin inhibitor (SPTI) exhibited thioltransferase-like and glutathione S-transferase (Huang et al. 2009). Trypsin inhibitor with molecular weight 23 kDa was purified from sweet potato (Jaw et al. 2007). A proteinaceous invertase inhibitor, designated ITI-L, with molecular weight of 10 kDa was purified from sweet potato leaves. It was thermostable (90 % of the activity remained after incubation at 100 °C for 20 min) (Wang et al. 2003).

Sweet potato tuberous roots contained large quantities of two proteins named sporamins A and B, were monomeric forms with similar M_rs (25,000) (Maeshima et al. 1985). They were very similar to each other with respect to amino acid composition, peptide map and immunological properties. Sporamin was found to account for more than 80 % of the total soluble proteins of tuberous roots of sweet potato, but very little, if any, in other tissues of the same plant (Hattori et al. 1985; Maeshima et al. 1985). Northern blot analysis showed that sporamin mRNA is approx-

imately 950 nucleotides in length and is specifically present in tuberous roots and very little, if any, in leaves, petioles and non-tuberous roots. Nucleotide sequence of the cDNA predicted a 37 amino acid extension in the precursor at the amino-terminus of the mature protein. Sporamin consisted of two polypeptide classes, A and B (Maeshima et al. 1985; Murakami et al. 1986). The sporamin cDNA clones could also be classified into sporamin A and B sub-families based on their sequence homologies, with intra-sub-family homologies being much higher than inter-sub-family homologies. Sporamin, a vacuolar storage protein of tuberous roots of sweet potato, was synthesized by membrane-bound polysomes as a precursor which contained a 16 amino acids-long propeptide that follows the N-terminal signal peptide (Nakamura et al. 1993). A precursor to sporamin, expressed in transformed cells of tobacco suspension-cultured cell line BY-2, was sequentially processed from the N-terminus and correctly targeted to the vacuole. Sporamin comprised two distinct homology groups, subfamilies A and B (Sun et al. 2009). Sporamin B Q40091|Q40091_IPOBA was isolated as the major sporamin B from sweet potato *cv.* 55-2 tuber and found to have potent trypsin inhibitory activity. Sporamin accounts for about 60–80 % of total soluble protein in sweet potato tubers, and possessed significant amino acid sequence identity with some Kunitz-type trypsin inhibitors (Yeh et al. 1997). It was suggested that sporamin may have a defense role as a protease inhibitor, in addition to its role as a storage protein. Sporamin was reported to be constitutively expressed in the tuberous root and not normally expressed in the stem or leaves, but this protein was expressed systemically in response to wounding and other abiotic stresses (Senthilkumar and Yeh 2012). These dual expression patterns at the transcriptional level revealed that the complex regulatory mechanism of sporamin was modulated by environmental stresses.

Three full-length cDNA clones, designated TRX1, TRX2 and TRX3 encoding different but similar thioredoxin *h* polypeptides, were isolated from sweet potato storage roots (Huang et al. 2004a). All three thioredoxin genes were found to have the highest level in the storage roots; those corresponding to TRX2 and TRX3 were detected at the next higher level in flowers. In Western blot analysis, the thioredoxins were found to have the highest level in the storage roots and veins, higher level in leaves, and very low levels in sprouts of storage root and roots. The three thioredoxin *h* genes of sweet potato storage roots displayed differential gene expression patterns, which may be associated with the diverse roles and functions. Sweet potato storage root thioredoxin *h*2 was found to have dehydroascorbate (DHA) reductase and monodehydroascorbate (MDA) reductase activities (Huang et al. 2008a).

An arabinogalactan-protein (WSSP-AGP) with weight-average molecular weight of 126,800 g/mol was isolated from the tuberous cortex of the white-skinned sweet potato (Ozaki et al. 2010). It was found to consist of 95 % (w/w) carbohydrate and 5 % (w/w) protein with high contents of hydroxyproline, alanine and serine and sugar composition of *α*-L-Rha:*α*-L-Ara:*β*-D-Gal:*β*-D-GlcA in a molar ratio of 1.0:4.1:7.6:1.3. Structural analysis indicated that WSSP-AGP is a (1→3)-*β*-D-galactan highly branched at O-6 with (1→6)-*β*-D-galactan, in which the branched chains are substituted at the O-3 position with *α*-L-Araf-(1→ and *α*-L-Araf-(1→5)-*α*-L-Araf-(1→ and at the O-6 position typically with *α*-L-Rhap-(1→4)-*β*-D-GlcAp-(1→ as terminating groups.

Potentially valuable proteins could be extracted from sweet potato peel, a waste product of sweet potato processing (Maloney et al. 2012). The highest yield of protein extraction was obtained by mixing blanched sweet potato peelings with 59.7 mL of 0.025 mM NaCl per g peel and then precipitating with 6.8 mM $CaCl_2$. More than 370 protein spots in sweet potato tuberous roots were reproducibly detected by two-dimensional gel electrophoresis, in which 35 spots were upregulated (orange-fleshed cv Yulmi vs. purple-fleshed cv. Shinjami) or uniquely expressed (only Yulmi or Shinjami) in either of the two cultivars (Lee et al. 2012). Of these 35 protein spots, 23 were expressed in Yulmi and 12 were expressed in Shinjami. Fifteen proteins were identified in Yulmi and eight proteins in

Shinjami. The proteins identified in Yulmi were catechol oxidase I, putative oxalyl-CoA decarboxylase, α-amylase, 2 semialdehyde dehydrogenase family protein, disulfide-isomerase precursor-like protein, anionic peroxidase swpa, putative beta-1,3-glucanse precursor, cysteine proteinase inhibitor (phytocystatin), amino acid transporter, sporamin A precursor, sporamin B, unnamed protein product and 3 unknown protein. The proteins identified in cv Shilmi were PSG-RGH7 resistance protein, 2 sporamin B precursor, flavanone 3-hydroxyrase, aldo-ketose reductase, peroxidase precursor and two unknown protein. In Yulmi, α-amylase and isomerase precursor-like protein were uniquely expressed or upregulated and activities of α-amylase, monodehydroascorbate reductase and dehydroascorbate reductase were higher than in Shinjami. In Shinjami, peroxidase precursor and aldo-keto reductase were uniquely expressed or upregulated and peroxidase and aldo-keto reductase activities were higher than in Yulmi. PSG-RGH7 resistance protein uniquely expressed only in cv. Shinjami was evaluated more resistant than cv. Yulmi against the root-knot nematode, *Meloidogyne incognita* on the basis of shoot and root growth. Egg mass formation was 14.9-fold less in Shinjami than in Yulmi. A cDNA encoding a small cysteine-rich protein designated defensin (SPD1) was isolated from sweet potato storage roots (Huang et al. 2008c). It was found that defensin (SPD1) had both dehydroascorbate (DHA) reductase and monodehydroascorbate (MDA) reductase activities. SPD1 was also shown to inhibit the growth of both fungi and bacteria. A defensin protein (SPD1) with antioxidant activities in-vitro and ex-vivo was isolated from sweet potato storage roots (Huang et al. 2012).

A novel cyclophilin-type peptidylprolyl isomerase protein (SPPPI) was isolated from sweet potato storage roots (Liao et al. 2012). This cDNA encoded a pro-protein of 260 amino acids with a predicted molecular mass of 27,658 Da. Genomic Southern blot analyses using the full-length SPPPI cDNA probe revealed a multi-gene family in the sweet potato genome. Both the corresponding mRNA and protein level were found the highest in the storage roots, followed by that in sprout. In the Japanese purple-fleshed sweet-potato cultivar, Ayamurasaki, which accumulated anthocyanin in the storage roots, three types of IbMYB1 gene sequences, named IbMYB1-1, IbMYB1-2a and IbMYB1-2b were found whereas in the spontaneous mutant, AYM96, lacking anthocyanin, only IbMYB1-1 was found (Tanaka et al. 2012). An IbGrx cDNA encoding a putative dithiol glutaredoxin was cloned from sweet potato (Chi et al. 2012). Glutaredoxins play an important role in the reduction of protein glutathione-mixed disulphides

In response to fungal infection, protein increased 10–30 % in diseases tissues and proteins of supernatant, mitochondrial and microsomal fractions (Uritani and Stahmann 1961). In diseased tissues, free amino acids namely leucine, isoleucine, proline, valine, tyrosine and lysine were increased with a concomitant decrease of amides and alanine. New antigenic components were detected in the infection site of resistant cultivar. One such compound was identified as a peroxidase. Microsomes, especially 73S units, were increased following infection as compared to healthy tissues. The IPO (ipomoelin) wound-inducible gene was isolated from sweet potato cv Tainung 57 (Jih et al. 2003). When sweet potato was wounded, both hydrogen peroxide and nitric oxide were produced to regulate the expression of the IPO gene and enhance the plant's defense system. Ipomoelin, one of the wound-inducible proteins of sweet potato, was found to be a Jacalin-related lectin that possessed carbohydrate-binding properties and may play a role in plant defense (Chang et al. 2012). IPO showed high binding ability to methyl α-D-mannopyranoside (Me-Man), methyl α-D-glucopyranoside (Me-Glc), and methyl α-D-galactopyranoside (Me-Gal) forming carbohydrate complexes.

Cytokinins

The cytokinins, 9-β-D-glucopyranosyl-6-(3-methyl-2-butenylamino)purine (IPG) (Hashizume et al. 1982a) and *cis*-zeatin riboside (236 ng/kg) were isolated and identified from sweet potato tubers (Hashizume et al. 1982b). Three major

cytokinins were isolated from sweet potato ribosyl 9-*β*-D-ribosyl *trans*-zeatin (*trans*- RZ), 9-*β*-D-ribosyl *cis* zeatin (*cis*-RZ) and 9-*β*-D-glucopyranosyl-6-(3-methyl-2-butenylamino) purine (IPG) (Suye et al. 1983). RZ = 9-*β*-D-ribofuranosyl-6-(4-hydroxy-3-methyl-2--butenylamino)purine. The presence and levels of 14 cytokinin species were determined in the developing tuberous roots of sweet potato (Sugiyama and Hashizume 1989). The following structures were assigned: *trans*- and *cis*-zeatin, *trans*- and *cis*-ribosylzeatin, *trans*- and *cis*-glucosylzeatin, *trans*- and *cis*-2-methylthioribosylzeatin, dihydrozeatin riboside, N^6-isopentenyladenine, N^6-isopentenyladenosine, 9-glucosyl-N^6-isopentenyladenine and *trans*- and *cis*-zeatin riboside. Trans-zeatin, *trans-zeatin* riboside and N^6-(δ^2-isopentenyl)adenosine (i^6Ado) were found to be the major cytokinins in the small tubers (≤5 mm diameter) of sweet potato (Matsuo et al. 1983). During tuber development, cytokinin levels were high in tubers having a diameter below 12 mm, minimal in tubers with a diameter of 22.5 mm and then gradually increased as the tuber developed. The presence and levels of *trans*- and *cis*-ribosyl zeatin, *trans*- and cis-zeatin, N^6-isopentenyladenosine and 6-(3-methyl-2-butenylamino)-9-*β*-D-glucopyranosylpurine were determined in the root, and the stem and leaf of young sweet potato plants (Sugiyama et al. 1983). The level of 6-(3-methyl-2-butenylamino)-9-*β*-D-glucopyranosylpurine was found to be the highest in both the root, and the stem and leaf. *trans*-ribosyl zeatin were present at a high level in the root. The ratio of the *trans*- and *cis*-isomers of zeatin in the root, and in the stem and leaf were determined to be 10 and 2.4, respectively. Matsuo et al. (1988) found the levels of (i^6Ado) and abscisic acid (ABA) were much lower than that of *trans*-ZR throughout tuber development. The longitudinal distribution of the cytokinins and ABA in the developing sweet potato tubers showed that the level of IPG was higher than *trans*-RZ and the levels of *trans*-ZR were higher in parts of the proximal side of the stem than in other lower parts of the tubers. The level of IPG was higher than *trans*-RZ with maximum level of 270 μg/kg fresh weight.

Phenolic Compounds and Anthocyanins

Two major anthocyanins in purple sweet potato were identified as cyanidin 3-caffeylferulysophor oside-5-glucoside of cyanidin and peonidin 3-caffeylferulysophoroside-5-glucoside (Odake et al. 1992). Two major pigments and two minor pigments were found in fresh sweet potato tissue slices, the major pigments were identified as peonidin-3-diglucoside-5-glucoside with three molecules of ferulic acid and one molecule of caffeic acid and peonidin-3-diglucoside-5-glucoside with two molecules of ferulic acid and one caffeic acid (Shi et al. 1992). Two new acylated anthocyanins isolated from sweet potato storage roots were identified as the 3-*O*-(6-*O*-*trans*-caffeyl-2-*O*-*β*-glucopyranosyl-*β*-glucopyranoside)-5-*O*-*β*-glucoside of cyanidin and peonidin (Goda et al. 1997). Among nine anthocyanins isolated from purple sweet potatoes, the structures of three major anthocyanins were elucidated as 3-*O*-(6-*O*-*trans*-caffeyl)-2-*O*-(6-*O*-*trans*-caffeylglucopyransoyl-*β*-D-glucopyranosyl-5-*O*-(*β*-D-glucopyranosyl)-peonidin; 3-*O*-(6-*O*-*trans*-caffeyl)-2-*O*-(6-*O*-*trans*-feruloylglucopyransoyl-*β*-D-glucopyranosyl-5-*O*-(*β*-D-glucopyranosyl)-peonidin; and 3-*O*-(6-*O*-*trans*-caffeyl)-2-*O*-(6-*O*-*p*-hydroxylbenzoylglucopyranosyl-*β*-D-glucopyranosyl-5-*O*-(*β*-D-glucopyranosyl)-peonidin (Lee et al. 1997). Eight acylated anthocyanins were isolated from purple sweet potato storage roots; of these, six pigments were identified as diacylated anthocyanins of cyanidin and peonidin: cyanidin 3-*O*-(2-*O*-(6-*O*-*p*-hydroxybenzoyl-*β*-D-glucopyranosyl)-6-*O*-(*E*)-caffeyl-*β*-D-glucopyranoside)-5-*O*-*β*-D-glucopyranoside; cyanidin 3-*O*-(2-*O*-(6-*O*-*(E)*-caffeyl-*β*-D-glucopyranosyl)-6-*O*-(*E*)-caffeyl-*β*-D-glucopyranoside)-5-*O*-*B*-D-glucopyranoside; cyanidin 3-O-(2-*O*-(6-*O*-*(E)*-ferulyl-*β*-D-glucopyranosyl)-6-*O*-(*E*)-caffeyl-*β*-D-glucopyranoside)-5-*O*-*B*-D-glucopyranoside; and peonidin 3-*O*-(2-*O*-(6-*O*-*(E)*-ferulyl-*β*-D-glucopyranosyl)-6-*O*-(*E*)-caffeyl-*β*-D-glucopyranoside-5-*O*-*B*-D-glucopyranoside; peonidin 3-*O*-(2-*O*-(6-*O*-*(E)*-caffeyl-*β*-D-glucopyranosyl)-6-*O*-(*E*)-caffeyl-*β*-D-glucopyranoside)-5-*O*-*B*-D-

glucopyranoside; and peonidin 3-*O*-(2-*O*-(6-*O*-*p*-hydroxybenzoyl-*β*-D-glucopyranosyl)-6-*O*-(*E*)-caffeyl-*β*-D-glucopyranoside)-5-*O*-*B*-D-glucopyranoside (Terahara et al. 1999). The following anthocyanins were isolated and identified from a highly pigmented callus induced from the storage root of purple-fleshed sweet potato cultivar Ayamurasaki: cyanidin 3-*O*-sophoroside-5-*O*-glucoside; 3-*O*-(2-*O*-(6-*O*-(*E*)-*p*-coumaroyl-*β*-D-glucopyranosyl)-*β*-D-glucopyranoside)-5-*O*-*β*-D-glucopyranoside (Terahara et al. 2000); cyanidin 3-*O*-(2-*O*-(6-*O*-(*E*)-caffeoyl-*β*-D-glucopyranosyl)-*β*-D-glucopyranoside)-5-*O*-*β*-D-glucopyranoside; cyanidin 3-*O*-(2-*O*-(6-*O*-(*E*)-*p*-coumaroyl-*β*-D-glucopyranosyl)-6-*O*-(*E*)-caffeoyl-*β*-D-glucopyranoside)-5-*O*-*β*-D-glucopyranoside; cyanidin 3-*O*-(2-*O*-(6-*O*-(*E*)-*p*-coumaroyl-*β*-D-glucopyranosyl)-6-*O*-(*E*)-*p*-coumaroyl-*β*-D-glucopyranoside)-5-*O*-*β*-D-glucopyranoside; and peonidin 3-*O*-(2-*O*-(6-*O*-(*E*)-*p*-coumaroyl-*β*-D-glucopyranosyl)-6-*O*-(*E*)-*p*-coumaroyl-*β*-D-glucopyranoside)-5-*O*-*β*-D-glucopyranoside (Terahara et al. 2004).

Twenty-six anthocyanins were detected and characterized in the aqueous extract of a purple line cell line generated from the storage root of purple-fleshed sweet potato (*Ipomoea batatas* cv. Ayamurasaki): cyanidin 3-sophoroside-5-glucoside; cyanidin 3,5-diglucoside; pelargonidin 3-sophoroside-5-glucoside; peonidin 3-sophoroside-5-glucoside; cyanidin 3-*p*-hydroxybenzoylsophoroside-5-glucoside; cyanidin 3-(6″-caffeoylsophoroside)-5-glucoside; peonidin 3-*p*-hydroxybenzoylsophoroside-5-glucoside; peonidin 3-caffeoylsophoroside-5-glucoside; cyanidin 3-(6″-*p*-coumarylsophoroside)-5-glucoside; cyanidin 3-(6″-feruloylsophoroside)-5-glucoside; peonidin 3-(6″-*p*-coumarylsophoroside)-5-glucoside;peonidin3-(6″-feruloylsophoroside)-5-glucoside; pelargonidin 3-feruloylsophoroside-5-glucoside;cyanidin3-(6″,6‴-dicaffeoylsophoroside)-5-glucoside; cyanidin 3-caffeoylsophoroside-5-glucoside; cyanidin 3-(6″-caffeoyl-6‴-feruloylsophoroside)-5-glucoside; cyanidin 3-caffeoyl-*p*-coumarylsophoroside-5-glucoside; peonidin3-caffeoyl-*p*-hydroxybenzoyl-sophoroside-5-glucoside; peonidin 3-caffeoylsophoroside-5-g lucoside; peonidin 3-feruloyl-*p*-caffeoylsophoroside-5-glucoside; cyanidin 3-feruloyl-*p*-co umarylsophoroside-5-glucoside; cyanidin 3-(6″, 6‴-dicoumarylsophoroside)-5-glucoside, peonidin 3-feruloyl-*p*-coumarylsophoroside-5-glucoside; peonidin 3-(6″, 6‴-dicoumarylsophoroside)-5-glucoside and two unidentified compounds (Tian et al. 2005). Four Japanese purple sweet potato cultivars were categorized into two groups based on variation in two major pigments, cyanidin-3-(6′′-caffeoylsophoroside)-5-glucoside and peonidin-3-(6′′-caffeoylsophoroside)-5-glucoside; the cultivars Chiran Murasaki and Purple Sweet showed a high content of peonidin derivatives (peonidin type), whereas the varieties Tanegashima Murasaki and Naka Murasaki were classified as cyanidin types (Montilla et al. 2010). The non-acylated 3-sophoroside-5-glucoside of cyanidin, the monoacylated cyanidin-3-(6′′-caffeoylsophoroside)-5-glucoside as well as three diacylated major pigments,cyanidin-3-(6′′,6′′′-dicaffeoylsophoroside)-5-glucoside, cyanidin-3-(6′′-caffeoyl-6′′′-*p*-hydroxy-benzoylsophoroside)-5-glucoside and peonidin-3-(6′′-caffeoyl-6′′′-*p*-hydroxybenzoylsophoroside)-5-glucoside were also isolated.

Four anthocyanins were isolated and purified from purple sweet potato: peonidin 3-*O*-(6-*O*-(*E*)-caffeoyl-2-*O*-*β*-D-glucopyranosyl-*β*-D-glucopyranoside)-5-*O*-*β*-D-glucoside (1), cyanidin 3-*O*-(6-*O*-*p*-coumaroyl)-*β*-D-glucopyranoside (2), peonidin 3-*O*-(2-*O*-(6-*O*-(*E*)-caffeoyl-*β*-D-glucopyranosyl)-6-*O*-(*E*)-caffeoyl-*β*-D-glucopyranoside)-5-*O*-*β*-D-glucopyranoside (3), peonidin 3-*O*-(2-*O*-(6-*O*-(*E*)-feruloyl-*β*-D-glucopyranosyl)-6-*O*-(*E*)-caffeoyl-*β*-D-glucopyranoside)-5-*O*-*β*-D-glucopyranoside (4) (Qiu et al. 2009). The purities of 1–4 were 95.5, 95.0, 97.8 and 96.3 %, respectively. The following anthocyanins were found in purple sweet potato: cyanidin-3-sophoroside-5-glucoside; peonidin-3-sophoroside-5-glucoside; cyanidin-3-(6″,6‴-dicaffeoylsophoroside)-5-glucoside; cyanidin-3-(6″-caffeoyl-6‴-*p*-hydroxybenzoylsophoroside)-5-glucoside; cyanidin-3-(6″-caffeoylsophoroside)-5-glucoside; cyanidin-3-(6″-caffeoyl-6‴-feruloylsophoroside)-5-glucoside; peonidin-3-(6″,6‴-dicaffeoylsophoroside)-5-glucoside;

peonidin-3-(6″-caffeoylsophoroside)-5-glucoside; peonidin-3-(6″-caffeoyl-6‴-*p*-hydroxybenzoylsophoroside)-5-glucoside; and peonidin-3-(6″-caffeoyl-6‴-feruloylsophoroside)-5-glucoside (Montilla et al. 2011). Seventeen anthocyanins were identified in purple-fleshed Stokes Purple and NC 415 varieties with five major compounds: cyanidin 3-caffeoylsophorosi de-5-glucoside; peonidin 3-caffeoylsophoroside-5-glucoside; cyanidin 3-caffeoyl-*p*-hydroxybenzoylsophoroside-5-glucoside; peonidin 3-caffeoyl-*p*-hydroxybenzoyl-sophoroside-5-glucoside and peonidin-caffeoyl-feruloylsophoroside-5-glucoside (Truong et al. 2010). The other anthocyanin compounds were cyanidin 3-sophoroside-5-glucoside, peonidin 3-sophoroside-5-glucoside, cyanidin 3-*p*-hydroxybenzoylsophoroside-5-glucoside, pelargonidin compound, cyanidin 3-(6″-feruloylsophoroside)-5-glucoside, cyanidin 3-caffeoylsophoroside-5-glucoside, cyanidin 3-(6″-feruloylsophoroside)-5-glucoside, cyanidin 3-(6″,6‴-dicaffeoylsophoroside)-5-glucoside, cyanidin 3-(6″-caffeoyl-6‴-feruloylsophoroside)-5-glucoside, peonidin-dicaffeoylsophoroside-5-glucoside,cyanidin 3-caffeoyl-*p*-coumarylsophoroside-5-glucoside. Okinawa variety showed 12 pigments with 3 major peaks identified as cyanidin3-caffeoylsophoroside-5-glucoside,cyanidin3-(6″,6‴-dicaffeoylsophoroside)-5-glucoside and cyanidin 3-(6″-caffeoyl-6‴-feruloylsophoroside)-5-glucoside. Steam cooking had no significant effect on total anthocyanin content or the anthocyanin pigments. Cyanidin and peonidin were the major anthocyanidins in the acid hydrolyzed extracts. Thirteen anthocyanins were identified in the purple-fleshed sweet potato cultivar Jihei No. 1 (Li et al. 2013a). The main anthocyanins were 3-sophoroside-5-glucoside derivatives from cyanidin and peonidin, acylated with *p*-hydroxybenzoic acid, ferulic acid or caffeic acid. A unique anthocyanin, delphinidin-3,5-diglucoside, was also found. Eight kinds of anthocyanins with a yield of 90.02 % were extracted from purple sweet potatoes by aqueous two-phase extraction, and the major anthocyanins were cyanidin-caffeoy-fumaroysophoroside-3-*O*-glucoside, peonidin-caffeoyl-hydroxybenzoyl-3-*O*-glucoside, peonidin-caffeoyl-sophoroside-3-*O*-glucoside and peonidin-caffeoyl-fumaroyl-sophoroside-3-*O*-glucoside (Liu et al. 2013). A dimer of galloyl procyanidin was also found. Purple-fleshed sweet potato P40 cultivar contained anthocyanins up to 13 mg/g dry matter and a total 12 acylated anthocyanins were identified: cyanidin 3-caffeoyl-*p*-hydroxybenzoyl sophoroside-5-glucoside (16.43 %), peonidin 3-caffeoyl sophoroside-5-glucoside (19.72 %), cyanidin 3-(6″-caffeoyl-6″-feruloylsophoroside)-5-glucoside (12.10 %), cyanidin 3-sophoroside-5-glucoside (9.25 %), cyanidin 3-*p*-hydroxybenzoyl sophoroside-5-glucoside (8.83 %), cyanidin 3-(6″-caffeoyl sophoroside)-5-glucoside (4.74 %), peonidin 3-*p*-hydroxybenzoyl sophoroside-5-glucoside (1.29 %), cyanidin 3-(6″-feruloyl sophoroside)-5-glucoside (8.09 %), peonidin 3-(6″-feruloyl sophoroside)-5-glucoside (1.72 %), cyanidin 3-(6″,6″-dicaffeoyl sophoroside-5-glucoside (9.56 %), peonidin 3-caffeoyl-*p*-hydroxybenzoyl sophoroside-5-glucoside (6.32 %) and peonidin 3-(6″-caffeoyl-6″-feruloyl sophoroside)-5-glucoside (1.53 %) (Xu et al. 2013). Baking did not impact overall anthocyanins, but steaming, high pressure cooking, microwaving and frying significantly reduced 20 % of total anthocyanins. Monoacylated anthocyanins showed a higher resistance against heat than di- and non-acylated. Among which, cyanidin 3-*p*-hydroxybenzoylsophoroside-5-glucoside exhibited the best thermal stability. Studies found that X-ray irradiation treatment at doses up to 1000 Gy could reduce microbial populations while maintaining the physical quality and anthocyanin content of purple-fleshed sweet potato cubes up to 14 days of storage (Oner and Wall 2013).

Optimal conditions for anthocyanin and phenolic content extraction from purple sweet potato using response surface methodology were drying temperature 62.91, 60.94 °C; citric acid concentration 1.38, 1.04 %; and soaking time 2.53, 2.24 min, respectively (Ahmed et al. 2011). The experimental value of anthocyanin content was

19.78 mg/100 g and total phenolic content was 61.55 mg/g. Twenty-seven different anthocyanins were tentatively identified in the sweet potatoes in four Korean purple-fleshed sweet potato varieties (Borami, Mokpo 62, Shinzami, and Zami) (Lee et al. 2013). Borami was found to be a rare sweet potato variety with an exceptionally high quantity of pelargonidin-based anthocyanins. Major anthocyanins in the crude extracts of peel, flesh and whole roots of 10 Chinese purple-fleshed sweet potato genotypes were identified as peonidin or cyanidin 3-sophoroside-5-glucoside and their acylated derivatives, e.g., peonidin 3-sophoroside-5-glucoside, peonidin 3-(6″-*p*-feruloylsophoroside)-5-glucoside and cyanidin 3-(6″-*p*-feruloylsophoroside)-5-glucoside (Zhu et al 2010). The main hydroxycinnamic acid derivatives were identified as mono- and dicaffeoylquinic acids (e.g., 5-*O*-caffeoylquinic acid and 3,5-di-*O*-caffeoylquinic acid) and caffeoyl-hexoside. These main phenolic compounds identified were important contributors to the total antioxidant capacity of the tested sweet potato samples.

The periderm cork cells (skin), but not those of the adjacent parenchyma cells, of sweet potato tubers were found to contain high concentrations of anthocyanins (Philpott et al. 2009). Acid hydrolysis of the periderm extract followed by HPLC indicated the presence of the anthocyanidins cyanidin and peonidin. The pattern of anthocyanin accumulation in sweet potato roots was characterized into three distinctive phases: (1) an initial rapid increase during the 3rd to 6th week, (2) no change or a slight decrease during the 6th to 12th week and (3) a slight increase during the 12th to 17th week (Yoshinaga et al. 2000). mRNA levels of dihydroflavonol 4-reductase (DFR), one of the key enzymes of the anthocyanin biosynthetic pathway, was expressed throughout the stage of storage root development and appeared to be the most abundant at the 6th week, and declined at the 9th week, which coincided with the change in anthocyanin content. The rate of increase in anthocyanin content during the 3rd to 6th week was significantly higher in clones with a high anthocyanin content than in clones with a low anthocyanin content. Seven unknown aminoacyl sugars were isolated from the polar extracts of sweet potatoes (Dini et al. 2006b). Their structures were elucidated as *β*-D-fructofuranosyl-(2 →1)-*α*-D-[2-*O*-valyl]-glucopyranoside; *β*-D-fructofuranosyl-(2→1)-*α*-D-[2-*O*-tyrosyl]-glucopyranoside; *β*-D-fructofuranosyl-(2→1)-*α*-D-[2-*O*-threonyl]-glucopyranoside; *β*-D-fructofuranosyl-(2→1)-*α*-D-[2-*O*-hystidyl]-glucopyranoside; 2-*β*-D-fructofuranosyl-(2→1)-*α*-D-[2-*O*-alanyl]-glucopyranoside; *β*-D-fructofuranosyl-(2→1)-*α*-D-[2-*O*-tryptophanyl]-glucopyranoside and *β*-D-fructofuranosyl-(2→1)-*α*-D-[2-*O*-glycyl]-glucopyranoside.

Chlorogenic acid and related components were isolated from sound sweet potato, yielding caffeic acid by alkaline hydrolysis (Rudkin and Nelson 1947). Quinic acid was found in sweet potato root in response to cutting (Minamikawa 1967).

Four isomers of caffeoylquinic acid and an unidentified phenolic acid compound were found in 14 sweet potato cultivars (Thompson 1981). A koji (*Aspergillus awamori*) extract hydrolyzed caffeoylquinic acid derivatives from sweet potato, chlorogenic acid, 3,4-di-*O*-caffeoylquinic acid, 3,5-di-*O*-caffeoylquinic acid, 4,5-di-*O*-caffeoylquinic acid and 3,4,5-tri-*O*-caffeoylquinic acid to caffeic acid (Yoshimoto et al. 2005). Chlorogenic acid, isochlorogenic acid (several isomers possible), caffeic acid, neochlorogenic acid and 'Band510' (4-[[4-(3,4-dihydroxyphenyl)-1-oxo-2-propenyl]oxy]-1,3,5-trihydroxycyclohexanecarboxylic acid) were identified in sweet potato peelings (Sondheimer 1958). The levels were 335, 603, 11.53 and 548 mg/100 g dw respectively. Chlorogenic (7.92–20.27 mg/100 g fw), isochlorogenic acid-1 (0.74–4.29 mg), isochlorogenic acid-2 (5.52–23.21 mg) were the most abundant phenolics comprising 80 % of the total in sweet potato. The phenolic contents ranged from 14.18–51.24 mg/100 g fw depending on cultivar. 4-*O*-Caffeoylquinic (1.29–3.66 mg) was found in three cultivars.

Trans-cinnamic acid-2-^{14}C, *p*-coumaric acid-2-^{14}C and caffeic acid-2-^{14}C were administered to discs of sweet potato roots and incorporation of each radioactive compound into chlorogenic acid was compared (Kojima et al. 1969). The data suggested that chlorogenic acid was synthesized through either or both of two major pathways, phenylalanine → *t*-cinnamate → *t*-cinnamoyl derivative → *p*-coumaroyl derivative → chlorogenic acid and phenylalanine → *t*-cinnamate → *p*-coumarate → *p*-coumaroyl derivative → chlorogenic acid. The intermediate compound V was found to be the first intermediate after *trans*-cinnamic acid and to be a conjugate of *t*-cinnamic acid and some sugar different from quinic acid or shikimic acid (Kojima and Uritani 1972a, 1972b; 1973). Also, it was found that chlorogenic acid was not the final product, but was metabolized to isochlorogenic acid in sweet potato tissues. Compound V was determined to be *β*-1-cinnamoyl-D-glucose (Kojima and uritani 1972a). *Trans*-cinnamic-2-^{14}C and quinic acid –G-^{3}H^{1} were selectively incorporated into the aromatic and non-aromatic moieties of chlorogenic acid, respectively (Kojima and uritani 1972b). Slicing of sweet potato released enzymes involved in the biosynthesis of phenolics leading to the production of phenolic compounds, mainly chlorogenic and isochlorogenic acids, and isomers of dicaffeoylquinic acid.

Sweet potato root periderm contained 0.008 to 7.97 mg/g dry weight caffeic acid and the highest content was 0.047 mg/g in the cortex tissues (Harrison et al. 2003). Chlorogenic acid contents were determined in periderm and cortex tissues of sweet potato roots (Peterson et al. 2005). On a dry weight basis, contents of the chlorogenic content in the periderm tissues ranged from 33 to 214 μg/g tissue and in the cortex from 1416 to 4213 μg/g tissue (181 to 1384 μg/g FW). Sweet potato root surface and epidermal extracts showed significant variation in phenolic compound concentration of hexadecylcaffeic acid, hexadecylcoumaric acid, heptadecylcaffeic acid, octadecylcaffeic acid, octadecylcormaric acid and 5-*O*-caffeoylquinic acid (chlorogenic acid), with higher concentrations correlated with weevil resistance (Anyanga et al. 2013). 2,4-Di-*tert*-butylphenol was isolated from sweet potato (Choi et al. 2013). Two constituents were isolated from purple sweet potato, 6,7-dimethoxycoumarin and 5-hydroxymethyl-2-furfural by combination of silica gel column and high-speed counter-current chromatography (He et al. 2012).

Phenolics (chlorogenic acid, neochlorogenic acid, isochlorogenic acid isomers A, B and C and an unknown compound in sweet potato decreased in the order: outer tissues (6.29 mg/100 g fw, 43.47 %) > skin (4.81 mg, 34.71 %) > inner (3.05 mg, 21.44 %) (Walter and Schadel 1981). Phenols in the inner tissues were uniform throughout the root while the outer tissues of the stem end and root end were found to contain more phenolics than the mid-root outer tissues. Thus, about 78 % of the phenolics were found localized in the skin and outer 5 mm of root tissues. Walter and Purcell (1980) showed that the amount of darkening in homogenized sweet potato was directly proportional to the concentrations of phenolics and that the majority of the phenolics were esters formed between quinic acid (1,3,4,5-tetrahydroxycyclohexanecarboxyl icacid)andcaffeicacid([3-(3,4-dihydroxyphenyl)-2-propenoic acid], i.e., chlorogenic acids. Significant differences were found in the distribution of carbohydrates, organic acids and phenolics among four sweet potato cultivars (Son et al. 1991). Cultivars more resistant to weevils i.e., Regal and Resisto had higher concentration of malic acid than the most susceptible cv Centennial. Concentration (mg/g FW) of carbohydrates and organic acids in the root periderm tissues (outer 3 mm) of four sweet potato cultivars were : *α*-glucose 0.51–2.26 mg, *β*-glucose 0.89–3.98 mg, sucrose 19.47–46.6 mg, inositol 27.16–59.27 mg, total sugars 27.16–59.27 mg, malic acid 0.25–1.49 mg, fructose and citric acid 2.52–10.41 mg, quinic acid 0.69–1.51 mg; and phenolics; chlorogenic acid 0.69–1.72 mg, caffeic acid 0.34–1.38 mg, dicaffeoylquinic acid (1) 0.89–3.44 mg, dicaffeoylquinic acid (2) 0.74–2.94 mg, rutin 0.13–0.58 mg and total phenolics

3.79–9.17 mg. Sucrose was the major water-soluble carbohydrate. In periderm tissues (outer) of cv Centennial stored for 3 months, 3-*O*-CQA (chlorogenic acid), 5-*O*-CQA (neochlorogenic acid), 4-*O*-CQA, dicaffeoylquinic acids and rutin (quercetin-3-*β*-D-rutinoside) were detected.

Four different polyphenolic compounds were isolated from methanolic and hydromethanolic extracts of *Ipomoea batatas* tuber flour (Dini et al. 2006a). Their structures were determined as 4,5-di-*O*-caffeoyldaucic acid; 4-*O*-caffeoylquinic acid; 3,5-di-*O*-caffeoylquinic acid and 1,3-di-*O*-caffeoylquinic acid). Six major phenolic compounds in raw sweet potato were identified as *β*-D-fructofuranosyl 6-*O*-caffeoyl-*α*-D--glucopyranoside (FCG), chlorogenic acid (5-*O*-caffeoylquinic acid or 5-CQA), caffeic acid, 4,5-CQA (4,5-di-*O*-caffeoylquinic acid), 3,5-CQA (3, 5-di-O-caffeoylquinic acid) and 3,4-CQA (3,4-di-*O*-caffeoylquinic acid) (Takenaka et al. 2006). Two further compounds from heated sweet potato were identified as 3-CQA (3-*O*-caffeoylquinic acid) and 4-CQA (4-*O*-caffeoylquinic acid). There was an obvious decrease in caffeic acid derivatives during the boiling of cube-shaped blocks of sweet potatoes. They also decreased in a mixture of freeze-dried sweet-potato powder and water maintained at room temperature. When the mixture of powdered sweet potato and water was heated at 100 °C, there was only a negligible decrease in the total amount of phenolic compounds, and portions of 5-CQA and 3,5-CQA were found to be isomerized to 3-CQA, 4-CQA, 3,4-CQA and 4,5-CQA. The content and composition of the phenolic compounds in sweet potatoes differed between fresh and long-stored ones, as did their response to heating. The main polyphenolic components found in four Japanese cultivars were chlorogenic acid (ChA) and 3,5-di-*O*-caffeoylquinic acid (3,5-diCQA) (Ishiguro et al. 2007). ChA level increased more at 5 °C than at 15 °C, whereas that of 3,5-diCQA was greater at 15 °C. Caffeoylquinic acids and radical scavenging activity in cv. 'Murasakimasari', which contained a large amount of anthocyanin in flesh tissue, were extremely high at the beginning of storage and remained nearly constant or decreased over time. Also, a non-caffeoylquinic acid component, identified as caffeoyl sucrose [CSu, 6-*O*-caffeoyl-(*β*-D-fructofuranosyl-(2→1))-*α*-D-glucopyranoside] increased during storage, especially in cv. 'J-Red' at 15 °C. Of five phenolic acids, caffeic acid, chlorogenic acid, 4,5-di-*O*-caffeoylquinic acid, 3,5-di-*O*-caffeoylquinic acid and 3,4-di-*O*-caffeoylquinic acid, identified in sweet potato storage roots, chlorogenic acid was the most abundant (Truong et al. 2007). Steam cooking resulted in statistically non-significant increases in the concentration of total phenolics and all the individual phenolic acids identified. Five colourless caffeoyl monomers were isolated from purple-fleshed sweet potato cv. Ayamurasaki and identified as: 5-caffeoylquinic acid; 6-*O*-caffeoyl-*β*-D-fructofuranosyl-(2–1)-*α*-D-glucopyranoside; *trans*-4,5-dicaffeoylquinic acid; 3,5-dicaffeoylquinic acid and 4,5-dicaffeoylquinic acid (Zhao et al. 2014).

Total phenolic content in purple-fleshed sweet potato (PFSP) ranged from 313.6 to 1483.7 mg chlorogenic acid equivalent/100 g fresh weight (fw), and anthocyanin contents between 51.5 and 174.7 mg anthocyanins/100 g fw (Steed and Truong 2008). Unlike orange-fleshed sweet potatoes (OFSP), the steamed roots of PFSP formed a thick paste, which required a process modification to produce flowable purees. Rheological testing indicated that adjusting the dry matter of PFSP to 18–21 % produced purees with flow properties similar to the OFSP purees.

Miscellaneous Compounds including Alkaloids, Saponins, Terpenoids

Polyhydroxylated nortropane alkaloids calystegines A_3, B_1, B_2 and C_1 were detected in sweet potato (Asano et al. 1997). Calystegine A_3, B_1, B_3 were isolated from sweet potato root sample from Panana (Schimming et al. 1998). From Japan, sweet potato aerial part samples calystegine B_1 and B_2 were found but no biogenetic precursors were determined; from Panama, sweet potato root calystegine A_3 and B_1 were found and biogenetic precursors found in the biosynthesis of calystegines were 3-oxotropane (tropinone),

3-β-hydroxytropane (pseudotropine) and 3-β-hydroxynortropane (norpseudotropine) (Schimming et al. 2005). From Mexico, sweet potato aerial part samples, calystegines B_1, B_2, B_3 were found and from roots calystegines A_5, B_1, B_2, were found and biogenetic precursors found were 3-oxotropane (tropinone), 3-β-hydroxytropane (pseudotropine), 3-oxonortropane (nortropinone) and 3-β-hydroxynortropane (norpseudotropine). The dihydroxynortropane alkaloid was isolated from sweet potato (Asano et al. 2001).

A novel indole-type alkaloid, named ipomine A, was isolated from sweet potato hairy roots (Yuan et al. 2004). Triterpene saponins were isolated from sweet potato tuber flour and identified as oleanolic acid-3-*O*-[β-D-glucopyranosyl-(1→2)-β-D-galactopyranosyl-(1→2)-β-D-glucuronopyranosyl]-28-*O*-β-D-glucopyranoside (sandrosaponin IX) (161.20 mg/100 g dry weight) and oleanolic acid-3-*O*-[β-D-galactopyranosyl-(1→3)-β-D-glucuronopyranosyl]-28-*O*-β-D-glucopyranoside (14.67 mg/100 g dry weight) (Dini et al. 2009). Both saponins were moderate in relation to commercial standards when tested by DPPH and FRAP assay. Five sesquiterpenes in the volatiles from sweet potato storage roots and leaves that attracted sweet potato weevils were identified as copaene, *trans*-caryophyllene, γ-humulene, γ-cadinene and γ-elemene (Nottingham et al. 1989a) and triterpenol acetate from root surface (Nottingham et al. 1989b). Two pentacyclic triterpenoids were isolated from sweet potato tubers: beohmerol and boehmeryl acetate (Son et al. 1990). Boehmeryl acetate from sweet potato tuber was found to be an ovipositional stimulant for sweet potato weevil, *Cylas formicarius elegantulus*.

Alkaloids, saponin, tannins, steroids, anthocyanins, flavonoids, anthraquinones, oxalate and trypsin inhibitors, phytate (35 mg/100 g) and coumarin was found in sweet potato but not phlobatannin and cyanogenic glycosides (Anbuselvi and Muthumani 2014).

A bioprocess for L- and D-lactic acid production from raw sweet potato through simultaneous saccharification and fermentation by *Lactobacillus paracasei* and *Lactobacillus coryniformis*, respectively, was developed by Nguyen et al. (2013). The method produced 122.91–198.32 g/L of L-lactic acid with yields of 90.11–84.92 % (w/w) and 122.49–186.40 g/L D-lactic acid with yields of 90.11–84.92 % (w/w). Sweet potato roots (non-boiled/fully-boiled) were fermented with *Lactobacillus plantarum* MTCC 1407 at 28 °C for 48 h to make lacto-juice (Panda and Ray 2007). There were no significant variations in biochemical constituents (pH, 2.2–3.3; lactic acid, 1.19–1.27 g/kg root; titratable acidity, 1.23–1.46 g/kg root, etc.) of lacto-juices prepared from non-boiled and fully-boiled sweet potato roots except β-carotene concentration [130 mg/kg (fully boiled roots) and 165 mg/kg (non-boiled roots)]. The panelist evaluation scores ranged from 3–4.8 (in a hedonic scale of 1–5) from moderate liking to very much liking of sweet potato lacto-juice.

Volatiles

Thirty volatile compounds were identified from baked 'Jewel' sweet potatoes: methanol, ethanol, acetone; diethylether; dichloromethane; 2,3-butanedione (doacetyl); 3-methylpentane; hexane, tetrahydrofuran; methylcyclopentane; 2,3-pentaedione; methylbenzene (toluene); 2-methyltetrahydrofuran-3-one; furfuraldehyde; dimethylbenzene (xylene); isobutyronitrile; 2-pyrone; heptanal; 2-furyl methyl ketone; benzaldehyde; 5-methyl-2-furaldehyde; trimethylbenzene (mesitylene); octanal; 2-pentylfuran; phenylacetaldehyde; nonanal; linalool; decanal; β-ionone; and 4-(2,2,3,3-tetramethylbutyl)phenol (Purcell et al. 1980). During heat treatment of sweet potatoes, maltose concentration increased from 0.03 % fresh weight at 25 °C to 4.33 % at 80 °C, at which temperature maximum synthesis occurred (Sun et al. 1994). Microwave pre-treatment (2 or 4 min) resulted in a significant decrease in amounts of maltose and volatiles formed. At 80 °C, approximately 80 % of maltose synthesis was inhibited when pre-treated with microwave. It was found that maltose represented a primary precursor for many of the volatile compounds emanating from baked 'Jewel' sweet potatoes and the formation of these volatiles appeared to involve both enzymatic and

thermal reactions. Twelve experimental lines of sweet potatoes could be classified on the basis of 27 volatiles (Tiu et al. 1985). Five volatiles were associated with good flavour and eight volatiles with cultivars having poor flavour. Twenty-one volatile compounds consisting of aldehydes, ketones, furans, pyridine, alcohols, terpenes and palmitic acid were identified in the steam distillates from Jewel, Tainung 57 and a breeding line (No. 99) (Horvat et al. 1991). Maltose, sucrose, glucose and fructose were isolated and identified. Maltose, the principal sugar formed during cooking, ranged from 0.07 % in the breeding line No. 99 to 5.3 % in Tainung 57. Twenty-three volatiles emanated from baked Jewel sweet potatoes: 2,3-pentadione, acetol; 2,3-butandione; methylpropanoate; isobutyl alcohol; 2-methyl-2-hexene; oxabicyclo(2,2,1) heptanes; 2-furancarboxaldehyde; 2-furmethyl ketone; 2-furmethanol; isomaltol; benzaldehyde; 5-methyl-furfural; phenylacetaldehyde; β-cyclocitral; 2-acetyl pyrrole; maltol; 2-methyl-furaote; 3,5-dihydroxy-2-methyl, 5,6-dihydrogen-4-pyrone; 5-hydroxymethyl furaldehyde; C_{17} hydrocarbon; C_{18} hydrocarbon; and C_{19} hydrocarbon (Sun et al. 1993). The major components identified from baked Jewel and Centennial sweet potatoes were similar: acetol; 2-furancarboxaldehyde; 2-furmethyl ketone; 2-furanmethanol; 5-methyl-furfural; phenylacetaldehyde; maltol; 2,3-dihydro-3,5-dihydroxy-6-methyl-4*H*-pyran-4-one; 5-hydroxy-methyl-2-furancarboxaldehyde.

Volatile compounds identified from conventionally baked and thermolyzed sweet potatoes were similar except for the presence of acetic acid (thermolyzed) and phenylacetaldehyde (baked) sweet potatoes (Sun et al. 1995). The common compounds to both baked and thermolyzed sweet potatoes were: acetol; furyl aldehyde; 2-acetylfuran; benzaldehyde; 5-methyl-2-furfural; furfuryl alcohol; 3,4-dihydropyran; 3-hydroxy-2-methyl-4-pyrone (maltol); 2-hydroxyacetyl furan; 5-hydroxymethyl-2-furfural and an unknown. All the aforementioned compounds except for phenylacetaldehyde and β-ionone were found in the insoluble sweet potato fraction. Acetol, furyl aldehyde, 5-methyl-2-furfural, furfuryl alcohol and β-ionone were found in the non-polar fraction. Acetol, furyl aldehyde, 5-methyl-2-furfural, furfuryl alcohol and 5-hydroxymethyl-2-furfural were found in the polar fraction.

Volatile emanating from baked roots of sweet potato cultivars 'Jewel' and 'GA90-16' were: pyridine; 1,2,4-cyclopentanetriol; 1,2,4-trimethyl benzene; 3-furaldehyde; xylene; 2-furmethanol; pyridine; 1,2,4-cyclopentanetriol; 1,2,4-trimethyl benzene; 3-furaldehyde; xylene; 2-furmethanol; furfuryl alcohol; 2-acetyl furan; benzaldehyde; 5-methyl-2-furfural; 2-pentyl furan; 2,3-pentanedione; phenylacetaldehyde; limonene; 3,4-dihydropyran; 2-acetyl pyrrole; maltol; linalool; isopulegone; 4,5-dimethyl-4-hexen-3-one; geraniol; 2,4-nonadienal; 2-napthalenone; cyclohexanol; n-decanal; 2,2-dimethyl-1,3-cyclohexanediol; 2,3-nonadecanediol; 2,4-decadienal; octyl ketone; germacrene D; caryophyllene; cyperene; β-farnesene; α-copaene; α-bisabolene; bohlmann 176; 2(4*H*)-benzofuranone; β-ionone; nerol, 4-decanolide; unknown; tetradecanoic acid; 10-heneicosene; palmitic acid; octadecanol; 1-nonadecanol; and 9,12-octadecadienoic acid (Wang et al. 1998). Of 46 compounds identified, 38 were odour-active. Curing had a pronounced effect on aroma intensity and quality. Uncured roots yielded only 37 % (cv. 'Jewel') and 12 % ('GA90–16') of the total amount of odour-active volatiles of cured roots and had substantially depressed aromas. Certain compounds were not obtained from freshly harvested roots (i.e., only 22 of 38 compounds were present in 'GA90–16' and 34 of 37 in 'Jewel'). Curing appeared to enhance the synthesis of α- and β-amylase, which in turn facilitates starch hydrolysis during baking and formation of monosaccharides that acted as precursors for critical volatile flavour components.

Volatiles (49) emanating from baked 'Jewel' sweet potatoes were: pyridine; 1,2,4-cyclopentanetriol; 1,2,4-trimethyl benzene; 3-furaldehyde; xylene; 2-furmethanol; 2-furancarboxaldehyde; 2-acetyl furan; benzaldehyde; 5-methyl-2-furfural; 2-pentyl furan; 2,3-pentanedione; phenylacetaldehyde; limonene; 3,4-dihydropyran; 2-acetyl

pyrrole; maltol; linalool; isopulegone; 4,5-dimethyl-4-hexen-3-one; geraniol; 2,4-nonadienal; 2-napthalenone; cyclohexanol; n-decanal; 2,2-dimethyl-1,3-cyclohexanediol; 2,3-nonadecanediol; 2,4-decadienal; octyl ketone; methyl geranate; germacrene D; β-caryophyllene; β-farnesene; α-copaene; α-bisabolene; bohlmann 176; 2(4*H*)-benzofuranone; β-ionone; nerolidol; 4-decanolide; unknown; tetradecanoic acid; 10-heneicosene; palmitic acid; octadecanol; 1-nonadecanol; 9,12-octadecadienoic acid, silane and squalene (Wang and Kays 2000). Three compounds, phenylacetaldehyde (perfume), maltol (caramel) and methyl geranate (2,6-octadienoic acid, 3,7-dimethyl-, methyl ester) (sweet candy) possessed the highest flavour dilution (FD) values (1500) via AEDA. 2-Acetyl furan (baked potato), 2-pentyl furan (floral), 2-acetyl pyrrole (sweet, caramel), geraniol (sweet floral) and β-ionone (violet) had FD values of 1000. These compounds were deemed to be the most potent odourants in baked 'Jewel' sweet potatoes. Additionally, 1,2,4-trimethyl benzene, 2-furmethanol, benzaldehyde, 5-methyl-2-furfural, linalool, isopulegone, *n*-decanal, 2,4-decadienal, octyl ketone, α-copaene, 4-decanolide and one unidentified compound were also contributors to the aroma.

Thirty-seven odour-active compounds of which 36 were identified from 'Jewel' sweet potatoes cooked by baking, boiling and microwaving (Wang and kays 2001). These compounds were: pyridine; 1,2,4-trimethyl benzene; 3-furaldehyde; xylene; 2-furmethanol; 2-acetylfuran; benzaldehyde; 5-methyl-2-furfural; 2-pentyl furan; 2,3-pentanedione; phenylacetaldehyde; limonene; 3,4-dihydropyran; 2-acetyl pyrrole; maltol; linalool; isopulegone; geraniol; 2,4-nonadienal; cyclohexanol; *n*-decanal; 2,2-dimethyl-1,3-cyclohexanediol; 2,3-nonadecanediol; 2,4-decadienal; octyl ketone; methyl geranate; β-caryophyllene; β-farnesene; α-copaene; α-bisabolene; bohlmann 176; 2(4*H*)-benzofuranone; β-ionone; nerolidol; 4-decanolide, unknown and tetradecanoic acid. Compared with conventional baking, boiling and microwaving yielded only 54.26 % and 6.43 % of the relative total yield of aroma-active compounds, respectively. From aroma extract dilution analysis, four compounds contributed to the aroma from each cooking method. In addition, cooking method specific aroma contributing compounds accounted for the unique aroma of each of the cooked products.

Wang and Kays (2003) described an analytical means of assessing flavour which is amenable to large numbers of sweet potato lines to be screened and allows initial selection decisions to be made without the use of sensory panels. For this, they found 19 critical aroma compounds and their relative weighting factors in six baked sweet potato clones: Maillard/caramelization products – maltol 138.9; 5-methyl-2-furfural 444.4; 2-acetyl furan 227.3; 3-furaldehyde 0.7; 2-furmethanol 7.1; Strecker degradation products – benzaldehyde 380.9; phenylacetaldehyde 50.5; β-carotene degradation products – β-ionone 625; 1,2,4-trimethyl benzene 296; lipid degradation products – 2-pentyl furan 833.3; 2,2-decadienal 666.7; 2,4-nonadienal 333.3; *n*-decanal trace; terpenoids – linalool 1000; geraniol 2500; methyl geranate trace; cyperene 1.0; α-copaene 266.7 and sesquiterpene (MW204) 100 (Wang and Kays 2003)

Flavour volatile components identified in Kansho-shochu (sweet potato spirit) included: ethanol, ethyl hexanoate, ethyl octanoate; 1-octen-3-ol; furfural; linalool; ethyl decanoate; diethyl succinate; α-terpineol; citronellol; ethyl phenylacetate; nerol; β-phenylethyl acetate; ethyl laurate; geraniol; β-phenylethyl alcohol; nerolidol; ethyl myristate; ethyl cinnamate; ethyl palmitate; ipomeamarone; farnesol; cetyl alcohol; ethyl stearate; ethyl oleate, ethyl linoleate (Ohta et al. 1990). The monoterpene alcohols linalool, α-terpineol, citronellol, nerol and geraniol, together with ethyl phenyl acetate and ethyl cinnamate qualitatively contributed to the sensory property of Kansho-shochu flavour. They also found terpene alcohols linalool, α-terpineol, nerol, geraniol and *p*-menth-2-en-7ol in raw grated sweet potato and nerol, geraniol and *p*-menth-2-en-7ol in steamed sweet potato.

The following volatiles emanated from storage roots of four sweet potato cultivars over 24 h: 1,2-dimethylbenzene; 6-methyl-5-hepten-2-one;

1,4-dichlorobenzene; *p*-cymene; *dl*-limonene; diethylbenzene; undecane; neroloxide; naphthalene; nerol; *Z*-citral; *E*-citral; 1-methylnaphthalene; 2-methylnaphthalene; methylgeranate; alkazene; δ-cadinene; *α*-copaene; *β*-elemene; cyperene; *trans*-caryophyllene; germacrene D; *α*-humulene; *cis*-*α*-bisabolene; ylangene; *β*-selinene, *α*-muurolene; α-gurjunene, 1*S*,*cis*-calamenene (Wang and Kays 2002). The following volatiles emanated from aerial parts of four sweet potato cultivars over 24 h: 1,2-dimethylbenzene; 1,4-dichlorobenzene; *p*-cymene; *dl*-limonene; undecane; naphthalene; nerol; 1-methylnaphthalene; 2-methylnaphthalene; alkazene; *α*-copaene; *β*-elemene; cyperene; *trans*-caryophyllene; germacrene D; *cis*-*α*-bisabolene; ylangene; *β*-selinene, *α*-muurolene; 1*S*,*cis*-calamenene; *β*-ocimene; (*E*)-4,8-dimethyl-1,3,7-nonatriene; zingiberene; and (*E*)-*β*-farnesene. The sesquiterpene volatile fraction was repellent to female sweet potato weevil (SPW, *Cylas formicarius elegantulus*) with *α*-gurjunene, *α*-humulene and ylangene active in the concentration range emanating from storage roots. The aerial plant parts emanated a higher composite concentration of sesquiterpenes than storage roots. Differences in the relative attraction among four sweet potato cultivars to female SPW was inversely correlated with the composite concentration of headspace sesquiterpenes.

Suberin and Waxes

Suberin, isolated from sweet potato, was finely powdered and depolymerized with 14 % boron trifluoride in methanol and the soluble monomers were fractionated into phenolic fraction 17 %, aliphatic fraction 6 % and soluble fraction 26 % (Kolattukudy et al. 1975). The aliphatic fraction consisted of 36 % *ω*-hydroxy acids, 21 % dicarboxylic acids, 5 % fatty acids, 4 % fatty alcohols and 3 % polar compounds. Among the fatty acids, very long chain acids (>C_{20}) were the dominant components. Chain length of fatty acids of root suberin were C_{16} 0.9 %, C_{17} 0.2 %, $C_{18:U}$ 3.9 %, C_{18} 1.1 %, C_{19} 0.09 %, C_{20} 0.9 %, C_{21} 0.08 %, C_{22} 3.3 %, C_{23} 0.3 %, C_{24} 4.7 %, C_{25} 0.5 %, C_{26} 16.6 %, C_{27} 1.6 %, C_{28} 20.2 %, C_{29} 7.5 %, C_{30} 34.3 %, C_{31} 3.7 % and C_{32} 0.4 %. In the alcohol fraction C_{18}, C_{20}, C_{22} and C_{24}, saturated primary alcohols were the major components. Chain lengths of fatty alcohol of root suberin were C_{18} 68.3 %, C_{19} 0.5 %, C_{20} 15.5 %, C_{21} 1.7 %, C_{22} 6.1 %, C_{23} 0.5 %, C_{24} 3.0 %, C_{25} 0.4 %, C_{26} 3 %, C_{27} 0.1 %, C_{28} 0.6 %, C_{29} 0.2 % and C_{30} 0.1 %. C_{16} and C_{18} dicarboxylic acids were the major dicarboxylic acids of the suberin. Chain length of dicarboxylic acids of root suberin were C_{15} 0.2 %, C_{16} 6.6 %, C_{17} 2.4 %, $C_{18:1}$ 80.5 %, C_{19} 0.7 %, C_{20} 0.2 % and C_{21} 0.4 %. The composition of the *ω*-hydroxy acid fraction was quite similar to that of the dicarboxylic acids; 18-hydroxy-octadec-9-enoic acid was the major component. Chain lengths of *ω*-hydroxy acids of root suberin were C_{16}: 4.9 %, $C_{18:1}$ 90.7 %, C_{18} 1.3 %, C_{20} 0.3 %, C_{22} 0.3 %, C_{24} 0.5 %, C_{26} 0.9 % and C_{28} 0.7 %. The amount of wax extracted from the periderm of the storage organs including sweet potato ranged from 2 to 32 μg/cm^2 (Espelie et al. 1980). The hydrocarbons from the suberin layer had a broader chain-length distribution, a predominance of shorter carbon chains and a higher proportion of even-numbered carbon chains than the leaf alkanes from the same plants. The major components of the free and esterified fatty alcohols and fatty acids had an even number of carbon atoms, and were similar in chain-length distribution to their counterparts found covalently attached to the suberin polymers. Also extracted from the suberin associated waxes were polar components which included fatty alcohols and fatty acids in a conjugated form, and *ω*-hydroxy acids and dicarboxylic acids.

Ceramides and glucocerebrosides containing the three different long-chain bases 4,8-sphingadienine (d18:$2^{\Delta4,\Delta8}$), 4-hydroxy-8-sphingenine (t18:$1^{\Delta8}$), and 8-sphingenine (d18:$1^{\Delta8}$) acylated to saturated and unsaturated hydroxy- and non-hydroxy fatty acids with 16–26 carbon atoms were detected in sweet potatoes and potatoes (Bartke et al. 2006). For ceramides and glucocerebrosides 4,8-sphingadienine (d18:$2^{\Delta4,\Delta8}$) was found as the major long-chain base, with lesser amounts of 4-hydroxy-8-sphingenine (t18:$1^{\Delta8}$) and 8-sphingenine (d18:$1^{\Delta8}$). 2-(*α*-)hydroxypalmitic acid (C16:0 h)

was the major fatty acid, which was found to be acylated to the long-chain bases. The analyzed samples of potatoes and sweet potatoes showed amounts of approximately 0.1–8 μg/kg single ceramides and amounts up to 500 μg/kg glucocerebrosides, with C16:0 h-glucosyl-4,8-sphingadienine as the major component.

Toxins and Phytoalexins

Ipomeamarone, a phytoalexin, was first reported as a stress metabolite of unknown structures in old damaged sweet potatoes in 1943 (Hiura 1943). The bitter substance produced in black rot-infected sweet potato was identified as ipomoeamarone, an open-chain ketone with two oxide rings and two ethylenic linkages (Ohno 1952). It yielded a liquid keto-acid and oxalic acid by oxidation with potassium permanganate at room temperature, and acetone, acetic acid, oxalic acid and a liquid acid by oxidation with the same reagent at 100 °C. Also, similarities were noted between ipomeamarone and ngaione. The correct accepted structure of ipomeamarone was elucidated through a series of oxidative degradation by Kubota and co-workers (Kubota and Ichikawa 1954; Kubota 1958; Kubota and Matsuura 1958; Kubota et al. 1965). Subsequent studies by Kubota and Matsuura (1958) established ipomeamarone and ngaione to be enantiomers, both having a *cis*-configuration. When sweet potato root tissue was infected by *Ceratocystis fimbriata*, terpenes like ipomeamarone was synthesized in the adjacent non-infected region and accumulated in the infected region; also chlorogenic acid increased in non-infected tissues adjacent to infected tissues (Akazawa and Wada 1961). Akazawa and Uritani (1962) hypothesized that the biosynthesis of ipomeamarone in the sweet potato root infected by *Ceratocystis fimbriata* might be induced by the alteration of the tricarboxylic acid cycle in the host tissues. Tracer studies showed although there was no significant change in lipid ester groups in both infected and non-infected tissues, increase in phospholipids was found in diseased tissue (Imaseki et al. 1964). Sterol isolated from fresh material was identified with *β*-sitosterol. Chromatographic patterns of non-phospholipid fraction of diseased tissue suggested that some metabolic alteration of this fraction might occur in response to infection. Tracer studies with the use of 2-^{14}C-acetate revealed that the infected tissue of diseased sweet potato roots with black rot was incapable of synthesizing ipomeamarone from acetate, but was capable of synthesizing it from some intermediate(s) which was produced from acetate by the non-infected tissue in a short period. The incorporation of mevalonate-2-^{14}C into ipomeamarone in sweet potato root tissue infected by *Ceratocystis fimbriata* was demonstrated, but the rate was low when compared with acetate-2-^{14}C (Oshima and Uritani 1968). The data supported the participation of mevalonate in ipomeamarone synthesis as an intermediate. Acetate-2-^{14}C, pyruvate-3-^{14}C and citrate-2,4-^{14}C were incorporated into ipomeamarone in sweet potato root tissues infected by *Ceratocystis fimbriata* (Oba et al. 1970). Rates of incorporation of ^{14}C, from these three substances, into the $CHCl^3$-CH_3OH-soluble lipid fraction and ipomeamarone were of the following order: acetate > pyruvate > citrate. Labelled studies showed that farnesol-2-^{14}C was incorporated into ipomeamarone (Oguni and Uritani 1970). Incorporation of ethanol-2-^{14}C into furanoterpenoids such as ipomeamarone in sweet potato infected with the black rot fungus, *Ceratocystis fimbriata*, was demonstrated by Oguni and Uritani (1971). The rate of incorporation of ethanol-2-^{14}C into ipomeamarone was about twofold more efficient than for acetate-2-^{14}C. The results suggested that ethanol was utilized for lipid synthesis after being directly converted to acetyl CoA via acetaldehyde, and it appeared likely that a CoA-linked aldehyde dehydrogenase operated in sweet potato root tissue infected with *C. fimbriata.*

3-hydroxy-3-methylglutaryl coenzyme, a reductase, was detected in the non-infected root tissues and found to participate in the synthesis of ipomeamarone (Suzuki et al. 1974). Fresh sweet potato root tissue had a very low activity of 3-hydroxy-3-methylglutaryl coenzyme A reductase; however, when infected by *Ceratocystis fimbriata*, the enzyme activity increased rapidly, and reached a maximum in 2 days, thereafter, the activity decreased rapidly (Suzuki et al. 1975).

Optimal activity of 3-hydroxy-3-methylglutaryl coenzyme A reductase occurred at pH 7·3 to 7·5.

An abnormal semicarbazone was formed by removal of two molecules of water when ipomeanine, one of the constituents of black rot-infected sweet potato, was treated with semicarbazide (Kubota et al. 1965). Structural formulae were assigned to semicarbazone and an abnormal semicarbazone of an ester, an intermediate in the synthesis of ipomeanine in fungal-rotted sweet potato (Kubota and Ichikawa 1954). From the volatile component of sweet potatoes infected by *Ceratostomella fimbriata* (black rot disease), four *β*-substituted furans, namely ipomeamarone, batatic acid, ipomeanine and *β*-furan carboxylic acid were isolated (Kubota 1958).

Sweet potato roots infected by the black rot fungus, *Ceratocystis fimbriata*, produced umbelliferon (7-hydroxy coumarin) and scopletin (6-methoxy-7-hydroxy coumarin) (Uritani and Hoshiya 1953; Minamikawa et al. 1962), esculetin (Minamikawa et al. 1962), along with polyphenols chlorogenic acid, isochlorogenic acid, caffeic acid and their derivatives were produced in the tissue adjacent to the wounded surface or infected region of sweet potato storage roots (Uritani and Muramatsu 1953; Uritani and Miyano 1955). The formation of these coumarins was very low in the uninfected or uninoculated cut tissues, indicating that coumarin synthesis occurred after infection (Minamikawa et al. 1963).

A new hepatotoxic metabolite, ipomeamaronol, was isolated and described from mold-damaged sweet potatoes (Kato et al. 1971; Yang et al. 1971). The toxic furan ipomeamarone was detected in damaged sweet potato (Wood and Huang 1975). A phytoalexin dehydroipomeamarone, a sesquiterpenoid, was isolated from sweet potato root tissues infected by the black rot fungus, *Ceratocystis fimbriata* (Oguni and Uritani 1973; 1974). Dehydroipomeamarone was found to be an intermediate precursor in the biosynthetic pathway of the phytoalexin, ipomeamarone (Oguni 1974). The hepatotoxin ipomeamarone and lung-oedema toxins 4, ipomeanol and ipomeanine, were found in several sweet potato samples in the United Kingdom by Coxon et al. (1975). Two glucosides of coumarin derivatives, skimmin (umbelliferone-7-*β*-glucoside) and scopolin (scopoletin-7-*β*-glucoside) were isolated from sweet potato roots infected with black rot (Minamikawa et al. 1964). *β*-hydroxy-*β*-methylglutaric acid derivative was synthesized from acetyl CoA by a cell-free system from non-infected tissues adjacent to the infected region of sweet potato infected with black rot (Oshima and Uritani 1967). A phytoalexin-like compound was isolated from sweet potato root tissue infected by the black-rot fungus, *Ceratocystis fimbriata*, and identified as 14-hydroxy-ipomeamarone and called ipomeamaronol (Kato et al. 1973). 4-hydroxydehydromyoporone was mainly converted to 4-hydroxymyoporone and a little to dehydroipomeamarone, and ipomeamarone, in *Ceratocystis fimbriata*-infected sweet potato tissues (Inoue and Uritani 1980a). Furano-sesquiterpene reductase reduced dehydroipomeamarone and 4-hydroxydehydromyoporone to ipomeamarone and 4-hydroxymyoporone, respectively, in fungal-inoculated sweet potato tissues (Inoue and Uritani 1980b). The enzyme was located in the microsomal fraction and required NADPH but not NADH.

A new stress toxic furano-sesquiterpene metabolite of sweet potato was isolated and identified as 1-(3-furoyl)-4,8-dimethyl-7-hydroxy-1,6-nonanedione (7-hydroxymyoporone) (Burka et al. 1974). Its toxicity was similar to that of the well known phytoalexin, ipomeamarone. This compound was produced along with other toxic furanoterpenoids in response to certain exogenous stimuli. The lowest limit of detection of ipomeamarone in damaged sweet potatoes was 2 ng (Wood and Huang 1975). Four lung-toxic furanoterpenoids produced by sweet potatoes following microbial infection were identified as 4-ipomeanol (1-(3-furyl)-4-hydroxy-1-pentanone), the isomeric 1-ipomeanol (1-(3-furyl)-1-hydroxy-4-pentanone), the corresponding diketone, ipomeanine (1-(3-furyl)-1,4-pentanedione), and the diol, 1,4-ipomeadiol (1-(3-furyl)-1,4-pentanediol) (Boyd et al. 1974). In addition to ipomeamarone and other hepatotoxins, a series of 1-(3-furyl)-1,4-dioxygenated pentanes were isolated from sweet potatoes

infected with *Fusarium solani* (Burka and Wilson 1976). Burka and Kuhnert (1977) demonstrated that the tetrahydrofuran ring was cleaved so that ipomeamarone was converted to 4-hydroxymyoporone. The following phytoalexins (μg/g fresh weight) were isolated from mercuric chloride stressed sweet potatoes: myoporone 39 μg, 6-hydroxymyoporol 109 μg and 1-hydroxymyoporol 53 μg (Burka and Iles 1979). 6-myoporol was two to three times more toxic than ipomeamarone.

A new sesquiterpenoid, 4-hydroxydehydromyoporone, was isolated from *Ceratocystis fimbriata*-infected sweet potato root tissue (Inoue et al. 1977). Injury to sweet potato tissue by a poison such as $HgCl_2$ caused changes very similar to those induced by inoculation with *Ceratocystis fimbriata*, namely faster respiration, increase of polyphenols and coumarins in adjacent, sound tissue and production of ipomeamarone (Uritani et al. 1960). Myoporone and 6-myoporol (Burka and Iles 1979) and 1-(3′-furyl)-6,7-dihydroxy-4,8-dimethylnonan-1-one (Burka 1978) were isolated from stress sweet potatoes. Treatment of sweet potato slices with 0.1 % $HgCl_2$ three times within 24 h caused production of ipomeamarone in both the injured and the adjacent tissue which did not brown. The following phytoalexins dehydroipomeamarone, ipomeamaronol, 4-hydroxydehydromyoporone, 4-hydroxymyoporoine were detected in sweet potato root tissues infected by *Ceratocystis fimbriata* (Inoue and Uritani 1979). Four sesquiterpene stress metabolites, 6-oxodendrolasin (8.5 μg/g), 6-hydroxydendrolasin (1.2 μg/g), 9-oxofarnesol (0.7 μg/g) and 9-hydroxyfarnesol (12.9 μg/g) were isolated from mercuric chloride-treated sweet potatoes (Burka et al. 1981). Oba and Uritani (1981) found that furano-terpene production in sweet potato root tissue treated with chemicals, such as $HgCl_2$, l-alanine and $_C$AMP was inhibited by antibiotics, such as cycloheximide, blasticidin S, puromycin and chloramphenicol when the antibiotics were administered to the tissue immediately after cutting. However, furano-terpene production was not inhibited by antibiotics when they were administered to the 18-h incubated cut tissue together with chemical elicitors of furano-terpene production. Two new compounds, 7-hydroxycostal and 7-hydroxycostal, were isolated from infected sweet potatoes as members of a new class of sweet potato phytoalexins (Schneider and Nakanishi 1983). Nine new sesquiterpenes related biosynthetically to ipomeamarone, the well-known sweet potato phytoalexin, were isolated from *Ceratocystis fimbriata*-infected sweet potato root tissue (Schneider et al. 1984). They were identified as 9-hydroxyfarnesoic acid; ipomeatetrahydrofuran; (*Z*)-1,6-dioxoisodendrolasin; (*E*)-1,6-dioxoisodendrolasin;10-hydroxyipomeabisfuran; 4-hydroxymyoporonol; 4-hydroxymyoporonol ketal and two butenolides 6-oxodendrolasinolide and ipomeamaronolide.

Furanoterpenoids including ipomeamarone and ipomeamaronol were isolated from sweet potato root tissues infected with *Ceratocystis fimbriata* (Shen 1997). *Plenodomus destruens, Diaporthe batatatis, Diplodia tubericola, Fusarium solani* and *Ceratocystis fimbriata* induced accumulation of relatively high concentrations of ipomeamarone (63–16, 523 μg/g), 4-ipomeanol (5–236 μg/g) and 1,4-ipomeadiol (ND–1406 μg/g) in sweet potatoes (Clark et al. 1981). *Macrophomina phaseoli* and *Sclerotium rolfsii* induced accumulation of relatively high concentrations of ipomeamarone (ND–23.346 μg/g) and 4-ipomeanol (4–227 μg/g), but did not induce accumulation of 1,4-ipomeadiol. *Rhizopus stolonifer* and *Erwinia carotovora* induced accumulation of relatively low concentrations of ipomeamarone (58–2675 μg/g), 4-ipomeanol (from not detectable [ND] to 112 μg/g) and 1,4-ipomeadiol (ND–16 μg/g). *Streptomyces ipomoeae, Monilochaetes infuscan,* and internal cork virus did not induce accumulation of detectable levels of ipomeamarone, 4-ipomeanol or 1,4-ipomeadiol. Mercuric acetate induced accumulation of low concentrations of total furanoterpenoids, ipomeamarone, 1-ipomeanol, 1,4-ipomeadiol. *Fusarium oxysporum* f. sp. *batatas* did not induce accumulation of furanoterpenoids in sweet potato vines. Concentrations of 4-ipomeanol and

1,4-ipomeadiol were highest in tissue infected with certain isolates of *Diplodia tubericola* and *Fusarium solani.* Sweet potato disks treated with 50 mM mercuric acetate and incubated for 2 weeks at 30 °C contained 1657 μg total furanoterpenoids, 1075 μg ipomeamarone, 3 μg 4-ipomeanol and 5 μg 1,4-ipomeadiol perg tissue.

The bitter principle of sweet potato roots destroyed by the sweet potato weevil, *Cylas formicarius elegantulus*, was isolated and proved to be ipomeamarone, which had originally been isolated from fungus-infected sweet potatoes (Akazawa et al. 1960). When sweet potato root tissues were infested by the larvae of sweet potato weevil, *Cylas formicarius* and West Indian sweet potato weevil, *Euscepes postfasciatus*, furanoterpenoids namely ipomeamarone, dehydroipomeamarone, 4-hydroxymyoporone, ipomeamaronol and component A and coumarins umbelliferone and scopoletin were produced in brown necrotic layer formed during the infestation (Uritani et al. 1975).

Sweet potato had been found to accumulate umbelliferone and scopoletin after biotic and abiotic stresses (Matsumoto et al. 2012) In the biosynthesis of the coumarins, they found that Ib1 proteins exhibited ortho-hydroxylation activity toward feruloyl coenzyme A (CoA) to form scopoletin. Ib2 proteins catalyzed ortho-hydroxylation of feruloyl-CoA and also of *p*-coumaroyl-CoA to form scopoletin and umbelliferone, respectively. Fungal and chitosan treatments increased levels of umbelliferone and its glucoside (skimmin) in the tubers, and expression of the Ib2 gene was induced concomitantly.

Two sesquiterpenoids phytoalexins, called components A_1 and A_2, were isolated from sweet potato root tissue either infected by *Ceratocystis fimbriata* or injured by $HgCl_2$ (Ito et al. 1984). Both components inhibited germination and term-tube growth of *Ceratocystis fimbriata*-oak strain. When ^{14}C-component A_2 was supplied to diseased tissue discs, the label was efficiently incorporated into dehydroipomeamarone, ipomeamarone, ipomeamaronol and component B_1, indicating it to be a close precursor of these components. In the case of ^{14}C-component A_1 administration, the label was mainly incorporated into the two unknown fractions.

Leaf/Stem/Nutrients/Phytochemicals

Sweet potato leaves provide a dietary source of vitamins, minerals, antioxidants, dietary fibre and essential fatty acids (Johnson and Pace 2010). Bioactive compounds contained in this vegetable play a role in health promotion by improving immune function, reducing oxidative stress and free radical damage, reducing cardiovascular disease risk and suppressing cancer cell growth. Research had affirmed the potential cardioprotective and chemopreventive advantages of consuming sweet potato leaves, thus indicating that increased consumption of this vegetable should be advocated. The levels of Fe, Ca and Mg essential trace elements of Cr, Co, Ni, Cu and Zn in 11 lines of sweet potato leaves were similar to those of common green leafy vegetables (Taira et al. 2013). The ratio of K and Na for the seven lines of sweet potato leaves was higher than that of spinach, indicating that sweet potato leaves may be used in antihypertensive diet. The selenium and manganese contents were higher in all the sweet potato leaves than in other green leafy vegetables, such as spinach and water spinach.

Crude protein ranged from 16.78 to 25.39 %; crude fibre from 9.75 to 12.14 %; crude fat from 0.38 to 1.91 %; ash content from 8.71 to 11.60 %; moisture content (fwb) ranged from 80.16 to 88.20 %; carbohydrate values from 53.29 to 59.01 % and calorific values ranged from 1344.00 to 1399.00 kJ/g (316.66–329.76 cal/g) for the sweet potato leaves (Oduro et al. 2008). Elemental analysis of the leaves in mg/100 g dry matter (DM) indicates the sweet potato leaves contained appreciable levels of calcium (1310.52–1402.27) and iron (9.62–23.02).

Fresh sweet potato leaves contained the following nutrient value per 100 g edible portion proximate: water 86.81 g, energy 42 kcal (175 kJ), protein 2.49 g, total lipid 0.51 g, ash 1.36 g, carbohydrate 8.82 g, total dietary fibre 5.3 g, Ca 78 mg, Fe 0.97 mg, Mg 70 mg, P 81 mg, K 508 mg, Na 6 mg, Se 0.9 mcg; vitamin C

11 mg, thiamin 0.156 mg, riboflavin 0.345 mg, niacin 1.130 mg, pantothenic acid 0.225 mg, vitamin B-6 0.190 mg, folate DFE 1 μg, vitamin A RAE 189 μg, vitamin A 3778 IU, β-carotene 2217 μg, α-carotene 42 μg, *β*-cryptoxanthin 58 μg, lutein-zeaxanthin 14,720 μg, vitamin K (phylloquinone) 302.2 μg, total saturated fatty acids 0.110 g, 16:0 (palmitic acid) 0.1 g, 18:0 (stearic acid) 0.01 g, total monounsaturated fatty acids 0.020 g, 18:1 undifferentiated (oleic acid) 0.020 g, total polyunsaturated fatty acids 0.228 g, 18:2 undifferentiated (linoleic acid) 0.192 g, 18:3 undifferentiated (linolenic acid) 0.036 g; amino acids – tryptophan 0.035 g, lysine 0.228 g, methionine 0.086 g, cystine 0.047 g, flavones – apigenin0.1mg,luteolin0.1mg,flavonols–kaempferol 2.1 mg, myricetin 4.4 mg and quercetin 16.9 mg (USDA-ARS 2014). Sweet potato leaves of ten sweet potato varieties were found to contain the following flavonoid (% DW) quercetin 0.98 %, myricetin 0.12 %, luteolin 0.23 %, apigenin 0.38 %, kaempferol 0.07 %, total flavonoids 1.79 % (Ojong et al. 2008). Sweet potato purple and green leaves were found to contain the flavonols quercetin (852.63; 83.22 μg/g dw), and morin (3266.11; 9376.21 μg/g dw) and anthocyanidins cyanidin (1437.32 μg/g Dw, not detected (nd)) and malvidin (30.37 μg/g DW, nd), respectively (Chao et al. 2014). Innami et al. (1998) reported the nutrient composition of freeze-dried sweet potato leaf powder as follows: moisture 3.9 %, protein 29.5 %, fat 5.5 %, ash 10.4 %, carbohydrates 10 % and total dietary fibre 40.7 %. A soluble dietary fibre (SDF) was extracted from the sweet potato leaf powder (Ishida et al. 2004). This substance was found to show high viscosity and to be mainly composed of xylose (34.7 %) and uronic acid (38.8 % as galacturonic acid).

The average contents of minerals and vitamin in the leaves of cv. 'Suioh' were 115 mg Ca, 1.8 mg Fe, 3.5 mg Carotene, 7.2 mg vitamin C, 1.6 mg vitamin E and 0.56 mg vitamin K per 100 g fresh weight leaves (Islam 2006). The nutrient and antinutrient (oxalate and polyphenol) content of green midrib sweet potato leaves and purple midrib sweet potato leaves were determined to be respectively as: moisture content 85.63, 86.28 %; dry matter 14.37, 13.72 %; crude protein 26.37, 37.06 %; total ash 12.87, 20.41 %, crude fibre 16.01, 21.48 %, total lipid 3.11, 3.27 %; soluble carbohydrates 41.62, 17.76 %; ascorbic acid 305.58, 273.17 mg/100 g, carotenoids 44.18, 53.32 mg/100 g, Fe 15.22, 17.48 mg/100 g; Ca 3457, 4255 mg/100 g, oxalates 3730.50, 2901.50 mg/100 g and polyphenols 5.28, 22.16 mg/100 g (Mwanri et al. 2011). Drying with salt and cooking with lemon reduced polyphenols significantly, with retention of 42 and 56 % respectively, while cooking with lemon lowered significantly the oxalate levels.

Two major anthocyanins in purple sweet potato were identified as cyanidin 3-caffeylferuly sophoroside-5-glucoside of cyanidin and peonidin 3-caffeylferulysophoroside-5-glucoside (Odake et al. 1992).

Fifteen anthocyanins of acylated cyanidin and peonidin types were identified in sweetpotato leaves: cyanidin 3-sophoroside-5-glucoside; peonidin3-sophoroside-5-glucoside; *p*-hydroxybenzoylated (cyanidin 3-sophoroside-5-glucoside); caffeoylated (cyanidin3-sophoroside-5-glucoside); *p*-hydroxybenzoylated (peonidin-3-sophoroside-5-glucoside); caffeoylated (peonidin 3-sophoroside-5-glucoside); feruloylated (cyanidin 3-sophoroside-5-glucoside); cyanidin 3-(6,6′-caffeoyl-*p*-hydroxybenzoylsophoroside)-5-glucoside; cyanidin 3-(6,6′-dicaffeoylsophoroside)-5-glucoside; cyanidin 3-(6-caffeoylsophoroside)-5-glucoside; cyanidin 3-(6,6′-caffeoylferuloylsophoroside)-5-glucoside; peonidin 3-(6,6′-dicaffeoylsophoroside)-5-glucoside; peonidin 3-(6,6′-caffeoyl-*p*-hydroxybenzoylsophoroside)-5-glucoside; cyanidin 3-(6-caffeoylsophoroside)-5-glucoside and peonidin 3-(6,6′-caffeoylferuloylsophoroside)-5-glucoside (Islam et al. 2002a).

From sweet potato leaves five compounds were isolated and identified as *β*-sitosterol, friedelin, acetyl-*β*-amyrin, caffeic acid and quercetin (Tan et al. 1995). Eight compounds: *β*-amyrin acetate, friedelin, epifriedelanol, *n*-triacontanol, *β*-sitosterol (5), ethyl caffeate, scopoletin and daucosterol were isolated from sweet potato leaves (Luo and Kong 2005a). Six different polyphenolic compounds were identified and quantified in sweet potato leaves (Islam et al. 2002b,

2003a, b). The relative levels of polyphenolic acids were as follows: 3,5-di-*O*-caffeoylquinic acid > 4,5-di-*O*-caffeoylquinic acid > chlorogenic acid (3-*O*-caffeoylquinic acid) > 3,4-di-*O*-caffeoylquinic acid > 3,4,5-tri-*O*-caffeoylquinic acid > caffeic acid. The highest 3,4,5-tri-*O*-caffeoylquinic acid and 4,5-di-*O*-caffeoylquinic acid occurred at 221 and 1183.30 mg/100 g dry weight, respectively. Six caffeoylquinic acid derivatives were quantified and divided into three categories based on the leaf polyphenol content: high, medium and low polyphenol accumulator (Islam et al. 2003b). The caffeic acid and derivatives were positively correlated with the total polyphenol contents, and the correlation coefficients varied widely among the different categories. Most of the phenolic compounds were highest in leaves from plants grown at 20 °C without shading except 4,5-di-*O*-caffeoylquinic acid (Islam et al. 2003a). The results indicated that growing leaves under moderately high temperatures and in full sun enhanced the accumulation of phenolic components. These phenolics exhibited various kinds of biological activities – radical scavenging, antimutagenic, anticancerous, antidiabetic and antimicrobial activities (Islam 2006). Of five phenolic acids caffeic acid, chlorogenic acid, 4,5-di-*O*-caffeoylquinic acid, 3,5-di-*O*-caffeoylquinic acid and 3,4-di-*O*-caffeoylquinic acid identified in sweet potato storage leaves, 3,5-di-*O*-caffeoylquinic acid and/or 4,5-di-*O*-caffeoylquinic acid were predominant (Truong et al. 2007). Sweetpotato leaves had the highest phenolic acid content followed by the peel, whole root and flesh tissues.

Sweet potato leaves were found to contain a high content of polyphenolics in comparison with 12 kinds of the major commercial vegetables (Yoshimoto et al. 2006). The polyphenolics were composed of caffeic acid (CA) and five kinds of caffeoylquinic acid derivatives, 3-mono-*O*-caffeoylquinic acid (chlorogenic acid, ChA), 3,4-di-*O*-caffeoylquinic acid (3,4-diCQA), 3,5-di-*O*-caffeoylquinic acid (3,5-diCQA), 4,5-di-*O*-caffeoylquinic acid (4,5-diCQA) and 3,4,5-tri-*O*-caffeoylquinic acid (3,4,5-triCQA). These polyphenolics showed various kinds of physiological functions, radical scavenging activity, antimutagenic activity, anticancer, antidiabetes and antibacterial activity in-vitro or in-vivo which may be helpful for maintaining and promoting human health.

Methanol extracts of vine latex of four sweet potato cultivars yielded hexadecyl, octadecyl and eicosyl *p*-coumarates (Snook et al. 1994). Both *Z*- and *E*-isomers of the phenolic acid were found, with the latter predominating. Trace quantities of hexadecyl (*Z*)- and (*E*)-ferulates were also identified in ester concentrates. Levels of octadecyl (*E*)-*p*-coumarate ranged from 0.7 % fresh weight in cv. Resisto to almost 2 % in cv. Jewel, while the hexadecyl ester levels were only 1/4 to 1/3 these values. Levels of the *Z*-esters were 1/10 to 1/20 of the levels of the corresponding *E*-isomers. Levels of the esters in cv. Jewel sweetpotato root latex were 2–10-fold the levels in the vine latex, while the ratio of *E*-esters to *Z*-esters was found to be 7–14-fold.

Sweet potato leaves, petiole and stem were found to contain folic acid and polyphenols (Taira et al. 2007). The amounts of folic acid and polyphenol in eligible parts (leaf and petiole), except in cv *Okinawa* 100 species, contained higher amounts than stems and were similar to amounts of spinach. *Miyanou* 36 had the highest amounts of folic acid.

Sweet potato leaves were found to contain functional polyphenols, such as caffeic acid (CA), chlorogenic acid (ChA), 4,5-dicaffeoylquinic acid (4,5-diCQA), 3,5-diCQA, 3,4-diCQA and 3,4,5-triCQA (Taira and Ohmine 2011). High-speed counter-current chromatography was used to isolate and purify four caffeoylquinic acid derivatives and a mixture of two flavonoids from sweet potato leaves (Li et al. 2012). The caffeoylquinic acid derivatives were 3-*O*-caffeoylquinic acid (1), 4,5-di-*O*-caffeoylquinic acid (4), 3,5-di-*O*-caffeoylquinic acid (5) and 3,4-di-*O*-caffeoylquinic acid (6). The mixture of flavonoids was separated by preparative high-performance liquid chromatography into quercetin-3-*O*-*β*-D-galactopyranoside (2) and quercetin-3-*O*-*β*-D-glucoside (3). The purities of compounds 1–6 were 95.8 % (5.4 mg), 99.5 % (6.1 mg), 98.7 % (15.1 mg), 97.8 % (14.5 mg), 96.2 % (10.3 mg) and 96.8 % (7.8 mg),

respectively. The following phenolic compounds were identified in the leaves of 20 sweet potato cultivars: quinic acid; hydroxybenzoic acids; hydroxycinnamic acids; caffeic acid; *p*-coumaric acid; ferulic acid; sinapic acid; chlorogenic acid; 5-CQA, 5-*O*-caffeoylquinic acid; 3-CQA, 3-*O*-caffeoylquinic acid; 4-CQA, 4-*O*-caffeoylquinic acid; 3,4-diCQA, 3,4-di-*O*-caffeoylquinic acid; 3,5-diCQA, 3,5-di-*O*-caffeoylquinic acid; 4,5-diCQA, 4,5-di-*O*-caffeoylquinic acid; 3,4,5-triCQA, 3,4,5-tri-*O*-caffeoylquinic acid; FQA, feruloylquinic acid; CFQA, caffeoyl-feruloylquinic acid (Luo et al. 2013). Two dihexosides (such as quercetin-*O*-dihexoside) and one hexoside of quercetin (quercetin-3-*O*-hexose-hexoside) were also characterized.

Three chemical compounds fumaric acid, succinic acid and 7,3′,4′-trimethylquercetin were isolated from sweet potato leaves and stems (Liu et al. 1991). Five flavonoid compounds were isolated from sweet potato leaves and identified as: tiliroside, astragalin, rhamnocitrin, rhamnetin and kaempferol (Luo and Kong 2005b). Six compounds were isolated from 90 % ethanol leaf extract and identified as tetracosane, myristic acid, *β*-sitosterol, *β*-carotene, daucosterol and quercetin (Lv et al. 2009). Sweet potato leaves were found to contain per 100 g dry weight, 345.65 mg alkaloids, 328.44 anthraquinones and phenolic compounds 662.02 mg (Pochapski et al. 2011).

Ipomoea batatas stems were found to contain sapons, flavonoids, phlobatannins, tannins and alkaloids (Essiett and Ukpong 2014). The stem had the following proximate composition: 10 % moisture, 7.2 % ash, 5.05 % crude fibre, 9.4 % crude protein, 4.65 % lipid, 73.7 % carbohydrate and 374.25 kcal energy value. It also contained antinutrients (mg/100 g) – hydrogen cyanide 0.103 mg, total oxalate 90.8 mg, soluble oxalate 66 mg, phytate 1.25 mg and 4.51 mg tannins.

Caffeoylquinic acids were quantitatively the major sub-group of chlorogenic acids detected in the stem and the only sub-group detected in the leaves (Zheng and Clifford 2008). This sub-group was dominated by 5-caffeoylquinic acid. The stem also contained three feruloylquinic acids, 3,5- and 4,5-dicaffeoylquinic acid, and small amounts of at least four caffeoyl-feruloylquinic acids. Chlorogenic acids were not detected in sweet potato root.

Young leafy shoots were reported to be rich in proteins and a good source of vitamin B1 thiamine, riboflavin B2, folic acid and ascorbic acid and a good source of *β*-carotene (Villareal et al 1979a, b, 1985, Woolfe 1992). The leaves had higher levels of Fe, P, K, Mg and Se than the tubers. Yields of sweet potato tips varied from 10 to 16 t/ha in weight and 61 to 352/m^2 in number, whilst yields of marketable roots varied from 1 to 16 t/ha (Villareal et al. 1979b). The tips appeared to provide an excellent source of vitamin B2, a vitamin often deficient in Asian diets. Cultivar differences in dry matter, fibre, ash, vitamin B2 and oxalate were also observed. Wide variation was found in the morphological characteristics of sweet potato leaf tips which were acceptable as a vegetable for human consumption in terms of yield, palatability, tenderness and nutritive traits (Villareal et al. 1979a). Among the leaf types evaluated, Kinangkong (a fine leaf cultivar) produced the highest yield, and had the highest amount of protein, lowest oxalate and fairly high dry matter, making it a very desirable leafy vegetable.

Antia et al. (2006) reported sweet potato leaves to contain an appreciable amount of nutrients, vitamins and mineral elements and low levels of toxicants and should be included in diets to supplement our daily allowance needed by the body. Their analyses reported that sweet potato leaves had the following nutrient composition (per 100 g dm): crude protein 24.85 %, crude fat 4.90 %, crude fibre 7.20 %, ash 11.10 %, carbohydrate 51.95 %, moisture content 82.21 %, calorific value 351.30 kcal, vitamin A 0.672 mg/100 g, vitamin C 15.20 mg/100 g. The elemental analysis of the leaves in mg/100 g, zinc 0.08 mg, potassium 4.05 mg, sodium 4.23 mg, manganese (4.64 mg, calcium 28.44 mg, magnesium 340 mg and iron 16 mg. The antinutrient composition for phytic acid, cyanide, tannins and total oxalate were 1.44, 30.24, 0.21 and 308.00 mg/100 g, respectively. Studies in Japan (Ishida et al. 2000) reported that sweet potato leaves contain a large amount of protein, showing high amino acid

score. Any part of sweet potato was rich in dietary fibre and in particular, leaves were soluble dietary fibre and stems were insoluble dietary fibre, respectively. Mineral content, particularly iron, and vitamin content such as carotene, vitamin B2, vitamin C and vitamin E were high in leaves in comparison with other vegetables. Furthermore, polyphenol and flavonoid content in leaves are comparatively high. These results suggest that the whole parts of sweet potatoes should be utilized as valuable foodstuffs to cope with future changes in food supply and demand, particularly in developing countries. Sweet potato leaves are excellent source of lutein with levels from 34–68 mg/100 mg depending on varieties (Khachatryan et al. 2003). This places sweet potato leaves second in lutein content after marigold flowers, and number one among edible vegetables. Lutein, an antioxidant carotenoid (3,3′-dihydroxy-D-carotene), has been identified as a dietary strategy that can delay the onset of age-related macular degeneration (AMD). AMD is a medical condition predominantly found in elderly adults in which the centre of the inner lining of the eye, known as the macula area of the retina, suffers thinning, atrophy and in some cases, bleeding. This can result in loss of central vision, which entails the inability to see fine details, to read or to recognize faces. Sweet potato leaves may help in the fight against age-related macular degeneration (AMD). The crude protein, crude fibre, crude fat, carbohydrate and ash contents of leaves from 40 sweet potato cultivars ranged between 16.69–31.08, 9.15–14.26, 2.08–5.28, 42.03–61.36 and 7.39–14.66 g/100 g dry weight (DW), respectively (Sun et al 2014b). According to the index of nutritional quality, sweet potato leaves were good sources of protein, fibre and minerals, especially K, P, Ca, Mg, Fe, Mn and Cu. The correlation coefficient between antioxidant activity and total polyphenol content was the highest (R^2 = 0.76032), indicating that polyphenols were important antioxidants in sweet potato leaves.

Sweet potato was reported to be a particularly rich source of ferulic acid (Min et al. 2006). The tuber contained 0.54 % per g dry weight; the non-tuber portion of the plant contained more than > 3.0 % per g by dry weight. Through enzymatic hydrolysis 45 % ferulic acid, 29 % vanillin, 16 % vanillic acid and negligible (about 0.5 %) cinnamic acid were obtained from the stems. The amount of vanillic acid and vanillin released were 11.04 and 14.69 mg/g, respectively, when incubated with 1.0 % Viscozyme L for an hour, compared with only 7.47 and 8.30 mg/g, respectively, when incubated for an hour with 1.0 % Ultraflo L.

The aerial parts of *Ipomoea batatas* were found to produce four new resin glycosides, designated as ipomotaosides A, B, C and D were isolated from sweet potato dried aerial parts (Yoshikawa et al. 2010).

Three kinds of pure polysaccharide, named PSPV I, PSPV II and PSPVIII, respectively, were obtained from sweet potato vines (SPV) (Luo and Gao 2008). PSPV I had a molecular weight of 6.278×10^4 D and was mainly composed of xylose, mannose and glucose. PSPV II and PSPVIII had molecular weights of 3.801×10^4 D and 1.418×10^4 D, respectively. PSPVII was mainly composed of mannose and galactan and PSPVIII mainly composed of glucose, xylose and rhamnose. Sweet potato leaves were found to contain galactolipids: 53,940 mg/kg monogalactosyldiacylglycerols (MGDG) and 22,640 mg/kg digalactosyldiacylglycerols (DGDG) (Napolitano et al. 2007).

Two forms of trypsin inhibitors with molecular weights of 31 and 14 kDa were found in sweet potato leaves and they were different from those found in the roots (Wang and Yeh 1996). A proteinaceous invertase inhibitor, designated ITI-L, with molecular weight of 10 kDa was purified from sweet potato leaves. It was thermostable (90 % of the activity remained after incubation at 100 °C for 20 min) (Wang et al. 2003). High-quality and intact total RNA with a yield of 0.2 mg/g fresh weight was isolated from leaf blade, petiole, stem, fibrous root, thick root and storage root purple-fleshed sweet potato (Zhou et al. 2009).

Chen et al. (2008b) found that elevation of cytosolic Ca^{2+} by ethylene may stimulate protein

phosphatase and mitogen-activated protein kinase kinase activation (MAPKK), which finally activated ipomoelin gene expression in sweet potato. They also found that cyclic guanosine monophosphate (cGMP) was necessary for expression of wounding-responsive microRNAs miR828, which repressed the expression of IbMYB and IbTLD, leading to accumulation of lignin and hydrogen peroxide to participate in the plant defense mechanisms (Lin et al. 2012a). In a recent subsequent study, they found wounding in leaves repressed IbHO expression and carbon monoxide production, induced hydrogen peroxide generation and extracellular signal-regulated kinase (ERK) phosphorylation and then stimulated ipomoelin expression (Lin et al. 2014). Three sweet potato metallothionein genes, cysteine-rich, low molecular weight, metal-binding proteins, type 1 (IbMT1), type 2 (IbMT2) and type 3 (IbMT3) were found to respond differentially to abiotic stress and heavy metal toxicity (Kim et al. 2014). IbMT1 was predominantly expressed in leaves, roots and callus. IbMT2 transcript was detected only in stems and fibrous roots, whereas IbMT3 was strongly expressed in leaves and stems. The levels of IbMT1 expression were strongly elevated in response to Cd and Fe, and moderately higher in response to Cu. The IbMT3 expression pattern in response to heavy metals was similar to that of IbMT1. Exposure to abiotic stresses such as methyl viologen (MV; paraquat), NaCl, polyethylene glycol (PEG) and H_2O_2 up-regulated IbMT expression; IbMT1 responded strongly to MV and NaCl, whereas IbMT3 was induced by low temperature and PEG. A 16-amino-acid peptide, IbACP (*Ipomoea batatas* anticancer peptide), was isolated from sweet potato leaves (Chang et al. 2013).

The presence and levels of zeatin, ribosylzeatin, 9-glucosylzeatin ([9G] Z), N^6-isopentenyladenosine and 9-glucosyl-N^6-isopentenyladenine ([9G] iP) were determined in calluses of sweet potato plant (Sugiyama et al. 1988). The levels of [9G]iP and [9G]Z were, respectively, 3.6 and 2.7 μg per 100 g fw of the callus derived from the tuberous root of sweet potato.

Antioxidant Activity

Of 21 sweet potato cultivars with different flesh colours, purple-fleshed cultivars which contained anthocyanins, had the highest *tert*-butylperoxyl radical (*t*-BuOO·)· scavenging activities (Furuta et al. 1998). Those cultivars with purple flesh also had the highest antioxidative activities against lipid peroxidation induced by auto-oxidation of linoleic acid. Most of the sweet potato cultivars with white, white-yellow, yellow and orange flesh had low *t*-BuOO· scavenging and antioxidative activities; however, some of them had higher activities. In all sweet potato cultivars tested, the *t*-BuOO· scavenging activities were higher with an increase in the total phenolic content. Due to its high phenolic and anthocyanin content, purple sweet potato exhibited high antioxidant activity (Cevallos-Casals and Cisneros-Zevallos 2002). Antioxidant activity and total phenolic content were 3.2 and 2.5 times higher, respectively, than that of a blueberry variety assayed. Antioxidant activity in purple sweet potato skin was found to be almost three times higher than in the rest of the tissue. Acylated anthocyanins isolated from a highly pigmented callus induced from the storage root of purple-fleshed sweet potato cultivar Ayamurasaki exhibited both higher stability and higher DPPH radical scavenging activity than corresponding nonacylated cyanidin and peonidin 3-*O*-sophoroside-5-*O*-glucosides (Terahara et al. 2004).

Purple-fleshed sweet potato containing anthocyanin pigments displayed a potent antioxidative activity (Oki et al. 2002; 2003) An 80 % ethanol extract from purple-fleshed sweet potato cultivars harvested exhibited a 1,1-diphenyl-2-picrylhydrazyl (DPPH) radical-scavenging activity of 8.6–49.0 μmol expressed as Trolox equivalent/g of fresh weight. Oki et al. (2002) found that the dominant DPPH radical-scavengers in sweet potato purple-fleshed cultivars 'Ayamurasaki' and 'Kyushu-132' were anthocyanins rather than phenolic compounds, while those in 'Miyanou-36' and 'Bise' were phenolic compounds, such as chlorogenic acid. 'Ayamurasaki' and 'Kyushu-132' were rich in anthocyanins with

peonidin aglycon, whereas 'Miyanou-36', 'Bise' and 'Tanegashimamurasaki' contained cyanidin aglycon. Steed and Truong (2008) reported that the DPPH radical-scavenging activities of purple-fleshed sweet potatoes were 47.0–87.4 μmol trolox equivalent (TE)/g fw, and the oxygen radical absorbance capacity (ORAC) values were between 26.4 and 78.2 μmol TE/g fw. Oki et al. (2003) developed a simple and rapid spectrophotometric method for selecting breeding lines of breeding lines of purple-fleshed sweet potato cultivars with a high DPPH radical-scavenging activity. In-vitro antioxidant assay *trans*-4,5-dicaffeoylquinic acid from purple sweet potato cv Ayamurasaki showed significant antioxidant activities (Zhao et al. 2014).

Purple sweet potato anthocyanin exhibited 1,1-diphenyl-2-picrylhydrazyl (DPPH) radical-scavenging activity and effectively inhibited lipid peroxidation initiated by Fe^{2+} and ascorbic acid in rat brain homogenates (Cho et al. 2003). Philpott et al. (2004) demonstrate in-vitro antioxidant activity by a mottled purple-fleshed sweet potato anthocyanins, where an additive effect with hydroxycinnamic acids was observed. They asserted that sweetpotato could be eaten several hundred grams at a time and as a staple could confer superior health protection against a variety of degenerative disease processes by anthocyanic varieties of sweet potato in comparison to most common fruits and vegetables. Anthocyanins from purple sweet potato (PSP) showed stronger 1,1-diphenyl-2-picrylhydrazyl (DPPH) radical-scavenging activity than anthocyanins from red cabbage, grape skin, elderberry or purple corn, and eight major components of the anthocyanins from PSP showed higher levels of activity than ascorbic acid (Kano et al. 2005). In PSP anthocyanin-injected rats and PSP beverage-administered human volunteers, DPPH radical-scavenging activity in the urine increased. The elevation of plasma transaminase activities induced by carbon tetrachloride was depressed in rats administered PSP anthocyanin solution. Results of in-vitro studies indicated that purple sweet potato anthocyanins maintained the intracellular redox balance of heat-shocked bovine embryos by reducing intracellular oxidative stress and increasing the glutathione (GSH) levels (Sakatani et al. 2007). Embryos exposed to heat shock without anthocyanins showed a significant decrease in blastocyst formation rate and GSH content and an increase in intracellular reactive oxygen species (ROS) compared with non-heat-shocked embryos.

The polyphenolic compound 4,5-di-*O*-caffeoyldaucic acid from sweet potato tuber exhibited higher antioxidant activity in both DPPH and FRAP methods compared to that of all antioxidant standards, l-ascorbic acid, *tert*-butyl-4-hydroxy toluene (BHT) and gallic acid used at the same molar concentration (Dini et al. 2006a, 2006b). Two components, cyanidin 3-*O*-(2-*O*-(6-*O*-(*E*)-caffeoyl-*β*-D-glucopyranosyl)-*β*-D-glucopyranoide)-5-*O*-*β*-D-glucopyranoside and peonidin 3-*O*-(2-*O*-(6-*O*-(*E*)-caffeoyl-*β*-D-glucopyranocyl)-*β*-D-glucopyranoide)-5-*O*-*β*-D-glucopyranoside, which were detected in the plasma, protected low-density lipoprotein from oxidation at a physiological concentration. At a concentration of 0.5 mg/mL, the reducing power of purple sweet potato (PSP) anthocyanins, *L*-ascorbic acid (*L*-AA) and butylated hydroxytoluene (BHT) reached 0.572, 0.460 and 0.121, respectively (Jiao et al. 2012; Jiao et al. 2014). PSP anthocyanins exhibited high DPPH and superoxide anion radicals-scavenging activities; for DPPH. IC_{50} values were PSP anthocyanins 6.94 μg/mL, *L*-AA 6.10 μg and BHT 123.46 μg/mL; for superoxide anion radical-scavenging activity IC_{50} values were PSP anthocyanin 3.68 μg/mL, *L*-AA 10.01 μg/mL and BHT 50 μg/mL. Sixteen kinds of anthocyanins cyanidin, peonidin and mono- and diacylated forms of cyanidin and peonidin were detected.

The total antioxidant activity (hydrophilic + lipophilic ORAC (oxygen radical absorbance capacity)) was highest (14.7–29.2 μmol TE/g fresh weight (fw)) for NC415 (purple-fleshed), orange-fleshed (5.89–10.3 μmol TE/g fw) and lowest (2.72–3.33 μmol TE/g fw) for Xushu 18 (white-fleshed) (Teow et al. 2007). The hydrophilic-ORAC values were significantly correlated with the DPPH (R^2=0.859) and 2,2′-azinobis(3-ethyl-benzothiazoline-6-sulfonic acid) (ABTS) (R^2=0.761) values. However, the

lipophilic-ORAC values were poorly correlated with the b-carotene contents (R^2=0.480). The total phenolic contents (0.011–0.949 mg chlorogenic acid equivalent/g fw) were highly correlated with the hydrophilic-ORAC (R^2=0.937) and DPPH (R^2=0.820) values. Purple sweet potato (PSP) ethanol extract (100-fold diluted) showed stronger radical (2,2-diphenyl-1-picrylhydrazyl radical) scavenging activity than the water extract of PSP and the ethanol extract of YSP (up to a sixfold higher activity) (Park et al. 2010). The ethanol extract of PSP also exhibited the highest increase in ferric reducing ability among all extracts. Cupric ion-mediated LDL oxidation was strongly inhibited by the ethanol extract of PSP, with similar potency to vitamin C treatment.

Small sweet potato roots (≈4 g root weight) had a higher antioxidant activity and phenolic content compared with full-sized marketable roots (≈300 g root weight) (Padda and Picha 2007). Phenolic content in marketable roots was significantly higher in the cortex tissue than in the internal pith tissue. The highest total phenolic content [chlorogenic acid equivalents (10.3 mg/g dry weight)] and antioxidant activity [Trolox equivalents (9.7 mg/g dry weight)] was found in cortex tissue of small-sized roots. Antioxidant activity of different sweet potato genotypes roots ranged from 1.3 to 4.6 mg/g dry weight (DW) Trolox equivalent (Padda and Picha 2008). Total phenolic content, expressed in terms of chlorogenic acid equivalent, in different sweet potato genotypes ranged from 1.4 to 4.7 mg/g DW. The highest total phenolic content and antioxidant activity were observed in a purple-fleshed genotype. Chlorogenic acid and 3,5-dicaffeoylquinic acid were the predominant phenolic acids, while caffeic acid was the least abundant in most genotypes. The highest content of chlorogenic acid (422.4 μg/g DW) was present in a white-fleshed cultivar 'Quarter Million' imported from Jamaica. However, a purple-fleshed genotype had the highest amounts of 3,5-dicaffeoylquinic (485.6 μg/g DW), 3,4-dicaffeoylquinic (125.6 μg/g DW), 4,5-dicaffeoylquinic (284.4 μg/g DW) and caffeic (20.5 μg/g DW) acids. *Ipomoea batatas* aqueous tuber extract showed strong DPPH radical-scavenging activity of about 83 % (Arockiamary et al. 2014). In ferric reducing antioxidant power (FRAP) assay, the extract showed increased absorbance which was directly related to the combined or total reducing power of the electron-donating antioxidants. Nitric oxide radical inhibition was 41 %. About 73 % of superoxide anion was inhibited, followed by 72 % of hydrogen peroxide radical-scavenging activity.

El Far and Taie (2009) found that the dominant DPPH radical-scavengers of Abees, the Egyptian orange-fleshed sweet potato cultivar, was due to the presence of anthocyanins and phenolic compounds rather than flavonoids, while in the genotype 199062.1 it was attributed flavonoids and in genotype 199004.2 was due to phenolic compounds. The DPPH and ABTS radical-scavenging activities of Egyptian sweet potato cultivars varied from 1.10 to 1.72 and 0.85 to 1.51 μmol trolox equivalent (TE)/g dw, respectively (Bellail et al. 2012). The reducing power between 0.1 and 0.25 mg chlorogenic acid equivalent (mg ChAE)/g dry weight basis (dw) and total phenolic content ranged from 0.53 to 0.87 mg ChAE/g dw. The most abundant individual phenolic acids in processed flesh roots tissues were chlorogenic acid followed by 3,5-dicaffeoylquinic acid. Total phenolic contents were highly correlated with RP, DPPH and ABTS, also the correlation between DPPH and ABTS values were significantly high. Thermal processing significantly increased the total phenolic content, as well as individual phenolic acids and antioxidant capacity of all the cultivars under study. Purple sweet potato cv. NCPuR02-020 was found to contain the highest levels of all phenolic components (Grace et al. 2014). A decrease in phenolic components was observed after curing and storage. Levels of carotenoids were significantly increased over curing and storage times. In contrast, antioxidant activity and ascorbic acid gradually decreased with storage. The DPPH scavenging activity of purple sweet potato wine was 58.95 % at a dose of 250 μg/mL (Ray et al. 2012).

The thioredoxin *h* protein from sweet potato storage roots at a concentration of 12.5 mg/mL

exhibited the highest activity (expressed as 0.37 mM ABTS* radical cation being cleared) in a total antioxidant status test (Huang et al. 2004b). In the DPPH staining thioredoxin h appeared as white spots when it was diluted to 50 mg/mL (a final amount of 15 μg). The reducing power, Fe^{2+}-chelating ability, FTC activity and protection against hydroxyl radical-induced calf thymus DNA damage were also found with the thioredoxin h protein. It was suggested that thioredoxin h might contribute to its antioxidant activities against hydroxyl and peroxyl radicals.

Antioxidant activities of the raw, boiled, steamed and fried sweet potato was 78.76, 89.67, 97.92 and 57.89 % (as DPPH radical inhibition percent), respectively (Tokusoglu and Yildirim 2012). With steaming process, radical inhibition percent increased 1.24-fold. Anthocyanin level of Turkish sweet potato cv Hatay Kirmizi (as cyanidin-3-glucoside (C3G) equivalent) was determined as 11,992 mg/100 g) (Tokusoglu and Yildirim 2012). Anthocyanins were detected as 13,767 mg/100 g; 24,756 mg/100 g; 6755 mg/100 g in boiled, steamed and fried sweet potatoes, respectively. The total anthocyanins increased 1.14-fold after boiling process and increased 3.22-fold after steaming process and decreased 1.78-fold after frying process. It was determined that steaming process was the most effective among the heat-treated sweet potatoes.

Dark green sweet potato leaves exhibited DPPH radical-scavenging activity with an EC_{50} of 4 nmol and polyphenol antioxidant index in free radical scavenging of 595.2 (Thu et al. 2004). Total polyphenol in the leaves was determined as 30.3 μmol catechin/g and free polyphenols as 23.6 μmol catechin/g. In the DPPH assay, it was found that sweet potato leaf ethanol extract had the highest radical-scavenging activity, followed by leaf vein water extract (Huang et al. 2004c). In the reducing power activity assay, it was found that the leaf water extract had the highest reducing power activity, followed by ethanol vein extract. The highest FTC (ferric thiocyanate) activity was found in the ethanol vein extract. Among all the extracts, the highest amount of total phenolic and flavonoid compounds was found in the ethanol vein extract. The highest TBARS values were obtained from root samples of sweet potato, and followed by stems and leaves, indicating that leaf sample showed the strongest antioxidant activity (Boo et al. 2005). Sweet potato leaves had a significantly higher phenolic content and antioxidant activity than roots (Padda and Picha 2007). Young, immature unfolded leaves had the highest total phenolic content (88.5 mg/g dry weight) and antioxidant activity (99.6 mg/g dry weight). Chlorogenic acid was the major phenolic acid in root and leaf tissues with the exception of young immature leaves in which the predominant phenolic acid was 3,5-dicaffeoylquinic acid. Sweet potato cultivars with yellow flesh and leaf part exhibited strong antioxidant activities. Studies showed that flavonoids of sweet potato vines had a strong scavenging effect on DPPH (Ding et al. 2010). Their scavenging activity on superoxide radicals was stronger than that of rutin, and the scavenging activity on hydroxyl radical was higher than that of rutin and vitamin C. Additionally, they exhibited strong antioxidant activities in the linoleic acid model system. Of six sweet potato varieties, leaves of the Indon variety showed the highest level of total phenolic contents at 5.35 g GAE/100 g DW (Hue et al. 2012). The flavonoid contents in the leaves ranged from 96±47.6 μg/g in Indon variety to 263.5±43.5 μg/g in Batu Biasa variety. In DPPH radical scavenging activity in leaves, the Indon and Biru Putih variety had the highest and lowest scavenging activities of 372.4 μg/mL (IC_{50}) and 597.61 μg/mL (IC_{50}), respectively. All varieties, except Biru Putih, showed high radical scavenging activity compared to the ascorbic acid standard. Besides, all the leaf varieties also showed increment in their reducing power with increasing concentrations. The total antioxidant capacity of sweet potato leaves was 42.94 % as compared to ascorbic acid. Purple-leaved sweet potato, an indigenous Taiwanese vegetable, was found to have higher antioxidant activity than green sweet potato leaves, which could be attributable to its higher contents of anthocyanidin content 275.60 unit/g DW, polyphenols 22.80 mg GAE (gallic acid equivalent)/g DW, 67.66 mg Que (quercetin)/g DW, flavonols 23.82 mg Que/g DW. In contrast,

green leaves had anthocyanidin content not detected, polyphenols 18.70 mg GAE/g DW, 42.45 mg Que/g DW, flavonols 7.80 mg Que/g DW. (Chao et al. 2014). The IC_{50} value for DPPH scavenging activity of purple and green sweet potato leaves were determined respectively as 803.13, 74.67 μg/mL; TEAC value in the leaf methanolic hydrolysate were 162.66; 92.76 μmol Trolox/g DW, TEAC value in the ethanolic hydrolysate were 25.58; 10.66 μmol Trolox/g DW). The ORAC antioxidant activity in the methanolic leaf hydrolysate were hydrophilic ORAC value 1174.98; 786.26 μmol Trolox/g DW and lipophilic ORAC value 92.43; 143.65 μmol Trolox/g DW.

The inhibition of LDL-oxidation of the sweet potato leaf extracts was correlated with the total amounts of caffeic acid derivatives, suggesting that sweet potato leaves may prevent atherosclerosis via reducing early atherogenesis (Taira and Ohmine 2011). caffeic acid derivatives, 4,5-dicaffeoylquinic acid (4,5-diCQA), 3,5-diCQA, 3,4-diCQA and 3,4,5-triCQA were the main polyphenol constituents of sweet potato leaf and total amounts of caffeic acid derivatives correlated with the polyphenol contents ($R^2=0.94$). All the caffeoylquinic acid (CQA) derivatives from sweet potato leaves indicated an anti-LDL (low-density lipoprotein) oxidation activity at low concentrations (1–5 μM) and particularly, the activity of 3,4,5-triCQA was remarkably higher than those of 5-CQA and diCQA, such as 4,5-diCQA, 3,5-diCQA and 3,4-diCQA (Taira et al. 2013). The antioxidant activity of sweet potato leaves was correlated with the amounts of CQA derivatives in the range of $R^2=0.69$–0.75. The results suggested that sweet potato leaves may prevent that developing atherosclerosis causes the oxidative modification of LDL. DPPH scavenging activity % in *n*-hexane leaf extracts of sweet potato cultivars with varying tuber colours were determined as: B (purple tuber cv) 23.35 %, C (yellow tuber cv) 23.18 %, E (orange tuber cv) 19.38 %, A (red-purple tuber cv 8.42 %, D (red-yellow tuber cv) 7.73 % (Fidrianny et al. 2013). The DPPH scavenging activity % in ethyl acetate leaf extracts were: E 93.26 %, D 91.47 %, C 90.01 %, A 77.69 %, B 65.47 %. The DPPH scavenging activity % in ethanol leaf extracts were: A 97.63 %, B 93.94 %, E 56.68 %, D 50.04 %, C 34.28 %. The highest antioxidant activity 97.63 % was given by the ethanolic extract of red-purple tuber cv. The ethyl acetate extract of B contained the highest total flavonoid (59.79 g QE/100 g). The ethanolic extract from B had the highest phenolic contents (19.64 g GAE/100 g), while the highest carotenoid 24.17 g BET/100 g was given by the ethyl acetate extract of C. The total phenolic contents were significantly correlated with DPPH scavenging activity in A (leaves of red-purple tubers) with $R^2=0.951$ and sample B (leaves of purples tubers) with $R^2=0.792$, but no correlation in sample C, D and E. The DPPH scavenging activity in sample A was negatively correlated with total flavonoid contents ($R^2=-0.772$), while sample B, C, D and E had no correlation. The total carotenoid content in sample C (leaves of yellow tubers) had correlation with DPPH scavenging activity ($R^2=0.778$) and no correlation in sample A, B, D and E.

The 33 kDa trypsin inhibitor, root storage protein, purified from sweet potato root, exhibited scavenging activity against 1,1-diphenyl-2-picrylhydrazyl (DPPH) radical (Hou et al. 2001). There was positive correlation between scavenging effects against DPPH (2 to 22 %) and amounts of 33 kDa TI (1.92 to 46 pmol). The sporamin B protein from sweet potato at a concentration of 100 μg/mL exhibited highest activity (expressed as 4.21 mM Trolox equivalent antioxidative value, TEAC) in total antioxidant status test (Huang et al. 2007a). In the DPPH staining sporamin B appeared as a white spot when the concentration was diluted to 25 mg sporamin B/mL (with an absolute amount of 75 μg). Like total antioxidant status, the reducing power, Fe^{2+} -chelating ability, FTC activity and protection calf thymus DNA against hydroxyl radical-induced damage all showed that sporamin B polypeptides had significant antioxidant activities. It was found that antioxidant activities of sporamin B increased from 19 % (0 h) to about 29 % (24 h) after 24 h hydrolysis by pepsin.

The cyclophilin-type peptidylprolyl isomerase protein (SPPPI) isolated from sweet potato storage roots and CP (calf thymus cytophilin, a

positive control) displayed the highest ABTS (2,2-azino-bis-(3-ethylbenzothiazoline-6-sulfonic acid) scavenging ability (15.36 and 17.79 %, respectively) at 100 μg/mL (Liao et al. 2012). In the DPPH assay, SPPPI and CP were found to have the highest radical-scavenging activity (5.78 and 4.05 %, respectively) at 100 μg/mL. The Fe^{2+}-chelating ability of SPPPI and CP was found to be the highest (12.47 and 14.576 %) at 100 μg/mL, respectively. The results suggested SPPPI to be an excellent candidate as a lead compound for the development of reductant agent. Sweet potato protein (SPP) hydrolysates prepared by six enzymes (alcalase, proleather FG- F, AS1.398, neutrase, papain and pepsin) exhibited antioxidant activity and protective effects on oxidative DNA damage (Zhang et al. 2012). Alcalase hydrolysates exhibited the highest hydroxyl radical-scavenging activity (IC_{50} 1.74 mg/mL) and Fe^{2+}-chelating ability (IC_{50} 1.54 mg/L). Compared with other five hydrolysates, the alcalase hydrolysates had the most abundant <3 kDa fractions. In addition, below 3 kDa fractions of alcalase hydrolysates showed the highest antioxidant activities and protective effects against DNA damage through both scavenging hydroxyl radicals and chelating Fe^{2+}, which was probably because of the increase in several antioxidant amino acids, such as histidine, methionine, cystein, tyrosine and phenylalanine, as well as the hydrophobic amino acids.

Sweet potato defensin (SPD1), a cystein-rich protein, was found to decrease the production of intracellular peroxide in HepG2 cells (Huang et al. 2012). Four of its peptides, namely GFR, GPCSR, CFCTKPC and MCESASSK, synthesized by hydrolysis was also tested for antioxidant activity. In the TEAC assay CFCTKPC performed the best (13.5TE/g dw), even better than reduced glutathione (7.3 μmol TE/g dw). In the DPPH radical assay, CFCTKPC again had the highest antioxidant activity (IC_{50} = 11.3 μM) even better than reduced glutathione (IC_{50} = 74.3 = μM). In the lipid peroxidation assay, once again CFCTKPC performed the best, with an IC_{50} value of 0.5 μM better than reduced glutathione (1.2 μM). The findings demonstrated that SPD1 might contribute various antioxidant properties through its hydrolytic peptides.

Clinical Studies

In a randomized, cross-over clinical study included 16 healthy adults (7 M, 9 F; aged 20–22 years), 2 weeks consumption of polyphenol-rich purple sweet potato leaves enhanced urinary total phenol excretion by 24.5 % at day 14 as compared to day 0, while the low polyphenol diet (LPD) decreased total phenol content in plasma and urine by 3.3 and 16.3 %, respectively (Chen et al. 2008a). Low-density lipoprotein lag time and glutathione concentration in erythrocytes at day 14 was significantly enhanced by 15.0 and 33.3 % by PSPL as compared to day 0, respectively. Urinary 8-hydoxy-deoxyguanosine (8-OHdG) excretion decreased significantly by PSPL consumption. The results suggested that polyphenols in 200 g purple sweet potato leaves were bioavailable and could enhance antioxidant defense and decrease oxidative stress in young healthy people.

Consumption of purple sweet potato leaves (PSPL) for 2 weeks by basketball players led to a significant increase of plasma polyphenol concentration and vitamin E and C levels (Chang et al. 2007). Low-density lipoprotein (LDL) lag time was significantly longer in the PSPLs group. A significant decrease of urinary 8-hydroxy-2-deoxyguanosine (8-OHdG) was noted; however, there was no significant change in plasma glutathione (GSH), total antioxidant status (TAS) and malondialdehyde + 4-hydroxy-2(E)-nonenal level after consuming the PSPLs diet. It was concluded that consumption of PSPLs diet for 2 weeks may reduce lipid and DNA oxidation and modulate the antioxidative status of basketball players during training period.

Anticancer Activity

In-Vitro Studies

Anthocyanin-rich aqueous extracts from cell suspension cultures of a high anthocyanin-producing sweet potato purple line grown under two differ-

ent media conditions, MM (multiplication medium) and APM (high anthocyanin-producing medium) exhibited higher radical-scavenging activities, 3.8- and 1.4-fold, respectively, than the field grown sweet potato storage root (SR) (Konczak-Islam et al. 2003). The antimutagenic activity of all extracts was found to be dose-dependent. At a dose of 1 mg/plate, the highest activity exhibited APM (73 % inhibition of Trp-P-1-induced reverse mutation of *Salmonella typhimurium* TA98), followed by MM (54 % inhibition) and SR (36 % inhibition). MM extract was the strongest inhibitor of the proliferation of human promyelocytic leukaemia HL-60 cells. At a concentration of 1.6 mg/mL medium during 24 h, it suppressed the growth of 47 % of HL-60 cells. A significantly lower growth suppression effect displayed APM and SR extracts (21 and 25 %, respectively). The MM extract, which exhibited the highest RSA and antiproliferation activities, contained the highest level of anthocyanins. Among them, nonacylated cyanidin 3-sophoroside-5-glucoside predominated. Sweet potato extract caused marked dose-dependent growth inhibition in several human colon carcinoma cell lines with IC_{50} values in the range of 20–50 μg/mL for HCT 116, SW480, HT29 and SW837 cell lines (Kaneshiro et al. 2005). However, the IC_{50} value was more than 100 μg/mL when CaCo2 cells were tested.

The water leaf vein extract of sweet potato had the highest antiproliferative activity in-vitro against human lymphoma NB4 cells with an EC_{50} of 449.6 μg/mL, followed by water extract of storage root, water extract of leaf, ethanol extract of storage root and ethanol extract of leaf (Huang et al. 2004c). Although the ethanol extract of vein showed strong antioxidant activity, it exhibited no antiproliferative activity under the experimental conditions tested. Studies showed that trypsin inhibitor (TI) from sweet potato tubers inhibited cellular growth of NB4 promyelocytic leukaemia cells in a time-dependent and dose-dependent manner, and treatment for 72 h induced a marked inhibition of cellular growth, showing an IC_{50} of 57.1 μg/mL (Huang et al. 2007b). TI caused cell cycle arrest at the G1 phase and induced apoptosis in NB4 cells through a mitochondria-dependent pathway, which was associated with the activation of caspase-3 and -8.

Caffeic acid, chlorogenic acid, 3,4-di-*O*-caffeoylquinic acid, 3,5-di-*O*-caffeoylquinic acid, 4,5-di-*O*-caffeoylquinic acid and 3,4,5-tri-*O*-caffeoylquinic acid, isolated from sweet potato leaves, dose-dependently depressed cancer cell proliferation, and the difference in sensitivity between caffeoylquinic acid derivatives and each kind of cancer cell was observed (Kurata et al. 2007). Specifically, 3,4,5-tri-*O*-caffeoylquinic acid effectively depressed the growth of human stomach cancer (Kato III), colon cancer (DLD-1) and a promyelocytic leukaemia cell (HL-60), and caffeic acid had an exceptionally higher effect against HL-60 cells than other di- and tricaffeoylquinic acids. Growth suppression of HL-60 cells by 3,4,5-tri-*O*-caffeoylquinic acid was determined to be the result of apoptotic death of the cells. IbACP (*Ipomoea batatas* anti-cancer peptide), isolated from sweet potato leaves, was found to dose-dependently inhibit Panc-1, a pancreatic cancer line, cell proliferation and induced cell death by apoptosis (Chang et al. 2013).

Batatosides L and O showed a weak inhibitory effect on the growth of Hep-2 cells, while the other batatosides proved to be inactive (Yin et al. 2009). The extract from baked sweet potato (cv. Koganesengan) showed potential cancer-preventing effects (Rabah et al. 2004). Fractions II-a and III suppressed strongly the proliferation of human myelocytic leukaemia HL-60 cells with apoptosis induction in a dose-dependent manner. Moreover, the two fractions markedly blocked TPA-induced cell transformation in the mouse skin JB6 cell line. Both fractions showed markedly strong radical scavenging effects on the DPPH radical, coinciding with the high content of total phenolic compounds in the fractions. Sporamin, the major soluble protein with a kunitz-type trypsin inhibitory activity, from sweet potato tuber, exhibited antiproliferative effects of human tongue cancer Tca8113 cells partly by induction of apoptosis by downregulating Akt/GSK-3 signaling pathway (Yao and Qian 2011). Of four polysaccharide components of purple sweet potato named as PPSP, PPSPII, PPSPIII and PPSPIV, PPSP II and PPSPIII inhib-

ited Hela and HepG2 tumour cells (Zhao et al. 2011). Treatment of human colonic SW480 cancer cells with sweet potato P40 anthocyanin-rich extracts at 0–40 μM of peonidin-3-glucoside equivalent resulted in a dose-dependent decrease in cell number due to cytostatic arrest of cell cycle at G1 phase but not cytotoxicity (Lim et al 2013). Further, dietary P40 at 10–30 % significantly suppressed azoxymethane-induced formation of aberrant crypt foci in the colons of CF-1 mice partially in conjunction with a lesser proliferative PCNA (proliferating cell nuclear antigen) and a greater apoptotic caspase-3 expression in the colon mucosal epithelial cells. Purified sweet potato root protein (SPP) exerted significant antiproliferative and antimetastatic effects on human colorectal cancer cell lines, both in-vitro and in-vivo (Li et al. 2013b). SPP inhibited the proliferation of human colorectal cancer SW480 cells in a dose-dependent manner in-vitro, with an IC_{50} value of 38.732 μmol/L. Both intraperitoneal (ip) and intragastric (ig) administration of SPP significantly suppressed growth of intraperitoneally inoculated human colorectal cancer HCT-8 cells in nude mice to 58.0 and 43.5 % of the controls, respectively, after 9 days treatment. Ig and ip administration of SPP markedly induced a significant decrease in spontaneous pulmonary metastatic nodule formation in C57 BL/6 mice (21.0 and 27.3 nodules/lung vs 42.5 5 nodules/lung in controls, respectively), after 25 days treatment. Moreover, the average weight of primary tumour nodules in the hind leg of mice decreased from 8.2 g/mice in the control to 6.1 g/mice in the ip group.

Sweet potato greens extract (SPGE) rich in polyphenols exerted significant antiproliferative activity in a panel of prostate cancer cell lines while sparing normal prostate epithelial cells (Karna et al. 2011). Mechanistically, SPGE disrupted cell cycle progression, reduced clonogenic survival, modulated cell cycle and apoptosis regulatory molecules and induced apoptosis in human prostate cancer PC-3 cells both in-vitro and in-vivo. Oral administration of 400 mg/kg SPGE markedly inhibited growth and progression of prostate tumour xenografts by ~69 % in nude mice, as shown by tumour volume measurements and non-invasive real-time bioluminescent imaging. SPGE did not cause any detectable toxicity to rapidly dividing normal tissues such as gut and bone marrow. In another study, a remarkably active polyphenol-enriched fraction, F5, of sweet potato greens extract was found to be ~100-fold more potent in anticancerous activity than the parent extract as shown by IC_{50} measurements in human prostate cancer cells (Gundala et al. 2013). F5 fraction was found to be rich in quinic acid (QA), caffeic acid, its ester chlorogenic acid and isochlorogenic acids, 4,5-di-CQA, 3,5-di-CQA and 3,4-di-CQA, especially in QA and chlorogenic acid. Sub-fractionation of F5 resulted in loss of bioactivity, suggesting synergistic interactions among the constituent phytochemicals. Daily oral administration of 400 mg/kg body wt of F5 inhibited growth and progression of prostate tumour xenografts by ~75 % in nude mice.

Studies found that inclusion of natural food anthocyanins, purple sweet potato colour and red cabbage colour (5 %) to the rat diet could reduce 2-amino-1-methyl-6-phenylimidazo[4,5-b]pyridine (PhIP)-associated colorectal carcinogenesis in rats initiated with 1,2-dimethylhydrazine (Hagiwara et al. 2002). Liu et al. (2008) found that sweet potato anthocyanins could have inhibitory effect on transplantation tumour of mice, and had no obvious toxicity and mutagenicity. At doses of 150 mg and 75 mg, the inhibition rates of mice carcinoma 180 were 45.04 % and 36.64 %, respectively, while the inhibition rate of mice liver cancer H22 was 33.33 % at 150 mg dose.

Antiviral/Antimicrobial Activity

Caffeoylquinic acid derivatives, 3,4,5-tri-*O*-caffeoylquinic acid exhibited a greater selective inhibition of HIV replication than 4,5-di-*O*-caffeoylquinic acid (Mahmood et al. 1993).

Purple sweetpotato extract at the tested concentration caused 40 % inhibition on the growth of *Salmonella enteritidis*, but no inhibition against *Escherichia coli* (Cevallos-Casals and Cisneros-Zevallos 2002).

Antimutagenic Activity

Analysis of anthocyanin in sweet potato tubers revealed a large distribution of anthocyanin pigments in the outer portion and showed that the content of cyanidin in the outer portion was higher than in the inner tissues (Yoshimoto et al. (1999a). The strong antimutagenicity of the purple-fleshed variety Ayamurasaki outer portion was attributed chiefly to the high concentration of cyanidin. The extracts from the outer portions of Koganesengan, Sunny Red and Joy White varieties showed an antimutagenic activity unlike the inner ones, suggesting that the antimutagenic component in the outer portions of these varieties was mainly associated with phenolics. Aqueous extract from the whole roots of the purple-coloured sweet potato Ayamurasaki variety effectively decreased the reverse mutation induced not only by Trp-P-1, Trp-P-2, IQ, B[a]P and 4-NQO but also by dimethyl sulphoxide extracts of grilled beef on *Salmonella typhimurium* TA 98 (Yoshimoto et al. 1999b). Two anthocyanin pigments purified from purple-coloured sweet-potato extract, 3-(6,6′-caffeylferulylsophoroside)-5-glucoside of cyanidin (YGM-3) and peonidin (YGM-6) effectively inhibited the reverse mutation induced by heterocyclic amines, Trp-P-1, Trp-P-2 and IQ in the presence of rat liver microsomal activation systems.

Anthocyanins purified from purple-fleshed sweet potato 3-sophoroside-5-glucoside of cyanidin and 3-sophoroside-5-glucoside of peonidin, the anthocyanin derivatives deacylated from the 3-(6,6′-caffeylferulylsophoroside)-5-glucoside of cyanidin (YGM-3) and 3-(6,6′-caffeylferulylsophoroside)-5-glucoside of peonidin (YGM-6) exhibited antimutagenicity activity using *Salmonella typhimurium* TA 98 (Yoshimoto et al 2001). A comparison of the antimutagenicity between YGM-3 and YGM-6 and the deacylated derivatives showed that the activity of cyanidin was stronger than that of peonidin. Deacylation of the peonidin-type pigment markedly decreased this antimutagenicity. Caffeic acid showed the strongest antimutagenicity of the constituent organic acids of the anthocyanin pigments, caffeic acid, ferulic acid and *p*-hydroxybenzoic acid. The results suggested that the catechol structure played an important role in the potent antimutagenicity of anthocyanin pigments. The caffeoylquinic acid derivatives, 3-mono-*O*-caffeoylquinic acid (chlorogenic acid, ChA), 3,4-di-*O*-caffeoylquinic acid (3,4-diCQA), 3,5-di-*O*-caffeoylquinic acid (3,5-diCQA), 4,5-di-*O*-caffeoylquinic acid (4,5-diCQA) and 3,4,5-tri-*O*-caffeoylquinic acid (3,4,5-triCQA) and caffeic acid (CA) isolated from the sweet potato leaf effectively inhibited the reverse mutation induced by Trp-P-1 on *Salmonella typhimurium* TA 98 (Yoshimoto et al. 2002). The antimutagenicity of these derivatives was 3,4,5-triCQA > 3,4-diCQA = 3,5-diCQA = 4,5-diCQA > ChA in this order. It was found that the number of caffeoyl groups bound to quinic acid played a role in the antimutagenicity of the caffeoylquinic acid derivatives. The caffeoylquinic acid derivatives isolated from sweet potato leaves showed high DPPH radical-scavenging activity (%), and high antimutagenicity by effectively inhibiting the reverse mutation induced by Trp-P-1 on *Salmonella typhimurium* TA 98 (Islam et al. 2003b).

Antidiabetic Activity

In-Vitro Studies

Sweet potato acylated anthocyanin (YGM) extract exhibited potent maltase inhibitory activity with IC_{50} of 0.36 mg/mL (Matsui et al. 2001). It also inhibited α-amylase action, indicating that anthocyanins would have a potential function to suppress the increase in post-prandial glucose level from starch.

Animal Studies

White-skinned sweet potato cortex (WSSP-cortex) was found to have potential hypoglycaemic activity in streptozotocin (STZ)-induced diabetic rats and hereditary diabetic mice (Type II, KK-A^Y/Ta Jcl, C57BL/KsJ db/db) by oral administration (Kusano et al. 1998). However, in normal rats, serum glucose levels were not changed by WSSP-cortex treatment. In glucose tolerance tests, treatment of normal and STZ dia-

betic rats by WSSP-cortex showed increased glucose tolerance and increased serum insulin levels like the tolbutamide treatment, whereas the treatment of KK-A^Y and db/db mice showed increased glucose tolerance and decreased serum insulin levels different from the tolbutamide treatment in response to oral glucose load. Results of further studies suggested that WSSP showed remarkable antidiabetic activity and ameliorated the abnormality of glucose and lipid metabolism by reducing insulin resistance (Kusano and Abe 2000). Hyperinsulinaemia in Zucker fatty rats was reduced by 23, 26, 60 and 50 %, respectively, 3, 4, 6 and 8 weeks after starting the oral administration of WSSP. Similar results were obtained with troglitazone. In the glucose tolerance test after 7 weeks of treatment, increases in blood glucose levels after glucose loading were inhibited by the administration of WSSP. Glucose tolerance was also improved. Blood triacylglyceride (TG) and free fatty acid (FFA) lactate levels were lowered by the oral administration of WSSP. Similar effects on blood insulin, lipid and lactate levels were observed after the administration of troglitazone. Body weight gain increased in the troglitazone group, but not in the WSSP group, compared to the control group. In histological examinations of the pancreas of Zucker fatty rats, remarkable re-granulation of pancreatic islet B-cells was observed in the WSSP and troglitazone groups after 8 weeks of treatment. The antidiabetic component of WSSP was found to be an acidic glycoprotein with molecular weight of 22,000 (Kusano et al. 2001). A single oral administration of sweet potato tuber diacylated anthocyanin peonidin 3-*O*-[2-*O*-(6-*O*-*E*-feruloyl-*β*-D-glucopyranosyl)-6-*O*-*E*-caffeoyl-*β*--D-glucopyranoside]-5-*O*-*β*-D-glucopyranoside exerted a potent maltase inhibitory activity with an IC_{50} value of 200 μM over sucrase inhibition in male Sprague–Dawley rats (Matsui et al. 2002). When the diacylated anthocyanin (100 mg/kg) was administered following maltose (2 g/kg), a maximal blood glucose level (BGL) at 30 min was significantly decreased by 16.5 % compared to vehicle. A minimum 10 mg/kg dose of the anthocyanin was necessary for the suppression of glycaemic rise, and the ED_{20} (69 mg/kg) was estimated to be approximately 30-fold lower than that of the therapeutic drug acarbose (ED_{20} = 2.2 mg/kg). A reduction of serum insulin secretion was also observed corresponding to the decrease in blood glucose level (BGL). No significant change in BGL was observed when sucrose or glucose was ingested, suggesting that the antihyperglycaemic effect of the anthocyanin was achieved by maltase inhibition, not by sucrase or glucose transport inhibition at the intestinal membrane. After 6 weeks feeding started, blood glucose and insulin levels in GK rats fed the 5 % sweet potato leaf (SPL) diet group were significantly reduced compared with those of the control diet group (Ishida et al. 2004). Effectiveness of SPL on modulation of glucose metabolism in the GK rats was accompanied by a significant reduction of blood cholesterol, especially LDL cholesterol. A soluble mucilaginous, dietary fibre (SDF) was extracted from the sweet potato leaf powder. An elevation of post-prandial blood glucose level in rats given 5.0 ml of 20 % glucose solution containing 1 % SPL-SDF orally was significantly suppressed as compared with that in the control rats given 20 % glucose solution only. The results suggested that the main effective substance of SPL for suppressing blood glucose elevation was a kind of mucilage SDF.

Treatment with sweet potato leaf flavones extract for 2 weeks resulted in a significant decrease in the concentration of plasma triglyceride (TG), plasma cholesterol (TC) and weight in non-insulin-dependent diabetes mellitus (NIDDM) rats (Zhao et al. 2006). Further, the extract markedly decreased fasting plasma insulin level, blood glucose (FBG) level, low-density lipoprotein cholesterol (LDL-C) and malondialdehyde (MDA) levels and significantly increased the Insulin Sensitive Index (ISI) and superoxide dismutase (SOD) level in NIDDM rats. The results suggested that sweet potato leaf flavone extract could control blood glucose and modulate the metabolism of glucose and blood lipid, and decrease outputs of lipid peroxidation and scavenge the free radicals in non-insulin-dependent diabetic rats. Li et al. (2009) demonstrated that sweet potato leaf flavonoids at a dose of

100 mg/kg bw exhibited the optimal antidiabetic effect on alloxan-induced diabetic mice. The leaf flavonoid treatment for 28 days resulted in a significant decrease in the concentration of fasting blood glucose, total cholesterol and triglyceride in diabetes mellitus mice and significantly increased body weight and serum high-density lipoprotein cholesterol. Studies by Ijaola et al. (2014) showed that oral treatment with 300 mg/kg/day of sweet potato leaf extract for 2 weeks produced the best hypoglycaemic effect (69.67 %) in alloxan-induced diabetic rats. The highest percentage blood sugar reductions of 69.67 % was recorded in rats treated with 300 mg/kg/day, followed by 59.24 % (400 mg/kg/day extract) while the least percentage sugar reduction of 52.18 % was observed in 200 mg/kg/day extract. The non-diabetic-induced rats exhibited steady increase (8.03 %) in their normal glucose level. It was revealed that alloxan induced rats treated with 200, 300 and 400 mg/kg/day sustained percentage weight loss of 48.91, 28.66 and 31.11 %, respectively, compared with non-diabetic-induced rats.

Administration of an arabinogalactan-protein (WSSP-AGP) from white-skinned sweet potato to KKAy mice significantly lowered fasting plasma glucose levels (Ozaki et al. 2010). This indicated that WSSP-AGP played an important role in the hypoglycaemic effects of white-skinned sweet potato. Treatment of hyperglycaemic db/db mice with sweet potato arabinogalactan protein (WSSP-AGP) decreased plasma glucose levels and ameliorated insulin resistance leading to hypoglycaemic effects (Oki et al. 2011). Feeding streptozotocin-induced diabetic rats with purple sweet potato flavonoids ameliorated diabetic symptoms (Jiang et al. 2011). Fasting blood glucose (FBG), GSP (glycated serum protein), total cholesterol, triglycerides (TG), LDL-C were decreased and serum HDL-C levels were increased in high-, medium-dose groups; while FBG, serum GSP, TG, LDL-C were also improved in low dose group. In another study, administration of dietary sweet potato leaf extract powder (SP) for 5 weeks significantly lowered hyperglycaemia in type 2 diabetic mice (Nagamine et al. 2014). Also, pre-administration of SP significantly stimulated glucagon-like peptide-1 (GLP-1) secretion and was accompanied by enhanced insulin secretion in rats, which resulted in a reduced glycaemic response after glucose injection. In in-vitro studies, it was found that SP and its caffeoylquinic acid derivatives significantly enhanced GLP-1 secretion. Studies by Chen et al. (2013) found that sweet potato starch (low GI) feeding for 4 weeks could improve insulin sensitivity in insulin-resistant rats, possibly by improving the adipocytokine levels, pro-inflammatory status and insulin signalling.

Clinical Studies

Ludvik et al. (2002) demonstrated in a randomized study that ingestion of 4 g *Ipomoea batatas* (Caiapo)/day for 6 weeks reduced fasting blood glucose and total as well as LDL cholesterol in male Caucasian type 2 diabetic patients previously treated by diet alone. The improvement of insulin sensitivity in the fasting serum intravenous glucose tolerance test FSIGT indicated that caiapo exerts its beneficial effects via reducing insulin resistance. The treatment was well tolerated, with no apparent side effects. After treatment with Caiapo, glycated haemoglobin HbA_{1c} decreased significantly from 7.21 to 6.68 %, whereas it remained unchanged in subjects given placebo (7.04 vs. 7.10 %). Fasting blood glucose levels decreased in the Caiapo group (143.7 vs. 128.5 mg/dL) and did not change in the placebo group (144.3 vs. 138.2 mg/dL). Mean cholesterol at the end of the treatment was significantly lower in the Caiapo group (214.6 mg/dL) than in the placebo group. A decrease in body weight was observed in both the placebo group and in the Caiapo group, probably due to a better-controlled lifestyle. In another study, they found that short-term treatment with 4 g/day of the nutraceutical Caiapo consistently improved metabolic control in type 2 diabetic patients by decreasing insulin resistance without affecting body weight, glucose effectiveness or insulin dynamics (Ludvik et al. 2003). No side effects related to the treatment were observed. They also confirmed that the long-term treatment of Caiapo had beneficial on glucose control as evidenced by the observed

decrease in HbA_{1c} in type 2 diabetic subjects (Ludvik et al. 2004).

Matured sweet potato tubers were cooked by roasting, baking, frying or boiling then immediately consumed by the ten non-diabetic test subjects (5 males and 5 females; mean age of 27 ± 2 years) (Bahado-Singh et al. 2011). Samples prepared by boiling had the lowest GI (41–50), while those processed by baking (82–94) and roasting (79–93) had the highest GI values. The study indicated that the glycaemic index of Jamaican sweet potatoes varied significantly with the method of preparation and to a lesser extent on intravarietal differences. Consumption of boiled sweet potatoes could minimize post-prandial blood glucose spikes and therefore, may prove to be more efficacious in the management of type 2 diabetes mellitus. In another study, fasted participants were measured blood glucose levels at 0, 30, 60, 90 and 120 min after consuming 25 g of available carbohydrate from 'Beauregard' sweet potato skin and flesh separately that were subjected to conventional cooking methods: baking at 163 °C for 1 h; microwaving for 5 min in a 1000 W microwave; dehydrating at 60 °C for 16 h; and steaming at 100 °C for 45 min (Allen et al. 2012). Glycaemic indices calculated from these methods for steamed, baked and microwaved sweet potato flesh were 63, 64 and 66, respectively, indicative of a moderate glycaemic index food. However, dehydrated and raw sweet potato flesh had a low glycaemic index (41 and 32, respectively). Steamed skin and baked skin, and raw flesh also had a low glycaemic index of 30, 34 and 19, respectively. A second experiment confirmed the low glycaemic index of raw sweet potato, especially the skin, and showed that a commercial extract of the sweet potato cortex, Caiapo, tended to lower the glycaemic index of white potato to a level that was not different from the raw sweet potato peel. The physiological mechanism for the lower glycaemic index was not due to a greater release or a greater clearance of insulin during the glycaemic response. Depending on cooking methods, 'Beauregard' sweet potato flesh and skin may be considered low and medium glycaemic index foods, which may prove beneficial for diabetic- or insulin-resistant consumers.

Reviews/Meta-analyses

Suksomboon et al. (2011) conducted a meta-analysis of the effect of herbal supplement on glycaemic control in type 2 diabetes and nine randomized trials (487 patients) that met the criteria of (1) randomized placebo-controlled trial of single herb aimed at assessing glycaemic control in type 2 diabetes, (2) of at least 8 weeks duration and (3) reporting HbA(1c). They found that supplementation with *Ipomoea batatas, Silybum marianum* and *Trigonella foenum-graecum* significantly improved glycaemic control, whereas *Cinnamomum cassia* did not. Ooi and Loke (2013) reviewed three randomized controlled trials ranged from 6 weeks to 5 months in duration and involving 140 participants in regards to the efficacy of sweet potato vs. placebo intervention for type 2 diabetes mellitus. There was a statistically significant improvement in glycosylated haemoglobin A1c (HbA1c) at 3–5 months with 4 g/day sweet potato preparation compared to placebo. However, it was concluded that there was insufficient evidence about the use of sweet potato for type 2 diabetes mellitus.

Antihyperlipidemic/ Antiatherosclerotic Activity

In-Vitro Studies

The purple sweet potato (PSP) ethanol extract (100-fold diluted) showed stronger (up to a six-fold higher) DPPH radical-scavenging activity than the water PSP extract and the ethanol extract of yellow sweet potato (Park et al. 2010). The PSP ethanol extract also exhibited the highest increase in ferric-reducing ability among all extracts. Cupric ion-mediated LDL oxidation was strongly inhibited by the PSP ethanol extract with similar potency to vitamin C treatment (final concentration, 10 mM). The PSP extract strongly inhibited fructose-mediated protein glycation and also inhibited the uptake of oxidized LDL into human macrophage cells with suppression of

malondialdehyde production in the cell culture medium. The data suggested that PSP extract could be used as a putative antiatherosclerotic and antidiabetic agent with strong antioxidant functions. Purple sweet potato extract exerted antiobesity, antioxidative and anti-inflammatory in 3 T3-L1 adipocytes in-vitro (Ju et al. 2011). It diminished leptin secretion, suppressed the expression of mRNAs of lipogenic and inflammatory factors and promoted lipolytic action. It exhibited DPPH radical-scavenging and ferric-reducing activity.

Animal Studies

Studies found feeding that male Sprague–Dawley rats for about 1 month feeding cholesterol-free diet containing dried powder of sweet potato leaves at 5 % level as dietary fibre significantly decreased hepatic cholesterol level (Innami et al. 1998). A significant increase in faecal weight was observed in the group fed the green leaf samples.

All the dried green leaves increased faecal excretion of bile acids per gram or per day compared with the control group. The results suggest that lowering of hepatic cholesterol by powdered green leaves was not necessarily due to the same factor, but to the increased faecal excretion of bile acids due to inhibited enterohepatic circulation in animals given the sample. Johnson et al. (2013) demonstrated that consumption of novel green leafy vegetables like sweet potato leaves improved liver fatty acid profiles of spontaneously hypertensive rats (SHRs) and protected against elevations in atherogenic fatty acids, which may be involved in cardiovascular disease pathogenesis. SHRs consuming diets containing such vegetables had significantly greater liver concentrations of γ-linolenic, docosahexaenoic and docosahexaenoic acids, as well as lower levels of lauric, palmitic and arachidonic acids.

Purple sweet potato anthocyanin fraction (200 mg/kg per day) reduced weight gain and hepatic triglyceride accumulation and improved serum lipid parameters in mice fed an high-fat diet (HFD) for 4 weeks (Hwang et al. 2011b). The fraction attenuated hepatic lipid accumulation through activating adenosine monophosphate-activated protein kinase signalling pathways in human HepG2 cells and obese mice. Chen et al. (2011) found that feeding hyperlipidaemic rats for 6 weeks with purple sweet potato could decrease serum lipids (TC, TG and LDL-C) and reduce hepatic oxidative stress. Serum SOD was significantly higher in high- and low-dosage group than in high-fat control group, whereas serum MDA was significantly lower than that in high-fat control group.

Consumption of purple sweet potato root affected post-translational modification of plasma proteins in male Syrian hamsters (Liao et al. 2013). The results indicated that 95 plasma proteins were identified and 28 post-translational modifications sites on 26 of these 95 proteins were affected by consumption of purple sweet potato. Methylation accounted for the largest percentage of affected modifications (35.71 %). Also, incorporation of purple sweet potato into the diet significantly lowered blood and liver lipids. Consumption of purple sweet potato root affected post-translational modification of plasma proteins in male Syrian hamsters (Liao et al. 2013). The results indicated that 95 plasma proteins were identified and 28 post-translational modifications sites on 26 of these 95 proteins were affected by consumption of purple sweet potato. Methylation accounted for the largest percentage of affected modifications (35.71 %). Also, incorporation of purple sweet potato into the diet significantly lowered blood and liver lipids.

Studies showed that rats administered sweet potato aqueous extract showed significant reduction in food intake, blood glucose level and body weight when compared with the control group (Olubobokun et al. 2013). It was suggested that consumption of sweet potato caused a reduction in food intake probably by increasing satiety and reduction in weight gain by using up the body's reserve of fat as a result of the low blood glucose.

APSP (anthocyanins from purple sweet potato) protected low-density lipoprotein against oxidation more potently than other anthocyanins and *L*-ascorbic acid in-vitro (Miyazaki et al. 2008). In apolipoprotein E-deficient mice, APSP

significantly lowered the atherosclerotic plaque area to about half of the control, the liver level of thiobarbituric acid-reactive substances as an oxidative stress marker and the plasma level of soluble vascular cell adhesion molecule-1 (sVCAM-1), but showed no effects on body weight and cholesterol and lipid levels in the plasma. Shin et al. (2013) demonstrated that mice fed a high-fat diet containing purple sweet potato extract (PSPE) presented lower increases in body and adipose tissue weights and reduced occurrences of hepatic steatosis than mice that were fed a high-fat diet without PSPE. The decreased adiposity induced by PSPE accounted for lower serum levels of leptin and a higher adiponectin/leptin ratio. PSPE administration also resulted in a significant decrease in serum and hepatic triglyceride and cholesterol levels and a significant increase in faecal triglyceride and cholesterol levels when compared to the high-fat group. PSPE suppressed the expression of sterol regulatory element-binding protein (SREBP)-1, acyl-CoA synthase (ACS), glycerol-3-phosphate acyltransferase (GPAT), HMG-CoA reductase (HMGR) and fatty acid synthase (FAS) in liver tissue in mice provided the high-fat diet. The results suggested that the antiobesity effect of PSPE in high-fat-fed mice occurred through its modulation of lipogenesis in the liver and inhibition of dietary lipid absorption.

In-vitro and clinical studies of 13 healthy volunteers by Nagai et al. (2011) found that sweet potato leaves had antioxidant activity leading to the suppression of low-density lipoprotein oxidation. Ingestion of sweet potato leaves prolonged the lag time for starting low-density lipoprotein oxidation and decreased low-density lipoprotein mobility.

Anti-inflammatory Activity

Leaves of nine Okinawan sweet potato cultivars and two comparable sweet potato cultivars suppressed nitrite production, an index of NO in LPS-stimulated RAW264.7 macrophages (Taira et al 2012). In addition, the sweet potato leave extracts decreased the amount of nitrite ions generated from NOR3, an NO donor, indicating that they had NO-scavenging activity. This NO-scavenging activity of sweet potato leaves was correlated with the total amount of polyphenol and its main constituents of caffeoylquinic acid (CQA) derivatives, such as 5-monoCQA, 4,5-diCQA, 3,5-diCQA, 3,4-diCQA and 3,4,5-triCQA. Therefore, the CQA derivatives may be responsible for the NO-scavenging activity demonstrating that potato leaves may be promising functional food materials for preventing various inflammatory diseases that cause excess NO production.

Cognitive Enhancing Activity

Purple sweet potato colour (anthocyanin rich colour) markedly enhanced cognitive performance, assessed by passive avoidance test in ethanol-treated mice (Cho et al. 2003). Its memory enhancing effects may be associated with its antioxidant properties. Treatment of *d*-galactose-treated mice with purple sweet potato colour (PSPC), a class of naturally occurring anthocyanins used to colour food, improved spatial learning and memory impairment by reversing the loss of pre- and post-synaptic proteins induced by galactose (Wu et al. 2008). Oral administration of purple sweet potato colour (anthocyanin rich) extract to domoic acid-treated mice significantly improved their behavioural performance in a step-through passive avoidance task and a Morris water maze task (Lu et al. 2012). These improvements were mediated through multiple pathways, involving a stimulation of oestrogen receptor-α-mediated mitochondrial biogenesis signalling, decreases in the expression of p47phox and gp91phox, decreases in reactive oxygen species and protein carbonylation were also observed, along with a blockade of the endoplasmic reticulum stress pathway.

Studies showed that purple sweet potato (PSP) extract rich in caffeoylquinic acid derivatives with or without anthocyanin had a neuroprotective effect on mouse brain and could improve the spatial learning and memory of senescence-accelerated prone mouse strain (SAMP) 8 (Sasaki et al. 2013). Additionally, PSP increased brain cell viability by 141.6 and 133 % as compared to Aβ1-42-treated cells.

Antihypertensive Activity

The purified mucilage from sweet potato tubers inhibited angiotensin converting enzyme (ACE) in a dose-dependent manner (28.7 to 59.8 % ACE inhibition, respectively, at 50 to 400 μg/mL mucilage) with IC_{50} of 364.5 μg/mL while that of captopril was 10 nM (8.68 μg/mL) (Huang et al. 2006). The commercial polysaccharide pectin (50 to 400 μg/mL) showed no inhibitory activity against ACE. Trypsin inhibitor (TI), the root storage protein of sweet potato, inhibited angiotensin converting enzyme (ACE) in a dose-dependent manner (50–200 μg/mL, with 31.9–53.2 % inhibition) using *N*-[3-(2-furyl) acryloyl]-Phe-Gly-Gly (FAPGG) as a substrate (Huang et al. 2008b). The 50 % inhibition (IC_{50}) of ACE activity required 187.96 μg/mL TI compared to 10 nM (868 ng/mL) of Captopril. ACE inhibitory activity of TI was found to increase from 34 % to about 8 3 % after 24 h of hydrolysis by pepsin. Ten peptides – namely HDHM, LR, SNIP, VRL, TYCQ, GTEKC, RF, VKAGE, AH and KIEL – were synthesized and were also found to have ACE inhibitory activity. IC_{50} values of individual peptides were 276.2, 746.4, 228.3, 208.6, 2.3, 275.8, 392.2, 141.56, 523.5 and 849.7 μM, suggesting that TYCQ might represent the main active site for the ACE inhibition. It was concluded that sweet potato TI and its hydrolysates might be good for control of hypertension and other diseases when people consume sweet potato tuberous roots. Similarly, sweet potato storage root thioredoxin *h*2 and its four peptic hydrolysates, namely EVPK, VVGAK, FTDVDFIK and MMEPMVK, exhibited angiotensin converting enzyme inhibitory activity (Huang et al. 2011a). The IC_{50} values of individual peptides were 1.73, 1.14, 0.42 and 1.03 mM, respectively, suggesting that FTDVDFIK might be the main active site of ACE inhibition. The results for Trx h2 and its hydrolysates suggested that consumption of sweet potato storage roots can aid in the control of hypertension and other diseases. Sweet potato defensin (SPD1) inhibits angiotensin converting enzyme in a dose-dependent manner (Huang et al. 2011b). The 50 % inhibition (IC_{50}) of ACE activity required 190.47 μg/mL SPD1 while that of Captopril was 10 nM (868 ng/mL). ACE inhibitory activity increased from 52.47 to about 74.38 % after 24 h hydrolysis. Hydrolysis afforded six peptides, namely GFR, FK, IMVAEAR, GPCSR, CFCTKPC and MCESASSK, with ACE inhibitory activity and the IC_{50} values of individual peptides were 94.25, 265.43, 84.12, 61.67, 1.31 and 75.93 μM, respectively, suggesting that CFCTKPC might represent the main domain for the ACE inhibition.

Studies found that continuous administration of colours (anthocyanins) from purple sweet potato to spontaneously hypertensive rats for 15 weeks decreased blood pressure and the heart rate (Shindo et al. 2007).

Cardioprotective Activity

Studies found that administration of probiotic-fermented purple sweet potato yogurt (PSPY) with high γ-aminobutyric acid (GABA) content inhibited cardiac hypertrophy in spontaneously hypertensive rat hearts (Lin et al. 2012b). Abnormal myocardial architecture and enlarged interstitial spaces were decreased in PSPY treated rats. The elevated protein levels of cardiac hypertrophic-related pathways and hypertension were completely reversed by the administration of PSPY. Also, it was found that PSPY may repress the activation of atrial natriuretic peptide (ANP) and brain (B-type) natriuretic peptide (BNP) which subsequently may inhibit the dephosphorylation of the nuclear factor of activated T-cells, cytoplasmic 3 and ultimately prevent the progression of cardiac hypertrophy. They also found that oral administration of 10 % probiotic-fermented sweet potato yogurt PSPY was strong enough to lower cardiac fibrosis in spontaneously hypertensive rats through the suppression of toll-like receptor 4 (TLR-4)-related inflammatory pathway (Lin et al. 2013b). They suggested that PSPY may be included in diets to help prevent cardiac fibrosis in patients with hypertension. In another study, they found that oral administration of PSPY may attenuate cardiomyocyte apoptosis in spontaneously hyper-

tensive rats' hearts by activating insulin-like growth factor-I receptor (IGF-IR)-dependent survival signalling pathways (Lin et al. 2013a). Administration of 2,4-di-tert-butylphenol from sweet potato increased alternation behaviour in mice injected with amyloid-beta peptide $A\beta_{1-42}$ (Choi et al 2013). The results suggest that sweet potato extract could be protective against $A\beta$-induced neurotoxicity, possibly due to the anti-oxidative capacity of its constituent, 2,4-di-*tert*-butylphenol.

Neuroprotective Activity

Studies showed that purple sweet potato anthocyanins (PSPA) alleviated *D*-galactose-induced brain aging in old mice by promoting survival of neurons via PI3K pathway and inhibiting cytochrome C-mediated apoptosis (Lu et al. 2010). PSPA enhanced open-field activity, decreased step-through latency and improved spatial learning and memory ability in *D*-galactose-treated old mice by decreasing advanced glycation end-products' (AGEs) formation and the AGE receptor (RAGE) expression, and by elevating Cu,Zn-superoxide dismutase (Cu,Zn-SOD) and catalase (CAT) expression and activity. PSPA also inhibited cleavage of capase-3 and the increase in terminal deoxynucleotidyl transferase (TdT)-mediated deoxyuridine triphosphate (dUTP) nick-end-labelling (TUNEL)-positive cells in *D*-galactose-treated old mice. In a separate study, aging mice administrated with PSPA via oral gavage showed significantly improved behaviour performance in the open field and passive avoidance test compared with *D*-galactose-treated mice (Shan et al. 2009). It was found that PSPA decreased the expression level of glial fibrillary acidic protein, inducible nitric oxide synthase (iNOS), and cyclooxygenase-2, inhibited nuclear translocation of nuclear factor-kappaB (NF-kappaB), increased the activity of Cu/Zn-SOD and (CAT), and reduced the content of malondialdehyde in the mouse brain. Pre-treatment of PC12 cells with PSPA reduced amyloid-beta peptide (Abeta)-induced toxicity, intracellular reactive oxygen species (ROS) generation and lipid peroxidation dose-dependently (Ye et al. 2010). Concomitantly, cell apoptosis triggered by Abeta characterized with the DNA fragmentation and caspase-3 activity were also inhibited by PSPA. The results suggested that PSPA could protect the PC-12 cell from Abeta-induced injury through the inhibition of oxidative damage, intracellular calcium influx, mitochondria dysfunction and ultimately inhibition of cell apoptosis and may have potential in the treatment of Alzheimer's disease and other oxidative-stress-related neurodegenerative diseases.

Oral administration of purple sweet potato anthocyanins (PSPA) to mice significantly reversed the impairment of motor and exploration behaviour induced by lipopolysaccharide in the open field tasks, and also improve learning and memory ability in step-through tests (Wang et al. 2010). The results suggested that PSPA may be useful for mitigating inflammatory brain diseases by inhibition of proinflammatory tumour necrosis factor-alpha (TNF-alpha), interleukin-1beta (IL-1beta) and interleukin-6 (IL-6) in LPS-stimulated mouse brain, partially through inhibition of extracellular signal-regulated kinase (ERK) and phosphorylated c-Jun N-terminal kinase (JNK) expression and nuclear factor kappa B (NF-kappaB) signalling. Sweet potato extract significantly reversed amyloid β peptide ($A\beta$)-induced neurotoxicity in ICR mice as assessed by the passive avoidance test (Kim et al. 2011). Additionally, it reduced the level of lipid peroxidation and increased catalase activities in brain tissue of mice. The results indicated that *I. batatas* might be beneficial against Alzheimer's disease, especially by limiting oxidative stress in the brain.

Immunomodulating Activity

Studies showed that white-skinned sweet potato (AWSSP) increased phagocytic activity and phagosome–lysosome fusion in neutrophils and monocytes in a dose-dependent manner, but had no significant effect on superoxide anion release (O_2^-) from human neutrophils (Miyazaki et al. 2005). The results suggested that AWSSP would

be useful in the prevention and improvement of diabetic symptoms by stimulating human immunity. In-vivo studies showed that the polysaccharide PSPP, purified from sweet potato root, improved the immune system in mice and could be deemed a biological response modifier (Zhao et al. 2005). PSPP at the dose of 50 mg/kg, significant increments in proliferation of lymphocytes and serum IgG concentration were observed. At the dose of 150 and 250 mg/kg, significant increments were observed in all tested immunological indexes. A dose-dependent manner was demonstrated in phagocytic function, haemolytic activity and serum IgG concentration, but not in proliferation of lymphocytes and natural killer cell activity. In a randomized cross-over study (two periods, each lasting for 2 weeks) involving 16 healthy non-smoking adults of normal weight, consumption of purple sweet potato leaves modulated various immune functions including increased proliferation responsiveness of peripheral blood mononuclear cells secretion of cytokines IL-2 and IL-4 and the lytic activity of natural killer cells (Chen et al. 2005).

The methanolic extract of sweet potato roots at the concentration range of 10–100 μg/mL stimulated cell-mediated immune system by increasing neutrophil phagocytic function and intracellular killing potency of human neutrophils (Patil et al. 2007).

Anti-inflammatory Activity

The resin glycoside ipomotaoside A, isolated from sweet potato aerial parts, were found to have inhibitory activity on both cyclooxygenase Cox 1 and Cox 2 (Yoshikawa et al. 2010). Arantes et al. (2014) investigated conformational characterization of ipomotaosides A–D in aqueous and non-aqueous solvents. The most abundant conformation of ipomotaoside A in solution was employed in flexible docking studies, providing a structural basis for the compound's inhibition of COX enzymes, further supporting its potential as a new anti-inflammatory agent. Purple sweet potato leaf extract and its components, cyanidin and quercetin, inhibited cell adhesion and inflammatory response induced by TNF-α in human aortic endothelial cells by modulation of NFκB and MAPK signalling (Chao et al. 2013).

Antifatigue Activity

Oral administration of sweet potato leaf flavonoids to male Kunming mice for 4 weeks exerted significant antifatigue effects (Li and Zhang 2013). The leaf extract extended the exhaustive swimming time, effectively inhibited the increase of blood lactic acid, decreased the level of serum urea nitrogen and increased the hepatic and muscle glycogen content of mice.

Hepatoprotective Activity

Pre-treatment of rats with purple-coloured sweet potato juice orally for five consecutive days prior to carbon tetrachloride treatment effectively reduced glutamic-oxaloacetic transaminase (GOT), glutamate pyruvate transaminase (GPT), lactate dehydrogenase (LDH) and thiobarbituric acid-reactive substance (TBARS) in serum and liver TBA-RS and oxidized protein levels (Suda et al. 1997). The results demonstrated ameliorating effects of purple-coloured sweet potato juice against carbon tetrachloride-induced liver injury.

Pre-treatment of mice with an anthocyanin fraction obtained from purple-fleshed sweet potato protected against acetaminophen (paracetamol)-induced hepatotoxicity by blocking CYP2E1-mediated paracetamol bioactivation, by upregulating hepatic glutathione levels, and by acting as a free radical scavenger (Choi et al. 2009). Studies showed that anthocyanin-rich purple sweet potato colour could protect mouse liver from *d*-galactose-induced injury by attenuating lipid peroxidation, renewing the activities of antioxidant enzymes (Cu, Zn-SOD (superoxide dismutase), catalase and glutathione peroxidase) and suppressing inflammatory response by inhibiting the upregulation of the expression of NF-kappaB p65, COX-2 and iNOS (Zhang et al. 2009). Administration of purple-fleshed sweet potato fraction to rats effectively

ameliorated liver fibrosis caused by dimethylnitrosamine (DMN) (Choi et al. 2010). Additionally, the fraction inhibited DMN-induced reductions in rat body and liver weights in a dose-dependent manner and decreased DMN-induced expression levels platelet-derived growth factor receptors-beta, tumour necrosis factor-alpha and transforming growth factor-beta.

Studies showed that purple sweet potato anthocyanin could protect mouse liver against *d*-galactose-induced hepatocyte apoptosis via attenuating oxidative stress, inhibiting the activation of caspase-3 and enhancing cell survival signalling (enhancing the level of antiapoptotic protein Bcl-2 and the activation of PI3K/Akt pathway) in *d*-galactose- treated mice (Zhang et al. 2010). Oral pre-treatment of purple sweet potato anthocyanin fraction prior to *t-tert*-butyl hydroperoxide treatment significantly lowered the serum levels of the hepatic enzyme markers (ALT and AST), reduced the incidence of liver lesion and reduced oxidative stress of the liver by evaluation of malondialdehyde and glutathione (Hwang et al. 2011a). In HepG2 cell, the fraction significantly reduced *t*-BHP-induced oxidative injury, as determined by cell cytotoxicity, intracellular glutathione content, lipid peroxidation, reactive oxygen species (ROS) levels and caspases activation. The hepatoprotective effects may be partly attributed to its ability to scavenge ROS and to regulate the antioxidant enzyme HO-1 via the Akt and ERK1/2/Nrf2 signaling pathways. Studies demonstrated that anthocyanins of the purple sweet potato attenuated dimethylnitrosamine-induced liver injury in rats by inducing nuclear erythroid 2-related factor 2 (Nrf2)-mediated antioxidant enzymes, reducing cyclooxygenase-2 and inducible nitric oxide synthase expression and reducing inflammation via nuclear factor kappa B (NF-κB) inhibition (Hwang et al. 2011c). Purple sweet potato was found to have a preventive effect on acute and subacute alcoholic liver damage in mice (Sun et al. 2014a). All tested biochemical and histological parameters were ameliorated after intragastric administration of purple sweet potato.

In a phase II study of patients with advanced hepatocellular carcinoma, 4-ipomeanol, a naturally occurring alkylating furan from fungal-infected sweet potato, at a dose of either 826 or 1032 mg/m^2 administered every 3 weeks did not demonstrate a relevant degree of clinical activity against advanced hepatocellular carcinoma (Lakhanpal et al. 2001). In a randomized, double-blind, placebo-controlled, parallel study of healthy adult men (30–60 years) with borderline hepatitis, ingestion of purple sweet potato beverage significantly decreased the serum levels of hepatic biomarkers gamma-glutamyl transferase (GGT), aspartate aminotransferase and alanine aminotransferase, particularly the GGT level (Suda et al. 2008).

Renoprotective Activity

Purple sweet potato anthocyanin (PSPA) (700 mg/kg per day) reduced body weight, ratio of urine albumin to creatinine, inflammatory cell infiltration and collagen IV accumulation in mice fed a high-fat diet (HFD) (60 % fat food) for 20 weeks (Shan et al. 2014). PSPA attenuated oxidative stress and kidney tissue damage in the kidney of HFD-treated mice. PSPA inhibited the activation of kidney IKKβ/NF-κB signalling in HFD-treated mice. PSPA also reduced the activation of NLRP3 inflammasome and decreased the protein expression of kidney oxidative stress-associated AGE receptor (RAGE) and thioredoxin interacting protein (TXNIP) in HFD-treated mice.

Wound Healing Activity

In the incision wound model, high tensile strength of the wounded skin was observed in Wistar rats treated with sweet potato peel extract gels and the peel bandage when compared with wounded control rats (Panda et al. 2011). The increase in tensile strength indicated the promotion of collagen fibres and that the disrupted wound surfaces were being firmly knit by collagen. In the excision wound model, significant wound closure was observed on the 4th day in rats treated with all three peel gel formulations when compared with wounded control rats. A significant increase in

hydroxyproline (index for collagen turnover) and ascorbic acid content in the peel gel-treated animals and a significant decrease in malondialdehyde content in the animals treated with peel-gel as well as peel bandage was observed when compared with the wounded control animal. It was concluded that sweet potato peels possessed a potent wound-healing activity which may be associated to its antioxidant property. The peel extract showed the presence of high levels of polyphenols (anthocyanins and phenolic acids) and sesquiterpenoids (6-myporol, 4-hydroxydehydromyoporone and ipomeamarone).

Antimelanogenic Activity

The extract from steamed sweet potato was found to suppress the melanogenesis of mouse melanoma B 16 cells (Shimozono et al. 1996). The phenolic acids extracted from steamed sweet potato such as chlorogenic acid (ChlA), 3,5-dicaffeoylquinic acid (3,5-diCQA), 3,4-dicaffeoylquinic acid (3,4-diCQA) and 4,5-dicaffeoylquinic acid (4,5-diCQA) also suppressed melanogenesis in mice.

Antiulcerogenic Activity

Studies showed that methanol sweet potato extract possessed gastroprotective activity as evidenced by its significant inhibition of mean ulcer score and ulcer index and a marked increase in GSH, SOD, CAT, glutathione peroxidase(GPx) and glutathione reductase (GR) levels and reduction in lipid peroxidation in a dose-dependent manner in cold stress and aspirin-induced gastric ulcers in Wistar rats (Panda and Sonkamble 2012).

Gastroenterologic Activity

In a prospective, randomized controlled trial with a sample of 93 hospitalized patients with acute coronary syndromes (ACS), sweet potato/foot-bath/acupressure massage (SFA) intervention was found to be a more effective, safe, economical and practical than usual care alone in managing constipation and satisfaction with defaecation in patients hospitalized with ACS (Ren et al. 2012).

Radioprotective Activity

Administration of a freshly prepared aqueous extract of sweet potato tubers to rats, 1 week pre-irradiation and during the period of radiation exposure significantly ameliorated the oxidative stress in liver and kidney tissues (Darwish et al. 2010). The significant amelioration in oxidative stress was substantiated by improvement of liver and kidney enzymes. Treatment of rats with sweet potato has significantly reduced the increase in serum alanine amino transferase (ALT), aspartate amino transferase (AST) and lactate dehydrogenase (LDH) activity, serum creatinine and urea levels. Furthermore, hyperglycaemia and alteration in lipid profile manifested by a significant increase in triglycerides (TG), total cholesterol (TC) and low-density lipoprotein cholesterol (LDL-C) and a significant decrease in high-density lipoprotein cholesterol (HDL-C) were improved in sweet potato-treated irradiated rats compared to those only irradiated. Pre-treatment of murine thymocytes with purple sweet potato pigments significantly inhibited ^{60}Co γ-ray-induced mitochondria-mediated apoptosis (Xie et al. 2010). The radioprotective effect might be related to reactive oxygen species scavenging, the enhancement of the activity of antioxidant enzymes, the maintenance of mitochondrial transmembrane potential and the sequential inhibition of cytochrome c release and downstream caspase and poly ADP-ribose polymerase (PARP) cleavage. The cosmetic cream with 0.61 mg of total anthocyanins (per 100 g cream) from TNG73 purple sweet potato absorbed approximately 46 % of the incident UV radiation (Chan et al. 2010). Although the anthocyanins absorbed both UV-A and UV-B radiation, they were particularly effective against UV-B rays. Acidic ethanol-extracted anthocya-

nins had better radical-scavenging ability, higher total phenolic content and stronger reducing ability than acidic water-extracted anthocyanins. The study demonstrated that the addition of anthocyanin extracts of purple sweet potato to a cosmetic cream improved the cream's UV absorption ability.

Purple sweet potato (PSP) pigments treatment prior to 4 Gy (60)Co γ-ray irradiation had a cytoprotective activity against γ radiation by increasing murine thymocytes viability and decreasing apoptosis (han et al. 2011). The protective effect of PSP pigments may be involving ROS scavenging, p53 depression and Bcl-2/Bax modulation in a caspase-dependent mitochondrial way.

Adaptogenic Activity

In a cross-over designed study of 15 healthy, non-trained, young male subjects, consumption of purple sweet potato leaves (PSPL) for 7 days significantly increased plasma total polyphenols concentration and total antioxidant power (i.e., the ferric-reducing ability of plasma) and decreased exercise-induced oxidative damage and pro-inflammatory cytokine secretion (Chang et al. 2010). However, no significant difference was found in heat shock protein HSP72 levels between PSPL and the control groups.

Antimicrobial Activity

Two antifungal (*Rhizopus stolonifer*) fractions were isolated from the periderm and outer cotex of sweet potato tubers; one active fraction comprised predominantly caffeic acid and the second more active fraction contained 3,5-dicaffeoylquinic acid (3,5-DCQA) with an EC_{50} of 2.2 g/L (Stange et al. 2001).

Vasorelaxant Activity

Ipomoea batatas plant extract exhibited more than 50 % relaxing effect on aortic ring preparations (Runnie et al. 2004). The vascular effects on the aortic ring preparations were mainly endothelium-dependent, and mediated by nitric oxide.

Vitamin A and Health Enhancement

In a study of 90 primary school children aged 5–10 years, consumption of β-carotene-rich orange-fleshed sweet potato was found to improve vitamin A status and could play a significant role in developing countries as a viable long-term food-based strategy for controlling vitamin A deficiency in children (van Jaarsveld et al. 2005). Jamil et al. (2012) found that daily consumption of orange-fleshed sweet potato for 60 days increased plasma β-carotene concentration, but did not increase total body vitamin A pool size in Bangladeshi women residing in a resource-poor community.

Amagloh et al. (2012) found that sweet potato-based formulations were superior to enriched Weanimix (maize-soybean blend) as complementary foods for infants in low-income countries, based on its higher fructose level (which makes the porridge naturally sweet) and lower phytate levels compared to the enriched Weanimix.

Removal of Trypsin Inhibitory Activity

Trypsin inhibitor activity (TIA) decreased during ensilage in sweet potato/maize powder samples of all treatments while the sweet-potato strips (SPS) mixed with maize powder (CP) mixture (7:3, w/w) ensiled for 3 months contained the lowest TIA (Lin et al. 1988). Rats fed on diets containing dried SPS-CP (8:2, w/w) showed significantly lower body-weight gain than rats fed on the control diet or ensiled SPS diets, at the end of the 8th week. They also showed enlargement of the pancreas. The adverse effect of SPS was associated with TIA which appeared to be prevented to some extent by ensilage. Among the four cultivars of sweet potatoes, RS-III-2 trypsin inhibitors were more heat-labile (Kiran and

Padmaja 2003). Heating at 100 °C led to rapid inactivation of TI of sweet potatoes. Microwave baking and flour preparation were the best methods to eliminate TI from sweet potatoes. Q40091|Q40091_IPOBA was isolated as the major sporamin B from sweet potato *cv.* 55-2 tuber and found to have potent trypsin inhibitory activity (Sun et al. 2009). There was a linear relationship between trypsin inhibitor activity (Ti activity) and amounts of this sporamin B (3–18 μg/mL). Heat treatment at more than 90 °C led to a dramatic decrease of trypsin inhibitor efficiency.

Pharmacokinetic Studies

Two major anthocyanin components of a beverage prepared from an extract of the tuber of purple sweet potato, cyanidin 3-*O*-(2-*O*-(6-*O*-(*E*)-caffeoyl-*β*-D-glucopyranosyl)-*β*-D-glucopyranoside)-5-*O*-*β*-D-glucopyranoside) and peonidin 3-*O*-(2-*O*-(6-*O*-(*E*)-caffeoyl-*β*-D-glucopyranosyl)-*β*-D-glucopyranoside)-5-*O*-*β*-D-glucopyranoside), were detected in the plasma and urine of both rats and humans (Harada et al. 2004). The plasma concentration of anthocyanins in humans reached a maximum 90 min after ingestion, and the recovery of anthocyanins in the urine was estimated as 0.01–0.03 %. An acylated anthocyanin, peonidin 3-caffeoylsophoroside-5-glucoside, was detected in the urine of 87 healthy volunteers 2 h after ingestion of purple-fleshed sweet potato beverage with various contents of anthocyanin (beverage A; 22.1 mg/250 mL, B; 107.8, C; 84.9) (Oki et al. 2006). The mean concentrations were 15.1 μg/L of urine, 46.6 and 53.3 for beverages A, B and C, respectively.

The tuber and leaves have also been used in traditional medicine. The leafy shoots are used as a galactagogue and as poultice. The leaves are used to treat diabetes, hookworm, haemorrhage and abscesses and reported to be used as a maturative cataplasm. The tuber is used to treat asthma and are laxative. Tubers are sliced, dried to make a tea to allay thirst.

Recent researches support their use for type 2 diabetes. Sweet potato was found to have beneficial effects on plasma glucose and total as well as LDL cholesterol levels in patients with type 2 diabetes. These effects related to a decrease in insulin resistance. Another study reported that increases in blood glucose levels after glucose loading in test animals were inhibited after oral administration of white skinned sweet potato (WSSP). WSSP shows remarkable antidiabetic activity and improves the abnormality of glucose and lipid metabolism by reducing insulin resistance. Almost all antidiabetic activity was found in the cortex of WSSP. This active component was presumed to be an acidic glycoprotein because it contained protein and sugar.

The phenolic compounds caffeic acid and di- and tricaffeoylquinic acids from sweet potato leaves were also reported to significantly depressed cancer cell proliferation. Specifically, 3,4,5-tri-O-caffeoylquinic acid effectively depressed the growth of three kinds of cancer cells: stomach cancer (Kato III), a colon cancer (DLD-1) and a promyelocytic leukaemia cell (HL-60). Caffeic acid had an exceptionally higher effect against HL-60 cells than other di- and tricaffeoylquinic acids. The findings indicate that 3,4,5-tri-O-caffeoylquinic acid may have the potential for cancer prevention.

One paper reported that flavone extracted from sweet potato leaf could control blood glucose and modulate the metabolism of glucose and blood lipid, and decrease outputs of lipid peroxidation and scavenge the free radicals in non-insulin-dependent diabetic rats.

Toxicity Issues

A lung-toxic furanoterpenoid, 4-ipomeanol, was isolated from moldy sweet potato (Boyd and Wilson 1972). The toxin was shown to produce a respiratory disease in mice characterized by severe pulmonary oedema, pleural effusion and death. Toxicity manifestations of mould-damaged sweet potatoes namely pulmonary oedema, emphysema and adenomatosis had been described as characteristic disease signs in cattle (Wilson et al. 1970; Wilson 1973). The experimental infection of viable sweet potato slices or

of the intact tubers by fungi resulted in the production of many furanoterpenoid compounds, some of which were markedly toxic for laboratory and commercial animals (Wilson et al. 1971). The respiratory tract toxin, 4-ipomeanol (1-[3-furyl]-4-hydroxy-1-pentanone) may be responsible for the typical lung oedema of cattle that were fed mouldy sweet potatoes as well as a similar pathological state in mice fed or injected with the pure compound (Boyd et al. 1972).

Four furanoterpenoids produced by sweet potatoes following microbial infection 4-ipomeanol, 1-ipomeanol, tipomeanine and 1,4-ipomeadiol were found to be acutely toxic to the lungs of experimental animals, characteristically producing pulmonary oedema and congestion, following a latent period of several hours after dosing (Boyd et al. 1974). Mice receiving lethal doses of the toxins usually die within 24 h, and pathological findings appeared most often only in the lungs. However, mice initially surviving near-lethal doses of the toxins, particularly 1-ipomeanol and 1,4-ipomeadiol, may show evidence of nephrotoxicity within 1–3 days. In addition to ipomeamarone and other hepatotoxins, a series of 1-(3-furyl)-1,4-dioxygenated pentanes were isolated from sweet potatoes infected with *Fusarium solani* (Burka and Wilson 1976). These compounds, especially 1-(3-furyl)-4-hydroxy-1-pentanone (4-ipomeanol) show marked pulmonary toxicity in laboratory animals. Cattle given intraruminal administration of 4-ipomeanol, a furanoterpenoid originally obtained from sweet potatoes infected with *Fusarium solani (F. javanicum)*, developed a respiratory syndrome clinically and histologically indistinguishable from atypical interstitial pneumonia (Doster et al. 1978). There were oedema and emphysema in the lungs and mediastinum. The maximum non-lethal oral dose of 4-ipomeanol was estimated to be between 7.5 and 9 mg/kg of body weight. Studies by Li and Castleman (1990) found significant elevations in numbers of neutrophils and macrophages were recovered by bronchoalveolar lavage at times from 24 to 96 h after 4-ipomeanol-treatment in calves. Hyperplasia of non-ciliated bronchiolar epithelial cells and of type II alveolar epithelial cells were observed at 72 and 96 h after treatment. The results indicated that type I alveolar epithelial cells and non-ciliated bronchiolar epithelial cells are most susceptible to 4-ipomeanol-induced damage and necrosis in calves. 4-Ipomeanol-induced pulmonary oedema in calves occurred prior to ultrastructurally demonstrable, mild, alveolar capillary endothelial cell damage. Also they found that 4-ipomeanol exacerbated interstitial pneumonia in calves induced by bovine parainfluenza type 3 virus (Li and Castleman 1991). Four-ipomeanol-enhanced viral pneumonia was characterized in part by extensive hyperplasia of type II alveolar epithelial cells and by dense aggregates of macrophages and neutrophils in alveolar spaces and interalveolar septa.

The pulmonary toxin, 4-ipomeanol, selectively alkylated the lungs of rats (Boyd and Burka 1978). Time-course and dose–response studies demonstrated a close correlation between the pulmonary alkylation and the lung toxicity of the compound. The LD_{50} values (μg/g) of the lung-toxic furanoterpenoids produced by sweet potatoes following microbial infection in mice were determined as follows: 4-ipomeanol 38 μg by oral administration, 36 μg by intraperitoneal and 21 μg by intravenous; 1-ipomeanol 79 μg by oral, 49 μg by intraperitoneal, 34 μg by intravenous; ipoemanine 26 μg by oral, 25 μg by intraperitoneal, 14 μg by intravenous and 1,4-ipomeadiol 104 μg by oral, 67 μg by intraperitoneal, 66 μg by intravenous administration (Boyd et al. 1974). Durham et al. (1987) found that the histologic severity of Sendai viral pneumonia in young adult female C57BL/6 J mice was closely correlated with increasing doses of the pulmonary toxicant, 4-ipomeanol.

Injection of five three-substituted furans, isolated from stressed sweet potato root tissue, into mice produced temporary neurological effects followed by development of extensive necrosis in the liver (Wilson and Burka 1979). The most toxic was 6-myoporol, with LD_{50} of 84 mg/kg, comparable to ipomeamarone. The other toxic compounds were ipomeamaronol LD_{50} 266 mg/kg, 4-hydroxymyoporone LD_{50} 235 mg/kg, 7-hydroxymyoporone LD_{50} 200 mg/kg and dihydro-7-hydroxymyoporone LD_{50} 184 mg/kg.

Fermentation of 6 weeks duration was observed to inadequately eliminate the lung, liver and kidney toxicity caused by mold-damaged sweet potatoes (Thibodeau et al. 2004). In fact, fermentation exacerbated the hepatotoxicity of mold-damaged sweet potatoes. Also, it was demonstrated that sweet potato regions lacking visible mold damage could induce lung and kidney injury, which, however, was preventable by fermentation.

Microwave and bake cooking operations destroyed approximately 90 % of the ipomeamarone in sweet potato roots (Cody and haard 1976). 4-Ipomeanol was more heat stable than ipomeamarone, although it also decreased substantially as a result of normal cooking. Catalano et al. (1977) found that ipomeamarone was primarily concentrated in the blemished and diseased sweet potato tissues. Neither baking nor boiling appeared to promote diffusion of ipomeamarone into the healthy tissues. Also, baking appeared to reduce the concentration of this hepatotoxin.

Sweet potato polyhydroxylated nortropane alkaloids calystegines A_3, B_1, B_2 and C_1 exhibited inhibitory activity on mammalian liver glucosidases (Asano et al. 1997). Calystegines B_1 and C_1 were potent competitive inhibitors of the bovine, human and rat β-glucosidase activities, with *Ki* values of 150, 10 and 1.9 μM, respectively, for B_1 and 15, 1.5 and 1 μM, respectively, for C_1. Calystegine B_2 was a strong competitive inhibitor of the α-galactosidase activity in all the livers. Human β-xylosidase was inhibited by all four nortropanes, with calystegine C_1 having a *Ki* of 0.13 μM. Calystegines A_3 and B2 selectively inhibited the rat liver β-glucosidase activity. The potent inhibition of mammalian beta-glucosidase and alpha-galactosidase activities in-vitro raises the possibility of toxicity in humans consuming large amounts of plants that contain these compounds.

Studies found that Wistar rats fed phytic acid extracted from sweet potato or commercial phytic acid-supplemented diets displayed reduced bone calcium levels and had significantly thinner bone in the trabecular region, compared to the groups fed formulated diet or zinc-supplemented formulated diet (Dilworth et al 2008). The results suggested that the consumption of foods high in phytic acid may contribute to a reduction in the minerals available for essential metabolic processes in rats.

Traditional Medicinal Uses

In Burkina Faso and the Ivory Coast, juice from the leaves are used for gum gargle and gum massage (Kerharo and Bouquet 1950), and leaf sap for burns and grounded leaves as enema to prevent miscarriage (Bouquet and Debray 1974). In the Popular Republic of Congo (Brazzaville), an infusion of leaves sweet potato and *Cassia occidentalis*, barks stem, branch, trunk of *Bridelia ferruginea*, are used as purgative (Bouquet 1969). In Casamance, Senegal, leaves are boiled and used to treat abscess and boils and the hot leaves are used as cataflam in cosmetics (Thomas 1972). In the Comoros, sweet potato is used for wound healing and analgesic and the leaves sap is used to treat serious sunburn (Adjanohoun et al. 1982). In Togo, a decoction of the roots and leaves of sweet potato and *Cassia occidentalis* is taken orally as a remedy for intercostal and rib pain and the powered mixture of the same is used to treat scarification pain (Adjanohoun et al. 1986). Leaves of *Ipomoea batatas, Vernonia amygdalina, Plumbago zeylanica* are burnt, and the ashes licked with palm oil to treat small pox in Benin (Verger 1995). In Gabon, crushed, macerated leaves of sweet potato is taken orally to facilitate child birth (Raponda-Walker and Sillans 1995). In the Democratic Republic of Congo (ex. Zaïre) region of Kisangani, the leaves are used to treat erythroderma (exfoliative dermatitis), measles and chicken pox (Kalanda and Bolamba 1994). In Cameroon Yaounde region, leaf pieces are added in food to treat diabetes (Tsabang et al. 2001). In Bulamogi, Uganda, the plant is used for insect stings and leaves used in steam bath as a restorative for lameness (Tabuti et al. 2003). In Uganda (Northern sector of Kibale National Park), an aqueous extract of dry leaves pounded with *Passiflora edulis*, *Coffea canephora*, is taken orally to treat diarrhoea (Namukobe et al. 2011).

An infusion of the leaves is used by women in the Sango Bay area, Uganda, for relaxation of the pelvic region during child birth (Ssegawa and Kasenene 2007). In Ethiopia, the leaves are used topically for boils (Giday et al. 2009). In Nigeria (Ndokwa Delta State), the leaves are squeezed and the sap taken orally as a therapy for stomach problem by the Abbi people (Ogie-Odia and Oluowo 2009). In Ogun state of Nigeria, peels of sweet potato are macerated and used to treat tuberculosis (Ogbole and Ajaiyeoba 2010). In northern Maputaland, KwaZulu-Natal Province, South Africa, sweet potato and *Tabernaemontana elegans* leaves are boiled in water and a decoction taken thrice daily to treat gonorrhoea (De Wet et al. 2012). In the Democratic Republic of Congo, pounded leaves are used as poultice for breast cancer (Mbuta et al. 2012). The leaves are used to treat insect stings and used by the women of Agnalazaha littoral forest (Southeastern Madagascar) to evacuate the placenta during pregnancy (Razafindraibe et al. 2013).

Other Uses

All parts of the plant are used for an animal feed and supplement especially in developing countries. In Papua New Guinea pigs are primarily raised on sweet potatoes. In the Canete valley in Peru sweet potato supports a modern dairy industry. There is growing interest in its potential use as a component in chicken feed. The vines are also used as mulches and compost. In Taiwan, companies are making alcohol biofuel from sweet potato. In South America, the juice of red sweet potatoes is combined with lime juice to make a dye for cloth. By varying the proportions of the juices, every shade from pink to purple to black can be obtained.

Sweet potato tubers and leafy shoots are used as animal (especially pigs and cattle) feed in developing countries (Scott 1992). Studies found that ensiled sweet potato leaves could replace fishmeal and groundnut cake in traditional Vietnamese diets for growing pigs (Van et al. 2005). There was a significant stimulatory impact of the intake of sweet potato leaves on growth performance of the growing pigs (Nguyen et al. 2004). Sweet potato tuber is used for livestock feed and for the production of starch in South Korea (Min et al. 2006). Studies found sweet potato vine pellet to be a good source of dietary supplement, which resulted in significant improvement in apparent digestibility, rumen fermentation and milk yield in lactating dairy cows fed on urea-treated rice straw (Phesatcha and Wanapat 2013). Megersa et al. (2013) found supplementation with sweet potato vine could replace the conventional concentrate and could be incorporated with poor quality hay to prevent body weight loss of goats in the absence of other feed supplements.

More than 50 million tons of sweet potato are used for starch production annually around the world (Cheng et al. 2014). A procedure was developed for separately recovering polyphenol oxidase (PPO), β-amylase, sporamins and small molecular nutrients (SMNs) from sweet potato wastewater in starch production. Purified powders of 4.3×10^5 units of PPO, 4.0×10^6 units of β-amylase, 8.70 g sporamins and 20.2 g SMNs were obtained from the wastewater of 1 kg sweet potato. A sweet potato medium derived from baked sweet potato supplemented with 0, 4 or 8 g/L of a nitrogen source such as yeast extract was found to be a suitable and low-cost medium for the cultivation of *Lactobacillus* (Hayek et al. 2013).

Cowdung was found to be a good seed for direct fermentation of sweet potato to produce biofuels (hydrogen and ethanol) (Chu et al. 2010). Also, acetate and butyrate with small quantities of propionate were produced at all pH values. Sweet potato was identified as a sustainable crop for fuel bioethanol production based on both its favourable energy balance and the net GHG emission reduction (Carrasco-Letelier et al. 2013). Studies found sweet potato to be an attractive raw material for fuel ethanol, since up to 4800 L ethanol per hectare can be obtained (Lareo et al. 2013). An energy-saving ethanol fermentation technology was developed using uncooked fresh sweet potato as raw material for fuel ethanol (Zhang et al. 2013). A mutant strain of *Aspergillus niger* isolated from mildewed sweet potato was used to produce abundant raw starch saccharification enzymes for treating

uncooked sweet potato storage roots. The ethanol fermentation was carried out by *Zymomonas mobilis*, and 14.4 g of ethanol (87.2 % of the theoretical yield) was produced from 100 g of fresh sweet potato storage roots. Studies found that sweet potato starch residue (SPSR) could be used as starting material to prepare an eco-friendly adsorbent to control heavy metal pollution (Hao et al. 2014). Life-cycle assessment (LCA) of the energy efficiency of sweet potato-based bioethanol production found the net energy ratio of sweet potato-based bioethanol to be 1.48 and the net energy gain was 6.55 MJ/L (Wang et al. 2013). Studies by Cai et al. (2010) demonstrated that sweet potato fuel ethanol wastewater could be used for electricity generation in microbial fuel cell while at the same time achieving wastewater treatment.

Seed germination inhibitors were isolated from sweet potato root periderm tissue as assessed using a proso millet seed germination bioassay, indicating it could have a weedicidal effect (Peterson and Harrison 1991). The highest levels of chlorogenic acid occurring in sweetpotato cortex tissue exceeded the minimum inhibitory concentrations for larval growth and survival of diamondback larvae, and growth of three out of the four sweetpotato pathogenic fungi tested (Peterson et al. 2005). Caffeic acid from sweet potato root inhibited the growth of four sweet potato pathogenic fungi and germination of proso millet seeds in bioassays (Harrison et al. 2003). Inhibitory activity in the bioassays suggests that high periderm caffeic acid levels contributed to the storage root defense chemistry of some sweet potato genotypes.

In Kenya, sap from the aerial parts is applied on the nail surface against tick (Wanzala et al. 2012), or the aerial part is boiled and the extract used as a dip to control the tick-borne cattle disease, theileriosis (Njoroge and Bussmann 2006).

Comments

According to 2012 FAO statistics, the then world's leading sweet potato producing countries are: China 77,375,000 MT (metric tonne), Nigeria 3,400,000 MT, United Republic of Tanzania 3,018,175 MT, Uganda 2,645,700 MT, Indonesia 2,483,467 MT, Vietnam 1,422,501 MT, the USA 1,201,203 MT, Ethiopia 1,185,050 MT, Madagascar 1,144,000 MT and India 1,072,800 MT (FAO 2014).

Patents that have been lodged over the past two decades related to alternative functional use of the sweet potato consisted largely under the category of ornamental products and alternative products such as sweet potato chips and fries, and a few fuel ethanol products (Barnes and Sanders 2012).

Selected References

Abegunde OK, Mu TH, Chen JW, Deng FM (2013) Physicochemical characterization of sweet potato starches popularly used in Chinese starch industry. Food Hydrocoll 33(2):169–177

Adjanohoun EJ, Ahyi MRA, Ahmed A, Eymê J, Guinko S, Kayonga A, Keita A, Lebras M (1982) Contribution aux Études Ethnobotaniques et Floristiques aux Comores. Agence de Coopération Culturelle et Technique. France, Paris, 216 pp

Adjanohoun EJ, Ahyi MRA, Aké Assi L, Akpagana K, Chibon P, El-Adji A, Eymé J, Garba M, Gassita JN, Gbeassor M, Goudote E, Guinko S, Hodouto KK, Houngnon P, Keita A, Keoula Y, Hodouto WP, Issa Lo A, Siamevi KM, Taffame KK (1986) Contributions aux Études Ethnobotaniques et Floristiques au Togo. Médecine Traditionelle et Pharmacopée Agence de Coopération Culturelle et Technique. France, Paris, 671 pp

Ahmed M, Akter MS, Eun JB (2011) Optimization conditions for anthocyanin and phenolic content extraction form purple sweet potato using response surface methodology. Int J Food Sci Nutr 62(1):91–96

Aina AJ, Falade KO, Akingbala JO, Titus P (2012) Physicochemical properties of Caribbean sweet potato (*Ipomoea batatas* (L) Lam) starches. Food Bioprocess Technol 5(2):576–583

Akazawa T, Uritani I (1962) Biosynthesis of Ipomeamarone: the incorporation of acetate-2-C^{14} into ipomeamarone. Agric Biol Chem 26(2):131–133

Akazawa T, Wada K (1961) Analytical study of ipomeamarone and chlorogenic acid alterations sweet potato roots infected by *Ceratocystis fimbriata*. Plant Physiol 36(2):139–144

Akazawa T, Uritani I, Kubota H (1960) Isolation of ipomeamarone and two coumarin derivatives from sweet potato roots injured by the weevil, *Cylas formicarius elegantulus*. Arch Biochem Biophys 88(1):150–156

Allen JC, Corbitt AD, Maloney KP, Butt MS, Truong VD (2012) Glycemic index of sweet potato as affected by cooking methods. Open Nutr J 6:1–11

Almeida-Muradian LB, Penteado MVC (1993) Carotenoids and provitamin A value of some Brazilian sweet potato cultivars (*Ipomoea batatas* Lam.). Rev Farm Bioquim Univ Sao Paulo 28(2):145–154

Amagloh FK, Coad J (2014) Orange-fleshed sweet potato-based infant food is a better source of dietary vitamin A than a maize-legume blend as complementary food. Food Nutr Bull 35(1):51–59

Anbuselvi S, Muthumani S (2014) Phytochemical and antinutritional constituents of sweet potato. J Chem Pharm Res 6(2):380–383

Antia BS, Akpan EJ, Okon PA, Umoren IU (2006) Nutritive and anti-nutritive evaluation of sweet potatoes *(Ipomoea batatas*) leaves. Pak J Nutr 5(2):166–168

Anyanga MO, Muyinza H, Talwana H, Hall DR, Farman DI, Ssemakula GN, Mwanga RO, Stevenson PC (2013) Resistance to the weevils *Cylas puncticollis* and *Cylas brunneus* conferred by sweetpotato root surface compounds. J Agric Food Chem 61(34):8141–8147

Arantes PR, Sachett LG, Graebin CS, Verli H (2014) Conformational characterization of ipomotaosides and their recognition by COX-1 and 2. Molecules 19(4):5421–5433

Arockiamary SN, Nadhiya K, Vijayalakshmi K (2014) Phytochemical screening and antioxidant study of *Ipomoea batatas* Lam (sweet potatoes). Int J Biol Pharm Sci 5(7):566–570

Asahi T, Sakano K, Nomura T, Ogawa H, Uritani I (1967) Isolation and properties of acid phosphatases of sweet potato roots. Agric Biol Chem 31(12):1394–1400

Asano N, Kato A, Matsui K, Watson AA, Nash RJ, Molyneux RJ, Hackett L, Topping J, Winchester B (1997) The effects of calystegines isolated from edible fruits and vegetables on mammalian liver glycosidases. Glycobiology 7(8):1085–1088

Asano N, Yokoyama K, Sakurai M, Ikeda K, Kizu H, Kato A, Arisawa M, Höke D, Dräger B, Watson AA, Nash RJ (2001) Dihydroxynortropane alkaloids from calystegine-producing plants. Phytochemistry 57(5):721–726

Bahado-Singh PS, Riley CK, Wheatley AO, Lowe HIC (2011) Relationship between processing method and the glycemic indices of ten sweet potato (*Ipomoea batatas)* cultivars commonly consumed in Jamaica. J Nutr Metab. Article ID 584832

Barnes SL, Sanders SA (2012) Advances in functional use of sweet potato, [*Ipomoea batatas* (L.) Lam]. Recent Pat Food Nutr Agric 4(2):148–154

Bartke N, Fischbeck A, Humpf HU (2006) Analysis of sphingolipids in potatoes (*Solanum tuberosum* L.) and sweet potatoes (*Ipomoea batatas* (L.) Lam.) by reversed phase high-performance liquid chromatography electrospray ionization tandem mass spectrometry (HPLC-ESI-MS/MS). Mol Nutr Food Res 50(12):1201–1211

Bechoff A, Westby A, Owori C, Menya G, Dhuique-Mayer C, Dufour D, Tomlins K (2010) Effect of drying and storage on the degradation of total carotenoids in orange-fleshed sweetpotato cultivars. J Sci Food Agric 90(4):622–629

Bellail AA, Shaltout OE, Youssef MM, El Gamal AMA (2012) Effect of home-cooking methods on phenolic composition and antioxidant activity of sweetpotato (*Ipomoea batatas* (L.) Lam.) cultivars grown in Egypt. Food Nutr Sci 3(4):490–499

Bengtsson A, Namutebi A, Alminger ML, Svanverg U (2008) Effects of various traditional processing methods on the all-trans-b-carotene content of orange-fleshed sweet potato. J Food Comp Anal 21:134–143

Boggess TS Jr, Marion JE, Woodroof JG, Dempsey AH (1967) Changes in lipid composition of sweet potatoes as affected by controlled storage. J Food Sci 32:554–558

Boggess TS Jr, Marion JE, Dempsey AH (1970) Lipid and other compositional changes in 9 varieties of sweet potatoes during storage. J Food Sci 35:306–309

Boo HO, Chon SU, Kim SM, Pyo BS (2005) Antioxidant activities of colored sweet potato cultivars by plant parts. Food Sci Biotechnol 14(2):177–180

Bossard E (1996) Quelques notes sur l'alimentation et les apports nutritionnels occultes en Angola. Garcia de Orta, Sér Bot Lisboa 13(1):7–41

Bouquet A (1969) Féticheurs et Médecines Traditionnelles du Congo (Brazzaville). Mém. O.R.S.T.O.M., 36, 282 pp

Bouquet A, Debray M (1974) Plantes Médicinales de la Côte d'Ivoire. Mémoires O.R.S.T.O.M., 32. Paris, 231 pp

Boyd MR, Burka LT (1978) In vivo studies on the relationship between target organ alkylation and the pulmonary toxicity of a chemically reactive metabolite of 4-ipomeanol. J Pharmacol Exp Ther 207:687–697

Boyd MR, Wilson BJ (1972) Isolation and characterization of 4-ipomeanol, a lung-toxic furanoterpenoid by sweet potatoes (*Ipomoea batatas*). J Agric Food Chem 20(2):428–430

Boyd MR, Wilson BJ, Harris TM (1972) Confirmation by chemical synthesis of the structure of 4-ipomeanol, a lung-toxic metabolite of the sweet potato, *Ipomoea batatas*. Nat New Biol 236(66):158–159

Boyd MR, Burka LT, Harris TM, Wilson BJ (1974) Lung-toxic furanoterpenoids produced by sweet potatoes (*Ipomoea batatas*) following microbial infection. Biochim Biophys Acta 337(2):184–195

Burka LT (1978) 1-(3'-furyl)-6,7-dihydroxy-4,8-dimethylnonan-1-one, a stress metabolite from sweet potatoes, *Ipomoea batatas*. Phytochemistry 17:317–318

Burka LT, Iles J (1979) Myoporone and related keto alcohols from stressed sweet potatoes. Phytochemistry 18(5):873–874

Burka LT, Kuhnert L (1977) Biosynthesis of furanosesquiterpenoid stress metabolites in sweet potatoes (*Ipomoea batatas*). Oxidation of ipomeamarone to 4-hydroxymyoporone. Phytochemistry 16(12):2022–2023

Burka LT, Wilson BJ (1976) Toxic furanosesquiterpenoids from mold-damaged sweet potatoes (*Ipomoea batatas*). In: Gould RF (ed) Mycotoxins and other fungal related food problems, vol 149, Advances in Chemistry., pp 387–399

Burka LT, Bowen RM, Wilson BJ, Harris TM (1974) 7-hydroxymyoporone, a new toxic furanosesquiterpene from mold-damaged sweet potatoes. J Org Chem 39(22):3241–3244

Burka LT, Felice LJ, Jackson SW (1981) 6-oxodendrolasin, 6-hydroxydendrolasin, 9- oxofarnesol and 9- hydroxy farnesol, stress metabolites of the sweet potato. Phytochemistry 20(4):647–652

Cai XB, Yang Y, Sun YP, Zhang L, Xiao Y, Zhao H (2010) Electricity generation from sweet potato fuel ethanol wastewater using microbial fuel cell technology. Huan Jing Ke Xue 31(10):2512–2517 (In Chinese)

Carrasco-Letelier L, Vázquez D, D'Ottone F, Resquín F, Scoz R, Vilaró F, Rodríguez G, Vicente E, Terra J (2013) Balance energético de cadenas agro-industriales de interés para la producción de bioenergías. Revista INIA 32:46–50

Catalano EA, Hasling VC, Dupuy HP, Constantin RJ (1977) Ipomeamarone in blemished and diseased sweet potatoes (*Ipomoea batatas*). J Agric Food Chem 25(1):94–96

Cevallos-Casals BA, Cisneros-Zevallos LA (2002) Bioactive and functional properties of purple sweetpotato (*Ipomoea batatas* (L.) Lam). Acta Hort (ISHS) 583:195–203

Chan CF, Lien CY, Lai YC, Huang CL, Liao WC (2010) Influence of purple sweet potato extracts on the UV absorption properties of a cosmetic cream. J Cosmet Sci 61(5):333–341

Chang WH, Chen CM, Hu SP, Kan NW, Chiu CC, Liu JF (2007) Effect of purple sweet potato leaf consumption on the modulation of the antioxidative status in basketball players during training. Asia Pac J Clin Nutr 16(3):455–461

Chang WH, Hu SP, Huang YF, Yeh TS, Liu JF (2010) Effect of purple sweet potato leaves consumption on exercise-induced oxidative stress and IL-6 and HSP72 levels. J Appl Physiol 109(6):1710–5

Chang WC, Liu KL, Hsu FC, Jeng ST, Cheng YS (2012) Ipomoelin, a jacalin-related lectin with a compact tetrameric association and versatile carbohydrate binding properties regulated by its N terminus. PLoS One 7(7), e40618

Chang VH, Yang DH, Lin HH, Pearce G, Ryan CA, Chen YC (2013) IbACP, a sixteen-amino-acid peptide isolated from *Ipomoea batatas* leaves, induces carcinoma cell apoptosis. Peptides 47:148–156

Chao PY, Huang YP, Hsieh WB (2013) Inhibitive effect of purple sweet potato leaf extract and its components on cell adhesion and inflammatory response in human aortic endothelial cells. Cell Adh Migr 7(2):237–245

Chao PY, Lin SY, Lin KH, Liu YF, Hsu JI, Yang CM, Lai JY (2014) Antioxidant activity in extracts of 27 indigenous Taiwanese vegetables. Nutrients 6(5):2115–2130

Chen CM, Li SC, Lin YL, Hsu CY, Shieh MJ, Liu JF (2005) Consumption of purple sweet potato leaves modulates human immune response: T-lymphocyte functions, lytic activity of natural killer cell and antibody production. World J Gastroenterol 11(37):5777–5781

Chen CM, Lin YL, Chen CY, Hsu CY, Shieh MJ, Liu JF (2008a) Consumption of purple sweet potato leaves decreases lipid peroxidation and DNA damage in humans. Asia Pac J Clin Nutr 17(3):408–414

Chen YC, Lin HH, Jeng ST (2008b) Calcium influxes and mitogen-activated protein kinase kinase activation mediate ethylene inducing ipomoelin gene expression in sweet potato. Plant Cell Environ 31(1):62–72

Chen WP, Mao TJ, Fan L, Zhou YH, Yu J, Jin Y, Hou PC (2011) Effect of purple sweet potato on lipid metabolism and oxidative stress in hyperlipidemic rats. Zhejiang Da Xue Xue Bao Yi Xue Ban 40(4):360–364 (In Chinese)

Chen YY, Lai MH, Hung HY, Liu JF (2013) Sweet potato [*Ipomoea batatas* (L.) Lam. "Tainong 57"] starch improves insulin sensitivity in high-fructose diet-fed rats by ameliorating adipocytokine levels, pro-inflammatory status, and insulin signaling. J Nutr Sci Vitaminol (Tokyo) 59(4):272–280

Cheng S, Zhang YF, Zeng ZQ, Lin J, Zhang YW, Ni H, Li HH (2014) Screening, separating, and completely recovering polyphenol oxidases and other biochemicals from sweet potato wastewater in starch production. Appl Microbiol Biotechnol (In Press)

Chi XW, Lin CT, Jiang YC, Wen L, Lin CT (2012) A dithiol glutaredoxin cDNA from sweet potato (*Ipomoea batatas* [L.] Lam): enzyme properties and kinetic studies. Plant Biol (Stuttg) 14(4):65–665

Cho J, Kang JS, Long PH, Jing J, Back Y, Chung KS (2003) Antioxidant and memory enhancing effects of purple sweet potato anthocyanin and cordyceps mushroom extract. Arch Pharm Res 26(10):821–825

Choi JH, Choi CY, Lee KJ, Hwang YP, Chung YC, Jeong HG (2009) Hepatoprotective effects of an anthocyanin fraction from purple-fleshed sweet potato against acetaminophen-induced liver damage in mice. J Med Food 12(2):320–326

Choi JH, Hwang YP, Choi CY, Chung YC, Jeong HG (2010) Anti-fibrotic effects of the anthocyanins isolated from the purple-fleshed sweet potato on hepatic fibrosis induced by dimethylnitrosamine administration in rats. Food Chem Toxicol 48(11):3137–3143

Choi SJ, Kim JK, Kim HK, Harris K, Kim CJ, Park GG, Park CS, Shin DH (2013) 2,4-Di-tert-butylphenol from sweet potato protects against oxidative stress in PC12 cells and in mice. J Med Food 16(11):977–983

Chu CY, Sen B, Lay CH, Lin YC, Lin CY (2012) Direct fermentation of sweet potato to produce maximal hydrogen and ethanol. Appl Energy 100:10–18

Clark CA, Lawrence A, Martin AF (1981) Accumulation of furanoterpenoids in sweet potato tissue following inoculation with different pathogens. Phys Chem Chem Phys 71:708–711

Cody M, Haard NF (1976) Influence of cooking on toxic stress metabolites in sweet potato root. J Food Sci 41(2):469–470

Cordeiro N, Freitas N, Faria M, Gouveia M (2013) *Ipomoea batatas* (L.) Lam.: a rich source of lipophilic phytochemicals. J Agric Food Chem 61(50):12380–12384

Coxon DT, Curtis RF, Howard B (1975) Ipomeamarone a toxic furanoterpenoid in sweet potatoes *(Ipomoea batatas)* in the United Kingdom. Food Cosmet Toxicol 13(1):87–90

Dansby MY, Bovell-Benjamin AC (2003) Physical properties and sixth graders' acceptance of an extruded ready-to-eat sweetpotato breakfast cereal. J Food Sci 68(8):2607–2612

Darwish MM, Farag MFS, Osman NN (2010) Sweet potato *Ipomoea batatas* modulates radiation-induced oxidative damage in liver and kidney of male albino rats. J Rad Res Appl Sci 3(3A):733–746

De Wet H, Nzama VN, Van Vuuren SF (2012) Medicinal plants used for the treatment of sexually transmitted infections by lay people in northern Maputaland, KwaZulu–Natal Province, South Africa. South Afr J Bot 78:12–20

Dilworth L, Omoruyi FO, Reid W, Asemota HN (2008) Bone and faecal minerals and scanning electron microscopic assessments of femur in rats fed phytic acid extract from sweet potato (*Ipomoea batatas*). Biometals 21(2):133–141

Dincer C, Karaoglan M, Erden F, Tetik N, Topuz A, Ozdemir F (2011) Effects of baking and boiling on the nutritional and antioxidant properties of sweet potato [*Ipomoea batatas* (L.) Lam.] cultivars. Plant Foods Hum Nutr 66(4):341–347

Ding HX, Gao YY, Lei HW, Luo LP, Chao HJ, Roger Ruan R (2010) In vitro antioxidant effects of flavonoids of sweet potato vines. Int J Foo Prop 13(2):360–368

Dini I, Tenore GC, Dini A (2006a) New polyphenol derivative in *Ipomoea batatas* tubers and its antioxidant activity. J Agric Food Chem 54(23):8733–8737

Dini I, Tenore GC, Trimarco E, Dini A (2006b) Seven new aminoacyl sugars in *Ipomoea batatas*. J Agric Food Chem 54(16):6089–6093

Dini I, Tenore GC, Dini A (2009) Saponins in *Ipomoea batatas* tubers: isolation, characterization, quantification and antioxidant properties. Food Chem 113(2):411–419

Donado-Pestana CM, Salgado JM, de Oliveira Rios A, dos Santos PR, Jablonski A (2012) Stability of carotenoids, total phenolics and in vitro antioxidant capacity in the thermal processing of orange-fleshed sweet potato (*Ipomoea batatas* Lam.) cultivars grown in Brazil. Plant Foods Hum Nutr 67(3):262–270

Doster AR, Mitchell FE, Farrell RL, Wilson BJ (1978) Effects of 4-ipomeanol, a product from mold-damaged sweet potatoes, on the bovine lung. Vet Pathol Online 15(3):367–375

Durham SK, Babish JG, Castleman WL (1987) 4-Ipomeanol-induced effects on Sendai viral pneumonia in mice. Am J Pathol 126(2):364–375

El Far MM, Taie HA (2009) Antioxidant activities, total anthocyanins, phenolics and flavonoids contents of some sweetpotato genotypes under stress of different concentrations of sucrose and sorbitol. Aus J Basic Appl Sci 3(4):3609–3616

Escalante-Sánchez E, Pereda-Miranda R (2007) Batatins I and II, ester-type dimers of acylated pentasaccharides from the resin glycosides of sweet potato. J Nat Prod 70(6):1029–1034

Escalante-Sánchez E, Rosas-Ramírez D, Linares E, Bye R, Pereda-Miranda R (2008) Batatinosides II-VI, acylated lipooligosaccharides from the resin glycosides of sweet potato. J Agric Food Chem 56(20):9423–9428

Espelie KE, Sadek NZ, Kolattukudy PE (1980) Composition of suberin-associated waxes from the subterranean storage organs of seven plants: parsnip, carrot, rutabaga, turnip, red beet, sweet potato and potato. Planta 148(5):468–476

Essiett UA, Ukpong UJ (2014) Comparative phytochemical, nutrient and anti-nutrient of stems of *Ipomoea Involucrata* Beauv, *Ipomoea. Triloba* L. and *Ipomoea Batatas* Lam. Am J Food Nutr 2(4):71–76

FAO (2014) FAO STAT. Food and Agricultural Organization of United Nations: Economic And Social Department: The Statistical Division. http://faostat.fao.org/site/567/DesktopDefault.aspx?PageID=567

Fidrianny I, Windyaswari AS, Wirasutisna KR (2013) DPPH scavenging activity of various extracts of sweet potatoes leaves with varying tubers colors. Int J Res Pharm Sci 3(2):133–145

Fujita M, Oba K, Uritani I (1982) Properties of a mixed function oxygenase catalyzing ipomeamarone 15-hydroxylation in microsomes from cut-injured and *Ceratocystis fimbriata*-infected sweet potato root tissues. Plant Physiol 70(2):573–578

Furuta S, Suda I, Nishiba Y, Yamakawa O (1998) High tert-butylperoxyl radical scavenging activities of sweet potato cultivars with purple flesh. Food Sci Technol Int Tokyo 4(1):33–35

Gichuki ST, BerenyiM ZDP, Hermann SJ, Glossl J, Burg K (2003) Genetic diversity in sweetpotato [*Ipomoea batatas* (L.) Lam.] in relationship to geographic sources as assessed with RAPD markers. Genet Resour Crop Evol 50(4):429–437

Giday M, Asfaw Z, Woldu Z, Teklehaymanot T (2009) Medicinal plant knowledge of the Bench ethnic group of Ethiopia: an ethnobotanical investigation. J Ethnobiol Ethnomed 5:34

Goda Y, Shimizu T, Kato Y, Nakamura M, Maitani T, Yamada T, Terahara N, Yamaguchi M (1997) Two acylated anthocyanins from purple sweet potato. Phytochemistry 44(1):183–186

Grace MH, Yousef GG, Gustafson SJ, Truong VD, Yencho GC, Lila MA (2014) Phytochemical changes in phenolics, anthocyanins, ascorbic acid, and carotenoids associated with sweet potato storage and impacts on bioactive properties. Food Chem 145:717–724

Gundala SR, Yang C, Lakshminarayana N, Asif G, Gupta MV, Shamsi S, Aneja R (2013) Polar biophenolics in sweet potato greens extract synergize to inhibit pros-

tate cancer cell proliferation and in vivo tumor growth. Carcinogenesis 34(9):2039–2049
Hagenimana V, Simard RE, Vézina LP (1994) Amylolytic activity in germinating sweetpotato (*Ipomoea batatas* L.) roots. J Am Soc Hort Sci 119(2):313–320
Hagiwara A, Yoshino H, Ichihara T, Kawabe M, Tamano S, Aoki H, Koda T, Nakamura M, Imaida K, Ito N, Shirai T (2002) Prevention by natural food anthocyanins, purple sweet potato color and red cabbage color, of 2-amino-1-methyl-6-phenylimidazo[4,5-b]pyridine (PhIP)-associated colorectal carcinogenesis in rats initiated with 1,2-dimethylhydrazine. J Toxicol Sci 27(1):57–68
Han YT, Chen XH, Xie J, Zhan SM, Wang CB, Wang LX (2011) Purple sweet potato pigments scavenge ROS, reduce p53 and modulate Bcl-2/Bax to inhibit irradiation-induced apoptosis in murine thymocytes. Cell Physiol Biochem 28(5):865–872
Hao ZH, Wang DJ, Chen H, Sun JM, Xu YY (2014) Sweet potato starch residue as starting material to prepare polyacrylonitrile adsorbent via SI-SET-LRP. J Agric Food Chem 62(8):1765–1770
Harada K, Kano M, Takayanagi T, Yamakawa O, Ishikawa F (2004) Absorption of acylated anthocyanins in rats and humans after ingesting an extract of *Ipomoea batatas* purple sweet potato tuber. Biosci Biotechnol Biochem 68(7):1500–1507
Harrison HF, Peterson JK, Snook ME, Bohac JR, Jackson DM (2003) Quantity and potential biological activity of caffeic acid in sweet potato [*Ipomoea batatas* (L.) Lam.] storage root periderm. J Agric Food Chem 51(10):2943–2948
Hashizume T, Suye S, Soeda T, Sugiyama T (1982a) Isolation and characterization of a new glucopyranosyl derivative of 6-(3-methyl-2-butenylamino)purine from sweet potato tubers. FEBS Lett 144:25–28
Hashizume T, Suye S, Sugiyama T (1982b) Isolation and identification of *cis*-zeatin riboside from tubers of sweet potato (*Ipomoea batatas* L.). Agric Biol Chem 46(3):663–665
Hattori T, Nakagawa T, Maeshima M, Nakamura K, Asahi T (1985) Molecular cloning and nucleotide sequence of cDNA for sporamin, the major protein of sweet potato tuberous roots. Plant Mol Biol 5:313–320
Hayek SA, Shahbazi A, Awaisheh SS, Shah NP, Ibrahim SA (2013) Sweet potatoes as a basic component in developing a medium for the cultivation of lactobacilli. Biosci Biotechnol Biochem 77(11):2248–2254
He K, Ye XL, Li XG, Chen HY, Yuan LJ, Deng YF, Chen X, Li XD (2012) Separation of two constituents from purple sweet potato by combination of silica gel column and high-speed counter-current chromatography. J Chromatogr B Analyt Technol Biomed Life Sci 881–882:49–54
Heinze PH, Appleman CO (1943) A biochemical study of curing processes in sweet potatoes. Plant Physiol 18(4):548–555
Hiura M (1943) Studies on storage and rot of sweet potato (2). Rep Gifu Agric Coll 50:1–5 (In Japanese)
Horvat RJ, Arrendale RF, Dull GG, Chapman GW, Kays SJ (1991) Volatile constituents and sugars of three diverse cultivars of sweet potatoes [*Ipomoea batatas* (L.) Lam.]. J Food Sci 56(3):714–715
Hou WC, Lin YH (1997a) Dehydroascorbate reductase and monodehydroascorbate reductase activities of trypsin inhibitors, the major sweet potato (*Ipomoea batatas* [L.] Lam) root storage protein. Plant Sci 128(2):151–158
Hou WC, Lin YH (1997b) Polyamine-bound trypsin inhibitors in sweet potato (*Ipomoea batatas* [L.] Lam cv. Tainong 57) storage roots, sprouted roots and sprouts. Plant Sci 126(1):11–19
Hou WC, Chen YC, Chen HJ, Lin YH, Yang LL, Lee MH (2001) Antioxidant activities of trypsin inhibitor, a 33 KDa root storage protein of sweet potato (*Ipomoea batatas* (L.) Lam cv. Tainong 57). J Agric Food Chem 49(6):2978–2981
Huang DJ, Chen HJ, Hou WC, Lin CD, Lin YH (2004a) Active recombinant thioredoxin h protein with antioxidant activities from sweet potato (*Ipomoea batatas* [L.] Lam 'Tainong 57') storage roots. J Agric Food Chem 52(15):4720–4724
Huang DJ, Lin CD, Chen HJ, Lin YH (2004b) Antioxidant and antiproliferative activities of sweet potato (*Ipomoea batatas* [L.] Lam 'Tainong 57') constituents. Bot Bull Acad Sin 45:179–186
Huang GJ, Chen HJ, Hou WC, Lin YH (2004c) Isolation and characterization of thioredoxin h cDNA from sweet potato (Ipomoea batatas [L.] Lam 'Tainong 57') storage roots. Plant Sci 166(2):515–523
Huang DJ, Hou WC, Chen HJ, Lin YH (2006) Sweet potato (*Ipomoea batatas* (L.) Lam. 'Tainong 57') storage root mucilage exhibited angiotensin converting enzyme inhibitory activity in vitro. Bot Stud 47:397–402
Huang GJ, Chen HJ, Chang YS, Sheu MJ, Lin YH (2007a) Recombinant sporamin and its synthesized peptides with antioxidant activities in-vitro. Bot Stud 48:133–140
Huang GJ, Sheu MJ, Chen HJ, Chang YS, Lin YH (2007b) Growth inhibition and induction of apoptosis in NB4 promyelocytic leukemia cells by trypsin inhibitor from sweet potato storage roots. J Agric Food Chem 55(7):2548–2553
Huang GJ, Chen HJ, Chang YS, Lu TL, Lin YH (2008a) Sweet potato storage root thioredoxin h2 with both dehydroascorbate reductase and monodehydroascorbate reductase activities. Bot Stud 49:1–7
Huang GJ, Chen HJ, Ho YL, Chen HJ, Chang YS, Huang SS, Hung HJ, Lin YH (2008b) Sweet potato storage root trypsin inhibitor and their peptic hydrolysates exhibited angiotensin converting enzyme inhibitory activity in vitro. Bot Stud 49:101–108
Huang GJ, Lai HC, Chang YS, Sheu MJ, Lu TL, Huang SS, Lin YH (2008c) Antimicrobial, dehydroascorbate reductase, and monodehydroascorbate reductase activities of defensin from sweet potato [*Ipomoea batatas* (L.) Lam. 'Tainong 57'] storage roots. J Agric Food Chem 56(9):2989–2995

Huang SS, Hung HJ, Chiu CS, Chen HJ, Lin SS, Lin YC, Chang HY, Huang GJ, Lin YH (2009) Sweet potato trypsin inhibitor with thioltransferase-like and glutathione S-transferase-like activities. Bot Stud 50:443–50

Huang GJ, Chen HJ, Susumu K, Wu JB, Hou WC, Wu CH, Sheu MJ, Huang SS, Lin YH (2011a) Sweet potato storage root thioredoxin h2 and their peptic hydrolysates exhibited angiotensin converting enzyme inhibitory activity in vitro. Bot Stud 52(1):15–22

Huang GJ, Lu TL, Chiu CS, Chen HJ, Wu CH, Lin YC, Hsiedh WT, Liao JC, Sheu MJ, Lin YH (2011b) Sweet potato storage root defensin and its tryptic hydrolysates exhibited angiotensin converting enzyme inhibitory activity in vitro. Bot Stud 52(3):257–264

Huang GJ, Deng JS, Chen HJ, Huang SS, Liao JC, Hou WC, Lin YH (2012) Defensin protein from sweet potato (*Ipomoea batatas* [L.] Lam 'Tainong 57') storage roots exhibits antioxidant activities in vitro and ex vivo. Food Chem 135(3):861–867

Hue SM, Boyce AN, Somasundram C (2012) Antioxidant activity, phenolic and flavonoid contents in the leaves of different varieties of sweet potato (*Ipomoea batatas*). Aus J Crop Sci 6(3):375–380

Hwang YP, Choi JH, Choi JM, Chung YC, Jeong HG (2011a) Protective mechanisms of anthocyanins from purple sweet potato against tert-butyl hydroperoxide-induced hepatotoxicity. Food Chem Toxicol 49(9):2081–2089

Hwang YP, Choi JH, Han EH, Kim HG, Wee JH, Jung KO, Jung KH, Kwon KI, Jeong TC, Chung YC, Jeong HG (2011b) Purple sweet potato anthocyanins attenuate hepatic lipid accumulation through activating adenosine monophosphate-activated protein kinase in human HepG2 cells and obese mice. Nutr Res 31(12):896–906

Hwang YP, Choi JH, Yun HJ, Han EH, Kim HG, Kim JY, Park BH, Khanal T, Choi JM, Chung YC, Jeong HG (2011c) Anthocyanins from purple sweet potato attenuate dimethylnitrosamine-induced liver injury in rats by inducing Nrf2-mediated antioxidant enzymes and reducing COX-2 and iNOS expression. Food Chem Toxicol 49(1):93–99

Ijaola TO, Osunkiyesi AA, Taiwo AA, Oseni OA, Lanreiyanda Y, Ajayi JO, Oyede RT (2014) Antidianetic effect of *Ipomoea batatas* in normal and alloxan-induced diabetic rats. IOSR J Appl Chem 7(5):16–25

Imaseki H, Uritani I (1964) Ipomeamarone accumulation and lipid metabolism in sweet potato infected by the black rot fungus II. Accumulation mechanism of ipomeamarone in the infected region with special regard to contribution of the non-infected tissue. Plant Cell Physiol 5(2):33–143

Imaseki H, Takei S, Uritani I (1964) Ipomeamarone accumulation and lipid metabolism in sweet potato infected by the black rot fungus I. Identification of sterol and changes in lipid metabolism during infection process. Plant Cell Physiol 5(2):119–132

Innami S, Tabata K, Shimizu J, Kusunoki K, Ishida H, Matsuguma M, Wada M, Sugiyama N, Kondo M (1998) Dried green leaf powders of Jew's mellow (*Corchorus*), persimmon (*Diospyros kaki*) and sweet potato (*Ipomoea batatas* Poir) lower hepatic cholesterol concentration and increase fecal bile acid excretion in rats fed a cholesterol-free diet. Plant Foods Hum Nutr 52(1):55–65

Inoue H, Uritani I (1979) Biosynthetic correlation of various phytoalexins in sweet potato root tissue infected by *Ceratocystisfimbriata*. Plant Cell Physiol 20:1307–1314

Inoue H, Uritani I (1980a) Conversion of 4-hydroxydehydromyoporone to other furanoterpenes in *Ceratocystis fimbriata*-infected sweetpotato. Agric Boil Chem 44(8):1935–1936

Inoue H, Uritani I (1980b) Furano-sesquiterpene reductase from fungal-inoculated sweetpotato root tissue. Agric Biol Chem 44(9):2245–2247

Inoue H, Kato N, Uritani I (1977) 4-Hydroxydehydromyoporone from infected *Ipomoea batatas* root tissue. Phytochemistry 16(7):1063–1065

Ishida H, Suzuno H, Sugiyama N, Innami S, Tadokoro T, Maekawa A (2000) Nutritive evaluation on chemical components of leaves, stalks and stems of sweet potatoes (*Ipomoea batatas* Poir). Food Chem 68(3):359–367

Ishida H, Furusyo T, Suzuno H, Yamamoto Y, Innami S, Tadokoro T (2004) Sweet potato (*Ipomoea batatas* Poiret) leaves suppress an elevation of blood glucose in rats. J Integr Study Diet Habit 15(2):111–117

Ishiguro K, Yahara S, Yoshimoto M (2007) Changes in polyphenolic content and radical-scavenging activity of sweet potato (*Ipomoea batatas* L.) during storage at optimal and low temperatures. J Agric Food Chem 55(26):10773–10778

Islam MS (2006) Sweetpotato (*Ipomoea batatas* L.) leaf: Its potential effect on human health and nutrition. J Food Sci 71:R13–R21

Islam MS, Yoshimoto M, Terahara N, Yamakawa O (2002a) Anthocyanin compositions in sweetpotato (*Ipomoea batatas* L.) leaves. Biosci Biotechnol Biochem 66(11):2483–2486

Islam SM, Yoshimoto M, Yahara S, Okuno S, Ishiguro K, Yamakawa O (2002b) Identification and characterization of foliar polyphenolic composition in sweetpotato (*Ipomoea batatas* L.) genotypes. J Agric Food Chem 50:3718–3722

Islam MS, Yoshimoto M, Ishiguro K, Okuno S, Yamakawa O (2003a) Effect of artificial shading and temperature on radical scavenging activity and polyphenolic composition in sweetpotato (*Ipomoea batatas* L.) leaves. J Am Soc Hort Sci 128:182–187

Islam MS, Yoshimoto M, Yamakawa O (2003b) Distribution and physiological functions of caffeoylquinic acid derivatives in leaves of sweetpotato genotypes. J Food Sci 68:111–116

Ito I, Kato N, Uritani I (1984) Biochemistry of two new sesquiterpenoid phytoalexins from sweet potato roots. Agric Biol Chem 48(1):159–164

Jamil KM, Brown KH, Jamil M, Peerson JM, Keenan AH, Newman JW, Haskell MJ (2012) Daily consumption of orange-fleshed sweet potato for 60 days increased plasma β-carotene concentration but did not increase total body vitamin A pool size in Bangladeshi women. J Nutr 142(10):1896–1902

Jaw KS, Chou LH, Chang SM, Duan KJ (2007) Purification of a trypsin inhibitor from sweet potato in an aqueous two phase system. Biotechnol Lett 29(1):137–140

Jiang HF, Li XR, Tang C (2011) Effect of purple sweet potato flavonoids on metabolism of glucose and lipids in diabetic rats. Zhejiang Da Xue Xue Bao Yi Xue Ban 40(4):374–379 (In Chinese)

Jiao YZ, Yang ZD, Jiang YJ, Zhai WW (2012) Study on chemical constituents and antioxidant activity of anthocyanins from purple sweet potato (*Ipomoea batatas* L.). Int J Food Eng 8(2):1–16

Jiao YZ, Jiang YJ, Zhai WW, Yang ZD (2014) Studies on antioxidant capacity of anthocyanin extract from purple sweet potato (*Ipomoea batatas* L.). Afr J Biotechnol 11(27):7046–7054

Jih PJ, Chen YC, Jeng ST (2003) Involvement of hydrogen peroxide and nitric oxide in expression of the ipomoelin gene from sweet potato. Plant Physiol 132(1):381–389

Johnson M, Pace RD (2010) Sweet potato leaves: properties and synergistic interactions that promote health and prevent disease. Nutr Rev 68(10):604–615

Johnson R, Moorthy SN, Padmaja G (2010) Production of high fructose syrup from cassava and sweet potato flours and their blends with cereal flours. Food Sci Technol Int 16(3):251–258

Johnson M, Pace RD, Dawkins NL, Willian KR (2013) Diets containing traditional and novel green leafy vegetables improve liver fatty acid profiles of spontaneously hypertensive rats. Lipids Health Dis 12:168

Jones DB, Gersdorff CE (1931) Ipomoein, A globulin from sweet potatoes, *Ipomoea batatas* isolation of a secondary protein derived from ipomoein by enzymic action. J Biol Chem 93(1):119–126

Ju JH, Yoon HS, Park HJ, Kim MY, Shin HK, Park KY, Yang JO, Sohn MS, Do MS (2011) Anti-obesity and antioxidative effects of purple sweet potato extract in 3T3-L1 adipocytes in vitro. J Med Food 14(10):1097–106

Kalanda K, Bolamba K (1994) Contribution à la connaissance des plantes médicinales du Haut Zaïre. Les plantes utilisées contre les maladies de la peau à Kisangani. Rev Méd Pharm Afr 8(2):179–188 (In French)

Kaneshiro T, Suzui M, Takamatsu R, Murakami A, Ohigashi H, Fujino T, Yoshimi N (2005) Growth inhibitory activities of crude extracts obtained from herbal plants in the Ryukyu Islands on several human colon carcinoma cell lines. Asian Pac J Cancer Prev 6:353–358

Kano M, Takayanagi T, Harada K (2005) Antioxidative activity of anthocyanins from purple sweet potato, *Ipomoea batatas* cultivar Ayamurasaki. Biosci Biotechnol Biochem 69(5):979–988

Karna P, Gundala SR, Gupta MV, Shamsi SA, Pace RD, Yates C, Narayan S, Aneja R (2011) Polyphenol-rich sweet potato greens extract inhibits proliferation and induces apoptosis in prostate cancer cells in vitro and in vivo. Carcinogenesis 32(12):1872–1880

Katayama K, Komaki K, Tamiyaeiyi S, Takayanagi K (1999) Varietal and annual variations in pasting properties of sweet potato. Breed Sci 49:173–178

Kato N, Imaseki H, Nakashima N, Uritani I (1971) Structure of a new sesquiterpenoid, ipomeamaronol, in diseased sweet potato root tissue. Tetrahedron Lett 12(13):843–846

Kato N, Imaseki H, Nakashima N, Akazawa T, Uritani I (1973) Isolation of a new phytoalexin-like compound, ipomeamaronol, from black-rot fungus infected sweet potato root tissue, and its structural elucidation. Plant Cell Physiol 14(3):597–606

Kerharo J, Bouquet A (1950) Plantes Médicinales et Toxiques de la Côte d'Ivoire - Haute-Volta. Vigot Frères, Paris, France, 291 pp

Khachatryan A, Bansode RR, Labonte DR, Losso JN (2003) Identification of sweet potato leaves (*Ipomoea batatas*) as an excellent source of lutein. Conference paper Institute of Food Technologists Annual meeting 2003, Chicago. http://ift.confex.com/ift/2003/techprogram/paper_20401.htm

Kim JK, Choi SJ, Cho HY, Kim YJ, Lim ST, Kim CJ, Kim EK, Kim HK, Peterson S, Shin DH (2011) *Ipomoea batatas* attenuates amyloid β peptide-induced neurotoxicity in ICR mice. J Med Food 14(3):304–309

Kim SH, Jeong JC, Ahn YO, Lee HS, Kwak SS (2014) Differential responses of three sweetpotato metallothionein genes to abiotic stress and heavy metals. Mol Biol Rep (In Press)

Kiran KS, Padmaja G (2003) Inactivation of trypsin inhibitors in sweet potato and taro tubers during processing. Plant Foods Hum Nutr 58(2):153–163

Kojima M, Uritani I (1972a) Elucidation of the structure of a possible intermediate in chlorogenic acid biosynthesis in sweet potato root tissue. Plant Cell Physiol 13(6):1075–1084

Kojima M, Uritani I (1972b) Studies on chlorogenic acid biosynthesis in sweet potato root tissue using trans-cinnamic acid-2-^{14}C and quinic acid-G-^{3}H. mediate in chlorogenic acid biosynthesis in sweet potato root tissue. PLoS One 13(2):311–319

Kojima M, Uritani I (1973) Studies on chlorogenic acid biosynthesis in sweet potato root tissue in special reference to the isolation of chlorogenic acid intermediate. Plant Physiol 51:768–711

Kojima M, Minamikawa T, Hyodo H, Uritani I (1969) Incorporation of some possible radioactive intermediates into chlorogenic acid in sliced sweet potato tissues. Plant Cell Physiol 10(2):471–474

Kolattukudy PE, Kronman K, Poulose AJ (1975) Determination of structure and composition of suberin

from the roots of carrot, parsnip, rutabaga, turnip, red beet, and sweet potato by combined gas–liquid chromatography and mass spectrometry. Plant Physiol 55(3):567–73

Konczak-Islam I, Yoshimoto M, Hou DX, Terahara N, Yamakawa O (2003) Potential chemopreventive properties of anthocyanin-rich aqueous extracts from in vitro produced tissue of sweetpotato (*Ipomoea batatas* L.). J Agric Food Chem 51(20):5916–5922

Kubota T (1958) Volatile constituents of black-rotted sweet potato and related substances. Tetrahedron 4:68–86

Kubota T, Ichikawa N (1954) Chemical studies on the black rot disease of sweet potato. VIII-IX. IX. On the chemical constitution of ipomeanine. Nippon kagaku zassi 75(4):450–456 (In Japanese)

Kubota T, Matsuura T (1958) The synthesis of (±)-ipomeamarone [(±)-ngaione] and its steric isomers. J Chem Soc 1958:3667–3677

Kubota T, Matsuura T, Kakuno Y (1965) Chemical studies on the black rot disease of sweet potatoes. XVII. Abnormal semicarbazone of ipomeanine and of related compounds. Bull Chem Soc Japan 38(7):1191–1194

Kurata R, Adachi M, Yamakawa O, Yoshimoto M (2007) Growth suppression of human cancer cells by polyphenolics from sweetpotato (*Ipomoea batatas* L.) leaves. J Agric Food Chem 55(1):185–90

Kusano S, Abe H (2000) Antidiabetic activity of white skinned sweet potato (*Ipomoea batatas* L.) in obese Zucker fatty rat. Biol Pharm Bull 23(1):23–26

Kusano S, Abe H, Okada A (1998) Study of antidiabetic activity of white skinned sweet potato (*Ipomoea batatas* L.): comparison of normal and streptozotocin induced diabetic rats and hereditary diabetic mice. Nippon Nougeikagaku Kaishi 72(9):1045–1052 (In Japanese)

Kusano S, Abe H, Tamura H (2001) Isolation of antidiabetic components from white-skinned sweet potato (*Ipomoea batatas* L.). Biosci Biotechnol Biochem 65(1):109–114

La Bronte DR, Picha DH, Johnson HA (2000) Carbohydrate-related changes in sweetpotato storage roots during development. J Am Soc Hort Sci 125(2):200–204

Lacroix D, Prado S, Kamoga D, Kasenene J, Namukobe J, Krief S, Dumontet V, Mouray E, Bodo B, Brunois F (2011) Antiplasmodial and cytotoxic activities of medicinal plants traditionally used in the village of Kiohima, Uganda. J Ethnopharmacol 133:850–855

Lai YC, Huang CL, Chan CF, Lien CY, Liao WC (2013) Studies of sugar composition and starch morphology of baked sweet potatoes (*Ipomoea batatas* (L.) Lam). J Food Sci Technol 50(6):1193–1199

Lakhanpal S, Donehower RC, Rowinsky EK (2001) Phase II study of 4-ipomeanol, a naturally occurring alkylating furan, in patients with advanced hepatocellular carcinoma. Invest New Drugs 19(1):69–76

Lareo C, Ferrari MD, Guigou M, Fajardo L, Larnaudie V, Ramirez MB, Martinez-Garreiro J (2013) Evaluation of sweet potato for fuel bioethanol production: hydrolysis and fermentation. Springerplus 2(1):493

Lee LS, Chang EJ, Rhim JW, Ko BS, Choi SW (1997) Isolation and identification of anthocyanins from purple sweet potatoes. J Food Sci Nutr 2(2):83–88

Lee JJ, Park KW, Kwak YS, Ahn JY, Jung YH, Lee BH, Jeong JC, Lee HS, Kwak SS (2012) Comparative proteomic study between tuberous roots of light orange- and purple-fleshed sweetpotato cultivars. Plant Sci 193–194:120–129

Lee MJ, Park JS, Choi DS, Jung MY (2013) Characterization and quantitation of anthocyanins in purple-fleshed sweet potatoes cultivated in Korea by HPLC-DAD and HPLC-ESI-QTOF-MS/MS. J Agric Food Chem 61(12):3148–3158

Lewthwaite SL, Sutton KH, Triggs CM (1997) Free sugar composition of sweetpotato cultivars after storage. NZ J Crop Hort Sci 25(1):33–41

Li X, Castleman WL (1990) Ultrastructural morphogenesis of 4-ipomeanol-induced bronchiolitis and interstitial pneumonia in calves. Vet Pathol 27(3):141–149

Li X, Castleman WL (1991) Effects of 4-ipomeanol on bovine parainfluenza type 3 virus-induced pneumonia in calves. Vet Pathol 28(5):428–437

Li HS, Oba K (1985) Major soluble proteins of sweet potato roots and changes in proteins after cutting, infection, or storage. Agric Biol Chem 49(3):737–744

Li CG, Zhang LY (2013) In vivo anti-fatigue activity of total flavonoids from sweetpotato [*Ipomoea batatas* (L.) Lam.] leaf in mice. Indian J Biochem Biophys 50(4):326–329

Li F, Li Q, Gao D, Peng Y (2009) The optimal extraction parameters and anti-diabetic activity of flavonoids from *Ipomoea batatas* leaf. Afr J Tradit Complement Altern Med 6(2):195–202

Li JY, Dong GP, Li ML, Liu ZH, Lu Y (2012) Efficient counter-current chromatographic isolation and structural identification of phenolic compounds from sweet potato leaves. J Liq Chromatogr Rel Technol 35(11):1517–1527

Li J, Li XD, Zhang Y, Zheng ZD, Qu ZY, Liu M, Zhu SH, Liu S, Wang M, Qu L (2013a) Identification and thermal stability of purple-fleshed sweet potato anthocyanins in aqueous solutions with various pH values and fruit juices. Food Chem 136(3–4):1429–1434

Li PG, Mu TH, Deng L (2013b) Anticancer effects of sweet potato protein on human colorectal cancer cells. World J Gastroenterol 19(21):3300–3308

Lian XJ, Li L, Zhang KS, Xu Y, Lu JX (2012) A new proposed sweet potato starch granule structure-pomegranate concept. Int J Biol Macromol 50(3):471–475

Liao JC, Chiu CS, Chen HJ, Huang SS, Hou WC, Lin WC, Lin YH, Huang GJ (2012) Characterization of a novel cyclophilin-type peptidylprolyl isomerase pro-

tein from sweet potato storage roots. Bot Stud 53(3):315–324

Liao CC, Chen YW, Jeng TL, Li CR, Kuo CF (2013) Consumption of purple sweet potato affects post-translational modification of plasma proteins in hamsters. J Agric Food Chem 61(50):12450–1248

Lien CY, Chan CF, Huang CL, Lai YC, Liao WC (2012) Studies of carotene extraction from sweet potato variety CYY95-26, *Ipomoea batatas*, L. Int J Food Eng 8(2):1–14

Lim SY, Xu JT, Kim JY, Chen TY, Su XY, Standard J, Carey E, Griffin J, Herndon B, Katz B, Tomich J, Wang WQ (2013) Role of anthocyanin-enriched purple-fleshed sweet potato p40 in colorectal cancer prevention. Mol Nutr Food Res 57(11):1908–1917

Lin YH, Huang TC, Huang C (1988) Quality improvement of sweet-potato (Ipomoea batatas L. Lam.) roots as feed by ensilage. Br J Nutr 60(1):173–184

Lin JS, Lin CC, Lin HH, Chen YC, Jeng ST (2012a) MicroR828 regulates lignin and H2O2 accumulation in sweet potato on wounding. New Phytol 196(2):427–440

Lin PP, Hsieh YM, Kuo WW, Lin CC, Tsai FJ, Tsai CH, Huang CY, Tsai CC (2012b) Inhibition of cardiac hypertrophy by probiotic-fermented purple sweet potato yogurt in spontaneously hypertensive rat hearts. Int J Mol Med 30(6):1365–1375

Lin PP, Hsieh YM, Kuo WW, Lin YM, Yeh YL, Lin CC, Tsai FJ, Tsai CH, Huang CY, Tsai CC (2013a) Probiotic-fermented purple sweet potato yogurt activates compensatory IGF-IR/PI3K/Akt survival pathways and attenuates cardiac apoptosis in the hearts of spontaneously hypertensive rats. Int J Mol Med 32(6):1319–1328

Lin PP, Hsieh YM, Kuo WW, Lin YM, Yeh YL, Lin CC, Tsai FJ, Tsai CH, Tsai CC, Huang CY (2013b) Suppression of TLR-4-related inflammatory pathway and anti-fibrosis effects of probiotic-fermented purple sweet potato yogurt in hearts of spontaneously hypertensive rats. Chin J Physiol 56(3):174–183

Lin JS, Lin HH, Li YC, King YC, Sung RJ, Kuo YW, Lin CC, Shen YH, Jeng ST (2014) Carbon monoxide regulates the expression of the wound-inducible gene ipomoelin through antioxidation and MAPK phosphorylation in sweet potato. J Exp Bot 65(18):5279–5290

Liu F, Jing Y, Peng Y, Huang Y (1991) Chemical constituents of *Ipomoea batatas* Lam. Zhongguo Zhong Yao Za Zhi 16(9):551–5522, 575 (In Chinese)

Liu N, Wang H, Wang C (2008) Study on antitumor effect and its toxicity of *Ipomoea batatas* Poir Cv anthocyanins. Wei Sheng Yan Jiu 37(4):489–491 (In Chinese)

Liu XL, Mu TH, Sun HN, Zhang M, Chen JW (2013) Optimisation of aqueous two-phase extraction of anthocyanins from purple sweet potatoes by response surface methodology. Food Chem 141(3):3034–3041

Lu J, Wu DM, Zheng YL, Hu B, Zhang ZF (2010) Purple sweet potato color alleviates D-galactose-induced brain aging in old mice by promoting survival of neurons via PI3K pathway and inhibiting cytochrome C-mediated apoptosis. Brain Pathol 20(3):598–612

Lu J, Wu DM, Zheng YL, Hu B, Cheng W, Zhang ZF (2012) Purple sweet potato color attenuates domoic acid-induced cognitive deficits by promoting estrogen receptor-α-mediated mitochondrial biogenesis signaling in mice. Free Radic Biol Med 52(3):646–659

Ludvik BH, Mahdjoobian K, Waldhaeusl W, Hofer A, Prager R, Kautzky-Willer A, Pacini G (2002) The effect of *Ipomoea batatas* (Caiapo) on glucose metabolism and serum cholesterol in patients with type 2 diabetes: a randomized study. Diabetes Care 25(1):239–240

Ludvik B, Waldhausl W, Prager R, Kautzky-Willer A, Pacini G (2003) Mode of action of *Ipomoea batatas* (Caiapo) in type 2 diabetic patients. Metabolism 52(7):875–880

Ludvik B, Neuffer B, Pacini G (2004) Efficacy of *Ipomoea batatas* (Caiapo) on diabetes control in type 2 diabetic subjects treated with diet. Diabetes Care 27:436–440

Luis G, Rubio C, Gutiérrez ÁJ, González-Weller D, Revert C, Hardisson A (2014) Evaluation of metals in several varieties of sweet potatoes (*Ipomoea batatas* L.): comparative study. Environ Monitor Assess 186(1):433–440

Luo L, Gao Y (2008) Isolation and identification of polysaccharide from sweet potato vines. Sheng Wu Gong Cheng Xue Bao 24(6):1056–1061 (In Chinese)

Luo JG, Kong LY (2005a) Lipophilic constituents from the leaves of *Ipomoea batatas* (cv.Simon). Nat Prod Res Dev 17(2):166–168 (In Chinese)

Luo JG, Kong LY (2005b) Study on flavonoids from leaf of *Ipomoea batatas*. Zhongguo Zhong Yao Za Zhi 30(7):516–518 (In Chinese)

Luo CY, Wang XX, Gao G, Wang L, Li YX, Sun CJ (2013) Identification and quantification of free, conjugate and total phenolic compounds in leaves of 20 sweetpotato cultivars by HPLC-DAD and HPLC-ESI-MS/MS. Food Chem 141(3):2697–2706

Lv LY, Shi GF, Li CL, Han XZ, Lv QN (2009) Study on the chemical constituents of the leaves of *Ipomoea batatas*. Zhong Yao Cai 32(6):896–897 (In Chinese)

Maeshima M, Sasaki T, Asahi T (1985) Characterization of major proteins in sweet potato tuberous roots. Phytochemistry 24:1899–1902

Mahmood N, Moore PS, Tommasi ND, Simone FD, Colman S, Hay AJ, Pizza C (1993) Inhibition of HIV infection by caffeoylquinic acid derivatives. Antiviral Chem Chemother 4:235–240

Maloney KP, Truong VD, Allen JC (2012) Chemical optimization of protein extraction from sweet potato (*Ipomoea batatas*) peel. J Food Sci 77(11):E307–E312

Maoka T, Akimoto N, Ishiguro K, Yoshinaga M, Yoshimoto M (2007) Carotenoids with a 5,6-dihydro-5,6-dihydroxy-beta-end group, from yellow sweet potato "Benimasari", *Ipomoea batatas* Lam. Phytochemistry 68(13):1740–1745

Matsui T, Ueda T, Oki T, Sugita K, Terahara N, Matsumoto K (2001) alpha-Glucosidase inhibitory action of natural acylated anthocyanins. 1. Survey of natural pigments with potent inhibitory activity. J Agric Food Chem 49(4):1948–1951

Matsui T, Ebuchi S, Kobayashi M, Fukui K, Sugita K, Terahara N, Matsumoto K (2002) Anti-hyperglycemic effect of diacylated anthocyanin derived from *Ipomoea batatas* cultivar ayamurasaki can be achieved through the α-glucosidase inhibitory action. J Agric Food Chem 50(25):7244–7248

Matsumoto S, Mizutani M, Sakata K, Shimizu B (2012) Molecular cloning and functional analysis of the ortho-hydroxylases of p-coumaroyl coenzyme A/feruloyl coenzyme A involved in formation of umbelliferone and scopoletin in sweet potato, *Ipomoea batatas* (L.) Lam. Phytochemistry 74:49–57

Matsuo T, Yoneda T, Itoo S (1983) Identification of free cytokinins and the changes in endogeneous levels during tuber development of sweet potato (*Ipomoea batatas* Lam.). Plant Cell Physiol 24:1305–1312

Matsuo T, Mitsuzono H, Okada R, Itoo S (1988) Variations in the levels of major free cytokines and free abscisic acid during tuber development of sweet potato. J Plant Gowrth Reg 7(4):249–258

Mbuta KK, Mwima K, Bitengeli M, Y'okolo I, Kavuna M, Mandanga M, Mpoyi K, Izamajole N, Kazembe K, Booto K, Vasaki N, Mwabonsika B, Lody D, Latham P (2012) Plantes medicinales de traditions – Province de l'Equateur. R.D. Congo I.R.S.S, Kinshasa, 419 pp

Megersa T, Urge M, Nurfeta A (2013) Effects of feeding sweet potato (*Ipomoea batatas*) vines as a supplement on feed intake, growth performance, digestibility and carcass characteristics of Sidama goats fed a basal diet of natural grass hay. Trop Anim Health Prod 45(2):593–601

Mei X, Mu TH, Han JJ (2010) Composition and physicochemical properties of dietary fiber extracted from residues of 10 varieties of sweet potato by a sieving method. J Agric Food Chem 58(12):7305–7310

Min JY, Kang SM, Park DJ, Kim YD, Jung HN, Yang JK, Seo WT, Kim SW, Karigar CS, Choi MS (2006) Enzymatic release of ferulic acid from *Ipomoea batatas* L. (sweet potato) stem. Biotechnol Bioprocess Eng 11(4):372–376

Minamikawa T (1967) The occurrence of quinic acid in sweet potato roots. Agric Biol Chem 31(1):124–126

Minamikawa T, Uritani I (1965) Phenylalanine ammonia-lyase in sliced sweet potato roots. Agric Biol Chem 29(11):1021–1026

Minamikawa T, Akazawa T, Uritani I (1962) Isolation of esculetin from sweet potato roots with black rot. Nature 195:726–727

Minamikawa T, Akazawa T, Uritani I (1963) Analytical study of umbelliferone and scopoletin synthesis in sweet potato roots infected by *Ceratocystis fimbriata*. Plant Physiol 38(5):493–497

Minamikawa T, Akazawa T, Uritani I (1964) Two glucosides of coumarin derivatives in sweet potato roots infected by *Ceratocystis fimbriata*. Agric Biol Chem 28(4):230–233

Miyazaki Y, Kusano S, Doi H, Aki O (2005) Effects on immune response of antidiabetic ingredients from white-skinned sweet potato (*Ipomoea batatas* L.). Nutrition 21(3):358–362

Miyazaki K, Makino K, Iwadate E, Deguchi Y, Ishikawa F (2008) Anthocyanins from purple sweet potato *Ipomoea batatas* cultivar Ayamurasaki suppress the development of atherosclerotic lesions and both enhancements of oxidative stress and soluble vascular cell adhesion molecule-1 in apolipoprotein E-deficient mice. J Agric Food Chem 56(23):11485–11492

Montilla EC, Hillebrand S, Butschbach D, Baldermann S, Watanabe N, Winterhalter P (2010) Preparative isolation of anthocyanins from Japanese purple sweet potato (*Ipomoea batatas* L.) varieties by high-speed countercurrent chromatography. J Agric Food Chem 58(18):9899–9904

Montilla EC, Hillebrand S, Winterhalter P (2011) Anthocyanins in purple sweet potato (*Ipomoea batatas* L.) varieties. Fruit Veg Cereal Sci Biotechnol 5(2):19–23

Moorthy SN, Naskar SK, Shanavas S, Radhika GS, Mukherjee A (2010) Physicochemical characterization of selected sweet potato cultivars and their starches. Int J Food Prop 13(6):1280–1289

Murakami S, Hattori T, Nakamura K (1986) Structural differences in full-length cDNAs for two classes of sporamin, the major soluble protein of sweet potato tuberous roots. Plant Mol Biol 7(5):343–355

Murata T (1971) Sucrose synthetase of sweet potato roots. Agric Biol Chem 35(9):1441–1448

Mwanri AW, Kogi-Makau W, Laswai HS (2011) Nutrients and antinutrients composition of raw, cooked and sun-dried sweet potato leaves. Afr J Food Agric Nutr Dev 11(3):1–15

Nagai M, Tani M, Kishimoto Y, Iizuka M, Saita E, Toyozaki M, Kamiya T, Ikeguchi M, Kondo K (2011) Sweet potato (*Ipomoea batatas* L.) leaves suppressed oxidation of low density lipoprotein (LDL) in vitro and in human subjects. J Clin Biochem Nutr 48(3):203–208

Nagamine R, Ueno S, Tsubata M, Yamaguchi K, Takagaki K, Hira T, Hara H, Tsuda T (2014) Dietary sweet potato (*Ipomoea batatas* L.) leaf extract attenuates hyperglycaemia by enhancing the secretion of glucagon-like peptide-1 (GLP-1). Food Funct 5(9):2309–16

Nakamura K, Matsuoka K, Mukumoto F, Watanabe N (1993) Processing and transport to the vacuole of a precursor to sweet potato sporamin in transformed tobacco cell line BY-2. J Exp Bot 44:331–338

Namukobe J, Kasenene JM, Kiremire BT, Byamukama R, Kamatenesi-Mugisha M, Krief S, Dumontet V, Kabasa JD (2011) Traditional plants used for medicinal purposes by local communities around the Northern sector of Kibale National Park, Uganda. J Ethnopharmacol 136(1):236–245

Napolitano A, Carbone V, Saggese P, Takagaki K, Pizza C (2007) Novel galactolipids from the leaves of *Ipomoea batatas* L.: characterization by liquid chromatography coupled with electrospray ionization-quadrupole time-of-flight tandem mass spectrometry. J Agric Food Chem 55(25):10289–10297

Nguyen LQ, Everts H, Hue HT, Beynen AC (2004) Feeding of spinach or sweet-potato leaves and growth performance of growing pigs kept on smallholder farms in Central Vietnam. Trop Anim Health Prod 36(8):815–822

Nguyen CM, Choi GJ, Choi YH, Jang KS, Kim JC (2013) D- and L-lactic acid production from fresh sweet potato through simultaneous saccharification and fermentation. Biochem Eng J 81:40–46

Njoroge GN, Bussmann RW (2006) Herbal usage and informant consensus in ethnoveterinary management of cattle diseases among Kikuyus (Central Kenya). J Ethnopharmacol 108:332–339

Noda N, Horiuchi Y (2008) The resin glycosides from the sweet potato (*Ipomoea batatas* L. Lam.). Chem Pharm Bull 56(11):1607–1610

Noda N, Yoda S, Kawasaki T, Miyahara K (1992) Resin glycosides. XV. Simonins I-V, ether-soluble resin glycosides (jalapins) from the roots of *Ipomoea batatas* (cv. Simon). Chem Pharm Bull 40(12):3163–3168

Nottingham SF, Son KC, Severson RF, Arrendale RF, Kays SJ (1989a) Attraction of adult sweet potato weevils, *Cylas formicarius elegantulus* (Summers), (Coleoptera: Curculionidae), to sweet potato leaf and root volatiles. J Chem Ecol 15(3):1095–1106

Nottingham SF, Son KC, Wilson DD, Severson RF, Kays SJ (1989b) Feeding and oviposition preferences of sweet potato weevil, *Cylas formicarius elegantulus* (Summers), on storage roots of sweet potato cultivars with differing surface chemistries. J Chem Ecol 15(3):895–903

Nwokocha LM, Aviria NA, Senan C, Williams PA (2014) A comparative study of properties of starches from Irish potato (*Solanum tuberosum*) and sweet potato (*Ipomea batatas*) grown in Nigeria. Starch-Stärke 66(7–8):714–723

Oba K, Uritani I (1981) Mechanism of furano-terpene production in sweet potato root tissue injured by chemical agents. Agric Biol Chem 45(7):1635–1639

Oba K, Shibata H, Uritani I (1970) The mechanism supplying acetyl-CoA for terpene biosynthesis in sweet potato with black rot: Incorporation of acetate-2-14C, pyruvate-3-14C and citrate-2, 4-14C into ipomeamarone. Plant Cell Physiol 11(3):507–510

Oba K, Makimoto N, Hattori T, Uritani I (1982) Induction of polysome formation in sweet potato root tissue in response to wounding. Agric Biol Chem 46(7):1929–1931

Odake K, Terahara N, Saito N, Toki K, Honda T (1992) Chemical structures of two anthocyanins from purple sweet potato, *Ipomoea batatas*. Phytochemistry 31(6):2127–2130

Oduro I, Ellis WO, Owusu D (2008) Nutritional potential of two leafy vegetables: *Moringa oleifera* and *Ipomoea batatas* leaves. Sci Res Essay 3(2):57–60

Ogbole OO, Ajaiyeoba EO (2010) Traditional management of tuberculosis in Ogun state of Nigeria: the practice and ethnobotanical survey. Afr J Tradit Complement Alternat Med 7(1):79–84

Ogie-Odia EA, Oluowo EF (2009) Assessment of some therapeutic plants of the Abbi People in Ndokwa West L.G.A of Delta State, Nigeria. Ethnobot Leaflets 13:989–1002

Oguni I (1974) Dehydroipomeamarone as an intermediate in the biosynthesis of ipomeamarone, a phytoalexin from sweet potato root infected with *Ceratocystis fimbriata*. Plant Physiol 53(4):649–652

Oguni I, Uritani I (1970) The incorporation of farnesol-2-14 C into ipomeamarone. Agric Biol Chem 34(1):156–158

Oguni I, Uritani I (1971) Utilization of ethanol-2-14C for the biosynthesis of ipomeamarone by sweet potato root tissue infected with *Ceratocystis fimbriata*. Agric Biol Chem 35(3):357–362

Oguni I, Uritani I (1973) Isolation of dehydro-ipomeamarone, a new sesquiterpenoid from the black-rot fungus infected sweet potato root tissue and its relation to the biosynthesis of ipomeamarone. Agric Biol Chem 37(10):2443–2444

Oguni I, Uritani I (1974) Dehydro-ipomeamarone from infected *Ipomoea batatas* root tissue. Phytochemistry 13(2):521–522

Oguni I, Oshima K, Imaseki H, Uritani I (1969) Biochemical studies on the terpene metabolism in sweet potato root tissue with black rot: effect of C_{10}- and C_{15}-terpenols on acetate-2-^{14}c incorporation into ipomeamarone. Agric Biol Chem 33(1):50–62

Ohno T (1952) The bitter substance, produced in black-rotten sweet potato. II. On the constitution of ipomoeamarone. Part 1. Bull Chem Soc Japan 25(3):222–225

Ohta T, Ikuta R, Nakashima M, Morimitsu Y, Samuta T, Saiki H (1990) Characteristic flavor of Kansho-shochu (sweetpotato spirit). Agric Biol Chem 54(6):1353–1357

Ohta T, Omori T, Shimojo H, Hashimoto K, Samuta T, Ohba T (1991) Identification of monoterpene alcohol β-glucosides in sweet potatoes and purification of a shiro-koji β-glucosidase. Agric Biol Chem 55(7):1811–1816

Ojong PB, Njiti V, Guo ZB, Gao M, Besong S, Barnes SL (2008) Variation of flavonoid content among sweetpotato accessions. J Am Soc Hort Sci 133(6):819–824

Oki T, Masuda M, Furuta S, Nishiba Y, Terahara N, Suda I (2002) Involvement of anthocyanins and other phenolic compounds in radical-scavenging activity of purple-fleshed sweet potato cultivars. J Agric Food Chem 67(5):1752–1756

Oki T, Osame M, Masuda M, Kobayashi M, Furuta S, Nishiba Y, Kumagai T, Sato T, Suda I (2003) Simple and rapid spectrophotometric method for selecting purple-fleshed sweet potato cultivars with a high radical-scavenging activity. Breed Sci 53(2):101–107

Oki T, Suda I, Terahara N, Sato M, Hatakeyama M (2006) Determination of acylated anthocyanin in human urine after ingesting a purple-fleshed sweet potato beverage

with various contents of anthocyanin by LC-ESI-MS/MS. Biosci Biotechnol Biochem 70(10):2540–2543

Oki N, Nonaka S, Ozaki S (2011) The effects of an arabinogalactan-protein from the white-skinned sweet potato (*Ipomoea batatas* L.) on blood glucose in spontaneous diabetic mice. Biosci Biotechnol Biochem 75(3):596–598

Olubobokun TH, Aluko EO, Iyare EE, Anyaehie UB, Atang DE (2013) *Ipomoea batatas* L. extract reduces food intake, fasting blood glucose levels and body weight. Eur J Med Plants 3(4):530–539

Oner ME, Wall MM (2013) Quality of fresh-cut purple-fleshed sweet potatoes after X-ray irradiation treatment and refrigerated storage. Int J Food Sci Technol 48:2064–2070

Ooi CP, Loke SC (2013) Sweet potato for type 2 diabetes mellitus. Cochrane Database Syst Rev 9, CD009128

Oshima K, Uritani I (1967) The enzymatic synthesis of a β-hydroxy-β-methylglutaric acid derivative by a cell-free system from sweet potato with black rot. Agric Biol Chem 31(9):1105–1107

Oshima K, Uritani I (1968) Participation of mevalonate in the biosynthetic pathway of ipomeamarone. Agric Biol Chem 32(9):1146–1152

Osundahunsi OF, Fagbemi TN, Kesselman E, Shimoni E (2003) Comparison of the physicochemical properties and pasting characteristics of flour and starch from red and white sweet potato cultivars. J Agric Food Chem 51(8):2232–2236

Ozaki S, Oki N, Suzuki S, Kitamura S (2010) Structural characterization and hypoglycemic effects of arabinogalactan-protein from the tuberous cortex of the white-skinned sweet potato (*Ipomoea batatas* L.). J Agric Food Chem 58(22):11593–11599

Pacheco S, Peixoto FM, Borguini RG, Nascimento LSM, Bobeda CRR, Santiago CPA, Godoy RLO (2014) Microscale extraction method for HPLC carotenoid analysis in vegetable matrices. Sci Agric (Piracicaba, Braz) 71(5):416–419

Padda MS, Picha DH (2007) Antioxidant activity and phenolic composition in 'Beauregard' sweetpotato are affected by root size and leaf age. J Am Soc Hort Sci 132(4):447–451

Padda MS, Picha DH (2008) Quantification of phenolic acids and antioxidant activity in sweetpotato genotypes. Sci Hort 119(1):17–20

Panda SH, Ray RC (2007) Lactic acid fermentation of beta-carotene rich sweet potato (*Ipomoea batatas* L.) into lacto-juice. Plant Foods Hum Nutr 62(2):65–70

Panda V, Sonkamble M (2012) Anti-ulcer activity of *Ipomoea batatas* tubers (sweet potato). Funct Foods Health Dis 2(3):48–61

Panda V, Sonkamble M, Patil S (2011) Wound healing activity of Ipomoea batatas tubers (sweet potato). Fucnt Foods Health Dis 10:403–415

Park KH, Kim JR, Lee JS, Lee H, Cho KH (2010) Ethanol and water extract of purple sweet potato exhibits anti-atherosclerotic activity and inhibits protein glycation. J Med Food 13(1):91–98

Patil KS, Chaudhari VV, Mahajan S (2007) Effect of *Ipomoea batatas* Linn. (Lam) root extracts on phagocytosis by human neutrophils. J Nat Remed 7(2):195–199

Peterson JK, Harrison HF Jr (1991) Isolation of substance from sweet potato (*Ipomoea batatas*) periderm tissue that inhibits seed germination. J Chem Ecol 17(5):943–951

Peterson JK, Harrison HF Jr, Snook ME, Jackson DM (2005) Content of chlorogenic acid in sweetpotato germplasm and possible roles in disease and pest resistance. Allelopathy J 16:239–250

Phesatcha K, Wanapat M (2013) Performance of lactating dairy cows fed a diet based on treated rice straw and supplemented with pelleted sweet potato vines. Trop Anim Health Prod 45(2):533–538

Philpott M, Gould KS, Lim C, Ferguson LR (2004) In situ and in vitro antioxidant activity of sweetpotato anthocyanins. J Agric Food Chem 52(6):1511–1513

Philpott M, Ferguson LR, Gould KS, Harris PJ (2009) Anthocyanidin-containing compounds occur in the periderm cell walls of the storage roots of sweet potato (*Ipomoea batatas*). J Plant Physiol 166(10):1112–1117

Picha DH (1985) HPLC determination of sugars in raw and baked sweet potatoes. J Food Sci 50:1189–1190

Picha DH (1986) Sugar content of baked sweet potatoes from different cultivars and lengths of storage. J Food Sci 51:845–848

Picha DH (1987) Carbohydrate changes in sweet potatoes during curing and storage. J Am Soc Hort Sci 112:89–89

Pochapski MT, Fosquiera EC, Esmerino LA, Dos Santos EB, Farago PV, Santos FA, Groppo FC (2011) Phytochemical screening, antioxidant, and antimicrobial activities of the crude leaves' extract from *Ipomoea batatas* (L.) Lam. Pharmacogn Mag 7(26):165–170

Purcell AE, Walter WM Jr (1968) Carotenoids of centennial variety sweet potato, *Ipomea batatas*. J Agric Food Chem 16(5):769–770

Purcell AE, Walter WM Jr (1980) Changes in composition of the nonprotein-nitrogen fraction of "Jewel" sweet potatoes (*Ipomoea batatas* (Lam.)) during storage. J Agric Food Chem 28(4):842–844

Purcell AE, Walter WM Jr, Giesbrecht FG (1978) Protein and amino acids of sweet potato (*Ipomoea batatas* (L.) Lam.) fractions. J Agric Food Chem 26(3):699–670

Purcell AE, Later DW, Lee ML (1980) Analysis of the volatile constituents of baked, "Jewel" sweet potatoes. J Agric Food Chem 28(5):939–941

Qiu F, Luo J, Yao S, Ma L, Kong L (2009) Preparative isolation and purification of anthocyanins from purple sweet potato by high-speed counter-current chromatography. J Sep Sci 32(12):2146–2151

Rabah IO, Hou DX, Komine S, Fujii M (2004) Potential chemopreventive properties of extract from baked sweet potato (*Ipomoea batatas* Lam. Cv.

Koganesengan). J Agric Food Chem 52(23):7152–7157

Raponda-Walker A, Sillans R (1995) Les Plantes Utiles du Gabon. Encyclopédie Biologique. Editions Sepia, 697 pp (First Published 1961)

Rautenbach F, Faber M, Laurie S, Laurie R (2010) Antioxidant capacity and antioxidant content in roots of 4 sweetpotato varieties. J Food Sci 75(5):C400–C405

Ravli Y, Da Silva P, Moreira RG (2013) Two-stage frying process for high-quality sweet-potato chips. J Food Eng 118(1):31–40

Ray RC, Panda SK, Swain MR, Sivakumar PS (2012) Proximate composition and sensory evaluation of anthocyanin-rich purple sweet potato (*Ipomoea batatas* L.) wine. Int J Food Sci Technol 47(3):452–458

Razafindraibe M, Kuhlman AR, Rabarison H, Rakotoarimanana V, Rajeriarison C, Rakotoarivelo N, Randrianarivony T, Rakotoarivony F, Ludovic R, Randrianasolo A, Bussmann RW (2013) Medicinal plants used by women from Agnalazaha littoral forest (Southeastern Madagascar). J Ethnobiol Ethnomed 9:73

Ren K, Qiu JB, Wang XH, Niu FL, Jiang TB (2012) The effect of a sweet potato, footbath, and acupressure intervention in preventing constipation in hospitalized patients with acute coronary syndromes. Gastroenterol Nurs 35(4):271–277

Rosas-Ramírez D, Pereda-Miranda R (2013) Resin glycosides from the yellow-skinned variety of sweet potato (*Ipomoea batatas*). J Agric Food Chem 61(39):9488–9494

Rosas-Ramírez D, Escalante-Sánchez E, Pereda-Miranda R (2011) Batatins III-VI, glycolipid ester-type dimers from *Ipomoea batatas*. Phytochemistry 72(8):773–780

Roullier C, Kamboupo R, Paofa J, Mckey D, Lebot V (2013) On the origin of sweet potato (*Ipomoea batatas* (L.) Lam.) genetic diversity in New Guinea, a secondary centre of diversity. Heredity 110:594–604

Roxas MJT, Fukuba H, Mendoza EMT (1985) The absence of oligosaccharides in storage roots of sweet potato (*Ipomoea batatas* L.) Philipp. J Crop Sci 10(3):161–163

Rudkin GO, Nelson JM (1947) Acid and respiration of sweet potatoes. J Am Chem Soc 69(6):470–1475

Runnie I, Salleh MN, Mohamed S, Head RJ, Abeywardena MY (2004) Vasorelaxation induced by common edible tropical plant extracts in isolated rat aorta and mesenteric vascular bed. J Ethnopharmacol 92(2–3):311–316

Sakatani M, Suda I, Oki T, Kobayashi S, Kobayashi S, Takahashi M (2007) Effects of purple sweet potato anthocyanins on development and intracellular redox status of bovine preimplantation embryos exposed to heat shock. J Reprod Dev 53(3):605–614

Salvador LD, Suganuma T, Kitahara K, Tanoue H, Ichiki M (2000) Monosaccharide composition of sweetpotato fiber and cell wall polysaccharides from sweetpotato, cassava, and potato analyzed by the high-performance anion exchange chromatography with pulsed amperometric detection method. J Agric Food Chem 48(8):3448–3454

Sasaki K, Han J, Shimozono H, Villareal MO, Isoda H (2013) Caffeoylquinic acid-rich purple sweet potato extract, with or without anthocyanin, imparts neuroprotection and contributes to the improvement of spatial learning and memory of SAMP8 mouse. J Agric Food Chem 61(21):5037–5045

Schimming T, Tofern B, Mann P, Richter A, Jenett-Siems K, Dräger B, Asano N, Gupta MP, Correa MD, Eich E (1998) Distribution and taxonomic significance of calystegines in the Convolvulaceae. Phytochemistry 49:1989–1995

Schimming T, Jenett-Siems K, Mann P, Tofern-Reblin B, Milson J, Johnson RW, Deroin T, Austin DF, Eich E (2005) Calystegines as chemotaxonomic markers in the Convolvulaceae. Phytochemistry 66:469–480

Schneider JA, Nakanishi K (1983) A new class of sweet potato phytoalexins. J Chem Soc Chem Comm 7:353–355

Schneider JA, Lee J, Naya Y, Nakanishi K, Oba K, Uritani I (1984) The fate of the phytoalexin ipomeamarone: furanoterpenes and butenolides from *Ceratocystis fimbriata*-infected sweet potatoes. Phytochemistry 23(4):759–764

Scott GJ (1992) Sweet potatoes as animal feed in developing countries: present patterns and future prospects. In: Machin D, Nyvold S (eds) Roots, tubers, plantains and bananas in animal feeding. FAO, Rome, Italy, pp 13–98

Senthilkumar R, Yeh KW (2012) Multiple biological functions of sporamin related to stress tolerance in sweet potato (*Ipomoea batatas* Lam). Biotechnol Adv 30(6):1309–1317

Shan Q, Lu J, Zheng YL, Li J, Zhou Z, Hu B, Zhang ZF, Fan SH, Mao Z, Wang YJ, Ma DF (2009) Purple sweet potato color ameliorates cognition deficits and attenuates oxidative damage and inflammation in aging mouse brain induced by d-galactose. J Biomed Biotechnol, 2009: article 564737

Shan Q, Zheng YL, Lu J, Zhang ZF, Wu DM, Fan SH, Hu B, Cai XJ, Cai H, Liu PL, Liu F (2014) Purple sweet potato color ameliorates kidney damage via inhibiting oxidative stress mediated NLRP3 inflammasome activation in high fat diet mice. Food Chem Toxicol 69:339–346

Shen Y (1997) Determination of furanoterpenoid toxins from sweet potato by thin-layer chromatography. Se Pu 15(4):328–330 (In Chinese)

Shi ZU, Bassa IA, Gabriel SL, Francis FJ (1992) Anthocyanin pigments of sweet potatoes – *Ipomoea batatas*. J Food Sci 57:755–757

Shimozono H, Kobori M, Shinmoto H, Tsushida T (1996) Suppression of the melanogenesis of mouse melanoma B16 cells by sweet potato extract. J Jpn Soc Food Sci Technol 43:313–317

Shin SJ, Bae UJ, Ahn M, Ka SO, Woo SJ, Noh SO, Kwon YS, Jung KH, Wee JH, Park BH (2013) Aqueous

extracts of purple sweet potato attenuate weight gain in high fat-fed mice. Int J Pharmacol 9(1):42–49

Shindo M, Kasai T, Abe A, Kondo Y (2007) Effects of dietary administration of plant-derived anthocyanin-rich colors to spontaneously hypertensive rats. J Nutr Sci Vitaminol (Tokyo) 53(1):90–93

Snook ME, Data ES, Kays SJ (1994) Characterization and quantitation of hexadecyl, octadecyl, and eicosyl esters of p-coumaric acid in the vine and root latex of sweetpotato [*Ipomoea batata*s (L.) Lam.]. J Agric Food Chem 42(11):2589–2595

Son KC, Severson RF, Arrendale RF, Kays SJ (1990) Isolation and characterization of pentacyclic triterpene ovipositional stimulant for the sweet potato weevil from *Ipomoea batatas* (L.) Lam. J Agric Food Chem 38(1):134–137

Son KC, Severson RF, Snook MF, Kays SJ (1991) Root carbohydrate, organic acids and phenolic chemistry in relation to sweet potato weevil resistance. Hort Sci 26(10):1305–1308

Sondheimer E (1958) On the distribution of caffeic acid and the chlorogenic acid isomers in plants. Arch Biochem Biophys 74:131–138

South Pacific Commission (1990) Sweet potato A store of goodness.. SPC South Pacific Foods Leaflet, No 13

Srisuwan S, Sihachakr D, Siljak-Yakovlev S (2006) The origin and evolution of sweet potato (*Ipomoea batata*s Lam.) and its wild relatives through the cytogenetic approaches. Plant Sci 171(3):424–433

Ssegawa P, Kasenene JM (2007) Medicinal plant diversity and uses in Sango bay area, Southern Uganda. J Ethnopharmacol 113:521–540

Stange RR Jr, Midland SL, Holmes GJ, Sims JJ, Mayer RT (2001) Constituents from the periderm and outer cortex of *Ipomoea batatas* with antifungal activity against *Rhizopus stolonifer*. Postharv Biol Technol 23:85–92

Steed LE, Truong VD (2008) Anthocyanin content, antioxidant activity, and selected physical properties of flowable purple-fleshed sweetpotato purees. J Food Sci 73:S215–S221

Suda I, Furuta S, Nishiba Y, Matsugano K, Sugita K (1997) Reduction of liver injury induced by carbon tetrachloride in rats administered purple-colored sweetpotato juice. Nippon Shokuhin Kagaku Kogaku Kaishi 44:315–318 (In Japanese)

Suda I, Yoshimoto M, Yamakawa O (1999) Sweet potato potentiality: prevention for life style-related disease induced by recent food habits in Japan. Food Food Ingred J Jpn 181:59–68 (In Japanese)

Suda I, Oki T, Masuda M, Kobayashi M, Nishiba Y, Furuta S (2003) Physiological functionality of purple-fleshed sweet potatoes containing anthocyanins and their utilization in foods. Jpn Agric Res Quart 37(3):167–174

Suda I, Ishikawa F, Hatakeyama M, Miyawaki M, Kudo T, Hirano K, Ito A, Yamakawa O, Horiuchi S (2008) Intake of purple sweet potato beverage affects on serum hepatic biomarker levels of healthy adult men with borderline hepatitis. Eur J Clin Nutr 62(1):60–67

Sugawara T, Miyazawa T (1999) Separation and determination of glycolipids from edible plant sources by high-performance liquid chromatography and evaporative light-scattering detection. Lipids 34(11):1231–1237

Sugiyama T, Hashizume T (1989) Cytokinins in developing tuberous roots of sweet potato. Agric Biol Chem 53(1):49–52

Sugiyama T, Suye S, Hashizume T (1983) Mass spectrometric determination of cytokinins in young sweet potato plants using deuterium-labeled standards. Agric Biol Chem 47(2):315–318

Sugiyama T, Nieda Y, Irie Y, Hashizume T (1988) Occurrence and levels of cytokinins in the calluses of *Ipomea* species plants. Plant Tissue Cult Lett 5(2):93–95

Suksomboon N, Poolsup N, Boonkaew S, Suthisisang CC (2011) Meta-analysis of the effect of herbal supplement on glycemic control in type 2 diabetes. J Ethnopharmacol 137(3):1328–1333

Sun JS, Severson RF, Kays SJ (1993) Quantitative technique for measurement of the volatile components of baked sweetpotatoes. Hortscience 28:1110–1113

Sun JB, Severson RF, Kays SJ (1994) Effect of heating temperature and microwave pretreatment on the formation of sugars and volatiles in jewel sweetpotato. J Food Qual 17(6):447–456

Sun JB, Severson RF, Schlotzhauer WS, Kays SJ (1995) Identifying critical volatiles in the flavor of baked Jewel' sweetpotatoes [*Ipomoea batatas* (L.) Lam.]. J Am Soc Hort Sci 120(3):468–474

Sun YL, Sun JM, Li QP (2009) Purification and trypsin inhibitor activity of a sporamin B from sweet potato (*Ipomoea batatas* Lam.). Agric Sci (China) 8(7):808–820

Sun HN, Mu TH, Liu XL, Zhang M, Chen JW (2014a) Purple sweet potato (*Ipomoea batatas* L.) anthocyanins: preventive effect on acute and subacute alcoholic liver damage and dealcoholic effect. J Agric Food Chem 62(11):2364–2373

Sun HN, Mu TH, Xi LS, Zhang M, Chen JW (2014b) Sweet potato (*Ipomoea batatas* L.) leaves as nutritional and functional foods. Food Chem 156:380–389

Suye S, Sugiyama T, Hashizume T (1983) Mass spectrometric determination of ribosyl trans-zeatin from sweet potato tubers (*Ipomoea batatas* L. cv. Kohkei No. 14). Agric Biol Chem 47(7):1665–1666

Suzuki H, Oba K, Uritani I (1974) Occurrence of 3-hydroxy-3-methylglutaryl coenzyme a reductase in sweet potato. Agric Biol Chem 38(10): 2053–2055

Suzuki H, Oba K, Uritani I (1975) The occurrence and some properties of 3-hydroxy-3-methylglutaryl coenzyme A reductase in sweet potato roots infected by *Ceratocystis fimbriata*. Physiol Plant Pathol 7(3):265–276

Tabuti JRS, Lye KA, Dhillion SS (2003) Traditional herbal drugs of Bulamogi, Uganda: plants, use and administration. J Ethnopharmacol 88:19–44

Taira J, Ohmine W (2011) Characteristics of caffeic acid derivatives in Okinawan sweet potato (*Ipomoea batatas* L.) leaves and the anti-LDL oxidation activity. Nippon Shokuhin Kagaku Kogaku Kaishi 58(1):16–20

Taira J, Ohmi N, Uechi K (2007) Characteristics of folic acid and polyphenol in Okinawan sweet potato (*Ipomoea batatas* L.) foliage. Nippon Shokuhin Kagaku Kogaku Kaishi 54(5):215–221

Taira J, Ohmine W, Ohmi N, Taira K, Nagata J (2012) Nitric oxide-scavenging activity of Okinawan sweet potato (*Ipomoea batatas* L.) leave extracts. Nippon Shokuhin Kagaku Kogaku Kaishi 59(6):295–300

Taira J, Taira K, Ohmine W, Nagata J (2013) Mineral determination and anti-LDL oxidation activity of sweet potato (*Ipomoea batatas* L.) leaves. J Food Comp Anal 29(2):117–125

Takagi H, Kuo CG, Sakamoto S (1996) *Ipomoea batatas* (L.) Lamk. In: Flach M, Rumawas F (eds) Plant Resources of South-East Asia No. 9, Plants yielding non-seed carbohydrates. Backhuys Publishers, Leiden, pp 102–107

Takahata Y, Noda T, Nagata T (1993) HPLC determination of β-carotene content of sweet potato cultivars and its relationship with color values. Jpn J Breed 43(3):421–427

Takahata Y, Noda T, Nagata T (1994) Effect of beta-amylase stability and starch gelatinization during heating on varietal differences in maltose content in sweetpotatoes. J Agric Food Chem 42(11):2564–2569

Takenaka M, Nanayama K, Isobe S, Murata M (2006) Changes in caffeic acid derivatives in sweet potato (*Ipomoea batatas* L.) during cooking and processing. Biosci Biotechnol Biochem 70(1):172–177

Tan GS, Xu PS, Dai ZY, Tang GC (1995) Studies on the chemical components of *Ipomoea batatas* Lam. Nat Prod Res Dev 7:44–46 (In Chinese)

Tanaka Y, Uritani I (1976) Immunochemical studies on fluctuation of phenylalanine ammonia-lyase activity in sweet potato in response to cut injury. J Biochem 79(1):217–219

Tanaka Y, Uritani I (1977) Synthesis and turnover of phenylalanine ammonia-lyase in root tissue of sweet potato injured by cutting. Eur J Biochem 73:255–260

Tanaka Y, Kojima M, Uritani I (1974) Properties, development and cellular- localization of cinnamic acid 4-hydroxylase in cut-injured sweet potato. Plant Cell Physiol 15:843–854

Tanaka M, Takahata Y, Kurata R, Nakayama H, Yoshinaga M (2012) Structural and functional characterization of IbMYB1 genes in recent Japanese purple-fleshed sweetpotato cultivars. Mol Breed 29(3):565–574

Teow CC, Truong VD, McFeeters RF, Thompson RL, Pecota KV, Yencho GC (2007) Antioxidant activities, phenolic and b-carotene contents of sweet potato genotypes with varying flesh colours. Food Chem 103:829–838

Terahara N, Shimizu T, Kato Y, Nakamura M, Maitani T, Yamaguchi M, Goda Y (1999) Six diacylated anthocyanins from the storage roots of purple sweet potato, *Ipomoea batatas*. Biosci Biotechnol Biochem 31(6):383–389

Terahara N, Konczak-Islam I, Nakatani M, Yamakawa O, Goda Y, Honda T (2000) Anthocyanins in callus induced from purple storage root of *Ipomoea batatas* L. Phytochemistry 54(8):919–922

Terahara N, Konczak-Islam I, Ono H, Yoshimoto M, Yamakawa O (2004) Characterization of acylated anthocyanins in callus induced from storage root of purple-fleshed sweet potato, *Ipomoea batatas* L. BioMed Res Int 5:279–286

Thao HM, Noomhorm A (2011) Physiochemical properties of sweet potato and mung bean starch and their blends for noodle production. J Food Process Technol 2:105

Thibodeau MS, Poore MH, Hagler WM Jr, Rogers GM (2004) Effect of fermentation on sweetpotato (*Ipomoea batatas*) toxicity in mice. J Agric Food Chem 52(2):380–384

Thomas LV (1972) De l'ethnobotanique à la médecine: l'exemple Diola. Notes Africaines 134:48–52

Thompson DP (1981) Chlorogenic acid and other phenolic compounds in fourteen sweet potato cultivars. J Food Sci 46:738–740

Thu NN, Sakurai C, Uto H, Chuyen NV, Lien DTK, Yamamoto S, Ohmori R, Kondo K (2004) The polyphenol content and antioxidant activities of the main edible vegetables in northern Vietnam. J Nutr Sci Vitaminol 50:203–210

Tian Q, Konczak I, Schwartz SJ (2005) Probing anthocyanin profiles in purple sweet potato cell line (*Ipomoea batatas* L. Cv. Ayamurasaki) by high-performance liquid chromatography and electrospray ionization tandem mass spectrometry. J Agric Food Chem 53(16):6503–6509

Tiu CS, Purcell AE, Collins WW (1985) Contribution of some volatile compounds to sweet potato aroma. J Agric Food Chem 33(2):223–226

Tokusoglu O, Yildirim Z (2012) Effects of cooking methods on the anthocyanin levels and antioxidant activity of a local Turkish sweetpotato [*Ipomoea batatas* (L.) Lam] cultivar Hatay Kirmizi: boiling, steaming and frying effects. Turk J Field Crops 17(1):87–90

Truong VD, Biermann CJ, Marlett JA (1986) Simple sugars, oligosaccharides and starch concentrations in raw and cooked sweet potato. J Agric Food Chem 34(3):421–425

Truong VD, McFeeters RF, Thompson RT, Dean LL, Shofran B (2007) Phenolic acid content and composition in leaves and roots of common commercial sweetpotato (*Ipomea batatas* L.) cultivars in the United States. J Food Sci 72(6):C343–C349

Truong VD, Deighton N, Thompson RT, McFeeters RF, Dean LO, Pecota KV, Yencho GC (2010) Characterization of anthocyanins and anthocyanidins in purple-fleshed sweetpotatoes by HPLC-DAD/ESI-MS/MS. J Agric Food Chem 58(1):404–410

Tsabang N, Nkongmeneck BA, Zapfack L, Dongmo Z, Nguenang GM, Lando G, Carlson TJ, Lowry PP II (2001) Inventaire des plantes à vertus antidiabétiques dans la région de Yaounde au Cameroun. Rev Méd Pharm Afr 15:87–94

Ukom AN, Ojimelukwe PC, Alamu EO (2011) All trans-cis β-carotene content of selected sweet potato (*Ipomea batatas* (L) Lam) varieties as influenced by different levels of nitrogen fertilizer application. Afr J Food Sci 5(3):131–137

Uritani I, Hoshiya I (1953) Phytopathological chemistry of the black-rotted sweet potato. Part VI. Isolation of coumarin substances from sweet potato and their physiology. J Agric Chem Soc Jpn 27:161–164

Uritani I, Miyano M (1955) Derivatives of caffeic acid in sweet potato attacked by black rot. Nature 175:812

Uritani I, Muramatsu K (1953) Phytopathological chemistry of black-rotted sweet potato. Part 4. Isolation and identification of polyphenols from the injured sweet potato (I). J Agric Biol Chem Jpn 27:29–33 (In Japanese)

Uritani I, Stahmann MA (1961) Changes in nitrogen metabolism in sweet potato with black rot. Plant Physiol 36:770–782

Uritani I, Uritani M, Yamada H (1960) Similar metabolic alterations induced in sweet potato by poisonous chemicals and by *Ceratocystis fimbriata*. Phytopathology 50(1):30–34

Uritani I, Saito T, Honda H, Kim WK (1975) Induction of furano-terpenoids in sweet potato roots by the larval components of the sweet potato weevils. Agric Biol Chem 39(9):1857–1862

US Department of Agriculture, Agricultural Research Service (USDA-ARS) (2014) USDA National Nutrient Database for Standard Reference, Release 27. Nutrient Data Laboratory Home Page: http://www.ars.usda.gov/ba/bhnrc/ndl

van Jaarsveld PJ, Faber M, Tanumihardjo SA, Nestel P, Lombard CJ, Benadé AJS (2005) β-Carotene–rich orange-fleshed sweet potato improves the vitamin A status of primary school children assessed with the modified-relative-dose-response test. Am J Clin Nutr 81(5):1080–1087

Van AL, Hong TT, Ogle B, Lindberg JE (2005) Utilization of ensiled sweet potato (*Ipomoea batatas* (L.) Lam.) leaves as a protein supplement in diets for growing pigs. Trop Anim Health Prod 37(1):77–88

Verger PF (1995) Ewé: the use of plants in Yoruba society. Odebrecht, Sao Paulo, 744p

Villareal RL, Lin SK, Chang LS, Lai SH (1979a) Use of sweet potato (*Ipomoea batatas*) leaf tips as vegetables. I. Evaluation of morphological traits. Exp Agric 15(2):113–116

Villareal RL, Tsou SCS, Lin SK, Chiu SC (1979b) Use of sweet potato (*Ipomoea batatas)* leaf tips as vegetables. II. Evaluation of yield and nutritive quality. Exp Agric 15(2):117–122

Villareal RL, Tsou SC, Lo HP, Chiu SC (1985) Sweet potato vine tips as vegetables. In: Bouwkamp JC (ed) Sweet potato products: a natural resource for the tropics. CRC Press, Boca Raton, pp 175–183

Vimala B, Nambisan B, Hariprakash B (2011) Retention of carotenoids in orange-fleshed sweet potato during processing. J Food Sci Technol 48(4):520–524

Walter WM Jr, Purcell AE (1980) Effect of substrate levels and polyphenol oxidase activity on darkening in sweet potato cultivars. J Agric Food Chem 28(5):941–944

Walter WM Jr, Schadel WE (1981) Distribution of phenols in "Jewel" sweet potato [*Ipomoea batatas* (L.) Lam.] roots. J Agric Food Chem 29(5):904–906

Walter WM Jr, Hansen AP, Purcell AE (1971) Lipids of cured centennial sweet potatoes. J Food Sci 36:795–797

Walter WM Jr, Purcell AE, McCollum GK (1979) Use of high-pressure liquid chromatography for analysis of sweet potato phenolics. J Food Chem 27(5):938–941

Wang Y, Kays SJ (2000) Contribution of volatile compounds to the characteristic aroma of baked 'Jewel' sweetpotatoes. J Am Soc Hort Sci 125(5):638–643

Wang Y, Kays SJ (2001) Effect of cooking method on the aroma constituents of sweet potatoes [*Ipomoea batatas* (L.) Lam.]. J Food Qual 24(1):67–78

Wang Y, Kays SJ (2002) Sweetpotato volatile chemistry in relation to sweetpotato weevil (*Cylas formicarius*) behavior. J Am Soc Hort Sci 127(4):656–662

Wang Y, Kays SJ (2003) Analytically directed flavor selection in breeding food crops. J Am Soc Hort Sci 128(5):711–720

Wang HY, Yeh KW (1996) Culvivar differences in trypsin inhibitory activities of sweet potato leaves and tuberous roots. Taiwania 41:27–34

Wang Y, Horvat RJ, White RA, Kays SJ (1998) Influence of postharvest curing treatment on the synthesis of the volatile flavor components in sweetpotato. Acta Hort (ISHS) 464:207–212

Wang YL, Wu CY, Chang CT, Sung HY (2003) Invertase inhibitors from sweet potato (*Ipomoea batatas*): purification and biochemical characterization. J Agric Food Chem 51(16):4804–4809

Wang YJ, Zheng YL, Lu J, Chen GQ, Wang XH, Feng J, Ruan J, Sun X, Li CX, Sun QJ (2010) Purple sweet potato color suppresses lipopolysaccharide-induced acute inflammatory response in mouse brain. Neurochem Int 56(3):424–430

Wang MX, Shi Y, Xia XF, Li DL, Chen Q (2013) Life-cycle energy efficiency and environmental impacts of bioethanol production from sweet potato. Bioresour Technol 133:285–292

Wanzala W, Takken W, Mukabana WR, Pala AO, Hassanali A (2012) Ethnoknowledge of Bukusu community on livestock tick prevention and control in Bungoma district, western Kenya. J Ethnopharmacol 140:298–324

Wilson BJ (1973) Toxicity of mold-damaged sweetpotatoes. Nutr Rev 31(3):73–78

Wilson BJ, Burka LT (1979) Toxicity of novel sequiterpenoids from the stressed sweet potato (*Ipomoea batatas*). Food Cosmet Toxicol 17(4):353–355

Wilson BJ, Yang DTC, Boyd MR (1970) Toxicity of mould- damaged sweet potatoes (*Ipomoea batatas*). Nature 227:521–522

Wilson BJ, Boyd MR, Harris TM, Yang DTC (1971) A lung oedema factor from mouldy sweet potatoes (*Ipomoea batatas*). Nature 231:52–53

Wireko-Manu FD, Ellis WO, Oduro I (2010) Production of a non-alcoholic beverage from sweet potato (*Ipomoea batatas* L.). Afr J Food Sci 4(4):180–183

Wood G, Huang A (1975) The detection and quantitative determination of ipomeamarone in damaged sweet potatoes (*Ipomoea batatas*). J Agric Food Chem J Agric Food Chem 23(2):239–241

Woolfe JA (1992) Sweet potato, an untapped food resource. Cambridge Univ. Press and the International Potato Center (CIP), New York, 643 pp

Wu DM, Lu J, Zheng YL, Zhou Z, Shan Q, Ma DF (2008) Purple sweet potato color repairs d-galactose-induced spatial learning and memory impairment by regulating the expression of synaptic proteins. Neurobiol Learn Mem 90(1):19–27

Xie J, Han YT, Wang CB, Yu WG (2010) Purple sweet potato pigments protect murine thymocytes from ^{60}Co γ-ray-induced mitochondria-mediated apoptosis. Int J Radiat Biol 6(12):1061–1069

Xu JT, Su XY, Lim SY, Griffin J, Carey E, Katz B, Tomich J, Wang WQ (2013) Identification of anthocyanins in purple-fleshed sweetpotato and stability during various cooking conditions. FASEB J 27:636.23

Yamakawa O, Suda I, Yoshimoto M (1998) Development and utilization of sweet potato cultivars with high anthocyanin content. Food Food Ingred J Jpn 178:69–77 (In Japanese)

Yang DTC, Wilson BJ, Harris TM (1971) The structure of ipomeamaronol: a new toxic furanosesquiterpene from moldy sweet potatoes. Phytochemistry 10(7):1653–1654

Yao J, Qian C (2011) Sporamin induce apoptosis in human tongue carcinoma cells by down-regulating Akt/GSK-3 signaling. Fundam Clin Pharmacol 25(2):229–236

Ye JL, Meng XJ, Yan CL, Wang CB (2010) Effect of purple sweet potato anthocyanins on beta-amyloid-mediated PC-12 cells death by inhibition of oxidative stress. Neurochem Res 35(3):357–365

Yeh KW, Chen JC, Lin MI, Chen YM, Lin CY (1997) Functional activity of sporamin from sweet potato (*Ipomoea batatas* Lam.): a tuber storage protein with trypsin inhibitory activity. Plant Mol Biol 33(3):565–570

Yen DE (1974) The Sweet Potato in Oceania: an Essay in Ethnobotany. Bernice P. Bishop Museum 236, Bishop Museum Press, Honolulu

Yin YQ, Kong LY (2008) Ether-soluble resin glycosides from the roots of *Ipomoea batatas*. J Asian Nat Prod Res 10(3–4):245–252

Yin YQ, Huang XF, Kong LY, Niwa M (2008a) Three new pentasaccharide resin glycosides from the roots of sweet potato (*Ipomoea batatas*). Chem Pharm Bull 56(12):1670–1674

Yin YQ, Li Y, Kong LY (2008b) Pentasaccharide glycosides from the tubers of sweet potato (*Ipomoea batatas*). J Agric Food Chem 56(7):2363–2368

Yin YQ, Shen ZB, Kong LY (2008c) Studies on chemical constituents from *Ipomoea batatas*. Zhong Yao Cai 31(10):1501–1503 (In Chinese)

Yin YQ, Wang JS, Luo JG, Kong LY (2009) Novel acylated lipo-oligosaccharides from the tubers of *Ipomoea batatas*. Carbohydr Res 344(4):466–473

Yin YQ, Shen ZB, Kong LY (2012) Chemical constituents from the tubers of *Ipomoea batatas*. Zhong Yao Cai 35(6):913–917 (In Chinese)

Yoshikawa K, Yagi C, Hama H, Tanaka M, Arihara S, Hashimoto T (2010) Ipomotaosides A-D, resin glycosides from the aerial parts of *Ipomoea batatas* and their inhibitory activity against COX-1 and COX-2. J Nat Prod 73:1763–1766

Yoshimoto M, Okuno S, Kumagai T, Yoshinaga M, Yamakawa O (1999a) Distribution of antimutagenic components in colored sweet potato. Jpn Agric Res Quart 33:143–148

Yoshimoto M, Okuno S, Yoshinaga M, Yamakawa O, Yamaguchi M, Yamada J (1999b) Antimutagenicity sweetpotato (*Ipomoea batatas*) roots. Biosci Biotechnol Biochem 63(3):537–541

Yoshimoto M, Okuno S, Yamaguchi M, Yamakawa O (2001) Antimutagenicity of deacylated anthocyanins in purple-fleshed sweetpotato. Biosci Biotechnol Biochem 65(7):1652–1655

Yoshimoto M, Yahara S, Okuno S, Islam MS, Ishiguro K, Yamakawa O (2002) Antimutagenicity of mono-, di-, and tricaffeoylquinic acid derivatives isolated from sweetpotato (*Ipomoea batatas*) leaf. Biosci Biotechnol Biochem 66(11):2336–2341

Yoshimoto M, Kurata-Azuma R, Fujii M, Hou DX, Ikeda K, Yoshidome T, Osako M (2005) Enzymatic production of caffeic acid by koji from plant resources containing caffeoylquinic acid derivatives. Biosci Biotechnol Biochem 69(9):1777–1781

Yoshimoto M, Kurata-Azuma R, Okuno S, Ishiguro K, Yamakawa O, Tsubata M, Mori S, Takagaki K (2006) Nutritional value and physiological functions of sweetpotato leaves. Acta Hort (ISHS) 703:107–116

Yoshinaga M, Tanaka M, Makoto N (2000) Changes in anthocyanin content and composition of developing storage root of purple-fleshed sweet potato (*Ipomoea batatas* (L.) Lam). Breed Sci 50(1):59–64

Yuan SQ, Zhao YM, You Y (2004) Alkaloids of the hairy roots of *Ipomoea batatas* Lam. Yao Xue Xue Bao 39(10):818–820 (In Chinese)

Zhang CP, Mu TH (2011) Optimisation of pectin extraction from sweet potato (*Ipomoea batatas*, Convolvulaceae) residues with disodium phosphate solution by response surface method. Int J food Sci Technol 46(11):2274–2280

Zhang D, Cervantes J, Huamán Z, Carey E, Ghislain M (2000) Assessing genetic diversity of sweet potato (*Ipomoea batatas* (L.) Lam.) cultivars from tropical

America using AFLP. Genet Resourc Crop Evol 47(6):659–665

Zhang Z, Wheatley CC, Corke H (2002) Biochemical changes during storage of sweet potato roots differing in dry matter content. Postharv Biol Technol 24(3):317–325

Zhang ZF, Fan SH, Zheng YL, Lu J, Wu DM, Shan Q, Hu B (2009) Purple sweet potato color attenuates oxidative stress and inflammatory response induced by d-galactose in mouse liver. Food Chem Toxicol 47(2):496–501

Zhang ZF, Lu J, Zheng YL, Hu B, Fan SH, Wu DM, Zheng ZH, Shan Q, Liu CM (2010) Purple sweet potato color protects mouse liver against d-galactose-induced apoptosis via inhibiting caspase-3 activation and enhancing PI3K/Akt pathway. Food Chem Toxicol 48(8–9):2500–2507

Zhang M, Mu TH, Sun MJ (2012) Sweet potato protein hydrolysates: antioxidant activity and protective effects on oxidative DNA damage. Int J Food Sci Technol 47(11):2304–2310

Zhang P, Chen CF, Shen YH, Ding TL, Ma DF, Hua ZC, Sun DX (2013) Starch saccharification and fermentation of uncooked sweet potato roots for fuel ethanol production. Bioresour Technol 128:835–883

Zhao GH, Kan JQ, Li ZX, Chen ZD (2005) Characterization and immunostimulatory activity of an (1→6)-α -D-glucan from the root of *Ipomoea batatas*. Int Immunopharmacol 5(9):1436–1445

Zhao R, Li QW, Long L, Li J, Yang RJ, Gao DW (2006) Antidiabetic activity of flavone from *Ipomoea Batatas* leaf in non-insulin dependent diabetic rats. Int J Food Sci Technol 42(1):80–85

Zhao J, Ruan H, Gao QP, Li MY, Tao YQ, Zheng Y (2011) Anti-tumor activity of components isolated from purple sweet potato polysaccharides. Zhejiang Da Xue Xue Bao Yi Xue Ban 40(4):365–373 (In Chinese)

Zhao JG, Yan QQ, Xue RY, Zhang J, Zhang YQ (2014) Isolation and identification of colourless caffeoyl compounds in purple sweet potato by HPLC-DAD-ESI/MS and their antioxidant activities. Food Chem 161:22–26

Zheng W, Clifford MN (2008) Profiling the chlorogenic acids of sweet potato (*Ipomoea batatas*) from China. Food Chem 106:147–152

Zhou W, Gong YF, Feng QL, Gao F (2009) An improved method of isolation of high quality total RNA from purple-fleshed sweet potato, *Ipomoea batatas* (L.) Lam. Prep Biochem Biotechnol 39(2):95–104

Zhu F, Cai YZ, Yang XS, Ke JX, Corke H (2010) Anthocyanins, hydroxycinnamic acid derivatives, and antioxidant activity in roots of different Chinese purple-fleshed sweetpotato genotypes. J Agric Food Chem 58(13):7588–7596

Zhu F, Corke H, Bertoft E (2011) Amylopectin internal molecular structure in relation to physical properties of sweetpotato starch. Carbohydr Polym 84(3):907–918

Cyperus papyrus

Scientific Name

***Cyperus papyrus* L.**

Synonyms

Chlorocyperus papyrus (L.) Rikli, *Cyperus antiquorum* (Willd.) Chiov., *Cyperus antiquorum* var. *palaestinae* Chiov., *Cyperus elapsus* Chiov., *Cyperus panormitanus* Chiov., *Cyperus papyraceus* Crantz, *Cyperus papyrus* subsp. *antiquorum* (Willd.) Kük., *Cyperus papyrus* var. *antiquorum* (Willd.) C.B. Clarke, *Cyperus papyrus* subsp. *antiquorum* (Willd.) Chiov., *Cyperus papyrus* subsp. *hadidii* Chrtek & Slavíková, *Cyperus papyrus* subsp. *niliacus* Tournay, *Cyperus papyrus* var. *niliacus* Tournay, *Cyperus papyrus* var. *palaestinae* (Chiov.) Kük., *Cyperus papyrus* subsp. *papyrus, Cyperus papyrus* subsp. *siculus* Kük., *Cyperus papyrus* subsp. *siculus* (Parl.) Chiov., *Cyperus papyrus* subsp. *ugandensis* (Chiov.) Kük., *Cyperus siculus* Chiov. [inval.], *Cyperus syriacus* Parl., *Cyperus ugandensis* Chiov., *Papyrus antiquorum* Willd., *Papyrus domesticus* Poir., *Papyrus mossambicensis* Parl., *Papyrus siculus* Parl. [inval.]

Family

Cyperaceae

Common Names

Bulrushes, Egyptian Paper Plant, Egyptian Paper Reed, Egyptian Papyrus, Mediterranean Sedge, Paper Reed, Papyrus, Papyrus Sedge

Vernacular Names

Brazil: Papiro;
Czech: Šáchor Papírodárný;
Danish: Ægte Papyrus, Papyrusfladaks, Papyrusplante;
Dutch: Papirus;
Eastonian: Papüürus-Lõikhein;
Finnish: Papyruskaisla;
French: Papier Du Nil, Souchet À Papier, Souchet Du Nil, Jonc Du Nil, Papyrus, Plante À Papier Du Nil, Papyrus Du Nil;
German: Papyrusstaude, Papyrus-Zypergras;
Iran: Bardi, Berdi, Burdi (Arabic);
Italian: Papiro;
Japanese: Papirusu;
Korean: Pa Pi Ru Su, P'a P'i Ru Seu;
Nepalese: Guda Mothe, Kagat Mothe;
Nigeria: Birdi, Burdi (Arabic), Umm Ganagan (Arabic-Shuwa), Fole, Kotolo (Kanuri);
Palestine: Babir, Kulan;
Polish: Papirus;
Portuguese: Papiro;
Spanish: Papiro;
Swedish: Papperssäv, Papyrus;

T.K. Lim, *Edible Medicinal and Non-Medicinal Plants: Volume 10, Modified Stems, Roots, Bulbs*,
DOI 10.1007/978-94-017-7276-1_6

Thailand: Kok I Yip (Bangkok);
Turkey: Bardi, Berdi, Burdi (Arabic)

Origin/Distribution

The species is native to Africa—East and Central tropical Africa, sometimes naturalized in the Mediterranean countries. It has been introduced to other areas in the tropics and subtropics.

Agroecology

In its natural range in Africa, Papyrus sedge forms vast colonies in swamps, shallow lakes, and along the shallow fringes of streams and rivers, but it has become rare in the Nile Delta. In deeper waters, it is the chief constituent of the floating, tangled masses of vegetation known as *sudd*. Optimum growth temperature ranges from 20 °C to 30 °C and optimum pH ranges from pH 6–8.5 (Duke 1983). It thrives best in full sun and is sensitive to frost.

Edible Plant Parts and Uses

Stem piths and rhizomes are edible and eaten in various parts of Africa (Burkill 1966; Lind and Morrison 1974; Burkill 1985; Duke 1983; Burmeister 2001). The pith of papyrus was recommended for food, while the starchy rhizomes and lowermost parts of the stem were cut off and consumed raw, boiled or roasted. They were also chewed, sucked and spat out, much as sugar cane is done today.

Botany

Tall, robust, aquatic, herbaceous, perennial sedge arising from a stout creeping rhizome with leaves reduced to sheaths (Plates 1 and 2). Culms trigonous, smooth, 2–5 m high, to 4 cm diameter, surrounded at base with coriaceous, large acuminate sheaths. Each culm is topped by a dense cluster of thin, bright green, linear, thread-like stems around 10–30 cm in length, resembling a feather duster when the plant is young (Plates 3, 4). Inflorescence large, compound to decompound, anthelate, comprising numerous primary branches to 40 cm long, with prominent brownish tubular prophylls to 6 cm long; spikes cylindrical, to 30 mm long and 10 mm diameter; involucral bracts brown not green, numerous, to 10 cm long. Spikelets flattened, numerous in each spike, 3–5 mm long, 1 mm wide. Glumes acute to obtuse, 2–2.5 mm long, straw-coloured to pale golden brown. Stamens 1–3. Style 3-fid. Nut trigonous, obovoid, 1 mm long by 0.5 mm across, grey-brown.

Nutritive/Medicinal Properties

The essential oil of *C. papyrus* was found to contain cyclosativene, α-copaene, sativene, cyperene, rotundene, cyprotene, cypera-2,4-diene, epoxycyperene and isopatchoula-3,5-diene (Sonwa 2000). The petroleum ether extract of *C. papyrus* tubers afforded a series of *n*-alkanes (87.7 %), campesterol (2.5 %), β-sitosterol (5.1 %), 7-stigmasterol (2.5 %), avenasterol (2.6 %) and 5-stigmasterol (2.9 %), and was rich in unsaturated fatty acid, predominated by oleic acid (Hassanein et al. 2011). The essential oil of *C. papyrus* was rich in monoterpenes (67 % and 61 % for tubers and stems, respectively) and only 33 % and 39 % sesquiterpenes in tubers and stems respectively; it also showed the absence of non-terpenoid compounds (Hassanein et al. 2014). The chemical constituents were identified to be pinene, eucalyptol, myrtenol, α-copaene, cyperene, caryophyllene, patchoulene and caryophyllene oxide.

An arabinoglucuronoxylan with a DPn of 57 was purified from *C. papyrus* stalk and found to consist of a main chain of $\beta(1 \rightarrow 4)$ linked D-xylopyranosyl residues to which are attached an average of 3·2 l-arabinofuranosyl residues and 1·7 D-glucopyranuronosyl residues (Buchala and Meier 1972).

The alkaloids tyramine and octopamine have been recorded present in the leaves (Wheaton and

Plate 1 Papyrus sedge in-situ habit

Plate 2 Papyrus sedge clump

Plate 3 Triangular green stems

Plate 4 Terminal clusters of feather-light linear green stems

Stewart 1970). Leaf flavonoids comprising glycosides of apigenin, luteolin, tricin, and quercetin were identified in 20 *Cyperus* species, including *C. papyrus* in Egypt (El-Habashy et al. 1989). Luteolin 5-0-methyl ether and the aurones, aureusidin and sulphuretin were found as aglycones.

Antioxidant & Cytoprotective Activity

Of three *Cyperus s*pecies, the 80 % ethanol extract of *C. papyru*s tubers was the most potent with direct antioxidant activity in both DPPH (EC_{50}: 5.1 μg/ml) and FRAP (FE (FRAP equivalent): 48.7 μg/ml) assays (Hamed et al. 2012). Pretreatment of hepa1c1c7 with 100 μg/ml of the ethanol extract produced significant full cytoprotection (100 % inhibition) against toxicity of t-butyl hydroperoxide.

Hepatoprotective Activity

*C. papyru*s tuber extracts exhibited hepatoprotective activity in rat hepatocyte (Hassanein et al. 2011). The 80 % ethanol extract of defatted tuber powder exhibited LC_{50} at 750 g/ml and petroleum ether extract showed LC_{50} at 1000 g/ml, while the 80 % ethanol extract of defatted powder exerted 95 % hepatoprotection at 200 g/ml but did not elicit100% protection from CCl_4-induced toxicity. The petroleum ether extract of *C. papyrus* did not show any hepatoprotective effect till a concentration of 2000 g/ml. DPPH was used as a method for antioxidant comparison screening of both species. The ethyl acetate fraction of *C. papyrus* tuber was a good scavenger of H_2O_2 and reacted rapidly with HO · radicals.

Traditional Medicinal Uses

Since antiquity, traditional folk medicine practitioners and pharmacologists in Egypt have included papyrus among medicinal plants (Duke 1983). The pith was recommended for widening and drying of fistula. Papyrus sheaths were burnt and the ash was employed for treating certain eye disease and malignant mouth ulcers. Ashes from burnt plant parts macerated in vinegar have been used to heal wounds. Europeans also listed this among their folk cancer cures. Decaying papyrus shoots are collected and used for medicinal uses in treating pregnant women (Katondo 2001).

Other Uses

Papyrus shoots have a wide array of nonedible uses (Lind and Morrison 1974; Duke 1983; Burmeister 2001; Katonda 2001). They are stripped, pressed, dried, and used for making mats used as beds, seats, toilet/baths; building business huts (as temporary local restaurants or pubs); for fencing of homes, and as ceiling and curtains in houses; packaging materials; and in broom making. Papyrus shoots are also used to make paper /parchments for writings, paintings, papyrus colouring book, bookmarks, sandals, boxes, fans, boats, funeral garlands, formal bou-

quets, cloth, medicine, cordage, ropes used in house construction and for tying luggage, thatch for roofs, food utensils—local bowls known as matangwa, makelejo, masonzo, and luungo. In Rwanda, papyrus shoots and rhizome are compressed into fuel briquette with high calorific values (Katonda 2001). The pith is used as support for cooking food such as fish. Papyrus is popularly used as an ornamental aquatic for many water features in gardens, parks and hotels. It is used in the manufacture of building boards in Uganda, and studies on the use of papyrus for hardboard have also been carried out (Steenberg 1965).

Papyrus is also used as animal forage. Analysis of the nutritive value of *C. papyrus* revealed that crude protein was higher in umbels than culms but ruminal dry matter digestibility of papyrus was higher in culms than umbels (Muthuri and Kinyamario 1989). Both the crude protein and ruminal dry matter digestibility decreased with increasing age of the plant. Values for crude protein and ruminal dry matter digestibility were similar to those reported for the range grasses that constituted the greatest percentage of forage in East Africa. In general, papyrus has some grazing potential and could be used as fodder, especially in the dry season when other forage is scarce and of low nutritive value.

Wetland aquatic plants such as *C. papyrus* had been used in the phytoremediation of submerged soil polluted by arsenic (As) (Jomjun et al. 2011). *C. papyrus* was found to be the largest biomass producer, which had arsenic accumulation capacity of 130–172 mg As/kg plant. Studies found that *C. papyrus* could be used to treat low-strength domestic waste water in constructed wetland (Perbangkhem and Polprasert 2010). Papyrus converted solar radiation to biomass of about 2200–3100 g dry weight/m(2) from the 2-month period of the experiments and the energy-capturing efficiencies were estimated to be in the range of 4.4–6.0 %, which were relatively high, compared with those of other plants. Earlier, Kyambadde et al. (2004) found that *C. papyrus* had good treatment efficiency for wastewater treatment in constructed wetlands. Papyrus showed higher ammonium-nitrogen and total reactive phosphorus (TRP) removal (75.3 % and 83.2 %) than *Miscanthidium violaceum* (61.5 % and 48.4 %) and unplanted controls (27.9 % ammonium-nitrogen). Papyrus root structures provided more microbial attachment sites, sufficient wastewater residence time, trapping and settlement of suspended particles, surface area for pollutant adsorption, uptake, assimilation in plant tissues and oxygen for organic and inorganic matter oxidation in the rhizosphere, accounting for its high treatment efficiency.

Comments

C. papyrus reproduces predominantly vegetatively by extension of its rhizomes, but seed production, normally sporadic, can increase in certain conditions, notably associated with water level changes (Terer et al. 2012).

Selected References

Buchala AJ, Meier H (1972) Hemicelluloses from the stalk of *Cyperus papyrus*. Phytochemistry 11(11):3275–3278

Burkill HM (1985) The useful plants of West Tropical Africa. Vol. 1. Families A-D. Royal Bot. Gardens, Kew, 960 pp

Burkill IH (1966) A dictionary of the economic products of the Malay Peninsula. Revised reprint of 1st ed 1935, 2 vols. Ministry of Agriculture and Co-operatives, Kuala Lumpur, Malaysia. Vol. 1 (A-H) pp. 1–1240, Vol. 2 (I-Z), pp 1241–2444

Burmeister M (2001) *Cyperus papyrus*: from the Nile to modern times. Ethnobot Leaflets 2001(1): Article 7

Duke JA (1983) *Cyperus papyrus* L. Handbook of energy crops. Unpublished. https://www.hort.purdue.edu/newcrop/duke

El-Habashy I, Mansour RMA, Zahran MA, El-Hadidi MN, Saleh NAM (1989) Leaf flavonoids of *Cyperus* species in Egypt. Biochem Syst Ecol 17(3):191–195

Haines RW, Lye KA (1983) The sedges and rushes of East Africa. East African Natural History Society, Nairobi, 404pp

Hamed A, Soltan M, Fry J, Hammouda F, Zaki A (2012) Antioxidant and cytoprotective properties of three Egyptian *Cyperus* species using cell-free and cell-based assays. Pharm Crops 3:88–93

Hassanein HD, Nazif NM, Shahat AA, Hammouda FM, Aboutable ESA, Saleh MA (2014) Chemical diversity of essential oils from *Cyperus articulatus, Cyperus*

esculentus and *Cyperus papyrus*. J Essent Oil Bear Plants 17(2):251–264

Hassanein HMDE, Nazif NM, Ehsan NA, Aboutabl EAH, Faiza M, Hammouda FM (2011) Lipid composition and evaluation of hepatoprotective activity of *Cyperus esculentus* L. and *Cyperus papyrus* tubers grown in Egypt. Adv Food Sci 33(2):65–72

Jomjun N, Siripen T, Maliwan S, Jintapat N, Prasak T, Somporn C, Petch P (2011) Phytoremediation of arsenic in submerged soil by wetland plants. Int J Phytorem 13(1):35–46

Katondo JM (2001) Traditional uses of *Cyperus papyrus* (L) and associated problems at Simiyu Fringing Wetland of Lake Victoria, Mwanza Region, Tanzania. In: Proceedings of Lake Victoria Environment Management Programme (LVEMP), first regional scientific conference 3rd–7th December 2001, Kisumu, Kenya, pp 227–245

Kyambadde J, Kansiime F, Gumaelius L, Dalhammar G (2004) A comparative study of *Cyperus papyrus* and *Miscanthidium violaceum*-based constructed wetlands for wastewater treatment in a tropical climate. Water Res 38(2):475–85

Lind EM, Morrison ES (1974) East African vegetation. Longman Group Ltd., London, 257 pp

Muthuri FM, Kinyamario JI (1989) Nutritive value of papyrus (*Cyperus papyrus*, Cyperaceae), a tropical emergent microphyte. Econ Bot 43(1):23–30

National Academy of Sciences (NAS) (1976) Making aquatic weeds useful: some perspectives for developing countries. National Academy of Sciences, Washington, DC

Perbangkhem T, Polprasert C (2010) Biomass production of papyrus (*Cyperus papyrus*) in constructed wetland treating low-strength domestic wastewater. Bioresour Technol 101(2):833–5

Purseglove JW (1972) Tropical crops monocotyledons. Longman, London

Reynders M, Huygh W, Larridon I, Muasya A, Govaerts R, Simpson D, Goetghebeur P (2011) Nomenclature and typification of names of genera and subdivisions of genera in the Cypereae (Cyperaceae): 3. Names in segregate genera of *Cyperus*. Taxon 60(3):885–895

Simpson DA, Inglis CA (2001) Cyperaceae of economic, ethnobotanical and horticultural importance: a checklist. Kew Bull 56:257–360

Sonwa MM (2000) Isolation and structure elucidation of essential oil constituents: comparative study of the oils of *Cyperus alopecuroides, Cyperus papyrus*, and *Cyperus rotundus*. Ph.D. thesis. University of Hamburg. http://ediss.sub.uni-hamburg.de/volltexte/2000/372/pdf/diss.pdf

Steenberg B (1965) Papyrus: problems in its utilization for pulp and paper making. In: Pulp and paper development in Africa and the Near East, vol 2. FAO, Rome, pp 865–876

Tackholm V, Drar M (1973) Flora of Egypt. Volume 2, Angiospermae, part monocotyledones: Cyperaceae-Juncaceae. Koeltz Antiquariat, Konigstein, West Germany, 547 p

Terer T, Triest L, Muasya AM (2012) Effects of harvesting *Cyperus papyrus* in undisturbed wetland Lake Naivasha Kenya. Hydrobiologia 680:135–148

The Plant List (2014) *Cyperus papyrus*. http://www.theplantlist.org/

Wheaton TA, Stewart I (1970) The distribution of tyramine, N-methyltyramine, hordenine, octopamine, and synephrine in higher plants. Lloydia 33(2):244–254

Cyperus rotundus

Scientific Name

Cyperus rotundus **L.**

Synonyms

Chlorocyperus rotundus (L.) Palla, *Chlorocyperus salaamensis* Palla, *Cyperus agrestis* Willd. ex Spreng. & Link, *Cyperus arabicus* Ehrenb. ex Boeckeler, *Cyperus badius* var. *inconspicuus* (Nyman) Nyman, *Cyperus bicolor* Vahl, *Cyperus bifax* C.B. Clarke, *Cyperus bulbosostoloniferus* Miq., *Cyperus comosus* Sm., *Cyperus disruptus* C.B. Clarke, *Cyperus elongatus* Sieber ex Kunth [Ileg.], *Cyperus herbicavus* Melliss, *Cyperus hexastachyos* Rottb., *Cyperus hildra* Poir., *Cyperus hydra* Michx., *Cyperus inconspicuus* Gennari [Ileg.], *Cyperus laevissimus* Steud., *Cyperus leptostachyus* Griff., *Cyperus longus* Boeckeler [Ileg.], *Cyperus merkeri* C.B. Clarke, *Cyperus micreilema* Steud., *Cyperus nubicus* C.B. Clarke, *Cyperus ochreoides* Steud., *Cyperus officinalis* Nees ex Godr. [Inval.], *Cyperus oliganthus* Gand., *Cyperus olivaris* O. Targ. Tozz., *Cyperus olivaris* var. *brevibracteatus* Le Grand, *Cyperus pallescens* Poir. [Ileg.], *Cyperus pallescens* Boiss. [Ileg.], *Cyperus patulus* M. Bieb. [Ileg.], *Cyperus platystachys* Cherm., *Cyperus procerulus* Nees, *Cyperus proteinolepis* Boeckeler [Ileg.], *Cyperus pseudovariegatus* Boeckeler, *Cyperus purpureovariegatus* Boeckeler, *Cyperus radicosus* Sm., *Cyperus retzii* Nees [Ileg.], *Cyperus rotundus* var. *acutus* Boeckeler, *Cyperus rotundus* var. *amaliae* C.B. Clarke, *Cyperus rotundus* subsp. *brevibracteatus* (Le Grand) M. Laínz, *Cyperus rotundus* var. *brevibracteatus* (Le Grand) Husn., *Cyperus rotundus* var. *carinalis* Benth., *Cyperus rotundus* var. *carinatus* F.M. Bailey, *Cyperus rotundus* var. *centiflorus* C.B. Clarke, *Cyperus rotundus* var. *comosus* (Sm.) Nyman, *Cyperus rotundus* f. *comosus* (Sm.) Kük., *Cyperus rotundus* subsp. *comosus* (Sm.) K. Richt., *Cyperus rotundus* f. *contractus* Kük., *Cyperus rotundus* f. *depallescens* Ekman & Kük., *Cyperus rotundus* var. *disruptus* (C.B. Clarke) Kük., *Cyperus rotundus* subsp. *divaricatus* Lye, *Cyperus rotundus* var. *elongatus* Boeckeler, *Cyperus rotundus* var. *hydra* (Michx.) A. Gray, *Cyperus rotundus* f. *inconspicuus* (Nyman) Kük., *Cyperus rotundus* var. *inconspicuus* Nyman, *Cyperus rotundus* subsp. *inconspicuus* (Nyman) K. Richt., *Cyperus rotundus* f. *latifolius* Kük., *Cyperus rotundus* f. *latimarginatus* Kük., *Cyperus rotundus* var. *macrostachyus* Boiss., *Cyperus rotundus* var. *major* Parl., *Cyperus rotundus* subsp. *merkeri* (C.B. Clarke) Kük., *Cyperus rotundus* var. *nubicus* (C.B. Clarke) Kük., *Cyperus rotundus* var. *pallidus* Benth., *Cyperus rotundus* var. *platystachys* Bojer ex C.B. Clarke, *Cyperus rotundus* var. *procerula* C.B. Clarke, *Cyperus rotundus*

T.K. Lim, *Edible Medicinal and Non-Medicinal Plants: Volume 10, Modified Stems, Roots, Bulbs*,
DOI 10.1007/978-94-017-7276-1_7

var. *quimoyensis* L.K. Dai, *Cyperus rotundus* subsp. *retzii* Kük., *Cyperus rotundus* var. *rotundus, Cyperus rotundus* subsp. *rotundus, Cyperus rotundus* var. *salsolus* C.B. Clarke, *Cyperus rotundus* var. *spadiceus* Boeckeler, *Cyperus rotundus* var. *taylorii* (C.B. Clarke) Kük., *Cyperus rotundus* var. *tetrastachyos* (Desf.) Trab., *Cyperus rotundus* subsp. *tuberosus* (Rottb.) Kük., *Cyperus rubicundus* Willd. ex Link [Ileg.], *Cyperus rudioi* Boeckeler, *Cyperus rudioi* var. *minor* Boeckeler, *Cyperus stoloniferus* var. *pallidus* Boeckeler, *Cyperus taylorii* C.B. Clarke, *Cyperus tenuifolius* Walp., *Cyperus tetrastachyos* Desf., *Cyperus tuberosus* Rottb., *Cyperus viridis* Roxb. ex C.B. Clarke [Inval.], *Cyperus weinlandii* Kük., *Cyperus yoshinagae* Ohwi, *Pycreus rotundus* (L.) Hayek, *Schoenus tuberosus* Burm. f.

Family

Cyperaceae

Common/English Names

Brown Nutsedge, Coco Grass, Java Grass, Ground Almond, Nutgrass, Nutsedge, Purple Nutsedge, Red Nutsedge

Vernacular Names

Arabic: Suad, Suadekufi;

Brazil: Alho-Bravo, Capim-Alho, Capim-Dandá, Juncinha, Te Mumute, Tiririca, Tiririca Do Brejo, Tiririca-Vermelha (Portuguese);

Burmese: Vorninniu;

Cambodia: Krâva:Nh Chru:K;

Chamorro: Chaguan Humatag;

Chinese: Suo Cao, Xiang Fu, Xiang Fu Zi, Xiang Tou Cao;

Cook Islands: Matie 'Oniani, Mauku 'Oniani, Oniani, 'Oniani Lau, 'Oniani Rau, 'Oniani Tita;

Czech: Šáchor Hlíznatý, Zelenošáchor Hlíznatý;

Danish: Cyperrod, Rund Fladaks;

Dutch: Notengras;

Fijian: Ivako, Malanga, Mot Ha, Soranakambani, Soro Ni Kabana, Soronakambani, Vucesa, Vuthesa;

French: Herbe-À-Oignon, Souchet À Tubercules, Souchet D'asie, Souchet En Forme D'olive, Souchet Rond;

German: Cyperngras, Knolliges Cypergras, Rundes Zypergras;

Ghana: Kulisaa (Dagbani), Ngɔi (Ga), Mbubule (Nzema);

Hawaiian: Kili'O'Opu, Mau'U Mokae, Pakopako;

I-Kiribati: Te mumute, Te Mutemute;

India: Keyabon (Assamese), Nagarmotha (Bengali), Bara-Nagar-Motha, Doongia, Doongla, Khal, Korehi-Jhar, Moth, Motha, Mothee, Mothi, Motho, Mutha, Nagarmotha (Hindu), Abda Hullu, Abdahallu, Abdahullu, Bhadra Hullu, Bhadra Mushti, Chandra Hullu, Gaekina Gedde, Gaeku, Gonarda, Gopura, Jaekina Gedde, Jalad, Jeykina Gadde, Konnaari Baeru, Konnaari Gedde, Konnari Gedde, Koranari Gedde, Koranari-Gadde, Tangahullu, Thunge Hullu, Tungegadde (Kannada), Karimuttan, Kora-Kizanna, Muttanna (Malayalam), Shembang Kouthum (Manipuri), Barik Motha, Barik-Moth, Bimbal, Bimbol, Lavola, Motha (Marathi), Mutha (Oriya), A, Abda, Abhrabheda, Ambhodhara, Ambuda, Arnoda, Bhadrakshi, Bhadramusta, Bhadramuste, Gangera, Gangeya, Ghana, Granthi, Gundra, Hima, Jalada, Jaladah, Kachhola, Kakshottha, Kasheru, Krodeshtha, Kuru, Kurubilva, Kurvinda, Kutannata, Megha, Mesa, Musta, Mustaka, Payoda, Rajkaseruka, Valya, Varahi, Varida, Varivaha, Vindakhya (Sanskrit), Accam, Araikkali, Araikkalippul, Avittam, Ayali, Ayil, Ayirpul, Campankorai, Cankam, Catatatikam, Celakam, Celam, Celatam, Celekam, Cevakam, Cevvetakam, Erumainkkuppul, Eruvai, Eruvaippul, Kaivarttam, Kancukam, Kankeyam, Karkkoli, Karkoli, Karkoli, Karkolippul, Karuvukatitam, Karuvukatitappul, Kerukam, Kolavunavu,

Kontankilanku, Kora, Korai, Korai Kilangu, Korai Kizhangu, Korai-K-Kilanku, Korai-Kizanghu, Koraikkilanku, Koraikkizhangu, Koraippul, Koraippurkilanku, Korankilanku, Korutan, Kotakilanku, Kotani, Kotanikkilanku, Kulamaccam, Kunram, Kupaiyatitam, Kupaiyatitappul, Kuruvintakan, Kuruvintak-kilanku, Kutiraivavikam, Kutiraivavikappul, Masta, Muthakasu, Muttakkacu, Muttati, Nanal, Netila, Paiyam, Panritonripputu, Panritontuputu, Panrittonri, Panrittonti, Panrittontipputu, Pathalamulam, Perun-kolikam, Perunkolitam, Talaikkorai, Tattaikkoraikkilanku, Tiraiyappul, Tiratkorai, Tirkkakantam, Tulam, Tunga-Gaddai, Tunka-muttu, Tunkamuttukkilanku, Tunkumustu, Ural, Vacanaippulkilanku, Vacciracalak-kilanku, Vacciracalam, Vacciracelatam, Vaccirakentam, Vacciratilatam, Varaki, Vicciral, Vicciram, Viccirappul, Visakkani (Tamil), Bhadra-Tunga-Muste, Bhadramuste, Bharamuste, Gandala, Kaivartakamuste, Kaivarthaka Muste, Mustakamu, Shakatunga, Shakha-Tunga-Veru, Shakhatungaveru, Thungamustha, Tunga-Muste, Tunga-Musthalu, Tungamuste, Tungamusti (Telugu), Habu-Ul-Zillam, Nagarmotha, Sad Kufi (Urdu);

Indonesia: Teki (General), Mota (Madura), Karelawai (Sumba);

Italian: Zigolo Infestante;

Japanese: Hamasuge, Kobushi;

Korean: Hyangbu, Hyangbuja, Jakpangdong Sani;

Laos: Hèwz Hmu:;

Malaysia: Teki, Rumput Haliya Hitan;

Mali: Numii Sami (Dogon), Gué, Hissel, Hisser (Fula-Pulaar), Digityeh, N-Tioko, N-Togon (Manding-Bambarra), Digai Sa (Songhai);

Marshallese: Tuteoneon;

Morocco: S-S'ad;

Myanmar: Monhnyin-Bin;

Naruan: Ibugiibugi;

Nepalese: Mothe;

New Caledonia: Herbe-À-Oignon;

Niger: Guiraguri (Huasa), Dúgú ɓì (Songhai);

Nigeria: Ayaare, Goye (Fula-Fulfulde), Àyaà-Áyaà, Giragiri, Girigiri, Gwaigwaya, Jigi, Jiji (Huasa), Nû, Ishọhọ I Toho (Kanuri);

Palauan: Tamanengi;

Philippines: Boto-Botones, Tarugug (Bikol), Balinsanga, Barsanga (Iloko), Galonalpas, Kusung, Mala-Apulid, Mota, Omadiung, Onoran, Sur-Sur (Pampangan), Mutha (Tagalog);

Polish: Cibora Okólkowa;

Portuguese: Junça;

Samoan: Mumuta;

Senegal: Gowé, Hissel, Hisser (Fula-Pulaar), N-Togon (Manding-Bambarra), Rôl (Sere), Ndidan, Ndiran, Tiehomtioli, Tiohamtiule (Wolof);

Sierra Leone: Njewɛ, Njewɛ Mumu, Njewɛ Wa, Wa, Tugbɛlɛ (Mende), Lela, Melai, Nelai (Susu), An-Roi-An-Rungi, An-Siri (Temne);

Spanish: Almendra De Tierra, Castanuela, Castañuela, Cebollín, Chufa, Coco, Coquille, Coquillo Purpura, Coquito, Cortadera, Juncia, Juncia Real;

Sri Lanka: Kalanduru;

Sudan: Eldeis;

Swedish: Nötag;

Thai: Ya Haeo Mu (Central Thailand), Ya Khon Mu;

Tibetan: Gla Sgan Gla-Gan;

Tokelauan: Mumuta;

Tongan: Pakopako;

Tuvaluan: Pakopako;

Vietnamese: Cỏ Gấu, Cỏ Cú, Cỏ Hương Phụ

Origin/Distribution

Purple nutsedge is a pantropical weed of doubtful origin (Parsons and Cuthbertson 2001; GRIN 2014). It is found as a weed in tropical, subtropical and warm-temperate countries including India, China, Taiwan, Korea, Philippines, Thailand, Vietnam, Malaysia, Indonesia, the Pacific Islands, Africa, south America, the Middle East, north America (eastern and southern USA), Mexico New Zealand and Australia.

Agroecology

Cyperus rotundus is found from near sea level to 1500 elevation in tropical, subtropical and temperate regions of the world. It occurs commonly in cultivated fields, farmlands, neglected areas, wastelands, grasslands, at the edges of forests, and on roadsides, sandy or gravelly shores, riverbanks, along rice fields, sugar fields, water course, irrigation canal banks, in disturbed areas and lawns/turf (Holm et al. 1977; Swarbrick 1997). It thrives well in almost every soil type, over a wide range of soil moisture, pH and elevation, fertile and infertile soils. Bendixen and Nandihalli (1987) reported that purple nutsedge grows best where soil moisture is high, such as in upland rice and sugarcane culture and consequently is not an important weed of arid regions, except on irrigated land (Bendixen and Nandihalli 1987). It is intolerant of cold temperatures and is restricted to latitudes where the average minimum air temperature for January is higher than −1 °C (Bendixen and Nandihalli 1987). In cool or waterlogged soils, it grows slower, flowers little and produces fewer tubers (Holm et al. 1977)

Plate 1 Clumps of nutsedge weeds (GF Chung)

Edible Plant Parts and Uses

C. rotundus nut-like tubers are edible raw or cooked (Hedrick 1972; Harrington 1974; Tanaka 1976; Facciola 1990; Cribb and Cribb 1987; Manandhar and Manandhar 2002; Deane 2014). Soaking the tuber will overcome the hardness, and the bitterness of the tuber can be reduced by drying for a few days before consumption, raw or cooked. The dried tubers can be ground into a powder and used as a cereal (Moerman 1998) and the tiny seeds are also edible (Kunkel 1984).

Botany

Rhizomatous clumping perennial (Plates 1 and 2), with slender rhizomes forming dark reddish-brown to black, woody, ovoid to ellipsoid tubers 5–10 mm diameter (Plate 3). Culms triquetrous, smooth, 15–30 (occasionally to 60) cm high, 1–2 mm diam. Leaves not septate-nodulose, equalling or shorter than culms, sheaths brown, leaf lamina flat, bluish-green. Inflorescence 10 cm long, compound or simple with 3–9 primary branches; spikes short, ovoid, to 2.5 cm long, 2–3 mm wide; involucral bracts leaf-like, 2–4 not or slightly exceeding inflorescence (Plate 4). Spikelets flattened, 3–10 per spike, 10–30-flowered, to 25 mm long, rachilla broadly winged, white, hyaline,persistent. Glumes blood-red to purplish brown on both surfaces but middle green, subdensely imbricate, ovate to oblong-ovate, about 3 mm, 5–7-veined, apex acute to obtuse and muticous. Stamens 3, anther linear, Style long, stigma 3 longer than style, exerted from glume. Nutlet (tuber) narrowly obovoid to ellipsoid, trigonous, dark-grey-brown, 1.5 cm by 0.5 cm.

Nutritive/Medicinal Properties

The chemical composition of *C. rotundus* dry roots was reported as follows (mg%): alkaloids 0.21–0.24, cardiac glycosides 0.62–0.74, flavo-

Plate 2 Foliage and inflorescence of nutsedge (GF Chung)

Plate 3 Nutsedge habit with nut-like tubers (GF Chung)

Plate 4 Nutsedge inflorescence (GF Chung)

noids 1.25, polyphenolic compounds 1.62, saccharides before hydrolysis 13.22, saccharides after hydrolysis 14.4, starch 9.2, pectins 3.72, ethereal oils 1.06, lipid compounds 2.98, resins 4.21, total acidity expressed as malic acid 3.25 and vitamin C 8.8 mg% (Akperbekova and Abdullaev (1966). Earlier, Asenjo (1942) reported in *Cyperus rotundus* tubers, reducing sugars amounted to 55 % of the molasses, non-reducing sugars 4 %, ketoses 9.3 % and aldoses 41.7 %. Wills (1972) found sucrose was the most concentrated sugar in mature rhizomes, glucose and fructose were the most concentrated sugars in the actively growing rhizomes and tubers. Phosphorus was also concentrated in the rhizomes while iron was concentrated in older tissues of the tubers and rhizomes.

Nutrient composition of the edible nut-like tubers had been reported per 100 g by Oladunni et al. (2011) as: moisture 24.73 %, fat 29.48 %, crude protein 9.04 %, ash 2.67 %, crude fibre 12.63 %, carbohydrate 21.47 %, Cu 28.11 mg, Mg 50.76 mg, Na 119.29 mg, Ca 16.40 mg and K 110.11 mg. The physicochemical properties were reported to be good with acid value 10.10 %, free fatty acid (oleic acid) 0.62 %, iodine value 24.15 %, Saponification value 37.71 %, peroxide value 0.65 %, refractive index 1.46 and specific gravity 0.93 (Oladunni et al. 2011). Earlier studies by Wang and Liu (2010) showed that K, Ca, Mg and Na were the most abundant of the major elements in Rhizoma cyperi (*C. rotundus* rhizome from 15 different zones in china) with average concentrations of K 26,221 μg/g; Ca 1097 μg/g; Mg 714 μg/g and Na 293 μg/g. Starch isolated from the subterraneous swollen tuberous bases and nodules of *Cyperus rotundus* afforded a yield of 24.1 % on a dry weight basis (Umerie and Ezeuzo 2000). The white starch granules were large-sized and comparable to the potato types, and had an appreciable amylose content, 26.73 %. Solutions of the starch showed a high pasting temperature, viscosity and adhesive strength. The *C. rotundus* starch noodles and the sized yarns both elicited fairly good tensile strengths, comparable to standards. *C. rotundus* was reported to contain phenols, tannins, glycoside, flavonoids and metals copper, lead, nickel and cadmium (Jebasingh et al. 2012). Physiochemical analysis revealed that the herb *C. rotundus* has low ash value and moderate water and alcohol solubility. A flavonol glycoside, rhamnetin 3-*O*-rhamnosyl-(1 → 4)-rhamnopyranoside (Singh and Singh 1986) was isolated from the mature tuber. An oleanolic-type saponin, 3-*O*-(2-rhamnosylglucosyl), was isolated from mature tubers of *Cyperus rotundus* (Singh and Singh 1980). Polyphenolic substances found in purple nutsedge tubers consisted primarily of catechol tannin of leucocyanidin, leucoyanidin-glucoside, catechine and chlorogenic acid at both stages, dormant and non-dormant (Ueki et al. 1974; Komai and Ueki 1975). Leucocyanidin was high in the mature tubers, seed heads and rhizomes. Phenolic acids detected in the hydrolysis of the phenols with HCl or NaOH were *p*-coumaric acid, ferulic acid, vanillic acid, *p*-hydroxybenzoic acid and protocatechuric acid (Komai and Ueki 1975). Ferulic, caffeic, hydroxyl benzoic, syringic, chlorogenic, p-coumaric acids and three unidentified compounds were found in C. rotundus tuber (Alsaadawi and Salih 2009b). Additional chromatographic analyses showed that the tuber also contained 11 volatile compounds, of which 10 were identified to be α-pinene, β-pinene, limonene, capsaicin, artemisinin, camphene, camphor, citronellal, farnesol and petalostemumol. Concentrations of all identified compounds except camphene, capsaicin and limonene were found to be higher in the tuber than in tops. Total volatile compound appeared to be considerably higher in the tuber. Ferulic acid content of the *C. rotundus* rhizomes, collected from 12 source areas in China, varied in a range of 0.027–0.0462 % (Li 2014). A new flavanone, 7,8-dihydroxy-5,6-methylenedioxyflavone and five known compounds, quercetin, kaempferol, luteolin, ginkgetin and isoginkgetin, were isolated from *Cyperus rotundu*s rhizomes (Zhou and Fu 2012). A novel enantiomeric and meso-stilbene trimers, i.e., (+)-(E)-cyperusphenol A, (−)-(E)-cyperusphenol A and (E)-meso-cyperusphenol A, a trimer bearing a novelhexacyclic ring system, cyperusphenol B, as well as known stilbenoids (cyperusphenols C and D, scirpusins A and B, and piceid) and luteolin were isolated from *C. rotundus* rhizomes (Ito et al. 2012). A new iridoid glycoside, rotunduside,

along with four known iridoid glycosides, 10-*O*-*p*-hydroxybenzoyltheviridoside, 10-*O*-vanilloyltheviridoside,6″-*O*-(*trans*-*p*-coumaroyl)-procumbide and loganic acid, were isolated from *Cyperus rotundus* rhizomes (Zhou et al. 2013). A new (2*RS*,3*SR*)-3,4′,5,6,7,8-hexahydroxyflavane, together with three known stilbene dimers cassigarol E, scirpusins A and B were isolated from *C. rotundus* rhizome (Tran et al. 2014).

Two sesquiterpenes, mustakone and copaene (Kapadia et al. 1963, 1965) and patchoulenone (Motl et al. 1963) were isolated from *C. rotundus*. Two sesquiterpenic alcohols, cyperol and isocyperol (Hikino et al. 1967b, 1968a); cyperolone (Hikino et al. 1967a, d); cyperotundone (Hikino et al. 1965, 1966, 1967b); cyperene, cyperotundone and patchoulenone (Hikino et al. 1968d); a sesquiterpenoid triol sugetriol (Hikino et al. 1967c, 1968c); a sesquiterpenic ketol sugeonol (Hikino et al. 1968b); cyperone, cyperotundone, and isopatchoul-4(5)-en-3-one (Neville et al. 1968); two norsesquiterpenoids, kobusone and isokobusone (Hikino et al. 1969b); two sesquiterpenic keto-alcohols, α-rotunol and β-rotunol (Hikino et al. 1969a; 1971) were isolated from nutgrass tubers. A sesquiterpenic oxido-alcohol was isolated from the rhizomes and identified as 4α-5α-oxidoeudesm-11-en--3α-ol (Hikino and Aota 1976). Paknikar et al. (1977) reported rotundene and rotudenol from *C. rotundus*. Four sesquiterpenes were isolated from tubers of purple nutsedge and were identified as cyperene, β-selinene, cyperenone and α-cyperone (Komai et al. 1977). Four sesquiterpenes were isolated from tubers of purple nutsedge and were identified as cyperene, β-selinene, cyperenone and α-cyperone (Iwamura et al. 1977). *Cyperus rotundus* tubers afforded the isolation of patchoulenone, caryophyllene α-oxide, 10,12-peroxycalamenene and 4,7-dimethyl-1-tetralone (Thebtaranonth et al. 1995). Kalsi et al. (1995) isolated β-caryophyllene, caryophyllene oxide and caryophylla-6-one from C. rotundus extract in Punjab, India. Ohira et al. (1998) identified 4,5-secoeudesmanolide, 10-epi-4,5-secoeudesmanolide, a cyclic acetal cyperolone, mustakone, α-cyperene, nootkatone, α-rotunol and β-rotunol in the hexane/water soluble fraction of C. rotundus methanol extract. Three sesquiterpene alkaloids: rotudines A, B and C were isolated from the rhizomes (Jeong et al. 2000). Five compounds cyperol, α-cyperone, cyperotundone, sugetriol triacetate and β-sitosterol were isolated from the methanol extract of C. rotundus tubers (Kim et al. 2000). Chhabra et al. (2002) reported cyperotundone, cyperone, isopatchoulenone, isophatchoul-3,5-diene in nutgrass. Two sesquiterpene ketones, cyperotundone (0.26 %) and α-cyperone (0.1 %), were isolated as major constituents from nutgrass-dried tubers (Morimoto and Komai 2005). A norsesquiterpene, named norcyperone (1), and three known compounds: (-)-clovane-2,9-diol; rosenonolactone and 5*a*,8*a*-epidioxy-(20*S*,22*E*,24*R*)-ergosta-6,22-dien-3*β*-ol were isolated from *C. rotundus* rhizomes (Xu et al. 2008). The structure of 1 was elucidated as 8,11,11-trimethylbicyclo [5.3.1] undecane-5*a*, 8*a*-epoxy-3-one. Two sesquiterpenes, epi-guaidiol A and sugebiol, together with four known sesquiterpenes, guaidiol A, sugetriol triacetate, cyperenoic acid and cyperotundone were isolated from the rhizomes (Xu et al. 2009). A new cerebroside, 1-O-(β-D-glucopyranosyloxy)-(2*S*, 3*R*, 4*E*, 8*Z*)-2-[(2′*R*)-2′-hydroxylignoceranoylamino]-4, 8-tetradecene-3-diol was isolated from the 60 % ethanol extract of *C. rotundus* (Liu et al. 2010). From the 70 % ethanol extract of nutgrass rhizomes, several major constituents including the sesquiterpene derivatives (valencene, nootkatone and caryophyllene α-oxide), monoterpenes (β-pinene, 1,8-cineole, and limonene) and 4-cymene were isolated (Jin et al. 2011). The content of (+)-nootkatone in *C. rotundus* rhizome was found to be higher in samples from India (30.47 μg/10 g) compared to samples from China (21.72 μg/10 g) (Jaiswal et al. 2014). Two compounds, 1α-methoxy-3β-hydroxy-4α-(3′,4′-dihydroxyphenyl)-1, 2,3,4-tetrahydronaphthalin and 1α,3β-dihydroxy-4α-(3′, 4′-dihydroxyphenyl)-1,2,3,4-tetrahydronaphthalin, along with six known compounds salicylic acid, caffeic acid, protocatechuic acid, *p*-coumaric acid, pongamone A and biochanin A were isolated from the rhizomes (Zhou and Yin 2012). From the n-hexane-soluble fraction of *C. rotundus* rhizomes, two new patchoulane-type sesquiterpenes (6S,9S)-patchoulan-4-ene-6,9-diol

diacetate {(6S,9S)-6,9-diacetoxy cyperene} and (6*S*)-patchoulan-4-ene-6-ol {(6*S*)-cyperene-6-ol} along with three known patchoulane-type sesquiterpenes (6S)-patchoulan-4-ene-6-ol acetate {(6S)-6-acetoxy cyperene}, sugetriol triacetate, cyperotunone and two known udesman-type sesquiterpenes α-cyperone and isocyperol were isolated (Kim et al. 2012). *Cyperus rotundus* rhizome extract afforded two new sesquiterpenoids with rearranged secoeudesmane and germacrene skeletons, and a new 9,10-seco-cycloartane triterpenoid, as well as seven previously reported terpenoids, including a monoterpenoid, five sesquiterpenoids with guaiane, patchoulane, and eudesmane skeletons, and a 3,4-seco-dammarane nortriterpenoid (Yang and Shi 2012). The following compounds were isolated from the n-hexane-soluble fraction of *C. rotundus* rhizomes: seven known eudesman-type sesquiterpenes α-rotunol; β-rotunol;(−)-eudesma-3,11-diene-5-ol; ligucyperonol; 14-hydroxy-α-cyperone; britanlin E and 1β,4β-dihydroxyeudesma-11-ene and two new patchoulane-type sesquiterpenes identified as sugetriol 6,9-diacetate and sugebiol 6-acetate (cyperene-3,6-diol 6-acetate) (Kim et al. 2013b).

Four major chemotypes H, M, K and O of *Cyperus rotundus* had been reported based on the sesquiterpene composition of essential oils in mature tubers (Komai and Ueki 1981; Komai and Tang 1989; Komai et al. 1991; Komai 1993; Jirovetz et al. 2004; Kilani et al. 2005a, b; Zoghbi et al. 2008). The O type was found to have the widest range of global distribution in Asia, USA, and Australia and was dominant in the Pacific Basin islands except for Hawaii, where the K-type was dominant. The K-type had higher concentrations of cyperene (28.7 %), cyperotundone (8.8 %), patchoulenyl acetate (8.0 %), patchouli acetate (8.0 %) and sugeonyl acetate (6.9 %) (Komai and Tang 1989). Other minor compounds were β-elemene (6.5 %), β-caryophyllene (5.0 %), α-humulene (3.9 %), δ-cadinene (3.8 %), α-copaene (3.6 %) and calamine (1.5 %). Sesquiterpenoids identified in the tuber essential oil of *C. rotundus* Type M and Type K were, respectively: cyperene (8 %, 31.3 %), cyperotundone (19.5 %, 12 %), patchoulenyl acetate (trace, 9.1 %), sugeonyl acetate (0.5 %, 8.5 %), β-elemene (2.5 %, 5.2 %), sugetriol acetate (2.5 %, 4.5 %), α-humulene (2.2 %, 4.1 %), α-copaene (trace, 0.7 %), β-caryophyllene (3 %, 3.8 %), δ-cadinene (3.1 %, 3.0 %), calamenene (trace, 1.5 %) and α-cyperone (31.4 %, not detected) (Komai et al. 1994). The K-type was also found in USA and Mexico (Komai et al. 1991). The H-type containing α-cyperone (36.6 %), β-selinene (18.5 %), cyperol (7.4 %) and β-caryophyllene (6.2 %) was found to dominate on the islands of Japan. Other minor components found were α-humulene (4.5 %), calamenene (3.9 %), δ-cadinene (1.0 %), β-elemene (0.8 %) and α-copaene (tr). The M-type from China, Hong Kong, Japan, Taiwan and Vietnam had α-cyperone (30.7 %), cyperotundone (19.4 %), β-selinene (17.8 %), cyperene (7.2 %) and cyperol (5.6 %). The minor components were β-caryophyllene (3.3 %), β-elemene (2.5 %), δ-cadinene (2.0 %), α-humulene (1.7 %), α-copaene (tr), calamenene (tr), patchouli acetate (tr) and sugeonol acetate (tr). The O-type from Japan, Taiwan, Thailand, Hawaii and the Philippines was characterised by cyperene (30.8 %), cyperotundone (13.1 %) and β-elemene (5.2 %). The minor components were α-humulene (4.0 %), β-caryophyllene (3.8 %), δ-cadinene (3.5 %), calamenene (2.6 %), α-copaene (2.0 %), β-selinene (tr), patchouli acetate (tr) and sugeonol acetate (tr). Additionally, the Hawaiian O-type had cyperotundone (25.0 %) and cyperene (20.7 %) as the major compounds. The minor components were α-humulene (4.7 %), δ-cadinene (3.4 %), β-caryophyllene (3.2 %), β-elemene (2.6 %), calamenene (1.7 %), α-copaene (tr), patchouli acetate (tr) and sugeonol acetate (tr).

The essential oil of *Cyperus rotundus* was found to contain at least 27 components comprising 10 sesquiterpene hydrocarbons (25 %), 3 sesquiterpene epoxides (12 %), 4 sesquiterpene ketones (20 %), 4 monoterpene and aliphatic alcohols (25 %) and unidentified components (Kapadia et al. 1967). Some new sesquiterpenoid compounds isolated were copadiene, rotundone and epoxy-quaine. Trivedi et al. (1964b) separated two sesquiterpenoids cyperence and patchoulenone from Chinese nutgrass oil. *C. rotundus* tuber oil was found to contain: cyperene, β-selinene, α-cyperone, β-cyperone, selin-

atriene, patchoulenone and monoterpenic compounds 1,8-cineol, limonene, β-pinene, *p*-cymene, camphene and unidentified α,β-unsaturated ketone and a monoterpenic alcohol (Trivedi et al. 1964a). Iwamura et al. (1977) reported kobusone, isokobusone, caryophyllene, caryophyllene-6,7-oxide and caryophylla-6-one in *C. rotundus* essential oil. Essential oil of nutgrass tubers was found to contain caryophyllene alcohol, caryophyllene ketone, ketone acetate and several terpenoids: kobusone, isokobusone, rotundenone, patchoulene, isopatchoulene, selinene, sacridione, cyperenol, rotundene, cyprene, *p*-cymene, camphene, 2-8-cineole, β-pinene, limonene (Dhillon et al. 1993). The oil and methylene chloride extract of *C. rotundus* from Chad was found to contain, respectively: cyperotundone (34.95, 15.6 %), eugenol acetate (8.7, 12.9 %), caryophyllene oxide (7.0, 6.5 %), mustakone (5.1, 5.9 %), viridiflorol (3.0, 1.6 %), cyperene (2.9, 15.9 %), cyperenal (4.6 %, 2.2 %); patchoulenone (4.5, 2.7 %), isomustakone (2.3,3.0 %), ylangadiene (1.0, 0.5 %), copadiene (0.9, 0.3 %), sugeonol (0.6,1.2 %), δ-cadinene (0.5, 0.2 %), verbenone (0.3, tr%), pinocarveol (0.2, tr%), α-ylangene (0.2, 3.9 %), myrtenol (0.1, tr%) and α-copaene (0.1, 0.4 %) (Mahmout et al. 1997).

Three new sesquiterpene hydrocarbons (−)-isorotundene, (−)-cypera-2,4(15)-diene, (−)-norrotundene and the ketone (+)-cyperadione were isolated from *C. rotundus* essential oil (Sonwa 2000; Sonwa and König 2001). The known sesquiterpenes isolated included cyprotene, cypera-2,4-diene, α-copaene, cyperene, α-selinene, rotundene, valencene, ylanga-2,4-diene, γ-gurjunene, *trans*-calamenene, δ-cadinene, γ-calacorene, epi-α-selinene, α-muurolene, γ-muurolene, cadalene and nootkatene. More than 90 compounds were identified in the tuber essential oil, and about 70 aroma volatiles were identified in the SPME-headspace of *C. rotundus* as follows: α-copaene (11.4 %), valerenal (9.8 %), caryophyllene oxide (9.7 %), cyperene (8.4 %), nootkatone (6.7 %) and *trans*-pinocarveol (5.2 %) were the main ones in the essential oil; α-copaene (12.1 %), cyperene (11.7 %), valerenal (8.7 %), caryophyllene oxide (7.8 %), *trans*-pinocarveol (7.4 %) and valencene (5.1 %) in the headspace volatiles (Jirovetz et al. 2004). Thirty-three compounds were identified in Tunisian nutgrass tuber oil, characterised by its high content of sesquiterpenes with cyperene (30.9 %) predominating (Kilani et al. 2005a). The other components were: rotundene (7.6 %), α-cyperone (4.5 %), β-pinene (3.9 %), mustakone (3.8 %), isorotundene (3.6 %), α-cubebene (3.4 %), α-cadinol (2.5 %), cypera-2,4-diene (2.4 %), isocyperol (2.0 %), *trans*-calamenene (2.0 %), τ-muurolol (1.9 %), τ-cadinol (1.4 %), α-pinene (1.4 %), α-calacorene (1.4 %), α-muurolene (1.1 %), *(E,E)*-farnesol (1.05), α-copaene (0.9 %), humulene epoxide (0.9 %), δ-cadinene (0.7 %), caryophyllene oxide (0.7 %), β-selinene (0.6 %), cadia-1,4-diene (0.6 %), α-humulene (0.5 %), borneol (0.3 %), β-caryophyllene (0.3 %), cyprotene (0.2 %) and elemol (0.2 %). More than 33 compounds were identified in *C. rotundus* tuber essential oil and cyperene, α-cyperone, isolongifolen-5-one, rotundene and cyperorotundene were the principal compounds comprising 62 % of the oil (Kilani et al. 2008b). Eight compounds were identified α-copaene, cyperene, β-selinene, β-cyperone and α-cyperone, selina-4,11-diene, aristol-9-en-8-one and aristol-9-en-3-one in nutgrass essential oil (Tam et al. 2007). Of three extraction methods, pressurized liquid extraction had the highest extraction efficiency for five components, including α-copaene, cyperene, β-selinene, β-cyperone and α-cyperone, while supercritical fluid extraction had the best selectivity for extraction of β-cyperone and α-cyperone compared to hydrodistillation.

Zoghbi et al. (2008) found α-cyperone (22.8 %), cyperotundone (12.1 %), caryophyllene oxide (6.9 %), isocyperol (6.7 %) and eudesma-3-11-dien-5-ol (6.2 %) as the main compounds from *C. rotundus* oil from Para state, Brazil. The other minor components (<6 %) included α-pinene, sabinene, β-pinene, *p*-menth-3-ene, limonene, 1,8-cineole, α-campholenal, *trans*-pinocarveol, *trans*-verbenol, pinocarvone, *p*-mentha-1,5-dien-8-ol, myrtenal + mrytenol, verbenone, *trans*-carveol, cyprotene, cypera-2,4-diene, cyperene, rotundene, eudesma-1,4,11-triene, β-selinene, α-selinene, epicubenol, δ-cadinene, *trans*-calamenene, cyperene epox-

ide, humulene epoxide ll, patchoulenone, cyperol, rotundone, eudesm-11-ene-4α,6α-diol and eudesma-3,11-diene-2-one. Sixteen compounds were identified constituting 67.55 % of the total oil (0.2 %) hydrodistilled from nutgrass tubers collected from Balawala, Dehradun, Uttarakhand (Bisht et al. 2011). The major components were 5-oxo-isolongifolene (16.268 %), α-gurjunene (10.22 %) and (*Z*)-valerenyl acetate (8.89 %) and α-selinene (4.48 %). Other components were valerenic acid (3.7 %), γ-cadinene (3.4 %), 9-H-cycloisolongifolene, 8-oxo (3.24 %), (+)-α-ylangene (2.9 %), isospathulenol (2.61 %), isolongifolene, 7,8-dihydro-8a-hydroxy (2.40 %), isocyclosyechellene (2.4 %), 9-aristolene-1-α-ol (2.16 %), β-elemene (1.33 %), (-) caryophyllene oxide (1.33 %), δ-cadinene (1.21 %) and *cis*-α-copane-8-ol (1.04 %).

Forty-one and 43 components were identified in the essential oil, representing 89.9 % and 92.0 % of sample A and sample B, from two different locations (Empangeni-A and KwaDlangezwa-B; in the Kwa Zulu-Natal Province of South Africa (Lawal and Oyedeji 2009). α-cyperone (11.0 %), myrtenol (7.9 %), caryophyllene oxide (5.4 %), β-pinene (5.3 %) and β-selinene (5.1 %) were major compounds in the oil of sample A. Minor constituents in sample A included α-pinene, camphene, myrcene, α-phellandrene, *p*-cymene, limonene, 1,8-cineol, terpineole, *trans*-pinocarveol, pinocarvone, *p*-mentha-1,5-diene-8-ol, terpinen-4-ol, verbenone, *trans*-carveol, cuminaldehyde, β-elemene, cyperene, β-caryophyllene, α-humulene, *allo*-aromadendrene, eudesma-2,4,11-triene, α-selinene, spathulenol, (*2R,5E*)-caryophyll-5-en-12-al, humulene epoxide II, oplopenone, patchenol, caryophylla-3,8(13)-dien-5-α-ol, caryophylla-3,8(13)-dien-5-β-ol, vulgarol B, caryophyllenol 11, aristolone, solavetivone, phytol, hexadecanoic acid and two unknowns. The main constituents of the oil of sample B were: β-pinene (11.3 %), α-pinene (10.8 %), α-cyperone (7.9 %), myrtenol (7.1 %), α-selinene (6.6 %) and limonene (5.7 %). Minor constituents from sample were: camphene, bicyclo[3.2.0] hept-6-ene, *p*-cymene, perillene, 3,3,5-trimethyl cyclohexene, fenchol, *trans*-pinocarveol, pinocarvone, camphene hydrate, borneol, terpinen-4-ol, verbenone, *trans*-carveol, cuminaldehyde, carvone, α-copaene, β-elemene, cyperene, β-caryophyllene, α-gurjunene, α -humulene, *allo*-aromadendrene, β-selinene, germacrene B, caryophyllene oxide, humulene epoxide II, globulol, patchenol, 2-cyclopropylthiophene, caryophylla-3,8(13)-dien-5-β-ol, vulgarol B, caryophyllenol 11, aristolone, aromadendrene epoxide, nootkatone, oxo-α-ylangene and one unknown.

Twenty-five components representing 91.2 % were found in the essential oil *C. rotundus* roots and rhizomes and the major components were α-cyperone (32.0 %) and spathulenol (11/0 %) (Mojab et al. 2009). The yield of essential oil from fresh *C. rotundus* rhizomes was 0.4 % (Chen et al. 2011). The major compounds were cyperene (41.03 %), β-caryophyllene oxide (5.32 %), α-selinene (4.37 %), α-copaene (4.36 %), naphthalene, 6-isoproenyl-4,8a-dimethyl-1, 2, 3, 5, 6, 7, 8, 8a-octahydro- (3.80 %) and α-cyperone (3.11 %). Sixty natural compounds consisting 95.8 % of the total components were identified from *C. rotundus* essential oil from Iran, with a yield of 0.2 % (w/w (Ghannadi et al. 2012). Among the oil constituents were cyperene (16.9 %), caryophyllene oxide (8.9 %) and α-longipinane (8.4 %). β-selinene (6.6 %), eugenol (4.7 %), aristolone (3.5 %), β-calacorene (3.3 %), α-copaene (3.2 %), *trans*-γ-bisabolene (3.1 %), α-cyperone (3.0 %), 1,8-cineole (2.4 %),α-ylangene (2.3 %), β-caryophyllene (2.3 %), aromadendrene (2.1 %), *cis*-γ-bisabolene (2.1) and α-guaiene (2.0 %) were the major components. Other components were present in amounts less than 2 % and included α-pinene, verbena, β-pinene, *o*-cymene, limonene, *p*-cymene, α-fenchol, *trans*-pinocarveol, *cis*-verbenol, pinocarvone, isopinocamphon, *p*-cymen-8-ol, α-terpineol, myrtenol, verbenone, *trans*-carveol, *trans*-myrtenyl acetate, cuminc aldehyde, carvone, cinnamaldehyde, *trans*-anethole, thymol, 2,4-decadienal, carvacrol, β-elemene, β-gurjunene, isoaromadendrene, α-humulene, α-caryophyllene, rotundene, γ-gurjunene, γ-muurolene, *n*-dodecanol, α-selinene, α-farnesene, α-calacorene, γ-elemene,

spathulenol, humulene epoxide, γ-gurjunene epoxide, *n*-hexadecanoic acid, phytol and methyl linoleate. Twenty-two compounds, mainly sesquiterpenes and also monoterpenes were identified from Iranian *C. rotundus* oil, of which cyperene (37.9 %) and cyperotundone (11.2 %) predominated (Aghassi et al. 2013).

Aerial Plant Part Phytochemicals

Cyperus rotundus was reported to contain essential oils, flavonoids, terpenoids, sesquiterpenes, cyprotene, cyperene, α-selinene, rotundene, valencene, cyperol, gurjunene, *trans*-calamenene, cadalene, cyperotundone, mustakone, isocyperol, α-cyperone, etc. (Imam et al. 2014). Ferredoxin was isolated from *C. rotundus* plant (Lee et al. 1970). The carboxyterminal amino acid of nutsedge ferredoxin was glycine and the amino-terminal amino acid was serine, which differed from spinach ferredoxin in which both terminal amino acids were alanine. Sucrose was the most concentrated sugar in the shoots, with the highest concentration in the younger leaves besides glucose and fructose (Wills 1972). Phosphorus was concentrated in the growing leaves while iron was most concentrated in older tissues, including rachises with seedheads and mature shoots. Three different aminopeptidases, one iminopeptidase, two or more carboxypeptidases and two or more different endopeptidases were detected in mature green leaves of *C. rotundus* (Fischer et al. 1998). The methanolic extract of *C. rotundus* aerial parts afforded the isolation of sitosterol, stigmasterol, sitosterol glucoside, stigmasterol glucoside, chrysoeriol, kaempferol, luteolin, quercetin, rutin and khellol-β-D-glucopyranoside (Sayed et al. 2001). A steroid glycoside named sitosteryl (6'-hentriacontanoyl)-β-D-galactopyranoside and three furochromones: khellin, visnagin and ammiol plus benzo-α-pyrone (coumarin), salicylic acid, caffeic acid, protocatechuic acid, *p*-coumaric acid, tricin and isorhamnetin were isolated from the aerial plant parts (Sayed et al. 2007). A fructose-amino acid conjugate, N-(1-deoxy-α-D-fructos-1-yl)-L-tryptophan and its tautomers, in addition to *n*-butyl-β-D-fructopyranoside; ethyl-α-D-glucopyranoside; adenosine; (-)-(*E*)-caffeoylmalic acid; vitexin; isovitexin; orientin; epiorientin; myricetin 3-*O*-β-D-galactopyranoside; luteolin 7-*O*-β-D-glucuronopyranoside-6″-methyl ester; chlorogenic acid; luteolin 4′-*O*-β-D-glucuronopyranoside; luteolin 7-*O*-β-D-glucuronopyranoside; uridine and ellagic acid were isolated from the aerial parts (Sayed et al. 2008). Rotundone (920 mg/kg) was found in the leaves of *C. rotundus* (Wood et al. 2008). Five major compounds (luteolin, ferulic acid, quercetin, 3-hydroxy, 4-methoxybenzoic acid and 6,7-dimethoxycoumarin) were isolated from ethyl acetate *C. rotundus* extract (Kilani-Jaziri et al. 2009). Ferulic, caffeic, hydroxyl benzoic, syringic, chlorogenic, p-coumaric acids and three unidentified compounds were found in *C. rotundus* shoot (Alsaadawi and Salih 2009b). Additional chromatographic analyses showed that the shoot also contained 13 volatile compounds, of which 10 were identified to be α-pinene, β-pinene, limonene, capsaicin, artemisinin, camphene, camphor, citronellal, farnesol and petalostemumol. Four flavonoids, namely, quercetin, kaempferol, catechin and myricetin were isolated from *in-vivo* (leaf and root) and *in-vitro* callus of *Cyperus rotundus* (Krishna and Renu 2013). The maximum amount of total flavonoid was found in 6-week-old callus tissue (1.96 mg/g.d.w.) and minimum (0.28 mg/g.d.w.) in 2-week-old callus tissue. Higher flavonoids content were found in leaf in free form (0.58 mg/g.d.w.) and bound form (0.48 mg/g.d.w.) when compared to root in free form (0.19 mg/g.d.w) and bound form (0.11 mg/g.d.w).

A number of pharmacological and biological activities including anti-*Candid*a, anti-inflammatory, antidiabetic, antidiarrhoeal, cytoprotective, antimutagenic, antimicrobial, antioxidant, cytotoxic as well as apoptotic, antipyretic and analgesic activities have been reported for this plant (Duarte et al. 2005). *Cyperus rotundus* has a wide range of medicinal and pharmacological applications. It is used in traditional system of medicine and exhibits anti-inflammatory, anti-arthritic, antipyretic, analgesic, antidiabetic, antidiarrhoeal, cytoprotective, antimutagenic, antimicrobial, antioxidant, cyto-

toxic, apoptotic and various other activities (Singh et al. 2012). Research studies had shown that it possesses various pharmacological activities such as diuretic, carminative, emmenagogue, anthelminthic, analgesic, anti-inflammatory, anti-dysenteric, antirheumatic activities (Imam et al. 2014).

Antioxidant Activity

Ethanol extract of *C. rotundus* root and rhizomes showed higher antioxidant activity than the petroleum ether (7.9 %) (35.5 %), chloroform (12 %) and aqueous extract (14.1 %) in the inhibition of haemoglobin glycosylation with a concentration of 1.0 mg/ml of each (Pal and Dutta 2006). The presence of flavonoids (1.25 %), ascorbic acid (0.01 %) and polyphenols (1.62 %) was suggested to be responsible for this activity. The hydroalcoholic extract of *C. rotundus* exhibited high reduction capability and potent free radical scavenging, especially against DPPH and superoxide anions as well as a moderate effect on NO (Yazdanparast and Ardestani 2007). The extract also showed inhibited lipid peroxidation in rat liver homogenate induced by Fe^{2+}/ascorbate and prevented deoxyribose degradation in both non-site-specific and site-specific assays, showing the hydroxyl radical scavenging and metal chelating activity of the hydroalcoholic extract. Further, the peroxidation inhibiting activity of the extract was demonstrated in the linoleic acid emulsion system.

Total oligomer flavonoids (TOF), ethyl acetate and methanol nutgrass extracts exhibited marked free radical scavenging activity towards the 1,1-diphenyl-2-picrylhydrazine (DPPH) free radical with IC50 values of 5, 20 and 65 μg/mL respectively (Kilani et al. 2005b). *C. rotundus* tuber essential oil was found to have antioxidant activity using DPPH, xanthine/xanthine oxidase assays, and the scavenging of superoxide radical assay generated by photo-reduction of riboflavin (Kilani et al. 2008b). The total oligomer flavonoid (TOF) and ethyl acetate extracts of *C. rotundus* tubers showed a significant ability to inhibit nitroblue tetrazolium reduction by the superoxide radical in a non-enzymatic superoxide generating system (Kilani et al. 2008a). In another study, TOF and ethyl acetate extracts of *C. rotundus* were found to be efficient in inhibiting xanthine oxidase with IC_{50} values of 240 and 185 μg/ml and superoxide anion with IC_{50} values of 150 and 215 μg/ml, respectively (Kilani-Jaziri et al. 2009). Also, the extracts tested were effective in reducing the production of thiobarbituric acid reactive substances (TBARS) and were able to protect against H_2O_2/UV-photolysis-induced DNA damage. The highest activity, measured as equivalents of MDA concentration, was observed in the ethyl acetate extract (MDA = 2.04 nM). Luteolin was the most active in reducing the production of TBARS (MDA = 1.5 nM), and in protecting against H_2O_2/UV-photolysis-induced DNA damage. The flavanone compound 7,8-dihydroxy-5,6-methylenedioxyflavon, isolated from the rhizome, showed inhibitory activity in superoxide radical scavenging assay with IC_{50} value of 3.1 μM (Zhou and Fu 2012). Compounds quercetin, kaempferol, luteolin, ginkgetin and isoginkgetin exhibited antioxidant activity in scavenging DPPH and superoxide radicals with IC_{50} values ranging from 3.1 to 23.7 μM.

Among the 70 % ethanolic, methanolic and water extracts of *C. rotundus* root the 70 % ethanolic extract was found to be most potent with the IC_{50} values of 64.64 μg/mL, 85.89 μg/mL and 8.42 mg/mL against DPPH, metal chelating and nitric oxide scavenging activities and the observed activity could be correlated with metabolites such as polyphenols, flavonoids and sesquiterpenes (Kumar et al. 2014). The extract showed 48 % protection against H_2O_2-induced DNA damage and inhibited 2, 2-azobis (2-amidino-propane) dihydrochloride (AAPH)- and 3-morpholinosydnonimine (*SIN-1*)-induced oxidation and nitration of bovine serum albumin. Moreover, *C. rotundus* pretreatment restored the antioxidant status in white blood cells treated with H_2O_2. *C. rotundus* showed acetylcholine esterase inhibitory activity and the extract was found to be non-toxic up to 100 μg/mL in SH-SY5Y human neuroblastoma cell line. The ethanolic extract (200 mg/kg) also proved to be a

potent anxiolyte as assessed by behavioural tests. Overall, the results suggested that *C. rotundus* extract may be useful in combating oxidative stress-related diseases through its antioxidant activity. The methanol and aqueous extracts of *C. rotundus* aerial parts showed, respectively, 88 % and 19 % inhibition of xanthine oxidase activity (Soumaya et al. 2014). Yet, the same extracts inhibited lipid peroxidation by 61.5 % and 42.0 %, respectively. Both extracts inhibited OH formation by 27.1 % and 25.3 %, respectively. Only the methanol extract of *C. rotundus* aerial parts induced DNA degradation towards K562 and L1210 cell lines (Soumaya et al. 2014). Orientin was determined as the major compound isolated from the butanol fraction of methanol extract.

Anti-tumour Activity

Visnagin, khellin and sitosteryl (6'-hentriacontanoyl)-β-D-galactopyranoside, isolated from the aerial parts, showed strong cytotoxic activity against L5178y mouse lymphoma cells and were also active in the brine shrimp lethality test (Sayed et al. 2007). *C. rotundus* tuber essential oil was found to be very effective against L1210 leukaemia cells line in in-vitro MTT cytotoxicity assay (Kilani et al. 2008b). The results correlated significantly with increased apoptotic DNA fragmentation. The total oligomers flavonoids and ethyl acetate extracts of *C. rotundus* tubers suppressed growth and proliferation of L1210 cells derived from murine lymphoblastic leukaemia (Kilani et al. 2008b). Morphological features of treated cells and characteristic DNA fragmentation revealed that the cytotoxicity was due to induction of apoptosis. Only the total oligomer flavonoid (TOF)-enriched extract of *C. rotundus* exerted growth inhibition on K562 human chronic myelogenous leukemia cells through apoptosis induction (Kilani-Jaziri et al. 2009). Luteolin was the most active in inhibiting significantly the proliferation of K562 cells (IC_{50}=25 μg/ml). Stilbene oligomers isolated from *C. rotundus* rhizome exhibited antiproliferative activity in the Jurkat cell line (human T-cell leukemia cells), with the IC_{50} potencies of a racemate of (+)-(*E*)-cyperusphenol A, (−)-(*E*)-cyperusphenol A and (*E*)-mesocyperusphenol A, cyperusphenol B, and cyperusphenol D were estimated as 27.4, 40.5, 26.4 and 26.3 μM, respectively (Soumaya et al. 2014). The suppression of cell growth by cyperusphenol D was due to the induction of apoptosis, which was characterised by nuclear changes and PARP-1 cleavage. Only the methanol extract of *C. rotundus* aerial parts induced DNA degradation towards K562 and L1210 cell lines. Total oligomeric flavonoids (TOF) extracted from dry *C. rotundus* rhizomes inhibited both human glioblastoma (AMGM) and mice mammary adenocarcinoma (AMN3) cancer cells proliferation by 67.09 % and 52.41 % at concentration of 350 μg/ml (IC_{50}) during incubation time of 24 h (Al-Saeedi 2013). The treatment with 1000 μg/ml increased the apoptotic cells to 89.5 % and 79.89 %, respectively. Both rat embryo fibroblast (REF) and African green monkey kidney (Vero) cell lines were totally resistant to TOF at 500 and even for 1000 μg/ml.

Antimutagenic Activity

C. rotundus oil acted as an antimutagen against Aflatoxin B1 (AFB 1) in both *Salmonella* strains (TA100 and TA98) and *Escherichia coli* strain (PQ37) and against nifuroxazide in *Escherichia coli* strain (PQ37) (Kilani et al. 2005a, b). The highest rates of AFB1 mutagenesis inhibition tested by Ames assay, ranged from about 82.56 % for TA100 strain to 85.47 % for TA98 strain at the same dose of 50 μg AFB1 per plate. Whereas, the mutagenic effect of nifuroxazide and AFB1 (50 μg/assay) were reduced by approximately 58.19 % and 81.67 %, respectively.

Anti-inflammatory Activity

The ethanol nutgrass rhizome extract was reported to exert hypotensive, anti-histaminic and anti-inflammatory activities in-vivo (Singh et al. 1970). Specifically, the extract exhibited an inhibitory effect against histamine-induced bron-

choconstriction in guinea pigs and strongly inhibited carrageenan-induced paw edema and formaldehyde-induced arthritis in rats, suggesting that it has anti-allergic and anti-inflammatory effects. The petroleum ether fraction of nutgrass extract showed the strongest anti-inflammatory action in-vivo (Gupta et al. 1971). Cyperol, α-cyperone and cyperotundone, isolated from the rhizome, exhibited inhibitory effects on LPS-induced nitric oxide production in RAW 264.7 cells in a dose-dependent manner (Kim et al. 2000). The methanol extract of *C. rotundus* rhizomes exhibited inhibition of nitric oxide (NO) production in a dose-dependent manner by RAW 264.7 cells stimulated with interferon-gamma plus lipopolysaccharide (Seo et al. 2001). The inhibition of NO production by the extract was due to the suppression of iNOS protein, as well as iNOS mRNA expression. Additionally, the methanol extract suppressed the production of superoxide O^{2-} by phorbol ester-stimulated RAW 264.7 cells in dose- and time-dependent manners. The results suggested that the methanol rhizome extract of *C. rotundus* could be developed as anti-inflammatory candidate for the treatment of inflammatory diseases mediated by overproduction of NO and O^{2-}. Nutgrass leaf extract dose-dependently inhibited rat paw edema induced by carrageenan (Sundaram et al. 2008). The extract also reduced significantly the wet weight of implanted granuloma in the subplantar region in the sub-acute inflammation model. The results indicated that the nutgrass leaf extract had anti-inflammatory property.

Among the four sesquiterpenes, isolated from nutgrass, α-cyperone and nootkatone, showed stronger anti-inflammatory and a potent NF-κB inhibitory effect on lipopolysaccharide (LPS)-stimulated RAW 264.7 cells than valencene and β-selinene (Khan and lee 2011). Molecular analysis revealed that various inflammatory enzymes (iNOS and COX-2) were reduced significantly and this correlated with downregulation of the NF-κB signaling pathway. Additionally, nootkatone and α-cyperone dramatically suppressed LPS-induced NF-κB-DNA binding activity. The *n*-hexane fraction of the 80 % ethanol nutgrass rhizome extract inhibited both NO and prostaglandin E_2 (PGE_2) production in RAW 264.7 cells (Jung et al. 2013). α-Cyperone isolated from the *n*-hexane fraction significantly inhibited PGE_2 production by suppressing the LPS-induced expression of inducible COX-2 at both the mRNA and the protein levels but had little effect on NO production and iNOS expression. Additionally, α-cyperone downregulated the production and mRNA expression of the inflammatory cytokine IL-6 and suppressed the transcriptional activity of NFκB and the nuclear translocation of the p65 NFκB subunit in LPS-induced RAW 264.7 cells. Of nine compounds isolated from the rhizome, sugebiol 6-acetate; β-rotunol; 14-hydroxy-α-cyperone; britanlin E and 1β,4β-dihydroxyeudesma-11-ene were found to have significant inhibitory effects on LPS-induced NO production (IC_{50} values were ≤ 100 μM) in RAW 264.7 cells (Kim et al. 2013b). The results indicated these compounds to have anti-inflammatory properties via inhibition of nitric oxide production. Aqueous, ethyl acetate, methanol and TOF-enriched extracts (300, 150 and 50 μg/ml) of *C. rotundus* aerial plant parts were able to decrease the mouse ear oedema induced by xylene revealing the anti-inflammatory of the extracts (Soumaya et al. 2013). Isocyperol from nutsedge rhizome significantly inhibited NO production by suppressing LPS-induced expression of iNOS at both mRNA and the protein levels in RAW 264.7 cell (Seo et al. 2014). Additionally, isocyperol downregulated the expression of inflammatory cytokine TNF-α and IL-1β, but not IL-6 and reduced STAT3 activation in LPS-stimulated RAW 264.7 cells. The data suggested that the anti-inflammatory activity of isocyperol was related to the downregulation of iNOS and pro-inflammatory cytokines by suppressing STAT 3 pathway in LPS-stimulated RAW 264.7 cells.

A polyherbal ayurvedic formulation containing four different plants viz., *Aegle marmelos, Coriandrum sativum, Cyperus rotundus* and *Vetiveria zinzanioids*, showed significant inhibitory activity against inflammatory bowel disease in acetic acid-induced colitis in mice and indomethacin-induced enterocolitis in rats. The results obtained established the efficacy of this polyherbal formulation against inflammatory bowel diseases (Jagtap et al. 2004). Compared

with the control, treatment with *C. rotundus* essential oil significantly reduced carrageenan paw edema in rats (Biradar et al. 2010). At 500 mg/kg, the oil produced a comparable anti-inflammatory effect (69.7 %) with indomethacin (72.1 %). The oil also significantly reduced the swelling in the formalin-injected (arthritic) left hind paw as compared to the diclofenac sodium-treated group. *Cyperus rotundus* exhibited anti-inflammatory effects in both carrageenan-induced rat paw edema and acetic acid-induced peritonitis in mice (Dang et al. 2011).

Among 12 compounds isolated from nutgrass rhizome extract, mostly sesquiterpenes induced stronger heme oxygenase expression than monoterpenes in macrophage cells (Tsoyi et al. 2011). The extract and its constituents, nootkatone and valencene (sesquiterpenes) significantly inhibited inducible nitric oxide synthase (iNOS) expression and NO production in LPS-simulated RAW264.7 cells. All three showed marked inhibition of high mobility group box-1 (HMGB1) in LPS-activated macrophages and increased survival rates in cecal ligation and puncture (CLP)-induced sepsis in mice. Aqueous, ethyl acetate, methanol and TOF-enriched extracts of *C. rotundus* aerial parts (300, 150 and 50 μg/ml) decrease mouse ear oedema induced by xylene (Kilani-Jaziri et al. 2013).

Studies by Choi et al. (2014) found that (+)-nootkatone, a major component of *Cyperus rotundus*, may suppress tumor necrosis factor α (TNF-α)/interferon γ (IFN-γ)-induced thymus and activation-regulated chemokine (TARC/CCL17) and macrophage-derived chemokine (MDC/CCL22) expression in HaCaT cells by inhibition of protein kinase Cζ and p38 mitogen-activated protein kinase (MAPK) signaling pathways that led to activation of nuclear factor kappa B (NF-κB). They proposed that (+)-nootkatone may be a useful therapeutic candidate for inflammatory skin diseases such as atopic dermatitis.

Analgesic Activity

Cyperus rotundus ethanol extract exhibited significant analgesic properties as evidenced by the significant reduction in the number of writhes and stretches induced in mice by 1.2 % acetic acid solution (Pal et al. 2009). It also potentiated analgesia induced by morphine and pethidine in mice. The behavioral studies on mice indicate CNS depressant activity of *C. rotundus* ethanol extract. The stretching times of mice induced by acetic acid were significantly reduced by *C. rotundus* rhizome essential oil at dosage of 0.1 g/kg, indicating it to have an analgesic effect (Chen et al. 2011).

Aqueous, ethyl acetate, methanol and TOF-enriched extracts of *C. rotundus* aerial parts (300, 150 and 50 μg/ml) reduced the number of abdominal contractions caused by acetic acid in mice, revealing the peripheral analgesic activity of these extracts (Kilani-Jaziri et al. 2013). Mice treated with doses up to 300 mg/kg b.w. of *Cyperus rotundus* extracts did not exhibit any toxicity. Aqueous, ethyl acetate, methanol and TOF-enriched extracts (300, 150 and 50 μg/ml) of *C. rotundus* aerial plant parts reduced the number of abdominal contractions caused by acetic acid in mice, revealing the peripheral analgesic activity of these extracts (Soumaya et al. 2013). Mice treated with doses up to 300 mg/kg b.w. of these extracts did not exhibit any toxicity.

Antidepressant Activity

The ethyl acetate and n-butanol fractions from *C. rotundus* ethanol extract significantly decrease the d immobility time in swimming and tail suspension tests (Zhou and Liu 2012). Both fractions markedly increased the content of serotonin (5-HT), dopamine with no action of noradrenalin. The study showed that both fractions could produce an antidepressant effect, the mechanism of which may be related to affecting the 5-HT and dopamine nervous system.

Immunomodulatory Activity

Aqueous, ethyl acetate, methanol and TOF-enriched extracts of *C. rotundus* aerial parts (300, 150 and 50 μg/ml) significantly enhanced lymphocyte proliferation at 1 mg/ml (Kilani-Jaziri et al. 2013). Rhizome compound *10-O-p-*

hydroxybenzoyltheviridoside exhibited considerable macrophages respiratory burst inhibitory activity with IC_{50} value of ~37 μM (Zhou et al. 2013). Aqueous, ethyl acetate, methanol and TOF-enriched extracts (300, 150 and 50 μg/ml) of *C. rotundus* aerial plant parts significantly enhance lymphocyte proliferation at 1 mg/ml (Soumaya et al. 2013).

Antiatherosclerotic Activity

A new cerebroside, 1-*O*-(β-D-glucopyranosyloxy)-(*2S*,*3R*,*4E*,*8Z*)-2-[(2′*R*)-2′-hydroxylignoceranoylamino]-4,8-tetradecene-3-diol 1-*O*-(β-D-glucopyranosyloxy)-(*2S*,*3R*,*4E*,*8Z*)-2-[(2′*R*)-2′-hydroxylignoceranoylamino]-4,8-tetradecene-3-diol isolated from the roots exhibited antiproliferative activity on vascular smooth muscle cells (VSMCs) (Liu et al. 2010).

Antimicrobial Activity

Essential oil of *C. rotundus* was reported to have anti-*Candida albicans* activity (Duarte et al. 2005). The antibacterial activity of nutgrass tuber oil, showed more important activity against Gram-positive bacteria, especially *Staphylococcus aureus* than Gram-negative bacteria (Kilani et al. 2005a). The growth and acid production of the cariogenic bacterium, *Streptococcus mutans,* were reduced by *C. rotundus* extract in a dose-dependent manner (Yu et al. 2007). The extract markedly inhibited the adherence of *S. mutans* to saliva-coated hydroxyapatite beads (HAs). The adherence was repressed by more than 50 % at the concentration of 0.5 mg/ml of the extract and complete inhibition was observed at the concentration of 4 mg/ml of the extract. On the activity of glucosyltransferase (GTFase), which synthesises water-insoluble glucan from sucrose, the extract displayed more than 10 % inhibition of at a concentration of 2 mg/ml. Kilani et al. (2008a) observed a marked inhibitory effect against *Salmonella enteritidis, Staphylococcus aureus* and *Enterococcus faecalis* with TOFs and ethyl acetate extracts of *C. rotundus* tubers. *Cyperus rotundus* oil exhibited inhibitory effect on in-vitro growth of *Staphylococcus aureus, Escherichia coli, Klebsiella pneumoniae* and *Streptococcus pyogenes* at high concentration of >25 % but not at concentration from 2.5 to 10 % (Nima et al. 2008). The cyperus oil exhibited no inhibitory effect on the growth of *Pseudomonas aeruginosa* and *Proteus vulgaris*.

Mojab et al. (2009) found the methanolic extracts of *C. rotundus* roots and rhizomes exerted antimicrobial effects on *Micrococcus luteus* and *Staphylococcus aureus*, but not on *Salmonella, Escherichia coli, Aspergillus niger* and *Cladosporium herebaceum.* Inhibitory concentrations for *Micrococcus luteus* and *Staphylococcus aureus* were 31.25 and 125 mg/ml. Nutgrass tuber oil was effective against *Escherichia coli, Staphylococcus aureus, Pseudomonas aeruginosa* and *Bacillus subtilis* but less so compared to chloramphenicol (Bisht et al. 2011). The oil was inhibitory against *Aspergillus fumigatus* and *Candida parapsilosis* and inhibited spore formation of *Aspergillus flavus* and *Fusarium oxysporum.* The crude extract of *Cyperus rotundus* exhibited broad-spectrum antibacterial activity against urinary tract infection (UTI) pathogens (Sharma et al. 2014). The minimum inhibitory concentrations of the ethanolic extract ranged from 2.5 to 10 mg/mL against different uropathogens. Phytochemical analysis of *C. rotundus* showed the presence of tannins and saponins.

Antidiarrhoeal Activity

The methanol extract of *Cyperus rotundus* rhizome, administered orally at the doses of 250 and 500 mg/kg b.w., showed significant antidiarrhoeal activity in castor oil-induced diarrhoea in mice (Uddin et al. 2006). Among the fractions, tested at 250 mg/kg, the petroleum ether fraction and residual methanol fraction were found to retain the activity, the latter being more active as compared to the control. The ethyl acetate fraction did not show any antidiarrhoeal activity. The decoction of *Cyperus rotundus* tubers exhibited antigiardial activity, reduced adherence of entero-

pathogenic *Escherichia coli* and invasion of enteroinvasive *E.coli* and *Shigella flexneri* to HEp-2 cells (Daswani et al. 2011). It also affected production of cholera toxin and action of heat labile toxin. The decoction of *C. rotundus* did not have marked antimicrobial activity and exerted its antidiarrheal action by mechanisms other than killing the pathogen.

In a prospective randomised clinical study carried out on 40 male and female patients 25 and 60 years of age who had been diagnosed with irritable bowel syndrome (IBS) for 5–10 years, 8-week treatment with a herbal mixture of *Mentha longifolia, Cyperus rotundus* and *Zingiber officinale*, resulted in improvements in all of their IBS symptoms after 8 weeks, as revealed by increase in their individual symptom scores and in their mean total improvement percentages (Sahib 2013). These results were comparable to those produced by the standard agent mebeverine.

Anti-hyperlipidemic/Antiobesity Activities

Administration of nutgrass ethanol rhizome extract to aged rats prevented the age-associated changes on glucose, total cholesterol, triglycerides, LDL cholesterol and VLDL cholesterol (Nagulendran et al. 2007). HDL cholesterol level was found to be increased significantly in both young and aged rats after treatment with nutgrass extract. The findings demonstrated that nutgrass extract normalised the age-associated altered levels of glucose and improved lipid profile status in aged rats, thereby decreasing the risk factors for diabetes mellitus and cardio vascular diseases associated with advancing age.

Studies showed that the powdered preparations of water and alcoholic extract of *C. rotundus* exhibited lipolytic action to mobilise fat from adipose tissues in rats and consequently helped in the reduction of obesity (Bambhole 1988). Lemaure et al. (2007) found that administration of 45 or 220 mg/kg/day of *C. rotundus* tubers hexane extract for 60 days in Zucker rats induced a significant reduction in weight gain without affecting food consumption or inducing toxicity. In-vitro, 250 microg/mL of this extract was able to stimulate lipolysis in 3 T3-F442 adipocytes, suggesting that this medicinal plant contained activators of beta-adrenoreceptors. The data suggested that the effect on weight loss exerted by *C. rotundus* tubers extract may be mediated, at least partially, through the activation of the beta3-adrenoreceptors. Aqueous extract of *Cyperus rotundus* exhibited promising activity against hyperlipidaemia in rats fed a high fat diet by attenuating the elevated serum lipid profile (Chandratre et al. 2012)

Anticonvulsant Activity

Pretreatment with *Cyperus rotundus* ethanol extract caused significant protection against strychnine and leptazol-induced convulsions in mice (Pal et al. 2009).

The ethanol nutgrass rhizome extract (100 mg / kg, p.o.) reduced hind limb extension in rats and duration of convulsion significantly, which was comparable to standard drug phenytoin (25 mg / kg, i.p.) and diazepam (4 mg /kg, i.p.), respectively (Shivakumar et al. 2009). The results suggested the potential of the extract for treatment of epilepsy and its anticonvulsant activity could be attributed to the pressmen of the flavonoids in the extract. *C. rotundus* essential oil at 500 mg/kg significantly decreased duration of clonus, stupor and phase of maximal electroshock –induced convulsion compared to the control rats (Biradar et al. 2010). Studies showed that the hydroalcoholic rhizome extract of *C. rotundus* could reduce intensity and duration of seizure in pentylenetetrazole (PTZ)-kindled mice (Khalili et al. 2011). Also, the extract could increase the level of superoxide dismutase and nitric oxide and decrease malondialdehyde level in mice brain. It was concluded that *C. rotundus* rhizome extract, probably via its antioxidant properties, could have exerted a potent antiepileptic effect. Studies showed that the hydroalcoholic rhizome extract of *C. rotundus* could reduce intensity and duration of seizure in pentylenetetrazole (PTZ)-kindled mice (Khalili et al. 2011). Also, the extract

could increase the level of superoxide dismutase and nitric oxide and decrease malondialdehyde level in mice brain. It was concluded that *C. rotundus* rhizome extract, probably via its antioxidant properties, could have exerted a potent antiepileptic effect.

Neuroprotective/CNS Activity

Among the four sesquiterpenes isolated from the rhizome, only isocurcumenol was found to inhibit [3H]Ro15-1788 binding and enhance [3H] flunitrazepam binding in the presence of gamma-aminobutyric acid ($GABA_A$) in rat cerebrocortical membranes, indicating that it may serve as a benzodiazepine receptor agonist and allosterically modulate GABAergic neurotransmission via enhancement of endogenous receptor ligand binding (Ha et al. 2002). Cyperi rhizome (*Cyperus rotundus* rhizome extract) exhibited neuroprotective effects against 6-hydroxydopamine (6-OHDA)-induced neuronal damage in an experimental in-vitro model of Parkinson's disease (Lee et al. 2010). In PC12 cells, the extract showed a significant protective effect on cell viability at 50 and 100 μg/mL. It inhibited generation of reactive oxygen species and nitric oxide, reduction of mitochondrial membrane potential, and caspase-3 activity, which were induced by 6-OHDA. It also showed a significant protective effect against damage to dopaminergic neurons in primary mesencephalic culture. Total oligomeric flavonoids (TOF) of *Cyperus rotundus* exhibited neuroprotective activity in the rat model of cerebral ischemia and reperfusion (IR) (Sunil et al. 2011). Neuroprotective effect of TOF was confirmed in terms of neurological deficits, excitotoxicity (glutamate, glutamine synthetase and Na^+K^+ATPase levels), oxidative stress (malondialdehyde, super oxide dismutase and glutathione) and neurobehavioral functions in the experimental animals. TOFs decreased glutamate, glutamine synthetase and increased Na^+K^+ATPase activity in a dose-dependent manner when compared to the IR rats. Treatment with TOF significantly reduced the neurological deficits and reversed the anxiogenic behavior in rats. Further, it also significantly decreased malondialdehyde and increased superoxide dismutas and glutathione content in brains of experimental rats. TOF also attenuated neuronal loss in stroke rats.

Studies by Kumar et al. (2013) showed that that pretreatment of neurons with *C. rotundus* rhizome extract ameliorated the mitochondrial and plasma membrane damage induced by 500 μM SIN-1. The extract inhibited NO generation by suppressing i-NOS expression and replenishing the SIN-1 induced depletion of antioxidant enzyme status. Also, it efficiently potentiated the SIN-1 induced apoptotic biomarkers such as bcl-2 and caspase-3, which orchestrate the proteolytic damage of the cell. The extract also attenuated peroxynitrite ($ONOO^-$) induced neurotoxicity of the human neuroblastoma SH-SY5Y cells. Further, pretreatment with the extract regulated 3-nitro tyrosine formation, indicating the potential of plant extract against tyrosine nitration. The results suggested that *C. rotundus* rhizome may be developed as a preventive agent against peroxynitrite-induced apoptosis. Studies showed pretreatment of human neuroblastoma SH-SY5Y cells with *Cyperus rotundus* rhizome extract for 2 h before administration of H_2O_2 for 24 h ameliorated the cytotoxicity induced by H_2O_2 as evidenced by MTT and LDH assays (Kumar and Khanum 2013). The extract exhibited potent antioxidant activity by regulating the enzymes/proteins levels such as SOD, CAT, GPx, GR, HSP-70, Caspase-3 and Bcl-2. The pretreatment restored H_2O_2-induced cellular, nuclear, and mitochondrial morphologies as well as increased the expression of brain-derived nerve growth factor (BDNF). The anti-oxidant and anti-apoptotic potentials of the plant extract may account for its high content of phenolics, flavonoids and other active principles. The findings suggested that *Cyperus rotundus* rhizome extract might be developed as an agent for neurodegeneration prevention or therapy.

Cyperi Rhizoma extract significantly induced the luciferase expression driven by an estrogen response element in PC12 cells, a dopaminergic cell line, in a dose-dependent manner (Kim et al. 2013a). In mice, 1-methyl-4-phenyl-1,2,3,6-

tetrahydropyridine (MPTP) significantly decreased the levels of dopamine in the striatum and behaviour performance; in contrast, both CRE and 17β-estradiol benzoate recovered these parameters to normal levels and also recovered TH immunopositive fibers and cells, respectively, from MPTP toxicity. The results suggested that the extract possessed estrogen-like and neuroprotective effects on dopaminergic neurons in estrogen-deprived mice treated with MPTP toxin. The results of animal studies suggested that *Cyperus rotundus* tubers ethanolic extract exhibited repairing effects on the memory and behavioral disorders produced by bilateral electric lesioning of the nucleus basalis of Meynert in Alzheimer's diseased rats as assessed by the shuttle box and Morris water maze task tests (Rabiei et al. 2013).

Antinociceptive Activity

In the hot-plate and tail-immersion tests, the hydromethanol extract of *C. rotundus* whole plant significantly increased the latency period to the thermal stimuli of mice at all the tested doses (50, 100 and 200 mg/kg) (Imam and Sumi 2014). In formalin-induced paw licking test, oral administration of the extract at 100 and 200 mg/kg doses decreased the licking of paw in early phase. All the tested doses (50, 100 and 200 mg/kg) significantly decreased the licking of paw in late phase of the test. The dose 200 mg/kg was most effective, showing maximum percentage of inhibition of licking in both early (61.60 %) and late phase (87.41 %). The results indicated the antinociceptive effect of *C. rotundus* and suggested that this effect was mediated by both peripheral and central mechanisms.

Wound Healing Activity

An alcoholic extract of nutgrass tubers in the form of ointment showed considerable difference in response in three types of wound models on rats: the excision, the incision and dead space wound model (Puratchikody et al. 2006). It was comparable to those of a standard drug nitrofurazone ointment (0.2 % w/w NFZ) in terms of wound contracting ability, wound closure time and tensile strength.

Gastroprotective Activity

Decoctions of *Rhizoma cyperi* nutgrass rhizome administered orally (1.25, 2.5, 4.0 g crude drug/kg) to rats 30 min before ethanol (40 % v/v, 10 mL/kg) exhibited gastric ulcer inhibitory effect in a dose-dependent manner (Zhu et al. 1997). The activity was also observed when the decoction was given subcutaneously (0.3–0.6 g/kg), suggesting that the herb possessed systemic effects on protecting the stomach. Compared with controls, gastric motility of the ethanol-treated rats was delayed significantly by either oral (2.5–4.0 g/kg) or subcutaneous (0.3 g/kg) administration of the decoction. *Cyperus rotundus* extract exhibited gastroprotective effect against acute gastric mucosal lesions induced by ischaemia/reperfusion in rats (Güldür et al. 2010). The mean ulcer index of rats treated with 200 and 100 mg/kg extract were significantly lower than that of control rats.

Antidiabetic Activity

Oral administration of 500 mg/kg of the *C. rotundus* extract (once a day for seven consecutive days) significantly lowered the blood glucose levels in alloxan-induced diabetic rats (Raut and Gaikwad 2006). This antihyperglycemic activity could be attributed to its antioxidant activity as it showed the strong DPPH radical scavenging action in-vitro. Studies showed that *C. rotundus* hydroalcoholic extract protected against protein oxidation and glycoxidation using an in-vitro model of fructose-mediated protein glycoxidation (Ardestani and Yazdanparast 2007). *C. rotundus* extract with glycation inhibitory activity also demonstrated antioxidant activity when a ferric reducing antioxidant power (FRAP) and Trolox equivalent antioxidant capacity (TEAC) assays as well as metal chelating activity were

applied. *C. rotundus* extract at different concentrations (25–250 μg/ml) significantly decreased the formation of AGEs (advanced glycation endproducts). Additionally, *C. rotundus* extract prevented oxidative protein damages, including effect on protein carbonyl formation and thiol oxidation associated with the glycoxidation process. Stilbene compound cassigarol E from the rhizome inhibited both α-glucosidase and α-amylase activities while (2*RS*,3*SR*)-3,4′,5,6,7,8-hexahydroxyflavane only showed effect on α-amylase, and compounds scirpusins A and B were active on α-glucosidase (Tran et al. 2014). All four compounds showed significant 2,2-diphenyl-1-picrylhydrazyl (DPPH) scavenging activity. The isolated compounds may contribute to the antidiabetic property of *C. rotundus*.

Anti-allergic Activity

Studies by Bae et al. (2010) found that Cyperi Rhizoma enhanced helper Th1 lineage development by increasing Th1-specific cytokine expression and secretion and reduced Th2 lineage development by repressing Th2-specific cytokine productions. The results suggested that Cyperi Rhizoma extract may be useful for preventing the onset of allergies or improving allergic symptoms. *C. rotundus* and its sesquiterpenic constituents, valencene, nootkatone and caryophyllene α-oxide, exerted anti-allergic activity in-vitro and in-vivo (Jin et al. 2011). In rat basophilic leukemia (RBL)-1cells, nutgrass sesquiterpenes strongly inhibited 5-lipoxygenase-catalyzed leukotrienes production and they inhibited β-hexosaminidase release by antigen-stimulated RBL-2H3 cells, with valencene having the highest inhibitory effect. Nutgrass rhizome ethanol extract inhibited leukotriene production and β-hexosaminidase release at 300 μg/mL. It was also found that the most active sesquiterpene (valencene) and the extract inhibited β-hexosaminidase degranulation by inhibiting the initial activation reaction, Lyn phosphorylation, in IgE-stimulated RBL-2H3 cells. Further, the extract valencene and nootkatone significantly inhibited the delayed-type hypersensitivity reaction in mice when administered orally at 50–300 mg/kg.

Anti-platelet Activity

Of *C. rotundus* rhizome's eight components: 4-cymene, (+)-nootkatone, b-pinene, 1,8-cienole, limonene, valencene, caryophyllene oxide, coumarin, (+)-nootkatone was found to have the most potent inhibitory effect on collagen-, thrombin- and AA-induced platelet aggregation (Seo et al. 2011). In addition, the rhizome extract and (+)-nootkatone-treated mice exhibited significantly prolonged bleeding times. Further, (+)-nootkatone had a significant inhibitory effect on rat platelet aggregation *ex vivo*.

Antiviral Activity

Cyperus rotundus hydroalcoholic extract was found to have virucidal effect against : herpes simplex-1 virus (Soltan and Zaki 2009).

Pre-menstrual Syndrome/ Dysmenorrhea Therapy

During 1998–2011, a total of 14,312 Chinese Herbal medicine (CHM) prescriptions for premenstrual syndrome (PMS) were provided in Taiwan (Chen et al. 2014a). Jia-Wei-Xiao-Yao-San (JWXYS) was the CHM, which had the highest prevalence (37.5 % of all prescriptions) and also the core of prescription network for PMS. For combination of two CHM, JWXYS with *Cyperus rotundus* was prescribed most frequently, 7.7 % of all prescriptions, followed by JWXYS with *Leonurus heterophyllus* 5.9 %, and *C. rotundus* with *L. heterophyllus* 5.6 %.

A total of 57,315 CHM prescriptions for primary dysmenorrhea women during 1998–2008 were analysed and, on average, 5.3 CHM was used in one prescription (Chen et al. 2014b). Dang-Gui-Shao-Yao-San (DGSYS) was the most commonly used herbal formula (27.2 %), fol-

lowed by Jia-Wei-Xiao-Yao-San (JWXYS) (20.7 %) and Wen-Jing-Tang (WJT) (20.5 %). *Corydalis yanhusuo* and *Cyperus rotundus* were the most commonly used single herb, found in 33.1 % and 29.2 %, respectively, of all prescriptions. Additionally, C. yanhusuo with *C. rotundus* was the most commonly used two CHM in combination, accounting for 14.24 % of all prescriptions, followed by DGSYS with *C. yanhusuo* (10.47 %). Multi-target effects on primary dysmenorrhea, such as analgesia, mood modifying and hormone adjustment, were found among commonly prescribed CHM in this study.

Na^+, K^+-ATPase Inhibitory Activity

The hexane extracts of *Cyperus rotundus* and *Orthosiphon aristatus* showed high potent inhibitory activity on crude enzyme Na^+, K^+-ATPase from rat brain (Ngamrojanavanich et al. 2006)

Antihypoxic Activity

Cyperus rotundus ethanol extract at doses of 200 and 400 mg/kg was able to protect against the cognitive impairments, and the locomotor activity and muscular coordination defects, induced by sodium nitrite-induced hypoxia injury in rats (Jebasingh et al. 2014). It was concluded that the protective effect of *C. rotundus* extract may help in designing better therapeutic regimes for hypoxia injury.

Hair Removal Therapy

In a study involving 91 female patients with androgenic hair (hirsutism and axillary hair), six months topical application of *Cyperus rotundus* oil was significantly more effective ($p<0.05$) than placebo in reducing hair growth without side effects (El-Kaream 2012). This result was confirmed by three assessment methods; difference in hair count, independent observer assessment and patients' self-assessment. In an open-label, prospective pilot study of 65 participants with unwanted axillary hair, topical treatment with *Cyperus rotundus* essential oil was found to be as effective as the Alexandrite laser photo-epilation for decreasing the growth of axillary hair (both dark and white) (Mohammed 2014).

Estrogenic Activity

Indira et al. (1956) reported that *C. rotundus* oil exhibited low order estrogenic activity. The hydrocarbon fraction was more active than other fractional distillates, but none of the components was found as active as the oil. The probability of these compounds being proestrogens was indicated by the ratio of systemically active to locally effective concentration.

Diuretic Activity

Akperbekova and Abdullaev (1966) reported the diuretic effect of drug from and galenicals from the roots of *Cyperus rotundus* growing in Azerbaijan.

Insecticidal Activity

The petroleum ether (PE) and ethyl alcohol (EA) extracts of *C. rotundus* exerted 98 % and 97 % mortality of *Aedes aegypti* larvae at 1,000 ppm, respectively (Imam et al. 2013). PE extracts exhibited LC_{50} 443.80 ppm and LC_{90} 882.98 ppm whereas EA extract exhibited LC_{50} 594.22 ppm and LC_{90} 936.25 ppm. Furochromones from the plant aerial parts, Khellin and visnagin, caused 99.5 % and 96.9 % inhibition of *Spodoptera littoralis* larval growth while ammiol and khellol β-D-glucopyranoside caused 33.1 % and 39.5 %, respectively, when incorporated in artificial diets of the insect (Sayed et al. 2007). The results of their study suggested that the presence of methoxyl group at positions C-5 and/or C-8 was essential for antifeedant activity.

Hexane tuber extract of *C. rotundus* exhibited repellent activity against mosquito vector *Anopheles culicifacies*, *Anopheles stephensi* and

Culex quinquefasciatus in the laboratory (Singh et al. 2009). Clear dose–response relationships were established with the highest dose of 10 % tuber extract evoking 100 % repellency. The consolidated data of the repellency observed in different species showed that the overall repellency rates varied between 80 and 100 % for different repellents concentrations (2.5 %, 5 %, and 10 %). *Cyperus rotundus* tuber oil exhibited ovicidal and larvicidal effect against eggs and fourth instar larvae of *Aedes albopictus* with EC_{50} 4.2 ppm for ovicidal effect and larval mortality with LC_{50} 12.2 ppm LC_{90} 18.8 ppm (Kempraj and Bhat 2008).

Solita and Castor (2011) found C. rotundus extract was more effective than carbamate and exhibted almost the same efficacy as that of organophosphate in killing ants. Constituents of *Cyperus rotundus* (L.) rhizome viz. *p*-cymene, nerol, linalool, *o*-cymene, (*S*)-(-)-citronellal, (1*S*)-(-)-camphor, terpinolene and *m*-cymene exhibited high contact fumigant toxicity to German cockroach, *Blattella germanica* KSS females (LD_{50}, 0.29–0.47 mg/cm^2) (Chang et al. 2012). The test compounds were effective in closed but not in open containers against field-collected SEL females, indicating that their route of insecticidal action was largely a result of vapor action.

Antimalarial Activity

Of 49 Tanzanian plants investigated for antimalarial activity, *Cyperus rotundus* tuber extract was found to be one of three most active with IC_{50} of 5–10 μg/mL (Weenen et al. 1990). The antimalarial activities of *C. rotundus* tuber sesquiterpenes patchoulenone, caryophyllene α-oxide, 10,12-peroxycalamenene and 4,7-dimethyl-1-tetralone were in the range of $EC_{50} = 10^{-4}–10^{-6}$ M, with the novel endoperoxide sesquiterpene, 10,12-peroxycalamenene, exhibiting the strongest antimalarial effect at EC_{50} 2.33×10^{-6} M (Thebtaranonth et al. 1995).

Molluscicidal Activity

Derris elliptica root and *C. rotundus* corm extracts showed the highest toxicity against 3-month-old *Pomacea canaliculata* snails with LC_{50} as 23.68 mg/l and 133.20 mg/l, respectively (Ruamthum et al. 2010). *C. rotundus* corm extracts inhibited esterase and glutathione S-transferase activity in the stomach, intestinal tracts and digestive glands of survival treated *P. canaliculata*.

Toxicity Studies

Acute toxicological studies on *C. rotundus* showed no mortality or morbidity up to 2000 mg/kg body weight in Wistar rats (Jebasingh et al. 2012). Subchronic toxicity study revealed that food, water consumption and body weight of animals did not vary significantly. But the hematological parameters showed an increase in WBC count and hemoglobin level. The kidney function and liver function did not change even after long-term exposure.

Traditional Medicinal Uses

Nutgrass roots and tubers are analgesic, antibacterial, antispasmodic, antitussive, aromatic, astringent, carminative, diaphoretic, diuretic, emmenagogue, litholytic, sedative, skin, stimulant, stomachic, tonic and vermifuge (Yeung 1985; Duke and Ayensu 1985; Chopra et al. 1986; Bown 1995). Nutgrass is a pungent bitter–sweet herb that relieves spasms and pain, acting mainly on the digestive system and uterus; used internally to treat digestive, menstrual problems and cervical cancer. It has been used in Indian Ayurvedic medicine of treating digestive disorders, fever, dysmenorrhea and other maladies. A decoction of the tubers is given in fever, diarrhoea, dyspepsia and stomach complaints. The fresh tubers are applied to the breast as a galactagogue and used also as an anthelmintic. The tubers are used in folkloric Chinese medicine as an antidiarrheal, antidepressant, anti-*Candida*,

antipyretic, analgesic, anti-inflammatory and anti-emetic remedy for dysentery and women's diseases (Lawal and Oyedeji 2009; Zhou and Liu 2012). Root paste is given to treat stomach trouble in traditional medicine in Macchegaun, Nepal (Joshi et al. 2011). The plant is ranked 8th amongst 250 potential antifertility plants in China (Duke and Ayensu 1985). *Cyperus rotundus* is used by the traditional medicine practitioners of ayurvedic medicine in India for CNS disorders like loss of memory, depression and epilepsy (Jebasingh et al. 2012). *Cyperus rotundus* was one of several plants, which had been successfully identified for the treatment of obesity in Ayurvedic, Unani, Siddha and Chinese, etc., systems of medicine (Vasudeva et al. 2012). The plant is used used in Tunisian folkloric medicine to treat stomach disorders and inflammatory diseases (Kilani-Jaziri et al. 2013). *Cyperus rotundus* rhizomes have been used in traditional Chinese medicine as an estrogenic and anti-inflammatory agent for the treatment of women's diseases and also used for treatment of stomach ache, bowel disorders and menstrual disorders (Kim et al. 2013b). *Cyperus rotundus* is a Tunisian medicinal plant used in folkloric (traditional) medicine to treat stomach disorders and inflammatory diseases (Soumaya et al. 2013). *Mentha longifolia, Cyperus rotundus* and *Zingiber officinale* are widely used in Iraqi traditional medicine for the treatment of multiple gastrointestinal diseases (Sahib 2013).

Other Uses

Studies had reported that nutgrass had insecticidal activity against crop pests, antifungal activity and also allelopathic properties, and may play a role in weed management.

Insecticidal (Crop Pest) Activity

Studies found that the active principle of nutgrass against diamondback moth (*Plutella xylostella*) was 4,11-selinnadien-3-one with LC_{50} against 2nd–3rd instar larvae of diamondback moth were 7–12 ppm (Visetson et al. 2001). The active principle was higher in summer than that in rainy season by ca. twofolds. This active principle varied according to geographical areas with Chanthaburi and Chaing Mai producing the highest amounts of 4,11-selinnadien-3-one (0.13–0.16 % ai. yield) compared with the other. Furthermore, synergists, Piperonyl butoxide (PB), triphenyl phosphate (TPP), could raise the effectiveness of the active principle up to ca. two–sixfold. At 2,000 ppm of 4,11-selinnadien-3-one, exposed mice showed no sign of acute dermal, acute oral or eye irritation effects. However, the active principle was toxic to other non-target organisms with LC_{50} of 28.01 ppm and 10.8 ppm to 1-month old guppies (*Poecilia reticulata*) and bee (*Apis florea*) larvae, respectively. Furochromones from the nutgrass aerial parts, Khellin and visnagin, caused 99.5 % and 96.9 % inhibition of *Spodoptera littoralis* larval growth while ammiol and khellol β-D-glucopyranoside caused 33.1 % and 39.5 %, respectively, when incorporated in artificial diets of the insect (Sayed et al. 2007). The results of their study suggested that the presence of methoxyl group at positions C-5 and/or C-8 was essential for antifeedant activity.

Antifungal Activity

Ethyl acetate fractions of *Cyperus rotundus* rhizome inhibited spore germination of *Alternaria alternata, Alternaria brassicicola, Alternaria solani, Alternaria chearanthi, Colletotrichum musae, Colletotrichum* sp., *Curvularia* lunata, *Curvularia* maculans, *Curvularia pallescens, Curvularia pennisetti, Helminthosporium pennissetti, Helminthosporium spiciferum, Helminthosporium echinoclova* and *Helminthosporium colocasiae* in-vitro (Singh et al. 2011). Germ tube elongation was also affected by 60–90 %. The germ tube branching and their elongation were affected in almost all species at 30–95 %. *A. brassicicola* was highly sensitive to all the fractions at all the concentrations.

Allelopathic Activity

The sesquiterpene from *C. rotundus* tuber, α-cyperone was the most inhibitory against elongation of wheat coleoptile segments in the presence of indole acetic acid and second leaf sheath growth of rice seedlings in the presence of gibberellin A_3 (GA_3) (Komai et al. 1977). α-Cyperone at 10^{-3} M strongly suppressed the elongation of hypocotyls and roots of lettuce seed in the presence of GA_3. This compound scarcely inhibited germination of lettuce itself. Biological inhibitory activities of 4β(H)-eudesm-11-ene-3-one, eudesm-4-ene-3-one and 4β(H)-eudesman-3-one derived from α-cyperone were clearly lower than those of α-cyperone. Essential oil and methanol extract of *C. rotundus* tubers precipitates inhibited the germination of lettuce and white clover at a concentration of 400 p.p.m., inhibited seedling growth of lettuce, *Digitaria adscendens* and white clover and inhibited the growth of nutsedge plants themselves (Komai and Ueki 1980). The essential oil content was high in tubers, rhizomes and roots but changed during the season, especially in roots from 30 to 60 days after planting. GLC-mass spectrometry indicated the presence of sesquiterpenoids in the steam distillate of rhizosphere soil, which might be the inhibitory agents. Inhibitory activity of the essential oils from *C. rotundus* tubers of four major chemotypes against the seedling growth of lettuce and oats was in the order of H>M>K>O (Komai et al. 1991). Studies suggested that *C. rotundus* of different chemotypes may have different allelopathic activity in the crop–weed interaction. Seven major sesquiterpenes were isolated from the oils, namely, α-cyperone, cyperotundone, cyperol, β-selinene, cyperene, sugeonyl acetate and patchoulene acetate and their inhibitory activities determined. Sesquiterpenes with ketone (i.e. a-cyperone and cyperotundone) or hydroxyl (i.e. cyperol) groups were more inhibitory than the acetates and hydrocarbons. α-cyperone at the 5 mmol level was nearly twice as inhibitory as /3-selinene, cyperene, sugeonol acetate or pachoulene acetate in both lettuce and oats. Two sesquiterpene ketone allelochemicals, cyperotundone (0.26 %) and α-cyperone (0.1 %), isolated from nutgrass, were postulated to undergo modification when released into the rhizosphere from the donor plant (Morimoto and Komai 2005). In their structure–activity relationship study, cyperotundone was oxidised with selenium dioxide in acetic acid to 4-patchoulene-2,3-dione and 4-patchoulene-2,3,6-trione. Subsequent hydrogenation of 4-patchoulene-2,3-dione and 4-patchoulene-2,3,6-trione gave hydroxylated derivatives, cyperotundon-2-ol and 3-hydroxy-4-patchoulene-2,6-dione, respectively. 4-Patchoulene-2,3-dione inhibited the hypocotyl growth of lettuce seedlings but promoted radicle elongation at 0.1–2 mg/L concentration without chlorosis. The effect of 4-patchoulene-2,3,6-trione showed a similar chlorosis, caused by 3,6,9-sugetriol triketone, against lettuce seedlings. These ketones did not show the radicle elongation.

Methanolic extract of nutgrass tubers strongly inhibited the activity of acetylcholinesterases from electric eel, wheat and tomato (Sharma and Gupta 2007). It also inhibited seed germination and seedling growth in wheat and tomato. The root exudates of *C. rotundus* significantly reduced the root and shoot growth of tomato and cucumber plants, while, its residues incorporated at 3 and 6 g per kg soil inhibited the seedling growth of sorghum, soybean and cowpea, and the reduction increased with the increased rates of residues (Alsaadawi and Salih 2009a). The volatile compounds released from its shoot and tubers significantly reduced the seedling growth of mungbean.

Phytoremediation and Decontamination

Studies found *C. rotundus* to be effective in phytoremediation of crude oil contaminated soil (Basumatary et al. 2012) and to degrade total petroleum hydrocarbon (TPH) in a petroleum sludge contaminated field (Basumatary et al. 2013). Studies found *C. rotundus* waste to be an effective and low-cost adsorbent for the removal of heavy metals, copper and zinc from aqueous solution (Ramesh et al. 2013).

Comments

Cyperus rotundus has been dubbed "the world's worst weed" as it is invasive as a weed in over 90 countries, and infests over 50 crops worldwide in tropical and temperate regions (Holm et al. 1977).

Selected References

Abbiw D (1990) Useful plants of Ghana. West African uses of wild and cultivated plants. Intermediate Technology and The Royal Botanic Gardens, Kew, 337pp

Aghassi A, Naeemy A, Feizbakhsh A (2013) Chemical composition of the essential oil of *Cyperus rotundus* L. from Iran. J Essent Oil Bear Plant 16(3):382–386

Akperbekova BA, Abdullaev RA (1966) Diuretic effects of drug from and galenicals from the roots of Cyperus rotundus growing in Azerbaidzhan. Izv Akad Nauk Azerb SSR Ser Biol Nauk 4:98–105

Al-Saeedi AT (2013) Total oligomeric flavonoids (TOF) of the herb tubers *Cyperus rotundus* induce growth inhibition and apoptosis in some cancer cell lines, a preliminary study. In: Proceedings of the AACR-NCI-EORTC International Conference: Molecular Targets and Cancer Therapeutics; 2013 Oct 19–23; Boston, MA. Philadelphia (PA): AACR; Mol Cancer Ther 12(11 Suppl):Abstract no A176

Alsaadawi IS, Salih NMM (2009a) Allelopathic potential of *Cyperus rotundas* L. I. Interference with crops. Allelopathy J 23(2):297–303

Alsaadawi IS, Salih NMM (2009b) Allelopathic potential of *Cyperus rotundus* L. II. Isolation and identification of phytotoxins. Allelopathy J 24(1):85–90

Ardestani A, Yazdanparast R (2007) *Cyperus rotundus* suppresses AGE formation and protein oxidation in a model of fructose-mediated protein glycoxidation. Int J Biol Macromol 41(5):572–578

Asenjo CF (1942) Some of the constituents of the tuber of coqui (*Cyperus rotundus* L.). III.—The sugars. J Pharm Sci 31:88–89

Bae H, Park N, Kim Y, Hong M, Shin M, Kim SH, Kim J (2010) The modulative effect of Cyperi Rhizoma on Th1/Th2 lineage development. Cytokine 51(3):259–265

Bambhole VD (1988) Effect of some medicinal plant preparations of adipose tissue metabolism. Anc Sci Life 8(2):117–124

Basumatary B, Saikia R, Bordoloi S (2012) Phytoremediation of crude oil contaminated soil using nut grass, *Cyperus rotundus*. J Environ Biol 33(5):891–896

Basumatary B, Saikia R, Das HC, Bordoloi S (2013) Field note: phytoremediation of petroleum sludge contaminated field using sedge species, *Cyperus rotundus* (Linn.) and *Cyperus brevifolius* (Rottb.) Hassk. Int J Phytoremediation 15(9):877–888

Bendixen LE, Nandihalli U (1987) Worldwide distribution of purple and yellow nutsedge (*Cyperus rotundus* and *C. esculentus*). Weed Technol 1(1):61–65

Biradar S, Kangralkar VA, Mandavkar Y, Thakur M, Chougule N (2010) Anti-inflammatory, anti-arthritic, analgesic and anticonvulsant activity of *Cyperus* essential oils. Int J Pharm Phar Sci 2(4):975–1491

Bisht A, Bisht GRS, Singh M, Gupta R, Singh V (2011) Chemical compsition and antimicrobial activity of essential oil of tubers of *Cyperus rotundus* Linn. collected from Dehradun (Uttarakhand). Int J Res Pharm Biomed Sci 2:661–665

Bown D (1995) Encyclopaedia of herbs and their uses. Dorling Kindersley, London, 424 pp

Burkill HM (1985) The useful plants of west tropical Africa. Vol.1. Families A-D. Gardens Kew, Kew, 960 pp

Chandratre RS, Chandarana S, Mengi SA (2012) Effect of aqueous extract of *Cyperus rotundus* on hyperlipidaemia in rat model. Int J Pharm Biol Arch 3(3):598–600

Chang KS, Shin EH, Park C, Ahn YJ (2012) Contact and fumigant toxicity of *Cyperus rotundus* steam distillate constituents and related compounds to insecticide-susceptible and-resistant *Blattella germanica*. J Med Entomol 49(3):631–639

Chen HY, Huang BS, Lin YH, Su IH, Yang SH, Chen JL, Huang JW, Chen YC (2014a) Identifying Chinese herbal medicine for premenstrual syndrome: implications from a nationwide database. BMC Complement Altern Med 14:206

Chen HY, Lin YH, Su IH, Chen YC, Yang SH, Chen JL (2014b) Investigation on Chinese herbal medicine for primary dysmenorrhea: implication from a nationwide prescription database in Taiwan. Complement Ther Med 22(1):116–125

Chen Y, Zhao YY, Wang XY, Liu JT, Huang LQ, Peng CS (2011) GC-MS analysis and analgesic activity of essential oil from fresh rhizoma of *Cyperus rotundus*. Zhong Yao Cai 34(8):1225–1229 (In Chinese)

Chhabra BR, Sharma A, Dhillon RS, Vij S (2002) Three new sesquiterpenes from nutgrass (*Cyperus rotundus* L.) and their evaluation as plant growth regulators. Res J Chem Environ 6(4):57–59

Choi HJ, Lee JH1, Jung YS (2014) (+)-Nootkatone inhibits tumor necrosis factor α/interferon γ-induced production of chemokines in HaCaT cells. Biochem Biophys Res Commun 447(2):278–284

Chopra RN, Nayar SL, Chopra IC (1986) Glossary of Indian Medicinal Plants. (Including the Supplement). Council Scientific Industrial Research, New Delhi, 330 pp

Cribb AB, Cribb JW (1987) Wild food in Australia, 2nd edn. Fontana Collins, Sydney, Australia, 240 pp

Dai LK, Tucker GC, Simpson DA (2010) *Cyperus* Linnaeus. In: Wu ZY, Raven PH, Hong DY (eds) Flora of China, Vol. 23 Acoraceae through Cyperaceae. Science Press/Missouri Botanical Garden Press, Beijing/St. Louis

Dang GK, Parekar RR, Kamat SK, Scindia AM, Rege NN (2011) Antiinflammatory activity of *Phyllanthus emblica, Plumbago zeylanica* and *Cyperus rotundus* in acute models of inflammation. Phytother Res 25(6):904–908

Daswani PG, Brijesh S, Tetali P, Birdi TJ (2011) Studies on the activity of *Cyperus rotundus* Linn. tubers against infectious diarrhea. Indian J Pharmacol 43(3):340–344

Deane G (2014) Chufa for two. http://www.eattheweeds.com/cyperus-esculentus-rotundus-for-lunch-2/

Dhillon RS, Singh S, Kundra S, Basra AS (1993) Studies on the chemical composition and biological activity of essential oil from *Cyperus rotundus* Linn. Plant Growth Regul 13(1):89–93

Duarte MC, Figueira GM, Sartoratto A, Rehder VL, Delarmelina C (2005) Anti-*Candida* activity of Brazilian medicinal plants. J Ethnopharmacol 97(2):305–311

Duke JA, Ayensu ES (1985) Medicinal Plants of China. Vols 1 & 2. Reference Publications, Inc. Algonac, 705 pp

El-Kaream GFA (2012) Role of *Cyperus rotundus* oil in decreasing hair growth. J Intercult Ethnopharmacol 1(2):111–118

Facciola S (1990) Cornucopia. A source book of edible plants. Kampong Publ, Vista. California, 677 pp

Fischer A, Brouquisse R, Raymond P (1998) Influence of senescence and of carbohydrate levels on the pattern of leaf proteases in purple nutsedge (*Cyperus rotundus*). Physiol Plant 102(3):385–395

Foundation for Revitalisation of Local Health Traditions (2008) FRLHT Database. http://envis.frlht.org

Ghannadi A, Rabbani M, Ghaemmaghami L, Malekian N (2012) Phytochemical screening and essential oil analysis of one of the Persian sedges; *Cyperus rotundus*. Int J Pharm Sci 3(2):424–427

Güldür ME, Ozgónül A, Kilic IH, Sögüt O, Ozaslan M, Bitiren M, Yalcin M, Musa D (2010) Gastroprotective effect of *Cyperus rotundus* extract against gastric mucosal injury induced by ischemia and reperfusion in rats. Int J Pharmacol 6(3):104–110

Gupta MB, Palit TK, Singh N, Bhargava KP (1971) Pharmacological studies to isolate the active constituents from *Cyperus rotundus* possessing anti-inflammatory, anti-pyretic and analgesic activities. Indian J Med Res 59(1):76–82

Ha JH, Lee KY, Choi HC, Cho JS, Kang BS, Lim JC, Lee DU (2002) Modulation of radioligand binding to the GABA(A)-benzodiazepine receptor complex by a new component from *Cyperus rotundus*. Biol Pharm Bull 25(1):128–130

Harrington HD (1974) Edible native plants of the rocky mountains. University New Mexico Press, Albuquerque 392 pp

Hedrick UP (1972) Sturtevant's edible plants of the world. Dover Publications, New York, 686 pp

Hikino H, Aota K (1976) 4α, 5α-Oxidoeudesm-11-en-3α-ol, sesquiterpenoid of Cyperus rotundus. Phytochemistry 15(8):1265–1266

Hikino H, Aota K, Kuwano D, Takemoto T (1969a) Structure of alpha-rotunol and beta-rotunol. Tetrahedron Lett 32:2741–2742

Hikino H, Aota K, Kuwano D, Takemoto T (1971) Structure and absolute configuration of α-rotunol and β-rotunol, sesquiterpenoids of *Cyperus rotundus*. Tetrahedron 27(19):4831–4836

Hikino H, Aota K, Maebayashi Y, Takemoto T (1967a) Structure and absolute configuration of cyperolone. Chem Pharm Bull 15(9):1349–1355

Hikino H, Aota K, Takemoto T (1965) Structure of cyperotundone. Chem Pharm Bull 13(5):628–630

Hikino H, Aota K, Takemoto T (1966) Structure and absolute configuration of cyperotundone. Chem Pharm Bull 14(8):890–896

Hikino H, Aota K, Takemoto T (1967b) Identification of ketones in *Cyperus*. Tetrahedron 23(5):2169–2172

Hikino H, Aota K, Takemoto T (1967c) Structure of sugetriol. Chem Pharm Bull 15(9):1433–1435

Hikino H, Aota K, Takemoto T (1968a) Structure and absolute configuration of cyperol and isocyperol. Chem Pharm Bull 15(12):1929–1933

Hikino H, Aota K, Takemoto T (1968b) Structure and absolute configuration of sugeonol. Chem Pharm Bull 16(1):52–55

Hikino H, Aota K, Takemoto T (1968c) Structure and configuration of sugetriol. Chem Pharm Bull 16(10):1900–1906

Hikino H, Aota K, Takemoto T (1969b) Structure and absolute configuration of kobusone and isokobusone. Chem Pharm Bull 17:1390–1394

Hikino H, Ito K, Aota K (1968d) Synthesis of cyperene, cyperotundone and patchoulenone. Chem Pharm Bull 16(1):43–51

Hikino H, Suzuki N, Takemoto T (1967d) Synthesis of cyperolone and 3-epi-cyperolone. Chem Pharm Bull 15(9):1395–1404

Holm LG, Plucknett DL, Pancho JV, Herberger JP (1977) The world's worst weeds: distribution and biology. University Press of Hawaii, East-West Center, 609 pp

Imam H, Riaz Z, Sofi G (2013) Evaluation of mosquito larvicidal effect of Nagarmotha (*Cyperus rotundus*) extracts against *Aedes aegypti* L. larvae. Int J Green Pharm 7(1):37–40

Imam H, Zarnigar S, Ghulamuddin A, Seikh LA (2014) The incredible benefits of Nagarmotha (*Cyperus rotundus*). Int J Nutr Pharmacol Nuerol Dis 4(1):23–27

Imam MZ, Sumi CD (2014) Evaluation of antinociceptive activity of hydromethanol extract of *Cyperus rotundus* in mice. BMC Complement Altern Med 14:83

Indira M, Sirsi M, Senich R, Dev S (1956) Occurrence of estrogenic substance in plants. I. Estrogenic activity of *Cyperus rotundus*. J Sci Ind Res 15C:202–204

Ito T, Endo H, Shinohara H, Oyama M, Akao Y, Iinuma M (2012) Occurrence of stilbene oligomers in *Cyperus rhizomes*. Fitoterapia 83(8):1420–1429

Iwamura JI, Kameda M, Komai K, Hirao N (1977) The constituents of essential oils of *Cyperus serotinus* Rottb. and *Cyperus rotundus* L. Nippon Kagaku Kaishi 7:1018–1020

Jagtap AG, Shirke SS, Phadke AS (2004) Effect of polyherbal formulation on experimental models of inflammatory bowel diseases. J Ethnopharmacol 90(2-3):195–204

Jaiswal Y, Liang Z, Guo P, Ho HM, Chen H, Zhao Z (2014) Tissue-specific metabolite profiling of *Cyperus rotundus* L. rhizomes and (+)-nootkatone quantitation by laser microdissection, ultra-high-performance liquid chromatography-quadrupole time-of-flight mass spectrometry, and gas chromatography-mass spectrometry techniques. J Agric Food Chem 62(29):7302–7316

Jansen PCM, Aguilar NO (1996) *Cyperus* L. In: Flach M, Rumawas F (eds) Plant resources of South-East Asia no 9. Plants yielding non-seed carbohydrates. Prosea Foundation, Bogor, pp 78–82

Jebasingh D, Jackson DD, Venkataraman S, Adeghate E, Emerald BS (2014) The protective effects of *Cyperus rotundus* on behavior and cognitive function in a rat model of hypoxia injury. Pharm Biol 15:1–12

Jebasingh D, Venkataraman S, Jackson DD, Emerald BS (2012) Physiochemical and toxicological studies of the medicinal plant *Cyperus rotundus* L (Cyperaceae). Int J Appl Res Nat Prod 5(4):1–8

Jeong SJ, Miyamoto T, Inagaki M, Kim YC, Higuchi R (2000) Rotundines AC, three novel sesquiterpene alkaloids from *Cyperus rotundus*. J Nat Prod 63(5):673–675

Jin JH, Lee DU, Kim YS, Kim HP (2011) Anti-allergic activity of sesquiterpenes from the rhizomes of *Cyperus rotundus*. Arch Pharm Res 34(2):223–228

Jirovetz L, Wobus A, Buchbauer G, Shafi MP, Thampi PT (2004) Comparative analysis of the essential oil and SPME-headspace aroma compounds of *Cyperus rotundus* L. roots/tubers from South-India using GC, GC-MS and olfactometry. J Essent Oil Bear Plant 7(2):100–106

Joshi K, Joshi R, Joshi AR (2011) Indigenous knowledge and uses of medicinal plants in Macchegaun, Nepal. Indian J Tradit Knowl 10(2):281–286

Jung SH, Jung Kim SJ, Jun BG, Lee KT, Hong SP, Oh MS, Jang DS, Choi JH (2013) α-Cyperone, isolated from the rhizomes of *Cyperus rotundus*, inhibits LPS-induced COX-2 expression and PGE_2 production through the negative regulation of NFκB signalling in RAW 264.7 cells. J Ethnopharmacol 147(1):208–214

Kalsi P, Sharma A, Singh A, Singh IP, Chhabra BR (1995) Biogenetically important sesquiterpenes from *Cyperus rotundus*. Fitoterapia 66(1):94

Kapadia VH, Nagasampagi BA, Naik VG, Dev S (1963) Structure of mustakone and copaen. Tetrahedron 21(24(28)):1933–1939

Kapadia VH, Nagasampagi BA, Naik VG, Dev S (1965) Studies in sesquiterpenes—XXII: structure of mustakone and copaen. Tetrahedron 21(2):607–618

Kapadia VH, Naik VG, Wadia MS, Dev S (1967) Sesquiterpenoids from the essential oil of *Cyperus rotundus*. Tetrahedron Lett 8(47):4661–4667

Kempraj V, Bhat SK (2008) Ovicidal and larvicidal activities of *Cyperus giganteus* Vahl and *Cyperus rotundus* Linn. essential oils against *Aedes albopictus* (Skuse). Nat Prod Radiance 7(5):416–419

Khalili M, Kiasalari Z, Roghani M, Azizi Y (2011) Anticonvulsant and antioxidant effect of hydroalcoholic extract of *Cyperus rotundus* rhizome on pentylentetrazole-induced kindling model in male mice. J Med Plant Res 5(7):1140–1146

Khan SM, Lee DU (2011) Sesquiterpene derivatives isolated from *Cyperus rotundus* L. inhibit inflammatory signaling mediated by NF-κB. Nat Prod Sci 17(3):250–255

Kemparaj V, Bhat SK (2008) Ovicidal and larvicidal activities of *Cyperus giganteus* Vahl and *Cyperus rotundus* Linn. essential oils against *Aedes albopictus* (Skuse). Nat Prod Rad 7:416–419

Kilani S, Abdelwahed A, Chraief I, Ben Ammar R, Hayder N, Hammami M, Ghedira K, Chekir-Ghedira L (2005a) Chemical composition, antibacterial and antimutagenic activities of essential oil from (Tunisian) *Cyperus rotundus*. J Essent Oil 17(6):695–700

Kilani S, Ben Sghaier M, Limem I, Bouhlel I, Boubaker J, Bhouri W, Skandrani I, Neffatti A, Ben Ammar R, Dijoux-Franca MG, Ghedira K, Chekir-Ghedira L (2008a) In vitro evaluation of antibacterial, antioxidant, cytotoxic and apoptotic activities of the tubers infusion and extracts of *Cyperus rotundus*. Bioresour Technol 99(18):9004–9008

Kilani S, Bouhlel I, Ben R, Abdelwahed A, Hayder N, Mahmoud A, Ghedira K, Chekir-Ghedira L (2005b) Evaluation of the antimutagenic and antiradical potentials of extracts from the tubers of (Tunisian) *Cyperus rotundus*. Toxicol Environ Chem 87(3):415–425

Kilani S, Ledauphin J, Bouhlel I, Ben Sghaier M, Boubaker J, Skandrani I, Mosrati R, Ghedira K, Barillier D, Chekir-Ghedira L (2008b) Comparative study of *Cyperus rotundus* essential oil by a modified GC/MS analysis method. Evaluation of its antioxidant, cytotoxic, and apoptotic effects. Chem Biodivers 5(5):729–742

Kilani-Jaziri S, Mhalla D, Châbane F, Ghedira Z, Limem I, Ghedira K, Chekir-Ghedira L (2013) Pharmacological, antioxidant, genotoxic studies and modulation of rat splenocyte functions by *Cyperus rotundus* extracts. BMC Complement Alternat Med 13(1):28

Kilani-Jaziri S, Neffati A, Limem I, Boubaker J, Skandrani I, Sghair MB, Bouhlel I, Bhouri W, Mariotte AM, Ghedira K, Dijoux Franca MG, Chekir-Ghedira L (2009) Relationship correlation of antioxidant and antiproliferative capacity of *Cyperus rotundus* products towards K562 erythroleukemia cells. Chem Biol Interact 181(1):85–94

Kim HG, Hong J, Huh Y, Park C, Hwang DS, Choi JH, Oh MS (2013a) Cyperi Rhizoma inhibits the 1-methyl-4-phenyl-1, 2, 3, 6-tetrahydropyridine-induced reduction in nigrostriatal dopaminergic neurons in estrogen-deprived mice. J Ethnopharmacol 148(1):322–328

Kim SJ, Kim HJ, Kim HJ, Jang YP, Oh MS, Jang DS (2012) New patchoulane-type sesquiterpenes from the

rhizomes of *Cyperus rotundus*. Bull Korean Chem 33(9):3115–3118

Kim SJ, Ryu B, Kim HY, Yang YI, Ham J, Choi JH, Jang DS (2013b) Sesquiterpenes from the rhizomes of *Cyperus rotundus* and their potential to inhibit LPS-induced nitric oxide production. Bull Korean Chem Soc 34(7):2207–2210

Kim SK, Hwang BY, Kang SJ, Lee JJ, Ro JS, Lee K (2000) Chemical components of *Cyperus rotundus* L. and inhibitory effects on nitric oxide production. Koran J Pharmacogn 31(1):1–6

Komai K (1993) Ecological and chemical studies on secondary metabolites in cyperaceous weed. Weed Res 38:69–75

Komai K, Ueki K (1975) Chemical properties and behaviour of polyphenolic substances in purple nutsedge (*Cyperus rotundus* L.). J Weed Sci Technol 20(2):66–71

Komai K, Iwamura J, Ueki K (1977) Isolation, Identification and physiological activities of sesquiterpenes in purple nutsedge tubers. J Weed Sci Technol 22(1):14–18

Komai K, Shimizu M, Tang CS, Tsutsui H (1994) Sesquiterpenoids of *Cyperus bulbosus, Cyperus tuberosus* and *Cyperus rotundus*. Mem Fac Agric Kinki Univ 27:39–45

Komai K, Tang CS (1989) A Chemotype of *Cyperus rotundus* in Hawaii. Phytochemistry 28(7):1883–1886

Komai K, Tang CS, Nishimoto RK (1991) Chemotypes of *Cyperus rotundus* in Pacific Rim and Basin: distribution and inhibitory activities of their essential oils. J Chem Ecol 17(1):1–8

Komai K, Ueki K (1980) Plant-growth inhibitors in purple nutsedge (*Cyperus rotundus* L.). Weed Res Japan 25(1):42–47

Komai K, Ueki K (1981) Geographical variation of essential oils in tubers of purple nutsedge. In: Proceedings, 8th Asian Pacific weed science society conference, Nov. 22–29, 1981, Bangalore, India, pp 387–389

Krishna S, Renu S (2013) Isolation and identification of flavonoids from *Cyperus rotundus* Linn. In vivo and in vitro. J Drug Del Ther 3(2):109–113

Kumar KH, Khanum F (2013) Hydroalcoholic extract of *Cyperus rotundus* ameliorates H(2)O(2)-induced human neuronal cell damage via its anti-oxidative and anti-apoptotic machinery. Cell Mol Neurobio 33(1):5–17

Kumar KH, Razack S, Nallamuthu I, Khanum F (2014) Phytochemical analysis and biological properties of *Cyperus rotundus* L. Ind. Crops Prod 52:815–826

Kumar KH, Tamatam A, PalA KF (2013) Neuroprotective effects of *Cyperus rotundus* on SIN-1 induced nitric oxide generation and protein nitration: Ameliorative effect against apoptosis mediated neuronal cell damage. Neurotoxicol 34:150–159

Kunkel G (1984) Plants for human consumption. Koeltz Scientific Books, An Annotated Checklist of the Edible Phanerogams and Ferns. Koenigstein

Lawal OA, Oyedeji AO (2009) Chemical composition of the essential oils of *Cyperus rotundus* L. from South Africa. Molecules 14(8):2909–2917

Lee CH, Hwang DS, Kim HG, Oh H, Park H, Cho JH, Lee JM, Jang JB, Lee KS, Oh MS (2010) Protective effect of Cyperi rhizoma against 6-hydroxydopamine-induced neuronal damage. J Med Food 13(3):564–571

Lee SS, Travis J, Black CC Jr (1970) Characterization of ferredoxin from nutsedge, *Cyperus rotundus* and other species with a high photosynthetic capacity. Arch Biochem Biophys 141(2):676–689

Lemaure B, Touché A, Zbinden I, Moulin J, Courtois D, Macé K, Darimont C (2007) Administration of *Cyperus rotundus* tubers extract prevents weight gain in obese Zucker rats. Phytother Res 21(8):724–730

Li YX (2014) Determination of ferulic acid content in *Cyperus rotundus* by HPLC. J Chem Pharm Res 6(3):1496–1500

Liu P, Liu L, Tang YP, Duan JA, Yang NY (2010) A new cerebroside and its anti-proliferation effect on VSMCs from the radix of *Cyperus rotundus* L. Chin Chem Lett 21(5):606–609

Mahmout Y, Bessiere JM, Dolmazon R (1997) Composants volatils des extraits de *Cyperus rotundus* L. du Tchad. J Soc Ouest Afr Chim 2(4):1–17 (In French)

Manandhar NP, Manandhar S (2002) Plants and people of Nepal. Timber Press, Oregon, 636 pp

Moerman DE (1998) Native American ethnobotany. Timber Press, Portland, 927 pp

Mohammed GF (2014) Topical *Cyperus rotundus* oil: a new therapeutic modality with comparable efficacy to Alexandrite laser photo-epilation. Aesthet Surg J 34(2):298–305

Mojab F, Vahidi H, Nickavar B, Kamali-Nejad M (2009) Chemical components of essential oil and antimicrobial effects of rhizomes from *Cyperus rotundus* L. J Med Plant 8(32):91–97, 186

Morimoto M, Komai K (2005) Plant growth inhibitors: patchoulane-type sesquiterpenes from *Cyperus rotundus* L. Weed Biol Manag 5:203–209

Motl O, Trevedi B, Herout V, Šorm F (1963) Sesquiterpenes in *Cyperus rotundus* L. oil. Chem Ind (Lond) 31:1284

Nagulendran KR, Mahesh R, Hazeena BV (2007) Preventive role of *Cyperus rotundus* rhizomes extract on age associated changes in glucose and lipids. Pharmacologyonline 2:318–325

Natural Products Research Institute, Seoul National University (1998) Medicinal Plants in the Republic of Korea, Western Pacific Series No 21. WHO Regional Publications, Manila, 316 pp

Neville GA, Nigam IC, Holmes JL (1968) Identification of ketones in *Cyperus*: NMR and mass spectral examination of the 2,4-dinitrophenylhydrazones. Tetrahedron 24(10):3891–3897

Ngamrojanavanich N, Manakit S, Pornpakakul S, Petsom A (2006) Inhibitory effects of selected Thai medicinal plants on Na+, K+-ATPase. Fitoterapia 77(6):481–483

Nima ZA, Jabier MS, Wagi RI, Hussain HA (2008) Extraction, identification and antibacterial activity of *Cyperus* oil from Iraqi *C. rotundus*. Eng Technol 26(10):1156–1159

Ohira S, Hasegawa T, Hayashi KI, Hoshino T, Takaoka D, Nozaki H (1998) Sesquiterpenoids from *Cyperus rotundus*. Phytochemistry 47(8):Sesquiterpenoids with 1577–1581

Oladunni OM, Abass OO, Adisa AI (2011) Studies on physicochemical properties of the oil, minerals and nutritional composition of nut of nut grass (*Cyperus rotundus*). Am J Food Technol 6:1061–1064

Pacific Island Ecosystems at Risk (PIER) (1999) *Cyperus rotundus* L., Cyperaceae. http://www.hear.org/pier/species/cyperus_rotundus.htm

Paknikar SK, Motl O, Chakravarti KK (1977) Structures of rotundene and rotundenol. Tetrahedron Lett 18(24):2121–2124

Pal DK, Dutta S (2006) Evaluation of the Antioxidant activity of the roots and rhizomes of *Cyperus rotundus* L. Indian J Pahrm Sci 68(2):256–258

Pal D, Dutta S, Sarkar A (2009) Evaluation of CNS activities of ethanol extract of roots and rhizomes of *Cyperus rotundus* in mice. Acta Pol Pharm 66(5):535–541

Parsons WT, Cuthbertson EG (2001) Noxious weeds of Australia, 2nd edn. CSIRO, Collingwood, Australia, p 712

Porcher MH et al. (1995–2020) Searchable World Wide Web multilingual multiscript plant name database. Published by The University of Melbourne, Australia. http://www.plantnames.unimelb.edu.au/Sorting/Frontpage.html

Priya RM, Padmakumari KP (2012) HPTLC and reverse phase HPLC methods for the simultaneous quantification and in vitro screening of antioxidant potential of isolated sesquiterpenoids from the rhizomes of *Cyperus rotundus*. J Chromatogr B, Anal Technol Biomed Life Sci 904:22–28

Puratchikody A, Devi CN, Nagalakshmi G (2006) Wound healing activity of *Cyperus rotundus* Linn. Indian J Pharm Sci 68(1):97–101

Rabiei Z, Hojjati M, Rafieian-Kopaeia M, Alibabaei Z (2013) Effect of *Cyperus rotundus* tubers ethanolic extract on learning and memory in animal model of Alzheimer. Biomed Aging Pathol 3(4):185–191

Ramesh ST, Gandhimathi R, Hamoneth Joesun J, Nidheesh PV (2013) Novel agricultural waste adsorbent, *Cyperus rotundus*, for removal of heavy metal mixtures from aqueous solutions. Environ Eng Sci 30(2):74–81

Raut NA, Gaikwad NJ (2006) Antidiabetic activity of hydro-ethanolic extract of *Cyperus rotundus* in alloxan induced diabetes in rats. Fitoterapia 77(7):585–588

Rogers HH, Runion GB, Prior SA, Price AJ, Torbert HA, Gjerstad DH (2008) Effects of elevated atmospheric CO_2 on invasive plants: comparison of purple and yellow nutsedge (*Cyperus rotundus* L. and *C. esculentus* L.). J Environ Qual 37(2):395–400

Ruamthum W, Visetson S, Milne JR, Bullangpoti V (2010) Toxicity of botanical insecticides on golden apple snail (*Pomacea canaliculata*). Commun Agric Appl Biol Sci 75(2):191–197

Sahib AS (2013) Treatment of irritable bowel syndrome using a selected herbal combination of Iraqi folk medicines. J Ethnopharmacol 148(3):1008–1012

Sayed HM, Mohamed MH, Farag SF, Mohamed GA (2001) Phytochemical and biological investigations of *Cyperus rotundus* L. Bull Fac Pharm Cairo Univ 39(3):195–203

Sayed HM, Mohamed MH, Farag SF, Mohamed GA, Omobuwajo OR, Proksch P (2008) Fructose-amino acid conjugate and other constituents from *Cyperus rotundus* L. Nat Prod Res 22(17):1487–1497

Sayed HM, Mohamed MH, Farag SF, Mohamed GA, Proksch P (2007) A new steroid glycoside and furochromones from *Cyperus rotundus* L. Nat Prod Res 21(4):343–350 (Refer again)

Seo EJ, Lee DU, Kwak JH, Lee SM, Kim YS, Jung YS (2011) Antiplatelet effects of *Cyperus rotundus* and its component (+)-nootkatone. J Ethnopharmacol 135(1):48–54

Seo JH, Yang YI, Jang DS, Choi JH (2014) Isocyperol isolated from rhizomes of *Cyperus rotundus*, inhibits iNOS and pro-inflammatory cytokines through suppressing STAT3 pathway in LPS-stimulated RAW 264.7 cells. Cytokine 70(1):68–68

Seo WG, Pae HO, Oh GS, Chai KY, Kwon TO, Yun YG, Kim NY, Chung HT (2001) Inhibitory effects of methanol extract of *Cyperus rotundus* rhizomes on nitric oxide and superoxide productions by murine macrophage cell line, RAW 264.7 cells. J Ethnopharmacol 76(1):59–64

Sharma A, Verma R, Ramteke P (2014) *Cyperus rotundus*: a potential novel source of therapeutic compound against urinary tract pathogens. J Herb Med 4(2):74–82

Sharma R, Gupta R (2007) *Cyperus rotundus* extract inhibits acetylcholinesterase activity from animal and plants as well as inhibits germination and seedling growth in wheat and tomato. Life Sci 80(24-25):2389–2392

Shivakumar SI, Suresh HM, Hallikeri CS, Hatapakki BC, Handiganur JS, Sankh K, Shivakumar B (2009) Anticonvulsant effect of *Cyperus rotundus* Linn rhizomes in rats. J Nat Remed 9(2):92–196

Singh A, Maurya S, Singh R, Singh UP (2011) Antifungal efficacy of some ethyl acetate extract fractions of *Cyperus rotundus* rhizomes against spore germination of some fungi. Arch Phytopathol Plant Prot 44(20):2004–2011

Singh N, Kulshrestha VK, Gupta MB, Bhargava KP (1970) A pharmacological study of *Cyperus rotundus*. Indian J Med Res 58(1):103–109

Singh N, Pandey BR, Verma P, Bhalla M, Gilca M (2012) Phyto-pharmacotherapeutics of *Cyperus rotundus* Linn. (Motha): an overview. Indian J Nat Prod Res 3(4):467–476

Singh NB, Singh PN (1986) A new flavonol glycoside from mature tubers of *Cyperus rotundus*. J Indian Chem Soc 64(4):450

Singh PN, Singh SB (1980) New saponin from mature tubers of *Cyperus rotundus*. Phytochemistry 19:2056

Singh SP, Raghavendra K, Dash AP (2009) Evaluation of hexane extract of tuber of root of *Cyperus rotundus* Linn (Cyperaceae) for repellency against mosquito vectors. J Parasitol Res 2009:Article ID 908085.

Solita ES, Castor L (2011) Phytochemical and pesticidal properties of barsanga (*Cyperus rotundus* Linn.). JPAIR Multidiscip Res 6(1):197–214

Soltan MM, Zaki AK (2009) Antiviral screening of forty-two Egyptian medicinal plants. J Ethnopharmacol 126(1):102–107

Sonwa MM (2000) Isolation and structure elucidation of essential oil constituents: comparative study of the oils of *Cyperus alopecuroides, Cyperus papyrus*, and *Cyperus rotundus*. Ph D. thesis. University of Hamburg. http://ediss.sub.uni-hamburg.de/volltexte/2000/372/pdf/diss.pdf

Sonwa MM, König WA (2001) Chemical study of the essential oil of *Cyperus rotundus*. Phytochemistry 58(5):799–810

Soumaya KJ, Dhekra M, Fadwa C, Zied G, Ilef L, Kamel G, Leila CG (2013) Pharmacological, antioxidant, genotoxic studies and modulation of rat splenocyte functions by *Cyperus rotundus* extracts. BMC Complement Altern Med 13:28

Soumaya KJ, Zied G, Nouha N, Mounira K, Kamel G, Genviève FD, Leila GC (2014) Evaluation of in vitro antioxidant and apoptotic activities of *Cyperus rotundus*. Asian Pac J Trop Med 7(2):105–112

Stuart GU (2013) Philippine alternative medicine. Manual of Some Philippine Medicinal Plants. http://www.stuartxchange.org/OtherHerbals.html

Sundaram MS, Sivakumar T, Balamurugan G (2008) Anti-inflammatory effect of *Cyperus rotundus* Linn. Leaves on acute and subacute inflammation in experimental rat models. Biomedicine 28:302–304

Sunil AG, Kesavanarayanan KS, Kalaivani P, Sathiya S, Ranju V, Priya RJ, Pramila B, Paul FD, Venkhtesh J, Babu CS (2011) Total oligomeric flavonoids of *Cyperus rotundus* ameliorates neurological deficits, excitotoxicity and behavioral alterations induced by cerebral ischemic–reperfusion injury in rats. Brain Res Bull 84(6):394–405

Swarbrick JT (1997) Weeds of the Pacific Islands. Technical paper no. 209. South Pacific Commission, Noumea, New Caledonia, 124 pp

Tam CU, Yang FQ, Zhang QW, Guan J, Li SP (2007) Optimization and comparison of three methods for extraction of volatile compounds from *Cyperus rotundus* evaluated by gas chromatography–mass spectrometry. J Pharm Biomed Anal 44(2):444–449

Tanaka T (1976) Tanaka's cyclopaedia of edible plants of the world. Keigaku Publishing, Tokyo, 924 pp

The Plant List (2014) *Cyperus rotundus* L. http://www.theplantlist.org/

Thebtaranonth C, Thebtaranonth Y, Wanauppathamkul S, Yuthavong Y (1995) Antimalarial sesquiterpenes from tubers of *Cyperus rotundus*: structure of 10,12-peroxycalamenene, a sesquiterpene endoperoxide. Phytochemistry 40(1):125–128

Tran HH, Nguyen MC, Le HT, Nguyen TL, Pham TB, Chau VM, Nguyen HN, Nguyen TD (2014) Inhibitors of α-glucosidase and α-amylase from *Cyperus rotundus*. Pharm Biol 52(1):74–77

Trivedi B, Motl O, Herout V, Šorm F (1964a) Composition of the oil from *Cyperus rotundus* L. structure of patchoulenone. Collect Czech Chem Commun 29(7):1675–1688

Trivedi B, Motl O, Smolíková J, Šorm F (1964b) Structure of the sesquiterpenic hydrocarbon cyperene. Tetrahedron Lett 5(19):1197–1201

Tsoyi K, Jang HJ, Lee YS, Kim YM, Kim HJ, Seo HG, Lee JH, Kwak JH, Lee DU, Chang KC (2011) (+)-Nootkatone and (+)-valencene from rhizomes of *Cyperus rotundus* increase survival rates in septic mice due to heme oxygenase-1 induction. J Ethnopharmacol 137(3):1311–1317

Uddin SJ, Mondal K, Shilpi JA, Rahman MT (2006) Antidiarrhoeal activity of *Cyperus rotundus*. Fitoterapia 77(2):134–136

Ueki K, Soga M, Komai K (1974) Studies on polyphenolic substances in tubers of Eleocharis kuroguwai Ohwi. *Cyperus serotinus Rottb*. and *Cyperus rotundus* L. J Weed Sci Technol 17:20–24

Umerie SC, Ezeuzo HO (2000) Physicochemical characterization and utilization of *Cyperus rotundus* starch. Bioresour Technol 72(2):193–196

USDA, ARS, National Genetic Resources Program Germplasm Resources Information Network – (GRIN) [Online Database] (2014) National Germplasm Resources Laboratory, Beltsville, Maryland. URL: http://www.ars-grin.gov/cgi-bin/npgs/html/index.pl

Wang H, Liu Y (2010) The mineral content of Rhizoma cyperi after microwave-assisted digestion and determination by AAS. Spectrosc Lett 43(3):149–154

Wood C, Siebert TE, Parker M, Capone DL, Elsey GM, Pollnitz AP, Eggers M, Meier M, Vössing T, Widder S, Krammer G, Sefton MA, Herderich MJ (2008) From wine to pepper: rotundone, an obscure sesquiterpene, is a potent spicy aroma compound. J Agric Food Chem 56(10):3738–3744

Vasudeva N, Yadav N, Sharma SK (2012) Natural products: a safest approach for obesity. Chin J Integr Med 18(6):473–480

Visetson S, Milne M, Milne J (2001) Toxicity of 4,11-Selinnadien-3-one from nutsedge (*Cyperus rotundus* L.) tuber extracts to diamondback moth larvae (*Plutella xylostella* L.), detoxification mechanisms and toxicity to non target species. Kasetsart J (Nat Sci) 35:284–292

Weenen H, Nkunya MH, Bray DH, Mwasumbi LB, Kinabo LS, Kilimali VA (1990) Antimalarial activity of Tanzanian medicinal plants. Planta Med 56(4):368–370

Wills GD (1972) Sugars, phosphorus and iron in purple nutsedge. Weed Sci 20(4):348–350

Xu Y, Zhang HW, Yu CY, Lu Y, Chang Y, Zou ZM (2008) Norcyperone, a novel skeleton norsesquiterpene from *Cyperus rotundus* L. Molecules 13(10):2474–2481

Xu Y, Zhang HW, Wan XC, Zou ZM (2009) Complete assignments of 1H and 13C NMR data for two new sesquiterpenes from *Cyperus rotundus* L. Magn Reson Chem 47(6):527–531

Yang JL, Shi YP (2012) Structurally diverse terpenoids from the rhizomes of *Cyperus rotundus* L. Planta Med 78(1):59–64

Yazdanparast R, Ardestani A (2007) In vitro antioxidant and free radical scavenging activity of *Cyperus rotundus*. J Med Food 10(4):667–674

Yeung HC (1985) Handbook of Chinese herbs and formulas, 2nd edn. Redwing Book, 431 pp

Yu HH, Lee DH, Seo SJ, You YO (2007) Anticariogenic properties of the extract of *Cyperus rotundus*. Am J Chin Med 35(3):497–505

Zhou ZL, Fu CY (2012) A new flavanone and other constituents from the rhizomes of *Cyperus rotundus* L. and their antioxidative activities. Chem Nat Comp 6:1–3

Zhou ZL, Yin WQ (2012) Two novel phenolic compounds from the rhizomes of *Cyperus rotundus* L. Molecules 17(11):12636–12641

Zhou ZL, Yin WQ, Zhang H, Feng Z, Xia J (2013) A new iridoid glycoside and potential MRB inhibitory activity of isolated compounds from the rhizomes of *Cyperus rotundus* L. Nat Prod Res 27(19):1732–1736

Zhou ZL, Liu YH (2012) Study on antidepressant effect and mechanism by *Cyperus rotundus* extracts. Chin J Tradit Med Formulae 7:191–193

Zhu M, Luk HH, Fung HS, Luk CT (1997) Cytoprotective effects of *Cyperus rotundus* against ethanol induced gastric ulceration in rats. Phytother Res 11(5):392–395

Zoghbi MDGB, Andrade EHA, Carreira LMM, Rocha EAS (2008) Comparison of the main components of the essential oils of “priprioca”: *Cyperus articulatus* var. *articulatus* L., *C. articulatus* var. *nodosus* L., *C. prolixus* Kunth and *C. rotundus* L. J Essent Oil Res 20:42–46

Eleocharis dulcis

Scientific Name

Eleocharis dulcis (Burm. f.) Trinius ex Henschel

Synonyms

Bolboschoenus maritimus subsp. *tuberosus* (Desf.) J. Sojak, *Carex tuberosa* (Schult.) Blanco (illeg.), *Eleocharis austrocaledonica* Vieill., *Eleocharis dulcis* var. *tuberosa* (Schult.) T. Koyama, *Eleocharis equisetina* J.Presl & C. Presl, *Eleocharis esculenta* Vieill., *Eleocharis indica* (Lour.) Druce, *Eleocharis plantaginea* (Retz.) Roem. & Schult., *Eleocharis plantaginea* var. *stolonifera* Boeckeler, *Eleocharis plantagineiformis* Tang & F.T.Wang, *Eleocharis plantaginoidea* W.Wight (illeg.), *Eleocharis plantaginoides* (Rottb.) Domin, *Eleocharis sphacelata* Boeck., *Eleocharis tuberosa* Schult., *Eleocharis tumida* (Roxb.) Schult., *Hippuris indica* Lour., *Limnochloa plantaginea* (Retz.) Nees, *Limnochloa tumida* (Roxb.) Nees, *Scirpus dubius* Roxb., *Scirpus plantagineus* Retz., *Scirpus plantaginoides* Rottb., *Scirpus tuberosus* Roxb. (illeg.), *Scirpus tumidus* Roxb.

Family

Cyperaceae

Common/English Names

Bush Nut, Chinese Water Chestnut, Edible Spike Rush, Ground Chestnut, Spike Rush, Water Chestnut, Water Nut.

Vernacular Names

Arabic: El-Kustnaa, El-Maee;
Chamorro: Uchaga-Lane;
Chinese: Bi-Qi, Pi Ch'i, Di-Li, Ti-Li, Matai, Ma Ti, Tian Cig U, T'ien Tzu Ku;
Danish: Kinavandkastanie;
Fijian: Kuta, Sasa, Taria;
French: Châtaigne D'eau, Châtaigne D'eau Vietnamienne, Eléocharide À Tubercules Comestibles, Cevuga Vula;
German: Chinesische Wassernuß, Wasserkastanie;
India Singhara (Hindu), Pani Phul;

T.K. Lim, *Edible Medicinal and Non-Medicinal Plants: Volume 10, Modified Stems, Roots, Bulbs*,
DOI 10.1007/978-94-017-7276-1_8

Indonesia: Teki, Teki-Tiki, Dekeng (Javanese), Peperetan, Teki (Malay) Chikai (Sumatra), Teke, Teki, Babawangan (Sudanese);
Khmer: Mëm Phlông;
Kwara'Ae: Ngwano;
Japanese: Okuro Guwai, Ohkuro Guai, Shinu Kuro Guwai, Shiro Guwai, Inu Kuro Guwai;
Malaysia: Tike, Tikitt, Ubi Puron;
Nepalese: Kasuur;
Palauan: Kerdikes;
Philippines: Apulid (Bikol), Buslig (Bisaya), Pagappo (Ibanag), Buslig (Iloko), Apulid, Kalangub (Tagalog); Portuguese: Castanheiro De Água;
Russian: Bolotnitsa Sladkaia;
Samoan: Utu Utu, Utuutu;
Spanish: Cabezas De Negrito, Nuez China;
Sri Lanka: Boru-Pan;
Taiwan: Shui Deng Xin Cao,
Thai: Yaa-Chongkatiam, Haeo, Haeo-Chin;
Tongan: Kuta;
Vietnamese: Củ Mã Thầy (North), Củ Năng (South).

Origin/Distribution

The plant is indigenous to the old world tropics—tropical West Africa and Madagascar to southern Asia, Indian Ocean Islands, south-east Asia, southern China, Taiwan and Japan, Malesia, northern Australia and eastward to the Pacific in Melanesia and Micronesia.

Agroecology

Water Chestnuts flourish in the mud of margins of shallow lakes, ponds, paddy fields, swamps and marshes. It prefers slightly acidic soil conditions and a sunny position. For cultivation, it requires a rich, friable soil, well manured or fertilised in shallow waters, creeks or marshes, with a fine loam top soil and clay base. Plants are frost tender; the tubers should be harvested at the end of the growing season. Plants perform best at temperatures between 30 and 35 °C during the leafy stage of growth, and about 5 °C lower during tuber formation.

Edible Plant Parts and Uses

The corm has a crispy, white flesh and can be eaten raw, slightly boiled, fried, cooked, grilled, pickled, candied or canned (Hodge 1956; Burkill 1966; Uphof 1968; Usher 1974; Ochse and van den Brink 1980; Facciola 1990; Low 1991; Larkcom 1991; Hu 2005). Even after cooking, which softens the flesh of most other vegetables, the flesh of water chestnuts, whether sliced, diced or chopped, still remains crisp. Canned and peeled water chestnuts, whole or sliced, are commonly sold in all Asian food stores around the world. The corm is a common ingredient in many Asian cuisines and delicacies. In China and south east Asia, they are mostly eaten raw and fresh as dessert, sometimes sweetened. Chinese water chestnut may be either used as a vegetable (*hon matai*) or in starch (flour) manufacture (*sui matai*). Because of its crisp, sweet, apple-like flesh, *hon matai* is widely eaten out-of-hand in China as a substitute for fresh fruits. Chinese water chestnut can be sliced, diced or grated with other vegetables in omelettes, soups, chop suey, salads, etc (Hodge 1956) or in stews, curry. It is often shredded for meat and fish dishes and is used as a common ingredient in stuffings and fillings for dim sum. Peeled *matai* on bamboo skewers is a popular and common item sold by street or train hawkers. Minced *matai* is made into puddings and fritters, including *matai ko*, a kind of cake, which can be bought from "portable made-shift kitchens" on the street corners or in tea houses. Matai is often served boiled or steamed as a winter vegetable, sometimes in a sauce of sugar, butter and flour. Two popular Chinese recipes are jellied water chestnut prepared from ground water chestnut, agar and sugar; and water chestnut meatballs made with minced water chestnut, beef or pork, scallions, ginger, corn starch, salt and soy sauce. The meat balls are first cooked and then celery cabbage is added together with onions, Sichuan pepper and the chicken broth, and brought to a boil.

The corms are also used to prepare sweetmeat and desserts, sometimes combined with coconut milk and palm sugar. In Vietnam, water chestnut is used as the main ingredient for *bánh củ năng hấp* (dimsum cake or bread) and *chè mã thầy* (sweet dessert soup or pudding).

Matai starch is used in China in the same way that cornstarch is used in food preparations.

A flour or starch can be made from the dried and ground up corm and this is used to thicken sauces and to give a crisp coating to various deep-fried foods, pastries and cakes, such as the water chestnut cake, which is common as part of the dim sum cuisine. In Asia, the corms are made into a drink by either blending raw water chestnuts in water or boiling peeled or unpeeled corms in water for 15–30 min and adding a little extra palm sugar to enhance the flavour. The starch (flour) can be similarly used to prepare a sweet morning drink. This drink is reputed to have cooling properties, popular on hot days in Asian cities.

According to Ochse and van den Brink (1980), the sweetish corms are usually cooked as a delicacy in Indonesia. They are made into *Kreepik*, the so-called *emping tekee*. Following cooking, the peels are removed and the corms are then crushed with a small hammer and dried in the sun to produce a meal called *emping*. *Emping* when fried in coconut oil is eaten as a delicacy or eaten with rice. *Emping* is commonly sold in local markets in Java.

Botany

Tufted, perennial partly submerged, with short subterranean rhizome and elongated stolons, most of these terminated by a depressed globose brownish to blackish corm 2–5 cm diameter with 3–5 distinct rings externally and flesh white within (Plates 1 and 2). Culms erect, glabrous, tufted, terete, 20–60 cm high, 1.5–4 mm thick, hollow, surface with transverse septa. Leaves reduced to thin tubular sheaths, glaucous or greenish-yellow, purplish or brown bearing a liguiform acute appendix at the apex. Inflorescence a single terminal spikelet, cylindric, 2–6 cm long, 3–6 mm broad, obtuse at apex, white to straw-coloured, many-flowered; Flowers bisexual, with perianth of 6–8 filiform unequal white to brown bristles, stamens 3, stigma 3, style bifid or trifid with enlarged deltoid-attenuate base persistent in the fruit. Fruit an obovoid nutlet, 1.5–2.3 mm by 1.2–1.8 mm, glossy yellow to brown when mature.

Nutritive/Medicinal Properties

Corm Phytochemicals

Nutrient composition of the raw corm per 100 g edible portion was reported as: energy 68 cal, moisture 81.1, protein 1.4 g, fat 0.32 g, total carbohydrates 16.1 g, dietary fibre 0.6 g, ash 1.1 g, Ca 5 mg, P 77 mg, Fe 0.7 mg, Na 10 mg, K 481 mg, thiamin 0.03 mg, riboflavin 0.02 mg, niacin 1.0 mg and ascorbic acid 6 mg (Leung et al. 1972). The nutritive value of raw water chestnut per 100 g edible portion is: energy 141 kJ, moisture 50.6 g nitrogen 0.59 g, protein 3.7 g, fat 0.6 g, ash 3.1 g, total dietary fibre 7 g, Ca 27 mg, Cu 0.8 mg, Fe 95 mg, Mg 35 mg, K 243 mg, Na 16 mg, Zn 1.9 mg, Niacin (derived from tryptophan or protein) 0.6 mg, niacin equivalent 0.6 mg and vitamin C 52 mg (Brand Miller et al. 1993). The corm also contains about 60 % starch. Mean particle size of *Eleocharis dulcis* corm starch was 12 ± 5 μm and the starch was A-type or C-type just close to A-type (Yu et al. 1999). The starch had an intermediate nature between amylose and short-chained amylopectin. Amylograms of the starch suggested that the gelatinized starch was difficult to retrograde. The cytokinin, N^6- (Δ^2-isopentenyl)-adenosine (iA), was found at a concentration of 60ug/kg of tissues in the corm (Tsui et al. 1983). This high cytokinin level suggested the corm to be a site of cytokinin biosynthesis and/or of accumulation of cytokinin produced elsewhere in the plant (e.g., the roots). Six compounds: hexacosanoic acid, 5α-stigmastane-3, 6-dione, β-sitosterol, stigmasterol, betulin and tricin were isolated from a methylene chloride extract of *E. dulcis* (Miles et al. 1994). Liu et al. (2006) isolated a functional

Plate 1 Chinese water chestnut

Plate 2 Close up of Chinese water chestnut

compound 24-ethylcholesta-Δ^7-cholesterol from Chinese water chestnut. The major phenolic compounds present in Chinese water chestnut tissues were identified as (−)-gallocatechin gallate, (−)-epicatechin gallate and (+)-catechin gallate (You et al. 2007).

Sugar content was higher in the parenchyma cell walls (931 μg/mg) than in the subepidermis (775 μg/mg) or epidermis (685 μg/mg) of Chinese water chestnut corm (Grassby et al. 2013) The alkali-extractable phenolic content was greater in the epidermal (32.4 μg/mg) and subepidermal cell walls (21.7 μg/mg) than in the cell walls of the parenchyma (12.3 μg/mg). The proportion of diferulic acids was higher in the parenchyma. The Klason lignin content of epidermal and sub-

epidermal cell walls was ~15 %. Methylation analysis of Chinese water chestnut cell-wall polysaccharides identified xyloglucan as the predominant hemicellulose in the parenchyma, and also a significant pectin component. Phenolic compounds: (−)-gallocatechin gallate, (−)-epicatechin gallate and (+)-catechin gallate were isolated from the corms and were presumed to be the potential endogenous polyphenoloxidase substrates due to their ortho-diphenolic or pyrogallolic structures (Sun et al. 2010). These polyphenols might be catalysed by polyphenoloxidase, resulting in the browning of corms after fresh-cut processing.

Of 24 varieties/lines of Chinese water chestnut studies, the corms of 'Guangdong Fenti (FT2)' and 'Guangdong Fenti (FT3)' were biggest, and 'Guangxi Zhongshanfenti' the smallest (Jiang et al. 2009). For total starch content, 'Guangdong Xiaomati (Fanyu)' was highest with 220.61 g/kg (FW), and 'Guangxi Guigangmati' the lowest with 16.07 g/kg (FW). For total soluble sugar content, 'Guangxi Linchuanmati' was highest with 82.47 g/kg (FW), and Guangxi Pinglefenti' was least with 10.52 g/kg (FW). 'Guangdong Xiaomati (Fanyu)' had the highest protein content, while 'Guangxi Beihaimati' and 'Guangxi Guipingmati' had the highest amino acids content. 'Guangdong Fenti (FT2)', 'Guangdong Fenti (FT3)' and 'Guangxi Pinglefenti' could be used for processing starch.

The optimum parameters for microwave extraction of flavonoids from the peel of *Eleocharis tuberosa* was determined at ethanol concentration of 50 %, solid to liquid 1:25 (W/V), microwave treating time 4 min at microwave power 200 W (Wei et al. 2009). Under the optimal conditions, the ultrasound-assisted extraction rate of natural brown pigment from *E. tuberosa* peel was 22.79 %, which increased by 16.99 % compared with conventional extract method (Luo et al. 2008b).

Kanes and Vines (1977) found that water chestnut corms stored for 8 months in a sodium hypochlorite solution could be successfully peeled by the standard method. After 8 months of storage in the solution, these corms showed a 20 % germination ability under favourable germination conditions. Low temperature storage (1.5 °C) of Chinese water chestnut corms over a 6-month period in aqueous solutions eliminated storage losses due to desiccation (Kays and Sanchez 1985). The use of an aqueous environment, however, resulted in permeation of sound corms with odoriferous metabolites from rotting corms within the container. Treatments such as dilute salt solutions (i.e. 10 % NaCl) gave superior results and prevented corm discolouration, rots and odour formation. Visual quality, soluble solids and texture were not significantly changed, although the concentration of sucrose was reduced while fructose and glucose increased. The sodium concentration within the tissue increased markedly but did not result in cellular death. Sodium could not be sufficiently removed from unpeeled corms prior of marketing by passive diffusion, indicating that the surface of the corm represented a major diffusive barrier. Major sugars found in the corms at harvest were sucrose (>90 % of total), glucose and fructose, the concentrations of which were altered by various storage treatments.

Chinese water chestnut corms can be harvested based on epidermis (peel) development, and mature corms stored at 1 or 5 °C had very little deterioration (Brecht et al. 1992). Initial soluble sugar levels varied from 1 to 3 % but increased by at least twofold at all temperatures during the first month of storage, but did not increase further during the second month of storage and did not differ significantly among harvests. Increases in sugar levels during storage were greatest in the earliest-harvested corms, the larger (2.5 and 3.2 cm) sizes and at 10 °C storage temperature. Initial starch levels were generally greater than 20 % FW (fresh weight) and as high as 35 % FW. Starch changes in storage mirrored sugar changes, decreasing especially in earlier-harvested corms, larger sizes and at 10 °C storage temperature. Incidence of decay (mainly due to *Fusarium* sp. and *Geotrichum* sp.) was otherwise greatest at 10 °C. Weight loss during storage was insignificant (less than 1 % FW) in all cases except those corms showing chilling injury. Texture measurements indicated that smaller and less mature corms were more tender than larger

and mature corms, respectively; however, texture changes during storage were small. With development of effective decay control, 10 °C storage may be desirable due to the greater sweetening that occurs at that temperature.

Heat treatment of Chinese water chestnut effectively prevented browning associated with ammonia lyase (PAL), polyphenol oxidase (PPO) and peroxidase (POD) activities and total phenolic content as well as delayed the decrease in eating quality, which was associated with reduced total soluble solids, titratable acidity and ascorbic acid, compared with fresh-cut Chinese water chestnut (Peng and Jiang (2004). Inhibition of browning by heat treatment without microbial growth was achieved for 12 days of storage at 4 °C. Among the browning inhibitors tested, 4-hexylresorcinol, at a concentration of 0.3 mmol/L, showed the strongest inhibition (70 %) against the polyphenol oxidase activity of Chinese water chestnut, followed by 3.0 mmol/L N-acetyl-L-cysteine with an inhibition of 53 % (Lu et al. 2006).

Plant Phytochemicals

Eleocharis dulcis plants yielded 1350 g/m^2 green biomass and 216 g/m^2 leaf protein concentrate (LPC) in two cuts within 100 days (Pandey and Srivastava 1991). The green biomass showed 14 % protein content and 65.50 % extractability of protein N in the LPC. The LPC contained 56.88 % protein, which was digestible in-vitro to 67.87 %. The amino acid composition of LPC was found to be nutritionally adequate, except for sulphur-containing amino acids.

Antioxidant Activity

Chinese water chestnut phenolic extract strongly inhibited linoleic acid oxidation and exhibited a dose-dependent free-radical scavenging activity against alpha,alpha-diphenyl-beta-picrylhydrazyl (DPPH) radicals, superoxide anions and hydroxyl radicals, which was superior to ascorbic acid and butylated hydroxytoluene (BHT), two commercially used antioxidants (You et al. 2007). Further, the extract was found to have a relatively higher reducing power, compared with BHT. *Eleocharis tuberosa* peel (0.1 g/L) was found to have strong antioxidant activity (Guo and Hu 2007). The inhibition rate of superoxide anion free radical was 48.45 % and the scavenging rate of hydroxyl free radical (OH) was 67.52 %. Adding 0.1 % (molar percentage) of the extract to lard, enhanced its antioxidant activity much better than that of 0.02 % BHT. The extracts from peel showed higher activities than from the sap. In separate studies, Luo et al. (2008a) found *E. tuberosa* peel aqueous extract had hydroxyl radical scavenging of 33.7 %. The antioxidant activity was enhanced after diluting the extract 5.7 times. Polysaccharides from *E. tuberosa* peels exhibited radical scavenging activity, which was much lower than that of tea polyphenols (Li et al. 2008). Polysaccharides without proteins demonstrated a higher ability to scavenge DPPH radicals than polysaccharides with proteins, with EC_{50} values of 0.26 and 76.22 μg/ml, respectively. They exhibited similar ability to scavenge hydroxyl (·OH) and superoxide anion (·O2-·), their EC_{50} being 0.118,0.124,10.87 and 9.53 mg/ml, respectively. A red-brown extract obtained from *E. tuberosa* peel was found to contain polyphenols (3.31 % w/w DM basis), flavonoids and other compounds (Jia et al. 2007). The extract had a DPPH radical scavenging activity of IC_{50} 130.37 ppm compared to BHT with IC_{50} value of 94.16 ppm.

Anoxia pretreatment (4-h exposure to pure nitrogen) significantly inhibited browning of fresh-cut Chinese water chestnut slices during storage, accompanied by lower contents of malondialdehyde, H_2O_2 and lipoxygenase activity (You et al. 2012). Further, anoxia stimulated the activities of superoxide dismutase and ascorbate peroxidase, scavenging reactive oxygen species and alleviating lipid peroxidation. Additionally, better maintenance of reducing power and free-radical-scavenging activities against DPPH, superoxide anions and hydroxyl was observed in N_2-treated Chinese water chestnut slices, with

higher phenolic compounds and ascorbic acid contents.

Antimicrobial Activity

The milky juice from *E. tuberosa* inhibited the growth of *Micrococcus pyogenes* var. *aureus*, *Escherichia coli* and *Aerobacter aerogenes* (Chen et al. 1945). The active substance, puchin, was insoluble in organic solvents, relatively stable and was inactivated by hydrogen sulphide and by ethanol (Korzybski et al. 1967).

A functional component of Chinese water chestnut was found to have bacteriostatic effect equivalent to 0.1 % potassium sorbate in prolonging shelf life of Chinese water chestnut cake (Zhao et al. 2005). It reduced the development of oxygen and microbial decay. Liu et al. (2006) isolated a functional compound from Chinese water chestnut, 24-ethylcholesta-Δ^7-cholesterol, which had in-vivo and in-vitro antimicrobial effects on bacteria, inhibited many kinds of inflammation and showed visible activity against pain caused by acetic acid.

Traditional Medicinal Uses

Water chestnut is described as cold and sweet and is used in Chinese traditional herbal remedy for diabetes, jaundice, urinary strains, pink eyes, sore throat and hypertension (Duke and Ayensu 1985; Lu 2005). To relieve hypertensions, a tea is prepared with fresh mandarin orange peel and drank thrice daily. Eating water chestnut or drinking its juice twice daily have been used to treat sore throat, haemorrhoids and mouth canker. Eating water chestnut steeped in rice wine and washing down with the steeped rice wine twice daily is used for diarrhoea. The plant is used also to treat abdominal pain, amenorrhoea, hernia and liver problems.

Chinese water chestnut has also been used as a folk medicine to treat hypertension, chronic nephritis, constipation, pharyngitis, laryngitis and enteritis (You et al. 2007).

Other Uses

The straw is used for weaving bags, sleeping mats, grass aprons, hats and as fodder for cattle and pigs. The straw is also used as mulch and compost. The plant was used for making salt in Zimbabwe.

Studies found that Chinese water chestnut could remove nutrients in sufficient quantities (108.06 kg of nitrogen, 6.90 kg of calcium and 37.46 kg of magnesium per hectare) to improve water quality and allow increases in feeding rates and greater catfish production (McCord Jr and Loyacano Jr 1978).

Eleocharis dulcis also contain compounds with insecticidal activity (Miles et al. 1994). 5α-Stigmastane-3,6-dione, betulin and tricin, isolated from a methylene chloride extract of *E. dulcis,* showed antifeedant activity against larvae of the boll weevil, *Anthonomus grandis*. Japanese and Vietnamese scientists found that the roots of *Ludwigia adscendens* and *Eleocharis dulcis* harbour ammonia-producing bacteria, which was identified as a new species, *Curtobacterium ammoniigenes* (Aizawa et al. 2007). These bacteria produced ammonia responsible for alkanisation of the soil. Thus, both plants that grow in highly acidic swamps have potential to be used as natural bioremediation agent to alkanise actual acid sulfate soils (AASS).

Comments

In China, a sweet dessert form '*hon matai*' and a starchy type '*sui matai*' of *E. dulcis* are being cultivated.

Selected References

Aizawa T, Ve NB, Kimoto K, Iwabuchi N, Sumida H, Hasegawa I, Sasaki S, Tamura T, Kudo T, Suzuki K, Nakajima M, Sunairi M (2007) *Curtobacterium ammoniigenes* sp. nov., an ammonia-producing bacterium isolated from plants inhabiting acidic swamps in actual acid sulfate soil areas of Vietnam. Int J Syst Evol Microbiol 57(7):1447–1452

Backer CA, Bakhuizen van den Brink RC Jr (1968) Flora of Java, vol 3. Wolters-Noordhoff, Groningen, 761 pp
Brand Miller J, James KW, Maggiore P (1993) Tables of composition of Australian aboriginal foods. Aboriginal Studies Press, Canberra
Brecht JK, Bergsma KA, Sanchez CA, Snyder GH (1992) Harvest maturity and storage temperature effects on quality of Chinese water chestnuts (*Eleocharis dulcis*). Acta Hort (ISHS) 318:313–320
Burkill IH (1966) A dictionary of the economic products of the Malay Peninsula. Revised reprint of 1st ed, 1935, 2 vols. Ministry of Agriculture and Co-operatives, Kuala Lumpur, Malaysia. Vol. 1 (A-H) pp. 1–1240, Vol. 2 (I-Z). pp. 1241–2444
Chen SL, Cheng BL, Cheng WK, Tang PS (1945) An antibiotic substance in the Chinese water chestnut, *Eleocharis tuberosa*. Nature 156:234–234
Duke JA, Ayensu ES (1985) Medicinal plants of China, vols 1 & 2. Reference Publications, Inc., Algonac, 705 pp
Facciola S (1990) Cornucopia. A source book of edible plants. Kampong Publ, Vista, 677 pp
Grassby T, Jay AJ, Merali Z, Parker ML, Parr AJ, Faulds CB, Waldron KW (2013) Compositional analysis of Chinese water chestnut (*Eleocharis dulcis*) cell-wall material from parenchyma, epidermis, and subepidermal tissues. J Agric Food Chem 61(40):9680–9688
Guo YH, Hu SQ (2007) Study on antioxidant activities of *Eleocharis tuberosa* peel. Food Ferment Indus 10:128–130
Hodge WH (1956) Chinese water chestnut or matai – a paddy crop of China. Econ Bot 10:49–65
Hu SY (2005) Food plants of China. The Chinese University Press, Hong Kong, 844 pp
Huxley AJ, Griffiths M, Levy M (eds) (1992) The new RHS dictionary of gardening, 4 vols, MacMillan, London
Jia DY, Cao DD, Yao L (2007) DPPH radical scavenging activity of *Eleocharis tuberosa* peel extract. Nat Prod Res Dev 19(5):745–747 (In Chinese)
Jiang W, Li YR, Yang LT, Chen LJ, M P (2009) Study on agronomic characters and nutrition of Chinese waterchest nut (*Eleocharis tuberosa*). China Veg 2:51–54 (In Chinese)
Kanes CA, Vines HM (1977) Storage conditions for Chinese water chestnuts, *Eleocharis dulcis* (Burm F.) Trin. Acta Hort (ISHS) 62:151–160
Kays SJ, Sanchez MGC (1985) Storage of Chinese water chestnut [*Eleocharis dulcis* (Burm. F.) Trin.] corms. Acta Hort (ISHS) 157:149–159
Korzybski T, Kowszyk-Gindifer Z, Kuryłowicz W (1967) Puchiin. Antibiotics origin. Nat Prop, p 1531
Larkcom J (1991) Oriental vegetables. John Murray Publishers Ltd., London, 232 pp
Leung WTW, Butrum RR, Huang Chang F, Narayana Rao M, Polacchi W (1972) Food composition table for use in East Asia. FAO, Rome, 347 pp
Li JS, Jia DY, Yao K, Liu XM (2008) In vitro radical-scavenging activity of polysaccharides from *Eleocharis tuberosa* peels. Amino Acid Biot Resour 4(7–9):20
Liu X, Zhao LC, Zhou AM (2006) Preliminary study on functional component and functional activities of waste slurry derived in processing water chestnut starch. Food Sci 27:251–256 (In Chinese)
Low T (1991) Wild food plants of Australia. Angus & Robertson, Sydney, 240 pp
Lu HC (2005) Chinese natural cures. Black Dog & Leventhal Publishers, New York, 512 pp
Lu SM, Tong GP, Long Y, Feng H (2006) Partial purification and characterization of polyphenol oxidase from fresh-cut Chinese water chestnut. J Food Biochem 30(2):123–137
Luo YH, Fan Y, Jiang ZL, Xie QL (2008a) Determination of antioxidant activity of *Eleocharis tuberosa* peel by spectrophotometry with acridine red. J Jiangxi Normal Univ (Nat Sci Ed) 3:277–280
Luo YH, Huang HY, Wei FM, Liu YQ, Mao XP (2008b) Ultrasound-assisted extraction of natural brown pigment from *Eleocharis tuberosa* peel. Appl Chem Ind 9:999–1001,1005
McCord CL Jr, Loyacano HA Jr (1978) Removal and utilization of nutrients by Chinese waterchest nut in catfish ponds. Aquaculture 13(2):143–155
Miles DH, Tunsuwan K, Chittawong V, Hedin PA, Kokpol U (1994) Boll weevil antifeedants from *Eleocharis dulcis* Trin. J Agric Food Chem 42:1561–1562
Ochse JJ, Bakhuizen van den Brink RC (1980) Vegetables of the Dutch Indies, 3rd edn. Ascher & Co., Amsterdam, 1016 pp
Pacific Island Ecosystems at Risk (PIER) (2002) *Eleocharis dulcis* (Burm.f.) Trin. ex Henschel, Cyperaceae. http://www.hear.org/pier/species/eleocharis_dulcis.htm
Paisooksantivatana Y (1996) Eleocharis dulcis (Burm. f.) Trinius ex Henschel. In: Flach M, Rumawas F (eds) Plants yielding non-seed carbohydrates, Plant resources of South East Asia no. 9. Prosea Foundation, Bogor, pp 97–100
Pandey VN, Srivastava AK (1991) Yield and nutritional quality of leaf protein concentrate from *Eleocharis dulcis* (Burm. f.) Hensch. Aquat Bot 41(4):369–374
Peng L, Jiang Y (2004) Effects of heat treatment on the quality of fresh-cut Chinese water chestnut. Int J Food Sci Technol 39(2):143–148
Sun J, You YL, García-García E, Long X, Wang JB (2010) Biochemical properties and potential endogenous substrates of polyphenoloxidase from chufa (*Eleocharis tuberosa*) corms. Food Chem 118(3):799–803
The Plant List (2013) *Eleocharis dulcis* (Burm. f.) Trinius ex Henschel. http://www.theplantlist.org/
Tsui C, Shao LM, Wang CM, Tao GQ, Letham DS, Parker CW, Summons RE, Hocart CH (1983) Identification of a cytokinin in water chestnuts (corms of *Eleocharis tuberosa*). Plant Sci Lett 32:225–231
Uphof JCT (1968) Dictionary of economic plants, 2nd ed. (1st ed. 1959). Cramer, Lehre, 591 pp
Usher G (1974) A dictionary of plants used by man. Constable, London, 619 pp

Wei XF, Luo YH, Deng NF, Luo J, Huang DX (2009) The microwave extraction techniques of flavonoid in *Eleocharis tuberosa* peel. Hubei Agric Sci 4:949–951

You YL, Duan XW, Wei XY, Su XG, Zhao MM, Sun J, Ruenroengklin N, Jiang YM (2007) Identification of major phenolic compounds of Chinese water chestnut and their antioxidant activity. Molecules 12(4):842–852

You YL, Jiang YM, Sun J, Liu H, Song LL, Duan XW (2012) Effects of short-term anoxia treatment on browning of fresh-cut Chinese water chestnut in relation to antioxidant activity. Food Chem 132(3):1191–1196

Yu B, Fujii S, Kishihara S, Wakiuchi N, Tamaki H (1999) Physicochemical property of Huaishan (*Rhizoma dioscorea*) and Matai (*Eleocharis dulcis*) starches. Starch-Starke 51(1):5–10

Zhao LC, Liu X, Chen YQ (2005) Preliminary study of chufa functional components in preserving water chestnut cake. Food Sci 26:422–424 (In Chinese)

Dioscorea alata

Scientific Name

***Dioscorea alata* L.**

Synonyms

Dioscorea alata var. *globosa* (Roxb.) Prain, *Dioscorea alata* var. *purpurea* (Roxb.) A. Pouchet, *Dioscorea alata* var. *tarri* Prain & Burkill, *Dioscorea alata* var. *vera* Prain & Burkill, *Dioscorea atropurpurea* Roxb., *Dioscorea colocasiifolia* Pax, *Dioscorea eburina* Lour., *Dioscorea eburnea* Lour., *Dioscorea globosa* Roxb., *Dioscorea javanica* Queva, *Dioscorea purpurea* Roxb., *Dioscorea rubella* Roxb., *Dioscorea sapinii* De Wild., *Dioscorea sapinii* De Wild., *Dioscorea vulgaris* Miq., *Elephantodon eburnea* (Lour.) Salisb., *Polynome alata* (L.) Salisb.

Family

Dioscoreaceae

Common/English Names

Asiatic yam, Greater Yam, Guyana Arrowroot, Manila Yam, Purple Yam, Ten Months Yam, Water Yam, Winged Yam

Vernacular Names

Arabic: Batata Maee;
Argentina: Batatilla;
Burmese: Myauk Uu Ni, Taw Myauk Uu, Mautinsong, Myauk-U, Taw-Myauk-U;
Cambodia: Damlong Dong, Damlong Chime Moan, Dmalong Phluk;
Cameroon: Joma;
Chinese: Man Bo, Shen Shu, Tai Shue, Da Shu;
Chuuk: Eep, Kááp;
Cuba: Ñame Peludo;
Czech: Jam Křídlatý;
Danish: Jams, Yam;
Estonian: Vesijamss;
Ethiopia: Boyye;
Fiji: Uvi, 'Uhi, The Iam;
French: Igname De Chine, Pacala, Grande Igname, Igname Ailee;
Gambia: Nyamba Ba;
German: Geflügelter Yam, Yamswurzel, Wasseryam Wasser, Yamswurzel;
Ghana: Adzugo, Droboli, Gaga;
Guam: Dago;
Guinea: Gbara-Gué;
Hawaiian: Uhi, Puku'i;
India: Bengo Nari, Chupri Alu, Kham Alu (Bengali), Chupri Alu, Kada Kanda, Kada-Kanda, Khamalu, Ratalu (Hindi), Dandaanu, Dappa Genasu, Hennu Genasu, Mudigenasu, Noorele Genasu, Shigenasu, Thoona Genasu, Tuna Genasu, Tuna-Genasu, Tunakereng

T.K. Lim, *Edible Medicinal and Non-Medicinal Plants: Volume 10, Modified Stems, Roots, Bulbs*,
DOI 10.1007/978-94-017-7276-1_9

(Kannada), Kaccil, Kacil, Katsjikelengu, Katsjil-Kelengu, Kavattu, Peruvallikkilannu, Peruvallikkizhannu (Malayalam), Chinem, Chipari-Aalu, Goradu, Khanphal, Pindaalu (Marathi), Desia Alu, Kambo Alu, Kham Alu (Oriya), Alukam, Dandalu, Kandaka, Kasthaluka, Raktaluka (Sanskrit), Cirakavalli, Iyamkilanku, Kappa-Kavali, Kappan Kaccil, Kayavalli, Kayvalli, Mullu Valli, Mullu-Valli, Perumvallikilanku, Siru-Valli, Siruvalli, Vettilai-Valli (Tamil), Daeshavaali Pendalam, Dukka Pendalam, Gadimidondapendalam, Gadimidondapendalamu, Gadinidonda Pendalamu, Guna Pendalamu, Gunapendalamu, Kavili-Gadda, Naarathega, Nelavupandalum, Niluvapendalamu, Niluvu Pendalam, Niluvupendalam, Pendalam, Pendalamu, Yadduthoka Dumpa (Telugu);

Indonesia: Uwi, Uwi Klapa, Uwi Legi, Uwi Manis (Javanese), Ubi (Madurese), Ubi, Ubi Kalapa, Ubi Manis (Malay), Huwi, Huwi Kalapa, Huwi Tehang (Sundanese);

Italian: Igname;

Japanese: Daisho, Daijo, Daijyo;

Kosrae: Muta;

Laotian: Man Man Hliemx, Houo;

Malaysia: Pokok Ubi, Ubi Tiyang, Ubi Kipas, Ubi Kemali;

Nepalese: Ghara Tarul Kukur Tarul;

Nigeria: Agadaga, Agbo, Sakata, and Mbala, Nvula (Igbo);

Papua New Guinea: Yam Tru, Nyaing, Kolpur;

Philippines: Knamap, Kinampai Ubi (Bisaya), Ubi (Iloko) Ubi (tagalog);

Pohnpei: Kehp;

Portuguese: Cará De Angola, Cará Branco, Inhame Bravo, Inhame Da India, Cará Da Terra;

Russian: Iams Krylatyi, Dioscoreia krylataia, Iams Belyi;

Satawal: Wot Omalu;

Spanish: Cabeza De Negro, Cará Branco, Ñame Blanco, Ñame Blanco Grande, Ñame Blanco De Agua, Ñame Branco, Ñame De Agua, Ñame Grande, Ñame De Gua, Tabena;

Sri Lanka: Hingurala, Raja-Ala

Tahitian: Uhi;

Thailand: Man-Sao (Central Thailand), Noi (Chiang Mai), Man Bak Hep (Don Daeng), Man liam (Northern Thailand), Man-Thu;

Ulithi: Ioth;

Venezuela: Ñame Asiatico;

Vietnamese: Cậm Kẹnh, Cň Sa, Củ Cái, Củ Cầm, Củ Canh, Củ đỏ, Củ lỗ, Củ Mỡ, Củ Ngŕ, Củ Nhŕ, Củ Tía, Củ Vạc, Khoai Bướu, Khoai Long, Khoai Mỡ, Khoai Ngŕ, Khoai Ngọt, Khoai Tía, Khoai Trắng, Khoai Trút, Khoai Vạc, Mắn Hăm;

Yapese: Du'og

Origin/Distribution

Dioscorea alata is native to Southeast Asia and has been distributed throughout the tropics worldwide. It is the most important yam for Southeast Asia and is also a staple food crop in New Guinea and is widely grown in tropical Asia. In Africa, it is second to white yam (*Dioscorea rotunda*) in popularity.

Agroecology

Greater yam thrives in the warm, humid tropics with annual precipitation of 1000–15,000 mm per year. It flourishes in the lowlands to an elevation of 2500 m. It tolerates soils of low fertility but is sensitive to aluminium toxicity.

Marcos et al. (2009) found that small changes in photoperiod and temperature, very usual in the tropics, had a big effect on the tested *D. alata* yam varieties. Emergence to tuber initiation was mainly affected by photoperiod and to a lesser extent by temperature. Both factors also affected the duration of tuber initiation to harvest but their effects were less noticeable.

Edible Plant Parts and Uses

The starchy tubers are eaten boiled, roasted, fried or pounded and eaten with various sauces (Burkill 1966; Ochse and van den Brink 1980; Udensi et al. 2008; South Pacific Commission 1990). Yams are eaten with other meat, shellfish, vegetables and green leaves. Yams can be mashed and added to other fruits and green leaves, fish and are good food for babies. Tubers of certain culti-

vars are suitable for production of chips and flakes. Purple-flesh varieties are used for ice cream, cakes and other confectioneries. Water yam can be processed into flour and reconstituted into *fufu* dough in Africa (Udensi et al. 2008). Young leaves are eaten in Congo. Yam can be converted into a meal and used as a substitute for wheat flour, although rarely used for this (Burkill 1966). In Indonesia, the tubers are often eaten cooked as a delicacy, although they may also be eaten raw (Ochse and van den Brink 1980). Tubers are cut into slices and fried in oil. Thinly cut slices of the yam dried in the sun can be made into *kripik* (chips). The tubers are also used for *sayur*. In Odisha, India the tubers, bulbils and leaves are consumed as vegetables (Kumar et al. 2012).

Some popular yam dishes in the South Pacific (South Pacific Commission 1990) are:

(a) Yam salad—cooked yam, chopped onions, salad cream lettuce, tomatoes and hard-boiled eggs
(b) Boiled yam in coconut cream—yam pieces, dilute coconut cream and aibika (*Abelmoschus manihot*) leaves
(c) Stuffed yam with cheese
(d) Small yam, cooked fish, chopped tomato, coconut milk or cream with grated cheese
(e) Yam and vegetable curry—chopped yam, sliced onion, chopped chillies, curry powder, cloves garlic crushed, chopped vegetables (e.g. beans, tomato, pumpkin carrots),
(f) Baked yam and pawpaw savoury—chopped yams, coconut cream, ripe pawpaw, onion, wrap in banana leaf and bake or steam
(g) Yam delicious—yam pieces mix with eggs, chopped onion, seasoning and fry
(h) Yam fritters
(i) Grated yams raw mixed with eggs, flour, baking powder and fry in oil.

Studies in Puerto Rico showed that *D. alata* yams can be used to produce instant flakes (Rodríguez-Sosa et al. 1972). Addition of 0–2 % of okra powder to reconstituted *D. alata* yam flake reduced the sensory impairments of *ojojo*—a fried yam (*D. alata*) snack, which compared favourably with those made from raw yam (*D. alata*) in terms of colour, flavour and taste (Shittu and Olaitan 2014).

Dioscorea alata yams can be used as a fat replacer in the manufacture of Chinese sausages up to a level of 5 %, resulting in the production of Chinese sausages with about 22 % less fat content (Tan et al. 2007). The products with 5 % yams added had no significant difference in colour, flavour, hardness, juiciness and overall acceptability when comparing with the control. Results of studies suggested that micronization by ball milling treatments could improve functional properties of the fibre components of micronized peels of yam (*Dioscorea alata*), taro (*Colocasia esculenta*) and sweet potato (*Ipomoea batatas*), thus providing a good source of dietary fibre in food applications (Huang et al. 2010). The micronization treatments decreased the bulk density but increased the solubility and water-holding capacities of the micronized peels.

Botany

A climbing, perennial herbaceous unisexual–dioecious vine. Tubers are subterranean, large, variable in shape, oblate, globose, conical, cylindric or branched; externally, the epidermis is corky brown or purplish and the internal flesh colour is white or purplish (Plates 1, 2, and 3). Stem twining to right, glabrous, ridged, with four narrow, membranous wings. Bulblets present at leaf axils, small and variable in shape subglobose to narrowly ovoid. Leaves alternate basally on stem, distally opposite on stem, on 9–18 cm long petioles, simple; petiole green or purplish red, 4–15 cm; leaf blade green or purplish red, ovate, 6–20×4–13 cm, papery, glabrous, with 7–9 distinct veins, base sagittate to deeply, narrowly cordate, apex shortly acuminate or caudate (Plate 4). Inflorescences glabrous. Male spikes solitary or a few together, 1.54 cm, sometimes forming an axillary terminal panicle from axils of bracts; rachis distinctly zigzag. Male flowers: outer perianth lobes broadly ovate, 1.52 mm, pale yellow to greenish yellow; fertile stamens 6. Female spikes solitary or 2 or 3 together, lax and unbranched. Female flowers: staminodes 6; perianth delate-subglobose, 5 mm across, yellow;

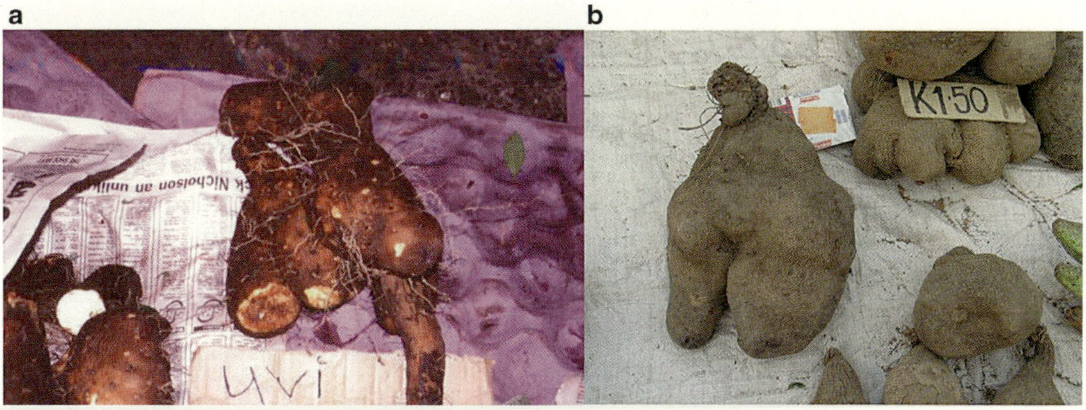

Plate 1 (**a**, **b**) variously and irregularly shaped, massive, white-fleshed tubers

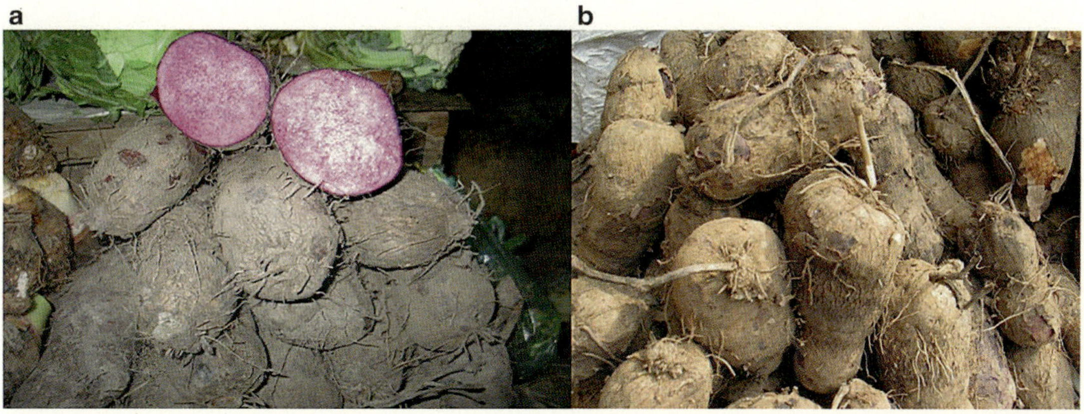

Plate 2 (**a**, **b**) variously and irregularly shaped, massive, purple-fleshed tubers

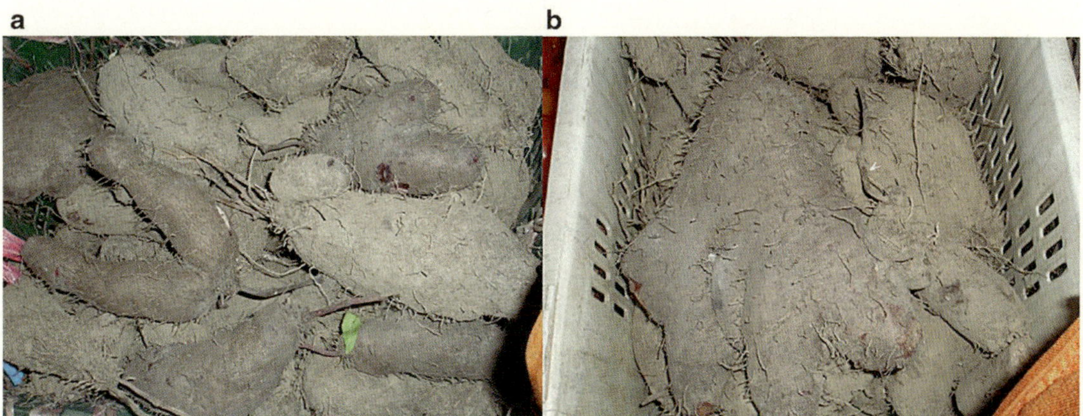

Plate 3 (**a**, **b**) variously and irregularly shaped, massive, purple- fleshed tubers

Plate 4 Narrowly cordate leaves with 7–9 veins

ovary glabrous. Capsule not reflexed, oblate, sometimes obcordate, 15–25 mm; wings 1.2–2.2 cm wide. Seeds winged all round.

Nutritive/Medicinal Properties

Tuber Nutrients/Phytochemicals

Nutrient composition of the raw tuber per 100 g edible portion was reported as: energy 87 cal, moisture 76.4 g, protein 1.9 g, fat 0.2 g, total carbohydrates 19.9 g, dietary fibre 0.6 g, ash 1.6 g, Ca 38 mg, P 28 mg, Fe 1.1 mg, Na 12 mg, K 397 mg, β-carotene equivalent 5 μg, thiamin 0.10 mg, riboflavin 0.04 mg, niacin 0.5 mg, and ascorbic acid 6 mg (Leung et al. 1972). Brand et al. (1993) reported the proximate value nutrient composition per 100 g edible portion of water yam as follows: water 72.4 g, energy 86 KJ, protein 1.4 g, fat 0.02 g, ash 0.9 g, available carbohydrate 0.0 g, total dietary fibre 6.8 g, minerals—Ca 15 mg, Fe 0.8 mg, Mg 15 mg, K 256 mg, Na 9 g, Zn 0.33 mg, Cu 0.29 mg, vitamins—vitamin C 62 mg, thiamine 0.02 mg, riboflavin 0.2 mg, niacin 0.2 mg. Shajeela et al. (2011) reported the following proximate nutrient composition of *D. alata* tubers (g/100 g): moisture 82.91 %, crude protein 7.57 g, crude lipid 5.28 g, crude fibre 3.96 g, ash 3.56 g, starch 49.13 g, NFE (nitrogen free extract) 79.63 g, energy 1655.30 kJ/100 g DM, niacin 36.20 mg, ascorbic acid 74.56 mg, and minerals mg/100 g Na 44.56 mg, K 786.30 mg, Ca 448.36 mg, Mg 656.31 mg, P 140.14 mg, Zn 2.26 mg, Mn 6.36 mg, Fe 34.30 mg and Cu 11.20 mg. The anti-nutritional factors were per 100 g: total free phenolics 0.68 g, tannins 0.41 g, hydrogen cyanide 0.17 mg, total oxalate 0.58, amylase inhibitor 6.21 AIU/mg soluble starch, and trypsin inhibitor 3.65 TIU/mg protein (Shajeela et al. 2011). The in-vitro protein digestibility was 5.23 % and the in-vitro starch digestibility was 39.40 %.

The total dietary fibre (TDF) content of 20 varieties of *Dioscorea alata* varied widely ranging from 4.10 to 11.00 % (Faustina Dufie et al. 2013). The dry matter composition ranged from 19.10 to 33.80 % and amylose was from 27.90 to 32.30 %. Mineral contents (mg/kg) were from 10.10 to 17.60 for Zn, 10,550 to 20,100 for K, 83 to 131 for Na, 260 to 535 for Ca and 390 to 595 for Mg. Physico-chemical characteristics of tubers of 48 *D. alata* accessions from Vanuatu in terms of mean and range were found respectively as follows: dry matter 23.44 % (13.64–31.42 %), starch 73.1 % (63.6–78.6 %), amylase 17.2 % (13.4–20.7 %), amylase/starch ratio 0.17 (0.13–0.21), minerals 3.3 % (2.5–4.9 %), lipids 0.3 % (0.2–0.5 %), proteins 11.95 % (8.8–17 %), sugars 1.85 % (0.6–5.71 %), and mean gelatinization temperature 74.9 °C to 84.2 °C (Lebot et al. 2006). The TDF content of 20 varieties of

Dioscorea alata varied widely ranging from 4.10 to 11 % (Dufie et al. 2013). The dry matter composition ranged from 19.10 to 33.80 % and amylose was from 27.90 to 32.30 %. Mineral contents (mg/kg) of the varieties were from 10.10 to 17.60 for Zn, 10,550 to 20,100 for K, 83 to 131 for Na, 260 to 535 for Ca, and 390 to 595 for Mg. Chemical composition of 18 *D. alata* varieties tubers were: moisture 56.47–79.31 %, mean 69.07; dry matter 20.7–43.53 %, mean 30.93 %; protein 5.07–.05 %, mean 6.51 %; sugar 2.43–6.91 %, mean 4.28 % (Wireko-Manu et al. 2011).

Studies conducted by Udensi et al. (2008) reported tubers of *D. alata* varieties contained 2.25–3.155 ash, 5.69–8.31 % crude protein, 81.6–87.6 % carbohydrates, 0.75–1.13 % crude fibre and 361-01-385.33 Kcal/100 g energy. The mineral contents mg/100 g ranged from 240 to 400 mg; with 240–400 mg K, 190–380 mg Na, 100–340 mg P, 20.04–80.16 mg Ca and 20.22–35.20 mg Mg. Vitamin C content of the yam tubers ranged from 16.72 to 35.20 mg/100 g, fresh weight. Functional property levels in the yam tubers were found to be in the range of 0.64–0.76 g/cm^3 (bulk density); 2.90–3.65 g/g (water absorption capacity); 27.0–3.5 s (wettability) and 30–50 % w/v (gelation capacity). Huang et al. (2007) found tubers of four *D. alata* cultivars had a substantial amount of protein 10.4–13.0 g/100 g (dry basis (db)) at time of harvest (day 260 post-emergence) when compared with other root and tuber crops. Starch content of the yam tubers increased as growth progressed and remained in the range of 70.5–85.3 g/100 g (db) during their growth period. The activity of yam-contained polyphenol oxidase (PPO) decreased markedly over the early period of harvest (day 155–225 post-emergence), and subsequently decreased only slightly as growth progressed to harvest. In contrast, the activity of α-amylase and dioscorin content of yam tuber increased significantly over the growth period for all cultivars. All the yam cultivars contained substantial levels of essential amino acids, all of which were superior to the FAO reference pattern for such amino acids except for sulphur-containing amino acids and lysine contents.

Wanasundera and Ravindran (1994) reported the following nutrient composition of tubers of several *D. alata* cultivars (mean values on percent DW basis): moisture 72.7 %, crude protein 7.4 %, crude fat 1.0 %, crude fibre 1.5 %, ash 3.4 %, starch 79.5 %, soluble sugars 1.2 %, total energy 365 Kcal/100 g; minerals (mg/100 g DM) K 1620 mg, Na 66.6 mg, P163 mg, Ca 68.6 mg, Mg 69.6 mg, Cu 6.6 mg, Fe 10.4 mg, Mn 3.6 mg, and Zn 3.9 mg. They also reported the presence of antinutrients (DM basis): low phytic acid 131 mg/100 g (range 58.6–198 mg/100 g), phytin as % of total phytin 383 %, water-soluble oxalates and total oxalates 334 mg and 585 mg/100 g. Phytic acid contributed 9.7–35.9 % of total phytin. They found that crude fat, crude fibre, starch and total sugar contents of tubers were unaffected by cooking but crude protein tended to decrease with cooking but not significantly (Wanasundera and Ravindran 1992). Water-soluble minerals leached out during boiling, thus causing a reduction in the ash content of boiled tubers. All cooking methods lowered the vitamin C content of the tubers. Phytate contents were unaffected, whereas total oxalate contents were significantly lowered by the cooking methods employed. The loss of oxalates was greater with boiling (40–50 %) compared to steaming (20–25 %) and baking (12–15 %). Ezeocha and Ojimelukwe (2012) found that the crude protein contents (10.27 %), ash (2.93 %) and lipid (0.15 %) were significantly lowered in the boiled tubers while the carbohydrate (76.57 %) significantly increased in the boiled tubers. The antinutrients; alkaloids (2.77 %), saponins (2.71 %), flavonoids (1.38 %) and tannins (0.21 %) were significantly reduced in the boiled tubers. They concluded that boiling had both positive and negative effect on water yam. A cooking time of between 30 and 60 min at 100 °C was recommended for *D. alata*.

Although yam proteins from *D. alata* and *D. alata* var. *purpurea* consisted of similar amino acid residues, they still exhibited significant differences in conformational arrangement (Liao et al. 2004). The secondary structure of *D. alata* was mainly an alpha-helix, while *D. alata* var. *purpurea* was mostly in antiparallel beta-sheets.

FT-Raman spectroscopy directly proved the existence of S-S in yam proteins, implying that oligomer formation in yam proteins might be due to disulfide linking of dioscorin (32 kDa).

Akissoe et al. (2005) found that polyphenoloxidase activity was 50 % higher in nonprocessed freeze-dried Florido (*Dioscorea alata*) than in nonprocessed freeze-dried Deba (*Dioscorea rotundata*). Polyphenoloxidase activity decreased progressively during blanching. Forty-five percent of polyphenoloxidase activity remained after 50 min of blanching at 60 or 65 °C, whereas the peroxidase activity declined sharply to less than 20 % of the initial activity after 10 min of blanching, whatever the blanching temperature. No anthocyanidins could be detected in nonprocessed freeze-dried yam. Flavanols and cinnamic acid compounds were detected. Catechin was identified as the major flavanol with concentrations ranging from 0.26 to 0.41 μM/g. One cinnamic compound, ferulic acid, was identified and assessed in both yams (0.03–0.04 μM/g). Total phenol, flavanol and cinnamic contents decreased during blanching independently of temperature.

D. alata tuber was reported to contain 0–0.25 % sapogenin (Anzaldo et al. 1956). The recoveries of furostanol and spirostanol glycosides were above 92 % in the three Taiwanese yam cultivars (two *D. alata* and one *D. pseudojaponica*), the contents of furostanol glycosides in the two *D. alata* cultivars were 34.81 and 97.58 μg/g dw, while the contents of spirostanol glycosides were 46 and 79.67 μg/g dw (Yang et al. 2003). The furostanol glycosides were: 26-*O*-β-D-glucopyranosyl-22α-methoxyl-25-(*R*)-furost-5-en-3β,26-diol3-*O*-α-L-rhamnopyranosyl-(1 → 2)-*O*-{[α-L-rhamnopyranosyl-(1 → 4)]-*O*-[β-L-rhamnopyranosyl-(1 → 4)]}-β-D-glucopyranoside; methyl protodioscin and methyl protogracillin. The spirostanol glycosides were 25(*R*)-spirost-5-en-3β-ol 3-*O*-α-L-rhamnopyranosyl-(1 → 2)-*O*-{[α-L-rhamnopyranosyl-(1 → 4)]-*O*-[α-L-rhamnopyranosyl-(1 → 4)]}-β-D-glucopyranoside; dioscin and gracillin. From the tubers, hydro-Q(9) chromene and γ-tocopherol-9, together with three known compounds, RRR-α-tocopherol, coenzyme Q(9) and 1-feruloylglycerol were silated (Cheng et al. 2007).

Two anthocyanins, cyanidin and peonidin 3-gentiobioside acylated with sinapic acid, were isolated from *Dioscorea alata* 'King yam' tuber from Sri Lanka (Shoyama et al. 1990). Three anthocyanins, alatanins A, B and C, were isolated from the tuber of purple yam *Dioscorea alata* (Yoshida et al. 1991b). Alatanin C was found to be an unusually stable monoacyalated anthocyanin in neutral aqueous solutions (Yoshida et al. 1991a). The stability was ascribed to the intramolecular stacking of sinapic acid and to the chiral self-association of anthocyanidin nuclei. The choline contents of Yangmingshan yam (*D. alata*) and Ming-Chien yam (*D. purpurea*) tubers determined using the original AACC method and the modified AACC method through coupling an additional bubble separation procedure, respectively, were 0.77 and 1.78 mg/g solid for *D. alata* and 0.44 and 1.35 mg/g solid for *D. purpurea* (Fu et al. 2005).

D. alata tuber peel was found to have high levels of macro-minerals (Na, K, Ca and P) compared to micro-minerals (Mg, Zn, Fe, Cu and Cr) (Yahaya et al. 2012). The micro-nutrients were found to be generally lower than the dietary mineral requirement for animal feeds. The peel contained 4.59–12.2 % protein, 9.71–41.7 % fibre, 37.5–45.5 % carbohydrate and 0.62–1.86 % lipid. The peels collected during dry season contained higher levels of phytate (2.41–4.18 %), hydrogen cyanide (4.69–5.05 %), soluble oxalate (1.15–1.34 %) and tannin (1.54–2.45 %) than the peels collected during wet season.

Two forms of phosphorylase were purified from *Dioscorea alata* (Oluoha and Ugochukwu 1995). The molecular masses obtained for fractions I and II of *D. alata* phosphorylase were 120,000 and 170,000, respectively. The catecholase enzyme was also found in *D. alata* (Adamson and Abigor 1980).

Ireland and Passam (1984) found a gradual decrease in growth inhibitory phenolics in tubers of *Dioscorea alata* and *D. esculenta* during dormancy and in *D. alata* this closely paralleled a decrease in batatasin content. It was found that batatasin-type growth inhibitory phenolics accumulated rapidly in developing tubers just prior to the onset of dormancy and were asymmetrically

distributed, being concentrated in the proximal (head) region and in the peripheral zone just beneath the periderm. Gibberellin A_3 treatment produced a promotion of the dormant period and a correlative rise in the growth inhibitory phenolic level. Effects of maleic hydrazide and ethylene chlorohydrin were also reported.

Tuber Starch/Flour

Jayakody et al. (2006) reported the chemical and physical composition and granule morphology of the Hingurala and Raja-ala varieties of *D. alata* native starches, respectively, as follows: starch yield (based on tuber weight) 14.25 %, 18.80 %, moisture 8.25 %, 8.75 %, ash 0.13 %, 0.17 %, nitrogen 0.02 %, 0.01 %, phosphorus 0.05 %, 0.04 %, lipids (extract by chloroform-methanol) 0.05 %, 0.08 %, lipids (extracted by *n*-propanol-water) 0.25 %, 0.20 %, total amylose content 26.98 %, 31.02 %, amylose complexed with lipids 8.34 %, 5.58 %, granule size range 30–40 μm, 35–45 μm, granule morphology, truncated oval, truncated spade, crystallinity 43 %, 43 % and crystalline type C-type, B-type. The gelatinization temperatures for Hingurala and Raja-ala varieties were, respectively, as follows: onset (To) 78.17, 75.45 °C, mid point (Tp) 85.13,78.49 °C and conclusion (Tc) 92.87,85.70 °C, gelatinization temperature range (Tc–To) 14.7, 12.25 °C and gelatinization enthalpy (ΔH) 18.98, 18.60 J/g. The hydrolysis percentage of Hingurala and Raja-ala starches by porcine pancreatic a-amylase were 56.14 %, 56.63 %, respectively. Both starches differed significantly from each other with respect to peak viscosity (Raja-ala > Hingurala), viscosity breakdown (Hingurala > Raja-ala) set-back (Hingurala > Raja-ala) and pasting temperature (Hingurala > Raja-ala). Percent amylose leaching was higher for Raja-ala than Hingurala. Melting enthalpies (ΔH_R) of amylopectin recrystallisation (reflecting the extent of retrogradation during the storage period of 40 °C for a week) was higher for Hingurala than Raja-ala. Huang et al. (2006) found the starch content of tubers of four *D. alata* cultivars tubers ranged from 70.5 to 85.3 % on a dry basis. The shapes of the starch granules were round to oval or angular. The size of starch granule increased, with growth time ranging from 10 to 40 μm. The X-ray diffraction patterns could be classified as typical of B-type starch for the four cultivars. The transition temperature of gelatinization of the four yam starches decreased during maturity The starch paste showed a lower breakdown at an early harvest time. It appeared to be thermo-stable during heating but had a high setback after cooling, which might result in a tendency towards high retrogradation. The results for pasting behaviours showed that higher amylose content was associated with a lower pasting temperature and a higher peak viscosity in these starches. Amylograms of *D. alata* tuber starch showed that starch pastes maintained a high viscosity under heat treatment and mechanical stirring in neutral to slightly acidic conditions (Mali et al. 2003). Brabender viscosity increased when gums were added; the effect of guar gum on viscosity was more marked than that of xanthan gum. Xanthan gum, at a concentration of 0.5/100 g suspension, showed higher effectiveness than guar gum in reducing exudate production during refrigerated storage. The results suggested the addition of hydrocolloids could allow yam starch to be used in foods requiring low temperatures. Fifteen test varieties of *D. alata* had lower starch content (68.3 %), swelling power (9.9), peak viscosity (283.9 RVU), trough (221.5 RVU), breakdown (20.2 RVU), final viscosity (283.9 RVU) and setback (62.4 RVU) but higher sugar (5.4 %), solubility (11.9 %), peak time (6.3 min) and pasting temperature (89.2 °C) than the control variety (Baah et al. 2009). Multiple comparison sensory tests by a trained panel showed poor quality of pounded yam from test varieties relative to the control, however, TDa 98–159 and TDa 291 compared well with the control. Starch characteristics of 18 *D. alata* varieties tuber flour were starch 60.42–77.56 %, mean 76.2 %, amylose 2.69–31.56 %, mean 26.41 %; swelling power 6.23–9.75 % mean 7.6 % (Wireko-Manu et al. 2011). The pasting characteristics of the starch were peak viscosity 74.80–284.60 RVU (rapid visco units), mean 157.66 RVU; final viscosity 112.25–317.20 RVU, mean 195.08 RVU; setback 27.45–308.10 RVU, mean 59.56 RVY, and pasting temperature 83.60–90.10 °C, mean 85.89 °C.

Significant associations were found, through canonical correlation analysis, between pasting characteristics of fresh yams from six varieties, each, of *Dioscorea rotundata* and *Dioscorea alata* and the textural quality of pounded yam samples prepared from them (Otegbayo et al. 2006). Good textural quality of pounded yam was associated with high peak viscosity, breakdown, final viscosity, holding strength and setback viscosity but with low pasting temperature in the fresh yam. Otegbayo et al. (2011) found that *D. alata* yam starch with high amylase content, water-binding capacity and low swelling power gave pounded yam samples, which were very soft, unstretchable, sticky and incohesive compared to *D. rotundata*.

Kpodo and Plahar (1992) found that *D. alata* yam flour (starch) could be successfully extruded with maximum expansion at a feed moisture range of 8–10 % using extrusion temperatures of 100–115 °C. Steam pressure treatment of *Dioscorea alata* and *Dioscorea rotundata* starches led to vast changes in physico-chemical properties content (Moorthy 1999). The treatment did not significantly affect the total amylose but the soluble amylose content decreased threefold to fivefold. Reducing values and lambda (max) of the iodine complexes were unaffected. Viscosity of the starch paste was reduced by the treatments, and at higher pressures and longer time of treatments, the peak viscosity values were reduced to very low values. Pasting temperatures were enhanced considerably. Swelling volumes underwent reduction, but no change in solubility occurred. Clarity and paste stability were markedly lowered. Studies revealed that carboxymethylation improved thermal stability of *Dioscorea alata* native starch (Lawal et al. 2008). The degree of substitution (DS) increased progressively as the steps of carboxymethylation increased from 2 to 9 and an optimal DS of 2.24 was obtained. Initial increases in carboxymethylation step increased the reaction efficiency progressively up to 82.1 % after the seventh carboxymethylation step but declined with further increases in the carboxymethylation step. Starch crystallinity reduced significantly after carboxymethylation. Thermogram of native starch showed a characteristic three-step decomposition with 13.16 %, 61.54 % and 24.79 % weight losses progressively, while carboxymethyl derivative showed four decomposition stages with 9.86 %, 36.57 %, 3.04 % and 23.07 % weight losses progressively.

Dioscorea alata purpurea yam contained starch granules mostly in the range of 10–80 μm, and about 1 % of starch granules was smaller than 1 μm (Yeh et al. 2009). Decreasing water content from 90 % to 40 % did not significantly alter the onset temperature (To) and peak temperature (Tp), but raised the conclusion temperature (Tc). Mucilage exhibited greater storage modulus (G′) and smaller loss modulus (G″) than the isolated yam starch at water content of 90 %. Water content also influenced the effect of mucilage on the rheological properties of starch–mucilage mixture, but did not significantly affect To and Tp.

After defatting, *Dioscorea alata* yam and cassava starches were found to have amylose contents of 36.2 and 24.2 %, respectively (Freitas et al. 2004). *D. alata* starch showed a more energetic gelatinization process when compared to cassava starch and also had a lower rate constant, indicating a relatively slow gelatinization process at higher temperatures. *D. alata* yam gels formed by autoclaving a suspension (50 g/L) showed after 24 h of refrigeration, a stronger structure than for a cassava gel.

Antioxidant Activity

Peel portions of *D. alata* yam were found to have a better effect on scavenging DPPH free radical than flesh portions, especially for the ethyl acetate partition of the peel portion of Tainung #2 yam (Chen et al. 2004). Its EC_{50} value (14.5 μg/mL) was even lower than that of ascorbic acid (21.4 μg/mL). Various extracts of *D. alata* rhizome, viz. aqueous, 30 % ethanol and boiled 30 % ethanolic extracts effectively inhibited the copper-driven Fenton reaction-induced damage of calf thymus DNA, while inhibition was less

pronounced in the case of X-ray induced strand breakage of plasmid DNA (Wang et al. 2004). While boiled aqueous extract potently inhibited X-ray induced strand breaks in plasmid pGL3 DNA, it failed to inhibit, and even greatly enhanced, copper-H2O2 induced damage of calf thymus DNA. The results demonstrated strong copper chelating and weak hydroxyl radical scavenging activities in *D. alata* rhizome extracts, and these activities may vary depending on the procedures used in preparing the extract.

Aqueous methanolic (50 % MeOH) extracts of the tubers (peel and flesh) of nine cultivars of *Dioscorea alata* were found to have relatively high antioxidant activities among which two cultivars (Ubong upo, purple, LA096, white) had activities as high as those of α-tocopherol and butylhydroxyanisole (BHA) (Lubag et al. 2008). Serial fractionation of the extract yielded two compounds P1 and P2, which showed antioxidant activities higher than those of BHA and α-tocopherol . P1 was established to be a purple anthocyanidin very similar to alatanin C. Initial results for P2 indicated its phenolic nature with a glucose moiety and a molecular weight of 306.

Dioscorea alata yam peel showed antioxidant activity in mouse liver cell lines (Hsu et al. 2011b). The peel water extract augmented tert-butylhydroperoxide (t-BHP)-induced cytotoxicity in mouse Hepa 1–6 cells, while the Yam peel ethanol extract reduced t-BHP -induced cytotoxicity in mouse FL83B cells. GPx activity was found to play important role on reducing t-BHP-induced oxidative stress. The methanol extract of *Dioscorea alata* tuber showed potent hydroxyl, superoxide, ABTS radical cation scavenging activities while the ethanol tuber extract showed strong DPPH radical scavenging activity (Sakthidevi and Mohan 2013). The maximum inhibitory concentration (IC50) in all models viz. DPPH, hydroxyl, superoxide and ABTs radical cation scavenging activity of tuber of *D. alata* were found to be 27.16, 26.12, 30.65 and 25.53 μg/mL, respectively, at 1 μg/mL concentration. The total phenolics and flavonoids in methanol extract were found to be 0.68 g/100 g and 1.21 g/100 g respectively.

Antidiabetic Activity

Ramdath et al. (2004) reported on the glycaemic index (GI) of eight staple foods eaten in the Caribbean: high GI food cassava (*Manihot esculenta*) 94, dasheen (*Colocasia esculenta*) 77, moderate GI food: breadfruit (*Artocarpus altilis*) 60, cooking green banana (*Musa* spp.) 65, 'sadha roti' 65, eddoes (*Colocasia esculenta* var. *antiquorum*) 61, Irish potato (*Solanum tuberosum*) 71, tannia (*Xanthosoma sagittifolium*) 60 and white yam (*Dioscorea alata*) 62. Crushing did not significantly affect the GI of dasheen, tannia or Irish potato. Studies by Bahado-Singh et al. (2006) found that 14 commonly eaten carbohydrate-rich foods, including *D. alata*, processed by roasting or baking may result in higher GI. Conversely, boiling of foods may contribute to a lower GI diet.

Treatment of glucose-loaded normal rats with *D. alata* tuber extract, at dose levels of 100 and 200 mg/kg, significantly reduced blood glucose levels (Maithili et al. 2011). The extract did not produce hypoglycemic activity at both dose levels in normal, fasted rats. In alloxan-induced diabetic rats treated with the extract, the body weight significantly increased after 21 days treatment; blood glucose level was reduced significantly by 47.48 % and 52.09 % after 21 days of treatment at dose levels 100 and 200 mg/kg, respectively. Serum lipid levels, total protein, albumin and creatinine were reversed towards near normal in treated rats as compared to diabetic control.

Dispo85E (*D. alata* rhizome extract) enhanced the endocytosis and degradation activity of advanced glycation end products (AGEs) in murine hepatic nonparenchymal cells (NPCs) (Peng et al. 2011b). Further, the hepatocyte growth factor (HGF) expression level was positively correlated with the clearance capacity of the AGEs in NPCs after Dispo85E treatment. It was also shown that recombinant mouse HGF could enhance the endocytosis and autophagic clearance of AGEs in NPCs. The in-vivo data indicated that Dispo85E increased hepatic HGF messenger RNA expression levels and decreased serum AGEs level in diabetic mice. Also, the function of retina and kidneys was improved by

Dispo85E treatment in AGEs-induced diabetic mice. The study suggested that Dispo85E enhanced the clearance of AGEs through HGF-induced autophagic-lysosomal pathway and could be a candidate drug for the treatment of diabetic vascular complications. Studies showed that administration of *Dioscorea alata* L. (*DA*) extract at 100, 200 and 300 mg/kg of body weight to male wistar rats significantly reduced food intake, fasting blood glucose level and body weight when compared with the control group (Helen et al. 2013). The results suggested that *Dioscorea alata* could serve as a great therapeutic diet in the management of diabetes.

Antiosteoporotic Activity

Extracts from *D. alata* roots and leaves were found to strongly stimulate proliferation of both bone marrow cells and splenocytes, significantly increasing cell concentrations (Tulin and Ecleo 2007). A cytokine mimetic with molecular weight of 35 kDa was isolated from greater yam root and found to be biologically active, stimulating a dose-dependent proliferative response. Studies showed that 2 weeks of feeding *D. alata* yam prevented loss of bone mineral density and improved bone calcium status without stimulating uterine hypertrophy in ovariectomised female BALB/c mice (Chen et al. 2009). Phyto rhizomes (Dispo85E) increased the activity of alkaline phosphatase (ALP) and bone nodule formation in primary bone marrow cultures (Peng et al. 2011a). The extract promoted osteoblastogenesis by increasing ALP activity and bone nodule formation in both intact and ovariectomised (OVX) mice. It ameliorated the deterioration of trabecular bone mineral density, trabecular bone volume/total volume, and trabecular bone number in OVX mice.

Antiallergic Activity

All the dioscorins from *D. alata* or *D. japonica* suppressed allergic reactions by decreasing the serum IgE and histamine levels in ovalbumin-induced allergy mice (Hsu et al. 2013). The IL-5 levels decreased to basal levels in dioscorin-treated mice and in most of the lymphoid cells of the dioscorin-treated mice in response to ConA stimulation. The spleen cells from the dioscorin-treated mice also exhibited an up-regulation of IFN-γ secretion in response to ConA stimulation The decrease of IgE and histamine levels was concomitant with an increase in IFN-γ and IgG2a levels and with a decrease in IL-5 levels, suggesting that dioscorins suppressed the ovalbumin -induced allergic reactions, possibly through modulating an imbalanced Th1/Th2 immune response.

Antihyperhomocysteinemia Activity

The results of studies indicated that hyperhomocysteinemia (HHcy) induced by methionine in rats could be reversed by *D. alata* feeding (Chang et al. 2004). *D. alata* powder feeding for 12 weeks significantly decreased plasma homocysteine levels and exhibited its antioxidative effects in HHcy. *D. alata* also alleviated thrombin-induced platelet aggregation, lipid peroxidation and oxidative stress, but did not induce activity of antioxidant enzymes, which had already adaptively increased by hyperhomocysteinemia.

Antihypertensive Activity

Dioscorin, the tuber storage protein of *Dioscorea alata* yam, inhibited dose-dependently 12.5–750 μg angiotensin converting enzyme (ACE), producing 20.83–62.5 % inhibition (Hsu et al. 2002). The 50 % inhibition (IC_{50}) of ACE activity was 6.404 μM dioscorin (250 μg corresponding to 7.81 nmol) compared to that of 0.00781 μM (0.0095 nmol) for captopril. The ACE inhibitory activity was increased from 51.32 % to about 75 % during 32 h hydrolysis with pepsin. The results suggested that dioscorin and its hydrolysates may have potential for hypertension control when people consume yam tuber.

Powdered yam product of *D. alata*, which included alcohol-insoluble solids of yam tuber,

hot air drying (HAD) of yam tuber slices, steam cooked once or twice followed by HAD, which were subsequently powdered, and liquid yam products of *D. alata* heated at 90 or 95 °C were found to effectively reduce the blood pressure of to spontaneously hypertensive rats (SHRs) and should be beneficial in food processing in the development of functional foods for blood pressure regulation (Liu et al. 2009).

Estrogenic Activity

Ethyl acetate extracts of various species/varieties of yam, including *D. alata*, were found to activate estrogen receptors alpha and beta to various extents (Cheng et al. 2007). Fractionation of *D. alata* cv. Tainung No. 2 tuber extract afforded two new compounds, hydro-Q(9) chromene and γ-tocopherol-9, together with three known compounds, RRR-α-tocopherol, coenzyme Q(9) and 1-feruloylglycerol; all were shown to activate human ERalpha and beta. These results confirmed the beneficial effect of yam for menopausal women.

Menopausal Symptoms Amelioration Activity

In a study of 22 apparently healthy postmenopausal women who completed the study, replacing two-thirds of staple food (rice for the most part) with 390 g of yam (*Dioscorea alata*) for 30 days improved the status of sex hormones, lipids and antioxidants (Wu et al. 2005). After yam ingestion, there were significant increases in serum concentrations of estrone (26 %), sex hormone binding globulin (SHBG) (9.5 %), and near significant increase in estradiol (27 %). Free androgen index estimated from the ratio of serum concentrations of total testosterone to SHBG decreased. Urinary concentrations of the genotoxic metabolite of estrogen, 16α-hydroxyestrone, decreased significantly by 37 %. Plasma cholesterol concentration decreased significantly by 5.9 %. Lag time of low-density lipoprotein oxidation prolonged significantly by 5.8 % and urinary isoprostane levels decreased significantly by 42 %.

In a two-centre, randomised, double-blind, placebo-controlled clinical investigation on 50 menopausal women, intake of *Dioscorea alata* improved menopausal symptoms, particularly the psychological parameters in menopausal women, compared with placebo (Hsu et al. 2011a). Apparent improvements were noted in the parameters 'feeling tense or nervous', 'insomnia', 'excitable' and 'musculoskeletal pain,' among those receiving *Dioscorea* yam. *Dioscorea* consumption also resulted in positive effects on blood hormone profiles. Safety monitoring indicated that standardised extracts of *D. alata* were safe during daily administration over a period of 12 months.

Antihypercholesterolemic Activity

Chen et al. (2003) found the *Dioscorea alata* 50 % yam diet consistently improved cholesterol profile in the plasma and liver in adult Balb/c mice. Also, faecal excretions of neutral steroid and bile acids were increased and fat absorption was decreased in mice fed on 50 % yam diet. The 25 % yam diet was sufficient to modulate intestinal enzyme activities, but not the plasma and hepatic cholesterol levels. The leucine-aminopeptidase activity in the small intestine was increased for 79 % and 102 % with 25 % and 50 % yam diet, respectively. In contrast, the sucrase activities were decreased by both yam diets. The results suggested that the reducing effects of 50 % yam diet on the plasma and hepatic cholesterol levels could be mediated through the inflated faecal fat and steroid excretion. Another study showed that male Wistar rats fed with high cholesterol (10 %) diet supplemented by 40 % *D. alata* yam significantly reduced plasma triglyceride and cholesterol (Yeh et al. 2007). Plasma aspartate transaminase and alanine transaminase activities were significantly increased. The results suggested that yam may inhibit hypertriglyceridemia induced by a high cholesterol diet in rats.

Anti-fibrosis Activity

D. alata aqueous extract treatment of murine fibroblast cells (NRK-49F) with cellular fibrosis induced by β-hydroxybutyrate, attenuated renal interstitial cellular fibrosis by suppressed β-hydroxybutyrate -induced expression of fibronectin concomitantly with the inhibition of Smad2/3, pSmad2/3 and Smad4 (Liu et al. 2012). The extract also caused a decrease in α-smooth muscle actin and MMP-2 levels, and an increase in E-cadherin expression. They proposed that d. alata extract might act as a novel fibrosis antagonist, which acts partly by down-regulating the transforming growth factor-beta (TGF-β)/smad signalling pathway and modulating epithelial-mesenchymal transition.

Immunomodulatory Activity

Dioscorin isolated from *Dioscorea alata* induced cytokine expression in macrophages by activating Toll-like receptor 4 (TLR4)- signalling pathways crucial for both innate and adaptive immunity (Fu et al. 2006). Dioscorin from *D. alata* tuber (5–100 μg/ml) was able to stimulate nitric oxide production (expressed as nitrite concentrations) in RAW264.7 cells (Liu et al. 2007). The stimulation index on the phagocytosis of RAW264.7 cells against *Escherichia coli* and the oxidative burst (determined by the intensity of rhodamine fluorescence) of RAW264.7 cells were both enhanced by different concentrations of dioscorin (5–100 μg/ml). Dioscorin (5–100 μg/ml) was found able to induce IL-6, TNF-alpha and IL-1beta production in RAW264.7 cells and human monocytes. The stimulated proliferation index of splenic cells from BALB/c mice, ranged from 1.38- to 1.48-fold for phytohemagglutinin alone or for phytohemagglutinin mixed with different concentrations of dioscorin (10, 25 and 50 μg/ml). The results suggested that the tuber storage protein of yam dioscorin functioned as an immunomodulatory substance. Another study found that yam storage protein dioscorins from *Dioscorea alata* and *D. japonica* exhibit distinct immunomodulatory activities in mice (Lin et al. 2009). Intraperitoneal injection of the *D. alata*-dioscorins was found to have a higher ability to stimulate the phagocytic activity of the lymphoid cells than *D. japonica* -dioscorins, whereas *D. japonica* -dioscorins possessed more abilities than *D. alata* -dioscorins to enhance the proliferation of the lymphoid cells.

The phytoextract from 50 to 75 % ethanol-precipitated fraction of *Dioscorea alata* var. *purpurea* Tainung no. 5 tuber, designated as DsII-TN5 was found to confer immunomodulatory activities ex-vivo and improve regeneration of bone marrow cells in- vivo (Chang et al. 2013). The extract showed a strong augmentation of tumour cell lysate- (TCL-) loaded dendritic cell-mediated activation of T-cell proliferation. It stimulated the expression of CD40, CD80, CD86 and IL-1 β in TCL-loaded DCs and down-regulated the expression of TGF- β 1. The extract as a dendritic cell vaccine adjuvant showed strong antimelanoma activity and reduced myeloid-derived suppressor cell population in tested mice. The extract could also activate dendritic cells to enhance Th1- and Th17-related cytokine expressions. Biochemical analysis showed that the extract consisted mainly of polysaccharides containing a high level (53 %) of mannose residues.

Hepatoprotective and Nephroprotective Activities

Feeding of crude water extract of *D. alata* yam to rats with acute hepato-nephrotoxicity induced by acetaminophen ameliorated renal tubular degranulation changes, necrosis and disintegration; and protected against the inflammation of central vein and necrosis of liver tissue (Lee et al. 2002).

In-vivo studies showed that ethanol extract *D. alata* (ethanol extract) treatment of Wistar rats aniline-induced spleen toxicity (oxidative and nitrosative stress) for a month resulted in significant recovery in aniline-induced splenic toxicity (Khan et al. 2014). The protection may be due to its antioxidant property and the presence of different phytochemicals. Earlier studies by Liu et al. (2012)

reported that *Dioscorea alata* attenuated renal interstitial cellular fibrosis by regulating Smad- and epithelial mesenchymal transition signaling pathway.

Traditional Medicinal Uses

Dioscorea alata is an important herb in Chinese medicine, widely used for the treatment of clinical diabetes mellitus (Liu et al. 2012). In Indian ethnomedicine, decoction of tubers is used in leprosy, piles and gonorrhoea.

Other Uses

Okunlola and Odeku (2011) found *D. alata* yam starch had high brittle fracture index and friability and could be useful in chloroquine phosphate tablet formulation when faster disintegration time of tablets is desired.

Studies showed that *D. alata* meal could replace up to 80 % of maize or constitute 40 % of a laying chicken diet, provided the rations are isocaloric and isonitrogenous (Agwunobi 1999).

Beta-sitosterol from *D. alata* tuber peel exhibited antifungal activity towards spore germination of two yam pathogens with an inhibition of less than 57 % at a concentration of 50 mg/L, while inhibition on the elongation of germ tubes of *Fusarium moniliforme* was as high as 82 % at the same concentration (Aderiye et al. 1996).

In Papua New Guinea, *D. alata* is also used for ceremonial purposes (Onwueme and Ganga 1996).

Comments

Dioscorea alata is a polyploid species with a ploidy level ranging from diploid (2n = 2x = 40) to tetraploid (2n = 4x = 80) (Nemorin et al. 2013). It was found that the polyploids of *D. alata* would have appeared through the formation of unreduced gametes. Triploids could be derived through the formation of 2n gametes in diploid females as the result of the non-viability of seeds resulting from the formation of 2n sperm and of the non-viability of inter-cytotype crosses. The tetraploids would have appeared through bilateral sexual polyploidization via the union of two unreduced gametes due to the sterility of triploids.

Selected References

Adamson I, Abigor R (1980) Transformation associated with catecholase in *Dioscorea alata* during storage. Phytochemistry 19(8):1593–1595

Aderiye BI, Ogundana SK, Adesanya SA, Roberts MF (1996) Antifungal properties of yam (*Dioscorea alata*) peel extract. Folia Microbiol (Praha) 41(5):407–412

Agwunobi LN (1999) *Dioscorea alata* (water yam) as a replacement for maize in diets for laying hens. Trop Anim Health Prod 31(6):391–396

Akissoe N, Mestres C, Hounhouigan J, Nago M (2005) Biochemical origin of browning during the processing of fresh Yam (*Dioscorea* spp.) into dried product. J Agric Food Chem 53(7):2552–2557

Anzaldo FE, Muranon J, Ancheta S (1956) Screening of Philippine plants for steroidal sapogenins. Philipp J Sci 85:305–314

Baah FD, Maziya-Dixon B, Asiedu R, Oduro I, Ellis WO (2009) Physicochemical and pasting characterisation of water yam (*Dioscorea* spp.) and relationship with eating quality of pounded yam. J Food Agric Environ 7:107–112

Bahado-Singh PS, Wheatley AO, Ahmad MH, Morrison EY, Asemota HN (2006) Food processing methods influence the glycaemic indices of some commonly eaten West Indian carbohydrate-rich foods. Br J Nutr 96(3):476–481

Brand Miller J, James KW, Maggiore P (1993) Tables of composition of Australian aboriginal foods. Aboriginal Studies Press, Canberra

Brown WH (1951–1957) Useful plants of the Philippines. Reprint of the 1941–1943 edition. 3 volumes. Technical Bulletin 10. Department of Agriculture and Natural Resources. Bureau of Printing, Manila, the Philippines. Vol. 1 (1951) 590 pp, Vol. 2 (1954) 513 pp, Vol. 3 (1957) 507 pp

Burkill IH (1951) Dioscoreaceae. In: van Steenis CGGJ (ed) Flora Malesiana; ser. I; Vol. 4. Noordhoff-Kolff, Djakarta, pp 293–335

Burkill IH (1966) A dictionary of the economic products of the Malay Peninsula. Revised reprint. 2 volumes. Ministry of Agriculture and Co-operatives, Kuala Lumpur, Malaysia. Vol. 1 (A-H) pp 1–1240, Vol. 2 (I-Z) pp 1241–2444

Chang SJ, Lee YC, Liu SY, Chang TW (2004) Chinese yam (*Dioscorea alata* cv. Tainung No. 2) feeding exhibited antioxidative effects in hyperhomocysteinemia rats. J Agric Food Chem 52(6):1720–1725

Chang WT, Chen HM, Yin SY, Chen YH, Wen CC, Wei WC, Lai P, Wang CH, Yang NS (2013) Specific *Dioscorea* phytoextracts enhance potency of TCL-loaded DC-based cancer vaccines. Evid Based Complement Alternat Med 2013:932040

Chen HL, Wang CH, Chang CT, Wang TC (2003) Effects of Taiwanese yam (*Dioscorea alata* L. cv. Tainung No. 2) on the mucosal hydrolase activities and lipid metabolism in Balb/c mice. Nutr Res 23(6):791–801

Chen PY, Tu YX, Wu CT, Jong TT, Chang CM (2004) Continuous hot pressurized solvent extraction of 1,1-diphenyl-2-picrylhydrazyl free radical scavenging compounds from Taiwan yams (*Dioscorea alata*). J Agric Food Chem 52(7):1945–1949

Chen HL, Hong LT, Kang Lee JK, Huang CJ (2009) The bone-protective effect of a Taiwanese yam (*Dioscorea alata* L. cv. Tainung No. 2) in ovariectomised female BALB/C mice. J Sci Food Agric 89:517–522

Cheng WY, Kuo YH, Huang CJ (2007) Isolation and identification of novel estrogenic compounds in yam tuber (*Dioscorea alata* Cv. Tainung No. 2). J Agric Food Chem 55(18):7350–7358

Dufie WMF, Oduro I, Ellis WO, Asiedu R, Maziya-Dixon B (2013) Potential health benefits of water yam (*Dioscorea alata*). Food Funct 4(10):1496–1501

Ezeocha VC, Ojimelukwe PC (2012) The impact of cooking on the proximate composition and anti-nutritional factors of water yam (*Dioscorea alata*). J Stored Prod Postharv Res 3(13):172–176

Faustina Dufie WM, Oduro I, Ellis WO, Asiedu R, Maziya-Dixon B (2013) Potential health benefits of water yam (*Dioscorea alata*). Food Funct 4(10):1496–1501

Foundation for Revitalisation of Local Health Traditions (2008) FRLHT Database. http://envis.frlht.org

Freitas RA, Paula RC, Feitosa JPA, Rocha S, Sierakowski MR (2004) Amylose contents, rheological properties and gelatinization kinetics of yam (*Dioscorea alata*) and cassava (*Manihot utilissima*) starches. Carbohydr Polym 55(1):3–8

Fu YC, Chen SH, Huang PY, Li YJ (2005) Application of bubble separation for quantitative analysis of choline in *Dioscorea* (yam) tubers. J Agric Food Chem 53(7):2392–2398

Fu SL, Hsu YH, Lee PY, Hou WC, Hung LC, Lin CH, Chen CM, Huang YJ (2006) Dioscorin isolated from *Dioscorea alata* activates TLR4-signaling pathways and induces cytokine expression in macrophages. Biochem Biophys Res Commun 339(1):137–144

Helen OT, Olusola AE, Bond AU (2013) *Dioscorea alata* L. reduces body weight by reducing food intake and fasting blood glucose level. Brit J Med Med Res 3(4):1871–1880

Hsu FL, Lin YH, Lee MH, Lin CL, Hou WC (2002) Both dioscorin, the tuber storage protein of yam (*Dioscorea alata* cv. Tainong No. 1), and its peptic hydrolysates exhibited angiotensin converting enzyme inhibitory activities. J Agric Food Chem 50(21):6109–6113

Hsu CC, Kuo HC, Chang SY, Wu TC, Huang KE (2011a) The assessment of efficacy of *Dioscorea alata* for menopausal symptom treatment in Taiwanese women. Climacteric 14(1):132–139

Hsu CK, Yeh JY, Wei JH (2011b) Protective effects of the crude extracts from yam (*Dioscorea alata*) peel on tert-butylhydroperoxide-induced oxidative stress in mouse liver cells. Food Chem 126(2):429–434

Hsu YJ, Weng CF, Lin KW, Lin KC (2013) Suppression of allergic reactions in ovalbumin-sensitized mice by yam storage proteins dioscorins. J Agric Food Chem 61(47):11460–11467

Huang CC, Lin MC, Wang CCR (2006) Changes in morphological, thermal and pasting properties of yam (*Dioscorea alata*) starch during growth. Carbohydr Polym 64(4):524–531

Huang CC, Chiang PY, Chen YY, Wang CC (2007) Chemical compositions and enzyme activity changes occurring in yam (*Dioscorea alata* L.) tubers during growth. LWT-Food Sci Technol 40(9):1498–1506

Huang CC, Chen YF, Wang CC (2010) Effects of micronization on the physico-chemical properties of peels of three root and tuber crops. J Sci Food Agric 90(5):759–763

Ireland CR, Passam HC (1984) The level and distribution of phenolic plant growth inhibitors in yam tubers during dormancy. New Phytol 97:233–242

Jayakody L, Hoover R, Liu Q, Donner E (2006) Studies on tuber starches. II. Molecular structure, composition and physicochemical properties of yam (*Dioscorea* sp.) starches grown in Sri Lanka. Carbohydr Polym 69(1):148–163

Khan R, Upaganlawar AB, Upasani C (2014) Protective effects of *Dioscorea alata* L. in aniline exposure-induced spleen toxicity in rats: a biochemical study. Toxicol Int 21(3):294–299

Kpodo KA, Plahar WA (1992) Effects of extrusion variables and starch modification on the puffing characteristics of yam (*Dioscorea alata*) flour. Food Control 3(4):200–204

Kumar S, Jena PK, Tripathy PK (2012) Study of wild edible plants among tribal groups of Simlipal Biosphere Reserve Forest, Odisha, India; with special reference to *Dioscorea* species. Int J Biol Technol 3(1):11–19

Lawal OS, Lechner MD, Kulicke WM (2008) Single and multi-step carboxymethylation of water yam (*Dioscorea alata*) starch: synthesis and characterization. Int J Biol Macromol 42(5):429–435

Lebot V, Malapa R, Molisale T, Marchand JL (2006) Physico-chemical characterisation of yam (*Dioscorea alata* L.) tubers from Vanuatu. Genet Resour Crop Evol 53(6):199–1208

Lee SC, Tsai CC, Chen JC, Lin JG, Lin CC, Hu ML, Lu S (2002) Effects of "Chinese yam" on hepato-nephrotoxicity of acetaminophen in rats. Acta Pharmacol Sin 23(6):503–508

Leung WTW, Butrum RR, Huang Chang F, Narayana Rao M, Polacchi W (1972) Food composition table for use in East Asia. FAO, Rome. 347 pp

Liao YH, Wang CH, Tseng CY, Chen HL, Lin LL, Chen W (2004) Compositional and conformational analysis

of yam proteins by near infrared fourier transform Raman spectroscopy. J Agric Food Chem 52(26):8190–8196

Lin PL, Lin KW, Weng CF, Lin KC (2009) Yam storage protein dioscorins from *Dioscorea alata* and *Dioscorea japonica* exhibit distinct immunomodulatory activities in mice. J Agric Food Chem 57(11):4606–4613

Liu YW, Shang HF, Wang CK, Hsu FL, Hou WC (2007) Immunomodulatory activity of dioscorin, the storage protein of yam (*Dioscorea alata* cv. Tainong No. 1) tuber. Food Chem Toxicol 45(11):2312–2318

Liu YH, Lin YS, Liu DZ, Han CH, Chen CT, Fan M, Hou WC (2009) Effects of different types of yam (*Dioscorea alata*) products on the blood pressure of spontaneously hypertensive rats. Biosci Biotechnol Biochem 73(6):1371–1376

Liu SF, Chang SY, Lee TC, Chuang LY, Guh JY, Hung CY, Hung TJ, Hung YJ, Chen PY, Hsieh PF, Yang YL (2012) *Dioscorea alata* attenuates renal interstitial cellular fibrosis by regulating Smad- and epithelial-mesenchymal transition signaling pathways. PLoS One 7(11):e47482

Lubag AJM Jr, Laurena AC, Mendoza EMT (2008) Antioxidants of purple and white greater yam (*Dioscorea alata* L.) varieties from the Philippines. Philipp J Sci 137(1):61–67

Maithili V, Dhanabal SP, Mahendran S, Vadivelan R (2011) Antidiabetic activity of ethanolic extract of tubers of *Dioscorea alata* in alloxan induced diabetic rats. Indian J Pharmacol 43(4):455–459

Mali S, Ferrero C, Redigonda V, Beleia AP, Grossmann MVE, Zaritzky NE (2003) Influence of pH and hydrocolloids addition on yam (*Dioscorea alata*) starch pastes stability. LWT-Food Sci Technol 36(5):475–481

Marcos J, Lacointe A, Tournebize R, Bonhomme R, Sierra J (2009) Water yam (*Dioscorea alata* L.) development as affected by photoperiod and temperature: experiment and modelling. Field Crops Res 111(3):262–268

Moorthy SN (1999) Effect of steam pressure treatment on the physicochemical properties of *Dioscorea* starches. J Agric Food Chem 47(4):1695–1699

Nemorin A, David J, Maledon E, Nudol E, Dalon J, Arnau G (2013) Microsatellite and flow cytometry analysis to help understand the origin of *Dioscorea alata* polyploids. Ann Bot 112(5):811–819

Ochse JJ, Bakhuizen van den Brink RC (1980) Vegetables of the Dutch Indies, 3rd edn. Ascher & Co., Amsterdam. 1016 pp

Okunlola A, Odeku OA (2011) Evaluation of starches obtained from four *Dioscorea* species as binding agent in chloroquine phosphate tablet formulations. Saudi Pharm J 19(2):95–105

Oluoha U, Ugochukwu EN (1995) Comparison of structure and kinetic mechanism of phosphorylase forms isolated from water yam (*Dioscorea alata*) and white yam (*Dioscorea rotundata*) tubers. Phytochemistry 40(3):677–684

Onwueme IC, Ganga ZN (1996) *Dioscorea alata* L. In: Flach M, Rumawas F (eds) Plant resources of South-East Asia 9. Plants yielding non-seed carbohydrates. Prosea Foundation, Bogor, pp 90–93

Otegbayo B, Aina J, Asiedu R, Bokanga M (2006) Pasting characteristics of fresh yams (*Dioscorea* spp.) as indicators of textural quality in a major food product – 'pounded yam'. Food Chem 99(4):663–669

Otegbayo B, Bokanga M, Asiedu R (2011) Physicochemical properties of yam starch: effect on textural quality of yam food product (pounded yam). J Food Agric Environ 9(1):145–150

Peng KY, Horng LY, Sung HC, Huang HC, Wu RT (2011a) Antiosteoporotic activity of *Dioscorea alata* L. cv. Phyto through driving mesenchymal stem cells differentiation for bone formation. Evid Based Complement Alternat Med 2011:712892

Peng KY, Horng LY, Sung HC, Huang HC, Wu RT (2011b) Hepatocyte growth factor has a role in the amelioration of diabetic vascular complications via autophagic clearance of advanced glycation end products: Dispo85E, an HGF inducer, as a potential botanical drug. Metabolism 60(6):888–892

Powell JM (1976) Ethnobotany. In: Paijmans K (ed) New Guinea vegetation. Elsevier, Amsterdam, pp 106–183. 212 pp

Ramdath DD, Isaacs RL, Teelucksingh S, Wolever T (2004) Glycaemic index of selected staples commonly eaten in the Caribbean and the effects of boiling v. crushing. Br J Nutr 91(6):971–977

Rodríguez-Sosa EJ, González MA, Martin FW (1972) Evaluation of ten varieties of yam (*Dioscorea* spp.) for production of instant flakes. J Agric Puerto Rico 56:235–243

Sakthidevi G, Mohan VR (2013) Total phenolic, flavonoid contents and in vitro antioxidant activity of *Dioscorea alata* L. tuber. J Pharm Sci Res 5(5):115–119

Shajeela PS, Mohan VR, Jesudas LL, Soris PT (2011) Nutritional and antinutritional evaluation of wild yam (*Dioscorea* spp.). Trop Subtrop Agroecosyt 14:723–730

Shittu TA, Olaitan OF (2014) Functional effects of dried okra powder on reconstituted dried yam flake and sensory properties of ojojo – a fried yam (*Dioscorea alata* L.) snack. J Food Sci Technol 51(2):359–364

Shoyama Y, Nishioka I, Herath W, Uemoto S, Fujieda K, Okubo H (1990) Two acylated anthocyanins from *Dioscorea alata*. Phytochemistry 29(9):2999–3001

South Pacific Commission (1990) Yam a status food. SPC South Pacific foods leaflet, No 14

Tan FJ, Liao FY, Jhan YJ, Liu DC (2007) Effect of replacing pork backfat with yams (*Dioscorea alata*) on quality characteristics of Chinese sausage. J Food Eng 79(3):858–863

The Plant List (2014) Version 1.1. Published on the internet. http://www.theplantlist.org/. Accessed 10 Aug 2014

Tulin EE, Ecleo ZT (2007) Cytokine-mimetic properties of some Philippine food and medicinal plants. J Med Food 10(2):290–299

Udensi EA, Oselebe HO, Iweala OO (2008) The Investigation of chemical composition and functional

properties of water yam (*Dioscorea alata*): effect of varietal differences. Pak J Nutr 7(2):342–344

Walter A, Lebot V (2007) Gardens of Oceania. ACIAR Monograph No. 122. 326 pp

Wanasundera JPD, Ravindran G (1992) Effects of cooking on the nutrient and antinutrient contents of yam tubers (*Dioscorea alata* and *Dioscorea esculenta*). Food Chem 45(4):247–250

Wanasundera JPD, Ravindran G (1994) Nutritional assessment of yam (*Dioscorea alata*) tubers. Plant Foods Hum Nutr 46(1):33–39

Wang TS, Liang SJ, Lii CK, Liu SY (2004) Protective effect of water yam (*Dioscorea alata* L.) extract on the copper-driven fenton reaction and X-ray induced DNA damage in vitro. Phytother Res 18(4):325–328

Wireko-Manu FD, Ellis WO, Oduro I, Asiedu R, Maziya-Dixon B (2011) Physicochemical and pasting characteristics of water yam (*D. alata*) in comparison with pona (*D. rotundata*) from Ghana. Eur J Food Res Rev 1(3):149–158

Wu WH, Liu LY, Chung CJ, Jou HJ, Wang TA (2005) Estrogenic effect of yam ingestion in healthy postmenopausal women. J Am Coll Nutr 24(4):235–243

Yahaya IA, Nok SSA, Achi SS, Bonire JJ (2012) Nutritional and antinutritional assessment of the peel of *Dioscorea alata* (water yam). Chem Class J 4:65–69

Yang DJ, Lu TJ, Hwang LS (2003) Simultaneous determination of furostanol and spirostanol glycosides in Taiwanese yam (*Dioscorea* spp.) cultivars by high performance liquid chromatography. J Food Drug Anal 11(4):271–276

Yeh YH, Lee YT, Hwang DF (2007) Yam (*Dioscorea alata*) inhibits hypertriglyceridemia and liver enlargement in rats with hypercholesterol diet. J Chin Med 18(1/2):65–74

Yeh AI, Chan TY, Chuang GCC (2009) Effect of water content and mucilage on physico-chemical characteristics of Yam (*Discorea alata* Purpurea) starch. J Food Eng 95(1):106–114

Yoshida K, Kondo T, Goto T (1991a) Unusually stable monoacylated anthocyanin from purple yam *Dioscorea alata*. Tetrahedron Lett 32(40):5579–5580

Yoshida K, Kondo T, Kameda K, Kawakishi S, Lubag AJM, Mendoza EMT, Goto T (1991b) Structures of alatanin A, B and C isolated from edible purple yam *Dioscorea alata*. Tetrahedron Lett 32(4):5575–5578

Dioscorea bulbifera

Scientific Name

Dioscorea bulbifera **L.**

Synonyms

Dioscorea anthropophagorum A.Chev., *Dioscorea bulbifera* var. *anthropophagorum* (A.Chev.) Summerh., *Dioscorea bulbifera* var. *crispata* (Roxb.) Prain, *Dioscorea bulbifera* var. *elongata* (F.M.Bailey) Prain & Burkill, *Dioscorea bulbifera* var. *pulchella* (Roxb.) Prain, *Dioscorea bulbifera* var. *sativa* Prain, *Dioscorea bulbifera* var. *suavia* Prain & Burkill, *Dioscorea bulbifera* var. *vera* Prain & Burkill, *Dioscorea crispata* Roxb., *Dioscorea heterophylla* Roxb., *Dioscorea hoffa* Cordem., *Dioscorea hofika* Jum. & H. Perrier, *Dioscorea korrorensis* R.Knuth, *Dioscorea latifolia* Benth., *Dioscorea longipetiolata* Baudon, *Dioscorea perrieri* R.Knuth, *Dioscorea pulchella* Roxb., *Dioscorea rogersii* Prain & Burkill, *Dioscorea sativa* f. *domestica* Makino, *Dioscorea sativa* var. *elongata* F.M.Bailey, *Dioscorea sativa* var. *rotunda* F.M.Bailey, *Dioscorea sylvestris* De Wild., *Dioscorea tamifolia* Salisb., *Dioscorea tenuiflora* Schltdl., *Dioscorea violacea* Baudon (illeg.), *Helmia bulbifera* (L.) Kunth, *Polynome bulbifera* (L.) Salisb.

Family

Dioscoreaceae

Common/English Names

Air Yam, Air Potato, Aerial Yam, Bubil Bitter Yam, Bearing Yam, Cheeky Yam, Malacca Yam, Otaheite Potato, Otaheite Yam, Potato Yam, Shoebutton Air Potato

Vernacular Names

Benin: Wérekou Dokoro (Bariba);
Burmese: Ah Lu Thi, Hpwut Sa Uu, Pat Sa Uu, Putsa U;
Central Africa: Nyitti;
Central African Republic: Motoko (Issongo); Koré (Sango);
Chinese: Huang Yao, Huang Yao Zi, Huang Du, Shan Ci Gu;
Colombia: Ńame De Aire;
Cuba: Ńame De Gunda, Ńame Volador;
Czech: Jam Bradavičnatý;
French: Igname Pousse En L'air, Igname Bulbifère, Igname Patate, Masako, Pomme De L'air, Pousse En L'Air;
Danish: Yams;

T.K. Lim, *Edible Medicinal and Non-Medicinal Plants: Volume 10, Modified Stems, Roots, Bulbs*,
DOI 10.1007/978-94-017-7276-1_10

Democratic Republic of Congo: Te'e (Balese, Efe)

German: Brotwurzel, Bulbenyams, Kartoffelyam, Kartoffel-Yam, Knollen-Yam, Luft-Kartoffel, Luft-Yams, Yamswurzel;

Gabon: Lèga (Apindji), Lisogo (Baduma, Bavili), Désomé (Bakèlè), Disogu (Balumbu, Bavarama), Disogu-di-duntsau (Bapumu, Bavungu, Eshira, Masangu), Ésogo (Bavové, Mitsogo), Lesogo (Banzabi), Ulèga (Benga), dirôga (Béséki, Ngowé), Abang, Apala, Apyala (Fang), Irôga (Galoa, Mpongwè, Myéné, Nkomi, Orungu), Lisoko (Loango), Léyiga (Mindumu);

Hawaiian: Hoi;

India: Ban Alu, Rat Alu, Roth Alu (Bengali), Bhirvolikanda, Gaithi, Genth, Genthi, Gethi, Karawakand, Karu Kunda, Karukanda, Khildri, Modi, Pitaalu, Ratalu, Suaralu, Zamin-Kand, Zaminkand (Hindi), Hansi Gedde, Heggenasu, Kaikaraande, Kaimoode Gedde, Kunta Genasu, Kuntagenasu (Kannada); Kattukaccil, Kattukachil, Katu-Katsjil, Katu-Kelengu, Katukatsjil (Malayalam), Dukarakanda, Dukarkkanda, Gathalu, Kadoo Karaandaa, Kadu Kand, Kadu-Kamdo, Karamdo, Konaphala, Konfa Goradu, Kukarkand, Mataru, Mibaelikand, Varaahi (Marathi), Pita Alu, Pita Kanda, Pita Lau (Oriya), Amrita, Badarakachha, Balya, Bilvamula, Brahmaputri, Brahmikanda, Charmakaraluka, Ghrishti, Grsti, Kandaka, Kanya, Kaumari, Krodakanya, Krodi, Kushthanashaka, Madhaveshtagrishtika, Magadhi, Mahaushadha, Mahavirya, Saukari, Shambarakanda, Shukari, Sukandaka, Trinetra, Vanavasi, Varahi, Varahikanda, Vishvaksenapriya, Vridhida, Vyadhihanta (Sanskrit), Cirakavalli, Kaayvalli, Kaattukkaayvalli, Kai Vallikkodi, Kaivalli Kodi, Karu-Karinda, Kattuccirakavalli, Kattuk Kayvalli, Kattukkilangu, Kattuvalli, Kayvalli, Kayvalli-Koti, Pannu Kilangu, Pannukilangu, Verrilai Valli (Tamil), Adavi Dumpa, Chedupaddudampa, Karu Kanda, Malaakaa-kaayapendalamu, Malakakayapendalam, Malakakayapendalamu, Malaka Kayependalamu (Telugu), Zaminekand (Urdu);

Indonesia: Jebubug Basu, Jebubug Endog, Gembolo, Uwi Gandul (Javanese), Kombulu (Madurese), Ubi Singapur (Malay), Huwi Blichik, Huwi Buwah, HUwi Gandul, Huwi Upas (Sundanese);

Italian: Igname;

Ivory Coast: Akaï (Ashanti);

Khmer: Damlong Duhs, Damlong Sdam Prei;

Japanese: Maruba-Dokoro, Kashû-Imo, Niga-Gashû;

Laotian: Manpauz, Houo I Mou, Hwai, Muz;

Malaysia: Ubi Atas, Ubi China, Ubi Kastela, Ubi Kemili Hutan, Memali Hutan, Hubi Kapor, Tolegn (in Semang);

Mexico: Papa Cimarrona;

Nepalese: Giitthaa, Gittha, Giitthe Tarul, Gitthe Tarul, Jada Bis, Kukur Tarul, Van Tarul;

Papua New Guinea: Poepoe Golagola (Tawala, Milne Bay), Puka (Kuanua, East New Britain Province), Kutu Kutu (Nupura, Eastern Highlands);

Persian: Zaminekanda;

Philippines: Dadakan (Bagio), Palugan (Bikol), Aribukbuk (Iloko), Utong-Utongan, Ubi-Ubihan, Bayag-Kabayo (Tagalog);

Popular Republic of Congo: Gambela (Beembe), Soko, Gamba (Laadi), Sola-nkiti (Laali), Vuba (Mbaamba), Ivuba Dumbala (Ndasa), Banga (Tiè), Makambi, Niengulingwe (Yoombe);

Portuguese: Inhame De Bolbos Aéreos, Inhame De Rama, Cará De Rama, Cará De Sapateiro;

Russian: Iams Lukovitsenosnyi;

Spanish: Ñame Del Aire, Ñame Congo, Ñame De Gunda, Papa De Aire, Papa Voladora;

Thailand: De Khwa, Lo Chae Mue, Man Nok, La Sa Mi, Man I Mo, Man Khamin, Wan Prachin, Hampao, Man-Soen, Wan Sam Phan Dtang, Wan Sam Phan Thueng;

Tibetan: Su Ka Ri;

Venezuela: Ńame Criollo;

Vietnamese: Củ Dại, Khoai Trời, Khoai Dái

Origin/Distribution

The species is indigenous to tropical Africa, Asia, the Pacific Islands and northern Australia. It is now naturalised throughout the West Indies and tropical America.

Agroecology

A tropical yam with high sensitivity to freezing temperatures that has limited its expansion into more temperate areas. In its native and naturalised range, it is found in mixed forest margins, river banks, valley sides; near sea level to 2300 m. It is an aggressive, herbaceous vine that can attain lengths of up to 20 m, scrambling up high into large trees and disperses itself easily via the bulbils.

Edible Plant Parts and Uses

Aerial bulbils, especially the large bulbils and starchy tubers, are eaten cooked. In Peninsular Malaysia, the tubers of wild varieties are eaten by the pagan tribes who grate the tubers and mix into a dough with ashes from burnt leaves (Burkill 1966). The Semang tribe made dough by rasping the tubers, kneading them with lime, wrapped in plantain leaf and roasted or buried in the ground for it to ferment. The dough is called 'kleb' when roasted and 'koyi' when fermented. In Odisha, India, the bulbils and tubers are used in curries (Kumar et al. 2012).

The absence of a viscosity peak and the high temperature stability of the ñame congo (*Dioscorea bulbifera*) flour make it an ideal ingredient for instant soup mixes (Rincón et al. 2000). *D. bulbifera* bulbils can be used to prepare cara-de-rama chips and French fries (Ferreira 1995).

Botany

A herbaceous, glabrous, perennial climber (Plate 1). Subterranean tubers, present or absent, usually solitary, globose, oblate, or lobed 10–20 cm long, 20–30 cm wide and 3–10 cm thick, reddish brown outside and white inside, corky outside with numerous fibrous roots. Axillary tubers (bulbils) sessile, at leaf axils, globose, reniform or oblate, greyish brown (Plate 2). Stem terete, twining to left, glabrous, smooth, without spines. Bulbils purplish brown with orbicular spots, globose or ovoid, variable in size, in axils of leaves. Leaves alternate, simple, broadly cordate, 8–26 × 2–26 cm, glabrous, margin entire or slightly undulate, apex caudate-acuminate, basal lobes rounded on long petioles 3–5.5 cm (Plate 3). Leaf veins (7–13) prominent radiate from single basal point. Male spikes usually clustered in leaf axils or along leafless, axillary shoots, pendent, sometimes branched. Male flowers yellowish-green, solitary with ovate bract and bracteole; perianth purple, lobes lanceolate; stamens 6, inserted at base of perianth, filaments nearly as long as anthers. Female spikes often two or more together, similar to male ones, 20–30 cm long. Female flowers with six staminodes. Capsule reflexed or drooping, straw-coloured, densely purplish dotted, oblong-globose, 1.5–3 cm, glabrous, base and apex rounded; wings 0.25–0.7 cm wide. Seeds oblong, 1.2–1.6 by 0.5 cm, dark brown with orbicular wings.

Nutritive/Medicinal Properties

Tuberous Rhizome Nutrients/ Phytochemicals

Proximate nutrient composition of *D. bulbifera* var. *vera* tubers (g/100 g) was reported by Shajeela et al. (2011) as: moisture 86.70 %, crude protein 7.28 g, crude lipid 6.14 g, crude fibre 3.48 g, ash 3.31 g, starch 38.10 g, NFE (nitrogen free extract) 79.79 g, energy 1685.55 kJ/100 g DM, niacin 33.74 mg, ascorbic acid 91.05 mg, and minerals mg/100 g Na 78.24 mg, K 1554.36 mg, Ca 338.15 mg, Mg 396.20 mg, P 154.42 mg, Zn 1.48 mg, Mn 9.40 mg, Fe 19.20 mg and Cu 2.14 mg. The anti-nutritional factors were per 100 g: total free phenolics 2.20 g, tannins 1.48 g, hydrogen cyanide 0.19 mg, total oxalate 0.78 g, amylase inhibitor 1.36 AIU/mg soluble

Plate 1 Herbaceous perennial twining vine with aerial axillary tubers (bulbils)

Plate 2 Axillary tubers—oblate, reniform

starch, and trypsin inhibitor 1.21 TIU/mg protein (Shajeela et al. 2011). The in-vitro protein digestibility was 4.61 % and the in-vitro starch digestibility was 56.84 %.

Alexis and Georges (2012) reported the following nutrient composition of *D. bulbifera* tubers as: moisture 69.80 % FW and the following per dry weight basis protein 6.29 %, lipid 3.53 %, soluble carbohydrate 2.29 %, starch 73.01 %, total carbohydrates 86.87 %, cellulose 3.71 %, ash 3.31 %, energy 383 cal/100 g; and anti-nutrient factors (mg/100 g DW) as oxalic acid 6.83 mg, tannins 421 mg, hydrocyanic acid 10.03^{-2} mg, alkaloid 248.23 mg and sapogenins 0.08 % (Among the different *Dioscorea* species found in Orissa, *D. bulbifera* was found to have the highest content of diosgenin, 1383 mg/100 g tuber and highest vitamin C content of 8.217 mg/100 g (Behera et al. 2010)).

Plate 3 Broadly cordate—rotund leaves

Dioscorea bulbifera bulb afforded a yield of 28.48 g/100 g DW starch, which contained 29.37 % amylase (Araujo de Vizcarrondo et al. 2004). Most of the starch granules were of irregular shape, similar to a pyramid with rounded vertices, and a smaller number were elongated with smooth surface. The starch showed a gelatinization temperature of 70.8 °C and maximum viscosity at 88.6 °C of 435 Brabender units (BU). It presented a relatively stable consistency to the cooking process and a low tendency to retrogradation, suggesting its potential use in food products that need a fast viscosity and a gel with a stable consistency.

Early phytochemical investigations on the root tubers of Japanese *D. bulbifera* had revealed no steroidal saponins but instead furanoid norditerpenoids or glycosides (Komori 1997; Ida et al. 1978a, b). Five furanoid norditerpene, designated as Diosbulbins-D, -E, -F, -G and -H were isolated from fresh tubers of *Dioscorea bulbifera* (Ida et al. 1978b). The structure of diosbulbin-G was resolved by crystal structure analysis (Ida et al. 1978a). Two furanoid norditerpene glucosides named diosbulbinosides D and F were isolated from the fresh root tubers *of Dioscorea bulbifera* forma *spontanea* (Ida et al. 1978c). They were assigned the structures, 6-*O*-β-D-glucopyranosides of the enol forms, 5-en-6-ols, of coexisting diosbulbins D and F, respectively. From *D. bulbifera* var. *sativa* tubers, a 18-nor-clerodane diterpenoid, 8-epidiosbulbin E acetate and the known norditerpenoid diosbulbin D were isolated (Murray et al. 1984). Six steroids were isolated from *D. bulbifera* tubers and identified as diosgenin, β-sitosterol, stigmasterol, daucosterol, diosgenin-3-*O*-α-L-rhamnopyranosyl (1 → 2)-β-D-glucopyranoside (prosapogenin A of dioscin) and diosgenin-3-*O*-[di-α-L-rhamnopyranosyl (1 → 2,1 → 3)]-β-D-glucopyranoside (taccaoside) (Li et al. 1998). A novel plasmid-curing compound identified as 8-epidiosbulbin E acetate (norditerpene) was isolated from an aqueous methanolic extract of *D. bulbifera* bulb (Shriram et al. 2008). Two clerodane diterpenoids, bafoudiosbulbins A 1, and B 2, together with five known compounds: tetracosanoic acid; 1-(tetracosanoyl)-glycerol; *trans*-tetracosanylferulate; β-sitosterol and 3-*O*-β-D-glucopyranosyl-β-sitosterol (Teponno et al. 2006b), and a pennogenin glycoside spiroconazole A (Teponno et al. 2006a) were isolated from the of *Dioscorea bulbifera* var *sativa* tubers. Two new furanoid norditerpenes named diosbulbins I and J were isolated from the tubers (Wang et al. 2009b). Fourteen compounds were isolated from *D. bulbifera* tubers and identified as stigmasterol; mono-arachidin; 1,7-*bis*-

(4-hydroxyphenyl)-1E,4E,6E-heptatrien-3-one; behenic acid; demethyl batatasin IV; 2,3′-di-hydroxy-4′,5′-dimethoxybibenzyl; diosbulbin B; diosbulbin D; docosyl ferulate; 7-*bis*-(4-hydroxyphenyl)-4E, 6E-heptadien-3-one; 5,3,4-trihydroxy-3,7-dimethoxyflavone; tristin; protocatechuic acid and adenosine (Wang et al. 2009a). Four steroidal sapogenins, named diosbulbisins A-D, two spirostane glycosides, diosbulbisides A and B, one cholestane glycoside, diosbulbiside C and known compounds 8–10 pennogenin; pennogenin-3-*O*-α-L--rhamnopyranosyl-(1 → 3)-[α-L-rhamnopyranosyl-(1 → 2)]-β-D-glucopyranoside; and pennogenin-3-*O*-α-L-rhamnopyranosyl-(1 → 4)-[α-L-rhamnopyranosyl- (1 → 2)]-β-D-glucopyranoside were isolated from *D. bulbifera* rhizomes (Liu et al. 2009). Three norclerodane diterpenoids, diosbulbins K, L, M, and one analogous enolglycoside, diosbulbinoside G, together with four norclerodane diterpenoids, diosbulbins B, E, F and G, were isolated from *Dioscorea bulbifera* rhizomes (Liu et al. 2010). One bibenzyl, and one new diarylheptanone, diobulbinone A together with 16 known compounds were isolated from *D. bulbifera* rhizomes (Liu et al. 2011b). Two new steroidal saponins, diosbulbisides D and E, along with five known saponins including two 3-*O*-trisaccharides of diosgenin spirostanes were isolated from *D. bulbifera* rhizomes (Liu et al. 2011a).

D. bulbifera tuber extract was found to be rich in flavonoid, phenolics, reducing sugars, starch, diosgenin, ascorbic acid and citric acid (Ghosh et al. 2012a). Two flavonoids 3,7-dimethoxy-5,4′-dihydroxyflavone; 3,7-dimethoxy-5,3′,4′-trihydroxyflavone and an anthraquinone, emodin was isolated from *D. bulbifera* (Li et al. 2000). Seven flavonoids kaempferol-3,5-dimethyl ether (1), caryatin (2), (+)-catechin (3), myricetin (4), quercetin-3-O-galactopyranoside (5), myricetin-3-O-galactopyranoside (6), myricetin-3-O-glucopyranoside (7) and diosbulbin B were isolated from the rhizomes (Gao et al. 2002). Twenty-eight compounds were isolated from the ethyl acetate and n-butanol soluble fractions of *D. bulbifera* rhizome and identified as diosbulbin-B; daucosterol; β-sitosterol; palmatic acid; succinic acid; shikimic acid; 3, 5-dimethoxykaempferol; 3, 5, 3′-trimethoxyquercetin; caryatin; (+)catechin; myricetin; myricetin-3-*O*-β-D-galactopyranoside (12), myricetin-3-*O*-β-D-glucopyranoside; hyperoside; kaempferol; kaempferol-3-*O*-β-D-galactopyranoside; kaempferol-3-*O*-β-D--glucopyranoside; protocatechuic acid; vanillic acid; isovanillic acid; (+)epicatechin; methyl-*O*-α-D-fructofuranoside; butyl-*O*-α-D-fructofuranoside; ethyl-*O*-β-D-fructopyranoside; butyl-*O*-β-D-fructopyranoside; 3-phenyl-2-propenyl-*O*-β-D-glucopyranoside; 2-(4-methoxyphenyl)-ethyl-*O*-β-D-glucopyranoside; phenyl-methyl-*O*-β-D--glucopyranoside (Gao et al. 2007). Three new apianen lactones 3α-hydroxy-13β-furan-11--ketoapian-9-en-(20,6)-olide; 13β-furan-11-ketoapian-3(4),8-dien-(20,6)-olide; and 7α-methoxy-13β-furan-11-keto-apian-3(4),8-dien-(20,6)-olide were isolated from the root together with three known compounds: 3,7-dimethoxy-5,3′,4′-trihydroxy-flavone; morin-7-*O*-β-D-glucoside; 5-hydroxy-7-methoxy-6-*O*-[α-L-rhamnopyranosyl(1 → 2)-β-D-fucopyranosyl]flavone (Zheng et al. 2003)

In case of petroleum ether extract of *D. bulbifera* bulbs, among identified compounds, ethyl ester of undecanoic acid; Z-1,9-dodecadiene and n-hexadecanoic acid were found to be predominant (Ghosh et al. 2013) Other compounds present were 3-oxo-androsta-1,4-dien-17α,spiro-2′3′-oxetane; apiol; 1,2-benzenedicarboxylic, butyl 8-methylnonylester; pentadecanoic acid, 14-methyl-, methyl ester; 9,12-octadecadienoic acid, methyl ester, (E,E)-; 9-octadecenoic acid, ethyl ester; methyl 17-methyl-octadecanoate; 2,4-hexadienedioic acid,3,4-diethyl-,dimethylester, (Z,Z)-; benzene, pentafluorol[(2-methylphenoxy)methyl]-; 3-methoxyestra-1,3,5(10),8,14-pentaen-17(+,–) -; and octadecanoic acid, ethyl ester. Notable amount 94.05 % of diosgenin was confirmed in ethyl acetate extract of *D. bulbifera* bulbs. Apart from diosgenin, ethyl ester of eicosanoic acid was also found. Diosgenin was prevalent as major phytochemical even in 70 % (v/v) ethanolic

extract of *D. bulbifera.* Also present were 1-(2-aminoethylamino)-2-propanol; ethyl acetate, diosgenin (3α, 25R) acetate; and squalene. Major compound present in the methanol extract were: 2-pyrrolidinone,1-methyl-; 2-pentaol,acetate; butylated hydroxytoluene; 9H-fluorene, 9-methylene-;5-(methylamino)-1,2,3,4-thiatraizole; acetic acid, [(1,1,-dimethylethyl)thio]-; 2-(1-methylcyclopentyloxy)-tetrahydropyran; pentanoicacid,1,1,-dimethylpropyl ester; decane, 2,4,6-trimethyl-; undecane; heptylcyclohexane.

Yellow pigment from *Dioscorea bulbifera* tubers were found to be xanthophylls identified as lutein, neaxanthin, violaxanthin, zeaxanthin, auroxanthin and cryptoxanthin, all present in small amounts (Martin et al. 1974)., Other pigments present included chlorophyll, an anthocyanin and an unidentified phenolic compound. An antifungal alkaloid dihydrodioscorine was isolated from the tubers (Adeleye and Ikotun 1989).

Four wild Nepalese yams, *D. bulbifera*, *D. versicolor, D. deltoidea* and *D. triphylla* were found to contain bitter principles (Bhandari and Kawabata 2005). The bitter components were identified as furanoid norditerpenes diosbulbins A and B, found in the range of 0.023–0.046 and 0.151–0.442 g/kg. They found that diosbulbin B (0.314 g/kg) was the principal bitter compound as compared to diosbulbin A (0.037 g/kg). The toxic alkaloid, dioscorine and histamine (an allergen) were not detected in these tubers, whereas cyanogens (as HCN equivalent) content were found ranging from 3.2 to 6.0 ppm satisfactorily below the safety limits. Domestic cooking methods were found to be very efficient in removing bitterness, thus making the bitter yams palatable.

Bulbil Nutrients/Phytochemicals

Nutrient composition of *D. bulbifera* bulbils was reported as: moisture 81.44 % FW and the following per dry weight basis protein 8.82 %, lipid 3.99 %, soluble carbohydrate 3.59 %, starch 57.15 %, total carbohydrates 82.44 %, cellulose 7.67 %, ash 4.74 %, energy 381 cal/100 g; and anti-nutrient factors (mg/100 g DW) as oxalic acid 9.33 mg, tannins 470.03 mg, hydrocyanic acid 1.973^{-2} mg, alkaloid 165.63 mg and sapogenins 0.22 % (Alexis and Georges 2012).

An antifungal alkaloid dihydrodioscorine was isolated from the bulbils (Adeleye and Ikotun 1989). The ethyl acetate extract of the bulbil infected with *Botryodiplodia theobromae* gave demethylbatatasin IV as the major phytoalexin (Adesanya et al. 1989). From *Dioscorea bulbifera* L. var. *sativa*, bulbils three new clerodane diterpenoids, bafoudiosbulbin C (= methyl (2β,8α,12*S*)-17-oxo-2,19:8,19:12,17:15,16-tetraepoxycleroda-3,13(16),14-triene-18-carboxylate; bafoudiosbulbin D (= methyl (2β,6β,12*R*)-17,19-dioxo-2,19:6,17:8,12:15,16-tetraepoxycleroda-13(16),14-diene-18-carboxylate; and bafoudiosbulbin E (= methyl (2β,3α,4α,6β,12*R*)-17,19-dioxo-2,19:3,4:6,17:8,12:15,16-pentaepoxycleroda-13(16),14-diene-18-carboxylate; were isolated, together with the known compounds bafoudiosbulbins A and B, 3-*O*-β-D-glucopyranosyl-β-sitosterol, and 6′-stearoyl-3-*O*-β-D-glucopyranosyl-β-sitosterol (Teponno et al. 2007); two clerodane diterpenoids, Bafoudiosbulbins F (1) and G (2), together with five known compounds: bafoudiosbulbins A–C; 3,5,40-trihydroxy-30-methoxybibenzyl, and kaempferol were isolated (Teponno et al. 2008).

A clerodane diterpenoid was isolated from the acetone extract of *D. bulbifera* bulbils and its structure established as 15,16-epoxy-6α-*O*-acetyl-8β-hydroxy-9-nor-clero-13(16),14-diene-17,12;18,2-diolide (Kidyu et al. 2011). Six compounds were isolated from the bulbils of *D. bulbifera*, namely, bafoudiosbulbins A, B, C, F, G and 2,7-dihydroxy-4-methoxyphenanthrene (Kuete et al. 2012).

Flower Phytochemicals

Eleven steroidal saponins, dioscoreanosides A-K, along with five known congeners, pennogenin 3-*O*-α-L-rhamnopyranosyl-(1→4)-α-L-rhamnopyranosyl-(1 → 4)-[α-L-rhamnopyranosyl-(1→2)]-β-D-glucopyranoside; 26-*O*-ß-D-glucopyranosyl-(25R)-5-en-furost-3ß,17α,22α,26-tetraol-3-*O*-α-L-rhamnopyranosyl-(1 → 4)-α-L-rhamnopyranosyl-(1→4)-[α-L-rhamnopyranosyl-(1→2)]-β-D-glucopyranoside; 23β,27-dihydroxy-pennogenin 3-*O*-α-L-rhamnopyranosyl-(1 → 4)-α-L-rhamnopyranosyl-(1 → 4)-[α-L-rhamnopyranosyl-(1→2)]-β-D-glucopyranoside; spiroconazole A; and floribundasaponin B (Tapondjou et al. 2013).

Other Plant Parts Phytochemicals

From the leaves and stems, diosbulbin B was isolated (Yonemitsu et al. 1993). The highest concentration of diosgenin, 2.94 and 2.14 % dry weight, was obtained in immobilised cell culture and cell suspension culture of *D. bulbifera* treated with 100 mg/L cholesterol, respectively (Jyothishwaran and Seetharam 2008). Diosgenin yield in *D. bulbifera* plantlets micropropagated through nodal segments and bulbils, reached a maximum after 20 weeks (Narula et al. 2003). The presence of Cu in the culture medium appeared to stimulate diosgenin production in regenerants of *D. bulbifera* (Narula et al. 2005). The regenerants also differentiated bulbils on lower concentrations of Cu. At $CuSO_4$ (100 μM); however, cultures showed poor growth as well as a low diosgenin yield. Increased proline and protein contents were recorded in cultures grown on Cu-enriched media. Genetic fidelity of in-vitro regenerants of *D. bulbifera* with high diosgenin content was achieved by encapsulation of shoot tips in 3 % (w/v) calcium alginate for storage and germplasm exchange (Narula et al. 2007). An increase in sucrose concentration from 2 % (w/v) to 8 % (w/v) and/or a decrease in cytokinin concentration in the basal culture medium induced tuberisation of stem nodes in *D. bulbifera* explants under non-inductive day lengths (Forsyth and van Staden 1984).

Antioxidant Activity

Dioscorea bulbifera, Eriobotrya japonica, Tussilago farfara and *Ephedra sinica* had the highest antioxidant capacities among the 56 Chinese plants based on a combinative consideration of the results obtained by Ferric reducing antioxidant power (FRAP) and Trolox equivalent antioxidant capacity (TEAC) assays (Song et al. 2010). All four could be potential rich sources of natural antioxidants. *D. bulbifera* extracts gave TEAC value of 708.73 μmol Trolox/g, FRAP value of 856.92 μmol Fe^{2+}/g and phenolic content of 59.43 mg GAE/g. The ethanolic extracts *Dioscorea bulbifera* tuber were to have good enzymatic and non-enzymatic antioxidant potential (Suriyavathana and Indupriya 2011). The level of enzymatic antioxidants, namely, glutathione peroxidase (GPx), catalase (CAT), superoxide dismutase (SOD) were high, while glucose-6-phosphate dehydrogenase (G6PD) and glucose-s-transferase (GST) were low in the ethanolic extracts. *Dioscorea bulbifera* also contained notable levels of vitamin C, vitamin E and reduced glutathione (GSH).

Chen et al. (2013) found that 70 % ethanol extract of *D. bulbifera* rhizome could scavenge DPPH radical at 2 mg/ml; 80 % ethanol extract could scavenge hydroxyl radicals (55.2 %) and also possessed the strongest reducing ability per gram of the extract equal to 49.3 μmol ferrous. The bibenzyl compound isolated from the rhizome showed high antioxidant capacity in FRAP assay and DPPH radical-scavenging activity (Liu et al. 2011b). The ethanolic plant extract of *D. bulbifera* showed high antioxidant activity (EC_{50} = 11.07 μg/ml) (Chunthorng-Orn et al. 2012). It had high total phenolic content of106.26 mg/g. The total phenolic content of the extract correlated with DPPH radical scavenging activity.

Ethyl acetate extract of *D. bulbifera* bulbs exhibited excellent scavenging of pulse radiolysis generated ABTS(•+) radical with a second-order rate constant of 1.72×10^6, while ascorbic acid failed to show any activity of the pulse radiolysis generated OH radical with a second-order rate constant of 4.46×10^6 compared with ascorbic acid (1.34×10^6) (Ghosh et al. 2013). In the hydroxyl radical scavenging activity, the ranking order of the extracts was methanol (76.11 % > ethyl acetate 66.67 % > ethanol 64.23 % > petroleum ether 44.51 %. In the DPPH radical scavenging, the ranking order of the extracts was methanol 84.94 % > ethyl acetate 82.79 % > ethanol 80.64 % > petroleum ether 61.82 %. In the superoxide anion scavenging activity, the ranking order of the extracts was methanol 59.75 % > ethyl acetate 57.60 % > ethanol 54.76 % > petroleum ether 26.88 %. In the superoxide radical scavenging activity, the ranking order of the extracts was methanol 59.65 % > ethyl acetate 59.24 % > ethanol 57.34 % > petroleum ether 28.30 %. In the nitric oxide scavenging activity, the ranking order of the extracts was methanol 57.59 % > ethyl acetate 54.55 % > ethanol 48.85 % > petroleum ether 20.57 %. In the ferric reducing antioxidant power activity, the order was ethyl acetate 123.03 GAEAC (gallic acid equivalent antioxidant capacity), methanol 97.88 GAEAC followed by ethanol and petroleum ether (40 GAEAC). Total phenolic and flavonoid contents were, respectively, in the ethyl extract, 98 μg/ml and 27.86 μg/ml; in the petroleum ether extract, 49.22 μg/ml and 4.95 μg/ml; in the methanol extract, 145.44 μg/ml and 12.76 μg/ml; in the ethanol extract, 85.89 μg/ml and 12.10 μg/ml.

Antitumour Activity

Seven flavonoids kaempferol-3,5-dimethyl ether (1), caryatin (2), (+)-catechin (3), myricetin (4), quercetin-3-O-galactopyranoside (5), myricetin-3-O-galactopyranoside (6), myricetin-3-O-glucopyranoside (7) isolated from the rhizome, exhibited potent inhibitory effect against the tumour promotion of JB6 (Cl 22 and Cl 41) cells induced by a promoter, 12-O-tetradecanoylphorbol-13-acetate (TPA) (Gao et al. 2002). Compared with (−)-epicatechin, (+)-catechin exhibited much stronger inhibitory activity, which suggested that chemical stereo structures of compounds affect the efficiency of inhibition. Anticancer active compounds were mainly extracted by petroleum ether fraction from hydrophobic constituents of *Dioscorea bulbifera* (Yu et al. 2004). Of all the fractions, the petroleum ether fraction was the most inhibitory and water extract the weakest. Life span of mice bearing HepA ascites was prolonged after exposed to 100 mg/k petroleum ether fraction and shortened significantly after exposed to water fraction. *Dioscorea bulbifera* was found to contain the biggest polysaccharide, 16. 509 % which exhibited antitumour activity (Zhang et al. 2007). The 75 % ethanol extract of *D. bulbifera* rhizome and its fractions showed different inhibitory activities against tumour promotion of mouse epidermal JB6 (Cl 22 and Cl 41) cells induced by the promoter of 12-*O*-tetradecanoylphorbol-13-acetate (TPA) (Gao et al. 2007). The ethyl acetate and n-butanol soluble fractions were found to be potent antitumour promoters. They almost completely inhibited the soft agar colony induction in JB6 cells at 30 μg/mL 100 % and 92.6 % and their 50 % inhibitory concentrations (IC_{50}) were 2.6 μg/mL and 4.4 μg/mL, respectively. Twenty-eight compounds were purified in the ethyl acetate and n-butanol fractions, amongst which the following exhibited inhibitory activities against tumour promotion of JB6 (Cl 22 and Cl 41) cell lines induced by TPA: 3, 5-dimethoxykaempferol at 10 μg/ml with an IC_{50} of 0.64 μg/ml; caryatin at 10 μg/ml with IC_{50} of 3 μg/ml, catechin at 30 μg/ml with IC_{50} 13.1 μg/ml, myricetin 10 μg/ml with IC_{50} 3.7 μg/ml, myricetin-3-O-galactoside 30 μg/ml with IC_{50} 9.2 μg/ml. Antitumour-promoting activities based on the inhibition of soft agar colony induction of TPA in the JB-6 Cl 22 C141 cell lines were 99.8 % for 3, 5-dimethoxykaempferol, 92.8 % for caryatin, 98.3 % for catechin, 93.8 % for myricetin, and 79.2 % for myricetin-3-*O*-galactoside. Diosbulbin B showed moderate activity. Of 10 compounds isolated

from the rhizomes, only pennogenin-3-O-α-L--rhamnopyranosyl-(1 → 3)-[α-L-rhamnopyranosyl-(1 → 2)]-β-D-glucopyranoside; andpennogenin-3-O-α-L-rhamnopyranosyl-(1 → 4)-[α-L-rhamnopyranosyl- (1 → 2)]-β-D-glucopyranoside at a concentration of 10 μM showed significant cytotoxic activity with 99.1 % and 92.6 % inhibition of human hepatocellular carcinoma cell line Bel-7402, respectively (Liu et al. 2009). Both compounds exhibited cell growth inhibition activity towards SMMC7721 human hepatocellular carcinoma cells, and with IC_{50} values of 4.54 μM and 4.85 μM, respectively. Kim et al. (2002) reported that boiled extracts of *D. bulbifera* killed stomach cancer AGS cells in a dose-dependent manner but not significantly.

Two 3-*O*-trisaccharides of diosgenin spirostanes, isolated from *D. bulbifera* rhizome, showed moderate cytotoxic activity against human hepatocellular carcinoma cells, with IC_{50} values of 3.89 μM and 7.47 μM on SMMC7721, and 10.87 μM and 19.10 μM on Bel-7402 cell lines, respectively (Liu et al. 2011a). The ethanol and ethyl acetate extracts of *Dioscorea bulbifera* decreased tumour weight in S180 and H22 tumour cells bearing mice, while the water and non-ethyl acetate extracts had no such effect (Wang et al. 2012). In addition, the ethyl acetate extract altered the weight of spleen and thymus, and the amounts of total leukocytes, lymphocytes and neutrophils in tumour-bearing mice. Also, diosbulbin B demonstrated antitumour effects in the dose-dependent manner at the dosage of 2–16 mg/kg without significant toxicity in- vivo. The results suggested *D. bulbifera* had potential antitumour effects, which may be related to influencing the immune system, and the compound diosbulbin B to be the major antitumour component of *D. bulbifera*. Various concentrations of ethanol extracts of *Dioscorea bulbifera* rhizome were found to inhibit the proliferation of SGC-7901 cancer line in a dose-dependent manner (Chen et al. 2013).

Administration of Antitumor B, a Chinese herbal mixture of six plants (*Sophora tonkinensis, Polygonum bistorta, Prunella vulgaris, Sonchus arvensis* L., *Dictamnus dasycarpus*, and *Dioscorea bulbifera*) inhibited 4-nitroquinoline-1-oxide (4NQO) induced oral squamous cell carcinoma development in A/J mice by 59.19 % (Wang et al. 2013).

Antimicrobial Activity

Two clerodane diterpenoids, bafoudiosbulbins A 1, and B 2 isolated from the tubers showed significant antibacterial activities in-vitro against *Pseudomonas aeruginosa*, *Salmonella typhi, Salmonella paratyphi A* and *Salmonella paratyphi B* (Teponno et al. 2006b). The ethanolic plant extract of *D. bulbifera* inhibited growth of *Escherichia coli* (MIC = 5 mg/ml) and Gram-positive bacteria such as *Streptococcus aureus* and *Bacillus subtilis* with the same MIC values (2.5 mg/ml) (Chunthorng-Orn et al. 2012). The crude methanol extract, fractions DBB1 and DBB2 as well as compounds bafoudiosbulbins B to F isolated from the bulbils, were able to prevent the growth in-vitro of all the fifteen microorganisms, namely, three strains of *Escherichia coli*, three strains of *Enterobacter aerogenes*, three strains of *Klebsiella pneumoniae*, two strains of *Pseudomonas aeruginosa*, one strain of *Mycobacterium smegmatis* and two strains of *M. tuberculosis*, within the concentration range of 8–256 μg/mL (Kuete et al. 2012). The lowest MIC value for the methanol extract and fractions (16 μg/mL) was obtained with DBB1 and DBB2 on *E. coli* AG100A and DBB2 on *Mycobacterium tuberculosis* MTCS2. The lowest value for individual compounds (8 μg/mL) was recorded with bafoudiosbulbin C on *M. smegmatis* and M. tuberculosis ATCC and MTCS2 strains, respectively. The activity of the samples on many multidrug-resistant (MDR) bacteria such as *Enterobacter aerogenes* EA289, CM64, *Klebsiella pneumoniae* KP63 and *Pseudomonas aeruginosa* PA124 was better than that of chloramphenicol. When tested in the presence of the efflux pump inhibitor against MDR Gram-negative bacteria, the activity of most of the samples increased. MBC values not greater than 512 μg/mL were recorded on all studied microor-

ganisms with fraction DBB2 and bafoudiosbulbins B to F.

A novel plasmid-curing compound identified as 8-epidiosbulbin E acetate (norditerpene) was isolated from an aqueous methanolic extract of *Dioscorea bulbifera* L. bulbs (Shriram et al. 2008) on the basis of modern spectroscopic analysis and X-ray crystallography. A norditerpene, 8-epidiosbulbin E acetate (EEA), isolated from *D. bulbifera* bulb, exhibited broad-spectrum plasmid-curing activity against multidrug-resistant (MDR) bacteria, including vancomycin-resistant enterococci (Shriram et al. 2008). EEA cured antibiotic resistance plasmids (R-plasmids) from clinical isolates of *Enterococcus faecalis, Escherichia coli, Shigella sonnei* and *Pseudomonas aeruginosa* with 12–48 % curing efficiency. The reference plasmids of *Bacillus subtilis* (pUB110), *E. coli* (RP4), *P. aeruginosa* (RIP64) and *Salmonella typhi* (R136) were cured with efficiency ranging from 16 to 64 %. EEA-mediated R-plasmid curing decreased the minimal inhibitory concentration of antibiotics against MDR bacteria, thus making antibiotic treatment more effective.

Silver nanoparticles synthesised by the reduction of aqueous Ag^+ ions using *D. bulbifera* tuber extract were found to possess potent antibacterial activity against both Gram-negative and Gram-positive bacteria (Ghosh et al. 2012b). Beta-lactam (piperacillin) and macrolide (erythromycin) antibiotics showed a 3.6-fold and 3-fold increase, respectively, in combination with silver nanoparticles selectively against multidrug-resistant *Acinetobacter baumannii*. Marked synergy was seen between silver nanoparticles and chloramphenicol or vancomycin against *Pseudomonas aeruginosa*, and was supported by a 4.9-fold and 4.2-fold increase in zone diameter, respectively. Similarly, a maximum 11.8-fold increase in zone diameter of streptomycin when combined with silver nanoparticles against *E. coli* was obtained, providing strong evidence for the synergistic action of a combination of antibiotics and silver nanoparticles.

Antidiabetic Activity

The aqueous extract of *Dioscorea bulbifera* tubers at 250, 500 and 1000 mg/kg doses administered for 7 weeks to streptozotocin-treated rats produced significant reduction in blood glucose level and increase in body weight (Ahmed et al. 2009).

Dioscorea bulbifera and *Gnidia glauca* extracts displayed significant in-vitro inhibition with porcine pancreatic amylase and crude murine glucosidases as well as pure α-glucosidase (Ghosh et al. 2012a). This may have beneficial effects in managing type II diabetes mellitus and could be used as effective herbal formulation in combinational therapy. Petroleum ether, ethyl acetate and methanol extracts of *D. bulbifera* bulb (tuber) inhibited porcine α-amylase activity in-vitro by 61.5 %, 73.39 %, 73.54 %, respectively, but the 70 % ethanol extract showed lower activity. Petroleum ether (22.3 %), ethyl acetate (23.59 %) and methanol 26.1 %) extracts of *D. bulbifera* showed higher inhibition of murine pancreatic glucosidase activity than extracts of G. glauca stem, leaf and flower (13.12–19.32 %). Petroleum ether extract of *D. bulbifera* showed a maximum inhibition of 74.36 % (IC_{50} 33.61 μg/mL) of murine small intestinal glucosidase activity that was found to be more potent compared to acarbose IC_{50} 48.79 μg/mL). Ethyl acetate extract of *D. bulbifera* bulb was found to be having an inhibition percentage of 51.41 %, which was significant as compared to both methanol (50.24 %) and 70 % ethanol extracts (43.54 %). The inhibition rates of *D. bulbifera* extracts against murine liver glucosidase were: petroleum ether 73.3 %, ethyl acetate 40 %, methanol 43.3 %, ethanol 53.3 %. Petroleum ether extract of *D. bulbifera* showed strong inhibitory potential of pure α-glucosidase with a percentage inhibition of 92.87 % as compared to acarbose. Ethyl acetate extract of *D. bulbifera* bulb was found to be the strongest inhibitor, showing an inhibition as high as 99.6 %. Similarly, methanol and 70 % ethanol extracts of *D. bulbifera* exhibited an inhibition of 98.81 and 79.27 %, which were more significant as compared to others. Stem, leaf and flower of

G. glauca showed moderate inhibition of 43.54, 21.77 and 51.23 %, respectively.

Ingestion of *Dioscorea bulbifera* flour attenuated the hyperglycaemia and the bone fragility in female diabetic rats (De Salgado Rêgo et al. 2014). The yam flour lowered glucose levels and augmented radio density of femoral head compared to the untreated diabetic rats fed a high fat diet. In a hospital-based, single-centre, prospective, open-label randomised, case–control interventional study of patient with diabetic nephropathy, *Dioscorea bulbifera* was found to be more effective than fosinopril in controlling blood pressure, glycaemia, cholesterolemia and inflammatory state in diabetic nephropathy. (Singh et al. 2013) Both agents decreased proteinuria. However, creatinine clearance significantly decreased with both the drugs, more so with *Dioscore*.

Antidyslipidemic Activity

The aqueous extract of *Dioscorea bulbifera* tubers at 250, 500 and 1000 mg/kg doses administered for 4 weeks to high fat diet-fed C57BL/6J mice showed significant antidyslipidemic effects (Ahmed et al. 2009). Serum glucose and lipid levels were reversed towards normal in *D. bulbifera* treated high fat diet-fed mice.

Antiinflammatory Activity

Animal studies showed that the aqueous and methanol extracts of *D. bulbifera* var *sativa* bulbils possessed potent analgesic and anti-inflammatory activities (Mbiantcha et al. 2011). Oral administration of aqueous and methanol extracts caused significant anti-inflammatory activity on paw oedema induced by histamine, serotonin and formalin. The aqueous extract showed a dose-dependent inhibition of pain and inflammation induced by acetic acid, formalin and pressure with a maximum effect of 56.38 %, 73.06 % and 42.79 %. Similarly, the methanol extract at the same dose, respectively, inhibited these models of pain by 62.70 %, 84.54 % and 47.70 %.

Antinociceptive Activity

The methanol *D. bulbifera* bulb extract showed significant antinociceptive effects in persistent pain induced by complete Freund's adjuvant and on neuropathic pain induced by partial ligation sciatic nerve (Nguelefack et al. 2010). The extract significantly inhibited acute lipopolysaccharides (LPS)-induced pain but failed to reduce thermal hypernociception and capsaicin-induced spontaneous nociception. The antinociceptive effects of the extract in prostaglandin-E(2) model was antagonised by either l-NAME or glibenclamide.

Cardioprotective Activity

Studies showed that *D. bulbifera* hydroalcoholic extract could ameliorate myocardial ischaemia and reperfusion injury in rats by improving ventricular function and inhibition of cardiomyocyte necrosis and apoptosis (Vasanthi et al. 2010). The extract also prevented ischaemic/reperfusion -mediated down-regulation of survival protein Akt and HO-1. They found that the cardioprotective effect of flavonoid rich fraction of *D. bulbifera* in isoproterenol-induced myocardial infarction rats could be elucidated by amelioration of lipid peroxidation, enhancement of the antioxidant status as evidenced by the increase in the reduced glutathione (GSH) content and the activity of antioxidant enzymes and up-regulation of energy-producing mitochondrial enzymes, including tricarboxylic acid cycle enzymes such isocitrate dehydrogenase (ICDH), succinate dehydrogenase (SDH), malate dehydrogenase (MDH) and α-ketoglutarate dehydrogenase (α-KGDH), NADH dehydrogenase and cytochrome-C-oxidase (Jayachandran et al. 2010).

Wound Healing Activity

Studies found that *D. bulbifera* tuber extract exhibited significant wound healing activity on experimentally induced excision wound model in rats for period of 22 days (Panduraju et al. 2010). The extract produced a high rate of wound contraction and decreased the period for epithelisation; its activity was comparable with standard ointment.

Anthelminthic Activity

Dioscorea bulbifera bulb aqueous extract at a high concentration (50 mg/ml) was found to be effective against the earthworm *Eisenia fetida*, the roundworm, *Ascaridia galli* and the tapeworm, *Raillietina spiralis,* causing paralysis and shortening the time of death (Kosalge and Fursule 2009).

Pharmacokinetic Studies

Gender exerted a significant influence on the pharmacokinetics and bioavailability of diosbulbin B main ingredient in *D. bulbifera* in rats (Yang et al. 2013). Female rats showed significantly better absorption (2 %) of diosbulbin B than males after oral administration.

Anorexic Activity

Primary studies showed that *D. bulbifera* extract exhibited anorexic activity in rats, depressing their appetite (1969).

Hepatotoxic Activity

The methanol extract and chloroform fraction of *D. bulbifera* rhizome caused significant liver toxicity in rats (Tan et al. 2003). Oral administration of *D. bulbifera* aqueous extract (200 %) caused toxicity in the liver and kidney of ICR mice (Su et al. 2003). The pathological changes showed fatty change and the increasing glycogen of liver cells; degeneration and necrosis of the epithelia of uriniferous tubules. The serum blood urea nitrogen and alanine aminotransferase of the extract-treated mice were higher than that of control group. The activity of the succinate dehydrogenase and glucose-6-phosphatase decreased and metabolism was affected.

The ethyl acetate extract of *D. bulbifera* rhizome (80–480 mg/kg) was found to be hepatotoxic in mice (Wang et al. 2010). Serum alanine transaminase (ALT) and aspartate transaminase (AST) were significantly elevated after 14 consecutive administrations of the extract. Liver lipid peroxidation (LPO) level increased, while glutathione amounts, glutathione-S-transferase (GST), superoxide dismutase (SOD) and catalase (CAT) activities all decreased in the time-dependent manner. Similar hepatotoxic effects were found after oral administration of *D. bulbifera* ethyl acetate extract (640 mg/kg) to mice (Wang et al. 2011). They found that ALT and AST activities in female mice were significantly lower than those in male but liver glutathione amounts and CAT activity in female mice were higher than those in males. They found diosbulbin B could be the main hepatotoxic chemical compound in *D. bulbifera* and that *D. bulbifera* could induce gender-related liver oxidative stress injury in mice. They also found that bile acids tauroursodeoxycholic acid (TUDCA), taurochenodeoxycholic acid (TCDCA), taurocholic acid (TCA), taurodeoxycholic acid (TDCA), cholic acid (CA) could be considered as sensitive biomarkers of ethanol *D. bulbifera* extract-induced and diosbulbin B-induced liver injury in mice (Xu et al. 2011). Good correlation could be observed between the bile acids and ALT, AST. Disobulbin-D, a hepatotoxic furano norclerodane diterpenoid, isolated from *D. bulbifera* rhizome, inhibited the viability of human liver cell line L-02 in a concentration and time-dependent manner (Ma et al. 2012). It induced cell apoptosis, which was shown to be caspase 3-dependent. Oral administration of diosbulbin B for 12 consecutive days was found to cause oxidative stress liver injury in mice (Ma et al. 2014). The level of malondialdehyde (MDA) in the liver

was increased in mice treated with diosbulbin B, while the glutathione amount and the enzymatic activity of glutathione peroxidase (GPx), glutathione-S-transferase (GST), copper/zinc-superoxide dismutase (CuZn-SOD), manganese-SOD (Mn-SOD) and catalase (CAT) were all decreased. Diosbulbin B also decreased the gene expression of CuZn-SOD and CAT. They found that *Grateloupia filicina* polysaccharide could attenuate the oxidative stress liver injury induced by *Dioscorea bulbifera* in mice (Ma et al. 2013).

Traditional Medicinal Uses

Dioscorea bulbifera is used medicinally in China, but, unlike in Africa, it is not much used as a food plant. In China, it is used as a remedy for sore throat, struma, diabetes, leprosy and various tumours (Gao et al. 2002; Yang et al. 2013). In India, *D. bulbifera* is used as tonic, expectorant, in asthma, diarrhoea, dysentery, syphilis, haemorrhoids, leucoderma as aphrodisiac, anthelmintic, to cure wounds, and is also an excellent dressing for boils (Chopra et al. 1986; Kirtikar and Basu 1999; Panduraju et al. 2010). Tribals of Satpuda Hills in India used *D. bulbifera* as an anthelminthic (Kosalge and Fursule 2009). Aqueous powdered tuber extract is taken orally to treat intestinal worm infestations. *Dioscorea bulbifera* had been reported to be used for the treatment of dysentery and diarrhoea by Bhoxa community in the district Dehradun, Uttarakhand, India (Gairola et al. 2013).

Dioscorea bulbifera var *sativa* is a medicinal plant commonly used in Cameroonian traditional medicine to treat pain and inflammation (Nguelefack et al. 2010). In Benin, the fresh sap of the leaves is used for dystocia (Adjanohoun et al. 1989). The pulp of the bulb is used for abscesses in Gabon (Adjanohoun et al. 1984). In Central African Republic, the bulbils are used to treat hypertension (Apema et al. 2011). In the Popular Republic of Congo, the sap *of D. bulbifera* is used for purulent ophthalmia and snake bite; the pounded bulb in palm oil is used as a massage for mycosis and dermatoses (Bouquet 1969). In the Democratic Republic of Congo, a raw slice of bulbil is applied to ringworm to cure it (Terashima et al. 1985).

Other Uses

Ghosh et al. (2011) reported on the synthesis of gold nanoparticles using *D. bulbifera* tuber extract as the reducing agent. The non-spherical gold nanoparticles possess unusual optical and electronic properties, improved mechanical properties and specific surface enhance spectroscopies that make them ideal structures for emerging applications in photonics, electronics, optical sensing and imaging, biomedical labelling and catalysis (Aizpurua et al. 2003; Ghosh et al. 2011). Ghosh et al. (2012b) also reported on rapid synthesis of silver nanoparticles by reduction of aqueous Ag^+ ions using *D. bulbifera* tuber extract.

An antifungal alkaloid dihydrodioscorine, isolated from the tubers and bulbils retarded the growth of five plant pathogenic fungi, namely: *Sclerotium rolfsii, Curvularia lunata, Fusarium moniliforme, Macrophomina phaseolina* and *Botryodiplodia theobromae* (Adeleye and Ikotun 1989). Also, sclerotia formation in *S. rolfsii* was delayed and the number was reduced. Conidia formation in *C. lunata* and *F. moniliforme* was also delayed and reduced but pycnidia formation in *M. phaseolina* and *B. theobromae* was not affected.

Comments

The chief means of propagation in *D. bulbifera* is asexual and is dependant on vegetative growth from underground tubers and aerial bulbils. Bulbils drop off of parent vines to the ground in abundance and give rise to sprouts.

Selected References

Adeleye A, Ikotun T (1989) Antifungal activity of dihydrodioscorine extracted from a wild variety of *Dioscorea bulbifera* L. J Basic Microbiol 29:265–267

Adesanya SA, Ogundana SK, Roberts MF (1989) Dihydrostilbene phytoalexins from *Dioscorea bulbifera* and *D. dumetorum*. Phytochemistry 28(3):773–774

Adjanohoun EJ, Adjakidjè V, Ahyi MRA, Aké Assi L, Akoègninou A, d'Almeida J, Apovo F, Boukef K, Chadare M, Cusset G, Dramane K, Eyme J, Gassita JN, Gbaguidi N, Goudote E, Guinko S, Houngnon P, Lo I, Keita A, Kiniffo HV, Kone-Bamba D, Musampa Nseyya A, Saadou M, Sodogandji T, De Souza S, Tchabi A, Zinsou Dossa C, Zohoun T (1989) Contribution aux Études Ethnobotaniques et Floristiques en République Populaire du Bénin. Médecine Traditionelle et Pharmacopée. Agence de Coopération Culturelle et Technique, Paris. 895 pp

Adjanohoun EJ, Ahyi AMR, Chibon P, de Vecchy H, Duboze E, Eymé J, Gassita JN, Goudote E, Guinko S, Keita A, Koudogbo B, Le Bras M, Mourambou I, Mve-Mengome E, Nguéma MG, Ollome JB, Posso P, Sita P (1984) Contribution aux études ethnobotaniques et floristiques au Gabon. Agence de coopération culturelle et technique, France. 294 pp

Ahmed Z, Chishti MZ, Johri RK, Bhagat A, Gupta KK, Ram G (2009) Antihyperglycemic and antidyslipidemic activity of aqueous extract of *Dioscorea bulbifera* tubers. Diabetol Croat 38(3):63–72

Aizpurua J, Hanarp P, Sutherland DS, Kall M, Bryant GW, de Abajo FJG (2003) Optical properties of gold nanorings. Phys Rev Lett 90(5):057401/1–057401/4

Alexis SD, Georges AN (2012) Classification of some wild yam species tubers of Ivory Coast forest zone. Int J Biochem Res Rev 2(4):137–151

Apema R, Mozouloua D, Kosh-Komba E, Ngoule Y (2011) Les plantes médicinales utilisées dans le traitement de l'hypertension artérielle par les tradipraticiens à Bangui. http://www.gfmer.ch/Activites_internationales_Fr/PDF/HTA-Apema-2011.pdf

Araujo de Vizcarrondo AC, Rincón AM, Padilla F (2004) Characterisation of *Dioscorea bulbifera* native starch. Arch Latinoam Nutr 54(2):241–245 (in Spanish)

Behera KK, Sahoo S, Prusti A (2010) Biochemical quantification of diosgenin and ascorbic acid from the tubers of different *Dioscorea* species found in Orissa. Libyan Agric Res Cent J Int 1(2):123–127

Bhandari MR, Kawabata F (2005) Bitterness and toxicity in wild yam (*Dioscorea* spp.) tubers of Nepal. Mater Veg 60(3):129–135

Bouquet A (1969) Féticheurs et Médecines Traditionnelles du Congo (Brazzaville). Memoires O.R.S.T.O.M., 36. Paris. 282 pp

Bouquet A, Debray M (1974) Plantes Médicinales de la Côte d'Ivoire. Memoires O.R.S.T.O.M., 32. Paris. 231 pp

Burkill IH (1966) A dictionary of the economic products of the Malay Peninsula. Revised reprint of 1st ed 1935. 2 volumes. Ministry of Agriculture and Co-operatives, Kuala Lumpur, Malaysia. Vol. 1 (A-H) pp 1–1240, Vol. 2 (I-Z) pp 1241–2444

Chen X, Wu SH, Zeng XB, Jiang XW, Chen XP, Yuan J, Lu BH, Li J (2013) Antioxidant and SGC-7901 cell inhibition activities of Rhizoma Dioscoreae bulbiferae. Ethanol extracts. Afr J Tradit Complement Altern Med 10(5):261–266

Chopra RN, Nayar SL, Chopra IC (1986) Glossary of Indian medicinal plants (Including the Supplement). Council Scientific Industrial Research, New Delhi, 330 pp

Chunthorng-Orn J, Panthong S, Itharat A (2012) Antimicrobial, antioxidant activities and total phenolic content of Thai medicinal plants used to treat HIV patients. J Med Assoc Thai 95(Suppl 1):S154–S158

Coursey DG (1967) Yams: an account of the nature, origins, cultivation, and utilization of the useful members of Dioscoreaceae. Longmans, Green and Co. Ltd., London, 230 pp

De Salgado Rêgo T, Da Silva Ash L, Pessoa L, Da Silva Feijó MB, Leite J, De Sousa Dos Santos A, Soares da Costa CA, Teles Boaventura G (2014) The intake of yam (*Dioscorea bulbifera* Linn) attenuated the hyperglycemia and the bone fragility in female diabetic rats. Nutr Hosp 29(2):370–375

Ferreira SM (1995) Cará-de-rama (*Dioscorea bulbifera*, L) fried processing. Arch Latinoam Nutr 45(2):128–133 (in Portuguese)

Forsyth C, van Staden J (1984) Tuberization of *Dioscorea bulbifera* stem nodes in culture. J Plant Physiol 115(1):79–83

Gairola S, Sharma J, Gaur RD, Siddiqi TO, Painuli RM (2013) Plants used for treatment of dysentery and diarrhoea by the Bhoxa community of district Dehradun, Uttarakhand, India. J Ethnopharmacol 150(3): 989–1006

Gao HY, Hou BL, Kuroyanagi M, Wu LJ (2007) Constituents from anti-tumor-promoting active part of *Dioscorea bulbifera* L. in JB6 mouse epidermal cell. J Asian J Trad Med 21(3):104–109

Gao HY, Kuroyanagi M, Wu LJ, Kawahara N, Yasuno T, Nakamura Y (2002) Antitumor-promoting constituents from *Dioscorea bulbifera* L. in JB6 mouse epidermal cells. Biol Pharm Bull 25(9):1241–1243

Ghosh S, Ahire M, Patil S, Jabgunde A, Bhat Dusane M, Joshi BN, Pardesi K, Jachak S, Dhavale DD, Chopade BA (2012a) Antidiabetic activity of *Gnidia glauca* and *Dioscorea bulbifera*: potent amylase and glucosidase inhibitors. Based Complement Altern Med 2012:929051

Ghosh S, Derle A, Ahire M, More P, Jagtap S, Phadatare SD, Patil AB, Jabgunde AM, Sharma GK, Shinde VS, Pardesi K, Dhavale DD, Chopade BA (2013) Phytochemical analysis and free radical scavenging activity of medicinal plants *Gnidia glauca* and *Dioscorea bulbifera*. PLoS One 8(12):e82529

Ghosh S, Patil S, Ahire M, Kitture R, Jabgunde AS, Kale S, Pardesi K, Bellare J, Dhavale DD, Chopade BA (2011) Synthesis of gold nanoanisotrops using *Dioscorea bulbifera* tuber extract. J Nanomaterials 2011:354793

Ghosh S, Patil S, Ahire M, Kitture R, Kale S, Pardesi K, Cameotra SS, Bellare J, Dhavale DD, Jabgunde A, Chopade BA (2012b) Synthesis of silver nanoparticles using *Dioscorea bulbifera* tuber extract and evaluation of its synergistic potential in combination with antimicrobial agents. Int J Nanomedicine 7:483–496

Ida Y, Komori T, Kawasaki T (1978a) Furanoid-norditerpene aus pflanzen der familie Dioscoreaceae, VI. Kristallstrukturanalyse von diosbulbin G. Liebigs Ann Chem 1978:834–838

Ida Y, Kubo S, Fujita M, Komori T, Kawasaki T (1978b) Furanoid-norditerpene aus pflanzen der familie Dioscoreaceae, V. Struktur der diosbulbine D, E, F, G und H. Liebigs Ann Chem 1978:818–833

Ida Y, Noda N, Kubo S, Komori T, Kawasaki T (1978c) Furanoid norditerpenes from *Dioscorea* plants. VII. Structures of diosbulbinosides D and F chem. Pharm Bull 26(2):435–439

Jayachandran KS, Vasanthi HR, Rajamanickama GV (2010) Flavonoid rich fraction of *Dioscorea bulbifera* Linn. (Yam) enhances mitochondrial enzymes and antioxidant status and thereby protects heart from isoproterenol induced myocardial infarction. Curr Pharm Biotechnol 11(8):887–894

Jindal MN, Kelkar VV, Doctor RB (1969) The anorexient activity of Kalio-kund (*Dioscorea bulbifera* Linn.), methylphenidate and cocaine in rats: a preliminary study. Indian J Med Res 57(6):1075–1080

Jyothishwaran G, Seetharam YN (2008) Biotransformation of cholesterol to diosgenin by freely suspended and immobilised cells of *Dioscorea bulbifera* L. J Asian Nat Prod Res 10(1–2):139–145

Kidyu K, Thaisuchat H, Meepowpan P, Sukdee S, Nuntasaen N, Punyanitya S, Pompimon W (2011) New clerodane diterpenoid from the bulbils of *Dioscorea bulbifera*. Nat Prod Commun 6(8):1069–1072

Kim SH, Ryu BH, Ryu KW, Yoon SH, Kim JS (2002) An anti-tumor effects of *Dioscorea bulbifera* L. on stomach cancer cells. J Korean Orient Int Med 23(1):99–106

Kirtikar KR, Basu BD (1999) Indian medicinal plants. Vol. IV. L.M Basu, Allahabad, p 2479

Kosalge SB, Fursule RA (2009) Investigation of anthelmintic potential of some plants claimed by tribals of Satpuda hills. Int J PharmTech Res 1(1):68–72

Komori T (1997) Glycosides from *Dioscorea bulbifera*. Toxicon 35(10):1531–1535

Kuete V, Betrandteponno R, Mbaveng AT, Tapondjou LA, Meyer JJ, Barboni L, Lall N (2012) Antibacterial activities of the extracts, fractions and compounds from *Dioscorea bulbifera*. BMC Complement Altern Med 12:228

Kumar S, Jena PK, Tripathy PK (2012) Study of wild edible plants among tribal groups of Simlipal Biosphere Reserve Forest, Odisha, India; with special reference to *Dioscorea* species. Int J Biol Technol 3(1):11–19

Li SS, Deng JZ, Zhao SX (1998) Steroids from tuber of *Dioscorea bulbifera* L. J Plant Resour Environ 8(2):61–62

Li SS, Iliya IA, Deng JZ, Zhao SX (2000) Flavonoids and anthraquinone from *Dioscorea bulbifera* L. Zhongguo Zhong Yao Za Zhi 25(3):159–160 (in Chinese)

Liu H, Chou GX, Guo YL, Ji LL, Wang JM, Wang ZT (2010) Norclerodane diterpenoids from rhizomes of *Dioscorea bulbifera*. Phytochemistry 71(10):1174–1180

Liu H, Chou GX, Wang JM, Ji LL, Wang ZT (2011a) Steroidal saponins from the rhizomes of *Dioscorea bulbifera* and their cytotoxic activity. Planta Med 77(8):845–848

Liu H, Chou GX, Wu T, Guo YL, Wang SC, Wang CH, Wang ZT (2009) Steroidal sapogenins and glycosides from the rhizomes of *Dioscorea bulbifera*. J Nat Prod 72(11):1964–1968

Liu H, Tsim KW, Chou GX, Wang JM, Ji LL, Wang ZT (2011b) Phenolic compounds from the rhizomes of *Dioscorea bulbifera*. Chem Biodivers 8(11):2110–2116

Ma M, Jiang ZZ, Ruan JL, Tan XQ, Liu J, Wang CF, Zha XM, Zhang LY (2012) The furano norclerodane diterpenoid disobulbin-D induces apoptosis in normal human liver L-02 cells. Exp Toxicol Pathol 64(6):611–618

Ma YB, Ji LL, Wang SC, Shi SS, Wang ZT (2013) Protection of *Grateloupia filicina* polysaccharide against hepatotoxicity induced by *Dioscorea bulbifera* L. Yao Xue Xue Bao 48(8):1253–1258 (in Chinese)

Ma YB, Niu CW, Wang JM, Ji LL, Wang ZT (2014) Diosbulbin B-induced liver injury in mice and its mechanism. Hum Exp Toxicol 33(7):729–736

Martin FW, Telek L, Ruberté RM (1974) Yellow pigments of *Dioscorea bulbifera*. J Agric Food Chem 22(2):335–337

Mbiantcha M, Kamanyi A, Teponno RB, Tapondjou AL, Watcho P, Nguelefack TB (2011) Analgesic and anti-inflammatory properties of extracts from the bulbils of *Dioscorea bulbifera* L. var sativa (Dioscoreaceae) in mice and rats. Evid Based Complement Altern Med 2011. pii:912935

Murray RDH, Jorge ZD, Khan NH, Shahjahan M, Quaisuddin M (1984) Diosbulbin D and 8-epidiosbulbin E acetate, norclerodane diterpenoids from *Dioscorea bulbifera* tubers. Phytochemistry 23:623–625

Narula A, Kumar S, Bansal KC, Srivastava PS (2003) In vitro micropropagation, differentiation of aerial bulbils and tubers and diosgenin content in *Dioscorea bulbifera*. Planta Med 69(8):778–779

Narula A, Kumar S, Srivastava PS (2005) Abiotic metal stress enhances diosgenin yield in *Dioscorea bulbifera* L. cultures. Plant Cell Rep 24(4):250–254

Narula A, Kumar S, Srivastava PS (2007) Genetic fidelity of in vitro regenerants, encapsulation of shoot tips and high diosgenin content in *Dioscorea bulbifera* L., a potential alternative source of diosgenin. Biotechnol Lett 29(4):623–629

Nguelefack TB, Dutra RC, Paszcuk AF, Andrade EL, Tapondjou LA, Calixto JB (2010) Antinociceptive activities of the methanol extract of the bulbs *of Dioscorea bulbifera* L. var sativa in mice is dependent

of NO-cGMP-ATP-sensitive-K(+) channel activation. J Ethnopharmacol 128(3):567–574

Panduraju T, Bitra VR, Vemula SK, Prabhakar Reddy VR (2010) Wound healing activity of *Dioscorea bulbifera* Linn. J Pharm Res 3(12):3138–3139

Raponda-Walker A, Sillans R (1995) Les Plantes Utiles du Gabon. Encyclopédie Biologique. Editions Sepia. 697 pp. (First Published 1961)

Rincón AM, Araujo de Vizcarrondo C, Carrillo de Padilla F, Martín E (2000) Evaluation of the possible technological uses of some *Dioscorea* tubers: name congo (*Dioscorea bulbifera*) and mapuey (*Dioscorea trifida*). Arch Latinoam Nutr 50(3):286–290 (in Spanish)

Shajeela PS, Mohan VR, Jesudas LL, Soris PT (2011) Nutritional and antinutritional evaluation of wild yam (*Dioscorea* spp.). Trop Subtrop Agroecosyt 14:723–730

Shriram V, Jahagirdar S, Latha C, Kumar V, Puranik V, Rojatkar S, Dhakephalkar PK, Shitole MG (2008) A potential plasmid-curing agent, 8-epidiosbulbin E acetate, from *Dioscorea bulbifera* L. against multidrug-resistant bacteria. Int J Antimicrob Agents 32(5):405–410

Singh RG, Rajak M, Ghosh B, Usha AA, Dubey GP (2013) Comparative evaluation of fosinopril and herbal drug *Dioscorea bulbifera* in patients of diabetic nephropathy. Saudi J Kidney Dis Transpl 24(4):737–742

Song FL, Gan RY, Zhang Y, Xiao Q, Kuang L, Li HB (2010) Total phenolic contents and antioxidant capacities of selected Chinese medicinal plants. Int J Mol Sci 11(6):2362–2372

Su L, Zhu JH, Cheng LB (2003) Experimental pathological study of subacute intoxication by *Dioscorea bulbifera* L. Fa Yi Xue Za Zhi 19(2):81–83 (in Chinese)

Suriyavathana M, Indupriya S (2011) Screening of antioxidant potentials in *Dioscorea bulbifera*. Int J Pharm Life Sci 2(4):661–664

Tan XQ, Ruan JL, Chen HS, Wang JY (2003) Studies on liver-toxicity in rhigoma of *Dioscorea bulbifera*. Zhongguo Zhong Yao Za Zhi 28(7):661–663 (in Chinese)

Tapondjou LA, Jenett-Siems K, Böttger S, Melzig MF (2013) Steroidal saponins from the flowers of *Dioscorea bulbifera* var. *sativa*. Phytochemistry 95:341–50

Teponno RB, Tapondjou AL, Abou-Mansour E, Stoekli-Evans H, Tane P, Barboni L (2008) Bafoudiosbulbins F and G, further clerodane diterpenoids from *Dioscorea bulbifera* L. var *sativa* and revised structure of bafoudiosbulbin B. Phytochemistry 68:2374–2379

Teponno RB, Tapondjou AL, Djoukeng JD, Abou-Mansour E, Tabacchi R, Tane P, Lontsi D, Park HJ (2006a) Isolation and NMR assignment of a pennogenin glycoside from *Dioscorea bulbifera* var *sativa*. Nat Prod Sci 12:62–66

Teponno RB, Tapondjou AL, Gatsing D, Djoukeng JD, Abou-Mansour E, Tabacchi R, Tane P, Stoekli-Evans H, Lontsi D, Park HJ (2006b) Bafoudiosbulbins A and B, two antisalmonellal clerodane diterpenoids from *Dioscorea bulbifera* var. *sativa*. Phytochemistry 67:1957–1963

Teponno RB, Tapondjou AL, Jung HJ, Nam JH, Tane P, Park HJ (2007) Three new clerodane diterpenoids from the bulbils of *Dioscorea bulbifera* var. *sativa*. Helv Chim Acta 90:1599–1605

Terashima H, Ichikawa M, Sawada S (1985) *Dioscorea bulbifera* L. AFlora The database of plant utilization in Africa. The Center for African Area Studies, Kyoto University. http://aflora.africa.kyoto-u.ac.jp/records

Vasanthi HR, Mukherjee S, Ray D, Pandian Jayachandran KS, Lekli I, Das DK (2010) Protective role of air potato (*Dioscorea bulbifera*) of yam family in myocardial ischemic reperfusion injury. Food Funct 1(3):278–283

Walter A, Lebot V (2007) Gardens of Oceania. ACIAR Monograph No. 122. 326 pp

Wang G, Lin BB, Liu JS, Wang GK, Wang F, Liu JK (2009a) Chemical constituents from tubers of *Dioscorea bulbifera*. Zhongguo Zhong Yao Za Zhi 34(13):1679–1682 (in Chinese)

Wang G, Liu JS, Lin BB, Wang GK, Liu JK (2009b) Two new furanoid norditerpenes from *Dioscorea bulbifera*. Chem Pharm Bull (Tokyo) 57:625–627

Wang JM, Ji LL, Branford-White CJ, Wang ZY, Shen KK, Liu H, Wang ZT (2012) Antitumor activity of *Dioscorea bulbifera* L. rhizome in vivo. Fitoterapia 83(2):388–394

Wang JM, Ji LL, Liu H, Wang ZT (2010) Study of the hepatotoxicity induced by *Dioscorea bulbifera* L. rhizome in mice. Biosci Trends 4(2):79–85

Wang JM, Liang QN, Ji LL, Liu H, Wang CH, Wang ZT (2011) Gender-related difference in liver injury induced by *Dioscorea bulbifera* L. rhizome in mice. Hum Exp Toxicol 30(9):1333–1341

Wang Y, Yao RS, Gao S, Wen WD, Du YQ, Szabo E, Hu M, Lubet RA, You M (2013) Chemopreventive effect of a mixture of Chinese herbs (antitumor B) on chemically induced oral carcinogenesis. Mol Carcinog 52(1):49–56

Watt JM, Breyer-Brandwijk MG (1962) The medicinal and poisonous plants of Southern and Eastern Africa, 2nd edn. E. and S. Livingstone, Edinburgh/London. 1457 pp

Xu Y, Chen CC, Yang L, Wang JM, Ji LL, Wang ZT, Hu ZB (2011) Evaluation on hepatotoxicity caused by *Dioscorea bulbifera* based on analysis of bile acids. Yao Xue Xue Bao 46(1):39–44 (in Chinese)

Yang BH, Wang XT, Liu W, Zhang Q, Chen KX, Ma YM, Wang CH, Wang ZT (2013) Gender-related pharmacokinetics and absolute bioavailability of diosbulbin B in rats determined by ultra-performance liquid chromatography-tandem mass spectrometry. J Ethnopharmacol 149(3):810–815

Yonemitsu M, Fukuda N, Kimura T, Komori T (1993) Diosbulbin-B from the leaves and stems of *Dioscorea bulbifera*: ^{1}H-^{1}H and 13C-^{1}H COSY NMR studies. Planta Med 59(6):57

Yu ZL, Liu XR, McCulloch M, Gao J (2004) Anticancer effects of various fractions extracted from *Dioscorea bulbifera* on mice bearing Hep A. Zhongguo Zhong Yao Za Zhi 29(6):563–567 (in Chinese)

Zhang SD, Fang YX, Chen MM, Huang JP, Zhang K (2007) Extraction and anti-tumor activity study of polysaccharide in several traditional Chinese medicine. Zhong Yao Cai 30(2):179–182 (In Chinese)

Zheng SZ, Guo Z, Shen T, Zhen XD, Shen XW (2003) Three new apianen lactones from *Dioscorea bulbifera* L. Indian J Chem Sect B 42(4):946–949

Dioscorea esculenta

Scientific Name

Dioscorea esculenta **(Lour.) Burkill**

Synonyms

Dioscorea aculeata Roxb. (illeg.), *Dioscorea aculeata* var. *spinosa* (Prain) Roxb. ex Prain & Burkill, *Dioscorea esculenta* var. *fasciculata* (Roxb.) Prain & Burkill, *Dioscorea esculenta* var. *fulvidotomentosa* R.Knuth, *Dioscorea esculenta* var. *spinosa* (Prain) R.Knuth, *Dioscorea esculenta* var. *tiliifolia* (Kunth) Fosberg & Sachet, *Dioscorea fasciculata* Roxb., *Dioscorea fasciculata* var. *spinosa* Prain, *Dioscorea papillaris* Blanco, *Dioscorea papuana* Warb., *Dioscorea spinosa* Roxb. ex Hook.f. (illeg.), *Dioscorea spinosa* Roxb. ex Wall., *Dioscorea tiliifolia* Kunth, *Dioscorea tugui* Blanco, *Oncorhiza esculentus* (Lour.) Pers., *Oncus esculentus* Lour.

Family

Dioscoreaceae

Common/English Names

Asiatic Yam, Birch Rind Yam, Chinese Yam, Goa Potato, Hausa Potato, Karen Potato, Lesser Yam, Sweet Yam, Spiny Chinese Yam

Vernacular Names

Antilles: Ñame Papa;
Brazil: Inhame De Sâo Tomé;
Burmese: Sadwe-U, Tadwe-U, Sadwe Uu, Tadwe Uu, Thadut Ni, Wet-Ka-Uu, Wakka Uu;
Chinese: Gan Shu, Pu Tong Shu Yu, Tian Shu,Ci Shu Yu, You Ci Gan Shu;
Danish:Yans;
Fiji: Kawai;
French: Igname Des Blancs, Igname De Chine, Igname Patate, Petite Igname, Igname Pas Possible, Igname De Chine Épineuse;
Germany: Chinesischer Yam, Kartoffel-Yams, Yamswurzel;
Ghana: Anagote, Blefo Hie;
Guatemala: Ñame Akam, Ñame Blanco;
India: Mou Alu, Sushni Alu, Susni Alu (Bengali), Kodikkilangu, Musilamvalli Kilangu, Siruvalli Kilangu (Tamil), Tivvi Tiga,Tippa Tiga (Telugu), Manalu, Kanga, Kharsang, Sasniali;
Indonesia: Gembolo, Gembili, Kombili, Sudo, (Java), Kaburan, Kamburan (Madurese), Ubi Aung, Ubi Gembili (Malay), Huwi Butul, Huwi Jahe, Huwi Kamayung, Huwi Kawayang, Huwi Landak, Huwi Taropong, Huwi Cheker (Sundanese);
Italian: Igname;
Japanese: Hari-Imo, Togedokoro, Toge Imo;
Khmer: Damlong Sya;
Laos: Hwa Kathad, Manonz;
Malaysia: Kembili, Kemili, Kemurang, Ubi Torak;

T.K. Lim, *Edible Medicinal and Non-Medicinal Plants: Volume 10, Modified Stems, Roots, Bulbs*,
DOI 10.1007/978-94-017-7276-1_11

Nepalese: Suthanii Tarul, Suthni Tarul;
New Caledonia: Ouale;
Nigeria: Ishu Alubosa;
Papua New Guinea: Mami, Taitu, Kalak, Diba;
Philippines: Kamiging (Bikol), Aneg, Luttu (Ibanag), Boga, Tugi,(Iloko), Dukai (Ivatan), Toñgo, Tugi, Tuñgo (Tagalog);
Portuguese: Inhame De São Tomé, Inhame-De-São;
Russian: Dioscoreia S'edobnaia, Iams S'edobnyi;
Samoa: Ufi Lei;
Solomon Islands: Pana;
Spanish: Ñame Papa, Ñame Pequeño, Batata De China;
Sri Lanka: Java Ala, Kukulala;
Swedish: Småjams;
Thailand: Man Chuak, Man-Musua, Man Mung (Central Thailand), Nam Chuak (Northern Thailand);
Tongan: Ufi Lei (Tonga);
Vanuatu: Sarsau;
Vietnamese: Cầm Ghim, Củ Bňng, Củ Mỡ, Củ Từ, Khoai Bướu, Khoai Từ, Từ Trơn, Từ Gai;
West Africa: Hausa Potato;

Origin/Distribution

It is indigenous to tropical Asia, occurs as wild types in South China, North Vietnam, North Thailand, Laos, Myanmar and India (Assam), Malaysia, the Philippines and New Guinea. It has long been cultivated in tropical Asia and is known to have been cultivated in South China for at least 1700 years. The thornless forms are probably selections from an original, thorny form. Recently, Papua New Guinea has been recognised as a major centre of diversity.

Agroecology

The species thrives in the warm and wet tropics. It grows in thickets, swamp forest and secondary forests at low altitudes up to 1500 m. It performs best in well-drained, friable, organic matter-rich soils with pH 5.5–6.5. It is cultivated in home gardens.

Edible Plant Parts and Uses

The oval to spindle-shaped, soft, mucilaginous tubers are rich in starch and are eaten like potatoes after cooking, boiling or roasting. They have a slightly sweet and agreeable taste. The starch is more easily digested than that of other yams; therefore, this species is used in special diets for persons with alimentary disorders. The tuber is a staple food in most parts of West Africa, including Nigeria.

Botany

A slender, pubescent, often spiny, herbaceous climbing annual, with left-handed twining terete stem (Plate 1), growing to 4–5 m with numerous shallow-rooted tubers in clusters of 4–20, arising downwards from a rhizome, on stolons 5–50 cm long. Mature tubers are yellowish brown, brown, or grey-brown, thin, smooth (Plate 2a, b) or often rough with indurated bases of protruding fragile rootlets (Plate 3a,b); flesh white. The tubers are oval to spindle-shaped to cylindrical, reaching lengths of 12–20 cm. Leaves are alternate, simple green, petiolated (petiole 5–8 cm), lamina broadly cordate to 15 by 17 cm with 9–13 basal veins, acute apex and cordate, rounded base. Very young leaves at the tips of the vines are pilose. Inflorescences unisexual, axillary, slender with green small flowers, 4 mm across. Male spike solitary, dense, 15 cm. Male flowers: usually solitary, rarely in cymules of 2–4, sessile or subsessile, with ovate bracts, shallowly cupular, puberulent perianth and 6 stamens inserted in the perianth tube. Female spike solitary, from upper leaf axils, pendent, to 40 cm. Fruit, a reflexed capsule, very seldom maturing, about 3 cm, base truncate, apex slightly emarginate. Seeds winged.

Plate 1 Dextrose twining climber with broadly cordate leaves

a b

Plate 2 (**a, b**) Brown/greyish brown, spindle-shaped, tubers smooth without protruding fragile rootlets

Nutritive/Medicinal Properties

Tuber Nutrients/Phytochemicals

Nutrient composition of the raw tuber per 100 g edible portion was reported as: energy 102 cal, moisture 73.6 g, protein 1.5 g, fat 0.2 g, total carbohydrates 23.9 g, dietary fibre 0.6 g, ash 0.8 g, Ca 12 mg, P 35 mg, Fe 0.8 mg, Na 8 mg, K 366 mg, thiamin 0.10 mg, riboflavin 0.01 mg, niacin 0.8 mg and ascorbic acid 15 mg (Leung et al. 1972). The nutrient composition of the tuber per 100 g edible portion was reported as: moisture 70–80 g, protein 1.3–2.1 g, fat 0.1–0.3 g, carbohydrates 26–36 g, starch 25 g, sugars 1–11 g, fibre 0.2–1.5 g, ash 0.5–1.2 g, vitamin A 0.017 mg, vitamin B1 0.08 mg, vitamin B2 0.02 mg and vitamin C 20.3 mg (JIRCAS 2010). Proximate nutrient composition of the tubers (g/100 g) was also reported by Shajeela et al. (2011b) as: moisture 83.37 %, crude protein 9.76 g, crude lipid 4.68 g, crude fibre 6.62 g, ash 5.17 g, starch 62.40 g, NFE (nitrogen free extract) 73.77 g, energy 1571.38 kJ/100 g DM, niacin

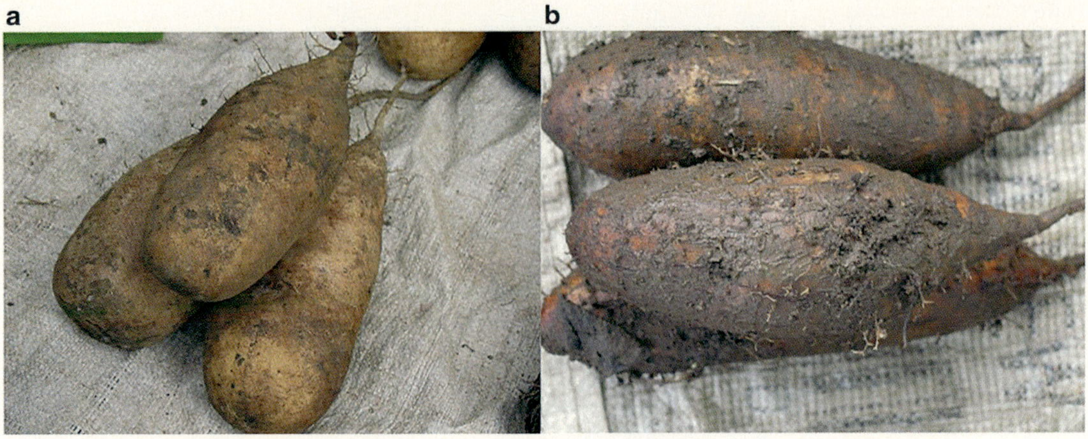

Plate 3 (**a, b**) *Brown*, spindle-shaped tubers smooth with sparsely protruding fragile rootlets

41.36 mg, ascorbic acid 84.06 mg, and minerals mg/100 g Na 86.40 mg, K 1594.31 mg, Ca 314.01 mg, Mg 436.06 mg, P 138.10 mg, Zn 1.76 mg, Mn 5.46 mg, Fe 11.48 mg, and Cu 3.40 mg. The anti-nutritional factors were per 100 g: total free phenolics 0.79 g, tannins 0.20 g, hydrogen cyanide 0.21 mg, total oxalate 0.33 g, amylase inhibitor 7.80 AIU/mg soluble starch and trypsin inhibitor 1.92 TIU/mg protein (Shajeela et al. 2011b). The in-vitro protein digestibility was 4.21 % and the in-vitro starch digestibility was 46.26 %. Harijon et al. (2013) reported the following nutrient composition of Lesser Yam water-soluble polysaccharides: water 7.39 %, ash 1.62 %, protein 4.47 %, fat 3.19 %, carbohydrate 83.44 %, crude fibre 3.07 % and starch 15.81 %.

The chemical composition (dry weight (dw) basis) of *D. esculenta* tuber flour and starch was reported, respectively, as: total starch 74.66 %, 86.3 %; protein 9.02 %,1.2 %; crude fat 1.55, 0.42 %; crude fibre 2.33 %, ash 2.1 %, 0.25 % (Senanayake et al. 2013). The tubers contained (mg/100 g dw) Ca 8.29 mg, Fe 6.12 mg, Mg 33.6 mg, K 190.91 mg, Zn 1.19 mg, saponins 20.01 mg, flavonoids 12.4 mg, and alkaloids 1.89 mg. The granular shape of starch was polygonal.

The presence of saponins, disgenin, β-sitosterol, stigmasterol, cardiac glycosides, fat and starch in the tubers was reported by (Olayemi and Ajaiyeoba 2007). Among the different *Dioscorea* species found in Orissa, *D. esculenta* was found to have a moderate content of diosgenin, 533.33 mg/100 g tuber and second lowest vitamin C content of 4.167 mg/100 g (Behera et al. 2010). Higher amounts of inulin could be extracted from *D. esculenta* compared to other tubers (Harmayani et al. 2011). They demonstrated the production of inulin powder from *D. esculenta* by the foam mat drying method using maltodextrin and egg white as filler and foaming agent.

In the fully matured tubers of *D. esculenta* and turmeric rhizomes starch degradation and the enzymes involved, viz. α-amylases and starch phosphorylase showed a lower level of activity during early period of dormancy, while sugar content and enzymes of carbohydrate metabolism increased rapidly during sprouting (Panneerselvam et al. 2007). The isoenzymic profiles of α-amylases showed marked variations in these two phases. The key enzymes of glycolysis, tricarboxylic acid (TCA) cycle and pentose phosphate pathway (PPP), viz. aldolase, succinic dehydrogenase, malic dehydrogenase, glucose-6-phosphate dehydrogenase and 6-phosphogluconate dehydrogenase were increased even before the visible appearance of sprouting and their activities were at their maximum during sprouting. The total loss in moisture content of *D. esculenta* tubers for the entire storage period (7 weeks) up to sprouting was 6.33 % (Panneerselvam and Abdul Jaleel 2008). There

was a total loss of 10.88 % of starch content from the period of harvest to sprouting. There was a sharp increase of non-reducing sugars from 5 to 8 WAS (weeks after storage). From start of storage till sprouting, the total increase of sucrose content was 33.75 %, the total increase of glucose content was 92.2 % and total increase of fructose content was 69.2 %. Glucose-1-phosphate and glucose-6-phosphate were observed till the 3 WAS. Fructose-6-phosphate was present from the 4 WAS and organic acids: succinic, malic and citric acids were observed. All the acids were present at the time of sprouting the tubers.

Jayakody et al. (2006) reported the chemical and physical composition and granule morphology of the Kukulala, Java-Ala and Nattala varieties of *D. esculenta* native starches, respectively, as follows: starch yield (based on tuber weight) 16.81 %, 10.21 %,12.22; moisture 10.61 %, 9.90 %, 10.97; ash 0.17 %, 0.22 %,0.32 %; nitrogen 0.01 %, 0.03 %, 0.01 %; phosphorus 0.05 %, 0.07 %, 0.10 %; lipids (extract by chloroform-methanol) 0.01 %, 0.01 %, 0.03 %; lipids (extracted by *n*-propanol-water) 0.39 %, 0.35 %,0.44 %; total amylose content 23.97 %, 20.07 %, 19.98 %; amylose complexed with lipids 14.98 %, 19.33 %, 22.02 %; granule size range 8–10 μm, 4–5 μm, 3–4 μm; granule morphology, all polygonal, crystallinity 49 %, 50 % 53 % and crystalline type all B-type. The gelatinization temperatures for Kukulala, Java-Ala and Nattala varieties were, respectively, as follows: onset (To) 72.30, 72.55, 72.45 °C, mid point (Tp) 75.73, 75, 75.60 °C and conclusion (Tc) 85.40, 81.65, 82.25 °C, gelatinization temperature range (Tc–To) 18.07, 17.32, 17.90 °C and gelatinization enthalpy (ΔH) 18.07, 17.32, 17.90 J/g. The hydrolysis percentage of the three varieties of starches by porcine pancreatic α-amylase were 68.84 %, 76.58 %, 66.08 %, respectively. The extent of hydrolysis followed the order: Java-Ala > Kukulala > Nattala. The *D. esculenta* starches differed only marginally (Kukulala > Java-Ala = Nattala) with respect to their pasting properties. However, they differed significantly with respect to peak viscosity (Nattala > Kukulala > Java-Ala), viscosity breakdown during the holding cycle (Nattala > Java-Ala > Kukulala) and degree of set-back (Kukulala > Nattala > Java-Ala). Melting enthalpies (ΔHR) of amylopectin recrystallization (reflecting the extent of retrogradation during the storage period of 40 °C for a week) was highest for Java-Ala > Kukulala > Nattala.

Crude protein contents of *D. esculenta* tubers tended to decrease with cooking, but the differences were not statistically significant and crude fat, crude fibre, starch and total sugar contents were unaffected by cooking (Wanasundera and Ravindran 1992) Water-soluble minerals leached out during boiling, thus causing a reduction in the ash content of boiled tubers. All cooking methods lowered the vitamin C content of the tubers.

D. esculenta yam flour contained 7.19 % protein and 1.10 % fat (Ukapi 2010). The maximum gelatinization temperature of the lesser yam flour paste was 91.7 °C and maximum viscosity of <700 Brabender units. Bread produced with 20 % lesser yam inclusion had 87 % specific loaf volume of the control sample and was not significantly scored lower than the sole wheat bread in overall acceptability, taste, appearance and softness by the sensory assessors. All the bread samples made with the experimental lesser yam flour had no observable cracks on their respective crusts.

Ireland and Passam (1984) found a gradual decrease in growth inhibitory phenolics in tubers of *Dioscorea alata* and *D. esculenta* during dormancy. It was found that batatasin-type growth inhibitory phenolics accumulated rapidly in developing tubers just prior to the onset of dormancy and were asymmetrically distributed, being concentrated in the proximal (head) region and in the peripheral zone just beneath the periderm. Gibberellin A_3 treatment produced a promotion of the dormant period and a correlative rise in the growth inhibitory phenolic level. The results suggested that batatasin-type phenolics were involved in the mechanism of tuber dormancy.

Antioxidant Activity

The methanol extracts of *D. esculenta* tuber was found to have potential in-vitro antioxidant activity

in a dose-dependent manner (Murugan and Mohan 2012). The DPPH radical scavenging activity was 79.33 % for 1000 μg/ml extract with I IC_{50} value of 38.33 μg/ml while the reference standard ascorbic acid had an IC_{50} value of 18.25 μg/ml. The hydroxyl scavenging activity ranged from 17.33 to 43.11 % at concentration 125–1000 μg/ml extract and the IC_{50} was about 15.5 μg/ml, while ascorbic acid IC_{50} value was 18.45 μg/ml. The superoxide anion radical scavenging activity increased from 39.40 to 78.55 at 125–1000 μg/ml concentrations and the IC_{50} value was 56.38 μg/ml and IC_{50} for ascorbic acid was 72.08 μg/ml. The ABTS radical cation scavenging activity increased from 46.14 to 64.11 % at concentration of 125–1000 μg/ml, the IC_{50} value was 40.50 μg/ml and IC_{50} value for Trolox was 20.67 μg/ml. Reducing power of the methanol extract also dose-dependently increased but was lower than ascorbic. Total phenolic content in methanol extract of *D. esculenta* was found to be 0.79 g/100 g and flavonoids content was found to be 0.26 g/100 g.

Antiinflammatory Activity

The defatted methanol extract of *D. esculenta* tuber exhibited significant dose-dependent inhibition of the carrageenan- induced oedema in rats at doses of 100 mg/kg and 150 mg/kg (Olayemi and Ajaiyeoba 2007). The observed activity was comparable to that of 150 mg/kg acetylsalicylic acid, the reference standard. The presence of saponins, disgenin, β-sitosterol, stigmasterol, cardiac glycosides, fat and starch was confirmed in the extract. The results supported the folkloric use for management of inflammation.

Hypoglycemic Activity

Administration of aqueous extract of *D. esculenta* tuber water-soluble polysaccharide (WSP), papain assisted-WSP extract, and tempeh inoculum assisted-WSP extract reduced blood glucose level in alloxan-induced hyperglycaemic rats (Estiasih et al. 2012). All the extracts inhibited glucose absorption and short chain fatty acids (SCFA) formation. Tempeh inoculum assisted-WSP extract exhibited the most significant hypoglycaemic activity. In another study, feeding biscuits (BIS) containing water-soluble polysaccharides from wild yam (*Dioscorea hispida*) (WY) or lesser yam (*Dioscorea esculenta*) (LY) tubers and commercial alginate (ALG) exerted a hypoglycaemic effect on alloxan-induced hyperglycaemic rats (Harijon et al. 2013). All hydrocolloid-containing biscuits affected the absorption of glucose into the blood and were able to maintain it at a normal level. After 4 weeks, the glucose blood levels were significantly reduced: ALG-BIS (57 % reduction) was the most effective, followed by WY/ALG-BIS (50 % reduction), LY/ALG-BIS (49 % reduction), WY-BIS (39 % reduction) and LY-BIS (35 %). The order of concentration of SCFAs profile from the caecum digesta at the end of feeding treatment was similar, i.e., acetic acid > propionic acid > butyric acid.

Antifertility Activity

The ethanol extract of *Dioscorea esculenta* tuber exhibited antifertility effects in male albino rats (Shajeela et al. 2011a). The relative weight of testes and epididymis were decreased significantly; epididymal sperm count, motility and sperm abnormality were reduced significantly. An increase in serum levels of follicle-stimulating hormone and estrogen but decrease in the serum levels of luteinizing hormone and testosterone were observed in treated rats compared to control. No significant changes were observed in the vas deferens, seminal vesicle and prostrate and in serum biochemical and liver marker enzymes SGOT (serum glutamic oxaloacetic transaminase), SGPT (serum glutamic-pyruvic transaminase) and ALP (alkaline phosphatase) in extract-treated rats.

Probiotic Activity

LY (*D. esculenta)* tuber inulin stimulated the growth of *Bifidobacterium breve*, *Bifidobacterium bifidum*, *Bifidobacterium longum* and *Lactobacillus casei*, but it did not have an effect on the growth of *Lactobacillus acidophilus* (Winarti et al. 2013). LY inulin inhibited the growth of *Escherichia coli*. Fermentation with LY inulin medium had higher pH values (from 6.8 to 5.46) compared with glucose (from 6.8 to 3.89). Lactic acid produced using *Bifidobacteria* and *Lactobacillus* in medium of LY inulin were lower (0.54–1.05 %) than that in glucose medium (2.24 %). Fermentation of LY inulin using *Bifidobacterium longum* produced the highest acetic acid (113.794 mMol) and propionic acid (9.217 mMol). High butyric acid was produced by *Bifidobacterium breve* (3.262 mMol). The results indicated that LY inulin had prebiotic effect and increased the amount of short-chain fatty acids (SCFAs), and probiotic bacteria better than commercial inulin.

Traditional Medicinal Uses

In Indo-China, a decoction of the tubers is given in the treatment of rheumatism and as a diuretic (Stuart 2013). In China, it is used for beriberi. Tubers are also used for piles and syphilis. In Africa, the tuber has been applied to ulcers, boils and abscesses (Olayemi and Ajaiyeoba, 2007). It contains allantoin, a cell-proliferant that speeds up the healing processes. It has been used traditionally as a contraceptive, in the treatment of menopausal symptoms and various disorders of the genital organs. It has been suggested for ethnomedicinal uses as an anti-fatigue, anti-inflammatory, anti-stress, anti-spasmodic and immune deficiency remedies in various ethnomedicines. In addition, it is used for spasmodic cough, diarrhoea and nausea of pregnant women. The peel has been reported to possess anticancer and antifungal properties.

Other Uses

Dioscorea esculenta tuber starch modified by carboxymethylation can be used as an excellent tablet disintegrant in low concentration (Nattapulwat et al. 2009). The tablets containing 1.0–4.0 % carboxymethyl yam starch disintegrated faster than 5 min.

The peelings are used for pig feed.

Comments

The plant is propagated by using healthy seed tubers and tubers divided into pieces of 250 g. These materials are established on mounds, ridges or on the flat.

Selected References

Aderiye BI, Ogundana SK, Adesanya SA, Roberts MF (1996) Antifungal properties of Yam (*Dioscorea alata*) peel extract. Folia Microbial (Praha) 41(5):407–442

Behera KK, Sahoo S, Prusti A (2010) Biochemical quantification of diosgenin and ascorbic acid from the tubers of different *Dioscorea* species found in Orissa. Libyan Agric Res Cent J Int 1(2):123–127

Burkill I H (1951) Dioscoreaceae. In: van Steenis CGGJ (ed) Flora Malesiana; ser. I; ol. 4. Noordhoff-Kolff, Djakarta, pp 293–335; 631 pp

Burkill IH (1966) A dictionary of the economic products of the Malay Peninsula. Revised reprint of 1st ed 1935. 2 volumes. Ministry of Agriculture and Co-operatives, Kuala Lumpur, Malaysia. Vol 1 (A–H) pp 1–1240, Vol 2 (I–Z). pp 1241–2444

Coursey DA (1968) The edible aroids. World Crops 20(4):25–30

Estiasih T, Sunarharum WB, Suwita IK (2012) Efek hipoglikemik polisakarida larut air gembili (*Dioscorea esculenta*) yang diekstrak dengan berbagai metode [Hypoglycaemic Effect of water soluble polysaccharides extracted from gembili (*Dioscorea esculenta*) by various methods]. J Teknol Industri Pangan 23(1):1–8 (in Indonesian)

Hahn SK (1995) Yams *Dioscorea* spp. (Dioscoreaceae). In: Smartt J, Simmonds NW (eds) Evolution of crop plants, 2nd edn. Longman Science & Technology, Harlow, pp 112–120; 531 pp

Harijono ET, Sunarharum WB, Hartono MD (2013) Hypoglycemic effect of biscuits containing water-

soluble polysaccharides from wild yam (*Dioscorea hispida* Dennts) or lesser yam (*Dioscorea esculenta*) tubers and alginate. Int Food Res J 20(5):2279–2285

Harmayani E, Winarti S, Nurismanto R (2011) Preparation of inulin powder from *Dioscorea esculenta* tuber with foam mat drying method. Paper present at the 121th Asean Food Conference, 16th -18th June 2011, BITEC Bangna, Bangkok, Thailand

Ireland CR, Passam HC (1984) The level and distribution of phenolic plant growth inhibitors in yam tubers during dormancy. New Phytol 97:233–242

Jayakody L, Hoover R, Liu Q, Donner E (2006) Studies on tuber starches. II. Molecular structure, composition and physicochemical properties of yam (*Dioscorea* sp.) starches grown in Sri Lanka. Carbohydr Polym 69(1):148–163

JIRCAS (2010) Local vegetables of Thailand. Japan International Research Center for Agricultural Sciences. http://www.jircas.affrc.go.jp/project/value_addition/Local_Vegetables_of_Thailand_home.html

Krochmal AC (1973) A guide to the medicinal plants of the United States. Quadrangle, the New York Times Book Co

Leung WTW, Butrum RR, Huang Chang F, Narayana Rao M, Polacchi W (1972) Food composition table for use in East Asia. FAO, Rome. 347 pp

Murugan M, Mohan VR (2012) *In vitro* antioxidant studies of *Dioscorea esculenta* (Lour). Burkill Asian Pac J Trop Biomed 2(3):S1620–S1624

Nattapulwat N, Purkkao N, Suwithayapan O (2009) Preparation and application of carboxymethyl yam (*Dioscorea esculenta*) starch. AAPS PharmSciTech 10(1):193–198

Olayemi JO, Ajaiyeoba EO (2007) Anti-inflammatory studies of yam (*Dioscorea esculenta*) extract on wistar rats. Afr J Biotechnol 6(16):1913–1915

Onwueme IC (1996) *Dioscorea esculenta* (Lour.) Burkill. In: Flach M, Rumawas F (eds) Plant Resources of South-East Asia, no. 9. Plants yielding non-seed carbohydrates. Prosea Foundation, Bogor, Indonesia, pp 93–95, 237 pp

Panneerselvam R, Abdul Jaleel C (2008) Starch and sugar conversion in *Dioscorea esculenta* tubers and *Curcuma longa* rhizomes during storage Caspian. J Environ Sci 6(2):151–160

Panneerselvam R, Abdul Jaleel C, Somasundaram R, Sridharan R, Gomathinayagam M (2007) Carbohydrate metabolism in *Dioscorea esculenta* (Lour.) Burk. tubers and *Curcuma longa* L. rhizomes during two phases of dormancy. Colloids Surf B Biointerfaces 59(1):59–66

Powell JM (1976) Ethnobotany. In: Paijmans K (ed) New Guinea vegetation. Elsevier, Amsterdam, pp 106–183, 212 pp

Senanayake SA, Ranaweera KKDS, Bamunuarachchi A, Gunaratne A (2013) Proximate analysis and phytochemical and mineral constituents in four cultivars of yams and tuber crops In Sri Lanka. Trop Agric Res Ext 15(1):32–36

Shajeela PS, Mohan VR, Jesudas LL, Soris PT (2011a) Antifertility activity of ethanol extract of *Dioscorea esculenta* (L.) Schott on male albino rats. Int J PharmTech Res 3(2):946–954

Shajeela PS, Mohan VR, Jesudas LL, Soris PT (2011b) Nutritional and antinutritional evaluation of wild yam (Dioscorea spp.). Trop Subtrop Agroecosyt 14:723–730

Sofowora A (1993) Medicinal plants and traditional medicine in Africa. Spectrum Books Limited, Ibada

Stuart GU (2013) Philippine alternative medicine. Manual of some Philippine medicinal plants. http://www.stuartxchange.org/OtherHerbals.html

The Plant List (2014) Version 1.1. Published on the Internet. http://www.theplantlist.org/. Accessed 20 Nov 2014

Ting CC, Gilbert MG (2000) Dioscoraceae R. Brown. In: Wu ZY, Raven PH (eds) Flora of China. Vol. 24 (Flagellariaceae through Marantaceae). Science Press, Beijing, and Missouri Botanical Garden Press, St. Louis

Ukpabi UJ (2010) Farmstead bread making potential of lesser yam (*Dioscorea esculenta*) flour in Nigeria. Aust J Crop Sci 4(2):68–73

Walter A, Lebot V (2007) Gardens of Oceania. ACIAR monograph no. 122. Canberra, Australia, 326 pp

Wanasundera JPD, Ravindran G (1992) Effects of cooking on the nutrient and antinutrient contents of yam tubers (*Dioscorea alata* and *Dioscorea esculenta*). Food Chem 45(4):247–250

Winarti S, Harmayani E, Marsono Y, Pranoto Y (2013) Effect of inulin isolated from lesser yam (*Dioscorea esculenta*) on the growth of probiotics bacteria and SCFA formation during fermentation. Int Res J Microbiol 4(2):53–63

Dioscorea hamiltonii

Scientific Name

Dioscorea hamiltonii **Hook. f.**

Synonyms

Dioscorea persimilis Prain & Burkill (Plate 2), *Dioscorea persimilis* var. *pubescens* C.T. Ting & M.C. Chang, *Dioscorea raishaensis* Hayata, *Dioscorea persimilis* var. *wukangensis* Hand.-Mazz.

Family

Dioscoreaceae

Common/English Names

Mountain Yam, Yam

Vernacular Names

Bangladesh: Murmujja Amiala (Chakma), Brara (Bawm), Thakan Budo, Tan Pawa (Marma), Tha Lang Ta (Tripura);
Chinese: Guan Shan Yao, He Bao Shu Yu, Yuan Bian Zhong;
French: Igname;
India: Naagar-kanda (Bihar), Moodavenni, Sandikilangu, Thalikkizhangu, Venni (Malayalam), Merumtua Sanga, Palru, Piska Sanga, Sika Kanda, Suta Alu (Oriya);
Vietnamese: Cù Mai, Khoai Mai, Son Duoc, Man Chèn (Tày), Hoài Son, Co Man Kep (Thái), Man On (Nung), Gò Lôn (K'dong), Hia Dòi (Dao).

Origin/Distribution

The species is found in Himalayas (Nepal, Sikkim, Bhutan), Western Ghats, Assam, Orissa and Bengal in northeast India, Peninsular Thailand, Vietnam and Laos (south China) and Taiwan. It grows wild in the lowlands and mountainous regions and is also cultivated for its edible and medicinal tubers.

Agroecology

The plant occurs wild in mixed forests, scrub forests, mountain slopes, valleys, roadsides from 100 to 2000 m elevation. It thrives in a hot, humid or moderate climate, with average temperature of 20–30 °C and annual rainfall of over 2000 mm. It does well in the mountains in the

T.K. Lim, *Edible Medicinal and Non-Medicinal Plants: Volume 10, Modified Stems, Roots, Bulbs*,
DOI 10.1007/978-94-017-7276-1_12

Plate 1 Fresh young planting of mountain Yam

midland and plains. It is hygrophilous, shade-tolerant and does best on fertile, well-drained sandy loam soils. It is intolerant of water-logged soils.

Edible Plant Parts and Uses

The tubers are edible. The rhizome is commonly consumed by the native ethnic minorities and is also cultivated for medicinal purpose in Vietnam. *Chè củ mài* is made from the rhizomes of *D. persimilis. Chè* is a traditional Vietnamese thick, sweet dessert soup or pudding. In Odisha, India, tubers are eaten as chips in the morning (Kumar et al. 2012).

In Bangladesh, tubers are boiled and eaten by the Chakma, Tripura, Marma, Bawm ethnic communities (Uddin 2014). The tuber is eaten with meat in curries by the Bawm community.

Botany

Glabrous climber. Tuber, single or paired, stout, cylindrical, vertical and slightly flat; the tip is round, penetrating deep into the soil. Tubers, yellowish-brown externally and have white flesh. Stems twinning to right, slender, hard, often 4–8 ridges usually bearing axillary bulbils (aerial tubers). Leaves, alternate or opposite, broadly ovate-cordate to elliptic-ovate 4–15 by 2–13 cm, papery, apex acute, acuminate or caudate, base sagittate or hastate, margin entire, main veins 5–7 radiating from base (Plates 1 to 3). Inflorescence in axillary racemes; flowers small, yellow, unisexual, dioecious, perianth of six equal segments. Male spikes, 2–4 together, 40 cm long and rachis distinctly zig-zagged. Male flowers with purplish-brown spotted, ovate bracts and 6 stamens. Female spikes solitary or paired, curved rachis reaching 20 cm long. Female flowers with ovate outer perianth lobes and small staminodes. Fruit, an oblate capsule, 1.5–2.5 cm across, 3- winged, membranous. Seeds with basal wings, greyish-brown.

Nutritive/Medicinal Properties

The levels of the following mineral elements Mn, Fe, Zn, Ca, Cu, Mg, P, Se, Cu and Zn in Guangshanyao (*D. persimilis*) rhizomes were much lower than that in serum of normal people

Plate 2 Plant label in Vietnamese

Plate 3 Mature foliage

(Huang et al. 2002). Traces of As and Pb were also present.

The rhizome contained hydrocarbons 63.25 %, proteins 6.75 %; lipids 0.45 %; mucilaginous substances 2–2.8 %, saponins: dioscin, sapotoxin, allantoin and dioscorin; and amino acids (arginine, choline) (Nguyen and Doan 1989; NIMM 1999).

Rhizomes from various localities in southern China were reported to have the following proximate nutrient composition of the rhizomes moisture 13.96–18.52 %, protein 7.06–9.79 %, starch

79.13–86.51 % and amino acids (18) 5.23–6.83 % (He et al. 2002). The tubers contained 85.5 % carbohydrates and 8.30 % albuminoids (Khare 2004). Diosgenin was identified in leaf samples of all the 12 *Dioscorea* species from Southern Western Ghats of India while tuber samples of only 4 species viz., *Dioscorea hispida, D. hamiltonii, D. spicata* and *D. pubera* revealed its presence (Asha and Nair 2005). Of these, the maximum diosgenin yield was recorded in *D. pubera* (1220 μg/g d. wt.) followed by *D. spicata* (305 μg/g d. wt.), *D. hispida* (57 μg/g d. wt) and *D. hamiltonii* (3 μg/g d. wt.). Among the different *Dioscorea* species found in Orissa, *D. hamiltonii* was found to have a low content of diosgenin, 138.83 mg/100 g tuber and lowest vitamin C content of 4.133 mg/100 g (Behera et al. 2010).

Moisture content of starches isolated from ten different *Dioscorea* plants, including *Dioscorea persimilis* ranged from 7.52 to 15.75 % and protein content varied between 0.010 and 0.028 % (Jiang et al. 2014). All of the starches gave a typical C-type X-ray diffraction pattern except D.BY (C_A-type pattern) and *Dioscorea persimilis* (B-type pattern) starches. The relative crystallinity of them varied from 12.02 to 51.68 %. The starches displayed significant variability in thermal transition temperatures and susceptibility to *in-vitro* digestion, and varied in rapidly digestible starch (RDS), slowly digestible starch (SDS), resistant starch (RS) contents, hydrolysis index (HI) and glycemic index (GI).

The amylose content of *Dioscorea zingiberensis*, *Dioscorea persimilis* and *Dioscorea opposita* starches decreased after 15 days of alkaline treatment and then increased after 30 days of alkaline treatment (Jiang et al. 2014). There were similar changes in relative crystallinity for *D. zingiberensis* and *D. persimilis* starches. It was observed that the three starches displayed a reduction trend in swelling power with a significant increase in solubility. Adhesion among some of the starch granules was observed after alkali treatment for 30 days in *D. zingiberensis* and *D. opposita* starches, while *D. persimilis* starch showed some hollows on the granule surface. The rise in the absorbance ratio was observed during alkali treatment for *D. persimilis* and *D. opposita* starches. Alkali treatment elevated the *in-vitro* digestibility with resistant starch values climbing up from 50.16 % to 64.95 % and from 66.14 % to 70.74 % for *D. zingiberensis* and *D. persimilis* starches, respectively, but there was no significant change in resistant starch value for *D. opposita* starch.

Spleen Deficiency Amelioration

In-vivo studies showed that administration of *D. persimilis* ameliorated syndrome of spleen asthenia in mice (Qin et al. 2003). Body weight, body temperature, the weight of thymus glands and spleen and the level of xylose in the serum were increased remarkably by *D. persimilis*. The effect with *D. opposita* was weaker.

Clinical observation study of 60 cases of spleen deficiency syndrome showed that oral administration for 15 days of *Dioscorea persimilis* (Guang Shan Yao) and *Dioscorea* opposita (Huai Shan Yao) aqueous paste extract ameliorated symptoms of the spleen deficiency and strengthened the spleen and nourished the stomach (Fang et al. 2002).

Antimicrobial Activity

The methanol, ethanol and ethyl acetate extracts of *D. hamiltonii* leaves showed good antimicrobial activity in-vitro against Gram-negative and Gram-positive bacteria *Escherichia coli, Lactobacillus delbrueckii subsp lactis, Vibrio parahaemolyticus, Pseudomonas aeruginosa, Proteus Vulgaris, Streptococcus pyogenes, Streptococcus thermophilus, Bacillus subtilis, Enterococcus faecalis* and the yeast *Saccharomyces cerevisiae* (Kaladhar et al. 2010). The aqueous leaf extract showed comparatively weaker activity. All the leaf extracts showed comparatively weaker activity against fungi tested, *Aspergillus niger* and *Penicillium chrysogenum*.

Hormonal Activity

Studies reported that *D. persimilis* may have anabolic and gonadotrophic activities (NIMM 1999). Oral administration of the rhizome powder for 28 consecutive days resulted in increase in uterus weight of female albino rats and weight increase of vesicles, prostate and seminal vesicle and anus levator muscle in male albino rats.

Traditional Medicinal Uses

According to traditional Vietnamese medicine, *D. persimilis* tones up the spleen and stomach, invigorates the lung and kidneys, and promotes the production of body fluid and removes thirst (Nguyen and Doan 1989; NIMM 1999). The rhizomes provide a restorative and antifebrile remedy. They have a beneficial effect in dyspepsia, general debility, chronic enteritis, chronic diarrhoea and dysentery, spermatorrhoea, night sweats, diabetes mellitus, polyuria, metrorrhoea, lumbago, vertigo and photopsia. They are also employed externally for poulticing boils.

In Bangladesh, the leaf paste is employed by the Chakma ethnic people for the treatment of jaundice and mumps (Uddin 2014).

Other Uses

In north Vietnam, some varieties are mainly cultivated as a medicinal crop.

Comments

The plant is readily propagated from crown cuttings or aerial tubers.

Selected References

Asha KI, Nair GM (2005) Screening of *Dioscorea* species for Diosgenin from Southern Western Ghats of India. Indian J Plant Gent Resour 18(2):227–230

Behera KK, Sahoo S, Prusti A (2010) Biochemical quantification of diosgenin and ascorbic acid from the tubers of different *Dioscorea* species found in Orissa. Libyan Agric Res Cent J Int 1(2):123–127

Fang XM, Lin SN, Huang B, Feng XZ, Luo ZL (2002) Clinical observation on the syndrome of spleen deficiency treated by *Dioscorea persimilis* Prain. J Guangxi Trad Chin Med Univ 5(3):19–21 (in Chinese)

Govaerts R, Wilkin P, Raz L, Téllez-Valdés O (2006) World checklist of Dioscoreaceae. Facilitated by the Royal Botanic Gardens, Kew. Published on the Internet. http://apps.kew.org/wcsp/. Retrieved 22 Feb 2014

He BZ, Jiang JP, Mo JG, Huang DP, Zhou L, Ban HL, He YZ (2002) Study on the quality of rhizome of *Dioscorea persimilis*. Zhong Yao Cai 25(4):233–237 (in Chinese)

Huang DP, Mo JG, Lao YW, Huang YF, He YZ, He BZ (2002) Analysis and comparison of sixteen kinds of elements in Guangshanyao and Huaishanyao. Guangdong Trace Elem Sci 9(2):47–50 (in Chinese)

Jiang QQ, Gao WY, Li X, Man SL, Shi YP, Yang YX, Huang LQ, Liu CX (2014) Comparative susceptibilities to alkali-treatment of A-, B-and C-type starches of *Dioscorea zingiberensis, Dioscorea persimilis* and *Dioscorea opposita*. Food Hydrocoll 39:286–294

Kaladhar DSVGK, Rao VN, Barla S, Harasreeramulu S (2010) Comparative antimicrobial studies of *Dioscorea hamiltonii* Hook. f. tubers with *Azadirachta indica* stem. J Pharm Sci Technol 2:284–287

Khare CP (2004) Indian medicinal plants: an illustrated dictionary. Springer, Heidelberg, 900 pp

Kumar S, Jena PK, Tripathy PK (2012) Study of wild edible plants among tribal groups of Simlipal Biosphere Reserve Forest, Odisha, India; with special reference to *Dioscorea* species. Int J Biol Technol 3(1):11–19

National Institute of Materia Medica (NIMM) (1999) Selected medicinal plants in Vietnam, vol 1. Science and Technology Publishing House, Hanoi. 439 pp

Nguyen VD, Doan TN (1989) Medicinal plants in Vietnam World Health Organization (WHO), Regional Publications, Western Pacific Series No 3. WHO, Regional Office for the Western Pacific, Manila, The Philippines and Institute of Materia Medica, Hanoi, Vietnam

Qin JJ, Zhou F, Fang H, Lin HS, Liang DY, HE BZ (2003) Effect of *Dioscorea persimilis* on syndrome of splenasthenia in mice. Lishizhen Med Mat Med Res 14(3):F003–F004

Ting CC, Gilbert MG (2000) Dioscoreaceae R. Brown. In: Wu ZY, Raven PH (eds) Flora of China. Flagellariaceae through Marantaceae, vol 24. Science Press/Missouri Botanical Garden Press, Beijing/St. Louis

Uddin SB (2014) *Dioscorea hamiltonii*. Bangladesh Ethnobotany Online Database. http://www.ebbd.info/

Dioscorea hispida

Scientific Name

***Dioscorea hispida* Dennst.**

Synonyms

Dioscorea daemona Roxburgh, *Dioscorea daemona* var. *reticulata* Hook.f., *Dioscorea hirsuta* Blume, *Dioscorea hispida* var. *daemona* (Roxb.) Prain & Burkill, *Dioscorea hispida* var. *reticulata* (Hook.f.) Sanjappa, *Dioscorea lunata* Roth, *Dioscorea mollissima* Blume, *Dioscorea triphylla* var. *daemona* (Roxb.) Prain & Burkill, *Dioscorea triphylla* var. *mollissima* (Blume) Prain & Burkill, *Dioscorea triphylla* var. *reticulata* (Hook.f.) Prain & Burkill, *Helmia daemona* (Roxb.) Kunth, *Helmia hirsuta* (Blume) Kunth

Family

Dioscoreaceae

Common/English Names

Asiatic Bitter Yam, Intoxicating Yam, Spirit Yam, Wild Yam

Vernacular Names

Burmese: Kywe;
Chinese: Bai Shu Liang;
French: Igname Épineuse Amčre, Morsure De Cobra;
German: Bittere Yamswurzel, Bittere Yamswurzel, Giftige Yams;
India: Bolkande, Podava-Kizhangu, Podava Kelengu, Podukkilangu, Venni (Malayalam), Baichandi, Bhul Kand, Dukar Kand (Marathi),

T.K. Lim, *Edible Medicinal and Non-Medicinal Plants: Volume 10, Modified Stems, Roots, Bulbs*,
DOI 10.1007/978-94-017-7276-1_13

Baikanda, Banya Alu, Bainya Alu, Hasar Sanga, Khulu Sanga, Kolhua, Kolokanda, Kulia Kulia Kanda (Oriya), Hastyaluka, Marpa Shpoli (Sanskrit), Kavalakodi, Pei Perendai, Periperendai (Tamil), Chanda Gadda, Puli Dumpa, Tellaagini Geddalu, Thella Chanda Gadda, Thella Gadda (Telugu), Magasirigadda, Peccheruvu (Andhra Pradesh);

Indonesia: Ghadhung (Madurese), Gadung, Gadung Ketan, Gadung Kuning, Gadung Padi (Malay), Gadung (Javanese), Gadung, Huwi Gadung (Sundanese);

Japanese: Mitsuba Dokoro;

Khmer: Dâmlô:Ng K'duöch;

Laos: Houo Koi;

Malaysia: Gadog, Gadong, Gadong Lilin, Gadong Mabok, Taring Pelanduk, Tuba Ubi, Ubi Arak, Ubi Akas, Ubi Cerok, Ubi Kendudok, Ubi Kipas, Ubi Nasi (Malay), Ha, Ha-U (Pagan), Bigap, Gadongan, Gang, Gong, Kedut, Kuoi, Kuoe, Ki-E, Sulur Gadong (Sakai), Ha, Ha'u, Hubi Gak (Semang), Bekoi, Bekoya, Gakn, Ubi Bekoi (Tembe);

Philippines: Mamo (Bikol), Gigos, Kalot, Orkot (Bisaya), Karot (Iloko), Bagai (Manobo), Kahit (Pampangan), Korot (Samar-Leyte Bisaya), Kahit, Kulot (Sambali), Karoti (Sulu), Kahit, Kayos, Nami (Tagalog);

Taiwan: Bai, Da;

Thailand: Khli, Koi, Kloi, Kloi-Nok, Kloy Kao Niaw, Man Kloi;

Vietnamese: Củ Năn, Củ Nần, Củ Nê, Củ Nâu Trắng, Dây Năn, Mài Lông

Origin/Distribution

The species is indigenous to tropical Asia (India to southwest China, Taiwan, Andaman Islands to Papua New Guinea and the Philippines).

Agroecology

Its natural habitat is chiefly in tropical rain forests, thickets and scrub forests, forest margins, at low and medium altitudes; near sea level to 1500 m. The plant thrives best on soils rich in humus. The plant is seldom cultivated, and tubers are harvested from the wild.

Edible Plant Parts and Uses

The tuber of this species is the chief famine food of tropical Asia and has been used as a staple food during World War I. The tuber has to be thoroughly processed to remove the toxic alkaloid dioscorine before it is consumed. Generally, the tubers are peeled, sliced, washed and boiled in several changes of salt water to remove the toxic alkaloid before it is eaten. In Java, the tubers are sliced, coated with ashes for a day, steeped in sea water for a few days, washed and dried in the sun (Ochse and Bakhuizen van den Brink 1980). To make *kripik gadung*, the tubers are peeled, mixed with ashes and sun dried for several days, washed and dried and then fried in oil and eaten. Starch extracted from the tubers is used for culinary or industrial purposes like the manufacture of glucose. The tubers can be ground to flour for cakes and puddings. In Thailand, kloi tubers are used to make a dessert called *Kao Nuew Kloi* (Theerasin and Baker 2009). In Odisha, the tubers are sliced soaked in running water and boiled successively with leaves of Matha sag (*Antidesma acuminata*) or tamarind. After filtering excess water, the slices are cooked in curry or eaten as such during food scarcity (Kumar et al. 2012a; Misra et al. 2013). In Terengganu, Malaysia, *D. hispida* is used in making popular traditional desserts and cakes such as *kuih putri mandi, kuih onde-onde, kuih cik mek baru, kuih koleh, lempeng, pengat* and porridge (Nashriyah et al. 2012). Additionally, the villagers eat yam after removal of toxin with glutinous rice and grated coconut for especially breakfast during the rainy season.

Botany

A robust, herbaceous, perennial, dextrose (left twining) vine reaching several meters (Plate 1). Stem terete, stout, pubescent when young, glabrous and prickly when mature. Tubers subterra-

nean, numerous, densely crowded in clusters, irregularly shaped and variable in size (Plate 3), compressed-oblong, angular, ovoid-globular, lobed tuberculate or not, yellowish-brown to light grey, beset with short or long, stiff or susub-spinous rootlets, transverse section white or yellow. Leaves alternate, palmately 3-foliolate (Plate 2); petiole to 30 cm, hairy; leaflets subequal, middle leaflet ovate to elliptic, 6–12(−17.5)×4–12 cm, adaxially sparsely hispid, glabrescent, abaxially hispid, palmately veined, apex acuminate; lateral leaflets ovate-elliptic or nearly broadly oblong, oblique, smaller than middle leaflet, margin entire. Bulbils absent. Male inflorescence axillary panicles to 50 cm with two levels of branching, most parts densely tomentose. Male flowers sessile, in dense clusters; perianth. 1 mm, yellow, outer lobes smaller and thinner than inner ones; stamens 6. Female inflorescence solitary, to 40 cm. Female flowers sessile, with densely pubescent 3-celled ovary. Fruit woody oblong to ellipsoid capsule, 3.5–7 cm, leathery, densely pubescent; honey coloured, three wings 1.2–1.5 cm wide. Seeds inserted near apex of capsule; wing pointing towards capsule base.

Nutritive/Medicinal Properties

Harijono et al. (2013) reported the following nutrient composition of *D. hispida* tuber water-soluble polysaccharides: water 7.96 %, ash 0.49 %, protein 5.59 %, fat 2.55 %, carbohydrate 83.40 %, crude fibre 3.19 % and starch 14.61 %. Burkill (1966) reported that the tuber contained 58 % moisture and 24 % starch or 57 % starch on a dry basis.

The alkaloid dioscorine was first isolated from the tuber by Leyva and Gutierrez (1937). Numerous studies had been reported on the biosynthesis of dioscorine in *D. hispida* using radiolabelled compounds. Reductive degradation of the lactone ring of dioscorine of *D. hispida* yielded a saturated tertiary alcohol (Pinder 1957). From infra-red studies on the above tertiary alcohol dioscorine was found too be derived from 7α-hydroxytropane. The administration of [1-^{14}C]-acetate of *Dioscorea hispida* plants afforded radioactive dioscorine (0·2 % incorporation) (Leete and Pinder 1972). Dioscorine obtained from *Dioscorea hispida* plants, which had been fed [6-^{14}C]-Δ^1-piperideine, had low activity (0·03 % incorporation). The administration of nicotinic-[2-^{14}C] acid to *Dioscorea his-*

Plate 1 Robust, dextrose twining habit

Plate 2 Digitately trifoliate leaves

Plate 3 Irregularly shaped subterranean tubers covered with stiff rootlets

pida plants afforded radioactive dioscorine (1.9 % absolute incorporation) and a systematic degradation of the alkaloid indicated that essentially all the activity was located at C-3 (Leete 1977). [3-^{14}C]- and [3,3½-$^{13}C_2$,3-^{14}C]-3-Hydroxy-3-methylglutaric acid were administered to *Dioscorea hispida* plants, resulting in the formation of labelled dioscorine (0.2 % absolute incorporation). A chemical degradation indicated that there was a non-random incorporation of ^{14}C into the lactone ring of the alkaloid, the major part of the radioactivity being at its C-10 position (Leete and Michelson 1989). The results suggested 3-hydroxy-3-methylglutaric acid, or more likely,

its monocoenzyme A ester, was an intermediate between acetate and the branched eight-carbon unit required for the biosynthesis of dioscorine. They also reported the biosynthesis of disocorine from trigonelline in *D. hispida* (Leete and Michelson 1988). [6-^{14}C,2-^{3}H]Nicotinic acid, [6-^{14}C,6-^{3}H]nicotinic acid, [methyl-^{14}C,2-^{2}H,^{3}H] trigonelline, and [methyl-^{14}C,6-^{2}H,^{3}H] trigonelline were all incorporated into the isoquinuclidine moiety of the alkaloid dioscorine found in *Dioscorea hispida* with complete retention of ^{3}H relative to ^{14}C. A chemical degradation on the dioscorine derived from the labelled trigonellines indicated that all the ^{14}C was located on its *N*-methyl group.

Phenolic compounds found in *D. hispida* tuber peel were caffeic acid, chlorogenic acid, p-hydroxybenzaldehyde and methylester of protocatechuic acid (methoxyprotocatechuate); only methylester of protocatechuic acid was found in the tuber flesh (Theerasin and Baker 2009).

A saponin glycoside prazerigenin A (3-*O*-β-D-glucopyranoside) was isolated from *D. hispida* (Nguyen et al. 1987). Diosgenin was identified in leaf samples of all the 12 *Dioscorea* species from Southern Western Ghats of India while tuber samples of only 4 species viz., *Dioscorea hispida, D. hamiltonii, D. spicata* and *D. pubera* revealed its presence (Asha and Nair 2005). Of these, the maximum diosgenin yield was recorded in *D. pubera* (1220 μg/g d. wt.) followed by *D. spicata* (305 μg/g d. wt.), *D. hispida* (57 μg/g d. wt) and *D. hamiltonii* (3 μg/g d. wt.). Among the different *Dioscorea* species found in Orissa, *D. hispida* was found to have the third highest content of diosgenin, 825 mg/100 g tuber and second highest vitamin C content 7.43 mg/100 g (Behera et al. 2010).

Dioscorea hispida yam starch when hydrothermally modified at 13 g water/100 g starch (wb) gave starch gel with higher complex modulus (G^*) when compared to its native starch gel (Tattiyakul et al. 2006). The X-ray diffraction pattern revealed that starch granule's crystalline structure changed from B to C-type when modified at 13 g water/100 g starch (wb) and stayed unchanged when modified at 18–30 g water/100 g starch (wb). They also found that the crystallographic pattern of the starch was altered from type B in native starch to type C in the starch modified at 90 °C and to type A in those modified at 100–130 °C (Tattiyakul et al. 2012). Along with the change in crystallographic pattern, decrease in granule swelling, starch solubility and amylose leaching was observed. Up to the modification temperature of 100 °C, a reduction in the estimated degree of crystallinity and an increase in peak viscosity were observed. Seven percent starch gel of all modified starches, except for that modified at 130 °C, showed higher complex moduli over 0.001 to 10 Hz in dynamic shear test when compared with 7 % native starch gel.

Ginger oil-modified *D. hispida* flour was found to have similar water solubility, swelling and gelatinization properties as American wheat flour, with values of 7.28 (g/100 g), 7.9 (g/g) and 56.2 °C, respectively (Kumoro et al. 2012). However, a setback was the presence of the remaining ginger aroma in the flour.

Antioxidant Activity

D. hispida tuber was found to have antioxidant potential (Theerasin and Baker 2009). The DPPH value for 20 % methanol peel and flesh fractions were EC_{50} 1.05×10^2 μg/ml and EC_{50} 2.04×10^2 μg/ml, respectively. FRAP values were 3.26×10^2 μmol Fe^{2+}/mg for peel and 2.93×10^2 μmol Fe^{2+}/mg for flesh. The total phenol content of peel was 0.068 mg GAE/mg and 0.085 mg GAE/mg for flesh.

Hypoglycaemic Activity

Administration of aqueous extract of *D. hispida* tuber water-soluble polysaccharide (WSP), papain assisted-WSP extract, and tempeh inoculum assisted-WSP extract reduced blood glucose level in alloxan-induced hyperglycaemic rats (Estiasih et al. 2012). All the extracts inhibited glucose absorption and short chain fatty acids (SCFAs) formation. The order of hypoglycaemic activity was tempeh assisted-WSP > papain

assisted WSP>aqueous WSP. The order of glucose absorption inhibition was papain assisted WSP>tempeh assisted-WSP>aqueous WSP. Tempeh inoculum assisted-WSP extract exhibited the most significant hypoglycaemic activity. In another study, feeding biscuits (BIS) containing water-soluble polysaccharides from wild yam (*Dioscorea hispida*) (WY) or lesser yam (*Dioscorea esculenta*) (LY) tubers and commercial alginate (ALG) exerted a hypoglycaemic effect on alloxan-induced hyperglycaemic rats (Harijono et al. 2013). All hydrocolloid containing biscuits affected the absorption of glucose into the blood and were able to maintain it at a normal level. After 4 weeks, the glucose blood levels were significantly reduced: ALG-BIS (57 % reduction) was the most effective, followed by WY/ALG-BIS (50 % reduction), LY/ALG-BIS (49 % reduction), WY-BIS (39 % reduction) and LY-BIS (35 %). The order of concentration of the SCFAs profile from the caecum digesta at the end of feeding treatment was similar, i.e. acetic acid>propionic acid>butyric acid.

Analgesic Activity

Leaf extract of *D. hispida* dose-dependently inhibited writhing response induced by acetic acid in mice; 200 mg body weight of the extract caused 78.85 % writhing inhibition whereas the standard drug indomethacin produced 69.42 % inhibition (Panduranga Murthy et al. 2011). The extract also showed significant antinociceptive action in the hot plate reaction time assay in mice; the effect was comparable to the standard drug pentazocine.

Antiinflammatory Activity

Leaf extract of *D. hispida* at a dose of 200 mg/kg body weight exerted 57.4 % inhibition in the carrageenan-induced paw oedema in rats, whereas the standard drug indomethacin produced 64.56 % inhibition (Panduranga Murthy et al. 2011).

Anticancer Activity

Studies showed that the ethanolic leaf extract of *Dioscorea hispida* possessed significant anticancer activity at the dose varying from 100 to 200 mg/kg body weight (Punithkumar et al. 2011). It increased the survival of Swiss albino mice with ascites tumour; decreased the body weight induced by the tumour burden; reduced packed cell volume and viable tissue cell count and also reduced elevated level of lipid peroxidation due to higher content of phenolic compounds.

Central Nervous System Activity

In rat clonal phaeochlomocytoma (PC12) cells, dioscorine from *D. hispida* tubers, accelerated the desensitization of current induced by 100 μMM acetylcholine at concentrations of 0.45–450 μMM and suppressed the current in a dose-dependent manner (Nagata et al. 1999). The suppressive effect of dioscorine on the acetylcholine-induced current was suggested to play an important role in its toxic actions in animals. Also, dioscorine had been reported to have insecticidal and antifeedant activities.

Miscellaneous Pharmacological Activities

Studies showed that the alkaloid dioscorine from *D. hispida* exhibited convulsant activity, toxicity, analeptic action, local anaesthetic activity, adrenaline potentiating action, antidiuretic effect and anti-acetylcholine activity (Broadbent and Schneiden 1958). In rats and mice, dioscorine caused convulsions similar to those produced by picrotoxin. The LD_{50} value in mice was 60 mg/kg. In isolated guinea pig ileum, disocorine showed slight antiacetylcholine activity at a concentration of 10^{-6}. Dioscorine 2 mg had no effect on isolated rabbit heart but diminished the responses of the heart to subsequent injection of acetylcholine. Forty mg/kg disocorine, when administered intravenously, increased the respi-

ratory rates of anaesthetized rats. A 5 % dioscorine when injected intradermally into guinea pigs had local anaesthetic activity almost equivalent to 0.05 % cocaine. When injected intravenously at doses of 10–20 mg/kg, dioscorine did not elicit any significant changes in the cat blood pressure, but the blood response to acetylcholine and adrenaline were altered. Mydriasis occurred during intravenous injection of the alkaloid. Dioscorine exhibited antidiuretic activity, 1 mg had the activity of about 100 μU of pitressin.

D. hispida extracts exhibited pharmacological properties similar to those of nicotine (Dejatowongse et al. 1980). Animals injected with the extract exhibited hyperpnea, tachycardia, increase in blood pressure, restlessness and convulsions. The rise in blood pressure, contraction of nictitating membrane and of smooth muscle induced by *D. hispida* were inhibited by previous treatment with hexamethonium. *D. hispida* stimulated the contraction of striated muscle; the effect was inhibited by pretreatment with d-tubocurarine.

Tuber Toxin Removal Studies

Hudzari et al. (2011) developed an automatic alkaloid removal system for the toxic poison dioscorine from *D. hispida* tuber, using the survival of *Cyprinus carpio* fish as a tool for detection of dioscorine removal. Kumoro et al. (2011) found that processing water flow rate, leaching time and steaming affected the efficiency of cyanides removal from *D. hispida* tubers in the leaching process. Best processing condition was at leaching using 5.00×10^{-5} m^3/s for 1 h, followed by steaming for 1 h to obtain cyanides content of 29.9 mg/kg. The yielded tuber chips were considered as safe for consumption. Sasongko (2012) used *Mucor* fermentation in the detoxification process to decrease the concentration of cyanide in *D. hispida* tuber so as to have safe yam such as gadung flour. The best treatment was the combination of 5 % mold and 72 h fermentation. The characteristics of this gadung flour were 4.01 % of moisture content; 21.74 ppm of cyanide content; 46.66 % of starch content; 3.6 % of fiber content; 17.00 % of total sugar content; 9.52 % reducing sugar content; L* colour of 60.97; a* of 13.27; b* of 18.45; 3.27 % of hygroscopicity, and starch gelatinization temperature of 83 °C.

Traditional Medicinal Uses

In the Philippines, tubers are used as an anodyne and maturative in cases of tumours and buboes, and also against arthritic and rheumatic pains and similar afflictions (Stuart 2013). In Bangladesh, tubers are used to kill worms in wounds. Various plant parts are used in whitlow, sores, boils and bites of rabbit, jackal or dog. According to Burkill (1966), in Peninsular Malaysia, the tubers are used medicinally for external application on puru sores, sometimes with lime or pounded with turmeric and benzoin and applied to sore feet; the leaves boiled in a decoction is applied on the feet. In Johore, a decoction of the tuber is used as an alternative and diuretic in chronic rheumatism. In Terengganu, the villagers used *D. hispida* as a deworming medicine (Nashriyah et al. 2012). The leaves are pounded and applied on the stomach. Pounded leaves are also used to cure stomach bloating, hernia and asthma. The Temuan tribe in Selangor, Malaysia, uses the pounded leaves from intoxicating yam for healing sores yaw sores (Hanum and Hamzah 1999)

Tubers of *Dioscorea hispida* are used for birth control among women in Odisha, India, as they contain dioscorine, a lactonic alkaloid, which is also used in the manufacture of birth control pills (Kumar et al. 2012b)

Other Uses

D. hispida tuber has insecticidal property. The methanol tuber extract of *D. hispida* showed strong feeding deterrent activity to diamondback moth (DBM), *Plutella xylostella* larvae (Banaag et al. 1997). Two alkaloids, A and B, were isolated as active components from alkaloid fraction. Alkaloid A was more active than alkaloid B at 100–250 μg/ml, and a binary mixture (1:1) of these two alkaloids inhibited larval feeding at

lower concentrations (50–100 μg/ml) than either alkaloid separately. Both alkaloids significantly retarded larval molting and reduced larval weight gain, and high mortality at the larval stage (70–86 %) and during emergence (98–100 %) resulted when DBM larvae were reared on treated radish seedlings.

Dioscorea hispida tuber extracts were found to have allelopathic effects (Nornasuha et al. 2012). The methanol crude extracts significantly inhibited the radicle length of all bioassay species; mustard (*Brassica* sp.), cucumber (*Cucumis* sp.), spinach (*Amaranthus* sp.) and radish (*Raphanus* sp.) at all concentrations. The water extract strongly inhibited the germination and radicle length of mustard at 12.5 g/L and 50.0 g/L concentrations, and at 25.0 g/L inhibited the germination and radicle length of cucumber. However, the concentration at 12.5 g/L stimulated the germination and radicle length of maize (*Zea mays*).

The juice of the tubers has been reported to be used in criminal poisoning by thieves with the intention of causing drowsiness to cover their visit (Burkill 1966). Tuber sap is used with *Antiaris toxicaria* poison in the preparation of poison darts for blowpipes by the Benua Jakuns, Sakai and Semang natives. The tuber has been used as fish poison and as bait to catch prawns in Terengganu, Malaysia (Nashriyah et al. 2012). In the Philippines, the flesh and sap of tubers is used for bleaching clothes and abaca fibres (Stuart 2013). Tubers are used as cure for myiasis of the scrotum in carabao. The people in Jeypore tract, Orissa India, reportedly use the intoxicating effect of *D. hispida* to forget their sorrows, as they get an effect similar to drinking beer (Mishra et al. 2011). In Bangladesh, tubers are used to kill worms in wounds. Various plant parts are used for whitlow, sores, boils and bites of rabbit, jackal or dog (Yusuf et al. 2009).

Comments

The plant can be propagated by planting pieces of tuber in prepared mounds.

Selected References

Asha KI, Nair GM (2005) Screening of *Dioscorea* Species for Diosgenin from Southern Western Ghats of India. Indian J Plant Gent Resour 18(2):227–230

Banaag A, Honda H, Shono T (1997) Effects of alkaloids from yam, *Dioscorea hispida* Schlussel, on feeding and development of larvae of the diamondback moth, *Plutella xylostella* (Lepidoptera: Yponomeutidae). Appl Entomol Zool 32(1):119–126

Behera KK, Sahoo S, Prusti A (2010) Biochemical quantification of diosgenin and ascorbic acid from the tubers of different *Dioscorea* species found in Orissa. Libyan Agric Res Cent J Int 1(2):123–127

Broadbent JL, Schneiden H (1958) A comparison of some pharmacological properties of dioscorine and dioscine. Brit J Pharmacol 12:213–215

Brown WH (1951) Useful plants of the Philippines, vol 1, Technical bulletin 10. Department of Agriculture and Natural Resources Rep, Manila, 590 pp

Bureau of Plant Industry (2005) Medicinal plants of the Philippines. Department of Agriculture Republic of Philippines. http://www.bpi.da.gov.ph/Publications/mp/mplants.html

Burkill IH (1951) Dioscoreaceae (pp 293–335). In: van Steenis CGGJ (ed) Flora Malesiana; ser. I; vol 4. Noordhoff-Kolff, Djakarta, 631 pp

Burkill IH (1966) A dictionary of the economic products of the Malay Peninsula. Revised reprint of 1st ed 1935. 2 volumes. Ministry of Agriculture and Co-operatives, Kuala Lumpur, Malaysia. Vol 1 (A–H) pp 1–1240, Vol 2 (I–Z) pp 1241–2444

Dejatowongse N, Ayudhya Q, Ketusinh O (1980) Pharmacological effects of the extract of *Dioscorea hispida*. Siriraj Med J 32(6):30–34

Estiasih T, Harijono SWB, Rahmawati A (2012) Hypoglycemic activity of water soluble polysaccharides of yam (*Dioscorea hispida* Dents) prepared by aqueous, papain, and tempeh inoculum assisted extractions. World Acad Sci Eng Technol 70:311–317

Hanum F, Hamzah N (1999) The use of medicinal plant species by the Temuan Tribe of Ayer Hitam Forest, Selangor, Peninsular Malaysia. Pertanika J Trop Agric Sci 22(2):85–94

Harijono ET, Sunarharum WB, Hartono MD (2013) Hypoglycemic effect of biscuits containing water-soluble polysaccharides from wild yam (*Dioscorea hispida* Dennts) or lesser yam (*Dioscorea esculenta*) tubers and alginate. Int Food RES J 20(5):2279–2285

Hudzari RM, Somad MAHA, Rizuwan YM, Asimi MNN, Abdullah ABC (2011) Development of automatic alkaloid removal system for *Dioscorea hispida*. Front Sci 1(1):16–20

Kumar S, Jena PK, Tripathy PK (2012a) Study of wild edible plants among tribal groups of Simlipal Biosphere Reserve Forest, Odisha, India; with special reference to *Dioscorea* species. Int J Biol Technol 3(1):11–19

Kumar S, Tripathy PK, Jena PK (2012b) Evaluation of nutritional and medicinal values of wild tuber crops (*Dioscorea* spp.) from Simlipal Biosphere Reserve Forest, Odisha, India. In: Proceeding of 37th annual conference and *P. Parija* memorial national conference on "Recent advances in plant biotechnology", 22–23 December 2012, pp 41–42

Kumoro AC, Retnowati DS, Budiyati CS (2011) Removal of cyanides from gadung (*Dioscorea hispida* Dennst.) tuber chips using leaching and steaming techniques. J Appl Sci Res 7(12):2140–2147

Kumoro AC, Retnowati DS, Budiyati CS (2012) Water solubility, swelling and gelatinization properties of raw and ginger oil modified gadung (*Dioscorea hispida* Dennst) flour. Res J Appl Sci Eng Technol 4(1):2854–2860

Leete E (1977) Biosynthesis of dioscorine: incorporation of nicotinic acid into the isoquinuclidine moiety. Phytochemistry 16(11):1705–1709

Leete E, Michelson RH (1988) Biosynthesis of dioscorine from trigonelline in *Dioscorea hispida*. Phytochemistry 27(12):3793–3798

Leete E, Michelson RH (1989) The incorporation of 3-hydroxy-3-methylglutaric acid into the lactone ring of dioscorine in *Dioscorea hispida*. Phytochemistry 28(12):3325–3330

Leete E, Pinder AR (1972) Biosynthesis of dioscorine. Phytochemistry 11(11):3219–3224

Leyva JF, Gutierrez (1937) Toxicological studies of nami, *Dioscorea hispida*. J Philipp Med Ass 17:349–356

Mishra S, Swaih S, Chaudhary SS, Ray S (2011) Wild edible tubers (*Dioscorea* spp.) and their contribution to the food security of tribes of Jeypore tract, Orissa, India. PGR Newsl FAO-Biodiv Issue 56:63–67

Misra RC, Sahoo HK, Pani DR, Bhandari DC (2013) Genetic resources of wild tuberous food plants traditionally used in Similipal Biosphere Reserve, Odisha, India. Genet Resour Crop Evol 60(7):2033–2054

Nagata K, Aistrup GL, Honda H, Shono T, Narahashi T (1999) Modulation of the nicotinic acetylcholine receptor by dioscorine in clonal rat phaeochlomocytoma (PC12) cells. Pestic Biochem Physiol 64(3):157–165

Nashriyah M, Nornasuha Y, Salmah T, Norhayati N, Mohd. Rohaizad (2010) *Dioscorea hispida* Dennst. (Dioscoreaceae): an overview. Buletin UniSZA, No. 4, Universiti Sultan Zainal Abidin, Kuala Trengganu, Malaysia

Nashriyah M, Salmah T, Nur Atiqah MY, Siti Nor Indah O, Muhamad Azhar AW, Munirah S, Nornasuha Y, Abdul Manaf A (2012) Ethnobotany and distribution of *Dioscorea hispida* Dennst. (Dioscoreaceae) in Besut, Marang and Setiu Districts of Terengganu, Peninsular Malaysia. World Acd Sci Eng Technol 72:240–243

Nguyen XC, Zeigan D, Voigt G, Hiller K (1987) Steroidal saponins from *Dioscorea hispida* L. Pharmazie 42(2):143

Nornasuha Y, Nashriyah M, Hasbullah M, Ain K, Mohd Noor Z, Adilah N, Azhar M (2012) Allelopathic effects of *Dioscorea hispida* Dennst. (Dioscoreaceae) tuber from Terengganu, Peninsular Malaysia. J Agric Technol 8(5):1855–1865

Ochse JJ, Bakhuizen van den Brink RC (1980) Vegetables of the Dutch Indies, 3rd edn. Ascher & Co., Amsterdam. 1016 pp

Onwueme IC (1996) *Dioscorea hispida* Dennst. L. In: Flach M, Rumawas F (eds) Plant resources of South-East Asia, No. 9. Plants yielding non-seed carbohydrates. Prosea Foundation, Bogor, pp 95–97, 237 pp

Panduranga Murthy G, Punith Kumar TG, Suresh A, Raviashankar HG, Chandrasekhar KB, Lokesh S (2011) Evaluation of ethanolic extract of *Dioscorea hispida* Dennst. for anti-inflammatory and analgesic activities. Int J Pharm Ind Res 1(2):83–87

Pinder AR (1957) An alkaloid of *Dioscorea hispida*, Dennstedt–IV: further investigations on the lactose ring. Tetrahedron 1(4):301–309

Powell JM (1976) Ethnobotany. In: Paijmans K (ed) New Guinea vegetation. Elsevier, Amsterdam, pp 106–183, 212 pp

Punithkumar TG, Panduranga MG, Suresh A, Suresh V, Senthilkumar N, Raviashankar HG (2011) Evaluation of antitumour activity and antioxidant status in *Dioscorea hispida* Dennst. leaves on Ehrlich ascites carcinoma in Swiss albino mice. Int J Drug Dev Res 3(2):203–2

Rao PS, Beri RM (1952) Tubers of *Dioscorea hispida* Dennst. Indian Forester 78(3):146–152

Sasongko P (2012) Detoksifikasi umbi gadung (*Dioscorea hispida* Dennst.) melalui proses fermentasi menggunakan kapang *Mucor* sp [Detoxification of gadung (*Dioscorea hispida* Dennst.) by *Muco*r sp. fermentation.]. J Teknol Pert 10(3):205–215 (in Indonesian)

Stuart GU (2013) Philippine alternative medicine. Manual of some Philippine medicinal plants. http://www.stuartxchange.org/OtherHerbals.html

Tattiyakul J, Naksriarporn T, Pradipasena P, Miyawaki O (2006) Effect of moisture on hydrothermal modification of yam *Dioscorea hispida* Dennst starch. Starch-Starke 58:170–176

Tattiyakul J, Naksriarporn T, Pradipasena P (2012) X-ray diffraction pattern and functional properties of *Dioscorea hispida* Dennst starch hydrothermally modified at different temperatures. Food Bioprocess Technol 5(3):964–971

The Plant List (2014) Version 1.1. Published on the Internet. http://www.theplantlist.org/. Accessed 10 Aug 2014

Theraseen S, Baker AT (2009) Analysis and identification of phenolic compounds in *Dioscorea hispida* Dennst. Asian J Food Agro-Ind 2(4):547–560

Yusuf M, Begum J, Hoque MN, Chowdhury JU (2009) Medicinal plants of Bangladesh, 2nd edn. BCSIR Laboratories, Chittagong, p 599

Dioscorea oppositifolia

Scientific Name

Dioscorea oppositifolia L.

Synonyms

Dioscorea opposita Thunb., *Dioscorea oppositifolia* var. *dukhunensis* Prain & Burkill, *Dioscorea oppositifolia* var. *linnae* Prain & Burkill, *Dioscorea oppositifolia* var. *thwaitesii* Prain & Burkill

Family

Dioscoreaceae

Common/English Names

Air Potato, Betel Yam, Chinese Potato, Chinese Yam, Cinnamon Vine, Cinnamon Yam, Korean Yam, Common Yam, Japanese Yam, Long Chinese Yam, Wild Yam

Vernacular Names

Chinese: Huai Shan Yao, Jia-Shan-Yao, Shan Yao, Shu Yu
Czech: Jam Čínský;
Danish: Jams, Kinyams;
Estonian: Hiina Jamss;
French: Igname De Chine, Igname Khmer;
German: Chinesische Yams, Chinesische Yamswurzel, Echte Yamswurzel, Japanische Berg-Yams, Koreanische Yams;
India: Aenna Soora, Bellaraai, Bellarai, Bellare, Bellaroy, Ennasara, Hakki Genasu, Imsara, Inaasara, Inasara, Kaadu Genasu, Neerabatte, Neerbatte, Nirballe, Nirbatte, Nirbatti, Tabinere, Taveneve, Thaabi Nere (Kannada), Kachil (Malayalam), Choratogu, Gilandru, Mar-Paspoli, Paspoli (Marathi), Pani Alu, Pit Kanda, Pitli Kanda, Pittalu (Oriya), Amlardraka, Sarpakhya (Sanskrit), Alakamatitakkilanku, Alakamatitam, Avatengatige, Kavala-Kodi, Kayvalli2, Kayvallikkilanku, Maruvalli, Maruvallikkilanku, Thavai-Kachchu, Varivalli, Varivallikkoti, Venil Valli, Verrilaivalli, Verrilaivallikkilankukkoti, Verrolaivalli, Vettilaivalli (Tamil), Adavidumpathige, Adavidumpatige, Adaviyatagathige, Adaviyatagatige, Adividumpatheega, Adiviyatagatige, Aretega, Aretegalu, Arethegalu, Aretige, Arithige, Athyaga, Atyaga, Avatengatige, Avathengathige, Cencudumpa, Chenchudampa, Chenchudumpa, Ganuga, Karrapendalamu, Naarabadlu, Tella Gadda, Tsuntsugadda, Yella Gadda (Telugu);
Indonesia: Ubi China;

T.K. Lim, *Edible Medicinal and Non-Medicinal Plants: Volume 10, Modified Stems, Roots, Bulbs*,
DOI 10.1007/978-94-017-7276-1_14

Japanese: Nagaimo, Ichōimo, Tororo Imo, Tsukune Imo, Yamatoimo;
Korean: Ma, Sanwu, Sanyak, Seoyeo;
Malaysia: Ubi Utan;
Philippines: Tuge (Iloko);
Polish: Pochrzyn Chiński;
Russian: Iams Kitaiskii, Kitaiskii Iams;
Slovašcina: Kitajski Krompir;
Thai: Huai Sua;
Vietnamese: Củ Mài, Khoai Mài

Origin/Distribution

Dioscorea oppositifolia is found in South India, Sri Lanka, Eastern Himalaya to Myanmar (Govaerts et al. 2006). Rubatzky and Yamaguchi (1997) reported *D. opposita* to have a Chinese origin.

Agroecology

In its native range, the plant is commonly found at natural forest margins, riparian zones, ruderal/disturbed, urban areas, hill slopes wetlands, along stream banks and drainage ways and near fencerows (Tu 2002). It thrives best in partial shade but tolerates full sun and full shade. It flourishes in silty loam and alluvial soils relatively rich in nitrogen. It is tolerant to frost and can be grown in much cooler conditions than other yams.

Edible Plant Parts and Uses

Nagaimo is a highly prized food ingredient all over Japan. Grated Nagaimo is called '*Toror*o' (Nagai and Nagashima 2006). It is commonly used in Japanese noodle dishes *tororo udon, totoro soba* and/or added to soups or bowl of warm steamed rice. It is slimy and is used as a binding/thickening agent in the batter of dishes like *Okonomiyaki. Toror*o is also mixed with other ingredients that typically include *tsuyu* broth (*dashi*), wasabi and green onions. Nagaimo can also be diced into cubes and added to salads to give it an interesting, almost water chestnut like texture. Nagaimo can also be fried, roasted, baked or stewed. It can be grind and blended with vanilla ice cream and milk to make a thick healthy milkshake. Other popular recipes Cheese *Goha*n (nagaimo and rice with melted cheese), *Korori Ishikoro* (nagaimo sauteed in butter with croquette), and Hatsukoi—thin slices of nagaimo smeared with a clear, light, sweet sauce. In China, some common recipes of *D. opposita* or Shan yao include *Shan yao rou tang* (yam pork broth), yam gruel for the elderly and *Ba si shan yao* (candied yam) (Hu 2005). It is used as an ingredient in *bupin* (home-made herbal, tonic food, cooked with other ingredients like rice, red beans, chicken or spare ribs).

In Odisha, India, the tubers are boiled and used in curries (Kumar et al. 2012).

Supplementation of Chinese yam was found to have a positive effect on the physicochemical, functional, nutraceutical and sensory characteristics of yogurt (Kim et al. 2011). It was found that the concentrations (0.2–0.6 %, wt/vol) of yam powder could be used to produce yam-supplemented yogurt without significant adverse effects on physicochemical, microbial and sensory properties, and with enhanced functional components from the supplementation.

Botany

A deciduous perennial creeping and twinning vine reaching 3–4 m high with large, elongated sparsely hairy tubers (Plate 1). Stem slender, terete, glabrous, purplish, clockwise (left to right) bearing bulbils in the leaf axils. tuber root-like, cylindrical 50–90 cm long. Leaves alternate in basal stem, opposite above the middle, rarely 3 in a whorl, on long petioles, simple, entire margin, broad ovate-triangular, with cordate base, and acuminate apex, 7–9 nerved 3–9 cm long, 2–7 cm wide, flowers are small, white (greenish-yellow), 4 mm, unisexual with a cinnamon fragrance in paniculate or spicate inflorescences. Male spike, 2–8 cm long, and nearly erect; 2–8 of them are located in leaf axils and occasionally in conical arrangement; rachis is clearly in the shape of zig-zag; male flowers 6 perianth lobes in two whorls, the outer are broadly ovate while the inner are

Plate 1 Chinese yam

oval; stamens 6. Female spikes 1–3 in the leaf axils; female flower ovary inferior, 3-loculed, ovules usually 2 per locule arranged in axile placentation, style 3. Capsule trigonous, membranous 1.2–2.0 cm long, 1.5–3.0 cm wide, and with white powder outside. Seed with membranous wings.

Nutritive/Medicinal Properties

Tuberous Rhizome Nutrients/ Phytochemicals

Proximate nutrient composition of *D. oppositifolia* var. *oppositifolia* tubers (g/100 g) was reported by Shajeela et al. (2011) as: moisture 89.39 %, crude protein 8.44 g, crude lipid 4.40 g, crude fibre 7.69 g, ash 5.39 g, starch 46.04 g, NFE (nitrogen free extract) 74.08 g, energy 1543.96 kJ/100 g DM, niacin 44.30 mg, ascorbic acid 90.51 mg, and minerals mg/100 g, Na 124 mg, K 1534.21 mg, Ca 646.20 mg, Mg 634.14 mg, P 124.12 mg, Zn 6.26 mg, Mn 9.04 mg, Fe 40.76 mg, Cu 7.62 mg. The antinutritional factors were per 100 g: total free phenolics 0.56 g, tannins 0.36 g, hydrogen cyanide 0.33 mg, total oxalate 0.46 g, amylase inhibitor 2.10 AIU/mg soluble starch, and trypsin inhibitor 11.26 TIU/mg protein (Shajeela et al. 2011). The in-vitro protein digestibility was 6.74 % and the in-vitro starch digestibility was 43.25 %.

Chinese yam rhizome had been reported to contain many chemical components such as mannan, allantoin, diosgenin, dopamine, batatasine, phytic acid, abscisin II, amino acids, glucoprotein, choline, ergosterol, campesterol, dioscorin, saponins, starch, non-starch polysaccharides, flavonoids, polyphenols minerals (K, S, Ca, Mg, Fe, Zn, Cu, Mn) and others (Tang 1987; Nie et al. 1993; Liao et al. 2003; Zhao et al. 2003, 2005; Shu et al. 2006; Zhang et al. 2006; Nagai and Nagashima 2006; Yuan 2008; Zhou et al. 2008; Park et al. 2010; Jiang 2011; Lee et al. 2011). Twelve compounds were isolated from *D. opposita* and identified as palmitic acid, β-sitosterol, oleic acid, β-sitosterol acetate, 5-(hydroxymethyl)furfural, nonanedioic acid, β-daucosterol, cyclo-(Phe-Tyr), cyclo-(Tyr-Tyr), 6-methyl citrate, 1, 5-dimethyl citrate and trimethyl citrate (Bai et al. 2008).

A phenanthrene glycoside, 3,4,6-trihydroxyphenanthrene-3-O-β-D-glucopyranoside, and five known compounds, soyacerebroside I, adenosine, β-sitosterol, palmitic acid and palmitoyloleoylphosphatidylcholine were isolated from *D. opposita* rhizome (Sautour et al. 2004). Two new phenanthrene glycosides, dioscopposide

A and dioscopposide B were isolated from *Dioscorea opposita* rhizomes (Zheng et al. 2014).

The following phenolic compounds were isolated from the chloroform-soluble fraction of *Dioscorea opposita* rhizome, including four new compounds, 3,5-dihydroxy-4-methoxybibenzyl; 3,3′,5-trihydroxy-2′-methoxybibenzyl; 10,11-dihydro-dibenz[*b*,*f*]oxepin-2,4-diol and 10,11-dihydro-4-methoxy-dibenz[*b*,*f*]oxepin-2-ol; together with an additional fifteen known compounds—six hydrostilbenes including batatasin III; batatasin IV; tristin; 2′,3,5-trihydroxybibenzyl; 2′4-dihydroxy-3,5-dimethoxybibenzyl and 3,4-dimethoxy-2′-hydroxybibenzyl; three phenanthrenes including 3,5-dimethoxy-2-7-phenanthrenediol; hircinol and 9,10-dihydro-7-methoxy-2,5-phenanthrenediol and five diarylheptanoids including (1*E*,4*E*,6*E*)-1,7-bis(4-hydroxyphenyl)1,4,6-hepta-trien-3-one; (4*E*,6*E*)-7-(4-hydroxy-3-methoxyphenyl)-1-(4-hydroxyphenyl)-4,6-heptadien-3-one; (4*E*,6*E*)-1,7-bis(4-hydroxyphenyl)-4,6-heptadien-3-one; (3R,5R)-3,5-dihydroxy-1, 7-bis(4-hydroxyphenyl)-3,5-heptanediol;and (3*R*,5*R*)-1,7-bis(4-hydroxy-3-methoxyphenyl)-3,5-heptanediol; and apigenin (Yang et al. 2009a, b). Only the bulbils of *D. opposita* were found to contain all five batatasins I-V (Ireland et al. 1981). Phenanthrene derivatives, 3,5-imethoxyph enanthrene-2,7-diol and batatasin-I could be using as non-polar standard marker compounds for quality control of various *Dioscorea* rhizome preparations using conventional HPLC-UV-DAD method (Yoon et al. 2007). The amounts of both markers were lower than those of saponins and allantoin in *Dioscorea* rhizomes.

Phenanthrene derivatives, 3,5-imethoxyphena nthrene-2,7-diol and batatasin-I could be using as non-polar standard marker compounds for quality control of various *Dioscorea* rhizome preparations using conventional HPLC-UV-DAD method (Yoon et al. 2007). The amounts of both markers were lower than those of saponins and allantoin in *Dioscorea* rhizomes.

Eleven gibberellins (GAs) were identified and quantified in tuber extracts of *D. opposita* (Kim et al. 2003). Five of these gibberellins were members of the early-13-hydroxylation pathway (GA53 → GA44 → GA19 → GA20 → GA1), and six were members of the non-13-hydroxylation pathway (GA12 → GA15 → GA24 → GA36/ GA9 → GA4) in the biosynthesis of GA1 and GA4.

The polysaccharide RDPS-I, isolated from Chinese yam tuber, was found to compose of Glc, Man and Gal, in a molar ratio of 1:0.4:0.1, with a backbone composed of α- D -(1,3)-Glcp and, a short branch of α- D -(1 → 2)-Manp-β- D -(1)-Galp attached to the main chain (Zhao et al. 2003). It had a molecular weight of 41,000. Another polysaccharide (YP-1) from *Dioscorea opposita* with molecular weight of 42 kDa was found to contain c glucose, mannose and galactose in the molar ratio of 1:0.37:0.11 (Zhao et al. 2005). The polysaccharide had a backbone of (1 → 3)-linked α-D-glucopyranosyl residues, with occasional branches at O-6. A homogeneous polysaccharide RP with beta configurations and composed of glucose, D-mannose, D-galactose was isolated from *D. opposita* (Xu et al. 2006). Chinese yam polysaccharide isolated from *Dioscorea opposita* rhizome was mainly composed of mannose, glucose, galactose and glucuronic acid in the ratio of 0.5:1.2:0.3:0.3 (Ju et al. 2014). A higher yield of polysaccharides from *Dioscorea opposita* was achieved at 55 °C, pH 5.5, with a load of alpha-amylase 10 mg for 1.0 h, the extraction rate was increased by in compare of sonolysis treatment alone (Zhang et al. 2008). The optimum conditions for microwave-assisted extraction of polysaccharides from Chinese yam were determined as: microwave power 464 W, material:water ratio 1:20, extracting temperature of 60 °C and using ethanol-concentrated liquid system (Xu et al. 2007). The optimum extracting conditions for water-soluble crude polysaccharide from *Dioscorea opposita* cv. Tiegun were 1:20 material: water ratio, extracting under 100 °C for 3 h and settling under 90 % ethanol system and the total polysaccharide content obtained was 1.40 % (Li et al. 2008). The optimal conditions for ultrasonic-assisted extraction of water-soluble crude polysaccharides from *Dioscorea opposita* cv. Tiegun were as follows: ultrasonic power 800 W, extraction temperature

60 °C, ultrasonic time 30 min and solid–liquid ratio 1 : 30;and the yield of the crude polysaccharides obtained was 4.6 % (Wei et al. 2010).

A 70-kDa galactose-specific dimeric lectin was purified from *Dioscorea opposita* cv. nagaimo tubers (Chan and Ng 2013). Nagaimo lectin displayed moderate thermostability, retaining full hemagglutinating activity after heating up to 62 °C for 30 min. It also manifested stability over a wide pH range from pH 2 to 13. Nagaimo lectin was a galactose-specific lectin, as evidenced by binding with galactose and galactose-containing sugars such as lactose and raffinose. The minimum concentration of galactose, lactose and raffinose required to exert an inhibitory effect on hemagglutinating activity of nagaimo lectin was 20 mM, 5 mM and 40 mM, respectively.

Chemical composition of raw and processed (mucilage removed) *D. opposita* tuber powder were reported, respectively, as: moisture (3.14,4.29 %), crude protein (14.3, 2.34 %), crude fat (0.01,0.01 %), crude carbohydrate (71.17,68.20 %), ash (3.25,1.13 %), allantoin (0.69, 0.44 %) and diosgenin (3.19, 3.17 %) (Park et al. 2010). The following minerals were detected in *D. opposita* rhizome: potassium 0.43 %, calcium 1.10 %, sodium 4.41 %, magnesium 0.68 %,, copper 1.44 %, zinc 1.88 %, iron 1.29 %, manganese 0.03 %, and also strontium and nickel (Zhang et al. 2006).

Dioscorea opposita tuber mucilage totoro aqueous extract contained about 280ug/ml protein and the main protein bands with an MW of about 33 kDa without 2-mercaptoethanol (2-ME) and about 31 kDa with 2-ME (Nagai et al. 2006, 2007). The yam tuber also contained relatively high contents of vitamins, different micro- and macroelements, enzymes and dietary fibres, and could increasingly become health-promoting food. The molecular weight of the purified viscous protein was estimated to be about 200 kDa and was identified as dioscorin, the major storage protein in yam tubers (Nagai and Nagashima 2006; Nagai et al. 2007). The purified protein showed carbonic anhydrase activity. Optimum pH in viscosity was about 5.8 and optimum temperature 20–30 °C. Heat denaturation of the protein occurred at about 40 °C, but the protein retained more than 66 % of the original viscosity at 70 °C for 30 min.

Removal of yam mucilage rich in allantoin content during extraction process resulted in significant loss of allantoin. Lee et al. (2011) found that raw yam with highly viscous manna-protein molecules, but not processed yam, could enhance the growth of *Lactobacillus acidophilus*, which may have been beneficial to health. Yam powder also contained polyphenols, flavonoids, mannans and other functional material. They found that fermentation of raw yam (6 % concentration for 12 h) with *L. acidophilus* to be most desirable as a functional food with abundant viable *L. acidophilus* cells and nutraceutical component of allantoin and diosgenin. Similar results were achieved with fermentation of raw yam powder with *L. bulgaricus* (Park et al. 2010). Chang et al. (2005) found that *D. opposita* tubers contained higher diosgenin content (3.32 %), than *D. japonica* tubers (2.61 %). Diosgenin content in *D. opposita* tubers (0.016 4 %) was slightly lower than in the bulbils (0.021 3 %) (Teng et al. 2012). The average recovery of diosgenin was 98.46 %. Among the different *Dioscorea* species found in Orissa, *D. oppositifolia* was found to have the second highest content of diosgenin, 958.33 mg/100 g tuber and a vitamin C content of 4.817 mg/100 g (Behera et al. 2010).

Fresh tubers of *Dioscorea opposita* 'Tsukuneimo', 'Yamatoimo' and 'Nagaimo' were found to contain 2.6 mg, 2.3 mg, 0.47 mg per gram FW of allantoin and 7.4 mg,6.6 mg, 4.7 mg per gram DW allantoin, respectively (Ninomiya et al. 2003). Tubers contained negligible amounts of allantoic acid as total ureides (containing allantoin and allantoic acid) contents were 7.4 mg,6.6 mg, 4.7 mg per gram DW, respectively. Leaves or stems had undetectable of allantoin. Much earlier, Ueda and Sasaki (1956) determined the weight of crystallised allantoin of 'Tuskuneimo' after removal of starch and fibres and precipitating mucous substances with 60 % ethanol. The ureide, allantoin, is thought to prevent inflammation and ulcers in humans

(Sagara et al. 1988) as well as to play an important role for storage and translocation of nitrogen in higher plants (Thomas and Schrader 1981). Studies by Ninomiya et al. (2004) found the content of allantoin in the leaves of *Dioscorea opposita* 'Tsukuneimo' decreased after any fertilization, while that in the stems significantly increased after the second fertilization, 26 days later, i.e. 96 days after planting. Allantoin content in tubers remained constant during tuber development. Arginine content in leaves and stems during the growth of 'Tsukuneimo' were lower than 0.15 and 4.8 μmol/gFW, respectively. The time course of arginine in stems content was similar to that of allantoin. In tubers, arginine content at 170 days after planting increased to reach 14 times that at 142 days but then decreased, demonstrating that *D. opposita* transiently accumulated arginine in stems as well as allantoin in leaves and stems but only accumulated the latter in developing tubers.

Allantoin content was found to be highest in *D. opposita* processed product stir-fried with wheat bran and lowest in crude *D. opposita* stir-fried with no added material (Liu et al. 1997). Of seven *Dioscorea opposita* processed samples, the sample processed with honey bran had the highest total phospholipid content (180.31 mg/100 g) (Wang et al. 1993).

During storage at ambient conditions, moisture content of Chinese yam tuber decreased from 68 to 56.5 %; starch decreased from 45 to 42 %; total sugars increased from 6.5 to 9.8 %; reducing sugars increased from 1.55 to 2.2 %; protein increased from 13 to14.6 %, and allantoin content increased from 2 mg/g to 4.7 mg/g (Zhang et al. 2014). Amylase activity increased from 1.3 mg/min/g to1.6 mg/min/g. The changes were more significant at cold temperatures and packaged conditions than at ambient conditions. The data suggested that after-ripening occurred in the early stages of Chinese yam tubers, which positively affected the nutritional potential of the tubers by a marked increase in nutrients. Low-temperature sweetening greatly affected the nutritional potential of tubers by a series of complicated interactions between starch and sugars at 4 °C.

Tuber Starch

Native starch of *D. opposita* was reported to have an amylose content of 24.5 %, granules of about 20 μm and to be the typical B-type with relative crystallinity of 26.9 % (Xie et al. 2011). In contrast, investigations by Wang and co-workers found C-type starch crystal in most *D. opposita* cultivars (Wang et al. 2006a, b, 2007, 2008a, b, c, d; Jiang, et al. 2011, 2012) The shape of starch granules separated from different *D. opposita* cultivars varied from round to oval or irregular (Wang et al. 2008a). The surface of the starch granules appeared to be smooth without any fissures. The average particle diameter of starches from different *D. opposita* Thunb. cultivars was 40.3 and 38.7 μm for *D*. 47 and *D*. SXY starch, respectively. The crystal type of starches separated from two different *D. opposita* cultivars was a typical C-type pattern. The degree of crystallinity of two *D. opposita* cultivars starches was about 45.9 % and 31.5 %, respectively. *Dioscorea opposita* cultivars (Taigu, Ribenbai, Wenxi and Zhongbowen) had amylose contents ranging from 21.17 to 25.00 % (Wang et al. 2006a). The shape of starch granules separated from different *D. opposita* Thunb. cultivars varied from round or oval to elliptic or caky, with smooth surface without any fissures. The average particle diameter of starches from different cultivars ranged from 25.90 to 28.06 μm. Starch crystal type from all cultivars was a typical C-type pattern and the relative degree of crystallinity was about 38.79 %, 39.88 %, 41.67 % and 49.03 %, respectively. The transition temperatures (T_o, T_p and T_c) ranged from 73.1 to 73.9, 77.6 to 80.4, 82.1 to 85.9 °C. The enthalpy of gelatinization (ΔH_{gel}) ranged from 6.548 to 12.13 J/g. Amylose content of *D. opposita* starches from 4 different cultivars (Yongji, Anguo, Jinpingyao and Shandongniutuimi) were in the range of 19.38–22.02 % (Wang et al. 2008b). Starch granules were smooth or rough surface, oval to spherical-shaped granules, with mean particle size in the range of 29.2–36.96 μm. Crystal type was a typical C-type X-ray diffraction pattern and the relative crystallinity of the starches ranged from 34.3 to 43.1 % The gelatinization transition temperatures (T_o, T_p and T_c) ranged from 70.2 to 75.8,

77.5 to 81.1 and 82.8 to 86.9 °C, respectively. Anguo starch showed the highest enthalpy of gelatinization (ΔH_{gel}), while cv. Shandongniutuimi starch showed the lowest values (10.54 J/g). Jiang et al. (2012) also found that *D. opposita* starch exhibited C-type crystals with a degree of crystallinity of 33.90 %.

The crystal type of the starch separated from *Dioscorea opposita* var. Zhongbowen changed from typical C-type X-ray diffraction pattern to the representative A-type pattern in the process of 8 days of acid hydrolysis (Wang et al. 2007, 2008c). It was found that B-type polymorphs present in C-type starch granule was preferentially degraded at the first stage of hydrolysis followed by A-type polymorphs (Wang et al. 2008d). The acid-thinned starch granules displayed a gradual reduction in the particle diameter followed the acid thinning. The most notable phenomenon was the conversion of thick starch granules (oval or spherical) to thin starch granules (caky or sheet). The crystallinity level increased with an increase in the hydrolysis time. In addition, the particle diameter of *Dioscorea* starch decreased from 26.8 to 18.2 μm with an increase in the hydrolysis time. Jiang et al. (2011) found that during hydrolysis, the glucoamylase primarily attacked the interior of the C-type starch granules and then the exterior of starch granules. FT-IR confirmed that the amorphous regions in the starch granules were firstly hydrolysed and could be hydrolysed completely as long as the hydrolysis time was sufficient. Chinese yam starch treated with alkali (starch-A) or enzyme (starch-E) were compared with Chinese yam starch isolated using ordinary method (starch-O) (Wang et al. 2011). The amylose content of three starches ranged from 19.47 to 22.17 % and granule surfaces were all smooth. The crystalline pattern of the three starches was a C-type. The gelatinization transition temperatures (T_o, T_p and T_c) varied from 70.11 to 73.64, 79.23 to 81.74 and 84.30 to 86.65 °C, respectively. The starch-E showed the highest enthalpy of gelatinization ΔH_{gel}, followed by the starch-A, while it was lowest for the starch-O. According to the viscosity measurement, starch-O had the lowest pasting temperature, highest peak viscosity and breakdown viscosity, which were contrary to those of starch-E.

The amylose content of *Dioscorea opposita*, *Dioscorea zingiberensis* and *Dioscorea persimilis* starches decreased after 15 days of alkaline treatment and then increased after 30 days of alkaline treatment (Jiang et al. 2014). All three starches displayed a reduction trend in swelling power with a significant increase in solubility. Adhesion among some of the starch granules was observed after alkali treatment for 30 days in *D. zingiberensis* and *D. opposita* starches, while *D. persimilis* starch showed some hollows on the granule surface. Alkali treatment elevated the in-vitro digestibility, with resistant starch values increasing from 50.16 to 64.95 % and from 66.14 to 70.74 % for *D. zingiberensis* and *D. persimilis* starches, respectively, but there was no significant change in resistant starch value for *D. opposita* starch.

Bulbil Phytochemicals

Dioscorea opposita bulbils were found to be rich in polysaccharides (0.66 %), proteins (4.61 %), amino acids, minerals but low in crude fat (0.27 %) (Sheng et al. 2009). The essential amino acid content was about 34.89 % of the total amino acids; taurine content was 15.7 mg/100 g, which accounted for 3.68 % of the total free amino acids.

The following endogenous gibberellins were purified from dormant *D. opposita* bulbil extract: GA_4, GA_9, GA_{12}, GA_{19}, GA_{20}, GA_{24}, GA_{36} and GA_{53} (Tanno et al. 1992). Their presence suggested the occurrence of two biosynthetic pathways in *D. opposita* bulbils, the early 13-hydroxylation pathway and the non-13-hydroxylation pathway.

Leaf/Aerial Plant Parts Phytochemicals

Petroleum ether extract of *D. oppositifolia* plant was found to contain steroids, triterpene, sugar,

tannin and amino acid; the methanol plant extract contained sugar, alkaloid, phenolic group, flavone, catechin, tannin and amino acid (Felix et al. 2009). Benzene and chloroform plant extracts showed the presence of steroid, sugar and steroid, triterpene, sugar, respectively. The distilled water plant extract contained steroid, triterpene, sugar and tannin. In all the extracts, saponin was absent. The total ash content of the leaf was 7.7 %; the acid insoluble ash content was 1.9 %; water insoluble ash content was (23.2 %) and sulphate ash content is 24 %.

The following compounds were isolated from the ethanol extract of the aerial parts of *Dioscorea opposita*: 6,7-dihydroxy-2-methoxy-1,4-phenanthrenedione, and four known compounds, chrysoeriol 4′-*O*-β-D-glucopyranoside, chrysoeriol 7-*O*-β-D-glucopyranoside, alternanthin and daucosterol (Ma et al. 2005). A class IV chitinase composed of eight amino acids (a C-terminal extension) at the C-terminal was found in *D. opposita* leaf (Mitsunaga et al. 2004; Karasuda et al. 2003). It was found that the C-terminal extension of yam class IV chitinase played a role as a targeting signal for plant vacuoles. Eleven gibberellins GA53, GA44, GA19, GA20, GA1, GA12, GA15, GA24, GA9, GA36 and GA4 were identified and quantified in leaf extracts of *D. opposita* (Kim et al. 2003). It was suggested that the higher level of GA4 in the leaves and tubers may be closely related to tuber enlargement.

Yuan (2008) reported *D. opposita* to have antioxidant, immunomodulatory, hypoglycaemic, antihyperlipidaemic, antitumour, antimutagenic and assimilation modulatory activities.

Antioxidant Activity

The water extract of *Dioscorea opposita* tuber mucilage tororo possessed high antioxidative activity and scavenging activities against superoxide anion and hydroxyl radicals (Nagai et al. 2006). Dioscorin, the soluble viscous protein from *D. opposita* tuber mucilage tororo, exhibited high scavenging activities against hydroxyl radicals (IC_{50}=195.1 μg/ml) and superoxide anion radicals (IC_{50}=92.7 μg/ml) (Nagai and Nagashima 2006). Shu et al. (2006) found that protein-bound polysaccharides from Chinese yam had high antioxidant activity scavenging hydrogen peroxide, superoxide and hydroxyl radicals.

The autolysate and three enzymatic hydrolysates (pepsin, trypsin and papain) prepared from yam tsukuneimo (*Dioscorea opposita*) tuber mucilage tororo were found to have the following properties μg/mg sample powder, respectively: yield—4.0,5.0, 4.5,5.0 μg; protein—36.7, 64.9, 60.2,36.3 μg; total phenol 6.4, 15, 3, 11.2, 7.4 μg (Nagai et al. 2014). The autolysate and enzymatic hydrolysates (pepsin, trypsin and papain) at 100 mg/ml significantly prolonged the induction period of auto-oxidation of linoleic acid, which was similar to 5 mM ascorbic acid. These hydrolysates also possessed high scavenging activities such as superoxide anion radicals, hydroxyl radicals and DPPH radicals. Except for pepsin hydrolysate, tororo papain and trypsin hydrolysates and autolysate were perfectly digested.

Of 19 phenolic compounds isolated from the chloroform fraction of *D. opposita* rhizome tristin; 2′,4-dihydroxy-3,5-dimethoxybibenzyl; 3,5-dimethoxy-2,7,-phenanthrenediol; hircinol; 9,10-dihydro-7-methoxy-2,5-phenanthrenediol; (4*E*,6*E*)-7-(4-hydroxy-3-methoxyphenyl)-1-(4-hydroxyphenyl)-4,6-heptadien-3-one; and (3*R*,5*R*)-1,7-bis(4-hydroxy-3-methoxyphenyl)-3,5-heptanediol exhibited radical DPPH and superoxide scavenging activities in-vitro (Yang et al. 2009b). The crude polysaccharides of *D. opposita* exhibited antioxidant activity in-vitro, scavenging superoxide anion and hydroxyl free radicals, and superoxide anion scavenging rate reached 93.75 % (Wei et al. 2010). In-vitro, Chinese yam polysaccharide isolated from *Dioscorea opposita* rhizome exhibited a potent scavenging activity on the DPPH, hydroxyl and superoxide radicals (Ju et al. 2014).

Antiinflammatory Activity

Of 19 phenolic compounds isolated from the chloroform-soluble fraction of *D. opposita* rhizome,3,3′,5-trihydroxy-2′-methoxybibenzyl;10,11-

dihydro-dibenz[*b,f*]oxepin-2,4-diol; 2′,3,5-trihydroxybibenzyl; 9,10-dihydro-7-methoxy-2,5-phenanthrenediol; (4*E*,6*E*)-7-(4-hydroxy-3-methoxyphenyl)-1-(4-hydroxyphenyl)-4,6-heptadien-3-one; (4*E*,6*E*)-1,7-bis(4-hydroxyphenyl)-4,6-heptadien-3-one showed selective inhibitory activities against cyclooxygenase-2 (COX-2) in-vitro (Yang et al. 2009b). Two phenanthrene glycosides, dioscopposide A and dioscopposide B, isolated from *Dioscorea opposita* rhizomes, exhibited inhibitory effects on the lipopolysaccharide-induced nitric oxide production using murine macrophage RAW 264.7 cells (Zheng et al. 2014). The IC_{50} values of dioscopposide A and dioscopposide B were 5.8 and 7.2 μM, respectively.

Anticancer Activity

Nagaimo tuber lectin inhibited the growth of some cancer cell lines, including breast cancer MCF7 cells, hepatoma HepG2 cells and nasopharyngeal carcinoma CNE2 cells, with IC_{50} values of 3.71 μM, 7.12 μM and 19.79 μM, respectively, after 24-h treatment with nagaimo lectin (Chan and Ng 2013). The induction of phosphatidylserine externalisation and mitochondrial depolarisation indicated that nagaimo lectin evoked apoptosis in MCF7 cells. However, the anti-proliferative activity of nagaimo lectin was not blocked by application of galactose, signifying that the activity was not related to the carbohydrate-binding specificity of the lectin.

The polysaccharide RDPS-I, isolated from Chinese yam tuber, significantly inhibited the cancer cell line of melanoma B16 and Lewis lung cancer in mice in-vivo (Zhao et al. 2003).

Prebiotic Activity

Lee et al. (2011) found that raw yam (*D. opposita*) with highly viscous manna-protein molecules, but not processed yam, could enhance the growth of *Lactobacillus acidophilus*, which may have beneficial to health. Oral administration of mice with diosgenin markedly restored the diminished density of faecal lactic acid bacteria associated with allergic reactions (Huang et al. 2012). The presence of diosgenin significantly enhanced the growth of *Lactobacillus murinus* and *Lactobacillus reuteri*, but not enterococci. The results indicated that steroidal sapogenins like diosgenin may be a novel class of prebiotics to lactic acid bacteria.

Antidiabetic Activity

D. opposita extract reduced significantly blood insulin and glucose levels in dexamethasone-induced diabetic rats (Gao et al. 2007). In-vitro, the yam extract significantly enhanced insulin-stimulated glucose uptake in 3T3-L1 adipocytes and increased mRNA expression of GLUT4 glucose transporter in 3T3-L 1 adipocytes. Studies showed that Chinese yam polysaccharides had significantly hypoglycaemic effect in streptozotocin diabetic rats compared with the model group (Yang et al. 2010). Hexokinase, succinate dehydrogenase and malate dehydrogenase activities increased significantly.

Four compounds isolated from *Dioscorea opposita* tuberous rhizomes : *trans-N-p*-coumaroyltyramine (1) (IC_{50}=0.40 μM), 1,7-bis(4-hydroxyphenyl)heptane-3,5-diol (2) (IC_{50}=0.38 mM), 6-hydroxy-2,4,7-trimethoxyphenanthrene (3) (IC_{50}=0.77 mM) were found to be active α-glucosidase inhibitors, and *cis-N-p*-coumaroyltyramine (4), an isomer of compound 1, showed no inhibitory activity against α-glucosidase (Zhang et al. 2011).

Neuroprotective Activity

Compounds isolated from the aerial parts, namely, 6,7-dihydroxy-2-methoxy-1,4-phenanthrenedione, and four known compounds, chrysoeriol 4′-*O*-β-D-glucopyranoside, chrysoeriol 7-*O*-β-D-glucopyranoside, alternanthin and daucosterol exhibited both promising neuroprotective effects and discernible to moderate antioxidant activities in-vitro (Ma et al. 2005).

Acute treatment (200 mg/kg body weight, p.o.) and 10 days daily administration (50 mg/kg body weight, p.o.) of chloroform-soluble extract of *D. opposita* showed significant spatial learning and memory improvement on mice (Yang et al. 2009a). Pre-treatment of primary cultured cortical neurons of rats with the extract protected against glutamate- and H_2O_2-induced neurotoxicity. Liu et al. (2013a) reported that treatment with Liuwei Dihuang decoction, a well-known herbal prescription of traditional Chinese medicine comprising *Dioscorea oppositifolia, Rehmannia glutinosa, Cornus officinalis, Paeonia ostii, Alisma orientale* and *Poria cocos*, exerted neuroprotective effect in streptozotocin-induced diabetic encephalopathy rats (STZ-DE) and may have potential in treating diabetic encephalopathy. Treatment with the decoction (1 and 2 g/kg, p.o., once daily, 30 days) significantly reduced the escape latency time and path length, and markedly enhanced the spent time in the target quadrant and platform crossings in Morris water maze test compared with the control group. The decoction significantly decreased the level of fasting blood glucose, acetylcholinesterase and iNOS (inducible nitrate synthase) activities in hippocampus and increased Na^+-K^+-ATP enzyme and choline acetyltransferase activities, and glutathione level. The decoction also significantly improved the expressions of IGF-1(insulin-like growth factor-1) and BDNF (brain-derived neurotrophic factor), attenuated the neural apoptosis, over-expression of caspase-3 and Aβ (amyloid beta) deposition in the hippocampus and cerebral cortex of STZ-DE rats.

Gastroprotective Activity

Jeon et al. (2006) reported that 40 % of ethanol extract of Chinese yam flour inhibited the secretion of gastric acid and showed the improvement of gut functions as observed by gastrointestinal transit and lactose-fermenting bacteria in faeces. They found that Chinese yam extract not only induced an improvement in digestive capability, but also affected the conversion of some intestinal flora to helpful bacteria. They found that serum glucose, neutral lipid and total cholesterol levels were reduced to some degree by long-term feeding on Chinese yam extract and may prove useful as a digestion-aiding agent for patients suffering from hyperglycaemia or hyperlipidaemia. Lee et al. (2011) found that feeding rats with *Lactobacillus acidophilus* fermented raw yam powder inhibited gastric lesions. When 200 mg/kg BW of both raw powder and fermented raw yam was injected, the gastric lesion was reduced apparently with 48.29 % and 53.41 % inhibition ratio, respectively, compared with that in control rats.

Park et al. (2010) also observed that yam powder fermented with *Lactobacillus bulgaricus* had higher preventive effect of gastric lesion in mice than the raw yam powder and recommend its inclusion as an ingredient in yogurt manufacture.

Antihyperlipidaemic Activity

Chinese yam starch significantly decreased the serum total cholesterol, triglyceride and LDL-cholesterol levels in hyperlipidaemic rats (Wang et al. 2008a). The serum total cholesterol, triglyceride and LDL-cholesterol decreased by approximately 33.8 %, 46.2 % and 27.5 %, respectively. The HDL-cholesterol level was not modified significantly; potato starch also reduced total cholesterol, triglyceride and LDL-cholesterol, but the results were not significant. Nishimura et al. (2011) found that raw *Dioscorea opposita* yam (RY) containing 33.9 % resistant starch as opposed to boiled yam (BY) with 6.9 % resistant starch, promoted caecal fermentation and reduced plasma non-HDL cholesterol concentration in rats. Plasma total cholesterol concentrations in the tail vein and non-HDL concentrations in arterial plasma of rats fed the 30 % RY diet were significantly lower than in the C (cholesterol-free diet) group throughout the 3-week feeding period. Liver cholesterol concentration in rats fed the 30 % RY diet was significantly higher compared with those fed the C diet. Hepatic cholesterol 7α-hydroxylase mRNA and faecal bile acid excretion were significantly higher in the BY, but

not the RY group, compared with the C group. Faecal cholesterol excretion in the 30 % RY group was greater compared with the C group. Hepatic microsomal triacylglycerol transfer protein mRNA was significantly lower in the 30 % RY group compared with the C group. Caecal pools of acetate, propionate and butyrate were 113–257 %, 181–476 % and 410–789 % greater in the RY group compared with the C group. The results suggested raw yam was effective as a source of resistant starch and production of short chain fatty acid (SCFA), especially butyrate, in the rat caecum. Additionally, raw yam exerted a plasma-cholesterol lowering effect, possibly due to the inhibited release of VLDL.

A total of 15 phenolic compounds isolated from *Dioscorea opposita* reduced pancreatic lipase activity at IC_{50} values of less than 50 μM and 3,3′,5-trihydroxy-2′-methoxybibenzyl showed the highest inhibition with an IC_{50} value of 8.8 μM (Yang et al. 2014).

Adipocyte Aquaglyceroporin Modulatory Activity

D. opposita was found to modulate aquaglyceroporin-7 channel expression and activation in adipocytes (Cals-Grierson 2007). The plasma membrane protein, aquaglyceroporin-7 (AQP7), exclusively expressed in adipocytes and as a channel for glycerol entry and exit, could be stimulated to open and to release intracellular glycerol a thus reducing the size of the lipid reservoir. Of several plant extracts with potential anti-cellulite properties tested, addition of *D. opposita* to culture medium of human preadipocytes and mouse 3 T3-L1 preadipocytes, stimulated the release of glycerol in a dose-dependent manner.

Antihypertensive Activity

Dioscorin, the soluble viscous protein from *D. opposita* tuber mucilage tororo, had antihypertensive activity, and it exhibited extremely high angiotensin I-converting enzyme inhibitory activity (IC_{50} = 41.1 μg/ml) (Nagai and Nagashima 2006). Treatment of renovascular hypertensive rats with *Dioscorea opposita* aqueous extract significantly reduced mean systolic and diastolic blood pressure (Amat et al. 2014). The extract also significantly increased plasma superoxide dismutase activity but decreased plasma malondialdehyde concentration. The extract reduced plasma angiotensin-II activity and plasma endothelin-1 concentration. Renal function was improved with captopril and the extract. They could also significantly reduce the left ventricular hypertrophy and cardiac mass index. The results suggested that *D. opposita* may have an antihypertensive effect on hypertension by inhibition of endothelin-converting enzyme and antioxidant activity.

Three enzymatic hydrolysates (pepsin, trypsin and papain) prepared from yam tsukuneimo (*Dioscorea opposita*) tuber mucilage tororo were found to have antihypertensive activity and may be useful for preventing diseases associated with reactive oxygen species and blood pressure in the body system (Nagai et al. 2014). The angiotensin-converting enzyme (ACE) inhibitory IC_{50} values (μg/mg sample powder) were 5.64 μg, 1.67 μg and 2.78 μg for pepsin, trypsin and papain hydrolysates, respectively. In contrast, ACE inhibitory in the autolysate was not detected at all.

Immunomodulatory Activity

A polysaccharide (YP-1) from *Dioscorea opposita* was found to stimulate ConA-induced T lymphocyte proliferation and its branches were extremely important for the expression of the enhancement of the immunological activity (Zhao et al. 2005).

Fertility-Enhancing Activity

Chinese yam polysaccharide CYP isolated from *Dioscorea opposita* rhizome promoted the proliferation of human endometrial epithelial cells, especially beyond the concentration of 100 μg/ml

after 36-h exposure (Ju et al. 2014). The anti-apoptotic protein Bcl-2 was up-regulated after endometrial epithelial cells were treated with CYP, while the protein level of Bax was attenuated, thus leading to the down-regulation of Bax/Bcl-2 ratio. The findings suggested that CYP may prove to be a potential candidate of the natural antioxidants as a therapeutic agent for female infertility.

Antimutagenic Activity

Water and ethanol extracts of nagaimo (*D. opposita*) exhibited antimutagenic effects against Trp-P 2-induced mutagenicity in *Salmonella typhimurium* TA 98 (Okabe et al. 1996).

Traditional Medicinal Uses

The leaves, flowers, tender shoots and tubers of *Dioscorea oppositifolia* are used for cooling and demulcent; they are used in the form of decoction for leprosy and cancerous lesions (Felix et al. 2009). The leaves are antiseptic; the paste is applied on ulcers and abscesses. The root is chewed to cure toothache and aphthae. The whole plant is used in application for oedematous tumours and the ash extract of flowering twigs along with tender leaves cure cancer and leprosy. The whole plant extract is used for secondary syphilis and psoriasis.

In China, the tuber is used in tonic food for those physically ill and debilitated due to age; it is also a major ingredient in prescriptions for diabetes (Hu 2005). Hua Shan has been reported to be beneficial to the lungs, spleen, pancreas and kidneys and is frequently used to treat pulmonary tuberculosis, fatigue and diseases associated with the spleen, pancreas and kidneys (Lu 2005). Hua Shan was found to be effective against a number of disorders, including diabetes, diarrhoea and premature ejaculation. The tuber has been eaten for the treatment of poor appetite, chronic diarrhoea, asthma, dry coughs, frequent or uncontrollable urination, diabetes and emotional instability (Tu 2002). Externally, the tuber has also been applied to ulcers, boils and abscesses. It has also been used traditionally as a contraceptive and in the treatment of various disorders of the genitary organs as well as for asthma and arthritis, and the leaf juice can be used to treat snakebites and scorpion stings.

The commonly prescribed TCM (Traditional Chinese medicine) herbs for malignant tumours included Huangqi (*Astragalus*), Nüzhenzi (Fructus Ligustri Lucid), Lingzhi (*Ganoderma lucidum*), Huaishan (*Dioscorea opposita*), Xiakucao (*Prunella vulgaris*) and Baihuasheshecao (Herba Hedyotidis) (Liu et al. 2013b). These herbs were mainly used to treat deficiency of both Qi and Yin and internal accumulation of toxic heat.

Huai Shan Yao, a common staple food in China, has been used for more than 2000 years in TCM to treat different systemic diseases, including hypertension (Amat et al. 2014). Chinese yam (Rhizoma dioscoreae, *Dioscorea opposita*) has been used in traditional Chinese medicine for many years, to strengthen stomach function, alleviate anorexia, and cure diarrhoea, and used as a delicious food in Chinese diets (Zhou et al. 2008).

Other Uses

The plant is also planted as an ornamental. The flowers smell like cinnamon and the twining vine is attractive for arbors, trellises and porches.

A class IV chitinase E from *D. opposita* was found to have fungicidal activity (Karasuda et al. 2003). Spraying strawberry fruit and leaves with the chitinase alone or β-1,3-glucanase degraded the powdery mildew fungus and the disease disappeared for more than weeks. The chitinase could be used as a safe and biodegradable biocontrol agent instead of conventional fungicides.

Dioscorea opposita is rich in starch and can be used to prepare the plasticized starch, which can be composited with halloysite nanotube (HNT) by the casting process (Xie et al. 2011). HNT could improve the pasting viscosity, mechanical properties, thermal stability and water vapour barrier of the composites.

Acid modified *D. oppositifolia* starch tablets showed higher crushing force and acceptable disintegration time and could therefore be useful as

directly compressible excipient in pharmaceutical tablets (Odeku and Picker-Freyer 2009a). Modification of *D. oppositifolia* starch by cross-linking, hydroxypropylation and dual modification-cross-linking followed by hydroxypropylation increased the swelling power (Odeku and Picker-Freyer 2009b). Hydroxypropylation and cross-linking did not significantly improve the flowability and compressibility but improved bonding, which resulted in an increased compaction and higher tablet crushing force, even though they all disintegrated rapidly. Thus, the modified Dioscorea starch showed potential for development as new excipients in solid dosage form design, and could be useful as disintegrant or for soft tableting. Okunlola and Odeku (2011) found Chinese yam starch to be useful in enhancing bond strength and in minimising the problems of lamination and capping in chloroquine phosphate tablet formulation.

Comments

In China, Huai Shan Yao is produced in Hebei, Shanxi and Shandong, with the best from Xinxiang county of Henan Province (including Wenxian) (Zhou et al. 2008). To date, Chinese yam has become not only an international medicinal and edible crop but also important special vegetable that China exports.

Selected References

Amat N, Amat R, Abdureyim S, Hoxur P, Osman Z, Mamut D, Kijjoa A (2014) Aqueous extract of *Dioscorea opposita* Thunb. normalizes the hypertension in 2K1C hypertensive rats. BMC Complement Altern Med 14:36

Bai B, Li MJ, Wang Y, Liu XH (2008) Studies on chemical constituents of *Dioscorea opposita*. Zhongguo Zhong Yao Za Zhi 33(11):1272–1274 (in Chinese)

Behera KK, Sahoo S, Prusti A (2010) Biochemical quantification of diosgenin and ascorbic acid from the tubers of different *Dioscorea* species found in Orissa. Libyan Agric Res Cent J Int 1(2):123–127

Cals-Grierson MM (2007) Modulation of activity of the adipocyte aquaglyceroporin channel by plant extracts. Int J Cosmet Sci 29(1):7–14

Chan YS, Ng TB (2013) A lectin with highly potent inhibitory activity toward breast cancer cells from edible tubers of *Dioscorea opposita* cv. nagaimo. PLoS One 8(1):e54212

Chang KJ, Park BJ, Paek JH, Kim SL, Park CH (2005) Tuber enlargement and chemical components of yams (*Dioscorea opposita* Thunb.). Korean J Plant Res 18:161–168

Felix R, Kumar NN, Raj TLS (2009) Pharmacognostical study of *Dioscorea oppositifolia* L. Ethnobot Leaflet 13:77–82

Foundation for Revitalisation of Local Health Traditions (2008) FRLHT database. http://envis.frlht.org

Gao XP, Li BG, Jaing HW, Liu FL, Xu DY, Liu ZR (2007) *Dioscorea opposita* reverses dexamethasone induced insulin resistance. Fitoterapia 78(1):12–15

Govaerts R, Wilkin P, Raz L, Téllez-Valdés O (2006) World checklist of Dioscoreaceae. Facilitated by the Royal Botanic Gardens, Kew. Published on the Internet; http://apps.kew.org/wcsp/. Retrieved 22 Feb 2014

Hu SY (2005) Food plants of China. The Chinese University Press, Hong Kong. 844 pp

Huang CH, Cheng JY, Deng MC, Chou CH, Jan TR (2012) Prebiotic effect of diosgenin, an immunoactive steroidal sapogenin of the Chinese yam. Food Chem 132(1):428–432

Ireland CR, Schwabe WW, Coursey DG (1981) The occurrence of batatasins in the Dioscoreaceae. Phytochemistry 20(7):1569–1571

Jeon JR, Lee JS, Lee CH, Kim JY, Kim SD, Nam DH (2006) Effect of ethanol extract of dried Chinese yam (*Dioscorea batatas*) flour containing dioscin on gastrointestinal function in rat model. Arch Pharm Res 29(5):348–353

Jiang HB (2011) Study on pharmacological activity and development for *Dioscorea opposita* Thunb. Chem Bioeng 4:9–12 (in Chinese)

Jiang QQ, Gao WY, Li X, Zhang JZ (2011) Characteristics of native and enzymatically hydrolyzed *Zea mays* L., *Fritillaria ussuriensis* Maxim. and *Dioscorea opposita* Thunb. starches. Food Hydrocoll 25(3):521–528

Jiang QQ, Gao WY, Li X, Xia YZ, Wang HY, Wu SS, Huang LQ, Liu CX, Xiao PG (2012) Characterizations of starches isolated from five different *Dioscorea* L. species. Food Hydrocoll 29(1):35–41

Jiang QQ, Gao WY, Li X, Man SL, Shi YP, Yang YX, Huang LQ, Liu CX (2014) Comparative susceptibilities to alkali-treatment of A-, B- and C-type starches of *Dioscorea zingiberensis, Dioscorea persimilis* and *Dioscorea opposita*. Food Hydrocoll 39:286–294

Ju Y, Zhai QZ, Wang XH (2014) Antioxidant Chinese yam polysaccharides and its pro-proliferative effect on endometrial epithelial cells. Int J Biol Macromol 66:81–85

Karasuda S, Tanaka S, Kajihara H, Yamamoto Y, Koga D (2003) Plant chitinase as a possible biocontrol agent for use instead of chemical fungicides. Biosci Biotechnol Biochem 67(1):221–224

Kim SK, Lee S, Shin DH, Jang SW, Nam JW, Park TS, Lee IJ (2003) Quantification of endogenous gibberellins in leaves and tubers of Chinese yam, *Dioscorea opposita* Thunb. cv. Tsukune during tuber enlargement. Plant Growth Reg 39(2):125–130

Kim SH, Lee SY, Palanivel G, Kwak HS (2011) Effect of *Dioscorea opposita* Thunb. (yam) supplementation on physicochemical and sensory characteristics of yogurt. J Dairy Sci 94(4):1705–1712

Kumar S, Jena PK, Tripathy PK (2012) Study of wild edible plants among tribal groups of Simlipal Biosphere Reserve Forest, Odisha, India; with special reference to *Dioscorea* species. Int J Biol Technol 3(1):11–19

Lee SY, Ganesan P, Ahn J, Kwak HS (2011) *Lactobacillus acidophilus* fermented yam (*Dioscorea opposita* Thunb.) and its preventive effects on gastric lesion. Food Sci Biotechnol 20(4):927–932

Li F, Cheng DM, Jin H, Shu XY (2008) Study on extraction technology of water-soluble crude polysaccharide from *Dioscorea opposita* Thunb. cv. Tiegun. J Anhui Agric Sci 11:4556–4557 (in Chinese)

Liao ZH, Zhu BF, Liu AL, Liu Z, Wu CG (2003) Analysis of major biochemical components in Chinese yam. J Shaoguan Univ 6:67–69 (in Chinese)

Liu B, Ma X, Li F (1997) A comparative study on chemical constituents in different processed products of *Dioscorea opposita*. Zhong Yao Cai 20(4):185–187 (in Chinese)

Liu JP, Feng L, Zhang MH, Ma DY, Wang SY, Gu J, Fu Q, Qu R, Ma SP (2013a) Neuroprotective effect of Liuwei Dihuang decoction on cognition deficits of diabetic encephalopathy in streptozotocin-induced diabetic rat. J Ethnopharmacol 150(1):371–381

Liu ZZ, Chen SY, Cai J, Zhang ED, Lan L, Zheng JT, Liao LM, Yang XM, Zhou CG, Du J (2013b) Traditional Chinese medicine syndrome-related herbal prescriptions in treatment of malignant tumors. J Tradit Chin Med 33(1):19–26

Lu HC (2005) Chinese natural cures. Black Dog & Leventhal Publishers, New York. 512 pp

Ma C, Wang W, Chen YY, Liu RN, Wang RF, Du LJ (2005) Neuroprotective and antioxidant activity of compounds from the aerial parts of *Dioscorea opposita*. J Nat Prod 68(8):1259–1261

Mitsunaga T1, Iwase M, Yuki D, Koga D (2004) Intracellular localization of a class IV chitinase from yam. Biosci Biotechnol Biochem 68(7):1518–1524

Nagai T, Nagashima T (2006) Functional properties of dioscorin, a soluble viscous protein from Japanese yam (*Dioscorea opposita* thunb.) tuber mucilage Tororo. Z Naturforsch C 61(11–12):792–798

Nagai T, Suzuki N, Nagashima T (2006) Antioxidative activity of water extracts from the yam (*Dioscorea opposita* Thunb.) tuber mucilage tororo. Eur J Lipid Sci Technol 108(6):526–531

Nagai T, Nagashima T, Suzuki N (2007) Purification and partial characterization of major viscous protein from yam (*Dioscorea opposita* Thunb.) tuber mucilage tororo. Int J Food Prop 10(3):515–526

Nagai T, Suzuki N, Kai N, Tanoue Y (2014) Functional properties of autolysate and enzymatic hydrolysates from yam tsukuneimo (*Dioscorea opposita* Thunb.) tuber mucilage tororo: antioxidative activity and antihypertensive activity. J Food Sci Technol 51(12):3838–3845

Nie GH, Zhou KF, Dong XH, Zhang C (1993) Research advances of yam. Chinese Trad Herbal Drugs 24:158–160

Ninomiya A, Murata Y, Tada M, Shimoishi Y (2003) Quantitative analysis of allantoin in fresh tubers of *Dioscorea opposita* 'Tsukuneimo'. J Japan Soc Hort Sci 72:321–323

Ninomiya A, Murata Y, Tada M, Shimoishi Y (2004) Changes in allantoin and arginine contents in *Dioscorea opposita* 'Tsukuneimo' during the growth. J Japan Soc Hort Sci 73(6):546–551

Nishimura N, Tanabe H, Yamamoto T, Fukushima M (2011) Raw Chinese yam (*Dioscorea opposita*) promotes cecal fermentation and reduces plasma non-HDL cholesterol concentration in rats. J Nutr Sci Vitaminol (Tokyo) 57(5):340–347

Odeku OA, Picker-Freyer KM (2009a) Characterization of acid modified *Dioscorea* starches as direct compression excipient. Pharm Dev Technol 14(3):259–270

Odeku OA, Picker-Freyer KM (2009b) Evaluation of the material and tablet formation properties of modified forms of *Dioscorea* starches. Drug Dev Ind Pharm 35(11):1389–1406

Okabe Y, Shinmoto H, Tsushida T, Tokuda S (1996) Antimutagenicity of the extracts from four root crops on the Trp-P2-induced mutagenicity to *Salmonella typhimurium* TA 98. J Japan Soc Food Sci Technol 43(1):36–39

Okunlola A, Odeku OA (2011) Evaluation of starches obtained from four *Dioscorea* species as binding agent in chloroquine phosphate tablet formulations. Saudi Pharm J 19(2):95–105

Park JH, Ahn J, Kwak HS (2010) Physicochemical properties of *Dioscorea opposita* Thunb. solution fermented by *Lactobacillus bulgaricus*. Korean J Food Sci An 30:773–779

Porcher MH et al (1995–2020) Searchable World Wide Web Multilingual Multiscript Plant Name Database. Published by The University of Melbourne, Australia. http://www.plantnames.unimelb.edu.au/Sorting/Frontpage.html

Rubatzky VE, Yamaguchi M (1997) Yams, *Dioscorea*. In: World Vegetables Principles, Production, and Nutritive Values. Chapman and Hall, London, pp 162–182

Sagara K, Suto K, Kawaura M, Yoshida T (1988) Analysis of *Dioscorea* rhizome. Pharm Tech Japan 4:811–813

Sautour M, Mitaine-Offer AC, Miyamoto T, Wagner H, Lacaille-Dubois MA (2004) A new phenanthrene glycoside and other constituents from *Dioscorea opposita*. Chem Pharm Bull (Tokyo) 52(10):1235–1237

Shajeela PS, Mohan VR, Jesudas LL, Soris PT (2011) Nutritional and Antinutritional evaluation of wild yam

(*Dioscorea* spp.). Trop Subtrop Agroecosyt 14:723–730

Sheng W, Xue JP, Xie BJ (2009) Chemical constituents in the bulbil of *Dioscorea opposita* and its nutrition evaluation. Food Sci Technol 8:76–79

Shu Y, Liu AJ, Wang LX (2006) The effects of protein-bound polysaccharides from Chinese yam on the antioxidant activities. Food Res Dev 11:39–42 (in Chinese)

Tang SR (1987) Analysis of nutrient constituents in *Dioscorea opposita*. Zhong Yao Tong Bao 12(4):36–38 (in Chinese)

Tanno N, Yokota T, Abe M, Okagami N (1992) Identification of endogenous gibberellins in dormant bulbils of Chinese yam, *Dioscorea opposita*. Plant Physiol 100(4):1823–1826

Teng JT, Gao X, Xue JP, Li ZW, Zhang AM, Sheng W (2012) Determination of the diosgenin content of *Dioscorea opposita* Thunb. tubers and bulbils. Med Plant 3(10):89–91 (in Chinese)

The Plant List (2014) *Dioscorea oppositifolia* L. http://www.theplantlist.org/

Thomas RJ, Schrader LE (1981) Ureide metabolism in higher plants. Phytochemistry 20:361–371

Tu M (2002) Element stewardship abstract for *Dioscorea oppositifolia* L. The Nature Conservancy. http://wiki.bugwood.org/Dioscorea_oppositifolia

Ueda H, Sasaki T (1956) Chemical studies on the plants of *Dioscorea* species. I J Pharm Soc Japan 76:745–746

Wang SL, Lu JC, Xu YM, Ye DJ (1993) Effect of processing on phospholipid of *Dioscorea opposita* Thunb. Zhongguo Zhong Yao Za Zhi 18(6):340–341, 381

Wang SJ, Gao WY, Liu HY, Chen HX, Yu JG, Xiao PG (2006a) Studies on the physicochemical, morphological, thermal and crystalline properties of starches separated from different *Dioscorea opposita* cultivars. Food Chem 99(1):38–44

Wang SJ, Yu JL, Gao WY, Liu HY, Xiao PG (2006b) New starches from traditional Chinese medicine (TCM) - Chinese yam (*Dioscorea opposita* Thunb.) cultivars. Carbohydr Res 341(2):289–293

Wang SJ, Yu JL, Gao WY, Pang JP, Liu HY, Yu JG (2007) Granule structural changes in native Chinese Yam (*Dioscorea opposita* Thunb var. Anguo) starch during acid hydrolysis. Carbohydr Polym 69(2):286–292

Wang SJ, Yu JL, Liu HY, Chen WP (2008a) Characterisation and preliminary lipid-lowering evaluation of starch from Chinese yam. Food Chem 108(1):176–181

Wang SJ, Yu JL, Yu JG, Chen HX, Pang JP, Liu HY (2008b) Partial characterization of starches from *Dioscorea opposita* Thunb. cultivars. J Food Eng 88(3):287–293

Wang SJ, Yu JL, Yu JG, Liu HY (2008c) Granule structure of C-type Chinese Yam (*Dioscorea opposita* Thunb var. Zhongbowen) starch by acid hydrolysis. Food Hydrocoll 22(4):538–542

Wang SJ, Yu JL, Yu JG, Liu HY (2008d) The partial characterization of C-type rhizoma Dioscorea starch granule during acid hydrolysis. Food Hydrocoll 22(4):531–537

Wang YG, Zhang LM, Li XL, Gao WY (2011) Physicochemical properties of starches from two different yam (*Dioscorea opposita* Thunb.) residues. Braz Arch Biol Technol 54(2):243–251

Wei T, He PX, Zheng JL (2010) Ultrasonic extraction technology of water-soluble polysaccharides from *Dioscorea opposita* Thunb. cv. tiegun and in vitro antioxidant activity. J Henan Univ Technol 6:25–28 (in Chinese)

Xie YF, Chang PR, Wang SJ, Yu JG, Ma XF (2011) Preparation and properties of halloysite nanotubes/plasticized *Dioscorea opposita* Thunb. starch composites. Carbohydr Polym 83(1):186–191

Xu Q, Xu ZL, Shen ZG, Shi YY (2006) Research of polysaccharide from *Dioscorea opposita* Thunb. Zhong Yao Cai 29(9):909–912 (in Chinese)

Xu BB, Zhang SJ, Jiang HB (2007) Study on the microwave assisted extraction technology of polysaccharide from Chinese yam. Anhui Agric Sci Bull 12:34–35, 60

Yang MH, Yoon KD, Chin YW, Park JH, Kim SH, Kim YC, Kim JW (2009a) Neuroprotective effects of *Dioscorea opposita* on scopolamine-induced memory impairment in in vivo behavioral tests and in vitro assays. J Ethnopharmacol 121(1):130–134

Yang MH, Yoon KD, Chin YW, Park JH, Kim JW (2009b) Phenolic compounds with radical scavenging and cyclooxygenase-2 (COX-2) inhibitory activities from *Dioscorea opposita*. Bioorg Med Chem 17(7):2689–2694

Yang HL, Zhang HX, Li LH, Wang Y, Zhang WW, Jiang X (2010) Study of the hypoglycemic mechanism of Chinese yam polysaccharide in type 2 diabetic rats. J Agric Univ Hebei 3:100–103

Yang MH, Chin YW, Yoon KD, Kim J (2014) Phenolic compounds with pancreatic lipase inhibitory activity from Korean yam (*Dioscorea opposita*). J Enzyme Inhib Med Chem 29(1):1–6

Yoon KD, Yang MH, Nam SI, Park JH, Kim YC, Kim JW (2007) Phenanthrene derivatives, 3,5-imethoxyphenan threne-2,7-diol and batatasin-I, as non-polar standard marker compounds for Dioscorea Rhizoma. Nat Prod Sci 13(4):378–383

Yuan SL (2008) Research advances on chemical compositions and bioactivities of *Dioscorea opposita* Thunb. (Chinese yam). Food Res Dev 3:176–179 (in Chinese)

Zhang W, Zhang ZY, Shi YZ, Fan GQ (2006) Determination of nine trace elements in *Dioscorea opposita* Thunb by flame atomic absorption spectrophotometry. Guang Pu Xue Yu Guang Pu Fen Xi 26(5):963–965 (in Chinese)

Zhang Y, Lin Q, Wei JN, Zhu HJ (2008) Study on enzyme-assisted extraction of polysaccharides from *Dioscorea opposita*. Zhongguo Zhong Yao Za Zhi 33(4):374–377

Zhang L, Bai B, Liu XH, Wang Y, Li MJ, Zhao DB (2011) α-Glucosidase inhibitors from Chinese Yam (*Dioscorea opposita* Thunb.). Food Chem 126(1):203–206

Zhang ZD, Gao WY, Wang RF, Huang LQ (2014) Changes in main nutrients and medicinal composition of Chinese yam (*Dioscorea opposita*) tubers during storage. J food Sci Technol 51(10):2535–2543

Zhao GH, Li XZ, Chen ZD (2003) Structural analysis and antitumour activity of RDPS-I polysaccharide from Chinese yam. Acta Pharm Sinica 38:37–41

Zhao GH, Kan JQ, Li ZX, Chen ZD (2005) Structural features and immunological activity of a polysaccharide from *Dioscorea opposita* Thunb roots. Carbohydr Polym 61(2):125–131

Zheng KY, Zhang ZX, Zhou W, Cao H, Xiang FJ (2014) New phenanthrene glycosides from *Dioscorea opposita*. J Asian Nat Prod Res 16(2):148–152

Zhou YQ, Zhou CN, Yao HL, Liu YJ, Tu RT (2008) Application of ISSR markers in detection of genetic variation among Chinese yam (*Dioscorea opposita* Thunb) cultivars. Life Sci J 5(4):6–12

Dioscorea rotundata

Scientific Name

***Dioscorea cayennensis* subsp. *rotundata* (Poir.) J. Miège.**

Synonyms

Dioscorea rotundata Poir.

Family

Dioscoreaceae

Common/English Names

African Yam, Connie Yam, Eboe Yam, Eight Months Guinea Yam, Eight Months Yam, Half-A-Yam, White Guinea Yam, White Yam

Vernacular Names

Arabic: Batata Beda;
Brazil: Cará Branco (Portuguese);
Bulgarian: Iams;
Chinese: Bai Shu Yu, Yuan Xing Shu Yu;
Cote D'ivoire: Kiri Kiri, Kuana, Sopéré (Baule);
Danish: Yams;
Dutch: Yam;
Finnish: Jamssi;
French: Igname Blanc, Igname De Guinée, Igname De Guinée Blanche;
German: Guinea-Yams, Guinea Yamswurzel, Weißer Guinea Yam;
Ghana: Yɛ (Adangme), Ade, Akim Bali, Akwakor, Akwakor Pechiwa, Akwakwaleekwa, Atief, Deeka, Hie, Jeow, Kate, Kentem, Kumiyo, Made, Mali, Obobi, Odjam, Odomor, Oho Hiɛ, Okoryeownye, Osiecutchie, Osuban (Adangme Krobo), Bayerɛ, E-Dwo, Σdeɛ (Akma-Akyem), Bayerɛ (Asante), Bamegu, Dakpan, Gungunsale, Gunuukple, Kpasajσ, Kpeney, Larbakσ, Larebakσ, Lili, Lilia, Perenga, Kprenga, Sagalanga, Sanaraje, Ziglanbo, Zon, Zugulangbσ, Zulanbo (Dagbani), Ayebir, Pona (Fante), Dana, Male, Tonto, Yɛlɛ (Ga), Ade, Akuku, Dreeka, Dze-Tɛ, Ge-Tɛ, Klewu, Kokolimakoe, Nkam, Sonka, Sunka, Tɛ, Tɛ-Dze, Te-Gba, Te-La (Gbe-Vhe), Kújʮ (Guang-Gonja), Gbelenga, Ade, Adipa, Akimbayeray, Akuku, Akwakor, Aso Bayerɛ, Ayebir, Bayerɛ, Jeow, Kokolimakoe, Krukupã, Kumiyow, Nana-Ntσ, Obobi, Odannã, Ode, Ode Bayerɛ, Σde Kwasea, Σde Pa, Osibã, Pona, Tonto (Konkomba), Ade, Djafuto, Dzσgσli, Klewu, Kpasa, Tɛkpuri (Vhe);
Greek: Glycopatata;
Guinea: Khabi (Susu), Dian Fasaka, Ku-Gué, Uraka (Manding-Maninka);
Italian: Igname;

T.K. Lim, *Edible Medicinal and Non-Medicinal Plants: Volume 10, Modified Stems, Roots, Bulbs*,
DOI 10.1007/978-94-017-7276-1_15

Japanese: Shiro Ginia Yama;
Nigeria: Èkà Èdiá, Èkárá Édiā (Anaang), Kit Ryáŋ (Berom), Èmọ́wè, Ígìerù, Ómí Ómí Funfun, Òrí (Edo), Afia Oko, Àfia Òkò, Òkò, Èkà Biâ, Ekefìà, Èkò, Ṇdìahà, Oko, Okpo, Okpo Uman, Okpo Úmàn Íwā, Okpuru, Otuk Okpo (Efik), Àḍịà (Epie), Dóóya (Huasa), Èkò, Ńdiágà, Okpo (Ibibio), Àbị̀, Jí Àbí, Àgà, Jí Àgà, Ala, Alafolo, Jí Alafolo Alẹfulu, Asokolo, Awudo, Jí Awudo, Asuku, ẹkpẹ, Jí Àyọ̀bè, Jí Íkē, Jí Ọ́chá, Jí Òkò Jí-Akero, Jiaga, Ólū, Omi (Igbo), Bùrú (Ijo), Ọ̀lẹ́ (Isoko), Nwángē (Tiv), Àbẹ́, Okpuru, Ọ̀lẹ́, Olumuda, Ómí, Ùjèrú, Ukpokoro, Urevbos (Urhobo), Ààlàoko, Àbàjẹ, Aga, Agake, Agẹmọkun, Aginni, Agogo, Aladẹ, Alo, Àlọ̀lọ, Ayin, Apepe, Awure, Bọki, Bùnbun Dodoro, Efùrù, Ehuru, ẹlẹyintu, Esinméérin, Fodu, Gàmù Gámù, Gbẹ̀hinrà, Gùdùgbú, Igun, Ihobia, Ilọlọ, Isu, Isu Ewara, Isu Funfun, Iyawo, Iyawo Alaajì, Iyawo ọlọ́run, Janyin-Janyin, Joyin-Joyin, Kangi, Kangì-Ojúnlaja, Kọ̀tisàn, Kukundu, Lásanrin, Layinbo, ọdọ, Olóféèré, Olonkoọlọtùn, ọlọtùn Iyaagbà, Omi Funfun, Omifun, Oparaga, Oputu, Ṣaja, Sofini, Wawaji (Yoruba);
Norwegian: Jams;
Polish: Yam;
Portuguese: Inhame Da Guiné Branco;
Russian: Belyi Iams, Dioskoreia Okruglaia;
Senegal: Kape (Fula-Pulaar);
Sierra Leone:
Sierra Leone: Mbole-Gowe, Mbole-Gowuli (Mende), A-Yams-A-Fɛra An-Won (Temne);
South America: Cará Branco, Inhame Da Guine;
Spanish: Ñame Blanco, Ñame Guineo Blanco;
Swedish: Jams;
Togo: Lili (Bassari)';
Vanuatu: White Yam, African Yam, Martinique Yam, 6-Month Yam

Origin/Distribution

D. rotundata and *D. cayenensis* both known as Guinea yam have been reported to be yams domesticated from wild Dioscoreaceae of the Enantiophyllum Uline section that have speciated in Africa. There is still considerable confusion on the taxonomy of both species (see Comments). Both species coexist in West Africa and is now being cultivated in the other African countries from Ivory Coast to southern Nigeria, and to some extent in East Africa. It has been introduced into the Caribbean Region (Jamaica, Puerto Rico, Cuba) as a result of import by the slave trade, also in Brazil (São Paulo) and in the Philippines. *D. rotundata* has been recently introduced into Oceania, spreading from New Caledonia and other French territories within the last 50 years. It is now found in New Caledonia, Vanuatu, Fiji, Papua New Guinea and the Solomon Islands.

Agroecology

White Guinea yam is a tropical species requiring a temperature of 25–30 °C, full sun and evenly distributed rainfall of 1500 mm/year for optimal growth. It is normally found from sea level to 900 m altitude. Water availability must be adequate for 6–7 months of the plant's growth phases. Temperatures below 20 °C retard vine growth and soil temperature above 35 °C retards sprouting of planted sets. The yam is tolerant of dry conditions and is adaptable to savannah regions with long dry season. White Guinea yam thrives on fertile, well-drained, friable soil devoid of coarse gravel or stones or a hard pan. Optimum pH is 5.5–6.5. Aluminium toxicity is problematic at pH less than 5.5. Nitrogen, phosphorus and potassium deficiencies are frequently encountered.

Edible Plant Parts and Uses

D. rotundata yam is consumed as boiled slices, pounded yam (*fufu* or *futu*), fried yam, roasted yam, yam flour, yam pottage, cakes, instant yam flakes and chips prepared for consumption in a variety of ways, including boiling, drying, blanching, fermentation, frying, milling, pounding, roasting and steaming (Achi 1991; Degras

and Coste 1993; Iwuoha 2004; Onwuka and Ihuma 2007; Otoo and Asiedu 2009; Taiga 2012). Besides being the choicest food for festive occasions, it has also become a common street or fast food in most urban areas in West Africa. Novel products such as lager beer, ice cream, jellies, candies and chips for snacks can be made from the tubers, and yam flour is utilised as dough conditioner in bread making, as stabiliser in ice cream, as well as a thickener in soups (Degras and Coste 1993) In the Niger-Delta of Nigeria, it is cooked into a delicacy known as 'ukodo' (yam and pepper soup), which is usually used for marriage and burial ceremonies or as breakfast, particularly during the cold season (Taiga 2012). Studies by Onwuka and Ihuma (2007) found that *D. alata* produced more acceptable yam chips, whereas *D. rotundata* gave more acceptable yam fufu and cake.

Countries from Cote d' Ivoire to Nigeria account for over 90 % of *D. rotundata* production in Africa and the yams are popularly consumed in pounded *fufu*, while north of this area and throughout Central Africa, the tubers are just boiled and served with various stews (Dumont et al. 2000). Inhabitants of urban areas of West Africa also consumed boiled or fried yams, often as a snack away from home. The use of yam flour (produced by milling dried chips) is another emerging habit. Yam flour is very well adapted to urban cooking requirements and is used to prepare a dough called *amala*, a staple or occasional food for about 50 % of the population of Cotonou (Benin) and towns of south-western Nigeria.

Botany

A large, perennial right-handed twining, glabrous, dioecious (rarely monecious) vine, with a terete stem, 10–12 m long, prickly at the base, which arises from subterranean tuber (Plate 1). Tuber is solitary, cylindrical to irregularly shaped and branched (Plate 2), up to 10–25 kg in weight, brown skinned and with firm white or yellow flesh. Leaves alternate in basal part of stem and opposite distally. Leaves simple, a stipulate on 5–12 cm long petiole; lamina entire, broadly ovate to suborbicular, 5–12 cm by 5–10 cm, base broadly cordate, apex acuminate (Plate 3) and 5–7-veined. Inflorescence an axillary unisexual spike, male spike1–3 together and 4–6 cm long, female spike 1–2 together and 10–12 cm long (Plate 3). Flowers actinomorphic with 6-lobed perianth. Male flowers small (1–2 mm across), sessile to subsessile, with 6 stamens; female flowers with inferior, 3-loculed ovary, and 3 short styles. Fruit a capsule wider than long, 20–25 mm by 30–35 mm, with up to 6 seeds. Seeds 10–15 mm across, with large circular wing.

Nutritive/Medicinal Properties

The proximate nutrient composition of the tuber on a 100 g fresh weight basis was reported as 58–80 g water, 15–23 g carbohydrate, 1–2 g crude protein, 0.05–0.12 g lipids, 0.35–0.79 g crude cellulose and 0.68–2.56 g ash (Onwueme and Hamon 2002). Carbohydrates, predominantly starch (50–80 %), comprise most of the dry matter. The starch granules of white Guinea yam are oval and measure 5–45 μm in diameter. The protein content, though generally low, is highest close to the skin. The protein fraction is high in aspartic and glutamic acids, and low in tryptophan and cystine. Some cultivars have significant amounts of vitamin C and thiamine.

Proximate composition of *D. rotundata* flour sample (% Dw) was reported by Alinnor and Akalezi (2010) as moisture 54.50 %, ash 1.40 %, crude fat 2.70 %, crude protein 0.087 %, crude fibre 0.70 %, available carbohydrate 40.61 %, energy 731.75 kJ and minerals mg/100 g Na 185.15 mg, K 209.13 %, Ca 132.02 mg, Mg 45.90 mg, Fe 81.85 mg, Cu 10.05 mg, Zn 5.48 mg, P 54.00 mg. Ihediohanma et al. (2012) reported the following proximate composition of white yam: moisture 80.8 %, ash 6.3 %, protein 7.8 %, fat 0.5 %, total carbohydrate 85.4 % and dietary fibre 31.4 %. Onwuka and Ihuma (2007) found that the moisture content, ash, fibre, protein, fat, amylose fraction and total sugar were higher in *D. alata* yam while carbohydrate, starch content and amylopectin fraction were higher in *D. rotundata*. The *D. rotundata* also maintained

Plate 1 A large, perennial right-handed twining vine

Plate 2 Leaves of white Guinea yam

higher values in most of the functional properties except in whippability, emulsion capacity and wettability.

Proximate and anti-nutrient composition (dry matter basis) of different white Guinea yam diets, viz. fresh dried yam (RY), boiled yam (BY), pounded yam, soup and stew (PYSS), boiled yam and stew (BYS) and fried yam and stew (FYS) and boiled yam with rat chow (BYPo) were compared by Lawal et al. (2012). RY contained

Plate 3 Massive, irregular-shaped, white-fleshed yam

7.82 % crude protein, 2.84 % ash, 0.42 % crude fibre, 19.5 % moisture, 0.84 % crude fat, 69.50 % carbohydrate, energy 316.87 kcal/100 g and antinutrients (mg/100 g)-hydrocynate 0.23 mg, phytate 7.85 mg, total oxalate 1.07 mg and soluble oxalate 0.88 mg. Crude protein ranged from 6.53 % for boiled yam to 17.13 % for PYSS; crude fat ranged from 1.00 % for BY to 12.72 % for FYS. Carbohydrate ranged from 41.86 % for PYSS to 68.57 % for BY. Crude fibre ranged from 0.39 % to 1.01 % for PYSS. Crude fat ranged from 1 % for BY to 18.32 % for BTPo. PYSS had significantly higher protein level but lower carbohydrate and phytate levels. BY contained the highest hydrocyanic level at 0.19 mg/100 g. Phytate levels ranged from 1.91 mg/100 g for PYSS to 2.37 mg/100 g for FYS. There were significant increases in the levels of total and soluble oxalate in all the diets compared to the boiled yam. The results showed that the various processing methods adopted reduced the hydrocyanic and phytate levels.

Major storage proteins (about 85 % total protein content) of *Dioscorea rotundata* tuber were found to consist mainly of subunits of molecular weight 31,000 and *N*-terminal amino acid glutamine/glutamic acid, of which there were a number of charge isomers; these usually contain one intra-chain disulphide bond (Harvey and Boulter 1983). They were found to be intracellularly located as protein 'aggregates' within cellular protein vacuoles, and also within the cytoplasm.

Various domestic processing methods, namely, drying, roasting, boiling and frying affected the amino acid (g/100 g crude protein) profile of *D. rotundata* yam (Ogunlade et al. 2011). The total amino acid ranged between 28.27 and 91.47 g/100 g crude protein. Glutamic acid was the most abundant amino acid, with values ranging from 3.13 to 7.15 g/100 g crude protein, with the drying process recording the highest value. Histidine (1.19 g fried to 5.20 g dried) showed higher value than the FAO standard for the four methods. Values for isoleucine (1.47–6.22 g) were higher when compared to the FAO standards. The lysine contents (3.0–7.10 g/100 g) were comparable with the reference egg protein. The phenylalanine and tyrosine (Phe + Tyr) levels ranged from 1.99 to 9.40 g/100 g crude protein, showing that the boiling and drying techniques gave values that were comparable with 6.3 FAO/WHO/UNU standards, suggesting that *D. rotundata* could be exploited to enhance protein quality of weaning/complimentary feeding, especially in dry form. Another amino acid found to be lower than FAO standards for preschool children was valine 2.26–3.33 g/100 g crude protein. In general, the amino acid content

was significantly higher when the yam was subjected to drying process, suggesting that *D. rotundata* flour has potential as a high quality protein source and can be exploited to enhance protein quality of diet for adults and weaning/ complimentary feeding for children. The frying method was the most limiting in terms of values of the amino acid present after processing.

Two membrane-bound acid phosphatases (B) and (C) were purified from *Dioscorea cayenensis rotundata* and their physicochemical properties compared to those of a cytoplasmic acid phosphatase (A) (Kamenan and Diopoh 1983). The apparent molecular weights were: 55,000 (enzyme B), 65,000 (enzyme C) and 98,000 (enzyme A). Phosphorylated sugars were hydrolysed only by the membrane-bound acid phosphatases, whereas the cytoplasmic enzyme showed no effect on four of the five phosphorylated sugars tested. The enzymatic activities of all three enzymes were inhibited by Ca^{2+}, Hg^{2+}, Mo, EDTA, but stimulated by several ions, particularly by Mg^{2+}. In healthy *D. rotundata* tubers stored for 8 months under local storage conditions, all enzymes monitored showed low and maximum activities before and at onset of sprouting, respectively, except for polyphenol oxidase (Ikediobi and Oti 1983). The activities of these enzymes after sprouting remained higher than those in the pre-sprouting period. Hexokinase, alcohol dehydrogenase and phosphorylase showed an approximately twofold increase in activity at sprouting while glucose-6-phosphate dehydrogenase showed more than threefold increase in activity. Polyphenol oxidase (PPO) activity and starch content decreased steadily, with storage with PPO activity decreasing to one-half by the end of storage. L-ascorbate, total polyphenols, carotenoids and lipids increased with storage and peaked at sprouting. Phosphorylase from *D. rotundata* tuber with molecular mass 188,000 and two subunits was different to the two forms of phosphorylase from *Dioscorea alata* (Oluoha and Ugochukwu 1995). Dansi et al. (2000) reported the following enzymes in *D. rotunda-D. cayenensis*: aspartate aminotransferase, esterase, glucose-6-phosphate, isocitrate dehydrogenase, phosphoglucomutase, phosphoglucoisomerase and shikimate dehydrogenase. The head region of *Dioscorea rotundata* tuber that imparted bitter taste when eaten was found to be the most productive region for propagation when compared to the other two physiological tuber regions; middle and tail (Isamah et al. 2000). This was elucidated by the high o-diphenolase activity in the head region, followed by the tail region and lastly, the middle region while the level of lipid peroxidation was highest in the middle region, followed by the tail and lastly the head region. The activities of superoxide dismutase and catalase exhibited a similar pattern of distribution with the highest activities recorded in the tail region and followed by the head region. The middle region had the least antioxidant enzyme activities.

Losses of moisture, dry matter, crude protein and ascorbic acid were observed in white yam (*D. rotundata*) and yellow yam (*D. cayenensis*), while the total reducing sugars increased concomitantly during the first 120 days and thereafter changes in sprouting and rotting became apparent (Onayemi and Idowu 1988). The levels of polyphenolic and glycoalkaloid substances in the stored tubers increased and became concentrated at the head regions, which were attributed to their tendency towards discolouration. The improved taste of the cooked stored tuber compared to the fresh tubers was related to the preponderance of sugars in the stored tubers, masking the bitter effect of the residual polyphenol and glycoalkaloids such as isomers A and B of dihydrodipyridopyrazine.

The starch content of six clonal selections *D. rotundata* was 23–25 % on a fresh weight basis (Moorthy and Nair 1989). The shape of the starch granules was the same, round to oval and mean dimension 27–34 μm. The reducing values of the starches (0.3–0.6) were below 1.0 for all the cultivars, indicating that the average molecular weight too be high and similar to other tuber crop starches. The total amylase content was found to vary between 20.9 and 24.6 %. The 2 % viscosity of starch of the different cultivars showed variations between 37 and 46 s. Peak viscosity was

observed between different cultivars, with values ranging from 325 B.U. (Barbender units) for starch of 184, to 550 B.U. for starch of U-195 at 5 % concentration. The data on viscosity values at 97 °C showed that the complete gelatinization occurred in most cases after 97°. The gel strength was high and desirable for many food applications. Such high viscosity stability was not observed for cassava and sweet potato starches. Blanching and fermentation of *D. rotundata* yam was found to improve colour and texture, producing flours with acceptable sensory qualities after 48 h (Achi 1991). Fermentation offered potential for preservation and improvement in quality of yam flour. Fermented products, including samples inoculated with yeasts, had higher protein values. Blanching amala, a traditional thick paste obtained from dry *Dioscorea cayenensis-rotundata* yam flour, reduced peroxidase activity and drying reduced polyphenoloxidase activity, but total phenol content and the brown index of flour and of amala increased dramatically during the latter operations (Akissoé et al. 2003). The brown index of amala was significantly correlated with the total phenol content of the flour (R^2=0.84) and the peroxidase activity of the fresh tubers (R^2=0.75). Amylose content and starch gelatinization enthalpy remained stable. Studies by Iwuoha (2004) found that water retention capacity (WRC), swelling index (SI), solubility (TSS) and iodine affinity of starch (IAS) correlated very much better and significantly with flour particle size than with tuber steeping duration. The study also observed that white yam, steeped for up to 4 days at tropical ambient temperatures, and the resultant flour classified / pulverised into ≤125 m flour particle size will yield the optimum physicochemical features in the paste. Good textural quality of pounded *Dioscorea rotundata* yam was associated with high peak viscosity, breakdown, final viscosity, holding strength and setback viscosity but with low pasting temperature in the fresh yam (Otegbayo et al. 2006). *Dioscorea rotundata* starch granules exhibited sizes <40 μm, accounting for 42.2 % and 12 % in local and hybrid cultivars, respectively (Malomo and Jayeola 2010). Swelling power was relatively higher in the hybrid, than in the local cultivars. Cultivars with lower solubility values could be important to diabetics and other health-conscious individuals while those that displayed higher solubility and swelling power may be important for dietary improvement and uses in pharmaceutical formulations. Aishat et al. (2007) found that starches obtained from *D. rotundata* tubers at 3 months after emergence could be better suited for certain products than those obtained when yam is normally harvested (6 months after emergence) due to high values for holding strength and cold paste viscosity. In normal conditions (control), sprouting of *D. rotundata* tubers occurred on 70–80 days after storage under 30 °C in the dark (Jaleel et al. 2007). The starch content decreased, while protein, amino acid, sugar contents and protease and α-amylase activities increased due to triadimefon treatment and led to early sprouting.

In a study of five normal (healthy) volunteers given white yam (*Dioscorea rotundata*), water yam (*Dioscorea alata*) and three-leaved yam (*Dioscorea dumetorum*), the mean IAUCs (incremental areas under the curve) (glycemic responses) for the test foods and glucose ranged from 863.5 to 3642.84 and were significantly different (Ihediohanma et al. 2012). Expectedly, glucose had the highest glycaemic response of 3642.84. The mean blood glucose IAUCs for white yam, water yam and three-leaved yam were 2386.50, 863.50 and 1929.08, respectively. The GI (glycemic index) of white yam (67) and three-leaved yam (56) was significantly higher than that of water yam (24). The results indicted white yam and three-leaved yam to be intermediate or medium glycemic food, and water yam a low GI food.

Among *D. rotundata* cultivars, the highest diosgenin content was 0.9263 % and the lowest 0.45 % (Vendl et al. 2006). 2′,3-Dihydroxy-5-methoxybibenzyl (Batatasin IV), its demethyl derivative and 3,5-dihydroxybibenzyl (dihydropinosylvin) were isolated only from flesh of *Dioscorea rotundata* tuber infected with *Botryodiplodia theobromae* and were therefore deemed phytoalexins (Fagboun et al. 1987).

Antimicrobial Activity

2',3-Dihydroxy-5-methoxybibenzyl (Batatasin IV), its demethyl derivative and 3,5-dihydroxybibenzyl (dihydropinosylvin) isolated from the tubers were found to be antifungal using bioassays with *Cladosporium cladosporioides*, *Botryodiplodia theobromae*, *Aspergillus niger* and *Penicillium schlerotgenum* (Fagboun et al. 1987). Dihydro-pinosylvinin also exhibited strong antibacterial activity against *Bacillus cereus*, *Staphylococcus aureus*, *Pseudomonas aeruginosa* and *Escherichia coli*.

Other Uses

The yam peelings from household waste are used to feed livestock (Degras and Coste 1993).

Starches obtained from *Dioscorea dumetorum* (Bitter yam), *Dioscorea oppositifolia* (Chinese yam), *Dioscorea alata* (Water yam) and *Dioscorea rotundata* (White yam) varied considerably in their physicochemical properties and suitability as binding agents in chloroquine phosphate tablet formulations (Okunlola and Odeku 2011). The ranking for the tensile strength and the disintegration and dissolution times for the formulations was Chinese > Bitter > Corn > White > Water, while the ranking was reversed for brittle fracture index and friability. The results suggest that Water, White and Corn could be useful when faster disintegration time of tablets is desired, while Chinese and Bitter could be more useful when bond strength is of concern and in minimising the problems of lamination and capping in tablet formulation.

Comments

Considerable confusion still exists on the taxonomy of *D. rotundata* (White Guinea yam) and *D. cayenensis* (Yellow Guinea yam). Some taxonomists treat them as two distinct species *D. rotundata* and *D. cayenensis* on various botanical and genetic traits (Hutchinson and Dalziel 1931; Burkill 1960; Akoroda and Chheda 1983; Onyilagha and Lowe 1986). In 1864, *D. rotunda* was reduced to subspecific status within *D cayenensis* by Grisebach. This view was endorsed by Prain and Burkill (1919), Chevalier (1936) and Miège (1968, 1979, 1982). Because of the confused taxonomic situation, the concept of *Dioscorea cayenensis-Dioscorea rotundata* species complex was proposed at the 1978 seminar on Yams in Cameroon (Dumont et al. 2000). This was endorsed by Hamon and Toure (1990). Using morphological descriptors and starch gel electrophoresis of 393 west African cultivated yam accessions (*D. cayenensis-rotundata* complex), and representatives of 20 varietal groups, Hamon and Toure (1990) proposed to treat this species complex as *D. cayenensis-rotundata.* Other workers like Martin and Rhodes 1978; Miège 1979; Terauchi et al. 1992; Ramser et al. 1997; Dansi et al. 2000; Chaïr et al. 2005 deemed them as two subspecies or varieties within the species *D. cayenensis.* Dansi et al. (2000) using isozyme polymorphism analysis of seven enzyme systems: aspartate aminotransferase (AAT), esterase (EST), glucose-6-phosphate dehydrogenase (G6PDH), isocitrate dehydrogenase (IDH), phosphoglucomutase (PGM), phosphoglucoisomerase (PGI) and shikimate dehydrogenase (SKDH), divided 467 accessions of cultivated Guinea yam (*Dioscorea cayenensis/Dioscorea rotundata* complex) into two groups corresponding to *D. rotundata* Poir. and *D. cayenensis* Lam., supporting the concept that the two forms of Guinea yam represented different genetic entities. More recently, studies on chloroplast DNA revealed that *D. cayenensis* and *D. rotundata* bore the same chloroplast DNA, which would make them the same species (Terauchi et al. 1992; Chaïr et al. 2005). Restriction fragment length polymorphisms (RFLP) analysis of chloroplast DNA and nuclear ribosomal DNA revealed that *Dioscorea rotundata* (white yam) was domesticated from either *D. abyssinica, D. liebrechtsiana* or *D. praehensilis* or their hybrid, and that *D. cayenensis* (yellow yam) was derived from hybridization between a male plant of either *D. burkilliana, D. minutiflora* or *D. smilacifolia* and a female plant of either *D. rotundata, D. abyssinica, D. liebrechtsiana* or *D. praehensilis*

(Terauchi et al. 1992). Thus, they proposed that the previous nomenclature of white yam should be retained as *D. rotundata* Poir. nomen nudum, and that yellow yam should be treated as a variety of *D. rotundata*, denoted as *D. rotundata* var. x '*cayenensis*'. In another study, using random amplified polymorphic DNA (RAPD) and microsatellite-primed PCR (MP-PCR), analysis of 21 accessions of Guinea yam (*Dioscorea rotundat*a - *Dioscorea cayenensis*), Ramser et al. (1997) found that all investigated species fell into two main clusters, one comprising *D. rotundata, D. cayenensis, Dioscorea abyssinica, Dioscorea liebrechtsiana* and *Dioscorea praehensilis*; the other comprising *Dioscorea smilacifolia, Dioscorea minutiflora, Dioscorea burkilliana* and *Dioscorea togoensis*. A series of diagnostic RAMP (random amplified microsatellite polymorphism) bands was identified, which clearly distinguished between *D. rotundata* and *D. cayenensis*. Studies on changes in chloroplast DNA simple sequence repeats (cpSSR) in 148 yam accessions revealed that *Dioscorea cayenensis* and *D. rotundata* shared the same haplotype (Chaïr et al. 2005).

Selected References

Achi OK (1991) Effect of natural fermentation of yams (*Dioscorea rotundata*) on characteristics of processed flour. J Food Sci 56(1):272

Aishat AB, Adebayo AI, Busie MD, Robert A (2007) Effect of tuber harvest time and storage period on the pasting properties of yam (*Dioscorea rotundata*) starch. World J Agric Sci 3(6):781–787

Akissoé N, Hounhouigan J, Mestres C, Nag M (2003) How blanching and drying affect the colour and functional characteristics of yam (*Dioscorea cayenensis-rotundata*) flour. Food Chem 82(2):257–264

Akoroda MO, Chheda HR (1983) Agro-botanical and species relationships of Guinea yams. Trop Agric 60:242–248

Alinnor IJ, Akalezi CO (2010) Proximate and mineral compositions of *Dioscorea rotundata* (white yam) and *Colocasia esculenta* (white cocoyam). Pak J Nutr 9(10):998–1001

Ayensu ES, Coursey DG (1972) Guinea yams. The botany, ethnobotany, use and possible future of yams in West Africa. Econ Bot 26:301–318

Brown WH (1954) Useful plants of the Philippines, vol 2. Department of Agriculture and Natural Resources, Bureau of Printing, Manila, 513 pp

Burkill IH (1960) The organography and the evolution of the Dioscoreaceae, the family of the yams. J Linn Soc London Bot 56:319–412

Burkill HM (1985) The useful plants of West Tropical Africa. 2nd edn, vol 1, Families A–D. Royal Botanic Gardens, Kew, 960 pp

Burkill IH (1966) A Dictionary of the economic products of the Malay Peninsula. Revised reprint of 1st ed 1935. 2 volumes. Ministry of Agriculture and Co-operatives, Kuala Lumpur, Malaysia. Vol 1 (A–H) pp 1–1240, Vol 2 (I–Z). pp 1241–2444

Chaïr H, Perrier X, Agbangla C, Marchand JL, Dainou O, Noye JL (2005) Use of cpSSRs for the characterisation of yam phylogeny in Benin. Genome 48(4):674–684

Chevalier A (1936) Contribution à l'étude de quelques espèces africaines du genre *Dioscorea*. Bull Mus Natl Hist Nat Bot 8:520–525

Coursey DG (1967) Yams. Tropical agriculture series. Longmans, London, 230 pp

Coursey DG, Booth RH (1977) Root and tuber crops. In: Leakey CLA, Wills JB (eds) Food crops of the lowland tropics. Oxford University Press, Oxford, pp 75–96), 345 pp

Dansi A, Mignouna HD, Zoundjihekpon J, Sangare A, Asiedu R, Ahoussou N (2000) Using isozyme polymorphism to assess genetic variation within cultivated yams (*Dioscorea cayenensis/Dioscorea rotundata* complex) of the Republic of Benin. Genet Resour Crop Evol 47(4):371–383

Degras L, Coste R (eds) (1993) The yam: a tropical root crop. Macmillan Press, London and Basingstoke. 408 pp

Dumont R, Dansi A, Vernier P, Zoundjihekpon J (2000) Biodiversity and domestication of yams in West Africa: traditional practices leading to *Dioscorea rotundata* Poir. CIRAD-IPGR Montpellier, Rome. 95 pp

Fagboun DE, Ogundana SK, Adesanya SA, Roberts MF (1987) Dihydrostilbene phytoalexins from *Dioscorea rotundata*. Phytochemistry 26(12):3187–3189

Grisebach AHR (1854) Grundriss der Systematischen Botanik für Akademische Vorlesungen. Verlag der Dieterichschen Buchhandlung, Gottingen, p 167

Hahn SK (1995) Yams *Dioscorea* spp. (Dioscoreaceae) (pp 112–120). In: Smartt J, Simmonds NW (eds) Evolution of crop plants, 2nd edn. Longman Science and Technology, London

Hamon P, Toure B (1990) The classification of the cultivated yam (*Dioscorea cayenensis-rotundata* complex) of West Africa. Euphytica 47:179–187

Harvey PJ, Boulter D (1983) Isolation and characterization of the storage protein of yam tubers (*Dioscorea rotundata*). Phytochemistry 22(8):1687–1693

Hutchinson J, Dalziel JM (1931) Flora of West Tropical Africa, vol II, 1st edn. Crown Agents, London, pp 379–382

Ihediohanma NC, Peter-Ikechukwu AI, Ojimba NC (2012) A comparative study and determination of glycemic indices of three yam cultivars (*Dioscorea rotundata, Dioscorea alata* and *Dioscorea dumetorum*). Pak J Nutr 11(6):547–552

Ikediobi CO, Oti E (1983) Some biochemical changes associated with post-harvest storage of white yam (*Dioscorea rotundata*) tubers. J Sci Food Agric 34:1123–1129

Isamah GK, Asagba SO, Thomas AE (2000) Lipid peroxidation, o-diphenolase, superoxide dismutase and catalase profile along the three physiological regions of *Dioscorea rotundata* Poir cv Omi. Food Chem 69(1):1–4

Iwuoha CI (2004) Comparative evaluation of physicochemical characteristics of flours from steeped tubers of white yam (*Dioscorea rotundata* Poir), water yam (*Dioscorea alata* L.) and yellow yam (*Dioscorea cayenensis* Lam). Tropicultura 22(2):56–63

Jaleel AC, Gopi R, Manivannan P, Kishorekumar A, Gomathinayagam M, Panneerselvam R (2007) Changes in biochemical constituents and induction of early sprouting by triadimefon treatment in white yam (*Dioscorea rotundata* Poir.) tubers during storage. J Zhejiang Univ Sci B 8(4):283–288

Kamenan A, Diopoh J (1983) Properties of two membrane-bound acid phosphatases compared with those of a cytoplasmic acid phosphatase from *Dioscorea cayenensis rotundata*. Plant Sci Lett 32(3):305–312

Lawal OO, Agiang MA, Eteng MU (2012) Proximate and anti-nutrient composition of white Guinea yam (*Dioscorea rotundata*) diets consumed in Ibarapa, South West region of Nigeria. J Nat Prod Plant Resour 2(2):256–260

Malomo O, Jayeola AA (2010) Micromorphological and chemical characterization of starches in *Dioscorea rotundata* L. (white yam). Not Bot Horti Agrobot Cluj-Napoca 38(1):14–20

Martin FW, Rhodes AM (1978) The relationship of *Dioscorea cayenensis* and *D. rotundata*. Trop Agric 55:193–206

Miège J (1968) Dioscoreaceae in: Flora of Tropical West Africa ed. 2 Hepper éd. London 3–1:144–154

Miège J (1979) Sur quelques problemes taxonomiques posés par *Dioscorea cayenensis* et *D. rotundata*. In: Kunkel G (ed) Taxonomic aspects of African economic botany. Proceedings of the IXth plenary meeting of AETFAT. Las Palmas de Gran Canaria, pp 109–113

Miège J (1982) Appendice. Notes sur les espèces *Dioscorea cayenensis* Lamk. et *D. rotundata* Poir. In: Miège J, Lyonga SN (eds) Yams ignames. Oxford University Press, Oxford, pp 367–375

Moorthy SN, Nair SG (1989) Studies on *Dioscorea rotundata* starch properties. Starch/Stärke 41:81–83

Ogunlade I, Olaifa O, Adeniran OA, Ogunlade AO (2011) Effect of domestic processing on the amino acid profile of *Dioscorea rotundata* (White yam). Afr J Food Sci 5(1):36–39

Okunlola A, Odeku OA (2011) Evaluation of starches obtained from four *Dioscorea* species as binding agent in chloroquine phosphate tablet formulations. Saudi Pharm J 19(2):95–105

Oluoha U, Ugochukwu EN (1995) Comparison of structure and kinetic mechanism of phosphorylase forms isolated from water yam (*Dioscorea alata*) and white yam (*Dioscorea rotundata*) tubers. Phytochemistry 40(3):677–684

Onayemi O, Idowu A (1988) Physical and chemical changes in traditionally stored yam tubers (*Dioscorea rotundata* Poir and *Dioscorea cayenensis* Lam). J Agric Food Chem 36(3):588–591

Onwueme IC, Hamon P (2002) *Dioscorea cayenensis* Lam. Record from Protabase. In: Oyen LPA, Lemmens RHMJ (eds) PROTA (Plant Resources of Tropical Africa/Ressources végétales de l'Afrique tropicale). The University, Wageningen

Onwuka GI, Ihuma C (2007) Physico-chemical composition and product behaviour of flour and chips from two yam sp. (*Dioscorea rotundata* and *Dioscorea alata*). Res J Appl Sci 2(1):35–38

Onyilagha JC, Lowe J (1986) Studies on the relationships of *Dioscorea cayenensis* and *Dioscorea rotundata* cultivars. Euphytica 35:733–739

Otegbayo B, Aina J, Asiedu R, Bokanga M (2006) Pasting characteristics of fresh yams (*Dioscorea* spp.) as indicators of textural quality in a major food product – 'pounded yam'. Food Chem 99(4):663–669

Otoo E, Asiedu R (2009) Sensory evaluation: The last hurdle in varietal development of yams (*Dioscorea rotundata*, Poir) in Ghana. Afr J Biotechnol 8(21):5747–5754

Prain D, Burkill IH (1919) *Dioscorea sativa*. Bull Misc Inform 9:339–375

Ramser J, Weising K, Terauchi R, Kahl G, Lopez-Peralta C, Terhalle W (1997) Molecular marker based taxonomy and phylogeny of Guinea yam (*Dioscorea rotundata - D. cayenensis*). Genome 40(6):903–915

Taiga A (2012) Differential rate of dry rot in *Dioscorea rotundata* (white yam) along the tuber length due to rot causing fungi. Adv Microbiol 2:451–455

Terauchi R, Chikaleke VA, Thottappilly G, Hahn SK (1992) Origin and phylogeny of Guinea yams as revealed by RFLP analysis of chloroplast DNA and nuclear ribosomal DNA. Theor Appl Genet 83(6–7):743–751

Vendl O, Wawrosch C, Noe C, Molina C, Kahl G, Kopp B (2006) Diosgenin contents and DNA fingerprint screening of various yam (*Dioscorea* sp.) genotypes. Zeitschrift Fur Naturforschung C 61(11–12):847–855

Tacca leontopetaloides

Scientific Name

Tacca leontopetaloides **(L.) Kuntze**

Synonyms

Chaitaea tacca Sol. ex Seem., *Chaitea tacca* Solander ex Parkinson, *Leontice leontopetaloides* L., *Tacca abyssinica* Hochst. ex Baker, *Tacca artocarpifolia* Seem., *Tacca browni* var. *paeoniifolia* Limpr., *Tacca brownii* Seem., *Tacca dubia* Schult. & Schult.f., *Tacca gaogao* Blanco, *Tacca guineensis* G. Don ex Loudon, *Tacca hawaiiensis* H.Limpr., *Tacca involucrata* Schumach. & Thonn., *Tacca involucrata* var. *acutifolia* (H.Limpr.) H.Limpr., *Tacca maculata* Zipp. ex Span. (inval.), *Tacca madagascariensis* Bojer, *Tacca madagascariensis* (H.Limpr.) H.Limpr., *Tacca oceanica* Seem., *Tacca phallifera* Schult. & Schult.f., *Tacca pinnatifida* J.R.Forst. & G. Forst., *Tacca pinnatifida* var. *acutifolia* H.Limpr., *Tacca pinnatifida* var. *brownii* (Seem.) Bailey, *Tacca pinnatifida* subsp. *eupinnatifida* Limpr., *Tacca pinnatifida* subsp. *interrupta* Warb. ex H. Limpr., *Tacca pinnatifida* subsp. *involucrata* (Schumach. & Thonn.) H.Limpr., *Tacca pinnatifida* var. *maculata* Limpr., *Tacca pinnatifida* subsp. *madagascariensis* H.Limpr., *Tacca pinnatifida* subsp. *minor* Limpr., *Tacca pinnatifida* var. *obtusata* Limpr., *Tacca pinnatifida* f. *obtusata* Limpr., *Tacca pinnatifida* var. *paeoniifolia* Domin, *Tacca pinnatifida* var. *permagna* Domin, *Tacca pinnatifolia* Gaertn., *Tacca quanzensis* Welw., *Tacca umbrarum* Jum. & H.Perrier, *Tacca viridis* Hemsl.,

Family

Dioscoreaceae

Common/English Names

Arrowroot, East Indian Arrowroot, Fiji Arrowroot, Indian Arrowroot, Polynesian Arrowroot, Tacca, Tahiti Arrowroot, Williams Arrowroot

Vernacular Names

Chinese: Ju Ruo Shu;
Chuuk: Mwéék;
Cook Islands: Pia;
Fais: Tacca Stärkewurzel;
French: Arrowroot De Tahiti, Fécule De Tavoul, Pia, Tacca;
French Polynesia: Pia;
Fiji: Yabia;
German: Ostin, Ostindisches Arrowroot, Ostindische Pfeilwurz, Ostindisches Teufelsblüte, Tahiti-Pfeilwurz, Takka;

T.K. Lim, *Edible Medicinal and Non-Medicinal Plants: Volume 10, Modified Stems, Roots, Bulbs*,
DOI 10.1007/978-94-017-7276-1_16

Guam: Gabgab, Gapgap, Gaogao;
Hawaii: Pia;
Ifaluk: Mogmog;
India: Suurana (Ayurvedic), Bagh-Moochh, Devkanda (Hindi), Devkanda (Marathi), Surna (Sanskrit), Karachunai (Siddha), Cenai, Kakanam, Karachunai, Kattu-K-Karunai (Tamil), adavidumpa (Telugu);
Indonesia: Chodang, Labing, Keker, Likir, Kachunda, Kechondang (Java), Telo (Sulawesi), Taka Laut (Sumatra), Gadung Tikus (General);
Japanese: Porineshiakuzuukon;
Kiribati: Te Makemake;
Korean: Polli Ne Sian Chilg;
Malaysia: Lukeh, Poko Lukeh;
Marshall Islands: Makmok, Makemok, Mogumok, Mok Mok;
Murulo: Mukmuk;
Myanmar: Pembroau, Toukta;
Namoluk: Mókumók;
Namonweito: Mwakamwak;
New Caledonia: Hâolaa;
Nigeria: Gignya Biri, Gaatarin Zoomoo (Huasa), Aduro Susu, Akana Maigbo (Yoruba);
Niue: Pia;
Nomwin: Mukmuk;
Nukuoro: Pié;
Palau: Seboseb, Ubechub;
Papua New Guinea: Masoa, Pia;
Philippines: Kanobong, Tayobong (Bisaya), Panarien (Iloko), Tambobon (Sambali), Gau-Gau, Yabyagan (Tagalog);
Pohnpei: Mwekimwek;
Pulusuk: Mwakumwak;
Samoa: Masoa;
Spanish: Arrowroot De Taiti, Yabia;
Swedish: Fliktacca;
Tahiti: Pia, Vitian;
Thailand: Buk-Ro (Eastern), Thaoyaimom (Central);
Tokelau: Masoa;
Tonga: Mahoa'a;
Ulithi: Mogmog;
Vietnamese: Bạch Tinh, Củ Nưa;
Woleai: Mogmog;
Yap: Chabchab

Origin/Distribution

The exact origin of tacca is still uncertain and is believed to be in Malaysia; it is naturally distributed from western Africa through southeast Asia to northern Australia, across to New Guinea and Polynesia. It was intentionally brought to tropical Pacific islands where it has naturalised.

Agroecology

Tacca is tropical plant species. It is occasionally cultivated but long naturalised in disturbed areas, behind the seashore often associated with beach vegetation, alang-alang fields, savannas, margins of secondary forests, grassy slopes and mesic valley floors and coconut plantations, from 3 to 330 m, occasionally to 100 m.

Edible Plant Parts and Uses

The tubers of *Tacca leontopetaloides* contain starch, which was an important food source for many Pacific Island cultures, primarily for the inhabitants of low islands and atolls. Fresh tubers are inedible because of toxic substance and is only eaten after some preparation by boiling, roasting, baking and fermentation (Murai et al. 1958; Burkill 1966; Kay 1973; Spennemann 1992,1994; Stuart 2013). Edible starch can be derived from the tubers that are used for baking bread, pastries, pudding, porridge, also mixed with other ingredients like coconut juice or milk, fruit pulp, wheat flour and other flavourings. The starch is easily digestible and suitable for the sick, old and infants. In Fiji, unprocessed starch is wrapped in leaves, buried under ground to ferment before being eaten. In Polynesia, tacca flour was mixed with mashed taro, breadfruit or *Pandanus* fruit extract and mixed with coconut cream to prepare puddings. Some of the many traditional arrowroot dishes in Marshall islands reported by Spennemann (1994) include: *aikui*—soup made with spongy coconut embryo and

arrowroot flour; *auiik*—arrowroot flour boiled or rolled in with grated coconut; *benben ni mokmok*—arrowroot flour boiled in water with coconut sap to a jelly-like consistency, shaped into small balls and rolled in grated coconut; *bwiro iiok*—preserved breadfruit mixed with arrowroot flour and coconut sap in breadfruit leaves and baked; *bobo*—arrowroot flour mixed with coconut water and cooked in coconut shells with added sugar, when jelled, cut into squares and rolled in grated coconut, food mainly used for the sick, the old and infants; *kebieltak*—arrowroot flour, crackers and coconut sap; *jup ni mokmok*—arrowroot flour, fish and coconut milk; *wagakgak*—meal prepared from arrowroot flour with grated coconut. In Tahiti, tacca starch used to make '*poi*' ('*poke*' in the Cook Islands), a traditional food, which consists of a mixture of fruit pulp and starch flavoured with vanilla and lemon and cooked in an oven (Kay 1973). In Hawaii, fresh tacca mixed with coconut milk is wrapped in the leaves baked in oven (Ihara 1971). Grated arrowroot grated, boiled in water to form a spongy ball is then covered with freshly grated coconut meat and eaten. In Hawaii, a local favorite is *haupia*, which was originally made with *pia* flour, coconut cream and *kō* (cane sugar) (Brennan 2000). The tuber has been used in curry in Peninsular Malaysia, boiled or roasted for food in Java and the starch prepared from the tuber used for sweetmeats in the Philippines (Burkill 1966).

The fruits are edible in Nigeria (Borokini and Ayidele 2012) and eaten by children in Gabon; the leaves are also used as vegetable (Jukema and Paisooksantivatana 1996).

Botany

An erect, perennial herb 1–2 m high with annual leaves and flowers arising from a tuberous rhizome (Plate 1). Tuber oblate, obovate, ellipsoid-globose or broadly ellipsoidal, usually potato-sized but up to 25 cm in diameter and sometimes weighing up to 500 g, tubers lighter colour when young turning to dark grey or brown with age, thin skinned and white fleshed. Leaves 1–3(−13) on solid or hollow, cylindrical, longitudinally ribbed, glabrous, 30–100 cm long, pale green petioles with a sheathing base; lamina large, broadly obovate in outline, palmately three-lobed, lobes pinnately dissected into orbicular to linear segments (Plate 2), upper surface with depressed veins. Flowers perfect, actinomorphic in 10–40 flowered umbellate cymes on a long scape (longer than petiole), drooping, subtended by an involucre of 4–12 bracts, individual flowers often subtended by filiform, purplish or black-brown bracteoles and on ribbed pedicels; tepals 6, basically petaloid, pale yellow, pale yellowish-green, or dark purplish-green, usually connate into a short tube, lobes in 2 alternate whorls, outer lobes elliptic to ovoid, inner ones broadly to narrowly ovate; stamens 6 white, yellow, brown or purple, adnate to tepal segment with short flattened and petaloid filaments and a connective forming a hood-like structure over the dorsifixed anthers; pistil on a hairy, glandular disk, ovary inferior, obconic to obovoid, unilocular, with short style with three-lobed stigma inflexed like an umbrella over the style. Fruit subglobose, ovoid or ellipsoid, berry-like, 3.5 × 2–2.5 cm, pale orange, pendulous, ribbed, many seeded. Seed flattened ovoid to ellipsoid, 5–8 × 4–6 mm, yellowish brown encircled by spongy white aril.

Nutritive/Medicinal Properties

Tuber Phytochemicals

The bitter brown skinned tacca tuber was found to have 28.25–29.00 % dry matter content, 25.00–27.25 % starch content, 1.67 g mL^{-1} density, 40–43 mg/100 g ascorbic acid and 3.15–3.58 % crude flavonoid extract (Ukpabi et al. 2009). The proximate composition of the tuber flesh was 1.10–1.50 % protein, 2.70–2.73 % ash, 0.28–0.68 % fibre, 0.08–0.10 % fat and 95.02–95.42 total carbohydrate on dry matter basis. The white coloured starch samples had 10.00 % moisture content, 0.71–0.77 g/mL packed bulk density, 6.55–6.75 g/g water absorption capacity, 6.90–7.30 oil absorption capacity, while the paste

Plate 1 Arrowroot plant habit

clarity (percent transmittance) generally tended to increase with storage.

Nutritive composition (per 100 g edible portion) of tacca arrowroot flour in the Pacific Islands was reported as: water 12.1 %, energy 346 cal, protein 0.18 g, fat 0.05 g, total carbohydrate 85.74 g, ash 1.89 g, Ca 58 mg, P 7.2 mg, Fe 0.55 mg (Murai et al. 1958). Dignan et al. (2004) reported the food composition (per 100 g) of Polynesian arrowroot flour as: water 12 g, energy 336 kcal (1404 kJ), protein 0.1 g, fat 0.2 g, available carbohydrate 84.5 g, total dietary fibre 0.1 g, Na 2 mg, Mg 4 mg, K 12 mg, Ca 35 mg, Fe 0.5 mg, Zn 0.6 mg, thiamine 0.1 mg, riboflavin 0.02 mg, niacin 0.5 mg, vitamin E traces

An analysis of African tubers on a dry weight basis had been reported as: protein 5.1 %; ether extract 0.2 %; carbohydrate 89.4 %; cellulose 2.1 %; fibre 8.8 %; ash 3.2 %; calcium 0.27 %; phosphorus 0.2 % (Kay 1973). The principal amino acids present in the protein were reported as arginine, glutamic and aspartic acids, leucine, lysine and valine.

Tacca starch was found to have a higher amylose content than maize starch but a lower content than potato starch (Kunle et al. 2003). The starch granules were small (average particle size 3.5 μm) relative to maize and potato starches and were predominantly polyhedral with edges. The gelatinisation characteristics except the temperature were similar to those of maize starch but much higher than those of potato starch. Tacca starch had relatively higher swelling power and solubility than the other starches. Tacca starch exhibited a monomodal distribution of irregularly shaped granules with a mean particle size of 2.64 μm (Manek et al. 2005). The spherulites of both samples indicated an A-type pattern, with the degree of crystallinity estimated to be 35 % for tacca starch and 38 % for maize starch. The gelatinisation parameters of tacca starch were found to be 65.57 – 68.56 – 73.10 °C while that of maize starch were 67.30 – 70.97 – 76.25 °C.

β-sitosterol, ceryl alcohol and a bitter principle, taccalin were isolated from tacca tubers (Scheuer et al. 1963). Tacca peels were found to

Plate 2 Close-up of arrowroot leaf

contain both nutrients and anti-nutritional factors but found toxic to animals (Ubwa et al. 2011). The peels contained moisture 15.40–28.38 %, protein 0.07–0.21 %, lipids 1.10–3.80 %, carbohydrates 62.94–71.20 %, fibre 1.10–2.07 %, ash 4.13–9.60 % and antinutrients (mg/kg)—cyanogenic glycosides 43–45 mg, saponins 31.50–35.00 mg, phytate 28.50–29.50 mg, haemagglutinin 20–23 mg and oxalate 15.50–19.00 mg (Ubwa et al. 2011)

Leaf Phytochemicals

Tacca leaves were found to contain alkaloids, saponins, tannins, anthraquinones and cardiac glycosides, and tacca tubers were found to contain cardiac glycosides and alkaloids (Borokini and Ayidele 2012). The following compounds were isolated from tacca leaves: taccagenin [furost-5-ene-3,26,27-triol-22,25-epoxy(3β,22*R*,26,27)] and (25*R*)- and (25*S*)-spirotaccagenins [spirost-5-ene-3,25,27-triol(3β,25,27)] (Abdel-Aziz et al. 1990a); Diosgenin and its ring-F-hydroxylated derivatives isonarthogenin [spirost-5-ene-3,27-diol (3β, 22*R*, 25*S*)] and isonuatigenin [spirost-5-ene-3,25-diol (3β, 22*R*, 25*S*)], together with the 22,25-epoxyfurost-5-ene isomer nuatigenin [furost-5-ene-3,26-diol-22,25-epoxy (3β, 22*R*, 25*S*)](Abdel-Aziz et al. 1990b); leontogenin,25(*R*)-B-*nor*(7)-6β-formyl-spirostane-3β,5β-diol (Abdel-Aziz et al. 1990c).

Antioxidant Activity

The percentage antioxidant (AA %) activity of the plants extract, including tacca, showed a dose-dependent increase, *Cissampelos owariensis* had the highest percentage antioxidant activity at 125 μg/mL (91 %), while the lowest results were recorded for *T. leontopetaloides* at 125 μg/mL (86 %) (Habila et al. 2011). The results of the reducing potential and total phenolic content expressed in terms of gallic acid equivalent (GAE) showed that *C. owariensis* possessed the highest reducing potential and total phenolic content (0.703 nm, 12.4 mg), and *T. leontopetaloides* the least reducing potential (0.217 nm, 6.90 mg), when compared with GAE standard (1.268 nm).

Antimicrobial Activity

Tacca plant extract exhibited antimicrobial activities in-vitro against *Staphylococcus aureus, Streptococcus pyogenes, Salmonella typhi, Escherichia coli, Shigella dysenteriae, Proteus vulgaris* and *Candida albicans*, with zones of inhibition ranging from 18 to 27 mm. The result of the minimum inhibition concentration (MIC) and minimum bactericidal concentration (MBC) showed that all the plants' extract, including tacca, inhibited and completely killed *S. aureus* and *E. coli* at a concentration range of 6.25–50 mg/mL and 25–100 mg/mL, respectively (Habila et al. 2011).

Toxicity Studies

Animal studies showed toxicity of unprocessed tacca depended on the quantity ingested (Ndouyang et al. 2014). At low levels in the diet, ingestion of unprocessed tacca reduced the level of LDL-cholesterol and limit digestion of lipids by increasing the faecal lipids. No abnormality in biochemical markers of toxicity was observed in rats fed unprocessed tacca. They found that unprocessed tacca could be recommended for ingestion a day at a quantity not higher than 1.34 g unprocessed tacca per kg body weight of rat per day.

Traditional Medicinal Uses

In India, tuber is considered acrid, astringent carminative, anthelmintic, and employed in treatment of piles, haemophilic conditions, internal abscessed, colic, enlargement of spleen, vomiting, asthma, bronchitis, elephantiasis and intestinal worms (Khare 2007).

In Hawaii, the plant is consumed as a secondary ingredient to treat *ho'opapailua lua'i* (nausea) and *lua'i mau* (continual vomiting) (Chun 1994). In the Polynesian Islands, the raw tubers mixed with water and red clay are consumed for diarrhoea and dysentery, and to stop stomach haemorrhages (Kay 1973). The grounded root is applied on guinea worm infected area to stop the infection, and is also taken in infusion to treat hepatitis in Nigeria (Borokini and Ayidele 2012). In Plateau state of Nigeria, a root preparation is used for snake bite and some other ailments and the flowers are rubbed on a snake bite. In cote d'ivoire, leaf decoction is taken orally for scrotal elephantiasis and stomach edema. In Australia, during the early days, the tuber starch was much used and highly recommended for dysentery and diarrhoea in the same way as the more familiar arrowroot has been used (Cribb and Cribb 1981).

Other Uses

The water in which the tuber gratings have been washed is used as a detergent in Plateau state of Nigeria (Borokini and Ayidele 2012); the plant finds relevance in traditional worship and sacrifices. The petioles afford good braiding material for the manufacture of hats, which is marketable in Europe and fishing utensils (Burkill 1966). Polynesian arrowroot starch used as laundry starch for starching clothes, tacca petioles and flower stalks used as cigarette holders, seed used as spear-like projectiles by children, flower stalks afford thin fibres woven into hats (Spennemann 1994). Starch is also used as glue between thin layers of paper mulberry to make tapa cloth in Samoa. In Yap, leaves are used for making skirts.

Tacca starch can be used with natural rubber to make natural biopolymers. Blending of olein oil, glycerol and crude palm oil with tacca starch-based (thermoplastic elastomer 3 TPE3) and natural rubber biopolymer imparted higher thermal resistance towards high temperature up to 310 °C (Mohd Makhtar et al. 2013, 2014). Increased volume of olein oil and glycerol tended to increase strength of hydrogen bonding in the intermolecular TPEs, imparting high thermal resistant towards high temperature. Conversely, increased volume of crude palm oil decreased the strength of hydrogen bonding in structural starch and crude palm oil; therefore, TPE with crude palm oil imparted high degradability with up to 100 % weight reduction at 500 °C.

Comments

The plant is propagated by seeds or tubers.

Selected References

Abdel-Aziz AM, Brain KR, Blunden G, Crabb T, Bashir AK (1990a) Isomeric F-ring dihydroxylated 22, 25-epoxyfurostene and spirostene sapogenins from *Tacca leontopetaloides*. Phytochemistry 29(5):1643–1646

Abdel-Aziz AM, Brain KR, Blunden G, Crabb T, Bashir AK (1990b) Steroidal sapogenins from *Tacca leontopetaloides*. Planta Med 56(2):218–221

Abdel-Aziz AM, Brain KR, Shatalebi MA, Blunden G, Patel A, Crabb TA, Bashir AK (1990c) A B-ring contracted spirostane from *Tacca leontopetaloides*. Phytochemistry 29(8):2623–2627

Backer CA, Bakhuizen van den Brink RC Jr (1968) Flora of Java, (Spermatophytes only), vol 3. Wolters-Noordhoff, Groningen. 61 pp

Borokini TI, Ayidele AE (2012) Phytochemical screening of *Tacca leontopetaloides* (L.) Kuntze collected from four geographical locations in Nigeria. Int J Modern Bot 2(4):97–102

Brennan J (2000) Tradewinds & coconuts: a reminiscence & recipes from the Pacific Islands, Periplus, p 344

Burkill IH (1966) A dictionary of the economic products of the Malay Peninsula (Revised reprint of 1st ed 1935), 2 vols. Ministry of Agriculture and Co-operatives, Kuala Lumpur, vol 1 (A–H) pp 1–1240, vol 2 (I–Z), pp 1241–2444

Caddick LR, Wilkin P, Rudall PJ, Hedderson TAJ, Chase MW (2002) Yams reclassified: a recircumscription of Dioscoreaceae and Dioscoreales. Taxon 51:103–114

Chun MN (Translator) (1994) Native Hawaiian medicine, vol 1. First People's Productions, Honolulu

Cribb AB, Cribb JW (1981) Wild medicine in Australia. Collins, Sydney. 228 pp

Dignan C, Burlingame B, Kumar S, Aalbersberg W (2004) The Pacific Islands food composition tables, 2nd edn. FAO, Rome. 135 pp

Drenth E (1972) A revision of the family Taccaceae. Blumea 20(2):367–406

Habila JD, Bello IA, Dzikwe AA, Ladan Z, Sabiu M (2011) Comparative evaluation of phytochemicals, antioxidant and antimicrobial activity of four medicinal plants native to northern Nigeria. Aus J Basic Appl Sci 5(5):537–543

Ihara VK (1971) Life in ancient Hawai'i – foods. A supplement. Research materials, Issue 15. State of Hawaii, Department of Education, Honolulu, 19 pp

Jukema J, Paisooksantivatana Y (1996) *Tacca leontopetaloides* (L.) O. Kuntze. In: Flach M, Rumawas F (eds) Plant resources of South-East Asia No 9. Plants yielding non-seed carbohydrates. PROSEA Foundation, Bogor, pp 156–159

Kay DE (1973) Root crops. Crops and products digest 2-root crops, 2nd edn. Tropical Products Institute, London. 245 pp

Khare CP (ed) (2007) Indian medicinal plants: an illustrated dictionary. Springer, Heidelberg. 900 pp

Kunle OO, Ibrahim YE, Emeje MO, Shaba S, Kunle Y (2003) Extraction, physicochemical and compaction properties of tacca starch: a potential pharmaceutical excipient. Starch/Stärke 55(7):319–325

Manek RV, Kunle OO, Emeje MO, Builders P, Rao GVR, Lopez GP, Kolling WM (2005) Physical, thermal and sorption profile of starch obtained from *Tacca leontopetaloides*. Starch/Stärke 57:55–61

Massal E, Barrau J (1956) Polynesian arrowroot. Food plants of the south sea islands. South Pacific Commission technical paper, no. 94, pp 11–12. South Pacific Commission, Noumea, 51 pp

Mohd Makhtar NS, Hamdan N, Muhd Rodhi MN, Mohibah Musa M, Ku Hamid KH (2014) The potential use of Tacca leontopetaloides as green material in manufacturing automotive part. App Mech Mat 575:65–72

Mohd Makhtar NS, Muhd Rodhi MN, Musa M, Ku Hamid KH (2013) Thermal behavior of Tacca leontopetaloides starch-based biopolymer. Int J Polym Sci 2013. Article ID 37854

Murai M, Pen F, Miller CD (1958) Some tropical south pacific island foods. Description, history, use, composition, and nutritional value. University of Hawaii Press, Honolulu, p 104

Ndouyang CJ, Nguibou RM, Njintang YN, Scher J, Facho B, Mbofung CMF (2014) In vivo assessment of the nutritional and subchronic toxicity of *Tacca leontopetaloides* (L.) tubers. Scholar J Agric Sci 4(1):5–13

Nurul Shuhada MM, Miradatul Najwa MR, Mohibah M, Ku Halim KH (2013) Thermal behavior of *Tacca leontopetaloides* starch-based biopolymer. Int J Polym Sci 2013:Article ID 373854

Nurul Shuhada MM, Ainatul Mardhiah MA, Nur Farzana AG, Nur Shahidah AA, Miradatul Najwa MR, Ku Halim KH (2014) Volume plasticizers on the thermal performance of *Tacca leontopetaloides* starch biopolymer. Key Eng Mat 594–595:882–886

People and Plants of Micronesia (2013) *Tacca leontopetaloides*. University of Hawaii, Manoa. http://manoa.hawaii.edu/botany/plants_of_micronesia/

Purseglove JW (1972) Taccaceae. Tropical crops: monocotyledons, vol 2, pp 517–518. Longman Group Ltd, London, 607 pp (2 vols)

Scheuer PJ, Swanholm CE, Madamba LA, Hudgins WR (1963) The constituents of *Tacca leontopetaloides*. Lloydia 26(3):133–140

Spennemann DHR (1992) Arrowroot production in the Marshall Islands: past, present, and future. NZ J Crop Hort Sci 20:97

Spennemann DHR (1994) Traditional arrowroot production and utilization in the Marshall Islands. J Ethnobiol 14(2):211–234

Stone EL (1951) Polynesian arrowroot, 'Makmok'. The agriculture of Arno Atoll, Marshall Islands. Atoll research bulletin, no. 6, pp 24–25. Pacific Science Board, Washington, DC, 46 pp

Stuart GU (2013) Philippine alternative medicine. Manual of some Philippine medicinal plants. http://www.stuartxchange.org/OtherHerbals.html

The Plant List (2014) Version 1.1. Published on the Internet; http://www.theplantlist.org/. Accessed 10 Oct 2014

Ting CC, Larsen K (2000) Taccaceae Dumortier. In: Wu ZY, Raven PH (eds) Flora of China. Vol. 24 (Flagellariaceae through Marantaceae). Science Press/Missouri Botanical Garden Press, Beijing/St. Louis

Ubwa ST, Anhwange BA, Chia JT (2011) Chemical analysis of *Tacca leontopetaloides* peels. Am J Food Technol 6:932–938

Ukpabi UJ, Ukenye E, Olojede AO (2009) Raw-material potentials of Nigerian wild polynesian arrowroot (*Tacca leontopetaloides*) tubers and starch. J Food Technol 7(4):135–138

Wagner WL, Herbst DR, Sohmer SH (1999) Manual of the flowering plants of Hawaii. Revised edition. Bernice P. Bishop Museum special publication. University of Hawai'i Press/Bishop Museum Press, Honolulu, 1919 pp (two volumes)

Manihot esculenta

Scientific Name

Manihot esculenta **Crantze**

Synonyms

Janipha aipi (Pohl) J.Presl, *Janipha manihot* (L.) Kunth, *Jatropha aipi* (Pohl) A.Moller, *Jatropha diffusa* (Pohl) Steud., *Jatropha digitiformis* (Pohl) Steud., *Jatropha dulcis* J.F.Gmel., *Jatropha flabellifolia* (Pohl) Steud., *Jatropha glauca* A.Rich. (Illeg.), *Jatropha janipha* Lour. (Illeg.), *Jatropha lobata* var. *richardiana* Müll. Arg., *Jatropha loureiroi* (Pohl) Steud., *Jatropha manihot* L., *Jatropha mitis* Sessé & Moc. (Illeg.), *Jatropha mitis* Rottb., *Jatropha paniculata* Ruiz & Pav. ex Pax, *Jatropha silvestris* Vell., *Jatropha stipulata* Vell., *Mandioca aipi* (Pohl) Link, *Mandioca dulcis* (J.F.Gmel.) D.Parodi, *Mandioca utilissima* (Pohl) Link, *Manihot aipi* Pohl, *Manihot aipi* var. *lanceolata* Pohl, *Manihot aipi* var. *latifolia* Pohl, *Manihot aipi* var. *lutescens* Pohl, *Manihot aypi* Spruce, *Manihot cannabina* Sweet, *Manihot cassava* Cook & Collins (Inval.), *Manihot diffusa* Pohl, *Manihot digitiformis* Pohl, *Manihot dulcis* (J.F.Gmel.) Baill., *Manihot dulcis* (J.F. Gmel.) Pax, *Manihot dulcis* var. *aipi* (Pohl) Pax, *Manihot dulcis* var. *diffusa* (Pohl) Pax, *Manihot dulcis* var. *flabellifolia* (Pohl) Pax, *Manihot edule* A.Rich., *Manihot edulis* A. Rich., *Manihot esculenta* var. *argentea* Cif., *Manihot esculenta* var. *coalescens* Cif., *Manihot esculenta* var. *debilis* Cif., *Manihot esculenta* var. *digitifolia* Cif., *Manihot esculenta* subsp. *flabellifolia* (Pohl) Cif., *Manihot esculenta* var. *flavicaulis* Cif., *Manihot esculenta* var. *fuscescens* Cif., *Manihot esculenta* subsp. *grandifolia* Cif., *Manihot esculenta* var. *grandifolia* Cif., *Manihot esculenta* var. *nodosa* Cif., *Manihot esculenta* var. *sprucei* Lanj., *Manihot flabellifolia* Pohl, *Manihot flexuosa* Pax & K.Hoffm., *Manihot guyanensis* Klotzsch ex Pax (Illeg.), *Manihot loureiroi* Pohl, *Manihot manihot* (L.) H.Karst. (Inval.), *Manihot melanobasis* Müll.Arg., *Manihot palmata* var. *aipi* (Pohl) Müll.Arg., *Manihot palmata* var. *diffusa* (Pohl) Müll.Arg., *Manihot palmata* var. *digitiformis* (Pohl) Müll.Arg., *Manihot palmata* var. *flabellifolia* (Pohl) Müll.Arg., *Manihot sprucei* Pax, *Manihot utilissima* Pohl, *Manihot utilissima* var. *castellana* Pohl, *Manihot utilissima* var. *sutinga* Pohl,

Family

Euphorbiaceae

T.K. Lim, *Edible Medicinal and Non-Medicinal Plants: Volume 10, Modified Stems, Roots, Bulbs*,
DOI 10.1007/978-94-017-7276-1_17

Common/English Names

Bitter Brazilian Arrowroot, Bitter Cassava, Bitter Tapioca Meal, Brazilian Arrowroot, Cassava, Manicot, Manidoca, Manihot, Manioc, Sweet Brazilian Arrowroot, Sweet Cassava, Sweet-Potato Tree, Sweet Tapioca Meal, Tapioca, Tapioca Manihot, Yuca

Vernacular Names

Afrikaans: Maniok;
Angola: Mbowe, Tombo (Umbundu), Mandioca, Mandioqueira, Mandioqueira-Amarga (Portuguese);
Benin: Koutou (Adja), Fènyèn (Fon), Fènyèn (Goun);
Brazil: Aipim, Macaxeira, Mandioca, Mandiya, Manioca;
Burmese : Palaw-pinan-u-pin;
Cambodia: Damlong Chhe;
Cameroon: Kasinga (Bangangte), Nkwamba (Douala), Mbom (Sangmelima), Mbom (Yaoundé);
Chinese: Mu Shu, Shu Ge, Shu Shu;
Croatian: Tropska Biljka;
Czech: Maniok Jedlý;
Danish: Maniok, Kassava;
Democratic Republic of Congo: Anaoseagasa (Batiabetuwa), Okpwa (Ngbandi), Muhogo (Swahili);
Dutch: Cassave, Maniok;
Fiji: Kasera, Katafaga, Sakarkanda, Sokobale;
Finnish: Kassava, Maniokki;
French: Cassave, Manio, Maniok, Tapioca;
Gabon: Moguma (Apindji), Muyondo (Baduma), Mayaka (Bahoumbou), N'wondo (Bakèlè), N'wondo (Bakota), Ovondo (Baléngi), Diyaga (Balumbu), Pita (Banzabi), Didjaga (Bapunu), Gigongu (Bavarama), Gigongo (Bavili), Gégongo (Bavové), Diyaga (Bavungu), Uvondo (Benga), N'wondo (Béséki), Gigongu (Eshira), Mboe (Fang), Iloti (Galoa), Gégongo (Ivéa), Yaka (Loango), Gékwo (Mindumu), Gégongo (Mitsogo), Oguma (Mpongwgwé), Iloti (Myéné), Diyaga (Ngowé), Iloti (Nkomi), Iloti (Orungu);
German: Bittere Cassava, Cassavastrauch, Kassave, Maniok, Maniokstrauch, Tapiokapflanze;
Ghana: Bankye (Asante-Twi);
Guyana: Cassava ;
Hungarian: Kasszáva, Manióka, Tápióka;
India : Kuri Aloo, Shimolu Aalu, Ximolu Alu (Assamese), Tabolchu (Garo), Sabudana (Pearls) (Gujarati), Kasāvā, Marachini, Mara Valle Kilangu, Maravalli, Simla Aloo, Simul Alu, Ṭaipi'ōkā, Ṭaipī'ōka, Tikhoors Maravalli (Hindu), Kolli, Maragenasu, Sabakki (pearls), Sabba Akki (Kannada), Cheeni, Kappa, Maraccīni (Pearls), Maracheeni, Kolli, Marakizhangu (Malayalam), Sabu Dana (Pearls) (Marathi), Karrapendalamu (Sanskrit), Javvarisi (Pearls), Kuchikezhangu (Roots), Maravallikilangu, Maravallikizhang (Tamil), Kanda, Karrapendalam, Karrapendalamu, Pendalamu, Saggu Biyyaṁ (Telugu), Sābūdānā (Pearls) (Urdu);
Indonesia: Bodin, Katel Budin, Katela Mantri, Kasawe, Kaspe, Sikong, Singkong, Katela Puhun, Puhung, Tela Cabut (Javanese), Balandong, Menjuk, Pohung, Sabbhrang Balanda, Sawe, Sawi, Tenggang (Madurese), Ketela Pohon, Ubi Kayu, Singkong (Malay), Huwi Dangdeur, Kasapen, Sampeu (Sundanese), Ubi Paranchih (Sumatra), Ubi Goa (Timor);
Italian: Cassàve, Manioca, Mandiòca, Iucca;
Ivory Coast: Bédé (Ashanti), Gbou Kwé (Shien);
Japanese: Imo Noki, Kyassaba, Maniokku, Tapioka Noki;
Korean: Kasaba;
Laos: Mantonz;
Madagascar: Kazaha (Agnalazaha Forest);
Malaysia: Ubi Kayu, Ubi Benggala, Ubi Belanda, Amlei (Sakai);
Nepalese: Simal Tarul;
Nicaragua: Aikavitu, Anaha, Belaselika, Cassava, Itk, Kasaleka, Kasera, Yauhra, Yuca;
Nigeria: Iwa (Ibibio), Akpu, Jigbo, Ugboro, Jiaphu (Igbo), Ege (Ekiti State), Gbaguda (SouthWestern Nigeria);
Norwegian: Kassava, Maniok, Tapioca, Tapiok;
Papua New Guinea: Cassava, Mandioca, Nao Harnaka, Noumea, Sakarkanda;

People's republic of Congo: Ayaka, Eyala (Akwa), Saka-Saka (Lingala), Manioc (French);
Philippines: Kamoteng-Kahoi, Kanggos (Bikol), Balangai, Balinghoy, Kamoteng-Kahoi (Bisaya), Padpadi (Bontok); Kamote Ti Moro (Ibanag), Kamonting-Kahoy, Kamote Ti Moro, Katimoro (Illoko), Balangai (Samar-Leyte Bisaya); Kamoteng-Kahoi (Sambali), Kamote-Kahoi, Panggi-Kahui (Sulu) Balinghoy, Kamoteng Kahoi (Tagalog), Malambonga (Tagbanua);
Rodrigues Islands: Manioc;
Russian: Maniok S'edobnyj;
Portuguese: Aipim, Macaxeira, Mandioca, Mandioca Brava, Maniba;
Samoa: Tapioka;
Puerto Rico: Yuca;
Serbian: Maniok;
Sierra Leone: Tangae (Kpaa Mende);
Slovašcina: Kasava, Maniok;
Spanish: Caxamote, Farinha, Guacamote, Huacamote, Mandioca, Mañoco, Yuca;
Sri Lanka: Mangonokka, Manioc (Sinhalese);
Sudan: Bafra, Bavra (Arabic);
Swahili: Mhogo;
Swedish: Brasiliansk Arrowroot, Maniok, Tapioka;
Tanzania: Muhoko, Mhogo;
Thai: Man-sampalang, Man-Samrong, Mam-Mai;
Togo: Agbéli, Akutéti, Katawoli (Ewé), Kuta (Mina);
Turkish: Manyok;
Uganda: Chombe (Alor and Jonam), Mattu Gamwogo (Bugisi), Muhogo (Bulamogi Country), Soutigbanda (Kakwa), Potmogo (Langi), Gbandabi (Madi), Muhogo (Rukiga), Muhogo (Runyankore), Muhogo (Runyoro), Muhogo (Rutooro);
Venezuela: Cassava, Maniota, Mannyok, Merelesita, Tapioka, Tavioka, Vula'tolu, Yabia, Yabia Damu;
Vietnamese: Sắn, Mỳ, Khoai Mì, Củ Mì (Root), Bột Năng (Flour), Bột Sắn (Starch);
Zaire: Cassava, Coci

Origin/Distribution

About 98 species of *Manihot* are known. All of them are native to the New World and are concentrated in four regions in Brazil and Mexico in Central America (Nassar 2002). Nassar (1978) identified four centres of diversity for these species: Mexico, and Northeast, central and southwest Brazil. Nassar (1978) postulated cassava to be an interspecific hybrid that appeared by domestication within the last 2000 years. Cassava may have originated by hybridisation between two wild *Manihot* species, followed by vegetative reproduction of the hybrid. Contrary to views that cassava (*Manihot esculenta*) was only known in cultivation, an argument was presented by Allem (1994) that wild accessions of the species grew over much of the American neotropics, in Brazil, Bolivia, Peru, Venezuela, Guyana and Surinam. Three subspecies were recognised: (1) *M. esculenta* subsp. *esculenta* as the domesticate and includes all cultivars known in cultivation, (2) the wild *M. esculenta* subsp. *peruviana* occurring in eastern Peru and western Brazil and (3) the wild *M. esculenta* subsp. *flabellifolia*, which showed a wider distribution and ranges from the central Brazilian state of Goiás northward to Venezuelan Amazonia. Allem (1987, 1994, 1999) asserted that *M. esculenta* subsp. *peruviana* and *M. esculenta* subsp. *flabellifolia* were the likely original stocks from which cassava domesticate descended from. Five Brazilian *Manihot* species were suggested as the closest wild relatives of cassava (Allem 1994). One of them (*M. esculenta* ssp. *flabellifolia*) was s regarded as the wild progenitor of modern cultivars and thus part of the primary genepool of the root crop. Another species (*M. pruinosa*) was regarded as the nearest species to the GP1 of cassava and could hardly be separated from the wild strain *M. esculenta* ssp. *flabellifolia* on morphological grounds. This was in sharp contrast to earlier views held by Rogers (1963, 1965) and Rogers and Appan (1973) that cassava had no

known ancestry and was a by-product of indiscriminate introgression among some wild relatives. Also Rogers singled out a species native to Mexico and Meso America as the closest morphologically to cassava on the basis of computer analyses, the closest wild relative of *M. esculenta* to be *M. aesculifolia*. Using AFLP (amplified fragment length polymorphism) analyses, Roa et al. (1997) found *Manihot esculenta* subsp. *flabellifolia* and *M. esculenta* subsp. *peruviana* most similar to cassava, while *M. aesculifolia*, *M. brachyloba* and *M. carthaginensis* were more distant. The demonstration of unique genetic diversity in the two *M. esculenta* subspecies and their genetic similarity to the crop supported the hypothesis that these materials may be the ancestors of cassava.

Cassava, *Manihot esculenta* Crantz (Euphorbiaceae), is one of the leading food and feed plants of the world. It ranks fourth among staple crops, with a global production of about 160 million tons (Nassar 2002). Most of this is grown in three regions: West Africa and the adjoining Congo basin, tropical South America and south and southeast Asia. Cassava do not grow wild (Nassar 2000)

Agroecology

Cassava cultivation is limited to areas between 30° north and 30° south of the equator, from sea level to an altitude of 2,000 m. It thrives best in areas with a mean temperature of 25–29 °C, and a soil temperature of about 30 °C; below 10 °C the plant stops growing. The crop grows best in areas with an annual uniformly distributed precipitation of 1000–1500 mm, but can tolerate semi-arid conditions with rainfall as low as 500 mm and may have a competitive advantage over other crops under those conditions. It is one of the most efficient producers of carbohydrates and energy among all the food crops (De Vries et al. 1976). Being a hardy and fairly drought resistant crop, cassava will adapt to a wide range of growing conditions, soil types, and fertility levels and is often grown in marginal areas with poor soils, and/or high risk of drought where other crops will not survive. Cassava does best on well-drained, light-textured, deep soils of intermediate fertility. Under high fertility conditions, top growth may be stimulated at the expense of root growth. Optimum soil pH is between 4.5 and 6.5.

Edible Plant Parts and Uses

Cassava is the third most important carbohydrate food crop after rice and maize in the tropics. Cassava is an important staple food crop in most regions of Africa, Asia and Latin America (Fasuyi 2005; Kubo et al. 2006). *Manihot esculenta* is economically the most important, because it is the basic green vegetable for people in many parts of sub-Saharan Africa such as Nigeria, Cameroon, Gabon, Democratic Republic of Congo (DRC), Uganda, Angola, etc., and is also consumed in Latin America, the Philippines, Indonesia, Malaysia and other Asian countries (Almazan and Theberge 1989; Kubo et al. 2006). Cassava leaf is the principal ingredient for maniçoba, a local dish in Pará, Brazil (Kubo et al. 2006).

Cassava storage roots and young leaves are normally cooked to remove the hydrocyanic acid before consumption. Cassava roots are processed into a wide variety of granules, paste, flour, starch, etc., or consumed freshly boiled or raw (Suresh et al. 2011). Cassava starch or flour is used for thickening soups, sauces, gravies and puddings and cakes, or mixed with wheat flour for bread, biscuit or other snacks. In Malaysia, cassava starch is used in sweetened and unsweetened biscuits and in cream sandwiches at the rate of 5–10 % to soften the texture, add taste and render the biscuit non-sticky (Burkill 1966). Cassava starch and molasses are the major raw materials used in the manufacture of monosodium glutamate (MSG) used extensively in many parts of the world in powder or crystal form as a flavouring agent in foods such as meats, vegetables, soups, sauces and gravies.

Cassava juice or 'yari' is used to prepare both alcoholic and non-alcoholic drinks (Lancaster et al. 1982). A thick drink known as 'karato' is

prepared by the Karinya people from cassava bread. The Macusi prepare a similar drink from cassava flour. The preparation of cassava 'beers' is widespread in tropical America although in many areas the cassava-based beverages are of secondary importance to those prepared from maize. Fermented cassava beverages are commonly referred to as either 'kashiri' or 'chicha' although the name chicha is also frequently applied to maize beers. In the Caribbean, juice extracted from cassava roots is flavoured with cinnamon, cloves and sugar and called cassareep; it is used for preserving and flavouring meats and is an essential ingredient in pepperpot stew. *Sago*, often called *tapioca*, is a rolled, pre-cooked cassava starch. It has a 'pearl-like' appearance and is used for desserts and baby foods. In Brazil, sour cassava starch is a traditional fermented (LAB and yeast) food used in the preparation of fried foods and baked goods such as traditional cheese breads (Lacerda et al. 2005). Farinha is a yellowish coarse cassava meal used in many Brazilian dishes, especially in the north-east region (Diop 1998). The central-south region of Brazil produced *farinha de raspas* (dried, ground and sieved cassava chips) for the bread-making market and the northeast region produced *farinha de mandioca* (table flour) for consumption (Chuze 2001).

In the South pacific Islands, some popular cassava dishes include 'laplap' (grated raw cassava, coconut cream, corned beef, chopped green leaves); cassava soufflé (cooked cassava, eggs, milk); cassava 'bibinka' (melted butter or margarine, grated raw cassava, sugar, coconut cream, scrapped coconut chopped nuts, cheese, eggs) and cassava soup with chicken (SPC 1986). In the South Pacific, cassava is preserved by fermentation to make 'bila' – fermentation increases the vitamin B content.

In Indonesia, the roots are eaten raw or cooked, roasted in ashes or steamed (Ochse and Bakhuizen van den Brink 1980). All kinds of delicacies are processed from the roots such as 'dage', 'kolek', 'kripik krupuk', 'opak', 'tape', 'chendol' (a refreshing beverage). Even the rind of thick roots can be eaten in 'sayor' or as 'sambal goreng'. The roots are grated, mixed with palm sugar and roasted grated coconut and wrapped in banana leaf and steam to prepare the delicacy 'katimus'. Young roots, young leaves and shoots are eaten in sayur, urap or sepan. The young leaves are cut fine, mixed with grated coconut, wrapped in banana leaf and fried to made the dish 'gembrot'. Some processed cassava food products in Java, Indonesia, include starch, cassava flour, chip and galpek (peeled dried cassava) (Djazuli and Bradbury 1999).

In the Philippines cassava for food use is still at a semi-commercial or subsistence level (Loreto and Orias 2000). Cassava is traditionally eaten as a staple or a staple supplement when cereals are not adequately available. It is boiled, steamed or fried (e.g., 'kabkab'), or processed into local delicacies of various procedures, forms and taste. Among the local delicacies, cassava pie, pudding and cake are gaining popularity in the urban areas. These products are traditionally prepared using fresh grated cassava. Also cassava grates is the main component of high-value food products like cassava cake, 'pitsi-pitsi' and cookies. However, fresh cassava and its grated form have high perishability; hence, market reach is constrained. On the other hand, the market for cassava cakes, pies and pudding is slowly developing. The author listed the following substitution level of cassava flour in food products : 'paborita' 50 %, cheese crackers 50 %, coconut cookies 50 %, doughnuts 50 %, 'gollorias' 50 %, 'polvoron' 100 %, 'pandesal' 20 %, fried cheese sticks 50 %, cinnamon roll 50 %, muffins 50 %, chiffon cake 100 %, butter cake 100 %, 'cacharon' 100 %, hot roll 20 % and loaf bread 10 %.

Extrusion-cooked cassava/soy flour porridges were found to have good potential for use as high-energy/high-protein complementary foods and to have acceptable sensory properties (Muoki et al. 2012). In Nigeria, cassava roots are processed into a mash (40 % moisture content) which is then mixed into different blends to produce traditional cassava fried snack 'Ajogun', and fried and baked extrudates (modified 'Ajogun') snacks (Obadina et al. 2013). Other food products include 'gari', a roasted fermented cassava meal, 'agbelima', which is a fermented cassava mash, and the dried cassava chips known

as 'kokonte', which is further processed into cassava flour. Tapioca is a minor product or by-product from cassava processing. For industrial use, cassava is processed to obtain starch.

Microbial fermented food products are commonly consumed in Africa. 'Lafun' is a microbial (aerobic bacteria, lactic acid bacteria (LAB) and yeast) fermented cassava food product consumed in parts of West Africa (Padonou et al. 2009). 'Fufu' a common indigenous food of Nigeria is a yeast-fermented wet paste made from cassava (Oyewole 2001). Kivunde is a LAB fermented traditional Tanzanian food product (Kimaryo et al. 2000). 'Akyeke' is a steamed sour cassava meal prepared by fermentation of cassava with LAB (Obilie et al. 2004). LAB fermentation of cassava dough yields the fermented cassava meal 'agbelima' (Amoa-Awua et al. 1996). 'Attiéké' is a microbial fermented cassava product consumed mainly in Cote d'Ivoire (Coulin et al. 2006). The fermented, pre-gelatinised cassava meal is generally consumed with milk or meat and vegetables (Diop 1998). 'Gari' is a fermented (LAB and yeast) and gelatinised dry coarse flour, very popular in West Africa and a staple food in Nigeria, Ghana, Benin and Togo (Diop 1998). Baton is a fermented and pounded cassava mash eaten alone or with a side dish in Cameroon, Congo, Zaire and Gabon. 'Chicouangue' is a pre-gelatinised cassava paste usually in the form of balls wrapped in leaves. In Congo and Zaire, 'chicouangue' is steamed before being sold. Chips are another cassava food product consumed in Nigeria, Cameroon, Benin, Togo and Ghana (Diop 1998). Chips are small pieces of sun-dried cassava sometimes fermented, and marketed before being ground into flour (Diop 1998). The flour is mixed into paste with hot water to form a thick, sticky mass known as 'fufu' in West Africa or 'ugali' in East Africa. It is consumed in Nigeria, Cameroon, Benin, Togo and Ghana. In Central Africa, two main products associated with LAB fermented cassava are 'fu-fu' and 'chikwangue' (Ampe et al. 1994). The former is a flour obtained from crushed, sun-dried cassava mash. This flour may be mixed with boiling water, and served with sauce and fish or meat. 'Chikwangue' (cassava bread) is obtained after multiple post-retting (LAB fermentation) steps, including defibring, pugging, kneading and several cooking steps. 'Bikedi' is a fermented cassava root food obtained by retting of cassava in Congo (Dunican 1990; Lancaster et al. 1982; Ongusua et al. 1983), while 'ntoba mbodi' is a vegetable obtained by semi-solid fermentation of cassava leaves. In Haiti, Dominican Republic and Venezuela, cassava bread called 'casaba' in Spanish is consumed. It is a white, flat, circular, light textured bread baked from moist cassava pulp (Diop 1998).

Cassava starch used as edible coatings on fresh-cut pineapple in slices prolonged shelf-life at 5 °C without changing the quality parameters of fresh fruit and reduced microbial spoilage compared to other antibrowning agents (citric acid, ascorbic acid, calcium lactate) used as control (Bierhals et al 2011). Cassava starch was found to have use as a stabiliser of soy-based beverages for maintaining the nutritional value and good sensory quality of the products as functional foods (Drunkler et al. 2012).

Botany

Slender lactiferous shrub 2–3 m to a small tree 3–6 m high, sparingly branched or branching towards the apex. Stems robust, thick with prominent leaf scars, glabrescent. Roots long, tapered, large tuberous with a firm, homogeneous, chalk-white or yellowish flesh encased in a detachable rind, about 1 mm thick, rough and brown on the outside (Plate 1a–d). Latex watery and white. Young stems glaucous, green or yellowish green or tinged violet. Petiole 4–25 cm long, yellowish-green often reddish, terete, glabrous (Plates 3). Leaves (Plates 2 and 3) alternate, long-petiolate, palmately 3–9-partite, shallowly cordate, sometimes very slightly peltate, lobes obovate-lanceolate, tapering at base, apex acuminate to acute, margin entire, glaucous, shining above, lighter colour and dull under, 5–20 cm long by 206 cm wide. Stipules triangular-lanceolate 4–5 × 2 mm. Flowers in lax panicles, pedicelled, apetalous. Male flowers calyx campanulate, 5-partite, lobes triangular, greenish with crimson

Plate 1 a-d Harvested cassava roots

Plate 2 Cassava foliage

Plate 3 Cassava leaves with green–petioles and red petioles

tinge, 5 long and 5 short stamens, anthers yellow and tufted, disk 10-lobed, lobes concave and acute. Female flowers, calyx 5-partite, lobes triangular-ovate; disk shallowly 5-lobed ovary -3-celled style connate; stigma broad, many-lobed. Fruit obovoid-globose with 6 longitudinal, undulate wings, 1.3–1.7 by 1.3–1.5 mm, greenish. Seeds ellipsoid shiny, pale grey with 3 mm wide caruncle.

Nutritive/Medicinal Properties

Root Phytonutrients

Cassava tuber contained 54.6 g% fresh weight moisture, 1.5 g% dry weight protein and 0.671 g% fresh weight total amino acid content (Yeoh and Chew 1977). Nutrient composition of the raw tuberous root per 100 g edible portion was reported as: energy 135 cal, moisture 65.5 g, protein 1.0 g, fat 0.2 g, total carbohydrates 32.4 g, dietary fibre 1.0 g, ash 0.9 g, Ca 26 mg, P 32 mg, Fe 0.9 mg, Na 2 g, K 394 g, thiamin 0.05 mg, riboflavin 0.04 mg, niacin 0.6 mg and ascorbic acid 34 mg (Leung et al. 1972). Proximate nutrient composition of the raw cassava per 100 g edible portion was reported as: energy 160 kcal (667 kJ), moisture 59.68 g, protein 1.36 g, fat 0.28 g, carbohydrates 38.06 g, total dietary fibre 1.8 g, total sugars 1.70 g, ash 0.62 g, Ca 16 mg, P 27 mg, Fe 0.27 mg, Mg 21 g, Na 14 g, K 271 g, Zn 0.34 mg, Cu 0.1 mg, Mn 0.384 mg, Se 0.7 μg, ascorbic acid 20.6 mg, thiamin 0.087 mg, riboflavin 0.048 mg, niacin 0.854 mg, pantothenic acid 107 mg, vitamin B-6 0.088 mg, total folate 27 μg, total choline 23.7 mg, betaine 0.4 mg, vitamin A 13 IU, vitamin A 1 μg RAE, β-carotene 8 μg, vitamin E (α-tocopherol) 0.19 mg, vitamin K (phylloquinone 1.9 μg), total saturated fatty acids 0.074 g, 12:0 0.001 g, 16:0 0.069 g, 18:00.005 g, total monounsaturated fatty acids 0.075 g, 18:1 undifferentiated 0.075 g, total polyunsaturated 0.048 g, 18:2 undifferentiated 0.032 g, 18:3 undifferentiated 0.017 g, tryptophan 0.019 g, threonine 0.028 g, isoleucine 0.027 g, leucine 0.039 g, lysine 0.044 g, methionine 0.011 g, cysteine 0.028 g, phenylalanine 0.026 g, tyrosine 0.017 g, valine 0.035 g, arginine 0.137 g, histidine 0.020 g, alanine 0.038 g, aspartic acid 0.079 g, glutamic acid 0.206 g, glycine 0.028 g, proline 0.033 g and serine 0.033 g (USDA-ARS 2014).

Cassava roots had been found to have a low-protein content (0.7–2 %) (Nassar and Sousa 2007). Amino acids such as lysine and methionine were also low. In comparing the amino acid profiles of a common cassava cultivar and an interspecific hybrid, namely, ICB 300, they found that the interspecific hybrid (*M. esculenta* x *M. oligantha*) had 10 times more lysine and 3 times more methionine than the common cassava cultivar. The amino acid content (g/100 g) of the common cultivar and the hybrid ICB300 was determined, respectively, as: alanine 0.02, 0.093 g, arginine 0.037, 0.261 g, aspartic acid

0.016, 0.146 g, cysteine 0.027, 0.029 g, glutamic acid 0.039, 0.222 g, glycine 0.012, 0.078 g, histidine 0.0, 0.038 g, isoleucine 0.008, 0.068 g, leucine 0.016, 0.131, lysine 0.010, 0.098, methionine 0.014, 0.041 g, phenylalanine 0.016, 0.129 g, proline 0.0, 0.054 g, serine 0.012, 0.088, threonine 0.008, 0.061 g, tyrosine 0.0, 0.0, valine 0.019, 0.115 g, and total amino acids 0.254, 1.654 g. Among the cassava clones studied, the most impressive was the indigenous Brazilian UnB-400 cultivar known popularly by Amarela having 236 mg/kg of lutein compared to zero in other cultivars (Nassar et al. 2005). ICB-300 contained 5 % protein in the roots compared to 1.5–2 % in common cassava and 20 mg/g *trans*-β-carotene and 9108 mg/g of lutein in the leaves, an excellent source of these important components. The following quality traits (mean values) of roots from 2457 CIAT cassava genotypes comprising landraces and improved clones were found by Chávez et al. (2005): root dry matter 34.27 %, HCN 263.7 ppm, carotene 0.246 mg/100 g FT carotene, PPD 24.47 %, total sugars 2.87 %, reducing sugars 0.753 %, minerals (mg/kg) Fe 17.1 mg, Mn 1.4 mg, B 2.0 mg, Cu 5.8 mg Zn 7.5 mg, NA 129.2 mg, Al 11.5 mg, Ca 0.076 %, Mg 0.1.05 %, K 1.172 %, P 0.165 %, S 0.027 mg and crude protein content 3.063 %.

Cassava tuber contained 86.1 % starch, 2.4 % of 80 % ethanol-soluble sugars (1.3 % fructose, 0.9 % glucose, 0.2 % sucrose), 3.4 % uronic acid, 0.5 % lignin, 0.9 % ash, 6.7 % others and 4.5 % cell wall material (CWM) (Kajiwara and Maeda 1983). Monosaccharide composition of CWM by Saeman hydrolysis yielded 1.9 % rhamnose, 1.2 % fucose, 2.6 % arabinose, 4.2 % xylose, 2 % mannose, 12.8 % galactose, 52.7 % glucose, total 77.4 %. *Aspergillus niger* cellulose preparation was the most effective in CWM degradation and 57.1 % of CWM was degraded by the enzyme preparation. About 50 % of the hemicellulose parts as pentosan and hexosan, apart from cellulose, were extracted from CWM by hot water only and the extract appeared to contain pectin-like material precipitated by ethanol.

Considerable variation in β-carotene content was found in cassava germplasm (Moorthy et al. 1990). The values varied from 0·04 to 0·79 mg per 100 g edible portion. Cassava tubers contained 0.1–3 mg/kg FW of β-carotene and 0.05–0.6 mg/kg FW of lutein (Adewusi and Bradbury 1993). Carotenoids present in small amounts included α-, γ- and §-carotenes and β-cryptoxanthin and others were separated and partially identified. Yellow tubers contained much more β-carotene than white tubers and cassava was also a good source of pro-vitamin A carotenoids compared with other root crops. In cassava roots, the following carotenoids were found: *trans*-β-carotene 47.17 %, 9-*cis*-β-carotene 36.59, 13-*cis*-β-carotene 36.59 %, violaxanthin 0.73 %, 5,8-epoxy-lutein 0.77 %, lutein 8.07 %, β-cryptoxanthin 0.87 %, *cis*-β-cryptoxanthin 1.64 %, *trans*-α-carotene 1.43 %, phytofluene 1.72 % (Nassar et al. 2005). Nassar et al. (2007) reported that one of the many cassava clones grown by indigenous Brazilian farmers showed a high level of lycopene content (5 mg/kg versus zero in common cultivars, and 12–20 mg/kg in tomato, a lycopene-rich vegetable). A second landrace called UnB 400 had a high content of β-carotene, which reached 4 mg/kg. The total carotenoid in raw roots of seven sweet cassava cultivars varied from 2.64 to 14.15 μg/g and total β-carotene from 1.99 to 10.32 μg/g (Carvalho et al. 2012b). The β-carotene predominated in all the roots. The all-*E*- β-carotene isomer was the most abundant ranging from 1.0 to 7.27 μg/g. The 13-Z-β-carotene isomer varied from 0.52 to 0.52 μg/g and that of 9-Z-β-carotene from 0.25 to 1.12 μg/g. The 9-Z-β-carotene percentages were higher than those of the 13-Zisomer in five sweet yellow cassava cultivars. Evaluating the real retention percentage (RR%) in sweet yellow cassava after home cooking methods showed differences that could be attributed to the total initial carotenoid contents. However, no cooking method uniformly provided a higher total carotenoid or β-carotene retention in all the cultivars. Fried yellow-fleshed cassava roots showed the highest micellarisation efficiency for total carotenoids and all-*trans*-β-carotene (14.1 % and 14.37 %, respectively), compared with boiled and raw samples (Gomes et al. 2013).

Glutamic acid rich proteins, Pt2L4 and C54 proteins were found in cassava storage roots (De Souza et al. 2002, 2004, 2006; Zhang et al. 2003). The deduced amino acid composition of the Pt2L4 protein revealed that the most abundant amino acids were glutamic acid (31.6 %), alanine (16.94 %), valine (13.55 %) and proline (11.29 %) (De Souza et al. 2006). Pt2L4 and C54 proteins were 60 % identical with similar molecular weights (16.7 and 18.0 kDa, respectively) and isoelectric points (3.70 and 3.97) (Zhang et al. 2003; De Souza et al. 2006). Two or more homologous genes coding for glutamic acid-rich proteins in the cassava genome were reported (Zhang et al. 2003; De Souza et al. 2004, 2006). Studies by De Souza et al. (2004) revealed five genes with higher expression levels in secondary xylem of storage roots than adventitious roots. Among them, the Mec1 gene coding for Pt2L4 glutamic acid-rich protein and a putative RING zinc finger and LEA protein genes were strongly induced in secondary xylem tissue. They isolated and characterised a cDNA sequence (Mec1) coding for a glutamic acid-rich protein (Pt2L4) from cassava storage roots (De souza et al. 2006). Comparative sequence analysis showed a high identity of Pt2L4 with cassava protein C54, which was expressed in vascular tissues of storage roots. Studies revealed that their transcripts were most strongly expressed in vascular tissues and in parenchyma cells of storage roots, indicating an important role in storage root formation (Zhang et al. 2003; De Souza et al. 2004, 2006). In addition, Zhang et al. (2003) reported greatest C54 promoter activity in vascular cambium and starch-rich parenchyma cells of storage roots from transgenic cassava plants containing this promoter fused to the β-glucuronidase (GUS) reporter gene. De Souza et al. (2009) isolated and characterised the promoter sequence of the cassava gene Mec1 coding for Pt2L4 differentially expressed in cassava storage roots. Carotenoid-protein complex (CPC) was isolated from chromoplast-enriched suspensions of cassava storage root (Carvalho et al. 2012a). Small heat shock proteins (sHSPs) were the most abundant proteins identified in the CPC. Western blot analysis showed that fibrillin and Or-protein were present in chromoplast-enriched suspensions of yellow root but not in the complex or white root. Results from qRT-PCR helped identify an isoform of HSP21 possessing four single point mutations in the intense yellow CSR that may be responsible for increased sequestration of β-carotene.

The DNA sequence that they identified was a new promoter that could be a candidate for genetic engineering of cassava roots. β-Cyanoalanine synthase, the major enzyme involved in the conversion of cyanide to β-cyanoalanine, was partially purified from cassava leaf, rind and tuber (Elias et al. 1997). Its specific activity in the tuber was significantly greater than that in the leaf and rind.

Sucrose formed the bulk of the sugars in cassava root-tubers, accounting for more than 69 % of the total sugars (Ketiku and Oyenuga 1972). Other sugars included fructose, glucose and maltose. Maltose was consistently present as the lowest amount. The highest concentration of sugars (5.7 %) was attained 9 months after planting. Starch accounted for the highest proportion of the carbohydrates. A peak value of 81 % was observed 8 months after planting. The decrease to 78 % at 9 months was accompanied by an increase in sugar concentration from 3.5 to 5.7 %. The sum of cellulose and hemicellulose constituted the non-available carbohydrate fraction and was less than 7 % of total carbohydrates. The neutralised hydrolysate of the extracted hemicellulose revealed the presence of glucose and xylose only. The amylose content of cassava starch varied between 16.2 and 17.4 % during growth. The separated amylose had an iodine affinity of 17.0 % while amylopectin had 0.1 %.

The starch content in cassava tubers varied from 73.7 to 84.9 % on dry weight basis (Rickard et al. 1991). The gelatisation temperature for cassava starch ranged from 62 to 73 °C (Rickard et al. 1991). Cassava starch granules are round structures, flat on one side and containing a conical pit with a size range of 5–40 μ_m (Moorthy 2001). Starch granule size increased in all six cassava varieties up to the 6th month from the time of tuber initiation and thereafter remained almost constant (Moorthy and Ramanujan 1986).

The amylose content and reducing values did not vary much at different stages of growth. The swelling volume and swelling power of starch showed large variations particularly after the 10th month. Cassava starch can be fractionated into two types of polymers, namely, amylose and amylopectin (Munyikwa et al. 1997). Amylose consists essentially of linear chains of 100–10000 $\alpha/(1 \rightarrow 4)$ linked glucose residues. There is a low degree of $\alpha/(1 \rightarrow 6)$ branching within the amylose chains. The amylose content of cassava starch ranged from 13.6 to 23.8 % (Ketiku and Oyenuga 1972; Kawabata et al. 1984; Moorthy and Ramanujan 1986). Cassava starch was found to contain 9.20–15.30 % moisture, 17.9–18.9 % amylose content, 3.50–3.69 I_2 mg/100 mg iodine affinity, ash 44.4–886.1 mg/100 g dm and minerals mg/100 g dm phosphorus 7.4–14 mg, sodium trace to 12.5 mg, potassium 1–23.5 mg, calcium 6–280 mg, magnesium 0.7–4 mg (Kawabata et al. 1984). Cassava starch A washed in demineralised water and starch B washed in 0.5 m $CaCl_2$ in demineralised water had peaked distribution at 12.7–20.2 μm in granular size while cassava starch B washed in tap water had peak at 10.1–16.0 μm. Cassava had C X-ray type diffraction pattern (close to A). The swelling power of cassava starches prepared in various washing water was in the following sequence: demineralised water>tap water>0.5 m $CaCl_2$ in demineralised water. Viscosity rising temperature for cassava starches B and C was 65.8 °C and 64.9 °C for cassava starch A. The maximum viscosity of cassava starch A and B was 750B.U. (Brabender units) while starch C gave 682 B.U. Cassava starch C showed minimum breakdown at 437 B.U. and the largest setback at 215 B.U. For cassava starch A, the degree of gelatinisation at 5 °C was 96.2 % and at −20 °C was 95 %. The soluble amylose (which is thought to be mainly responsible for cohesiveness in cooked starch) content of cassava was determined to range from 10 to 40 % of total amylose (Moorthy 2001). Amylopectin is made up of much shorter chains of /-D glucopyranose units (Munyikwa et al. 1997). These are primarily linked by $\alpha/(1 \rightarrow 4)$ bonds with $\alpha/(1 \rightarrow 6)$ branches. The outer regions of the amylopectin molecules which are short and unbranched are called the A chains. Whereas the B chains exhibit multiple branching, the C chains have a single non-reducing end (Guilbot and Mercier 1985). The amylopectin molecules are highly organised and have a definite crystalline nature. Three main types of X-ray diffraction crystalline patterns in starch have been described namely A, B and C starch patterns (Gallant et al. 1982; Guilbot and Mercier 1985). Cassava had been found to consist largely of the A type pattern. Type A starches showed a higher susceptibility to amylase attack (Gallant et al. 1982). Extracted amylopectin was more stable than amylase due to limited hydrogen bonding, and this enabled it to remain fluid conferring high viscosity and elasticity to pastes and thickeners. Cassava starch also contain crude fat (0.08–1.54 %), crude protein (0.03–0.6 %), ash (0.02–0.33 %) and very low phosphorous levels (0.75–4 %) (Rosenthal et al. 1974). The genes coding for the main enzymes involved in cassava starch biosynthesis had been cloned and characterised such as genes coding for ADP glucose pyrophosphorylase B and S subunits, branching enzyme, granule bound starch synthase and their isoforms (Munyikwa et al. 1997). These genes together with the appropriate storage specific promoters are the tools with which to genetically modify starch biosynthesis in cassava (Munyikwa et al. 1997). Two starch biosynthetic genes, granule-bound starch synthase (GBSS) and branching enzyme (BE), were found to be much more expressed in in-vitro grown plants than in the corresponding organs of greenhouse grown plants (Salchuzzaman et al 1994). GBSS was more abundantly expressed in leaf than in stem, while the reverse was true for BE. Transcript levels of both genes were strongly induced by the exogenous supply of metabolisable carbohydrates such as sucrose, glucose and fructose. A tuber-specific cDNA library of cassava was constructed and a full-length cDNA for granule-bound starch synthase (GBSS, also known as waxy protein), the enzyme responsible for the synthesis of amylose in reserve starch, was cloned (Salehuzzaman). Sequencing of the cloned cDNA showed that it had 74 % identity with potato GBSS and 60–72 % identity with

GBSS from other plant species. The cDNA encoded a 608 amino acid protein of which 78 amino acids formed a chloroplast/amyloplast transit peptide of 8.37 kDa. The mature protein had a predicted molecular mass of 58.61 kDa (530 amino acids).

Root quality traits (minimum, maximum, average) from more than 4000 cassava genotypes expressed on dry basis (db) were reported respectively by Sánchez et al. (2009) as follows: dry matter content 14.3, 48.1, 33.6 %; cyanogenic potential 14, 3247, 324.7 ppm; total sugar content 0.2, 18.8, 3.8 %; reducing sugars content 0, 15.7, 1.3 %; starch content 65, 91, 84.9 %. The starch quality traits (minimum, maximum, average) of the > 4000 cassava genotypes were: amylose contents 15.2, 26.5, 20.7 %; water solubility 0.2, 16.6, 2.2 % db; swelling power o.8, 15.5, 4.6 %; paste clarity 12.5, 96.6, 45.2 %. Pasting properties (minimum, maximum, average) of the starches from > 4000 cassava genotypes were: pasting temperature 58.8, 71.2, 65.3 °C; maximum viscosity 146, 1505, 777.5 mPa s; breakdown 28.1, 859, 298.1 mPa s; consistency 0, 626, 155.8 mPa s; setback −702, 273, −144.5 mPa s; ease of cooking 1.1, 5.6, 2.8 min.

Proximate composition (dry basis) of starch from Ugandan cassava parental varietal lines and their F1 progenies were reported respectively as: moisture content 14.04–16.66, 13.82–15.37 %; protein 0.29–0.52, 0.28–0.35 %; fibre 0.02–0.56, 0.330.93 %; lipid 0.12–0.38, 0.19–0.35 % (Nuwamanya et al. 2010). Starch content of the cassava parental lines ranged from 70.36 to 89.90 % and 73.48-93.85 % in the progenies. Amylose content ranged from 23.01 to 26.98 % in the parents and 19.69-26.63 % in progenies. The amylose content was negatively correlated to swelling power and solubility but positively correlated to starch content. Average starch granule sizes ranged from 7 to 12 μm, smaller granules 2–6.9 μm and large granules 13–20 μm. Granules were mostly truncated- polygonal to truncated-oval and truncated oval, rounded in shape. Peak viscosity ranged from 170.79 to 344.96 RVU, hot paste viscosity 80.46–127.45 RVU, breakdown viscosity 76.26–256.21RVU, final viscosity 130.07–227.17 RVU, setback 57.99–89.17 RVU, pasting temperature ranged from 64.75 to 70.4 °C, peak time 4.24–7.77 min. The parental lines had higher peak viscosity, set back viscosity and breakdown viscosity than their progenies. Swelling powers increased with increasing temperatures.

The centesimal composition of the cassava starch indicated: 12.2 % of moisture, 88.43 % of starch, 0.15 % of fibre, 0.14 % of total sugar, 0.26 % of lipids, 0.07 % of protein and 0.1 % of ash (Leonel et al. 2009). The expansion index of cassava starch expanded products ranged from 2.79 to 3.69. There were no effects of extrusion temperature, feed moisture and screw speed. The WSI (water solubility index) obtained in the different extrusion treatments ranged from 12.3 to 30.74 %. Considering the WSI of cassava starch before extrusion (0.69 %), there was a pronounced increase in solubility with the extrusion process. The WAI (water absorption index) of extruded cassava starch varied from 4.19 to 6.41 g/g; such values were higher than those observed for non-extruded starch (1.64 g/g). The viscosity peak (VP) of cassava starch was 439.67 RVU (rapid viscosity units), breakdown (B) was 319.75 RVU, final viscosity (FV) was 217.83 RVU, retrogradation tendency (R) was 97.92 RVU, and paste temperature was 67.20 °C. The breakdown of the extruded products varied from 2.83 to 12.83 RVU. The final viscosity (FV) of extruded starch varied from 7.25 to 13.1 RVU. The retrogradation tendency values of the extruded products varied from 2.50 to 12.50 RVU. Extruded starch had less luminosity, higher chroma a* level and greater intensity of chroma b* than native starch. There is a pronounced increase of the water solubility and water absorption indexes with the extrusion process. High moisture, low screw speed and intermediate temperature provide lower starch degradation which is desirable in pre-cooked starch. Out of the extrusion parameters, barrel temperature has the most pronounced effect on expansion, paste properties and colour of extruded starch, followed by feed moisture and screw speed. Twin screw extrusion of cassava starch resulted in gelatinisation, partial or complete destruction of the crystalline structure and macromolecular

degradation of both amylose and amylopectin, as well as protein denaturation, and formation of complexes between starch and lipids, and between protein and lipids (Colonna and Mercier 1983). The formation of lower molecular weight material was observed by a decrease of intrinsic viscosities of both starch components.

Gelatinised starches have numerous industrial, nonfood uses such as in drilling oil wells, sizing textiles, paper making, charcoal briquetting, and in obtaining water-based paints (Powell 1965). In food, gelatinised starches can be used almost any time thickening is desired. The gelatinisation of starch also significantly affects the characteristics and quality of food such as loaf volume and crumb bread, the elasticity and softness of paste products, digestibility and palatability, the tolerance of batter properties in cakes, frostings and doughnut mixes, the sugar crystal growth in foods, and the texture, volume, shelf-life and freeze-thaw stability of bread and cakes (Powell 1965; Williams and Lesseelleur 1970; Michael and Brown 1968).

Proximate composition of cassava starch moisture 10.4 %, fat 1 %, protein 0.8 %, ash 0.1 %, carbohydrate 87.8 %, energy 362.9 kcal/100 g., starch 84.5 %, starch as amylase 19.5 %, cyanide 0.05 ppm, sugar 1.5 %, gluten 0 % (Oladunmoye et al. 2014). Functional properties of cassava starch were determined as follows: diastatic activity 96 mg/10 g, a-amylase activity 260 s; swelling power 9 %, solubility 2.2 %, WAC (water absorption capacity) 1647 %, WBC (water binding capacity) 147.5 %, starch damage AACC (American Association of Cereal Chemists) method 7.2 %, absorption rate 97.1 % absorption speed 28 s. Pasting properties of cassava starch found were: peak viscosity 364.3 RVU, holding strength 153.5 RVU breakdown 210.8RVU, final viscosity 216.0 RVU, setback value 63.5RVU, peak time 4 min, pasting temperature 50.3 °C.

Root Phytochemicals

Low levels of tannins (proanthocyanidins) have been detected in dried/processed cassava products and may pose a factor limiting their nutritional value (Rickard 1986). The aqueous and ethanolic extracts of raw cassava tuber were found to contain alkaloids, flavonoids, tannins, reducing sugars and anthocyanosides, but do not contain cardiac glycosides, anthraquinone, phlobatannins and saponins (Ebuehi et al. 2006). The proximate composition of the raw and boiled cassava tubers was not significantly different, except in the carbohydrate, fat and moisture. Vitamins A, C and E and minerals, namely, calcium, magnesium, phosphorus, iron, sodium and chloride ions were identified in the raw and boiled cassava tubers and their levels were significantly reduced by boiling. Raw and boiled cassava tubers are good sources of water and carbohydrate. All cassava cultivars were reported to contain cyanogenic glucosides (mainly linamarin) which liberate the toxic hydrogen cyanide by enzymatic hydrolysis (Veltkamp and De Bruijn 1996).

The cyanogenic glucoside linamarin was obtained in good yield from manioc tubers (Clapp et al. 1966). From the reaction of acetobromoglucose and acetone cyanohydrin both the α- and β-anomers of the glucoside of acetone cyanohydrin were obtained, and comparison of the properties of natural linamarin with those of the synthetic samples established the identity of the natural glucoside as the β-anomer. A glucoside of 2-hydroxy-2-methylbutyronitrile was shown to be present, in low concentration compared with linamarin, in three samples of relatively bitter tubers of manioc (Bisset et al 1969). The less levorotatory of the synthetic glucosides, having the (R) configuration at the asymmetric centre of the aglycone, was shown to be identical with the compound lotaustralin. In addition to lotaustralin and linamarin, a novel cyanogenic glycoside, 2-((6-*O*-(β-D-apiofuranosyl)-β-D-glucopyranosyl)oxy)-2-methylbutanenitrile, two novel non-cyanogenic glycosides, (2*S*)-((6-*O*-(β-D-apiofuranosyl)-β-D-glucopyranosyl)oxy) butane and 2-((6-*O*-(β-D-apiofuranosyl)-β-D-glucopyranosyl)oxy)propane, and a simple non-cyanogenic glycoside, ethyl β-D-glucopyranoside, were isolated from an ethanolic extract of the fresh root cortex of *Manihot esculenta* (Prawat et al. 1995). The methanol extract of cassava

root parenchyma and cortex yielded a new compound isopropyl-β-D-apiofuranosyl-(16)-β-D-glucopyranoside (IAG), and small amounts of phenylalanine and tryptophan (King and Bradbury 1995). Also present were the sugars: glucose, fructose, sucrose, maltose and organic anions oxalate, malate and citrate and the cyanogenic glucosides linamarin and lotaustralin. The following compounds were found to contribute to the level of bitterness in cassava: linamarin, lotaustralin, IAG, phenylalanine and tryptophan. The occurrence of linamarin [2-β-D-glucopyranosyl-2-methylpropiononitrile] and lotaustralin [(2*R*)-2-β-D-glucopyranosyloxy-2-methylbutyronitrile] in roots and leaves with the linamarase-resistant cyanogenic disaccharides linustatin [2-(6-*O*-β-D-glucopyranosyloxy)-2methylpropiononitrile] and neolinustatin [(2*R*)-2-(6-*O*-β-D-glucopyranosyl-β-D-glucopyranosyloxy)-2-methylbutyronitrile] strongly suggested the operation of the linustatin pathway in cassava when the cyanogenic glucosides linamarin and lotaustralin were transported from the leaves to the roots (Lykkesfeldt and Moller 1994). Large amount of hydrogen cyanide were released from fractions known to contain linamarin and lotaustralin which occurred at levels of 10 μmol/g and 0.7 μg/mol, respectively, in cassava seedlings.

All cassava tissues, with the exception of seeds, contain the cyanogenic glycosides linamarin (>90 % total cyanogen) and lotaustralin (<10 % total cyanogens) (McMahon et al. 1995). Leaves contained the highest cyanogenic glycoside levels (5.0 g linamarin/kg fresh weight), whereas roots had approximately 20-fold lower linamarin levels (White et al. 1998). In addition to tissue-specific differences, there were cultivar-dependent differences in root cyanogen levels. Total root linamarin levels ranged between 100 and 500 mg linamarin/kg fresh weight for low- and high-cyanogenic cultivars, respectively. Linamarin, the predominant cyanogenic glycoside in cassava, could accumulate to concentrations as high as 500 mg/kg fresh weight in roots and to higher levels in leaves (McMahon et al. 1995). De novo synthesis of linamarin in cassava roots was demonstrated in-vivo by feeding [14C] valine to excited segments of phelloderm (Du et al. 1995). In vitro, a microsomal enzyme system isolated from cassava roots was shown to catalyse the conversion of valine to acetone cyanohydrin, the aglycone of linamarin. Cyanogenic glucosides were found to accumulate in cassava roots, but hitherto de novo synthesis had only been demonstrated in the leaves, suggesting translocation of cyanogenic glucosides from leaves to roots. The results showed that at least part of the cyanogenic glucosides were synthesised in the roots. The highest linamarin was found in both root cortex and parenchyma tissues extracted with H_2SO_4 (Sornyotha et al. 2007). However, the concentration of linamarin in the extract of the root cortex was higher than that of the root parenchyma.

Recently, the pathway of linamarin synthesis and the cellular site of linamarin storage was reported by Siritunga and Sayre (2004) as follows: valine → *N*-hydroxy valine → 2-methyl proponal oxime → acetone cyanohydrin (plus UDP-glucose) → linamarin → acetone cyanohydrin and glucose → acetone and HCN. The enzymes involved were valine → *N*-hydroxy valine (CYPD79D1), *N*-hydroxy valine → 2-methyl proponal oxime (CYPD79D2), 2-methyl proponal oxime → acetone cyanohydrin (putative CYPE1E), acetone cyanohydrin (plus UDP-glucose) → linamarin (UDP-glucosyl transferase), linamarin → acetone cyanohydrin and glucose (linamarase), acetone cyanohydrin and glucose → acetone and HCN (hydroxynitrile lyase) or spontaneously. In addition, the cyanogenic enzymes, linamarase and hydroxynitrile lyase, had been characterised and their genes cloned. Studies by Jørgensen et al. (2011) concluded that CYP71E7 was the oxime-metabolising enzyme in cyanogenic glucoside (linamarin and lotaustralin) biosynthesis in cassava. Heterologous expression of CYP71E7 in yeast afforded microsomes converting 2-methylpropanal oxime (valine-derived oxime) and 2-methylbutanal oxime (isoleucine-derived oxime) to the corresponding cyanohydrins, which dissociated into acetone and 2-butanone, respectively, and hydrogen cyanide.

Hughes (1992) reported cyanogenesis (release of hydrogen cyanide) to be initiated in cassava when the plant tissue was damaged and depended on the sequential action of a β-glucosidase (linamarase) and an α-hydroxynitrilase on cyanoglucosides. High levels of linamarase activity were demonstrated in the latex of leaf petioles and this activity was shown to be dependent on the presence of attached leaflets. In contrast α-hydroxynitrilase activity occurred at low levels in the latex and must be located elsewhere in the leaf. Earlier, Mkpong et al. (1990) identified three isozymes of limarase in cassava leaves based on differences in isoelectric points. The enzyme was capable of hydrolysing a number of beta-glycosides in addition to limarin. They also found that the steady state levels of linamarin and linamarase in leaves of high and low cyanogenic varieties were not correlated with the varietal differences in the steady state levels of linamarin in roots. Rupture of the vacuole releases linamarin, which is hydrolysed by linamarase (McMahon et al. 1995). Hydrolysis of linamarin was found to yield an unstable hydroxynitrile intermediate, acetone cyanohydrin, plus glucose. Acetone cyanohydrin spontaneously decomposed to acetone and HCN at pH >5.0 or temperatures >35 °C and could be broken down enzymatically by a-hydroxynitrile lyase (White et al. 1998). Studies by White et al. (1994; White and Sayre 1995; McMahon et al. 1995) demonstrated that root tissues were capable of synthesising linamarin at rates comparable to those in leaves and that linamarin was stored in vacuoles. They also demonstrated that hydroxynitrile lyase, like linamarase, was localised in the apoplast. Recent studies by Santana et al. (2002) demonstrated that linamarase was found mainly in laticifer cells of petioles and roots of both cultivars of low and high cyanide with no significant differences between them. At the subcellular level, there were sharp differences because linamarase was found mainly in the cell walls of the high cyanide cultivar, whereas in the low-cyanide cultivar, the enzyme was present in vacuoles and cell wall of laticifer cells. Reverse transcriptase-PCR on cassava tissues showed no expression of linamarase in cassava roots, thus, the transport of linamarase from shoots to roots through laticifers was proposed. More recent studies by Narayanan et al. (2011) found that overexpression of hydroxynitrile lyase (HNL) in cassava roots elevated protein and free amino acids while reducing residual cyanogen levels. They hypothesised that the over-expression of HNL in cassava roots was under the control of a root-specific, patatin promoter that would not only accelerate cyanogenesis during food processing, resulting in a safer food product, but also lead to increased root protein levels since HNL was sequestered in the cell wall.

Sakai et al. (1986) identified methyl palmitate, palmitic acid, methyl linoleate, methyloleate, linoleic acid, oleic acid, R(−)-1-glyceryl monopalmitate, R(−)-1-glyceryl monolineate, R(−)-1-glycerylmonooleate, cholesterol, campesterol, stigmasterol and β-sitosterol in healthy cassava root tissues. They found that diterpenic and other stress metabolites were formed after the production of coumarins and phenols in discoloured tissues. They classified the stress metabolite into two groups: 4 steroids campest-4-ene-3-one, stigmast-4-en-3-one, stigmasta-4,22-dien-3-one and stigmasta-4,6-dien-3-one, and more than 20 diterpenoids. Two diterpene hydrocarbons were isolated (+)-stachene; (−)-ent-pirara-8(14),15-diene were identified as major stress metabolites in damaged tissues. The minor ent-kaurene was not isolated but its structure was confirmed. Twenty-two diterpenic stress metabolites were isolated and identified from cassava (*Manihot esculenta*) root tissues damaged by cutting or fungal-infection (Sakai and Nakagawa 1988). The metabolites were classified into four families: *en*t-beyerane (10 components), *ent*-pimarane (9 components), *ent*-atisane (2 components) and *en*t-kaurane (1 component). Apart from three previously known alkenes, (+)-stachene (*en*t-beyerene, yucalexin B-1), *ent*-kaurene (yucalexin K-3, the sole component of the *ent*-kaurane class), and (−)-*en*tpimara-8(14),15-diene, all showed oxygen substitution in the A and C rings. The three major components were: yucalexin B-9 (*ent*-3β-hydroxybeyer-15-en-2,12-dione), the most abundant diterpenoid component in deteriorated cassava root (10 mg/kg fresh weight), yucalexin

P-8 (*ent*-8a,14α-epoxy-3β-hydroxypimara-9(11), 15-dien-2,12-dione), an *ent*-pimarane, was the second most abundant diterpenoid (4 mg/kg fresh weight), and yucalexin A-19 (*en*t-3β,16α-dihydroxyatis-13-en-2-one), an *ent*-atisane, also a novel compound was the third most abundant diterpenoid (1.4 mg/kg fresh weight). The known compounds isolated were yucalexin B-9; yucalexin A-19 together with the *ent*-beyeranes, yucalexin B-5 (*ent*-2-hydroxybeyera-1,15-dien-3,12-dione) and B-6(*en*t-beyer-1,15-dien-3, 12-dione) and an atisane, yucalexin A-16 (*ent*-16α-hydroxyatis-13-en-3-one); yucalexin P-4 (*ent*-pimara-8(14),15-dien-3,12-dione); yucalexin B-22 (*en*t-2α,3β-dihydroxybeyer-15-en-12-one); yucalexin P-15 (*ent*-3β,14α-dihydroxypimara-7,9(11),15-trien-2,12-dione); andyucalexinP-17(*ent*-3β,12β-dihydroxypimara-8(14),15-dien-2-one).

A rapid accumulation of bluish fluorescent compounds and phenolic components in the parenchyma of cassava root which suffered from cut injury, post-harvest physiological deterioration (PPD) and microbial deterioration was reported by various authors (Tanaka et al. 1983; Uritani et al.1983; 1984a; b). These compounds were identified as the hydroxycoumarins, scopoline (6-methoxy-7-hydroxycoumaroyl-7-β-D-glucoside) scopoletin (6-methoxy-7-hydroxycoumarin), and esculin (6,7-dihydroxycoumaroyl-6-β-D-glucoside) and two conjugates containing scopoletin and esculetin, respectively. A main phenolic component of the periderm found was (+) catechin and a minor component (+)-gallocatechin. Also some enzymes related to the production of the secondary metabolites such as acid invertase, phenylalanine ammonia lyase (PAL) and peroxidase were found in cut-injured and on-injured tissues adjacent to the soft rotten parts (Uritani et al. 1983; Tanaka et al. 1983). Further, polyphenol oxidase was present in both fresh and deteriorated tissues. Wheatley (1982) found a 150–200-fold increase of scopoletin during the first 24–48 h after wounding. Wheatley and Schwabe (1985) showed that rapid PPD of cassava roots required the presence of oxygen and scopoletin, the latter acting in some autocatalytic fashion. Other compounds that were identified from cassava roots and that may play a role in PPD included leucoanthocyanins, cyanidin and delphinidin (Akinrele 1964), the flavan-3-ols,(+)-catechin and (+)-gallocatechin (Uritani et al. 1983; 1984a), and 4 steroids (Sakai et al. 1986) and 22 diterpenoid compounds (Sakai and Nakagawa 1988). H_2O_2 was quantified and localised histochemically at the tissue and cell level in deteriorating cassava roots during PPD (Buschmann et al. 2000a). This reactive oxygen species accumulated during the first 24 h after harvest, especially in the inner parenchymatic tissue. Three flavan-3-ols, (+)-catechin, (+)-catechin gallate, and (+)-gallocatechin accumulated during the storage of cassava roots. Four hydroxycoumarins (esculin, esculetin, scopolin and scopoletin), compounds derived from the phenylpropanoid pathway, were found to accumulate in cassava tuberous roots during the time course of PPD (Buschmann et al. 2000b). Scopoletin and scopolin showed the most dramatic increases in concentration, peaking by day 2 after harvesting. Also evidence for the metabolism of scopoletin to an insoluble coloured product by means of a peroxidase was presented. A cassava catalase (MecCAT1) expressed during PPD was isolated from cassava root (Reilly et al. 2001).

Besides the accumulation of secondary metabolites, the increasing activities of enzymes like phenylalanine ammonia lyase (PAL) (Tanaka et al. 1983), the key entry enzyme of the phenylpropanoid pathway, had been described. Evidence from cycloheximide treatment of root discs (which partially inhibited symptoms), in-vivo labelling of proteins that were newly synthesised during deterioration time courses, and cDNA cloning and Northern hybridisation experiments clearly demonstrated PPD to be an active process involving the increase and de novo synthesis of proteins, and changes in gene expression (Beeching et al. 1998; Reilly et al. 2007). A cDNA clone (cMeHRGP1) for a hydroxyproline-rich glycoprotein expressed during the deterioration response was isolated and characterised in cassava roots suffering PPD (Han et al. 2001). Using elicitor (yeast)-challenged cassava suspension cells and leaves, phenylalanine ammonia-

lyase (PAL), phenylpropanoids and peroxidases (POD) were found to play a role in cassava innate defence against microbial infection (Gómez-Vásquez et al. 2004). Fungitoxicity was also determined against the cassava pathogens *Fusarium solani, F. oxysporum* and the saprotroph *Trichoderma harzianum*. Phenolic levels in elicited cells were not enhanced and were, theoretically, too low to be inhibitory. However, in combination and when oxidised they may contribute to defence, because oxidation of esculetin and scopoletin by peroxidase and of esculetin by tyrosinase enhanced their fungitoxicity up to 20-fold.

Using the cDNA-AFLP method, Huang et al. (2001) systematically screened 70 genes expressed during PPD in cassava tuberous roots: (EST) inosine 5'phosphate dehydrogenase 1, DAD1 (defender against apoptotic death 1), isovalery-CoA dehydrogenase precursor, flavonol synthase, similar to expansin, allergen like protein C_2H_2-type zinc finger protein, peroxidase, 40s ribosomal protein, CPRD 12 protein, EST similar to hypothetical protein, Ca^{2+}-binding protein, no apical meristem like protein, beta-COP like protein, 40s ribosomal protein, similar to peptide transport proteins, HVA22 homologs, (EST) putative glucotransferase, (EST) putative 21kD protein precursor, R2R3-MYB transcription factor, elongation factor 1-alpha, ZPT2-14 zinc finger protein, cytochrome P450 monooxygenase, UDP-glucose pyrophosphorylase, putative serine/threonine kinase, WIZZ (transcription factor upon wounding), chlorophyllase, cytochrome P450, heat shock 70 protein, eukaryotic release factor 1, putative receptor-like protein kinase, 26S ribosomal protein, Golgi associated protein se-wrap41, ATP synthase delta chain, Ca^{2+}- binding protein, putative pectin acetyl esterase protein, Ca^{2+-}binding protein and (EST) translation factor EF-1-alpha-like protein. Reilly et al. (2007) identified the full array of genes expressed during cassava post-harvest physiological deterioration (PPD). Many of the up-regulated and PPD-specific expressed sequence tags (EST) were predicted to play a role in cellular processes including reactive oxygen species turnover, cell wall repair, programmed cell death, ion, water or metabolite transport, signal transduction or perception, stress response, metabolism and biosynthesis, and activation of protein synthesis. These PDP EST found were as follows: PPD (postharvest physiological deterioration) upregulated genes included: genes with roles in ROS (reactive oxygen species) turnover: catalase CAT 1, catalase Cat2, ascorbate peroxidase, secretory peroxidase, thioredoxin peroxidase, thioredoxin-like protein, glutathtone S-transferase, metallothionein, quinine-oxidoreductase, aldo/keto reductase, early light induced protein; roles in signal transduction or perception: ACC (1-aminocyclopropane-1-carboxylate) oxidase, phospholipase, immunophilin; genes with potential rolls in programmed cell death: cysteine protease, class IV chitinase; roles in stress response: dehydrin, heat shock protein cognate 70; genes with roles in ion or metabolite transport: PIP (plasma membrane intrinsic protein) 1 type aquaporin, PIP2 type aquaporin, gamma adaptin, H^+-PPase, ATP/ADP translocase precursor; genes with potential roles in cell wall metabolism and remodelling: xyloglucan endotransglycosylase, UDP-glucose dehydrogenase, hydroxyproline-rich glycoprotein, extension, germin-like protein; genes with roles in biosynthesis and metabolism: cytochrome P_{450} CYP79D1, cytochrome P_{450} CYP79D2, cytochrome P_{450} CYP71E, cytochrome P_{450}, cytochrome b_5, UDP (uridine diphosphate) glycosyltransferase, L-asparaginase, ketol acid reductoisomerase, transaldolase protein, arginine decarboxylase, sulphite reductase precursor; genes with roles in transcription or translation: ribosomal protein L5, ribosomal protein S10, ribosomal protein S16, ribosomal protein S29, eukaryotic initiation factor eIF4, ATP-dependent RNA helicase, seryl tRNA synthase, elongation factor eF1a, oligouridylate-binding protein, alternative splicing factor SF2a, pentatricopeptide repeat (PPR) protein; genes with role unknown or uncharacterised: putative endopeptidase; MtN19, hypothetical protein, putative protein, three unknown proteins, three no protein match; and PPD downregulated genes: cystatin-like protein, late embryogenesis-abundant protein LEA, translationally controlled tumour protein (TCTP), auxin repressed protein ARP1, auxin repressed protein ARP2, PWWP

domain protein, two expressed proteins, and unknown protein. Twenty-one storage root-specific genes were found: genes with potential roles in signal transduction or perception: Auxin-repressed protein-like protein ARP1, GTP-binding protein, nuclear transport factor, putative; gene with potential roles in cell wall metabolism and remodelling: cell wall-plasma membrane linker protein; genes with roles in transcription or translation: translation initiation factor, putative ribosomal protein L10a, structural constituent of ribosome, nucleic acid binding; genes with roles in protein modification: electron transporter/thiol-disulphide exchange, polyubiquitin; genes with role unknown or uncharacterised: cystatin-like protein, translationally controlled tumour protein (TCTP), *M. esculenta* allergenic-related protein Pt2L4I, *Arabidopsis* expressed, four unknown proteins, conserved hypothetical protein, two no protein matches. Of the group of 21 storage root-specific genes, none were common to the up-regulated PPD genes, though four were identical to the smaller group of down-regulated genes: cystatin-like protein, TCTP, ARP1 and an unidentified expressed protein.

Bayoumi et al. (2008b) showed that the major biosynthetic pathway to scopoletin and its glucoside, scopolin, in cassava roots during post-harvest physiological deterioration (PPD) was through p-coumaric, caffeic and then ferulic acids. An alternate pathway through 2′,4′-dihydroxycinnamate and umbelliferone led to esculetin and esculin. They demonstrated that the major pathway was through o-hydroxylation and not via a proposed spirolactone-dienone intermediate. Further they found that the E-Z-isomerisation step in the biosynthesis of scopoletin and scopolin, in cassava roots during PPD, was not photochemical, but could be catalysed by an isomerase which was independent of light.

Galactosyl diacylglycerides (2 mg/kg FW), β-carotene (<0.2 mg/kg), linamarin, and β-sitosterol glucopyranoside (4 mg/kg) were identified in fresh cassava roots (Bayoumi et al. 2010). The hydroxycoumarin scopoletin and its glucoside scopolin were identified from cassava roots during post-harvest physiological deterioration (PPD), as well as trace quantities of esculetin and its glucoside esculin. There was no isoscopoletin in cassava roots during PPD.

Fermentation of cassava roots in cold water in *gari* preparation was found to impart an objectionable odour to the fermented mass and the cooked 'fufu' (Ohochuku and Ballantine 1983). The compounds responsible for this objectionable odour were isolated from the acidic fraction of fermentation liquor and identified as butanoic acid, propanoic acid, acetic acid and butanoic acid. During fermentation of cassava tuber, there was a decrease of 32 % in the crude protein in sour flour and 69 % in sweet flour (Padmaja et al. 1994). The ether extractive fraction increased by 26 % in sour flour while there was tremendous reduction in ash content in fermented flours. Although there was a decrease in all of the amino acids in the sour and sweet flours, the decrease in arginine, histidine and glutamic acid was quite noticeable. Despite the decrease in all amino acids in sweet flour, the protein quality based on the essential amino acid scores in sweet flour appeared to be good. Sucrose, glucose and fructose were the prominent sugars in nonfermented flour but fermentation reduced the sucrose content significantly. There was no significant difference in in-vivo digestibility of fermented and non-fermented flours. During cassava fermentation (72 h) with a mixed culture inoculums, there was a slight decrease in starch content from 216 to 331.3 g/kg to 200–302.2 g/kg (George et al. 1995). A sharp decrease in pH was observed after 24 h fermentation itself, resulting from the production of organic acids. The starchy flour had higher fibre content than that obtained from the non-fermented tubers. Optical and scanning electron microscopy revealed clustering of starch granules in the starchy flour from fermented tubers.

Aerial Parts Phytochemicals

The proximate nutrient and fibre composition (% dry weight (dw)) of very young expanding and mature cassava leaves were determined respectively as follows: moisture content (wet weight) 89.1, 79.4 %, food energy 452, 478 kcal, crude

protein 38.1, 19.7 %; crude fat 3.8, 6.8 %, crude fibre 8.3, 27.4 %, ash 4.0, 7.9 %, carbohydrate 45.8, 38.2 %, neutral detergent fibre 18.1, 46.3 %, acid detergent fibre 9.0, 30.3 %, hemicelluloses 9.1, 16 %, cellulose 8.4, 22.1 % lignin 0.9, 8.4 %; minerals (g/100gdw) – K 2.26, 1.38 g, Ca 0.43,1.14 g, Mg 0.37, 0.26 g, P 0.23, 0.18, Na 0.08, 0.12 g, Zn 20.9, 16.4 g, Mn 24, 15.9 g, Fe 15.2, 26.6, Cu 3.9, 4 g; amino acid (g/16gN) aspartic acid 10.9, 7.6 g, threonine 5, 3.2, serine 5.7, 3.3, glutamic acid 10.1, 13.2, proline 3.7, 5.8 g, glycine 4.7, 12.1 g, alanine 6.3, 3.2, valine 5.7, 5.1 g, cysteine 1.2, 0.7 g, methionine 2, 1.3 g, isoleucine 5, 3.9 g, tyrosine 4.6, 2.8 g, phenylalanine 5.3, 5.4, lysine 7.5, 3.8 g, histidine 2.5, 1.1 g, and arginine 5.7, 4.0 g (Ravindran and Ravindran 1988). The crude protein and carbohydrate contents decreased with maturity, whereas all other proximal and fibre components increased. The mineral profile showed cassava leaves to be good sources of most minerals, particularly of calcium and trace minerals. The P and Na contents, however, were low. The values for K, Mg, P, Zn and Mn decreased with leaf maturity, while those for Ca, Na and Fe increased. Cassava leaves were found to be rich in all essential amino acids, except methionine and phenylalanine. As the leaves matured, the tendency was for the amino acid concentrations to decrease. Only glutamic acid, proline and glycine contents increased, while those of valine and phenylalanine were unaffected.

Cassava leaves had been reported to be a good source of supplementary protein and also vitamins A and C and minerals (Lancaster and Brooks 1983). Proximate nutrient composition of cassava leaves (mean of 4 cultivars in g/kg DM (dry matter)): gross energy 469.3 mJ/kg, crude protein 34.8 g, crude fibre 12.1 g, ether extract 7 g, ash 6.9 g, N-free extract 39 g, Ca 2.1 g, Mg 0.26 g, Zn 2.6 g, Ni 2 g, Na 0.65 g, K 1.6 g, P 600 mg, Fe 105 mg (Fasuyi 2005). The range and mean content (per kg) of hydrocyanic acid, tannin and phytin in the leaves were 40.2–60.6 mg (52.9 mg), 6.9-15 g (9.7 g) and 107.3–249.1 mg (192 mg), respectively. Compositions of cassava leaf protein concentrate (g/kg DM) were reported in the following range (mean): crude protein 42.4–50 g (47 g), crude fibre 1.4–2.6 g (2 g), ether extract 19.4–22.8 (21.6 g), ash (6.9–9.3 (7.8 g) N-free extract 11.5–23.5 g (mean 15.9 g), gross energy 500.8–539.9 MJ/kg (524.3 MJ/kg), digestible energy 433.6–476.3 (461 MJ/kg) (Fasuyi and Aletor 2005). Amino acid profile (g/16 g N) was: alanine 6.12–6.51a, aspartic acid 7.10–9.98 g, arginine 5.96–6.38 g, glutamic acid 10.85–11.62 g, his 2.63–2.90 g, isoleucine 5.50–5.74 g, lysine 6.69–6.88 g, methionine 2.41–2.53 g, cysteine 1.08–1.33 g, leucine 9.56–9.78, serine 4.87–5.30 g, threonine 4.72–5.4 g, phenylalanine 6.01–6.52, val 5.98–6.49 g, tyrosine 4.71–5.00 g, tryptophan 2.21–2.36 g, methionine+cysteine 3.74–3.84 g. Functional properties of cassava leaf protein concentrate were :water absorption capacity 118–225.5 % (181.5 %), oil absorption capacity19.2–40.8 % (mean 33.4 %), fat emulsion capacity 25–40.8 % (mean 32.5 %), fat emulsion stability 25–40.8 % (mean 32.5 %), foaming capacity 21–38.9 % (mean 32.1 %), least gelation concentration 8–16 % (mean 12.5 %), foaming stability (after 30 min) 8–16.4 cm^3 (mean 10.2 cm^3). Functional properties of cassava leaf meal were: water absorption capacity 400–417 % (409.6 %), oil absorption capacity 48.3–60.8 % (mean 56.8 %), fat emulsion capacity 26.3–28.9 % (mean27.4 %), fat emulsion stability 27.3–46.9 % (mean 41.2 %), foaming capacity 15.8–21 % (mean 17.7 %), least gelation concentration 8–12 % (mean 9 %), foaming stability (after 30 min) 4.1–4.4 cm^3 (mean 4.3 cm^3).

Rogers and Milner (1963) reported the proximate composition and cyanide of leaves of Jamaican and Brazilian cassava cultivars (mean values, fresh weight) respectively as follows: moisture content 79.2, 71.2 %; protein 5.4, 7.8 %; ether extract 1.6 %, 3 %; ash 1.8, 1.6 %; crude fibre 1.7, 1.4 %; carbohydrate 10.4, 14.9 %; energy 77.4, 118 cal/100 g; cyanide 46, 50.5 ppm. The amino acid content was respectively: alanine 5.98, 6.19 %; arginine 5.28, 6.12 %; aspartic acid 10.14, 9.63 %; cysteine 1.37, 1.04 %, glutamic acid 10.22, 10.12 %; glycine 5.39, 5.32 %; histidine 2.23, 2.56 %; isoleucine 5.012, 4.84 %; leucine 2.29 g8.89, 8.85 %; lysine 7.20, 6.33 %; methionine 1.65,1.71 %; phenylalanine 5.82,

5.53 %; proline 4.64, 5.40 %; serine 5.16, 4.60 %; threonine 4.92, 4.73 %; tryptophan 1.47, 2.07 %, tyrosine 4.18, 3.93%g; valine 5.73, 5.58 %.

The leaf protein content of six cassava varieties ranged from 29.3 to 38.6 % dry weight basis and estimated leaf protein production varied from 242 to 953 kg/ha (Yeoh and Chew 1976a;b). Total amino acids varied from 8.42 to 9.4 % while the essential amino acids averaged 4.21 % and the sulphur containing amino acids only 0.25 %. The amino acid content (range % fresh leaf weight) of fresh young cassava leaves was: isoleucine 0.43-o.61 %, leucine 0.81-a.00 %, lysine 0.54–0.70, methionine 0.10–0.25 %, cysteine 0.07–0.09 %, phenylalanine 0.36–0.51 %, tyrosine 0.23–0.42 %, threonine 0.32–0.51 %, tryptophan 0.12–0.35 %, valine 0.33–0.61 %, arginine 0.34–0.48 %, histidine 0.07–0.35 %, alanine 0.40–0.61 %, aspartic acid 0.93–1.36 %, glutamic acid 0.95–1.14 %, glycine 0.35–0.61 %, proline 0.26–0.46 %, serine 0.28–0.46 %, total essential amino acids 3.87–4.38 %, total amino acids 8.34–9.40 %.

In cassava leaves the following carotenoids were found: *trans*-β-carotene 76.57 %, 9-*cis*-β-carotene 10.42 %, 13-cis-β-carotene 10.42 %, violaxanthin 0.5 %, 5,8-epoxy-lutien 0.53 %, lutein 10.16 %, and *trans*-α-carotene 0.27 % (Nassar et al. 2005). The content of both *trans*-β-carotene and lutein in leaves of cassava clones UnB-400 and ICB-300 were high 27.40 mg/g in the former and 20 mg/g in the latter (Nassar et al. 2005). The major carotenoids of cassava leaves were the non-vitamin A carotenoid lutein (86–290 mg/kg fresh weight (FW)) and the pro-vitamin A carotenoid β-carotene (13–78 mg/kg FW) (Adewusi and Bradbury 1993). Immature leaves contained less than mature leaves.

Of the three Nigerian cassava varieties, the protein content of the young leaves (oven-dried) was highest in IITA white (32.4 %) followed by oko iyawo (31.3), IITA red (29.3 %) (Awoyinka et al. 1995). Ash content was the lowest (4.6 %) in oko iyawo variety, 6.0 and 6.4 % for IITA white and red varieties, respectively. Calcium content was 429 mg/kg in oko iyawo, 500 mg in IITA white and 551 mg in IITA red. 429–551 mg/kg iron content was highest in oko iyawo 429 mg/kg, 235 mg/kg in IITA white and 218 mg in IITA red. Phosphorus content was 780 mg/kg in oko iyawo, 710 mg in IITA white and 800 mg in IITA red. Dietary fibre was 26.9 % for oko iyawo, 33.7 % for IITA white and 35.5 % for IITA red. Tannin content was low in 'okoiyawo' variety (2.1 mg/100 g) but high 8.3 and 29.7 mg/100 g in IITA white and red cassava varieties, respectively. In-vitro digestibility of oven-dried cassava leaves was very low (15.6 for IITA white, 19.2 % for IITA red and -20.5 % for oko iyawo. HCN-potential in fresh cassava leaves was 8.1–12.6 mg HCN/100 g fresh weight. Blanching increased protein content and in vitro protein digestibility but decreased ash, minerals, dietary fibre and tannin, while HCN-potential was unchanged. Grinding reduced both HCN-potential and tannin by 84 and 71 %, respectively, while oven drying only reduced the HCN content marginally.

Young cassava leaves were found to have low content of lipids (3.02 %) of which 22.4, 25.1 and 48.2 were non-polar lipids, glycolipids and phospholipids, respectively (Khor and Tan 2006). Pigments (11.5 %), wax and hydrocarbons (1.2 %), steryl esters (2.9 %), methyl esters of fatty acids (2.0 %), triglycerides (1.5 %), fatty acids (2.1 %), diglycerides (1.1 %) and sterols (0.1 %) constituted the leaf non-polar lipids. The leaf glycolipids were made up of esterified steryl glycosides (2.1 %), monogalactosyl diglycerides (12.5 %), steryl glycosides (1.1 %), cerebrosides (4.2 %) and digalactosyl diglycerides (5.2 %). The leaf phospholipids were found to include cardiolipin (3.6 %), phosphatidylglycerol (21.5 %), phosphatidylserine (0.7 %), phosphatidylethanolamine (16.4 %), phosphatidylinositol (4.0 %) and other unidentified phospholipids (2.5 %). Phosphatidylcholine was present only in trace quantity. Analysis of the fatty acid composition of each of the leaf lipids revealed that, with the exception of steryl esters, all leaf lipids possessed high content of polyunsaturated fatty acids.

Cassava leaf meals from 12 month old plants of five Brazilian cultivars contained the highest levels of crude protein (CP) (29.17–35.90 g/100 g), β-carotene (113.83–137.38 mg/100 g), iron (202.90–225.60 mg/kg),

magnesium (0.19–0.29 mg/100 g), phosphorus (0.28–0.33 g/100gdm), potassium (1.45–12.63 g/100 g) and sulphur 0.33–0.42 g/100 g per 100 g compared to 15 and 17 month old plants (Wobeto et al. 2006). The IAC 289–70 cv. showed the highest levels of magnesium, as well as considerable contents of CP, β-carotene, iron, zinc and sulphur.

Cassava leaves were found to contain significant amount of rutin (Subramanian et al. 1971). The phenolic extract obtained from fresh cassava leaves yielded 2 biflavones characterised as amentoflavone and podocarpus-flavone a (4‴-*O*-methyl amentoflavone) (Kamil et al 1994). From the methanolic extract of the fresh cassava leaves lotaustralin and linamarin, and two flavonoid glycosides, kaempferol-3-*O*-rutinoside and quercetin-3-*O*-rutinoside (rutin) were isolated (Prawat et al. 1995). Cassava leaf was found to contain a large amount of rutin or quercetin 3-*O*-rutinoside which was completely hydrolyzed to quercetin during cooking processes (Kubo et al. 2006). In the case of total phenol content in cassava leaf stalks, the acetone extract was found to have maximum phenol content followed by acidified methanol and methanol (Suresh et al. 2011). For anthocyanin content, acidified methanol extract gave maximum yield followed by methanol and acetone extracts.

The content of some antinutrients in cassava leaf meal from 12 to 17 month old plants of five Brazilian cultivars was: oxalate 1.36–2.49, 1.37–2.86 g/100 g DM (dry matter); cyanide 10.80–19.33, 12.38–35.02 mg/100 g DM; saponin 1.74–4.41, 3.62–4.43 g/100 g DM; and trypsin inhibitor 0.57–2.75, 2.46–3.28 mg ITU (inhibited trypsin unit) and polyphenol content 43.37–61.49, 71.15–106.43 mg/g DM (Wobeto et al. 2007). The nitrate levels decreased with the maturity of the plant, whereas those of cyanide increased. The polyphenol, saponin and trypsin inhibitor contents increased with the maturity of the plant. The hemagglutinative activity decreased with the maturity of the plant, but this tendency was not found in the cultivars Ouro do Vale. The cultivar IAC 289–70 had the lowest antinutrient levels, except for saponin and oxalate.

Cassava leaves showed a wide variation of cyanide content, ranging from 12.5 mg to 85.4 mg/100 g fresh weight (fw), and in the tubers the pulp had lower cyanide content 2.2 mg-16.6 mg/100 g fw than in the peel (11.1–70 mg/100 g fw) (Yeoh and Oh 1979). Cassava leaves were found to contain <20 % crude protein and the neutral-detergent fibre (NDF) was found to be rich prodelphinidin condensed tannin which may pose an important factor limiting the nutritive value of cassava forage (Reed et al. 1982). The raw cassava leaves had more presence of alkaloids, flavonoids, tannins, anthraquinone, phlobatannins, saponins, reducing sugars and anthocyanosides, but do not contain cardiac glycosides (Ebuehi et al. 2006). Raw cassava leaves contained more crude fibre and protein as compared to the raw or boiled cassava tubers. Vitamins A, C and E and minerals, namely, calcium, magnesium, phosphorus, iron, sodium and chloride ions were identified in raw leaves and their levels were significantly reduced by boiling. The raw leaves are rich in crude fibre, protein and water. Raw cassava leaves contained more antinutrients, such as tannins, oxalate, phytate and higher trypsin inhibitor activity than in the raw tuber.

Antinutritional factors, condensed tannins and cyanide, found in cassava leaves, reduce the nutritional quality of the leaf meal (Padmaja 1989). It was found that wilting the whole branches under shade for 16 h followed by drying the detached leaves at 60 °C was even more advantageous in reducing the levels of these toxic principles. Chopping of the wilted leaves retained a higher percentage of cyanide on drying compared with the drying of wilted whole leaf blades. Highly significant reduction in assayable tannin levels could be brought about when the leaves were sprayed with either sodium hydroxide or ammonia. While sun-drying and shredding + sun-drying reduced cyanide to innocuous levels, the processing techniques were less efficient with regard to tannin and phytin removal. Phytic acid or phytate, the free-acid form of myo-inositol hexakisphosphate was detected in cassava leaves (Hadi Alkarawi and Zotz 2014). Cyanide levels in leaves of five Brazilian cassava cultivars

ranged from 62.41 to 152.41 mg/100 g DM and in cassava leaf flour from 12.38 to 35.02 mg/100 g DM, presenting a loss of 62.09–80.16 % cyanide in the leaf flour (Wobeto et al. 2004). Hue et al. (2012) found dry matter (DM), neutral detergent fibre (NDF), acid detergent fibre (ADF) and total tannin contents were higher in fully matured cassava leaves, while crude protein (CP) and hydrogen cyanide (HCN) were lower in developing leaves. Increasing the harvesting frequency increased DM and CP production in cassava foliage. HCN content was lower at the first foliage harvest than at later harvests for all cutting cycles. At subsequent harvests the content of total tannins tended to decline, while HCN content increased.

Rhodanese (cyanide sulphur transferase) activity was detected in a crude extract of tapioca leaves (Chew and Boey 1972). Optimal activity was found at a high pH (10·2–11·0) and temperature (57–59°). Under these conditions, rhodanese from 0·5 ml of the crude extract (75 mg leaf fresh weight) catalysed the formation of ten·2 μmoles thiocyanate per 15 min. Rhodanese was purified 7.8-fold from young cassava leaves and found to be a hetero-dimer with two subunits of molecular weights of 16000 and 17000 (Boey et al. 1976). Superoxide dismutase (SOD) cDNA, mSOD2, encoding cytosolic copper/zinc SOD (CuZnSOD) cDNA was isolated from suspension-cultured cells of cassava (Shin et al. 2005). mSOD2 was 774 bp in length with an open reading frame (ORF) of 152 amino acids, corresponding to a protein of predicted molecular mass 15 kDa and a pI of 5.22. mSOD2 was highly induced in leaf tissue by several abiotic stresses, including high temperature (37 °C), chilling (4 °C), methyl viologen (MV) exposure and wounding treatment. The results indicated the mSOD2 gene to be involved in the antioxidative process triggered by oxidative stress induced by environmental change.

The ethyl acetate extract of cassava leaves, stems and twigs yielded known triterpenoids including: taraxerol, 3-epi-taraxerol, taraxerone, β-amyrin, β-amyrone, lupenone, and two novel triterpenoid acids, 3α-hydroxytaraxer-14-en-29--oic acid (esculentoic acid A) and 3-oxo-taraxer-14-en-29-oic acid (esculentoic acid B) (Chaturvedula et al. 2003).

Twenty-five components accounting for 83.53 % of cassava stem oil, and 22 components accounting for 83.53 % of cassava leaf oil were identified (Hu et al. 2010). Major compounds in the stem oil were palmitic acid (31.39 %), oleic acid (8.94 %) and linoleic acid (5.57 %), and in the leaf oil palmitic acid (16.85 %), phytol (15.02 %) and isophytol (11.21 %).

The ethyl acetate fraction of cassava stem afforded 10 phenolic compounds: coniferaldehyde (1), isovanillin (2), 6-deoxyjacareubin (3), scopoletin (4), syringaldehyde (5), pinoresinol (6), *p*-coumaric acid (7), ficusol (8), balanophonin (9) and ethamivan (10) (Yi et al. 2011). A new diterpene named yucalexin P-23 together with known compounds yucalexin P-15, protocatechuic acid and catalpinic acid were isolated from cassava stems (Li et al. 2011). A new eremophilane sesquiterpene, sporogen AO-2, and a new beyerane diterpene, thecacorin C, together with two known compounds, longifoamide-B and methylcholestane-3β,5α,6β-triol, were isolated from cassava stems (Zeng et al. 2014). Two new compounds, maniesculentins A and B, together with four known ones yucalexin P-21, cleistanthe-type sonderianol, calliterpenone and *ent*-kauran-3α,16α,17triol were isolated from cassava stems (Pan et al. 2015). The structures of maniesculentin A was elucidated as 3*a*,12*a*-dihydroxy-pimara-8(14),15-dien and maniesculentin B as tetrahydro-2*a*-(4′-hydroxy-3′-methoxyphenyl)- 4*a*-carbonxyl-3*b*-hydroxymethyl furan. Maniesculentins A and B were assayed for antibacterial activity against four bacteria lines (*Staphylococcus aureus*, *Streptococcus*, *Escherichia coli*, *Pseudomonas aeruginosa*) but were inactive.

Chemical composition of cassava stems was reported by Chen et al. (2011) as: ashes 4.97 %, cold water extraction 12.04 %, hot water extraction 12.57 %, 1 % sodium hydroxide solution extraction 34.16 %, benzene-alcohol solution extraction 4.20 %, nitric acid-alcohol cellulose 35.86 %, holo-cellulose 72.62 %, pentosan 19.20 %, acid-soluble lignin 2.51 %, acid-insoluble lignin 26.10 %, organic solvent-soluble

lignin 1.07 %, pectin 0.02 %. Phenolic hydroxyl content of organic solvent-soluble lignin and acid-insoluble lignin were 1.245 mmol/g/ and 0.261 mmol/g, respectively. The authors stated that cassava stems could be used as a non-wood renewable source of natural products. Cassava stems were found to contain high α-cellulose 56.82 %, lignin 21.72 %, acid detergent fibre (ADF) 21.45 % and fibre length 0.05–0.5 cm (Sumada et al. 2011). Stages to extract α-cellulose included cutting and grating stems, oven drying, prehydrolysis, delignification, bleaching and analysis of α-cellulose. Also found were β-cellulose and γ-cellulose.

Seed/Seedling Phytochemicals

Cassava seed kernel contained 6.6 g% fresh weight moisture, 31.6 g% dry weight protein and 23.53 g% fresh weight (Yeoh and Chew 1977). Cassava seed kernel was found to constitute 57 % of the whole seed (Nartey et al. 1974). The endosperm, comprising 96 % of the seed kernel, was found to contain the bulk of storage lipids and proteins, although the embryonic tissues showed a similar distribution of storage materials. The major seed storage reserve was composed of lipids, which occurred to the extent of 47 % of the dry kernel weight (Nartey et al. 1973, 1974). Kernels were found to contain relatively little carbohydrate. Starch occurred to the extent of only 0.3 %, while soluble carbohydrates occurred in a much higher concentration, 3.8 %. The latter was almost entirely composed of sucrose. Total seed proteins accounted for 34 %, while total soluble nitrogenous compounds accounted for only 0.13 %. Inorganic phosphate accounted for only 0.08 %, whereas organic-bound phosphate occurred in a much higher concentration, 1.36 %. Seed storage lipids comprised mainly triglycerides (98 %), with trace amounts of di- and mono-glycerides, phospholipids, glycolipids and a steroid derivative. No free fatty acids could be detected. The fatty acid composition of the storage lipids of cassava seeds comprised 61.6 % linoleic acid (22.4 % oleic acid and 10.3 % palmitic acid as major components), with myristic acid, palmitoleic acid, stearic and linolenic acid as minor components (Nartey and Møller 1973). A trace of arachidic acid occurred during early germination. The overall fatty acid composition of total lipids in dark- and light-grown seedlings remained relatively constant and indicated that no specific fatty acids were preferentially metabolised during seed germination and growth. Total nitrogen accounted for 5.31 % of the dry cassava kernel weight and free amino acids constituted 2.0 % of the total nitrogen in cassava seed (Nartey and Møller 1976). Total, free and protein bound amino acid contents of cassava seed kernels expressed as mg/100 g dried matter were determined respectively as follows: aspartic acid 3600, 95, 3505 mg; threonine 1345, 9, 1336 mg; serine 1340,13.4, 1327 mg; glutamic acid 5815, 109.4, 5706 mg; proline 1595, 29.4, 1566 mg; glycine 1405, 12.5,1392 mg; alanine 1460, 18.7, 1441 mg; valine 2770, 7, 2763 mg; isoleucine 1010, 6.7, 1003 mg; leucine 2110, 9.8, 2100 mg; tyrosine 1000, 6.7, 993 mg; phenylalanine 1410, 6.1, 1404 mg; lysine 1040, 17.3, 1023 mg; histidine 765, 5.1, 760 mg; arginine 4590, 32.9, 4557 mg; methionine 570, 2.9, 567 mg; cysteine 540, 7.4, 533, and tryptophan 540, 0, 0 mg.

The extracted cassava seed oil was liquid at room temperature (25 °C), pale yellow, clear and transparent, odourless and tasteless with specific gravity of 0.94 at room temperature and free of sediments (Popola and Yangomodou 2006). The physicochemical profile of cassava seed oil was reported as moisture 0.73 %, ash 0.5 %, crude fibre 15.20 %, crude protein 20.20 %, carbohydrate 20.10 %, lipid 25.02 %, free fatty acids 0.39 %, saponification value 270 mg KOH/g, acid value 0.7 mg KOH/g, unsaponifiable fraction 2 g/kg, specific weight at room temperature 0.94, peroxide value 2.5 g O2 mg/g, cyanide 0.027 %, refractive index 1.444 and iodine 90 Wij's.

No HCN could be detected in seeds of one cassava cultivar whereas seeds of two other cultivars contained 5.2–8.5 μg HCN/g fresh weight (Nartey 1968). However, 10–14-day-old seedlings of all three cultivars contained 156–260 μg HCN/g fresh weight indicating a rapid biosynthe-

sis of cyanogenic material occurred during germination of cassava seeds. Linamarin, 2(β-D-glucopyranosyloxy) isobutyronitrile, accounted for 93 %, while lotaustralin, 2(β-D-glucopyranosyloxy) 2-methylbutyronitrile, accounted for 7 % of the total HCN evolved by autolysing etiolated cassava seedlings. L-[U-^{14}C]-Valine and L-[U-^{14}C]-isoleucine were incorporated by etiolated seedlings into the aglycone moieties of linamarin and lotaustralin, respectively, indicating that the amino acids were effective precursors of these glucosides in cassava. Seedlings of all three cultivars contained linamarase, the β-glucosidase which catalyses the hydrolysis of both glucosides. A crude preparation of the enzyme from leaves showed strong activity against linamarin and lotaustralin, mild activity against salicin and weak activity against β-methyl glucoside and amygdalin. A microsomal system catalysing the in-vitro synthesis of the aglycones of the two cyanogenic glucosides linamarin and lotaustralin was isolated from young etiolated cassava seedlings (Koch et al. 1992). The rates of conversion of the parent amino acids valine and isoleucine to their cyanohydrins were 19 and 6 nmol/h/mg protein, respectively. The conversion rates for the corresponding oximes (2-methylpropanal oxime and 2-methylbutanal oxime) were 475 and 440 nmol/h/mg protein and for the nitrites (2-methylpropionitrile and 2-methylbutyronitrile) 45 and 75 nmol/h/mg protein. With the exception of 2-cyclopentenylglycine, none of the additionally tested amino acids were metabolised, whereas a broad substrate specificity was observed using oximes and nitriles as substrates. All tissues of the cassava seedling contained cyanogenic glucosides. The microsomal enzyme system responsible for their synthesis was restricted to the cotyledons and their petioles. This demonstrated that the cyanogenic glucosides were actively transported to other parts of the seedling. The enzyme activity decreased with the height of the etiolated seedling and was barely detectable in seedlings above 75 mm.

Antioxidant Activity

Aqueous cassava leaf extract exhibited antioxidant activity with DPPH radical scavenging activity of 78.89 μmol Trolox equivalent per g dry plant material, Fe^{3+} reducing antioxidant potential (FRAP) activity 144.85 μmol Trolox equivalent per g dry plant material, Cu^{2+} ion chelating activity 66.56 % and total phenolic content of 18.77 mg gallic acid equivalent per g dry plant material (Wong et al. 2006). The total antioxidant activity values of aqueous extract of shoots as evaluated by the ferric thiocyanate (FTC) and thiobarbituric acid (TBA) with absorbance of 500 nm and 532 nm, respectively, in descending order were as follows: *M. utilissima* (fresh) > *Diplazium esculentum* (fresh) > *Sauropus androgynus* (fresh) > *M. utilissima* (boiled) > *D. esculentum* (boiled) > *S. androgynus* (boiled) (Rahmat et al. 2004). Quercetin, derived from cassava leaf, scavenged the superoxide anion generated by xanthine oxidase and inhibited uric acid formation on xanthine oxidase, while rutin scavenged the superoxide anion but without inhibiting the enzyme (Kubo et al. 2006). 3,4-Dihydroxyphenylacetic acid, a microfloral metabolite of quercetin, also scavenged the superoxide anion generated by xanthine oxidase and inhibited uric acid formation on xanthine oxidase. In addition, quercetin inhibited the oxidation of linoleic acid catalysed by soybean lipoxygenase-1.

The following inhibitory efficiency of ethylene production (markers of advanced lipid peroxidation) was observed with aqueous leaf extracts: *Manihot esculenta* > *Pteridium aquilinum* > *Abelmoschus esculentus* > *Hibiscus acetosella* (Tsumbu et al. 2011). The addition of the extracts at different final concentrations inhibited significantly the ethylene production in a concentration-dependent relationship. For the same extract at the same concentration, the levels of the residual ethylene production were moderately higher for the boiled than for the not heated samples. For production of lipid hydroperoxides,

the addition of final extract concentrations higher than 5 μg/mL decreased significantly the intensity of the absorbance in a dose-dependent manner. The IC_{50} values for boiled and not heated samples showed the following order of inhibitory efficiency of lipid hydroperoxides: *Manihot*>*Pteridium*>*Abelmoschus*>*Hibiscus* without great differences from one plant to another. All plant extract showed extremely significant inhibitory effects on free radical production vs. control depending on the concentration and the plant, except *Abelmoschus* and *Hibiscus* at 10 μg/mL, which increased the formation of free radicals, compared to the control. All plant extract showed no toxicity at the highest concentration of 100 μg/mL on HL60 (human promyelocytic leukaemia) cells, and there was no significant difference of viability between cells incubated with plant extracts and those without extract solutions (control cells). On the inhibition of ROS (reactive oxygen species) produced by phorbol-12-myristate-13-acetate (PMA)-activated HL60 monocytes, *Manihot* was the most efficient, followed by *Pteridium, Abelmoschus* and *Hibiscus*. The not heated samples displayed slightly more inhibitory efficiency than the boiled ones.

The ethyl acetate and *n*-butanol fractions of cassava stems showed greater DPPH and ABTS·+scavenging activities than other fractions. The ethyl acetate fraction contained the highest content of total phenolics 29.82 g GAE/g extract in comparison to *n*-butanol fraction 15.06 g, petroleum ether fraction 4.58 g and aqueous fraction 6.02 g. The ethyl acetate fraction afforded 10 phenolic compounds: coniferaldehyde (1), isovanillin (2), 6-deoxyjacareubin (3), scopoletin (4), syringaldehyde (5), pinoresinol (6), *p*-coumaric acid (7), ficusol (8), balanophonin (9) and ethamivan (10), which exhibited significant antioxidant activities. The relative order of DPPH· scavenging capacity for these compounds was: ascorbic acid (reference)>6>1>8>10>9>3>4>7>5>2, and that of ABTS·+scavenging capacity was 5>7>1>10>4>6>8>2>Trolox (reference compound)>3>9. The results showed that these phenolic compounds contributed to the antioxidant activity of cassava. The presence of phenolic compound and anthocyanin contributed to the free radical scavenging activity of cassava leaf stalk extracts (Suresh et al. 2011).

Anticancer Activity

Esculentoic acids A and B, isolated from the aerial parts of cassava, exhibited moderate in-vitro cytotoxicity against the A2780 human ovarian cancer cell line (Chaturvedula et al. 2003). Cassava stem oil showed significant cytotoxicity against K562 (human erythromyeloblastoid leukaemia) cell lines with IC_{50} value of 7 μg/ml (Hu et al. 2010).

Antihyperglycemic/Antidiabetic Activity

Among the Malaysian plant extracts cassava shoot (*Manihot esculenta*), papaya shoot (*Carica papaya*), ulam raja (*Cosmos caudatus*), pegaga (*Centella asiatica*) and kacang botol (*Psophocarpus tetragonolobus*) tested for inhibitory activity against two key enzymes involved in hyperglycaemia, cassava shoot extract was the most inhibitory (Loh and Hadira 2011). In the α-amylase inhibition assay, the inhibitory potential was highest in cassava shoot for both hexane (59.22 %) and dichloromethane extract (54.15 %). Cassava shoot hexane extract showed the highest inhibitory potential (95.01 %) on α-glucosidase activity.

Wang et al. (2014) found that administration of cassava cross-linked octenyl succinic maltodextrin (CCOMD) to streptozotocin-induced diabetic mice may be helpful for treating and preventing hyperlipidaemia and hyperglycaemia in diabetes. It was also found that the blood glucose and insulin levels were ameliorated in the diabetic mice by the CCOMD diet. Moreover, the CCOMD diet decreased the plasma total cholesterol level (8.1–9.1 %) and LDL cholesterol level (28.9–39.4 %), and improved the plasma HDL cholesterol level (13.8–15.3 %) and intestine short chain fatty acid content.

Hypotensive Activity

Cassava shoot extract also exhibited angiotensin I-converting enzyme (ACE) inhibitory activity (Loh and Hadira 2011). The dichloromethane extract (28 %) exhibited higher ACE inhibition than the hexane extract (3 %).

Antimicrobial Activity

Chloroform cassava leaf extract exerted positive antibacterial activity against *Listeria monocytogenes, Vibrio cholerae, Shigella flexneri* and *Salmonella typhi* while the ethanol leaf extract was effective against *Pseudomonas aeruginosa, Corynebacterium diphtheria* and *V. cholerae* (Zakaria et al. 2006). Cassava seed oil exhibited inhibitory effect in-vitro on the growth of clinical isolates of *Staphylococcus aureus, Propionibacterium acnes, Escherichia coli, Pityrosporum ovale* and *Candida albicans* (Popola et al. 2007). The most pronounced inhibition as confirmed by the zones of inhibition around growing colonies was on *S. aureus*, *P. acnes* was moderately inhibited, while inhibition of growth of *E. coli* was mild. Growth inhibition by the oil was not significant between *P. ovale* and *C. albicans*. Cassava stem and leaf volatile oils exhibited inhibitory activity against *Staphylococcus aureus* and methicillin-resistant *Staphylococcus aureus* MRSA (Hu et al. 2010). Protocatechuic acid and catalpinic acid isolated from cassava stems exhibited antimicrobial activity (Li et al. 2011). Longifoamide-B from cassava stem exhibited modest inhibitory effects on *Staphylococcus aureus* and methicillin-resistant *S. aureus*, and methylcholestane-3β,5α,6β-triol exhibited modest inhibitory effect on *S. aureus* (Zeng et al. 2014).

Analgesic Activity

Oral administration of ethanol extract of cassava leaves to mice reduced the number of writhings as a response to pain induced by intraperitoneal injection of 0.6 % acetic acid compared to control (Miladiyah et al. 2011). Cassava leaf ethanol extract at concentrations of 100, 250, 500 mg/kg inhibited acid writhing in rats (Okechukwu and Bokanisereme 2013). Terpernoids, tannins, flavonoids and carotenoids were found present in the cassava leaf extract. Aqueous cassava leaf extract (100–400 mg/kg, orally and 1–4 %, w/w, topically), like aspirin (100 mg/kg, i.p.), exhibited significant inhibition of acetic acid and acetylcholine-induced mouse writhing tests, compared to untreated control (Adeyemi et al. 2008). Effects produced by cassava leaf extract were significantly lower than those produced by aspirin (100 mg/kg, i.p.) in the analgesic models, except for the topically administered extract on acetylcholine-induced pain.

Antiinflammatory Activity

Cassava leaf aqueous extract at 100–400 mg/kg, p.o. and 1–4 % (w/w), topically, produced significant inhibitions of carrageenan-induced rat paw oedema and xylene-induced ear swelling in mice (Adeyemi et al. 2008). Effects produced by the extract were significantly higher than those produced by indomethacin (10 mg/kg, s.c. or 1 %, w/w in petroleum jelly) in the anti-inflammatory models. Cassava leaf ethanol extract at concentrations of 100, 250, 500 mg/kg inhibited acid carrageenan and histamine induced oedema in rats (Okechukwu and Bokanisereme 2013). Terpernoids, tannins, flavonoids and carotenoids were found present in the cassava leaf extract.

Antipyretic Activity

Cassava leaf ethanol extract at concentrations of 100, 250, 500 mg/kg inhibited yeast induced pyrexia in rats (Okechukwu and Bokanisereme 2013).

Hepatoprotective Activity

Co-administration of a cassava rich diet and alcohol in rats reduced the alcohol induced toxicity

which was evidenced by the lower activities of GOT (glutamic-oxaloacetic transaminase), glutamic pyruvic transaminase (*GPT*), γ-glutamyltransferase (*GGT*), acid phosphatase and alkaline phosphatase in the liver and serum (Boby and Indira 2004). The pyruvate content in the blood increased while the lactate content, lactate/pyruvate ratio and the activity of lactate dehydrogenase decreased in the blood due to co-administration. The blood cyanide content, serum thiocyanate content and the activities of rhodanese and beta-glucuronidase increased on co-administration. The histopathological studies also revealed that co-administration reduced the alcohol-induced toxicity.

Anthelminthic Activity

A study on fresh cassava leaves incorporated into the diets of West African goats in Cameroon showed decrease in helmintic and coccidia infections (Pamo et al. 2006). The methanol cassava leaf extract (25 mg/ml) was the most efficient compared to hexane, chloroform and ethyl acetate leaf extracts at the same concentration causing 59.33 % paralysis of *Trichostrongylus* nematode larvae extracted from goat faeces but was less potent than Ivermectin 0.01 mg/l1 causing 99.67 % paralysis (Al-Rofaai et al. 2010).

Prebiotic Activity

Cassava fibre, a waste product formed in starch production, when incorporated into wheat to give composite flour ratios of 60/40 and 50/50 afforded a cracker like biscuit with prebiotic effects as evaluated in a rat assay (Osundahunsi et al. 2012). The protein content of the cracker-like product based on the 50/50 and 60/40 (fibre/wheat flour) ratios were 15.0 % and 10.0 %, respectively. Crude fibre ranged from 14.1 to 17.1 % while ash ranged from 3.0 to 5.0 %. Low cholesterol levels of 28.75 mg/dL and 18.75 mg/dL were recorded for the 50/50 and 60/40 composite ratios, respectively. The result of liver function test showed that the rats that were fed the fibre-based cracker product had an average value of 44.00 IU/L of aspartate amino transferase (AST), which is lower than the 67.75 IU/L recorded for the control. There was a significant increase in the packed cell volume (PCV) of the rats fed a fibre-based diet, relative to those fed 'Ogi' (fermented maize). Data obtained from the faecal analysis showed that the rats fed with the composite ratios and other diets had an increased *Lactobacillus* count. However, by increasing the number of days that the rats were fed the fibre-based diet, the *Escherichia coli* count in the rat faeces reduced significantly. The data obtained showed that cassava fibre-based crackers had good nutraceutic effects, with reduction in the *E. coli* count found in the rat faeces and healthy performances in terms of weight gain.

Amoebicidal Activity

Cassava plant extract exhibited antiamoebic activity in-vitro against the trophozoites of *Entamoeba histolytica* isolated from dysenteric patient faeces (Moundipa et al. 2005). The extract exhibited antiamoebic activity of <50 % at 48 h incubation but showed >50 % activity at 96 h.

Detoxification of Cassava Antinutrient

Hydroxynitrile lyase (HNL), which catalyses the conversion of acetone cyanohydrin to cyanide, was found to be expressed predominantly in the cell walls and laticifers of leaves (Siritunga et al. 2004). They showed that HNL activity increased more than twofold in leaves and 13-fold in roots of transgenic plants relative to wild-type plants. Unlike acyanogenic cassava, transgenic plants over-expressing HNL in roots retained the herbivore deterrence of cyanogens while providing a safer food product. Of four diets, namely, sun-dried cassava peel meal (S), ensiled cassava peel meal (E) and retted cassava peel meal (R), and the control administered to rabbits, ensiled cassava peel meal depressed feed and digestible nutrient intakes, weight gain, nutrient digestibil-

ity, digestible protein (DP), digestible energy (DE) and DP/DE ratio but increased protein intake to gain and feed to gain ratios compared to other diets (Olafadehan et al. 2012). Intake of HCN decreased successively in the order: R, S and E. The results indicated that retting and sun-drying were more effective in cassava peel detoxification than ensiling, and dietary HCN concentration and intake of 56 and 4 mg/kg BW, respectively, were not toxic under the conditions of the experiment.

Pharmacokinetic/Toxicity Studies

No intact linamarin was identified in the faeces or blood but 5.65 mg was excreted in the urine along with 0.823 mg of thiocyanate ion after oral intubation of linamarin in the rat (Barret et al. 1977). A 50-mg dose of linamarin was lethal to 7 of 10 rats receiving this dose. The amount of thiocyanate ion excreted in the urine by rats that received linamarin in their diet was higher than that excreted by rats that did not receive linamarin at both levels of dietary methionine (Barret et al. 1978). The plasma thiocyanate concentration was also higher for linamarin-fed animals than for animals that did not receive linamarin. Higher plasma thiocyanate levels were associated with a lower percentage uptake of radioiodine by the thyroid. In Mozambican subjects consuming insufficiently processed cassava the mean urinary linamarin levels were 211 μmol/l, indicating for the first time that substantial amounts of the main cyanogenic glycoside in cassava may be absorbed from the human gut and excreted intact in the urine (Brimer and Rosling 1993). In 75 Tanzanian subjects consuming insufficiently processed cassava, the mean urinary linamarin concentration was 104 μmol/l (range 0 – 644 μmol/L), while that for thiocyanate was 486 μmol/l (range 10–2,940 μmol/l), giving an approximate 1:5 M concentration ratio between urinary linamarin and thiocyanate (Carlsson et al. 1995). In two separate studies of healthy adults, after consumption of cassava porridge meal prepared from cassava flour, it was found that one-half of orally ingested linamarin was converted to cyanide and hence thiocyanate, about one-quarter was excreted unchanged and another quarter was metabolised into an as yet unknown compound (Carlsson et al. 1999).

In the brine shrimp toxicity assay among the chloroform cassava extracts, CME4 (chloroform root) (LC_{50}=413.9 ppm) possessed significantly high toxicity followed by CME39 (chloroform young leaf extract) (LC_{50}=496.2 ppm) and CME2 (chloroform old leaf extract) (LC_{50}=532.9 ppm) while among the ethanol cassava extracts, EME3 (ethanol young leaf extract) (LC_{50}=344.7 ppm) was significantly more toxic followed by EME2 (ethanol young leaf extract) (LC_{50}=534.3 ppm) and EME4 (ethanol root extract) (LC_{50}=609.6 ppm) (Zakaria et al. 2006). Overall, EME3 and CME4 were more toxic than their counterpart (CME3 and EME4). Overall, the results indicated that cassava possessed low toxicity level.

Cassava, a staple food in many tropical countries, had been suspected as a cause of human congenital defects (Frakes et al. 1985). Ingestion of cassava during pregnancy had been reported to induce limb defects, microcephaly, open eye and growth retardation in rats. An oral dose of 120 or 140 mg/kg of the glycoside, linamarin, was associated with an increased incidence of vertebral and rib anomalies as well as the production of encephaloceles in the offspring of pregnant hamsters. These larger doses of linamarin also resulted in obvious maternal toxicity. Linamarin treatment had no effect on fetal body weight, ossification of fetal skeletons, embryonic mortality or litter size.

It was suspected that traditional processing of cassava flour was insufficient to eliminate all the toxic cyanogenic glycosides (Shama and Wasma 2011). The authors demonstrated that oral administration of aqueous and methanolic extracts of cassava roots to Wistar rats resulted in alteration in aspartate transaminase (AST), alanine transaminase (ALT) and alkaline phosphatase (ALP) activities, changes in the concentration of urea, cholesterol and other serobiochemical parameters, pathological changes in vital organs, including necrosis and shrinkage of the glomeruli and aggregates of lymphocytes in the renal cortex

was observed; these changes were accompanied with cytoplasmic fatty vacuolation of entrilobula hepatocytes and cerebral neurons. They concluded that cassava was toxic causing alteration on various serobiochemical and hematological parameters and that this toxicity was correlated to dysfunction of vital organs and attributable to the presence of cyanogenic glycosides linamarin and lotaustralin.

Adverse Toxicity Issues

Cassava leaf flour (FFM) contained high content of proteins, vitamins and minerals but also cassava leaves contained substances deemed as antinutritive and/or toxic, such as cyanide, polyphenols, nitrate, oxalic acid, hemagglutinin, saponins and trypsin inhibitors (Pereira, et al. 2008). The highest protein content and hemagglutinating activity were obtained in the extraction of cassava leaf with distilled water at the 1:20 ratio (p/v) followed by the precipitation with ammonium sulphate at 80 % of saturation. Out of the four purification fractions (I, II, III and IV), I and II activities presented higher specific hemagglutinating activity. The same fractions were injected intraperitoneal in mice of 20 g weight with doses of 2 mg (fraction I), 3 mg (fraction II), 54 mg (fraction III) and 52 mg (fraction IV). No deaths or any adverse effects were observed after 120 h.

Akintonwa and Tunwashe (1992) reported that three patients admitted to the Accident and Emergency Unit of Lagos University Teaching Hospital (LUTH) after eating a cassava based meal 'Gari' died shortly after admission. The patients vomited and complained of abdominal pain immediately after the meal. They were unconscious with renal failure and died of cardiopulmonary arrest. The cyanide levels in the blood and urine averaged 1.12 and 0.54 mg/L, respectively.

Scopoletin was isolated and identified in gari, a cassava food consumed in Nigeria (Obidoa and Obasi 1991). Its levels in gari and cassava flour were not altered by post processing treatments such as sun-drying, refrigeration and storage. Scopoletin had been found to be a potent hypotensive and non-specific spasmolytic agent. These pharmacological effects of Scopoletin were probably the underlying factors in the slowly developing tropical neuropathy characterised by optic atrophy, nerve deafness and ataxia endemic among populations subsisting on cassava diets such as gari. The mean total cyanogen content of the test samples was 25.4, 20.00 and 17.5 mg/kg sample for garri, fufu and tapioca, respectively (Adindu et al. 2003). The mean total cyanogen levels for all the samples were higher than the FAO/WHO safe level for cassava flour (10 mg/kg). The implication of the findings was that consumers of these improperly detoxified products in the locality may be exposed to health problems through intake of cyanide.

Kwashiorkor is characterised by low serum albumin concentration, abnormal plasma amino acid pattern and ratios, oedema, potassium depletion, increase in subcutaneous fat with muscle wasting, low insulin levels, pancreatic dysfunction and atrophy and nephrosis (Kamalu 1993). Cassava (gari) diets were reported to be associated caused with all these conditions suggesting a linkage and possibly a direct causative relationship between cassava consumption and kwashiorkor and that link was suggested by the author to be intact linamarin. The author hypothesised that intact linamarin absorbed from cassava diets, acting as a glucoside, caused the inhibition of Na+–K+'-ATPase, giving rise to electrolyte imbalance with potassium depletion. Potassium depletion caused cellular swelling (oedema), vacuolation and rupture in tissues, but at rates that differed from tissue to tissue. Swelling, vacuolation and rupture of the epithelial cells of the proximal convoluted tubules constituted nephrosis, which resulted in proteinuria and caused low serum albumin. Distortion of plasma amino acid pattern and ratios was related to cellular damage in the liver and pancreas and to gluconeogenesis. These characteristics – potassium depletion, oedema, nephrosis, low serum albumin concentration, distorted plasma amino acid pattern and ratios, increase in subcutaneous fat with muscle wasting, low insulin levels and pancreatic dysfunction and atrophy – were all characteristic fea-

tures of kwashiorkor. All these substantiated a direct role of intact linamarin, in the linkage of kwashiorkor and cassava consumption. Such a linkage between cassava consumption and kwashiorkor was also supported both by geographical occurrence and by animal studies. The effect of intact linamarin appeared to be exacerbated by low protein levels.

Delange et al. (1994) found that dietary cyanide in the human body resulting from cassava consumption may be converted to a potential goitrogenic agent thiocyanate that may aggravate iodine deficiency disorders such as goiter and cretinism, a severe form of mental retardation, as their main effects. Therefore the goitrogenic action of cassava depended on the glucoside levels in fresh roots, the effectiveness of processing, the frequency of cassava consumption and the iodine intake. The goitrogenic effect of cassava could be corrected entirely by increasing the intake of iodine. Hernández et al. (1995) found that consumption of 1–4 kg boiled fresh sweet cassava roots by adult, nonsmoking subjects over 2 days caused negligible cyanide exposure. The main cyanide metabolite, thiocyanate, and 28 % of linamarin cyanogenic glycoside were excreted in the urine. The results indicated that the dominant cyanogen in boiled cassava was glycosides that pass through the human body without causing cyanide exposure.

In a study of 86 patients with tropical chronic pancreatic (TCP) and 90 healthy control, there was significant reduction in serum cyanogens detoxifying enzyme rhodanese activity in both cassava consumer- and non-consumer TCP patients as compared to controls but no significant difference between cassava consumer- and non-consumer TCP patients was observed (Girish et al. 2011). Serum thiocyanate was significantly lower in cassava consumer TCP patients as compared to cassava consumer controls but not significantly different from cassava non-consumer TCP patients. Plasma methionine, cysteine and urinary inorganic sulphate/creatinine ratio was significantly lower in both cassava consumer and non-consumer TCP patients as compared to controls but were comparable among cassava consumers and non-consumers. The results suggested that TCP patients were at higher risk of defective detoxification of cyanogens. However, there was no difference between cassava consumers and non-consumers. Low levels of sulphur amino acids may contribute to the development of pancreatitis.

In a study of two populations of children from the northern and southern zones of the Bandundu region, Democratic Republic of Congo (former Zaire), Banea-Mayambu et al. (2000) found that dietary cyanide exposure from cyanogenic glucosides in insufficiently processed cassava may be a factor aggravating growth retardation in children in Bandundu.

Neurological disorders had been reported from parts of Africa with protein-deficient populations and attributed to cyanide (CN-) exposure from prolonged dietary use of cassava, a cyanophoric plant (Tor-Agbidye et al. 1999). Cyanide is normally metabolised to thiocyanate (SCN-) by the sulphur-dependent enzyme rhodanese. However, in protein-deficient subjects where sulphur amino acids (SAA) are low, CN may be converted to cyanate (OCN-), which is known to cause neurodegenerative disease in humans and animals. Tor-Agbidye et al. (1999) found a strongly positive linear relationship between blood CN- and plasma OCN-concentrations in rats. Their data were consistent with the hypothesis indicating cyanate to be an important mediator of chronic cyanide neurotoxicity during protein-calorie deficiency. Konzo is a distinct form of tropical myelopathy characterised by abrupt onset of spastic paraparesis (Tylleskär et al. 1992) and presents an irreversible upper-motor neuron disorder affecting children (Boivin et al. 2013). Neurocognitive deficits in children with konzo were documented by Boivin et al. (2013). Konzo is an irreversible paralysis of the legs that occurs mainly among children and young women in remote villages in tropical Africa and is associated with a monotonous diet of bitter cassava (Banea et al. 2014). Konzo was first identified by Dr. Trolli in 1938 in Popokabaka Health Zone (Banea et al. 2012). High cyanide intake from consumption of insufficiently processed cassava had been advanced as a possible aetiology of the upper motor neurone disease

konzo (Banea-Mayambu et al. 1997). Epidemics in East Africa had been attributed to dietary cyanide exposure from insufficiently processed cassava but a study done in Zaire disputed such an aetiology. Studies by Tylleskär et al. (1992) indicated a causal role in konzo of sustained high blood cyanide concentrations maintained by a deficient sulphur intake impairing cyanide to thiocyanate conversion. The underlying causes of konzo were poverty and food shortage, but a minor improvement of food processing may be preventive. Nine patients with konzo were reexamined after 14 years (Cliff and Nicala 1997). This long-term follow-up showed that the neurological signs in konzo patients remained constant. Four severely affected patients, however, showed functional improvement. One child, originally classified as a konzo case, showed signs of cretinism at follow-up.

The results of an ecological study involving 22 cases of konzo, 57 unaffected household members and 116 members from unaffected households, a total of 195 subjects, in konzo-affected savanna villages in Zaire showed that the mean value of urinary thiocyanate, the main cyanide metabolite, was higher in the three groups in konzo-affected villages than in unaffected villages. Their findings supported an aetiological role for cyanide in konzo (Banea-Mayambu et al. 1997). However they found that urinary linamarin, the cyanogenic glucoside and source of cyanide in cassava, was more closely associated with the occurrence of konzo. The mean value of urinary linamarin in the konzo cases was 632 μmol/l and in their household members 657 μmol/l, which was significantly higher than in members of control households in the same village (351 μmol/l) and in unaffected villages (147 μmol/l). This suggested that a specific neurotoxic effect of linamarin, rather than the associated general cyanide exposure resulting from glucoside breakdown in the gut, may be the cause of konzo. Fifty konzo cases were identified in four villages in Popokabaka Health Zone where one-third of people had only one meal per day, mainly of cassava flour consumed as a thick porridge (fufu) and pounded, boiled cassava leaves (Banea et al. 2012). After 1.5 years intervention with wetting method (soaking for 3–4 days) of cassava flour the total cyanide content of cassava flour was reduced to below 10 ppm. No new konzo cases occurred, which included two dry seasons when konzo peaks and it has now been prevented for the first time in the same area. The wetting methodology of cassava was adopted in three villages in Boko Health Zone, Bandundu Province, Democratic Republic of Congo (DRC), with 61 konzo cases and konzo prevalences of 2.5 %, 4.1 % and 7.5 %, respectively (Banea et al. 2013). The high mean cyanide content of cassava flour of 50 ppm was due to short soaking of cassava roots for 1–2 days instead of 3–4 days. Following 1 year intervention with the wetting method, no new cases of konzo were encountered. This represented the second time that konzo had been controlled and success depended on regular use of the wetting method by village women. The methodology is now being used in other villages in DRC with financial support of AusAID. The wetting methodology was adopted in Kay Kalenge village, DRC, in 2011 and till 2014 no new cases of konzo were found (Banea et al. 2014). The wetting method had been readily accepted by rural women as a simple and useful method to control konzo by removing cyanide from cassava flour, and its use had spread to nearby villages. Prevention measures such as proper cassava processing, that is, detoxification of cassava roots before their consumption, and promotion of genetically engineered low-toxin strains of the plant were reported to be of paramount importance as they may help eradicate konzo disease (Tshala-Katumbay et al. 2013). Siritunga and Sayre (2003) generated cyanogen-free transgenic cassava plants in which the expression of the cytochrome P450 genes (CYP79D1 and CYP79D2), which catalyse the first-dedicated step in linamarin synthesis, was inhibited. They observed 94 % reduction in leaf linamarin content associated with an inhibition of CYP79D1 and CYP79D2 expression. Importantly, the linamarin content of roots also was reduced by 99 % in transgenic plants having between 60 and 94 % reduction in leaf linamarin content. The results suggested that linamarin was transported from leaves to roots and that a thresh-

old level of leaf linamarin production was required for transport. To produce cultivars that promote rapid cyanide volatilisation, hydroxynitrile lyase (HNL), which catalyses the last step in cyanogenesis, the conversion of acetone cyanohydrin to cyanide, was overexpressed in roots (Siritunga et al. 2004; Siritunga and Sayre 2007). HNL was found to be expressed predominantly in the cell walls and laticifers of leaves (Siritunga et al. 2004). They showed that HNL activity increased more than 2-fold in leaves and 13-fold in roots of transgenic plants relative to wild-type plants. Unlike acyanogenic cassava, transgenic plants over-expressing HNL in roots retained the herbivore deterrence of cyanogens while providing a safer food product.

The toxicity of cyanogenic glycosides in the context of their metabolic involvement with goitre and tropical ataxic neuropathy was reviewed by Oke (1980). Ataxic polyneuropathy, occurring in endemic form in an area in southwest Nigeria, was attributed to exposure to cyanide from consumption of cassava foods; however, studies carried out by Oluwole et al. (2002) found that the occurrence of ataxic polyneuropathy was low in a community where exposure to cyanide was high suggesting that exposure to cyanide was not a direct cause of ataxic polyneuropathy.

Man e5, the first purified allergen from cassava demonstrated IgE cross-reactivity with Hev b 5 from rubber latex (Santos et al. 2013). Data suggested Hev b 5 might act as primary sensitiser and could therefore lead to allergic manifestations upon manioc consumption without prior exposition.

Prolonged ingestion of cassava leaves by goats were toxic (Soto-Blanco and Górniak 2010). A mild increase in the number of resorption vacuoles in the thyroid follicular colloid, slight vacuolation of periportal hepatocytes and spongiosis of the mesencephalon were found in goats treated with cassava. The toxic could be attributed to the action of cyanide released from cyanogenic glycosides, and none of the effects was promoted by these glycosides directly.

Traditional Medicinal Uses

The nourishing root is used as an appetiser, aperients, vulvenrry and tonic (HMRC 2002). The roots are useful for the treatment of anorexia, dyspepsia, constipation, wounds, ulcers and general debility. The plant is also used by Malays to treat ulcers, body aches and rheumatism in Malaysia. Cassava is used in Chinese traditional medicine to cure furuncle and ascariasis (Hu et al 2010). In Brazil, cassava leaves flour (CLF) has been used to combat undernourishment, because it is a source of vitamins and minerals (Wobeto et al. 2004). In Indonesia, cassava leaves (and seeds) are used in folk medicine to alleviate fever, headache, rheumatism and haemorrhoids (Miladiyah et al. 2011).

In south western Nigeria, a decoction of the leaves are taken orally for diabetes (Abo et al. 2008) and in Ekiti state, Nigeria, cassava leaves and tuberous roots are used to treat sexually transmitted diseases, including gonorrhoea, trichomoniasis, chlamydial infection, syphilis, and HIV and AIDS (Kayode and Kayode 2008). In south eastern Nigeria, premature roots are used to treat eye ailments (Obute 2005). In Nigeria, they are also utilised in the treatment of ringworms, tumour, conjunctivitis, sores and abscesses (Miladiyah et al. 2011). Aqueous cassava leaf extract is being used orally and topically in traditional African medicine for the treatment of inflammation and pain (Adeyemi et al. 2008). In Nigeria, Akwa Ibom State, sap from crushed cassava stems is used as eye-drops to treat conjunctivitis and sap from crushed leaves used as local application for skin disease (Ajibesin et al. 2008). In the People's Republic of Congo, pounded cassava leaves are used for measles and chicken pox; roots together with *Abrus precatorius* leaves and peanut seeds are pounded, macerated and taken orally to treat azoospermia; cassava roots and *Voandzeia subterranea* seeds are pounded and used to treat metrorrhagia; stem and leaves with other plant ingredients are used to treat fractures and headache and migraine

(Adjanohoun et al. 1988). In the Democratic Republic of Congo, pounded leaves are used to treat measles and filariasis and the macerated roots taken in orally to cure lactation disorder (Nyakabwa and Dibaluka 1990); pounded leaves are used to stop bleeding nose (Kawukpa and Angoyo 1994). In Togo, petioles are chewed and applied to snakebite and the leaf pulp is used for the same (Adjanohoun et al. 1986). In Benin, cassava leaves are used to treat snake bites and pounded tuberous roots decoction taken orally for azoospermia, metrorrhagia and dysuric (Adjanohoun et al. 1989). In Ghana, district of Gosomtwi-Atwima-Kwanwoma, a decoction of cassava leaves is used as poultice for wounds, stomach ulcers and sores (Agyare et al. 2009). In Tanzania, Kimboza forest reserve in Morogoro, a leaf infusion is taken orally for stomach ache (Amri and Kisangau 2012). In Uganda, Bulamogi country, the roots are taken orally to treat for baby infection and the leaves are used to treat lameness (Tabuti et al. 2003); dry leaves are boiled and used as bath for fever (Namukobe et al. 2011). In south eastern Madagascar, Agnalazaha Forest, cassava leaves are used to treat boils, pneumonia, gonorrhoea and painful spasma (Razafindraibe et al. 2006). In Gabon, the leaves are used to treat diarrhoea and pectoral/cardiac pain (Adjanohoun et al. 1984); the roots are used to remedy lactation disorder and miscarriage, the leaves are used for wounds, cuts, chicken pox, skin diseases and painful menstruation, and leaf petioles are used as purgative (Raponda-Walker and Sillans 1995). In Cameroon Southern region of Sangmelima, the leaf latex is used as an abortifacient (Noumi and Tchakonang 2001). The enlarged end of a fresh leaf petiole of the bitter variety is inserted into the vagina such that the white latex comes in contact with the cervix to cause bleeding. The preparation is maintained until the foetus has been expelled. In Sierra Leone, Moyamba District, a leaf infusion of *Microdesmis puberula* together with cassava leaves is rubbed on the head and given orally to victims of snake bite (Lebbie and Guries 1995). The leaves of the two species are ground with clay and the poultice rubbed around the snake bite wound. In Angola, manioc is used as a proodge for dysentery and the macerated extract is taken orally for haematuria; the pounded leaves are used to treat otitis and bee sting (Bossard 1996). In Ivory coast, cassava leaves are used to treat conjunctivitis, jaundice, haemoglobinuria, haematuria and non-occurrence of menses (Bouquet and Debray 1974).

Other Uses

Cassava roots are an important source of dietary and industrial carbohydrates, mainly eaten as a source of starch, forming the staple food to over 500 million; additionally, the roots have value as a raw material for industrial starch production and for animal feed giving the crop high economic value (Blagbrough et al. 2010). More than two-thirds of the total production of cassava is used as food for humans, with lesser amounts being used for animal feed and industrial uses (Tonukari 2004). Cassava starch can replace maize, rice and wheat starch in processed foods. Starch is utilised in sizing and dyeing in the textile industries to increase brightness and weight of the cloth. In the pharmaceutical industries, starch serves as a filler material and bonding agent for making tablets. Cassava starch also have several other numerous uses such as an additive in cement to improve the setting time, and it is used to improve the viscosity of drilling muds in oil wells. It is also used to seal the walls of bore holes and prevent fluid loss. Starch is also the main raw material in glue and adhesive industries. In paper production, cassava starch is currently used as glue to achieve brightness and strength. Starch is also an important raw material for powder in the cosmetics industries. In detergent soap manufacture, starch is used to get better recovery and to improve the shelf life of detergents. While in the rubber and foam industries, starch is employed for getting better foaming and colour.

Premkumar et al. (2000) listed the potential markets for cassava starch products such as pregelatinised instant and convenience foods, namely, 'yucca rava' and 'yucca' porridge, cassava 'papads' an important snack food item pre-

pared from cassava flour consumed by frying in oil, cassava wafers from cassava starch and fried cassava chips; extruded and fermented food products; animal feed products using by-product utilisation for poultry, and value-addition through microbial enrichment and ensiling technology; modified starch products like sweeteners cassava starch used as raw materials for production of glucose, dextrose, fructose and maltose syrup, cold water-soluble starch, cassava gums, adhesives, commodity chemicals like citric acid, ethanol, biodegradable polymers and plastics incorporating cassava starch, biogas from starch factory wastes. In the Philippines, cassava can be processed into various products and replace imported raw materials such as substitute for maize in animal feed, molasses for the production of sweetener or alcohol and wheat flour in various bakery products (Loreto and Orias 2000).

Cassava starch can be converted by *Aspergillus niger* to biomass, carbohydrates such as oligosaccharides, maltotriose, maltose and glucose as well as to other modified sugars and organic acids – citric, malic, gluconic, succinic and fumaric acids (Tan et al. 1984). Cassava starch can be used to make fructose syrups (Vuilleumier 1993) and formulate gelatin capsules (Nduele et al. 1993). Roble et al. (2003) demonstrated the production L-lactic acid from raw cassava starch in a bioreactor using *Aspergillus awamori* (fungus) and *Lactococcus lactis* spp. *lactis* (bacteria). Cassava dregs can be employed for phytase production after the addition of a nitrogen source (ammonium nitrate) and mineral salts (e.g., sodium dodecyl sulphate) (Hong et al. 2001). A maximum phytase yield of 6.73 U/g of dry mass was obtained. Activated carbons were prepared from waste. Cassava oil has good values as edible oil and may also find industrial applications in soap, shampoo and other related cosmetic industries (Popola and Yangomodou 2006). The properties of lipid extracted from cassava seeds compare well with those of processed cottonseed oil, soybean oil, corn oil and sunflower-seed oil. Cassava seed oil has promising potentials which could be tapped for domestic and industrial uses. In India, nearly 60 % of cassava is used industrially in the production of sago, starch and dry chips (Srinivas 2007). The projected demand for diversified uses of cassava is predominantly in the adhesive sector, especially in the corrugation gums and paper conversion industry, and in the paper industry. Biocomposites made from cassava starch and kenaf fibres were found to have improved tensile strength and moduli (Zainuddin et al. 2013). Novel multifunctional excipients were prepared by coprocessing tapioca starch with mannitol (Adeoye and Alebiowu 2014). They found that coprocessing tapioca starch and mannitol would enhance the flow, packing and compaction properties of the novel excipient and that the co-fusion method of coprocessing would produce novel excipients with enhanced direct compression potential compared to the co-grinding method. Studies by Sugumaran et al. (2014) found cassava bagasse to be a cheap and novel substrate for pullulan production in solid state fermentation. Esterified tapioca maltodextrin could be used as an emulsifier to make n-hexadecane oil/water emulsions (Udomrati and Gohtani 2014). Edible coatings produced using cassava starch (2 % and 3 % w/v) containing cinnamon bark (0.05 % to 0.30 % v/v) or fennel (0.05% to 0.30 % v/v) essential oils showed barrier properties, an antioxidant capacity and microbial inhibition (Oriani et al. 2014). Antimicrobial tests showed that the addition of 0.30 % (v/v) cinnamon bark essential oil to the cassava starch edible coating inhibited the growth of *Staphylococcus aureus* and *Salmonella choleraesuis*, and 0.30 % fennel essential oil inhibited just *S. aureus*. Cassava starch is also used to make liquid starch adhesives, pre-gel starch adhesives and dextrin-based adhesives. The fruit is used as a fish poison in California and Brazil.

Bioethanol/Butanol/Acetone Production

The use of cassava as a source of ethanol for fuel is already being exploited and very promising. Cassava compared favourably to other crops such as maize, sugarcane and sweet sorghum in biofuel conversions performance and bioethanol yield (Wang 2002). Sugarcane affords a yield of

70 tonne/ha/year, conversion rate to bioethanol 70 L/ tonne, and bioethanol a yield of 4,900 l/ha/year. Cassava affords a yield of 40 tonne/ha/year, a conversion rate to bioethanol 150 L/tonne, and bioethanol a yield of 6,000 l/ha/year. For maize a yield of 5tonne/ha/year, a conversion rate to bioethanol 150 L/tonne, and bioethanol a yield of 2,800 l/ha/year. For sweet sorghum a yield of 35 tonne/ha/year, a conversion rate to bioethanol 80 L/tonne, and bioethanol a yield of 2,800 l/ha/year. For wheat a yield of 4 tonne/ha/year, a conversion rate to bioethanol 390 L/tonne, and bioethanol a yield of 1560 l/ha/year. Recently, cassava-derived bioethanol production in China has been increasing due to its economic benefits compared to other bioethanol-producing crops in the country (Jansson et al 2009). The sequential co-culture biphasic fermentation approach provided a consolidated bio-processing means to produce ethanol and hydrogen from cassava pulp (Li and Zhu 2011). The ethanol level reached 8.83 g/L with a fermentation efficiency of 64.95 %. Hydrogen production of 4.06 mmol by the co-culture system was 1.54 and 2.09-fold greater than that produced by mono-cultures of *Clostridium thermocellum* and *Thermoanaerobacterium aotearoense*, respectively. Studies by Lépiz-Aguilar et al. (2013) suggested that the use of cassava as a substrate in acetone-butanol-ethanol fermentation by *Clostridium beijerinckii* could be a cost-effective way of producing butanol in tropical regions.

Liquid biofuel (bio-oil) was produced by pyrolising cassava peel in a fixed-bed tubular reactor at temperatures ranging from 400 to 600 °C with a heating rate of 20 °C/min (Ki et al. 2013). Fed-batch fermentation of concentrated cassava bagasse hydrolysate containing 584.4 g/L glucose in a fibrous bed bioreactor with continuous gas stripping generated a product containing 10–16 % (w/v) of butanol, ~4 % (w/v) of acetone, a small amount of ethanol (<0.8 %) and almost no acids, resulting in a highly concentrated butanol solution of ~64 % (w/v) after phase separation (Lu et al. 2012).

Animal Feed

Studies by Liu and Zhaung (2000) found that compound feed made from dry cassava roots and leaf silage has an advantage over maize with respect to taste, daily gain, feed conversion rate and economic returns. Maize can be partially or completely replaced by cassava to feed various livestock and poultry. Cassava leaf silage has enormous value and can compensate for the lack of protein in cassava root powder. Cassava roots and leaves contain hydrocyanic acid, but this will be reduced to non-toxic levels by either drying or ensiling. Nguyen et al. (2000) found that ensiled cassava leaves can be used as a protein source for feeding pigs under village conditions. Ensiling cassava leaves supplemented with cassava root meal, rice bran or molasses at 5 or 10 % (fresh basis) produced good quality silage that could be stored for up to 5 months. Oni et al. (2010) found that leaves of Nigerian cassava cultivars could be ranked for their potential feeding value as TMS 30572 (71.5 %)>MS6 (65.8 %)>Idileruwa (63.0 %)>TMS 30555 (50.4 %). TMS 30572 and MS 6 exhibited good potential as feed resources for ruminant animals and could be used in ruminant feeding as protein source ingredient. They also found that leaves of the varieties MS 6 and TMS 30555 were superior to the others in terms of crude protein and gas production indicating a higher digestibility and energy content and thus nutritive potential (Oni et al. 2011). Studies by Ly et al. (2012) found that supplementation of diet containing ensiled cassava leaves as the main protein source with synthetic amino acids, DL-methionine alone or with L-lysine, improved growth performance and protein gain of cross-bred (Large White × Mong Cai) pigs. Cassava roots are made into pellets which are used in compounding animal feeds for cattle, sheep, goats, pigs, poultry, and farmed fish (Jackson et al. 2014).

Chicks fed with diet containing 100 g/kg cassava root meal had higher final live-weight and weight gain and reduced HCN intake than chicks

fed with diet containing 200 g/kg cassava root meal (Akapo et al. 2014). Dietary inclusion of peeled cassava root meal (PCRM) for broiler chicks resulted in increased final live-weight, weight gain and feed intake when compared with chicks fed with diet containing unpeeled cassava root meal (UCRM). The least final live-weight and weight gain and worst feed-to-gain ratio were obtained with chicks fed with a diet containing 200 g/kg UCRM. In comparison with chicks fed with a diet containing UCRM, dietary inclusion of PCRM resulted in increased red blood cell (RBC) count and haemoglobin (Hb) concentration and reduced white blood cell (WBC) count and serum thiocyanate concentration. Although inclusion of 100 g/kg PCRM showed some economic sense, dietary inclusion of either peeled or unpeeled cassava root posed a threat on growth and health status of broiler chicks. Oso et al. (2014) found that supplementation of broiler chicken diet containing up to 100 g/kg unpeeled cassava root meal with 6 g/kg charcoal showed improved weight gain without any deleterious effect on serum metabolites.

Biodegradable Plastics

Cassava starch is the cheapest carbon source in the region and can be applied for the production of biodegradable plastics in two different ways (Sriroth et al. 2000): (a) as polymers: cassava starch can serve as a carbon source in the fermentation process leading to the formation of high molecular weight polymers, polyhydroxybutyrate (PHB) poly(3-hydroxybutyrate-co-3-hydroxyvalerate) (PHBV), use, or organic acids such as succinic acid and lactic acid, which can subsequently undergo direct condensation to high molecular weight polymers; (b) as the blending material cassava starch can be modified in different ways for use in various starch-polymer blends (PCL, polylactide (PLA), polyvinyl acetate (PVA) and polyhydroxyalkanoates (PHA), so that the properties of the starch in the polymer blend are the best possible. Luk et al. (2013) found that at low moisture (<11 %) the addition of 1–8 % amylose complexing fatty acids (CFA), such as linoleic and oleic acids, to cassava starch caused an anti-plasticisation effect, while at higher moisture contents it produced plasticisation. The anti-plasticising effect of CFA on cassava starch was attributed to amylose-lipid complex formation.

Pest Control

Arirob et al. (2013) found that spraying with 1,000 mg/l cassava leaf tannin extract could repel cassava mealy bug (*Pseudococcus jackbeardsleyi*) infestations of cassava plants.

Biomedical Application

Chalapathi et al. (2010) found that Paracetamol tablets manufactured by using cassava starch was better in friability and hardness than those made of industrial starch (maize). Results showed increased disintegration time and binding capacity. Also it presented a potential as a cheaper alternative to the tablet manufacturing industry.

Comments

The top 10 cassava producing countries (2012) are: Nigeria 54,000,000 MT, Thailand 29,848,000 MT, Indonesia 24,177, 372 MT, Brazil 23,044,557 MT, Democratic Republic of the Congo 16,000,000 MT, Ghana 14,547,279 MT, Angola 10,636,400 MT, Mozambique 10,0513,364 MT, Vietnam 9,745,545MT and India 8,746,500 MT (FAO 2014).

Selected References

Abo KA, Fred-Jaiyesimi AA, Jaiyessimi AEA (2008) Ethnobotanical studies of medicinal plants used in the management of diabetes mellitus in South Western Nigeria. J Ethnopharmacol 115:67–71

Adeoye O, Alebiowu G (2014) Flow, packing and compaction properties of novel coprocessed multifunctional directly compressible excipients prepared from

tapioca starch and mannitol. Pharm Dev Technol 19(8):901–910

Adewusi SRA, Bradbury JH (1993) Carotenoids in cassava – comparison of open column and HPLC methods of analysis. J Sci Food Agric 62(4):375–383

Adeyemi O, Yemitan O, Afolabi L (2008) Inhibition of chemically induced inflammation and pain by orally and topically administered leaf extract of *Manihot esculenta* Crantz in rodents. J Ethnopharmacol 119:6–11

Adindu MN, Olayemi FF, Nze-Dike OU (2003) Cyanogenic potential of some cassava products in Port Harcourt markets in Nigeria. J Food Comp Anal 16(1):21–24

Adjanohoun EJ, Ahyi AMR, Chibon P, de Vecchy H, Duboze E, Eymé J, Gassita JN, Goudote E, Guinko S, Keita A, Koudogbo B, Le Bras M, Mourambou I, Mve-Mengome E, Nguéma MG, Ollome JB, Posso P, Sita P (1984) Contribution aux études ethnobotaniques et floristiques au Gabon. Agence de coopération culturelle et technique, France, 294 pp

Adjanohoun EJ, Ahyi MRA, Aké Assi L, Akpagana K, Chibon P, El-Adji A, Eymé J, Garba M, Gassita JN, Gbeassor M, Goudote E, Guinko S, Hodouto KK, Houngnon P, Keita A, Keoula Y, Hodouto WP, Issa Lo A, Siamevi KM, Taffame KK (1986) Contributions aux Études Ethnobotaniques et Floristiques au Togo. Médecine Traditionelle et Pharmacopée Agence de Coopération Culturelle et Technique, Paris, 671 p

Adjanohoun EJ, Ahyi AMR, Aké Assi L, Baniakina J, Chibon P, Cusset G, Doulou V. Enzanza A, Eymé J, Goudoté E, Keita A, Mbemba C, Mollet J, Moutsamboté J-M, Mpati J, Sita P (1988) Contribution aux Études Ethnobotaniques et Floristiques en République Populaire du Congo. Médecine Traditionelle et Pharmacopée Agence de Coopération Culturelle et Technique, Paris, France, 606 p

Adjanohoun EJ, Adjakidjè V, Ahyi MRA, Aké Assi L, Akoègninou A, d'Almeida J, Apovo F, Boukef K, Chadare M, Cusset G, Dramane K, Eyme J, Gassita JN, Gbaguidi N, Goudote E, Guinko S, Houngnon P, Lo I, Keita A, Kiniffo HV, Kone-Bamba D, Musampa Nseyya A, Saadou M, Sodogandji T, De Souza S, Tchabi A, Zinsou Dossa C, Zohoun T (1989) Contribution aux Études Ethnobotaniques et Floristiques en République Populaire du Bénin. Médecine Traditionelle et Pharmacopée. Agence de Coopération Culturelle et Technique, Paris, 895 p

Agyare C, Asase A, Lechtenberg M, Niehues M, Deters A, Hensl A (2009) An Ethnopharmacological survey and in vitro confirmation of ethnopharmacological use of medicinal plants used for wound healing in Gosomtwi-Atwima-Kwanwoma area, Ghana. J Ethnopharmacol 125:393–403

Ajibesin KK, Ekpo BA, Bala DN, Essien EE, Adesanya SA (2008) Ethnobotanical survey of Akwa Ibom State of Nigeria. J Ethnopharmacol 115(3):387–408

Akapo AO, Oso AO, Bamgbose AM, Sanwo KA, Jegede AV, Sobayo RA, Idowu OM, Fan J, Li L, Olorunsola RA (2014) Effect of feeding cassava (*Manihot esculenta* Crantz) root meal on growth performance, hydrocyanide intake and haematological parameters of broiler chicks. Trop Anim Health Prod 46(7):1167–1172

Akinrele IA (1964) Fermentation of cassava. J Sci Food Agric 15:589–594

Akintonwa A, Tunwashe OL (1992) Fatal cyanide poisoning from cassava-based meal. Hum Exp Toxicol 11(1):47–49

Allem AC (1987) *Manihot esculenta* is a native of the neotropics. Plant Genet Resourc Newsl 71:22–24

Allem AC (1994) The origin of *Manihot esculenta* Crantz (Euphorbiaceae). Genet Resour Crop Evol 41(3):133–150

Allem AC (1999) The closest wild relatives of cassava (*Manihot esculent*a Crantz). Euphytica 107(2):123–133

Almazan AM, Theberge RL (1989) Influence of cassava mosaic virus on cassava leaf-vegetable quality. Trop Agric 66:305–308

Al-Rofaai A, Rahman WA, Sulaiman SF (2010) In vitro anthelmintic activity of cassava (*Manihot esculenta*) extract on Trichostongyloid larvae. In: Proceedings of the 7th IMT-GT UNINET and the 3rd international PSU-UNS conferences on bioscience for the future 2010, Prince of Songkla University Thailand 7–8 October 2010, pp 233–236

Amoa-Awua WK, Appoh FE, Jakobsen M (1996) Lactic acid fermentation of cassava dough into agbelima. Int J Food Microbiol 31(1–3):87–98

Ampe F, Brauman A, Trèche S, Agossou A (1994) Cassava retting: optimisation of a traditional fermentation by an experimental research methodology. J Sci Food Agric 65:355–361

Amri E, Kisangau DP (2012) Ethnomedicinal study of plants used in villages around Kimboza forest reserve in Morogoro, Tanzania. J Ethnobio Ethnomed 8:1

Arirob W, Insung A, Pumnuan J, Won-In K, Dararutana P (2013) Investigation of tannin crude extract from cassava leaves for mealy bug control. Adv Sci Lett 19(12):3579–3581

Awoyinka AF, Abegunde VO, Adewusi SR (1995) Nutrient content of young cassava leaves and assessment of their acceptance as a green vegetable in Nigeria. Plant Foods Human Nutr 47(1):21–28

Banea JP, Nahimana G, Mandombi C, Bradbury JH, Denton IC, Kuwa N (2012) Control of konzo in DRC using the wetting method on cassava flour. Food Chem Toxicol 50(5):1517–1523

Banea JP, Bradbury JH, Mandombi C, Nahimana D, Denton IC, Kuwa N, Tshala Katumbay D (2013) Control of konzo by detoxification of cassava flour in three villages in the Democratic Republic of Congo. Food Chem Toxicol 60:506–513

Banea JP, Bradbury JH, Mandombi C, Nahimana D, Denton IC, Kuwa N, Tshala Katumbay D (2014) Effectiveness of wetting method for control of konzo and reduction of cyanide poisoning by removal of cyanogens from cassava flour. Food Nutr Bull 35(1):28–32

Banea-Mayambu JP, Tylleskär T, Gitebo N, Matadi N, Gebre-Medhin M, Rosling H (1997) Geographical and seasonal association between linamarin and cyanide exposure from cassava and the upper motor neurone disease konzo in former Zaire. Trop Med Int Health 2(12):1143–1151

Banea-Mayambu JP, Tylleskar T, Tylleskar K, Gebre-Medhin M, Rosling H (2000) Dietary cyanide from insufficiently processed cassava and growth retardation in children in the Democratic Republic of Congo (formerly Zaire). Ann Trop Paediatr 20(1):34–40

Barret MD, Hill DC, Alexander JC, Zitnak A (1977) Fate of orally dosed linamarin in the rat. Canadian J Physiol Pharmacol 55(1):134–136

Barret MD, Alexander JC, Hill DC (1978) Effect of linamarin on thiocyanate production and thyroid activity in rats. J Toxicol Environ Health 4(5–6):735–740

Bayoumi SAL, Rowan MG, Blagbrough IS, Beeching JR (2008a) Biosynthesis of scopoletin and scopolin in cassava roots during post-harvest physiological deterioration: the E-Z-isomerisation stage. Phytochemistry 69(17):2928–2936

Bayoumi SAL, Rowan MG, Blagbrough IS, Beeching JR (2008b) Investigation of biosynthetic pathways to hydroxycoumarins during post-harvest physiological deterioration in cassava roots by using stable isotope labelling. ChemBio Chem 9(18):3013–3022

Bayoumi SAL, Rowan MG, Beeching JR, Blagbrough IS (2010) Constituents and secondary metabolite natural products in fresh and deteriorated cassava roots. Phytochemistry 71(5–6):598–604

Beeching JR, Han Y, Gomez-Vasquez R, Day RC, Cooper RM (1998) Wound and defense responses in cassava as related to post-harvest physiological deterioration. Recent Adv Phytochem 32:231–248

Bierhals VS, Chiumarelli M, Hubinger MD (2011) Effect of cassava starch coating on quality and shelf life of fresh-cut pineapple (Ananas comosus L. Merril cv "Pérola"). J Food Sci 76(1):E62–E72

Bisset FH, Clapp RC, Coburn RA, Ettlinger MG, Long L (1969) Cyanogenesis in manioc: concerning lotaustralin. Phytochemistry 8(11):2235–2247

Blagbrough IS, Bayoumi SAL, Rowan MG, Beeching JR (2010) Cassava: an appraisal of its phytochemistry and its biotechnological prospects. Phytochemistry 71:1940–1951

Boby RG, Indira M (2004) Effect of co-administration of cassava (*Manihot esculenta* Crantz) rich diet and alcohol in rats. Indian J Physiol Pharmacol 48(1):41–50

Boey CG, Yeoh HH, Chew MY (1976) Purification of tapioca leaf rhodanese. Phytochemistry 15:1343–1344

Boivin MJ, Okitundu D, Makila-Mabe Bumoko G, Sombo MT, Mumba D, Tylleskar T, Page CF, Tamfum Muyembe JJ, Tshala-Katumbay D (2013) Neuropsychological effects of konzo: a neuromotor disease associated with poorly processed cassava. Pediatrics 131(4):e1231–e1239

Bossard E (1996) La médecine traditionnelle au centre et à l'ouest de l'Angola. Ministerio da Ciencia e da Tecnologia. Instituto de Investigacao Cientifica Tropical, Lisboa, 531 p. (In French)

Bouquet A, Debray M (1974) Plantes Médicinales de la Côte d'Ivoire. Mémoires O.R.S.T.O.M., 32. Paris, 231 p

Brimer L, Rosling H (1993) Microdiffusion method with solid state detection of cyanogenic glycosides from cassava in human urine. Food Chem Toxicol 31(8):599–603

Burkill IH (1966) A dictionary of the economic products of the Malay Peninsula. Revised reprint of 1st ed 1935. 2 volumes. Ministry of Agriculture and Co-operatives, Kuala Lumpur, Malaysia. Vol. 1 (A-H) pp. 1–1240, Vol. 2 (I-Z), pp 1241–2444

Buschmann H, Reilly K, Rodriguez MX, Tohme J, Beeching JR (2000a) Hydrogen peroxide and flavan-3-ols in storage roots of cassava (*Manihot esculenta* Crantz) during postharvest deterioration. J Agric Food Chem 48(11):5522–5529

Buschmann H, Rodriguez MX, Tohme J, Beeching JR (2000b) Accumulation of hydroxycoumarins during post-harvest deterioration of tuberous roots of cassava (*Manihot esculenta* Crantz). Ann Bot 86(6):1153–1160

Carlsson L, Mlingi N, Ronquist G, Rosling H (1995) A specific and sensitive method for the determination of linamarin in urine. Nat Toxins 3(5):378–382

Carlsson L, Mlingi N, Juma A, Ronquist G, Rosling H (1999) Metabolic fates in humans of linamarin in cassava flour ingested as stiff porridge. Food Chem Toxicol 37(4):307–312

Carvalho LJ, Lippolis J, Chen S, Batista de Souza CR, Vieira EA, Anderson JV (2012a) Characterization of carotenoid-protein complexes and gene expression analysis associated with carotenoid sequestration in pigmented cassava (*Manihot esculenta* Crantz) storage root.Open. Biochem J 6:116–130

Carvalho LMJ, Oliveira ARG, Godoy RLO, Pacheco S, Nutti MR, de Carvalho JLV, Pereira EJ, Fukuda WG (2012b) Retention of total carotenoid and β-carotene in yellow sweet cassava (*Manihot esculenta* Crantz) after domestic cooking. Food Nutr Res 56:15788

Chalapathi V, Yuvaraj TV, Jaganathan A (2010) Formulation of paracetamol tablets using a novel binder isolated from *Manihot esculenta* L. and its evaluation. Int J ChemTech Res 2(1):406–411

Chaturvedula VSP, Schilling JK, Malone S, Wisse JH, Werkhoven MCM, Kingston DGI (2003) New cytotoxic triterpene acids from the above ground parts of

Manihot esculenta from the Suriname rainforest. Planta Med 69(3):271–274

Chávez AL, Sánchez T, Jaramillo G, Bedoya JM, Echeverry J, Bolanos EA, Ceballos H, Iglesias CA (2005) Variation of quality traits in cassava roots evaluated in landraces and improved clones. Euphytica 143(1–2):125–133

Chen CJ, Li Y, Su QM, Qu LJ, Chen D (2011) Study on chemical compositions of *Manihot esculenta* Crantz (*M. utilissima* Pohl) stalks. Adv Mat Res 236:394–398

Chew MY, Boey CG (1972) Rhodanese of tapioca leaf. Phytochemistry 11:167–169

Chiang BY, Johnson JA (1977) Gelatinization of starch in extruded products. Cereal Chem 54:436–443

Chuze G (2001) The cassava processing industry in Brazil: Traditional techniques, technological developments, innovations and new markets. Afr J Food Nutr Sec 1(1):46–59

Clapp RC, Bissett FH, Coburn RA, Long L (1966) Cyanogenesis in manioc: Linamarin and isolinamarin. Phytochemistry 5(6):1323–1326

Cliff J, Nicala D (1997) Long term follow-up of konzo patients. Trans Royal Soc Trop Med Hyg 91(4):447–449

Colonna P, Mercier C (1983) Macromolecular modifications of manioc starch components by extrusion-cooking with and without lipids. Cabohydr Polym 3:87–108

Coulin P, Farah Z, Assanvo J, Spillmann H, Puhan Z (2006) Characterisation of the microflora of attiéké, a fermented cassava product, during traditional small-scale preparation. Int J Food Microbiol 106(2):131–136

De Souza CR, Carvalho LJ, De Almeida ER, Gander ES (2002) Towards the identification of cassava root protein genes. Plant Foods Hum Nutr 57(3–4): 353–363

De Souza CR, Carvalho LJ, de Mattos Cascardo JC (2004) Comparative gene expression study to identify genes possibly related to storage root formation in cassava. Protein Pept Lett 11(6):577–582

De Souza CR, Carvalho LJ, De Almeida ER, Gander ES (2006) A cDNA sequence coding for a glutamic acid-rich protein is differentially expressed in cassava storage roots. Protein Pept Lett 13:653–657

De Souza CR, Aragão FJ, Moreira ECO, Costa CNM, Nascimento SB, Carvalho LJ (2009) Isolation and characterization of the promoter sequence of a cassava gene coding for Pt2L4, a glutamic acid-rich protein differentially expressed in storage roots. Genet Mol Res 8(1):334–344

De Vries CA, Ferweds JD, Flash M (1976) Choice of crops in relation to actual and potential production in tropics. Neth J Agric Sci 15:241–246

Delange F, Ekpechi LO, Rosling H (1994) Cassava cyanogenesis and iodine deficiency disorders. Acta Hort (ISHS) 375:289–294

Diop A (1998) Storage and processing of roots and tubers in the tropics. (ed) Calverley DJB. Food and Agriculture Organization of the United Nation, Rome

Djazuli M, Bradbury JH (1999) Cyanogen content of cassava roots and flour in Indonesia. Food Chem 65:523–525

Drunkler NL, Leite RS, Mandarino JM, Ida EI, Demiate IM (2012) Cassava starch as a stabilizer of soy-based beverages. Food Sci Technol Int 18(5):489–499

Du LC, Bokanga M, Møller BL, Halkier BA (1995) The biosynthesis of cyanogenic glucosides in roots of cassava. Phytochemistry 39(2):323–326

Dunican LK (1990) Strategies for developing the cassava industry.Nuclear and related techniques in the improvement of traditional fermentation processing of cassava. International Atomic Energy Agency, Vienna, pp 9–14

Ebuehi OAT, Babalola O, Ahmed Z (2006) Phytochemical, nutritive and anti-nutritive composition of cassava (*Manihot esculenta* L) tubers and leaves. Nigerian Food J 23(1):40–46

Elias M, Sudhakaran PR, Nambisan B (1997) Purification and characterisation of β-cyanoalanine synthase from cassava tissues. Phytochemistry 46(3):469–472

El-Kamali HH (2009) Ethnopharmacology of medicinal plants used in North Kordofan (Western Sudan). Ethnobot Leaflets 13:89–97

FAO (2014) FAO STAT3. Food and agricultural organization of United Nations: economic and social department: The Statistical Division. http://faostat3.fao.org/home/E

Fasuyi AO (2005) Nutrient composition and processing effects on cassava leaf (*Manihot esculenta* Crantz) antinutrients. Pakistan J Nutr 4(1):37–42

Fasuyi AO, Aletor VA (2005) Varietal composition and functional properties of cassava (*Manihot esculenta*, Crantz) leaf meal and leaf protein concentrates. Pak J Nutr 4(1):43–49

Frakes RA, Sharma RP, Willhite CC (1985) Developmental toxicity of the cyanogenic glycoside linamarin in the golden hamster. Teratology 31(2):241–246

Gallant DJ, Bewa M, Buy QH, Bouchet B, Szylit O, Sealy L (1982) On ultrastructural and nutritional aspects of some tropical tuber starches. Starch-Starke 34:255–262

George M, Moorthy SN, Padmaja G (1995) Biochemical changes in cassava tuber during fermentation and its effect on extracted starch and residue. J Sci Food Agric 69(3):367–371

Girish BN, Rajesh G, Vaidyanathan K, Balakrishnan V (2011) Assessment of cassava toxicity in patients with tropical chronic pancreatitis. Trop Gastroenterol 32(2):112–116

Gomes S, Torres AG, Godoy R, Pacheco S, Carvalho J, Nutti M (2013) Effects of boiling and frying on the bioaccessibility of beta-carotene in yellow-fleshed cassava roots (*Manihot esculenta* Crantz cv. BRS Jari). Food Nutr Bull 34(1):65–74

Gómez-Vásquez R, Day R, Buschmann H, Randles S, Beeching JR, Cooper RM (2004) Phenylpropanoids, phenylalanine ammonia lyase and peroxidases in elicitor-challenged cassava (*Manihot esculenta*) suspension cells and leaves. Ann Bot 94(1):87–97

Guilbot A, Mercier C (1985) Starch. In: Aspinall GO (ed) The polysaccharide, vol 3. Academic Press, London, pp 209–282

Hadi Alkarawi H, Zotz G (2014) Phytic acid in green leaves. Plant Biol (Stuttg) 16(4):697–701

Han YH, Gómez-Vásquez R, Reilly K, Li HY, Tohme J, Cooper RM, Beeching JR (2001) Hydroxyproline-rich glycoproteins expressed during stress responses in cassava. Euphytica 120:59–70

Herbal Medicine Research Center (HMRC) (2002) Compendium of medicinal plants used in Malaysia, vol 1 & 2. Institute for Medical Research, Kuala Lumpur

Hernández T, Lundquist P, Oliveira L, Pérez Cristiá R, Rodriguez E, Rosling H (1995) Fate in humans of dietary intake of cyanogenic glycosides from roots of sweet cassava consumed in Cuba. Nat Toxins 3(2):114–117

Hong K, Ma Y, Li MQ (2001) Solid-state fermentation of phytase from cassava dregs. Appl Biochem Biotechnol 91–93:777–785

Hu LF, Mei WL, Wu J, Wang WQ, Peng M, Dai HF (2010) The chemical constituents and bioactivities of volatile oils from stems and leaves of cassava (*Manihot esculenta*) in Hainan Chin. J Trop Crops 31(1):126–130 (In Chinese)

Huang J, Bachem C, Jacobsen E, Visser RG (2001) Molecular analysis of differentially expressed genes during postharvest deterioration in cassava (*Manihot esculenta* Crantz) tuberous roots. Euphytica 120(1):85–93

Hue KT, Thanh VDT, Ledin I, Wredle E, Spörndly E (2012) Effect of harvesting frequency, variety and leaf maturity on nutrient composition, hydrogen cyanide content and cassava foliage yield. Asian-Australas J Anim Sci 25(12):1691–1700

Hughes AP (1992) In-situ localization of cyanogenic β-glucosidase (linamarase) gene expression in leaves of cassava (*Manihot esculenta* Cranz) using non-isotopic riboprobes. Plant J 2(5):821–827

Jackson BA, Oladipo NO, Agaja MO (2014) Cassava: a potential crop for industrial raw material in Nigeria. Int J Life Sci 3(3):105–112

Jansson C, Westerbergh A, Zhang JM (2009) Cassava, a potential biofuel crop in the People's Republic of China. Appl Energy 86(1):S95–S99

Jørgensen K, Morant AV, Morant M, Jensen NB, Olsen CE, Kannangara R, Motawia MS, Møller BL, Bak S (2011) Biosynthesis of the cyanogenic glucosides linamarin and lotaustralin in cassava: isolation, biochemical characterization, and expression pattern of CYP71E7, the oxime-metabolizing cytochrome P450 enzyme. Plant Physiol 155(1):282–292

Kajiwara S, Maeda H (1983) The monosaccharide composition of cell wall material in cassava tuber (*Manihot utilissima*). Agric Biol Chem 47(10):2335–2340

Kamalu BP (1993) Cassava (*Manihot esculenta* Crantz) in the aetiology of kwashiorkor. Nutr Res Rev 6:121–135

Kamil M, Ilyas W, Okigawa M, Kawano N (1994) Biflavones from *Manihot utilissima*. Phytochemistry 13(12):2619–2620

Kawabata A, Sawayama S, Nagashima N, del Rosaria RR, Nakamura M (1984) Some physico-chemical properties of starches from cassava, arrow-root and sago. J Jpn Soc Starch Sci 31(4):24–232

Kawukpa UU, Angoyo MM (1994) Plantes utiles chez les Batiabetuwa de l'ile de Mbie, Kisangani, Zaïre. Afr Study Monographs 15(2):49–68 (In French)

Kayode J, Kayode GM (2008) Ethnomedicinal survey of botanicals used in treating sexually transmitted diseases in Ekiti State, Nigeria. Ethnobot Leaflets 12:44–55

Kerharo J, Bouquet A (1950) Plantes Médicinales et Toxiques de la Côte d'Ivoire – Haute-Volta. Vigot Frères, Paris, 291 pp

Ketiku AO, Oyenuga VA (1972) Changes in the carbohydrate constituent of cassava root tuber (*Manihot utilissima* Pohl) during growth. J Sci Food Agric 23(12):1451–1456

Khor HT, Tan HL (2006) The lipids of young cassava (*Manihot esculenta,* Crantz) leaves. J Sci Food Agric 32(4):399–402

Ki OL, Kurniawan A, Lin CX, Ju YH, Ismadji S (2013) Bio-oil from cassava peel: a potential renewable energy source. Bioresour Technol 145:157–161

Kimaryo VM, Massawe GA, Olasupo NA, Holzapfel WH (2000) The use of a starter culture in the fermentation of cassava for the production of "kivunde", a traditional Tanzanian food product. Int J Food Microbiol 56(2):179–190

King NLR, Bradbury JH (1995) Bitterness of cassava: identification of a new apiosyl glucoside & other compounds that effect its bitter taste. J Sci Food Agric 68(2):223–2230

Koch B, Nielsen VS, Halkier BA, Olsen CE, Møller BL (1992) The biosynthesis of cyanogenic glucosides in seedlings of cassava (*Manihot esculenta* Crantz). Arch Biochem Biophys 292(1):141–150

Kubo I, Masuoka N, Nihei KI, Burgheim B (2006) Maniçoba, a quercetin-rich Amazona dish. J Food Compos Anal 19:579–588

Lacerda IC, Miranda RL, Borelli BM, Nunes ÁC, Nardi RM, Lachance MA, Rosa CA (2005) Lactic acid bacteria and yeasts associated with spontaneous fermentations during the production of sour cassava starch in Brazil. Int J Food Microbiol 105(2):213–219

Lancaster P, Brooks J (1983) *Cassava* leaves as human food. Econ Bot 37:331–348

Lancaster PA, Ingram JS, Lin HY, Coursey DG (1982) Traditional cassava-based foods: survey of processing techniques. Econ Bot 36:12–25

Lebbie AR, Guries RP (1995) Ethnobotanical value and conservation of sacred groves of the Kpaa Mende in Sierra Leone. Econ Bot 49(3):297–308

Leonel M, de Freitas TS, Mischan MM (2009) Physical characteristics of extruded cassava starch. Sci Agric (Piracicaba Braz) 66(4):486–493

Lépiz-Aguilar L, Rodríguez-Rodríguez CE, Arias ML, Lutz G (2013) Acetone-Butanol-Ethanol (ABE) production in fermentation of enzymatically hydrolyzed cassava flour by *Clostridium beijerinckii* BA101 and solvent separation. J Microbiol Biotechnol 23(8):1092–1098

Leung WTW, Butrum RR, Huang Chang F, Narayana Rao M, Polacchi W (1972) Food composition table for use in East Asia. FAO, Rome, 347 pp

Li P, Zhu MJ (2011) A consolidated bio-processing of ethanol from cassava pulp accompanied by hydrogen production. Bioresour Technol 102(22):10471–10479

Li SS, Hu LF, Zhao YX, Zuo WJ, Zeng YB, Li XN, Mei WL, Dai HF (2011) A new diterpene from the stems of *Manihot esculenta*. J Asian Nat Prod Res 13(10):961–964

Liu JP, Zhaung ZT (2000) The use of dry cassava roots and silage from leaves for pig feeding in Yunnan Province of China. In: Howeler RH, Tan SL (eds) Cassava's potential in Asia in the 21st century: present situation and future research and development needs. Proceedings of the Sixth Regional Workshop held in Ho Chi Minh City, Vietnam. Feb 21–25, 2000, pp 527–537

Loh SP, Hadira O (2011) In vitro inhibitory potential of selected Malaysian plants against key enzymes involved in hyperglycemia and hypertension. Mal J Nutr 17(1):77–86

Loreto AB, Orias RR (2000) New products of future potential in the Philippines: cassava flour and grates. In: Howeler RH, Tan SL (eds) Cassava's potential in Asia in the 21st century: Present situation and future research and development needs. Proceedings of the sixth regional workshop held in Ho Chi Minh City, Vietnam. Feb 21–25, 2000, pp 587–592

Lu CG, Zhao JB, Yang ST, Wei D (2012) Fed-batch fermentation for n-butanol production from cassava bagasse hydrolysate in a fibrous bed bioreactor with continuous gas stripping. Bioresour Technol 104:380–387

Luk E, Sandoval AJ, Cova A, Müller AJ (2013) Antiplasticization of cassava starch by complexing fatty acids. Carbohydr Polym 98(1):659–664

Ly NT, Ngoan LD, Verstegen MW, Hendriks WH (2012) Pig performance increases with the addition of DL-methionine and L-lysine to ensiled cassava leaf protein diets. Trop Anim Health Prod 44(1):165–172

Lykkesfeldt J, Moller BL (1994) Cyanogenic glycosides in cassava, *Manihot esculenta* Crantz. Acta Chem Scand 48:178–180

McMahon JM, White WL, Sayre RT (1995) Cyanogenesis in cassava (*Manihot esculenta* Crantz). J Exp Bot 46(7):731–741

Michael E, Brown JA (1968) Improve texture, consistency, shelf-life and freeze thaw stability at reduced cost. Food Proc 28(6):2436–2443

Miladiyah I, Dayi F, Desrini S (2011) Analgesic activity of ethanolic extract of *Manihot esculenta* Crantz leaves in mice. Universa Medicina 30(1):3–10

Mkpong OE, Yan H, Chism G, Sayre RT (1990) Purification, characterization and localization of linamarase in cassava. Plant Physiol 93(1):176–181

Moorthy SN (2001) Tuber crop starches, Tech bull no 18. Central Tuber Crops Research Institute, Thiruvananthapuram

Moorthy SN, Ramanujan T (1986) Variation in properties of cassava starch in relation to age of crop. Starch-Starke 38(2):58–61

Moorthy SN, Jos JS, Nair RB, Sreekumari MT (1990) Variability of β-carotene content in cassava germplasm. Food Chem 36(3):233–236

Moundipa PF, Flore KGM, BilongC F, Bruchhau I (2005) In vitro amoebicidal activity of some medicinal plants of the Bamun region (Cameroon). Afr J Trad Complement Altern Med 2(2):113–121

Munyikwa TRI, Langeveld S, Salehuzzaman SNIM, Jacobsen E, Visser RGF (1997) Cassava starch biosynthesis: new avenues for modifying starch quantity and quality. Euphytica 96(1):65–75

Muoki PN, de Kock HL, Emmambux MN (2012) Effect of soy flour addition and heat-processing method on nutritional quality and consumer acceptability of cassava complementary porridges. J Sci Food Agric 92(8):1771–1779

Namukobe J, Kasenene JM, Kiremire BT, Byamukama R, Kamatenesi-Mugisha M, Krief S, Dumontet V, Kabasa JD (2011) Traditional plants used for medicinal purposes by local communities around the Northern sector of Kibale National Park, Uganda. J Ethnopharmacol 136:236–245

Narayanan NN, Ihemere U, Ellery C, Sayre RT (2011) Overexpression of hydroxynitrile lyase in cassava roots elevates protein and free amino acids while reducing residual cyanogen levels. PLoS ONE 6(7), e21996

Nartey F (1968) Studies on cassava, *Manihot utilissima* Pohl—I. Cyanogenesis: the biosynthesis of linamarin and lotaustralin in etiolated seedling. Phytochemistry 7(8):1307–1312

Nartey F, Møller BL (1973) Fatty acid profiles in germinating *Manihot esculenta*. Phytochemistry 12(12):2909–2911

Nartey F, Møller BL (1976) Amino acid profiles of cassava seeds (*Manihot esculenta*). Econ Bot 30(4):419–423

Nartey F, Møller BL, Andersen MR (1973) The major constituents of *Manihot esculenta* Crantz seeds. Trop Sci 15:273–273

Nartey F, Møller BL, Andersen MR (1974) Changes in the major constituents of *Manihot esculenta* Crantz seeds during germination and growth. Econ Bot 28:145–145

Nassar NMA (1978) Conservation of the genetic resources of cassava *Manihot esculenta*; determination of wild species localities with emphasis on probable origin. Econ Bot 32:311–320

Nassar NMA (2000) Cytogenetics and evolution of cassava (*Manihot esculenta* Crantz). Genet MolBiol 23(4):1003–1014

Nassar NMA (2002) Cassava, *Manihot esculent*a Crantz, genetic resources: origin of the crop, its evolution and relationships with wild relatives. Genet Mol Res 1(4):298–305

Nassar NMA, Sousa MV (2007) Amino acid profile in cassava and its interspecific hybrid. Genet Mol Res 6(2):192–197

Nassar NMA, Vizzotto CS, da Silva HL, Schwartz CA, Pires Júnior OR (2005) Potentiality of cassava cultivars as a source of carotenoids. J Food Agric Environ 3(3–4):33–35

Nassar NMA, Vizzotto CS, Schwartz CA, Pires Júnior OR (2007) Cassava diversity in Brazil: the case of carotenoid-rich landraces. Genet Mol Res 6(1):116–121

National Germplasm Resources Laboratory, Beltsville, Maryland. URL: http://www.ars-grin.gov.4/cgi-bin/npgs/html/taxon.pl?300073. 09 April 2015

Nduele M, Ludwig A, Van Ooteghem M (1993) The use of cassava starch in the formulation of gelatin capsules. J Pharm Belgique 48(5):325–334 (In French)

Nguyen THL, Nguyen TL, Du TH, Le VA (2000) The use of cassava leaf silage for feeding growing pigs and sows in Central Vietnam. In: Howeler RH, Tan SL (eds) Cassava's potential in Asia in the 21st century: present situation and future research and development needs. Proceedings of the Sixth Regional Workshop held in Ho Chi Minh City, Vietnam. Feb 21–25, 2000, pp 517–526

Noumi E, Tchakonang NYC (2001) Plants used as abortifacients in the Sangmelima region of Southern Cameroon. J Ethnopharmacol 76:263–268

Nuwamanya E, Baguma Y, Emmanbux N, Taylor J, Patrick R (2010) Physciochemical and functional characteristics of cassava starch in Ugandan varieties and their progenies. J Breeding Crop Sci 2(1):1–11

Nyakabwa M, Dibaluka W (1990) Plantes médicinales cultivées dans la zone de Kabondo à Kisangani (Zaïre). Afr Study Monogr 11(2):87–99 (In French)

Obadina AO, Oyewole OB, Williams OE (2013) Improvement in the traditional processing method and nutritional quality of traditional extruded cassava-based snack (modified Ajogun). Food Sci Nutr 1(4):350–356

Obidoa O, Obasi SC (1991) Coumarin compounds in cassava diets: 2 health implications of scopoletin in gari. Plant Food Human Nutr 41(3):283–289

Obilie EM, Tano-Debrah K, Amoa-Awua WK (2004) Souring and breakdown of cyanogenic glucosides during the processing of cassava into akyeke. Int J Food Microbiol 93(1):115–121

Obute GC (2005) Ethnomedicinal plant resources of south eastern Nigeria. Ethnobot Leaflets 2(5):78–79. http://www.ethnoleaflets.com/leaflets/obute.htm

Ochse JJ, Bakhuizen van den Brink RC (1980) Vegetables of the Dutch Indies, 3rd edn. Ascher & Co., Amsterdam, 1016 p

Ohochuku NS, Ballantine JA (1983) Fermented cassava: odor active components. J Agric Food Chem 31(6):1386–1387

Oke OL (1980) Toxicity of cyanogenic glycosides. Food Chem 6:97–109

Okechukwu PN, Bokanisereme YUF (2013) Anti-inflammatory, analgesic and anti – pyretic activity of cassava leaves extract. Asian J Pharm Clin res 6(4):89–92

Oladunmoye OO, Aworh OC, Maziya-Dixon B, Erukainure OL, Elemo GN (2014) Chemical and functional properties of cassava starch, durum wheat semolina flour, and their blends. Food Sci Nutr 2(2):132–138

Olafadehan OA, Olafadehan OO, Obun CO, Yusuf AM, Adewumi MK, Omotugba SK, Daniel NE (2012) Influence of processing cassava peels on the hydrogen cyanide concentration, nutritive value and performance of growing rabbits. Trop Anim Health Prod 44(2):285–291

Oluwole OSA, Onabolu AO, Cotgreave IA, Rosling H, Persson A, Link H (2002) Low prevalence of ataxic polyneuropathy in a community with high exposure to cyanide from cassava foods. J Neurol 249(8):1034–1040

Ongusua O, Okafor N, Onyekwere OO, Akinrele IA (1983) Nigerian garri. In: Steinkraus KH (ed) Handbook of indigenous fermented foods, vol 9. Marcel Dekker, Balsen, New York, pp 208–220

Oni AO, Onwuka CFI, Arigbede OM, Oni OO, Anele UY, Yusuf KO, Oduguwa BO, Onifade OS (2010) Chemical composition and in sacco degradability of four varieties of cassava leaves grown in Southwestern Nigeria in the rumen of sheep. Trop Anim Health Prod 42(7):1385–1393

Oni AO, Onwuka CF, Arigbede OM, Anele UY, Oduguwa OO, Onifade OS, Tan ZL (2011) Chemical composition and nutritive value of four varieties of cassava leaves grown in South-Western Nigeria. J Anim Physiol Anim Nutr (Berl) 95(5):583–590

Oriani VB, Molina G, Chiumarelli M, Pastore GM, Hubinger MD (2014) Properties of cassava starch-based edible coating containing essential oils. J Food Sci 79(2):E189–E194

Oso AO, Akapo O, Sanwo KA, Bamgbose AM (2014) Utilization of unpeeled cassava (*Manihot esculenta* Crantz) root meal supplemented with or without charcoal by broiler chickens. J Anim Physiol Anim Nutr 98(3):431–438

Osundahunsi OF, Williams AO, Oluwalana IB (2012) Prebiotic effects of cassava fibre as an ingredient in cracker-like products. Food Funct 3(2):159–163

Oyewole OB (2001) Characteristics and significance of yeasts' involvement in cassava fermentation for 'fufu'production. Int J Food Microbiol 65(3):213–218

Padmaja G (1989) Evaluation of techniques to reduce assayable tannin and cyanide in cassava leaves. J Agric Food Chem 37(3):712–716

Padmaja G, George M, Moorthy SN, Bainbridge Z, Plumb V, Wood JF, Powell CJ (1994) Nutritional evaluation of the starchy flour obtained from cassava tubers on fermentation with a mixed-culture inoculum. J Agric Food Chem 42(3):766–770

Padonou SW, Nielsen DS, Hounhouigan JD, Thorsen L, Nago MC, Jakobsen M (2009) The microbiota of Lafun, an African traditional cassava food product. Int J Food Microbiol 133(1):22–30

Pamo ET, Awah-Ndukum J, Boukila B, Kana JR, Tendonken F, Essie FMN (2006) A study of anthelmintic property of fresh cassava (*Manihot esculenta*) leaves incorporated in the diet of West African dwarf goats. Bull Anim Health Prod Africa 54(3):230–233

Pan YM, Zhng Y, Wang XN, Chen HP, Li SL, Di YT, Chne DZ, Guo LL, Hao XJ, He HP (2015) Chemical constituents from the stems of *Manihot esculenta*. Nat Prod Bioprospect 5(1):55–59

Pereira CA, Corrêa AD, Santos CDD, Abreu CMPD, Sousa RVD, Magalhães MM (2008) Hemaglutinin of cassava leaves (*Manihot esculenta* Crantz): partial purification and toxicity. Ciênc Agrotec 32(3):900–907

Popola TOS, Yangomodou OD (2006) Extraction, properties and utilization potentials of cassava seed oil. Bioechnology 5:38–41

Popola TOS, Yangomodou OD, Akintokun AK (2007) Antimicrobial activity of cassava seed oil on skin pathogenic microorganisms. Res J Med Plant 1:60–64

Powell EL (1965) Product and use of pregelatinized starch. In: Whistler RL, Paschall EF (eds) Starch: chemistry and technology. Academic Press, New York, pp 253–536

Prawat H, Mahidol C, Ruchirawat S, Prawat U, Tuntiwachwut-tikul P, Tooptakong U, Taylor WC, Pakawatchai C, Skeleton BW, White AH (1995) Cyanogenic and non-cyanogenic glycosides from *Manihot esculenta*. Phytochemistry 40(4):1167–1173

Premkumar T, Padmaja G, Moorthy SN, Nanda SK, George M, Balagopalan C (2000) New cassava products of future potential in India. In: Howeler RH, Tan SL (eds) Cassava's potential in Asia in the 21st century: present situation and future research and development needs. Proceedings of the Sixth Regional Workshop held in Ho Chi Minh City, Vietnam, 21–25 Feb 2000, pp 564–577

Rahmat A, Kumar V, Fong LM, Endrini S, Sani HA (2004) Determination of total antioxidant activity in three types of local vegetables shoots and the cytotoxic effect of their ethanolic extracts against different cancer cell lines. Asia Pac J Clin Nutr 13(3):308–311

Rajeshwarisivaraj R, Sivakumar S, Senthilkumar P, Subburam V (2001) Carbon from cassava peel, an agricultural waste, as an adsorbent in the removal of dyes and metal ions from aqueous solution. Bioresour Technol 80(3):233–235

Raponda-Walker A, Sillans R (1995) Les Plantes Utiles du Gabon. Encyclopédie Biologique. Editions Sepia, 697 p. (First Published 1961)

Ravindran G, Ravindran V (1988) Changes in the nutritional composition of cassava (*Manihot esculenta* Crantz) leaves during maturity. Food Chem 27(4):299–309

Razafindraibe M, Kuhlman AR, Rabarison H, Rakotoarimanana V, Rajeriarison C, Rakotoarivelo N, Randrianarivony T, Rakotoarivony F, Ludovic R, Randrianasolo A, Bussmann RW (2006) Medicinal plants used by women from Agnalazaha littoral forest (Southeastern Madagascar). J Ethnobiol Ethnomed 2:22

Reed JD, McDowell RE, Van Soest PJ, Horvath PJ (1982) Condensed tannins: a factor limiting the use of cassava foliage. J Sci Food Agric 33:213–220

Reilly K, Han Y, Tohme J, Beeching JR (2001) Isolation and characterisation of a cassava catalase expressed during post-harvest physiological deterioration. Biochim Biophys Acta -Gene Struct Express 1518(3):317–323

Reilly K, Bernal D, Cortés DF, Gómez-Vásquez R, Tohme J, Beeching JR (2007) Towards identifying the full set of genes expressed during cassava post-harvest physiological deterioration. Plant Mol Biol 64(1–2):187–203

Rickard JE (1986) Tannin levels in cassava, a comparison of methods of analysis. J Sci Food Agric 37(1):37–42

Rickard JE, Asaoka M, Blanshard JMV (1991) The physicochemical properties of cassava starch. Trop Sci 31:189–207

Roa AC, Maya MM, Duque MC, Tohme J, Allem AC, Bonierbale MW (1997) AFLP analysis of relationships among cassava and other *Manihot* species. Theor Appl Genet 95(5–6):741–750

Roble ND, Ogbonna JC, Tanaka H (2003) L-Lactic acid production from raw cassava starch in a circulating loop bioreactor with cells immobilized in loofa (*Luffa cylindrica*). Biotechnol Lett 25(13):1093–1098

Rogers DJ (1963) Studies of *Manihot esculenta* Crantz and related species. Bull Torrey Bot Club 90:43–54

Rogers DJ (1965) Some botanical and ethnological considerations of *Manihot esculenta*. Econ Bot 19:369–377

Rogers DJ, Appan SG (1973) *Manihot* and Manihotoides (Euphorbiaceae), a computer-assisted study. Flora Neotropica, monograph. Hafner Press, New York

Rogers DJ, Milner M (1963) Amino acid profile of manioc leaf protein in relation to nutritive value. Econ Bot 17:211–217

Rosenthal FRT, Nakamura T, Espindola AMC, Jochimek MR (1974) Structure of starch granules. Part 3. Some

considerations on leguminosae and tuberosae. Starch-Starke 26(2):50–55

Sakai T, Nakagawa Y (1988) Diterpenic stress metabolites from Cassava roots. Phytochemistry 27:3769–3779

Sakai T, Nakagawa Y, Uritani I, Data ES (1986) Occurrence of various kinds of metabolites in physiologically and microbially damaged cassava (*Manihot esculenta* Crantz) roots. Agric Biol Chem 50(11):2905–2907

Salchuzzaman SNIM, Jacobsen E, Visser RGF (1994) Expression patterns of two starch biosynthetic genes in vitro cultured cassava plants and their induction by sugars. Plant Sci 98:53–62

Salehuzzaman SNIM, Jacobsen E, Visser RGF (1993) Isolation and characterization of cDNA encoding granule-bound starch synthase in cassava (*Manihot esculenta* Crantz) and its antisense expression in potato. Plant Mol Biol 23:947–962

Sánchez T, Salcedo E, Ceballos H, Dufour D, Mafla G, Morante N, Calle F, Pérez JC, Debouck D, Jaramillo G, Moreno IX (2009) Screening of starch quality traits in cassava (*Manihot esculenta* Crantz). Starch-Starke 61(1):12–19

Santana MA, Vásquez V, Matehus J, Aldao RR (2002) Linamarase expression in cassava cultivars with roots of low- and high-cyanide content. Plant Physiol 129(4):1686–1694

Santos KS, Gadermaier G, Vejvar E, Arcuri HA, Galvão CE, Yang AC, Resende VM, Martins Cde O, Himly M, Mari A, Liso M, Pomponi D, Breiteneder H, Wagner S, Kalil J, Ferreira F, Castro FF (2013) Novel allergens from ancient foods: Man e 5 from manioc (*Manihot esculenta* Crantz) cross reacts with Hev b 5 from latex. Mol Nutr Food Res 57(6):1100–1109

Shama IYA, Wasma AAA (2011) Evaluation of the toxicity of Manihot esculenta on Wistar rats after traditional sudanese processing. J Pharmacol Toxicol 6(4):418–426

Shin SY, Lee HS, Kwon SY, Kwon ST, Kwak SS (2005) Molecular characterization of a cDNA encoding copper/zinc superoxide dismutase from cultured cells of *Manihot esculenta*. Plant Physiol Biochem 43:55–60

Siritunga D, Sayre RT (2003) Generation of cyanogen-free transgenic cassava. Planta 217:367–373

Siritunga D, Sayre R (2004) Engineering cyanogen synthesis and turnover in cassava (*Manihot esculenta*). Plant Molr Biol 56(4):661–669

Siritunga D, Sayre R (2007) Transgenic approaches for cyanogen reduction in cassava. J AOAC Int 90:1450–1455

Siritunga D, Arias-Garzon D, White W, Sayre RT (2004) Over-expression of hydroxynitrile lyase in transgenic cassava roots accelerates cyanogenesis and food detoxification. Plant Biotechnol J 2(1):37–43

Sornyotha S, Khin Lay Kyu KL, Ratanakhanokchai K (2007) Purification and detection of linamarin from cassava root cortex by high performance liquid chromatography. Food Chem 104(4):1750–1754

Soto-Blanco B, Górniak SL (2010) Toxic effects of prolonged administration of leaves of cassava (*Manihot esculenta* Crantz) to goats. Exp Toxicol Pathol 62(4):361–366

South Pacific Commission (SPC) (1986) Cassava a popular plant. SPC South Pacific foods leaflet, No 5

Srinivas T (2007) Industrial demand for cassava starch in India. Starch-Starke 59(10):477–481

Sriroth K, Chollakup R, Piyachomkwan K, Oates CG (2000) Biodegradable plastics from cassava starch in Thailand. In: Howeler RH, Tan SL (eds) Cassava's potential in Asia in the 21st century: present situation and future research and development needs. Proceedings of the Sixth Regional Workshop held in Ho Chi Minh City, Vietnam, 21–25 Feb 2000, pp 538–552

Subramanian SS, Nagarajan S, Sulochana N (1971) Flavonoids of some Euphorbiaceous plants. Phytochemistry 10:2548–2549

Sugumaran KR, Jothi P, Ponnusami V (2014) Bioconversion of industrial solid waste–cassava bagasse for pullulan production in solid state fermentation. Carbohydr Polym 99:22–30

Sumada K, Tamara PE, Alqani F (2011) Isolation study of efficient α-cellulose from waste plant stem *Manihot esculenta* Crantz. J Teknil Kimia 5(2):434–438

Suresh R, Saravanakumar M, Suganyadevi P (2011) Anthocyanins from Indian cassava (*Manihot esculenta* Crantz) and its antioxidant properties. Int J Pharm Sci Res 2:1819–1828

Tabuti JRS, Lye KA, Dhillion SS (2003) Traditional herbal drugs of Bulamogi, Uganda: plants, use and administration. J Ethnopharmacol 88:19–44

Tan KH, Ferguson LB, Carlton C (1984) Conversion of cassava starch to biomass, carbohydrates, and acids by *Aspergillus niger*. J Appl Biochem 6(1–2):80–90

Tanaka Y, Data ES, Hirose S, Taniguchi T, Uritani I (1983) Biochemical changes in secondary metabolites in wounded and deteriorated cassava roots. Agric Biol Chem 47(4):693–700

The Plant List (2015) Version 1.1. Published on the Internet. http://www.theplantlist.org/. Accessed 1st February

Tonukari NJ (2004) Cassava and the future of starch. Electron J Biotech 7(1):5–8

Tor-Agbidye J, Palmer VS, Lasarev MR, Craig AM, Blythe LL, Sabri MI, Spencer PS (1999) Bioactivation of cyanide to cyanate in sulfur amino acid deficiency: relevance to neurological disease in humans subsisting on cassava. Toxicol Sci 50(2):228–235

Tshala-Katumbay D, Mumba N, Okitundu L, Kazadi K, Banea M, Tylleskär T, Boivin M, Muyembe-Tamfum JJ (2013) Cassava food toxins, konzo disease, and neurodegeneration in sub-Sahara Africans. Neurology 80(10):949–951

Tsumbu CN, Deby-Dupont G, Tits M, Angenot L, Franck T, Serteyn D, Mouithys-Mickalad A (2011) Antioxidant and antiradical activities of *Manihot esculenta* Crantz (Euphorbiaceae) leaves and other selected tropical green vegetables investigated on lipoperoxidation and phorbol-12-myristate-13-acetate (PMA) activated monocytes. Nutrients 3(9):818–838

Tylleskär T, Banea M, Bikangi N, Cooke RD, Poulter NH, Rosling H (1992) Cassava cyanogens and konzo, an upper motoneuron disease found in Africa. Lancet 339(8787):208–211

U.S. Department of Agriculture, Agricultural Research Service (USDA-ARS) (2014) USDA National Nutrient Database for Standard Reference, Release 27. Nutrient Data Laboratory Home Page. http://www.ars.usda.gov/ba

Udomrati S, Gohtani S (2014) Enzymatic esterification of tapioca maltodextrin fatty acid ester. Carbohydr Polym 99:379–384

Uritani I, Data ES, Villegas RJ, Flores P, Hirose S (1983) Relationship between secondary metabolism changes in cassava root tissue and physiological deterioration. Agric Biol Chem 47(7):1591–1598

Uritani I, Data ES, Tanaka Y (1984a) Biochemistry of post harvest deterioration of cassava and sweet potato roots. In: Uritani I, Reyes ED (eds) Tropical root crops: post harvest physiology and processing. Japan Scientific Societies Press, Tokyo, pp 61–65

Uritani I, Data ES, Villegas RI, Flores P (1984b) Changes in secondary metabolism in cassava roots, in relation to physiological deterioration. In: Uritani I, Reyes ED (eds) Tropical root crops: post harvest physiology and processing. Japan Scientific Societies Press, Tokyo, pp 109–118

USDA-ARS (2015) National genetic resources program. Germplasm resources information network – (GRIN) [Online Database]. National Germplasm Resources Laboratory, Beltsville, Maryland. https://npgsweb.ars-grin.gov/gringlobal/taxonomydetail.aspx?23373

Veltkamp HJ, De Bruijn GH (1996) Manihot esculenta Crantz. In: Flach M, Rumawas F (eds) Plant resources of south east Asia no. 9. Plants yielding non-seed carbohydrates. Prosea Foundation, Bogor, pp 107–113

Vuilleumier S (1993) Worldwide production of high-fructose syrup and crystalline fructose. Am J Clin Nutr 58(5):733S–736S

Wang W (2002) Cassava production for industrial utilization in China –present and future perspective. In: Howeler RH (ed) Cassava research and development in Asia: exploring new opportunities for an ancient crop. Proceedings of the 7th Regional workshop, Bangkok, Thailand 28 Oct–1 Nov 2002, pp 33–38

Wang L, Zheng M, Wang Y, Zhang Y, Qian H, Zhang H, Qi X (2014) Anti-diabetic activity of cassava cross-linked octenyl succinic maltodextrin in STZ-induced diabetic mice. Int J Biol Macromol 64:247–251

Wheatley CC (1982) Studies on cassava (Manihot esculenta Crantz) root physiological deterioration. Ph.D. thesis. University of London, p 246

Wheatley CC, Schwabe WW (1985) Scopoletin involvement in post-harvest physiological deterioration of cassava root (*Manihot esculenta* Crantz). J Exp Bot 36:783–791

White W, Sayre RT (1995) The characterization of hydroxynitrile lyase for the production of safe food products from cassava (*Manihot esculent*a, Crantz). In: Gustine DL, Flores HE (eds) Phytochemicals and health: current topics in plant physiology, vol 15, American Society of Plant Physiologists. Rockville, MD, pp 303–304

White WL, McMahon JM, Sayre RT (1994) Regulation of cyanogenesis in cassava. Acta Hort (ISHS) 375:69–78

White WL, Arias-Garzon DI, McMahon JM, Sayre RT (1998) Cyanogenesis in cassava the role of hydroxynitrile lyase in root cyanide production. Plant Physiol 116(4):1219–1225

Widowati S, Hartojo K (2000) Production and use of cassava flour: a new product of future potential in Indonesia. In: Howeler RH, Tan SL (eds) Cassava's potential in Asia in the 21st century: present situation and future research and development needs. Proceedings of the Sixth Regional Workshop held in Ho Chi Minh City, Vietnam. Feb 21–25, 2000, pp 578–586

Williams PC, Lesseelleur GC (1970) Determination of damaged starch in flour. A comparitve study of present day procedures. Cereal Sci Today 15:4–19

Wobeto C, Corrêa AD, de Abreu CM, Santos CDD (2004) Cyanide in the leaves and cassava leaves flour (*Manihot esculenta* Crantz). Ciênc Agrotec 28(5):1115–1118 (In Portuguese)

Wobeto C, Corrêa AD, Abreu CMPD, Santos CDD, Abreu JRD (2006) Nutrients in the cassava (*Manihot esculenta* Crantz) leaf meal at three ages of the plant. Food Sci Technol (Campinas) 26(4):865–869

Wobeto C, Corrêa AD, Abreu CMPD, Santos CDD, Pereira HV (2007) Antinutrients in the cassava (*Manihot esculenta* Crantz) leaf powder at three ages of the plant. Food Sci Technol (Campinas) 27(1):108–112

Wong SP, Leong LP, Koh JHW (2006) Antioxidant activities of aqueous extracts of selected plants. Food Chem 99:775–783

Yeoh HH, Chew MY (1976a) Free amino acids of cassava leaves. Mal Agric J 50:435–441

Yeoh HH, Chew MY (1976b) Protein content and amino acid composition of cassava leaf. Phytochemistry 15:1597–1599

Yeoh HH, Chew MY (1977) Protein content and amino composition of cassava seed and tuber. Mal Agric J 51(1):1–6

Yeoh HH, Oh HY (1979) Cyanide content of cassava. Mal Agric J 52(1):24–28

Yi B, Hu LF, Mei W, Zhou KB, Wang H, Luo Y, Wei XY, Dai HF (2011) Antioxidant phenolic compounds of cassava (*Manihot esculenta*) from Hainan. Molecules 16(12):10157–10167

Zainuddin SY, Ahmad I, Kargarzadeh H, Abdullah I, Dufresne A (2013) Potential of using multiscale kenaf fibers as reinforcing filler in cassava starch-kenaf biocomposites. Carbohydr Polym 92(2):2299–2305

Zakaria ZA, Hussain MK, Somchit MN, Sulaiman MR, Mat Jais AM, Rezal I, Mat Zaid NN, Abdul Wahab SNZ, Fadzil NS, Abdullah M, Fatimah CA (2006) The in vitro antibacterial activity and brine shrimp toxicity of *Manihot esculenta* extract. Int J Pharmacol 2(2):218–220

Zeng YB, Li SS, Mei WL, Dong WH, Li KM (2014) Two new terpenoids from the stems of *Manihot esculenta*. J Asian Nat Prod 17(3):1–5

Zhang P, Bohl-Zenger S, Puonti-Kaerlas J, Potrykus I, Gruissem W (2003) Two cassava promoters related to vascular expression and storage root formation. Planta 218(2):192–203

Glycyrrhiza glabra

Scientific Name

Glycyrrhiza glabra L.

Synonyms

Glycyrrhiza brachycarpa Boiss., *Glycyrrhiza glabra* var. *caduca* X.Y. Li, *Glycyrrhiza glabra* var. *glabra*, *Glycyrrhiza glabra* subsp. *glandulifera* (Waldst. & Kit.) Ponert, *Glycyrrhiza glabra* var. *glandulifera* (Waldst. & Kit.) Regel & Herder, *Glycyrrhiza glabra* var. *glandulifera* (Waldst. & Kit.) Boiss., *Glycyrrhiza glabra* var. *glandulosa* X.Y. Li, *Glycyrrhiza glabra* var. *laxifoliolata* X.Y. Li, *Glycyrrhiza glabra* var. *typica* L., *Glycyrrhiza glabra* var. *violacea* (Boiss. & Noe) Boiss., *Glycyrrhiza glandulifera* Waldst. & Kit., *Glycyrrhiza hirsuta* Pall., *Glycyrrhiza pallida* Boiss. & Noe, *Glycyrrhiza pallida* Boiss., *Glycyrrhiza violacea* Boiss. & Noe

Family

Fabaceae

Common/English Names

Black Sugar, Common Liquorice, Licorice, Licorice-Root, Liquorice, Liquorice Root, Persian Licorice, Rhizoma Glycyrrhizae, Russian Licorice, Russian Liquorice, Si-Pei Licorice, Sinkiang Licorice, Spanish Juice, Spanish Licorice, Spanish Liquorice Sweet Root, Sweet Wood, Sweet Wood Liquorice, True Licorice, True Liquorice

Vernacular Names

Afrikaans: Drop
Albanian: Gliciriza E Shogët, Glicirizë
Arabic: Irq As-Sus, Irqu As-Sus, Irqu Al-Sus, Sous, Sus
Armenian: Madoodag, Matutak
Azeri: Biyanlıq
Basque: Erregaliz, Gotxerro, Makilgoxo
Belarusian: Lakryčnik
Brazil: Alcaçuz (Portuguese)
Breton: Regalis
Bulgarian: Sladnik, Sladuk Koren
Burmese: Noekiyu
Catalan: Regaléssia
Chinese: Gān Cǎo, Gam Chou, Guangguo–Gancao, Kan Tsau, Xi-Bei, Yang Gan Cao
Croatian: Sladki Korijen, Slatki Sladić
Czech: Lékořice, Lékořice Lysá, Sladký Dřevo
Danish: Glat Lakrids, Lakrids, Lakrids Rod, Lakridsrod, Lakridsplante
Dutch: Zoethout
Esperanto: Glicirizo

T.K. Lim, *Edible Medicinal and Non-Medicinal Plants: Volume 10, Modified Stems, Roots, Bulbs*,
DOI 10.1007/978-94-017-7276-1_18

Estonian: Lagrits, Lagritsa-Magusjuur, Magusjuur
Farsi: Shirin Bayan
Finnish: Lakritsi, Lakritsijuuri, Lakritsikasvi
French: Bois Sucré, Régalisse, Reglisse, Réglisse Glabre
Gaelic: Carra-Meille, Carrchan, Maide-Milis
German: Gemeines Süssholz, Kahles Süssholz, Lacrisse, Lakritz, Lakritze, Lakritzeholz, Lakritzenwurzel, Lakritzpflanze, Lakrizenwurzel, Spanisches Süssholz, Süsscholz, Süssholz, Süsswurz
Greek: Glikoriza, Glykoriza, Glykyrrhiza
Hebrew: Shush, Shush Kireah
Hungarian: Édesfa, Igazi Édesgyökér
Icelandic: Lakkrís, Lakkrísrót
India: Jesthimadhu, Yeshtmadhu, Zostimodhu (Assamese), Jaishbomodhu, Yashthimadhu (Bengali), Veymui (Dhivehi), Mulethi (Dogri), Jethimadha, Jethimard, Jethimadh, Jethimadhu (Gujarati), Jetimadh, Jethimadhu, Jethimadh, Jothi-madh, Mulathi, Mulethi, Mulaithi, Mulhathi, Yashtimadhu (Hindi), Atimadhura, Jestamadu, Jyeshtamadhu, Madhuka, Yasthimadhu (Kannada), Multhi (Kashimiri), Mauddh (Maithili), Etthimadhiram, Iradimadhuram, Irattimadhuram, Madhugam, Yashtimadhugam (Malayalam), Jestamadha (Marathi), Jatimadhu, Jasthimadhu (Oriya), Jethimadh, Malathi, Mulathi (Punjabi), Madhuuka, Madhuyasti, Yashti, Yastika, Yashtimadhu, Yashtomadhu (Sanskrit), Athimadhuram, Atimaduram (Tamil), Atimadhuramane, Atimadhuramu, Yastimadhu, Yashtimadhukkam (Telugu), Asl-us-sus, Mulhati, Mulethi (Urdu)
Indonesian: Akar Manis, Licorice
Irish: Liocras
Italian: Liquirizia, Liquirizia Commune, Regalizia, Regolizia
Japanese: Nankin-Kanzō, Nankin-Kanzo, Rikorisu, Kanzō, Kanzo
Kazakh: Mïya, Qızılmïya
Korean: Kamcho, Rikeorisu, Rikorisu
Laotian: Sa-Em, Sa-Em Thet
Latvian: Lakrica
Lithuanian: Saldymedis, Paprastasis Saldymedis
Macedonian: Sladok Koren
Malaysia: Akar Likuoris
Mongolian: Chiher Övs
Nepal: Istami, Jethimadhu (Nepali), Istivi (Nepalbhasa)
Norwegian: Lakrisplante, Lakrisrot
Pashto: Shireen Buya
Polish: Korzeń Lukrecji, Lukrecja Gładka
Portuguese: Alcaçuz
Provençal: Recalicé, Recalissi
Romanian: Lemn Dulce, Rădăcină Dulce, Reglisă
Russian: Koren Solodki, Lakrichnik, Lakritsa, Solodka
Serbian: Konjeda, Slacić, Sladić, Slatki Koren, Slatko Drvce
Slovak: Sladké Drievko, Sladovka Hladkoplodá
Slovenian: Golostebelni Sladki Koren, Sladki Koren, Sladki Koren Golostebelni
Spanish: Alcazuz, Orozus, Orozuz, Paloduz, Regaliz
Sri Lanka: Atimaduram, Valmi (Sinhala)
Swahili: Susu
Swedish: Äkta Lakrits, Lakrits, Lakritsrot, Lakritsväxt, Sötlakrits
Thai: Cha-Em Thet
Tibetan: Shi Na Ma Ngar, Shina ngar
Turkish: Biyam, Meyan, Meyankökü, Piyan, Tatlı Kök, Tatlı Meyan
Ukrainian: Lokrytsya, Solodkyj Korin, Solodka Hola
Uzbek: Miya, Qizilmiya
Vietnamese: Cam Thảo
Yiddish: Lakrets, Ziswortsl

Origin/Distribution

Glycyrrhiza glabra is native to Eurasia, in central and south-western Asia and the Mediterranean region (Plate 2). According to Hayashi (2009), Hayashi and Sudo (2009), *G. glabra* is found in South Europe (Spain, Italy), Turkey, Iran, Iraq, Central Asia and the north-western part of China, while *G. uralensis* is found in Central Asia, Mongolia and north-western and north-eastern parts of China, and *G. inflata* is found only in the north-eastern part, the Xinjiang Uygur Autonomous Region of China. *G. glabra* is divided into two varieties: *G. glabra* var. *typica* (Spanish licorice) and *G. glabra* var. *glandulifera*

(Russian licorice) (Hayashi and Sudo 2009). Three varieties of *G. glabra* have been reported; the Spanish and Italian licorice, assigned to *G. glabra* var. *typica*; Russian licorice to *G. glabra* var. *glandulifera*; Persian and Turkish licorice to *G. glabra* var. *violacea* (Nomura et al. 2002).

Countries producing liquorice include Iran, Afghanistan, the People's Republic of China, Pakistan, Iraq, Azerbaijan, Uzbekistan, Turkmenistan and Turkey. In China, commercial licorice is produced from the three aforementioned species.

Agroecology

Licorice grows well in temperate, warm and subtropical climate. It thrives best in well-limed, well-drained, composted, loose, friable, deep soil, preferably in full sun. Licorice is not bothered by frosts, as it is dormant in winter, and actually benefits by the defined cold period, which induces the translocation of properties to the underground rhizomes. They are easily grown from divisions or root cuttings.

Edible Plant Parts and Uses

Fresh liquorice root when washed is externally of a bright yellowish-brown colour and is chewed fresh or dried as a mouth freshener, for teething in children and also as a tooth cleaner (Chiej 1984). Dried liquorice root can be chewed as a sweet. The extract of liquorice in rolls is glossy black in colour, often used in cough lozenges and pastilles. In Calabria a popular liqueur is made from pure liquorice extract. Liquorice is also very popular in Syria and Egypt, where it is sold as a drink, in shops as well as by the street vendors. Liquorice is used by brewers to flavour and colour porter classes of beers, and the enzymes in the root also stabilize the foam heads produced by beers brewed with it. Licorice powder used in sweets, baked goods, ice cream, soft drinks, etc., and the powdered root is also used as a sweetener in other herbal teas (Facciola 1990). The leaves are used as a tea substitute in Mongolia (Facciola 1990).

The licorice root contains glycyrrhizin, a substance that is 50 times sweeter than sucrose (Hill 1952; Facciola 1990; Bown 1995). Glycyrrhizin imparts a sweet taste to foods; moreover, it has salt-softening and flavour-enhancing properties and is also heat stable (Hayashi and Sudo 2009). Most Japanese people do not like the long-lasting sweet taste of glycyrrhizin; however, a more acceptable sweetness can be created by using a combination of glycyrrhizin and natural sugars or other sweeteners. Therefore, glycyrrhizin and licorice extracts are used as food additives in a variety of foods such as snacks, instant noodles, sausages and sauces. Glycyrrhizin is used in sweet foods such as sweet snacks, sweets and candies, ice creams and sherbets to enhance their sweetness. It is also used to reduce the saltiness of salty foods such as soy sauce, other sauces, savoury snacks, 'kamaboko' (boiled fish paste), tsukudani (fish boiled in soy sauce), 'tsukemono' (Japanese pickles) and sausages in Japan. In Japan, enzymatically modified licorice extract (a-glycosyl-glycyrrhizin) and enzymatically hydrolysed licorice extract (glycyrrhetinic acid 3-*O*-glucuronide) are also used as sweeteners (Hayashi and Sudo 2009). Most liquorice is used as a flavouring agent for tobacco. Licorice not only imparts a sweet taste but also an aroma of tobacco, which makes it mild (Nieman 1959). It also prevents the desiccation of tobacco. The licorice extracts used in the tobacco industry are supplied by an American company, namely, MAFCO. Licorice extracts were first used for flavouring confectionery products in England during the eighteenth century in Pontefract in Yorkshire; it was blended with sugar, flour and other ingredients to make Pontefract cakes (Nieman 1959). Nowadays, licorice confectionery is widely available in western countries, and large quantities of licorice are used in the confectionery industry. In the Netherlands, where liquorice candy ('drop') is one of the most popular forms of sweets, only a few of the many forms that are sold contain aniseed, although mixing it with mint, menthol or with laurel is quite popular. *Glycyrrhiza glabra* root is one of the common traditional Chinese medicines and used as flavouring and sweetening agents for tobaccos, chewing gums, candies, toothpaste and beverages (Dong et al. 2007).

Botany

A herbaceous perennial, with stem 0.5–1.5 m high, woody at base, densely scaly glandular punctuate with stoloniferous roots. Leaves imparipinnate, 7–15 cm long with 9–17 ovate-oblong, oblong-lanceolate, or elliptic leaflets 1.7–4.0 by 0.8–2.0 cm, abaxially densely scaly glandular punctate and pubescent on veins, adaxially glabrescent or pilose (Plate 1). Stipules caducous, linear, 1–2 mm. Inflorescence open, racemose, many flowered. Flowers 0.8–1.2 cm long. Calyx campanulate, 5–7 mm, 5-toothed, upper 2 teeth mostly joined; corolla purple or pale whitish blue, 9–12 mm, standard ovate or oblong, 1–1.1 cm, base clawed, wings 8–9 mm, keel straight, 7–8 mm; ovary glabrous. Fruit oblong, flat, glabrous or sparsely hairy legume, 2–3 cm long, containing 2–8, dark green, smooth seeds, 2 mm across.

Plate 1 Liquorice foliage

Plate 2 Liquorice plant label

Nutritive/Medicinal Properties

Root/Stolon Phytochemicals

Licorice is a powerful natural sweetener, 50–170 times sweeter than sucrose (Mukhopadhyay and Panja 2008). The chemical constituents of the roots include several bioactive compounds such as glycyrrhizin (~16 %), different sugars (up to 18 %), flavonoids, saponoids, sterols, starches, amino acids, gums and essential oils. Licorice roots were reported to contain 25–30 % starch, 3–10 % D-glucose and sucrose, 3–5 % glycyrrhizin and traces of flavonoids, saponoids, sterols, amino acids, etc. (Fenwick et al. 1990). Licorice root contained phenolic constituents (such as coumarin compounds, glycerol, glycerine, glycycoumarin, herniarin, umbelliferone, licopyranocoumarin, licoarylcoumarin and licocoumarone), amines (1–2 % asparagines, betaine and choline), amino acids, sterols (stigmasterol and β-sitosterol) and sugars (5–15 % as glucose, sucrose and mannitol), and starch about 20 % of dried root (Blumenthal 2000). Flavonoids, saponins and sugars were found in the methanol root extract, and sterols in the crude petroleum ether extract (Chopra et al. 2013). Alkaloids, proteins and tannins were not detected. The mineral elements found in the roots included K (0.66 %), Ca (1.87 %), S (0.09 %), Fe (0.14 %), P (0.06 %), Mg (0.17 %), Na (0.04 %), Si (0.12 %), Al (0.05 %), Sr (0.06 %), Mn (tr), Ti (tr) and AS (tr). The amount of total phenolics in Turkish *G. glabra* roots was 12.88 μgGAE/mg DW (Ercisli et al. 2008). The average composition (mg/100 g) of N, P, K, Ca, Mg, Fe, Mn, Zn, Na and Cu in licorice roots was 2.80 %, 175 mg, 1400 mg, 147 mg, 120 mg, 20 mg, 6 mg, 4.4 mg, 2.1 mg and 0.1 mg, respectively. Eight commercial licorice extracts used as food additive (sweetener) in Japan were found to contain 0.3–12.% ash, 10.9–77.4 % glycyrrhizin, 0.1–1.2 % sodium, 0.3–5 % potassium and 0.03–2.5 % ammonium nitrogen and pH of 4.1–6.8 (Iida et al. 2007).

The roots of *G. glabra* were reported to contain 1.6 % of water soluble polysaccharides consisting of rhamnose, arabinose, mannose, glucose and galactose, and also 9.7 % of total polysaccharides (Dzhumamuratova et al. 1978). Denisova et al. (2003) found that >50 % of the ethanol extract of licorice root consisted of monosaccharides and disaccharides (7–8 mass % of the dry raw material). The principal component was saccharose (46.78 %). Significant quantities of D-mannopyranose (9.06 %), β-D-glucopyranose (7.06 %) and 2-*O*-hydroxyethylglucose (12.84 %) and smaller quantities of sorbose (4.12 %), α-D-fructose (2.01 %), β-D-fructose (2.56 %) and β-D-galactofuranose (1.88 %) were observed among the identified sugars. The sugar alcohols mannopyranosyl-D-glucitol (3.49 %), ribitol (0.95 %), mannitol (1.33 %), and myo-inositol (0.33 %) were present in insignificant amounts. Rhizomes were reported to contain alkaloids, triterpenes, saponins, flavonoids, polysaccharides, steroids and tannins (Meena et al. 2010). An acidic polysaccharide, named glycyrrhizan GA, was isolated from the stolon of *Glycyrrhiza glabra* var. *glandulifera* (Shimizu et al. 1991). Its molecular mass was estimated to be 85,000 and it comprised L-arabinose: D-galactose: L-rhamnose: D-galacturonic acid: D-glucuronic acid in the molar ratio of 22:10:1:2:1, in addition to small amounts of *O*-acetyl groups. Glycyrrhizan GA, a representative polysaccharide with remarkable phagocytosis-enhancing activity, was isolated from the stolon of *Glycyrrhiza glabra* var. *glandulifera* (Takada et al. 1992). The core structural features of glycyrrhizan GA included a backbone chain composed of β-1,3-linked D-galactose residues. Three-fifths of the galactose units in the backbone carry side chains composed of β-1,3- and β-1,6-linked D-galactosyl residues at position 6. Monosaccharide composition of the polysaccharides isolated, measured as alditol acetates, of Chinese and Hungarian *G. glabra* roots was very similar, but the Lithuanian *G. glabra* and the *G. echinata* were quite different (Gyémánt et al. 2001). All investigated samples contained glucuronic acid, the *G. echinata* contained also galacturonic acid. Although the yield of Hungarian origin species was found lower than the yield of eastern species, the uronic acid content was similar (Kiss et al. 1998). The water-extracted arabinogalactan protein enriched fraction of *Glycyrrhiza glabra* was found to con-

sist mainly of 3- and 3,6-linked galactopyranosyl, and 5- and 3,5-linked arabinofuranosyl residues (Saha et al. 2011). The hexane extract of *G. glabra* was found to contain 70 % neutral and 30 % polar lipids (Denisova et al. 2007). Among neutral lipids, the main components were sterol esters (SEs), which accounted for about half of this fraction. Triacylglycerides (TAGs), free fatty acids (FFAs) and free sterols (FSs) comprised this fraction in approximately equal proportions, amounting to 10.0, 10.5 and 11.5 %.

The contents (mg/g) of total phenols, total flavonoids and total tannins in licorice extracts of *G. glabr*a roots at different harvest times were determined to vary a from 72.10 to 107.93 mg/g, 18.42 to 44.2 mg/g and 4.8 to 12.78 mg/g, respectively (Cheel et al. 2013). Liquiritin and glycyrrhizin, the major components of licorice extract, varied in the range of 28.65–62.80 and 41.84–114.33 mg/g, respectively. The relative proportion of glycyrrhizin derivative, glabridin, glabrene and liquiritigenin derivative, varied in the range of 0.88–11.38 %, 1.86–10.03 %, 1.80–18.40 % and 5.53–16.31 %, respectively. Treatment of in-vitro cultured 65-day-old *G. glabra* plantlets with 0.1–2 mM methyl jasmonate and 0.1 and 1 mM salicylic acid enhanced the production of glycyrrhizin by 3.8 and 4.1 times, respectively, as compared to the controls (Shabani et al. 2009). Increasing amounts of glycyrrhizin in the roots treated with methyl jasmonate inhibited root growth, while salicylic acid increased the amount of glycyrrhizin without negative effects on growth.

Thirteen terpenoids, minor sapogenins were isolated from *G. glabra* roots by Canonica and co-workers (Canonica et al. 1966a, b, c, 1967a, b, c, 1968) glabrolide, isoglabrolide, 11-deoxoglabrolide, liquiritic acid, 11-deoxoglycyrrhetinic acid, 3β-hydroxy-ll, 13(18)-oleanadien-30-oic acid, glypallidifloric acid, glycyrretol (glycyrrhetol), 21α-hydroxyisoglabrolide, 24-hydroxyglycyrrhetic acid, 24-hydroxy-11-deoxoglycyrrhetic acid, 18α-hydroxyglycyrrhetic acid, liquiridiolic acid (glyyunnansapogenin B_1) and 24-hydroxyliquiritic acid. Glabric acid was isolated by Beaton and Spring (1956). Bogatkina et al. (1975) isolated 3,24-dihydroxy-ll, 13(18)-oleanadien-30-oic acid as a methyl ester. The following triterpenoid compounds were isolated from *G. glabra* roots, a pentacyclic triterpenoid, liquoric acid, was isolated as the methyl ester (Elgamal et al. 1965); another triterpenoid glabric acid (Elgamal and Fayez 1968), 7-hydroxy-4′-methoxy-isoflavone (formononetin) (Reiners 1966; Elgamal and Fayez 1972), another triterpenoid, 28-hydroxyglycyrrhetic acid (Elgamal and El-Tawil 1975); a carboxyliedialcoholic triterpenoid was isolated as the diacetate-methyl ester (Elgamal and Fayez 1975); five new pentacyclic triterpenoids 11-desoxoglycyrrhetic acid acetate methyl ester; 24-acetoxy-l 1-desoxoglycyrrhetic acid acetate methyl ester; 11-desoxo-glabrolide acetate; glabrolide acetate and 3 ß-acetyl-18 ß-hydroxy-1 l-keto-olean-12-en-30-oic acid, 30,18 ß-lactone were isolated as minor constituents together with 24-hydroxy-11-desoxoglycyrrhetic acid methyl ester, 3 ß, 18 ß-dihydroxy-11 -keto-olean-12-en-30-oic acid, 30,18 ß-lactone and glabrolide (Elgamal et al. 1990). *Glycyrrhiza glabra* root yielded two saponins named glabranin-A, a pentaglycoside of glycyrrhetic acid, and glabranin-B, both a heptaglycoside of glycyrrhetic acid (Varshney et al. 1983). Glabranin A on hydrolysis gave glucose and rhamnose (4:1) whereas glabranin-B gave glucose, xylose and rhamnose in the molar ratio (4:1:2).

About 70 phenolic compounds were isolated from the subterranean parts of *G. glabra* (Nomura and Fukai 1998). The flavanone glycoside 7,4′-dihydroxyflavanone, better known as liquiritigenin and liquiritin its aglycone were first isolated from *G. glabra* roots by Shinoda and Ueeda (1934a, b). Liquiritin and the corresponding chalcone, isoliquiritin were isolated from the dried root (Puri and Seshadri 1954); they also isolated isoliquiritin from fresh roots but not liquiritin. Litvinenko et al. (1963a, b) isolated liquiritigenin [7,4′-dihydroxyflavanone], liquiritin [liquiritigenin 4′-(β-D-glucopyranoside)], neoliquiritin [liquiritigenin 7-(β-D-glucopyranoside)], [lacroside [liquiritigenin7-(β-D-glucopyranosyl(1 → 2)-β-d-apiofuranoside)] and uraloside [liquiritigenin 4′-(β-d-glucopyranosyl(1 → 4)-β-d-apiofuranoside)]

from *G. glabra roots*. Liquiritoside, a flavonoside, was isolated from root of licorice, *G. glabra* (Paris and Guillot 1955).

Licuroside was first isolated from the roots of *Glycyrrhiza glabra* by Litvinenko (1964), Litvinenko and Obolentseva (1964), and was found not to be a homogenous compound (Miething and Speicher-Brinker 1989). It was separated into two isomeric glycosides: isoliquiritigenin-4-β-D-apiofuranosyl-2″-β-D-glucopyranoside with the proposed name neolicuroside and isoliquiritigenin-4′-β-D-apiofuranosyl-2″-β-D-glucopyranoside. Licoricidin was isolated from *G. glabra* root and its structure elucidated as 3′,6-diisopentenyl-2′,4′,5-trihydroxy-7-methoxyisoflavan (Shibata and Saitoh 1968). An isoflavan named glabridin and glabrol, a flavanone, were isolated from Russian *G. glabra* root (Saitoh et al. 1976b); glabranin and two flavonoids identified as 5,7-dihydroxyflavanone (= pinocembrin) and 4′,5-dihydroxy-7-methoxyisoflavone (= prunetin) (Kattaev and Nikonov 1972, 1974); flavonoids 7-hydroxy-2-methylisoflavone; 7-acetoxy-2-methylisoflavone and quercetin, kaempferol, apigenin, liquiritigenin isoliquiritigenin (Bhardwaj et al. 1976b), liqcoumarin with the assigned structure 6-acetyl-5-hydroxy-4-methylcoumarin (Bhardwaj et al. 1976a); and an isoflavone glyzarin (2-methyl-7-hydroxy-8-acetyl-isoflavone) (Bhardwaj et al. 1977). From Chinese licorice, Sipei or Xi-bei (Seihoku Kanzo in Japanese) and assigned to *G. glabra* var. *glandulifera* by Hattori et al. (1986), a new flavonol licoflavonol together with known compounds kumatakenin, glycerol and licoricone with the structure 6-γ, γ-dimethylallyl-kaempferol were isolated (Saitoh et al. 1976a). The Chinese licorice *G. uralensis* had been given the Japanese name Tohoku Kanzo (Saitoh et al. 1976a). Isoliquiritin, rhamnoisoliquiritin, liquiritin, liquiritin apioside were isolated from the roots of *Glycyrrhiza glabra* var. *glandulifera* grown in Iran (Afchar et al. 1980).

From the root of *Glycyrrhiza glabra* var. *typica* two new flavonoids were isolated (van Hulle et al. 1971). The flavanone was identified as 7,4′ dihydroxy–flavanone with a glucose–rhamnose moiety at the 4′–position. The other flavonoid was the corresponding chalcone. The structure of a new 3-arylcoumarin, glycerin, isolated from the root of *Glycyrrhiza* sp. (si-pei licorice = Seihoku Kanzo) was determined as 2′, 4′-dihydroxy-5, 7-dimethoxy-6-γ, γ-dimethylallyl-3-arylcoumarin (Kinoshita et al. 1978). Two new flavanone glycosides, liquiritigenin 4′-apiosyl(1 → 2)-glucoside and liquiritigenin 7,4′-diglucoside together with a known flavone, apigenin 6,8-di-*C*-glucoside, were isolated from licorice (Yahara and Nishioka 1984). The following isoflavonoids and related substances glabridin (1), glabrol (2), glabrene (3), 3-hydroxyglabrol (4), 4′-*O*-methylglabridin (5), 3′-methoxyglabridin (6), formononetin (7), phaseollinisoflavan (8), hispaglabridin A (9), hispaglabridin B (13), salicylic acid and O-acetyl salicylic acid were isolated from *Glycyrrhiza glabra* var. *typica* (Mitscher et al. 1980). Two compounds 9,12,13-trihydroxy-(10*E*)-octadecenoic and 9,12,13-trihydroxy-10,11-epoxy-octadecanoic acid were isolated from licorice (Panossian et al. 1988). A new prenylated isoflavan derivative, kanzonol R, was isolated from *G. glabra* (Fukai et al. 1994). Two new pyrano-2-arylbenzofuran derivatives named glabrocoumarones A and B were isolated from commercially available licorice of *Glycyrrhiza glabra* origin, and their structures were elucidated as 4′-6-dihydroxy-[6″, 6″-dimethylpyrano(2″,3″:2′,3′)]-2-arylbenzofuran and 2′, 6-dihydroxy-[6″, 6″-dimethylpyrano (2″,3″:4′,3′)]-2-aryl-benzofuran, respectively (Kinoshita et al. 1996b). Six known compounds were also obtained and identified as glabrol, 3-hydroxygrabrol, shinflavanone, [6″, 6″-dimethylpyrano(2″,3″:7,8)]-[6″,6″-dimethylpyrano (2″,3″:4′,3′)]-flavanone (xambioona), 3, 3′-di-γ, γ-dimethylallyl-2′-4,4′-trihydroxychalcone and [6″,6″-dimethylpyrano(2″, 3″:4,5)]-3′-γ,γ-dimethylallyl-2′,3,4′-trihydroxychalcone. A new isoflavan 8-prenyl-phaseollinisoflavan was isolated from *Glycyrrhiza glabra* root, together with five known isoflavans identified as glabridin, 4′-*O*-methylglabridin, hispaglabridins A, B and 3′-hydroxy-4′-*O*-methylglabridin (Kinoshita et al. 1996a). 2, 2′, 4′-

Trihydroxychalcone was reported from *G. glabra* (Zhu et al. 2010).

High-performance liquid chromatography (HPLC) profiles of ethyl acetate extract of underground parts *G. glabra* : Type A glabrene, glabridin; 3,4-dihydroglabridin; 3-hydroxyglabrol; glabrol, 4′-*O*-methylglabridin; shinflavanone, hispaglabradin A and hispaglabridin B; and Type B glabrene, parvisoflavone B, glabridin; 3,4-dihydroglabridin; 3-hydroxyglabrol; glabrol, 4′-*O*-methylglabridin; shinflavanone, hispaglabradin A and hispaglabridin B (Kusano et al. 2003). Two new prenylated isoflavanones were isolated from licorice roots along with the known compounds cetoleic acid, β-sitosterol, stigmasterol, lanast-5,24-dien-3 β-D-glucuronopyranoside and glucuronic acid (Suman et al. 2009). The structures of the prenylated isoflavanones were established as 8-isoprenyl-7,4′-dihydroxylicoisoflavanone (glabraisoflavanone A) and 7,3′-dihydroxy-8-isoprenyl-4′-cyclogeranioloxyisoflavanone (glabraisoflavanone B). The following flavonoids were isolated from the ethyl acetate extract of licorice root: glabridin, the principal component, 4′-*O*-methylglabridin, hispaglabridin B, the isoflavene glabrin, unsilylated glabridin and monotrimethylsilyl derivative of 4′-*O*-methylglabridin (Denisova et al. 2006). Twelve flavonoids, hispaglabridin A, glabrol, 4′-*O*-methoxy glabridin, glabridin, 4′,7-dihydroxy flavones, 7-hydroxy-4′-methoxy flavones, isoliquiritigenin, 3,3′,4,4′-tetrahydroxy-2-methoxychalcone, liquiritigenin, licuroside, isoliquiritoside and isoononin were isolated from *Glycyrrhiza glabra* roots (Birari et al. 2011). The ethyl acetate extract of the rhizome afforded 7-hydroxycoumarin (umbelliferone) (Kaur et al. 2012a).

One hundred and twenty six compounds including flavonoids, terpenoids, saponins, essential oils, amino acids, other nitrogen containing compounds, hydrocarbons, fatty acids and their esters were found in the ethanolic extract of licorice (*G. glabra*) root (Vijayalakshmi and Shourie 2013). The most abundant phytoconstituents identified were: 5-(hydroxymethyl)-2-furancarboxaldehyde (23.15 %), *N*-methyl-4-(4-methyl-1-phthalazinylamino)-benzamide (7.18 %), 2-phenyl-furo[b]benzopyran-4(4H)-one (5.69 %), 1, 2-benzenedicarboxylic acid (5.31 %), 4H-pyan-4-one, 2,6-dimethyl- (4.48 %), dicycloctanopyridazine (2.7 %), 4-methyl-herniarin (2.67 %), cyclotetracosane (2.28 %), glabridin (2.02 %), tetracosan-1-ol (1.95 %), 5,6,6a,6b,7,12,12a,12b-dodecahydrodicyclopent(a,c)-anthracene-7 (1.82 %), cycolongifolene, 9,10-dehydro- (1.08 %), cycloserine (1.08 %) and 3,4-furandimethanol (1.01 %). Other compounds (>1 %) included: 4,5-dimethyl-1,3-dioxol-2-one; 4-hydroxy-*N*-methylpiperidine; 2,5-dimethyl-3(2*H*)furanone; 2,4,5-trimethyl-1,3-dioxolane; *N*, 1-dimethyl-4-piperidinamine; glycerine monoacetate; 3-isopropoxy alanine; 2,8,4,6 (epoxyethanediylidenoxy)[1,3]dioxino [5,4-d]-1,3-dioxin; 2-methoxy-4-vinylphenol; *cis*-dimethyl morpholine; 2-chloro-4-formyl-6-methoxyphenyl 4-morpholine carboxylate; 3-methyl-3-[2-oxopropyl]amino-2-butanone; *N*-methyl- 3-hdroxymethylpyrrolidin-2-one; sarreroside, 6-methoxy-8-methyl-8-azabicyclo [3.2.1]octan-3-ol; 5-ethylfuran-2-carboxylic acid; 1-dodecanol; 4-allyl-4-hydroxyprolin; 3,7-dimethylimidazo[1,2-a]pyrimidine-2,5(1H,3H)-dione; 3,5-ditert-butylphenol; dodecanoic acid; acetic acid, (2-isopropenylcycl opentylidene)-, methyl ester; 7-methoxy-2-benzofuranyl methyl ketone; isonicotinic acid, 2-tetrahydrofurylmethyl ester; (3*Z*)-3-ethyl-2-methyl-1,3-hexadiene; 1-(4-amino-furazan-3-yl)-5-pyrrolidin-1-ylmethyl-1H-[1,2,3]triazole-4-carboxylic acid ethyl ester; 6-methoxy-2-hydroxyquinoxaline-4-oxide; 2-furan propionic acid; β-methyl-4-methoxycinnamic acid; 2-pyridine carboxylic acid; benzoic acid; tetradecanoic acid, 4-((1*E*)-3-hydroxy-1-propenyl)-2-methoxyphenol; 4-(7-methoxy-3,3,7-trimethyl-oxepan-2-ylidene)-butan-2-one; linalool; stypticin; 2,6-nonadienoic acid; eicosanoic acid, 1-tert-butyl-2-methoxy-4-methyl-3,5-dinitorbenzene; *o*-nitrocumene,2-isopropylnitrobenzene; 1,2-benzenedicarboxylic acid; salicylic acid; pseduosarsasapogenin-5,20-dien; 3-doxy-D-mannonic acid; ascorbic acid; mome inositol; benzenepropanoic acid, 2,5-dimethoxy-; hexadecanoic acid, 1-methylethyl ester; 9-octadecadienoic

acid, methyl ester; 9-octadecanoic acid; octadecanoic acid, methyl ester; linoleic acid, methyl ester; 22-tricosanoic acid; hystrene S-97; *N*-[3-[6-hydroxyhexyl]aminopropyl]aziridine; hydrazine carboxaldehyde, 2,2-diethyl-, diethyl hydrazone; undecanoic acid, 11-amino-; 1-phenanthrenecarboxylic acid; 5α-andro-7-ene; tetrahydropyran-4-carboxylic acid; 2-hydroxybenzoic acid (1-methyl-2-oxo-1,2-dihydroindol-3-ylidene)-hydrazide; *N*-(4-hydroxybutyl) phthalimide; 2-methyl-6-phenyl-2,3,4,5-tetrahydro-3-pyridazinone;4-(4-methoxyphenyl)-6-phenyl-2-pyrimidinol; 3(2H)-benzofuranone-3-18ol,6-methoxy-2-(3-phenyl-2-propenylidene); 1,2,3,4-tetrahydroisoquinolin; cis-9-tricosane;2-phenyl-furo[b]benzopyran-4(4H)-one-flavone; *N*-(2,4-dimethyl-phenyl)-3-oxobutyramide; 2,3-dihydro-7-methoxy-8-(3-methylbut-2-enyl)-2-phenyl-4H-1-benzopyran-4-one; licoisoflavone B; β-methylumbelliferone (hymecromone); 7-acetoxy-4-methylcoumarin; 7-hydroxy-8-(γ,γ-dimethylally); benzene, 1,3,5-trimethyl-2-(2-phenylethenyl)-,(*Z*)-;4-(3,4-dimethoxyphenyl)-6-phenyl-2-pyrimidinol; 5,5-dimethyl-4-phenylcarbonyl-1,3,4-oxadoazoline; 3-quinoline carboxylic acid; 1-hexadecanesulfonyl chloride; 7-(ethylamino)-4,6-dimethyl-2H-chromenone; squalene, 2,6,10,14,18,22-tetracosahexaene; isocordoin; pyrimidine-2,4(1H,3H)-dione, 6-amino-1-(2-methylphenyl)-3-(2-phenylethyl)-; 4H,8H-benzo(1,2b:3,4b′)dipyran-4-one; 1-(4-[6-(4-acetylphenyl)hexyl]phenyl)ethanone; 2-(5-oxo-1,5-diphenyl-3-p-tolyl-pent-2-enylidene)-indan-1-one; 1-triacontanol; ethanone, 1-(4-cyclohexylphenyl)-, 4-cyclohexylacetophenone; liquiritigenin; coumarine, 8-allyl-7-hydroxy-6-ethy l-4-methyl-; 1,3-dioxolo[4,5-g]isoquinolin-5-ol,5,6,7,8-tetrahydro-6-methyl-; glycyrrhiza chalcone (licochalcone A); quinazolin-4(1H)-one,2,3-dihydro-2-methyl-3-(4-dimethylaminobenzylidenamino)-; cholest-5-en-3-ol(3β)-; olean-12-ene-28 al; dihydrocoumarin, 5,7,8-trimethyl-; 2H,8H-benzol [1,2-b:5,4-b′]dipyran-2-one; β-sitosterol; 3,4-heptadien-2-one, 3,5-dicyclopentyl-6-methyl-; 2,2-dimethyl-7-hydroxy-6-(2,4-dimethoxycinnamoyl)chromene; stigmasterol; 3-[2-(2-chloro-benzyloxy)-ethyl]-1H quinazolin 2H-dione; 2H,8H-benzo[1,2-b:5,4-b′]dipyran-2-one,4-hydroxy-5-methoxy-3-(4-methoxyphenyl)-8,8-dimethyl-; octadeamethyl-cyclononasiloxane; 4-(7-8- dihydro-tetrazolo[1,5-b][1,2,4]triazin-7-yl)-2,6-dimethyl-phenol; 2-tert-butyl-4,6-*bis*(3,5-di-tert-butyl-4-hydroxybenzyl)phenol; 7-methyl-1,2,3,4,4a,9a-hexahydro-9H-fluoren-9-one.

Glycyrol, glycyrin, isoglycyrol and glycycoumarin were isolated from the methanolic extract of licorice roots from northwest China, *G. glabra* var. *glandulifera* (Hattori et al. 1986). A new 2-arylbenzofuran derivative named licocoumarone with the structure 2-(2, 4-dihydroxyphenyl)-6-hydroxy-4-methoxy-5-(3-methyl-2-butenyl) coumarone was isolated from commercially available xibei licorice (seihoku kanzo) along with a known 3-arylcoumarin derivative, glycycoumarin (Demizu et al. 1988). An anti-HIV (human immunodeficiency virus) phenolic constituent, licopyranocoumarin, and two other new phenolics named licoarylcoumarin and glisoflavone together with glycyrrhisoflavanone, kaempferol 3-*O*-methyl ether and licocoumarone were isolated from Si-pei licorice (a commercial licorice; root and stolon of *Glycyrrhiza* sp. from the north-western region of China) (Hatano et al. 1989). Licuraside was isolated from *G. glabra* and on hydrolysis afforded liquiritigenin and isoliquiritigenin (Gorecki et al. 1991). From *G. glabra* (Xinjiang) roots the following were isolated five flavonoids and four coumarins, namely, liquiritin, isoliquiritin, liquiritigenin, isoliquiritigenin, glycycoumarin, isoglycycoumarin, licochalcone A, glycyrol and isoglycyrol (Zeng et al. 1990); glycyrrhizic acid, isoliquiritin, liquiritin, liquiritigenin, glycoumarin, isoglycoumarin and uralsaponin B (Zeng et al. 1991a); and three saponins glycyrrhizic acid (S-I), uralsaponin B and uralsaponin A (Zeng et al. 1991b). Five new flavonoid compounds named glucoliquiritin apioside (a flavonone bisdesmoside), prenyllicoflavone A (a bisprenylflavone), shinflavone (a prenylated pyranoflavanone), shinpterocarpin and 1-methoxyphaseollin (both pyranopterocarpans) were isolated together with eight known saponins (glycyrrhizin, licorice-saponins A3, C2, E2, G2, H2, apioglycyrrhizin and araboglycyrrhi-

zin), and seven known flavonoid glycosides (ononin, liquiritin, liquiritin apioside, isoliquiritin, neoisoliquiritin, luciraside and isoliquiritin apioside) from the aqueous fraction of the methanol extract of *G. glabra* dried roots ('Shinkyo-Kanzo' in Japanese) collected in Xinjiang province, China (Kitagawa et al. 1994). The ethyl acetate fraction yielded three known chalcones (licochalcone A and B, and echinatin), two new prenylated flavonoids (prenyllicoflavone A and shinflavanone) together with two known prenylated flavonoids (glabrol, licoflavone A), three known isoflavonoids (hispaglabridin A and B, and methylhispaglabridin B), two new pyranopterocarpans (shinpterocarpin and 1-methoxyphaseollin) and two known pterocarpans (medicarpin and *ent*-(−)-hemileiocarpin). A new prenylated isoflavan derivative, kanzonol R, was isolated from *G. glabra* (Fukai et al. 1994). A large amount of glabridin, a prenylated flavonoid, was detected exclusively in the cork layer and the decayed part of the thickening roots while large amounts of flavonoid glycosides, liquiritigenin glycosides and isoliquiritigenin glycosides were mainly distributed in the woody part of the thickening roots (Hayashi et al. 1996b). The isoflavans hispaglabridin A, hispaglabridin B, glabridin, and 4′-*O*-methylglabridin, the two chalcones, isoprenylchalcone derivative and isoliquiritigenin, and the isoflavone, formononetin were isolated from *G. glabra* roots (Vaya et al. 1997). Two new 3-arylcoumarin derivatives were isolated from *Glycyrrhiza glabra* root and their structures were elucidated as 2′,4′-dihydroxy-[6″, 6″-dimethylpyrano(2″,3″:7,8)]-3-arylcoumarin and [6′,6″-dimethylpyrano(2″,3″:7,8)]-2′-hydroxy-4′-methoxy-3-arylcoumarin (Kinoshita et al. 1997). Two known isoflavonoids glabrene and glabrone were also isolated. From the ether soluble fraction of the crude licorice root, licoricidin, 1-methoxyphaseollin, 6,8-diprenylgenistein and 1-methoxyphaseollidin were isolated (Nagumo et al. 1999). Isoflavan derivatives, glabridin, hispaglabridin A, hispaglabridin B, 4′-*O*-methylglabridin and 3′-hydroxy-4′-*O*-methylglabridin were isolated from *Glycyrrhiza glabra* (Haraguchi et al. 2000). Yuldashev et al. (2000b) isolated the following flavonoids from above ground parts of *G. glabra* grown in Uzbekistan: 7-*O*-methylglabranin, glabranin (8-C-prenylpinocembrin) (5,7-dihydroxy-8-(γ,γ-dimethylallyl)-flavanone), pinocembrin, galangin and a new isoflavanoid glabrisoflavone with the structure (*E*)-5,7,3′-trihydroxy-6-(3-hydroxymethyl-2-butenyl)-isoflavone.

The main phenolic compounds of licorice were reported as glycosides of liquiritigenin (4′,7-dihydroxyflavanone) and isoliquiritigenin (2′,4,4′-trihydroxychalcone), e.g., liquiritin, isoliquiritin, liquiritin apioside, etc. (Nomura et al. 2002). Minor phenolic compounds comprising many isoprenoid-substituted flavonoids, chromenes, coumarins, dihydrostilbenes and dihydrophenanthrenes were isolated from *Glycyrrhiza* species. Ninety phenolic compounds were isolated from *G. glabra*. Fifty were substituted isoprenoid groups, for example, in Type I licorice Spanish and Russian, the main isoprenoid-substituted flavonoid was a pyronoisoflavan, glabridin. The 5-position of most flavonoids was unsubstituted, for example, glabrene, glabrol and 3-hyrdoxyglabrol. Type 2 licorice comprising Chinese and Kirghiz *G. glabra*, both 5-unsubstituted flavonoids and 5-oxygenated flavonoids, for example, 3,8′-diprenylated dalbergioidin was isolated. Most flavonoids were 5-hyrdoxy-flavonoids or 5-methoxy-flavonoids. The main isoprenoid-substituted flavonoid of Kirghiz licorice was 3,8′-diprenylated dalbergioidin but glabridin had not been isolated. Fourteen flavonoids were isolated and purified from the benzene extract (33 g) of the roots of Kirghiz licorice (*G. glabra*): 3′,8-diprenyldalbergioidin (630 mg) 3′,6-diprenyldalbergioidin (67 mg), licoisoflavanone (19 mg), glyasperin A (25 mg), glyasperin C (20 mg), glyasperin D (49 mg), isoderone (1 mg), semilicoisoflavone B (3 mg), 8-(γ,γ-dimethylallyl)-wighteone (13 mg), gancaonin G (2 mg), gancaonin H (27 mg), 1-methoxyphaseollidin (53 mg), edudiol (3,9-dihydroxy-1-methoxy-2-prenypterocarpan) (12 mg) and glabrene (5 mg), 3′(γ,γ-dimethylallyl) kievitone, glisoflavanone (3′,6-diprenyl-2,4′,5,7-tetrahydroxyisoflavanone), isoderone (2,2-dimethylpyranol [4′,3′]-5,7-dihydroxyisofla

vone), 1-methoxyphaseollidin (Fukai et al. 2001). From European *G. glabra* cultivated in Japan the following flavonoids were isolated: glabrene, glabridin, 4′-*O*-methylglabridin, hispaglabridin A (3′-prenylglabridin), glabrol, 3-hydroxyglabrol, glabrone, medicarpin (3-hydroxy-9-methoxypterocarpan), shinpterocarpin, euchrenone a_5, glyinflanin K, glyinflanin G, two pyrano-2-arylbenzofurans, kanzonols U and V, a pyrano-3-arylcoumarin, kanzonol W, a diprenylated isoflavan, kanzonol X, a diprenylated α-hydroxydihydrochalcone kanzol Y ((*αR*)-3,5′-diprenyl-*α*,2′,4,4′-tetrahydroxydihydrochalcone), kanzol Z (prenylated 3-hydroxypyranoflavanone) and 3-hydroxyparatocarpin (Fukai et al. 1996a, 1998). A new isoprenoid-substituted isoflavone, kanzonol T, was isolated from Chinese licorice, *Glycyrrhiza glabra*, along with eight known flavonoids (Fukai et al. 1996b). The structure of glabrene was revised. Seven constituents, isolated from *Glycyrrhiza glabra*, were identified as the isoflavans hispaglabridin A, hispaglabridin B, glabridin and 4′-*O*-methylglabridin, the two chalcones, isoprenylchalcone derivative and isoliquiritigenin, and the isoflavone, formononetin (Vaya et al. 1997). Two new isoflavones named glabroisoflavones A and B, and a 3-arylcoumarin derivative glabrocurmarin and a known isoflavene derivative 3,4-didehdyroglabridin were isolated from the dichloromethane extract of commercially available licorice root (Kinoshita et al. 2005). Licochalcone-A, a novel flavonoid, was isolated from licorice root (*Glycyrrhiza glabra*) (Fu et al. 2004). Licocalchone-C was isolated from *Glycyrrhiza glabra* (Franceschelli et al. 2011). Two new flavonosides were isolated from *Glycyrrhiza glabra* roots and identified as 5,8-dihydroxy-flavone-7-*O*-β-D-glucuronide, glychionide A and 5-hydroxy-8-methoxyl-flavone-7-*O*-β-D-glucuronide, glychionide B (Li et al. 2005).

Licorice root was found to contain an estrogenic hormone in appreciable quantity (Costello and Lynn 1950). Eight phytoestrogenic compounds were found and assessed from the roots of *G. glabra* from Syria: daidzein (4′,7-dihydroxyisoflavone), daidzin (diadzein-7-glucoside), genistein (4′,5,7-trihydroisoflavone), genistin (genistein-7-glucoside), formononetin (7-hydroxy-4′-methoxyisoflavone), ononin (formononetin-7-glucoside), glycitein (4′,7-dihydroxy-6-methoxyisoflavone) and coumestrol (Khalaf et al. 2010). Phytosterols β-sitosterol, stigmasterol, campesterol and ergosterol were isolated from *G. glabra* roots from Syria (Khalaf et al. 2011).

Glycyrrhizin, a major bioactive compound in liquorice root, had been reported to be 50 times sweeter than sugar (Nitalikar et al. 2010). Licorice extracts had been widely used in pharmaceutical and confectionery industries because of the presence of glycyrrhizin. Glycyrrhizin was postulated to be most likely derived from the triterpene β-amyrin, an initial product of the cyclization of 2,3-oxidosqualene (Seki et al. 2008). *CYP88D6*, a cytochrome P450 monooxygenase (P450) gene, was successfully identified as a glycyrrhizin-biosynthetic gene. *CYP88D6* was characterized by in-vitro enzymatic activity assays and shown to catalyse the sequential two-step oxidation of β-amyrin at C-11 to produce 11-oxo-β-amyrin, a possible biosynthetic intermediate between β-amyrin and glycyrrhizin. *CYP88D6* expression was detected in the roots and stolons but not in the stem and leaves. A second relevant P450 (CYP72A154) was identified and shown to be responsible for C-30 oxidation in the glycyrrhizin pathway (Seki et al. 2011). CYP72A154 expressed in an engineered yeast strain that endogenously produced 11-oxo-β-amyrin (a possible biosynthetic intermediate between β-amyrin and glycyrrhizin) catalysed three sequential oxidation steps at C-30 of 11-oxo-β-amyrin supplied in situ to produce glycyrrhetinic acid, a glycyrrhizin aglycone.

The ethanol licorice (*G. glabra*) root extract afforded the following phenolic compounds: 5′-formylglabridin; (2*R*,3*R*)-3,4′,7-trihydroxy-3′-prenylflavane; echinatin; (3*R*)-2′,3′,7-trihydroxy-4′-methoxyisoflavan; kanzonol X; kanzonol W; shinpterocarpin; licoflavanone A; glabrol; shinflavanone; gancaonin L; glabrone; licochalcone B; morachalcone A; 2′3′,4′-trihydroxy- 3′γ,γ- dimethylallyl- 6″, 6″- dimethylpyrano[2″,3″:4,5]chalcone;1-(2′,4′-dihydroxphenyl)-

2-hydroxy-3-(4″-hydroxyphenyl)-1-propanone; kanzonol Y; (3*R*)- vestitol; glabridin; 4′-*O*-methylglabridin; 3′-hydroxyl-4′-*O*-methylgalbridin; hispaglabridin A; hispaglabridin B; glabrene; kanzonol W; kanzonol U; *O*-methylshinpterocarpin; licoagrocarpin; xambioona; 8′,8-dimethyl-3,4-dihydro-2*H*,8*H*-pyrano [2,3-*f*]chromone-3-ol; 3,3′,4,4′-tetrahydroxy-2′-methoxy-5- prenylchalcone; 3′,4,4′-tetrahydroxy- 3,5′-diprenylchalcone; 2′,3,4,4′, α-pentahydroxy- 3′,5-diprenyl-dihydrochalcone; 5′-formyl glabridin; (2*R*,3*R*)-3,4′,7-trihydroxy-3′-prenylflavanone; and 7,8-dihydroxy-4′-methoxy-6-prenylisoflavanone8-hydroxymethyl-8-methyl-3,4-dihydro-2*H*,8*H*-pyrano[2,3-*f*]-chromon-3-ol (Kuroda et al. 2010).

The essential oil yield from *G. glabra* root was 0.047 % by steam distillation, and 82 components of the volatile compounds were identified (Kameoka and Nakai 1987). The predominant constituent was hexanoic acid (31.57 %), followed by other components hexadecanoic acid (3.3 %), heptanoic acid (2.54 %), hexanol (1.7 %), octanoic acid (1.44 %), γ-nonalactone (1.33 %), 4-methyl-1-isopropyl-3-cyclohexen-1-ol (1.30 %) and *N*-methyl-2-pyrrolidone (1.07 %). Other compounds (<1 %) included methyl hexanoate, 2,3-dihydro-4-methyl-furan, 1-methoxy-4-isopropyl cyclohexane, 2-hexanal, 2-pentylfuran, 2-ethyl-1,4-dimethyl benzene, 3-methoxy-2-methyl propane, tridecane, 6-methyl-5-hepten-2-one, hexyl formate, camphor, 6-methyl-3-undecene, 3-methyl-hepten-2-one, tetradecane, acetic acid, furfural, 5-methyl-2-undecene, 9-methyl-3-undecene, linalool, 3,5-octadien -2-one, dihydro-5,5-dimethyl-2(3*H*)-furanone, myrtenal, 6-methyl-1-isopropyl-3-cyclohexen-1-ol, butanoic acid, estragole (methyl chavicol), 2-methyl-6-methylen-7-octen-2-ol, pentanoic acid, 2-methyl-5-isopropyl-2-cyclohexen-1-one, *o*-cresol, 2-methalyl phenol, 3-methyl-6-propryl phenol, anethole, cumic alcohol, pseudoionone, phenethyl alcohol, 5-pentylpyran-2-one, *o*-tolunitrile, 2-methyl-3-decen-5-one, 7-methoxy-3,7-dimethyl-octanal, 1-pentadecanol, 2-hydroxy-4-methyl benzaldehyde, 1-methoxy-4-isopropyl benzene, isobutyl adipate, nonanoic acid, eugenol, 2-methyl-5-isopropyl phenol, methyl hexadecanoate, decanoic acid, hexadecyl acetate, 2,3-dihydro benzofuran, indole dodecanoic acid and pentacosane. Those found in traces were dodecane decane, undecane, 2-methyl propanoic acid, pentadecane, phenyl acetaldehyde, acetophenone, heptadecane, octadecane, guaiacol, nonadecane, eicosane, heneicosane, docosane, tricosane, undecanoic acid, tridecanoic acid, tetradecanoic acid and pentadecanoic acid. Volatile compounds (%) found in *G. glabra* root essential samples from Egypt, Afghanistan and Syria were, respectively, 5-methyl furfural (3.6, 10.1, 9.4 %), *o*-guaiacol (2.2 %, tr, tr), 2-phenylethanol (0.5, 2.1 %, tr), *Z*-pinene hydrate (tr, 2.1 %, -), tetrahydro-lavandulol (-, 7.6, 4.1 %), terpinene-4-ol (tr, -, 3.6 %), (*E*)-linalool oxide (2.1 %, -, -), *p*-cymen-8-ol (-, 2.7, 3.0 %), α-terpineol (tr, 0, 2.4 %), methyl chavicol (-, - 2.4 %), (4*E*)-decenal (5.3, 2.8, 5.4 %), cumin aldehyde (4.7, 1.8, 3.3 %), carvone (2.1, 0.2, 3.1 %), piperitone (9.4, 13.1, 7.2 %), (*E*)-cinnamaldehyde (3.6, 4.5, 6.2 %), (*E*)-anethole (1.3 %, - , -), thymol (27.2, 6.0, 5.5 %), indole (-, 7.4, 1.8 %), carvacrol (11.1, 1.4, 5.8 %), *p*-vinyl-guaiacol (8.5, 8.5, 9.5 %), unknown aldehyde (-, 3.7, 5.7 %), eugenol (9.4, 7.5, 8.8 %), γ-nonalactone (1.8, 2.5, 7.4 %), methyl eugenol (3.5, 0.2, 3.8 %), β-caryophyllene (0.5, 0.5, 1.1 %), himachalene epoxide (-, 8.8, 1.6 %) and β-caryophyllene oxide (1.1, -, 1, 1 %) (Farag and Wessjohann 2012). Total phenols (23.4–58.5 %) were most dominant in *G. glabra* followed by aldehydes (17.2–305), ketones (10.3–13.3 %), ether/epoxides (6.5–11.4 %) and alcohols (0.5–9.2 %). Two phenols, thymol and carvacrol, were found exclusively in essential oil and headspace samples of *G. glabra*, and with highest amounts for samples that originated from Egypt. Principal component and hierarchical cluster analyses clearly separated *G. echinata* and *G. inflata* from *G. glabr*a, with phenolics and aliphatic aldehydes contributing mostly for species segregation.

Compounds (and aglycone class) isolated from *G. glab*ra methanol root extract by LC-MS included: rhamanoliquiritin, isoviolanthin, liquiritin apioside, liquiritin, choerospondin, 5,7-dihydroxyflavone, licorice D2/D1,

3-hydroxyglabrol, glabrol (flavanone); neolicuroside, neolicuroside isomer, isoliquiritin, licochalcone B, licorice glycoside A, isoliquiritigenin (chalcone); glycyrrhizin, 22-acetoxyglyccyrrhizin, licorice saponin A3, licorice saponin G2, licorice saponin J2, glycyrrhizin isomer, licorice saponin C2, 11-deoxoglycyrrhetinic, glycyrrhetic acid (triterpene), glabrene (isoflavene), glabridin (isoflavan), kanzonol Y (dihydrochalcone) and kanzonol F (pterocarpan) (Farag et al. 2012). Chemical shift of constituents identified in *G. glabra* root included: glycyrrhizin, sucrose, liquiritin, isoliquiritin, liquiritigenin, iso liquiritigenin, 4-hydroxyphenyl acetic acid, fatty acids, licochalcone A and rhamnose (glycosides). Three new oleanane-type triterpene saponins, namely, licorice-saponin M3, licorice-saponin N4 and licorice-saponin O4, an artificial product, namely, ester of licorice-saponin G2, as well as five known triterpene glucuronides, namely, 24-hydroxylicorice saponin A3, licorice saponin G2, 22B-acetoxyl-glycyrrhizin, licorice saponin A3 and glycyrrhizin were isolated from *G. glabra* roots (Wei et al. 2014). Dibenzoylmethane (DBM; 1, 3-diphenyl-1, 3-propadinedione), a beta-diketone analogue of curcumin, had been reported to be constituent of licorice (Jackson et al. 2002; Thimmulappa et al. 2008; Khor et al. 2009; Liao et al. 2015) and was confirmed by Mancia et al. (2014) to be a natural constituent of Licorice root (*G. glabra*).

Based on nucleotide sequences of ribulose-1,5-bisphosphate carboxylase/oxygenase large subunit gene (rbcL) sequences, six *Glycyrrhiza* species were divided into two groups: three, *G. glabra*, *G. uralensis* and *G. inflata*, producing glycyrrhizin as a major saponin, and the others, *G. echinata*, *G. macedonica* and *G. pallidiflora*, producing macedonoside C as a major saponin (Hayashi et al. 2000b). Among the three glycyrrhizin-producing species, only two nucleotide substitutions were observed between the rbcL sequences of *G. glabra* and *G. uralensis*, and the sequence of *G. uralensis* was identical to that of *G. inflata*, indicating *G. uralensis* and *G. inflata* to be closely related. Regarding six main constituents of licorice, glycyrrhizin, liquiritin, liquiritin apioside, isoliquiritin, isoliquiritin apioside and liquiritigenin, the constituent property of *G. glabra* was similar to that of *G. inflata*; both were dissimilar to *G. uralensis* which was characterized by a wide content variation of the six constituents compared to those of *G. glabra* and/or *G. inflata* (Kondo et al. 2007). The mean contents of liquiritin, isoliquiritin or liquiritigenin in *G. uralensis* were significantly higher than those of *G. glabra* or *G. inflata*. Additionally, glycycoumarin, glabridin and licochalcone A were reconfirmed as the species-specific typical constituents of *G. uralensis*, *G. glabra* and *G. inflata*, respectively. Hatano et al. (1991a) found that the root and rhizome of *G. glabra*, and the licorice specimens imported from the Soviet Union and Afghanistan (Type B) contained glabridin and glabrene; the roots and rhizomes of *G. uralensis*, commercial licorice specimens from the north-western region of China (Seihoku-kanzo) and from the north-eastern region of China (Tohoku-kanzo) in Japanese markets, and also several licorice specimens from Chinese markets (Type A) contained licopyranocoumarin, glycycoumarin and/or licocoumarone, which were not found in *G. glabra* and *G. inflata*. Root sample of *G. inflata* (Type C) contained licochalcones A and B, which were not found in the samples of the other two *Glycyrrhiza* species. Extracts of some licorice specimens of types A and B, and all of the licorice specimens of type C inhibited 40–56 % of the xanthine oxidase activity at the concentration of 30 μg/ml. Extracts of some licorice specimens of types A and B also showed inhibitory effects on monoamine oxidase (44–64 % inhibition, at the concentration of 30 μg/ml), which were slightly weaker than that of harmane hydrochloride. Hatano et al. (1991b) isolated two new phenolic compounds glicoricone and licofuranone together with echinatin, genistein and licopyranocoumarin from a licorice species imported from the north-western region of China. Chinese *G. glabra* root was found to contain the highest levels of glycyrrhizic acid, followed by those from Italy (Calabria) (Montoro et al. 2011). *G. uralensis* was found to contain quercetin but *G. glabra* did not (Liao et al. 2012). Fifteen bioactive components were found in both

species. *G. glabra* provided by Brion Research Institute of Taiwan contained uralsaponin B, but *G. glabra* brought from herbal shops did not. The following mono- and di-prenylated flavonoids were extracted from the 70 % aqueous- ethanol, ethanol and ethyl acetate extracts of *Glycyrrhiza glabra* roots (Simons et al. 2009): liquiritin apioside, liquiritin, glabrol, 3-hydroxy-glabrol, isoliquiritin apioside, lichochalone B, glycyrrhizinic acid, glabrene, formononetin, glabrone, glabridin, 3′-hydroxy-4′-*O*-methylglabridin, 4′*O*-methylglabridin, hispaglabridin A and hispaglabridin B. Khalaf et al. (2012) found phenol carboxylic acids were present in larger quantities than flavonoids in *G. glabra* tincture. *p*-Coumaric acid, ferulic acid, luteolin and apigenin were found and quantified in both hydrolysed and unhydrolysed samples. Kaempferol, fisetin, myricetin, hyperoside, quercitrin, isoquercitrin and rutoside could not be found in the analysed samples. Regarding the phytoestrogens content, significant quantities of ononin and its aglycone formononetin were detected. Among sterols, the largest amount recovered in the tincture was β-sitosterol and the smallest ergosterol.

New esters of 18α-glycyrrhizic acid (18α-GA) namely 18 α-glycyrrhizic acid and its monopotassium salt; 18 α-glycyrrhizic acid pentasulphate sodium disodium salt; 18 α-glycyrrhizic acid di-*O*-nicotinate; 18 α-glycyrrhizic acid penta-*O*-nicotinate and penta-*O*-4-methoxycinnamate ester of 18 α-glycyrrhizic acid were synthesized; these were D/E-trans-isomers of natural 18α-GA, the major triterpene glycoside in roots of Spanish licorice and Urals licorice roots (Baltina et al. 2010). Drought stress was found to enhance the levels of secondary metabolites and key gene expression involved in the biosynthesis of triterpenoid saponins in liquorice (Nasrollahi et al. 2014). Due to osmotic stress, the gene expression levels of squalene synthase (SQS) and β-amyrin synthase (bAS) were increased, whereas those of cycloartenol synthase (CAS) were relatively unchanged at the seedling stage. At the adult plant stage, the expression levels of SQS and bAS were increased under drought stress conditions, whereas the gene expression level of CAS remained relatively constant. The glycyrrhizin content in stolons was increased only under severe drought stress conditions (28 days).

Derivatives were synthesized from isoliquiritigenin, a chalcone and liquiritigenin, a flavonoid found in *G. glabra* rhizomes (Gaur et al. 2014). 4,4′-Diacetoxy-2′-hydroxy chalcone; 2′,4′dimethoxy-4-hydroxychalcone; 4-acetoxy-2′,4′-dimethoxychalcone; 4-benzoyloxy-2′,4′-dimethoxychalcone and 2′,4′-dimethoxychalcone from isoliquiritigenin and liquiritigenin 7,4′-diacetate; liquiritigenin 4′-acetate; liquiritigenin 7,4′-dibenzoate and liquiritigenin-oxime from liquiritigenin.

Glycyrrhizin, 18β-glycyrrhetic acid and 18α-glycyrrhetic acid in liquorice roots were separated and quantified by means of capillary zone electrophoresis (Sabbioni et al. 2005). Linearity was found over the 5–200 and 2.5–100 μg/mL concentration ranges for glycyrrhizin and glycyrrhetic acid, respectively. Glycyrrhizin, glycyrrhetic acid, glabridin, liquiritin and licochalcone A and liquiritin apioside were found in samples of *Glycyrrhiza glabra*, *G. uralensis*, *G. inflata* and commercial licorice from Europe and China by means of capillary-zone electrophoresis (Rauchensteiner et al. 2005). The maximum recovery of mono-ammonium glycyrrhizate from licorice roots was achieved at 110 °C and 5 atm pressure with the ratio of 40 ml/g of 0.01 % (w/v) ammonia solution to powdered feed after 90 min of extraction (Mukhopadhyay and Panja 2008). Under the optimum extraction condition comprising a mixture of ethanol/water (30:70, v/v) and extraction time 60 min under 50 °C, 2.39 mg/g of glycyrrhizic acid and 0.92 mg/g of glabridin were extracted from Chinese licorice and the recoveries were 89.7 % and 72.5 %, respectively (Tian et al. 2008). The simultaneous HPTLC quantification of glycyrrhetinic acid and apigenin from *G. glabra* was 0.65 % and 0.0074 %, respectively (Rathee et al. 2010). The HPLC recoveries of 18β-glycyrrhetinic acid from *G. glabra* were 99.60–102.81 % (Esmaeili et al. 2010).

Highest concentrations of glycyrrhizinic acid were found in the main roots, lower concentrations in the lateral roots (Fuggersberger-Heinz and Franz 1984). The green parts of the plant

were shown to contain no glycyrrhizinic acid. Enzymatic hydrolysis of glycyrrhizinic acid with β-glucuronidase afforded the monoglucuronide of β-glycyrrhetinic acid as intermediate. Biosynthetic studies with licorice roots showed that acetate was specifically incorporated into the aglycone moiety of the triterpene saponin and glucuronic acid mainly in the sugar moiety of the diglucuronide, respectively. Glycyrrhizin was found to be localized exclusively in the woody parts of thickened roots but not in the leaf, seed, stem, rootlets, or root nodules while soyasaponins were detected in all parts of the plants examined, and the contents were higher in the seeds, rootlets and root nodules than in other parts (Hayashi et al. 1988, 1993a, 2004, Hayashi 2009). The contents of soyasaponins were higher in younger parts of a growing stolon, whereas those of glycyrrhizin tended to be higher in the older parts. As the primary roots grew and became thicker, the soyasaponin content tended to decrease, while the glycyrrhizin content increased. On the other hand, betulinic acid was localized to the rootlets, root nodules and the cork layer of thickening roots (Hayashi et al. 1988, 2004). Since both soyasaponins and betulinic acid were produced in the rootlets, root nodules and cultured cells, the triterpenoid metabolism of the cultured licorice cells was similar to that of the rootlets and root nodules, whereas glycyrrhizin was detected exclusively in the thickening root and the stolon (Hayashi et al. 1993a). An inverse relationship was found between the soyasaponins content and the glycyrrhizin content in the growing stolon and in the roots at different stages of growth. The glycyrrhizin content in 1-year-old roots rapidly increased from October to November, whereas the isoliquiritigenin glycoside content increased up to October (Hayashi et al. 1998a). In 3-year-old plants, although the isoliquiritigenin glycoside content rapidly increased from June to July, the glycyrrhizin content did not show any significant increase from May to August. The glycyrrhizin content increased during the senescence of the aerial parts as well as during the early stage of shoot elongation. The incorporation of [^{14}C]mevalonic acid into the glycyrrhizin fraction by the root segments was high in May, June and September, and low in August and winter

Leaf Aerial Plant Parts Phytochemicals

Nomura and Fukai (1998) listed the phenolic compounds found in all parts of *G. glabra* plant and suspension cell/hairy root cultures and callus tissues under the following categories: chalcone, flavanone, flavones, flavonol, isoflavanone, isoflavone, pterocarpan, coumestan, isoflavan, isoflav-3-ene, 2-arylbenzofuran, 3-arylcoumarin and miscellaneous phenolic compounds and also listed the saponins found. Ammosov and Litvinenko (2007) listed>250 phenolic compounds in all parts of plants of the genera *Glycyrrhiza* in their review paper. The classes of phenolic compounds reported in *G. glabra* plants included: phenols (2), hydroxycinnamic acids (5), coumarins (5), chalcones (7), chalcone glycosides (6), flavanones (9), flavan glycosides (6), hydroxyflavanones or flavanonols (2), flavones (4), glycoflavonoids or C-glycosides (4), hydroxyflavones or flavonols (6), flavonol glycosides (8), isoflavanones (3), isoflavones (17), 2-methylisoflavones (4), isoflavone glycosides (1), isoflavans (15), isoflavan-4-enes (isoflavenes) (2), ketoisofla-4-enes (3-arylcoumarins) (2), pterocarpans (7) and benzofurans (2-arylbenzofurans) (3).

Litvinenko (1966) isolated a new flavonoid glycoside named glyphoside with the chemical structure quercetin 3-β-D-glucopyranosyl-2′-acetate, kaempferol-3-*O*-diglucoside, and astragalin (kaempferol-3-*O*-glucoside) monoacetate from the aerial parts of Spanish *G. glabra.* Litvinenko and Kovalev (1967) isolated glycoflavonoids vitexin with the structure 5,7,4′-trihydroxyflavone 8-C-β-D-glucopyranoside (8-*syn*-isomer) and isovitexin (saponaretin) with the structure 5,7,4′-trihydroxyflavone 8-C-β-D-glucopyranoside (8-*anti*-isomer) from the aerial parts. Litvinenko and Nadezhina (1972) isolated *cis*-3-hydroxyflavanone folerogenin from the above ground parts. Kir'yalov et al. (1970) isolated methyl glycyrrhetate, two terpenoid com-

pounds not containing a conjugated keto group and presumed to be homo- and heteroannular dienes and uralenic acid which was identified by Belous et al. (1965) as 18α-glycyrrhetic acid. From the aerial parts, pinocembrin, 6-prenylpinocembrin and naringenin were isolated (Batirov et al. 1986).

An isoflavonoid phytoalexin isolated from *Glycyrrhiza glabra* leaves was characterized as 7,2′-dihydroxy-3′,4′-dimethoxyisoflavan (isomucronulatol) (Ingham 1977). Isomucronulatol was produced on inoculation with *Helminthosporium carbonum*. A new prenylated flavanone named licoflavanone was isolated, together with pinocembrin, from *Glycyrrhiza glabra* var. *typica* leaves (Fukui et al. 1988). Upon treatment with methanolic HCl, it was converted to cyclolicoflavanone. Coumarins xanthotoxin (8-methoxy(furano-7,6:2′,3′)coumarin) and bergapten (5-methoxy(furano-7,6:2′,3′)coumarin) were isolated from *G. glabra* leaves (Saleh et al. 1989). Flavonoids genistein, pinocembrin, prunetin, 6-prenylnaringenin, licoflavanone and wighteone were isolated from the leaves of *Glycyrrhiza glabra* collected on the west coast of Anatolia, whereas lupiwighteone was found only in the leaves of *G. glabra* growing in middle or east Anatolia (Hayashi et al. 1996d). Large amounts of flavonoids such as pinocembrin and licoflavanone were present on the outer surface of the young leaves, whereas isoquercitrin, a common flavonoid glycoside, was detected inside the leaves (Hayashi et al. 1996a). 7-*O*-Methylglabranin, 6-*C*-prenylpinocembrin, glabranin, pinocembrin, galangin and a novel isoflavonoid, (*E*)-5,7,4′-trihydroxy-6-(3-hydroxymethyl-2-butenyl)isoflavone (glabrisoflavone) were isolated from the aerial parts of *Glycyrrhiza glabra* (Yuldashev et al. 2000a). A new flavanonglycoside pinocembroside, 2(S)-7-*O*-*β*-D-glucopyranosyl-5-hydroxyflavanone, was isolated from the aerial part of *Glycyrrhiza glabra* (Yuldashev 2001). Five new prenylated dihydrostilbenes, α,α′-dihydro-3,5,4′-trihydroxy-4,5′-diisopentenylstilbene (1), α,α′-dihydro-3,5,3′,4′-tetrahydroxy-4,5′-diisopentenylstilbene (2), α,α′-dihydro-3,5,4′-trihydroxy-5′-isopentenylstilbene (3), α,α′-dihydro-3,5,3′-trihydroxy-4′-methoxy-5′-isopentenylstilbene (4) and α,α′-dihydro-3,5,3′,4′-tetrahydroxy-5′-isopentenyl stilbene (5), along with four known flavonoids, glabranin (6), pinocembrin, (7), licoflavone (8) and wighteone (9), were isolated from a lipid extract of the leaves of Sicilian *Glycyrrhiza glabra* (Biondi et al. 2003). HPLC profile of methanol extract from aerial parts of *G. glabra* (1–4) samples: *G. glabra*-1 schaftoside, isoquercitrin, astragalin, genistein, gancaonin C and pinocembrin; *G glabra*-2 vicenin 2, isoquercitrin, astragalin, ononin, genistein, gancaonin C, pinocembrin, licoflavanone, lupiwighteone and wighteone; *G. glabra*-3 vicenin 2, schaftoside, isoquercitrin, astragalin, genistein, gancaonin C, pinocembrin, licoflavanone, lupiwighteone and wighteone; *G. glabra*-4 schaftoside, rutin, genistein, gancaonin C, pinocembrin, licoflavanone and wighteone (Kusano et al. 2003).

Licorice leaves of *G. glabra* var. *glandulifera* and *G. glabra* var. *glabra* were found to have the following proximate and monosaccharide compositions (% w/w), respectively: ash 10.3, 7 %; total lipid 10, 7.3 %; total carbohydrate 24.6, 11.8 %; protein 24.1, 19.8 %; monosaccharide – glucose 20, 50.4 %; galactose 21.7, 13.2 %; mannose traces 7.4 %; arabinose 15.8, 7 %; xylose 16.5, 6 %; and uronic acid 26, 16 % (Helmy et al. 2013).

Four new dihydrostilbenes – α,α′-dihydro-3,5-dihydroxy-4′-acetoxy-5′-isopentenylstilbene; α,α′-dihydro-3,3′,4′-trihydroxy-5-*O*-isopentenyl-6-isopentenylstilbene; α,α′-dihydro-3,5,3′-trihydroxy-4′-methoxystilbene; and α,α′-dihydro-3,3′-dihydroxy-5β-D-*O*-glucopyranosyloxy-4′-methoxystilbene – together with seven known flavonoids, glabranin isomer, naringenin, lupiwighteone, pinocembrin 7-*O*-glucoside, astragalin, isoquercitrin, vicenin II, and the inositol, pinitol, were isolated from the leaves of *Glycyrrhiza glabra* grown in Sicily (Biondi et al. 2005). Thirty compounds were isolated from *G. glabra* leaf ethyl acetate, *n*-hexane and methanol extracts: lutein, β-carotene; naringenin; α-α′-dihydro-3,5,3′,4′-tetrahydroxy-5′-isopentenylstilbene; α-α′-dihydro-3,5,3′-trihydroxy-4′-methoxystilbene; α-α′-dihydro-3,5,3′-trihydroxy-4′-methoxy -5′-isopentenylstilbene; pinocembrin; α-α′-dihydro-3,5-dihydroxy-4′-acethoxy-5′-

isopentenylstilbene; licoflavanone; α-α′-dihydro-3,5,4′-trihydroxy-5′-isopentenylstilbene; acetoxy derivative of α-α′-dihydro-3,5,4′-trihydroxy-5′-isopentenylstilbene; unknown, α-α′-dihydro-3,5,3′, 4′-tetrahydroxy-4,5′-diisopentenylstilbene; α-α′-dihydro-3,5,4′-trihydroxy-4,5′-diisopentenylstilbene; α-α′-dihydro-3,5,4′-trihydroxy-5-*O*-isopentenyl-6-isopentenylstilbene; glabranin; glabranin isomer; kaempferol dihexoside; apigenin 6,8 di-C-glucoside (vicenin-2); aromadendrin dihexoside; pinocembrin hexoside-deoxyhexoside; apigenin di-C-hexoside-pentoside; quercetin rhamno-glucoside (rutin), quercetin hexoside (glucoside)-pentoside; isoquercitrin; quercetin 3-*O*-glucoside 6′acetate; α-α′-dihydro-3,3′-dihydroxy-5β-D-*O*′-glucopyranosyl-4′-methoxystilbene; kaempferol 3-*O*-glucoside (astragalin); kaempferol 7-*O*-glucoside and pinocembrin 7-*O*-glucoside (Siracusa et al. 2011).

Studies showed that at least three unknown ingredients were detected in rough bark (Cortex Glycyrrhizae) which were not in Fen Gancao (barked licorice root), and glycyrrhizic acid content in the Cortex Glycyrrhizae was higher than that in Fen Gancao (Rong et al. 2006). The results suggested that Cortex Glycyrrhizae could be used as the material not only to extract glycyrrhizic acid but also for making additives. Three oleanane-type monoglycosides along with glycyrrhizin and 3-*O*-[β-D-glucuronopyranosyl-(1 → 2)-β-D-glucuronopyranosyl]-18β-liquiritic acid were isolated from enzymatically hydrolysed licorice extract (EHLE) (Liut et al. 2001). The structures of the three compounds were determined to be 3-*O*-β-D-glucuronopyranosyl-24-hydroxy-18β-glycyrrhetinic acid; 3-*O*-β-D-glucuronopyranosyl-18β-glycyrrhetinic acid and 3-*O*-β-D-glucuronopyranosyl-18β-liquiritic acid. Six major constituents were isolated from enzymatically modified licorice extract: glycyrrhizin (major sweet constituent, 1), 3-*O*-[β-D-glucuronopyranosyl-(1 → 2)-β-D-glucuronopyranosyl] liquiritic acid (minor sweet component 2) and their derivatives glucosylated at the C-4 position of the terminal glucuronopyranose with additional one (compounds 3 and 4, respectively) and two (compounds 5 and 6, respectively) glucose moieties (Liu et al. 2000). Compound 2 was sweeter than compound 1. Compound 3, a monoglucosylated derivative of compound 1, was sweeter than compound 4. Compounds 5 and 6, with two additional glucose moieties, showed only slight sweetness.

Hayashi and co-workers (1988, 1995, 1998b, 2003a, b) conducted field surveys of *G. glabra* in Turkey, Italy, Spain, Kazakhstan and found that it could be divided into two types according to their major flavonol glycosides in leaves (Shibano et al. 1996; Hayashi et al. 2003a). The major leaf flavonol glycoside of *G. glabra* collected in Turkey, Italy and Spain was isoquercitrin (IQ), but that collected in Kazakhstan was rutin (RT) (Hayashi et al. 2003b). In addition to these, the three flavanones, pinocembrin, licoflavanone and glabranin, were recognized as major compounds common to both types. *G. glabra* and *G. uralensis* were found growing together forming a mixed population, and intermediate-type plants between them were at three sites in Kazakhstan (Hayashi et al. 2003b). Although two nucleotide substitutions of the chloroplast rbcL gene were observed between *G. uralensis* and *G. glabra*, rbcL sequences of the intermediate-types were divided into *G. uralensis* -type (G-A type) and *G. glabra*-type (A-T type). Roots of *G. glabra* contained glabridin and no glycoumarin while that of *G. uralensis* glycycoumarin and no glabridin, but neither flavonoid was detected in underground parts of the intermediate-types. All three types contained glycyrrhizin, the rbcL genes for *G. glabra* plants having linear-oblong leaflets, and straight fruits from the three collection sites were identified as the A-T type sequence (*G. glabra*-type), and those of *G. uralensis* plants, having ovate leaflets and falcated fruits, were the G-A type sequence (*G. uralensis*-type). Leaves of *G. glabra* contained rutin, isoquercitrin, licoflavanone and pinocembrin, while *G. uralensis* contained rutin, isoquercitrin, 2 unidentified compounds and no licoflavanone and pinocembrin. Both *G. glabra*-specific and *G. uralensis*-specific compounds were detected in the leaves of the intermediate-type, thus suggesting that the intermediate plants were hybrids of *G. glabra* and *G. uralensis*.

G. glabra plants collected in Uzbekistan could be divided into two types, RT-type and IQ-type, according to their major flavonol glycosides,

rutin or isoquercitrin, respectively (Hayashi et al. 2003a). In Uzbekistan, HPLC analysis of the underground parts (root, stolon) of *G. glabra* indicated that glycyrrhizin contents varied from 3.3 to 6.1 % of dry weight, and that glabridin, a species-specific flavonoid for *G. glabra*, was detected in all underground samples (0.08–0.35 % of dry weight). All the *G. glabra* plants in Sicily and Spain were morphologically similar to each other, and their fruits had no glandular hairs on the capsule (Hayashi et al. 1998b). The glycyrrhizin content in the stolons ranged from 4.4 to 0.7 % of dry weight, and the stolons contained glabridin. A survey of the habitat of *Glycyrrhiza glabra* in the Mus district of East Anatolia, Turkey, revealed that *G. glabra* var. *glandulifera* and *G. glabra* var. *glabra* grew together forming a mixed population (Tabata et al. 1988). No significant morphological difference was observed between these two varieties except that the former had glandular hairs on the capsule. The glycyrrhizin content of the roots of these plants bearing fruits in early August was found to be from 0.6 to 3.5 % dw (dry weight), while that in the roots collected in spring in the same region was 3.0–6.1 % dw. In another study, HPLC analysis of leaves indicated a significant difference in the chemical composition between the *G. glabra* plants growing in the west and those in the other regions in Turkey (Hayashi et al. 1995). Pinocembrin (o.18–1.5 % dw), licoflavone 0.07–0.79 % dw and three unidentified compounds were found in the leaves. Root and stolon of all samples contained glabridin (0.15–0.70 % dw) and glycyrrhizin contents (1.1 to 8.0 % of dw). Nine samples of *Glycyrrhiza glabra* were collected in various sites of Calabria, Italy, which showed remarkable differences in chemical composition and biological activity (Statti et al. 2004).

Phytochemicals in Callus/Cell Suspension Cultures

The metabolites detected in licorice (*G. glabra*) single cell suspension culture included a volatile apple aroma, a polysaccharide pectin-like material, steroids and triterpenoids (Wu et al. 1974). The analyses of the licorice cell volatile apple aroma indicated the presence of ethanol and some related esters. The monosaccharides found in the pectin-like polysaccharide hydrolysate were glucose, fructose, galactose, arabinose, xylose, galacturonic acid and glucuronic acid. Glycyrrhizinic acid, the common licorice constituent found in the root, could not be detected in the suspension cultures. However, several other related compounds which gave typical steroid and triterpenoid reactions were found. Sorbitol, glucose and fructose were found to be the three major sugars which accumulated in free form in the licorice cell medium. Callus and cell suspension cultures of *Glycyrrhiza glabra* failed to produce detectable amounts of glycyrrhizin, the major oleanane-type triterpene glycoside of the thickening root, or of its 11-deoxoderivative (Hayashi et al. 1988). However, betulinic acid, a lupane-type triterpene, which was found in the root bark, and a small amount of β-amyrin, a possible precursor of oleanane-type triterpenes, were detected in cell suspension cultures in addition to lupeol, a fundamental form of lupane-type triterpenes. The biotransformation of papaverine with cell suspension cultures of G. *glabra* was not reported by Dorisse et al. (1988). The main metabolite of the transformation was a hydroxylated compound, papaverinol. *G. glabra* suspension cultures did not accumulate either glycyrrhizin of glycyrrhetic acid but did produce the isoflavonoid formononetin (Arias-Castro et al. 1993). The initial pH, sucrose concentrations, medium composition, auxin and cytokinins were found to affect the accumulation of formononetin. Mousa et al. (2007) reported regenerative callus and cell suspension system of licorice (*Glycyrrhiza glabra*) to be a prerequisite for the production of the sweetener glycyrrhizin in cell suspension.

Soyasaponins I and II, oleanane-type triterpene glycosides, were shown to be produced by licorice cell suspension cultures (Hayashi et al. 1990b). The soyasaponin content in the licorice cell cultures varied from 0.017 to 1.1 % of the dry weight of cells depending on culture strains and was also greatly influenced by plant growth hormones.

Two biotransformation products formed from 18β-glycyrrhetinic acid by cell suspension cultures of *Glycyrrhiza glabra* were isolated and their structures determined as 3-*O*-[α-L-arabinopyranosyl-(1 → 2)-β-D-glucuronopyranosyl]-24-hydroxy-18β-glycyrrhetinic acid and 30-*O*-β-D-glycopyranosyl-18β-glycyrrhetinic acid (Hayashi et al. 1990a). The structures of seven triterpenoid metabolites including new compounds, 3-*O*-β-D-glucuronopyranosyl-24-hydroxy-18β-glycyrrhetinic acid and 24-hydroxy-18β-glycyrrhetinic acid 30-β-D-glucopyranosyl ester, derived from exogenous 18β-glycyrrhetinic acid administered to glycyrrhizin-free cell suspension cultures of *Glycyrrhiza glabra*, were determined (Hayashi et al. 1992a). 18β-Glycyrrhetinic acid 24-hydroxylase activity was detected in the microsomal fraction prepared from *Glycyrrhiza glabra* cell suspension cultures (Hayashi et al. 1993b). Cultured cells of *G. glabra* produced no detectable amount of glabridin, liquiritigenin glycosides, isoliquiritigenin glycosides, pinocembrin, licoflavanone and isoquercitrin but produced formononetin, an isoflavonoid (Hayashi et al. 1996b). Distribution of soyasaponin and betulinic acid triterpenoids was reported to be different in the intact plant of *G. glabra* (Hayashi et al. 1993a). Betulinic acid (0.01–0.17 %) was detected in all the callus cultures derived from various organs (hypocotyl, root, stem and leaf) of 1-month-old *G. glabra* seedlings. The presence of stigmasterol (0.04–0.07 % of cell dry weight), sitosterol (0.06–0.12 %) and small amounts of β-amyrin and lupeol were also detected. Both substrates [1- ^{14}C]acetate and [2-^{14}C]mevalonate labelled β-amyrin, an oleanane-type triterpene, and cycloartenol and 24-methylenecycloartanol, both intermediates of phytosterol biosynthesis in plant organs and tissue cultures of *Glycyrrhiza glabra* var. *glandulifera* (Ayabe et al. 1990). The label in esterified triterpenes was distributed mainly in phytosterol intermediates, but not in β-amyrin. The ratio of β-amyrin formation among the three triterpenes from [2- ^{14}C]mevalonate was relatively high in stolon segments and in root cultures, but negligible in callus cultures. Administration of a specific inhibitor of squalene-2, 3-epoxide:cycloartenol (lanosterol) cyclase caused a marked increase of β-amyrin synthesis in root suspension cultures, and of 24-methylenecycloartanol synthesis in cell suspension cultures, from [2- ^{14}C]mevalonate. Echinatin and glabridin were isolated from *G. glabra* callus tissues (Oda et al. 1995)

UDP-glucuronic acid:triterpene glucuronosyltransferase activities for soyasapogenol B, soyasapogenol C, 24-hydroxyglycyrrhetinic acid and 24-hydroxyglycyrrhetinic acid methyl ester were detected in cultured licorice cells (Hayashi et al. 1996c). The pH dependency of the activity for soyasapogenol B was different from that for 24-hydroxyglycyrrhetinic acid methyl ester. Hayashi et al. (1996a) isolated two cDNAs (GgSQS1 and GgSQS2) encoding squalene synthase of *Glycyrrhiza glabra* by cross-hybridization with that of *Arabidopsis thaliana* squalene synthase. Their nucleotide sequences contained an open reading frame for a polypeptide of 413 amino acids (GgSQS1) and 412 amino acids (GgSQS2), respectively. The deduced amino acid sequence of GgSQS1 exhibits 88 %, 81 %, 78 %, 45–44 % and 45–41 % identity to those of the GgSQS2, *Nicotiana benthamiana*, *Arabidopsis thaliana*, mammal, and yeast squalene synthases, respectively. The cell-free extracts of *E.coli* transformed with the respective plasmid converted 3H-farnesyl diphosphate into squalene. Two cDNAs (GgSQS1 and GgSQS2) encoding squalene synthase were isolated from licorice, *Glycyrrhiza glabra* (Hayashi et al. 1999). Squalene synthase activity was found in the cell-free extracts of *Escherichia coli* transformed with the recombinant plasmids for GgSQS1 and GgSQS2, respectively. Northern blot analysis showed that GgSQS2 mRNA was mainly expressed during the exponential growth phase of the cultured licorice cells. A cDNA clone (GgCAS1) encoding cycloartenol synthase (CAS) was isolated from *Glycyrrhiza glabra* by cross-hybridization with that of *Pisum sativum* CAS as a probe (Hayashi et al. 2000a). Southern blot analysis suggested that at least two CAS genes exist in the licorice genome. An oxidosqualene cyclase cDNA, termed GgbAS1, was isolated from cultured cells of *Glycyrrhiza glabra* by heterolo-

gous hybridization with cDNA of *Arabidopsis thaliana* LUP1 lupeol synthase (Hayashi et al. 2001). It was shown that the level of β-amyrin synthase mRNA was drastically changed in the cultured licorice cells, whereas the mRNA level of cycloartenol synthase was relatively constant. Exogenously applied methyl jasmonate (MeJA) stimulated soyasaponin biosynthesis in cultured *Glycyrrhiza glabr*a cells (Hayashi et al. 2003c). mRNA level and enzyme activity of β-amyrin synthase (bAS), an oxidosqualene cyclase (OSC) situated at the branching point for oleanane-type triterpene saponin biosynthesis, were upregulated by MeJA, whereas those of cycloartenol synthase, an OSC involved in sterol biosynthesis, were relatively constant. Two mRNAs of squalene synthase (SQS), an enzyme common to both triterpene and sterol biosyntheses, were also upregulated by MeJA. In addition, enzyme activity of UDP-glucuronic acid: soyasapogenol B glucuronosyltransferase, an enzyme situated at a later step of soyasaponin biosynthesis, was also upregulated by MeJA. Accumulations of bAS and two SQS mRNAs were not transient but lasted for 7 days after exposure to MeJA, resulting in the high-level accumulation (more than 2 % of dry weight cells) of soyasaponins in cultured licorice cells. The cultured cells and intact plants of *Glycyrrhiza glabra* produced betulinic acid and oleanane-type triterpene saponins (soyasaponins and glycyrrhizin) (Hayashi et al. 2004). They found that the mRNA expression levels of lupeol synthase and β-amyrin synthase were consistent with the accumulation of betulinic acid and oleanane-type triterpene saponins, respectively. The transcript of lupeol synthase was highly expressed in the cultured cells and root nodules. The transcript of β-amyrin synthase, an OSC responsible for oleanane-type triterpene biosynthesis, was highly expressed in the cultured cells, root nodules and germinating seeds, where soyasaponin accumulated, and in the thickened roots where glycyrrhizin accumulated. Yeast extract promoted betulinic acid accumulation in cultured cells of *G. glabra* whereas soyasaponin accumulation was suppressed (Hayashi et al. 2005). The results indicated that soyasaponin and betulinic acid production were differently regulated in cultured cells of *G. glabra*. Differential pathway of soyasaponin and betulinic acid production in *G. glabra* were proposed as follows:

acetyl-CoA → mevalonic acid → squalene → squalene-2,3-epoxide → β-amyrin → glycyrrhetinic acid → glycyrrhizin;

acetyl-CoA → mevalonic acid → squalene → squalene-2,3-epoxide → β-amyrin → soyasapogenol B → Soyasaponin I (R = galactose) and soyasaponin II (R = arabinose)

acetyl-CoA → mevalonic acid → squalene → squalene-2,3-epoxide → lupeol → betulinic acid;

acetyl-CoA → mevalonic acid → squalene → squalene-2,3-epoxide → cycloartenol → β-sitosterol.

Kanzonol Y, 4-hydroxylonchocarpin, xambioona, glyinflanin K and several other unnamed compounds were isolated from hairy root cultures (Asada et al. 1997). A new compound named licoagrodione was isolated from *Glycyrrhiza glabra* hairy root cultures together with five known prenylated flavonoids (Li et al. 1998). Two new prenylated flavonoids, licoagrochalcone A and licoagrocarpin, were isolated from the hairy root cultures of *Glycyrrhiza glabra* along with eight known flavonoids (Asada et al. 1998). The structures of the new compounds were elucidated as 3-prenyl-2′,4,4′-trihydroxychalcone and (6a*R*, 11a*R*)-4-prenyl-3-hydroxy-9-methoxypterocarpan, respectively. A new prenylated biaurone, licoagrone, was isolated from the hairy root cultures of *Glycyrrhiza glabra* together with five known flavonoids, kanzonol D, afrormosin, odoratin, phaseol and echinatin (Asada et al. 1999). Asada et al. (2000) found that the biosynthesis of the hemiterpene moiety of glabrol, the main prenylated flavanone in the *Glycyrrhiza glabra* hairy root cultures proceeded via a glyceraldehyde/pyruvate non-mevalonate pathway.

An unusual biflavonoid named licoagrodin was isolated from *G. glabra* hairy root cultures along with three prenylated retrochalcones, licoagrochalcones B, C, D, a prenylated aurone, lico-

agroaurone and four known prenylated flavonoids, licochalcone C, kanzonol Y, glyinflanin B and glycyrdione A (Li et al. 2000). From the glycosidic fraction, a new isoflavone glycoside, licoagroside A, and a new maltol glycoside, licoagroside B were isolated together with known isoflavone glycosides onionin, calycosin 7-*O*-glucoside, wistin, afrormosin 7-*O*-(6″-malonylglucoside), vicenin-2, and isoschaftoside and three other known glycosides tachioside, isotachioside and dimethylallyl 6-*O*-α-L-arabinopyranosyl-β-D-glucopyanoside.

Pharmacological Activities

Licorice rhizome are considered to possess an expectorant and carminative, flavouring agent, depressant, antimicrobial, hypolipidemic, antianthersclerotic, antiviral, antiulcerogenic, hypotensive, hepatoprotective, spasmolytic, antidiuretic, antimutagenic, antipyretic, anti-inflammatory (Isbrucker and Burdock 2006; Meena et al. 2010). Licorice had been reported to possesses many therapeutic properties as to potentiate the action of cortisol, to reduce testosterone synthesis, especially in women, to exert an estrogen-like activity and to reduce body fat mass (Armanini et al. 2002). Licorice flavonoid constituents mainly comprised flavones, flavonals, isoflavones, chalcones, bihydroflavones and bihydrochalcones (Xing et al. 2003). Pharmacological investigation concluded that they had antioxidant, antibacterial, anti-tumour and anti-HIV activities. Glycyrrhiza Radix is a commonly used Chinese herbal medicine, derived from the dried roots and rhizomes of Glycyrrhiza uralensis, *G. glabra* and *G. inflata* (Gao et al. 2009). The main bioactive constituents of licorice comprised triterpene saponins and flavonoids. Various pharmacological properties of liquorice had been reported including anti-ulcer, anti-inflammation, spasmolysis, anti-oxidative, contravariance, antiviral, anticancer activities, hepatoprotective, expectorant and memory enhancing effects. Glabridin, a prenylated isoflavonoid of *G. glabra* roots had been associated with a wide range of biological properties such as antioxidant, anti-inflammatory, anti-atherogenic, regulation of energy metabolism, estrogenic, neuroprotective, anti-osteoporotic and skin-whitening (Simmler et al. 2013). While glabridin is one of the most studied licorice flavonoids, both glabridin and standardized licorice extracts have significant impact on food, dietary supplements (DSs) and cosmetic markets, as evidenced by the amount of available patents and scientific articles since 1976, when glabridin was first described.

Antioxidant Activity

The highest antioxidant activity (β-carotene bleaching assay) of *G. glabra* root extract was 88.7 % at a concentration of 800 μg/mL (Ercisli et al. 2008). Aqueous and ethanol extract of Turkish *G. glabra* aerial parts and roots inhibited 87.9, 83.6, 88.6 and 80.1 % lipid peroxidation of linoleic acid emulsion at 30 μg/mL concentration, respectively (Tohma and Gulçin 2010). In contrast, α-tocopherol and trolox had inhibition of 68.1 and 81.3 %, respectively. The hydromethanolic *G. glabra* root extract displayed possessed potent hydroxyl radical scavenging activity with IC_{50} value of 80 μg/ml against the positive control standard ascorbic acid with IC_{50} value of 50 μg/ml (Varsha et al. 2013).

The organic extracts of two licorices, known in commerce as Russian licorice (*G. glabra* var. *glandulifera*) and Xinjiang licorice (*G. inflata*) exhibited potent antimicrobial and antioxidant activity (Okada et al. 1989). The bioassay-directed chemical investigation of both licorices revealed glabrene, glabridin and licochalcones A and B as active principles. Glabrene showed most potent antioxidant, three times as potent as vitamin E, glabridin showed no significant activity. The DPPH radical scavenging activity of *G. uralensis* and *G. glabra* achieved approximately 72–75 % if 10 mg/mL or more licorice extract was used (Liao et al. 2012). *G. uralensis* had slightly better radical scavenging activity than *G. glabra*. *G. uralensis* also showed higher reducing ability than *G. glabra*. Cheel et al. (2013) found that the chemical profile of licorice quantitatively varied at different harvest times and these fluctu-

ations determined changes in its bioactivities such as antioxidant and free radical scavenging activities. In general, the samples from May and November showed the most favourable free radical scavenging and antioxidant effects. Liquiritin and glycyrrhizin, the major constituents in the February and May licorice extract, appeared to contribute to the superoxide radical scavenging activity. Glabridin and glabrene, the compounds with the highest relative proportion in the November licorice extract, accounted for the antioxidant and DPPH scavenging activities of licorice.

Seven constituents with antioxidant capacity were isolated from *Glycyrrhiza glabra* (Vaya et al. 1997). The isolated compounds were identified as the isoflavans hispaglabridin A (1), hispaglabridin B (4), glabridin (3) and 4′-*O*-methylglabridin (2), the two chalcones, isoprenylchalcone derivative (5) and isoliquiritigenin (6), and the isoflavone, formononetin (7). The isoflavans (1–4) at a concentration of 50 μM inhibited *β*-carotene consumption, following 90 min of incubation at 50 °C, similar to the inhibitory effect of the whole licorice crude extract (at 16 mg/l). The chalcones (5 and 6) exhibited moderate inhibition and the isoflavone 7 was almost inactive, whereas vitamin E (50 μM) completely inhibited *β*-carotene consumption. Compounds 1–6 exhibited high inhibitory activity at a concentration of 30 μM on 2,2′-azobis (2-amidinopropane) dihydrochloride (AAPH)-induced LDL oxidation, but compound 7 and vitamin E were not active. A dose-dependent inhibitory effect of glabridin, on the formation of cholesteryl linoleate hydroperoxide (CLOOH), in an AAPH-induced LDL oxidation system was also shown. Glabridin, at 5 or 40–60 μM concentration, inhibited the CLOOH formation by 62 % and 90 %, respectively. The results suggested that constituents 1–6 were very potent antioxidants toward LDL oxidation with glabridin (11.6 % wet weight) being the most abundant and potent antioxidant.

Licorice root extract was tested for antioxidative activity in comparison with antioxidants (sodium metabisulphite and BHT) at 0.1 %, 0.5 %, 1.0 % and 2.0 % w/w in 2 % w/w hydroquinone cosmetic cream (Morteza-Semnani et al. 2002). After 3 months, at 25 °C and 45, the extract demonstrated more antioxidant activity than from two other commercial antioxidants at all concentrations, with about 43–53 % and 34–46 %, respectively, more hydroquinone remaining than in the control system. In the third month, the preparation containing 0.1 %, 0.5 %, 1.0 % and 2.0 % extract gave good physical formulation stability with about 72 %, 76 %, 78 % and 81 % hydroquinone remaining at 25 °C and 51 %, 55 %, 60 % and 63 % hydroquinone remaining at 45 °C respectively. Aqueous and ethanolic extracts of *Glycyrrhiza glabra* root demonstrated the dose-dependent scavenging activity against nitric oxide (IC_{50} = 72 and 62.1 μg/ml, respectively), superoxide (IC_{50} = 64.2 and 38.4 μg/ml, respectively), hydroxyl (IC_{50} = 81.9 and 63 μg/ml, respectively), DPPH (IC_{50} = 43.6 and 28.3 μg/ml, respectively) and ABTS•+ (IC_{50} = 77.3 and 57.2 μg/ml, respectively) radicals (Visavadiya et al. 2009). Further, both extracts showed strong reducing power and iron-chelating capacities. In the Fe^{2+}/ascorbate system, both extracts were found to inhibit mitochondrial fraction lipid peroxidation. In copper-catalysed human serum and low-density lipoprotein oxidation models, both extracts significantly lengthened the lag phase along with a decline in the oxidation rate, conjugated dienes, lipid hydroperoxides and thiobarbituric acid reactive substance formation. The findings showed *G. glabra* possessed considerable antioxidant activity and protective effect against the human lipoprotein oxidative system.

Isoflavans from *G. glabra* were shown to be effective in protecting mitochondrial function against oxidative stresses (Haraguchi et al. 2000). Mitochondrial lipid peroxidation linked to respiratory electron transport and that induced non-enzymatically were inhibited by these isoflavans. Hispaglabridin A strongly inhibited both peroxidations and 3′-hydroxy-4′-*O*-methylglabridin was the most effective at preventing NADH-dependent peroxidation. 3′-Hydroxy-4′-*O*-methylglabridin protected mitochondrial respiratory enzyme activities against NADPH-dependent peroxidation injury.

Dihydroxyfumarate-induced mitochondrial peroxidation was also prevented by this isoflavan. At the concentration of 0.10, 0.25 and 0. 5 mg/mL, licorice glabridin inhibited microsomal free radical (ROS) formation by 67 %, 73 % and 83 %, respectively (Ablise et al. 2007). At lower concentration, the ROS inhibitory activity of glabridin was the same with those of *Ginkgo biloba* extract EGB761. Administration of *Glycyrrhiza glabra* polysaccharides (GGP) dose-dependently and significantly enhanced immune and antioxidant enzyme activities in the GGP-treated high-fat mice (Hong et al. 2009). *Glycyrrhiza glabra* root samples irradiated with 20 and 25 kGy doses gamma irradiation as a method of decontamination for food and herbal materials, increased phenolic contents and DPPH scavenging activity (Khattak and Simpson 2010). Licorice infusion weakly scavenged DPPH, and its compounds liquiritin and glycyrrhizin showed negligible effects (Cheel et al. 2010). Both licorice infusion and glycyrrhizin substantially scavenged superoxide radicals. The β-carotene bleaching was inhibited by licorice infusion, but liquiritin and glycyrrhizin showed no effect. The licorice infusion, liquiritin and glycyrrhizin exhibited no meaningful activities against hypochlorous acid, and they showed pro-oxidant effects in the myeloperoxidase-chlorinating system.

The DPPH radical scavenging activity of *G. glabra* leaf extracts (SC_{50}) were determined as: methanol extract 35.01 μg/ml, ethyl acetate extract 52.02 μg/ml and n-hexane extract 92.01 μg/ml and the total phenols were 104.09 μg/mg, 297.25 μg/mg and 111.53 μg/gm, respectively (Siracusa et al. 2011). Chloroform fraction of licorice methanol extract was the most effective antioxidant in scavenging 1, 1-diphenyl-2-picrylhydrazyl radical (DPPH) with 87.7 % activity but the activity was less than the crude methanolic extract, that is, 90 % (Lateef et al. 2012). Chloroform fraction showed the same trend in reducing power as that in radical scavenging activity. Significant anti-urease activity, that is, 72 % was observed in the ethyl acetate fraction with respect to standard inhibitor thiourea. Umbelliferone isolated from *Glycyrrhiza glabr*a rhizome exhibited DPPH free radical scavenging potential of 43.88 % at 616.75 μM (Kaur et al. 2012a). The methanol extract of *G. glabra* roots was found to have good antioxidant activity of 67.22 % at 500 μg/mL with the IC_{50} value of 359.45 μg/mL (Chopra et al. 2013).

Incorporation of licorice flavonoids liquiritigenin (LQG) and liquiritin (LQ) into ceramide liposome-in-cellulose hydrogel complex system enhanced their skin permeability (Kim et al. 2014b). Encapsulation efficiencies for liquiritigenin and liquiritin-loaded liposome-in-hydrogel were 69.39 % and 64.71 %, respectively. Liposome-in-hydrogel complex systems (LQG: 56.55 %, LQ: 66.99 %) had greater skin permeability than control (LQG: 4.92 %, LQ: 5.30 %) or a single liposome systems (LQG: 43.34 %, LQ: 48.97 %) and hydrogel systems (LQG: 38.21 %, LQ: 55.07 %). Liposome-in-hydrogel system could be a potential drug delivery system for topical delivery of antioxidants such as licorice flavonoids to construct antioxidative skin barrier. Licorice had been shown to have antioxidant properties and may play a role in the treatment and prevention of photo-ageing and a natural ingredient used in cosmetics (Bowe and Pugliese 2014). When licorice extract (LE) was also used in emulsion preparation, a remarkable synergistic oxidation inhibition was observed with pea protein hydrolysates (PPH) (Zhang et al. 2013, 2014b). Remarkable synergistic effects were observed on both Flavourzyme (Fla-PPH) or Protamex (Pro-PPH) with licorice extract (LE) (Zhang et al. 2013). The presence of LE enhanced the antioxidant potential by producing a more compact network apparently via PPH-LE complexation. LE adsorbed onto oil droplets either directly or through associating with PPH to produce a thick and compact interfacial membrane enabling the defence against oxygen species (Zhang et al. 2014b). Liquiritin apioside, neolicuroside, glabrene and 18β-glycyrrhetic acid were the predominant phenolic derivatives partitioning at the interface and most likely the major contributors to the notable synergistic antioxidant activity when coupled with pea protein hydrolysates.

Five macroporous resins showed similar and effective adsorption and desorption properties for enriching flavonoids from licorice leaf (Dong

et al. 2015). Further column chromatography of two representative resins XAD-16 and SP825 showed that the total flavonoids (from 16.8 to 55.6 % by XAD-16 and to 53.9 % by SP825) and pinocembrin (from 5.49 to 15.2 % by XAD-16 and to 19.8 % by SP825) were enriched in 90 % ethanol fractions. Meanwhile, the antioxidant capacities and nitrite-scavenging capacities were 2–3 times higher than those of the crude extract. The fractions with high flavonoid and pinocembrin contents could be used as biologically active ingredients in functional food.

Anticancer Activity

In-Vitro Studies

Studies by Kobayashi et al. (1995) showed that the anti-angiogenic effect of licorice extract depended on the anti-tube formation effect of isoliquiritin. Isoliquiritin (0.31–3.1 mg/kg), a licorice-derived flavonoid, inhibited the carmine content of granuloma tissue 50-fold greater than licorice extract (Kobayashi et al. 1995). Glycyrrhizin (20–80 mg/kg), a licorice-derived saponin, inhibited carmine content with a weak potency. The licorice extract (0.01–1 mg/ml) also inhibited tube formation from vascular endothelial cells in a concentration-dependent manner. From the chemical structure-activities of used licorice-derived flavonoids (0.1–100 μM), their potencies for anti-tube formation were in the order isoliquiritigenin > isoliquiritin > liquiritigenin >> isoliquiritin-apioside. Glycyrrhizin (0.1–100 μM) and glycyrrhetinic acid (0.1–10 μM) increased tube formation. A glycyrrhizin (82 μg/ml)-induced increase in tube formation was inhibited by isoliquiritin. The combined effect of a mixture of 82 μg/ml glycyrrhizin and 4.2 μg/ml isoliquiritin, a similar concentration ratio to their yield ratio in the licorice extract, corresponded to the effect of 100 μg/ml extract. Various extracts of *G. glabra* and its constituents were found to be cytotoxic in-vitro (Rathi et al. 2009). The IC_{50} values of standard 18 β-glycyrrhetic acid was 0.412 μM and those for the three different extracts (chloroform, methanol and water) of *G. glabra* on MCF7 cancerous cell line were 0.4485, 0.9906 and 1.288 μM, respectively. HPTLC study indicated that the amount of 18 β-glycyrrhetic acid in three different extracts (chloroform, methanol and water extract) was 26.6, 12.5 and 5.6 μg/g, respectively. *Glycyrrhiza* extract exhibited antiangiogenic activity in the zebrafish antiangiogenic model (Li 2012). Of seven fraction form the ethyl acetate extract, Fr5 and Fr6 showed antiangiogenic activity. Results of studies suggested that licorice root extract could mitigate the tumorigenic effects of 2,3,7,8-tetrachlorodibenzo-p-dioxin (TCDD), a endocrine disrupting chemical in MCF-7 cell breast cancer cells by suppression of aryl hydrocarbon receptor (AhR), AhR nuclear translocator and cytochrome P450 1A1 in a dose-dependent manner and cell cycle arrest (Chu et al. 2014). Thus, licorice could be used as a potential toxicity-alleviating agent against endocrine disrupting chemical-mediated diseases.

Glycyrrhetic acid inhibited the action of tumour promoter in-vitro and in-vivo (Nishino et al. 1984). Glycyrrhetic acid inhibited the increased phospholipid metabolism of cultured cells induced by tumour promoters, 12-*O*-tetradecanoylphorbol-13-acetate or teleocidin, and it markedly suppressed the promoting effect of teleocidin on skin tumour formation in mice initiated with 7,12-dimethylbenz[a]anthracene. Application of glycyrrhetic acid, to Swiss 3 T3 mouse fibroblasts prior to the treatment with 12-*O*-tetradecanoylphorbol-13-acetate (TPA), a potent tumour promoter, showed a time- and dose-dependent inhibitory effect on the TPA-stimulated 3-*O*-methyl-glucose transport (Kitagawa et al. 1984). Glycyrrhetic acid inhibited the specific binding of TPA to mouse epidermal membrane fractions in a dose- and time-dependent manner (Kitagawa et al. 1986). Glycyrrhizic acid, a glycoside of glycyrrhetic acid, exerted no inhibitory effect on TPA binding. The results of kinetic analysis suggested that glycyrrhetic acid directly bound to the TPA receptor, resulting in the competitive inhibition of the binding of TPA to its receptor without affecting the number of binding sites. The authors suggested that the inhibitory effect of glycyrrhetic acid on TPA binding to the membrane

receptor may play a role in its antitumor-promoting activity in-vivo. Glycyrrhetic acid an anti-inflammatory agent isolated from licorice root inhibited 12-*O*-tetradecanoylphorbol-13-acetate (TPA)-mediated tumour promotion in mouse skin (O'Brian et al. 1990). It was demonstrated that glycyrrhetic acid inhibited the Ca^{2+}- and phospholipid-dependent phosphotransferase activity of protein kinase C (PKC), the phorbol ester tumour promoter receptor. Therefore, inhibition of PKC may play a role in the anti-promoting activity of glycyrrhetic acid.

Studies showed that 18α-glycyrrhetinic acid (AGA) inhibited proliferation and growth of prostate cancer cell line DU-145 cells by inducing apoptosis (Shetty et al. 2011). Also it was shown that HUVEC tube formation was drastically reduced when cultured in conditioned medium of AGA-treated DU-145 cells. Additionally, AGA treatment prevented the invasion of DU-145 prostate cancer cells on matrigel coated transwells via downregulation of NF-κB (p65), VEGF and MMP-9 expression. Further, AGA treatment also downregulated the expression of pro-inflammatory cytokine/growth factor genes HMGB1, IL-6 and IL-8 in DU-145 cells. The results suggested that AGA may be a promising anticancer agent for the chemoprevention and treatment of prostate cancer. Non-polar compounds in the ethanol extract of roasted licorice (EERL) exerted potent anti-carcinogenic effects in inhibiting the growth of DU145 and MLL prostate cancer cells, as well as HT-29 colon cancer cells (Park et al. 2014a, b). In contrast aqueous extracts of un-roasted and roasted licorice showed minimal effects on cell growth. EERL potently inhibited growth of MCF-7 and MDA-MB-231 breast, B16-F10 melanoma, and A375 and A2058 skin cancer cells, whereas EERL slightly stimulated the growth of normal IEC-6 intestinal epithelial cells and CCD118SK fibroblasts.

18-α-glycyrrhetinic acid (AGA) caused more than 95 % rapid and reversible inhibition of intercellular gap-junctional communication at concentrations as low as 2 μM (Davidson et al. 1986). The related compounds 18-β-glycyrrhetinic acid and carbenoxolone also caused communication inhibition. Glycyrrhetinic acid was shown previously to inhibit intercellular gap-junctional communication between human fibroblasts (Davidson and Baumgarten 1988). Of 31 derivatives of glycyrrhetinic acid tested for their ability to inhibit communication, eight of the compounds inhibited communication with high potency (IC_{50} less than 3 μM) and showed low toxicity, properties which suggested they may be useful pharmacological probes for studies of gap-junction function. Endogenous and exogenous factors which modulate intercellular gap-junctional communication may be efficiently used to prevent potential cancer cells deviating from normal cell society and homeostasis (Yamasaki 1990). 2-cyano substituted analogues of glycyrrhetinic acid, namely, methyl 2-cyano-3,11-dioxo-18β-olean-1,12-dien-30-oate (β-CDODA-Me) and methyl 2-cyano-3,11-dioxo-18α-olean-1,12-dien-30-oate (α-CDODA-Me) were found to be more cytotoxic to colon cancer cells (SW480, HT-29, HCT-15) than their des-cyano analogues and to have selective peroxisome proliferator-activated receptor gamma (PPARgamma) agonist activity (Chintharlapalli et al. 2007). This selectivity in colon cancer cells was shown for the induction of two proapoptotic proteins, namely, caveolin-1 and the tumour-suppressor gene Krüppel-like factor-4 (KLF-4). The data suggested that PPARgamma agonists derived from glycyrrhetinic acid induced cell-dependent caveolin-1 and KLF-4 expression through receptor-dependent pathways.

Glycyrrhetinic acid, the metabolite of the natural product glycyrrhizin, had been found to be a nonselective inhibitor of 11β-hydroxysteroid dehydrogenase (11β-HSD) type 1 and type 2 (Stanetty et al. 2010). 11β-HSD2 inhibitors may find therapeutic applications in chronic inflammatory diseases and certain forms of cancer, whereas 11β-HSD1 inhibitors may find use for treatment of metabolic diseases, such as obesity and diabetes. Several 3-amino and 29-hydroxamic acid derivatives of glycyrrhetinic acid (metabolite of glycyrrhizin) were synthesised and showed high selectivity for 11β-hydroxysteroid dehydrogenase 2 (11β-HSD2) (Stanetty et al. 2010). The most potent and selective compound was active

against human 11β-HSD2 in the low nanomolar range with a 350-fold selectivity over human 11β-HSD1. Kratschmar et al. (2011) found selected 18β-glycyrrhetinic acid derivatives potently inhibited 11β-HSD2 in intact SW-620 colon cancer cells, although the rank order of inhibitory potential differed from that obtained in cell lysates. The biological activity of compounds was further demonstrated in glucocorticoid receptor (GR) transactivation assays in cells co-expressing GR and 11β-HSD1 or 11β-HSD2. Earlier novel 18βa-glycyrrhetinic acid analogues, 7 *N*-(2-hydroxyethyl)-3β-hydroxy-11-oxo-18β--olean-12-en-30-oic acid amide (Su et al. 2004) and 5, *N*-(2-hydroxyethyl)-3β-hydroxy-11-oxo-18β-olean-12-en-30-oic acid amide (Vicker et al. 2004) were found to be very potent selective inhibitors of 11β-hydroxysteroid dehydrogenase 2 with IC_{50} values of 4pM. Gaware et al. (2011) described novel 29-urea- and 29-hydroxamic acid derivatives of glycyrrhetinic acid as well as derivatives with the Beckman rearrangement of the 3-oxime to a seven-membered ring, and the rearrangement of the C-ring from 11-keto-12-ene to 12-keto-9(11)-ene to have improved selective inhibition of 11β-HSD2 in the lower nanomolar range with up to 3600-fold selectivity. Glycyrrhetinic acid (GA) significantly suppressed the viability of NCI-H460 and A549 non-small lung cancer cells in-vitro (Song et al. 2014). Also, GA significantly increased the sub G1 population by cell cycle analysis and terminal deoxynucleotidyl transferase dUTP nick end labelling (TUNEL) positive cells in a concentration dependent manner in NCI-H460 non-small lung cancer cells. The results suggested that GA induced apoptosis via inhibition of PKC α/βII and activation of JNK in NCI-H460 non-small lung cancer cells. 18β-Glycyrrhetinic acid (18β-GA) suppressed cell proliferation of non-small cell lung cancer by inhibiting thromboxane synthase and its ERK/CREB signalling (Huang et al. 2014).

Glycyrrhetinic acid (GA) was found to have potential as a chemotherapeutic agent for human non-small cell lung cancer (Zhu et al. 2015). GA suppressed the proliferation of A549 and NCI-H460 cells by inducing G1-phase cell cycle arrest by upregulation of cyclin-dependent kinase inhibitors (CKIs) (p18, p16, p27 and p21) and inhibition of cyclins (cyclin-D1, -D3 and -E) and cyclin-dependent kinases (CDKs) (CDK4, 6 and 2). GA also upregulated the unfolded proteins, Bip, PERK and ERP72 in the endoplasmic reticulum.

Isoliquiritigenin, one of the components in the root of *Glycyrrhiza glabra*, significantly inhibited the proliferation of DU145 and LNCaP prostate cancer cell lines in a dose-dependent and time-dependent manner (Kanazawa et al. 2003). Isoliquiritigenin induced S and G2/M phase arrest and enhanced the expression of GADD153 mRNA and protein associated with cell cycle arrest. Licochalcone from *G. glabra* roots inhibited growth and induced modest apoptosis of androgen-independent p53-null PC-3 prostate cancer cells but had more pronounced effect on cell cycle progression arresting cells in G2/M, accompanied by suppression of cyclin B1 and cdc2 (Fu et al. 2004). It also inhibited phosphorylation of Rb and decreased expression of transcription factor E2F concurrent with reduction of cyclin D1, downregulation of CDKs 4 and 6, but increased cyclin E expression. Kwon et al. (2009) found that isoliquiritigenin inhibited migration, and the metastatic and invasive capacity of human prostate cancer DU 145 cells possibly by decreasing Jun N-terminal kinase (JNK)/activator protein (AP)-1 signalling. Isoliquiritigenin (ISL), a flavonoid isolated from licorice, inhibited human promyelocytic leukaemia (HL-60) cell proliferation and decreased the iROS levels in a dose-dependent manner, while the treatment did not increase the lethality rate (Li et al. 2009). Isoliquiritigenin (ISL) triggered the mammalian target of rapamycin (mTOR)-dependent autophagic and apoptotic cell death in adenoid cystic carcinoma (ACC) (Chen et al. 2012). The results suggested that induction of mTOR-dependent autophagic and apoptotic cell death may be an important mechanism in cancer chemotherapy by ISL. Wang et al. (2013c) found that isoliquiritigenin (ISL) significantly inhibited the VEGF-induced proliferation of human umbilical vein endothelial cells (HUVECs) at non-toxic concentration. A series of angiogene-

sis processes including tube formation, invasion and migration abilities of HUVECs were also interrupted by ISL in-vitro (Wang et al. 2013b). Additionally, ISL suppressed sprout formation from VEGF-treated aortic rings in an ex-vivo model. In-vivo studies further showed that ISL administration could inhibit breast cancer growth and neoangiogenesis accompanying with suppressed VEGF/VEGFR-2 signalling, elevated apoptosis ratio and little toxicity effects. Studies by Kang et al. (2010) demonstrated that isoliquiritigenin blocked JNK- or p38 MAPK-responsive pathways leading to direct matrix metalloproteinases (MMPs) activation of PMA-exposed endothelial cells. The results suggested that isoliquiritigenin inhibition of MMP may boost a therapeutic efficacy during angiogenesis. Li et al. (2013) found that isoliquiritigenin induced growth inhibition and apoptosis through downregulating multiple key enzymes in arachidonic acid (AA) metabolic network and the deactivation of PI3K/Akt in human breast cancer cells. Isoliquiritigenin diminished cell viability, 5-bromo-2′-deoxyuridine (BrdU) incorporation, and clonogenic ability in both MCF-7 and MDA-MB-231 cells, and induced apoptosis. Furthermore, isoliquiritigenin inhibited mRNA expression of multiple forms of AA-metabolizing enzymes, including phospholipase A2 (PLA2), cyclooxygenases (COX)-2 and cytochrome P450 (CYP) 4A, and decreased secretion of their products, including prostaglandin E2 (PGE2) and 20-hydroxyeicosatetraenoic acid (20-HETE), without affecting COX-1, 5-lipoxygenase (5-LOX), 5-lipoxygenase activating protein (FLAP) and leukotriene B4 (LTB4). Isoliquiritigenin suppressed the migration of MDA-MB-231 cells by inhibiting upstream signalling pathways (Wang et al. 2013c). It reduced the secretions and protein levels of vascular endothelial growth factor (VEGF), hypoxia-inducible factor 1-alpha (HIF-1α) and also inhibited the expression and gelatinolytic activity of matrix metalloproteinase MMP-2 and MMP-9. In another study, isoliquiritigenin inhibited breast cancer MDA-MB-231 cell metastasis by preventing anoikis resistance, migration and invasion through downregulating COX-2 and CYP 4A signalling (Zheng et al. 2014). The results suggested that isoliquiritigenin could be a promising multi-target agent for preventing breast cancer metastasis, and anoikis could represent a novel mechanism through which flavonoids may exert the anti-metastatic activities. Licorice flavonoid isoliquiritigenin diminished 7,12-dimethylbenz[α]anthracene (DMBA)-induced DNA (xenobiotic response element – XRE) binding of aryl hydrocarbon receptor (AhR) in MCF-7 breast cancer cells (Wong et al. 2014). Quantitative reverse transcriptase-polymerase chain reaction (RT-PCR) assay demonstrated that expressions of genes with XRE-containing promoters, including CYP1A1, 1A2 and 1B1, followed the same pattern of XRE transactivation. The present study illustrated that isoliquiritigenin might downregulate polycyclic aromatic hydrocarbon (PAH)-induced expressions through antagonizing AhR translocation. Isoliquiritigenin was found to be a promising chemopreventive agent against oral cancer (Hsia et al. 2015). It caused DNA damage and inhibited ataxia telangiectasia mutated (ATM) expression leading to G2/M phase arrest and apoptosis in oral squamous cell carcinoma.

Studies showed that licorice (*Glycyrrhiza glabra*) could induce caspase-dependent and autophagy-related cell death in human LNCaP prostate cancer cells (Yo et al. 2009). Licorice and its constituent licochalcone-A induced autophagy in LNCaP prostate cancer cells by suppression of Bcl-2 expression and the mTOR pathway. Licochalcone A showed the most cytotoxic activity (IC_{50} values of 40–92 μM) among all seven licorice compounds (lichochalone A, glycyrrhizic acid, glycyrrhetinic acid, isoliquiritigenin, glabridin, liquiritigenin, glycyrrhizic acid ammonium salt) tested against gastric cancer cell lines GES-1, MKN-28, SGC7901 AGS and MKn-45 (Xiao et al. 2011). Licochalcone A induced apoptosis of gastric cancer cell lines via the caspase-dependent mitochondrial pathway and induced G2 cell cycle arrest through regulation of G2/M phase check-point proteins. Studies by Kim et al. (2014a) demonstrated that licorice compound licochalcone-A-induced apoptosis in KB oral cancer cells was mediated by the extrinsic apoptotic signalling pathway, which involved

a caspase-dependent FasL-mediated death receptor pathway.

Animal Studies

The aqueous licorice root extract inhibited the in-vivo (mice) and in-vitro proliferation of Ehrlich ascites tumour cells (Sheela et al. 2006). The angio-inhibitory activity of *G. glabra* was confirmed by its inhibition of angiogenesis in in-vivo assays, peritoneal and chorioallantoic membrane assay. Licorice extract decreased VEGF production and the cytokine-induced neovascularization. The results suggested that licorice root extract may be a potential supplemental source for cancer therapy. Administration of the licorice extract inhibited the growth of mouse colon carcinoma in BALB/C mice inoculated with CT-26 colon cancer cells, without any adverse effects, and reduced the cisplatin-induced toxicity (Lee et al. 2007). In addition, the administration of the licorice extract significantly reduced the cisplatin-induced oxidative stress.

Glycyrrhetic acid suppressed tumour promoter-induced effects in-vitro, such as stimulation of 32Pi-incorporation into phospholipids of cultured cells and downregulation of the epidermal growth factor receptor (Nishino et al. 1986). Glycyrrhetic acid inhibited the promoting activity of both 12-*O*-tetradecanoylphorbol-13-acetate (TPA) and teleocidin on skin tumour formation in mice initiated with 7,12-dimethylbenz[a]anthracene (DMBA). The percentage of tumour-bearing mice in the group treated with DMBA plus teleocidin was 88 % at week 18, whereas that in the group treated with DMBA plus teleocidin and glycyrrhetic acid (10 mumol/painting) was 6 %. Similarly, the percentage of tumour-bearing mice of the group treated with DMBA plus TPA was 97 % at week 20, whereas that of the group treated with DMBA plus TPA and glycyrrhetic acid was 40 %. Thus, glycyrrhetic acid was confirmed to inhibit the activity of two different tumour promoters, teleocidin and TPA, in mouse skin.

Stereoisomeric forms of glycyrrhetinic acid (GA) α-GA and β-GA 18α (α-GA) and 18β (β-GA) were found to inhibit the mutagenicity of benzo[*a*]pyrene (B[*a*] 2-aminoflourene) and aflatoxin B_1 in *Salmonella typhimurium* TA98 and TA100 (Wang et al. 1991). β-GA was more effective than α-GA as an antimutagen. In the two-stage skin tumorigenesis protocol using 7,12-dimethylbenz[*a*]anthracene (DMBA) as the tumour initiating agent followed by twice weekly applications of 12-*O*-tetradecanoylphorbol-13-acetate as tumour promoter, pretreatment of SENCAR mice with α-GA or β-GA resulted in significant protection against tumour initiation as well as tumour promotion. As an anti-tumour initiating agent, β-GA was found to be more effective than α-GA. Similarly, topical application of β-GA was found to be more effective than α-GA in inhibiting the binding of both [^{3}H]B[*a*]P and [^{3}H]DMBA to epidermal DNA.

Glycyrrhizin and caffeine inhibited the action of tumour promoter in mouse skin which was comparable to the effect of quercetin (Yasukawa et al. 1988). Glycyrrhizin and caffeine inhibited the 12-*O*-tetradecanoylphorbol-13-acetate (TPA)-induced inflammation, and these markedly suppressed the promoting effect of TPA on skin tumour formation in mice initiated with 7, 12-dimethylbenz[a]anthracene (DMBA). Agarwal et al. (1991) showed that oral feeding of licorice glycyrrhizin to Sencar mice resulted in substantial protection against skin tumorigenesis caused by DMBA initiation and TPA promotion. The latent period prior to the onset of tumour development was considerably prolonged in glycyrrhizin-fed animals compared with non-glycyrrhizin fed and resulted in significant decrease in the number of tumours per mouse, during and at the termination of the experiment. Oral feeding of glycyrrhizin in drinking water also resulted in inhibition in the binding of topically applied [3H]benzo[a]pyrene and [3H] DMBA to epidermal DNA.

A topical application of a chalcone derivative, 4,2′,4′-trihydroxychalcone (isoliquiritigenin) inhibited epidermal ornithine decarboxylase (ODC) induction and ear oedema formation, that is, inflammation, caused by a topical application of TPA in CD-1 mice (Yamamoto et al. 1991). Also, isoliquiritigenin potently inhibited DMBA-initiated and TPA-promoted skin papilloma formation and epidermal ODC induction and

skin tumour promotion caused by 7-bromomethylbenz[alpha]anthracene (BrMBA), a non-TPA type of tumour-promoting agent, in DMBA-initiated mice. It was found that isoliquiritigenin exerted its anti-tumour-promoting action through the lipoxygenase inhibition by acting on cells other than the target epidermal cells. Intragastric administration of isoliquiritigenin (ISL) for 12 weeks in mice significantly decreased azoxymethane (AOM) induced colon cancer incidence, multiplicity and tumour size by 60 %, 55.4 % and 42.6 %, respectively (Zhao et al. 2014). Moreover, ISL inhibited M2 macrophage polarization in colitis-associated tumorigenesis through downregulating PGE2 and IL-6.

Licorice liquiritigenin effectively inhibited the growth of tumours xenografted in nude mice from human cervical cancer cell line HeLa cells (Liu et al. 2012). Also microvascular density (MVD) of the tumour exposed to liquiritigenin was reduced in a dose dependent manner, especially in the high dose group. Moreover, the expression and secretion of VEGF were down-regulated by liquiritigenin in-vivo and in-vitro. Liquiritigenin markedly reduced cell viability, enhanced apoptotic rate, induced lactate dehydrogenase over-release, and increased intracellular reactive oxygen species (ROS) level and caspase 3 activity in both hepatocellular carcinoma PLC/PRL/5 and HepG2 cells (Wang et al. 2014b). It was found that liquiritigenin induced tumour cell death through mitogen-activated protein kinase- (MPAKs-) mediated pathway. This antitumor activity was further confirmed in PLC/PRL/5-xenografted mice model. In another study, liquiritigenin inhibited cell viability, caused G1 phase arrest and initiated apoptosis in both pituitary adenoma MMQ and GH3 cells via Ras/ERKs and ROS-dependent mitochondrial pathways, suggesting it to be a potential suppressor of pituitary adenoma (Wang et al. 2014c). In mice with GH3 cells xenograft, liquiritigenin markedly reduced tumour size without affecting bodyweight.

Dibenzoylmethane (DBM), a minor β-diketone constituent of licorice and sunscreens, had been shown to exhibit anti-neoplastic effects in chemically induced skin and mammary cancers in several animal models (Jackson et al. 2002). They found that DBM inhibited the growth of LNCaP, DU145 and PC-3 prostate carcinoma cell lines by induction of cell cycle deregulation. Frazier et al. (2004) investigated the proteomic changes induced by treatment of human prostatic LNCaP cancer cells with DBM in order to develop a model for the mechanism to elucidate how DBM induced cell cycle arrest and repressed androgen receptor expression. Dibenzoylmethane was reported to be a promising chemopreventive agent for colon, breast and skin cancer; it mediated the induction of phase II enzymes by Nrf2 activation and inhibited benzo[a]pyrene induced DNA adducts by enhancing its detoxification in lungs (Thimmulappa et al. 2008). Khor et al. (2009) demonstrated that DBM-fed TRAMP mice had a lower incidence of palpable tumour and high-grade prostatic intraepithelial neoplasia. Their findings suggested that DBM blocked the growth and progression of prostate cancer in TRAMP mice via modulation of tumour cell cycle regulation. Dibenzoylmethane (DB), a minor constituent of the root extract of licorice, hydroxydibenzoylmethane and hydroxymethyl-dibenzoylmethane with an identical structure to DB, inhibited phorbol-12-myristate 13-acetate-induced breast carcinoma cell migration and invasion through the protein kinase PI3K/PKCδ-mediated matrix metalloproteinase (MMP)-9 pathway (Liao et al. 2015).

Antimutagenic Activity

For all the compounds tested namely *Glycyrrhiza glabra* extract, glycyrrhizinic acid, 18 α- and 18 β-glycyrrhetinic acids, no desmutagenic activity was observed against ethyl methanesulfonate and N-methyl-N′-nitro-N-nitrosoguanidine; only *Glycyrrhiza glabra* extract showed antimutagenic activity against ethyl methanesulfonate (Zani et al. 1993). On using the ribose-lysine mutagenic browning mixture, the desmutagenic activities of the *Glycyrrhiza glabra* extract, glycyrrhizinic acid, 18 α- and 18 β-glycyrrhetinic

acids were observed. 18 β-Glycyrrhetinic acid was the most active compound. *Glycyrrhiza glabra* extract also exhibited antimutagenic activity against ribose-lysine. *Glycyrrhiza* extract and one of its components, glycyrrhizin, inhibited the mutagenicities of 3-amino-1,4-dimethyl-5 H-pyrido[4,3-b]-indole (Trp-P-1) and 3-amino-1-methyl-5 H-pyrido[4,3-b]indole (Tanaka et al. 1987). *Glycyrrhiza* extract and glycyrrhizin also inhibited the mutagenicities of benzo[a]pyrene, 3-methylcholanthrene, 2-naphthylamine, 2-amino-6-methyldipyrido [1,2-a:3′,2′-d]-imidazole, dimethylnitrosamine and dimethylaminoazobenzene. The mutagenicity of 2-(2-furyl)-3-(5-nitro-2-furyl)acrylamide (AF-2) was inhibited by the *Glycyrrhiza* extract but not by glycyrrhizin. A standardized licorice extract GutGard™ did not show significant increase in number of revertant colonies in *Salmonella typhimurium* strains (TA98 and TAMix) with/without S9 fraction (Chandrasekaran et al. 2011b). In chromosome aberration and micronucleus studies, GutGard™ did not show clastogenic effect at 4 and 18 h treatments with and without S9 fraction. The results indicated that GutGard™ was not mutagenic in a battery of genotoxicity tests.

Antigenotoxic Activity

The antimutagenic and genoprotective activities of *Glycyrrhiza glabra* root extracts were demonstrated both on plant test systems – *Allium fistulosum*, *Allium cepa*, *Vicia faba* and on animals – Wistar rats (Agabeĭli 2012). Various genotoxic studies had indicated that glycyrrhizin was neither teratogenic nor mutagenic, and may possess anti-genotoxic properties under certain conditions (Isbrucker and Burdock 2006). Studies showed that licorice flavonoid oil (LFO) a functional food ingredient appeared not to present any genotoxic hazard to humans on dietary consumption (Nakagawa et al. 2008a) in a reverse mutation assay using four *Salmonella typhimurium* strains and *Escherichia coli*, LFO did not increase the number of revertant colonies in any tester strain with or without metabolic activation by rat liver S9 mix. In a chromosomal aberration test using Chinese hamster lung (CHL/IU) cells, LFO did not induce any chromosomal aberrations either in the short period test without rat liver S9 mix or in the continuous treatment (24 or 48 h) test. No significant or dose-dependent increases in the frequency of micronucleated polychromatic erythrocytes (MNPCE) were observed and the high dose suppressed the ratio of polychromatic erythrocytes (PCE) to total erythrocytes in the bone marrow micronucleus test using male F344 rats. No micronuclei induction either in hepatocytes or PCE was observed even at the highest dose of 5000 mg/kg/day. In-vitro studies found that isoliquiritin apioside, a chalcone oligoglucose isolated from *G. glabra* exhibited marked potential to combat oxidative stress-induced genotoxicity (Kaur et al. 2009). It exhibited modulatory effect against hydrogen peroxide and 4-nitroquinoline-N-oxide induced genotoxicity in *Escherichia coli* PQ37 using SOS chromotest and in human peripheral blood lymphocytes using the Comet assay. Umbelliferone isolated from *Glycyrrhiza glabra* rhizome exhibited moderate response by reducing the induction factor of mutagens hydrogen peroxide by 68.99 % (IC_{50} 223.44 μM) and that of 4NQO (4-nitroquinoline-N-oxide) by 59.71 % (IC_{50} 280.74 μM) in the SOS chromotest using PQ37 strain of *Escherichia coli* (Kaur et al. 2012a). In comet assay in human blood lymphocytes, it exhibited good activity by inhibiting the genotoxicity of both hydrogen peroxide and 4NQO by 61.64 % (IC_{50} 330.02 μM) and 50.66 % (IC_{50} 577.83 μM), respectively. None of licorice leaf (methanol, ethyl acetate and n-hexane) extracts exhibited genotoxic effects in the SOS (short term bacterial) chromotest assay used at doses up to 100ug/assay both in the presence and absence of enzymatic metabolization (Siracusa et al. 2011). Licorice fraction comprising glycyrrhizic acid inhibited the genotoxicity of oxidative mutagens, namely, H_2O_2 and 4NQO quite efficiently (Kaur et al. 2012b). In SOS chromotest, using *Escherichia coli* PQ37 tester strain, it inhibited induction factor induced by H_2O_2 and 4NQO by 75.54 % and 71.69 % at the concentration of 121.46 μM, respectively. In Comet assay, it reduced the tail moment induced by H_2O_2 and

4NQO by 70.21 % and 69.04 %, respectively, at the same concentration in human blood lymphocytes. The isolated fraction also exhibited DPPH free radical scavenging activity and was able to scavenge 85.95 % radicals at a concentration of 120 μM. The results showed glycyrrhizic acid to be a potential modulator of genotoxins as well as efficient scavenger of free radicals. In-vitro study demonstrated that *G. glabra* extracts protected against cadmium-induced genetic and oxidative damage in human lymphocytes (Dirican and Turkez 2014). Co-application of *G. glabra* extract (5, 10 and 20 ppm) and $CdCl_2$ resulted in decreases of micronucleus and sister chromatid exchange formations as compared to the group treated with CdCl2 alone.

Antimicrobial Activity

In-Vitro Studies

The methanol extracts of *G. glabra* roots showed antibacterial activities against *Agrobacterium tumefaciens*, *Bacillus cereus*, *Bacillus subtilis* and *Pseudomonas syringae* pv. *tomato*, but none of the water extracts showed any antibacterial activity against the microorganisms (Ercisli et al. 2008).

Of the isoflavanoids and related substances isolated from *Glycyrrhiza glabra* var. *typica* hispaglabridin A, hispaglabridin B, 4′-*O*-methylglabridin, glabridin, glabrol and 3-hydroxyglabrol exhibited significant antimicrobial activity in-vitro (Mitscher et al. 1980). Both glycoumarin and licocoumarone inhibited the growth of Gram-positive bacteria *Staphylococcus aureus* and *Bacillus subtilis* as strongly as streptomycin. Both also inhibited yeasts *Saccharomyces cerevisiae*, *Candida utilis* and *Pichia nakazawae*, while streptomycin was inactive (Demizu et al. 1988). Pinocembrin, licoflavanone (from licorice leaves) and its converted product cyclolicoflavanone were weakly active against *Bacillus subtilis*, *Staphylococcus aureus* and *Candida albicans*, but were inactive against *Escherichia coli* (Fukui et al. 1988). Glabrene and glabridin, from Russian licorice (*G. glabra* var. *glandulifera*) showed potent antimicrobial activity against Gram positive bacteria, *Staphylococcus aureus* and *Bacillus subtilis* with IC_{50} values of 1.95–7.81 μg/ml, yeast – *Saccharomyces cerevisiae* and *Candida utilis* (IC_{50} values of 7.81–31.3 μg/ml) and the fungus, *Mucor pusillus* (IC_{50} values of 3.91–15.6 μg/ml) (Okada et al. 1989). Seven licorice phenolic compounds showed induction activities of DNA damage in a recombinationless mutant of *Bacillus subtilis* M45 (Fukai et al. 1998). *Glycyrrhiza glabra* rhizome extracts exhibited antifungal activity against *Candida albicans* in-vitro with MIC values of 1.56 mg/ml (Motsei et al. 2003). *Glycyrrhiza glabra* rhizome extract exhibited antifungal activity against *Candida albicans* in-vitro with MIC values of 1.56 mg/ml (Motsei et al. 2003). *G. glabra e*xtracts from samples collected in various sites of Calabria, Italy, exhibited antibacterial activity against *Staphylococcus aureus*, *Enterococcus faecalis*, *Micrococcus luteus* and the fungus *Trichophyton mentagrophytes*, some samples inhibited *Pythium ultimum* (Statti et al. 2004). The hydroalcoholic extract of *G. glabra* exhibited antifungal activity in-vitro against *Candida albicans* and *Aspergillus niger* (Tharkar et al. 2010). The ethanolic extract of *Glycyrrhiza glabra* showed good antifungal activity in-vitro against *Aspergillus niger*, *Aspergillus fumigatus*, *Candida albicans*, *Mucor* sp. and *Penicillium marneffei* (Geetha and Roy 2013). The in-vitro growth of the *Candida albicans* strains was markedly reduced, in a pH-dependent manner, by relatively low doses (6.2 μg/mL) of 18-β glycyrrhetinic acid, from *Glycyrrhiza glabra* root (Pellati et al. 2009). The results demonstrated 18-β glycyrrhetinic acid to be a promising biological alternative for the topical treatment of recurrent vulvovaginal candidiasis.

Of various oriental herb extracts tested only *Glycyrrhiza glabra* showed a remarkable antibacterial activity against *Propionibacterium acnes*, resulting in negligible induction of resistance, in comparison with a marked development of resistance in the bacteria treated with erythromycin (Nam et al. 2003). The ethanol extract (0.01 %) of *Angelica dahurica* markedly suppressed neutrophil chemotaxis, comparable to

the effect of erythromycin (0.01 %), whereas a strong antilipogenic effect was obtained with rhizoma coptidis (*Coptis chinensis*) extract (0.01 %), leading to a higher efficacy than that of retinoic acid (0.01 %). The authors suggested that an appropriate formulation containing *A. dahurica*, *Coptis chinensis* rhizome and *G. glabra* could be helpful for the prevention and treatment of acne lesions. The mixture of *Capsella bursa-pastoris* and *Glycyrrhiza glabra* extracts was more effective against all the oral pathogenic bacteria *Streptococcus mutans*, *S. sanguis*, *Actinomyces viscosus*, *Enterococcus faecalis* than the separate individual extracts indicating synergistic effects between the two plant extracts (Soleimanpour et al. 2013).

Fukai et al. (2002b) found that glabridin exhibited antibacterial activity in-vitro against methicillin sensitive *Staphylococcus aure*us, methicillin resistant *S. aure*us (MRSA) and licochalcone A exhibited antibacterial activity against MRSA with MIC values ranging from 6.25 to 16 μg/mL depending on the strain of microorganism. *G. glabr*a extract at concentration higher than 7.5 % exhibited inhibitory effects in-vitro against *Salmonella typhi*, *S. paratyphi B*, *Shigella sonnei*, *S. flexneri* and enterotoxigenic *Escherichia coli* (Shirazi et al. 2007). Antimycobacterial activity of *Glycyrrhiza glabra* root was found at 500 μg/mL concentration (Gupta et al. 2008). Bioactivity guided phytochemical analysis identified glabridin as potentially active against both *Mycobacterium tuberculosis* H_{37}Ra and H_{37}Rv strains at 29.16 μg/mL concentration. It exhibited antimicrobial activity against both Gram-positive and Gram-negative bacteria. Raw polysaccharides from *Glycyrrhiza glabra* were shown to have strong anti-adhesive effects against *Porphyromonas gingivalis*, a major pathogen for induction of periodontal inflammations (Wittschier et al. 2009). Glabridin, from licorice roots, was found to be active against both yeast and filamentous fungi (Fatima et al. 2009). Glabridin also showed resistance modifying activity against drug-resistant mutants of *Candida albicans* at a minimum inhibitory concentration of 31.25–250 μg/mL. The vegetative cell growth of *Bacillus subtilis* was inhibited dose-dependently by licochalcone A and was completely inhibited at a concentration of 3 μg/ml (Tsukiyama et al. 2002). Licochalcone A did not inhibit the germination of heat-treated spores of *B. subtilis* induced by L-alanine. Licochalcone A showed effects against all Gram-positive bacteria tested and was especially effective against all *Bacillus* spp. tested, with MICs of 2–3 μg/ml, but was not effective against Gram-negative bacteria or eukaryotes at 50 μg/ml. Glabridin and licochalcone A showed antifungal activity on *C. albicans* while glycyrrhizic acid had no effect (Messier and Grenier 2011). Biofilm formation was inhibited by 35–60 % in the presence of licochalcone A (0.2 μ/ml/). A strong inhibitory effect (>80 %) on hyphal formation was observed with licochalcone A or glabridin (100 μg/ml).

Licorice root and leaf extracts showed activity against *Candida albican*s, and tested Gram-positive bacteria *Bacillus subtilis*, *Enterococcus faecalis* and *Staphylococcus aureus* in a dose dependent manner (Irani et al. 2010). The ethanolic extract of the leaves was the most active extract against Gram-positive bacteria. Ether, chloroform, acetone extracts of liquorice roots showed significant antibacterial activities against two Gram-positive (*Bacillus subtilis* and *Staphylococcus aureus*) and two Gram-negative (*Escherichia coli* and *Pseudomonas aeruginosa*) bacteria (Nitalikar et al. 2010). The hydro-methanolic *G. glabra* root extract displayed antibacterial activities in-vitro against *Pseudomonas aeruginosa*, *Escherichia coli*, *Shigella flexneri*, *Staphylococcus aureus*, *Staphylococcus epidermidis* and *Bacillus subtilis* (Varsha et al. 2013). *Shigella flexneri* was the most sensitive. Aqueous and ethanolic licorice extracts were found to have antimicrobial activity (Ajagannanavar et al. 2014). MIC of aqueous and ethnolic licorice root extract against oral pathogens, *Streptococcus mutans* and *Lactobacillus acidophilus* were 25 % and 12.5 %, respectively. The methanol licorice root extract exhibited moderate antimicrobial activity (Chopra et al. 2013). The extract was most potent against *Staphylococcus aureus* at 500 μg/mL (inhibition zone 13 mm) among bacteria and showed maximum potency against

Rhizopus spp. at 500 μg/ml (inhibition zone 11 mm) among fungi. It was least active against *Aspergillus awamori*. Diethyl carbonate extracts of *Glycyrrhiza glabra* root from Astrakhan region (Russia) exhibited maximum activity against test bacterial strains *Staphylococcus aureus*, *Escherichia coli* and *Bacillus subtilis*; activity of Astrakhan licorice was superior among 50 % ethanol extracts from Astrakhan (Russia) and Calabria (Italy) (Astaf'eva and Sukhenko 2014). Antibacterial activity was directly proportional to the content of glycyrrhizin and 18β-glycyrrhetinic acid in the extracts. The content of these chemical components in *Glycyrrhiza glabra* root from Astrakhan region was higher than in licorice growing in Italy.

Clinical Studies

Recent research suggested that licorice and its bioactive ingredients such as glycyrrhizin, glabridin, licochalcone A, licoricidin and licorisoflavan A possessed potential beneficial effects against oral microbial pathogens and the host immune response involved in common ora-dental diseases (dental caries, periodontitis, candidiasis and recurrent aphthous ulcers) (Messier et al. 2012). A 1989 split-mouth clinical study on 21 dental subjects found that glycyrrhizin had a tendency toward a statistically significant effect for controlling dental plaque formation caused by cariogenic bacteria after just a few days (Steinberg et al. 1989). However, another pilot study involving 40 male and female volunteers showed that tooth brushing for up to 42 days with a toothpaste containing glycyrrhizin had no effect on the plaque index compared with using a control toothpaste (Goultschin et al. 1991). Possible explanations for the lack of efficiency in improvement of plaque, gingival and bleeding indices may have been an insufficient glycyrrhizin concentration and/or chemical incompatibility in a toothpaste containing a mixture of an anionic detergent and an organic antibacterial surface agent. In a more recent pilot study of pre-school children, clinical reduction of *Streptococcus mutans* was obtained by using a licorice root lollipop intervention twice daily for 3 weeks (Peters et al. 2010). High-risk children showed the steepest early decrease in mean log-*Streptococcus mutan*. At the end of a follow-up (9 week) period, the log *Streptococcus mutans* decrease moved the high-risk group down to moderate-risk level. Glycyrrhizol A extracted from licorice roots was found to have strong antimicrobial activity against cariogenic bacteria like *Streptococcus mutans* (Hu et al. 2011). In two pilot human studies they found that a brief application of sugar-free licorice lollipops (twice a day for 10 days) led to a marked reduction of cariogenic bacteria in oral cavity among most human subjects tested. This herbal lollipop could be a novel tool to promote oral health through functional foods. Jain et al. (2013) conducted as a double-blind pilot study of paediatric patients ($N=60$), aged 7–14 years, equally divided by randomization into three groups, namely, Group 1 using aqueous liquorice mouthwash (15 %), Group 2 using ethanolic liquorice mouthwash (3.75 %) and Group 3 using chlorhexidine gluconate (0.156 %) as positive control. They found the mean *Streptococcus mutans* colony counts in all three groups decreased significantly immediately after the oral rinsing. The reduction in colony counts was significant in ethanolic liquorice group as compared to the control. Liquorice extracts also led to an immediate rise in salivary pH. The study affirmed that both aqueous and ethanolic liquorice extracts were potent cariostatic agents and were palatable by child patients.

Antiviral Activity

Glycyrrhizin exhibited a concentration-dependent inhibition of the replication of hepatitis A virus (HAV) in PLC/PRF/5 cells (Crance et al. 1994). It was shown to inhibit an early stage of the HAV replication. In in-vitro studies, glycyrrhizin suppressed the secretion of hepatitis B surface antigen (HBsAg) dose-dependently in PLC/PRF/5 cells (Takahara et al. 1994). It was found that glycyrrhizin suppressed the intracellular transport of HBsAg at the trans-Golgi area after O-linked glycosylation and before its sialylation. Glycyrrhizin administered intravenously in guinea pigs might bind to hepatocytes at the con-

centration at which glycyrrhizin could modify the expression of chronic hepatitis B virus (HBV)-related antigens on the hepatocytes and suppress sialylation of HBV surface antigen (HBsAg) (Sato et al. 1996). Glycyrrhizin and its diastereoisomers were found to be inhibitory against two new human herpes virus: HHV-6 and HHV-7 (Ceremelli et al. 1996). Randomized controlled trials confirmed that the *Glycyrrhiza glabra* derived compound glycyrrhizin and its derivatives reduced hepatocellular damage in chronic hepatitis B and C (Fiore et al. 2008). In hepatitis C virus-induced cirrhosis the risk of hepatocellular carcinoma was reduced. Animal studies demonstrated a reduction of mortality and viral activity in herpes simplex virus encephalitis and influenza A virus pneumonia. In-vitro studies revealed antiviral activity against HIV-1, SARS related coronavirus, respiratory syncytial virus, arboviruses, vaccinia virus and vesicular stomatitis virus. Mechanisms for antiviral activity of *Glycyrrhiza* spp. include reduced transport to the membrane and sialylation of hepatitis B virus surface antigen, reduction of membrane fluidity leading to inhibition of fusion of the viral membrane of HIV-1 with the cell, induction of interferon gamma in T-cells, inhibition of phosphorylating enzymes in vesicular stomatitis virus infection and reduction of viral latency (Fiore et al. 2008). Matsumoto et al. (2013) demonstrated that glycyrrhizin treatment of hepatitis C virus (HCV)-infected Huh7 cells caused a reduction of infectious HCV production. The results suggested that glycyrrhizin inhibited the release of infectious HCV particles via its inhibitory effect on 1B phospholipase A2. Also combination treatment with glycyrrhizin augmented IFN-induced reduction of virus in the cell culture-produced HCV system.

Ito et al. (1987) found that glycyrrhizin completely inhibited human immunodeficiency virus (HIV)-induced plaque formation in molt-4 (MT-4) cells at a concentration of 0.6 mM, the 50 % inhibitory dose being 0.15 mM. Glycyrrhizin completely inhibited the cytopathic effect of HIV and the HIV-specific antigen expression in MT-4 cells at a concentration of 0.3 and 0.6 mM, respectively. Further, glycyrrhizin inhibited giant cell formation of HIV-infected Molt-4 clone No. 8 cells. The anti-HIV-1 activity of glycyrrhizin may be attributed to its inhibition of protein kinase activity and partly to an interference with virus-cell binding (Ito et al. 1988). Five phenolics licochalcone A, isolicoflavonol, glycycoumarin, glycyrrhisoflavone and licopyranocoumarin isolated from licorice inhibited the cytopathic activity of a human immunodeficiency virus (Hatano et al. 1988). An anti-HIV (human immunodeficiency virus) phenolic constituent, licopyranocoumarin was isolated from Si-pei licorice (a commercial licorice; root and stolon of *Glycyrrhiza* sp. from the north-western region of China) (Hatano et al. 1989). Betulinic acid and dihydrobetulinic acid derivatives were reported to be potent anti-HIV agents (Kashiwada et al. 1996). The results of in-vitro studies indicated that glycyrrhizin had the potential to inhibit a non-syncytium-inducing variant of HIV (NSI-HIV) replication in HIV-infected patients peripheral blood mononuclear cell cultures by inducing the production of β-chemokines CCL4 and CCL5 (Sasaki et al. 2002–2003). 18α-glycyrrhizic acid exhibited anti-HIV activity based on primary infection of MT-4 lymphoid cells with HIV (an acute HIV infection model) using strain HIV/EVK but its esters 18α-glycyrrhizic acid pentasulphate sodium disodium salt (III), 18α-glycyrrhizic acid di-*O*-nicotinate (IV) exhibited lower activity (Baltina et al. 2010). Changes in the stereochemistry of the18α-glycyrrhizic acid aglycone led to significant decreases in its anti-HIV-1 activity.

Glycyrrhizic acid inactivated herpes simplex virus particles irreversibly (Pompei et al. 1979). Studies by Utsunomiya et al. (1995) found that glycyrrhizin improved the resistance of thermally injured mice to opportunistic infection of herpes simplex virus type 1 through the induction of CD4+ contrasuppressor T cells. Glycyrrhizin was found to reduce morbidity and mortality of mice infected with lethal doses of influenza virus through stimulation of IFN-gamma production by T cells (Utsunomiya et al. 1997). In-vitro studies showed that soyasaponin II was more potent than soyasaponin I against herpes simplex virus type 1 (HSV-1) as shown by reduction of

HSV-1 production (Hayashi et al. 1997). When acyclovir and soyasaponin II were evaluated in combination for anti-HSV-1 activity, additive antiviral effects were observed for this virus. Soyasaponin II was also found to inhibit the replication of human cytomegalovirus, influenza virus and human immunodeficiency virus type 1. The action was not due to inhibition of virus penetration and protein synthesis, but might involve a virucidal effect. Ghannad et al. (2014) found that aqueous licorice extract pretreatment of Vero cells infected with HSV-1 was efficacious. There was significant difference in the efficacy of the extract with regard to incubation period between 1 and 4 h, 1 and 8 h, 4 and 12 h and 8 and 12 h. When glycyrrhizic acid and rapamycin were added to HeLa cells together with the viruses, glycyrrhizic acid demonstrated a strong anti-herpes simplex virus type 1 (HSV1) activity, whereas rapamycin had no activity (Laconi et al. 2014). However, if the compounds were added to the cells 24 h before the viruses, glycyrrhizic acid induced the production of an even higher amount of Beclin 1 and showed an improved antiviral effect; under these conditions, rapamycin was also able to exert a significant anti-HSV1 activity. The results suggested glycyrrhizic acid to be a strong inducer of the autophagy activator Beclin 1.

The study of Omer et al. (2014) showed that *Glycyrrhiza glabra* extract (60 mg/100 ml) inhibited replication of Newcastle disease virus and was non-toxic in the embryonated eggs. Glycyrrhizic acid from *Glycyrrhiza glabra* root inhibited growth and cytopathology of several unrelated DNA and RNA viruses: Vaccinia, Newcastle disease, vesicular stomatitis, herpes simplex type 1, influenza, etc., in-vitro, while not affecting cell activity and the ability to replicate (Pompei et al. 1979). Glycyrrhizic acid was also found to have antiviral activity against enterovirus 71 (EV71) of foot and mouth disease and coxsackievirus A16 (CVA16) on Vero cells (Wang et al. 2013a). Treatment with glycyrrhizin enhanced the production of interferon-γ in human peripheral lymphocyte-macrophage cultures by concanavalin A (Con A) (Shinada et al. 1986). Collaboration between enriched T-lymphocytes and macrophages, both treated with glycyrrhizin, was needed for the enhancement of interferon-γ production. A smaller increase in interferon production was also observed in the glycyrrhizin-treated peripheral lymphocyte-macrophage cultures derived from an asymptomatic carrier of hepatitis B virus, in response to Con A and surface antigen of hepatitis B virus.

Glycyrrhizin exhibited antiviral activity against varicella-zoster virus (VZV) in-vitro (Baba and Shigeta 1987). When human embryonic fibroblast (HEF) cells were treated with glycyrrhizin after inoculation of virus (post-treatment), the average 50 %-inhibitory dose (ID_{50}) for five VZV strains was 0.71 mM, and the selectivity index (ratio of ID50 for host-cell DNA synthesis to ID50 for VZV replication) was 30. Glycyrrhizin was also effective against VZV replication when HEF cells were treated 24 h before the inoculation (pretreatment). In addition, at a concentration of 2.4 mM glycyrrhizin inactivated more than 99 % of virus particles within 30 min at 37 °C. In combination with other anti-herpes drugs (acyclovir, adenine arabinoside, bromovinyldeoxyuridine and phosphonoformate) or human native β-interferon, glycyrrhizin had an additive or slightly synergistic effect on VZV replication. Glycyrrhizin, licorice and ammonium salt of glycyrrhizic acid exhibited antiviral activity on three strains of Japanese encephalitis virus (JEV), Nakayama, P-20778 and 821564XY48 (Badam 1997). Purified glycyrrhizin inhibited plaque formation in all three strains of JEV at a concentration of 500 μg/ml at 96 h. Similar effect was observed at 1000 μg/ml concentration with licorice and ammonium salt of glycyrrhizic acid. The minimal inhibitory concentrations were not toxic to porcine stable kidney (PS) and human cervical carcinoma (HeLa) cell lines. Glycyrrhizin inhibited the viral antigen expression of human cytomegalovirus (HCMV) in human monocytic cell line U-937 and human embryonic lung cell line MRC-5 in-vitro (Numazaki et al. 1994). Glycyrrhizin (GA) and primary metabolite 18β-glycyrrhetinic acid (GRA) pharmacologically active components of the medicinal licorice root had both been shown to have antiviral and immunomodulatory proper-

ties (Hardy et al. 2012). However, GRA, but not GA, exhibited significant antiviral activity against rotavirus replication in-vitro. GRA treatment reduced rotavirus yields by 99 % when added to infected cultures post-virus adsorption. In in-vivo studies, they showed that GRA delivered orally to mice with rotavirus infection augmented lymphocyte recruitment to the intestinal mucosa and induced maturation of B cell-rich ILF independently of ectopic antigenic stimulus (Hendricks et al. 2012). GRA reduced the duration of viral antigen shedding, and endpoint serum antibody titers were higher in GRA-treated animals.

Glycyrrhizic acid, a component of licorice root was active against Epstein Barr virus replication in superinfected Raji cells in a dose-dependent fashion (Lin 2003). The IC_{50} values for viral inhibition and cell growth were 0.04 and 4.8 mM, respectively. The selectivity index (ratio of IC_{50} for cell growth to IC_{50} for viral DNA synthesis) was 120. Of the antiviral compounds ribavirin, 6-azauridine, pyrazofurin, mycophenolic acid and glycyrrhizin assessed against two clinical isolates of SARS-associated coronavirus (FFM-1 and FFM-2) from SARS patients, glycyrrhizin was the most active in inhibiting replication of the SARS-associated virus (Cinatl et al. 2003). Aqueous *Glycyrrhiza glabra* root extracts were found to have strong significant antiviral activity against rhesus rotavirus with a 50 % inhibitory concentration (IC_{50}) < 300 μg/ml; its pure constituent, 18ß-glycyrrhetinic acid, was found to have the strongest antiviral activity (IC_{50} 46 μM) (Knipping et al. 2012).

Antihyperglycaemic/Antidiabetic Activity

11β-Hydroxysteroid dehydrogenase type 1 (11β-HSD1) inhibitors had been investigated as potential treatments for metabolic diseases, such as diabetes mellitus type 2 or obesity (Su et al. 2007; Classen-Houben et al. 2009; Stanetty et al. 2010). Su et al. (2007) reported the inhibition of human and rat 11β-hydroxysteroid dehydrogenase type 1 by 18β-glycyrrhetinic acid (18β-GA) derivatives. The 11-modified 18β-GA derivatives 2 and 3 with apparent selectivity for rat 11β-HSD1 showed a high percentage inhibition for human microsomal 11β-HSD1 at 10 μM and exhibited IC_{50} values of 400 and 1100 nM, respectively. Classen-Houben et al. (2009) compared the biological activity of 18β-GA and its diastereomer 18α-GA against the two enzymes in lysates of transfected HEK-293 cells and showed that 18α-GA selectively inhibited 11β-HSD1 but not 11β-HSD2. This was in contrast to 18β-GA, which preferentially inhibited 11β-HSD2. The side chain modified 18β-GA derivatives 4 and 5, although showing selectivity for rat 11β-HSD1 inhibited human microsomal 11β-HSD1 with IC_{50} values in the low micromolar range.

Administration of 18β-glycyrrhetinic acid to streptozotocin-induced diabetic rats reduced hyperglycaemia and hyperlipidaemia related to the risk of diabetes mellitus (Kalaiarasi et al. 2009). All elevated levels of total cholesterol, triglyceride, free fatty acid and phospholipids and reduced HDL cholesterol level in the diabetic rats were reverted back to normalcy. Results of the study by Sawada et al. (2010) indicated that glabridin, a prenylated isoflavone in licorice, may possess a therapeutic effect on metabolic disorders, such as diabetes and hyperglycaemia, by modulating glucose metabolism through adenosine monophosphate-activated protein kinase (AMPK)-dependent GLUT4 translocation pathway in the plasma membrane of mice skeletal muscle cells. Licorice rhizome isoliquiritigenin and its derivatives 4,4′-diacetoxy-2′-hydroxy chalcone; 2′,4′dimethoxy-4-hydroxychalcone; 4-acetoxy-2′,4′-dimethoxychalcone; 2′,4′-dimethoxychalcone and liquiritigenin derivatives liquiritigenin 4′-acetate; and liquiritigenin 7,4′-dibenzoate showed significant blood glucose lowering effect in normal Swiss albino male mice (Gaur et al. 2014). Isoliquiritigenin, 2′,4′dimethoxy-4-hydroxychalcone and liquiritigenin 7,4′-dibenzoate were selected for in-vivo antidiabetic activity and found to be potential candidates for treatment of diabetes

Studies found that oral administration of male Sprague–Dawley rats with high-fat/high-sucrose (HF/HS) diet elevated the fasting blood glucose level and insulin resistance index which was pre-

vented by licorice root glycyrrhizic acid (GA) supplementation (Cheng et al. 2014). GA treatment significantly lowered the circulating advanced glycation end product (AGE) independent of its glucose-lowering effect. HF/HS diet also triggered receptor for advanced glycation end product (RAGE) upregulation in the abdominal muscles while GA administration downregulated RAGE expression in the abdominal muscles, aorta and subcutaneous adipose tissues. It was concluded that HF/HS diet could cause glucose intolerance, insulin resistance and upregulation of RAGE expression while GA ameliorated the metabolic dysregulation besides exhibiting inhibitory effects on the AGE-RAGE axis.

Antihyperlipidemic/Hypocholesterolaemic/Anti-obesity Activity

Animal Studies

Administration of licorice root extract to hypercholesterolaemic rats reduced the elevated levels of triglycerides and total lipids and augmented the low level of phospholipids in the rat serum and most organs (Sitohy et al. 1991). Licorice exhibited hypocholesterolaemic action and improved the impaired function of both liver and kidney. Administration of *Glycyrrhiza glabra* root extract to male rats at oral doses of 200, 400 and 800 mg/kg for 4 weeks induced a significant decrease in food intake and significant increases in body weight gain and feed efficiency ratio as compared to the control (Shalaby et al. 2004). At doses of 400 and 800 mg/kg, the extract caused significant decreases in total cholesterol and triglycerides associated with non-significant reductions in HDLc, LDLc and VLDLc concentrations in the serum. The extract at all doses produced significant decreases in the levels of serum liver enzymes (AST and ALT) and urea nitrogen, while the creatinine concentration significantly decreased by the high dose only. Studies showed that feeding obese diabetic KK-A(y) mice with diets containing licorice flavonoid oil (LFO) suppressed body weight gain, weights of abdominal adipose tissues and blood glucose levels (Nakagawa et al. 2004). Furthermore, LFO and licorice ethanolic extract stimulated human adipocyte differentiation in-vitro. The results indicated that licorice hydrophobic flavonoids possessed abdominal fat-lowering and hypoglycaemic effects, possibly mediated via activation of peroxisome proliferator-activated receptor-gamma (PPAR-gamma).

A 4-week administration of licorice root powder (5 and 10 gm% in diet) to hypercholesterolaemic rats resulted in significant reduction in plasma, hepatic total lipids, cholesterol, triglycerides and plasma low-density lipoprotein and VLDL-cholesterol accompanied by significant increases in HDL-cholesterol levels (Visavadiya and Narasimhacharya 2006). Furthermore, significant increases in faecal cholesterol, neutral sterols and bile acid excretion along with an increase in hepatic HMG-CoA reductase activity and bile acid production were observed in these animals. The root powder administration to hypercholesterolaemic rats also decreased hepatic lipid peroxidation with a concomitant increase in superoxide dismutase (SOD) and catalase activities and total ascorbic acid content. The antioxidant status of these animals also was improved upon treatment. Animal studies showed that *G. glabra* extract significantly decreased total cholesterol, LDL cholesterol and triglycerides levels, increased HDL cholesterol and reduced atereosclerotic lesion in the aorta of hypercholesterolemic rabbits (Asgary et al. 2007). This effect was ascribed to the effect of licorice on plasma lipoproteins and its antioxidant and anti-inflammatory properties.

Administration of ethanolic extract and its ethyl acetate soluble, water soluble and hexane soluble fractions to dyslipidaemic hamsters fed a high-fat diet (containing g/kg fructose 500 g, casein 190 g, Dalda Vansapati Ghee 110 g, wheat/corn/gram flour 150 g, cholesterol 5 g, methionine 3 g, vitamin mix 3 g, mineral mixture 40 g), decreased serum level of total cholesterol by 25.9, 38.0, 39.0 and 26.3 %, respectively (Maurya et al. 2009). The ethanolic extract, ethyl acetate soluble, water soluble and hexane soluble fraction increased the serum HDL-cholesterol level

by 14.8, 34.3, 27.3 and 17.2 %, respectively. The ethanolic extract, ethyl acetate fraction, aqueous fraction and hexane fraction decreased the triglyceride level by 31.3, 37.2, 41.2 and 28.9 %, respectively. The reduction in LDL-cholesterol level by ethanolic extract, ethyl acetate soluble fraction and water soluble fraction were 43.9, 31.0, 33.4 and 24.6 %, respectively. Compared with obese mice in the control group, those fed a high-fat diet containing1% and 2 % licorice flavonoid oil (LFO) presented reductions in the weight of abdominal white adipose tissues and body weight gain (Aoki et al. 2007a). A histological examination revealed that the adipocytes became smaller and the fatty degenerative state of the hepatocytes was improved in the 2 % LFO group. The findings suggested that LFO prevented and ameliorated diet-induced obesity via the regulation of lipid metabolism-related gene expression in the liver. Their findings suggested that the decrease in abdominal adipose tissue weight by LFO was mediated by the transcriptional regulation of sterol regulatory element-binding protein-1c (SREBP-1c) (a transcription factor promoting hepatic fatty acid synthesis), and peroxisome proliferator-activated receptor-alpha (PPAR-alpha) (a transcription factor promoting hepatic fatty acid oxidation) in the liver (Honda et al. 2009). Among the isolated phenolic compounds from *G. glabra* root ethanol extract, 5′-formylglabridin; (2*R*,3*R*)-3,4′,7-trihydroxy-3′-prenylflavane; echinatin; (3*R*)-2′,3′,7-trihydroxy-4′-methoxyisoflavan; kanzonol X; kanzonol W; shinpterocarpin; licoflavanone A; glabrol; shinflavanone; gancaonin L; and glabrone all exhibited significant PPAR-γ ligand-binding activity (Kuroda et al. 2010). The activity of these compounds at a sample concentration of 10 μg/mL was three times more potent than that of 0.5 μM troglitazone. Among 12 flavonoids isolated from *G. glabra* roots, isoliquiritigenin; 3,3′,4,4′-tetrahydroxy-2-methoxychalcone; licuroside and isoliquiritoside showed strong inhibition against pancreatic lipase in-vitro with IC_{50} values of 7.3 μM, 35.5 μM, 14.9 μM and 37.6 μM, respectively (Birari et al. 2011). In high fat diet (HFD) fed rats supplemented with isoliquiritigenin, the body weight increase was only 23.2 g as compared to 64.2 g in the HFD control group while rats treated with licuroside showed 23.2 g weight gain only. Isoliquiritigenin decreased the levels of plasma total cholesterol (TC) to 84.6 mg/dl and plasma total triglycerides (TG) to 128.8 mg/dl. Licuroside also lowered the plasma TC and TG levels considerably. The results indicated the potential of the chalcone scaffold as a source of pancreatic lipase inhibitors for preventing obesity.

Glabridin, an isoflavan isolated from licorice, effectively inhibited adipogenesis in 3T3-L1 cells and glabridin-rich supercritical fluid extract of licorice (LSC) showed inhibitory effect on adipogenesis in a dose-dependent manner (Ahn et al. 2013). LSC significantly reduced weight gain by high-fat diet mice in a dose-dependent manner. The reductions of the hypertrophy of white adipose tissue and of fat cell size were also observed. In the liver, LSC supplementation effectively inhibited high-fat diet-induced hepatic steatosis through downregulation of gluconeogenesis related phosphoenolpyruvate carboxykinase and glucose 6-phosphatase and upregulation of the β-oxidation related carnitine palmitoyltransferase 1. The results suggested that glabridin and glabridin-rich licorice extract would be effective anti-obesity agents.

Clinical Studies

In a study of 15 normal-weight subjects (7 males, age 22–26 years, and 8 females, age 21–26 years), consumption of 3.5 g a day of a commercial preparation of licorice for 2 months was found to reduce body fat mass (BFM) and plasma renin activity and aldosterone were suppressed, without any change in body mass index (BMI) (Armanini et al. 2003). It was suggested that licorice could reduce fat by inhibiting 11β-hydroxysteroid dehydrogenase Type 1 at the level of fat cells. In another study of 18 healthy women (age range 20–33 years) with normal BMI, topical application of a cream containing 2.5 % glycyrrhetinic acid to the thigh for a month reduced the thickness of subcutaneous layer of thigh fat in comparison to the contralateral untreated thigh and to control subjects treated with the placebo cream (Armanini et al. 2005).

The effect of glycyrrhetinic acid on the thickness of subcutaneous fat was likely related to a block of 11βa-hydroxysteroid dehydrogenase type 1 at the level of fat cells; therefore, glycyrrhetinic acid could be effectively used in the reduction of unwanted local fat accumulation.

Hepatoprotective Activity

In-Vitro Studies

Fractionated extracts of *Glycyrrhiza glabra* and *Schisandra chinensis* inhibited the action of acetaminophen or D-galactosamine in small scale rat hepatocyte primary culture model (Nakagiri et al. 2003). It was concluded that the hepatocyte microculture system presented was suitable for the screening of hepatoprotective substances. Glycyrrhizic acid (GA) exhibited protective effects against aflatoxin B_1 (AFB_1)-induced cytotoxicity in human hepatoma cell line (HepG2) (Chan et al. 2003). Both CYP1A1 and glutathione *S*-transferase (GST) activities were increased in cells after treatment with the GA. For cells without GA pretreatment, cell injury was implicated as indicated by the decrease in cell viability. It was found that GA pretreatment provided protective effects in terms of the enzyme activity and increase in cell viability. GA also protected against aflatoxin-induced oxidative stress. In a hepatocyte model of cholestatic liver injury in a hepatocyte model of cholestatic liver injury, pre-incubation with licorice glycyrrhizin exerted pro-apoptotic properties, whereas pre-incubation with its metabolite 18β-glycyrrhetinic acid potently inhibited bile acid-induced apoptosis and necrosis in a manner consistent with its antioxidative effect (Gumpricht et al. 2005). Pretreatment with glycyrrhizic acid (GA) (4 μg) protected the hepatocytes against t-BHP induced oxidative injury and the results were comparable to the pretreatment with positive control, that is, silymarin (Tripathi et al. 2009). The protective potential against cell death was achieved mainly by preventing intracellular GSH depletion, decrease in ROS formation as well as inhibition of mitochondrial membrane depolarization. GA was found to modulate critical end points of oxidative stress-induced apoptosis and could be beneficial against liver diseases.

Animal Studies

Oral administration of *G. glabra* extract (200 mg/kg, bodyweight) protected against paracetamol induced liver damage in rats (taju et al. 2011). All altered levels of biochemical markers were restored to the near normal levels in the dose dependent manner. Histological examination of the liver tissues confirmed the hepatoprotective effect of *G. glabra.*

Studies by Jeong et al. (2002) showed that protective effects of 18β-glycyrrhetinic acid (GA) against the carbon tetrachloride-induced hepatotoxicity in mice may be due to its ability to block the bio-activation of carbon tetrachloride, primarily by inhibiting the expression and activity of P450 2E1, and its free radical scavenging effects. Biotransformation of 18β-glycyrrhetinic acid by *Absidia pseudocylinderospora*, *Gliocladium viride* and *Cunninghamella echinulata* afforded seven metabolites, including three new ones 15α-hydroxy-18α-glycyrrhetinic acid; 13β-hydroxy-7α,27-oxy-12-dihydro-18β--glycyrrhetinic acid and 1α-hydroxy-18β-glycyrrhetinic acid and known metabolites 7β, 15α-dihydroxy-18β-glycyrrhetinic acid; 7β-hydroxy-18β-glycyrrhetinic acid; 5α-hydroxy-18 β-glycyrrhetinic acid and 3-oxo derivative of glycyrrhetinic acid (Maatooq et al. 2010). Two major metabolites 7β, 15α-dihydroxy-18β-glycyrrhetinic acid and 13β-hydroxy-7α,27-oxy-12-dihydro-18β-glycyrrhetinic acid displayed significant hepatoprotective activity against CCl_4-induced hepatotoxicity in albino mice. Pretreatment of Wistar rats with 18-β Glycyrrhetinic acid (18β-GA) at two different doses (45 and 75 mg kg^{-1} b.w.) significantly ameliorated 2-acetylaminofluorene-induced increased lipid peroxidation, alanine transaminase and aspartate transaminase, xanthine oxidase activities and activities of phase-II detoxifying enzymes along with the levels of glutathione content (Hasan et al. 2015). Administration of 18β-GA also significantly restored the expressions of proliferating cell nuclear antigen, cyclooxygenase 2, inducible

nitric oxide synthase and nuclear factor κB. Furthermore, histological observations also supported the preventive effects of 18β-GA. Their findings suggested that pretreatment with 18β-GA showed potential hepatoprotective effects via attenuation of oxidative stress, inflammation and hyperproliferation.

In-vivo studies demonstrated that liquiritigenin, an aglycone of liquiritin in licorice root, efficaciously protected the liver from acute injuries induced by acetaminophen-induced or from acetaminophen plus buthionine sulfoximine-induced severe injuries in rats (Kim et al. 2006). Liquiritigenin pretreatments significantly reduced the potentiated liver necrosis, decreasing mortality. Another animal study demonstrated that liquiritigenin had a choleretic effect and the ability to induce transporters and phase-II enzymes in the rat liver, which may be associated with a hepatoprotective effect against galactosamine/LPS (Kim et al. 2009). Liquiritigenin treatments attenuated galactosamine/LPS-induced hepatitis in rats, as supported by decreases in the plasma alanine aminotransferase, liver necrosis and plasma TNF-alpha. Intragastric administration of liquiritigenin (20 mg/kg) for 15 days effectively inhibited the growth of transplanted H(22) hepatocarcinoma in mice (Zhou et al. 2010). It also decreased malondialdehyde content and increased thymus weight. The study by Kim et al. (2010) demonstrated that isoliquiritin, a licorice antioxidant flavonoid, had the ability to repress liver X receptor-α (LXRα)-dependent hepatic steatosis through JNK1 inhibition and protected hepatocytes from oxidative injury inflicted by fat accumulation. In mice fed a high-fat diet, isoliquiritin treatment inhibited hepatic steatosis, as shown by a decrease in fat accumulation and repression of lipogenic genes. The results of blood biochemistry and histopathology confirmed attenuation of high-fat diet-induced liver injury by isoliquiritin. Moreover, isoliquiritin inhibited oxidative stress, as indicated by decreases in thiobarbituric acid-reactive substance formation, iNOS and COX2 induction, and nitrotyrosinylation.

Studies showed that compared to glycyrrhizin (Gly) and matrine (Mat) alone, a combination of Gly + Mat reduced the mortality of acetaminophen overdosed mice more effectively, attenuated acetaminophen-induced hepatotoxicity, and reduced the number and area of γ-GT positive foci, thus protecting liver function and preventing hepatocellular carcinoma from occurring (Wan et al. 2009). Further, Gly + Mat had a protective effect on immunosuppression, a strong non-specific anti-inflammatory effect, and an effect of reducing the incidence of sodium and water retention. Glycyrrhizin (Gly) in combination with matrine (Mat) had a better effect in inhibiting the proliferation of activated rat hepatic stellate cell (HSC) line and diminishing collagen I and HA levels secreted by HSC as compared with Gly or Mat alone (Zhao et al. 2012a). The combination significantly reduced serum hexadecenoic acid and laminin and procollgan type-III levels in the rat model of CCl_4-induced liver fibrosis compared with Gly or Mat alone. Hepatic histological analysis also confirmed that Gly + Mat administration exhibited the recovery effect remarkably and could improve liver fibrosis in-vitro and in-vivo more effectively. El-Tahawy et al. (2011) found that pretreatment with glycyrrhizin protected against lipopolysaccharide/D-galactosamine-induced acute hepatitis in albino rats by its anti-inflammatory and anti-apoptotic effects. Studies by Tu et al. (2012) found that mice treated with glycyrrhizin prevented concanavalin A-induced liver inflammation and fibrosis. Glycyrrhizin was found to alleviate liver injury and fibrosis progression via regulation of $CD4^+T$ cell response in JNK, ERK and PI3K/AKT-dependent pathways. Results of studies by Tsai et al. (2013) suggested that glycyrrhizin pretreatment decreased total parenteral nutrition-associated acute liver injury factors in rats by suppressing endoplasmic reticulum stress and reactive nitrogen stress. Results of animal studies by Kuroda et al. (2014) revealed that a single injection of LPS/GalN (lipopolysaccharide/D-galactosamine) might stimulate apoptosis of mouse hepatocytes through the binding of HMGB1 (high mobility group box 1) protein to Gsto1(Glutathione S-transferase omega-1) promoter region and that glycyrrhizin-treatment might prevent the apoptosis and inflammatory

infiltrates caused by LPS/GalN-injection in mouse liver by disturbing the binding of HMGB1 protein to Gsto1 promoter sequence. Intraperitoneal (i.p.) administration of 200 mg/kg glycyrrhizin from licorice significantly protected lithocholic acid (LCA)-induced liver damage in mice, indicated by alleviated histology alteration and prevention of the alanine transaminase elevation (Han et al. 2014). Glycyrrhizin treatment significantly prevented LCA-induced reduction of the three phospholipid compounds, lysophosphatidylcholine LPC 16:0, LPC 18:0 and LPC 18:2. Glycyrrhizin and omega-3 fatty acids (ω-3) alone or in combination protected rat liver from thioacetamide-induced hepatotoxic effects as they significantly decreased serum aspartate aminotransferase activity and serum total bilirubin level; they also significantly increased serum albumin and total protein levels (El Magd et al. 2015). The hepatoprotective effects of glycyrrhizin and ω-3 were confirmed by the histopathological analysis as they significantly reduced the necroinflammatory scores and the extent of fibrosis. GL and ω-3 significantly decreased liver malondialdehyde level.

Results of in-vivo studies indicated that glycyrrhizic acid effectively protected against titanium dioxide nanoparticles (NTiO2) in rats (Orazizadeh et al. 2014). Pretreatment of glycyrrhizic acid significantly decreased alanine aminotransferase (ALT), aspartate aminotransferase (AST) and alkaline phosphatase (ALP), attenuated the histopathology of hepatic injury, decreased apoptotic index, ameliorated oxidative stress in hepatic tissue and increased the activities of superoxide dismutase (SOD) and glutathione peroxidase (GPx).

Clinical Studies

In a clinical trial of 18 patients with subacute hepatic failure due to viral hepatitis, thrice weekly intravenously treatment of the interferon stimulator named Stronger Neo Minophagen-C (SNMC) derived from the plant *G. glabra* for 8 weeks increased survival rate (Acharya et al. 1993). The survival rate among these patients was 72.2 %, as compared to the earlier reported rate of 31.1 % in 98 patients who received supportive therapy. In a retrospective study, long-term administration of the Japanese medicine 'Stronger Neo-Minophagen C' (SNMC), which contained 0.2 % glycyrrhizin (GL), 0.1 % cysteine and 2.0 % glycine in physiologic saline solution, used for the treatment of chronic hepatitis C, was effective in preventing liver carcinogenesis (Arase et al. 1997). The 10th-year rates of cumulative hepatocellular carcinoma (HCC) incidence for Group A (SNMC-treated) and Group B (vitamin K-treated) were 7 % and 12 %, respectively, and the 15th-year rates were 12 % and 25 %. HCC is caused by hepatitis C virus. By Cox regression analysis, the relative risk of HCC incidence in patients not treated with SNMC (Group B) was 2.49 compared with that of patients treated with SNMC (Group A). SNMC was also reported to be particularly helpful in patients who failed to respond to interferon and in patients who could not be treated with it for various reasons (Kumada 2002). In a prospective randomized controlled trial of 170 patients glycyrrhizin and glycyrrhizin plus ursodeoxycholic acid were compared for efficacy against chronic hepatitis C virus infection (Tsubota et al. 1999). It was found that the combined therapy with ursodeoxycholic acid and glycyrrhizin was safe and effective in improving liver-specific enzyme abnormalities and may be an alternative to interferon in chronic hepatitis C virus infection, especially for interferon-resistant or unstable patients. In a double-blind, randomized, placebo-controlled phase I/II trial of 57 patients with chronic hepatitis C, intravenous glycyrrhizin administration at a dose of 240 mg, thrice weekly for 4 weeks, lowered serum alanine aminotransferase (ALT) during treatment, but had no effect on hepatitis C virus (HCV)-RNA levels (van Rossum et al. 1999). The drug appeared to be safe and is well tolerated. The mechanism by which glycyrrhizin improved liver biochemistry and histology are not well elucidated (van Rossum and De Man 1998). In another clinical study of 69 European patients with chronic hepatitis C, glycyrrhizin treatment induced a significant decrease in ALT. Six times per week glycyrrhizin treatment for 4 weeks appeared more effective than three times per week. In another randomized phase II

trial, HCV-RNA-positive patients with elevated ALT and marked fibrosis or necro-inflammation who were not eligible for interferon therapy were treated for 4 weeks with six infusions weekly of glycyrrhizin (Orlent et al. 2006). Seventy two patients with an ALT response at week 4 were randomized to continue treatment for 22 weeks in three dose frequency groups: 6×, 3× or once weekly. They found that ALT responses induced by 4 weeks glycyrrhizin therapy could be maintained in a subset of chronic hepatitis C patients receiving at least three injections weekly. The observed ALT response did not translate in a significant histological improvement after 6 months treatment.

Ikeda (2007) retrospectively analysed 1249 patients with chronic hepatitis with or without cirrhosis and found that long-term glycyrrhizin injection therapy significantly decreased the incidence of hepatocellular carcinoma in patients with interferon-resistant active chronic hepatitis C, whose average aminotransferase value was twice or more of the upper limit of normal after interferon.

Yasui et al. (2011) treated 17 patients defined with acute onset autoimmune hepatitis with intravenous glycyrrhizin and 17 patients of severe disease with intravenous glycyrrhizin and corticosteroids. The alanine aminotransferase level could be controlled at an early stage using glycyrrhizin with no significant difference compared with the combined treatment. Recovery rate was higher in the glycyrrhizin group than in the glycyrrhizin + corticosteroid group. Glycyrrhizin could be used safely and be useful for patients with difficult-to-diagnose acute liver disease as an 'initial' treatment tool to improve liver inflammation before starting disease-specific treatment. In a randomized, double-blind, placebo-controlled, intravenous administration of glycyrrhizin, 5×/or 3×/week, and 5×/week placebo for 12 weeks to 379 patients, followed by a randomized, open comparison of glycyrrhizin i.v. 5×/versus 3×/week for 40 weeks was found to be effective in treating chronic hepatitis C patients who failed to respond to interferon-based therapies (Manns et al. 2012). Glycyrrhizin exhibited a significantly higher ALT reduction compared to placebo after 12 weeks of therapy and an improvement of necro-inflammation and fibrosis after 52-weeks treatment. Generally, glycyrrhizin treatment was well tolerated.

In a study of 38 patients with non-severe aplastic anaemia (NSAA), the combination therapy of glycyrrhizin and cyclosporine was found to be an effective treatment for NSAA in terms of improvement of response rate, reduction in cyclosporine-related liver injury, and attenuation of severity of nausea and other adverse events (Ren et al. 2013).

Neuroprotective Activity

In-Vitro Studies

The results of studies by Kao et al. (2009) suggested that glycyrrhizic acid may protect PC12 cells from ischemic injury caused by 6-hydroxydopamine (6-OHDA)-induced cytotoxicity via modulation of the intracellular antioxidant system and mitochondria-induced apoptosis. Moreover, glycyrrhizic acid and 18β-glycyrrhetinic acid may modulate the ratio of the mitochondrial Bcl-2 family and influence PI3K/Akt signalling. Glycyrrhizin and its metabolite 18β-glycyrrhetinic acid in *Glycyrrhiza*, a constituent herb of yokukansan, a traditional Japanese medicine, ameliorated thiamine deficiency-induced dysfunction of glutamate transport in cultured rat cortical astrocytes in a dose-dependent manner (Kawakami et al. 2010). Studies indicated that extracellular signal-regulated kinases (ERKs) and mitochondria-related pathways were essential for the neuroprotective effect of glycyrrhizic acid against glutamate-induced toxicity in DPC12 cells (Wang et al. 2014a). Glycyrrhizic acid pretreatment enhanced activation of ERKs but not AKT (protein kinase B). This was further confirmed by Teng et al. (2014a) who demonstrated the involvement of the ERK pathway in the neuroprotective effects of glycyrrhizic acid against the 1-methyl-4-phenylpyridinium (MPP+)-induced apoptosis of dopaminergic neuronal cells. Pretreatment with glycyrrhizic acid had no

effects on the expression of phosphorylated AKT (p-AKT) and total AKT (t-AKT).

Studies found that liquiritin exhibited neuroprotective effect against glutamate toxicity in differentiated PC12 (DPC12) rat pheochromocytoma cells, predominantly through the extracellular signal-regulated kinase (ERK) and protein kinase B (AKT)/glycogen synthase kinase-3β (GSK-3β) pathways, indicating the potential of liquiritin for the treatment of neurodegenerative diseases (Teng et al. 2014b).

Animal Studies

In in-vivo studies, gerbils treated with roasted licorice but not raw licorice exhibited significant neuroprotection against ischemic damage in the hippocampus after transient forebrain ischemia (Hwang et al. 2006). It was found that non-polar compounds containing glycyrrhizin-degraded products, such as glycyrrhetinic acid and glycyrrhetinic acid monoglucuronide, were increased in roasted licorice. In an in-vitro study, both raw and roasted licorice significantly reduced acetate dehydrogenase release from PC12 cells exposed to hypoxia. *G. glabra* root extract (250 and 500 mg/kg) promoted the locomotor activity and spatial behaviour significantly, which was impaired in hypoxic rats (Muralidharan et al. 2009). The extract administration restored the decreased levels of brain enzymes such as glutamate and dopamine and decreased acetylcholinesterase (AchE) activity significantly. Levels of antioxidant enzymes such as superoxide dismutase, glutathione peroxidase, glutathione reductase and catalase were reduced due to hypoxia and were restored to near normalcy by administration of *G. glabra* extract. Oral administration of aqueous licorice root extract in the dose of 150 and 225 mg/kg to 1-month-old male Wistar albino rats showed a significant enhancement of dendritic arborization (dendritic branching points) and dendritic intersections along the length of both apical and basal dendrites in hippocampal (CA3) pyramidal neurons (Chakravarthi and Avadhani 2014). The results indicated that constituents present in aqueous licorice root extract had neuronal dendritic growth stimulating properties.

Pretreatment of rats with isoliquiritigenin (ISL), flavonoid constituent of *G. glabra* root, significantly reduced the cerebral infarct volume and oedema and produced significant reduction in neurological deficits (Zhan and Yang 2006). ISL pretreatment increased the brain ATP content, energy charge and total adenine nucleotides in a dose-dependent manner. ISL significantly inhibited the increases of brain malondialdehyde content and prevented the decline in activities of brain superoxide dismutase (SOD), catalase (CAT) and glutathione peroxidase (GSH-Px) caused by cerebral ischemia–reperfusion.

Yu et al. (2008) showed that glabridin, a major flavonoid of *Glycyrrhiza glabra*, at 25 mg/kg by intraperitoneal injection, significantly decreased the focal infarct volume, cerebral histological damage and apoptosis in stroked rats induced by cerebral artery occlusion middle (MCAO) compared to sham-operated rats. Glabridin significantly attenuated the level of brain malonyldialdehyde (MDA) in MACO rats while it elevated the level of two endogenous antioxidants in the brain, that is, superoxide dismutase (SOD) and reduced glutathione (GSH). Co-treatment with glabridin significantly inhibited the staurosporine-induced cytotoxicity and apoptosis of cultured rat cortical neurons and DNA laddering caused by staurosporine in a concentration-dependent manner. Glabridin also suppressed the elevated Bax protein and caspase–3 proenzyme and decreased bcl–2 induced by staurosporine in cultured rat cortical neurons, facilitating cell survival. Glabridin also inhibited superoxide production in cultured cortical neurons exposed to staurosporine. The findings indicated that glabridin had a neuroprotective effect via modulation of multiple pathways associated with apoptosis.

In-vivo studies by Zhang et al. (2014a) showed that glycyrrhizin had neuroprotective efficacy against brain ischemia–reperfusion injury in mice through high mobility group box 1 (HMGB1)-Toll-like receptor 4 (TLR4) signalling pathway. Administration of glycyrrhizin (10 mg/kg) intravenously in rats at 3 or 6 h after middle cerebral artery occlusion reduced infarct volumes to 12.9 % and 46.2 %, respectively (Kim

et al. 2012). This neuroprotective effect was accompanied by improvements in motor impairment and neurological deficits and suppressions of microglia activation and pro-inflammatory cytokine induction. The results indicated that glycyrrhizin had neuroprotective efficacy in the post-ischemic brain via its anti-inflammatory, anti-excitotoxic and anti-oxidative effects and in particular, it exerted anti-inflammatory effect, at least in part, by inhibiting HMGB1 (high mobility group box 1) secretion. Song et al. (2013) found that oral administration of glycyrrhizin alleviated neuro-inflammation and memory deficit induced by systemic lipopolysaccharide (LPS) treatment in mice. Glycyrrhizin significantly reduced TNF-α and IL-1β mRNA, COX-2 and iNOS protein expressions and the elevated Iba1 protein expression and the average cell size of Iba1-expressing microglia induced by LPS. In the Morris water maze test, glycyrrhizin significantly prolonged the swimming time spent in the target and peri-target zones. Glycyrrhizin also significantly increased the target heading and memory score numbers. The neuroprotective effect of glycyrrhizin was found to be mediated through the inhibition of pro-inflammatory mediators and microglial activation in the brain tissue. Glycyrrhizin exhibited neuroprotective effect on ischemia-reperfusion injury in rat brains through the inhibition of inflammation, oxidative stress and apoptotic injury by antagonizing the cytokine activity of high mobility group box 1 (HMGB1) (Gong et al. 2014). Pretreatment with glycyrrhizin significantly reduced infarct volume and improved the accompanying neurological deficits in locomotor function. The expression levels of inflammation- and oxidative stress-related molecules including TNF-α, iNOS, IL-1β and IL-6, which were over-expressed in I/R, were decreased by glycyrrhizin. Glycyrrhizin was found to reduce secondary brain injury and improved outcomes in rat following traumatic brain injury (TBI) by downregulation of high-mobility group box 1 (HMGB1)/HMGB1 receptors (toll-like receptor 4 (TLR4) and receptor for advanced glycation end products (RAGE))/NF-κB-mediated inflammatory responses in the injured rat brain (Gu et al. 2014). Beam walking performance impairment and brain oedema were significantly reduced in TBI + glycyrrhizin group compared with TBI group. In separate studies, Okuma et al. (2014) found glycyrrhizin to be a novel for TBI through its interference with HMGB1 and RAGE interaction in rats. The beneficial effects of glycyrrhizin on motor and cognitive functions persisted for 7 days after injury.

The anti-inflammatory and anti-excitotoxic effects of glycyrrhizic acid were verified in LPS-treated primary microglial cultures and in N-methyl-D-aspartate (NMDA)-treated or kainic acid-treated primary cortical cultures (Luo et al. 2013). Also they found that the neuroprotective effect of glycyrrhizic acid in the kainic acid-injected mouse brain might be attributable to the inhibitions of HMGB1 induction and release, which in turn mitigated the inflammatory process.

Results of studies by Shi et al. (2011) suggested that pinocembrin provided neuroprotection against global cerebral ischemic injury in rats with a wide therapeutic time window, which may be attributed to its anti-oxidative, anti-inflammatory and anti-excitotoxic effects. Studies by Meng et al. (2011) demonstrated that pinocembrin alleviated blood-brain barrier injury induced by global cerebral ischemia/reperfusion (GCI/R) in rats. Pinocembrin decreased neurological score and lessened brain oedema induced by GCI/R. Pinocembrin also alleviated the ultrastructural changes of cerebral microvessels, astrocyte end-feet and neurons, and improved cerebral blood flow in the GCI/R rats.

Central Nervous System (CNS) Activity

In in-vivo studies in dogs, *G. glabra* extract was found to have anti-cholinergic action as it blocked the stimulatory effect of acetylcholine; the extract produced inhibition of the intestinal movements (Shihata and Elghamry 1963b). Studies with licorice-derived enzyme inhibitors indicated functional effects for 11β-hydroxysteroid dehydrogenase (11β-HSD) in the adult brain, notably in the periventricular hypothalamus and limbic

system (Seckl 1997). 11β-HSD catalyses the conversion of the active glucocorticoids corticosterone and cortisol to inert 11 keto-products (11-dehydrocorticosterone, cortisone), thus regulating access of glucocorticoids to receptors. Thus, 11beta-HSD represents a novel and potentially important level of control of glucocorticoid action in the CNS. Enzyme modulation by pharmacological or other agents may provide a useful means to target increased or attenuated glucocorticoid action to specific sites in the brain. Administration of *Glycyrrhiza glabra* aqueous extract (150 mg/kg) significantly improved learning and memory of mice in the elevated plus-maze and passive avoidance test (Dhingra et al. 2004). Furthermore, this dose significantly reversed the amnesia induced by diazepam (1 mg/kg i.p.) and scopolamine (0.4 mg/kg i.p.). Anti-inflammatory and antioxidant properties of liquorice may be contributing favourably to the memory enhancement effect (Parle et al. 2004). As scopolamine-induced amnesia was reversed by liquorice, it was suggested that the beneficial effect on learning and memory may be because of facilitation of cholinergic transmission in brain. The results showed *G. glabra* to have promise as a memory enhancer in both exteroceptive and interoceptive behavioural models of memory. Administration of 150 mg/kg aqueous licorice extract significantly reduced the immobility times of mice in both forced swim test (FST) and tail suspension test (TST), without any significant effect on locomotor activity of mice (Dhingra and Sharma 2006). The efficacy of extract was found to be comparable to that of imipramine (15 mg/kg i.p.) and fluoxetine (20 mg/kg i.p.). Liquorice extract reversed reserpine-induced extension of immobility period of mice in FST and TST. It was found that the antidepressant-like effect of liquorice extract appeared to be mediated by increase of brain norepinephrine and dopamine, but not by increase of serotonin (Cui et al. 2008). Oral administration of the higher doses (2 and 4 mg/kg) of glabridin and piracetam to mice significantly antagonized the amnesia induced by scopolamine in both the elevated plus maze test and passive avoidance test. Furthermore, glabridin (2 and 4 mg/kg per os) and metrifonate (50 mg/kg intraperitoneally), used as a standard drug, both remarkably reduced the brain cholinesterase activity in mice compared to the control group. The results suggested glabridin to be a promising candidate for memory improvement.

Zhu et al. (2010) reported that the natural product 2,2′,4′-trihydroxychalcone (TDC) from *Glycyrrhiza glabra* functioned as a specific non-competitive inhibitor against β-site amyloid precursor protein (APP)-cleaving enzyme 1 (BACE1) enzyme, and potently repressed β-cleavage of APP and production of amyloid-β (Aβ) in human embryo kidney cells-APPswe cells. In APP-PS1 double transgenic mice, treatment with 9 mg/kg/day of TDC markedly decreased Aβ production and Aβ plaque formation, while efficiently improving the memory impairment based on Morris water maze test. Their findings demonstrated that the natural product TDC as a new BACE1 inhibitor could ameliorate memory impairment in mice and could have potential as a lead compound for further anti-Alzheimer's disease reagent development. Studies showed that liquiritigenin treatment improved the behavioural performance of Abeta (25–35) hippocampal-injected rats and attenuated neuronal loss in the brain (Liu et al. 2010a). More importantly, liquiritigenin treatment decreased mRNA levels and protein expression of Notch-2, an effect that could promote the generation of new neurons. They found that treatment of brain-derived progenitor cell cultures with liquiritigenin increased the number of cells that differentiated into neurons; but the treatment did not alter the growth of astrocytes (Liu et al. 2010b). In addition, treatment with liquiritigenin decreased Notch-2 mRNA and protein expression, which could promote the growth of new neurons. In a subsequent study, they found that treatment with liquiritigenin improved the behavioural performance of transgenic mice and it attenuated the protein expression of oligomeric form of amyloid β-peptide (Aβ) (Liu et al. 2011). Furthermore, treatment with liquiritigenin inhibited astrocytosis in the hippocampus, through its inhibitory activities on Notch-2, an important molecular regulating neural proliferation and dif-

ferentiation. These findings provided evidence for beneficial activity of liquiritigenin in a mouse model of Alzheimer's disease.

Glycyrrhiza glabra ethanol extract (GGE) dose-dependently potentiated pentobarbital-induced sleep and increased the amount of non-rapid eye movement sleep in mice without decreasing delta activity (Cho et al. 2012). The major flavonoid glabrol was isolated from the flavonoid-rich fraction of GGE; it inhibited [^{3}H] flumazenil binding to the $GABA_A$-BZD receptors in rat cerebral cortex membrane. Glabrol increased sleep duration and decreased sleep latency in a dose-dependent manner (5, 10, 25 and 50 mg/kg); its hypnotic effect was also blocked by flumazenil a well-known γ-aminobutyric acid type A-benzodiazepine ($GABA_A$-BZD) receptor antagonist. The results implied that GGE and its flavonoid glabrol induced sleep via a positive allosteric modulation of $GABA_A$-BZD receptors. Results of animal studies demonstrated that the anti-amnestic dose of *Glycyrrhiza glabra* extract (150 mg/kg for 7 days) resulted in prepulse inhibition (PPI) disruption; it augmented cortical, hippocampal and striatal monoamine levels in mice (Michel et al. 2013). It was concluded that liquorice extract (150 mg/kg)-induced PPI deficit was mediated through augmenting monoaminergic transmission in the cortex, hippocampus and striatum.

A panchagavya Ayurvedic formulation (300, 500 mg/kg, po) containing *Emblica officinalis*, *G. glabra*, and cow's ghee produced a significant prolongation of pentobarbital-induced sleeping time and reduced spontaneous locomotor activity (Achliya et al. 2004). The formulation also significantly antagonized the amphetamine induced hyper-locomotor activity (500, 750 mg/kg, po) and protected mice against tonic convulsions induced by maximal electroshock (500, 750 mg/kg, po). The formulation slightly prolonged the phases of seizure activity but did not protect mice against lethality induced by pentylenetetrazole. The formulation did not show neurotoxicity. Combination drug therapy of *G. glabra* and *Piper nigrum* in rats exerted better antidepressant effects than the individual dosage as evaluated by the force swim and tail suspension tests (Sohi et al. 2013).

Anti-inflammatory/Anti-allergic Activities

In-Vitro Studies

Liquiritigenin and 18β-glycyrrhetinic acid most potently inhibited the degranulation of RBL-2H3 cells induced by IgE with the antigen (DNP-HSA) and rat peritoneal mast cells induced by compound 48/80 (Shin et al. 2007). Liquiritigenin and 18β-glycyrrhetinic acid potently inhibited the passive cutaneous anaphylactic reaction as well as the scratching behaviour in mice induced by compound 48/80. These components inhibited the production of IgE in ovalbumin-induced asthma mice but liquiritigenin had little effect. The results suggested that the anti-allergic effects of licorice were mainly due to glycyrrhizin, 18β-glycyrrhetinic acid and liquiritigenin, which could relieve IgE-induced allergic diseases such as dermatitis and asthma. At non-toxic ≥10 μM concentration, isoliquiritigenin blocked the induction of vascular cell adhesion molecule-1 (VCAM-1) and E-selectin on activated human umbilical vein endothelial cells (HUVEC) and markedly interfered with THP-1 monocyte adhesion to TNF-alpha-activated endothelial cells (Kwon et al. 2007). Isoliquiritigenin abolished TNF-alpha-induced mRNA accumulation of VCAM-1 and E-selectin. Additionally, isoliquiritigenin attenuated platelet endothelial cell adhesion molecule-1 (PECAM-1) expression induced by TNF-alpha. In contrast, other components found in licorice, 18β-glycyrrhetinic acid, glycyrrhizin, formononetin and ononin did not downregulate the expression of VCAM-1 and/or PECAM-1 activated by TNF-alpha, implying that these components are inactive in modulating adhesion of leukocytes to stimulated endothelial cells. Isoliquiritigenin downregulated CAM proteins in TNF-alpha-activated HUVEC at the transcriptional levels by blocking degradation of IkappaBalpha and nuclear translocation of NF-kappaB. The results demonstrated that the induction blockade of VCAM-1 and E-selectin

by isoliquiritigenin was directly mediated by its interference with the cell adhesion molecules (CAM) mRNA transcription through NF-kappaB-dependent mechanisms under inflammatory conditions. In lipopolysaccharide (LPS)-treated RAW 264.7 macrophages, isoliquiritigenin (ILG), more potently inhibited LPS-induced nitric oxide (NO) and prostaglandin E(2) (PGE(2)) production than isoliquiritin (ILT) (Kim et al. 2008a). ILG dose-dependently reduced the LPS-induced expressions of inducible nitric oxide synthase (iNOS) and cyclooxygenase-2 (COX-2) at the protein and mRNA levels via suppression of the transcription activity of nuclear factor-kappa B (NF-kappaB). The results suggested that the anti-inflammatory properties of ILG are caused by iNOS, COX-2, TNF-alpha and IL-6 downregulation due to NF-kappaB inhibition via the suppression of IKK, ERK1/2 and p38 phosphorylation in RAW 264.7 cells. Isoliquiritigenin and glycyrrhizin inhibited lipopolysaccharide (LPS)-induced NF-κB activation and interleukin IL-6 production in dose-dependent manners in RAW264.7 cells (Honda et al. 2012). They both modulated LPS sensor toll-like receptor 4/MD-2 complex at the receptor level, leading to suppress LPS-induced activation of signalling cascades and cytokine production, but their effects were exerted at different steps of TLR4/MD-2 signalling.

G. glabra crude extract displayed remarkable reactivity with free stable 1,1′-diphenyl-2-picrylhydrazyl (DPPH) radical, inhibitory efficacy in peroxidatively damaged unilamellar dioleoyl phosphatidylcholine (DOPC) liposomes and inhibition of ROS chemiluminescence, generated by whole blood, induced by both receptor-bypassing stimuli (PMA) and receptor operating stimuli (Opz) in the ranking order of stimuli PMA > Opz (Račková et al. 2007). These activities were postulated to be attributed to phenolic antioxidants involving isoflavan derivatives, coumarins and chalcones. Nonetheless, triterpene saponin glycyrrhizin exhibited no efficacy in the system of DPPH reaction and peroxidation of liposomal membrane, and negligible inhibition of chemiluminescence generated by inflammatory cells. These results indicate that the mechanism of anti-inflammatory effect of glycyrrhizin most probably did not involve ROS and this major constituent was not responsible for the inhibition effects of liquorice extract on neutrophil functions. The results of studies in a lipopolysaccharide (LPS)-stimulated macrophage model indicated that licorice glycyrrhizic acid and 18β-glycyrrhetinic acid may provide an anti-inflammatory effect by attenuating the generation of excessive NO, PGE(2) and ROS and by suppressing the expression of pro-inflammatory genes through the inhibition of NF-κB and PI3K activity (Wang et al. 2011). Kao et al. (2010) demonstrated that both glycyrrhizic acid (GA) and 18β-glycyrrhetinic acid (18βGA) reduced inflammatory cytokine production and its resulting anti-inflammation (Kao et al. 2010). GA acted via PI3K/Akt/GSK3β to reduce cytokine production, while 18βGA caused the dissociation of a glucocorticoid receptor (GR)-HSP90 complex to block inflammation. They proposed that GA and 18βGA may be valuable biological inhibitors of lung inflammation. In-vitro studies using murine macrophages (J774A.1) and human neutrophil (HL-60) cells revealed that *G. glabra* and glabridin significantly inhibited prostaglandin E_2 (PGE_2), thromboxane (TXB_2), cyclooxygenase (COX), leukotriene (LTB_4), leukotriene (LTB_4) and lipoxygenase (LOX), while isoliquiritigenin exerted inhibitory effect only against COX products but failed to suppress LOX products (Chandrasekaran et al. 2011a). However, glycyrrhizin at the tested concentrations failed to exhibit inhibitory effect on both COX and LOX products. Umbelliferone isolated from *Glycyrrhiza glabra* rhizome exhibited 95.68 % inhibition of COX-2 at 10 μM concentration with $IC_{50} < 1$ μM (Kaur et al. 2012a). In the xylene-induced ear oedema and ovalbumin-induced mouse paw oedema assay, extracts of the cell culture of *Glycyrrhiza* exhibited similar anti-inflammatory effects to those of its field cultivated equivalent, through the enhancement of the SOD activity of plasma and liver tissues (Man et al. 2013).

Studies by Thiyagarajan et al. (2011) showed that the inhibitory effect of *G. glabra* extract on

lipopolysaccharide (LPS)-induced pro-inflammatory mediators was influenced by glabridin and isoliquiritigenin and was not contributed by glycyrrhizin. *G. glabra* and isoliquiritigenin significantly inhibited LPS stimulated NO, IL-1 β and IL-6 production. Glabridin showed significant inhibition of NO and interleukin IL-1 β release, but failed to attenuate IL-6 levels at the tested concentrations. In addition, glycyrrhizin did not exhibit inhibitory response towards any of the LPS-induced pro-inflammatory mediators at the tested concentrations. Treatment of THP-1 (human myelomonocytic leukaemia) cells with licochalcone C attenuated lipopolysaccharide (LPS)-IFN-γ-induced inflammatory response by significantly decreasing the expression and activity of inducible nitrate synthase (iNOS) via NFκB (nuclear factor kappa-B), by influencing extracellular O_2^- production, and by modulating the antioxidant network activity of SOD (superoxide dismutase), CAT (catalase) and GPx (glutathione peroxidase) activity. It was hypothesized that licochalcone C had antioxidant properties since it reduced the production of superoxide radicals and consequently reduced the activity of iNOS (Franceschelli et al. 2011). Licorice extract inhibited nitric oxide (NO) production and inducible NO synthase (iNOS) expression in lipopolysaccharide (LPS)-stimulated RAW264 murine macrophage cells (Uto et al. 2012). However, treatment of glycyrrhizin alone could not show the suppression of NO production and iNOS expression. The combined treatment with glycyrrhizin and glycyrrhizin-removed extract (GC–KO) extract enhanced the attenuated inhibition. The ethyl acetate leaf extract showed good dose-dependent ability to inhibit release of both thromboxane B_2 (TxB_2) and prostaglandin E_2 (PGE_2) in whole blood (Siracusa et al. 2011). The methanol leaf extract appeared to inhibit only the PGE2 release, suggesting a selective action on the cyclooxy genase COX-2 pathway. No effect on TxB2 and PGE_2 was observed with *n*-hexane leaf extract.

Studies by Bhattacharjee et al. (2012) showed that glycyrrhizic acid (GA) treatment caused an enhanced expression of iNOS2 along with inhibition of Cox-2 in *Leishmania donovani*-infected macrophages. GA treatment in infected macrophages enhanced the expression of interleukin IL-12 and tumour necrosis factor TNF-α, concomitant with a downregulation of interleukin IL-10 and transforming growth factor TGF-β. GA increased macrophage effector responses via inhibition of Cox-2-mediated prostaglandin E2 release in *L. donovani*-infected macrophages. Studies showed that monoammonium glycyrrhizate (MAG) derived from licorice suppressed TNF-α-induced chemokine (including CXCL8, CX3CL1 and CXCL16) mRNA expression in human dermal microvascular endothelial cell line (HMEC-1) cells, in a dose-dependent manner, and reduced the secretion of these chemokines in culture supernatant (Cao et al. 2014). MAG also suppressed TNF-α-induced chemokine production in HMEC-1 cells. The results revealed MAG to be a potential anti-inflammatory agent capable of improving inflammatory skin diseases.

Treatment of licochalcone A, from *G. glabra* root, suppressed polyinosinic-polycytidylic acid (poly-IC)-induced thymic stromal lymphopoietin (TSLP) in BEAS 2B cells and primary bronchial epithelial cells in a dose dependent manner (Kim et al. 2015). The poly-IC-induced mRNA expression of other pro-inflammatory mediators such as MCP-1, RANTES and interleukin IL-8 was suppressed by licochalcone A. Licochalcone A inhibited the Iκβ kinase (IKK)/nuclear factor kappa (NF-κB) signalling pathway, which might be involved in the pathogenesis of virus-exacerbated asthma.

Animal Studies

The anti-inflammatory activity of aqueous licorice extract in rat foot arthritis model was ascribed to glycyrrhizin and glycyrrhetic acid (Gujral et al. 1959). Effect at doses tested was comparable to that of hydrocortisone and butazolidin. The *Streptococcus* LJ-22-transformed product, 18β-glycyrrhetinic acid-3-*O*-β-D-glucuronide (GAMG), of glycyrrhizin (18β-glycyrrhetinic acid-3-*O*-β-D-glucuronopyranosyl-(1 → 2)-β-D-glucuronide, GL) exhibited anti-allergic activity with IC_{50} values of 0.28 mM (Park et al. 2004). GAMG, which was sweeter than glycyrrhizin, and 18β-glycyrrhetinic acid, a GAMG metabolite

by human intestinal bacteria, also inhibited the passive cutaneous anaphylaxis and skin contact inflammation. Licorice root extract lipids produced statistically significant suppression of inflammatory oedema growth induced by 1 % carrageenan and 3 % formalin solutions in mice compared to that in the untreated control (Denisova et al. 2007) and their antiphlogistic action was comparable with that of the reference drug ortophen. In animal studies, glyderinine, a derivative of glycyrrhizic acid isolated from *Glycyrrhiza glabr*a, was found to exert a pronounced anti-inflammatory effect exceeding that of hydrocortisone and amidopyrine (Azimov et al. 1988). Oral administration of *Glycyrrhiza glabra* root extract to rats caused a significant reduction in pedal inflammation and swelling induced by formalin compared to the control group (Shalaby et al. 2004). However, the anti-inflammatory effect was less marked than that produced by tenoxicam (a standard anti-inflammatory drug). Glycyrrhizin, a major constituent of *G. glabra* (5 mg/kg) markedly inhibited ovalbumin (OVA)-induced immediate airway constriction, airway hyper-reactivity to methacholine, lung inflammation and infiltration of eosinophils in the peribronchial and perivascular areas in BALB/c mice (Ram et al. 2006). It prevented the reduction of interferon IFN-γ and decreased interleukin IL-4, IL-5 and eosinophils in the bronchoalveolar lavage fluid. Also, it reduced OVA-specific IgE levels and prevented the reduction of total IgG2a in serum. Studies by Ma et al. (2013) demonstrated that glycyrrhizic acid exerted anti-asthmatic effects via modulation of Th1/Th2 cytokines and enhancement of CD4+CD25+Foxp3+ regulatory T cells in ovalbumin (OVA)-sensitized mice. Glycyrrhizic acid inhibited OVA-induced increases in Raw and eosinophil count; interleukin (IL)-4, IL-5, IL-13 levels were recovered in bronchoalveolar lavage fluid; and increased IFN-γ level in bronchoalveolar lavage fluid. Separate studies demonstrated that liquiritigenin exerted anti-inflammatory effects, through inhibition of NF-kappaB activation in Raw264.7 macrophages, thereby decreasing production of iNOS and pro-inflammatory cytokines (Kim et al. 2008b). In rats, liquiritigenin treatment inhibited the formation of paw oedema induced by carrageenan. Administration of glycyrrhizin at 10 mg/kg i.p. 5 min prior to carrageenan exerted potent anti-inflammatory effects in mice (Menegazzi et al. 2008). Injection of carrageenan into the pleural cavity of mice elicited an acute inflammatory response and carrageenan-induced pleurisy which were attenuated by glycyrrhizin. It was found that prevention of the activation of NF-κB and STAT-3 by glycyrrhizin reduced the development of acute inflammation. 18β-Glycyrrhetinic acid ameliorated acute *Propionibacterium acnes*-induced liver injury in C57BL/6 mice through reduced macrophage inflammatory protein (MIP)-1α expression in Kupffer cells by downregulating MyD88 expression and inhibiting NF-κB activation (Xiao et al. 2010).

Isoliquiritigenin (ILG) was found to be a potent inhibitor of NLRP3 inflammasome activation (Honda et al. 2014). In-vivo, analyses revealed that ILG potently attenuated high-fat diet (HFD)-induced obesity, hypercholesterolemia and insulin resistance. Further ILG treatment improved HFD-induced macrovesicular steatosis in the liver. Additionally, ILG markedly inhibited diet-induced adipose tissue inflammation and IL-1β and caspase-1 production in white adipose tissue in ex-vivo culture. The results suggested ILG to be a potential drug target for treatment of NLRP3 inflammasome-associated inflammatory diseases.

Oral administration of Saiboku-To, a herbal medicine comprising *G. glabra*, *Magnolia officinalis* and *Suctellaria baicalensis* and their constituents medicarpin (*G. glabra*), baicalein, magnolol (*M. officinalis*) and baicalin (*S. baicalensis*) (100 mg/kg), inhibited picryl chloride-induced ear swelling significantly by 23.5, 40.1, 30.5, 23.6 and 20.9 %, respectively, though the effects were weaker than that of 5 mg/kg of prednisolone (52.9 %) (Taniguchi et al. 2000). Medicarpin derived from *Glycyrrhiza glabra*, magnolol and 8,9-dihydroxydihydromagnolol from *Magnolia officinalis*, baicalein, wogonin and oroxylin A from *Suctellaria baicalensis* inhibited concanavalin A-induced human lymphocyte blastogenesis in dose-dependent fashion

with IC_{50} values ranging from 3.0 to 7.7 μg/mL. The results suggested that flavonoids and lignans tested in the present study were implicated in anti-asthmatic effect of Saiboku-To through suppression of type IV allergic reaction.

Clinical Studies

In double-blind clinical trial of 30 patients with atopic dermatitis, treatment with licorice extract prepared as a 2 % licorice topical gel was more effective than 1 % in reducing the scores for erythema, oedema and itching over 2 weeks (Saeedi et al. 2003). The quantity of glycyrrhizinic acid was determined 20.3 % in the extract and 19.6 % in the topical preparation. Kolbe et al. (2006) conducted a prospective randomized vehicle-controlled clinical trial to assess the anti-irritative efficacy of cosmetic formulations containing licochalcone A in a post-shaving skin irritation model and on UV-induced erythema formation. Topical licochalcone A caused a highly significant reduction in erythema relative to the vehicle control in both the shave- and UV-induced erythema tests, demonstrating the anti-irritative properties of licochalcone A. further, licochalcone A was found to be a potent inhibitor of pro-inflammatory in-vitro responses, including N-formyl-MET-LEU-PHE (fMLP)- or zymosan-induced oxidative burst of granulocytes, UVB-induced PGE(2) release by keratinocytes, lipopolysaccharide (LPS)-induced PGE(2) release by adult dermal fibroblasts, fMLP-induced LTB(4) release by granulocytes, and LPS-induced IL-6/TNF-alpha secretion by monocyte-derived dendritic cells.

Cardiovascular Protective/Anti-atherosclerotic and Haemodynamic Activities

In-vitro studies showed that glycyrrhizin inhibited saponin-induced haemolysis in washed erythrocytes (Segal et al. 1977). The saponins inhibited included digitonin, tomatin, saponin A and cscin. Licorice extract and its major polyphenol glabridin protected low-density lipoprotein (LDL) against lipid peroxidation: in-vitro and in ex-vivo dietary supplementation studies in humans and in atherosclerotic apolipoprotein E-deficient mice (Fuhrman et al. 1997). These results could be related to the absorption and binding of glabridin to the LDL particle and subsequent protection of the LDL from oxidation by multiple modes as shown in humans and in E zero mice. They also found that glabridin or quercetin consumption by atherosclerotic apolipoprotein E-deficient mice resulted in a 53 and 54 % reduction in copper ion induced oxidation, respectively, and a 95 and 83 % reduction in 2,2′-azobis′2-amidino propane hydrochloride (AAPH) induced LDL oxidation, respectively (Belinky et al. 1998a). In the in-vitro oxidation of LDL induced by AAPH (5 mM), glabridin inhibited the formation of TBARS, lipid peroxides and cholesteryl linoleate hydroperoxide (CLOOH) at all the concentrations tested (5–60 μM), while in oxidation induced by copper ions (10 μM), glabridin exhibited a pro-oxidant activity at concentrations lower than 20 μM, and a clear antioxidant activity at concentrations greater than 20 μM. Glabridin (30 μM) inhibited the formation of cholest-5-ene-3,7-diol (7-hydroxycholesterol), cholest-5-ene-3-ol-7-one (7-ketocholesterol) and cholestan-5,6-epoxy-3-ol (5,6-epoxycholesterol) after 6 h of AAPH-induced LDL oxidation, by 55, 80 and 40 %, respectively, and after 6 h of copper ion induced LDL oxidation, by 73, 94 and 52 %, respectively. Glabridin also inhibited the consumption of β-carotene and lycopene by 38 and 52 %, respectively, after 0.5 h of LDL oxidation with AAPH, but failed to protect vitamin E. The in-vivo and in-vitro reduction of the susceptibility of LDL to oxidation obtained with glabridin may be related to the absorption or binding of glabridin to the LDL particle and subsequent protection of LDL from oxidation by inhibiting the formation of lipid peroxides and oxysterols, and by protecting LDL associated carotenoids. Results of further studies suggested that the antioxidant effect of glabridin on LDL oxidation appeared to reside mainly in the 2′ hydroxyl, and that the hydrophobic moiety of the isoflavan was essential to obtain this effect. It was also shown that the position of the hydroxyl group at B ring

significantly affected the inhibitory efficiency of the isoflavan derivatives on LDL oxidation, but did not influence their ability to donate an electron to DPPH or their peak potential values (Belinky et al. 1998b). Both mouse peritoneal macrophages (MPMs) and the J-774 A.1 macrophage-like cells accumulated up to 1.5 μg of glabridin/mg of cell protein after 2 h (Rosenblat et al. 1999). In parallel, in glabridin-enriched cells, macrophage-mediated oxidation of LDL was inhibited by up to 80 % in comparison with control cells. In glabridin-enriched macrophages, protein kinase C activity was reduced by approximately 70 % and in in-vivo studies, using the atherosclerotic apolipoprotein E-deficient (E0) mice, glabridin reduced capability to oxidize LDL by 80 % in comparison with placebo-treated mice. This latter phenomenon was associated with a reduction in the lesion oxysterols and a 50 % reduction in the aortic lesion size. It was concluded that glabridin accumulation in macrophages was associated with reduced cell-mediated oxidation of LDL and decreased activation of the NADPH oxidase system and these phenomena could be responsible for the attenuation of atherosclerosis in E0 mice, induced by glabridin.

Animal studies demonstrated that glycyrrhizin protected rat heart against myocardial ischemia-reperfusion-induced injury via directly inhibiting extracellular HMGB1 cytokine activity and blocking the phospho-JNK/Bax pathway (Zhai et al. 2012). Intravenous administration of glycyrrhizin (10 mg/kg) significantly reduced the infarct size, but did not change the hemodynamic parameters at different time points during reperfusion. Glycyrrhizin significantly decreased the levels of serum HMGB1, TNF-α and IL-6. Glycyrrhizin changed the distribution of Bax and cytochrome c expression between the mitochondrial and cytosolic fractions in the heart tissue, resulting in inhibition of myocardial apoptosis. Studies showed that *Glycyrrhiza glabra* protected rats from myocardial ischemia–reperfusion injury by improving hemodynamic, biochemical, histopathological and ventricular function (Ojha et al. 2013). Pretreatment with *G. glabra* significantly prevented the depletion of the antioxidant enzymes: superoxide dismutase (SOD), catalase (CAT) and glutathione peroxidase (GSH-Px) and myocyte injury marker enzymes: creatine phosphokinase-MB isoenzyme and lactate dehydrogenase. It also prevented depletion of glutathione (GSH) and inhibited lipid peroxidation in the heart. In addition to improving biochemical indices of myocardial function, *G. glabra* also significantly reinstated mean arterial pressure, heart rate, (±) LVdP/dtmax and attenuated abrupt rise in left ventricular end diastolic pressure. Histopathological preservation evidenced by reduced infiltration of cells and myonecrosis depicted the myocardial salvaging effect of *G. glabra*. Basso et al. (1994) reported a case of a 63-year-old patient with severe postural hypotension caused by autonomic diabetic neuropathy who recovered after licorice (equivalent to 3 g/day of glycyrrhizic acid) treatment for 7 days. The results suggested that licorice could be used for the therapy of postural hypotension attributable to autonomic diabetic neuropathy.

Haemodynamic changes induced by liquorice consumption in 20 subjects versus 30 controls with average blood pressures of 120/68 and 116/64 mmHg, respectively, were investigated by Leskinen et al. (2014). Two weeks of daily liquorice consumption increased extracellular volume, amplified pressure wave reflection from the periphery and elevated central systolic and diastolic blood pressure. Heart rate, systemic vascular resistance, cardiac output and pulse wave velocity did not differ between the groups. Licorice active constituent, glycyrrhetinic acid (GA) effectively restored vascular contractility in the model of lipopolysaccharide (LPS)-treated rat aorta (Muller et al. 2014). GA was as effective as the NO synthase inhibitor N(G)-nitroarginine methylester. The results suggested GA to be an interesting alternative to NO synthase inhibitors in sepsis-associated vascular dysfunction.

The polyphenols glabridin (derived from licorice), rosmarinic acid or carnosic acid (derived from rosemary), as well as garlic (which contains a mixture of natural antioxidants) inhibited LDL oxidation in a dose-dependent manner (Fuhrman et al. 2000). In-vivo, a meal of lycopene in com-

bination with other natural antioxidants reduced susceptibility to oxidation by 21 % in four healthy subjects. It was concluded that lycopene acted synergistically, as an effective antioxidant against LDL oxidation, with several natural antioxidants such as vitamin E, the flavonoid glabridin, the phenolics rosmarinic acid and carnosic acid, and garlic. The results suggested a superior anti-atherogenic characteristic to a combination of different natural antioxidants over that of an individual one.

Anticoagulation/Fibrinolytic Activities

Aqueous licorice leaf extracts exhibited no anticoagulation activities at 2000 μg/ml concentration but addition of sulphate group to the aqueous extracts enhanced the anticoagulation activity (Helmy et al. 2013). The highest activity was found by using both the sulphated alkaline and neutral extract of *G. glabra* var. *glabra* at a concentration of 400 μg/ml. Most of the aqueous extracts and their sulphated extracts of licorice leaves (*G. glabra* var. *glabra* and var. *glandulifera*) exhibited fibrinolytic activities higher than standard (Hemoclar) preparation at concentration of 2000 μg/ml.

Anti-ulcer/Gastroprotective Activity

In-Vitro Studies

Among the chemical constituents of licorice, glabridin and glabrene, licochalcone A, licoricidin and licoisoflavone B exhibited inhibitory activity against the growth of *Helicobacter pylori* in-vitro (Fukai et al. 2002a). These flavonoids also showed anti-*H. pylori* activity against a clarithromycin (CLAR) and amoxicillin (AMOX)-resistant strains. Glycyrrhetinic acid was as the most potent compound (MIC_{50} = 50 mg/L, MIC_{90} = 100 mg/L) inhibiting 79.3 % *Helicobacter pylori* strains at 50 mg/L or <. Krausse et al. (2004) exhibited rapid, concentration and strain-dependent bactericidal activity. Clarithromycin-resistant strains of *H. pylori* were susceptible at 12.5 and 25 mg/L, and metronidazole-resistant strains at 25–50 and at 200 mg/L. The MIC distribution (mg/L) of the lipophilic acetylated derivative of glycyrrhetinic acid monoglucuronide was ≤6.25 (29.2 %), 50 (4.2 %), 100–200 (12.5 %) and ≥400 (54.1 %). Extractum liquiritiae and glycyrrhizic acid were less active (MICs >400 mg/L). Studies by Malek Jafarian and Ghazvini (2007) showed that therapeutically administered concentrations of licorice extract (100–400 mg/ml) could have growth-inhibiting effect on *H. pylori* in-vitro. Cheel et al. (2013) found that the chemical profile of licorice quantitatively varied at different harvest times and these fluctuations determined changes in its bioactivities including gastroprotective activity. Licorice samples in May gave the best gastroprotective effect. Liquiritin and glycyrrhizin, the major constituents in the February and May licorice extract, appeared to contribute to the superoxide radical scavenging and gastroprotective effects. GutGard®, a flavonoid rich extract of *Glycyrrhiza glabra*, exhibited anti-*Helicobacter pylori* activity in both agar dilution and microbroth dilution methods possibly by inhibiting protein synthesis, DNA gyrase and dihydrofolate reductase (Asha et al. 2013). Glabridin, the major flavonoid present in GutGard® exhibited superior activity against *Helicobacter pylori* while glycyrrhizin did not show activity even at 250 μg/ml concentration.

Animal Studies

Gastric mucosal damage induced by giving 60 mg aspirin orally to rats was reduced by simultaneous administration of 100–500 mg deglycyrrhizinated liquorice (Rees et al. 1979). Studies demonstrated that deglycyrrhizined liquorice (DGL) exerted a protective effect against aspirin and aspirin plus bile acid-induced gastric mucosal damage, and aspirin absorption in rats (Russell et al. 1984). In the first study, lesion scores were increased from 6(3;10) with aspirin alone to 12(5;16) by taurodeoxycholic acid (TDC) and reduced by DGL to 1(0;3.5) for aspirin alone and 3.5(0;6) for aspirin plus TDC. In the second study a slight reduction of aspirin absorption was found only at 20 min with

a median level of 0.9 mmol/l for the DGL treated rats and 1.2 mmol/l for the aspirin alone group. Although DGL diminished aspirin (128 mg/kg)-induced gastric mucosal damage from 17(12;25) to 8(3;14) when the two were given together it did not do so significantly when DGL was given before aspirin -15(20;22). Intraperitoneal deglycyrrhinized licorice reduced lesion scores from aspirin (128 mg/kg) from 14(11;24) to 7(1;19), thus indicating a systemic as well as a local effect of deglycyrrhinized licorice. Bennett et al. (1985) concluded that in rats, carbenoxolone and deglycyrrhized liquorice may exert their anti-ulcer effect by a non-prostaglandin mechanism. This contrasted with the mechanism occurring in man with carbenoxolone. Aspirin coated with liquorice was found to reduce the number and size of gastric ulcers, reducing the ulcer index from 1.5 to 0.5 and the incidence from 96 to 46 % in rats (Dehpour et al. 1994). Coating with derivatives including the deglycyrrhized form, a high glycyrrhized form, carbenoxolone and enoxolone was less effective (ulcer index, 0.70–0.94; incidence 62–76 %). GutGard dose dependently decreased gastric content, total acidity, ulcer index and increased the pH of gastric fluid in pylorus ligated ulcer rats (Mukherjee et al. 2010). In cold-restraint stress and indomethacin induced ulcer rats all the doses of GutGard decreased the ulcer index and increased the pH of gastric fluid. GutGard exhibited potent antioxidant activity with high hydrophilic and lipophilic oxygen radical absorbance capacity (ORAC) value.

Oral administration of *Glycyrrhiza glabra* root extract caused a significant reduction in the length of gastric ulcer induced by ethanol in rats (Shalaby et al. 2004). The curative ratios from gastric ulceration were 40.0, 65.9 and 67.3 % in groups of rats given the extract at 200, 400 and 800 mg/kg, respectively. Licorice root extract lipids markedly reduced the degree of damage stomach ulceration in mice caused by indomethacin and voltaren and was more efficacious than the well-known reference drugs – licurazide and liquiriton (Denisova et al. 2007). Special licorice extracts (containing low glycyrrhizin and high licochalcone A) afforded significant attenuation of either *Helicobacter pylori*-induced gastritis or tumorigenesis in interleukin-deficient mice through its potent antioxidative, anti-inflammatory and antimutagenic actions (Park et al. 2014a, b).

Clinical Studies

The double-blind, cross-over design study involving patients with chronic ulcer disease failed to demonstrate any healing effect of deglycyrrhizinated liquorice extract (Caved-S) 760 mg three times daily for 4 weeks on gastric ulcer (Engqvist et al. 1973). In another clinical trial of 96 patients with gastric ulcer, after 4 weeks no differences were found between placebo or treatment with deglycyrrhizinated liquorice whether assessed by gastroscopy or radiology, or in the percentage reduction in ulcer area, or in clinical improvement (Bardhan et al. 1978). In a clinical trial, retrospective endoscopic examination in 32 patients of chronic duodenal ulceration treated with deglycyrrhizinated liquorice tablets showed that healing of the ulceration had occurred and in the majority the mucosa appeared normal (Larkworthy and Holgate 1975). The authors asserted that for optimum effect the preparation in adequate dosage should be well chewed and swallowed on an empty stomach in the ambulant patient. Human faecal blood loss from gastric mucosal damage induced by 975 mg aspirin orally three times a day was less when 350 mg deglycyrrhizinated liquorice was given with each dose of aspirin (Rees et al. 1979).

Aqueous extract (1 mg/mL) of *Glycyrrhiza glabra* significantly inhibited the adhesion of *Helicobacter pylori* to human stomach tissue (Wittschier et al. 2009). This effect was related to the polysaccharides isolated from the extract, with one purified acidic fraction (0.25 SPB) as main active polymer. Purified polysaccharides did not exhibit direct cytotoxic effects against *Helicobacter pylori* and did not influence hemagglutination. In a double-blind clinical trial study of 60 patients with peptic ulcer disease, 4 weeks treatment with licorice was found to be as effective as bismuth in *Helicobacter pylori* eradication (Momeni et al. 2014). They suggested that in patients whom bismuth was contraindicated, licorice could be used safely instead. In a double-

blind study of 40 patients with peptic ulcer, licorice was found to be a good a replacement for bismuth sub-nitrate in treatment of peptic ulcer in a quadruple therapy comprising amoxicillin (500 mg, 3 times/day after diet for 15 days), metronidazole (250 mg, 4 times/day after diet for 15 days), omeprazole (20 mg, 2 times/day ½h before the diet for 30 days) and licorice (250 mg, 3 times/day ½h before the diet for 30 days). In a randomized, double-blind clinical trial, the use of an over-the-counter licorice medicated intraoral adhesive patch for treatment of recurrent aphthous ulcers significantly reduced ulcer size and pre-stimulus pain in treated subjects compared with placebo (Martin et al. 2008). In a double-blind, randomized prospective trial of 60 patients, Ghalayani et al. (2014) found both triamcinolone (30 patients) and licorice (30 patients) mucoadhesive films were effective in the management of oral mucositis during head and neck cancer radiotherapy. Furthermore, comparison of the pain scores between two groups demonstrated no meaningful difference, although an overall trend to reduced oral discomfort was seen in the licorice group.

Anti-dyspepsia Activity

In a randomized, double-blind, placebo-controlled study of patients with functional dyspepsia, administration of GutGard, an extract of *Glycyrrhiza glabra*, exerted a significant decrease in total symptom scores on day 15 and day 30, respectively, compared to placebo (Raveendra et al. 2012). The GutGard group also showed a significant decrease in the Nepean dyspepsia index on day 15 and 30, respectively, when compared to placebo. GutGard was generally found to be safe and well tolerated by all patients.

Immunomodulatory Activity

Glycyrrhizan GA, an acidic polysaccharide isolated from the stolon of *Glycyrrhiza glabra* var. *glandulifera* showed remarkable reticuloendothelial system-potentiating activity in a carbon clearance test (Shimizu et al. 1991). Glycyrrhizan GA, a representative polysaccharide with remarkable phagocytosis-enhancing activity was isolated from the stolon of *Glycyrrhiza glabra* var. *glandulifera* (Takada et al. 1992). Smith degradation product obtained from glycyrrhizan GA showed markedly higher values of anti-complementary and alkaline phosphatase-inducing activities than that from glycyrrhizin UA, the main polysaccharide from *G. uralensis* root. Kroes et al. (1997) found β-glycyrrhetinic acid to a potent inhibitor of the classical complement pathway (IC_{50}=35 μM), whereas non-inhibitory activity was observed towards the alternative pathway (IC_{50}>2500 μM). The anti-complementary activity of β-glycyrrhetinic acid was dependent on its conformation, since the α-form was not active. Detailed mechanistic studies revealed that β-glycyrrhetinic acid acted at the level of complement component C2. Crude polysaccharide fraction from *G. glabra* shoot was found to have immunomodulatory action; the fraction induced nitric oxide production by murine peritoneal macrophages in-vitro (Nose et al. 1998).

Studies demonstrated that glycyrrhizin could promote phenotypic and functional maturation of murine dendritic cells and this adjuvant-like activity may have potential therapeutic value (Hua et al. 2012). Also glycyrrhizin increased the production of interleukins IL-12, IL-10 and decreased the production of tumour necrosis factor alpha (TNF-α). Glycyrrhizin was able to upregulate the expression of CD40, CD86 and MHC-II maturation markers on dendritic cells (DCs) (Bordbar et al. 2012). DCs treated with glycyrrhizin enhanced proliferation of allogeneic T cells along with the production of IFN-γ and IL-10 cytokines and reduced IL-4 production. The data indicated that glycyrrhizin had the capacity to upregulate allostimulatory activity of professional antigen presenting DCs and conduct immune responses toward a T helper 1 response.

Granulocytes and NK cells were markedly activated by licorice infusion, whereas liquiritin and glycyrrhizin were inactive (Cheel et al.

2010). The results suggested that licorice infusion could be used as a potential non-specific immune stimulator. Leukocyte count and phagocytic index (carbon clearance) in sheep red blood cells was increased significantly with the treatment of aqueous liquorice extract (1.5 g/kg) compared to control (Mazumder et al. 2012). Zinc (45 mg/kg) in combination with licorice extract (0.75 g/kg) showed highly significant increase of leukocyte count and phagocytic index compared to control. The combination of zinc (45 mg/kg) and licorice (0.750 g/kg) showed significant increase in haemagglutinin titre and antibody secreting cells of mouse spleen value compared to vehicle control. In systemic anaphylaxis reaction test, results indicate a positive effect on anaphylaxis with the treatment of licorice in both doses and in combination with zinc. The results indicated that *G. glabra* in combination with zinc had shown potentiation of immunomodulatory activity.

In the in-vitro phagocytosis test with human granulocytes, Revitonil® (a phytopharmaceutical containing an extract of *Echinacea purpurea* and *Glycyrrhiza glabra* root) showed a 44–53 % immunostimulating effect at a concentration of 100 μg/ml (Wagner and Jurcic 2002). Whereas in the chemiluminescence test at a concentration of 1.25 μg/ml, Revitonil® exhibited a moderate enhancing effect only, a remarkable stimulating activity (30–50 %) was observed in the T-lymphocyte CD69 bioassay at a concentration of 100 μg–1 μg/ml. The highest immunological efficacy could be assigned to Revitonil® as revealed by the in-vivo carbon-clearance model in mice. With RCt/RCc-values of 2.0, Revitonil® exhibited a very high carbon elimination rate at oral administration. Because the *Echinacea* and *Glycyrrhiza* mono-extracts alone showed lower RCt/RCc-values (1.3–1.7) than Revitonil®, a potentiating synergistic effect of the extract mixture in Revitonil® could be postulated. Vikhe et al. (2013) reported that neutrophils when treated with *G. glabra* and *Tinospora cordifolia* plant extracts showed increase in phagocytic activity.

Adaptogenic Activity

Two anti-stress compounds were isolated from licorice 9,12,13-trihydroxy-(10*E*)-octadecenoic and 9,12,13-trihydroxy-10,11-epoxy-octadecanoic acid (Panossian et al. 1988). Rabbits were treated (orally) with a preparation of *Glycyrrhiza glabra* for 30 days and concurrently were exposed to vibration stress (30 days) (Oganesyan 2002). The licorice preparation reduced catalase activity in the peripheral blood and increased animal resistance to vibration stress. Active substances of licorice root accelerated metabolism in rabbit peripheral blood red cells of the bone marrow erythroid stem, enhanced compensatory reserve of the organism and increased animal's resistance to stress (Adamyan et al. 2005). Minasian et al. (2007) found that biological active substances of licorice accelerated metabolism processes of rabbit bone marrow stem cells, enlarged the animal compensatory abilities, thus providing its resistance to vibration.

Incorporation of *Glycyrrhiza glabra* powder to the feed media of *Drosophila melanogaster* was found to reduce stress in *D. melanogaster* induced by methotrexate at different concentration (Sowmya and Sathish Kumar 2010). Stress-related enzymes like catalase (CAT) and super oxide dismutase (SOD) were reduced by licorice powder. The exposure of mice to chronic fatigue stress for 15 days demonstrated an increased immobility time, increased anxiety, impaired memory, reduction in muscle co-ordination, reduced activity and increased pain perception (Trivedi and Sharma 2011). These altered behavioural parameters were attenuated significantly by the treatment of *Glycyrrhiza glabra* (100 and 200 mg/kg p.o.) comparable to fluoxetine (10 mg/kg, i.p.).

Estrogenic Activity

Glycyrrhiza glabra was confirmed to have a high estrogenic activity as proven by the effects of its alcoholic extract (25 mg dose) on mouse uterine response and vaginal opening which was compa-

rable to estradiol monobenzoate (Shihata and Elghamry 1963a). A higher dose of 50 mg daily for 3 days proved to possess a lower estrogenic activity when compared with estradiol monobenzoate. Upon uterine motility, licorice extract manifested an inhibitory influence upon the spontaneous movement of the organ at di-oestrus, oestrus and pregnancy. In in-vivo studies in dogs, *G. glabra* extract caused relaxation of the uterine musculature of both pregnant and non-pregnant bitches (Shihata and Elghamry 1963b).

Beta-sitosterol was isolated as an estrogenic principle from Egyptian *G. glabra* (Zayed et al. 1964). Studies demonstrated glabridin to be a phytoestrogen, binding to the human estrogen receptor and stimulating creatine kinase activity in rat uterus, epiphyseal cartilage, diaphyseal bone, aorta and left ventricle of the heart (Tamir et al. 2000). The stimulatory effects of 2.5–25 μ g/animal glabridin were similar to those of 5 μg/animal estradiol. Glabridin was found to be three to four times more active than 2′-*O*-methylglabridin and 4′-*O*-methylglabridin, and both derivatives were more active than 2′,4′-*O*-methylglabridin. The effect of increasing concentrations of glabridin on the growth of breast tumour cells was biphasic. Glabridin showed an estrogen receptor-dependent, growth-promoting effect at low concentrations (10 nm–10 μm) and estrogen receptor-independent antiproliferative activity at concentrations of >15 μm. Glabridin and its derivatives exhibited varying degrees of estrogen receptor agonism in different tests and demonstrated growth-inhibitory actions on breast cancer cells.

Animals fed with licorice extract, compared with estradiol and glabridin, showed an increase in creatine kinase (CK) activity, a known marker for estrogen responsive genes, which was higher than expected from the levels of glabridin in the extract indicating the possible presence of other components that may contribute to this strong estrogen agonist activity (Tamir et al. 2001). Results indicated that glabrene and isoliquiritigenin, (2′,4′,4-3 hydroxy chalcone) (ILC) in the licorice extract could bind to the human estrogen receptor with higher affinity (IC_{50}, 1 and 0.5 μM) than glabridin (IC_{50}, 5 μM). The stimulatory effects of glabrene in-vivo were tissue specific and similar to those of estradiol. The effect of increasing concentrations of glabrene and ILC on the growth of breast tumour cell were biphasic. Both showed an estrogen receptor-dependent growth-promoting effect at low concentrations (10 nM–10 μM), and estrogen receptor-independent antiproliferative activity at concentrations >15 μM.

Licorice root constituents: glabridin and glabrene exhibited estrogen-like activity in-vitro and in-vivo (Somjen et al. 2004b). Similar to estradiol-17β (E2), glabridin stimulated DNA synthesis in human endothelial cells (ECV-304; E304) and had a bi-phasic effect on proliferation of human vascular smooth muscle cells. In animal studies, in intact females or after ovariectomy, glabridin similar to E2 stimulated the specific activity of creatine kinase in aorta and in left ventricle of the heart. Raloxifene inhibited glabridin as well as E2 activities. Glabrene, on the other hand, had only the stimulatory effect on DNA synthesis in vascular cells, with no inhibition by raloxifene, suggesting a different mechanism of action. The authors suggested the use of glabrene with or without E2 as a new agent for modulation of vascular injury and atherogenesis for the prevention of cardiovascular diseases in post-menopausal women. In pre-menopausal human bone cells, the response to estradiol-17β and glabridin (at higher concentration) was higher than in post-menopausal cells; whereas, glabrene (at higher concentration) was more effective in post-menopausal cells (Somjen et al. 2004a). At both ages, the response to estradiol-17β and glabridin was enhanced by pretreatment with the less-calcemic Vitamin D analogue CB 1093 (CB) and the demonstrably non-calcemic analogue JK 1624 F(2)-2 (JKF). The response to glabrene was reduced by this pretreatment. Both glabridin and glabrene stimulated creatine kinase specific activity in diaphyseal bone and epiphyseal cartilage of prepubertal female rats. Daily feeding (3–14 days) of prepubertal female rats with glabridin, estradiol-17β or their combination also stimulated creatine kinase specific activity. Glabridin, similarly to estradiol-17β, also stimulated creatine kinase specific activity in

ovariectomized female rats. Raloxifene, in combination with glabridin or estradiol-17β, demonstrated the phenomenon of mutual annihilation of stimulation of creatine kinase specific activity in both epiphysis and diaphysis. Glabrene activity was not inhibited by raloxifene. They found that glabridin showed greater similarity to estradiol-17β and thus greater potential, with or without Vitamin D, to modulate bone disorders in postmenopausal women. In a study, nine healthy women 22–26 years old, in the luteal phase of the cycle were given 3.5 g of a commercial preparation of licorice (containing 7.6 % w.w. of glycyrrhizic acid) daily for two cycles (Armanini et al. 2004). It was found that licorice could reduce total serum testosterone probably due to the block of 17-hydroxysteroid dehydrogenase and 17–20 lyase. It was concluded that licorice could be considered an adjuvant therapy of hirsutism and polycystic ovary syndrome.

Studies by Dong et al. (2007) found that activation of rapid signalling pathways, including Erk1/2 and Akt, and the subsequent transcriptional regulation were involved in the proliferation of MCF-7 cells induced by the extract of *G. glabra* root. The extract had similar activity to that induced by 17β-estradiol (E_2), although glycyrrhizin did not show such an activity. Furthermore, the extract had estrogenic activity and a distinguishable profile of gene expression, suggesting the presence of potentially useful components other than glycyrrhizin in *G. glabra* root for hormone and anti-cancer therapies. In in-vivo studies, administration of alcoholic *G. glabra* extract at doses of 150 and 300 mg/kg exerted significant reduction of prostate weight, total serum testosterone and ventral prostate epithelium/stroma ratio in immature male rats (Zamansoltani et al. 2009). Increase in testosterone metabolism, downregulation of androgen receptors or activation of estrogen receptors could be the mechanisms involved. Most fraction of the ethyl acetate *G. glabra* root extract showed some estrogenicity on both human estrogen receptors (ER) α and β (Simons et al. 2011). A compound was considered a phytoestrogen when it activates the ER at concentrations $\leq 10^4$ times than that of estradiol (E_2). The main flavonoids in the fractions were glabrene, glabrone, glabridin, glabrol 4′-mehtyl-glabridin, 3′-hydroxy-4-*O*-methyl-glabridin, hispaglabridin A and hispaglabridin B. Fractions F16-20, rich in glabrene, showed a predominant estrogenic activity on the ERα. Several fractions displayed higher responses than the maximum response obtained with the reference compound, the natural hormone 17β-estradiol (E_2). In addition to glabrene, the estrogenic activity of licorice roots extract had been ascribed to the presence of glabridin and its derivatives. Glabridin did not exert agonistic activity to both ER subtypes. Prenylation of isoflavonoids had been suggested to induce antagonism towards the ERα. The estrogenic activities of all fractions, including this so-called superinduction, were clearly ER-mediated, as the estrogenic response was inhibited by 20–60 % known ER antagonists, and no activity was found in yeast cells that did not express the ERα or ERβ subtype. Prolonged exposure of the yeast to the estrogenic fractions that showed superinduction did not, contrary to E_2, result in a decrease of the fluorescent response. Glabridin displayed ERα-selective antagonism, similar to the ERα-selective antagonist RU 58668. Whereas glabridin was able to reduce the estrogenic response of E_2 by approximately 80 % at 6×10^{-6} M, glabrene-rich fractions only exhibited agonistic responses, preferentially on ERα.

Pulmonoprotective Activity

In-vivo studies showed that ovalbumin-induced asthmatic mice treated with glycyrrhizin ameliorated all established chronic histopathologic lung parameters (Hocaoglu et al. 2011). When the glycyrrhizin and dexamethasone groups were compared, there was no statistically significant difference between the two groups in the histopathologic parameters, including thickness of basement membrane, subepithelial smooth muscle, and epithelium and number of mast and goblet cells. Oral administration of glycyrrhizic acid (GA) (50 and 100 mg/kg b.wt.) significantly protected against benzo(a)pyrene-induced debilities in lungs of Wistar rats against B(a)P induced

debilities in lungs of Wistar rats (Qamar et al. 2012). GA protected lung epithelium by suppression of caspases activities in lung tissue and reduction of total protein, total cells, elastase activity, lactate dehydrogenase, alkaline phosphatase activities along with fortification of phospholipids in bronchoalveolar lavage fluid. Studies by Dai et al. (2013) found that liquiritigenin could protect human lung cells (A549) from *Staphylococcus aureus* α-haemolysin-mediated injury. Low concentrations of liquiritigenin remarkably decreased *Staphylococcus aureus* α-haemolysin production in a dose-dependent manner.

Antiplatelet/Antithrombotic Activity

Butanolic extract of *G. glabra* inhibited human platelet aggregation induced by adrenaline with an IC_{50} of 1.66 mg/mL (Sajid et al. 1991). The ether soluble fraction of the crude licorice root produced a 27.3 % inhibition of lysoPAF (platelet-activating factor) acetyltransferase in-vitro at a concentration 10 μg/ml (Nagumo et al. 1999). From this fraction, licoricidin, 1-methoxyphaseollin, 6,8-diprenylgenistein and 1-methoxyphaseollidin were isolated as active components, with IC_{50} values of 7.7, 57, 19 and 48 μM, respectively. Glycyrrhizin, isolated from *G. glabra*, was identified as a new thrombin inhibitor: (a) It prolonged plasma recalcification and thrombin and fibrinogen clotting times, and (b) it inhibited thrombin-induced, but not collagen-, PAF- or convulxin-induced platelet aggregation; but glycyrrhizin did not block thrombin's amidolytic activity upon S-2238 (Francischetti et al. 1997). Furthermore, the fluorescence emission intensity of dansyl-thrombin was increased upon glycyrrhizin binding. Moreover, glycyrrhizin displaced hirudin as an inhibitor of thrombin-catalysed hydrolysis of S-2238. The data indicated that glycyrrhizin was a selective inhibitor of thrombin and that it was able to exert its anti-thrombin action by interacting with the enzyme's anion binding exosite 1.

Intravenous administration of rats with glycyrrhizin caused a dose-dependent reduction in thrombus size on a venous thrombosis model that combines stasis and hypercoagulability (Mendes-Silva et al. 2003). It was observed that glycyrrhizin doses of 180 mg/kg body weight produced 93 % decrease on thrombus weight. Glycyrrhizin was also able to prevent thrombosis using an arteriovenous shunt model. Glycyrrhizin doses of 180 and 360 mg/kg decreased the thrombus weight by 35 and 90 %, respectively. In addition, glycyrrhizin doses above 90 mg/kg caused significant haemorrhagic effect. In contrast with heparin, glycyrrhizin did not potentiate the inhibitory activity of anti-thrombin III or heparin cofactor II towards thrombin.

GU-7, a 3-arylcoumarin derivative, from glycyrrhizae radix, inhibited platelet aggregation, phosphorylation of 40 K and 20 K dalton proteins, inositol 1,4,5-trisphosphate production, intraplatelet calcium increase and phosphodiesterase activity in-vitro (Tawata et al. 1990). The data indicated that GU-7 inhibited platelet aggregation by increasing intraplatelet cAMP concentration. Isoliquiritigenin, a new aldose reductase inhibitor purified from licorice (Glycyrrhizae radix), inhibited platelet aggregation (Tawata et al. 1992). It significantly inhibited the phosphorylation of 40,000- and 20,000-Da proteins, and inhibited the formation of 12 (S)-hydroxy-5,8,10-heptadecatrienoic acid, 12-hydroxyeicosatetraenoic acid and thromboxane B2. The inhibitory effect of isoliquiritigenin on platelet aggregation in-vitro was comparable to that of aspirin. It was suggested that isoliquiritigenin elicited an anti-platelet action by inhibiting not only cyclooxygenase but also lipoxygenase or peroxidase activity in platelets. Since the hyperaggregability of platelets had been implicated in the pathogenesis of diabetic complications, isoliquiritigenin may offer a unique benefit as an aldose reductase inhibitor. Isoliquiritigenin, glabridin, licoarylcoumarin, glycycoumarin, glycyrol, licoricone and licoricidin were identified as strong inhibitors of adenosine 3′, 5′-cyclic monophosphate (cAMP) phosphodiesterase in waste materials which were obtained during the industrial extraction of glycyrrhizin from licorice roots (Kusano et al. 1991). Isoliquiritigenin-4′-*O*-apioglucoside and liquiritigenin were weakly

inhibitory while liquiritin, glycyrin and isoglycerol were not inhibitory. (cAMP) phosphodiesterase inhibitors had also been reported to be cardiotonic.

Xanthine Oxidase Inhibitory/ Antigout Activity

Of 10 licorice phenolics, sinkiang licorice (licochalcone and licochalcone A) and si pei (licorice glycyrrhisoflavone, glycyrrhiza flavanone, 3-arylcoumarin, licopyranocoumarin, licoarylcoumarin, glisoflavone, kaempferol 3-*O*-methyl ether, 2-arylbenzofuran, licocoumarone and glycycoumarin), 4 phenolics licochalcone B, glycyrrhisoflavone, licocoumarone and licochalcone A showed 50 % inhibition on xanthine oxidase at the concentration of $1.3–5.6 \times 10^{-5}$ M (Hatano et al. 1989). However, all were weaker than allopurinol a remedy for gout. Among 12 licorice constituents examined, six compounds namely glicoricone, licofuranone, genistein, licopyranocoumarin, licocoumarone and glycyrrhisoflavone, inhibited the enzyme with the IC_{50} (concentration required for 50 % inhibition of the enzyme activity) values of 6.0×10^{-5}–1.4×10^{-4} M. Glycyrrhizin also inhibited monoamine oxidase with the IC_{50} value of 1.6×10^{-4} M (Hatano et al. 1991b).

Antiangiogenic Activity

Using various experimental models of ocular neovascularization, namely (1) silver nitrate cauterization-induced corneal neovascularization in BALB/c mice, followed by topical isoliquiritigenin (ISL) (0.2–50 μM) and CD31 immunofluorescence of corneal blood vessels; (2) argon laser photocoagulation-induced choroidal neovascularization in C57BL/6 mice, followed by intravitreal ISL (10–200 μM) and fundus fluorescein angiography and immunofluorescence with *Griffonia simplicifolia* isolectin-B4 (GSA I-B4); and (3) oxygen-induced retinopathy in C57BL/6 J mice pups, followed by intravitreal ISL (1–100 μM) and GSA I-B4 immunofluorescence mice. Jhanji et al. (2011) demonstrated that ISL from licorice extract had an antiangiogenic effect. The authors suggested that ISL may be a potential antiangiogenic agent in the development of therapy for neovascularization diseases.

Anti-tyrosinase Activity

Studies showed that glabridin, from licorice extract, at concentrations of 0.1–1.0 μg/ml, inhibited tyrosinase activity of cultured B16 murine melanoma cells and guinea pig skins and had no detectable effect on their DNA synthesis (Yokota et al. 1998). It was also shown that UVB-induced pigmentation and erythema in the skins of guinea pigs were inhibited by topical applications of 0.5 % glabridin. Anti-inflammatory effects of glabridin in-vitro were also shown by its inhibition of superoxide anion productions and cyclooxygenase activities. By replacing each of the hydroxyl groups of glabridin with others, it was revealed that the inhibitory effect of 2′-*O*-ethyl glabridin was significantly stronger than that of 4′-*O*-ethyl-glabridin on melanin synthesis in cultured B16 cells at a concentration of 1.0 mg/ml. The 50 % tyrosinase-inhibitory concentration of the *Glycyrrhiza glabra* methanol extract was 21.2 μg/ml (Khanom et al. 2000). A glabridin derivative, 3″,4″-dihydroglabridin exhibited higher tyrosinase inhibitory activity than glabridin (IC_{50} value = 11.40 μM), which was probably due to the 4-substituted resorcinol skeleton and the lacking of double bond between carbon atom 3″ and 4″ on its structure giving more conformational flexibility to interact with the enzyme more effectively (Jirawattanapong et al. 2009). In addition, various acylated derivatives were synthesized as glabridin prodrugs. The chemical and enzymatic hydrolysis of prodrugs revealed that the diacetate ester was rapidly hydrolysed by porcine liver esterase with the half-life of 2.36 min, while that of the dihexanoate was 14.8 h.

The cellular levels of tyrosinase mRNA, protein, enzyme activities and melanin contents in B16 murine melanoma cells were increased by glycyrrhizin in a dose-dependent manner (Jung et al. 2001). Expression of tyrosinase-related pro-

tein-2 (TRP-2) mRNA was also increased by glycyrrhizin, however, no significant change was observed on TRP-1. Glycyrrhetinic acid showed no effect on melanogenesis at the equivalent non-toxic concentrations, indicating that glycoside structure is important in the stimulatory effect of glycyrrhizin on melanogenesis. Liquiritin was the main compound in licorice root extract and exhibited strong inhibitory effect on mushroom tyrosinase (Dong et al. 2014). Pinocembrin, the main compound in the leaf extract showed good antioxidant activity and nitrite scavenging capacity, but moderate inhibitory effect on mushroom tyrosinase. Both compounds exhibited significant protection effect on H_2O_2-injured PC12 cells at a low concentration.

Wound/Burn Healing Activity

Licorice root extract (LRE) lipid fraction effectively stimulated the reparative regeneration of damaged burnt skin caused by hydrochloric acid (I), skin layer removal (II), thermal burns (III) and in guinea pigs and rats (Denisova et al. 2007). An analysis of the results of these tests showed that the healing of model wounds I and II treated with LRE lipids was comparable to the action of rosehip oil and more effective than the natural healing in the control. In series burn model III RE lipid fraction also reliably reduced the area of burned skin and this effect was more pronounced than the action of rosehip oil. Intraperitoneal administration of glycyrrhizin (Gly) (60 mg/kg) significantly reduced the levels of elevated serum TNF-α and IL-1β caused by severe skin scald burn injury in rats (Shen et al. 2015). Gly treatment reduced these biochemical indices accompanied by lower level of HMGB1 protein and mRNA expression.

Antitussive Activity

Glycyrrhetinic acid and its derivatives were active in antitussive activity in experiments using chemical stimulation in the un-anaesthetized guinea-pig and electrical stimulation in the lightly anaesthetized cat indicating a central antitussive effect (Anderson and Smith 1961). Several derivatives had approximately the same potency as codeine when given subcutaneously to guinea-pigs; one of these, dicholine glycyrrhetinic acid hydrogen succinate, exhibited the same degree of activity after oral administration. The water-extracted arabinogalactan protein enriched fraction of *Glycyrrhiza glabra*, when administered orally in a dose of 50 mg/kg body weight decreased the number of citric acid induced cough efforts in guinea pigs more effectively than codeine (Saha et al. 2011). It did not induce significant change in the values of specific airway resistance or provoked any observable adverse effects.

Renoprotective Activity

Malekinejad et al. (2011a) found that liquorice plant extract could reduce ochratoxin A-induced nephrotoxicity in rats. Liquorice plant extract like melatonin could alleviate an ochratoxin A-reduced antioxidant power of serum and lower the toxin-induced malondialdehyde generation. They also found that *Glycyrrhiza glabra* extract and melatonin could protect against ochratoxin A-induced damages on testes in mature rats through their antioxidative actions (Malekinejad et al. 2011b).

Oral administration of glabridin, a pyranoisoflavan isolated from *Glycyrrhiza glabra* (30 mg/kg/day) for 10 days to mice with glomerular disease (Masugi-nephritis) reduced the amount of urinary protein excretion from control level (100 mg/day) to a significantly lower level (47 mg/day) (Fukai et al. 2003). Pretreatment with DHC-1, a herbal formulation derived from *Bacopa monnieri*, *Emblica officinalis*, *Glycyrrhiza glabra*, *Mangifera indica* and *Syzygium aromaticum* protected arts against isoproterenol-induced myocardial infarction and cisplatin-induced renal damage (Bafna and Balaraman 2005). This beneficial effect may be attributed, at least in part, to its antioxidant activity. Glycyrrhizin was demonstrated to attenuate renal I/R injury in mice via administration of

glycyrrhizin, which suppressed the serum levels of creatinine and blood urea nitrogen 6 h following reperfusion; furthermore, the superoxide anions as well as the activity of superoxide dismutase within renal tissues was reduced by glycyrrhizin pretreatment (Ye et al. 2014). Further, the protein level of cleaved caspase-3, as well as its activity in renal tissue, was suppressed as a result of the glycyrrhizin pretreatment, indicating that glycyrrhizin inhibited I/R-induced renal cell apoptosis. Additionally, glycyrrhizin pretreatment appeared to ameliorate I/R-induced renal injury via inhibition of inflammatory cell infiltration, as well as the production of pro-inflammatory cytokines, including tumour necrosis factor-α, interferon-γ, interleukin (IL)-1β and IL-6. The results suggested that glycyrrhizin provided significant protection against I/R-induced renal injury in mice by inhibiting inflammatory responses and renal cell apoptosis.

Licorice and its compounds had been found to have anti-hyperkalaemia action. Hyperkalaemia is a common life-threatening problem in haemodialysis patients. Licorice treatment could provide an important tool to maintain predialysis [K(+)] within safe limits in some dialysis patients at risk of hyperkalaemic arrhythmias (Ferrari 2009). In a prospective, double-blind, cross-over, placebo study of seven patients with anuria or chronic haemodialysis, administration of glycyrrhetinic acid, active compound of licorice (1 g/day for 2 weeks) decreased plasma potassium concentration (Serra et al. 2002). The decline in potassium level was paralleled by an increase in plasma cortisol/cortisone ratio indicating the inhibition of renal 11beta-hydroxysteroid dehydrogenase type 2 (11beta-HSD2). The data indicated that extra-renal 11beta-HSD activity influenced serum potassium concentrations but did not regulate blood pressure independently of renal sodium retention. In another prospective, double-blind, placebo-controlled crossover study of 10 haemodialysis patients, administration of cookies or bread rolls supplemented with glycyrrhetinic acid for 6 months significantly lowered serum potassium levels (Farese et al. 2009). No differences were found in parameters reflecting sodium retention.

Hair Promoting Activity

Studies showed that petroleum ether extract of *G. glabra* roots had potentials as a hair growth promoting agent for female rats (Upadhyay et al. 2013). Female rats treated with *G. glabra* had longer hair than those treated with either minoxidil or control. A maximum of 76 % of hair follicles were in anagenic stage (active growth phase of hair) in licorice extract-treated animals, compared to 66 and 45 % in minoxidil-treated and control groups, respectively. In another study in male Wistar albino rats, they found that petroleum ether extract of *G. glabra* possessed anti-androgenic alopecia activity which was comparable to that of standard drug finasteride (Upadhyay and Singh 2013).

Antiosteoporotic Activity

In a clinical study, it was found that licorice 3.5 g of a commercial preparation of licorice (containing 7.6 %, w/w of glycyrrhizic acid) administration daily for 2 months could increase serum parathyroid hormone and urinary calcium levels from baseline value in nine healthy women (22–26 years old, in the luteal phase of the cycle) after only 2 months of treatment (Mattarello et al. 2006). The effect of licorice on calcium metabolism was postulated to be influenced by several components of the root, which showed aldosterone-like, estrogen-like and antiandrogen activity.

Treatment with glabridin (1–10 μM) prevented apoptosis induced by TNF-alpha in murine MC3T3-E1 osteoblastic cells (Choi 2005). Moreover, glabridin (50 μM) decreased the TNF-alpha-induced production of prostaglandin E2 and nitric oxide in osteoblasts. The data indicated that the enhancement of osteoblast function by glabridin may result in the prevention for osteoporosis and inflammatory bone diseases. Liquiritigenin protected osteoblastic MC3T3-E1 cells from antimycin A-induced cell death (Choi et al. 2014). It was found that modulation of PI3K, antioxidant effects and the attenuation of mitochondrial dysfunction by liquiritigenin rep-

resent an important mechanism for its protection of osteoblasts against cytotoxicity resulting from mitochondrial oxidative stress. Glabridin attenuated 2-deoxy-D-ribose(dRib)-induced oxidative cell damage in MC3T3-E1 mouse osteoblastic cells (Kim et al. 2013). Treatment with glabridin resulted in a significant elevation of alkaline phosphatase (ALP) activity, collagen contents and osteoblast differentiation genes [ALP, collagen, osteopontin (OPN), osteoprotegerin (OPG) and osteocalcin (OC)] and bone morphogenetic protein (BMP) genes (BMP2, BMP4 and BMP7). Glabridin activated dRib-induced decreased expression of phosphatidylinositol 3′-kinase (PI3K) and protein kinase B 2 (AKT2) genes, master regulators of survival-related signalling pathways. Glabridin also upregulated the gene expression of antioxidant enzymes, superoxide dismutase 1 (SOD1) and glutathione peroxidase 4 (GPX4), which were inhibited by dRib. The results suggested that glabridin may be useful for the treatment of diabetes-related bone disease.

Ming et al. (2014) found that 0.35 mg/L licochalcone A (L-A) had a strong effect in increasing the osteogenic differentiation and mineralization of bone marrow-derived mesenchymal stem cells (BMSCs) both in-vivo and in-vitro by upregulating FasL and further playing a role in regulating the ERK and GSK-3β-catenin systems. It was also demonstrated that the administration of L-A could restore the biological function of the damaged BMSCs differentiation by recovering or protecting bone mass in a disease state through activating the endosteal bone formation and partially inhibiting bone resorption in acute estrogen deficiency model.

Radioprotective Activity

G. glabra root extract was found to protect microsomal membranes from gamma irradiation, as evident from reduction in lipid peroxidation, and could also protect plasmid DNA from γ-radiation-induced strand breaks (Shetty et al. 2002).

Cytoprotective Activity

Incubation of H4IIE murine hepatoma cells with *Glycyrrhiza* radix extract (GRE) inhibited cell death induced by 10 μM cadmium (Kim et al. 2004). The results demonstrated that GRE blocked Cd-induced cell death by inhibiting the apoptotic processes involving translocation of Bad into mitochondria, decreases in mitochondrial Bcl_{xL} and cytochrome *c*, and poly(ADP-ribose)polymerase cleavage. Among the major components present in GRE, liquiritigenin, but not liquiritin, isoliquiritigenin or glycyrrhizin exerted cytoprotective effect.

Choleretic Activity

Recent studies indicated that licorice extract, when administered per os or i.v., caused a marked choleretic effect in rats (Raggi et al. 1995). Umbelliferone (7-hydroxycoumarin) and glycyrrhizin were found to be the bioactive constituents with choleretic effects. Unlike the glycyrrhizin, which was present in a fairly large amount, umbelliferone was present at a very low concentration (traces), both in licorice and in bile. Licorice extract presented a significant choleretic effect after both oral and i.v. administration in rats, which increased the excretion rate of glycyrrhizin (Cantelli-Forti et al. 1997). Intravenous administration of liquiritigenin, a flavonoid aglycone from licorice, was found to have a choleretic effect and the ability to induce hepatic transporters and phase-II enzymes in rats (Kim et al. 2009).

Anti-convulsant Activity

G. glabra var. *glandulifera* leaf ethanol extract and dichloromethane fraction showed anticonvulsant effect in pentylenetetrazol seizure test in mice (Yazdi et al. 2011). The ED_{50} value of 2.11 g/kg and 1.30 g/kg was obtained for the

crude extract and dichloromethane fraction, respectively. The anticonvulsant activity of the extract and dichloromethane fraction could be mainly attributed to the compounds of triterpenes/sterols class present in the leaves. *G. glabra* ethanol extract at three doses, namely, 100, 200 and 400 mg/kg i.p. delayed the onset of pentylenetetrazole (PTZ)-induced seizure in rat but the duration of convulsion was reduced only in higher dose level (200 and 400 mg/kg) (Chowdhury et al. 2013). Pretreatment with the extract attenuated lipid peroxidation due to increase in antioxidant enzymes in the rat brain tissues.

Aldose Reductase Inhibitory Activity

Isoliquiritigenin was found to have potent aldose reductase inhibiting activity (Aida et al. 1990). Isoliquiritigenin inhibited rat lens aldose reductase with an IC_{50} of 3.2 x 10^{-7} M, using DL-glyceraldehyde as a substrate. It inhibited sorbitol accumulation in human red blood cells in-vitro, with an IC_{50} of 2.0×10^{-6} M. Isoliquiritigenin, when administered via an intragastric tube to diabetic rats, suppressed sorbitol accumulation in the red blood cells, the sciatic nerve and the lens as effectively as ONO-2235. The results suggested that isoliquiritigenin may be effective in preventing diabetic complications.

Vasorelaxant Activity

Isoliquiritigenin was found to have vasorelaxant activity (Yu and Kuo 1995). Isoliquiritigenin caused endothelium-independent relaxation of phenylephrine-precontracted rat aortic rings. Relaxation of phenylephrine-precontracted rat aorta and carbachol-precontracted guinea-pig trachea by rolipram (phosphodiesterase, PDE IV inhibitor) was markedly enhanced by isoliquiritigenin, while response to cilostamide (PDE III inhibitor) was not significantly changed by isoliquiritigenin. It was concluded that isoliquiritigenin exerted a vasorelaxant effect by activating soluble guanylate cyclase and increasing cyclic GMP.

Anti-metrorrhagia Activity

In a study of 32 women with polycystic ovary syndrome (PCOS), mean blood pressure was significantly reduced during spironolactone treatment, while it was unchanged in women receiving spironolactone plus licorice. Twenty percent of women treated with spironolactone and none treated with the addition of licorice complained of symptoms related to volume depletion. Consistently, the activation of the renin-aldosterone system was significantly lower during spironolactone plus licorice than with spironolactone alone. The prevalence of metrorrhagia was lower in the combined therapy. The results suggested that in patients with PCOS the mineralocorticoid properties of licorice could reduce the prevalence of side effects related to the diuretic activity of spironolactone.

Cosmetic Application Property

Both α-glycyrrhizin and β-glycyrrhizin exhibited similar considerable interfacial activity for cosmetic applications (Kondo et al. 1986). However, the aqueous solution of β-glycyrrhizin formed an extremely rigid gel in acidic media, whereas α-glycyrrhizin showed no sign of gelation. β-glycyrrhizin could emulsify various oily materials over a wide range of required HLB (hydrophilic-lipophilic balance) values, while α-glycyrrhizin had solubilizing ability for several perfume materials. Results from a series of experiments on the solubilizing, emulsifying and gelling mechanisms of α- and β-glycyrrhizin suggested that the β-glycyrrhizin molecule which was found to be cyclically constructed constituted the micelies which in turn orient anisotropically to form a rigid gel and to stabilize the emulsion. Based on the already described properties, α-glycyrrhizin should be utilized as a solubilizer, and β-glycyrrhizin as an emulsifier and a

stabilizer in cosmetics. Baltina et al. (1996) separated *cis* (18β) and *trans* (18α) isomers of glycyrrhizic acid (GA) from *G. glabra* root. The two GA isomers possess close physicochemical compositions, but *trans*-GA differed from the *cis*-isomer by higher solubility in water, higher stability of the aqueous solutions and lack of gel formation in these solutions. *Trans*-GA was of interest as a surfactant and solubilizing agent for the production of cosmetics and stabilized water-soluble medication forms.

Postoperative Sore Throat (POST) Attenuation Activity

In a prospective, randomized, single blind study of 40 adults (18–60 year), ASA physical status I and II of either sex, undergoing elective lumber laminectomy, licorice gargle performed 5 min before anaesthesia was found effective in attenuating the incidence and severity of postoperative sore throat (Agarwal et al. 2009). In a randomized, double-blind study of 236 patients having elective thoracic surgery, it was found that preoperative gargling with licorice solution rather than sugar water prevented postoperative sore throat and post-extubation coughing in patients intubated with double-lumen tubes (Ruetzler et al. 2013). Licorice gargling halved the incidence of postoperative sore throat.

Antiparasitic Activity

In in-vivo studies in dogs, *G. glabra* extract was found to have an anthelmintic influence upon Taenia worms but not on Ascaris worms (Shihata and Elghamry 1963b). Glycyrrhizic acid (GA) decreased hepatic and splenic *Leishmania donovani* parasite burden and increased T-cell proliferation in *Leishmania*-infected BALB/c mice (Bhattacharjee et al. 2012). When treated with GA at 75 mg/kg/day, *L. donovani*-infected mice exhibited a 95 % and 92 % reduction of the parasite burden in the liver and spleen, respectively. GA significantly enhanced the cell-mediated immune response.

Licorice glycyrrhetinic acid (GA) was found to be effective against microfilariae in-vitro (LC_{100}: 12.5 μM; IC_{50}: 1.20 μM), but was inactive against adult *Brugia malayi* worms (Kalani et al. 2013b). Of six GA analogues, the benzyl amide analogue killed adults and microfilariae at 25 and 50 μM concentration, respectively, and inhibited 49 % MTT reduction potential of the adult parasites. The IC_{50} values were found to be 8.8 and 2.2 μM for adults and microfilariae, respectively. In contrast, the octylamide analogue required much higher concentration to adversely affect the parasites. In-vivo using *B. malayi*-jird model, the benzyl amide analogue exhibited promising macrofilaricidal activity at 100 mg/kg, s.c. ×5 days and around 40 % of the treated animals showed calcified masses of worm fragments in the peritoneal cavity of the animals. The in-vitro studies against *Plasmodium falciparum* showed significant (IC_{50} 1.69 μg/ml) anti-malarial potential for licorice 18β-glycyrrhetinic acid (Kalani et al. 2013a). In-vivo evaluation showed a dose dependent anti-malarial activity ranging from 68 to 100 % at doses of 62.5–250 mg/kg on day 8.

Pharmacokinetic Studies

The pharmacokinetics of glycyrrhizin had been described and indicated its reduced bioavailability when consumed as licorice and based on available in-vivo and clinical evidence. Isbrucker and Burdock (2006) proposed an acceptable daily intake of 0.015–0.229 mg glycyrrhizin/kg body weight/day.

It was found that the time required for a maximum concentration (Tmax) of glycyrrhizin (G) in the rat plasma was 8 h after oral administration of licorice extract (Ozaki et al. 1990). In contrast, glycyrrhizin reached a maximum plasma concentration at less than 6 h after administration of glycyrrhizin. The plasma level of glycyrrhizin fell slowly within 24 h after their oral administration, and it was still detected in the plasma even after 24 h. Glycyrrhizin in the rat plasma, bile and urine could be precisely determined in concentrations as low as 1, 1 and 2.5 μg/ml, respectively, in a 0.1-ml sample using selective high-performance

liquid chromatographic method (Yamamura et al. 1991). The equivalent values for the glycyrrhetic acid-3-*O*-glucuronide were 1, 2.5 and 2.5 μg/ml, respectively. After oral administration of glycyrrhizin (100 mg) to three normal subjects, the major metabolite of glycyrrhizin (glycyrrhetic acid) appeared in plasma (<200 ng/mL), but glycyrrhizin was not found (Yamamura et al. 1992). In contrast, glycyrrhizin was found in urine, and the amount excreted was 1.1–2.5 % of the dose. The finding suggested that glycyrrhizin was partly absorbed in the intact form from the gastrointestinal tract. The concentration of glycyrrhizin in plasma after intravenous administration of glycyrrhizin (40, 80 and 120 mg) showed bi-exponential profiles during the 24-h period after administration of each dose. The glycyrrhizin metabolites, glycyrrhetic acid and glycyrrhetic acid-3-*O*-glucuronide, were not detected in either plasma or urine. Glycyrrhizin was not detected in plasma after oral administration of the usual therapeutic dose of glycyrrhizin, and no dose dependency of the drug was observed in the dose range of 40–120 mg.

The pharmacokinetics of glycyrrhetinic acid and glycyrrhizic acid humans and experimental animals could be described by a biphasic elimination from the central compartment with a dose-dependent second elimination phase (Krähenbühl et al. 1994). Depending on the dose, the second elimination phase in humans had a half-life of 3.5 h for glycyrrhizic acid and between 10 and 30 h for glycyrrhetinic acid. The major part of both glycyrrhetinic acid or glycyrrhizic acid was eliminated by the bile. While glycyrrhizic acid could be eliminated unmetabolized and underwent enterohepatic cycling, glycyrrhetinic acid was conjugated to glycyrrhetinic acid glucuronide or sulphate prior to biliary excretion. Orally administered glycyrrhizic acid was almost completely hydrolysed by intestinal bacteria and reached the systematic circulation as glycyrrhetinic acid. HPLC methods were found suitable in terms of precision and accuracy for the glycyrrhizin and glycyrrhetic acid determination in plasma and urine of human volunteers and in bile, plasma and urine of rats (Raggi et al. 1994a). Raggi et al. (1994b) found that after oral administration of licorice extract or glycyrrhizin to rats and humans, glycyrrhizin showed significantly reduced bioavailability when administered as licorice extract compared to when administered as glycyrrhizin. Significantly lower glycyrrhizin and glycyrrhetic acid plasma levels were found in rats and humans treated with licorice extract compared to the levels obtained with those in which glycyrrhizin alone was administered (Cantelli-Forti et al. 1994). The pharmacokinetic curves showed significant differences in the areas under the plasma-time curve (AUC), Cmax, and Tmax parameters. The data obtained from urine samples are in agreement with the above results and confirm a reduced bioavailability of glycyrrhizin present in licorice extract compared to pure glycyrrhizin. Glycyrrhizic acid was mainly absorbed after presystemic hydrolysis as glycyrrhetic acid (Ploeger et al. 2000, 2001). Once absorbed, glycyrrhetic acid was transported, mainly taken up into the liver by capacity-limited carriers, where it was metabolized into glucuronide and sulphate conjugates. These conjugates were transported efficiently into the bile possibly by the hepatic transfer protein 3α-hydroxysteroid dehydrogenase. After outflow of the bile into the duodenum, the conjugates were hydrolysed to glycyrrhetic acid by commensal bacteria; glycyrrhetic acid was subsequently reabsorbed, causing a pronounced delay in the terminal plasma clearance.

After oral administration of *Glycyrrhiza* extract to rats, glycyrrhizin and glycyrrhetic acid were detected in the plasma (Wang et al. 1995). However, the plasma concentration-time curves of glycyrrhizin and glycyrrhetic acid after *Glycyrrhiza* extract oral administration were much lower than those of pure glycyrrhizin, indicating the marked reduction in bioavailability of glycyrrhizin and as glycyrrhetic acid after this administration. It was found that the lipophilic components of *Glycyrrhiza* extract reduced the gastric emptying rate and the absorption of glycyrrhizin from the small intestine, while these effects were not observed in the hydrophilic components. In contrast, the bioavailability of glycyr-

rhizin as glycyrrhetic acid was increased by the hydrophilic components, but not the lipophilic ones.

The results of toxicological and glycyrrhetic acid analyses showed significantly lower concentrations of glycyrrhizin in bile samples from rats treated with pure aqueous liquorice extract (LE) compared to pure glycyrrhizin (Cantelli-Forti et al. 1997). Furthermore, LE presented a significant choleretic effect after both oral and i.v. administration, which increased the excretion rate of glycyrrhizin. In case of glycyrrhetic acid, all the concentrations were very low, often below the detection limit.

Ruminococcus sp. PO1-3 and *Clostridium innocuum* ES24-06 intestinal bacteria isolated from human faeces were found capable of metabolizing glycyrrhizin (Hattori et al. 1985). *Ruminococcus* sp. had the ability to hydrolyse GL to glycyrrhetic acid (GA) and to reduce 3-dehydroglycyrrhetic acid (DGA) to GA while the *Clostridium innocuum* had the ability to reduce DGA to 3-epiglycyrrhetic acid (EGA). A mixture of the two strains could not only reduce DGA to both GA and EGA, but also epimerized GA to EGA and vice versa, possibly through a 3-dehydro intermediate. A bacterial strain capable of hydrolysing glycyrrhizin (GL) to glycyrrhetic acid (GA) was isolated from human faeces and identified as *Eubacterium* sp. (Akao et al. 1987). The GL-hydrolysing activity increased in parallel with the growth of this bacterium, which also produced β-D-glucuronidase acting on β-D-glucuronides of phenolic compounds such as phenolphthalein mono-β-D-glucuronide. Akao and Kobashi (1988) found that *Eubacterium* sp. strain GLH, isolated from human faeces, produced two kinds of β-D-glucuronidase, one enzyme specific for glycyrrhizin (GL) and the other for phenyl β-D-glucuronides. GL or p-nitrophenyl-mono-β-D-glucuronide (pNPG) stimulated the production of GL or pNPG β-glucuronidases and the growth of strain GLH in a basal medium lacking carbohydrate. D-Glucuronic acid also stimulated the growth of the bacterium, but glycyrrhetic acid did not. During mixed cultivation of the *Eubacterium* strain with *Streptococcus faecalis*, which does not produce GL β-glucuronidase, GL β-glucuronidase was also increased by GL or pNPG, but not by glycyrrhetic acid and p-nitrophenol. It was suggested that GL stimulated the growth of strain GLH even in the mixed culture. Glycyrrhizin (GL), a main constituent of liquorice, was hydrolysed to 18β-glycyrrhetic acid mono-β-D-glucuronide (GAMG, glycyrrhetyl monoglucuronide) by rat liver homogenate, and the hydrolytic activity was localized in the lysosomes among the same subcellular fractions as acid β-D-glucuronidase activity (p-nitrophenyl β-D-glucuronide (pNPG)-hydrolysing activity) (Akao et al. 1991). Rat liver lysosomes hydrolysed GAMG to 18β-glycyrrhetic acid (GA) at only 30 % rate compared with the rate of GL to GAMG. GA was also produced slowly from GL after time lag by the lysosomes. *Ruminococcus* sp. PO1-3, metabolized glycyrrhizin (GL) to glycyrrhetic acid (GA) and GA to 3-oxo-glycyrrhetic acid (3-oxo-GA) and possessed GL β-D-glucuronidase and 3β-hydroxysteroid dehydrogenase (3β-HSD) involved in the metabolism of GL (Akao 1999). This bacterial growth was enhanced by GL at a concentration of 0.4 mm and was suppressed by GA at a concentration of 1.0 mM. Chenodeoxycholic acid, deoxycholic acid and lithocholic acid among the bile acids added to this bacterium suppressed the growth and GL β-D-glucuronidase activity and 3β-HSD activity incident to it at a concentration of 1.0 mM, while cholic acid, hyodeoxycholic acid and glycine and taurine conjugates of cholic acid, chenodeoxycholic acid, deoxycholic acid and lithocholic acid had almost no effect on this bacterium at a concentration of 0.2–1.0 mm

Glycyrrhizin (18β-glycyrrhetic acid β-D-glucuronyl α-D-glucuronic acid, GL) and baicalin (baicalein β-D-glucuronic acid) were metabolized to glycyrrhetinic acid and baicalin, respectively, by human intestinal bacteria (Kim et al. 1996). However, α-glucuronidase of *Bacteroides* JY-6 isolated from human intestinal bacteria, hydrolysed GL or 18β-glycyrrhetinic acid α-D-glucuronic acid to 18β-glycyrrhetic acid but not to baicalin. However, *E. coli* β-glucuronidase from human intestinal bacteria hydrolysed baicalin to baicalein, but did not hydrolyse GL.

pressed. In contrast, 5α-reductase was markedly increased. In humans 5β-reductase is quantitatively the major enzyme and plays an important role in the regulation of cortisol and aldosterone metabolism. The authors suggested that the suppression of 5β-reductase activity by GA or GL administration may delay the clearance of corticosteroids and prolong the biological half-life of cortisol resulting in the synergism of these steroids and GA or GL. Latif et al. (1990) demonstrated that glycyrrhetinic acid (GA) did not affect either microsomal 5α-reductase or cytosolic 3α-hydroxysteroid dehydrogenase activity. However, GA was a potent inhibitor of cytosolic 5β-reductase. GA specifically inhibited microsomal 3β-HSD enzyme activity by apparently a competitive inhibition mechanism, causing a build-up of the intermediate, 5α-dihydroaldosterone (DHAldo). The results indicated that GA had a profound effect on hepatic ring A-reduction of aldosterone. Inhibition of 5β-reductase and 3β-HSD resulted in decreased synthesis of both 3α, 5β-tetrahydroaldosterone (THAldo) and 3β, 5α-THAldo and, hence, accumulation of aldosterone and 5α-DHAldo, both potent mineralocorticoids. The ingestion of liquorice, and/or its active metabolites, could sometimes produce an acquired form of apparent mineralocorticoid excess (AME) syndrome, expressed as sodium and fluid retention, potassium loss and suppression of the renin-angiotensin-aldosterone system, in addition to clinical consequences such as raised blood pressure (hypertension) and oedema (Sigurjónsdóttir et al. 1995; Olukoga and Donaldson 2000). Excessive ingestion of licorice had been known to induce a syndrome of hypokalaemia and hypertension reflecting increased activation of renal mineralocorticoid receptors by cortisol (Walker and Edwards 1994). A similar syndrome of cortisol-dependent mineralocorticoid excess had been reported to occur in congenital deficiency of the enzyme 11β-hydroxysteroid dehydrogenase. Liquorice had been reported to induce pseudohyperaldosteronism by inhibiting the 11β-hydroxysteroid dehydrogenase type 2 and was also known to inhibit the renin-angiotensin-aldosterone system (RAAS) (Sigurjónsdóttir et al. 2006). The continuous, high level exposure to glycyrrhizin compounds could produce hypermineralocorticoid-like effects in both animals and humans as biochemical studies indicated that glycyrrhizinate inhibited 11β-hydroxysteroid dehydrogenase, the enzyme responsible for inactivating cortisol (Isbrucker and Burdock 2006). These effects were reversible upon withdrawal of licorice or glycyrrhizin. Consumption of large amounts of liquorice could cause hypertension and hypokalaemia as liquorice contains glycyrrhetinic acid, which inhibits the enzyme 11β-hydroxysteroid dehydrogenase type 2, and ultimately leads to an apparent mineralocorticoid excess syndrome (Nielsen et al. 2012). Excessive intake of licorice may cause a hypermineralocorticoidism-like syndrome characterized by sodium and water retention, hypertension, hypokalaemia, metabolic alkalosis, low-renin activity and hypoaldosteronism (Celik et al. 2012), and generally the onset and severity of symptoms depend on the dose and duration of licorice intake, as well as individual susceptibility (Mumoli and Cei 2008). Glycyrrhizic acid, contained in licorice, possessed a mineralocorticoid-like effect and chronic excess intake of licorice could induce the rare syndrome of 'apparent mineralocorticoid excess', due to the inhibitory effect of glycyrrhizic acid on 11β-hydroxysteroid dehydrogenase type 2 determining clinical/biochemical manifestations as resistant hypertension, metabolic alkalosis and severe hypokalaemia (Bisoni et al. 2014). They emphasized the importance of anamnesis in the diagnosis so as to avoid unnecessary and expensive investigations, and reduce the duration of hospitalization. Liquorice intoxication could be confirmed by the shut-off of the renin-angiotensin-aldosterone axis, and by the increase of the urinary ratio of [cortisol metabolites (5α tetrahydrocortisol + 5β tetrahydrocortisol)]/[cortisone metabolite (5β tetrahydrocortisol)] together with increase of urinary free cortisol excretion (Luchon et al. 1993).

Case Reports

Gross et al. (1966) reported an obese patient striving to reduce weight ingested excessive

amount of licorice leading to severe potassium depletion and acute myopathy and myoglobinuria. Hypertension, initially normokalaemic, was eventually associated with hypokalaemia, alkalosis, suppressed plasma renin activity and aldosteronopenia in a 58-year-old man who had ingested two to three 36-gm licorice candy bars daily for 6 years to 7 years (Conn et al. 1968). Metabolic balance studies recorded complete recovery upon cessation of licorice ingestion. Holmes et al. (1970) reported a case of a 63-year-old man with pseudohyperaldosteronism induced by habitual ingestion of liquorice (3.5 lb licorice for the last 15–20 years). Chamberlain (1970) described a case of a previously healthy 53-year-old male patient who presented with fulminant congestive heart failure (CHF) after ingesting large quantities of licorice for a week. Kuriyama et al. (1975) described a patient who developed marked hypokalaemia due to chronic administration of glycyrrhizin (150 mg daily). Bannister et al. (1977) reported a case of a 58-year-old women with cardiac arrest associated with hypokalaemia caused by ingesting 1.8 kg of liquorice a week. Three months after stopping liquorice, she remained well and all laboratory values were normal. Takeda et al. (1979) reported the natural recovery from the aggravated hypertension, hypokalaemia and suppression of the renin-aldosterone axis after the glycyrrhizin discontinuation in two mild hypertensive women aged 71 and 68 years, who had been administered 273–546 mg glycyrrhizin daily for 1.5 and 6 months, respectively, for the treatment of liver disease. About one month after the glycyrrhizin discontinuation, acceleration of hypertension, hypokalaemia and suppression of the renin-aldosterone system still continued in both patients. About one and a half months later, the improvements of aggravated hypertension, hypokalaemia and suppressed renin-aldosterone system gradually occurred in both patients. The results demonstrated that both patients had a prolongation of the syndrome resembling primary aldosteronism except the low plasma aldosterone level about 1 month after the glycyrrhizin discontinuation. A man 85 years old, who habitually swallowed saliva while chewing tobacco leaf containing about 8.3 % (w/w) liquorice paste with a glycyrrhizinic acid content of 0.15 %, developed the classical syndrome of exogenous mineralocorticoid excess: hypokalaemia, hypertension, renal potassium wasting, metabolic acidosis, sodium retention and low plasma rennin (Blachley and Knochel 1980). Cuspidi et al. (1981) reported four female subjects with a form of severe systo-diastolic hypertension, recalcitrant to previous anti-hypertensive treatment, accompanied by marked hypokalaemia caused by habitual licorice ingestion. Abstinence from licorice led to normalization of kalemia in a period varying from 6 to 15 days, while arterial pressure values and all other essential parameters examined (plasma renin activity, etc.) recovered their balance more slowly. A 54-year-old man was admitted to hospital with acute rhabdomyolysis and myoglobinuria due to hypokalaemia caused by chronic licorice ingestion and diuretic treatment (Heidemann and Kreuzfelder 1983). The myoglobinaemia led to a glomerulopathy and tubulopathy, however there was no clinical evidence of acute renal failure. Four cases of pseudohyperaldosteronism due to chronic ingestion of liquorice-containing laxatives were described (Scali et al. 1990). All patients had hypertension and hypokalaemia with suppression of plasma renin activity and aldosterone.

Böcker and Breithardt (1991) reported two cases of licorice-induced arrhythmias. In both cases the ingestion of large amounts of licorice caused a marked hypokalaemia. Okada et al. (1987) reported a case of Sjögren's syndrome associated with hypokalaemic myopathy due to glycyrrhizin. Chubachi et al. (1992) reported the case of a 72-year-old man who developed acute renal failure (ARF) following severe hypokalaemic rhabdomyolysis; the hypokalaemia was due to chronic glycyrrhizin (glycyrrhizic acid) administration. Caradonna et al. (1992) reported a patient with acute rhabdomyolysis and absence of myoadenylate deaminase (MADA) associated with chronic licorice intoxication. These were completely reversed with potassium supplementation and licorice withdrawal. A case of a 55-year-old man of hypertension and hypokalaemia was reported by Kageyama (1992). He had been administered glycyrrhizin from 1 year

before admission for the treatment of chronic hepatitis. His blood pressure, potassium, plasma renin activity and plasma aldosterone concentration returned to normal within about 4 weeks after discontinuation of the glycyrrhizin. Re-administration of glycyrrhizin caused increases in plasma aldosterone concentration (PAC) and plasma renin activity (PRA). His urinary cortisol excretion was increased and urinary cortisone excretion decreased, while his serum cortisol level remained unchanged. The results suggested that increased renal cortisol as a result of decreased conversion to cortisone might play an important role in the development of pseudoaldosteronism as well as in its own mineralocorticoid activity. Corsi et al. (1983) reported a 35-year-old man who had been ingesting one or two bags of tablets of pure licorice daily (20–40 g/day) for about 2 years, developed an acute myopathy with high levels of serum muscle enzymes and the typical features of mineralocorticoid excess: serious hypokalaemia, hypertension and metabolic alkalosis. Both plasma renin and serum aldosterone were below the normal values. Van der Zwan (1993) reported a 15-year-old boy who developed a hypertension encephalopathy after ingestion of 0.5 kg liquorice candy. He recovered completely in the course of 5 months.

Luchon et al. (1993) reported a case of chronic intoxication with glycyrrhizinic acid, at a dosage of 1000–1500 mg per month over a period of 11 months, in a former alcoholic. This intoxication was revealed by profound hypokalaemia and rhabdomyolysis but blood pressure remained constantly normal. A 78-year-old Japanese man hospitalized because of muscular weakness and acute renal failure was diagnosed to suffer from licorice-induced pseudoaldosteronism that produced hypokalaemic rhabdomyolysis, resulting in acute renal failure and profound deposition of calcium into the damaged skeletal and cardiac muscles (Saito et al. 1994). Brayley and Jones (1994) described a 29-year-old female patient who presented with acute severe hypokalaemia after increasing her licorice consumption from 300 to 600 g/day. She had a history of anorexia nervosa with bulimia. Berlango Jiménez et al. (1995) presented the case of a 36-year-old patient who, as the result of the intake of five daily licorice sticks (25 g/day) for a month, developed analytical and clinical signs of acute rhabdomyolysis characterized by typical disorders of mineralocorticoid excess, such as severe hypokalaemia, arterial hypertension and metabolic alkalosis. Heikens et al. (1995) described of a 40-year-old female with severe hypertension and hypokalaemic metabolic alkalosis, due to prolonged liquorice ingestion. They suggested that glycyrrhetinic acid, the hydrolytic metabolite of glycyrrhizic acid, was the active component of liquorice which caused inhibition of the peripheral metabolism of cortisol. Cortisol had been found to bind with the same affinity as aldosterone to the mineralocorticoid receptor resulting in a hypermineralocorticoid condition. Ingestion of liquorice may therefore result in retention of sodium and water, hypertension, hypokalaemia, alkalosis and suppression of the renin-aldosterone system.

Seelen et al. (1996) reported a case of 38-year-old woman who was hospitalized because of hypertension and hypokalaemic alkalosis caused by the intake of liquorice (200 g per day). De Klerk et al. (1997) reported the case of a 21-year-old woman with hypertension associated with chewing of licorice-flavoured chewing gum. Chamberlain and Abolnik (1997) described a case of a 64-year-old previously healthy person who developed pulmonary oedema after a binge of black licorice sweet consumption. Cataldo et al. (1997) described a 6 1/2-year-old child with pseudohyperaldosteronism due to excessive and prolonged liquorice ingestion and its unusual association with haemorrhagic gastritis never observed in the course of liquorice intoxication. A case of a 69-year-old women with pseudoaldosteronism characterized by hypertension with hypokalaemia induced by a mouth refresher containing licorice, cinnamon, ginger and other spices was reported by Kageyama et al. (1997). Dilated cardiomyopathy during hypokalaemic myopathy resulting from excessive use of licorice and glycyrrhizin for gastritis had been reported by Hasegawa et al. (1998). Doeker and Andler (1999) reported an 11-year-old boy who had hypoparathyroidism and Addison's disease.

The boy reported an excessive daily intake of 300–400 g liquorice corresponding to 600–800 mg glycyrrhizic acid because of salt craving. After complete withdrawal of liquorice, all symptoms of hypermineralocorticoidism diminished and growth velocity increased. Ishikawa et al. (1999) reported a patient with a history of anorexia nervosa who developed licorice-induced hypokalaemic myopathy and hypokalaemic renal tubular damage. With potassium replacement, high creatinine phosphokinase blood level and myopathic signs returned to normal.

Lozano et al. (2000) reported a young woman with right upper limb ischemia induced by chronic licorice ingestion. The patient had a long lasting history, longer than 10 years, of continuous licorice ingestion. Blood samples showed severe hypokalaemia that caused EKG changes. Transoesophageal echocardiogram discovered mild mitral valve prolapse. Russo et al. (2000) described two cases of hypertension encephalopathy (in addition to the classical symptoms of hypertension, hypokalaemia and suppression of the renin-aldosterone system) which resulted in pseudohyperaldosteronism syndrome due to the regular daily intake of low doses of liquorice. Glycyrrhizic acid, a component of liquorice, had been known to produce both hypermineralocorticism and the onset of encephalopathy through the inhibition of 11βa-hydroxysteroid dehydrogenase. Harada et al. (2002) reported an 84-year-old man with congestive heart failure caused by digitalis toxicity after taking Chinese herbal laxative containing licorice. Cartier et al. (2002) described the case of a 33-year-old woman herbalist who developed occupational asthma due to liquorice roots as confirmed by specific inhalation challenges. Elinav and Chajek-Shaul (2003) described a patient who suffered life-threatening hypokalaemic paralysis caused by consumption of licorice in the form of a tea sweetener superimposed on long-term consumption of licorice candy. Aggressive fluid and potassium replenishment produced complete and lasting recovery. Lin et al. (2003) reported an elderly Asian man with hypokalaemic muscle paralysis cause by chronic licorice ingestion. Campana et al. (2003) described a case of cardiac arrest due to 'torsade de pointes' resulting from a marked hypokalaemia caused by the patient's habit of eating daily of an appreciable quantity of licorice. Elinav and Chajek-Shaul (2003) described a patient who suffered life-threatening hypokalaemic paralysis caused by consumption of licorice in the form of a tea sweetener superimposed on long-term consumption of licorice candy. Aggressive fluid and potassium replenishment produced complete and lasting recovery. Ishiguchi et al. (2004) described a 90-year-old woman with hypertension who developed metabolic alkalosis and myoclonus from ingestion of antacid medication containing licorice. A provisional diagnosis of licorice-induced metabolic alkalosis was established and the patient was successfully treated after correction of serum pH and cessation of the medications.

Cooper et al. (2007) described the case of a 42-year-old woman who self-treated undiagnosed Addison's disease for several years with soy sauce and liquorice sticks consuming around 46 g of salt per week. She presented with a 4-week history of decreased energy, malaise and postural dizziness. Mumoli and Cei (2008) described a patient with hypokalaemia caused by long-term consumption of natural licorice root after quitting smoking. Sontia et al. (2008) described a patient with long-standing hypokalaemia and uncontrolled hypertension related to excessive ingestion of liquorice. Yaguchi et al. (2008) reported a case of an 86-year-old female with licorice-induced hypokalaemic myopathy; she had been taking two kinds of Chinese medicines containing licorice for 4 years. The patient presented with difficulty in holding her head up and proximal-dominant tetraparesis with significant laterality. The general reflexes were decreased, and the bilateral Chaddock's reflexes were repeatedly positive. They found that the pathologic reflex was caused by the aggravation of cervical spondylotic myelopathy due to neck weakness and that tetraparesis with laterality was caused by hypokalaemic myopathy. Tacconi et al. (2009) reported one patient presented with carpal tunnel syndrome with nerve conduction studies revealing bilateral median neuropathies likely attributed to licorice-induced water

retention caused by excessive licorice consumption.

Johns (2009) reported a 49-year-old female physician who presented with peripheral oedema, weight gain and relative hypertension caused by the consumption of licorice candy cigars containing glycyrrhizic acid (GZA) found in natural licorice extract. Yorgun et al. (2010) reported a 50-year-old woman who was admitted to the emergency department with an aborted cardiac arrest due to ventricular fibrillation and electrocardiographic changes consistent with Brugada syndrome due to liquorice-induced hypokalaemia. Licorice consumption was found to be the cause of a posterior reversible encephalopathy syndrome (PRES) in a 49-year-old woman admitted to hospital because of thunderclap headache and blurred vision (van Beers et al. 2011). The combination of sequential computed tomography (CT) and the triad of hypertension, hypokalaemia and metabolic alkalosis in this patient suggested the diagnosis. Omar et al. (2012) reported a 35-year-old man from Egypt, with no past medical history, who presented to the emergency room with progressive weakness that started in his lower extremities and quickly progressed to involve the upper limbs. He was diagnosed with hypokalaemic myopathy due to excessive licorice ingestion. Celik et al. (2012) reported a case of an association of hypokalaemia, oedema and thrombocytopenia that was developed due to the excessive intake of licorice. Nielsen et al. (2012) described a 50-year-old woman with hypertension and hypokalaemia-induced limb paresis due to chronic liquorice ingestion. The patient was treated with potassium supplementation and spironolactone. Her blood pressure and electrolyte status normalized within a month after cessation of liquorice intake. A 47-year-old woman was admitted to the emergency department with a history of asthenia, periorbital and lower limbs oedema, associated with hypokalaemia and increased blood pressure levels (Robles et al. 2013). It was revealed that she had been consuming several sachets of raw liquorice lollies obtained from a herbalist a month ago and clinical tests established the cause to liquorice poisoning. During the patient's stay at the hospital, liquorice was stopped and potassium supplements were started. Subsequently, a week after, the patient fully recovered without any significant sequelae. O'Connell et al. (2014) reported a case of a 56-year-old lady who presented with thunderclap headache, visual disturbance and a generalized tonic–clonic seizure, diagnosed with hypokalaemia and posterior reversible encephalopathy syndrome (PRES) associated with regular liquorice consumption.

Nugmanova and Kalitina (1979) reported a case of contact dermatitis caused by licorice. Sailler et al. (1993) reported three cases of diffuse acute oedema caused by licorice. Dobbins and Saul (2000) reported five cases of transient visual loss after licorice ingestion. The visual symptoms were similar to one with ocular migraine without headache. According to the authors, the underlying pathogenesis may involve vasospasm of the optic nerve blood vessels leading to transient monocular or binocular visual loss/aberrations. The occurrence of adverse ocular side effects related to licorice ingestions were reported by Hall and Clemett (2004), Fraunfelder (2004) and Santaella and Fraunfelder (2007).

Preclinical and Clinical Studies

Epstein et al. (1977) studied the electrolyte status and renin-angiotensin-aldosterone axis after withdrawal of licorice in four sick women aged 35–55 years admitted with chronic licorice intoxication. They had been consuming 25–200 g licorice daily for 6 months to 5 years. All patients showed normal renin, angiotensin and aldosterone values 2–4 months later, however, indicating that long-term suppression of the renin-angiotensin-aldosterone axis was uncommon despite several years of liquorice ingestion. Ingestion of licorice, 100 g daily for 8 weeks, caused a rise of 81 % in plasma atrial natriuretic peptide (ANP) concentration in 12 healthy subjects (Forslund et al. 1989). The plasma concentrations of antidiuretic hormone, aldosterone and plasma renin activity decreased. Blood pressure increased transiently and two subjects developed reversible hypertension. The rise in plasma ANP concentration during ingestion of licorice may be

considered a physiological response to prevent fluid retention and development of hypertension.

Gomez-Sanchez and Gomez-Sanchez (1992) found that intracerebroventricular (icv) administration of the infusion of both glycyrrhizic acid, an active principle of licorice, and carbenoxolone, a synthetic analogue, into a lateral ventricle of the brain of a rat, at a dose less than that which had an effect when infused subcutaneously, produced hypertension. Oral administration of carbenoxolone or glycyrrhizic acid caused saline polydipsia and polyuria typical of chronic systemic mineralocorticoid excess, and the icv licorice derivatives produced hypertension without affecting saline appetite. The findings provided additional evidence for a central role in blood pressure control by mineralocorticoids that was distinct from their renal effects. They also suggested that more was involved in licorice-induced hypertension than only inhibition of 11β-hydroxysteroid dehydrogenase. Hayashi et al. (1992b) reported two patients with hypokalaemic myopathy induced by the administration of glycyrrhizin, 270–273 mg per day for a period of 2 and 8 months, respectively. Myotonic and repetitive discharges were observed when the serum chloride level fell below 90 mEq/l, and these discharges disappeared following administration of KCl. The findings supported the causal role of hypochloremia in myotonic discharges. Shintani et al. (1992) reported 59 cases of glycyrrhizin (licorice)-induced hypokalaemic myopathy (GIHM). In many cases, conditions leading to the onset of GIHM were habitual licorice ingestion, ingestion of antituberculosis agents containing licorice and long-term ingestion of licorice-containing agents for chronic gastritis, chronic hepatitis or chronic dermatitis. The main clinical symptom was flaccid quadriplegia in almost all cases, with muscle pain in 32.2 % and peripheral dysesthesia in the extremities, manifested mainly by numbness (27.1 %). Muscle biopsy was performed in 17 of the 59 cases with resultant findings of myopathic changes consisting mainly of phagocytosis, necrotic fibres, vacuolar degeneration, together with sporadic neurogenic changes. Complete cure was attained in 57 of the 59 cases of GIHM by discontinued ingestion of glycyrrhizin (licorice) and potassium supplement.

According to Schambelan (1994) studies had demonstrated that a paste prepared from succus liquiritiae, a dried watery extract of the roots of *Glycyrrhiza glabra*, could prevent the formation of gastric ulcers in experimental animals and confirmed the salutary effects in patients, but found that approximately 20 % of patients so treated developed facial and dependent oedema, often accompanied by headache, shortness of breath, stiffness and pain in the upper abdomen. Glycyrrhizic and glycyrrhetinic acids administration to rats in-vivo (75 mg/kg. day for 5 days) resulted in inhibition of 11β-hydroxysteroid dehydrogenase (11β HSD) activity, but also a significant reduction in steady state 11 β HSD mRNA levels in both predominantly mineralocorticoid (kidney and distal colon) and glucocorticoid (liver and pituitary) target tissues (Whorwood et al. 1993). In-vitro, 11βHSD mRNA and activity were present in rat pituitary GH3 cells (81 % conversion of corticosterone to 11-dehydrocorticosterone/4 × 10^6 cells after 24-h incubation) and inhibited by glycyrrhizic and glycyrrhetinic acids. Oral administration of a water freeze-dried extract of *Glycyrrhiza glabra* (liquorice) at doses of 100, 250 and 500 mg/kg in rats induced dose-dependent and mostly significant decreases in the plasma concentration of cortisol, adrenocorticotropic hormone, aldosterone and potassium (K) (Al-Qarawi et al. 2002). The results suggested a strong and dose-dependent suppression of the adrenal-pituitary axis, accompanied by stimulation of renin production from the kidney. Calò et al. (2004) reported a direct, mineralocorticoid-mediated effect on the protein expression of two markers of oxidative stress after incubation of mononuclear leukocytes with 1 × 10^{-8} M aldosterone (p22(phox)/β-actin = 1.38 and PAI-1/β--actin = 1.80). The same effect was also found with 3 × 10^{-5} M glycyrrhetinic acid, the principal constituent of licorice root (p22(phox)/β--actin = 1.37 and PAI-1/β-actin = 1.80). The

findings confirmed an involvement of mononuclear leukocytes in the pathogenesis of the oxidative stress induced by hyperaldosteronism.

Studies by Kato et al. (1995) suggested that licorice-induced pseudoaldosteronism was due to an increased concentration of 3 β-D-(monoglucuronyl)18β-glycyrrhetinic acid (3MGA), but not 18β-glycyrrhetinic acid (GA), in the circulating blood of these patients. They found an increased concentration of 3MGA in 10 patients with licorice-induced pseudoaldosteronism, but not in 11 patients without pseudoaldosteronism.

Ingestion of regular moderate liquorice consumption of 100 g of liquorice daily by 30 normotensive subjects caused a significant rise in systolic blood pressure (SBP) and a fall in plasma potassium (Sigurjónsdóttir et al. 1995). In a subgroup of 13 women the consumption of 50 g of liquorice daily also caused a significant rise in SBP of 5.6 mmHg ($P<0.001$) and diastolic blood pressure of 3.4 mmHg. A significant change in the cortisol/cortisone ratio in urine was observed during 100 g liquorice consumption indicating inhibition of 11β-hydroxysteroid dehydrogenase in kidneys. In another study of healthy Caucasian volunteers, consumption of licorice in various doses, 50–200 g/day, for 2–4 weeks, corresponding to a daily intake of 75–540 mg glycyrrhetinic acid, was found to increase systolic blood pressure in a dose–response relationship (Sigurjónsdóttir et al. 2001). They found even doses as low as 50 g of liquorice (75 mg glycyrrhetinic acid) consumed daily for 2 weeks can cause a significant rise in blood pressure. They also found that patients with essential hypertension were more sensitive to the inhibition of 11β-hydroxysteroid dehydrogenase (11 beta-HSD) by liquorice than normotensive subjects, and that this inhibition caused more clinical symptoms in women than in men although the difference in the effect on the blood pressure was not significant (Sigurjónsdóttir et al. 2003).

In another study of hypertensive patients (eight men and three women, mean age 40.7 years) and healthy controls (13 men and 12 women, mean age 31.2 years), licorice consumption of 100 g liquorice (containing 150 mg glycyrrhetinic acid) daily for 4 weeks, inhibited aldosterone secretion, the degree of licorice induced inhibition of aldosterone secretion differed between the genders and was not influenced by the blood pressure levels (Sigurjónsdóttir et al. 2006). In a clinical study, six male volunteers taking daily 7 g of a commercial preparation of licorice for 7 days, corresponding to an intake of 500 mg/day of glycyrrhizic acid developed pseudohyperaldosteronism during the treatment characterized by increase of body weight, suppression of plasma renin activity and plasma aldosterone and reduction of serum potassium (Armanini et al. 1996). The authors concluded that the pseudohyperaldosteronism from licorice was initially related to decreased activity of 11β-hydroxysteroid-dehydrogenase and afterwards also a direct effect of licorice derivatives on mineralocorticoid receptors becomes evident in some cases. Results of studies by Ohtake et al. (2007) suggested that accumulation of 3-monoglucuronyl-glycyrrhetinic acid (3MGA), a metabolite of glycyrrhizin, in the plasma may be involved in the pathogenesis of pseudoaldosteronism induced by chronic glycyrrhizin treatment. Studies by Makino et al. (2012) suggested that 3MGA was actively transported into tubules through organic anion transporters (OATs), resulting in the inhibition of type 2 11β-hydroxysteroid dehydrogenase (11β-HSD2). As the plasma level of 3MGA depended on the function of hepatic transporters, monitoring 3MGA levels in plasma or urine may be useful for preventing pseudoaldosteronism when licorice or GL was prescribed to patients. Celik et al. (2012) presented a case report on an association of hypokalaemia, oedema and thrombocytopenia that was developed due to the excessive intake of licorice.

In a study of a sample of 1049 Finnish women and their healthy singleton infants in 1998, Strandberg et al. (2001) found that heavy glycyrrhizin exposure during pregnancy did not significantly affect birth weight or maternal blood pressure, but it was significantly associated with shorter gestation. Heavy glycyrrhizin intake was sufficient to double the risk of being born before 38 weeks. In another study conducted in 2000–2001 of 95 Finnish women who delivered preterm singletons, heavy licorice consumption was found to be associated with a twofold to threefold increase in the risk of pre-

term (<37 weeks) birth (Strandberg et al. 2002). However, these results were later questioned due to the retrospective collection of data and the possibility of confounding factors that might have biased the results (Hughes et al. 2003).

In a study of nine healthy women 22–26 years old, in the luteal phase of the cycle, after 2 months of administration of licorice, serum parathyroid hormone, 25-hydroxycholecalciferol and urinary calcium were increased significantly from baseline values, while 1,25-dihydroxy Vitamin D and ALP did not change during treatment (Mattarello et al. 2006). All these parameters returned to pretreatment levels 1 month after discontinuation of licorice. Plasma renin activity and aldosterone were depressed during licorice therapy, while blood pressure and plasma cortisol remained unchanged. In a clinical study of 321 Finnish children 8.1 years of age born in 1998 as healthy singletons at 35–42 weeks of gestation, Räikkönen et al. (2009) found that prenatal exposure to licorice dose-dependently predicted poorer verbal and visuospatial abilities and narrative memory as well as increased risk of externalizing symptoms, attention, rule-breaking and aggression problems in children aged 8.1 years. Their findings supported adverse foetal 'programming' by overexposure to glucocorticoids and counsel concern against consuming excessive amounts of foodstuffs containing licorice during pregnancy. Further they found that maternal prenatal licorice consumption altered hypothalamic-pituitary-adrenocortical axis (HPAA) function in children (Räikkönen et al. 2010) and may increase risk of adult disease. Their findings lend support to prenatal 'programming' of HPAA function by overexposure to glucocorticoids.

Miscellaneous Adverse Issues

All analysed samples of licorice root and derived products (licorice-confectionery, licorice block and licorice extract) were found to contain ochratoxin A, and some of them showed extremely high concentrations up to 252.8 ng/g of ochratoxin A (Arino et al. 2007). Highest levels of ochratoxin A were found in dry licorice root, averaging 63.6 ng/g, while mean contents in fresh licorice root were 9.2 ng/g. Licorice-confectionery (sweets) contained 3.8 ng/g of ochratoxin A. Ochratoxin A was also abundant in two licorice derivatives, liquid licorice extract (16.0 ng/g) and solid licorice block (39.5 ng/g). The ochratoxin levels found in licorice and derived products are higher than those reported in the literature for other food commodities. The experiments of ochratoxin A transfer into the tea beverages showed that almost 5 % of the ochratoxin A present in dry licorice root is transferred to the corresponding decoction tea, whereas only 1 % of ochratoxin A remains in infusion tea.

Traditional Medicinal Uses

Liquorice is one of the most commonly used herbs in Western herbal medicine and has a very long history of use, both as a medicine and also as a flavouring to disguise the unpleasant flavour of other medications, in cough medicines and also in the treatment of catarrhal infections of the urinary tract (Grieve 1971). Licorice is deemed emollient, expectorant, laxative, moderately pectoral and tonic (Grieve 1971; Launert 1981; Uphof 1968). Liquorice root is taken internally in the treatment of Addison's disease, asthma, bronchitis, coughs, peptic ulcer, arthritis, allergic complaints and following steroidal therapy (Bown 1995). Liquorice should be used in moderation and should not be prescribed for pregnant women or people with high blood pressure, kidney disease or taking digoxin-based medication (Bown 1995). Prolonged usage raises the blood pressure and causes water retention (Chiej 1984; Bown 1995). Externally, the root is used in the treatment of herpes, eczema and shingles (Bown 1995).

Licorice root extracts have been used in traditional Chinese, Tibetan and Indian medicine for the treatment of pulmonary diseases and inflammatory processes (Kwon et al. 2007). It can be used in the folk medicine at different parts of the world to treat many diseases including bacterial infection, cough suppression, 4 gastric ulcer treatment 5, treatment of early Addison disease

6, 7, treatment of liver disease 8, 9 and as a laxative (Nitalikar et al. 2010). Licorice is clinically used for the treatment of stomach ulcers 10, 11. Its preparations are used as a conditioning and flavouring agent in tobacco products. Aqueous extracts from the roots of *Glycyrrhiza glabra* are widely used for treatment of stomach ulcer (Wittschier et al. 2009). Glycyrrhizin has long been used in China in the treatment of various liver diseases to lower transaminases (Ren et al. 2013). Glycyrrhizin, a major component of a herb (licorice), has been widely used to treat chronic hepatitis B in Japan (Takahara et al. 1994). Licorice is the most common ingredient of traditional Japanese Kampo medicines (Hayashi and Sudo 2009). The minimum content of glycyrrhizin in these medicines should be 2.5 % according to the standards of the Japanese Pharmacopeia. Glycyrrhizin is a prescription drug used in the treatment of liver and allergic diseases in Japan. It is manufactured as an injectable preparation (Stronger Neo-Minophagen® C) and in a tablet form (Glycyron®) by a Japanese company, namely, Minophagen Pharmaceutical Co. Ltd. Stronger Neo-Minophagen® C has been available in the Japanese market for over 60 years. Glycyrrhizin, glycyrrhetinic acid and licorice extracts are used in various over-the-counter drugs, including anti-allergic and anti-inflammatory drugs. In addition, in England, the glycyrrhetinic acid derivative glycyrrhetinic acid 3-β-*O*-hemisuccinate (carbenoxolone) is a prescription drug used in the treatment of peptic ulcers.

Licorice is used for many ailments including asthma, athlete's foot, baldness, body odour, bursitis, canker sores, chronic fatigue, depression, colds and flu, coughs, dandruff, emphysema, gingivitis and tooth decay, gout, heartburn, HIV, viral infections, fungal infections, ulcers, liver problems, Lyme disease, menopause, psoriasis, shingles, sore throat, tendinitis, tuberculosis, ulcers, yeast infections, prostate enlargement and arthritis (Khalaf et al. 2010). Glycyrrhizic acid coupled with glycyrrhetinic acid and 18-β-glycyrrhetic acid was developed in China or Japan as an anti-inflammatory, antiviral and anti-allergic drug for liver disease (Li et al. 2014).

Glycyrrhizin (GL) has been used in Japan to treat patients with chronic viral hepatitis (Matsumoto et al. 2013). In Japan, glycyrrhizin therapy is widely used for more than 20 years for chronic hepatitis C and reportedly reduces the progression of liver disease to hepatocellular carcinoma (van Rossum and De Man 1998; van Rossum et al. 1999).

The roots are sweet, refrigerant, emetic in large dose, tonic, mild, laxative, aphrodisiac, haemostatic (Meena et al. 2010). They are useful in hyperdipsia, cough, bronchitis, ulceration of urinary tract, pharyngitis, epilepsy and anaemia. In the Ayurvedic system of medicine it is used in the preparations of yashtyadi churna, Yashtimadhvadya taila, Brihatashwagandha hrita, Pippalyadi taila and Vridhihara lepa.

Other Uses

Liquorice extracts (in pharmacy called succus liquiritiae) are currently used mainly in the tobacco, pharmaceutical and confectionery industries (Fenwick et al. 1990). Licorice plant yields a substance that is used for etching steel sections in photomicrographic work (Hill 1952). Extracts from the root are used as a foaming agent in beers and fire extinguishers (Bown 1995). A fibre obtained from the roots is used for insulation, wallboard, boxboard, etc. (Hill 1952; Grieve 1971). The fibres can be used after the medicinal and flavouring constituents of the root have been extracted (Grieve 1971).

The world's leading manufacturer of liquorice products is M&F Worldwide, which manufactures more than 70 % of the worldwide liquorice flavours sold to end-users. Mafco Worldwide produces a variety of licorice products from licorice root, intermediary licorice extracts produced by others and certain other ingredients (M & F Worldwide Corp 2010). Approximately 63 % of Mafco Worldwide's licorice product sales are to the worldwide tobacco industry for use as tobacco flavour enhancing and moistening agents in the manufacture of American blend cigarettes, moist snuff, chewing tobacco and pipe tobacco. Mafco

Worldwide also sells licorice products to food processors, confectioners, cosmetic companies and pharmaceutical manufacturers for use as flavouring or masking agents, including its *Magnasweet* brand flavour enhancer, which is used in various brands of chewing gum, energy bars, non-carbonated beverages, lip balm, chewable vitamins, aspirin and other products. In addition, Mafco Worldwide sells licorice root residue as garden mulch under the name *Right Dress*. Mafco Worldwide also sells licorice products worldwide to food processors, confectioners, cosmetic companies and pharmaceutical manufacturers for use as flavouring and masking agents, including its *Magnasweet* brand flavour enhancer, which is used in various brands of chewing gum, lip balm, energy bars, non-carbonated beverages, chewable vitamins, aspirin and other products.

Licorice extracts and many glycyrrhizin derivatives are widely used in the preparation of cosmetics in Japan (Hayashi and Sudo 2009). Glycyrrhizin as well as powdered *Glycyrrhiza* roots, licorice extracts, glycyrrhetinic acid, stearyl glycyrrhetinate, pyridoxine glycyrrhetinate and glycyrrhetinic acid 3-β-*O*-hemisuccinate (carbenoxolone) are used in cosmetics for their anti-inflammatory action. Furthermore, glabridin-containing glycyrrhiza flavonoids isolated from *G. glabra* are used in cosmetic preparations owing to their skin-whitening, anti-sensitizing and anti-inflammatory properties (Yokota et al. 1998).

Glycyrrhiza glabra was liquefied by ethanol and acetone in an autoclave under high pressure using potassium hydroxide or sodium carbonate as the catalyst, as well as without catalyst at various temperatures (250, 270 and 290 °C) for producing bio-oil (Durak 2014). The maximum bio-oil yield was obtained in acetone (79 %) at 290 °C without catalyst. GC–MS identified 131 and 147 different compounds in the bio-oils obtained at 270 and 290 °C, respectively.

G. glabra was found to be a good carbon source in the biological denitrification of drinking water (Ovez et al. 2006). Complete denitrification was achieved with *G. glabra*. It was found that the nitrate removal rate of *G. glabra* was 6.96 mg/L/day.

Mohammadi et al. (2014) prepared high surface area-activated carbon from *Glycyrrhiza glabra* residue by ZnCl2 activation for removal of Pb(II) and Ni(II) from water samples. High values of intra-particle rate constants calculated show the high tendency of activated carbon for removal of lead and nickel ions. Studies showed that *G. glabra* root could be used as an adsorbent of toluene vapour from gaseous media (Mohammadi-Moghadam et al. 2013). Licorice adsorbent is a waste material with a sorption capacity of 2.2 mg/g. In comparison with other natural sorbents (e.g., compost, diatomaceous earth and chaff), licorice root appeared to be a cost-effective sorbent in the removal of toluene vapour.

G. glabra foliage has potential as animal feed. Kamalak (2006) found that *G. glabra* leaves harvested at the proper stage of maturity offered considerable potential as a high quality forage for ruminant during critical period in the semi-arid and arid regions. Neutral detergent fibre (NDF), acid detergent fibre (ADF) and condensed tannin (CT) contents of *G. glabra* leaves increased with increasing maturity whereas the crude protein decreased. Gas production, dry matter (DM) and crude protein (CP) disappearance and estimated parameters also decreased with increasing maturity. CP, ADF and CT contents ranged from 16.19 to 26.93 %, from 20.74 to 29.07 % and from 1.57 to 10.83 %, respectively. The potential gas production and metabolizable energy ranged from 65.34 to 72.12 ml/0.200 g of DM and from 10.14 to 12.12 MJ/kg DM, respectively. The effective DM degradability (EDMD) and effective CP degradability (ECPD) ranged from 58.70 to 70.59 % and from 57.32 to 73.72 %.

Comments

In Chinese Pharmacopoeia, three *Glycyrrhiza* species *Glycyrrhiza uralensis*, *G. glabra* and *G. inflata* are listed as licorice. While in Japanese Pharmacopoeia, two species *G. uralensis* and *G. glabra* are prescribed as licorice. Three varieties of *G. glabra* have been reported the Spanish and Italian licorice, assigned to *G. glabra var. typica*, Russian licorice to *G glabra var. glandulifera*, Persian and Turkish licorice to *G. glabra var. violacea* (Nomura et al. 2002).

Commercial licorice is derived from three *Glycyrrhiza* species, *G. glabra* L., *G. uralensis*, Fisch., and *G. inflata* Batal. in the family Fabaceae, which are indigenous to Asia and the Mediterranean region (Shibata 2000).

Selected References

Abe N, Ebina T, Ishida N (1982) Interferon induction by glycyrrhizin and glycyrrhetinic acid in mice. Microbiol Immunol 26(6):535–539

Ablise M, Kasimu R, Ma SY, Ioppinet V, Siest G (2007) Isolation and antioxidant property of glabridin. Nat Prod Res Dev 19(4):675–682

Acharya SK, Dasarathy S, Tandon A, Joshi YK, Tandon BN (1993) A preliminary open trial on interferon stimulator (SNMC) derived from *Glycyrrhiza glabra* in the treatment of subacute hepatic failure. Indian J Med Res 98:69–74

Achliya GS, Wadodkar SG, Avinash KD (2004) Neuropharmacological actions of panchagavya formulation containing *Emblica officinalis* Gaerth and *Glycyrrhiza glabra* Linn in mice. Indian J Exp Biol 42(5):499–503

Adamyan TI, Gevorkyan ES, Minasyan SM, Oganesyan KR, Kirakosyan KA (2005) Effect of licorice root on peripheral blood indexes upon vibration exposure. Bull Exp Biol Med 140(2):197–200

Afchar D, Cave A, Vaquette J (1980) Etudes des reglisses d'Iran. I. Flavonoides de *Glycyrrhiza glabra* var. *glandulifera*. Plant Med Phytother 14(1):46–50 (In French)

Agabeĭli RA (2012) Genetic effects of root extracts of *Glycyrrhiza glabra* L. in different test systems. Tsitol Genet 46(5):297–301 (In Russian)

Agarwal R, Wang ZY, Mukhtar H (1991) Inhibition of mouse skin tumor-initiating activity of DMBA by chronic oral feeding of glycyrrhizin in drinking water. Nutr Cancer 15(3–4):187–193

Agarwal A, Gupta D, Yadav G, Goyal P, Singh P, Singh U (2009) An evaluation of the efficacy of licorice gargle for attenuating postoperative sore throat: a prospective, randomized, single blind study. Anesth Analg 109:77–81

Ahn JY, Lee HJ, Jang JY, Kim SN, Ha TY (2013) Anti-obesity effects of glabridin-rich supercritical carbon dioxide extract of licorice in high-fat-fed obese mice. Food Chem Toxicol 51:439–445

Aida K, Tawata M, Shindo H, Onaya T, Sasaki H, Yamaguchi T, Chin M, Mitsuhashi H (1990) Isoliquiritigenin: a new aldose reductase inhibitor from glycyrrhizae radix. Planta Med 56(3):254–258

Ajagannanavar SL, Battur H, Shamarao S, Sivakumar V, Patil PU, Shanavas P (2014) Effect of aqueous and alcoholic licorice (*Glycyrrhiza glabra*) root extract against *Streptococcus mutans* and *Lactobacillus acidophilus* in comparison to chlorhexidine: an in vitro study. J Int Oral Health 6(4):29–34

Akao T (1999) Influence of various bile acids on the metabolism of glycyrrhizin and glycyrrhetic acid by *Ruminococcus* sp. PO1-3 of human intestinal bacteria. Chem Pharm Bull 22(8):787–793

Akao T (2000a) Competition in the metabolism of glycyrrhizin with glycyrrhetic acid mono-glucuronide by mixed *Eubacterium* sp. GLH and *Ruminococcus* sp. PO1-3. Biol Pharm Bull 23:149–154

Akao T (2000b) Effects of glycyrrhizin and glycyrrhetic acid on the growth, glycyrrhizin beta-D-glucuronidase and 3 beta-hydroxysteroid dehydrogenase of human intestinal bacteria. Biol Pharm Bull 23(1):104–107

Akao T, Kobashi K (1988) Glycyrrhizin stimulates growth of *Eubacterium* sp. strain GLH, a human intestinal anaerobe. Appl Environ Microbiol 54(8):2027–2030

Akao T, Akao T, Kobashi K (1987) Glycyrrhizin β-D-glucuronidase of *Eubacterium* sp. from human intestinal flora. Chem Pharm Bull 35(2):705–710

Akao T, Akao T, Hattori M, Kanaoka M, Yamamoto K, Namba T, Kobashi K (1991) Hydrolysis of glycyrrhizin to 18β-glycyrrhetyl monoglucuronide by lysosomal β-D-glucuronidase of animal livers. Biochem Pharmacol 41(6):1025–1029

Al-Qarawi AA, Abdel-Rahman HA, Ali BH, El Mougy SA (2002) Liquorice (*Glycyrrhiza glabra*) and the adrenal-kidney-pituitary axis in rats. Food Chem Toxicol 40(10):1525–1527

Ammosov AS, Litvinenko VI (2007) Phenolic compounds of the genera Glycyrrhiza L. and Meristotropis Fisch. et Mey. (review). Pharm Chem J 41(7):372–395

Anderson DM, Smith WG (1961) The antitussive activity of glycyrrhetinic acid and its derivatives. J Pharm Pharmacol 13(1):396–404

Aoki F, Nakagawa K, Tanaka A, Matsuzaki K, Arai N, Mae T (2005) Determination of glabridin in human plasma by solid-phase extraction and LC–MS/MS. J Chromatogr B 828(1–2):70–74

Aoki F, Honda S, Kishida H, Kitano M, Arai N, Tanaka H, Yokota S, Nakagawa K, Asakura T, Nakai Y, Mae T (2007a) Suppression by licorice flavonoids of abdominal fat accumulation and body weight gain in high-fat diet-induced obese C57BL/6J mice. Biosci Biotechnol Biochem 71(1):206–214

Aoki F, Nakagawa K, Kitano M, Ikematsu H, Nakamura K, Yokota S, Tominaga Y, Arai N, Mae T (2007b) Clinical safety of licorice flavonoid oil (LFO) and pharmacokinetics of glabridin in healthy humans. J Am Coll Nutr 26(3):209–218

Arase Y, Ikeda K, Murashima N, Chayama K, Tsubota A, Koida I, Suzuki Y, Saitoh S, Kobayashi M, Kumada H (1997) The long term efficacy of glycyrrhizin in chronic hepatitis C patients. Cancer 79:1494–1500

Arias-Castro C, Seragg AH, Rodriguez-Mendiola MA (1993) The effect of cultural conditions on the accumulation of formononetin by suspension cultures of *Glycyrrhiza glabra*. Plant Cell Tissue Organ Cult 34(1):63–70

Arino A, Herrera M, Estopanan G, Juan T (2007) High levels of ochratoxin A in licorice and derived products. Int J Food Microbiol 114(3):366–369

Armanini D, Lewicka S, Pratesi C, Scali M, Zennaro MC, Zovato S, Gottardo C, Simoncini M, Spigariol A, Zampollo V (1996) Further studies on the mechanism of the mineralocorticoid action of licorice in humans. J Endocrinol Invest 19(9):624–629

Armanini D, Fiore C, Mattarello MJ, Bielenberg J, Palermo M (2002) History of the endocrine effects of licorice. Exp Clin Endocrinol Diabetes 110(6):257–261

Armanini D, De Palo CB, Mattarello MJ, Spinella P, Zaccaria M, Ermolao A, Palermo M, Fiore C, Sartorato P, Francini-Pesenti F, Karbowiak I (2003) Effect of licorice on the reduction of body fat mass in healthy subjects. J Endocrinol Invest 26(7):646–650

Armanini D, Mattarello MJ, Fiore C, Bonanni G, Scaroni C, Sartorato P, Palermo M (2004) Licorice reduces serum testosterone in healthy women. Steroids 69(11–12):763–766

Armanini D, Nacamulli D, Francini-Pesenti F, Battagin G, Ragazzi E, Fiore C (2005) Glycyrrhetinic acid, the active principle of licorice, can reduce the thickness of subcutaneous thigh fat through topical application. Steroids 70(8):538–542

Armanini D, Castello R, Scaroni C, Bonanni G, Faccini G, Pellati D, Bertoldo A, Fiore C, Moghetti P (2007) Treatment of polycystic ovary syndrome with spironolactone plus licorice. Eur J Obstet Gynecol Reprod Biol 131:61–67

Asada Y, Li W, Yoshikawa T (1997) Constituents of hairy root cultures of licorice. In: 117th annual meeting of Pharmaceutical Society of Japan, Abstract Papers, part 2, p 117. Tokyo, March 1997

Asada Y, Li W, Yoshikawa T (1998) Isoprenylated flavonoids from hairy root cultures of *Glycyrrhiza glabra*. Phytochemistry 47(3):389–392

Asada Y, Li W, Yoshikawa T (1999) The first prenylated biaurone, licoagrone from hairy root cultures of *Glycyrrhiza glabra*. Phytochemistry 50(6):1015–1019

Asada Y, Li W, Yoshikawa T (2000) Biosynthesis of the dimethylallyl moiety of glabrol in *Glycyrrhiza glabra* hairy root cultures via a non-mevalonate pathway. Phytochemistry 55(4):323–326

Asano T, Ishihara K, Morota T, Takeda S, Aburada M (2003) Permeability of the flavonoids liquiritigenin and its glycosides in licorice roots and davidigenin, a hydrogenated metabolite of liquiritigenin, using human intestinal cell line Caco-2. J Ethnopharmacol 89(2–3):285–289

Asgary S, Jafari Dinani N, Madani H, Mahzoni P, Naderi GH (2007) Effect of *Glycyrrhiza glabra* extract on aorta wall atherosclerotic lesion in hypercholesterolemic rabbits. Pak J Nutr 6(4):313–317

Asha MK, Debraj D, Prashanth D, Edwin JR, Srikanth HS, Muruganantham N, Dethe SM, Anirban B, Jaya B, Deepak M, Agarwal A (2013) In vitro anti-*Helicobacter pylori* activity of a flavonoid rich extract of *Glycyrrhiza glabra* and its probable mechanisms of action. J Ethnopharmacol 145(2):581–586

Astaf'eva OV, Sukhenko LT (2014) Comparative analysis of antibacterial properties and chemical composition of *Glycyrrhiza glabra* L. from Astrakhan region (Russia) and Calabria region (Italy). Bull Exp Biol Med 156(6):829–832

Ayabe S, Takano H, Fujita T, Furuya T, Hirota H, Takahashi T (1990) Triterpenoid biosynthesis in tissue cultures of *Glycyrrhiza glabra* var. *glandulifera*. Plant Cell Rep 9(4):181–184

Azimov MM, Zakirov UB, Radzhapova SD (1988) Pharmacological study of the anti-inflammatory agent glyderinine. Farmakol Toksikol 51(4):90–93 (In Russian)

Baba M, Shigeta S (1987) Antiviral activity of glycyrrhizin against varicella-zoster virus in vitro. Antiviral Res 7(2):99–107

Badam L (1997) In vitro antiviral activity of indigenous glycyrrhizin, licorice and glycyrrhizic acid (Sigma) on Japanese encephalitis virus. J Commun Dis 29(2):91–99

Bafna PA, Balaraman R (2005) Antioxidant activity of DHC-1, an herbal formulation, in experimentally-induced cardiac and renal damage. Phytother Res 19(3):216–221

Baltina LA, Serdyuk NG, Flekhter OB, Krasnova LV, Davydova VA, Ismagilova AF, Zarudii FS, Tolstikov G (1996) Isomerization of glycyrrhizic acid: antiulcer activity. Pharm Chem J 30(10):613–616

Baltina LA, Stolyarova OV, Baltina LA, Kondratenko RM, Plyasunova OA, Pokrovskii AG (2010) Synthesis and antiviral activity of 18α-glycyrrhizic acid and its esters. Pharm Chem J 44(6):299–302

Bannister B, Ginsburg R, Shneerson J (1977) Cardiac arrest due to liquorice induced hypokalaemia. Br Med J 2:738–739

Bardhan KD, Cumberland DC, Dixon RA, Holdsworth CD (1978) Clinical trial of deglycyrrhizinised liquorice in gastric ulcers. Gut 19:779–782

Basso A, Dalla Paola L, Erle G, Boscaro M, Armanini D (1994) Licorice ameliorates postural hypotension caused by diabetic autonomic neuropathy. Diabetes Care 17:1356

Batirov EK, Kiyamitdinova F, Malikov VM (1986) Flavonoids of the epigeal part of *Glycyrrhiza glabra*. Chem Nat Comp 22(1):107–108

Beaton JM, Spring FS (1956) Triterpenoids Part LI. The isolation and characterisation of glabric acid, a new triterpenoid acid from liquorice root. J Chem Soc:2417–2419

Belinky PA, Aviram M, Fuhrman B, Rosenblat M, Vaya J (1998a) The antioxidative effects of the isoflavan glabridin on endogenous constituents of LDL during its oxidation. Atherosclerosis 137(1):49–61

Belinky PA, Aviram M, Mahmood S, Vaya J (1998b) Structural aspects of the inhibitory effect of glabridin on LDL oxidation. Free Radic Biol Med 24(9):1419–1429

Belous YN, Matyukhina LG, Ryabinik AA (1965) Uralenic acid. Zhurn Obshchei Khim (USSR) 35:401. (In Russian).

Bennett A, Melhuish PB, Stamford IF (1985) Carbenoxolone and deglycyrrhized liquorice have little or no effect on prostanoid synthesis by rat gastric mucosa ex vivo. Br J Pharmacol 86(3):693–695

Berlango Jiménez A, Jiménez Murillo L, Montero Pérez FJ, Muñoz Avila JA, Torres Murillo J, Calderón de la Barca Gázquez JM (1995) Acute rhabdomyolysis and tetraparesis secondary to hypokalemia due to ingested licorice. An Med Interna 12(1):33–35 (In Spanish)

Bernardi M, D'Intino PE, Trevisani F, Cantelli-Forti G, Raggi MA, Turchetto E, Gasbarrini G (1994) Effects of prolonged ingestion of graded doses of liquorice by healthy volunteers. Life Sci 55(11):863–872

Bhardwaj DK, Murari R, Seshadri TR, Singh R (1976a) Liqcoumarin, a novel coumarin from *Glycyrrhiza glabra*. Phytochemistry 15(7):1182–1183

Bhardwaj DK, Murari R, Seshadri TR, Singh R (1976b) Occurrence of 2-methylisoflavones in *Glycyrrhiza glabra*. Phytochemistry 15(2):352–353

Bhardwaj DK, Seshadri TR, Singh R (1977) Glyzarin, a new isoflavone form *Glycyrrhiza glabra*. Phytochemistry 16(3):402–403

Bhattacharjee S, Bhattacharjee A, Majumdar S, Majumdar SB, Majumdar S (2012) Glycyrrhizic acid suppresses Cox-2-mediated anti-inflammatory responses during *Leishmania donovani* infection. J Antimicrob Chemother 67(8):1905–1914

Bijlsma JA, Van Vloten P, Van Gelderen CEM, Mensinga TjT, Mout HA, Elvers LH, Van Leeuwen FXR, Stolker AAM, Van Ginkel LA, Looman CWN, Van der Maas PJ, Koomans HA, Savelkoul TJF (1996) Onderzoek naar de effecten van verschillende doseringen glycyrrhizine bij gezonde vrouwelijke vrijwilligers. [Study into the effects of different dosages of glycyrrhizin in health female volunteers; in Dutch]. RIVM report 348801004, RIVM, Bilthoven

Biondi DM, Rocco C, Ruberto G (2003) New dihydrostilbene derivatives from the leaves of *Glycyrrhiza glabra* and evaluation of their antioxidant activity. J Nat Prod 66(4):477–480

Biondi DM, Rocco C, Ruberto G (2005) Dihydrostilbene derivatives from *Glycyrrhiza glabra* leaves. J Nat Prod 68(7):1099–1102

Birari RB, Gupta S, Mohan CG, Bhutani KK (2011) Antiobesity and lipid lowering effects of *Glycyrrhiza* chalcones: experimental and computational studies. Phytomed 18(8-9):795–801

Bisoni V, Rossi GP, Calò LA (2014) Apparent mineralcorticoid excess syndrome, an often forgotten or unrecognized cause of hypokalemia and hypertension: case report and appraisal of the pathophysiology. Blood Press 23(3):189–192

Blachley JD, Knochel JP (1980) Tobacco chewer's hypokalemia: licorice revisited. N Engl J Med 302(14):784–785

Blumenthal M (2000) Licorice root herbal medicine. Expanded Commission E monographs. Am Bot Council, Austin, pp 233–239

Böcker D, Breithardt G (1991) Induction of arrhythmia by licorice abuse. Z Kardiol 80:389–391

Bogatkina VF, Murav'ev IA, Stepanova EF, Kir'yalov NP (1975) Triterpene compounds from the epigeal mass of *Glycyrrhiza glabra*. Chem Nat Comp 11(1):114–115

Bordbar N, Karimi MH, Amirghofran Z (2012) The effect of glycyrrhizin on maturation and T cell stimulating activity of dendritic cells. Cell Immunol 280(1):44–49

Bowe WP, Pugliese S (2014) Cosmetic benefits of natural ingredients. J Drugs Dermatol 13(9):1021–1025

Bown D (1995) Encyclopaedia of herbs and their uses. Dorling Kindersley, London, 424 pp

Brayley J, Jones J (1994) Life-threatening hypokalemia associated with excessive licorice ingestion. Am J Psychiatry 151(4):617–618

Calò L, Zaghetto F, Pagnin E, Davis P, De Mozzi P, Sartorato P, Martire G, Fiore C, Armanini D (2004) Effect of aldosterone and glycyrrhetinic acid on the protein expression of PAI-1 and p22(phox) in human mononuclear leukocytes. J Clin Endocrinol Metab 89:1973–1976

Campana A, Manzo M, Brigante M, Marrazzo N, Melchiorre G (2003) An unusual cause of cardiac arrest. Ital Heart J Suppl 4:510–513 (In Italian)

Canciu-Dobrea C, Kaycsa A, Dehelean C, Şoica C, Antal D (2012) *Glycyrrhiza glabra* and *Glycyrrhiza echinata* – sources of low hemotoxic saponins. Acta Med Marisiensis 58(3):150–152

Canonica L, Danielli B, Manitto P, Russo G (1966a) Triterpenes of *Glycyrrhiza glabra*. III. Structure of isoglabrolide. Gazz Chim Ital 96:843–851

Canonica L, Russo G, Bombardelli E (1966b) Triterpenes of *Glycyrrhiza glabra*. 2. Liquiritic acid. Gazz Chim Ital 96:833–842

Canonica L, Russo G, Bonati A (1966c) Triterpenes of *Glycyrrhiza glabra*. I. Two new lactones with an oleanane structure. Gazz Chim Ital 96:772–785

Canonica L, Danielli B, Manitto P, Russo G, Bombardelli E (1967a) *Glycyrrhiza glabra* triterpenes V. Glycyrrhetol and 21α-hydroxyisoglabrolide. Gazz Chim Ital 97:1347–1358

Canonica L, Danielli B, Manitto P, Russo G, Bonati A, Bombardelli E (1967b) *Glycyrrhiza glabra* triterpenes VI. 23-Hydroxyglycyrrhetic acid and 24-hydroxy-11-deoxyglycyrrhetic acid. Gazz Chim Ital 97:1359–1369

Canonica L, Danielli B, Russo G, Bonati A (1967c) Triterpenes from *Glycyrrhiza glabra* IV. 18α-Hydroxyglycyrrhetic acid. Gazz Chim Ital 97:769–786

Canonica L, Danielli B, Manitto P, Russo G, Bombardelli E, Bonati A (1968) *Glycyrrhiza glabra* triterpenes VIL 24-Hydroxyliquiritic acid (30,24-dihydroxy-11-oxo-olean-12-en-29-oic acid) and liquiridolic acid (3β, 21α,24-trihydroxy-olean-12-en-29-oic acid). Gazz Chim Ital 98:712–728

Cantelli-Forti G, Maffei F, Hrelia P, Bugamelli F, Bernardi M, D'Intino P, Maranesi M, Raggi MA (1994) Interaction of licorice on glycyrrhizin pharmacokinetics. Environ Health Perspect 102(Suppl 9):65–68

Cantelli-Forti G, Raggi MA, Bugamelli F, Maffei F, Villari A, Trieff NM (1997) Toxicological assessment of liquorice: biliary excretion in rats. Pharmacol Res 35(5):463–470

Cao J, Chen X, Liang J, Yu XQ, Xu AL, Chan E, Wei D, Huang M, Wen JY, Yu XY, Li XT, Sheu FS, Zhou SF (2007) Role of P-glycoprotein in the intestinal absorption of glabridin, an active flavonoid from the root of *Glycyrrhiza glabra*. Drug Metab Dispos 35(4):539–553

Cao N, Chen T, Guo ZP, Qin S, Li MM (2014) Monoammonium glycyrrhizate suppresses tumor necrosis factor-α induced chemokine production in HMEC-1 cells, possibly by blocking the translocation of nuclear factor-κB into the nucleus. Can J Physiol Pharmacol 92(10):859–865

Caradonna P, Gentiloni N, Servidei S, Perrone G, Greco A, Russo M (1992) Acute myopathy associated with chronic licorice ingestion: reversible loss of myoadenylate deaminase activity. Ultrastruct Pathol 16(5):529–535

Carmines EL, Lemus R, Gaworski CL (2005) Toxicologic evaluation of licorice extract as a cigarette ingredient. Food Chem Toxicol 43(9):1303–1322

Cartier A, Malo J, Labrecque M (2002) Occupational asthma due to liquorice roots. Allergy 57(9):863

Cataldo F, Di Stefano P, Violante M, Traverso G, Mulè M (1997) Pseudohyperaldosteronism secondary to licorice poisoning associated with hemorrhagic gastritis. Pediatr Med Chir 19(3):219–221 (In Italian)

Celik MM, Karakus A, Zeren C, Demir M, Bayarogullari H, Duru M, Al M (2012) Licorice induced hypokalemia, edema, and thrombocytopenia. Hum Exp Toxicol 31(12):1295–1298

Ceremelli C, Portolani M, Cotombari B, Castelli M, Baggio G, Galatulas I (1996) Activity of glycyrrhizin and its diasterioisomers against two new human herpes virus: HHV-6 and HHV-7. Phytochem Res 10:527–528

Chakravarthi KK, Avadhani R (2014) Enhancement of hippocampal CA3 neuronal dendritic arborization by *Glycyrrhiza glabra* root extract treatment in Wistar albino rats. J Nat Sci Biol Med 5(1):25–29

Chamberlain T (1970) Licorice poisoning, pseudoaldosteronism and heart failure. JAMA 213:1343

Chamberlain J, Abolnik I (1997) Pulmonary edema following a licorice binge. West J Med 167:184–185

Chan HT, Chan C, Ho JW (2003) Inhibition of glycyrrhizic acid on aflatoxin B1-induced cytotoxicity in hepatoma. Toxicology 188(2-3):211–217

Chandrasekaran CV, Deepak HB, Thiyagarajan P, Kathiresan S, Sangli GK, Deepak M, Agarwal A (2011a) Dual inhibitory effect of *Glycyrrhiza glabra* (GutGard™) on COX and LOX products. Phytomed 18(4):278–284

Chandrasekaran CV, Sundarajan K, Gupta A, Srikanth HS, Edwin J, Agarwal A (2011b) Evaluation of the genotoxic potential of standardized extract of *Glycyrrhiza glabra* (GutGard™). Reg Toxicol Pharmacol 61(3):373–380

Cheel J, Van Antwerpen P, Tůmová L, Onofre G, Vokurková D, Zouaoui-Boudjeltia K, Vanhaeverbeek M, JeanNève J (2010) Free radical-scavenging, antioxidant and immunostimulating effects of a licorice infusion (*Glycyrrhiza glabra* L.). Food Chem 122(3):508–517

Cheel J, Tůmová L, Areche C, Van Antwerpen P, Nève J, Zouaoui-Boudjeltia K, Martin AS, Vokřál I, Wsól V, Neugebauerová J (2013) Variations in the chemical profile and biological activities of licorice (*Glycyrrhiza glabra* L.), as influenced by harvest times. Acta Physiol Plant 35(4):1337–1349

Chen G, Hu X, Zhang W, Xu N, Wang FQ, Jia J, Zhang WF, Sun ZJ, Zhao YF (2012) Mammalian target of rapamycin regulates isoliquiritigenin-induced autophagic and apoptotic cell death in adenoid cystic carcinoma cells. Apoptosis 17(1):90–101

Cheng HS, Kong JM, Ng AX, Chan WK, Ton SH, Abdul Kadir K (2014) Novel inhibitory effects of glycyrrhizic acid on the accumulation of advanced glycation end product and its receptor expression. Nat Prod Bioprospect 4(6):325–333

Chiej R (1984) The Macdonald encyclopaedia of medicinal plants. Macdonald & Co, London, 447 pp

Chintharlapalli S, Papineni S, Jutooru I, McAlees A, Safe S (2007) Structure-dependent activity of glycyrrhetinic acid derivatives as peroxisome proliferator-activated receptor [gamma] agonists in colon cancer cells. Mol Cancer Ther 6(5):1588–1598

Cho SM, Park JH, Pae AN, Han DS, Kim DS, Cho NC, No KT, Yang HY, Yoon MS, Lee CH, Shimizu M, Baek NI (2012) Hypnotic effects and GABAergic mechanism of licorice (*Glycyrrhiza glabra*) ethanol extract and its major flavonoid constituent glabrol. Bioorg Med Chem 11(1):3493–3501

Choi EM (2005) The licorice root derived isoflavan glabridin increases the function of osteoblastic MC3T3-E1 cells. Biochem Pharmacol 70(3):363–368

Choi EM, Suh KS, Lee YS (2014) Liquiritigenin restores osteoblast damage through regulating oxidative stress and mitochondrial dysfunction. Phytother Res 28(6):880–886

Chopra PKPG, Saraf BD, Inam F, Deo SS (2013) Antimicrobial and antioxidant activities of methanol extract roots of *Glycyrrhiza glabra* and HPLC analysis. Int J Pharm Pharm Sci 5(S2):157–160

Chowdhury B, Bhattamisra SK, Das MC (2013) Anticonvulsant action and amelioration of oxidative stress by *Glycyrrhiza glabra* root extract in pentylenetetrazole-induced seizure in albino rats. Indian J Pharmacol 45(1):40–43

Chu XT, de la Cruz J, Hwang SG, Hong H (2014) Tumorigenic effects of endocrine-disrupting chemicals are alleviated by licorice (*Glycyrrhiza glabra*) root extract through suppression of AhR expression in mammalian cells. Asian Pac J Cancer Prev 15(12):4809–4813

Chubachi A, Wakui H, Asakura K, Nishimura S, Nakamoto Y, Miura AB (1992) Acute renal failure following hypokalemic rhabdomyolysis due to chronic glycyrrhizic acid administration. Intern Med 31(5):708–711

Cinatl J, Morgenstern B, Bauer G, Chandra P, Rabenau H, Doerr HW (2003) Glycyrrhizin, an active component of liquorice roots, and replication of SARS-associated coronavirus. Lancet 361(9374):2045–2046

Classen-Houben D, Schuster D, Da Cunha T, Odermatt A, Wolber G, Jordis U, Kueenburg B (2009) Selective inhibition of 11beta-hydroxysteroid dehydrogenase 1 by 18alpha-glycyrrhetinic acid but not 18beta-glycyrrhetinic acid. J Steroid Biochem Mol Biol 113(3–5):248–252

Conn JW, Rovner DR, Cohen EL (1968) Licorice-induced pseudoaldosteronism. Hypertension, hypokalemia, aldosteronopenia, and suppressed plasma renin activity. JAMA 205(7):492–496

Cooper H, Bhattacharya B, Verma V, McCulloch AJ, Smellie WSA, Heald AH (2007) Liquorice and soy sauce, a life-saving concoction in a patient with Addison's disease. Ann Clin Biochem 44(4):397–399

Corsi FM, Galgani S, Gasparini C, Giacanelli M, Piazza G (1983) Acute hypokalemic myopathy due to chronic licorice ingestion: report of a case. Ital J Neurol Sci 4(4):493–497

Costello CH, Lynn EV (1950) Estrogenic substances from plants: I. *Glycyrrhiza*. J Pharm Sci 39(3):177–180

Crance JM, Lévêque F, Biziagos E, van Cuyck-Gandré H, Jouan A, Deloince R (1994) Studies on mechanism of action of glycyrrhizin against hepatitis A virus replication in vitro. Antiviral Res 23(1):63–76

Cui YM, Ao MZ, Li W, Yu LJ (2008) Effect of glabridin from *Glycyrrhiza glabra* on learning and memory in mice. Planta Med 74(4):377–380

Cuspidi C, Gelosa M, Moroni E, Sampieri L (1981) Pseudo-Conn's syndrome after habitual ingestion of liquorice. Report on various clinical cases. Minerva Med 72(13):825–830 (In Italian)

Dai XH, Li HE, Lu CJ, Wang JF, Dong J, Wei JY, Zhang Y, Wang X, Tan W, Deng XM, Zhao SH, Zhang MJ (2013) Liquiritigenin prevents *Staphylococcus aureus*-mediated lung cell injury via inhibiting the production of α-hemolysin. J Asian Nat Prod Res 15(4):390–399

Datla R, Rao SR, Murthy KJ (1981) Excretion studies of nitrofurantoin and nitrofurantoin with deglycyrrhizinated liquorice. Indian J Physiol Pharmacol 25(1):59–63

Davidson JS, Baumgarten IM (1988) Glycyrrhetinic acid derivatives: a novel class of inhibitors of gap-junctional intercellular communication. Structure-activity relationships. J Pharmacol Exp Ther 246(3):1104–1107

Davidson JS, Baumgarten IM, Harley EH (1986) Reversible inhibition of intercellular junctional communication by glycyrrhetinic acid. Biochem Biophys Res Commun 134(1):29–36

De Klerk G, Nieuwenhuis M, Beutler J (1997) Hypokalaemia and hypertension and another 35 year old woman with profound hypokalaemia and hypertension ssociated with use of liquorice flavoured chewing gum. Br Med J 314:731–732

Dehpour AR, Zolfaghari ME, Samadian T, Vahedi Y (1994) The protective effect of liquorice components and their derivatives against gastric ulcer induced by aspirin in rats. J Pharm Pharmacol 46(2):148–149

Demizu S, Kajiyama K, Takahashi K, Hiraga Y, Yamamoto S, Tamura Y, Okada K, Kinoshita T (1988) Antioxidant and antimicrobial constituents of licorice: isolation and structure elucidation of a new benzofuran derivative. Chem Pharm Bull 36(9):3474–3479

Denisova SB, Galkin EG, Danilov VT, Murinov YI (2003) Mono- and dioses of *Glycyrrhiza glabra* root. Chem Nat Comp 39(3):237–239

Denisova SB, Galkin EG, Murinov YI (2006) Isolation and GC-MS determination of flavonoids from *Glycyrrhiza glabra* root. Chem Nat Comp 42(3):285–289

Denisova SB, Danilov VT, Yunusova SG, Davydova VA, Murinov YI, Zarudii FS (2007) Isolation and biological activity of lipids from licorice (*Glycyrrhiza glabra*) roots. Pharm Chem J 41(9):489–491

Dhingra D, Sharma A (2006) Antidepressant-like activity of *Glycyrrhiza glabra* L. in mouse models of immobility tests. Prog Neuropsychopharmacol Biol Psychiatry 30(3):449–454

Dhingra D, Parle M, Kulkarni SK (2004) Memory enhancing activity of *Glycyrrhiza glabra* in mice. J Ethnopharmacol 91(2–3):361–365

Dirican E, Turkez H (2014) In vitro studies on protective effect of *Glycyrrhiza glabra* root extracts against cadmium-induced genetic and oxidative damage in human lymphocytes. Cytotechnology 66(1):9–16

Dobbins KR, Saul RF (2000) Transient visual loss after licorice ingestion. J Neuroophthalmol 20:38–41

Doeker BM, Andler W (1999) Liquorice, growth retardation and Addison's disease. Horm Res 52(5):253–255

Dong SJ, Inoue A, Zhu Y, Tanji M, Kiyama R (2007) Activation of rapid signaling pathways and the subsequent transcriptional regulation for the proliferation of breast cancer MCF-7 cells by the treatment with an extract of *Glycyrrhiza glabra* root. Food Chem Toxicol 45(12):2470–2478

Dong Y, Zhao MM, Zhao TT, Feng MY, Chen HP, Zhuang MZ, Lin LZ (2014) Bioactive profiles, antioxidant activities, nitrite scavenging capacities and protective effects on H2O2-injured PC12 cells of *Glycyrrhiza glabra* L. leaf and root extracts. Molecules 19(7):9101–9113

Dong Y, Zhao MM, Sun-Waterhouse D, Zhuang MZ, Chen HP, Feng M, Lin LZ (2015) Absorption and desorption behaviour of the flavonoids from *Glycyrrhiza glabra* L. leaf on macroporous adsorption resins. Food Chem 168:538–545

Dorisse P, Gleye J, Loiseau P, Puig P, Edy AM, Henry M (1988) Papaverine biotransformation in plant cell suspension cultures. J Nat Prod 51(3):532–536

Durak H (2014) Bio-oil production from *Glycyrrhiza glabra* through supercritical fluid extraction. J Supercrit Fluids 95:373–386

Dzhumamuratova A, Seitmuratov E, Rakhimov DA, Ismailov ZF (1978) Polysaccharides of some species of *Glycyrrhiza*. Chem Nat Comp 14(4):437–437

El Magd NFA, El-Karel A, El-Shishtawy MM, Eissa LA (2015) Hepatoprotective effects of glycyrrhizin and omega-3 fatty acids on nuclear factor-kappa B pathway in thioacetamide-induced fibrosis in rats. Egyptian J Basic Appl Sci 2(2):65–74

Elgamal MHA, El-Tawil BA (1975) Constituents of local plants. XVIII. 28-hydroxyglycyrrhetic acid, a new triterpenoid isolated from the roots of *Glycyrrhiza glabra*. Planta Med 27(2):159–163

Elgamal MHA, Fayez MBE (1968) Structure of glabric acid a further triterpenoid constituent of *Glycyrrhiza glabra*. Acta Chim (Budapest) 58(1):75–84

Elgamal MHA, Fayez MBE (1972) Isolation of formononetin from the root of *Glycyrrhiza glabra* L. collected locally. Indian J Chem 10(1):128

Elgamal MHA, Fayez MBE (1975) A new triterpenoid from the roots of *Glycyrrhiza glabra* L. Naturwissenschaften 62(4):183–183

Elgamal MHA, Fayez MBE, Snatzke G (1965) Constituents of local plants – VI roots of *Glycyrrhiza glabra* L.: liquoric acid, a new triterpenoid from the roots of *Glycyrrhiza glabr*a L. Tetrahedron 21(8):2109–2115

Elgamal MHA, Hady FK, Hanna AG, Mahran GH, Duddeck H (1990) A further contribution to the triterpenoid constituents of *Glycyrrhiza glabra* L. Z Naturforsch C 45(9-10):937–941

Elinav E, Chajek-Shaul T (2003) Licorice consumption causing severe hypokalemic paralysis. Mayo Clinic Proc 78(6):767–768

El-Tahawy NF, Ali AH, Saied SR, Abdel-Wahab Z (2011) Effect of glycyrrhizin on lipopolysaccharide/D-galactosamine-induced acute hepatitis in albino rats: a histological and immunohistochemical study. Egypt J Histol 34(3):518–527

Engqvist A, von Feilitzen F, Pyk E, Reichard H (1973) Double-blind trial of deglycyrrhizinated liquorice in gastric ulcer. Gut 14(9):711–715

Epstein MT, Espiner EA, Donald RA, Hughes H (1977) Liquorice toxicity and the renin-angiotensin-aldosterone axis in man. Br Med J 1(6055):209–210

Ercisli S, Coruh I, Gormez A, Sengul M, Bilena S (2008) Total phenolics, mineral contents, antioxidant and antibacterial activities of *Glycyrrhiza glabra* L. roots grown wild in Turkey. Ital J Food Sci 20(1):91–99

Esmaeili S, Naghibi F, Mosaddegh M, Nader N (2010) Determination of 18β-glycyrrhetinic acid in *Glycyrrhiza glabra* L. extract by HPLC. Iran J Pharm Res 5(2):137–141

European Commission Scientific Committee on Food (2003) Opinion of the Scientific Committee on food on glycyrrhizinic acid and its ammonium salt. European Commission Heath and Consumer Protection Directorate General, Brussel, Belgium. http://ec.europa.eu/food/fs/sc/scf/out186_en.pdf

Facciola S (1990) Cornucopia. A source book of edible plants. Kampong Publ., Vista, 677 pp.

Farag MA, Wessjohann LA (2012) Volatiles profiling in medicinal licorice roots using steam distillation and solid-phase microextraction (SPME) coupled to chemometrics. J Food Sci 7(11):C1179–C1184

Farag MA, Porzela A, Wessjohann LA (2012) Comparative metabolite profiling and fingerprinting of medicinal licorice roots using a multiplex approach of GC–MS, LC–MS and 1D NMR techniques. Phytochemistry 76:60–72

Farese S, Kruse A, Pasch A, Dick B, Frey BM, Uehlinger DE, Frey FJ (2009) Glycyrrhetinic acid food supplementation lowers serum potassium concentration in chronic hemodialysis patients. Kidney Int 76(8):877–884

Fatima A, Gupta VK, Luqman S, Negi AS, Kumar JK, Shanker K, Saikia D, Srivastava S, Darokar MP, Khanuja SP (2009) Antifungal activity of *Glycyrrhiza glabra* extract and its active constituent glabridin. Phytother Res 23(8):1190–1193

Fenwick GR, Lutomski J, Nieman C (1990) Liquorice, *Glycyrrhiza glabra* L.- composition, uses and analysis. Food Chem 38(2):119–143

Ferrari P (2009) Licorice: a sweet alternative to prevent hyperkalemia in dialysis patients? Kidney Int 76(8):811–812

Ferrari P, Sansonnens A, Dick B, Frey FJ (2001) In vivo 11beta-HSD-2 activity: variability, salt-sensitivity, and effect of licorice. Hypertension 38(6):1330–1336

Fiore C, Eisenhut M, Krausse R, Ragazzi E, Pellati D, Armanini D, Bielenberg J (2008) Antiviral effects of *Glycyrrhiza* species. Phytother Res 22(2):141–148

Forslund T, Fyhrquist F, Frøseth B, Tikkanen I (1989) Effects of licorice on plasma atrial natriuretic peptide in healthy volunteers. J Intern Med 225(2):95–99

Franceschelli S, Pesce M, Vinciguerra I, Ferrone A, Riccioni G, Antonia P, Grilli A, Falaco M, Speranza L (2011) Licocalchone-C extracted from *Glycyrrhiza glabra* inhibits lipopolysaccharide-interferon-γ inflammation by improving antioxidant conditions and regulating inducible nitric oxide synthase expression. Molecules 16(7):5720–5734

Francischetti IM, Monteiro RQ, Guimaraes JA (1997) Identification of glycyrrhizin as a thrombin inhibitor. Biochem Biophys Res Commun 235(1):259–263

Fraunfelder F (2004) Ocular side effects from herbal medicines and nutritional supplements. Am J Ophthalmol 138:639–647

Frazier MC, Jackson KM, Jankowska-Stephens E, Anderson MG, Harris WB (2004) Proteomic analysis of proteins altered by dibenzoylmethane in human prostatic cancer LNCaP cells. Proteomics 4(9):2814–2821

Fu Y, Hsieh TC, Guo JQ, Kunicki J, Lee MYWT, Darzynkiewicz Z, Wu JM (2004) Licochalcone-A, a novel flavonoid isolated from licorice root (*Glycyrrhiza glabra*), causes G2 and late-G1 arrests in androgen-independent PC-3 prostate cancer cells. Biochem Biophys Res Commun 322(1):263–270

Fuggersberger-Heinz R, Franz G (1984) Formation of glycyrrhizinic acid in *Glycyrrhiza glabra* var. *typica*. Planta Med 50(5):409–413

Fuhrman B, Buch S, Vaya J, Belinky PA, Coleman R, Hayek T, Aviram M (1997) Licorice extract and its major polyphenol glabridin protect low-density lipoprotein against lipid peroxidation: in vitro and ex vivo studies in humans and in atherosclerotic apolipoprotein E-deficient mice. Am J Clin Nutr 66(2):267–275

Fuhrman B, Volkova N, Rosenblat M, Aviram M (2000) Lycopene synergistically inhibits LDL oxidation in combination with vitamin E, glabridin, rosmarinic acid, carnosic acid, or garlic. Antioxid Redox Signal 2(3):491–506

Fukai T, Nishizawa J, Yokoyama M, Tantai L, Nomura T (1994) Five new isoprenoid-substituted flavonoids, kanzonols M, P and R, from two *Glycyrrhiza* species. Heterocycles 38(5):1089–1098

Fukai T, Cai BS, Horikoshi T, Nomura T (1996a) Isoprenylated flavonoids from underground parts of *Glycyrrhiza glabra*. Phytochemistry 43(5):1119–1124

Fukai T, Tantai L, Nomura T (1996b) Isoprenoid-substituted flavonoids from *Glycyrrhiza glabra*. Phytochemistry 43(2):531–532

Fukai T, Cai BS, Maruno K, Miyakawa Y, Konishi M, Nomura T (1998) An isoprenylated flavanone from *Glycyrrhiza glabra* and rec-assay of licorice phenols. Phytochemistry 49(7):2005–2013

Fukai T, Cai BS, Nomura T (2001) Flavonoids from Kirghiz licorice (*Glycyrrhiza glabra*). Nat Med 55(6):311

Fukai T, Marumo A, Kaitou K, Kanda T, Terada S, Nomura T (2002a) Anti *Helicobacter pylori* flavonoids from licorice extract. Life Sci 71(12):1449–1463

Fukai T, Marumo A, Kaitou K, Kanda T, Terada S, Nomura T (2002b) Antimicrobial activity of licorice flavonoids against methicillin-resistant *Staphylococcus aureus*. Fitoterapia 73(6):536–539

Fukai T, Satoh K, Nomura T, Sakagami H (2003) Preliminary evaluation of antinephritis and radical scavenging activities of glabridin from *Glycyrrhiza glabra*. Fitoterapia 74(7):624–629

Fukui H, Goto K, Tabata M (1988) Two antimicrobial flavanones from the leaves of *Glycyrrhiza glabra*. Chem Pharm Bull 36:4174–4176

Gao XY, Wang WQ, Wei SL, Li WD (2009) Review of pharmacological effects of *Glycyrrhiza* radix and its bioactive compounds. Zhongguo Zhong Yao Za Zhi 34(21):2695–2700 (In Chinese)

Gaur R, Yadav KS, Verma RK, Yadav NP, Bhakuni RS (2014) In vivo anti-diabetic activity of derivatives of isoliquiritigenin and liquiritigenin. Phytomedicine 21(4):415–422

Gaware R, Khunt R, Czollner L, Stanetty C, Da Cunha T, Kratschmar DV, Odermatt A, Kosma P, Jordis U, Classen-Houben D (2011) Synthesis of new glycyrrhetinic acid derived ring A azepanone, 29-urea and 29-hydroxamic acid derivatives as selective 11β-hydroxysteroid dehydrogenase 2 inhibitors. Bioorg Med Chem 19(6):1866–1880

Geetha RV, Roy A (2013) In vitro evaluation of anti mycotic activity of ethanolic extract of *Glycyrrhiza glabra*. Asian J Pharm Clin Res 6(3):205–206

Ghalayani P, Emami H, Pakravan F, Nasr Isfahani M (2014) Comparison of triamcinolone acetonide mucoadhesive film with licorice mucoadhesive film on radiotherapy-induced oral mucositis: a randomized double-blinded clinical trial. Asia Pac J Clin Oncol (in press) DOI: 10.1111/ajco.12295

Ghannad MS, Mohammadi A, Safiallahy S, Faradmal J, Azizi M, Ahmadvand Z (2014) The effect of aqueous extract of *Glycyrrhiza glabra* on Herpes Simplex Virus 1. Jundishapur J Microbiol 7(7):1–6

Gomez-Sanchez EP, Gomez-Sanchez CE (1992) Central hypertensinogenic effects of glycyrrhizic acid and carbenoxolone. Am J Physiol 263(6 Pt 1):E1125–E1130

Gong G, Xiang L, Yuan LB, Hu L, Wu W, Cai L, Yin L, Dong HL (2014) Protective effect of glycyrrhizin, a direct HMGB1 inhibitor, on focal cerebral ischemia/reperfusion-induced inflammation, oxidative stress, and apoptosis in rats. PLoS One 9(3), e89450

Gorecki P, Drozdzynska M, Segiet-Kujawa E (1991) Isolation and determination of flavonoids and chalcones in *Glycyrrhiza glabra* roots. Planta Med 57:A118

Goultschin J, Palmon S, Shapira L, Brayer L, Gedalia I (1991) Effect of glycyrrhizin-containing toothpaste on dental plaque reduction and gingival health in humans. J Clin Periodontol 18:210–212

Grieve M (1971) A modern herbal, 2 vols. Penguin/Dover Publications, New York, 919 pp

Gross E, Dexter J, Roth R (1966) Hypokalemic myopathy with myoglobinuria associated with licorice ingestion. N Engl J Med 274:602–606

Gu XJ, Xu J, Ma BY, Chen G, Gu PY, Wei D, Hu WX (2014) Effect of glycyrrhizin on traumatic brain injury in rats and its mechanism. Chin J Traumatol 17(1):1–7

Gujral ML, Sareen KN, Tangri KK, Roy AK, Gupta GP, Amma MK (1959) Antiarthritic activity of *Glycyrrhiza glabra* Linn. Indian J Psychol 3(1):39–47

Gumpricht E, Dahl R, Devereaux MW, Sokol RJ (2005) Licorice compounds glycyrrhizin and 18beta-glycyrrhetinic acid are potent modulators of bile acid-induced cytotoxicity in rat hepatocytes. J Biol Chem 280(11):10556–10563

Gupta VK, Fatima A, Faridi U, Negi AS, Shanker K, Kumar JK, Rahuja N, Luqman S, Sisodia BS, Saikia D, Darokar MP, Khanuja SPS (2008) Antimicrobial potential of *Glycyrrhiza glabra* roots. J Ethnopharmacol 116(2):377–380

Gyémánt G, Lenkey B, Nánási P (2001) Comparative study of *Glycyrrhiza glabra* L. and *Glycyrrhiza echinata* L. of different origin. Medicinal plant polysaccharides III. Acta Pharm Hung 71(3):325–328 (In Hungarian)

Hall R, Clemett R (2004) Central retinal vein occlusion associated with liquorice ingestion. Clin Exp Ophthalmol 32:341

Han JC, Yu J, Gao YJ (2014) Lipidomics investigation of reversal effect of glycyrrhizin (GL) towards lithocholic acid (LCA)-induced alteration of phospholipid profiles. Pharm Biol 52(12):1624–1628

Harada T, Ohtaki E, Misu K, Sumiyoshi T, Hosoda S (2002) Congestive heart failure caused by digitalis toxicity in an elderly man taking a licorice-containing Chinese herbal laxative. Cardiology 98:218

Haraguchi H, Yoshida N, Ishikawa H, Tamura Y, Mizutani K, Kinoshita T (2000) Protection of mitochondrial functions against oxidative stresses by isoflavans from *Glycyrrhiza glabra*. J Pharm Pharmacol 52(2):219–223

Hardy ME, Hendricks JM, Paulson JM, Faunce NR (2012) 18β-glycyrrhetinic acid inhibits rotavirus replication in culture. Virol J 9:96

Hasan S, Khan R, Ali N, Khan A, Rehman M, Tahir M, Lateef A, Nafees S, Mehdi S, Rashid S, Shahid A, Sultana S (2015) 18-β Glycyrrhetinic acid alleviates 2-acetylaminofluorene-induced hepatotoxicity in Wistar rats: role in hyperproliferation, inflammation and oxidative stress. Hum Exp Toxicol 34(6):628-641.

Hasegawa J, Suyama Y, Kinugawa T, Morisawa T, Kishimoto Y (1998) Echocardiographic findings of the heart resembling dilated cardiomyopathy during hypokalemic myopathy due to licorice-induced pseudoaldosteronism. Cardiovasc Drugs Ther 12:599–600

Hatano T, Yasuhara T, Miyamoto K, Okuda T (1988) Anti-human immunodeficiency virus phenolics from licorice. Chem Pharm Bull 36(6):2286–2288

Hatano T, Yasuhara T, Fukuda T, Noro T, Okuda T (1989) Phenolic constituents of licorice. II. Structures of licopyranocoumarin, licoarylcoumarin and glisoflavone, and inhibitory effects of licorice phenolics on xanthine oxidase. Chem Pharm Bull 37(11):3005–3009

Hatano T, Fukuda T, Liu YZ, Nor T, Okuda T (1991a) Phenolic constituents of licorice. IV. Correlation of phenolic constituents and licorice specimens from various sources, and inhibitory effects of licorice extracts on xanthine oxidase and monoamine oxidase. Yakugaku Zasshi 111(6):311–321 (In Japanese)

Hatano T, Fukuda T, Miyase T, Noro T, Okuda T (1991b) Phenolic constituents of licorice. III. Structures of glicoricone and licofuranone, and inhibitory effects of licorice constituents of monoamine oxidase. Chem Pharm Bull 39(5):1238–1243

Hattori M, Sakamoto T, Yamagishi T, Sakamoto K, Konishi K, Kobashi K, Namba T (1985) Metabolism of glycyrrhizin by human intestinal flora. II. Isolation and characterization of human intestinal bacteria capable of metabolizing glycyrrhizin and related compounds. Chem Pharm Bull 33(1):210–217

Hattori M, Miyachi K, Shu YZ, Kaikiuchi N, Namba T (1986) Studies on dental caries prevention by traditional medicines. 9. Potent antibacterial action of coumarin derivatives from licorice roots against *Streptococcus mutans*. Shoyakugaku Zasshi 40:406–412

Hayashi H (2009) Molecular biology of secondary metabolism: case study for *Glycyrrhiza* plants. In: Kirakosyan A, Kaufman PB (eds) Recent advances in plant biotechnology. Springer, New York/Dordrecht, pp 89–103

Hayashi H, Sudo H (2009) Economic importance of licorice. Plant Biotechnol 26:101–104

Hayashi H, Fukui H, Tabata M (1988) Examination of triterpenoids produced by callus and cell suspension cultures of *Glycyrrhiza glabra*. Plant Cell Rep 7:508–511

Hayashi H, Fukui H, Tabata M (1990a) Biotransformation of 18β-glycyrrhetinic acid by cell suspension cultures of *Glycyrrhiza glabra*. Phytochemistry 29(7):2149–2152

Hayashi H, Sakai T, Fukui H, Tabata M (1990b) Formation of soyasaponins in licorice cell suspension cultures. Phytochem 29(10):3127–3129

Hayashi H, Yamada K, Fukui H, Tabata M (1992a) Metabolism of exogenous 18β-glycyrrhetinic acid in cultured cells of *Glycyrrhiza glabra*. Phytochemistry 31(8):2724–2733

Hayashi R, Maruyama T, Maruyama K, Yanagawa S, Tako K, Yanagisawa N (1992b) Myotonic and repetitive discharges in hypokalemic myopathy associated with glycyrrhizin-induced hypochloremia. J Neurol Sci 107(1):74–77

Hayashi H, Fukui H, Tabata M (1993a) Distribution pattern of saponins in different organs of *Glycyrrhiza glabra*. Planta Med 59:351–353

Hayashi H, Hanaoka S, Tanaka S, Fukui H, Tabata M (1993b) Glycyrrhetinic acid 24-hydroxylase activity in microsomes of cultured licorice cells. Phytochemistry 34:1303–1307

Hayashi H, Honda G, Tabata M, Yamamoto H, Yeşilada E, Sezik E (1995) A survey of distribution and characteristics of *Glycyrrhiza glabra* L. in Turkey. Nat Med 49(2):129–132

Hayashi H, Hiraoka N, Ikeshiro Y (1996a) Molecular cloning and functional expression of cDNAs for *Glycyrrhiza glabra* squalene synthase. Biol Pharm Bull 19(10):1387–1389

Hayashi H, Hiraoka N, Ikeshiro Y, Yamamoto H (1996b) Organ specific localization of flavonoids in *Glycyrrhiza glabra* L. Plant Sci 116(2):233–238

Hayashi H, Nishiyama Y, Tomizawa N, Hiraoka N, Ikeshiro Y (1996c) UDP-glucuronic acid: triterpene glucuronosyltransferase activity in cultured licorice cells. Phytochemistry 42:665–666

Hayashi H, Yasuma M, Hiraoka N, Ikeshiro Y, Yamamoto H, Yeşilada E, Sezik E, Honda G, Tabata M (1996d) Flavonoid variation in the leaves of *Glycyrrhiza glabra*. Phytochemistry 24(3):701–704

Hayashi K, Hayashi H, Hiraoka N, Ikeshiro Y (1997) Inhibitory activity of soyasaponin II on virus replication in vitro. Planta Med 63(2):102–105

Hayashi H, Hiraoka N, Ikeshiro Y, Yamamoto H, Yoshikawa T (1998a) Seasonal variation of glycyrrhizin and isoliquiritigenin glycosides in the root of *Glycyrrhiza glabra* L. Biol Pharm Bull 21(9):987–989

Hayashi H, Shibano M, Kusano G, Yamamoto H, Ikeshiro Y (1998b) A field survey of *Glycyrrhiza glabra* L in Sicily and Spain. Nat Med 52(3):259–264

Hayashi H, Hirota A, Hiraoka N, Ikeshiro Y (1999) Molecular cloning and characterization of two cDNAs for *Glycyrrhiza glabra* squalene synthase. Biol Pharm Bull 22:947–950

Hayashi H, Hiraoka N, Ikeshiro Y, Kushiro T, Morita M, Shibuya M, Ebizuka Y (2000a) Molecular cloning and characterization of a cDNA for *Glycyrrhiza glabra* cycloartenol synthase. Biol Pharm Bull 23(2):231–234

Hayashi H, Hosono N, Kondo M, Hiraoka N, Ikeshiro Y, Shibano M, Kusano G, Yamamoto H, Tanaka T, Inoue K (2000b) Phylogenetic relationship of six *Glycyrrhiza* species based on rbcL sequences and chemical constituents. Biol Pharm Bull 23(5):602–606

Hayashi H, Huang P, Kirakosyan A, Inoue K, Hiraoka N, Ikeshiro Y, Kushiro T, Shibuya M, Ebizuka Y (2001) Cloning and characterization of a cDNA encoding beta-amyrin synthase involved in glycyrrhizin and soyasaponin biosyntheses in licorice. Biol Pharm Bull 24(8):912–916

Hayashi H, Hattori S, Inoue K, Khodzhimatov O, Ashurmetov O, Ito M, Honda G (2003a) Field survey of *Glycyrrhiza* plants in Central Asia (3). Chemical characterization of *G. glabra* collected in Uzbekistan. Chem Pharm Bull 51(11):1338–1340

Hayashi H, Hattori S, Inoue K, Sarsenbaev K, Ito M, Honda G (2003b) Field survey of *Glycyrrhiza* plants in central Asia (1). Characterization of *G. uralensis*, *G. glabra* and the putative intermediate collected in Kazakhstan. Biol Pharm Bull 26(6):867–871

Hayashi H, Huang P, Inoue K (2003c) Up-regulation of soyasaponin biosynthesis by methyl jasmonate in cultured cells of *Glycyrrhiza glabra*. Plant Cell Physiol 44(4):401–411

Hayashi H, Huang P, Takada S, Obinata M, Inoue K, Shibuya M, Ebizuka Y (2004) Differential expression of three oxidosqualene cyclase mRNAs in *Glycyrrhiza glabra*. Biol Pharm Bull 27(7):1086–1092

Hayashi H, Hiraoka N, Ikeshiro Y (2005) Differential regulation of soyasaponin and betulinic acid production by yeast extract in cultured licorice cells. Plant Biotechnol 22(3):241–244

Heidemann HT, Kreuzfelder E (1983) Hypokalemic rhabdomyolysis with myoglobinuria due to licorice ingestion and diuretic treatment. Kline Wochenschr 61(6):303–305

Heikens J, Fliers E, Endert E, Ackermans M, van Montfrans G (1995) Liquorice-induced hypertension-a new understanding of an old disease: case report and brief review. A Neth J Med 47(5):230–234

Helmy WA, El-Kheir ZAA, AbdelHady MS, El-Hameid ARA (2013) Biological activities of aqueous extracts and their sulfated derivatives from licorice (*Glycyrrhiza glabra* L.). J Appl Res 9(6):3638–3645

Hendricks JM, Hoffman C, Pascual DW, Hardy ME (2012) 18β-glycyrrhetinic acid delivered orally induces isolated lymphoid follicle maturation at the intestinal mucosa and attenuates rotavirus shedding. PLoS One 7(11), e49491

Hill AF (1952) Economic botany, 2nd edn. McGraw-Hill Book Co., New York, 560 pp

Hocaoglu AB, Karaman O, Erge DO, Erbil G, Yilmaz O, Bagriyanik A, Uzuner N (2011) Glycyrrhizin and long-term histopathologic changes in a murine model of asthma. Curr Ther Res 72(6):250–261

Holmes A, Young J, Marrott P, Prentice E (1970) Pseudohyperaldosteronism induced by habitual ingestion of liquorice. Postgrad Med J 46(540):625–629

Honda K, Kamisoyama H, Tominaga Y, Yokota S, Hasegawa S (2009) The molecular mechanism underlying the reduction in abdominal fat accumulation by licorice flavonoid oil in high fat diet-induced obese rats. Anim Sci J 80(5):562–569

Honda H, Nagai Y, Matsunaga T, Saitoh S, Akashi-Takamura S, Hayashi H, Fujii I, Miyake K, Muraguchi A, Takatsu K (2012) Glycyrrhizin and isoliquiritigenin suppress the LPS sensor toll-like receptor 4/MD-2 complex signaling in a different manner. J Leukoc Biol 91(6):967–976

Honda H, Nagai Y, Matsunaga T, Okamoto N, Watanabe Y, Tsuneyama K, Hayashi H, Fujii I, Ikutani M, Hirai Y, Muraguchi A, Takatsu K (2014) Isoliquiritigenin is a potent inhibitor of NLRP3 inflammasome activation and diet-induced adipose tissue inflammation. J Leukoc Biol 96(6):1087–1100

Hong YK, Wu HT, Ma T, Liu WJ, He XJ (2009) Effects of *Glycyrrhiza glabra* polysaccharides on immune and antioxidant activities in high-fat mice. Int J Biol Macromol 45(1):61–64

Hou YC, Lin SP, Chao PD (2012) Liquorice reduced cyclosporine bioavailability by activating P-glycoprotein and CYP 3A. Food Chem 135(4):2307–2312

Hsia SM, Yu CC, Shih YH, Chen MY, Wang TH, Huang YT, Shieh TM (2015) Isoliquiritigenin causes DNA damage and inhibits ATM expression leading to G2/M phase arrest and apoptosis in oral squamous cell carcinoma. Head Neck doi: 10.1002/hed.24001. (in press)

Hu CH, He J, Eckert R, Wu XY, Li LN, Tian Y, Lux R, Shuffer JA, Gelman F, Mentes J, Spackman S, Bauer J, Anderson MH, Shi WY (2011) Development and evaluation of a safe and effective sugar-free herbal lollipop that kills cavity-causing bacteria. Int J Oral Sci 3(1):13–20

Hua H, Liang ZF, Li WW, Meng YM, Li X, Zhang ZJ, Lu CL, Meng JJ, Shan FP (2012) Phenotypic and functional maturation of murine dendritic cells (DCs) induced by purified glycyrrhizin (GL). Int Immunopharmacol 12(3):518–525

Huang RY, Chu YL, Huang QC, Chen XM, Jiang ZB, Zhang X, Zeng X (2014) 18β-Glycyrrhetinic acid suppresses cell proliferation through inhibiting thromboxane synthase in non-small cell lung cancer. PLoS One 9(4), e93690

Hughes J, Sellick S, King R, Robbé IJ (2003) Re: "preterm birth and licorice consumption during pregnancy". Am J Epidemiol 158(2):190–191

Hwang IK, Lim SS, Choi KH, Yoo KY, Shin HK, Kim EJ, Yoon-Park JH, Kang TC, Kim YS, Kwon DY, Kim DW, Moon WK, Won MH (2006) Neuroprotective effects of roasted licorice, not raw form, on neuronal injury in gerbil hippocampus after transient forebrain ischemia. Acta Pharmacol Sin 27(8):959–965

Iida K, Uematsu Y, Suzuki K, Yasuno T, Hirata K, Ito K (2007) Properties of commercial licorice extracts used as a food additive. Shokuhin Eiseigaku Zasshi 48(4):112–117 (In Japanese)

Ikeda K (2007) Glycyrrhizin injection therapy prevents hepatocellular carcinogenesis in patients with interferon-resistant active chronic hepatitis C. Hepatol Res 37(s2):S287–S293

Ikeda K, Arase Y, Kobayashi M, Saitoh S, Someya T, Hosaka T, Sezaki H, Akuta N, Suzuki Y, Suzuki F, Kumada H (2006) A long-term glycyrrhizin injection therapy reduces hepatocellular carcinogenesis rate in patients with interferon-resistant active chronic hepatitis C: a cohort study of 1249 patients. Dig Dis Sci 51(3):603–609

Ingham JL (1977) An isoflavan phytoalexin from leaves of *Glycyrrhiza glabra*. Phytochemistry 16(9):1457–1458

Irani M, Sarmadi M, Bernard F, Ebrahimi Pour GH, Shaker Bazarnov H (2010) Leaves antimicrobial activity of *Glycyrrhiza glabra* L. Iran J Pharm Res 9(4):425–428

Isbrucker RA, Burdock GA (2006) Risk and safety assessment on the consumption of Licorice root (*Glycyrrhiza* sp.), its extract and powder as a food ingredient, with emphasis on the pharmacology and toxicology of glycyrrhizin. Regul Toxicol Pharmacol 46(3):167–192

Ishiguchi T, Mikita N, Iwata T, Nakata H, Sato H, Higashimoto Y, Fujimoto H, Yoshida S, Itoh H (2004) Myoclonus and metabolic alkalosis from licorice in antacid. Intern Med 43(1):59–62

Ishikawa S, Kato M, Tokuda T, Momoi H, Sekijima Y, Higuchi M, Yanagisawa N (1999) Licorice-induced hypokalemic myopathy and hypokalemic renal tubular damage in anorexia nervosa. Int J Eat Disord 26(1):111–114

Ito M, Nakashima H, Baba M, Pauwels R, De Clercq E, Shigeta S, Yamamoto N (1987) Inhibitory effect of glycyrrhizin on the in vitro infectivity and cytopathic activity of the human immunodeficiency virus [HIV (HTLV-III/LAV)]. Antiviral Res 7(3):127–137

Ito M, Sato A, Hirabayashi K, Tanabe F, Shigeta S, Baba M, De Clercq E, Nakashima H, Yamamoto N (1988) Mechanism of inhibitory effect of glycyrrhizin on replication of human immunodeficiency virus (HIV). Antiviral Res 10(6):289–298

Ito C, Oi N, Hashimoto T, Nakabayashi H, Aoki F, Tominaga Y, Yokota S, Hosoe K, Kanazawa K (2007) Absorption of dietary licorice isoflavan glabridin to blood circulation in rats. J Nutr Sci Vitaminol (Tokyo) 53(4):358–365

Jackson KM, DeLeon M, Verret CR, Harris WB (2002) Dibenzoylmethane induces cell cycle deregulation in human prostate cancer cells. Cancer Lett 178(2):161–165

Jain E, Pandey RK, Khanna R (2013) Liquorice root extracts as potent cariostatic agents in pediatric practice. J Indian Soc Pedod Prev Dent 231:146–152

Jeong HG, You HJ, Park SJ, Moon AR, Chung YC, Kang SK, Chun HK (2002) Hepatoprotective effects of 18beta-glycyrrhetinic acid on carbon tetrachloride-induced liver injury: inhibition of cytochrome P450 2E1 expression. Pharmacol Res 46(3):221–227

Jhanji V, Liu HM, Law K, Lee VYW, Huang SH, Pang CP, Yam HHF (2011) Isoliquiritigenin from licorice root suppressed neovascularisation in experimental ocular angiogenesis models. Br J Ophthalmol 95:1309–1315

Jin SX, Fu SS, Han J, Jin SY, Lv QY, Lu Y, Qi JP, Wu W, Yuan HL (2012) Improvement of oral bioavailability of glycyrrhizin by sodium deoxycholate/phospholipid-mixed nanomicelles. J Drug Target 20(7):615–622

Jirawattanapong W, Saifah E, Patarapanich C (2009) Synthesis of glabridin derivatives as tyrosinase inhibitors. Arch Pharm Res 32(5):647–654

Johns C (2009) Glycyrrhizic acid toxicity caused by consumption of licorice candy cigars. Can J Emer Med 11:94–96

Jung GD, Yang JY, Song ES, Par JW (2001) Stimulation of melanogenesis by glycyrrhizin in B16 melanoma cells. Exp Mol Med 33(3):131–135

Kageyama Y (1992) A case of pseudoaldosteronism induced by glycyrrhizin. Nihon Jinzo Gakkai Shi 34(1):99–102

Kageyama K, Watanobe H, Nishie M, Imamura K, Suda T (1997) A case of pseudoaldosteronism induced by a mouth refresher containing licorice. Endocr J 44(4):631–632

Kalaiarasi P, Kaviarasan K, Pugalendi KV (2009) Hypolipidemic activity of 18β-glycyrrhetinic acid on streptozotocin-induced diabetic rats. Eur J Pharmacol 612(1-3):93–97

Kalani K, Agarwal J, Alam S, Khan F, Pal A, Srivastava SK (2013a) In silico and in vivo anti-malarial studies of 18β glycyrrhetinic acid from *Glycyrrhiza glabra*. PLoS One 8(9), e74761

Kalani K, Kushwaha V, Verma R, Murthy PK, Srivastava SK (2013b) Glycyrrhetinic acid and its analogs: a new class of antifilarial agents. Bioorg Med Chem Lett 23(9):2566–2570

Kamalak A (2006) Determination of nutritive value of leaves of a native grown shrub, *Glycyrrhiza glabra* L. using in vitro and in situ measurements. Small Ruminant Res 64(3):268–278

Kameoka H, Nakai K (1987) Components of the essential oil from the root of *Glycyrrhiza glabra*. Nippon Nogeikagaku Kaishi 61(9):1119–1121

Kanazawa M, Satomi Y, Mizutani Y, Ukimura O, Kawauchi A, Sakai T, Baba M, Okuyama T, Nishino H, Miki T (2003) Isoliquiritigenin inhibits the growth of prostate cancer. Eur Urol 43(5):580–586

Kang SW, Choi JS, Choi YJ, Bae JY, Li J, Kim DS, Kim JL, Shin SY, Lee YJ, Kwun IS, Kang YH (2010)

Licorice isoliquiritigenin dampens angiogenic activity via inhibition of MAPK-responsive signaling pathways leading to induction of matrix metalloproteinases. J Nutr Biochem 21(1):55–65

Kao TC, Shyu MH, Yen GC (2009) Neuroprotective effects of glycyrrhizic acid and 18beta-glycyrrhetinic acid in PC12 cells via modulation of the PI3K/Akt pathway. J Agric Food Chem 57(2):754–761

Kao TC, Shyu MH, Yen GC (2010) Glycyrrhizic acid and 18beta-glycyrrhetinic acid inhibit inflammation via PI3K/Akt/GSK3beta signaling and glucocorticoid receptor activation. J Agric Food Chem 58(15):8623–8629

Kashiwada Y, Hashimoto F, Cosentino LM, Chen CH, Garrett PE, Lee KH (1996) Betulinic acid and dihydrobetulinic acid derivatives as potent anti-HIV agents. J Med Chem 39:1016–1017

Kato H, Kanaoka M, Yano S, Kobayashi M (1995) 3-Monoglucuronyl-glycyrrhetinic acid is a major metabolite that causes licorice-induced pseudoaldosteronism. J Clin Endocrinol Metab 80(6):1929–1933

Kattaev NS, Nikonov GK (1972) Glabranin, a new flavanone from *Glycyrrhiza glabra*. Chem Nat Comp 8(6):790–791

Kattaev NS, Nikonov GK (1974) Flavonoids of *Glycyrrhiza glabra*. Chem Nat Comp 10(1):94–95

Kaur P, Kaur S, Kumar N, Singh B, Kumar S (2009) Evaluation of antigenotoxic activity of isoliquiritin apioside from *Glycyrrhiza glabra* L. Toxicol In Vitro 23(4):680–686

Kaur P, Kumar M, Singh B, Kumar S, Kaur S (2012a) Amelioration of oxidative stress induced by oxidative mutagens and COX-2 inhibitory activity of umbelliferone isolated from *Glycyrrhiza glabra* L. Asian Pac J Trop Biomed 2(1):S120–S126

Kaur P, Sharma N, Singh B, Kumar S, Kaur S (2012b) Modulation of genotoxicity of oxidative mutagens by glycyrrhizic acid from *Glycyrrhiza glabra* L. Pharmacogn Res 4(4):189–195

Kawakami Z, Ikarashi Y, Kase Y (2010) Glycyrrhizin and its metabolite 18β-glycyrrhetinic acid in *Glycyrrhiza*, a constituent herb of yokukansan, ameliorate thiamine deficiency-induced dysfunction of glutamate transport in cultured rat cortical astrocytes. Eur J Pharmacol 626(2–3):154–158

Kent UM, Aviram M, Rosenblat M, Hollenberg PF (2002) The licorice root-derived isoflavan glabridin inhibits the activities of human cytochrome P450s 3A4, 2B6, and 2C9. Drug Metab Dispos 30(6):709–715

Khalaf I, Vlase L, Lazar D, Corciovă A, Ivănescu B, Lazăr D (2010) HPLC-MS study of phytoestrogens from *Glycyrrhiza glabra*. Farmacia 58(1):89–94

Khalaf I, Corciovă A, Vlase L, Ivănescu B, Lazăr D (2011) LC/MS anlayis of sterolic compounds from *Glycyrrhiza glabra*. Studia Univ Babes-Bolyai Chem 56(3):97–102

Khalaf I, Vlase L, Lazar D, Ivănescu B, Lazăr D, Corciovă A (2012) HPLC analysis of polyphenolic compounds, phytoestrogens and sterols from *Glycyrrhiza glabra* L. tincture. Studia Univ Babes-Bolyai Chem 57(2):113–118

Khanom F, Kayahara H, Tadasa K (2000) Tyrosinase inhibitory activity of Bangladeshi indigenous medicinal plants. Biosci Biotechnol Biochem 64(9):1967–1969

Khattak KF, Simpson TJ (2010) Effect of gamma irradiation on the antimicrobial and free radical scavenging activities of *Glycyrrhiza glabra* root. Radiat Phys Chem 79(4):507–512

Khor TO, Yu S, Barve A, Hao X, Hong JL, Lin W, Foster B, Huang MT, Newmark HL, Kong AN (2009) Dietary feeding of dibenzoylmethane inhibits prostate cancer in transgenic adenocarcinoma of the mouse prostate model. Cancer Res 69(17):7096–7102

Kim DH, Jang IS, Lee HK, Jung EA, Lee KY (1996) Metabolism of glycyrrhizin and baicalin by human intestinal bacteria. Arch Pharm Res 19(4):292–296

Kim DH, Lee SW, Han MJ (1999) Biotransformation of glycyrrhizin to 18beta-glycyrrhetinic acid-3-O-beta-D-glucuronide by *Streptococcus* LJ-22, a human intestinal bacterium. Biol Pharm Bull 22(3):320–322

Kim SC, Byun SH, Yang CH, Kim CY, Kim JW, Kim SG (2004) Cytoprotective effects of Glycyrrhizae radix extract and its active component liquiritigenin against cadmium-induced toxicity (effects on bad translocation and cytochrome c-mediated PARP cleavage). Toxicology 197(3):239–251

Kim YW, Ki SH, Lee JR, Lee SJ, Kim CW, Kim SC, Kim SG (2006) Liquiritigenin, an aglycone of liquiritin in Glycyrrhizae radix, prevents acute liver injuries in rats induced by acetaminophen with or without buthionine sulfoximine. Chem Biol Interact 61(2):125–138

Kim JY, Park SJ, Yun KJ, Cho YW, Park HJ, Lee KT (2008a) Isoliquiritigenin isolated from the roots of *Glycyrrhiza uralensis* inhibits LPS-induced iNOS and COX-2 expression via the attenuation of NF-kappaB in RAW 264.7 macrophages. Eur J Pharmacol 584(1):175–184

Kim YW, Zhao RJ, Park SJ, Lee JR, Cho IJ, Yang CH, Kim SJ, Kim SC (2008b) Antiinflammatory effects of liquiritigenin as a consequence of the inhibition of NF-kappaB dependent iNOS and pro inflammatory cytokines production. Br J Pharmacol 154(1):165–173

Kim YW, Kang HE, Lee MG, Hwang SJ, Kim SC, Lee CH, Kim SG (2009) Liquiritigenin, a flavonoid aglycone from licorice, has a choleretic effect and the ability to induce hepatic transporters and phase-II enzymes. Am J Physiol Gastrointest Liver Physiol 296(2):G372–G381

Kim YM, Kim TH, Kim YW, Yang YM, Ryu DH, Hwang SJ, Lee JR, Kim SC, Kim SG (2010) Inhibition of liver X receptor-α-dependent hepatic steatosis by isoliquiritigenin, a licorice antioxidant flavonoid, as mediated by JNK1 inhibition. Free Rad Biol Med 49(11):1722–1734

Kim SW, Jin YC, Shin JH, Kim ID, Lee HK, Park SH, Han PYL, Lee JK (2012) Glycyrrhizic acid affords robust neuroprotection in the postischemic brain via anti-inflammatory effect by inhibiting HMGB1 phosphorylation and secretion. Neurobiol Dis 46(1):147–156

Kim HS, Suh KS, Ko A, Sul D, Choi D, Lee SK, Jung WW (2013) The flavonoid glabridin attenuates 2-deoxy-D-ribose-induced oxidative damage and cellular dysfunction in MC3T3-E1 osteoblastic cells. Int J Mol Med 31(1):243–251

Kim JS, Park MR, Lee SY, Kim DK, Moon SM, Kim CS, Cho SS, Yoon G, Im HJ, You JS, Oh JS, Kim SG (2014a) Licochalcone A induces apoptosis in KB human oral cancer cells via a caspase-dependent FasL signaling pathway. Oncol Rep 31(2):755–762

Kim SJ, Kwon SS, Jeon SH, Yu ER, Park SN (2014b) Enhanced skin delivery of liquiritigenin and liquiritin-loaded liposome-in-hydrogel complex system. Int J Cosmet Sci 36(6):553–560

Kim SH, Yang M, Xu JG, Yu X, Qian XJ (2015) Role of licochalcone A on thymic stromal lymphopoietin expression: implications for asthma. Exp Biol Med (Maywood) 240(1):26–33

Kinoshita T, Saitoh T, Shibata S (1978) A new 3 arylcoumarin from licorice root. Chem Pharm Bull 26(1):135–140

Kinoshita T, Kajiyama K, Hiraga Y, Takahashi K, Tamura Y, Mizutani K (1996a) Isoflavan derivatives from *Glycyrrhiza glabra* (licorice). Heterocyles 43:581–588

Kinoshita T, Kajiyama K, Hiraga Y, Takahashi K, Tamura Y, Mizutani K (1996b) The isolation of new pyrano-2-arylbenzofuran derivatives from the root of *Glycyrrhiza glabra*. Chem Pharm Bull 44(6):1218–1221

Kinoshita T, Tamura Y, Mizutani K (1997) Isolation and synthesis of two new 3-arylcoumarin derivatives from the root of *Glycyrrhiza glabra* (licorice), and structure revision of an antioxidant isoflavonoid glabrene. Nat Prod Res 9(4):289–296

Kinoshita T, Tamura Y, Mizutani K (2005) The isolation and structure elucidation of minor isoflavonoids from licorice of *Glycyrrhiza glabra* origin. Chem Pharm Bull 53(7):847–849

Kir'yalov NP, Murav'ev IA, Stepanova EF, Bogatkina VF (1970) Triterpene compounds of the herbage of *Glycyrrhiza glabra*. Chem Nat Comp 6(6):787–787

Kiss T, Gyémánt G, Dános B, Nánási P (1998) Polysaccharides of *Glycyrrhiza glabra* L. and *Glycyrrhiza echinata* L. Medicinal plants II. Acta Pharm Hung 68(5):263–268 (In Hungarian)

Kitagawa K, Nishino H, Iwashima A (1984) Inhibition of 12-O-tetradecanoylphorbol-13-acetate-stimulated 3-O-methyl-glucose transport in mouse Swiss 3T3 fibroblasts by glycyrrhetic acid. Cancer Lett 24(2):157–163

Kitagawa K, Nishino H, Iwashima A (1986) Inhibition of the specific binding of 12-O-tetradecanoylphorbol-13-acetate to mouse epidermal membrane fractions by glycyrrhetic acid. Oncology 43(2):127–130

Kitagawa I, Chen WZ, Hori K, Harada E, Yasuda N, Yoshikawa M, Ren J (1994) Chemical studies of Chinese licorice-roots. I. Elucidation of five new flavonoid constituents from the roots of *Glycyrrhiza glabra* L. collected in Xinjiang. Chem Pharm Bull 42(5):1056–1062

Knipping K, Garssen J, van't Land B (2012) An evaluation of the inhibitory effects against rotavirus infection of edible plant extracts. Virol J 9(1):137–144

Kobayashi S, Miyamoto T, Kimura I, Kimura M (1995) Inhibitory effect of isoliquiritin, a compound in licorice root, on angiogenesis in vivo and tube formation in vitro. Biol Pharm Bull 18(10):1382–1386

Kolbe L, Immeyer J, Batzer J, Wensorra U, tom Dieck K, Mundt C, Wolber R, Stäb F, Schönrock U, Ceilley RI, Wenck H (2006) Anti-inflammatory efficacy of Licochalcone A: correlation of clinical potency and in vitro effects. Arch Dermatol Res 298(1):23–30

Kondo M, Minamino H, Okuyama G, Honda K, Nagasawa H, Otani Y (1986) Physicochemical properties and applications of α and β-glycyrrhizins, natural surface active agents in licorice root extract. J Soc Cosmet Chem 37:177–189

Kondo K, Shiba M, Nakamura R, Morota T, Shoyama Y (2007) Constituent properties of licorices derived from *Glycyrrhiza uralensis*, *G. glabra*, or *G. inflata* identified by genetic information. Boil Pharm Bull 30(7):1271–1277

Krähenbühl S, Hasler F, Krap R (1994) Analysis and pharmacokinetics of glycyrrhizic acid and glycyrrhetinic acid in humans and experimental animals. Steroids 59(2):121–126

Kratschmar DV, Vuorinen A, Da Cunha T, Wolber G, Classen-Houben D, Doblhoff O, Schuster D, Odermatt A (2011) Characterization of activity and binding mode of glycyrrhetinic acid derivatives inhibiting 11β-hydroxysteroid dehydrogenase type 2. J Steroid Biochem Mol Biol 125(1–2):129–142

Krausse R, Bielenberg J, Blaschek W, Ullmann U (2004) In vitro anti-*Helicobacter pylori* activity of extractum liquiritiae, glycyrrhizin and its metabolites. J Antimicrob Chemother 54(1):243–246

Kroes BH, Beukelman CJ, van den Berg AJ, Wolbink GJ, van Dijk H, Labadie RP (1997) Inhibition of human complement by beta-glycyrrhetinic acid. Immunology 90(1):115–120

Kumada H (2002) Long-term treatment of chronic hepatitis C with glycyrrhzin [Stronger Neo-Minophagen C (SNMC)] for preventing liver cirrhosis and hepatocellular carcinoma. Oncology 62(S1):94–100

Kuo S, Shankel DM, Telikepalli H, Mitscher LA (1992) *Glycyrrhiza glabra* extract as an effector of interception in *Escherichia coli* K12+. Mut Res Lett 282(2):93–98

Kuriyama Y, Takano T, Okada F, Nukada T (1975) Hypersensitivity to glycyrrhizin. A case report. Med J Osaka Univ 26:75–78

Kuroda M, Mimaki Y, Honda S, Tanaka H, Yokota S, Mae T (2010) Phenolics from *Glycyrrhiza glabra* roots and their PPAR-γ ligand-binding activity. Bioorg Med Chem 18(2):962–970

Kuroda N, Inoue K, Ikeda T, Hara Y, Waka K, Sato T (2014) Apoptotic response through a High Mobility Box 1 protein-dependent mechanism in LPS/GalN-induced mouse liver failure and glycyrrhizin-mediated inhibition. PLoS One 9(4), e92884

Kusano A, Nikaido T, Kuge T, Ohmoto T, Monache G, Botta B, Botta M, Saitoh T (1991) Inhibition of adenosine 3′,5′ -cyclic monophosphate phosphodiesterase by flavonoids from licorice roots and 4-arylcoumarins. Chem Pharm Bull 39(4):930–933

Kusano G, Shibano M, Watanabe H, Ozaki K (2003) Pharmaceutical botanical studies on some *Glycyrrhiza* species. Yakugaku Zasshi 123(8):619–631 (In Japanese)

Kwon HM, Choi YJ, Choi JS, Kang SW, Bae JY, Kang IJ, Jun JG, Lee SS, Lim SS, Kang YH (2007) Blockade of cytokine-induced endothelial cell adhesion molecule expression by licorice isoliquiritigenin through NF-kappaB signal disruption. Exp Biol Med (Maywood) 232(2):235–245

Kwon GT, Cho HJ, Chung WY, Park KK, Moon A, Park JH (2009) Isoliquiritigenin inhibits migration and invasion of prostate cancer cells: possible mediation by decreased JNK/AP-1 signaling. J Nutr Biochem 20(9):663–676

Laconi S, Madeddu MA, Pompei R (2014) Autophagy activation and antiviral activity by a licorice triterpene. Phytother Res 28(12):1890–1892

Larkworthy W, Holgate PF (1975) Deglycyrrhizinized liquorice in the treatment of chronic duodenal ulcer. A retrospective endoscopic survey of 32 patients. Practitioner 215(1290):787–792

Lateef M, Iqbal L, Fatima N, Siddiqui K, Afza N, Zia-ul-Haq M, Ahmad M (2012) Evaluation of antioxidant and urease inhibition activities of roots of *Glycyrrhiza glabra*. Pak J Pharm Sci 25(1):99–102

Latif S, Conca T, Morris D (1990) The effects of the licorice derivative, glycyrrhetinic acid, on hepatic 3 alpha- and 3 beta-hydroxysteroid dehydrogenases and 5 alpha- and 5 beta-reductase pathways of metabolism of aldosterone in male rats. Steroids 55:52–58

Launert E (1981) Edible and medicinal plants. Hamlyn, London

Lee CK, Kk P, Lim SS, Park JH, Chung WY (2007) Effects of the licorice extract against tumor growth and cisplatin-induced toxicity in a mouse xenograft model of colon cancer. Bio Pharm Bull 30(11):2191–2195

Leskinen MH, Hautaniemi EJ, Tahvanainen AM, Koskela JK, Päällysaho M, Tikkakoski AJ, Kähönen M, Kööbi T, Niemelä O, Mustonen J, Pörsti IH (2014) Daily liquorice consumption for two weeks increases augmentation index and central systolic and diastolic blood pressure. PLoS One 9(8), e105607

Li YQ (2012) Antiangiogenic activity of *Glycyrrhiza* extract. Phys Procedia 33:144–149

Li W, Asada Y, Yoshikawa T (1998) Antimicrobial flavonoids from *Glycyrrhiza glabra* hairy root cultures. Planta Med 64(8):746–747

Li W, Asada Y, Yoshikawa T (2000) Flavonoid constituents from *Glycyrrhiza glabra* hairy root cultures. Phytochemistry 55(5):447–456

Li JR, Wang YQ, Deng ZZ (2005) Two new compounds from *Glycyrrhiza glabra*. J Asian Nat Prod Res 7(4):677–680

Li DF, Wang ZH, Chen HM, Wang JY, Zheng QS, Shang J, Li J (2009) Isoliquiritigenin induces monocytic differentiation of HL-60 cells. Free Rad Biol Med 46(6):731–736

Li HY, Xu W, Su J, Zhang X, Hu LW, Zhang WD (2010) In vitro and in vivo Inhibitory effects of glycyrrhetinic acid on cytochrome P450 3A activity. Pharmacology 86(5–6):287–292

Li Y, Zhao HX, Wang YZ, Zheng H, Yu W, Chai HY, Zhang J, Falck JR, Guo AM, Yue J, Peng RX, Yang J (2013) Isoliquiritigenin induces growth inhibition and apoptosis through downregulating arachidonic acid metabolic network and the deactivation of PI3K/Akt in human breast cancer. Toxicol Appl Pharmacol 272(1):37–48

Li JY, Cao HY, Liu P, Cheng GH, Sun MY (2014) Glycyrrhizic acid in the treatment of liver diseases: literature review. Biomed Res Int 2014:872139

Liao WC, Lin YH, Chang TM, Huang WY (2012) Identification of two licorice species, *Glycyrrhiza uralensis* and *Glycyrrhiza glabra*, based on separation and identification of their bioactive components. Food Chem 132(4):2188–2193

Liao YF, Tzeng YM, Hung HC, Liu GY (2015) Dibenzoylmethane, hydroxydibenzoylmethane and hydroxymethyldibenzoylmethane inhibit phorbol-12-myristate 13-acetate-induced breast carcinoma cell invasion. Mol Med Rep 11(6):4597-4604.

Lin JC (2003) Mechanism of action of glycyrrhizic acid in inhibition of Epstein-Barr virus replication in vitro. Antiviral Res 59(1):41–47

Lin SH, Yang SS, Chau T, Halperin ML (2003) An unusual cause of hypokalemic paralysis: chronic licorice ingestion. Am J Med Sci 325(3):153–156

Lin SP, Tsai SY, Hou YC, Chao PD (2009) Glycyrrhizin and licorice significantly affect the pharmacokinetics of methotrexate in rats. J Agric Food Chem 57(5):1854–1859

Litvinenko VI (1964) Chalcone glycosides from *Glycyrrhiza glabra* L. Dokl Akad Nauk SSSR155:600-2 (In Russian)

Litvinenko VI (1966) Glyphoside, a new flavonoid glycoside obtained from *Glycyrrhiza glabra*. Rast Resur 2:531–536 (In Russian)

Litvinenko VI, Kovalev IP (1967) Glycoflavonoids of *Glycyrrhiza glabra* I. Saponaretin and vitexin. Chem Nat Comp 3(1):47–48

Litvinenko VI, Nadezhina TP (1972) Flavonoids from the aboveground part of *Glycyrrhiza glabra*. Rastit Resur 8(1):35–42 (In Russian)

Litvinenko VI, Obolentseva GV (1964) Chemical and pharmacological investigation of flavonoids from licorice *Glycyrrhiza glabra* L. and *Glycyrrhiza uralensis* Fisch. Med Prom SSSR 18:20–23 (In Russian)

Litvinenko VI, Maksyutina NP, Kolesnikov IG (1963a) Flavonoid compounds of *Glycyrrhiza glabra*. Zhurn Obshchei Khimii (USSR) 33(1):296–299

Litvinenko VI, Maksyutina NP, Kolesnikov IG (1963b) Flavonoidal compounds of *Glycyrrhiza glabra*.

I. Flavonoid L-l. Zhurn Obshchei Khimii (USSR) 33(12):4014–4018

Liu HM, Sugimoto N, Akiyama T, Maitani T (2000) Constituents and their sweetness of food additive enzymatically modified licorice extract. J Agric Food Chem 48(12):6044–6047

Liu RT, Zou LB, Fu JY, Lu QJ (2010a) Effects of liquiritigenin treatment on the learning and memory deficits induced by amyloid beta-peptide (25–35) in rats. Behav Brain Res 210(1):24–31

Liu RT, Zou LB, Fu JY, Lu QJ (2010b) Promotion of rat brain-derived progenitor cell neurogenesis by liquiritigenin treatment: underlying mechanisms. Neurosci Lett 481(3):139–143

Liu RT, Tang JT, Zou LB, Fu JY, Lu QJ (2011) Liquiritigenin attenuates the learning and memory deficits in an amyloid protein precursor transgenic mouse model and the underlying mechanisms. Eur J Pharmacol 669(1-3):76–83

Liu Y, Sirou X, Wang Y, Luo K, Wang Y, Cai Y (2012) Liquiritigenin inhibits tumor growth and vascularization in a mouse model of HeLa cells. Molecules 17(6):7206–7216

Liut HM, Akiyama T, Sugimoto N, Maitani T (2001) Isolation and identification of main constituents in an enzymatically hydrolysed licorice extract sweetener. Food Addit Contam 18(4):281–284

Lozano P, Flores D, Martínez S, Artigues I, Rimbau EM, Gómez F (2000) Upper limb ischemia induced by chronic licorice ingestion. J Cardiovasc Surg (Torino) 41(4):631–632

Luchon L, Meyrier A, Paillard F (1993) Hypokalemia without arterial hypertension by licorice poisoning. Nephrologie 14(4):177–181 (In French)

Luo L, Jin YC, Kim ID, Lee JK (2013) Glycyrrhizin attenuates kainic acid-induced neuronal cell death in the mouse hippocampus. Exp Neurobiol 22(2):107–115

Luo L, Jin YC, Kim ID, Lee JK (2014) Glycyrrhizin suppresses HMGB1 inductions in the hippocampus and subsequent accumulation in serum of a kainic acid-induced seizure mouse model. Cell Mol Neurobiol 34(7):987–997

M & F Worldwide Corp (2010) M & F Worldwide Corp. Annual report on Form 10-K for the Year Ended December 31, 2010

Ma CH, Ma ZQ, Liao XL, Liu JP, Fu Q, Ma SP (2013) Immunoregulatory effects of glycyrrhizic acid exerts anti-asthmatic effects via modulation of Th1/Th2 cytokines and enhancement of CD4+CD25+Foxp3+ regulatory T cells in ovalbumin-sensitized mice. J Ethnopharmacol 148(3):755–762

Maatooq GT, Marzouk AM, Gray AI, Rosazza JP (2010) Bioactive microbial metabolites from glycyrrhetinic acid. Phytochemistry 71(2–3):262–270

Makino T, Okajima K, Uebayashi R, Ohtake N, Inoue K, Mizukami H (2012) 3-Monoglucuronyl-glycyrrhretinic acid is a substrate of organic anion transporters expressed in tubular epithelial cells and plays important roles in licorice-induced pseudoaldosteronism by inhibiting 11β-hydroxysteroid dehydrogenase 2. J Pharmacol Exp Ther 342(2):297–304

Malek Jafarian M, Ghazvini K (2007) In vitro susceptibility of *Helicobacter pylori* to licorice extract. Iran J Pharm Res 6:69–72

Malekinejad H, Farshid AA, Mirzakhani N (2011a) Liquorice plant extract reduces ochratoxin A-induced nephrotoxicity in rats. Exp Toxicol Pathol 63(1–2):125–130

Malekinejad H, Mirzakhani N, Razi M, Cheraghi H, Alizadeh A (2011b) Protective effects of melatonin and *Glycyrrhiza glabra* extract on ochratoxin A—induced damages on testes in mature rats. Hum Exp Toxicol 30(2):110–123

Man SL, Wang J, Gao WY, Guo SB, Li YY, Zhang LM, Xiao PG (2013) Chemical analysis and anti-inflammatory comparison of the cell culture of *Glycyrrhiza* with its field cultivated variety. Food Chem 136(2):513–517

Mancia MD, Reid ME, DuBose ES, Campbell JA, Jackson KM (2014) Qualitative identification of dibenzoylmethane in licorice root (*Glycyrrhiza glabra*) using gas chromatography-triple quadrupole mass spectrometry. Nat Prod Commun 9(1):91–94

Manns MP, Wedemeyer H, Singer A, Khomutjanskaja N, Dienes HP, Roskams T, Goldin R, Hehnke Y, Inoue H, European SNMC Study Group (2012) Glycyrrhizin in patients who failed previous interferon alpha-based therapies: biochemical and histological effects after 52 weeks. J Viral Hepat 19(8):537–546

Martin MD, Sherman J, van der Ven P, Burgess J (2008) A controlled trial of a dissolving oral patch concerning glycyrrhiza (licorice) herbal extract for the treatment of aphthous ulcers. Gen Dent 56(2):206–210

Matsumoto Y, Matsuura T, Aoyagi H, Matsuda M, Hmwe SS, Date T, Watanabe N, Watashi K, Suzuki R, Ichoinpose S, Wake K, Suzuki T, Miyamura T, Wakita T, Aizaki H (2013) Antiviral activity of glycyrrhizin against hepatitis C virus in vitro. PLoS One 8(7), e68992

Mattarello MJ, Benedini S, Fiore C, Camozzi V, Sartorato P, Luisetto G, Armanini D (2006) Effect of licorice on PTH levels in healthy women. Steroids 71(5):403–408

Maurya SK, Raj K, Srivastava AK (2009) Antidyslipidaemic activity of *Glycyrrhiza glabra* in high fructose diet induced dsyslipidaemic Syrian golden hamsters. Indian J Clin Biochem 24(4):404–409

Mazumder PM, Pattnayak S, Parvani H, Sasmal D, Rathinavelusamy P (2012) Evaluation of immunomodulatory activity of *Glycyrrhiza glabra* L roots in combination with zinc. Asian Pac J Trop Biomed 2(1):S15–S20

Meena AK, Singh A, Sharma K, Kumari S, Rao MM (2010) Physicochemical and preliminary phytochemical studies on the rhizomes of *Glycyrrhiza glabra* Linn. Int J Pharmacy Pharm Sci 2(2):48–50

Mendes-Silva W, Assafim M, Ruta B, Monteiro RQ, Guimaraes JA, Zingali RB (2003) Antithrombotic

effect of glycyrrhizin, a plant-derived thrombin inhibitor. Thromb Res 112(1–2):93–98

Menegazzi M, Di Paola R, Mazzon E, Genovese T, Crisafulli C, Dal Bosco M, Zou ZZ, Suzuki H, Cuzzocrea S (2008) Glycyrrhizin attenuates the development of carrageenan-induced lung injury in mice. Pharmacol Res 58(1):22–31

Meng FR, Liu R, Gao M, Wang YH, Yu XY, Xuan ZH, Sun JL, Yang F, Wu CF, Du GH (2011) Pinocembrin attenuates blood-brain barrier injury induced by global cerebral ischemia-reperfusion in rats. Brain Res 1391:93–101

Messier C, Grenier D (2011) Effect of licorice compounds licochalcone A, glabridin and glycyrrhizic acid on growth and virulence properties of *Candida albicans*. Mycoses 54(6):e801–e806

Messier C, Epifano F, Genovese S, Grenier D (2012) Licorice and its potential beneficial effects in common oro-dental diseases. Oral Dis 18(1):32–39

Methlie P, Husebye EES, Hustad S, Lien EA, Løvås K (2011) Grapefruit juice and licorice increase cortisol availability in patients with Addison's disease. Eur J Endocrinol 165(5):761–769

Michel HE, Tadros MG, Abdel-Naim AB, Khalifa AE (2013) Prepulse inhibition (PPI) disrupting effects of *Glycyrrhiza glabra* extract in mice: a possible role of monoamines. Neurosci Lett 544:110–114

Miething H, Speicher-Brinker A (1989) Neolicuroside a new chalcone glycoside from the roots of *Glycyrrhiza glabra*. Archiv Pharm (Weinheim, Ger) 322(3):141–143

Minasian SM, Adamian TI, Gevorkian ES (2007) Changes of morphological indicators of blood from vibration and liquorice effect. Ross Fiziol Zh Im I M Sechenova 93(9):1035–1042 (In Russian)

Ming LG, Jin F, Huang P, Luo HL, Liu WJ, Zhang LL, Yuan W, Zhang YJ, Jin Y (2014) Licochalcone A upregulates of FasL in mesenchymal stem cells to strengthen bone formation and increase bone mass. Sci Rep 4:7209

Mitscher LA, Park YH, Clark D, Beal JL (1980) Antimicrobial agents from higher plants. Antimicrobial isoflavanoids and related substances from *Glycyrrhiza glabra* L. var. *typica*. J Nat Prod 43(2):259–269

Mohammadi SZ, Hamidian H, Moeindini Z (2014) High surface area-activated carbon from *Glycyrrhiza glabra* residue by ZnCl2 activation for removal of Pb(II) and Ni(II) from water samples. J Ind Eng Chem 20(6):4112–4118

Mohammadi-Moghadam F, Amin MM, Khiadani M, Momenbeik F, Nourmoradi H, Hatamipour MS (2013) Application of *Glycyrrhiza glabra* root as a novel adsorbent in the removal of toluene vapors: equilibrium, kinetic, and thermodynamic study. J Environ Public Health 2013:1–7

Momeni A, Rahimian G, Kiasi A, Amiri M, Kheiri S (2014) Effect of licorice versus bismuth on eradication of *Helicobacter pylori* in patients with peptic ulcer disease. Pharmacogn Res 6(4):341–344

Montoro P, Maldini M, Russo M, Postorino S, Piacente S, Pizza CJ (2011) Metabolic profiling of roots of liquorice (*Glycyrrhiza glabra*) from different geographical areas by ESI/MS/MS/ and determination of major metabolites by LC-ESI/MS and LC-ESI/MS/MS. J Pharm Biomed Anal 54(3):535–544

Moon A, Kim SH (1997) Effect of *Glycyrrhiza glabra* roots and glycyrrhizin on the glucuronidation in rats. Planta Med 63(2):115–119

Morteza-Semnani K, Saeedi M, Shahnavaz B (2002) Comparison of antioxidant activity of extract from roots of licorice (*Glycyrrhiza glabra* L.) to commercial antioxidants in 2 % hydroquinone cream. J Cosmet Sci 54(6):551–558

Motsei ML, Lindsey KL, van Staden J, Jäger AK (2003) Screening of traditionally used South African plants for antifungal activity against *Candida albicans*. J Ethnopharmacol 86(2–3):235–241

Mousa NA, Siaguru P, Wiryowidagdo S, Wagih ME (2007) Establishment of regenerative callus and cell suspension system of licorice (*Glycyrrhiza glabra*) for the production of the sweetener glycyrrhizin in vitro. Sugar Technol 9(1):72–82

Mukherjee M, Bhaskaran N, Srinath R, Shivaprasad HN, Allan JJ, Shekhar D, Agarwal A (2010) Anti-ulcer and antioxidant activity of GutGard. Indian J Exp Biol 48(3):269–274

Mukhopadhyay M, Panja P (2008) A novel process for extraction of natural sweetener from licorice (*Glycyrrhiza glabra*) roots. Sep Purif Technol 63(3):539–545

Muller B, Aparin PG, Stoclet JC, Kleschyov AL (2014) Glycyrrhetinic acid reverses the lipopolysaccharide-induced hypocontractility to noradrenaline in rat aorta: implications to septic shock. J Pharmacol Sci 125(4):422–425

Mumoli N, Cei M (2008) Licorice-induced hypokalemia. Int J Cardiol 124(3):e42–e44

Muralidharan P, Balamurugan G, Babu V (2009) Cerebroprotective effect of *Glycyrrhiza glabra* Linn. root extract. Bangladesh J Pharmacol 4(1):60–64

Nagumo S, Fukuju A, Takayama M, Nagai M, Yanoshita R, Samejima Y (1999) Inhibition of lysoPAF acetyltransferase activity by components of licorice root. Biol Pharm Bull 22(10):1144–1146

Nakagawa K, Kishida H, Arai N, Nishiyama T, Mae T (2004) Licorice flavonoids suppress abdominal fat accumulation and increase in blood glucose level in obese diabetic KK-A(y) mice. Biol Pharm Bull 27(11):1775–1778

Nakagawa K, Hidaka T, Kitano M, Asakura M, Kamigaito T, Noguchi T, Hosoe K (2008a) Genotoxicity studies on licorice flavonoid oil (LFO). Food Chem Toxicol 46(7):2525–2532

Nakagawa K, Kitano M, Kishida H, Hidaka T, Nabae K, Kawabe M, Hosoe K (2008b) 90-Day repeated-dose toxicity study of licorice flavonoid oil (LFO) in rats. Food Chem Toxicol 46(7):2349–2357

Nakagiri R, Oda H, Kamiya T (2003) Small scale rat hepatocyte primary culture with applications for screening hepatoprotective substances. Biosci Biotechnol Biochem 67(8):1629–1635

Nam C, Kim S, Sim Y, Chang I (2003) Anti-acne effects of Oriental herb extracts: a novel screening method to select anti-acne agents. Skin Pharmacol Appl Skin Physiol 16(2):84–90

Nasrollahi V, Mirzaie-asl A, Piri K, Nazeri S, Mehrabi R (2014) The effect of drought stress on the expression of key genes involved in the biosynthesis of triterpenoid saponins in liquorice (*Glycyrrhiza glabra*). Phytochemistry 103:32–37

Nielsen ML, Pareek M, Andersen I (2012) Liquorice-induced hypertension and hypokalaemia. Ugeskr Laeger 174(15):1024–1025 (In Danish)

Nieman C (1959) Licorice. Adv Food Res 7:339–381

Nishino H, Kitagawa K, Iwashima A (1984) Antitumor-promoting activity of glycyrrhetic acid in mouse skin tumor formation induced by 7,12-dimethylbenz[a] anthracene plus teleocidin. Carcinogenesis 5(11):1529–1530

Nishino H, Yoshioka K, Iwashima A, Takizawa H, Konishi S, Okamoto H, Okabe H, Shibata S, Fujiki H, Sugimura T (1986) Glycyrrhetic acid inhibits tumor-promoting activity of teleocidin and 12-O-tetradecanoylphorbol-13-acetate in two-stage mouse skin carcinogenesis. Jpn J Cancer Res 77(1):33–38

Nitalikar MM, Munde KC, Dhore BV, Shikalgar SN (2010) Studies of antibacterial activities of *Glycyrrhiza glabra* root extract. Int J Pharm Tech Res 2(1):899–901

Nokhodchi A, Nazemiyeh H, Ghafourian T, Hassan-Zadeh D, Valizadeh H, Bahary LA (2002) The effect of glycyrrhizin on the release rate and skin penetration ofdiclofenac sodium from topical formulations. Farmaco 57(11):883–888

Nomura T, Fukai T (1998) Phenolic constituents of licorice (*Glycyrrhiza* species). In: Herz W, Kirby GW, Moore RE, Steglich W, Tamm C (eds) Fortschritte der Chemie organischer Naturstoffe/Progress in the chemistry of organic natural products. Springer, Vienna, pp 1–140

Nomura T, Fukai T, Akiyama T (2002) Chemistry of phenolic compounds of licorice (*Glycyrrhiza* species) and their estrogenic and cytotoxic activities. Pure Appl Chem 74(7):1199–1206

Nose M, Terawaki K, Oguri K, Ogihara Y, Yoshimatsu K, Shimomura K (1998) Activation of macrophages by crude polysaccharide fractions obtained from shoots of *Glycyrrhiza glabra* and hairy roots of *Glycyrrhiza uralensis* in vitro. Biol Pharm Bull 21(10):1110–1112

Nugmanova ML, Kalitina NF (1979) Case of contact dermatitis caused by licorice. Vestn Dermatol Venerol 11:62–64 (In Russian)

Numazaki K, Nagata N, Sato T, Chiba S (1994) Effect of glycyrrhizin, cyclosporin A, and tumor necrosis factor alpha on infection of U-937 and MRC-5 cells by human cytomegalovirus. J Leukoc Biol 55(1):24–28

O'Brian CA, Ward NE, Vogel VG (1990) Inhibition of protein kinase C by the 12-O-tetradecanoylphorbol-13-acetate antagonist glycyrrhetic acid. Cancer Lett 49(1):9–12

O'Connell K, Kinsella J, McMahon C, Holian J, O'Riordan S (2014) Posterior reversible encephalopathy syndrome (PRES) associated with liquorice consumption. Irish J Med Sci DOI: 10.1007/s11845-014-1214-0 (in press)

Oda M, Tamararaggi et Y, Mizutani K, Tanaka O (1995) Biological activity and constituents of cell cultures of licorice. In: 42nd Annual meeting of Japanese Society of Pharmacognosy, Abstract Papers, Fukuyama, September 1995, p 168

Oganesyan KR (2002) Antioxidant effect of licorice root on blood catalase activity in vibration stress. Bull Exp Biol Med 134(2):135–136

Ohtake N, Kido A, Kubota K, Tsuchiya N, Morita T, Kase Y, Takeda S (2007) A possible involvement of 3-monoglucuronyl-glycyrrhetinic acid, a metabolite of glycyrrhizin (GL), in GL-induced pseudoaldosteronism. Life Sci 80(17):1545–1552

Ojha S, Golechha M, Kumari S, Bhatia J, Arya DS (2013) *Glycyrrhiza glabra* protects from myocardial ischemia–reperfusion injury by improving hemodynamic, biochemical, histopathological and ventricular function. Exp Toxicol Pathol 65(1-2):219–227

Okada K, Kobayashi S, Tsunematsu T (1987) A case of Sjögren's syndrome associated with hypokalemic myopathy due to glycyrrhizin. Nippon Naika Gakkai Zasshi 76(5):744–745 (In Japanese)

Okada K, Tamura Y, Yamamoto M, Inoue Y, Takagaki R, Takahashi K, Demizu S, Kajiyama K, Hiraga Y, Kinoshita T (1989) Identification of antimicrobial and antioxidant constituents from licorice of Russian and Xinjiang origin. Chem Pharm Bull 37(9):2528–2530

Okuma Y, Liu K, Wake H, Liu R, Nishimura Y, Hui Z, Teshigawara K, Haruma J, Yamamoto Y, Yamamoto H, Date I, Takahashi HK, Mori S, Nishibori M (2014) Glycyrrhizin inhibits traumatic brain injury by reducing HMGB1-RAGE interaction. Neuropharmacol 85:18–26

Olukoga A, Donaldson D (2000) Liquorice and its health implications. J R Soche Promot Health 120(2):83–89

Omar HR, Komarova I, El-Ghonemi M (2012) Licorice abuse: time to send a warning message. Ther Adv Endocrinol Metab 3(4):125–138

Omer MO, AlMalki WH, Shahid I, Khuram S, Altaf I, Imran S (2014) Comparative study to evaluate the anti-viral efficacy of *Glycyrrhiza glabra* extract and ribavirin against the Newcastle disease virus. Pharmacogn Res 6(1):6–11

Orazizadeh M, Fakhredini F, Mansouri E, Khorsandi L (2014) Effect of glycyrrhizic acid on titanium dioxide nanoparticles-induced hepatotoxicity in rats. Chem Biol Interact 220:214–221

Orlent H, Hansen BE, Willems M, Brouwer JT, Huber R, Kullak-Ublick GA, Gerken G, Zeuzem S, Nevens F, Tielemans WCM, Zondervan PE, Lagging M, Westin J, Schalm SW (2006) Biochemical and histological effects of 26 weeks of glycyrrhizin treatment in chronic hepatitis C: a randomized phase II trial. J Hepatol 45(4):539–546

Orujei Y, Shabani L, Sharifi-Tehrani M (2013) Induction of glycyrrhizin and total phenolic compound production in licorice by using arbuscular mycorrhizal fungi. Russ J Plant Physiol 60(6):855–860

Ovez B, Ozgen S, Yuksel M (2006) Biological denitrification in drinking water using *Glycyrrhiza glabra* and *Arunda donax* as the carbon source. Process Biochem 41(7):1539–1544

Ozaki Y, Noguchi M, Kamakura H, Harada M (1990) Studies on concentration of glycyrrhizin in plasma and its absorption after oral administration of licorice extract and glycyrrhizin. Yakugaku Zasshi 110(1):77–81 (In Japanese)

Panossian AG, Shirinian EA, Barikian ML, Avakian OM (1988) 9,12,13-Trihydroxy-(10E)-octadecenoic and 9,12,13-trihydroxy-10,11-epoxy-octadecanoic acids – new anti-stress compounds from licorice. Isvestiya Academii Nauk USSR 6:932–935

Paris R, Guillot M (1955) Liquiritoside, flavonoside from root of licorice, *Glycyrrhiza glabra* L. Ann Pharm Fr 13(9–10):592–595 (In French)

Park HY, Park SH, Yoon HK, Han MJ, Kim DH (2004) Anti-allergic activity of 18beta-glycyrrhetinic acid-3-O-beta-D-glucuronide. Arch Pharm Res 27(1):57–60

Park JM, Park SH, Hong KS, Han YM, Jang SH, Kim EH, Hahm KB (2014a) Special licorice extracts containing lowered glycyrrhizin and enhanced licochalcone A prevented *Helicobacter pylori*-initiated, salt diet-promoted gastric tumorigenesis. Helicobacter 19(3):221–236

Park SY, Kim EJ, Choi HJ, Seon MR, Lim SS, Kang YH, Choi MS, Lee KW, Yoon Park JH (2014b) Anti-carcinogenic effects of non-polar components containing licochalcone A in roasted licorice root. Nutr Res Pract 8(3):257–266

Parle M, Dhingra D, Kulkarni SK (2004) Memory-strengthening activity of *Glycyrrhiza glabra* in exteroceptive and interoceptive behavioral models. J Med Food 7(4):462–466

Pellati D, Fiore C, Armanini D, Rassu M, Bertoloni G (2009) In vitro effects of glycyrrhetinic acid on the growth of clinical isolates of *Candida albicans*. Phytother Res 23(4):572–574

Peters MC, Tallman JA, Braun TM, Jacobson JJ (2010) Clinical reduction of *S. mutans* in pre-school children using a novel liquorice root extract lollipop: a pilot study. Eur Arch Paediatr Dent 11(6):274–278

Ploeger BA, Meulenbelt J, DeJongh J (2000) Physiologically based pharmacokinetic modeling of glycyrrhizic acid, a compound subject to presystemic metabolism and enterohepatic cycling. Toxicol Appl Pharmacol 162(3):177–188

Ploeger BA, Mensinga T, Sips A, Seinen W, Meulenbelt J, DeJongh J (2001) The pharmacokinetics of glycyrrhizic acid evaluated by physiologically based pharmacokinetic modeling. Drug Metab Rev 33(2):125–147

Pompei R, Flore O, Marcialis M, Pani A, Loddo B (1979) Glycyrrhizic acid inhibits virus growth and inactivates virus particles. Nature 281:689–690

Puri B, Seshadri TR (1954) Survey of anthoxanthins. V. Coloring matter of licorice roots. J Sci Ind Res 13B:475–480

Qamar W, Khan R, Khan AQ, Rehman MU, Lateef A, Tahir M, Ali F, Sultana S (2012) Alleviation of lung injury by glycyrrhizic acid in benzo(a)pyrene exposed rats: probable role of soluble epoxide hydrolase and thioredoxin reductase. Toxicology 291(1–3):25–31

Račková L, Jančinová V, Petríková M, Drábiková K, Nosáľ R, Štefek M, Košťálová D, Prónayová N, Kováčová M (2007) Mechanism of anti-inflammatory action of liquorice extract and glycyrrhizin. Nat Prod Res 21(14):1234–1241

Raggi MA, Bugamelli F, Nobile L, Schiavone P, Cantelli-Forti G (1994a) HPLC determination of glycyrrhizin and glycyrrhetic acid in biological fluids, after licorice extract administration to humans and rats. Boll Chim Farm 133(11):704–708

Raggi MA, Maffei F, Bugamelli F, Cantelli Forti G (1994b) Bioavailability of glycyrrhizin and licorice extract in rat and human plasma as detected by a HPLC method. Pharmazie 49(4):269–272

Raggi MA, Bugamelli F, Nobile L, Curcelli V, Mandrioli R, Rossetti A, CantelliForti G (1995) The choleretic effects of licorice: identification and determination of thepharmacologically active components of *Glycyrrhiza glabra*. Boll Chim Farm 134(11):634–638 (In Italian)

Rahnama M, Mehrabani D, Japoni S, Edjtehadi M, Firoozi MS (2013) The healing effect of licorice (*Glycyrrhiza glabra*) on *Helicobacter pylori* infected peptic ulcers. J Res Med Sci 18(6):532–533

Räikkönen K, Pesonen AK, Heinonen K, Lahti J, Komsi N, Eriksson JG, Seckl JR, Järvenpää AL, Strandberg TE (2009) Maternal licorice consumption and detrimental cognitive and psychiatric outcomes in children. Am J Edpidemiol 170(9):1137–1146

Räikkönen K, Seckl JR, Heinonen K, Pyhälä R, Feldt K, Jones A, Pesonen AK, Phillips DI, Lahti J, Järvenpää AL, Eriksson JG, Matthews KA, Strandberg TE, Kajantie E (2010) Maternal prenatal licorice consumption alters hypothalamic-pituitary-adrenocortical axis function in children. Psychoneuroendocrinology 35(10):1587–1593

Ram A, Mabalirajan U, Das M, Bhattacharya I, Dinda AK, Gangal SV, Ghosh B (2006) Glycyrrhizin alleviates experimental allergic asthma in mice. Int Immunopharmacol 6(9):1468–1477

Rathee P, Rathee S, Ahuja D (2010) Simultaneous quantification of glycyrrhetinic acid & apigenin using HPTLC from *Glycyrrhiza glab*ra Linn. Eurasian J Anal Chem 5(1):95–103

Rathi SG, Suthar M, Patel P, Bhaskar VH, Rajgor NB (2009) In-vitro cytotoxic screening of *Glycyrrhiza glabra* L. (Fabaceae): a natural anticancer drug. J Young Pharm 1(3):239–243

Rauchensteiner F, Matsumura Y, Yamamoto Y, Yamaji S, Tani T (2005) Analysis and comparison of Radix Glycyrrhizae (licorice) from Europe and China by capillary-zone electrophoresis (CZE). J Pharm Biomed Anal 38(4):594–600

Raveendra KR, Jayachandra, Srinivasa V, Sushma KR, Allan JJ, Goudar KS, Shivaprasad HN, Venkateshwarlu K, Geetharani P, Sushma G, Agarwal A (2012) An extract of *Glycyrrhiza glabra* (GutGard) alleviates symptoms of functional dyspepsia: a randomized, double-blind, placebo-controlled study. Evid Based Complement Alternat Med 2012:216970

Rees WD, Rhodes J, Wright JE, Stamford LF, Bennett A (1979) Effect of deglycyrrhizinated liquorice on gastric mucosal damage by aspirin. Scand J Gastroenterol 14(5):605–607

Reiners W (1966) 7-hydroxy-4′-methoxy-isoflavone (formononetin) from liquorice root. On substances contained in liquorice root. II. Experientia 22(6):359 (In German)

Ren CA, Li YX, Cui JY, Sheng ZX, Ran XH, Wang BH, Zhang MH (2013) Efficacy of glycyrrhizin combined with cyclosporine in the treatment of non-severe aplastic anemia. Chin Med J 126(11):2083–2086

Robles BJ, Sandoval AR, Dardon JD, Blas CA (2013) Lethal liquorice lollies (liquorice abuse causing pseudohyperaldosteronism). BMJ Case Rep bcr2013201007

Rong QX, Liu CS, Duan TX (2006) Content determination of glycyrrhizic acid in cortex glycyrrhizae. Zhong Yao Cai 29(9):891–892 (In Chinese)

Rosenblat M, Belinky P, Vaya J, Levy R, Hayek T, Coleman R, Merchav S, Aviram M (1999) Macrophage enrichment with the isoflavan glabridin inhibits NADPH oxidase-induced cell-mediated oxidation of low density lipoprotein. A possible role for protein kinase C. J Biol Chem 274(20):13790–13799

Ross IA (2001) *Glycyrrhiza glabra*. In: Medicinal plants of the world. Humana Press, Totowa, pp 191–240

Ruetzler K, Fleck M, Nabecker S, Pinter K, Landskron G, Lassnigg A, You J, Sessler DI (2013) A randomized, double-blind comparison of licorice versus sugar-water gargle for prevention of postoperative sore throat and postextubation coughing. Anesth Analg 117(3):614–621

Russell RI, Morgan RJ, Nelson LM (1984) Studies on the protective effect of deglycyrrhinized liquorice against aspirin (ASA) and ASA plus bile acid-induced gastric mucosal damage, and ASA absorption in rats. Scand J Gastroenterol Suppl 92:97–100

Russo S, Mastropasqua M, Mosetti MA, Persegani C, Paggi A (2000) Low doses of liquorice can induce hypertension encephalopathy. Am J Nephrol 20(2):145–148

Sabbioni C, Mandrioli R, Ferranti A, Bugamelli F, Saracino MA, Forti GC, Fanali S, Raggi MA (2005) Separation and analysis of glycyrrhizin, 18beta-glycyrrhetic acid and 18alpha-glycyrrhetic acid in liquorice roots by means of capillary zone electrophoresis. J Chromatogr A 1081(1):65–71

Saeedi M, Morteza-Semnani K, Ghoreishi MR (2003) The treatment of atopic dermatitis with licorice gel. J Dermatolog Treat 14(3):153–157

Saha S, Nosál'ová G, Ghosh D, Flešková D, Capek P, Ray B (2011) Structural features and in vivo antitussive activity of the water extracted polymer from *Glycyrrhiza glabra*. Int J Biol Macromol 48(4):634–638

Sailler L, Juchet H, Ollier S, Nicodème R, Arlet P (1993) Generalized edema caused by licorice: a new syndrome. Apropos of 3 cases. Rev Med Interne 14(10):984

Saito T, Tsuboi Y, Fujisawa G, Sakuma N, Honda K, Okada K, Saito K, Ishikawa S, Saito T (1994) An autopsy case of licorice-induced hypokalemic rhabdomyolysis associated with acute renal failure: special reference to profound calcium deposition in skeletal and cardiac muscle. Jpn J Nephrol 36(11):1308–1314

Saitoh T, Kinoshita T, Shibata S (1976a) Flavonols of licorice roots. Chem Pharm Bull 24(6):1242–1245

Saitoh T, Kinoshita T, Shibata S (1976b) New isoflavan and flavanone from licorice root. Chem Pharm Bull 24(4):752–755

Sajid TM, Rashid S, Saeed SA (1991) Inhibition of adrenaline-induced aggregation of human platelets by Pakistani medicinal plants. Pak J Pharm Sci 4(2):145–152

Saleh NAM, Elgamal MHA, Hanna AG (1989) Constituents of the leaves of *Glycyrrhiza glabra*. Fitoterapia 60(2):189

Santaella RM, Fraunfelder FW (2007) Ocular adverse effects associated with systemic medications: recognition and management. Drugs 67(1):75–93

Sasaki H, Takei M, Kobayashi M, Pollard RB, Suzuki F (2002–2003) Effect of glycyrrhizin, an active component of licorice roots, on HIV replication in cultures of peripheral blood mononuclear cells from HIV-seropositive patients. Pathobiology 70(4):229–236

Sato H, Goto W, Yamamura JI, Kurokawa M, Kageyama S, Takahara T, Wantanbe A, Shiraki K (1996) Therapeutic basis of glycyrrhizin on chronic hepatitis B. Antiviral Res 30(2–3):171–177

Sawada K, Yamashita Y, Zhang TS, Nakagawa K, Ashida H (2010) Glabridin induces glucose uptake via the AMP-activated protein kinase pathway in muscle cells. Mol Cell Endocrinol 393(1–2):99–108

Scali M, Pratesi C, Zennaro MC, Zampollo V, Armanini D (1990) Pseudohyperaldosteronism from liquorice-containing laxatives. J Endocrinol Invest 13(10):847–848

Schambelan M (1994) Licorice ingestion and blood pressure regulating hormones. Steroids 59(2):127–130

Seckl JR (1997) 11beta-Hydroxysteroid dehydrogenase in the brain: a novel regulator of glucocorticoid action? Front Neuroendocrinol 18(1):49–99

Seelen MA, de Meijer PH, Braun J, Swinkels LM, Waanders H, Meinders AE (1996) Hypertension caused by licorice consumption. Ned Tijdschr Geneeskd 140(52):2632–2635 (In Dutch)

Segal R, Milo-Goldzweig I, Kaplan G (1977) The protective action of glycyrrhizin against saponin toxicity. Biochem Pharmacol 26(7):643–645

Seki H, Ohyama K, Sawai S, Mizutani M, Ohnishi T, Sudo H, Akashi T, Aoki T, Saito K, Muranaka T (2008) licorice b-amyrin 11-oxidase, a cytochrome P450 with a key role in the biosynthesis of the triterpene sweetener glycyrrhizin. Proc Natl Acad Sci U S A 105(37):14204–14209

Seki H, Sawai S, Ohyama K, Mizutani M, Ohnishi T, Sudo H, Fukushima EO, Akashi T, Aoki T, Saito K, Muranaka T (2011) Triterpene functional genomics in licorice for identification of CYP72A154 involved in the biosynthesis of glycyrrhizin. Plant Cell 23(11):4112–4123

Serra A, Uehlinger DE, Ferrari P, Dick B, Frey BM, Frey FJ, Vogt B (2002) Glycyrrhetinic acid decreases plasma potassium concentrations in patients with anuria. J Am Soc Nephrol 13(1):191–196

Shabani L, Ehsanpour AA, Asghari G, Emami J (2009) Glycyrrhizin production by in vitro cultured *Glycyrrhiza glabra* elicited by methyl jasmonate and salicylic acid. Russ J Plant Physiol 56(5):621–626

Shalaby MA, Ibrahim HS, Mahmoud EM, Mahmound AF (2004) Some effects of *Glycyrrhiza glabra* (liquorice) roots extract on male rats. Egypt J Nat toxin 1:83–94

Sheela ML, Ramakrishna MK, Salimath BP (2006) Angiogenic and proliferative effects of the cytokine VEGF in Ehrlich ascites tumor cells is inhibited by *Glycyrrhiza glabra*. Int Immunopharmacol 6(3):494–498

Shen LY, Cui ZW, Lin Y, Wang SQ, Zheng DF, Tan Q (2015) Anti-inflammative effect of glycyrrhizin on rat thermal injury via inhibition of high-mobility group box 1 protein. Burns 41(2):372–378

Shetty TK, Satav JG, Nair CK (2002) Protection of DNA and microsomal membranes in vitro by *Glycyrrhiza glabra* L. against gamma irradiation. Phytother Res 16(6):576–578

Shetty AV, Thirugnanam S, Dakshinamoorthy G, Samykutty A, Zheng G, Chen A, Bosland MC, Kajdacsy-Balla A, Gnanasekar M (2011) 18α-glycyrrhetinic acid targets prostate cancer cells by down-regulating inflammation-related genes. Int J Oncol 39(3):635–640

Shi LL, Chen BN, Gao M, Zhang HA, Li YJ, Wang L, Du GH (2011) The characteristics of therapeutic effect of pinocembrin in transient global brain ischemia/reperfusion rats. Life Sci 88(11–12):521–552

Shibano M, Matsumoto Y, Kusano G, Shibata T (1996) Researches of *Glycyrrhiza* species grown at medicinal plant gardens in Japan and basic studies for selection of for selection of pharmaceutically fine races: 1. Comparative studies by HPLC patterns and constituents of aerial parts. Nat Med 50(4):273–283

Shibata S (2000) A drug over the millennia: pharmacognosy, chemistry, and pharmacology of licorice. Yakugaku Zasshi 120:849–862

Shibata S, Saitoh T (1968) The chemical studies on the oriental plant drugs. XIX. Some new constituents of licorice root. (1) The structure of licoricidin. Chem Pharm Bull 16(10):1932–1936

Shihata IM, Elghamry MI (1963a) Estrogenic activity of *Glycyrrhiza glabra* with its effect upon uterine motility at various stages of sex cycle. Zentralbl Veterinarmed A 10(2):155–162

Shihata MI, Elghamry MI (1963b) Experimental studies in the effect of *Glycyrrhiza glabra*. Planta Med 11(1):37–43

Shimizu N, Tomoda M, Satoh M, Gonda R, Ohara N (1991) Characterization of a polysaccharide having activity on the reticuloendothelial system from the stolon of *Glycyrrhiza glabra* var. *glandulifera*. Chem Pharm Bull 39(8):2082–2086

Shin YW, Bae EA, Lee B, Lee SH, Kim JA, Kim YS, Kim DH (2007) In vitro and in vivo antiallergic effects of *Glycyrrhiza glabra* and its components. Planta Med 73(3):257–261

Shinada M, Azuma M, Kawai H, Sazaki K, Yoshida I, Yoshida T, Suzutani T, Sakuma T (1986) Enhancement of interferon-γ production in glycyrrhizin-treated human peripheral lymphocytes in response to concanavalin A and to surface antigen of hepatitis B virus. Exp Biol Med 181(2):205–210

Shinoda J, Ueeda S (1934a) Über das flavanon-glucosid in *Glycyrrhiza glabra* L. var. *glandulifera* Regel et Herder. Chem Ber 67:434–440

Shinoda J, Ueeda S (1934b) Über das flavanon-glucosid in *Glycyrrhiza glabra* L. var. *glandulifera* Regel et Herder. Yakugaku Zasshi 54(8):707–714

Shintani S, Murase H, Tsukagoshi H, Shiigai T (1992) Glycyrrhizin (licorice)-induced hypokalemic myopathy. Report of 2 cases and review of the literature. Eur Neurol 32(1):44–51

Shirazi MH, Ranjbar R, Eshraghi S, Sadeghi G, Jonaidi N, Bazzaz N, Izadi M, Sadeghifard N (2007) An evaluation of antibacterial activity of *Glycyrrhiza glabra* extract on the growth of *Salmonella*, Shigella and ETEC E. coli. J Biol Sci 7(5):827–829

Sigurjónsdóttir HA, Ragnarsson J, Franzson L, Sigurdsson G (1995) Is blood pressure commonly raised by moderate consumption of liquorice? J Hum Hypertens 9(5):345–348

Sigurjónsdóttir HÁ, Franzson L, Manhem K, Ragnarsson J, Sigurdsson G, Wallerstedt S (2001) Liquorice-induced rise in blood pressure: a linear dose-response relationship. J Hum Hypertens 15(8):549–552

Sigurjónsdóttir HA, Manhem K, Axelson M, Wallerstedt S (2003) Subjects with essential hypertension are more sensitive to the inhibition of 11 beta-HSD by liquorice. J Hum Hypertens 17(2):125–131

Sigurjónsdóttir HA, Axelson M, Johannsson G, Manhem K, Nyström E, Wallerstedt S (2006) The liquorice effect on the RAAS differs between the genders. Blood Press 15(3):169–172

Simmler C, Pauli GF, Chen SN (2013) Phytochemistry and biological properties of glabridin. Fitoterapia 90:160–184

Simons R, Vincken JP, Bakx EJ, Verbruggen MA, Gruppen H (2009) A rapid screening method for prenylated flavonoids with ultra-high-performance liquid chromatography/electrospray ionisation mass spectrometry in licorice root extracts. Rapid Commun Mass Spectrom 23(19):3083–3093

Simons R, Vincken JP, Mol LA, Bovee TF, Luijendijk TJ, Verbruggen MA, Gruppen H (2011) Agonistic and antagonistic estrogens in licorice root (*Glycyrrhiza glabra*). Anal Bioanal Chem 401(1):305–313

Siracusa L, Saija A, Cristani M, Cimino F, D'Arrigo M, Trombetta D, Rao F, Ruberto G (2011) Phytocomplexes from liquorice (*Glycyrrhiza glabra* L.) leaves – Chemical characterization and evaluation of their antioxidant, anti-genotoxic and anti-inflammatory activity. Fitoterapia 82(4):546–556

Sitohy MZ, el-Massry RA, el-Saadany SS, Labib SM (1991) Metabolic effects of licorice roots (*Glycyrrhiza glabra*) on lipid distribution pattern, liver and renal functions of albino rats. Nahrung 35(8):799–806

Sohi PK, Parashar B, Kansal D (2013) Comparative evaluation of anti-depressant activity of *Glycyrrhiza glabra* and *Piper nigrum* combination with individual drugs effects and standard drug imipramine in mice. Int J Pharm Res Allied Sci 2(4):16–21

Soleimanpour S, Sedighinia FS, Afshar AS, Zarif R, Asili J, Ghazvini K (2013) Synergistic antibacterial activity of *Capsella bursa-pastoris* and *Glycyrrhiza glabra* against oral pathogens. Jundishapur J Microbiol 6(8), e7262

Somjen D, Katzburg S, Vaya J, Kaye AM, Hendel D, Posner GH, Tamir S (2004a) Estrogenic activity of glabridin and glabrene from licorice roots on human osteoblasts and prepubertal rat skeletal tissues. J Steroid Biochem Mol Biol 91(4–5):241–246

Somjen D, Knoll E, Vaya J, Stern N, Tamir S (2004b) Estrogen-like activity of licorice root constituents: glabridin and glabrene, in vascular tissues in vitro and in vivo. J Steroid Biochem Mol Biol 91(3):147–155

Song JH, Lee JW, Shim BS, Lee CY, Choi SY, Kang CH, Sohn NW, Shin JW (2013) Glycyrrhizin alleviates neuroinflammation and memory deficit induced by systemic lipopolysaccharide treatment in mice. Molecules 18(12):15788–15803

Song JH, Ko HS, Sohn EJ, Kim BL, Kim JH, Kim HJ, Kim CW, Kim JE, Kim SH (2014) Inhibition of protein kinase C α/βII and activation of c-Jun NH2-terminal kinase mediate glycyrrhetinic acid induced apoptosis in non-small cell lung cancer NCI-H460 cells. Bioorg Med Chem Lett 24(4):1188–1191

Sontia B, Mooney J, Gaudet L, Touyz RM (2008) Pseudohyperaldosteronism, liquorice, and hypertension. J Clin Hypertens (Greenwich) 10(2):153–157

Sowmya M, Sathish Kumar BY (2010) Antistress property of *Glycyrrhiza glabra* (Athimadhura) on stress induced *Drosophila melanogaster*. J Stress Physiol Biochem 6(4):18–27

Stanetty C, Czollner L, Koller I, Shah P, Gaware R, Cunha TD, Odermatt A, Jordis U, Kosma P, Classen-Houben D (2010) Synthesis of novel 3-amino and 29-hydroxamic acid derivatives of glycyrrhetinic acid as selective 11β-hydroxysteroid dehydrogenase 2 inhibitors. Bioorg Med Chem 18(21):7522–7541

Statti GA, Tundis R, Sacchetti G, Muzzoli M, Bianchi A, Menichini F (2004) Variability in the content of active constituents and biological activity of *Glycyrrhiza glabra*. Fitoterapia 75(3–4):371–374

Steinberg D, Sgan-Cohen HD, Stabholz A, Pizanty S, Segal R, Sela MN (1989) The anticariogenic activity of glycyrrhizin: preliminary clinical trials. Isr J Dent Sci 2(3):153–157

Strandberg T, Järvenpää A, Vanhanen H, McKeigue P (2001) Birth outcome in relation to licorice consumption during pregnancy. Am J Epidemiol 153(11):1085–1088

Strandberg TE, Andersson S, Järvenpää AL, McKeigue PM (2002) Preterm birth and licorice consumption during pregnancy. Am J Epidemiol 156(9):803–805

Su X, Lawrence H, Ganeshapillai D, Cruttenden A, Purohit A, Reed MJ, Vicker N, Potter BV (2004) Novel 18beta-glycyrrhetinic acid analogues as potent and selective inhibitors of 11beta-hydroxysteroid dehydrogenases. Bioorg Med Chem 12(16):4439–4457

Su X, Vicker N, Lawrence H, Smith A, Purohit A, Reed MJ, Potter BV (2007) Inhibition of human and rat 11beta-hydroxysteroid dehydrogenase type 1 by 18beta-glycyrrhetinic acid derivatives. J Steroid Biochem Mol Biol 104(3–5):312–320

Suman A, Ali M, Alam P (2009) New prenylated isoflavanones from the roots of *Glycyrrhiza glabra*. Chem Nat Comp 45(4):487–491

Tabata M, Honda G, Hayashi H, Gotoh K (1988) A field survey of *Glycyrrhiza glabra* in East Anatolia. Jpn J Pharmacogn 42(4):264–267 (In Japanese)

Tacconi P, Paribello A, Cannas A, Marrosu M (2009) Carpal tunnel syndrome triggered by excessive licorice consumption. J Peripher Nerv Syst 14:64–65

Taju G, Majeed SA, Nambi KSN, Jayanthi M (2011) Hepatoprotective and antioxidant effects of *Glycyrrhiza glabra* extract against paracetamol induced liver damage in rats. J Pharm Res 4(10):3798–3802

Takada K, Tomoda M, Shimizu N (1992) Core structure of glycyrrhizan GA, the main polysaccharide from the stolon of *Glycyrrhiza glabra* var. *glandulifera*; anti-complementary and alkaline phosphatase-inducing activities of the polysaccharide and its degradation product. Chem Pharm Bull 40(9):2487–2490

Takahara T, Watanabe A, Shiraki K (1994) Effects of glycyrrhizin on hepatitis B surface antigen: a biochemical and morphological study. J Hepatol 21(4):601–609

Takeda R, Morimoto S, Uchida K, Nakai T, Miyamoto M, Hashiba T, Yoshimitsu K, Kim KS, Miwa U (1979) Prolonged pseudoaldosteronism induced by glycyrrhizin. Endocrinol Jpn 26(5):541–547

Tamir S, Eizenberg M, Somjen D, Stern N, Shelach R, Kaye A, Vaya J (2000) Estrogenic and antiproliferative properties of glabridin from licorice in human breast cancer cells. Cancer Res 60(20):5704–5709

Tamir S, Eizenberg M, Somjen D, Izrael S, Vaya J (2001) Estrogen-like activity of glabrene and other constituents isolated from licorice root. J Steroid Biochem Mol Biol 78(3):291–298

Tamura Y, Nishikawa T, Yamada K, Yamamoto M, Kumagai A (1970) Effects of glycyrrhetinic acid and its derivatives on delta 4-5 alpha- and 5 beta-reductase in rat liver. Arzneimittelforschung 29(4):647–649

Tanaka M, Mano N, Akazai E, Narui Y, Kato F, Koyama Y (1987) Inhibition of mutagenicity by glycyrrhiza extract and glycyrrhizin. J Pharmacobiodyn 10(12):685–688

Taniguchi C, Homma M, Takano O, Hirano T, Oka K, Aoyagi Y, Niitsuma T, Hayashi T (2000) Pharmacological effects of urinary products obtained after treatment with saiboku-to, a herbal medicine for bronchial asthma, on type IV allergic reaction. Planta Med 66(7):607–611

Tawata M, Yoda Y, Aida K, Shindo H, Sasaki H, Chin M, Onaya T (1990) Anti-platelet action of GU-7, a 3-arylcoumarin derivative, purified from Glycyrrhizae radix. Planta Med 56(3):259–263

Tawata M, Aida K, Noguchi T, Ozaki Y, Kume S, Sasaki H, Chin M, Onaya T (1992) Anti-platelet action of isoliquiritigenin, an aldose reductase inhibitor in licorice. Eur J Pharmacol 212(1):87–92

Teng LS, Kou CJ, Lu CY, Xu JM, Xie J, Lu JH, Liu Y, Wang ZY, Wang D (2014a) Involvement of the ERK pathway in the protective effects of glycyrrhizic acid against the MPP+-induced apoptosis of dopaminergic neuronal cells. Int J Mol Med 34(3):742–748

Teng LS, Meng QF, Lu JH, Xie J, Wang ZZ, Liu Y, Wang D (2014b) Liquiritin modulates ERK- and AKT/GSK-3β-dependent pathways to protect against glutamate-induced cell damage in differentiated PC12 cells. Mol Med Rep 10(2):818–824

Tharkar PR, Tatiya AU, Shinde PR, Surana SJ, Patil UK (2010) Antifungal activity of *Glycyrrhiza glabra* Linn. and *Emblica officinalis* Gaertn by direct bioautography method. Int J Pharm Technol Res 2:1547–1549

The Pharmacopoeia Commission of People's Republic of China (ed) (2005) Pharmacopoeia of the Peoples's Republic of China, vol I. Chemical Industry Press, Beijing, pp 59–60

The Plant List (2013) Version 1.1. Published on the Internet; http://www.theplantlist.org/. Accessed 18 Sept 2014

The Society of Japanese Pharmacopoeia (ed) (2001) Japanese Pharmacopoeia, 15th edn. Ministry of Health, Labour and Welfare of Japan, Tokyo, p 1197

Thimmulappa RK, Rangasamy T, Alam J, Biswal S (2008) Dibenzoylmethane activates Nrf2-dependent detoxification pathway and inhibits benzo(a)pyrene induced DNA adducts in lungs. Med Chem 4(5):473–481

Thiyagarajan P, Chandrasekaran CV, Deepak HB, Agarwal A (2011) Modulation of lipopolysaccharide-induced pro-inflammatory mediators by an extract of *Glycyrrhiza glabra* and its phytoconstituents. Inflammopharmacology 19(4):235–241

Tian ML, Yan HY, Row KH (2008) Extraction of glycyrrhizic acid and glabridin from licorice. Int J Mol Sci 9(4):571–577

Tohma HS, Gulçin I (2010) Antioxidant and radical scavenging activity of aerial parts and roots of Turkish liquorice (*Glycyrrhiza glabra* L). Int J Food Prop 13(4):657–671

Tripathi M, Singh BK, Kakkar P (2009) Glycyrrhizic acid modulates t-BHP induced apoptosis in primary rat hepatocytes. Food Chem Toxicol 47(2):339–347

Trivedi R, Sharma K (2011) Hydroalcoholic extract of *Glycyrrhiza glabra* Linn. attenuates chronic fatigue stress induced behavioral alterations in mice. Int J Pharm Biol Arch 2(3):996–1001

Tsai JJ, Kuo HC, Lee KF, Tsai TH (2013) Glycyrrhizin represses total parenteral nutrition-associated acute liver injury in rats by suppressing endoplasmic reticulum stress. Int J Mol Sci 14(6):12563–12580

Tsubota A, Kumada H, Arase Y, Chayama K, Saitoh S, Ikeda K, Kobayashi M, Suzuki Y, Murashima N (1999) Combined ursodeoxycholic acid and glycyrrhizin therapy for chronic hepatitis C virus infection: a randomized controlled trial in 170 patients. Eur J Gastroenterol Hepatol 11(10):1077–1084

Tsukiyama RI, Katsura H, Tokuriki N, Kobayashi M (2002) Antibacterial activity of licochalcone a against spore-forming bacteria. Antimicrob Agents Chemother 46(5):1226–1230

Tu JH, He YJ, Chen Y, Fan L, Zhang W, Tan ZR, Huang YF, Guo D, Hu DL, Wang D, Zhou HH (2010a) Effect of glycyrrhizin on the activity of CYP3A enzyme in humans. Eur J Clin Pharmacol 66(8):805–810

Tu JH, Hu DL, Dai LL, Sun Y, Fan L, Zhang M, Tan ZR, Chen Y, Li Z, Zhou HH (2010b) Effect of glycyrrhizin on CYP2C19 and CYP3A4 activity in healthy volunteers with different CYP2C19 genotypes. Xenobiotica 40(6):393–399

Tu CT, Li J, Wang FP, Li L, Wang JY, Jiang W (2012) Glycyrrhizin regulates CD4+ T cell response during liver fibrogenesis via JNK, ERK and PI3K/AKT pathway. Int Immunopharmacol 14(4):410–421

Upadhyay S, Singh V (2013) Potentiality of petroleum ether (60–80)° C extract of *Glycyrrhiza glabra* on androgenic alopecia. Asian J Pharm Clin Res 6(3):52–55

Upadhyay S, Ghosh AK, Singh V (2013) Hair growth promotant activity of petroleum ether root extract of *Glycyrrhiza glabra* L (Fabaceae) in female rats. Trop J Pharm Res 11(5):753–758

Uphof JCT (1968) Dictionary of economic plants, 2nd edn (1st edn, 1959). Cramer, Lehre, 591 pp

Uto T, Morinaga O, Tanaka H, Shoyama Y (2012) Analysis of the synergistic effect of glycyrrhizin and other constituents in licorice extract on lipopolysaccharide-induced nitric oxide production

using knock-out extract. Biochem Biophys Res Commun 417(1):473–478

Utsunomiya T, Kobayashi M, Herndon DN, Pollard RB, Suzuki F (1995) Glycyrrhizin (20 beta-carboxy-11-oxo-30-norolean-12-en-3 beta-yl-2-O-beta-D-glucopyranuronosyl-alpha-D-glucopyranosiduronic acid) improves the resistance of thermally injured mice to opportunistic infection of herpes simplex virus type 1. Immunol Lett 44(1):59–66

Utsunomiya T, Kobayashi M, Pollard RB, Suzuki F (1997) Glycyrrhizin, an active component of licorice roots, reduces morbidity and mortality of mice infected with lethal doses of influenza virus. Antimicrob Agents Chemother 41(3):551–556

van Beers EJ, Stam J, van den Bergh WM (2011) Licorice consumption as a cause of posterior reversible encephalopathy syndrome: a case report. Crit Care 15(1):R64

Van der Zwan A (1993) Hypertension encephalopathy after liquorice ingestion. Clin Neurol Neurosurg 95(1):35–37

van Hulle C, Braeckman P, Vandewalle M (1971) Isolation of two new flavonoids from the root of *Glycyrrhiza glabra* var. *typica*. Planta Med 20(3):278–282

Van Rossum TG, De Man RA (1998) Glycyrrhizin as a potential treatment for chronic hepatitis C. Aliment Pharmacol Therap 12(3):199–205

Van Rossum TG, Vulto AG, Hop WC, Brouwer JT, Niesters HG, Schalm SW (1999) Intravenous glycyrrhizin for the treatment of chronic hepatitis C: a double-blind, randomized, placebo-controlled phase I/II trial. J Gastroenterol Hepatol 14(11):1093–1099

Van Rossum TG, Vulto AG, Hop WC, Schalm SW (2001) Glycyrrhizin-induced reduction of ALT in European patients with chronic hepatitis C. Am J Gastroenterol 96(8):2432–2437

Varsha S, Agrawal RC, Sonam P (2013) Phytochemical screening and determination of anti-bacterial and anti-oxidant potential of *Glycyrrhiza glabra* root extracts. J Environ Res Dev 7(4A):1552–1558

Varshney IP, Jain DC, Srivastava HC (1983) Study of saponins from *Glycyrrhiza glabra* root. Pharm Biol 21(4):169–172

Vaya J, Belinky PA, Aviram M (1997) Antioxidant constituents from licorice roots: isolation, structure elucidation and antioxidative capacity toward LDL oxidation. Free Radic Biol Med 23(2):302–313

Vicker N, Su X, Lawrence H, Cruttenden A, Purohit A, Reed MJ, Potter BV (2004) A novel 18 beta-glycyrrhetinic acid analogue as a potent and selective inhibitor of 11 beta-hydroxysteroid dehydrogenase 2. Bioorg Med Chem Lett 14(12):3263–3267

Vijayalakshmi U, Shourie A (2013) Gas chromatography-mass spectrometric analysis of ethanolic extracts of *Glycyrrhiza glabra* Linn. roots. Int J Pharm Bio Sci 4(4):741–755

Vikhe GP, Vikhe PP, Naik SS, Gavhane AJ, Gaikar RB (2013) In vitro effect of *G. glabra* and *T. cordifolia* plant extracts on phagocytosis by human neutrophils. Pravara Med Rev 5(2):12–15

Visavadiya NP, Narasimhacharya AV (2006) Hypocholesterolaemic and antioxidant effects of *Glycyrrhiza glabra* (Linn) in rats. Mol Nutr Food Res 50(11):1080–1086

Visavadiya NP, Soni B, Dalwadi N (2009) Evaluation of antioxidant and anti-atherogenic properties of *Glycyrrhiza glabra* root using in vitro models. Int J Food Sci Nutr Suppl 60(Sup 2):135–149

Wagner H, Jurcic K (2002) Immunological studies of Revitonil®, a phytopharmaceutical containing Echinacea purpurea and *Glycyrrhiza glabra* root extract. Phytomedicine 9(5):390–399

Walker BR, Edwards CR (1994) Licorice-induced hypertension and syndromes of apparent mineralocorticoid excess. Endocrinol Metab Clin North Am 23(2):359–377

Wan XY, Luo M, Li XD, He P (2009) Hepatoprotective and anti-hepatocarcinogenic effects of glycyrrhizin and matrine. Chem Biol Interact 181(1):15–19

Wang ZY, Agarwal R, Zhou ZC, Bickers DR, Mukhtar H (1991) Inhibition of mutagenicity in *Salmonella typhimurium* and skin tumor initiating and tumor promoting activities in SENCAR mice by glycyrrhetinic acid: comparison of 18α- and 18β-stereoisomers. Carcinogenesis 12(2):187–192

Wang ZH, Nishioka M, Kurosaki Y, Nakayama T, Kimura T (1995) Gastrointestinal absorption characteristics of glycyrrhizin from *Glycyrrhiza* extract. Biol Pharm Bull 18(9):1238–1241

Wang CY, Kao TC, Lo WH, Yen GC (2011) Glycyrrhizic acid and 18β-glycyrrhetinic acid modulate lipopolysaccharide-induced inflammatory response by suppression of NF-κB through PI3K p110δ and p110γ inhibitions. J Agric Food Chem 59(14):7726–7733

Wang JJ, Chen XQ, Wang W, Zhang YT, Yang ZY, Jin Y, Ge HM, Li EG, Yang G (2013a) Glycyrrhizic acid as the antiviral component of *Glycyrrhiza uralensis* Fisch. against coxsackievirus A16 and enterovirus 71 of hand foot and mouth disease. J Ethnopharmacol 147(1):114–121

Wang KL, Hsia SM, Chan CJ, Chang FY, Huang CY, Bau DT, Wang PS (2013b) Inhibitory effects of isoliquiritigenin on the migration and invasion of human breast cancer cells. Expert Opin Ther Targets 17(4):337–349

Wang ZY, Wang N, Han SW, Wang DM, Mo SL, Yu LZ, Huang H, Tsui KC, Shen JG, Chen JP (2013c) Dietary compound isoliquiritigenin inhibits breast cancer neoangiogenesis via VEGF/VEGFR-2 signaling pathway. PLoS One 8(7), e68566

Wang D, Guo TQ, Wang ZY, Lu JH, Liu DP, Meng QF, Xie J, Zhang XL, Liu Y, Teng LS (2014a) ERKs and mitochondria-related pathways are essential for glycyrrhizic acid-mediated neuroprotection against glutamate-induced toxicity in differentiated PC12 cells. Braz J Med Biol Res 47(9):773–779

Wang D, Lu JH, Liu Y, Meng QF, Xie J, Wang ZZ, Teng LS (2014b) Liquiritigenin induces tumor cell death through mitogen-activated protein kinase- (MPAKs-) mediated pathway in hepatocellular carcinoma cells. Biomed Res Int 2014:965316

Wang D, Wong HK, Feng YB, Zhang ZJ (2014c) Liquiritigenin exhibits antitumour action in pituitary adenoma cells via Ras/ERKs and ROS-dependent mitochondrial signalling pathways. J Pharm Pharmacol 66(3):408–417

Wei JH, Zheng YF, Li CY, Tang YP, Peng GP (2014) Bioactive constituents of oleanane-type triterpene saponins from the roots of *Glycyrrhiza glabra*. J Asian Nat Prod Res 16(11):1044–1053

Whorwood C, Sheppard M, Stewart P (1993) Licorice inhibits 11 betahydroxysteroid dehydrogenase messenger ribonucleic acid levels and potentiates glucocorticoid hormone action. Endocrinology 132:2287–2292

Wittschier N, Faller G, Hensel A (2009) Aqueous extracts and polysaccharides from liquorice roots (*Glycyrrhiza glabra* L.) inhibit adhesion of *Helicobacter pylori* to human gastric mucosa. J Ethnopharmacol 125(2):218–223

Wong TY, Lin SM, Poon CH, Leung LK (2014) The licorice flavonoid isoliquiritigenin reduces DNA-binding activity of AhR in MCF-7 cells. Chem Biol Interact 221:70–76

Wu CH, Zabawa EM, Townsley PM (1974) The single cell suspension culture of the licorice plant, *Glycyrrhiza glabra*. Can Inst Food Sci Technol J 7(2):105–109

Wu YP, Meng XS, Bao YR, Wang S (2013) Pharmacokinetic study of four flavones of *Glycyrrhiza* in rat plasma using HPLC–MS. J Ethnopharmacol 148(1):266–270

Xiang C, Qiao X, Wang Q, Li R, Miao WJ, Guo D, Ye M (2011) From single compounds to herbal extract: a strategy to systematically characterize the metabolites of licorice in rats. Drug Metab Dispos 39(9):1597–1608

Xiao YC, Wu JW, Mao C, Jin M, Wu Q, Zou J, Gu QL, Zhang Y, Zhang YY (2010) 18β-Glycyrrhetinic acid ameliorates acute *Propionibacterium acnes*-induced liver injury through inhibition of macrophage inflammatory protein-1α. J Biol Chem 285:1128–1137

Xiao XY, Hao M, Yang XY, Ba Q, Li M, Ni SJ, Wang LS, Du X (2011) Licochalcone A inhibits growth of gastric cancer cells by arresting cell cycle progression and inducing apoptosis. Cancer Lett 302(1):69–75

Xing GX, Li N, Wang T, Yao MY (2003) Advances in studies on flavonoids of licorice. Zhongguo Zhong Yao Za Zhi 28(7):593–597 (In Chinese)

Yaguchi M, Yaguchi H, Sakano M (2008) A case of hypokalemic myopathy mimicking hemiparesis. Brain Nerve 60(2):191–194

Yahara S, Nishioka I (1984) Flavonoid glucosides from licorice. Phytochemistry 23(9):2108–2109

Yamamoto S, Aizu E, Jiang H, Nakadate T, Kiyoto I, Wang JC, Kato R (1991) The potent anti-tumor-promoting agent isoliquiritigenin. Carcinogenesis 12(2):317–323

Yamamura Y, Kawakami J, Santa T, Kotaki H, Uchino K, Sawada Y, Iga T (1991) Selective high-performance liquid chromatographic method for the determination of glycyrrhizin and glycyrrhetic acid-3-O-glucuronide in biological fluids: application of ion-pair extraction and fluorescence labelling agent. J Chromatogr 567(1):151–160

Yamamura Y, Kawakami J, Santa T, Kotaki H, Uchino K, Sawada Y, Tanaka N, Iga T (1992) Pharmacokinetic profile of glycyrrhizin in healthy volunteers by a new high-performance liquid chromatographic method. J Pharm Sci 81(10):1042–1046

Yamasaki H (1990) Gap junctional intercellular communication and carcinogenesis. Carcinogenesis 11(7):1051–1058

Yasui S, Fujiwara K, Tawada A, Fukuda Y, Nakano M, Yokosuka O (2011) Efficacy of intravenous glycyrrhizin in the early stage of acute onset autoimmune hepatitis. Dig Dis Sci 56(12):3638–3647

Yasukawa K, Takido M, Takeuchi M, Nakagawa S (1988) Inhibitory effect of glycyrrhizin and caffeine on two-stage carcinogenesis in mice. Yakugaku Zasshi 108(8):794–796

Yazdi A, Sardari S, Sayyah M, Ezzati MH (2011) Evaluation of the anticonvulsant activity of the leaves of *Glycyrrhiza glabra* var. *glandulifera* grown in Iran, as a possible renewable source for anticonvulsant compounds. Iran J Pharm Res 10(1):75–81

Ye SJ, Zhu Y, Ming YZ, She XG, Liu H, Ye QF (2014) Glycyrrhizin protects mice against renal ischemia-reperfusion injury through inhibition of apoptosis and inflammation by downregulating p38 mitogen-activated protein kinase signaling. Exp Ther Med 7(5):1247–1252

Yo YT, Shieh GS, Hsu KF, Wu CL, Shiau AL (2009) Licorice and licochalcone-A induce autophagy in LNCaP prostate cancer cells by suppression of Bcl-2 expression and the mTOR pathway. J Agric Food Chem 57(18):8266–8273

Yokota T, Nishio H, Kubota Y, Mizoguchi M (1998) The inhibitory effect of glabridin from licorice extracts on melanogenesis and inflammation. Pigment Cell Res 11(6):355–361

Yorgun H, Aksoy H, Sendur MA, Ateş AH, Kaya EB, Aytemir K, Oto A (2010) Brugada syndrome with aborted sudden cardiac death related to liquorice-induced hypokalemia. Med Princ Pract 19(6):485–489

Yu SM, Kuo SC (1995) Vasorelaxant effect of isoliquiritigenin, a novel soluble guanylate cyclase activator, in rat aorta. Br J Pharmacol 114(8):1587–1594

Yu XY, Lin SG, Zhou ZW, Chen X, Liang J, Yu XQ, Chowbay B, Wen JY, Duan W, Chan E, Li XT, Cao J, Li CG, Xue CC, Zhou SF (2007) Role of P-glycoprotein in limiting the brain penetration of glabridin, an active isoflavan from the root of *Glycyrrhiza glabra*. Pharm Res 24(9):1668–1690

Yu XQ, Xue CCL, Zhou ZW, Li CG, Du YM, Liang J, Zhou SF (2008) In vitro and in vivo neuroprotective effect and mechanisms of glabridin, a major active isoflavan from *Glycyrrhiza glabra* (licorice). Life Sci 82(1–2):68–78

Yuldashev MP (2001) New flavanonglycoside from *Glycyrrhiza glabra*. Chem Nat Comp 37(3):224–245

Yuldashev MP, Batirov E, Vdovin AD, Abdullaev ND (2000a) Glabrizoflavone – a novel isoflavone from *Glycyrrhiza glabra* L. Bioorg Khim 26(11):873–876 (In Russian)

Yuldashev MP, Nigmatullajev AM, Sultanov S (2000b) Flavonoids of above-ground part of *Glycyrrhiza glabra* L. grown in Uzbekistan. Rastit Resur 36(4):56–59

Zamansoltani F, Nassiri-Asl M, Sarookhani MR, Jahani-Hashemi H, Zangivand AA (2009) Antiandrogenic activities of *Glycyrrhiza* glabra in male rats. Int J Androl 4:417–422

Zani F, Cuzzoni MT, Daglia M, Benvenuti S, Vampa G, Mazza P (1993) Inhibition of mutagenicity in *Salmonella typhimurium* by *Glycyrrhiza glabra* extract, glycyrrhizinic acid, 18 alpha- and 18 beta-glycyrrhetinic acids. Planta Med 59(6):502–507

Zayed SM, Hassan A, Elghamry MI (1964) Estrogenic substances from Egyptian *Glycyrrhiza glabra*. II. Beta-sitosterol as an estrogenic principle. Zentralbl Veterinarmed A 11(5):476–482

Zeng L, Zhang RY, Meng T, Lou ZC (1990) Determination of nine flavonoids and coumarins in licorice root by high-performance liquid chromatography. J Chromatogr 513:247–254

Zeng L, Lou ZC, Zhang RY (1991a) Quality evaluation of Chinese licorice. Yaoxue Xuebao 26(10):788–793 (In Chinese)

Zeng L, Zhang RY, Lou ZC (1991b) Quantitative determination of three saponins in licorice root by high-performance liquid chromatography. Yaoxue Xuebao 26(1):53–58 (In Chinese)

Zhai CL, Zhang MQ, Zhang Y, Xu HX, Wang JM, An GP, Wang YY, Li L (2012) Glycyrrhizin protects rat heart against ischemia-reperfusion injury through blockade of HMGB1-dependent phospho-JNK/Bax pathway. Acta Pharmacol Sin 33(12):1477–1487

Zhan C, Yang J (2006) Protective effects of isoliquiritigenin in transient middle cerebral artery occlusion-induced focal cerebral ischemia in rats. Pharmacol Res 53(3):303–309

Zhang X, Xiong YL, Chen J, Zhou LR (2013) Synergistic inhibition of lipid oxidation by pea protein hydrolysate coupled with licorice extract in a liposomal model system. J Agric Food Chem 61(35):8452–8461

Zhang JC, Wu Y, Weng ZL, Zhou T, Feng T, Lin Y (2014a) Glycyrrhizin protects brain against ischemia–reperfusion injury in mice through HMGB1-TLR4-IL-17A signaling pathway. Brain Res 1582:176–186

Zhang X, Xiong YL, Chen J, Zhou LR (2014b) Synergy of licorice extract and pea protein hydrolysate for oxidative stability of soybean oil-in-water emulsions. J Agric Food Chem 62(32):8204–8213

Zhao J, Wan XY, Luo M, Chen TS, He P (2012a) Antifibrotic effects of glycyrrhizin and matrine in vitro and in vivo. Biomed Prev Nutr 2(2):132–137

Zhao K, Ding M, Cao H, Cao ZX (2012b) In vitro metabolism of glycyrrhetinic acid by human and rat liver microsomes and its interactions with six CYP substrates. J Pharm Pharmacol 64(10):1445–1451

Zhao HX, Zhang XH, Chen XW, Li Y, Ke ZQ, Tang T, Chai HY, Guo AM, Chen HL, Yang J (2014) Isoliquiritigenin, a flavonoid from licorice, blocks M2 macrophage polarization in colitis-associated tumorigenesis through downregulating PGE2 and IL-6. Toxicol Appl Pharmacol 279(3):311–321

Zheng H, Li Y, Wang YZ, Zhao H, Zhang J, Chai HY, Tang T, Yue J, Guo AM, Yang J (2014) Downregulation of COX-2 and CYP 4A signaling by isoliquiritigenin inhibits human breast cancer metastasis through preventing anoikis resistance, migration and invasion. Toxicol Appl Pharmacol 280(1):10–20

Zhou M, Higo H, Cai Y (2010) Inhibition of hepatoma 22 tumor by liquiritigenin. Phytother Res 24(6):827–833

Zhu ZY, Li CJ, Wang X, Yang ZY, Chen J, Hu LH, Jiang HL, Shen X (2010) 2, 2′, 4′-Trihydroxychalcone from *Glycyrrhiza glabra* as a new specific BACE1 inhibitor efficiently ameliorates memory impairment in mice. J Neurochem 114(2):374–385

Zhu J, Mj C, Chen N, Ma AZ, Zhu CY, Zhao RL, Jiang M, Zhou J, Ye LH, Fu HA, Zhang X (2015) Glycyrrhetinic acid induces G1-phase cell cycle arrest in human non-small cell lung cancer cells through endoplasmic reticulum stress pathway. Int J Oncol 46(3):981–988

Pachyrhizus ahipa

Scientific Name

***Pachyrhizus ahipa* (Wedd.) Parodi**

Synonyms

Dolichos ahipa Wedd., *Pachyrhizus ahipa* var. *albiflora* Parodi, *Pachyrhizus ahipa* var. *violacea* Parodi

Family

Fabaceae

Common/English Names

Ahipa, Andean Bean, Andean Yam Bean, Yam Bean

Vernacular Names

Bolivia: Huitoto, Villu (Aymara), Frijol Chuncho (Spanish)
Brazil: Jacatupé
French: Dolique Tubereux d'Ande, L'Ahipa
German: Andine Jamsbohne, Andine Knollenbohne
Peru: Ahipa
Portuguese: Ahipa
Quechuan: Ajipa, Asipa
Spanish: Achipa, Ahipa, Ajipa, Judia Batata, Poroto Batata
Philippines: Singkamas (Tagalog)

Origin/Distribution

Pachyrhizus ahipa is indigenous to the Andes in South America (Grau 1997) and its area of origin is most likely in the 'ceja de montañas' Andean region (Ørting et al. 1996). *P. ahipa* is the only species without records of wild material (Sørensen et al. 1997). It is only found cultivated in Bolivia, some areas in the Jujuy and Salta provinces of Argentina, and in Andean valleys at altitudes between 1000 and 3000 m (Ørting et al. 1996; Sørensen et al. 1997). Indications of the *Pachyrhizus* species in general were also made at the southern coast of Peru, in the Nasca culture (Sørensen et al. 1997).

T.K. Lim, *Edible Medicinal and Non-Medicinal Plants: Volume 10, Modified Stems, Roots, Bulbs*,
DOI 10.1007/978-94-017-7276-1_19

Agroecology

P. ahipa is well adapted to cool subtropical conditions in the Andes; it tolerates low temperatures but is not frost hardy (Ørting et al. 1996; Grau 1997). It is mainly found in cool tropical and subtropical valleys, on sun-facing slopes, on the border between the warm and cold tropics in altitudes of 1800–300 m (Sørensen et al. 1997). The average temperature in this region is 16–18 °C, with precipitation of 400–700 mm, occurring within 4–6 months, with the rest being the dry season. Flowering occurs under decreasing day length and *P. ahipa* is regarded as a short day plant (Sørensen et al. 1997). *P. ahipa* thrive best a light rich well-drained sandy soil with pH of 6–8 (Ørting et al. 1996; Sørensen et al. 1997). Although it can tolerant long dry spell to increase tuber yield, an additional water supply is essential. Also flower removal has been reported to increase tuberous root yields.

Edible Plant Parts and Uses

Ahipa tuberous roots can be eaten raw or cooked. The roots are sweet and crispy, when eaten raw it can be peeled like banana (Grau 1997) or eaten like apple as snacks or in green and fruit salads or prepared as juice (Popenoe et al. 1989; Sørensen et al. 1997). The roots are thirst quenching and nutritious with an easily digestible starch (Ørting et al. 1996). Ahipa roots can be boiled and accompany dishes as a substitute for cassava or sweet potato, they retain their crunchy texture even after cooking (Sørensen et al. 1997). Young seed pods when thoroughly cooked to remove toxic rotenone (Huxley 1992) can be used like French beans (Uphof 1968; Usher 1974). Ahipa root and flour can be considered alternative food sources of gluten-free starch, or gluten free functional food ingredients with a considerable contribution of protein, fibre and minerals, such as potassium, calcium and iron (Doporto et al. 2011).

Botany

An erect, semi-erect to twining herbaceous perennial plant; erect type 15–40 cm high, semi-erect 30–60 cm and twining 60–200 cm long with tuberous roots. Tuberous roots 15 cm long, swollen at proximal end tapering distally like radish or turnip-shaped (Plate 1), 500–800 g, skin pale yellowish and flesh white. The leaves are trifoliate

Plate 1 Ahipa roots (Olivia Lopez)

Plate 2 Pod and leaf. (Frank Van Keirsbilck)

with stipules (Plate 2), leaflets entire with caduceus stipels. All plant types have short inflorescences 5–9 cm, with 0–6 lateral axes in the inflorescence with only 2–6 flowers per lateral axis. Flowers bisexual, papilionaceous, white or pale lavender and borne on short pedicels. Calyx tubular. Corolla wings and keel glabrous, wing petals curve outward post anthesis, stigma bent and in close contact with anthers. Legume pod 13–17 by 11–16 cm wide, cylindrical weakly dorsiventrally compressed (Plate 2). Seeds round, reniform black or white mottled, or lilac, 8–10 mm and 10 mm across.

Nutritive/Medicinal Properties

Tuberous Root Nutrient/ Phytochemicals

Ahipa tuberous roots and flour can be considered alternative food sources of gluten-free starch, with a considerable contribution of protein, fibre and minerals, such as potassium, calcium and iron (Doporto et al. 2011). Ahipa roots were reported to contain 15–30 % dry matter, 19–28 % sugars, <1 % fat, 48–54 % starch containing 99.9 % amylopectin, 8–18 % protein and about 80 % of the protein was reported to be water soluble and not extractable within a pH range of 2–10 (Sørensen et al. 1997). Popenoe et al. (1989) reported ahipa root to be low in sodium and calories (containing approximately 50 calories per cup raw) and a good source of potassium and vitamin C. Leidi et al. (2003) found ahipa root to contain 16.3–21.1 % dry matter, 38.6–54.4 % starch, 21.8–51.4 % sugars, and high free amino acids 0.26–0.41 % with asparagine as the main amino-N compound. Forsyth and Shewry (2002) found ahipa tuberous roots to contain between 0.77 and 1.34 % nitrogen on a dry weight basis equivalent to 4.8–8.4 % crude protein. It was calculated that salt-soluble proteins comprised about 60 % of the total tuber nitrogen, with low-molecular-mass nitrogenous components comprising a further 30 %. Five 'major' protein bands were found, which together accounted for about 19 % of the total salt-soluble protein fraction. These had similarities to α-amylases, chitinases and chitin binding proteins, cysteine proteinases (including major components from *P. erosus* tubers), a tuberization-specific protein from potato, and proteins induced in soybean and pea by stress or the plant hormone abscisic acid, respectively. Starch accounted for 56–58 % of ahipa root dry weight with granules occurring in a range of geometric forms and in sizes from below 5 μm to about 35 μm (mean about 10 μm in all accessions except two) (Forsyth et al. 2002). The amylose content ranged from 11.6 to 16.8 % and over 23 % in the seed starches. X- ray diffraction analysis showed A-type or C(A)-type diffraction patterns. The chain-length distribution of the amylopectin after enzyme debranching showed a peak at DP11 similar to that of wheat starch, but had a less marked shoulder at DP 21–22 and contained a higher proportion of longer chains. Differential scanning calorimetry showed an endothermic peak corresponding to gelatinization with T_{max} ranging from 59 to 63 °C, which was similar to the T_{max} of wheat (about 64 °C). The composition of the ahipa starch may mean that it is suitable for food applications that require low amylose content and low retrogradation after processing.

Leonel et al. (2003) found ahipa roots to have a moisture content of 82 %, starch 7.86 %, protein 1 %, fibre 0.74 %, reducing sugars 2.68 %m, total sugars 4.24 %, ash 0.4 %, lipid 0.15, pH 5.52 and titratable acidity 1.04. Ahipa starch extract had 12.3 % moisture, 84 % starch containing

12.84 % amylose. The starch granules had circular and polygonal shapes; the size ranged between 10 and 25 mm. The viscosity profile of the *P. ahipa* starch showed low pasting temperature (56 °C), peak at 272 RVU with high breakdown value. Leonel et al. (2005) found that planting dates and harvesting periods impacted on the nutrient composition of ahipa tuberous roots at harvesting. Nutrient composition of roots 5 months after planting in October and February were determined, respectively, as: moisture 83.38 %, 81.18 %; and per dry basis – ash 2.53, 2.65 %; reduced sugar 13, 9.68 %; total soluble sugar 26.85, 27.39 %; lipid 0.41, 0.83 %; protein (Nx6.25) 8.23, 7.18 %; starch 50.72, 47.28 %; and fibre 2.08, 8.36 %. Nutrient composition of roots 7 months after planting in October and February were determined, respectively, as: moisture 84.33 %, 83.91 %; and per dry basis – ash 1.99, 2.70 %; reduced sugar 8.99, 10.42 %; total soluble sugar 15.83, 28.31 %; lipid 0.70, 0.35 %; protein (N x 6.25) 6.14, 8.54 %; starch 57.16, 42.55 %; and fibre 8.93, 9.93 %. The average amylose values recorded (October planting 13.9 %, 13.3 % for 5 months, 7 months harvesting, respectively; February planting 15.8, 13.7 % for 5 months, 7 months harvesting, respectively) were close to those of Forsyth et al. (2002), who reported 11.6–16.8 % for six *P. ahipa* crops and much higher than those reported by Ørting et al. (1996) of 1–1.5 % amylose content. Plant age and planting periods were found to influence starch granule size distribution. Starch granule size was found to increase with plant development, 10–20 μm at 5 months, 15–25 μm at 7 months and 12–18 μm at 9 months in the October planting. At 7 months distribution was heterogeneous and at 9 months it was more homogenous. In the second planting period, starch granules also increased in size with plant development (13.79 μm and 15.48 μm average granule size at 5 and 7 months, respectively). However, at the end of 7 months, starch granule size distribution was not as homogeneous as for the first planting, at 9 months. They also found that pasting temperature, peak viscosity and hot paste stability decreased with tuber growth, while peak viscosity increased.

Planting date of ahipa impacted on productivity when delayed, because of growth season shortening (Leidi et al. 2004a). Reproductive pruning significantly augmented root yield. Increasing planting density affected root and fruit growth per plant but increased yield to a certain extent. Seed inoculation with effective rhizobia greatly enhanced root and seed production. Flower pruning was found to increase ahipa root yield without modifying the functional properties of the starch (López et al. 2010). Flower-pruning practice increased 1.6 times starch extraction yield. The nutrient composition determined for ahipa roots was moisture 81.9 %, crude protein (dry basis) 6.5 %, ash 3.7 %, fat 1.5 %, crude fibre 13.3 %, starch 56.54 % or 11.31 % w/w on wet basis and amylase 10.75 %. Starch diffractograms were assigned to a type C pattern. Ahipa starch granules were not associated as clusters; they exhibited round and polygonal shapes with irregular borders with a mean diameter of 7.95 μm or starch granules without flower pruning and 7.32 μm for those roots with flower pruning. Ahipa starch exhibited a monomodal straight distribution ranged from 2 to 18 μm, in both starch samples. Granular sizes were lower than those reported for potato starch, with an average size ranging between 30 and 70 μm, and those reported for cassava starch which exhibited average sizes of 12–15 μm Ahipa starch pastes gelatinized at relatively low temperature and showed low retrogradation tendency under refrigeration conditions. Rheological properties indicated that ahipa starch might be adequate as a food thickener. Thus, native ahipa starch appeared to be an alternative to other traditional starch source.

Harvested *P. ahipa* tuberous roots were found to have moisture 90.84 %, ash 3.97 %, lipids 0.44 %, proteins 11.57 %, carbohydrates 84.02 %, TSS (total soluble solids) 6.9°Brix, TTA (total titratable acidity) 0.22 % citric acid and pH 6.08 (Mussury et al. 2013). The lowest loss of root fresh weight and higher TTA were observed in the cold chamber and PVC bags. The lowest TSS contents were observed for roots stored in the cold chamber, and these did not vary among the packing materials. The average carbohydrate content percentage for all treatments was 84.9 %.

The percentage of lipids was highest in roots stored at room temperature while protein and ash contents were highest in tubers under refrigeration. The best storage conditions for roots were plastic bags packaging in a cold chamber, affording appropriate quality for commercialization for up to 30 days.

The grating process for ahipa flour production required a pressing step (AFGP) and the recovery of the starch leached (Doporto et al. 2011). The slicing procedure (AFS) was simpler and the resulting product showed higher contents of potassium, magnesium, calcium and protein than did AFGP, which showed lower sodium and higher acid detergent fibre contents, together with lower gelatinization temperature. Both flours differed in terms of α-amylase activity and swelling power, characteristics that may condition their specific applications, such as the incorporation of these flours as gluten-free functional food ingredients.

The carbohydrate-rich roots can be eaten raw and provide calories and vitamin K and C, as well as potassium to the diet (Leidi et al. 2003). Ahipa tubers are even considered to have a cleansing effect on the body. It is supposed to cure infections of the throat and the air passage (Sørensen et al. 1997). Its dry matter ranges from 15 to 30 % (Sørensen et al. 1997). Further the tuber contains 48–54 % starch, which consists of 96–99.9 % amylopectin. This is a very high value and especially interesting for food processing, where low solubilization and retrogradation are important. Because of the high starch and amylopectin content, ahipa is a good material for the starch industry (Sørensen et al. 1997). Further the suitability of the Andean bean depends on other factors such as extractability of the starch, the diameter of the starch granules or particles and their distribution. After 9 months, the starch granules are very homogenously distributed in the tuber. There is a high percentage of granules, which show the same size of about 12–18 μm diameter (Leonel et al. 2005). This indicates that the tuber growth is completed (Sørensen et al. 1997).

Seed Nutrients/Phytochemicals

P. ahipa seeds harvested in 1996 and 1997 were found to have the following oil, protein, fatty acid and tocopherol contents in mg/kg (range), respectively: oil content 24.1–28.1 %, 18.7–24.8 %; protein 23.2–276.6 %, 23.7–30.8 %; palmitic acid 27.5–31.05 %, 26.8–31.1 %; stearic acid 4.7–7.5 %, 4.7–6.4 %; oleic acid 21–24.9 %, 22.3–25.8 %; linoleic acid 35–40.7 %, 34.9–39.4 %; linolenic acid 1.3–2.5 %,1.3–2.3 %; total tocopherols 508.5–858 mg, 465.9–896.2 mg; α-tocopherol 0.1–0.7 mg, 0.1–0.9 mg; γ-tocopherol 97.7–99.8 mg, 97.7–99.3 mg; and δ-tocopherol 0.1–0.9 mg, 0.6–2 mg (Grüneberg et al. 1999). In all the samples, γ-tocopherol was predominant, accounting for more than 90 % of the total tocopherol content. The combination of high oil (18–28 %) and protein (23–31 %) contents, together with high palmitic acid (27–31 %), low linolenic acid (1.3–2.5 %), and high γ-tocopherol concentration (466–896 mg), makes *P. ahipa* an interesting alternative as sources of high palmitic acid oil for the food industry. At harvest, ahipa seeds showed high N accumulation whereas roots and pod-shells were the main recipients for P and K (Leidi et al. 2004b). The amount of nutrients removed by harvested roots and seeds was calculated in 67 kg N/ha, 13.7 kg P/ha and 80.2 kg K/h.

Ahipa seeds were found to have high protein (25.2–31.4 %) and high oil contents (18.7–22.4 %) and some accessions contained canavanine, amino acid structural analogue of arginine (Leidi et al. 2003). *Pachyrhizus erosus* and *P. ahipa* seeds were found to contain 1.13–2.76 mg/g dry material of rotenone using SPE HPLC-UV method (Lautié et al. 2012). Ahipa seeds were found to contain two toxic polyphenols, namely, pachyrrhizine 0.25–3.29 mg/g and rotenone 0.58–3.52 mg/g (Lautié et al. 2013). Different food processes, namely, drying, roasting, boiling, frying, alcohol extraction, tegument removal and traditional Beninese culinary recipes of yam bean seeds afforded a rotenone removal of up to 80 % (Catteau et al. 2013). The

most effective methods were the drying and roasting of the seeds and the maceration of their flour in local alcohol. Rotenone degradation and elimination were confirmed by cytotoxic assays, effectively inducing a decrease in sample toxicity.

Medicinal Uses

Ahipa tubers are considered to have a cleansing effect on the body and believed to cure infections of the throat and the air passage (Sørensen et al. 1997).

Other Uses

Ahipa is a legume crop of great potential for the production of raw materials (starch, sugar, oil and proteins) for industrial use (Leidi et al. 2004b). Ahipa's ability to fix atmospheric N_2 in association with rhizobia and the tolerance to insects makes it an attractive option for low input agriculture schemes. Ahipa was reported to have low environmental impact and low input requirement (fertilizer, pesticides), providing an attractive alternative to traditional sources of carbohydrates for conserving non-renewable resources and maintaining farmer profitability (Leidi et al. 2004a). Competitive yields for the simultaneous production of starch, feed protein and industrial oil may be obtained with low investment using available landraces.

The plant contains rotenone in the seeds and leaves (Grau et al. 1997), the active ingredient in the insecticide 'derris', and it has the potential to be used as an insecticide (Huxley et al. 1992).

Comments

Ahipa can be readily propagated from seeds or division of tubers.

Selected References

Catteau L, Lautié E, Koné O, Coppée M, Hell K, Pomalegni CB, Quetin-Leclercq J (2013) Degradation of rotenone in yam bean seeds (*Pachyrhizus* sp.) through food processing. J Agric Food Chem 61(46):11173–11179

Doporto MC, Mugridge A, García MA, Viña SZ (2011) *Pachyrhizus ahipa* (Wedd.) Parodi roots and flour: biochemical and functional characteristics. Food Chem 126(4):1670–1678

Forsyth JL, Shewry PR (2002) Characterization of the major proteins of tubers of yam bean (*Pachyrhizus ahipa*). J Agric Food Chem 50(7):1939–1944

Forsyth JL, Ring SG, Noel TR, Parker R, Cairns P, Findlay K, Shewry PR (2002) Characterization of starch from tubers of yam bean (*Pachyrhizus ahipa*). J Agric Food Chem 50(2):361–367

Grau A (1997) Ahipa, la legumbre tuberosa de los Andes. Ciencia Hoy 7:31–38

Grüneberg WJ, Goffman FD, Velasco L (1999) Characterization of yam bean (*Pachyrhizus* spp.) seeds as potential sources of high palmitic acid oil. J Am Oil Chem Soc 76(11):1309–1312

Huxley AJ, Griffiths M, Levy M (eds) (1992) The new RHS dictionary of gardening. 4 vols. MacMillan, New York

Lautié E, Rozet E, Hubert P, Quetin-Leclercq J (2012) Quantification of rotenone in seeds of different species of yam bean (*Pachyrhizus* sp.) by a SPE HPLC-UV method. Food Chem 131(4):1531–1538

Lautié E, Rozet E, Hubert P, Vandelaer N, Billard F, Zum Felde T, Grüneberg WJ, Quetin-Leclercq J (2013) Fast method for the simultaneous quantification of toxic polyphenols applied to the selection of genotypes of yam bean (*Pachyrhizus* sp.) seeds. Talanta 117:94–101

Leidi EO, Sarmiento R, Rodríguez-Navarro DN (2003) Ahipa (*Pachyrhizus ahipa* [Wedd.] Parodi): an alternative legume crop for sustainable production of starch, oil and protein. Ind Crop Prod 17:27–37

Leidi EO, Rodríguez-Navarro DN, Fernández M, Sarmiento R, Semedo J, Marques N, Matos A, Machado AP, Ørting B, Sørensen M, Matos MC (2004a) Factors affecting root and seed yield in ahipa (*Pachyrhizus ahipa* (Wedd.) Parodi), a multipurpose legume crop. Eur J Agron 20:395–403

Leidi EO, Vidueira JM, Cobo J, de Castro A, Rodríguez-Navarro DN (2004b) Nutrient requirements of ahipa, tuberous-root crop. J Plant Nutr 27:931–945

Leonel M, Sarmento SBS, Cereda MP (2003) Extração e caracterização de amido de jacatupé (*Pachyrhizus ahipa*). (Extraction and characterization of *Pachyrhizus ahipa* starch). Ciênc Tecnol Aliment 23(3):362–365 (in Portuguese)

Leonel M, Ferrari TB, Sarmento, SBS, de Oliveira MA (2005) Planting time, developmental stages and characteristics of roots and starch of *Pachyrhizus ahipa*. Sci Agric (Piracicaba, Braz.) 62(6):528–533

López OV, Viña SZ, Pachas ANA, Sisterna MN, Rohatsch PH, Mugridge A, Fassola HE, García MA (2010) Composition and food properties of *Pachyrhizus ahipa* roots and starch. Int J Food Sci Technol 45:223–233

Mussury RM, Scalon SP, Silva MA, Silva TF, Gomes H, Gassi R (2013) Postharvest conservation of the tuberous roots of *Pachyrhizus ahipa* (Wedd) Parodi. An Acad Bras Cienc 85(2):761–768

Ørting B, Grüneberg J, Sørensen M (1996) Ahipa (*Pachyrhizus ahipa* (Wedd.) Parodi) in Bolivia. Genet Resour Crop Evol 43:435–446

Popenoe H, King SR, Léon J, Kalinkowski LS, Vietmeyer ND, Dafforn M (1989) Lost crops of the Incas. Little known Plants of the Andes with Promise for Worldwide Cultivation. National Academy Press, Washington, DC, pp 37–46

Sørensen M, Gruneberg WJ, Ørting B (1997) Ahipa (*Pachyrhizus ahipa* (Wedd.) Parodi). In: Herman M, Heller J (eds) Andean roots and tubers: ahipa, arracacha, maca and yacon. Promoting the conservation and use of underutilized and neglected crops. International Plant Genetic Resources Institute, Rome, pp 13–73

Uphof JCT (1968) Dictionary of economic plants, 2nd edn (1st edn 1959). Cramer, Lehre, 591 pp

Usher G (1974) A dictionary of plants used by man. Constable, London. 619 pp

Pachyrhizus erosus

Scientific Name

Pachyrhizus erosus **(L.) Urban**

Synonyms

Cacara bulbous Thouars, *Cacara bulbosa* Rumphius ex Du Petit-Thouars, *Cacara erosa* (L.) Kuntze, *Cacara palmatiloba* (DC.) Kuntze, *Dolichos articulatus* Lam., *Dolichos bulbosus* L., *Dolichos erosus* L. (Basionym), *Dolichos palmatilobus* DC., *Pachyrhizus angulatus* Rich. ex DC., *Pachyrhizus articulatus* (Lam.) Duch. & Walp, *Pachyrhizus bulbosus* (L.) Kurz, *Pachyrhizus erosus* var. *erosus*, *Pachyrhizus erosus* var. *palmatilobus* (Moc. & Sessé ex DC.) R. T. Clausen, *Pachyrhizus erosus* var. *typicus* R.T.Clausen, *Pachyrhizus jicamas* Blanco, *Pachyrhizus palmatilobus* Benth. & Hook.f., *Pachyrhizus strigosus* R. T. Clausen, *Robynsia lobata* M.Martens & Galeotti, *Robynsia macrophylla* M. Martens & Galeotti, *Stizolobium bulbosum* (L.) Spreng., *Stizolobium domingense* Spreng., *Taeniocarpum articulatum* (Lam.) Desv.

Family

Fabaceae or Leguminosae

Common/English Names

Jicama, Yam Bean, Indian Potato; Chop Suey Bean, Mexican Jicama, Mexican Yam Bean, Chop-Suey Bean, Manioc Bean, Three-Lobed-Leaved Yam Bean, Four-Lobed-Root Yam Bean, Short-Podded Yam Bean, Mexican Potato, Mexican Turnip

Vernacular Names

Antilles: Patate-Cochon;
Argentina: Poroto Batata;
Brazil: Jacatupe;
Burmese: Pre Myit, Sane-Saar-U;
Chinese: Bai Tu Gua, Dou Shu, Fan Ge, Fan Ko, Liang Shu, Sha Ge, Sha Kot, Tu Gua;
Danish: Mexikansk Yamsbønne;
Dutch: Bengkoewang, Hoewi Iris, Hoewi Hiris;
French: Dolique Bulbeuse, Dolique Tubereux, Patate Cochon, Pois Batate, Pois; Manioc, Pois Patate;
German: Yambohne, Yamsbohne, Knollige Bohne;
Guinea: Pois Cachou, Pois Manioc:
Hawaii: Chopsui Potato;
India: Shankalu (Bengali), Kasaur (Bihar), Mishrikand, Sankalu (Hindi), Tani Uttan Kai (Tamil), Kandha (Telugu);

T.K. Lim, *Edible Medicinal and Non-Medicinal Plants: Volume 10, Modified Stems, Roots, Bulbs*,
DOI 10.1007/978-94-017-7276-1_20

Indonesia: Bengkawang, Besusu (Javanese), Benkuang, Bangkowang, Beto, Betok (Madurese), Sengkuang (Sumatra), Bankowang, Huwi Hiris (Sundanese);
Italian: Fagiolo Patata, Dolico Bulboso;
Japanese: Kuzuimo;
Khmer: Pe'kuĕk;
Laotian: Man Ph'au;
Malaysia: Bangkuan, Mengkuan, Sengkuang, Singkong;
Maya: Mehen-Chikam;
Mexico: Jicama, Jiquima;
Nepalese: Keshaura;
Peru: Jicama, Jiquima;
Philippines: Kamias (Bisaya), Kamas (Iloko), Sikamas (Pampangan), Lakamas (Pangasinan), Kamah (Sambali), Hinkamas, Iguama, Singkamas, Sinkamas (Tagalog);
Portuguese: Jacatupé, Jacutupé, Jocotupé;
Salvador: Frijol De Jicama;
Spanish: Achipa, Ajipa, Jícama, Judía Batata, Yuco De Bejuco;
Swedish: Jamsbönrot;
Thai: Man Kaeo, Huapaekua, Man Lao;
Turkish: Köklü Böyrüce;
Venezuela: Carota De Caballo, Nupe, Npera, Yuco De Bejuco;
Vietnamese: Cây Củ Đậu (Northern Vietnam), Củ Sắn, Sắn Nước (Southern)

Origin/Distribution

The species is indigenous to Central America – from South Mexico to Nicaragua and Costa Rica – and has naturalized elsewhere after introduction. It is cultivated as a tuber crop in the West Indies, in South and Southeast Asia from the islands of the Indian Ocean, India, Myanmar, Indo China, Southern China, Malaysia, Indonesia, the Philippines and Pacific islands and to a lesser extent in tropical South America, Africa and Australia.

Agroecology

The plant thrives in a warm, humid tropical climate from near sea level to 1400 m elevation with an optimum temperature range of 20–28 °C in zones with a moderate precipitation rate, that is, approximately 1500 mm mean annual rainfall. It prefers full sun and rich, moist, well-drained light, sandy-loamy, alluvial or volcanic soils; waterlogged soil is detrimental as it causes rotting of tubers. It is frost sensitive and hardy.

Day-length-short days were found to be necessary for tuberization of jicama (Cotter and Gomez 1979). When grown under a 14–15 h photoperiod, the vegetative growth was good, but there was little production of tuberous roots: short day length produced smaller, more bushy plants and good tuberization. Decreasing day length at time of planting initiated tuberous root development, whereas increasing day length inhibited tuberous root development and promoted vine and leaf growth (Paull et al. 1988). Once tuberous root formation and flowering began, stem and leaf growth ceased. At the time of planting, the critical day length for tuberous root formation and flowering was 11–12 h. Field studies in Senegal demonstrated *P. erosus* to be good drought avoider (Annerose and Diouf 1994). Leidi (2002) found that under sufficient water supply, the increase in air temperature and decrease in air humidity increased stomatal conductance and net photosynthetic rate in *P. erosus*. In a drying soil, yam bean showed greater sensitivity than ahipa (*P. ahipa*) to water restriction. High stomatal conductance at low humidity increased photosynthetic rate but resulted in lower water use efficiency. Flower and tuber formation were reported to occur almost simultaneously during *P. erosus* plant development (Arevalo 1998). Hence, removal of inflorescences would channel more assimilates to the tubers. Studies in P. erosus found that removal of flowers resulted in higher yield (Adjahossou and Ade 1998; Caro and Casillas 1998), biomass produc-

tion and sugar content (Arevalo 1998). Due to the cost of labour, it would be most efficient to prune flowers only once, rather than the four times necessary to eliminate flowers totally, as the single pruning increased tuber growth to almost that of multiple prunings (Caro and Casillas 1998). Use of chemicals such as NAA, chlormequat (1500 ppm) or 2,4-D (25 ppm) to remove flowers or manually by hand had been reported to be effective in increasing tuber yield (Panda and Sen 1995). Mora and Morera (1998) found hand removal of flowers gave higher tuber yield than spraying with 50 ppm 2,4-D.

Edible Plant Parts and Uses

Root tubers and immature pods (substitute for French beans) are eaten as vegetables in Southeast Asia. The crisp white flesh can be sliced, diced or cut into strips for use as a garnish, in salads, or with dips (Herklots 1972). It is frequently served as a snack, sprinkled with lime or lemon juice and a dash of chili powder. Jicama remains crisp after boiling and serves as a textural substitute for water chestnuts. Jicama is similar in food value to white potatoes, but with slightly fewer calories.

Seed pods – the young seed pods of *P. erosus* are sometimes eaten as a cooked vegetable, similarly to French beans, but cannot so be used as the seeds develop (Kay 1973). The young tubers are eaten raw in salads, or cooked as a vegetable, or in pickles and chutney.

The tubers are sliced and consumed fresh whole or in vegetable salads and chop suey. They are also eaten cooked, roasted, braised or simmered in soups, or cooked in stir-fried dishes with meat and seafood like shrimps and dried squid or cuttlefish or preserved in vinegar. The tuber is also cut into cubes and used as an ingredient for a mixed fruit cocktail. In Malaysia and Indonesia, slivers of the peeled tubers are mixed with other vegetables and fruit in a vegetarian dish called 'rujak' which is eaten with a spicy peanut sauce and prawn paste. Jicama is the predominant ingredient in the Malaysian speciality called 'popiah' and the Chinese salad, 'yusheng'. The flavour of the tuber is sweet, starchy, crispy and pleasant, reminiscent of apples. The most common way of eating jicama in Mexico is raw, cut into pieces (10–20 cm long by 0.5–2 cm wide) and topped with lemon juice and powdered chilli (Gómez-Aldapa et al. 2013). Frequently prepared in the home, fresh-cut jicama can also be purchased in restaurants, public markets and from street vendors. Several commercial products such as 'pikel' comprising fermented yam bean tuberous roots, vegetables and fruit in 15–20 % salt solution, yam bean syrup and juice, 'manisan bengkuang' in syrup (fresh root in syrup) or 'asinan bengkuang'(fresh root in salt solution) are sold in the local markets in Indonesia (Karuniawan 2004).

In Latin America, jicama is also a source of a starch used in custards and puddings.

Incorporation of the yam bean soluble fibre to stirred yogurt reduced significantly the syneresis and produced a more acceptable mouthfeel enriched yogurt in comparison to just stirred yogurt, indicating the viability of the process to obtain a commercial product (Ramirez-Santiago et al. 2010). *Pachyrhizus* spp. including *P. erosus* are important food crops in the tropics with a long history of cultivation and have shown promising potential in the production of oligosaccharides/monosaccharides, high glucose syrups, flour and wine and incorporated in yogurt or other milk products (Ramos-de-la-Peña, et al. 2013). Its importance in America with recent prospects for export, as well as its wide acceptance in western Africa and Southeast Asia, has stimulated interest in its conservation and processing.

Botany

Vigorous climbing herb, 2–5 m long (Plate 1) with a large, subglobose, large turnip-like shaped tuberous roots 10–20 cm or more across and can weigh up to 3 kg or more, yellowish-brown, coarse, membranous with creamy succulent white, crisp starchy flesh inside (Plates 2 and 3). Leaves pinnately trifoliate, coarse, stipules lanceolate or falcate, 0.5–1 cm long, petiole 10–15 cm long. Leaflets shortly stalked obliquely

Plate 1 Vigorous climbing jicama

ovate or rhomboid, base cuneate, apex acute, hispid on both surfaces, green, 6.5–13 cm long by 5–15 cm wide, margins entire or coarsely dentate in upper margins (Plate 4). Racemes axillary, fascicled, pentamerous, long peduncled, 20–60 cm long. Flowers shortly pedicelled, bluish-violet, glabrous, bracteoles at base of calyx, corolla glabrous, papilionaceous with standard, lateral wings and keel which is auricled at the base, stamen with elliptical, dorsifixed anthers, ovary, subsessile and multiovulate, style bearded, stigma oblique (Plate 4). Pods, subsessile, linear, acuminate, compressed, finely pubescent, 5–10 seeded, 7.5–15 cm long and 1.2–1.6 cm wide (Plate 5). Seeds flat rounded to squarish, olive-green to brown or reddish-brown.

Nutritive/Medicinal Properties

Tuberous Root Nutrients/ Phytochemicals

Analyses carried out in the United States report that raw, jicama tuber (excludes ends and skin) have the following proximate composition (per 100 g edible portion): water 90.07 g, energy 38 kcal (159 kj), protein 0.72 g, total lipid 0.09 g, ash 0.30 g, carbohydrates 58.29 g, total dietary fibre 4.9 g, total sugars 1.8 g, Ca 12 mg, Fe 0.6 mg, Mg 12 mg, P 18 mg, K 150 mg, Na 4 mg, Zn 0.16 mg, Cu 0.048 mg, Mn 0.060 mg, Se 0.7 μg, vitamin C 20.2 mg, thiamine 0.020 mg, riboflavin 0.029 mg, niacin 0.2 mg, pantothenic acid 0.1135 mg, vitamin B-6 0.042 mg, total folate 12 μg, choline 13.6 mg, vitamin A 21 IU, vitamin E (α-tocopherol) 0.46 mg, vitamin K (phylloquinone) 0.3 μg, total saturated fatty acids 0.021 g, 6:0 (palmitic acid) 0.018 g, 18:0 (stearic acid) 0.002 g, total monounsaturated fatty acids 0.005 g, 18:1 undifferentiated (oleic acid) 0.005 g, total polyunsaturated fatty acids 0.043 g, 18:2 undifferentiated (linoleic acid) 0.029 g, 18:3 undifferentiated (linolenic acid) 0.014 g, threonine 0.018 g, isoleucine 0.016 g, leucine 0.025 g, lysine 0.026 g, methionine 0.007 g, cysteine 0.0.006 g, phenylalanine 0.0017 g, tyrosine 0.012 g, valine 0.022 g, arginine 0.037 g, histidine 0.019 g, alanine 0.020 g, aspartic acid 0.2 g, glutamic acid 0.043 g, glycine 0.016 g, proline 0.025 g, serine 0.025 g and β-carotene 13 μg (USDA-ARS 2014). Evans et al. (1977) found 3.08 % N in the meal and 0.6 % N in extracted meal of the tuberous roots. Total amino acid nitrogen accounted for 57.5 % of total nitrogen in the tuberous roots. Yam bean contained the largest proportion of non-protein N in its meal. It was found that the free and small peptide amino-acid fraction probably contained all of the protein-type amino acid but most was accounted for by aspartic acid (55.14 g/100 g)

Plate 2 Harvested jicama

Plate 3 Close-up tuberous, large, turnip-like tuberous roots

and arginine (7.38 g/100 g). The composition (g/100 g recovered amino acid) of other amino acids in yam bean meal were: threonine 2.68 g, serine 3.28 g, glutamine 5.37 g, proline 2.53 g, glycine 1.86 g, alanine 2.46 g, valine 3.20, methionine 0.82 g, isoleucine 2.09 g, leucine 2.91 g, tyrosine 1.71 g, phenylalanine 2.31 g, histidine 2.53 g, lysine 3.28 g and cysteine 0.45 g. The tuberous roots of yam bean contained 0.52 % of 'protein' N, 1.25 % of free and small-peptide amino-acid N and 1.31 % of non-amino acid N, most of which was soluble in 70 % ethanol. Hence, 43 % of the total nitrogen in the tuberous roots of yam bean was non-amino acid nitrogen, which probably consisted of nitrogen in reduced forms.

Plate 4 Jicama leaves and flowers

Plate 5 Old jicama legume pod

The chemical analysis of *P. erosus* roots at different plant ages (20–36 weeks) gave range values for dry matter as 16.19–22.28 %, protein 1.11–1.62 %, fat 0.553–0.867 %, crude fibre 0.3048–0.3943 %, and ash 0.669–1.089 % (Fernandez et al. 1997). The chemical constituents fluctuated with age but without specific trends. After 21 days storage, the moisture content of tubers decreased from 88.5 to 86.5 %, total sugars (mg/100 g FW) increased from 2817 to 4361 mg, sucrose from 485 to 1073 mg, fructose from 1034 to 1435 mg, glucose from 1298 to 1853 mg (Kawabata et al. 1986). Also, the ratios of fructose and glucose to total sugar content increased, while the ratios of sucrose decreased. Heat treatment (roasting) increased water soluble fraction and water soluble pectic substances.

Two triterpenoid glycosides, kaikasaponin III and phaseoside IV, along with daidzin and (+)-abrin, were isolated from *Pachyrhizus erosus* root (Yahara et al. 1994). Six active compounds with antioxidant and skin whitening activities, namely, daidzein, daidzin, genistin, (8,9)-furanylpterocarpan-3-ol, 4-(2-(furane-2-yl)ethyl)-2-methyl-2,5-dihydro-furane-3-carbaldehyde and 2-butoxy-2,5-bis(hydroxymethyl)-tetrahydrofurane-3,4-diol were isolated from Bengkoang (*P. erosus*) (Lukitaningsih et al. 2013a).

The sweet flavour of the tuberous flesh comes from the oligofructose inulin (also called fructo-oligosaccharide) (Noman et al. 2007). The tuber was found to have a high level of moisture (82 %), appreciable amounts of carbohydrate (14.9 %), crude fibre (1.4 %), protein (1.2 %) and negligibly low amount of lipids (0.1 %). The total

caloric value corresponded to 39 kcal/100 g. The tuber is considered to be a potential source of potassium, sodium, phosphorus, calcium and magnesium and found to contain a fair amount of ascorbic acid. Thiamine, riboflavin, pyridoxine, niacin, adenine, choline, phytoestrogen and folic acid were also detected and very negligible contents of anti-nutrient components were observed (Noman et al. 2007; Nurrochmad et al. 2010).

Pectins were the largest alcohol insoluble solids (AIS) component of both jicama and Chinese water chestnut comprising approximately 50 % of AIS (Klockman et al. 1991). Further analysis showed a similar composition of water soluble, chelator soluble and alkaline soluble pectins in both jicama and water chestnut. Jicama contained much lower hemicelluloses. Studies found jicamas and water chestnuts contained less galactose but more arabinose and glucuronic acid and much more xylose than potatoes (Pressey 1993). Water chestnuts contained less galacturonic acid than jicamas and potatoes. The results suggested that resistance to thermal softening may be related to cell wall polysaccharides rich in arabinose, glucuronic acid and xylose rather than to pectic polysaccharides. The major differences in enzyme compositions were much higher β-amylase and exopolygalacturonase in jicamas and water chestnuts than in potatoes. Another study reported that the tuberous roots of the jicama contained large quantities of two acidic glycoproteins designated YBG1 and YBG2, with molecular mass of 28,000 and 26000, which accounted for more than 70 % of the total soluble proteins (about 3 g per 100 g of tuber on a dry weight basis) (Gomes et al. 1997). A third protein named YBP22 which accounted for 2–5 % of the total soluble proteins had an M(r) of 22,000. Both YBG1 and YBG2 showed strong homology with the papain class of cysteine proteases and exhibited cysteine proteolytic activity. In contrast, YBP22 showed sequence homology with several known protease inhibitors, and a polyclonal antibody raised against this protein cross reacted with soybean trypsin inhibitor. α-Amylase, a starch splitting enzyme, was purified with a yield of 22.8 % from post-harvest yam bean tuberous root (Noman et al. 2006). This glycoprotein had a molecular weight of 40 kDa and contained 2.6 % sugar. The enzyme displayed optimum activity at pH 7.3 and 37 °C with an apparent Km value of 0.29 % for starch as substrate. The enzyme was strongly inhibited by Cu^{2+}, Fe^{2+} and Zn^{2+}, moderately by Li^{2+}, Hg^{+} and Cd^{2+} and slightly by Ag^{+}, Mg^{2+} and K^{+}. Calcium ion almost doubled the activity whereas Fe^{3+}, Mn^{2+} and Na+enhanced it appreciably.

Yam bean starch paste (from the tuberous roots) was found to have a high viscosity profile, high retrogradation tendency and low stability on cooking (Mélo et al. 2003). Gelatinization temperature (53–63 °C) and the pasting temperature (64.5 °C) were lower than those of cereal starch; however, the swelling power was higher (54.4 g gel/g dried starch). Yam bean starch could be used as a potential new source of starch on the basis of similar functional properties as cassava starch. Jícama starch granules were spherical or polygonal with 1–15 μm diameters (Stevenson et al. 2007). The starch exhibited C_A-type X-ray diffraction pattern, an apparent amylose content of 28.1 % and absolute amylose content of 23.6 %. Jícama amylopectin weight-average molar mass (M_w) was 3.9×10^8 g/mol and gyration radius (R_z) was 363 nm. Average amylopectin branch chain-length was short (DP 22.7). Onset gelatinization temperature was very low (52.0 °C) and enthalpy change was 15.1 J/g. Peak (282 RVU), final (221 RVU) and breakdown (137 RVU) viscosity of 8 % jicama starch paste were high relative to other starches and pasting temperature was 72.3 °C. High paste viscosity and low gelatinization temperature could give jicama starch advantages in some industrial applications. High energy milling resulted in a partial fragmentation of jicama starch granules, increasing the water absorption index (WAI) and the water solubility index (WSI) (Martínez-Bustos et al. 2007). Thermal properties of jicama ball-milled starch were modified. The enthalpies were lower than native starch indicating that ball milling destroyed the crystallinity and double helical order arrangements. Ball-milled jicama starch showed some functional characteristics of gelatinization that allow their use in food systems as stabilizing, additives, moisture retainers and thickeners.

The extracted jicama root pectic polysaccharide (1.0 g/L) was used for the formation of polyelectrolyte complexes (PECs) with water-soluble chitosan (WSCh) (Ramos-de-la-Peña et al. 2011). The yield of the PECs was dependent on the concentration of WSCh used, it being 13.3, 26.7 and 18.3 % with WSCh at 0.5, 1.0 and 2.0 g/L, respectively. The PECs had higher thermostability compared with WSCh and had potential uses in a wide variety of specialized applications such as food technology and medicine.

Amylolytic enzymatic liquefaction of jicama tuberous roots increased the yields of juice (Ramos-de-la-Peña et al. 2012b). The highest volume of juice was 980 mL/kg tuberous root. The range for maximum glucose was 58.0–75.8 g/kg tuberous root and the lowest for residual AIS (alcohol insoluble solids) was in the range of 15.5–24.6 g/kg tuberous root after 10 h with 0.024 mL/g of enzyme per kg at 40 °C. Destarched jicama cell walls revealed a decrease in sugars after liquefaction except in xylose, glucose and mannose which increased as a result of this process. Hydrolysed jicama tuberous root was found to be a good source of high glucose syrups and destarched cell walls may need enzyme rich in xylanase to reduce residue further. Optimization of jicama liquefaction and saccharification was done using response surface methodology and response surface methodology-spline (Ramos-de-la-Peña et al. 2012a). At the optimum conditions (10 h and 0.024 mL/g) the pH was 4.24, total soluble solids (°Brix) amount was 9.00, weight loss percentage was 95.82 %, the total and reducing sugars liberated were 119.0 and 63.8 g/kg, respectively. Uronic acids obtained were 5.7 g/kg and polyphenols amount was 26.8 mg/kg.

When plants were fed with nitrate fertilizer, ^{14}C accumulated in leaf proteins and starch and less was partitioned to tubers or root nodules in contrast to control plants where the major photosynthate sinks were the shoot apices, root nodules and tubers (Vaillant et al. 1990). The negative effect of nitrate on nodules and tubers was consistent with the carbohydrate-deprivation hypothesis as a main cause of the nitrate inhibition of tuberous legume nodules. Tubers were strong storage sinks for carbon and accumulated more than 80 % of the total incorporated ^{14}C after a 72-h chase under short day (Vaillant and Desfontaines 1995). Although sucrose represented about 21 % of the tuber non-structural carbohydrate, that is, 15 % of the tuber dry weight, ^{14}C did not accumulate in sucrose but in glucose, fructose and starch. The data indicated the existence of a separate sucrose pool which was affected only very slowly by recent assimilates. The data also suggested the existence of two distinct pools of amino acids in the tuber, one utilized mostly for protein synthesis and the other probably stored in the vacuole. In the absence of tuber growth (because of long-day development), apex and roots were the main sinks for carbon (Robin et al. 1990). In the apex ^{14}C accumulated mostly in insoluble compounds. After 24 h, 27 % and 38 % of the apex ^{14}C was encountered in starch and proteins, respectively. It was concluded that carbon allocation to nodules was primarily used to sustain the assimilation of fixed nitrogen.

Tuberous roots sustained chilling injuries (internal discoloration, decay, loss of intact root firmness, weight loss, increased rate of water loss and ion leakage) when stored at temperatures < 10 °C (Paull and Chen 1988; Cantwell et al. 1992, 2002; Bergsma and Brecht 1992; Mercado-Silva et al. 1998). Roots stored at 12.5 °C for more than 4 weeks remained in excellent condition (Cantwell et al. 1992). Chilling injury symptoms occurred after 7 days at 10 °C or less, with an additional 2 days at 20 °C to allow symptoms to develop, but were not evident for up to 49 days at 15 °C (Bergsma and Brecht 1992). Roots stored at 20 °C for 6 weeks had no colour or textural changes although they lost about 40 % of their fresh weight (Cantwell et al. 2002). Chill-induced browning was associated with increased concentrations of soluble phenolic compounds and increased phenylalanine ammonia lyase activity. Phenolic compounds in chilled jicama root had UV spectra similar to those of catechins, though (+)-catechin and (−)-epicatechin were not present based on high-performance liquid chromatography (HPLC). Roots stored at

13 °C showed few internal quality changes over a 5-month period, although weight loss exceeded 35 % (Mercado-Silva et al. 1998). After 4 months storage at 22 °C, roots lost 14.5 % of their original weight; those stored at 12.5 °C lost 9.6 %; after 3 months storage, starch declined to one-sixth harvest level at 12.5 °C and two-thirds harvest level at 22 °C (Paull and Chen 1988). The decline in starch content at 12.5 °C was related to an increase in total sugars in the root, particularly sucrose. The sucrose content of roots stored at 12.5 °C tripled over a 3-month period. Glucose and fructose levels declined over the same period irrespective of storage temperature. Titratable acidity was very low, as was total phenols, and both did not change after harvest.

Studies found the browning of cut jicama at 20 °C to be related to the process of lignification in which the peroxidase enzyme played an important role (Aquino-Bolaños and Mercado-Silva 2004). After a week at 20 °C, the phenolic content, expressed as Gallic acid in fresh tissue, increased from 0.37 to 1.04 g/kg while the lignin content, expressed as Coumaric acid, increased from 16.50 to 52.22 mg/kg. Polyphenol oxidase (PPO) and peroxidase (POD) activities were induced by damage and were greater in the damaged external tissue than in the internal tissue; they were also influenced by temperature. Coumaric, caffeic and ferulic acids, coniferaldehyde and coniferyl alcohol (precursors in lignin synthesis) were found to be good substrates for POD.

Pod/Seeds Nutrients/Phytochemicals

Pachyrhizus erosus seeds (g/100 g) showed a high content of crude protein 28.27 g, lipids 26.80 g, Fe (16 mg/100 g) and Ca (356 mg/100 g), in comparison to other legumes (Santos et al. 1996). Its crude carbohydrate content was 26.85 g, crude fibre 6.20 g, ash 4.58 g and K 992 mg, P 286 mg, Na 6.8 mg, Zn 4 mg and Cu 1.2 mg. Glutelins constituted the highest protein fraction, followed by globulins. Antinutritional substances detected as tannins (10.20 mg/100 g), rotenoids (1.01 g/100 g), hemagglutinating activity and trypsin inhibitory activity (17.1 ITU (inhibited trypsin unit)) were in low concentrations. Seeds were also processed to obtain a flour which showed proper characteristics, good in vitro digestibility, significant rotenoid reduction level and amino acid composition rich in essential amino acids, except methionine. Proximate nutrient composition (g/100 g) of the flour was moisture 3.45 g, crude protein 51.20 g, lipid 1.51 g, crude fibre 4.42 g, ash 3.64 g and crude carbohydrate 35.78 g. The flour also contained 6 mg/100 g tannins, 0.06 mg/100 g rotenoids and had high in-vitro protein digestibility of 74.7 %. The amino acid pattern of *P. erosus* flour revealed its potential nutritional value as a protein source. All the dietary essential amino acids were present in amounts superior to the FAO/WHO (1985) standard (in parenthesis) except for sulphur amino acids which was also common in legumes like soya beans and *Phaseolus* beans. The amino acid profile of yam bean flour (g/100 g) was reported as isoleucine 4.4 g (2.8 g), threonine 3.5 g (3.4 g), phenylalanine 5.5 g (6.3 g for phenylalanine and tyrosine), tyrosine 3.2 g, valine 4.6 g (3.5 g), lysine 6.5 g (5.8 g), histidine 4.1 g (1.9 g), methionine 1.4 g (2.5 g recommended for 2–5-year children), leucine 7.7 g (1.1 g) and tryptophan (1.1 g). The non-essential amino acids were serine 4.3 g, aspartic acid 9.7 g, alanine 4 g, glycine 3.8 g, arginine 4.5 g, glutamic acid 15.8 g and proline 4.1 g.

The fatty acid composition of *P. erosus* seed oil was found to contain 26.7 % of palmitic, 5.7 % of stearic, 33.4 % of oleic and 34.2 % of linoleic acids (Broadbent and Shone 1963). The unsaponifiable matter of the oil contained 55 % of digitonide-forming sterols. The seed oil of *P. erosus* was found to have an oil content of 22.0–25.9 % and protein content of 28.9–31.8 % seed weight; fatty acids (expressed as % total fatty acids) palmitic acid 25.0–31.4 %, stearic acid 3.9–5.2 %, oleic acid 22.2–28.6 %, linoleic acid 35.9–38.1 %, linolenic acid 0.9–1,6 %; total tocopherols (mg/kg) 249.3–585.4 mg, α-tocopherol 2.3–9.7 mg, γ-tocopherol 90.1–97.5 mg and δ-tocopherol 0.1–2.5 mg (Grüneberg et al. 1999).

Proteins from jicama seeds comprised albumins as the major fraction (52.1–31.0 %), fol-

lowed by globulins (30.7–27.5 %) and prolamins (0.8 %) as the minor protein fraction (Morales-Arellano et al. 2001). Trypsin inhibitors were lower than those of other legumes. Overall, yam bean seed proteins showed an excellent balance of all essential amino acids and albumins contained the highest amount of essential amino acids.

Pachyrhizus erosus seeds were reported to contain rotenone (Hwang 1941; Norton 1943), saponins (Shangraw and Lynn 1955) and the glycoside pachyrhizid (Greshoff 1890, cited by Jones 1942). Pachyrhizid was found to be similar to derrid isolated from *Derris elliptica* and now known to be very impure rotenone (Van Sillevoldt 1899 cited by Jones 1942). Yam bean seeds were found to contain two toxic polyphenols, namely, pachyrrhizine 0.25–3.29 mg/g and rotenone 0.58–3.52 mg/g (Lautié et al. 2013). *P. erosus* seeds were also found to contain an isoflavone (dehydroneotenone), an isoflavanone (neotenone), 4 rotenoids and 2 furano-3-phenyl coumarins (Krishnamurthi and Seshadri 1966). Two new rotenoids 2a-hydroxydolineone and 12a-hydroxypachyrrhizone (Krishnamurti et al. 1970) and a new isoflavonone, erosenone were isolated from Indian yam bean (Kalra et al. 1977). Nine known components were isolated from *P. erosus* seeds: five rotenoids dolineone, pachyrrhizone, 12a-hydroxydolineone, 12a-hydroxypachyrrhizone and 12a-hydroxyrotenone; two isoflavonoids neotenone and dehydroneotenone; one phenylfuranocoumarin pachyrrhizine; and a monosaccharide (dulcitol) (Phrutivorapongkul et al. 2002). Rotenone, erosone, paquirrizone, dolineone and paquirrizine were isolated from the dichloromethane extract of *P. erosus* seeds and the isoflavone dehydroneotenone was isolated from the acetone extract (Barrera-Necha et al. 2004).

Extraction and analyses by the gravimetric and mercuric acetate methods showed the presence of rotenoids to the extent of 12.381 % and 10.431 %, respectively, in yam bean seeds (Sahu and Hameed 1983). When the seeds were subjected to rigorous extraction by Soxhlet apparatus, rotenoid oil content was found to be 22.2 %. Further extraction (after extraction of rotenoid oil) by polar solvent revealed the presence of rotenone and related compounds to the extent of 0.321 %. Rotenone in yam bean (*P. erosus* and *P. ahipa*) seeds was quantified to be 0.07–1.25 % (W/W) (Lautié et al. 2012). The results ranged from 1.13 to 2.76 mg/g dry material. Different food processes, namely, drying, roasting, boiling, frying, alcohol extraction, tegument removal and traditional Beninese culinary recipes of yam bean seeds afforded a rotenone removal of up to 80 % (Catteau et al. 2013). The most effective methods were the drying and roasting of the seeds and the maceration of their flour in local alcohol. Rotenone degradation and elimination were confirmed by cytotoxic assays, effectively inducing a decrease in sample toxicity. A novel ribosome-inactivating protein, pachyerosin, was purified from *P. erosus* seeds (Guo et al. 2014). Pachyerosin was found to be a type I ribosome-inactivating protein with a molecular mass of 29 kDa and an isoelectric point of 9.19. It strongly inhibited protein synthesis of rabbit reticulocyte lysate with an IC_{50} of 0.37 ng/mL and showed N-glycosidase activity on rat liver ribosomes with an EC_{50} of 85.9 pM. The N-terminal 27 amino acids of pachyerosin revealed a 60.71% sequence identity with abrin A from the seeds of *Abrus precatorius*.

An unnamed protein comprising one protein molecule of 33 kDa in an asymmetric unit was isolated from *P. erosus* seed (Lin et al. 1996). A 16 kDa protein SPE16 was purified from *P. erosus* seed (Wu et al. 2002, 2003). Its N-terminal amino acid sequence showed significant sequence homology to pathogenesis-related class 10 proteins. An activity assay indicated that SPE16 possessed ribonuclease activity in-vitro as do some other PR-10 proteins. A novel plant defensin protein SPE10 from *P. erosus* seeds was purified and partially sequenced (Song et al. 2004). Plant defensins are presumed to play an important role in the innate immunity of plants. Activity assays showed the recombinant SPE10 inhibited specifically the growth of several pathogenic fungi as native SPE10 (Song et al. 2005).

Plant Phytochemicals

Nodules, leaves and pods contained the major part of total nitrogen (Lamaze et al. 1985). Soluble nitrogen was found essentially in pods and stems. Ureide nitrogen content was low in leaves compared to other parts. In comparison to asparagine, glutamate, glutamine and aspartate, the ureides represented the main form of nitrogen transport. The ureides showed two peaks of accumulation, at first nodule appearance and at pod development. Xanthine dehydrogenase activity was detected only in the nodules, in which case, uricase activity was found to be 50–100 times higher than in other tissues. These two enzymes showed two peaks at the beginning of tuberization and at pod development. Allantoicase activity although present in all plant parts appeared to be greater in the stems. The furano pterocarpan, neodunol, was isolated as a major phytoalexin from the fungus, *Helminthosporium carbonum*-inoculated stems of *P. erosus* (Ingham 1979). Also isolated were small quantities of demethylmedicarpin and a prenylated pterocarpan, homoedudiol.

Titratable acidity was about 20 meq/g fresh weight if not induced to form tuberous roots, and approximately 15 meq/g fresh weight in plants forming tuberous roots (Paull et al. 1988). Total sugar was in the range of 30–50 mg/g fresh weight during tuberous root growth of induced plants. Total phenols in the tuberous root declined during development, whereas roots from plants uninduced to form tuberous roots had an increased level of total phenols.

Anticancer Activity

Yam bean seed extracts were highly cytotoxic to brine shrimps (Yongkhamcha and Indrapichate 2012). An immunotoxin was prepared by conjugating pachyerosin (from *P. erosus* seeds) with anti-human AFP monoclonal antibodies SM0736 (Guo et al. 2014). The immunotoxin pachyerosin-SM0736 efficiently inhibited the growth of the human hepatoma cell line HuH-7 with an IC_{50} of 0.050 nM, 2360 times lower than that of pachyerosin and 430 times lower than that of the immunotoxin against human gastric cancer cell line SGC7901. The results implied that pachyerosin may be used as a new promising anticancer agent. Rotenone, isolated from yam bean seeds, exhibited significant cytotoxic activity (IC_{50} = 13.05 μM), on K562 human leukaemia cells as determined by the MTT assay (Estrella-Parra et al. 2014). Three other isolated isoflavonoids were not cytotoxic.

Antiviral Activity

Among the compounds isolated from *P. erosus* seeds, 12a-hydroxydolineone and 12a-hydroxypachyrrhizone exhibited moderate anti-herpes simplex virus (HSV) Type 1 and II activity in-vitro (Phrutivorapongkul et al. 2002).

Immunomodulatory Activity

P. erosus powdered root fibre extract facilitated IgM production by the human hybridoma cell line HB4C5 cells (Kumalasari et al. 2014a). Additionally, production of IgM, IgG and IgA by mouse primary splenocytes was facilitated by the extract in a dose-dependent manner. The extract also significantly facilitated production of both interleukin-5 and interleukin-10 by splenocytes. In-vivo oral administration of the extract to mice for 14 days significantly enhanced serum Ig levels and activated production of Igs and cytokines (IL-6, IL-10, TNF-a, TGF-b and IFN-c) by lymphocytes from the spleen, mesenteric lymph nodes and Peyer's patch in mice. The results suggested the extract to have positive effects on the immune system in-vitro and in-vivo. They also found that the extract activated J774.1 cells via the MAPK and NF-κB signalling pathways (Kumalasari et al. 2014b). Treatment of J774.1 cells with the toll-like receptor e (TLR4) inhibitor significantly inhibited production of IL-6 and TNF-α, suggesting that TLR4 was the target membrane receptor for the extract. The immunostimulatory effect of the extract was nullified by the pectinase treatment, suggesting

the active ingredient in the extract to be pectin-like molecules.

CNS Depressant Activity

The result of animal study by Abid et al. (2006) found that ethanol extract of *P. erosus* seeds (150 mg/kg, p.o) decreased locomotor activity, produced muscle relaxation and showed antianxiety and anti-aggressive activity.

Antiosteoporotic Activity

Oral administration of ethyl acetate *P. erosus* root extract (EPE) at 200, 400 and 800 mg/kg BW significantly prevented bone loss in ovariectomized (ovx) rats whose effects were equivalent to that of estradiol (Nurrochmad et al. 2010). These effects were evident in increased length of femur and tibiae, bone density and mineral content of calcium and phosphorous in bone ash. EPE also significantly prevented OVX-induced uterine atrophy and increased body weight. The study suggested that phytoestrogen compound from *Pachyrhizus erosus* may offer a potential alternative therapy for the treatment of health problems such as osteoporosis in post-menopausal women.

Skin Whitening/Antityrosinase Activity

Six active compounds with antioxidant and skin whitening activities, namely, daidzein, daidzin, genistin, (8,9)-furanyl-pterocarpan-3-ol, 4-(2-(furane-2-yl)ethyl)-2-methyl-2,5-dihydrofurane-3-carbaldehyde and 2-butoxy-2,5-bis(hydroxymethyl)-tetrahydrofurane-3,4-diol were isolated from Bengkoang (*P. erosus*) (Lukitaningsih et al. 2013a). They also reported the following IC_{50} values for tyrosinase inhibition of the following compounds from bengkoang, daidzein 5.35 mM, Kojic acid 1.07 mM, daidzein-7-*O*-β-glucopyranose 22.20 mM, 5-hydroxyldaidzein-7-*O*-β-glucopyranose 4.38 mM, 4-(2-(furane-2-yl)ethyl)-2-methyl-2,5-dihydrofurane-3-carbaldehyde 0.198 mM, 2-butoxy-2,5-bis(hydroxymethyl)-tetrahydrofurane-3,4-diol 1.21 mM and (8,9)-furanyl-pterocarpan-3-ol 7.19 mM.

Photoprotective Activity

Cream and lotion formulations containing yam bean starch concentration of 15 %, 20 % and 25 % had sunscreen activity in-vivo in Swiss Webster female mice, with Sun Protecting Factor (SPF) value of 1.22, 1.52 and 2.38 (Zulkarnain et al. 2013).

Insecticidal Activity

All seed ethanolic extracts of yam bean (YSE), celery (CSE) and mint weed (*Hyptis suaveolens*) (MSE) were more toxic to *Aedes aegypti*, dengue virus vector, than water extracts (Yongkhamcha and Indrapichate 2012). The insecticidal efficacy of all extracts ranged as YSE/e > YSE/w > MSE/e > CSE/e > MSE/w > CSE/w. YSE/e was most toxic to *A. aegypti* second instar larvae and adults with LC_{50} of 16.22 μg/ml and 91.41 μg/ml, respectively, and much higher than MSE/s and CSE/e. YSE in combinations produced strong synergistic effects to the other extracts. The insecticidal activities of MSE and CSE were mild.

Adverse Issues

Yam bean seeds contained rotenone, a natural molecule previously used as an insecticide inhibiting the respiratory mitochondrial chain and was also proven to be toxic to mammals as chronic exposure led to the development of Parkinson-like symptoms in rats (Catteau et al. 2013). Rotenone genotoxicity was detected using the comet assay (Estrella-Parra et al. 2014). Rotenone induced cell death, and caspase-3 activation as indicated by TUNEL assay, and immunocytofluorescence. Plasmid nicking assay indicated that rotenone did not interact directly with DNA.

Narongchai et al. (2005) reported the case of a 59-year-old Taiwanese man who died within 2 h with respiratory failure after ingestion of yam bean seeds. Rotenone substance were found in yam bean seeds, gastric content and 72 ng/ml blood. Also generalized microscopic haemorrhage in the brain, lungs, liver and adrenal glands which were characteristic pathology were observed. The authors concluded that the cause of death was asphyxia from yam bean seed or rotenone toxicity. Hung et al. (2007) reported five patients presented with perioral numbness, nausea and vomiting after eating a same soup made from yam bean seeds. One of them, a 54-year-old woman, had difficulty breathing and lost consciousness. Physical examination showed dilated pupils and coma with no focal neurological signs. An initial diagnosis of cyanide intoxication was considered. Aggressive fluid and inotropic therapy were given and the patient eventually recovered. The other four patients suffered only minor gastrointestinal and neurological symptoms and received supportive treatment. Cyanide levels were negative in all five patients. Yam bean seed poisoning could cause acute metabolic acidosis and altered mental status, which could be confused with acute cyanide intoxication from a cyanogenic glycoside-containing plant.

Fu and Wang (2012) reported a 54-year-old Chinese woman who developed disturbed consciousness after eating 40 pieces of yam bean seeds. The diagnosis of this patient was toxic leukoencephalopathy by yam bean seeds intoxication.

Traditional Medicinal Uses

Seeds of *Pachyrhizus erosus* are used as folk medicine in treatment of insomnia (Abid et al. 2006). The crushed pod of *P. tuberosus*, mixed with lard, is used in China to cure itch; the dried roots are reported to be used as a cooling food for people with fever (Perry and Metzger 1980).

According to Burkill (1966), powdered seeds were used in Java for skin infections and prickly heat and seeds were regarded laxative and used as vermifuge in Indo-China. According to Stuart (2014), tincture from seeds is used for treatment of herpes. Warmed poultice of the stem pulp applied to painful areas in the leg decoction of the roots is used as a diuretic. In Taiwan, the roots are used for fever and haemorrhages

Other Uses

The seeds contain the toxin rotenone, which is used to poison insects and fish. Oil from the seeds is used like cotton seed (Kay 1973). The stems yield fibre for making fishing nets in Fiji (Burkill 1966). The plant is used as green manure and animal fodder. Old tubers are also used as feed for pigs and cattle. Yam bean leaves fermented using *Trichoderma koningii* were found to contain high crude protein and low crude fibre and could be used up to 12 % in broiler diet (Nuraini 2000; Nuraini et al. 2001).

Cosmetic products like yam bean face tonic, yam bean mask and yam bean powder, and a traditional cosmetic preparation comprising yam bean starch, rice meal, spices and flowers have been commercialized in Java and Sumatra (Karuniawan 2004).

On the basis of high toxicity in laboratory tests against the Mexican bean beetle, the silk worm and the bean aphid, it was suggested that the seeds of *P. erosus* and related species offered a possible substitute for rotenone insecticides (Hansberry and Lee 1943). Powdered yam bean contains insecticides known to work against insect pests. Finely grounded seeds were effective against third-instar larvae of *Epilachna varivestis* on bean leaves (Hansberry et al. 1947). Powdered yam bean added to wheat grains caused 83.33–96.66 mortality of rice weevil (Bhusan and Ghatak 1991). There was no egg hatching of rice moth *Corcyra cephalonica* at 6 days after treatment, *P. erosus* at 2.0 % (Ghatak and Bhusan 1995). Seed extract also worked against *Callosobruchus analis* (Kardinan and Wikardi 1997) and tobacco caterpillar *Spodoptera litura* but could not be used as a fumigant (Sahu and Hameed 1989). Yam bean seed extracts were also effective against black pepper stem borer,

Lophobaris piperis, in Indonesia (Rumbaina and Martono 1988; Deciyanto 1989).

P. erosus seeds contained antifungal compounds (Barrera-Necha et al. 2004). *P. erosus* seed powders inhibited *Colletotrichum gloeosporioides*, but stimulated the growth of *Fusarium oxysporum* and *Rhizopus stolonifer* at 0.5 and 5.0 mg/ml, respectively, in-vitro (Barrera-Necha et al. 2004). The hexane, dichloromethane and acetone extracts significantly inhibited the three fungi at 2.0, 5.0 and 10 mg/ml. The greatest antifungal activity was recorded for the dichloromethane extract on *R. stolonifer* (64.97 % inhibition), *F. oxysporum* (37.8 %) and *C. gloeosporioides* (36.4 %). Rotenone, erosone, paquirrizone, dolineone and paquirrizine were isolated from the dichloromethane extract of *P. erosus* seeds; the isoflavone dehydroneotenone was isolated from the acetone extract. These secondary metabolites significantly inhibited the three fungi at 250 μg/ml. Superior antifungal activity was recorded for rotenone on *R. stolonifer*, pachyrrizine on *F. oxysporum* and dehydroneotenone on *C. gloeosporioides*. SPE10, an antifungal protein, was isolated from *Pachyrhizus erosus* seeds (Song et al. 2005). Activity assays showed the recombinant SPE10 inhibited specifically the growth of several pathogenic fungi as native SPE10.

Comments

Cases of poisoning from ingestion of seeds and pods had been reported by Burkill (1966).

Selected References

Abid M, Hrishikeshavan HJ, Asad M (2006) Pharmacological evaluation of *Pachyrhizus erosus* (L) seeds for central nervous system depressant activity. Indian J Physiol Pharmacol 50(2):143–151

Adjahossou DF, Ade J (1998) Effects of inflorescence removal on tuber yield of some accessions of *Pachyrhizus erosus* (L.) Urban and *P. tuberosus* (Lam.) Sprengel. [French], pp 97–114. In: Sorensen M, Estrella Engelmann JE, Hamann OJ, Rios Ruiz SAMC (eds) Proceedings of the 2nd international symposium on tuberous legumes, Celaya, Guanajuato, Mexico, 5–8 August 1996, 566 pp

Annerose DJM, Diouf O (1994) Some aspects of response to drought in the genus *Pachyrhizus* Rich. ex DC, pp 199–214. In: Sørensen M (ed) Proceedings of the first international symposium on tuberous legumes, Guadeloupe, FWI, 21–24 Apr 1992. Jordbrugsforlaget, Copenhagen, Denmark

Aquino-Bolaños EN, Mercado-Silva E (2004) Effects of polyphenol oxidase and peroxidase activity, phenolics and lignin content on the browning of cut jicama. Postharv Biol Technol 33(3):275–283

Arevalo TA (1998) Effect of flower removal on the yield of root tubers of *Pachyrhizus erosus* L. Urban. [Spanish], pp 125–130. In: Sorensen M, Estrella Engelmann JE, Hamann OJ, Rios Ruiz SAMC (eds) Proceedings of the 2nd international symposium on tuberous legumes, Celaya, Guanajuato, Mexico, 5–8 August 1996, 566 pp

Barrera-Necha LL, Bautista-Banos S, Bravo-Luna L, García-Suárez FJ, Alavez-Solano D, Reyes-Chilpa R (2004) Antifungal activity of seed powders, extracts, and secondary metabolities of *P. erosus* L. against three postharvest fungi. Rev Mex Fitopatol 22(3):356–361

Bergsma KA, Brecht JK (1992) Postharvest respiration, moisture loss, sensory analysis and compositional changes in jicama (*Pachyrhizus erosus*) roots. Acta Horticult 318:325–332

Bhusan TK, Ghatak SS (1991) Evaluation of some plant powders as protectants against rice weevil *Sitophilus oryzae* Linn. (Curculionidae: Coleoptera). Environ Ecol 9(1):115–117

Broadbent JH, Shone G (1963) The composition of *Pachyrhizus erosus* (yam bean) seed oil. J Sci Food Agric 14(7):524–527

Burkill IH (1966) A Dictionary of the economic products of the Malay Peninsula. Revised reprint. 2 volumes. Ministry of Agriculture and Co-operatives, Kuala Lumpur, Malaysia. Vol. 1 (A–H) pp 1–1240, Vol. 2 (I–Z) pp 1241–2444

Cantwell M, Orozco W, Rubatzky V, Hernández L (1992) Postharvest handling and storage of jicama roots. Acta Hort (ISHS) 318:333–344

Cantwell MI, Peiser G, Mercado-Silva E (2002) Induction of chilling injury in jicama (*Pachyrhizus erosus*) roots: changes in texture, color and phenolics. Postharv Biol Technol 25(3):311–320

Caro VFJ, Casillas RD (1998) Effects of flower removal on yield of yam bean (*Pachyrhizus erosus* (L.) Urban) in Nayarit, Mexico. [Spanish], pp 131–137. In: Sorensen M, Estrella Engelmann JE, Hamann OJ, Rios Ruiz SAMC (eds) Proceedings of the 2nd international symposium on tuberous legumes, Celaya, Guanajuato, Mexico, 5–8 August 1996, 566 pp

Catteau L, Lautié E, Koné O, Coppée M, Hell K, Pomalegni CB, Quetin-Leclercq J (2013) Degradation of rotenone in yam bean seeds (*Pachyrhizus* sp.) through food processing. J Agric Food Chem 61(46):11173–11179

Cotter DJ, Gomez RE (1979) Daylength effect on root development of jicama (*Pachyrhizus erosus* Urban). HortSci 14(6):733–734

Deciyanto S (1989) Pengendalian hama utama lada secara terpadu di Indonesia. J Penelitian Pengembangan Pertanian 8(3):69–74 (in Indonesian)

Estrella-Parra EA, Gomez-Verjan JC, González-Sánchez I, Vázquez-Martínez ER, Vergara-Castañeda E, Cerbón MA, Alavez-Solano D, Reyes-Chilpa R (2014) Rotenone isolated from *Pachyrhizus erosus* displays cytotoxicity and genotoxicity in K562 cells. Nat Prod Res 23:1–6

Evans IM, Boulter D, Eaglesham ARJ, Dart PJ (1977) Protein content and protein quality of tuberous roots of some legumes determined by chemical methods. Qual Plant 27(3–4):275–285

Fernandez MV, Warid WA, Loaiza JM, Montiel A (1997) Developmental patterns of jicama (*Pachyrhizus erosus* (L.) Urban) plant and the chemical constituents of roots grown in Sonora, Mexico. Plant Foods Hum Nutr 50(4):279–286

Fu PK, Wang PY (2012) Toxic leukoencephalopathy due to yam bean seeds poisoning. Neurologist 18(4):199–201

Ghatak SS, Bhusan TK (1995) Evaluation on the ovicidal activity of some indigenous plant extracts on rice moth, *Corcyra cephalonica* Staint. (Galleriidae: Lepidoptera). Environ Ecol 13(2):284–286

Gomes AV, Sirju-Charran G, Barnes JA (1997) Major proteins of yam bean tubers. Phytochemistry 46(2):185–193

Gómez-Aldapa CA, Rangel-Vargas E, de León HB, Filardo-Kerstupp RS, Gordillo-Martínez AJ, Vázquez-Barrios ME, Castro-Rosas J (2013) Presence of diarrheagenic Escherichia coli pathotypes, Salmonella and indicator bacteria on fresh-cut jicama [*Pachyrhizus erosus* (L.) Urban] from public markets in Mexico. African J Microbiol Res 7(34):4323–4330

Greshoff M (1890) Cheraisch-pharmacologisch laboratorium eerste verslag van het onderzoek naar de plantenstof fen van Nederland- schindie. Meded. uit 's Lands Plantentuin VII. 127 pp. Batavia (in Dutch)

Grüneberg WJ, Goffman FD, Velasco L (1999) Characterization of yam bean (*Pachyrhizus* spp.) seeds as potential sources of high palmitic acid oil. J Am Oil Chem Soc 76(11):1309–1312

Guo JL, Cheng YL, Qiu Y, Shen CH, Yi B, Peng C (2014) Purification and characterization of a novel type I ribosome inactivating protein, pachyerosin, from *Pachyrhizus erosus* seeds, and preparation of its immunotoxin against human hepatoma cells. Planta Med 80(11):896–901

Hansberry R, Lee C (1943) The yam bean, *Pachyrhizus erosus* Urban, as a possible insecticide. J Econ Entomol 36(2):351–352

Hansberry R, Clausen RT, Norton LB (1947) Variations in the chemical composition and insecticidal properties of the yam bean (*Pachyrhizus*). J Agric Res 74(2):55–64

Herklots GAC (1972) Vegetables in South-East Asia. George Allen & Unwin, London. 525 pp

Hung YM, Hung SY, Olson KR, Chou KJ, Lin SL, Chung HM, Tung CN, Chang JC (2007) Yam bean seed poisoning mimicking cyanide intoxication. Intern Med J 37(2):130–132

Hwang SL (1941) A preliminary report on the chemical composition of yam bean (*Pachyrhizus erosus* Urban), a new Rotenone bearing plant. Kwangsi Agric 2(4):269–280 (in Chinese)

Ingham JL (1979) Isoflavonoid phytoalexins of yam bean *Pachyrhizus erosus*. Z Naturforsch 34c:683–688

Jones HA (1942) A list of plants reported to contain rotenone or rotenoids. U.S. Department of Agriculture, Bureau of Entomology and Plant Quarantine, Washington, DC. 14 pp

Kalra AJ, Krishnamurthi M, Nath M (1977) Chemical investigation of Indian yam beans (*Pachyrrhizus erosus*): Isolation and structures of two new rotenoids and a new isoflavonone, erosenone. Indian J Chem 15:1084–1086

Kardinan A, Wikardi EA (1997) The prospect of botanical insecticides on stored food insects management. Biotrop Spec Publ 59:199–208

Karuniawan A (2004) Cultivation status and genetic diversity of yam bean (*Pachyrhizus erosus* (L.) Urban) in Indonesia. Cuvillier Verlag, Göttingen. 81 pp

Kawabata A, Sawayama S, Del Rosario RR, Noel MG (1986) Effect of storage and heat treatments on the sugar constituents in cassava and yambean roots. Nippon Shokuhin Kogyo Gakkaishi (Japón) 33(6):441–449

Kay DE (1973) Root crop. Crops and products digest 2. Tropical Products Institute, London, 245 pp

Klockman DM, Pressey R, Jen JJ (1991) Characterization of cell polysaccharides of jicama *(Pachyrrhizus erosus)* and Chinese water chestnut *(Eleocharis dulcis)*. J Food Biochem 15:317–329

Krishnamurthi M, Seshadri TR (1966) Chemical components of yam beans: their evolution and interrelationship. Curr Sci 35(7):167–169

Krishnamurti M, Sambhy YR, Seshadri TR (1970) Chemical study of Indian yam beans (*Pachyrrhizus erosus*): Isolation of two new rotenoids: 12a-hydroxydolineone and 12a-hydroxypachyrrhizone. Tetrahedron 26:3023–3027

Kumalasari ID, Nishi K, Harmayani E, Raharjo S, Sugahara T (2014a) Immunomodulatory activity of Bengkoang (*Pachyrhizus erosus*) fiber extract in vitro and in vivo. Cytotechnology 66(1):75–85

Kumalasari ID, Nishi K, Putra AB, Sugahara T (2014b) Activation of macrophages stimulated by the bengkoang fiber extract through toll-like receptor 4. Food Funct 5(7):1403–1408

Lamaze T, Vansuyt G, Zinsou C (1985) Répartition des formes d'azote et catabolisme des purines chez le dolique tubéreux, *Pachyrhizus erosus* Urban. Agronomie (Francia) 5(10):933–938

Lautié E, Rozet E, Hubert P, Quetin Leclercq J (2012) Quantification of rotenone in seeds of different species

of yam bean (*Pachyrhizus* sp.) by a SPE HPLC–UV method. Food Chem 131(4):1531–1538

Lautié E, Rozet E, Hubert P, Vandelaer N, Billard F, Zum Felde T, Grüneberg WJ, Quetin-Leclercq J (2013) Fast method for the simultaneous quantification of toxic polyphenols applied to the selection of genotypes of yam bean (*Pachyrhizus* sp.) seeds. Talanta 117:94–101

Leidi EO (2002) Leaf gas exchange of *Pachyrhizus ahipa* and *P. erosus* under water and temperature stress. Photosynthetica 40(3):375–381

Lin Y, Hao B, Pan K (1996) Preliminary crystallographic studies of a protein from *Pachyrrhizus erosus*. Acta Crystallogr D Biol Crystallogr 52(Pt 2):414–415

Lukitaningsih E, Bahi M, Holzgrabe U (2013a) Tyrosinase inhibition type of isolated compounds obtained from *Pachyrhizus erosus*. Aceh Int J Sci Technol 2(3):98–102

Lukitaningsih E, Mustikawatty AA, Sulistyo B, Sudarmanto A (2013b) Homology modelling and molecular docking of active compounds from bengkoang (*Pachyrrhizus erosus*) as tyrosinase inhibitor in Homa Sapiens. J Ilmu Farmasi Indonesia 12(1):132–137

Martínez-Bustos F, López-Soto M, Martín-Martínez ES, Zazueta-Morales JJ, Velez-Medina JJ (2007) Effects of high energy milling on some functional properties of jicama starch (*Pachyrrhizus erosus* L. Urban) and cassava starch (*Manihot esculenta* Crantz). J Food Eng 78(4):1212–1220

Mélo EA, Stamford TLM, Silva MPC, Krieger N, Stamford NP (2003) Functional properties of yam bean (*Pachyrhizus erosus*) starch. Bioresour Technol 89(1):103–106

Mercado-Silva E, Garcia R, Heredia-Zepeda A, Cantwell M (1998) Development of chilling injury in five jicama cultivars. Postharv Biol Technol 13(1):37–43

Mora QA, Morera MJA (1998) Test of elimination of flowers in a cultivar of yam beans (*Pachyrhizus erosus* (L.) Urban) using 2,4-D. [Spanish], pp 221–229. In: Sorensen M, Estrella Engelmann JE, Hamann OJ, Rios Ruiz SAMC (eds) Proceedings of the 2nd international symposium on tuberous legumes, Celaya, Guanajuato, Mexico, 5–8 August 1996, 566 pp

Morales-Arellano GY, Chagolla-Lopez A, Paredes-Lopez O, Barba de la Rosa AP (2001) Characterization of yam bean (*Pachyrhizus erosus*) proteins. J Agric Food Chem 49(3):1512–1516

Narongchai P, Narongchai S, Thampituk S (2005) The first fatal case of yam bean and rotenone toxicity in Thailand. Chotmaihet Thangphaet (J Med Assoc Thailand) 88(7):984–987

National Academy of Sciences (NAS) (1979) Tropical legumes: resources for the future. Root crops: Yam bean, Washington, pp 22–27

Noman ASM, Hoque MA, Sen PK, Karim MR (2006) Purification and some properties of α-amylase from post-harvest *Pachyrhizus erosus* L. tuber. Food Cehm 99(3):444–449

Noman ASM, Hoque MA, Haque MM, Pervin F, Karim MR (2007) Nutritional and anti-nutritional components in *Pachyrhizus erosus* L. tuber. Food Chem 102(4):1112–1118

Norton LB (1943) Rotenone in the yam bean (*Pachyrhizus erosus*). J Am Chem Soc 61:2259–2260

Nuraini J (2000) The effect of inoculum percentage and duration of fermentation with *Trichoderma koningii* on water content, crude protein and crude fiber of bengkuang leaves. J Penelitian Andalas 12(32):132–136 (in Indonesian)

Nuraini J, Hellyward, Mirzah (2001) The effect of fermented bengkuang leaves with *Trichoderma koningii* on broiler performance and income over feed cost. J Penelitian Andalas 13(35):42–47 (in Indonesia)

Nurrochmad A, Leviana F, Wulancarsari CG, Lukitaningsih E (2010) Phytoestrogens of *Pachyrhizus erosus* prevent bone loss in an ovariectomized rat model of osteoporosis. Int J Phytomed 2:363–372

Nurrochmad A, Lukitaningsih E, Monikawati A, Septhea DB, Meiyanto E (2013) Combination of low-concentration of novel phytoestrogen (8,9)-furanyl-pterocarpan-3-ol from *Pachyrhizus erosus* attenuated tamoxifen-associated growth inhibition on breast cancer T47D cells. Asian Pac J Trop Biomed 3(11):847–852

Ochse JJ, Bakhuizen van den Brink RC (1980) Vegetables of the Dutch Indies, 3rd edn. Ascher & Co., Amsterdam. 1016 pp

Panda PK, Sen H (1995) Effect of growth regulators on flowering and tuber yield of yambean. J Root Crops 21(2):97–101

Paull RE, Chen NJ (1988) Compositional changes in Yam Bean during storage. HortSci 23(1):194–196

Paull RE, Chen NJ, Fukuda SK (1988) Planting dates related to tuberous root yield, vine length, and quality attributes of Yam Bean. HortSci 23(2):326–329

Perry LM, Metzger J (1980) Medicinal plants of East and Southeast Asia. MIT Press, Cambridge. 632 pp

Phrutivorapongkul A, Lipipun V, Ruangrungsi N, Watanabe T, Ishikawa T (2002) Studies on the constituents of seeds of *Pachyrrhizus erosus* and their anti herpes simplex virus (HSV) activities. Chem Pharm Bull 50(4):534–537

Pressey R (1993) Cell wall compositions and enzymes of potatoes, jicamas and Chinese water chestnuts. J Food Biochem 17:85–95

Purseglove JW (1968) Tropical crops: Dicotyledons I. Wiley, New York, pp 281–284

Ramirez-Santiago C, Ramos-Solis L, Lobato-Calleros C, Peña-Valdivia C, Vernon-Carter EJ, Alvarez-Ramírez J (2010) Enrichment of stirred yogurt with soluble dietary fiber from *Pachyrhizus erosus* L. Urban: Effect on syneresis, microstructure and rheological properties. J Food Eng 101(3):229–235

Ramos-de-la-Peña AM, Renard CMGC, Wicker L, Montañez J, Reyes-Vega MDLL, Contreras-Esquivel JC (2012a) Optimization of the liquefaction and saccharification of structural polysaccharides of jicama (*Pachyrhizus erosus* L.) tissue by enzymatic pulping. LWT Food Sci Technol 46(1):232–238

Ramos-de-la-Peña AM, Renard CMGC, Wicker L, Montañez J, Reyes-Vega MDLL, Voget C, Contreras-Esquivel JC (2012b) Enzymatic liquefaction of jicama (*Pachyrhizus erosus*) tuberous roots and characterization of the cell walls after processing. LWT Food Sci Technol 49(2):257–262

Ramos-de-la-Peña AM, Renard CMGC, Wicker L, Contreras-Esquivel JC (2013) Advances and perspectives of *Pachyrhizus* spp. in food science and biotechnology. Trends Food Sci Technol 29(1):44–54

Ramos-de-la-Peña AM, Rangel-Rodríguez AM, Casas-González MR, Balagurusamy N, Sañudo-Barajas JA, Carrillo-Castillo A, Contreras-Esquivel JC (2011) Analysis of polyelectrolyte complexes formed with jicama pectic polysaccharide and water-soluble chitosan. Polymer Int 60(6):916–919

Robin C, Vaillant V, Vansuyt G, Zinsou C (1990) Assimilate partitioning in *Pachyrhizus erosus* during long-day vegetative development. Plant Physiol Biochem 28(3):343–349

Rumbaina D, Martono (1988) Uji efikasi biji bengkuang (*Pachyrrhizus erosu*s Urb.) terhadap hama penggerek batang lada. Sub-Balittro, Natar, Lampung (in Indonesian)

Sahu RC, Hameed SF (1983) Assessment of the rotenoids of Indian yam bean seeds. J Inst Agric Anim Sci 4(1–2):57–63

Sahu RC, Hameed SF (1989) Effect of *Pachyrrhizus erosus* urban seed extracts against tobacco caterpillar, *Spodoptera litura* F. Tobacco Res 15(1):17–20

Santos AC, Cavalcanti MS, Coelho LC (1996) Chemical composition and nutritional potential of yam bean seeds (*Pachyrhizus erosus* L. Urban). Plant Foods Hum Nutr 49(1):35–41

Shangraw RF, Lynn EV (1955) The saponins of *Pachyrrhizus erosu*s (L) Urban. J Am Pharm Assoc (Baltim) 44(1):38–39

Song XM, Zhou ZC, Wang J, Wu F, Gong WM (2004) Purification, characterization and preliminary crystallographic studies of a novel plant defensin from *Pachyrrhizus erosus* seeds. Acta Crystallogr D Biol Crystallogr 60(Pt 6):1121–1124

Song XM, Wang J, Wu F, Li X, Teng MK, Gong WM (2005) cDNA cloning, functional expression and antifungal activities of a dimeric plant defensin SPE10 from *Pachyrrhizus erosus* seeds. Plant Mol Biol 57(1):13–20

Sørensen M (1996) Yam bean. *Pachyrrhizus* DC. Promoting the conservation and use of underutilized and neglected crops 2 IPK/IPGRI Gatersleben/Rome, 141 pp

Sørensen M, van Hoof WCH (1996) *Pachyrhizus erosus* (L.) Urban. In: Flach M, Rumawas F (eds) Plant resources of South-East Asia, No. 9. Plants yielding non-seed carbohydrates. Prosea Foundation, Bogor, pp 137–141

Sørensen M, Grum M, Paull RE, Vaillant V, Venthou-Dumaine A, Zinsou C (1993) Yam beans (*Pachyrhizus* species). In: Williams JT (ed) Pulses and vegetables, underutilized crop series. Chapman and Hall, London, pp 59–102

Stevenson D, Jane JL, Inglett GE (2007) Characterisation of jicama (Mexican potato) (*Pachyrhizus erosus* L. Urban) starch from taproots grown in USA and Mexico. Starch-Starke 59(3–4):132–140

Stuart GU (2014) Philippine alternative medicine. Manual of some Philippine medicinal plants. http://www.stuartxchange.org/OtherHerbals.html

The Plant List (2014) *Pachyrhizus erosus* (L.) Urb. http://www.theplantlist.org/

U.S. Department of Agriculture, Agricultural Research Service (USDA-ARS) (2014) USDA National Nutrient Database for Standard Reference, Release 26. Nutrient Data Laboratory Home Page. http://www.ars.usda.gov/ba/bhnrc/ndl

Vaillant V, Desfontaines L (1995) Assimilate partitioning in *Pachyrhizus erosus* tuber on short days. Physiol Plant 93:558–562

Vaillant V, Robin C, Zinsou C (1990) Effects of the inhibition of nodule metabolism by nitrate on the regulation of assimilate partitioning in *Pachyrhizus erosus*. Plant Physiol Biochem 28(1):131–136

van Sillevoldt HET (1899) Ueber das derrid und pachyrhizid, ein beitrag zur kenntnis der indischen fischgifte. Arch Pharm 237:595–616 (in Dutch)

WHO/FAO (1985) Energy and protein requirements, Report of Joint FAO/WHO Adhoc expert committee. FAO/WHO, Rome

Wu F, Li YK, Chang SJ, Zhou ZC, Wang F, Song XM, Lin Y, Gong WM (2002) Purification, characterization and preliminary crystallographic studies of a PR-10 protein from *Pachyrrhizus erosus* seeds. Acta Crystallogr D Biol Crystallogr 58(Pt 12):2165–2167

Wu F, Yan M, Li YK, Chang SJ, Song XM, Zhou ZC, Gong WM (2003) cDNA cloning, expression, and mutagenesis of a PR-10 protein SPE-16 from the seeds of *Pachyrrhizus erosus*. Biochem Biophys Res Commun 312(3):761–766

Yahara S, Irino N, Takaoka T, Nohara T (1994) Studies on the constituents of *Pachyrhizus erosus*. Nat Med 48(4):312–313

Yongkhamcha B, Indrapichate K (2012) Insecticidal efficacy of mintweed, yam bean and celery seed extracts on *Aedes aegypti* L. Int J Agric Sci 4(3):207–212

Zinsou C, Vansuyt G, Venthou-Dumaine A (1987) Croissance et développement du *Pachyrhizus erosus* Urban, 2: Effets de l'acide gibbérellique et du chlorure de chlorocholine sur la composition glucidique du tubercule. Agronomie (Francia) 7(10):821–825

Zulkarnain AK, Ernawati N, Sukardani NI (2013) Activities of yam starch (*Pachyrhizus erosus*) as sunscreen in mouse and the effect of its concentration on viscosity level. Majalah Obat Tradisional (Tradit Med J) 18(1):1–8 (in Indonesian)

Pueraria montana var. *lobata*

Scientific Name

***Pueraria montana* var. *lobata* (Willd.) Sanjappa & Pradeep**

Synonyms

Dolichos hirsutus Thunb., *Dolichos japonicus* hort., *Dolichos lobatus* Willd., *Neustanthus chinensis* Benth., *Pachyrhizus thunbergianus* Siebold & Zucc., *Phaseolus trilobus* (L.) Aiton, *Pueraria argyi* H. Lev. & Vaniot, *Pueraria bodinieri* H. Lev. & Vaniot, *Pueraria caerulea* H. Lev. & Vaniot, *Pueraria harmsii* Rech., *Pueraria hirsuta* (Thunb.) Matsum., *Pueraria koten* H. Lev. & Vaniot, *Pueraria lobata* (Willd.) Ohwi, *Pueraria lobata* subsp. *lobata* (Willd.) Ohwi, *Pueraria neo-caledonica* Harms, *Pueraria pseudo-hirsuta* T. Tang & Wang, *Pueraria pseudohirsuta* Tang & Wang, *Pueraria thunbergiana* (Siebold & Zucc.) Benth. (spelling variant), *Pueraria thunbergiana* (Siebold & Zucc.) Benth., *Pueraria triloba* (Houtt.) Makino, *Pueraria volkensii* Hosok.

Family

Fabaceae, Papilionaceae

Common/English Names

Foot-A-Night Vine, Ko-Hemp, Kudsu, Kudzu, Kudzu Bean, Kudzu Hemp, Kudzu Vine, Japanese Arrowroot, Taiwan Kudzu, Vine-That-Ate-The-South, Tropical Kudzu, Wild Kudzu

Vernacular Names

Afrikaans: Kudzuranker;
Arabic: Kaskt, Kisht Yâbânî;
Chinese: Bai Ge, Ge, Fen Ge, Fen Ge Teng, Ge Gen, Ge Hua, Ge Teng, Huang Ge Teng, Ma Ge Teng, Mao Ge Teng, Ri Ben Ge, Shan Get Eng, Tian Ge Teng;
Cook Islands: Kūtū;
Danish: Kudzibønne;
Dutch: Koedzoe;
Fiji: Nggariaka, Wa Yaka, Yaka;
French: Kudzu, Kudzu Du Japon, Koudzou, Dolique Kudzu, Nepalem, Puéraria, Puéraire Kudzu, Vigne Japonaise;
German: Knollengrische, Ko-Hemp, Kopoubohne, Kudzu-Bohne, Kudzu-Kletterwein;
India: Dab (Hindi), Anetli, Bepui, Kakmudga, Marjaragandhika, Mudgaparni, Re, Saha, Sheem Sisali, Surpaparni, Suting Rit, Tagek;

T.K. Lim, *Edible Medicinal and Non-Medicinal Plants: Volume 10, Modified Stems, Roots, Bulbs*,
DOI 10.1007/978-94-017-7276-1_21

Indonesia: Bitok, Tobi (Sundanese), Ngu Lok, Tebu;
Italian: Kudzo, Kuzu;
Japanese: Kakkon, Kuzu, Oykar, Tokiiro Kuzu;
Korean: Chilg, Chilk, Ch'iri Kkot;
Laotian: Chüa Tau Kung, Khauz Pied;
Niuean: Acha, Aka, Aka Fala, Fou Gau;
Papua New Guinea: Hgédafo, Hxgédafo, Kópitu, Ngkoqahi, Oka Moi, Owitu, Sifu;
Philippines: Baai, Tahaunon;
Portuguese: Kudzu, Kudzu Tropical, Pueraria, Puero
Russian: Kudzu;
Samoa: A'A, A'A;
Spanish: Hierba Kudzú; Kudzú Común, Kudzú Tropical;
Thai: Tamyakhrua;
Tongan: Aka, Akataha, Fue'Ae Puaka;
Vietnamese: Bach Cat, Ban Mam Keo, Cu Nang, Cu San Day, Khau Cat, San Day;
Wallis And Futuna: Aka, Deday Aka;
Yapese: Dedai

Origin/Distribution

The native range of *Pueraria montana* var. *lobata* is still obscure; one view is that it is native to eastern Asia (China, Japan, Korea, Thailand, Vietnam and Taiwan) and Malesia (Indonesia, Malaysia, Papua New Guinea and the Philippines), while some references state that the native range of kudzu extends across the western Pacific region (Fiji, New Caledonia, Solomon Islands, Tonga, Micronesia and Vanuatu (e.g. USDA-ARS 2015; Fosberg et al. 1979).

Agroecology

Kudzu grows well under a wide range of habitats, in woodland, forest margins, riparian habitats, thickets, abandoned fields, roadsides and disturbed areas and on low mountains up to 2700 m altitude. It thrives well under warm humid conditions, from 24 to 25 °C and rainfall of 1500–2000 mm in full sun or partial shade. It is intolerant of frost as above-ground parts are killed by frost. Although it grows on a wide range of soil types, it does not fare well on very light poor sand or on poorly drained heavy clay. It prefers fertile, friable and well-drained soil high in organic matter like humus with pH of 5–7. It is drought-tolerant as it has a long (1.8 m) and deep-rooted system. This nitrogen-fixing and fast-growing vine can quickly cover shrubs and trees with a dense tangle of stems, smothering them and shading out all other vegetation.

Edible Plant Parts and Uses

Kudzu produces an edible tuber, and its leaves, shoots and flowers can be used as vegetables.

The young leaves are cooked as a leafy vegetable and can be used for salad. The flowers are battered and fried (like squash flowers) and the starchy tuberous roots can be prepared as any root vegetable. The root contains about 10 % starch, which can be extracted and used as a crispy coating in deep fried foods, or for thickening soups (Facciola 1990). The starch is used especially in China, Japan and Papua New Guinea for sauces, soups, jelled salads, noodles, porridge, jelly puddings, confectionary and beverages (Groen et al. 1996). In China and Japan, the starchy roots are ground into fine-powdered flour called Ko-ken flour, which is used for varieties of Wagashi (Japanese confectionary served with tea) and herbal medicines. The flour is also used for noodles, soups and various other dishes in China and Japan. When added to water and heated, kudzu powder becomes clear and adds stickiness to the food. In China, the young shoots are peeled before eating; and the dried flowers are used as an ingredient of *Wu-Hua-Cha* (Five flowers tea), an herbal tea commonly sold in Guangzhou and Hong Kong; and the tuber is cooked with Chinese dates and sliced yams for soup. In Fiji, the leaves, tender shoots and flowers are eaten and the tubers are eaten in the uplands of PNG and New Caledonia. The purple flowers of Kudzu are also used to make a sweet jelly in the southern United States and its flowers are an excellent honey source. Flowers are

cooked or made into pickles and the leaves and shoots are eaten raw or cooked (Facciola 1990). Kudzu leaves, shoots and flowers are used in salads, soups, sauteed dishes and casseroles (Shurtleff and Aoyagi 1977). Kudzu flowers are useful in the production of an unusually fragrant, flavourful honey.

Dried root of *Pueraria lobata* is a botanical supplement widely used as a nutraceutical (Zou et al. 2014). Lotus and kudzu starches have been used as functional foods in East Asia for thousands of years (Zhong et al. 2007).

Botany

Kudzu is a robust, high-climbing, twining or trailing perennial, pubescent, deciduous vine with very large oblongoid tuberous roots up to 2 m long, 18–45 cm diameter and weighing up to 180 kg when old (Plates 1 and 2). Stems are semi-woody 2.5–10 cm thick, 10–30 m long covered with long yellowish-brown hairs and branched. Leaves are alternate, pinnately trifoliate, with a rachis 8–20 cm long and 4–10 mm petioles. The central (i.e. terminal) leaflet is usually slightly larger and three-lobed, while the two side (i.e. lateral) leaflets usually have two lobes. Leaflets ovate–orbicular, shallowly lobed, the lobes rounded, 5–19 cm by 14–18 cm, acuminate, margins wavy, leaf surfaces softly pubescent, stipules peltate. Flowers in many-flowered axillary racemes are 15–35 cm long, with three flowers per node. Calyx long-hairy, divided into five unequal lobes; corolla papilionaceous, 15–18 mm long, purplish to blue or pink and the standard petal with a basal yellow spot, stamens ten monadelphous with one free stamen. Fruit is a densely brown-hairy, flattened, oblong-linear legume and constricted between the seeds, 9–12 cm long with 8–12 flattened-ovoid small, 4–5 mm by 2 mm, red brown with black mosaic seeds.

Plate 1 Young kudzu root

Nutritive/Medicinal Properties

Root/Stem Nutrients/Phytochemicals

Over 70 phytochemicals had been identified in kudzu root, with isoflavonoids and triterpenoids as the major constituents (Wong et al. 2011). The tuberous roots were reported to be rich in starch and beneficial isoflavone compounds. Nutrient composition of the raw tuberous root per 100 g edible portion was reported as: energy 113 cal, moisture 68.6 g, protein 2.1 g, fat 0.1 g, total carbohydrates 27.8 g, dietary fibre 0.7 g, ash 1.4 g, Ca 15 mg, P 18 mg, Fe 0.6 mg (Leung et al. 1972). Duke et al. (2002) reported the tubers to contain (per 100 g edible portion) 68.6 % moisture, 2.1 g protein, 0.1 g fat, 27.1 g total carbohydrate, 0.7 g fibre, 1.4 % starch, 15 mg Ca, 18 mg P, and 0.6 mg Fe (Duke 1983). The starch of the roots contained (per 100 g) 340 cal, 16.5 % moisture, 0.2 g protein, 0.1 g fat, 83.1 g total carbohydrate, 0.1 g ash, 35 mg Ca, 18 mg P, 2.0 mg Fe and 2 mg Na. The water contents in the fresh roots of five wild kudzu accessions ranged from 57 to 66.58 %, starch 2.40–28.30 %DW, Ca 211.9–559.7 mg/kg, Zn 12.9–55.3 mg/kg, Mg 0.31–0.82 mg/kg, P 0.1–2.51 mg/kg (Li et al. 1998). The 17 amino acid content (% DW) determined were: aspartic acid 0.38–1.25 %, threonine 0.14–0.36 %, serine 0.20–0.47 %, glutamic acid 0.35–0.70 %, glycine 0.19–0.38 %, alanine 0.02–0.37 %, cystine 0.3–0.3 %, valine 0.31–0.60 %, methionine 0.16–0.22 %, isoleucine 0.23–0.45 %, leucine 0.29–0.57 %, tyrosine 0.16–0.36 %, phenylalanine 0.26–0.45 %, lysine 0.14–0.48 %, histidine 0.06–0.24 %, arginine 0.08–0.26 %, proline 0.22–0.69 %. In contrast, kudzu powder purchased from the market had a water content of 5.5 %, starch content of 39.10

Plate 2 Pieces of large old Kudzu roots

%Dw, Ca 0.03 %, Fe 60.5 %, Zn 0.5 %, Mg 0.01 % and P 0.02 % and very low amounts of aspartic acid 0.018 %, serine 0.01 %, glutamic acid 0.02 %, glycine 0.007 %, tyrosine 0.07 % and lysine 0.05 % and the other amino acids were not detected. Du et al. (2002) reported the following proximate composition (%DW) of *P. lobata* roots: moisture 62.5 %, protein 5.87 %, crude fat 1.09 %, dietary fibre 18.15 %, ash 3.89 %, isoflavones 10.41 % and starch 58.96 %. *P. lobata* starch was found to have 178 ppm phosphorus, 0.04 % db crude protein, 0.10 % crude fat, 0.17 % crude fibre, irregular shape, 7–40 μm in size, iodine affinity 4.12 %, 20.68 % amylose content and gelatinization temperature (initial, mid and endpoints) of 61–64–70.5 °C, respectively (Du et al. 2002). Starch granules were of A-crystalline, X-ray diffraction type. *P. lobata* starch showed negligible swelling up to 75 °C, while between 75 and 95 °C it exhibited swelling. The pasting temperature was 72 °C and the peak viscosity was 540 BU at 84 °C. The viscosity dropped to 525 BU after holding the temperature at 95 °C for 30 min. Hence, the breakdown value was 540–525 = 15 BU. The relatively smaller breakdown compared with corn starch indicated that the *P. lobata* starch paste had a higher stability of viscosity. The molecular weight of *P. lobata* amylopectin was greater than corn amylopectin and was at least 20×10^6. Nutritive value (% dry matter) of kudzu root as a feed for ruminants was reported as 8.6 % crude protein, 39.8 % neutral-detergent fibre, 53.3 % acid-detergent fibre, 4.3 % ash, 0.4 % Ca, 0.3 % K, 0.1 % Mg and 3600 mg/kg Fe (Corley et al. 1997).

The organic form of Se (selenium) accounted for 82.42 % of total content in selenium-enriched *P. lobata* root (Zou et al. 2014). Purification yielded three single fractions: RP-SeP-11, RP-SeP-22 and RP-SeP-33 with Se contents of 0.9562×10^{-3}, 0.6113×10^{-3} and 0.3827×10^{-3} g/kg, respectively. RP-SeP-11 (3.5 kDa) was composed of glucose, RP-SeP-22 (19.6 kDa) was composed of xylose and glucose, and RP-SeP-33 (97.9 kDa) comprised galactose, mannose and glucose. Two Se-containing proteins were obtained with Se content of 3.175×10^{-3} and 4.328×10^{-3} g/kg, respectively. One appeared as three subunits with molecular masses of 43.0, 29.0 and 17.8 kDa while the other appeared as two subunits with molecular masses of 43.0 and 26.3 kDa.

The average particle size (diameter) of kudzu starch was 24.08 μm for lotus and the amylose content was lower than lotus starch (Zhong et al. 2007). Kudzu starch exhibited a C-type diffraction pattern. Kudzu starch was characterized by a maximum viscosity immediately followed by a sharp decrease in viscosity. *P. lobata* starch granules were oval and polygonal in shape (Du et al. 2002). Centric birefringence was clearly observed when viewed under polarized light. The gelatini-

zation temperature range was 61–64–70.5 °C. The iodine affinity value (4.12 %) indicated an amylose content of 21.68 %. Kudzu starch showed a fairly high maximum viscosity and very low breakdown, indicating high hot stability of the viscosity. The starch underwent a single-stage swelling power pattern over one temperature range, and the solubility pattern paralleled the swelling power. The X-ray diffraction pattern of the starch showed a Ca-type crystallite. The kudzu starch from Vietnam had polygonal granules, whereas the kudzu starches from Japan and Korea contained both polygonal and spherical granules (Pham and Morita 2007). Total protein, lipid, ash and phosphorus contents present in these kudzu starches were less than 1 % (starch basis). Vietnamese and Korean kudzu starches contained both daidzein and daidzin, whereas the Japanese kudzu starch had only daidzein. These starches had similar actual amylose contents (22.2–22.9 %). However, λ_{max}, blue value and apparent amylose contents of the kudzu starch from Vietnam were lower than those from Japan and Korea. Amylose molecules of the kudzu starch from Vietnam had the largest average degree of polymerization (DP_n) and number of chains (NC), followed by the kudzu starches from Japan and Korea. Amylopectin molecules of the kudzu starch from Vietnam also had the largest DP_n and NC, followed by the kudzu starches from Korea and Japan. X-ray diffraction patterns of the kudzu starches from Vietnam, Japan and Korea were A-type, C-type and B-type, respectively. The kudzu starch from Vietnam was found to have the specific characteristics such as significantly high gelatinization temperature, transition enthalpy and degree of crystallinity as compared to the kudzu starch from Korea and Japan. The yellowness of kudzu starch–chitosan composite films was enhanced, whereas water content and water vapour permeability of the films declined with increasing storage time (Zhong and Li 2011). Tensile strength of films using acetic and lactic acid as solvents, and solubility of all the films increased within 30 days thereafter decreased. Storage temperature had no impact on X-ray diffraction, water content and solubility. Under the same storage conditions, the film using acetic acid as solvent presented the strongest mechanical property, the smallest solubility and the lightest colour. The film made from lactic acid solution was the most flexible and the most yellow. The film with malic acid solvent showed the highest ordering degree, the lowest water content and the best water barrier property.

Isoflavone derivatives daidzein, daidzin, daidzein diacetate, daidzein dimethyl ether, genistein triacetate and apigenin triacetate were isolated from the roots of *Pueraria thunbergiana, P. pseudo-hirsuta* and *P. thomsonii* (Shibata et al. 1959b). Puerarin, $C_{21}H_{20}O_9$, and unidentified substance f, were isolated from *Pueraria* root (Murakami et al. 1960). Puerarin was confirmed to be 8-D-glucopyranosyl-4′, 7-dihydroxyisoflavone (=8-D-glucopyranosyldaidzein). Ozonolysis of puerarin afforded *D*-glucose and *D*-arabinose. On hydrolysis with sulphuric acid, substance f yielded puerarin and *D*-xylose. A human intestinal bacterium, strain PUE, was found to cleave the *C*-glucosyl bond of puerarin to yield its aglycone daidzein and an intact glucose (Nakamura et al. 2011). ^{14}C-labelled isoliquiritigenin was efficiently incorporated into puerarin of *P. lobata*, but ^{14}C-labelled daidzein was not (Inoue and Fujita 1977). Further competitive feeding experiments with T-or ^{14}C-labelled isoliquiritigenin and liquiritigenin suggested that *C*-glycosylchalcone would be a possible intermediate for the biosynthesis of puerarin.

Daidzein, daidzin and puerarin were isolated from the methanol extracts of *P. lobata* roots (Hayakawa et al. 1984). Recoveries of these constituents were as follows; daidzein: 86 %, daidzin: 96 % and puerarin: 99 %. Puerarin, daidzin, daidzein, genistein, formononetin and their *O*- and *C*-glycosides were isolated from *P. lobata* by narrow-bore HPLC (Rong et al. 1998a). Puerarin, daidzein and rutin in *P. lobata* roots and vines were determined by capillary electrophoresis with electrochemical detection (Chen et al. 2001). The vines were found to have 11.378 mg/g puerarin, 0.5911 mg/g daidzein, rutin 0 mg/g; the roots were found to have 23.361–30.745 mg/g puerarin, 0.9784–1.6816 mg/g daidzein and 0.5665–2.6059 mg/g rutin. Chen et al. (2010b)

recovered five isoflavonoids from *Pueraria* root, the major component was puerarin, then daidzin and daidzein; genistin and genistein were the least abundant. The isoflavonoid recovery from Radix Puerariae was 90–113 %. The content of four isoflavones daidzein, daidzin, puerarin and daidzein-4′,7-diglucoside in *P. lobata* from eight areas were determined using TLC-densitometry (Zhao and Zhang 1985). Seven oleanane-sapogenols were obtained from the methanolysate of the crude saponin extract of *P. lobata* roots (Kinjo et al. 1985). Four of them, sapogenols 1–4, were identical with sophoradiol, cantoniensistriol and soyasapogenol B and soyasapogenol A. The structures of three new sapogenols were named kudzusapogenol C, kudzusapogenol A and kudzusapogenol B methyl ester. Two new isoflavone glycosides, the daidzein 8-*C*-apiosyl (1→6) glucosides and genistein 8-*C*-apiosyl (1→6) glucoside, were isolated from the roots and stems of *P. lobata* (Kinjo et al. 1987). From the Chinese drug Gegen, the root of *P. lobata*, several isoflavonoid compounds were isolated. Besides the known compounds, namely puerarin, daidzin, daidzein, formononetin, *Pueraria* glycosides (PG) 1-6 and puerarol (Ohshima et al. 1988), PG-1 (8-β-D-glucopyranosyl-7-hydroxy-3-(3,4-dihydroxyphenyl)-4*H*-1-benzopyran-4-one), PG-3 (8-β-D-glucopyranosyl-7-hydroxy-3-(4-hydroxy-3-methoxyphenyl)-4*H*-1-benzopyran-4-one) and daidzein (7-hydroxy-3(4-hydroxyphenyl)- 4*H*-1-benzopyran-4-one) were isolated from *P. lobata* root methanol extract (Sato et al. 1992).

Thirteen known isoflavonoids and related compounds were also found; daidzein, formononetin, genistein, daidzin, daidzein 4′,7-diglucoside, puerarin, PG-1, PG-3, puerarin xyloside, genistin, genistein-8-*C*-glucoside, coumestrol and isoliquiritigenin. The presence of 6″-*O*-malonyl esters of daidzin, genistin and 8-*C*-glucoside puerarin was also detected in the fresh roots and stems of *P. lobata* (Park et al. 1992). Isoflavones found in the roots of *P. lobata* included: puerarin-4′-*O*-D-glucoside, 3′-hydroxypuerarin, 3′-hydroxypuerarin-4′-*O*-deoxyhexoside, puerarin, 3′-methoxypuerarin, 6″-*O*-D-xylosylpuerarin, 3′-methoxy-6″-*O*-D-xylosylpuerarin, daidzin and 3′-methoxydaidzin, daidzen, 3′-methoxydaidzein, genistein, daidzein-7-*O*-methyl ether, 3′-methoxydaidzein-7-*O*-methyl ether or 3′-methoxyformononetin, formononetin, biochanin A, coumestrol and genistein-9-C-apiosyl (1→6) glucoside (Rong et al. 1998b). The major glycosides were derived from daidzein and most were 8-*C*-glycosides. The following isoflavones were identified in kudzu roots: puerarin, puerarin xyloside, daidzin, genistein 8-*C*-glycoside-xyloside, genistein 8-*C*-glycoside, genistein 8-*C*-glycoside, daidzein 4′-*O*-glycoside, genistin, *O*-glycoside of an unidentified compound, formononetin 7-*O*-glycoside, daidzein, genistein, formononetin, two unknown acidic compounds and an isoflavone with one rhamnose, one *C*-glycoside (Zhang et al. 2005). To improve the bioavailability of kudzu root isoflavones, crude beta-glycosidases prepared from microbes were used to hydrolyze the isoflavone glycosides. Twelve compounds were isolated from dried Pueraria roots and identified as 3′-hydroxypuerarin; puerarin; 3′-methoxypuerarin; 6′′-*O*-D-xylosylpuerarin; daidzin; genistin; 6,7-dimethoxycoumarin; daidzein; genistein; formononetin; isoliquiritigenin and biochanin A (Lin et al. 2005). A new 2-arylbenzofuran, puerariafuran and three known compounds, coumestrol, daidzein and genistein, were isolated from the methanol extract of *P. lobata* roots (Jang et al. 2006). The major isoflavonoids identified in *P. lobata* roots included puerarin, daidzin, genistein-6″-*O*-malonylester, 3′-methoxypuerarin and daidzein (Fang et al. 2006a). Isoflavonoids were separated from *P. lobata* (a rich source of isoflavonoids) by high-performance capillary electrophoresis (HPCE) (Fang et al. 2006b). The first-order polynomial regression between peak areas and isoflavone concentration was determined over the range 0.05–0.5 mg/mL for puerarin and 2.5–50 μg/mL for 3′-methoxypuerarin, daidzin and daidzein, respectively. Some common isoflavones identified in kudzu roots, puerarin, daidzin, daidzein, genistin, genistein, malonylgenistin, malonyldaidzin, formononetin and 4′,7-*O*-diglucoside of daidzein (Prasain et al.

2007). *P. lobata* root was found to contain hydroxyethylpuerarin (Guang et al. 2005; Wang et al. 2007b). Puerarin, rather than daidzin, was the most abundant component (8.44–0.60 mg/capsule) in commercially available kudzu dietary supplements (Prasain et al. 2003). Chen et al. (2007) found that 3-year-old *P. lobata* roots harvested in January had the highest yields of isoflavonoid compounds: 3′-hydroxypuerarin, puerarin, 3′-methoxypuerarin, daidzin, genistin, formononetin-7-glucoside and daidzein.

From *P. lobata* roots, four new oleanene-type triterpene glycosides, named kudzusaponins SA_1, SA_2, SA_3 and C_1 were isolated and their structures were determined to be 3-*O*-β-D-galactopyranosyl-(1 → 2)-β-D-glucuronopyranosyl soyasapogenol A; 3-*O*-β-D-galactopyranosyl-(1 → 2)-β-D-glucuronopyranosyl soyasapogenol A 22-*O*-α-L-arabinopyranoside; 3-*O*-α-L-rhamnopyranosyl-(1 → 2)-β-D-galactopyranosyl-(1 → 2)-β-D-glucuronopyranosyl soyasapogenol A 22-*O*-α-L-arabinopyranoside; and 3-*O*-α-L-rhamnopyranosyl-(1 → 2)-β-D-galactopyranosyl-(1 → 2)-β-D-glucuronopyranosyl kudzusapogenol C 21-*O*-β-D-glucopyranoside, respectively (Arao et al. 1995). From *P. lobata* root, six new oleanane-type triterpene glycosides, called kudzusaponins A_1 (1), A_2 (2), A_4 (3), A_5 (4), SA_4 (5), and SB_1 (6) were isolated together with kudzusaponin A_3, soyasaponin SA_3 and soyasaponin I (Arao et al. 1997a). The structures of 1–6 were determined to be 3-*O*-α-L-rhamnopyranosyl-(1 → 2)-β-D-arabinopyranosyl-(1 → 2)-β-glucuronopyranosyl kudzusapogenol A 22-*O*-β-D-xylopyranoside, 3-*O*-β-D-galactopyranosyl-(1 → 2)-β-D-glucuronopyranosyl kudzusapogenol A, 3-*O*-β-D-glucopyranosyl-(1 → 2)-β-D-glucuranopyranosyl kudzusapogenol A, 3-*O*-α-L-rhamnopyranosyl-(1 → 2)-β-D-glucopyranosyl-(1 → 2)-β-D-glucuronopyranosyl kudzusapogenol A, 3-*O*-β-D-glucuronopyranosyl-(1 → 2)-β-D-glucuronopyranosyl soyasapogenol A 22-*O*-α-L-arabinopyranoside and 3-*O*-α-L-rhamnopyranosyl-(1 → 2)-β-D-galactopyranosyl-(1 → 2)-β-D-glucuronopyranosyl (β-fabatriosyl) soyasapogenol B 22-*O*-α-L-arabinopyranoside, respectively. *P. lobata* root crude saponin extract contained soyasaponin 1and kudzusaponin SA3 (Arao et al. 1997b).

Two new isoflavone diglycosides, formononetin 8-*C*-[β-D-apiofuranosyl-(1 → 6)]-β-D-glucopyranoside and formononetin 8-*C*-[β-D-xylopyranosyl-(1 → 6)]-β-D-glucopyranoside, were isolated from *P. lobata* roots, together with four known compounds, 4′-methoxypuerarin, daidzin, genistin and daidzein (Sun et al. 2008). Two isoflavone C-glucosides, puerarin and PG-3; a but-2-enolide, (±)-puerol B; two isoflavone *O*-glucosides, daidzin and genistin; and three pterocarpans, (−)-medicarpin, (−)-glycinol and (−)-tuberosin, were isolated from a methanol extract of *P. lobata* roots (Kim et al. 2006a). Puerarin, daidzein 8-*C*-apiosyl-glucoside, daidzin and daidzein were extracted from Ge gen (*P. lobata*) and the recovery rates were in the range of 96.01–106.18 % (Chang et al. 2008). The purity of 3′-hydroxypuerarin, puerarin and daidzin of the product obtained from *P. lobata* root using optimum macroporous resins technology was 23.39, 52.09 and 19.21 %, respectively (Cui et al. 2013).

Spinasterol (stigmasta-7, 22-dien-3β-ol) was isolated from *P. lobata* roots (Jeon et al. 2005). Thirteen compounds were isolated from *P. lobata* roots and their structures were identified as lupeol (1), formononetin (2), salicylic acid (3), quercetin (4), genistein (5), 2,3-dihydroxypropyl palmitate (6), puerarol (7), daidzein (8), daidzin (9), puerarin (10), ononin (11), gallic acid (12) and 3′-methoxy puerarin (13) (Liu et al. 2009a). Twenty-two compounds were isolated from *P. lobata* roots and characterized as β-sitosterol palmitate (1), β-sitosterol (2), lupeol (3), lupeone (4), puerarol (5), diisobutyl phthalate (6), bis(2-ethylhexyl) phthalate (7), sophoracoumestan A (8), coumestrol (9), allantoin (10), daidzein (11), formononetin (12), 3′-methoxy daidzein (13), ononin (14), 3′-hydroxy dadzein (15), genistin (16), daidzein (17), 8-methoxy ononin (18), sissotrin (19), (−)-puerol B 2-*O*-glucopyranoside (20), (6*S*,9*R*)-roseoside (21) and sucrose (22) (Li et al. 2010a). A 50 % ethanol root extract of *P. lobata* was found to contain 10.42 % puerarin as the main constituent and smaller amounts of the related isoflavonoids 3′-hydroxypuerarin,

3′-methoxypuerarin, 6″-xylosylpuerarin, daidzin, genistin, daidzein and genistein (Bebrevska et al. 2010). Fourteen phenolic compounds were detected from *P. lobata* roots and identified as: puerarin-4′-*O*-glucoside (48.9 μg), puerarin-3′-methoxy-4′-*O*-glucoside (34.1 μg), diadzein-4′,7-*O*-glucoside (171.9 μg), puerarin (4353 μg), 6′′-*O*-xylosylpuerarin (1985.9 μg); mirificin (1299.7 μg), daidzin 763.2 μg); 3′-methoxypuerarin (332 μg), genistin (469.5 μg), sophoroside A (321.7 μg), ononin (417.9 μg), daidzein (1468.7 μg), genistein (75.6 μg) and formononetin (99.3 μg) (Du et al. 2010). The ethyl acetate-soluble extract of *P. lobata* afforded 12 phenolic compounds: puerarol, daidzein, formononetin, diazin, ononin, genistein, biochanin A, sissotrin, puerol B, puerol B 2-*O*-β-D-glucose, 4′,8-dimethoxyisoflavone 7-*O*-β-D-glucose and a new coumestan, 2-(α,α-dimethylallyl) coumestrol (Choi et al. 2010).

P. lobata root cultures were found to contain the following isoflavone glycosides: OH-acetylpuerarin, 2-OH-puerarin, OH-puerarin, 2×OH-puerarin, malonylpuerarin, licodiol hexoside, acetylpuerarin, acetyldaidzin, malonyldiaidzin, 2-OH-daidzin, daidzin, 2×OH-daidzin, daidzein-2-O-hexoside, OH-acetyldaidzin, genistin, OH-genistin, aglycones OH-daidzein, 6-OH-daidzein, 3′-OH-daidzein, genistein, biochanin A and methyl derivative of flavanone (Prasain et al. 2007).

The following radiolabelled isoflavones (mg/g extract) were recovered from kudzu (*P. lobata*) root cultures after incubation with uniformly labelled ^{14}C-sucrose in the culture medium for 21 days: puerarin 33.6 mg, daidzin 32.2 mg, malonyl-daidzin 29.1 mg, hydroxyl-puerarin 14.4 mg, malonyl-genistin 4.5 mg, genistin 1.3 mg/g, daidzein 0.9 mg, genistein 0.6 mg and total isoflavones 116.6 mg (Reppert et al. 2008). Two major isoflavonoids found in kudzu roots were 8-[α-D-glucopyranosyl-(1→6)-β-D-glucopyranosyl] daidzein and glucosyl-α-1,6-puerarin (2.3 mg/g) (Nguyen et al. 2009).

Seventeen compounds were identified in *P. lobata* roots: 3′-hydroxypuerarin; puerarin; isomer of daidzin; daidzein; genistein 8-*C*-glycoside-xyloside; 6′′-*O*-xylosylpuerarin; mirificin; puerarin-4′-*O*-glucoside; genistein; 3′-methoxypuerarin; 6′′-*O*-apiosylgenistin; daidzin; isodaidzein; daidzein; biochanin A; genistein; formononetin and sophoroside A (Zhang et al. 2011a). Five major isoflavonoids, including puerarin, daidzin, genistin, daidzein and genistein, were found in tubers of *Pueraria mirifica* from 28 provinces in Thailand and *P. lobata* from Guangzhou province, China (Cherdshewasart et al. 2007). In comparison with *P. lobata, P. mirifica* population exhibited differences only with a lower amount of daidzein. The total content of 13 isoflavones in *P. lobata* root (53.6–134.7 mg/g) was much higher than that in *P. thomsonii* root (5.36–12.5 mg/g) (Du et al. 2011). Among them, puerarin (21.1–54.3 mg/g in *P. lobata* root and 2.23–9.27 mg/g in *P. thomsonii* root) and daidzin (4.12–17.1 mg/g in *P. lobata* root, and 0.76–2.55 mg/g in *P. thomsonii* root) were the predominant components found in the roots. Additionally, five characteristic puerarin derivatives: 6-*O*-xylosylpuerarin, puerarin-4,7-*O*-glucoside, puerarin-4-*O*-glucoside, puerarin-3-methyoxy-4-*O*-glucoside and 3-methoxypuerarin were found in *P. lobata* root or *P. thomsonii* root. Genistein, genistin, daidzein ononin, formononetin and mirificin were also detected in the roots.

Five isoflavonoids, puerarin, daidzin, daidzein, genistin and genistein, were analyzed simultaneously from *Pueraria* root by high-performance liquid chromatography-diode array detection (HPLC-DAD) (Lau et al. 2009). Major isoflavonoids of *Pueraria* root (puerarin, daidzin, daidzein and genistein) were extracted by high-performance liquid chromatography with cyclodextrins as a mobile phase modifier (Zeng et al. 2012). Hydroxypropyl-β-cyclodextrin was found to be a very effective mobile phase additive; hence it could markedly reduce the retention of isoflavonoids, especially daidzein and genistein.

Optimization of ultrasound-assisted extraction coupled with response surface methodology afforded yields of 41 mg/g puerarin and 128 mg/g total isoflavones from *P. lobata* root (Wu et al. 2012). The methanol extract of *P. lobata* roots yielded the following: four known compounds from the anti-inflammatory and antioxidant EtOAc fraction (daidzein, genistein, puerarin, (+)-puerarol B-2-*O*-glucopyranoside); four known compounds from the anti-inflammatory *n*-hexane fraction (lupenone, lupeol, puerarol, coumestrol); seven known compounds from the antioxidant *n*-BuOH fraction (allantoin, 3′-hydroxypuerarin, daidzein 8-*C*-apiosyl-(1→6)-glucoside, puerarin, genistin, 3′-methoxypuerarin, daidzin) (Jin et al. 2012). The root outer bark of *P. lobata* root possessed higher isoflavonoids content than the whole root or kudzu root (Chen et al. 2012b). The main isoflavonoids in all sections of *P. lobata* roots were puerarin, daidzin, daidzein, genistin and genistein. Two new *C*-glucofuranosyl isoflavones 8-*C*-α-glucofuranosyl-7,4′-dihydroxyisoflavone and 8-*C*-β-glucofuranosyl-7,4′-dihydroxyisoflavone, named as neopuerarin A and neopuerarin B, were isolated from *P. lobata* roots (Zhang et al. 2010). Six isoflavones were isolated from kudzu root (*P. lobata*) and elucidated to be four isoflavone *C*-glycosides as 6″-*O*-α-D-glucopyranosylpuerarin, puerarin, 3′-methoxypuerarin, 6″-*O*-α-D-apiofranosylpuerarin, and two aglycones as biochanin A and formononetin (Kayano et al. 2012). Also related isoflavones daidzein, diazin, 3′,4′,7-trihydroxyisoflavone were also found.

Thirty-three components were identified in the volatile oil of *P. lobata* roots: 2 dimethyl esters of dibasic acids, 9 methylesters of fatty acids, 2 acetates, 3 aldehydes, 5 ketones, 1 acid, 1 hydrocarbon and 10 alcohols and phenols (Miyazawa and Kameoka 1988). The compounds were acetone, methyl propyl ketone, acetonyl acetone, 2-furfuryl methyl ketone, butanoic acid, 2-propanoyl acetate, 2-methoxy ethyl acetate, heptane, octanal, furfural, benzaldehyde, benzyl alcohol, thymol, *p*-cresol, *m*-cresol, *o*-cresol, paenol, phenol, β-phenyl ethyl alcohol, *p*-ethyl phenol, furfuryl alcohol, acetyl carbinol, methyl caprylate, methyl caprate, methyl laurate, methyl tridecanoate, methyl margarate, methyl stearate, methyl palmitate, methyl myristate, methyl pentadecanoate, dimethyl suberate, dimethyl azelate. Methyl palmitate 42.2 % was the major component. The two dimethyl esters of dibasic acids had a mildly fruit/winey odour, acetyl carbinol had a sweet-caremellic and choking ethereal odour and paenol had a warm, aromatic and botanpi-like odour. Furan compounds had a warm-oily, sweet, caramel odour. A mixture (acetyl carbinol, 42.0 %; paenol, 26.5 %; furfuryl alcohol, 23.5 %; dimethyl azelate, 6.5 %; dimethyl suberate, 0.5 %; furfural, 0.5 %; and 2-furfurylmethyl ketone, 0.5 %) had a slightly sweet odour as a volatile oil, and a similar odour to that of Kakkon.

A water-soluble linear glucan, PLB-2C, was isolated from the water extract of *P. lobata* (Cui et al. 2008). It was found to be composed of (1→6)-α-D-Glcp. Chain conformation study showed that the polysaccharide-possessed random coil compact conformation. *C*-glucosyltransferase, one of the key enzymes for the biosynthesis of puerarin, was extracted and partially purified from roots of *P. lobata* (Chen et al. 2010a). Higher activity was found in roots than in leaves and stems of *P. lobata*. It was the first time that the activity of *C*-glucosyltransferase, which transforms isoliquiritigenin to puerarin, was detected in *P. lobata*. A novel isoflavone 7-*O*-glucosyltransferase PlUGT1 was isolated from *P. lobata* (Li et al. 2014). PlUGT1 could convert daidzein to daidzin, genistein to genistin as well as formononetin to ononin. The PlUGT1 gene was highly expressed in *P. lobata* roots relative to other organs and strongly induced by methyl jasmonate signal in *P. lobata* cell suspension culture. The transcript abundance of PlUGT1 was correlated with the accumulation pattern of isoflavone glycosides such as daidzin in *P. lobata* plants or in cell suspension culture.

Flower Phytochemicals

From the steam distillate of the ether extract of flower of *Pueraria thunbergiana*, nine kinds of

alcohol namely *cis*-3-hexen-1-ol, benzyl alcohol, eugenol, isoamyl alcohol, ethyl acetate, octyl alcohol, 1-octen-3-ol, phenethyl alcohol, *l*-linalool and four esters of volatile acid viz. methyl propionate, methyl isovalerate, methyl capronate, methyl benzoate were isolated as the fragrant compositions (Kurihara and Kiruchi 1973). Irisolidone-7-*O*-glucoside was isolated from Japanese *Pueraria* flower (*P. lobata*), and tectoridin from Formosan *Pueraria* flower (*P. montana*). Irisolidone, genistein, daidzein, quercetin, biochanin A, *p*-coumaric acid and some paraffins were also identified in the residue of the steam distillation.

An isoflavone glycoside, kakkalide, was isolated from the methanol flower extract of *Pueraria thunbergiana* and its chemical structure was determined as irisolidone-7-β-D-xylopyranosyl-6-*O*-β-D-glucopyranoside (Kurihara and Kikuchi 1975). A new isoflavone, 6, 4′-dihydroxy-7-methoxy-isoflavone, was isolated from Chinese *Pueraria*, Ge hua (*P. lobata*) flowers (Kubo et al. 1975). Biochanin A, formononetin, ononin, sissotrin, genistin, β-sitosterol and β-sitosterol-3-*O*-β-D-glucoside were isolated from the methanolic flower extract of *Pueraria thunbergiana* (Kurihara and Kikuchi 1976). A new triterpenoidal saponin, 3-*O*-[α-L-rhamnopyranosyl(1 → 2)-α-L-arabinopyranosyl(1 → 2)-β-D-glucuronopyranosyl] sophoradiol, and a known oligoglycoside kaikasaponin III with the structure 3-*O*-[α-L-rhamnopyranosyl(1 → 2)β-D-galactopyranosyl(1 → 2) β-D-glucuronopyranosyl]sophoradiol together with kakkalide, daidzin, genistin, rutin, robinin (kaempferol 3-*O*-rhamnosyl (1 → 6)galactosyl-7-*O*-rhamnoside and nicotiflorin (kaempferol-3-*O*-rutinoside) were isolated from *P. lobata* flowers (Kinjo et al. 1988b). A tryptophan derivative, *N*-acyl-N_1-glucosyltryptophan, was isolated from *P. lobata* flower (Kinjo et al. 1988b). Oleanane glucuronides soyasaponin I, kaikasaponin III and kakkasaponin I were isolated from the flowers of *P. lobata and P. thomsonii* (Kinjo et al. 1999). Two new isoflavone glycosides 4′,7-dihydroxy-6-methoxyisoflavone-7-*O*-β-D-xylopyranosyl- (1 → 6)-β-D-glucopyranoside and 4′,5,7-trihydroxy-6-methoxyisoflavone-7-*O*-β-D-xylopyranosyl-(1 → 6)-β-D-glucopyranoside together with glycitein, tectoridin, irisolidone, kakkalide, kakkalidone, irisolidone-7-*O*-β-D-glucopyranoside and glycitin were isolated from *Pueraria thunbergiana* flowers (Park et al. 1999). Kim et al. (2003) reported the following compounds in *P. lobata* flowers: isolidone, genistein, daidzein, glycitein, glycitin, 6″-*O*-xylosyl-tectoridin, 6″-*O*-xylosyl-glycitin, tectorigenin, tectoridin, kakkalide, kakkatin, kaikasaponon III, soyasaponin I, soyasaponin b, soyasaponin Ab, glycyrrhizin, rutin, biocanin A, ononin, β-sitosterol, robinin, nicotiflorin and quercetin. *P. lobata* flowers were reported to contain isoflavones: irisolidone, irisolidone 7-glucoside, kakkalide, genistein, biochanin A and sissotrin and three saponins: soyasaponin I, kaikasaponin III and kakkasaponin I (Nohara 2004). Isoflavonoids like puerarin, daidzin and daidzein in *P. lobata* extract and its preparations were separated and identified by reversed-phase capillary liquid chromatography (RP-CapLC) coupled with photodiode array (PDA) detector and negative electrospray ionization quadrupole time of flight (Q-TOF) mass spectrometry (Tian et al. 2005). Puerarin was found to be the most abundant component in the extract (about 13 %, mass fraction) and its preparations (19.28–24.34 mg per tablet). Trace amount of unknown isoflavonoids were tentatively identified to be 3′-methoxypuerarin and 3′-methoxydaidzin.

Four isoflavones kakkalide, irisolidone, tectorigenin-7-*O*-xylosylglucoside and biochanin A were detected in *P. lobata* flowers (Zhang et al. 2009b). The calibration curve was linear within the range of 42.8–856 mg/L for kakkalide, 9.12–182.4 mg/L for irisolidone, 13.2–264 mg/L for tectorigenin-7-*O*-xylosylglucoside and 4.2–84 mg/L for biochanin A, respectively. The average recoveries for four marker compounds were from 98.3 to 100.7 %. Three major compounds, i.e. the isoflavones tectorigenin 7-*O*-[β-D-xylopyranosyl-(1 → 6)-β-D-glucopyranoside], tectorigenin 7-*O*-β-D-glucopyranoside and tectorigenin, were quantified in *P. lobata* flowers using HPLC (Bebrevska et al. 2007). Flowers of *P. lobata* and *P. thomsonii* were found to contain a

large amount of isoflavones and saponins (Niiho et al. 2010). *P. lobata* flowers contained 0.43–2.5 % total saponins, of which kaikasponin III, kaikasaponin I and soyasaponin were detected, and 1.84–2.86 % isoflavones comprising kakkalide detected in all six samples and irisolidone detected in four samples.

Rapid screening of antioxidants from *P. lobata* flowers based on high-performance liquid chromatography analysis combined with a 2,2′-diphenyl-1-picrylhydrazyl assay yielded 18 active isoflavones, puerarin-4′-*O*-β-D-glucopyranoside (1), glycitin (2), tectorigenin-7-*O*-β-D-xylosyl-(1 → 6)-β-D-glucopyranoside (3), tectoridin (4), genistein-8-*C*-β-D-glucopyranoside (5), irisolidone-7-*O*-β-D--glucopyranpsyl-(1 → 6)-β-D-glucopyranoside (6), puerarin (7), biochanin A-7-*O*-β-D--glucopyranoside (8), kakkalide (9), daidzin (10), 3′-methoxydaidzin (11), ononin (12), 3′-hydroxydaidzein (13), tectorigenin (14), daidzein (15), genistein (16), 3′-methoxydaidzein (17) and irisolidone (18) (Zhang et al. 2012b). Thirteen active isoflavones were isolated from the ethanolic extract of *P. lobata* flower and identified as glycitin (1), tectoridin (2), daidzin (3), 3′-methoxy daidzin (4), ononin (5), 3′-hydroxyl daidzein (6), tectorigenin (7), biochanin A (8), prunetin (9), genistein (10), 3′-methoxy daidzein (11), irisolidone (12) and 5,7-dihydroxy-3′,4′-methylenedioxyisoflavone (13), using bovine serum albumin (BSA) functionalized iron oxide magnetic nanoparticles (Fe3O4 MNPs) coupled with high-performance liquid chromatography–tandem mass spectrometry (HPLC–MS/MS) (Liu et al. 2012).

A total of 25 isoflavones, 13 saponins and 3 flavones were identified in the flowers of *P. lobata* and *P. thomsonii*: flavones: rutin, luteolin and dihydrokaempferol-*O*-hexoside; saponins: astragaloside VIII, soyasaponin I, irisolidone, soyasaponin IV, kaikasaponin III, kaikasaponin II, kaikasaponin I, kakkasaponin I, azukisaponin I, baptisiasaponin I, phaseoside IV, kakkasaponin II, abrisapogenol-F-*O*-rhamnosyl-pentosyl-glucuronide and kakkasaponin III; isoflavones: puerarin; 6-hydroxygenistein-6,7-di-*O*-glucoside; glycitein-*O*-pentosyl-hexoside; daidzin; glycitin 6-hydroxygenistein-7-*O*-glucoside, genistein-*O*-pentosyl-hexoside tectorigenin-7-*O*-xylosylglucosidegenistin tectoridin 6-hydroxybiochanin A-6, 7-di-*O*-glucoside; gehuain; isomer of tectorigenin; daidzein; glycitein; biochanin A-*O*-pentosyl-hexoside; kakkalide; isomer of kakkalide; genistein; biochanin A-*O*-hexoside; irisolidone-*O*-hexoside; tectorigenin; isomer of irisolidone; biochanin A and irisolidone (Lu et al. 2013). *P. lobata* had higher amounts of kakkalide (10.3–17.7 mg/g) and irisolidone (2.76–4.95 mg/g) than *P. thomsonii* flower (trace—4.28 mg/g for kakkalide, and 0.00–0.36 mg/g for irisolidone). In contrast, tectorigenin-7-*O*-xylosylglucoside, tectoridin and tectorigenin were more abundant in *P. thomsonii* flower (15.0–44.9 mg/g for tectorigenin-7-*O*-xylosylglucoside, 3.59–42.7 mg/g for tectoridin and 0.27–15.0 mg/g for tectorigenin) than in the *P. lobata* flower (1.06–2.57 mg/g for tectorigenin-7-*O*-xylosylglucoside and trace for tectoridin and tectorigenin). Thus, kakkalide and irisolidone were the major isoflavones of *P. lobata* flower, while tectorigenin-7-*O*-xylosylglucoside, tectoridin and tectorigenin of *P. thomsonii* flower. Some soy and red clover isoflavones were also detected in both *Pueraria* flowers. *P. lobata* flower had higher biochanin A (0.94–1.42 mg/g), but lower glycitin (not detected), genistin (trace—0.25 mg/g) and genistein (trace) than *P. thomsonii* flower (trace—0.05 mg/g for biochanin A, 0.21–6.24 mg/g for glycitin, 0.32–1.68 mg/g for genistin and 0.16–1.72 mg/g for genistein). However, daizin and daidzein, two important soy-isoflavones, were detected in traces or small amounts in both flowers. Total content of 15 isoflavones was much higher in the *P. thomsonii* flower (30.4–112.6 mg/g) than in the *P. lobata* flower (18.1–24.5 mg/g). Among the 12 saponins, kaikasaponin III and soyasaponin I were the major saponins found in both *P. lobata* and *P. thomsonii* flowers (1.26–15.2 mg/g for kaikasaponin III and 2.65–19.1 mg/g soyasaponin I). Kaikasaponin II and kakkasaponin I were higher in *P. lobata* flower (1.63–5.74 mg/g and 5.08–12.3 mg/g) than in *P. thomsonii* flower (trace—2.96 mg/g and 1.43–5.47 mg/g), suggesting that both to be the charac-

teristic saponins for *P. lobata* flower. In contrast, soyasaponin IV and baptisiasaponin I were more abundant in *P. thomsonii* flower (0.09–1.31 mg/g and 1.01–3.51 mg/g) compared to *P. lobata* flower (0.00–0.08 mg/g and 0.09–0.86 mg/g), suggesting both to be the characteristic saponins for *P. thomsonii* flower. The remaining six saponins, phaseoside IV, astragaloside VIII, kaikasaponin I, azukisaponin I, kakkasaponin II and kakkasaponin III were detected in low amounts in both *Pueraria* flowers. Total content of 12 saponins found in *P. lobata* flower was 23.2–60.6 mg/g, and that in *P. thomsonii* flower was 9.07–42.4 mg/g. Two flavones, rutin and luteolin, were found in small amounts or traces in *P. lobata* flower (trace—0.09 mg/g for rutin and 0.04–0.07 mg/g for luteolin) and *P. thomsonii* flower (0.13–4.64 mg/g for rutin and 0.05–0.42 mg/g for luteolin).

Tectorigenin-7-*O*-xylosylglucoside, tectoridin and tectorigenin were extracted from *P. lobata* flowers (Liu et al. 2010). The average recoveries of the three constituents were 99.5, 99.6 and 99.8 %, respectively. Tectoridin and 6″-*O*-xylosyltectoridin two major compounds were isolated from *P. lobata* flowers together with kakkalide, kakkalidone, puerarin, irisolidone, 6″-*O*-xylosylglycitin and genistin (Yao et al. 2010). *P. lobata* flowers from northern China contained 26.46–43.28 mg/g of tectoridin and 30.90–48.23 mg/g of 6″-*O*-xylosyl-tectoridin comparing to 10.00–19.81 mg/g of tectoridin and 11.08–37.03 mg/g of 6″-*O*-xylosyl-tectoridin in those from southern China. Two new isoflavone compounds were isolated, 5,6,7,4′-tetrahydroxyisoflavone-6,7-di-*O*-β-D-glucopyranoside and 5,6,7-trihydroxy-4′-methoxyisoflavone-6,7-di-*O*-β-D-glucopyranoside together with known ones 4′5,7-trihydroxy-6-methoxyisoflavone-7-*O*-β-D-xylopyranosyl-(1 → 6)-β-D-glucopyranoside and tectoridin from *P. lobata* flower extract (Yu et al. 2011).

Leaf/Stem Phytochemicals

Kudzu leaves were reported to be high in vitamins A and C, as well as calcium and protein. Cooked leaves contained (per 100 g) 36 cal, 89.0 % moisture, 0.4 g protein, 0.1 g fat, 9.7 g total carbohydrate. 7.7 g fibre, 0.8 fat, 34 mg Ca, 20 mg P, 4.9 mg Fe, 0.03 mg thiamin, 0.91 mg riboflavin, 0.8 mg niacin (Duke 1983). Nutritive value (% dry matter) of kudzu leaf as a feed for ruminants was reported as 17.5 % crude protein, 48.1 % neutral-detergent fibre, 38.2 % acid-detergent fibre, 8.3 % ash, 0.7 % Ca, 1 % K, 0.3 % Mg and 162.3 mg/kg Fe (Corley et al. 1997). Nutritive value (% dry matter) of kudzu stem as a feed for ruminants was reported as 10.3 % crude protein, 73.7 % neutral-detergent fibre, 44 % acid-detergent fibre, 7.9 % ash, 0.1 % Ca, 1 % K, <0.1 % Mg and 156.6 mg/kg Fe.

Robinin was isolated from *P. lobata* leaves (Saĭiad et al. 1979). From *P. lobata* leaf, kaikasaponin III, kakkalide, daidzin, genistin, rutin, robinin (kaempferol 3-*O*-rhamnosyl (1→6) galactosyl-7-*O*-rhamnoside and nicotiflorin (kaempferol-3-*O*-rutinoside) were isolated (Kinjo et al. 1988a). The concentrations of allelopathic substances, *cis,trans*-xanthoxin and *trans,trans*-xanthoxin, in *P. thunbergiana* leaves were 51.4 and 72.5 ng/g fresh weight, respectively (Kato-Noguchi 2003). A glycoside flavonoid robinin (kaempferol-3-*O*-robinoside-7-*O*-rhamnoside) was isolated from the methanolic kudzu foliage extract (Lau et al. 2005). Robinin accounted for 0.65 % (dry basis) of kudzu biomass.

Plant/Callus/Cell Culture Phytochemicals

Puerarin, daidzin, daidzein, genistein, coumestrol, methyl 2,4-dihydroxybenzoate, a mixture of methyl *p*-hydroxybenzoate and methyl *trans-p*-hydroxycinnamate, phytosterols and aliphatic acid esters were isolated from the methanol extract of callus tissues, which were induced from *P. lobata* stem (Takeya and Itokawa 1982). Isoflavone synthase activity was tested with a microsomal fraction of cell suspension cultures of *P. lobata* which had been treated with an endogenous elicitor prepared by hydrolysis of their own cell walls with a fungal endopolygalac-

turonase (Hakamatsuka et al. 1990). The deoxy types of flavanone and chalcone, liquiritigenin and isoliquiritigenin, were both converted into the corresponding isoflavone, daidzein, by a microsomal preparation (Hakamatsuka et al. 1990). From the methanol extracts of *P. lobata* cultured cells, it was found that most of the isoflavone 7-*O*-glucosides, daidzin, genistin and 8-C-glucoside puerarin, existed as their 6″-*O*-malonyl esters (Park et al. 1992). Chalcone synthase of *P. lobata* cell cultures treated with an endogenous elicitor afforded isoliquiritigenin, a deoxy-type chalcone, from *p*-coumaroyl and malonyl CoAs in the presence of NADPH (Hakamatsuka et al. 1988). But naringenin chalcone, a hydroxy-type chalcone, was the sole reaction product when NADPH was omitted from the reaction mixture. Hakamatsuka et al. (1994) found that tissue cultures of *P. lobata* grew rapidly and produced an almost equivalent amount of isoflavonoid as the mother plant. They reported that the cell line retained the ability to produce isoflavonoids even after subculture for 10 years. Also, the production of isoflavonoids in the cell suspension cultures was markedly enhanced by treatment with various elicitors such as endogenous elicitor (and yeast extracts) and the fungal glycoprotein elicitor (and $CuCl_2$). Four new isoflavonoids were isolated from kudzu cell cultures and were identified as daidzin-6″-*O*-malonylester, genistin-6″-*O*-malonylester, puerarin-6″-*O*-malonylester and genistein-8-C-glucoside-6″-*O*-malonylester (Kwon and Park 1995).

P. lobata roots, leaves and stem segments were found to be the best sources of callus tissue (Matkowski 2004). Callus from all organs contained isoflavonoid aglycones: genistein and daidzein, and daidzein glycosides: daidzin, puerarin and 3′-methoxypuerarin. The differences between each kind of explant were observed in both the total amount of isoflavonoids and in the proportion of individual compounds. The highest content was in root callus, followed by leaf and stem callus. The optimum medium for cell cultures of leaves of *P. lobata* seedlings was Gamborg B5 liquid medium supplemented with 2 % sucrose, 1.0 mg/L 2,4-D, 1.0 mg/L NAA, 0.5 mg/L kinetin and 20 mg/L casein hydrolysate (Li and Zhang 2006). The procedure use was found to be potentially useful for the production of isoflavones. Five major components of puerarin, daidzin-6″-*O*-acetylester, genistin-6″-*O*-malonylester, biochanin A-7-*O*-glucoside-6″-*O*-malonylester, and daidzein were detected and identified from the methanolic extract of *P. lobata* callus cultures (Fang et al. 2006a). The major isoflavonoid components of *P. lobat*a cell suspension cultures were identified as puerarin, daidzin, daidzin-6″-*O*-acetylester, genistin-6″-*O*-malonylester, biochanin A-7-O-glucoside-6″-*O*-malonylester, genistein-8-C-glucoside-6″-*O*-malonylester and daidzein, (Fang et al. 2006b). *P. lobata* suspension cells were successfully cultured in a bioreactor (Chen and Li 2007). An isoflavone yield of 326.9 μg/mL was obtained 73.4 μg/mL and daidzein 68 μg/mL after 12 days.

Five compounds were isolated from *P. lobata* (4*R*)-3-[2-hydroxy-4-methoxyphenyl]-4-(4-β-D-glucopyranosyloxybenzyl) but-2-en-4-olide, 4′,8-dimethoxyl-7-*O*-β-D-glucopyranosyl isoflavone, eicosanoic acid, hexadecanoic acid and tetracosanoic acid-2,3-dihydroxypropyl ester (Wang et al. 2007a). Twelve compounds were isolated from the vine of *P. lobata* and identified as: 3′-methoxydaidzein, formononetin, genistein, daidzein, daidzin, genistin, ononin, 5-hydroxyl ononin, calycosin, 6″-*O*-acetyl genistein, 6″-*O*-acetyl daidzin and puerarin (Zhang et al. 2009a).

The content sequence of mineral elements in various plant parts of *P. lobata* were as follows: copper: leaves > flowers > rattans (vines) > roots; zinc: leaves > flowers > rattans > roots; iron: roots > rattans > leaves > flowers; calcium: leaves > roots > flowers > rattans; magnesium: flowers > roots > leaves > rattans (Han et al. 2005b).

Aerial parts of kudzu plants (young shoots, leaf blades and leaf petioles) contained relatively low levels of isoflavonoids (daidzein, genistein, daidzin, genistin and puerarin) examined, whereas seeds and seedlings were intermediate in isoflavonoid levels and roots consistently had the highest levels, particularly puerarin and the glucosyl conjugates of genistein and daidzein,

namely, genistin and daidzin (Kirakosyan et al. 2003). Interestingly, commercially available kudzu root starch from Japan did not contain the isoflavonoids of interest, whereas homemade kudzu root starch contained all studied isoflavonoids in various amounts, and especially, high levels of puerarin. Shoots of light-grown kudzu seedlings, when compared with shoots of dark-grown seedlings, had higher levels of all isoflavonoids with the exception of daidzin. In contrast, for seedling roots, such differences were not greatly different between light-grown and dark-grown plants. Light-grown intact kudzu seedlings had significantly higher levels of soluble proteins than dark-grown seedlings. Mass spectrometer analyzes all kudzu samples for the toxic non-protein amino acid, *l*-canavanine, and indicated it to be absent using this method of detection and level of sensitivity.

Puerarin, the major bioactive ingredient isolated from *P. lobata* root had been widely used in the treatment of cardiovascular and cerebrovascular diseases, nervous system disorders, diabetes and diabetic complications, liver injury, osteoporosis and osteonecrosis, Parkinson's disease, Alzheimer's disease, endometriosis and cancer (Zhou et al. 2014; Wei et al. 2014). The beneficial effects of puerarin on the various medicinal purposes may be due to its wide spectrum of pharmacological properties such as inhibiting oxidative stress, vasodilation, cardioprotection, neuroprotection, antioxidant, anticancer, anti-inflammation, alleviating pain, promoting bone formation, inhibiting alcohol intake and attenuating insulin resistance.

Antioxidant Activity

Pueraria lobata tubers displayed higher DPPH antioxidant activity than *P. mirifica* tubers (Cherdshewasart and Sutjit 2008). Its isoflavonoids puerarin and daidzein exhibited the same level of antioxidant activity as alpha-tocopherol. The correlation analysis between antioxidant activity and major isoflavonoid contents of plant tubers indicated a significant correlation only with puerarin and a significant lack of correlation with daidzin, daidzein and genistein. Studies showed that *P. lobata* roots and its constituents exhibited in vitro scavenging activities against 1,1-diphenyl-2-picrylhydrazyl (DPPH), peroxynitrite (ONOO−), nitric oxide (NO·), superoxide anion (·O2−) and total ROS, and inhibitory activities against ONOO−-mediated tyrosine nitration (Jin et al. 2012). Among its constituents, lupeol showed significant inhibitory activity against intracellular ROS generation by t-BHP. Meanwhile, 3′-hydroxypuerarin showed marked ONOO−, NO·, total ROS scavenging activities, and weak·O2− scavenging activity, while 3′-methoxypuerarin showed ONOO− scavenging activity and weak NO·and O2− scavenging activities, suggesting that existence of the 3′-hydroxyl group in puerarin played an important role in the scavenging of ONOO−, NO·, and total ROS, as well as inhibiting the ONOO−-mediated tyrosine nitration mechanism.

PG-1 (8-β-D-glucopyranosyl-7-hydroxy-3-(3,4-dihydroxyphenyl)-4*H*-1-benzopyran-4-one) from *P. lobata* roots, rapidly scavenged 1, 1-diphenyl-2-picrylhydrazyl (DPPH) radical, and inhibited lipid peroxidation which was initiated enzymatically by reduced nicotinamide adenine dinucleotide phosphate (NADPH) or non-enzymatically be ascorbic acid or Fenton's reagent ($H_2O_2+Fe^{2+}$) in rat liver microsomes (Sato et al. 1992). PG-3 (8-β-D-glucopyranosyl-7-hydroxy-3-(4-hydroxy-3-methoxyphenyl)-4*H*-1-benzopyran-4-one) and daidzein (7-hydroxy-3(4-hydroxyphenyl)- 4*H*-1-benzopyran-4-one) also from *P. lobata* roots, did not show any of these effects Among the three isoflavones namely tectorigenin, glycitein and genistein isolated from *P. thunbergiana*, tectorigenin, glycitein, both bearing 6-methoxyl groups, showed significant free radical scavenging activities on 1,1-diphenyl-2-picrylhydrazyl (DPPH) radical and xanthine/xanthine oxidase (XOD) generating superoxide anion radical (Lee et al. 1999). Tectorigenin only showed a slight inhibitory effect on XOD. Each of them exhibited inhibitory effect on both ascorbic -acid/Fe^{2+}-and ADP/NADPH/Fe^{+3}-induced lipid peroxidation in rat liver microsomes. Moreover, tectorigenin exhibited the highest protection of hydrogen peroxide

damage on HepG2 and Vero cells among the three isoflavones, in the cytoprotective assay. Tectorigenin, a metabolite formed by transformation of tectoridin by intestinal microflora, was found to scavenge intracellular reactive oxygen species, and 1,1-diphenyl-2-picrylhydrazyl (DPPH) radical, and thus prevented lipid peroxidation (Kang et al. 2005b). The radical scavenging activity of tectorigenin protected the viability of Chinese hamster lung fibroblast (V79-4) cells exposed to hydrogen peroxide (H_2O_2) via activation of extracellular signal regulated kinase (ERK) pathway and by reducing the apoptotic cells formation and cell cycle arrest at G2/M phase induced by H_2O_2. Tectorigenin increased the activities of cellular antioxidant enzymes like superoxide dismutase, catalase, glutathione peroxidase, and also increased their protein level. Tectoridin and its aglycone, tectorigenin, from kudzu flowers, and modified tectorigenin were found to have antioxidant activity in-vitro including reducing power, superoxide anion radical scavenging activity, hydroxyl radical scavenging activity, 1,1-diphenyl-2-picryl-hydrazyl (DPPH) radical scavenging activity and anti-lipid peroxidation (Han et al. 2012). The results suggested that the antioxidant activity in all the experimental systems exhibited the same order as follows: tectorigenin sodium sulphonate > tectorigenin > tectoridin. Due to the high water-solubility and good antioxidant properties with tectorigenin sodium sulphonate, appropriate chemical modifications could greatly improve the biological activities of the naturally occurring products.

A glycoside flavonoid, robinin, isolated from the methanolic kudzu foliage extract, displayed antioxidant activity; one mg of fractionated robinin generated an ORAC value of 5.15 μmol/mg of Trolox, whereas 1 mg of pure robinin generated an ORAC value of 12.34 μmol/mg of Trolox (Lau et al. 2005). Zhang et al. (2011a) found that the antioxidant activity of *P. thomsonii* root was mainly attributed to 3′-hydroxypuerarin, puerarin, genistein 8-C-glycoside-xyloside, 6"-*O*-xylosylpuerarin, mirificin, isomer of daidzin and several unknown compounds. The main isoflavonoids in all sections of *P. lobata* roots were puerarin, daidzin, daidzein, genistin and genistein (Chen et al. 2012b). The root outer bark of *P. lobata* root possessed higher isoflavonoids content than whole root or kudzu root. The levels of antioxidant potential of the root outer bark assayed by total phenolic content, DPPH, ABTS and reducing power were significantly higher than those of the whole root or the kudzu root The major isoflavonoid puerarin in root outer bark exhibited the greatest antioxidant activity in *P. lobata* roots.

In vitro studies showed *Pueraria lobata* isoflavones (PLIs) could significantly inhibited lipid peroxidation (LPO) (Zhang and Fang 1997). The highest inhibition rate were found to be 79.7 %, 84.7 %, and 86.6 % in liver, kidney of mice and brain of rabbit respectively. When animals were injected with PLIs, LPO was significantly reduced and superoxide dismutase activity in animal blood and brain was increased in a dose-dependent manner. Isoflavones from kudzu roots: puerarin, daidzin, daidzein, biochanin A and genistein, acted as free-radical scavengers and blocked both linoleic acid peroxidation and lipoxygenase activity (Jun et al. 2003). Oral administration of ethanol kudzu root extract at a at a daily dose of 500 mg/kg root extract, corresponding to 50 mg/kg puerarin, during 3 weeks to diabetic rats significantly reduced plasma malondialdehyde, a marker of lipid peroxidation, to the same level as in healthy control animals, and as in the positive control group treated daily with 50 mg/kg alpha-tocopherol acetate 1 (Bebrevska et al. 2010). No obvious signs of toxicity were observed by administration of 10x the treatment dose. In-vitro studies showed that *P. thunbergiana* root ethanol extract inhibited cisplatin-induced damage of HEI-OC1 auditory cells through inhibition of lipid peroxidation and scavenging activities of superoxide radical, hydroxyl radical, hydrogen peroxide, and DPPH free radicals (Yu et al. 2010).

All three main methods based on high-performance liquid chromatography (HPLC) analysis combined with 2,2′-diphenyl-1-picrylhydrazyl (DPPH) assay, DPPH spiking HPLC analysis, on-line post-column HPLC-DPPH analysis, and HPLC-based DPPH activity

profiling could be used to screen the antioxidants from natural products successfully without the need of preparative isolation techniques, which then could speed up the identification of novel antioxidants from natural products such as *Pueraria* flowers (Zhang et al. 2012b). Among the three methods, DPPH spiking HPLC analysis method seemed to be more efficient and sensitive to the compounds with low HPLC separation resolution than the other two methods.

Anticancer Activity

Roots of *P. lobata* were found to contain 0.95 g/kg DW of the anticancer metabolite, daidzein and also a rich food source for both genistein and daidzein (Kaufman et al. 1997). Of six isoflavonoids, tectorigenin, glycitein, tectoridin, glycitin, 6″-O-xylosyltectoridin, and 6″-O-xylosylglycitin isolated from *P. thunbergiana* flower, tectorigenin and genistein exhibited cytotoxicity against various human cancer cells; glycitein showed only mild cytotoxicity (Lee et al. 2001b) Moreover, tectorigenin induced differentiation of human promyelocytic leukaemia HL-60 cells to granulocytes and monocytes/macrophages, and caused apoptotic changes of DNA in the cells, as did genistein. Tectorigenin also inhibited autophosphorylation of epidermal growth factor (EGF) receptor by EGF and decreased the expression of Bcl-2 protein, with less activity than genistein. From the results, tectorigenin may be a possible therapeutic agent for leukaemia.

The isoflavonoid genistein was found to be a potent inhibitor of the growth of human breast carcinoma cell lines, MDA-468 (oestrogen receptor negative), and MCF-7 and MCF-7-D-40 (oestrogen receptor positive) with IC_{50} values from 6.5 to 12.0 μg/ml, whereas biochanin A and daidzein were weaker growth inhibitors (IC_{50} values from 20 to 34 μg/ml) (Peterson and Barnes 1991). The isoflavone beta-glucosides, genistin and daidzin, have little effect on growth (IC_{50} values greater than 100 μg/ml). In addition, the effects of genistein and biochanin A were not attenuated by over-expression of the multi-drug resistance gene product (MCF-7-D40 vs MCF-7 cells). Fotsis et al. (1993) found genistein to potently inhibit endothelial cell proliferation and in-vitro angiogenesis at concentrations giving half-maximal inhibition of 5 and 150 μM, respectively. They found genistein concentrations in urine of subjects consuming a plant-based diet were in the micromolar range, while those of subjects consuming a traditional Western diet were lower by a factor of > 30. The high excretion of genistein in urine of vegetarians and results of their study suggested that genistein may contribute to the preventive effect of a plant-based diet on chronic diseases, including solid tumours, by inhibiting neovascularization. Thus, genistein may represent a member of a new class of dietary-derived anti-angiogenic compounds.

The ethanol-extracted fractions (PE1, PE4) of the dried powder from roots of *P. lobata* and *P. mirifica* exhibited significant antiproliferative effects on breast cancer cell lines, including MCF-7, ZR-75-1, MDA-MB-231, SK-BR-3, and Hs578T (Jeon et al. 2005). Of eight different components isolated, one of them affected the growth of some breast cancer cell lines (MCF-7, MDA- MB-231) in a dose- and time-dependent manner, as well as the growth of ovarian (2774) and cervical cancer cells (HeLa). The active cytotoxic component of *Pueraria* root was identified as spinasterol (stigmasta-7, 22-dien-3β-ol).

Studies found that after treatment with *Pueraria* root (PR) isoflavones (puerarin, daidzein, and genistein), a dose-dependent inhibition of cell growth and apoptosis in breast cancer HS578T, MDA-MB-231, and MCF-7 cell lines (Lin et al. 2009). The PR isoflavones induced cell apoptosis through a caspase-3-dependent pathway and mediated cell cycle arrest in the G2/M phase. They observed that the serum metabolites of PR (daidzein sulphates /glucuronides) inhibited proliferation of the breast cancer cells at a 50 % cell growth inhibition (GI_{50} value of 2.35 μM). The protein expression levels of the active forms of caspase-9, Bax, p53 and p21 in breast cancer cells were significantly increased by treatment with PR metabolites.

Puerarin, a natural isoflavonoid from *Pueraria lobata*, was found to dose-dependently reduced growth of colon cancer HT-29 cells in-vitro by

reducing cell viability and inducing apoptosis (Yu and Li 2006). Puerarin down-regulated MDR1 expression in MCF-7/adriamycin (MCF-7/adr), a human breast MDR cancer cell line (Hien et al. 2010). Puerarin treatment significantly inhibited MDR1 expression, MDR1 mRNA and MDR1 promoter activity in MCF-7/adr cells. Puerarin also inhibited nuclear factor kappa-B activity and IkappaB degradation. Puerarin stimulated AMP-activated protein kinase (AMPK), acetyl-CoA carboxylase and glycogen synthase kinase-3beta phosphorylation, but puerarin decreased cAMP-responsive element-binding protein phosphorylation. Both MDR1 protein expression and the transcriptional activity of cAMP-responsive element (CRE) were inhibited by puerarin and protein kinase A/CRE inhibitor (H89). Overall, the results suggested that puerarin down-regulated MDR1 expression via nuclear factor kappa-B and CRE transcriptional activity-dependent up-regulation of AMPK in MCF-7/adr cells.

Treatment of acute myeloid leukemia cell line Kasumi-1 cells with *Pueraria* root flavones caused apoptosis of Kasumi-1 cells via down-regulation of Bcl-2 protein expression and the activation of caspase-3 and caspase-8 protein (Shao et al. 2012). In-vitro studies showed that *P. lobata* root isoflavones could induce the apoptosis of human acute myeloid leukaemia SHI-1 cells, its molecular mechanism may be related to the activation of caspase hydrolase, activation of MAPK, downregulation of NF-κB, Bcl-2 and other signal molecules (Zhu et al. 2013).

Antimutagenic Activity

A methanol extract from *Pueraria lobata* showed a suppressive effect on umu gene expression of the SOS response in *Salmonella typhimurium* TA1535/pSK1002 against the mutagen 2-(2-furyl)-3-(5-nitro-2-furyl)acrylamide (furylfuramide) (Miyazawa et al. 2001). The active constituent was identified as tectorigenin. Tectorigenin and its methylated derivative [7,4′-di-O-methyltectorigenin] had the suppressive effects on umu gene expression of the SOS response in *Salmonella typhimurium* TA1535/pSK1002 against furylfuramide, 4-nitroquinoline-1-oxide, N-methyl-N′-nitrosoguanidine, and activated Trp-P-1, which did not require live metabolic activation by S9. These compounds also showed suppression of SOS-inducing activity against Trp-P-1 and AfB(1), which required liver metabolizing enzymes.

The ethyl acetate fraction (1 mg/plate) of *P. thunbergiana* flower methanol extract, decreased the number of revertants of *Salmonella typhimurium* TA100 induced by 95 % against aflatoxin B_1 (AFB_1) (Park et al. 2002). From the fraction four isoflavonoids (tectorigenin, glycitein, tectoridin and glycitin) and one saponin (kaikasaponin III) were isolated. Three isoflavonoids other than tectoridin showed significant antimutagenicity, the activity of kaikasaponin III was the most potent. Kaikasaponin III (1 mg/plate) decreased the number of revertants of S.*typhimurium* TA100 by 99 % against AFB_1 and by 75 % against N-methyl-N′-nitro-N-nitrosoguanidine (MNNG). Tectorigenin (1 mg/plate) inhibited the AFB_1-induced mutagenicity by 90 % and MNNG-induced mutagenicity by 76 %. Glycitein and glycitin were less active than tectorigenin and kaikasaponin III. The result suggested that kaikasponin III prevented the metabolic activation of AFB_1 and scavenged electrophilic intermediate capable of mutation. The two components with potent activities, tectorigenin and kaikasaponin III, significantly prevented malondialdehyde formation caused by bromobenzene in the rat.

Pueraria mirifica and *Pueraria lobata* plant extracts at the concentrations of 2.5, 5 and 10 mg/plate in the presence and absence of the S9 mixture were negative in the mutagenic Ames test (Cherdshewasart et al. 2009). In contrast, both extracts were positive in the antimutagenic Ames test towards either one or both of the tested mutagens 2-(2-furyl)-3-(5-nitro-2-furyl)-acrylamide and benzo(a)pyrene. In the micronucleus test both plant extracts at doses up to 300 mg/kg body weight (equivalent to 16 g/kg body weight plant tuberous powder) failed to exhibit significant micronucleus formation in rats. The tests con-

firmed the non-mutagenic but reasonably antimutagenic activities of the two plant extracts, supporting their current use as safe dietary supplements and cosmetics.

Hepatoprotective Activity

Supplementation of *Pueraria* root water extract (PRWE), resulted in a significant decrease in the plasma and liver total cholesterol concentrations in the ethanol-treated rats (Lee 2004). Ethanol administration significantly lowered the activities of hepatic superoxide dismutase (SOD) and catalase (CAT), whereas it increased the plasma and hepatic thiobarbituric acid reactive substances (TBARS) and the hepatic glutathione peroxidase (GSH-Px) activities. The results indicated that PRWE could alleviate the adverse effect of ethanol ingestion by enhancing the lipid metabolism as well as the hepatic antioxidant defence system.

Kakkalide, a major isoflavonoid of kudzu flower, at doses of 100 and 200 mg/kg, significantly reduced ethanol-induced lethality and acute hepatic injury in mice (Yamazaki et al. 1997). It prevented increased serum glutamic oxaloacetic transaminase and glutamic pyruvic transaminase activity and counteracted ethanol-induced elevation of glucose levels. Han et al. (2003) found that the hepatoprotective effects of kakkalide on ethanol-induced lethality and hepatic injury were dependent on its biotransformation into irisolidone via kakkalidone by human intestinal microflora. orally administered kakkalide and intraperitoneally administered irisolidone significantly reduced the mortality ceased by ethanol in mice but not kakkalidone. Orally administered kakkalide and intraperitoneally injected irisolidone greatly reduced serum alanine aminotransferase and aspartate aminotransferase activities and blood ethanol level in ethanol-intoxified mice. The results indicated kakkalide to be a prodrug of irisolidone in protecting against ethanol-induced lethality and hepatic injury. Subsequent studies by Lee et al. (2005a) also suggested irisolidone, a metabolite of kakkalide, to be hepatoprotective and kakkalide may be a prodrug transformed to irisolidone. Irisolidone protected HepG2 cells against cytotoxicity induced by tert-butyl hydroperoxide (t-BHP) but kakkalide did not. When kakkalide 100 mg/kg was orally administered to mice injured by t-BHP, it significantly inhibited the increase in plasma alanine aminotransferase and aspartate aminotransferase activities by 84 % and 85 % of t-BHP-treated control group, respectively. However, intraperitoneally administered kakkalide did not exhibit hepatoprotective activity. When irisolidone was intraperitoneally administered to mice, it exhibited potent hepatoprotective activity

Tectoridin and tectorigenin from Puerariae Flos exhibited hepatoprotective activity (Lee et al. 2005b). Tectorigenin at a dose of 50 mg/kg intraperitoneally administered to mice injured by *t*-BHP, significantly inhibited the increase the activities of plasma ALT and AST by 39 % and 41 %, respectively, in the *t*-BHP-treated group. The inhibitory effect of tectorigenin was much more potent than that of a commercially available dimethyl diphenyl bicarboxylate. Orally administered tectoridin showed hepatoprotective activity. However, when tectoridin was intraperitoneally administrated to mice, no hepatoprotective activity was observed. Tectorigenin also protected against the cytotoxicity of HepG2 cells induced by *t*-BHP. Tectorigenin may be hepatoprotective and tectoridin should be a prodrug that was transformed to tectorigenin. Intragastric treatment with tectoridin, from *P. lobat*a flowers, markedly attenuated the adverse effects of acute ethanol-induced liver steatosis in mice such as the significant increase of the alanine aminotransferase (ALT), aspartate aminotransferase (AST) and triglyceride (TG) levels and hepatic mitochondria dysfunction shown as the increase of MPT and the decrease of DeltaPsi(m) (Xiong et al. 2010). In addition, tectoridin notably alleviated the over-production of thiobarbituric acid-reactive substance and inhibited the decrease of PPARalpha expression and its target genes, including medium-chain acyl-CoA dehydrogenase (MCAD), acyl-CoA oxidase (ACO) and cytochrome P450 4A (CYP 4A) at mRNA and enzyme activity levels.

Beta-glucuronidase inhibitor tectorigenin isolated from kudzu flower protected mice from carbon tetrachloride-induced liver injury (Lee et al. 2003). tectorigenin significantly inhibited the increase of plasma ALT, AST and LDH activities and was found to be much more potent than that of dimethyl diphenyl bicarboxylate (DDB), a commercial hepatoprotective agent. When tectoridin transformed to tectorigenin by intestinal bacteria was orally administered to mice, it showed hepatoprotective activity but not when administered intraperitoneally. Also, orally administered tectoridin not only inhibited beta-glucuronidase but also increased GSH content and GST activity on CCl4-induced hepatotoxicity of mice. The results suggested tectorigenin may be hepatoprotective and tectoridin to be a prodrug transformed to tectorigenin. The main isoflavones of *Pueraria* flowers (7-O-xylosylglucoside, tectoridin and kakkalide) were metabolized by using 29 commercially available human intestinal bacterial strains and tested for hepatoprotective activity (Tsuchihashi et al. 2009). Tectoridin, which contained only one glucosyl moiety, was metabolized to its aglycone tectorigenin by various bacterial strains. On the other hand, the metabolism of 7-O-xylosylglucoside and kakkalide, which both contained disaccharide groups, was limited to specific bacterial strains. The glycosides 7-O-xylosylglucoside, tectoridin and kakkalide did not show any hepatoprotective activity, whereas aglycones tectorigenin and irisolidone showed moderate activity. Furthermore, the hepatoprotective activity of the demethylated metabolites 6-hydroxygenistein and 6-hydroxybiochanin A was extremely potent.

Kudzu flower (PF) and root (PR) water extract supplementations elicited a significant increase in the Cu/Zn SOD and/or CAT activities and a significant decrease in the GSH-Px activity in the ethanol-treated rats (Lee et al. 2001a). The mRNA levels of these antioxidant enzymes in the ethanol-treated rats were normalized to the control level and hepatic glutathione content, which was significantly lower in the ethanol-treated group than in the control group, were all normalized to the control level by supplementing with either PF or PR. Moreover, PF or PR supplement resulted in lowering the hepatic malondialdehyde to the control level in the ethanol-treated rats.

Eight weeks treatment with puerarin reverted alcohol plus carbon tetrachloride-induced liver fibrosis in rats (Zhang et al. 2006b). Puerarin significantly reversed the symptoms of liver fibrosis and other hepatic lesions. Serum alanine aminotransferase (ALT) and aspartate aminotransferase (AST), as indexes of hepatic cell disruption, were reduced with puerarin treatment, whereas no significant effect was found in the levels of alkaline phosphatase (ALP) and gamma-glutamyltransferase (GGT) activities. A significant increase in apoptosis of activated hepatic stellate cell (HSC) was found and the expression of bcl-2 mRNA was down-regulated after puerarin administration. Pre-treatment of mice with puerarin exhibited protective effects against carbon tetrachloride-induced hepatotoxicity (Hwang et al. 2007). The protective effect of puerarin was found to involve mechanisms related to its ability to block cytochrome CYP-mediated CCl_4 bioactivation, induction of glutathione S-transferase (GST) activity and free radical scavenging effects. They also found that pre-treatment of murine Hepa1c1c7 and human HepG2 cells with the phytoestrogen puerarin significantly reduced tert-butyl hydroperoxide (t-BHP)-induced caspase-3 activation and subsequent cell death (Hwang and Jeong 2008). Also, puerarin up-regulated HO-1 expression and this expression conferred cytoprotection against oxidative injury induced by t-BHP via an oestrogen-receptor-dependent Gbeta1/PI3K/Akt-Nrf2 and haeme oxygenase-1 (HO-1) pathways. Another study showed that puerarin significantly decreased the total cholesterol and triglyceride content in the liver of the non-alcoholic fatty disease rats, ameliorated steatosis in liver, lowered liver inflammatory reaction, decreased leptin level in serum, and enhanced the expression of leptin receptor mRNA and P-JAK2/P-STAT3 level (Zheng et al. 2009). All the results demonstrated that puerarin could exhibit therapeutic effect on non-alcoholic fatty liver disease by improving leptin signal transduction through JAK2/STAT3 pathways. Studies by Peng et al. (2013) found that oral administration

of puerarin ameliorated experimental alcoholic liver injury in mice by inhibition of liver tumour necrosis factor α, endotoxin gut leakage, Kupffer cell activation, and endotoxin receptors expression.

Separate studies demonstrated that puerarin, from *P. lobata*, successfully reversed hepatotoxicity in CCl_4-induced HF rats via the underlying mechanisms of lowering serum levels of alanine aminotransferase, aspartate aminotransferase, albumin, total protein and attenuating tumour necrosis factor alpha (TNF-α)/nuclear factor-kappa B (NF-κB) pathway for anti-inflammation response, as well as improving metabolic function in liver tissue (Li et al. 2013b). In-vivo studies demonstrated that puerarin exhibited hepatoprotection against chronic alcohol-induced liver injury in rats (Li et al. 2013a). The mechanisms underlying the cytoprotective effects of puerarin were associated with inhibiting immunotoxicity in hepatocytes and regulating the GSK-3β/NF-κB pathway, thereby maintaining metabolic homeostasis in the liver tissue. Another in-vivo study showed that puerarin could attenuate the CCl*4*-induced toxicity in the hepatocytes of hepatic fibrotic rats by regulating the peroxisome proliferator-activated receptor-gamma (PPAR-γ) expression and blocking the PI3K/Akt pathway to inhibit the excessive deposition of collagen (Guo et al. 2013). Shenge, a complex (1:1) preparation containing Dan shen (*Salvia miltiorrhiza*) and puerarin from *P. lobata* root, exerted significant cardioprotective effects against acute ischemic myocardial injury in rats, likely through its antioxidant and anti-lipid peroxidation properties, and thus may be an effective and promising medicine for both prophylaxis and treatment of ischemic heart disease (Wu et al. 2007).

Crude saponin from *P. lobata* root inhibited the elevation of alanine aminotransferase (ALT) activity at the dose of 90 μg/ml in rat primary hepatocyte cultures (Arao et al. 1997b). The inhibition was stronger than that of glycyrrhizin, a positive control drug. The representative saponins in the crude drug, soyasaponin I and kudzusaponin SA3, were also more effective than glycyrrhizin, although their effects were weaker than that of crude saponin at the lower doses (90, 200 μg/ml). At 500 μg/ml, kudzusaponin SA3 showed antihepatotoxic activity equal to that of crude saponin. Nine saponins kudzusaponins A_1, A_2, A_4, A_5, SA_4, SB_1 kudzusaponin A_3, soyasaponin SA_3, and soyasaponin I isolated from the roots of *Pueraria lobata* prevented in-vitro immunological liver injury in primary cultured rat hepatocytes (Arao et al. 1998). Although all tested saponins showed hepatoprotective action, the levels of activity of individual saponins differed due to differences in structure-activity relationships. The hepatoprotective activity of kaikasaponin III from *P. lobata/P. thomsonii* flowers, towards immunologically induced liver injury was significantly more effective than that of soyasaponin I (Kinjo et al. 1999).

Studies showed that alcohol-induced hepatic pathological changes, elevations in serum alanine aminotransferase (ALT) and aspartate aminotransferase (AST), and a decrease in superoxide dismutase (SOD) activity were significantly inhibited in *Pueraria* root extract- treated rats (Zhang et al. 2009c). Its inhibitory effect on alcohol-induced liver injury was associated with suppression of alcohol induced increase of intestinal permeability. Pre-treatment with intraperitoneal injection of daidzin, major isoflavone from *Pueraria* root (25, 50, 100 and 200 mg/kg) protected against D-galactosamine and lipopolysaccharide-induced hepatic failure in mice (Kim et al. 2009). Daidzin attenuated the apoptosis of hepatocytes. The liver protection of daidzin was attributed to reduced oxidative stress and its anti-apoptotic activity. Feeding of ovariectomized (OVX) rats with total isoflavones from *P. lobata* (PTIF) for 8 weeks significantly decreased the bone mineral density (BMD) loss in the femur and inhibited the increase in body weight and lipoprotein levels compared to the OVX-control group without elevating the serum levels of the liver enzymes, aspartate aminotransferase (AST) and alanine transaminase (ALT) (Lim et al. 2013a). PTIF exhibited a hepatoprotective effect in OVX-induced hepatic steatosis, indicated with reduced hepatic lipid contents.

Antidipsotropic (Anti-alcohol Abuse) Activity

Daidzin, an antioxidant isoflavonoid from *P. lobat*a, decreased blood alcohol levels after ethanol ingestion in both fasted and fed rats (Xie et al. 1994). Also, daidzin shortened sleep time for rats receiving ethanol intragastrically (7 g/kg) but not intraperitoneally. Studies demonstrated that a crude extract of kudzu root suppressed the free-choice ethanol intake of ethanol-preferring golden Syrian hamsters and two of its isoflavones, daidzin and daidzein, were found to account for this effect (Keung and Vallee 1998). Studies showed that kudzu root extract suppressed voluntary alcohol intake and alcohol withdrawal symptoms in alcohol preferring rats receiving free access to water and alcohol (Benlhabib et al. 2004). The 0.5 g/kg kudzu extract dose caused a 50–60 % reduction in alcohol consumption, abolished the development of alcohol withdrawal symptoms, but did not affect blood alcohol levels. The higher KdR doses did not further reduce alcohol consumption. The kudzu root extract contained 150 mg/g of puerarin, 13 mg/g of daidzin, 4 mg/g of daidzein, 3 mg/g of genistin, 0.2 mg/g of genistein, and 1 mg/g of glycetin. Rat's blood and liver samples contained mostly puerarin and a trace amount of daidzein that may have been formed by the hydrolysis of daidzin by liver enzymes. Rat brain samples obtained from kudzu-fed or alcohol+kudzu-fed rats did not contain any of the isoflavones. Animal studies showed that isoflavonoid compounds extracted from *Pueraria lobata* was effective in suppressing the appetite for alcohol when taken orally (Lin et al. 1996). When given orally to alcohol preferring rats at a dose of 100 mg/kg/day, daidzein, daidzin, and puerarin decreased ethanol intake by 75 %, 50 %, and 40 %, respectively. The decrease in alcohol consumption was accompanied by an increase in water intake, so that the total fluid volume consumed daily remained unchanged. The effects of these isoflavonoid compounds on alcohol and water intake were reversible. Isoflavones from P. *lobata* and *P. thomsonii* flowers namely genistein, tectorigenin and irisolidone substantially inhibited the lipopolysaccharide-induced nitric oxide release from primary cultured rat cortical microglia (IC_{50}: 1.3-9.3 μM) (Yuan et al. 2009). The inhibitory effects of isoflavones gehuain (from *P. lobata*), tectoridin, tectorigenin-7-*O*-β-D-xylosyl-(1 → 6)-beta-D-glucopyranoside and 6-hydroxygenistein-6,7-di-*O*-β-D-glucopyranoside (from *P. thomsonii*) (IC_{50}: 38–62 μM) were significant but weaker than the above aglycones. However, two compounds from *P. thomsonii*, 6-hydroxybiochanin A-6,7-di-*O*-β-D-glucopyranoside and 6-hydroxygenistein-7-*O*-β-D-glucopyranoside; genistin, and kakkalide (from *P. lobat*a) showed little inhibitory activity. With regard to the structure-activity relationships of the isoflavonoids for the inhibition of microglial activation, the glycosylation at the C-7 hydroxyl group was found to reduce the inhibitory activity. The methoxylation of 4′-hydroxyl group of 7-glycosylated isoflavonoids was found to reduce the inhibitory activity, while the methoxyl group at the 6-position enhance the activity. The results suggested that isoflavonoids of *Pueraria* flowers may be of therapeutic potential in alcoholism related to microglial activation.

According to Mc Gregor (2007) *Pueraria lobata* root appeared to be an inappropriate herb for use in herbal hangover remedies as it is an inhibitor of mitochondrial aldehyde dehydrogenase. The chronic usage of *Pueraria lobata* at times of high ethanol consumption, such as in hangover remedies, may predispose subjects to an increased risk of acetaldehyde-related neoplasm and pathology. Pueraria flos, which enhances acetaldehyde removal, is the traditional hangover remedy.

In a study of male and female "heavy" alcohol drinkers, kudzu treatment for 7 days resulted in significant reduction in the number of beers consumed that was paralleled by an increase in the number of sips and the time to consume each beer and a decrease in the volume of each sip (Lukas et al. 2005). These changes occurred in the absence of a significant effect on the urge to drink alcohol. There were no reported side effects of kudzu treatment. Data from a double-blind, placebo-controlled, crossover trial suggested that

the administration of kudzu root extract did not disturb sleep/wake cycles of moderate alcohol drinkers, and as such its utility as an adjunct treatment for alcohol dependence remained free of any potential side-effects on sleep (Bracken et al. 2011). In a week long double-blind, placebo-controlled, crossover design study of ten healthy adult volunteers, puerarin treatment reduced alcohol intake in heavy drinkers (Penetar et al. 2012). Participants consumed on average 3.5 (±0.55) beers when treated with placebo and 2.4 (±0.41) beers when treated with puerarin. Drinking topography also changed. When treated with puerarin, participants decreased sip size, took more sips to finish a beer, and took longer to consume each beer. Additionally, after finishing a beer, latency to opening the next beer was increased.

In another randomized between-subject, double-blind, placebo-controlled study of 17 male heavy alcohol drinkers aged 21–33 years (drinking 27.6 ± 6.5 drinks/week), treatment with a standardized kudzu extract significantly reduced the number of drinks consumed each week by 34–57 %, reduced the number of heavy drinking days, and significantly increased the percent of days abstinent and the number of consecutive days of abstinence (Lukas et al. 2013). The standardized formulation of kudzu extract produced minimal side effects, was well-tolerated, and resulted in a modest reduction in alcohol consumption in young non-treatment-seeking heavy drinkers. The results suggested that alcohol consumption patterns were influenced by puerarin administration and this botanical medication may be a useful adjunct in the treatment of excessive alcohol intake.

Anti-diabetic/Hypoglycemic Activity

6″-O-Xylosyltectoridin and tectoridin isolated from *Pueraria thunbergiana* flowers were metabolized to tectorigenin by a few human intestinal bacteria such as *Bifidobacterium breve* K-110 and *Eubacterium* A-44 (Bae et al. 1999). The metabolite, tectorigenin, had more potent hypoglycaemic activity as well as in-vitro cytotoxic activity against tumour cell lines than 6″-O-xylosyltectoridin and tectoridin. The results suggested that 6″-O-xylosyltectoridin and tectoridin were prodrugs which could be transformed to the active agents by certain human intestinal bacteria. Tectorigenin and kaikasaponin III from kudzu flowers exerted potent hypoglycaemic and hypolipidaemic effects in the streptozotocin-induced diabetic rats (Lee et al. 2000). Intraperitoneal administration of these two compounds with 5 and 10 mg/kg, respectively, for 7 days to streptozotocin-induced rats significantly reduced the blood glucose, total cholesterol, LDL- and VLDL-cholesterol and triglyceride levels when compared with those of control group. Additionally, tectorigenin showed in-vitro antioxidant effects on 1,1diphenyl-2-pirylhydrazyl (DPPH) radical, xanthine-xanthine oxidase superoxide anion radical, and lipid peroxidation in rat microsomes induced by enzymatic and non-enzymatic methods. Tectorigenin and kaikasaponin III also protected the Vero cell line (normal monkey kidney) from injury by hydrogen peroxide. Thus, the antioxidant action of tectorigenin and kaikasaponin III may alleviate the streptozotocin-induced toxicity and contribute to hypoglycaemic and hypolipidaemic effects. Studies found that kaikasaponin III (KS-III) may exhibit its hypoglycemic and hypolipidaemic effects by up-regulating or down-regulating antioxidant mechanisms via the changes in Phase I and II enzyme activities (Choi et al. 2004). It inhibited the formation of malondialdehyde (MDA) and hydroxy radicals in serum and liver, and increased Phase II enzyme activities such as SOD, glutathione peroxidase, and catalase, suggesting the activation of free radical-scavenging enzymes. Low MDA concentrations and low xanthine oxidase and aldehyde oxidase activities were observed in the KS-III-treated rats, suggesting that such Phase I enzyme activities were the major source of lipid peroxidation. *Pueraria* root ethanol extract protected the viability of Chinese hamster lung fibroblast (V79-4) cells exposed to H_2O_2 by reducing H_2O_2 –induced apoptosis and by scavenging intracellular reactive oxygen species (ROS), and 1,1-diphenyl-2-picrylhydrazyl (DPPH) radical,

and preventing lipid peroxidation (Kang et al. 2005a). The extract increased the activities of the cellular antioxidant enzymes, superoxide dismutase and catalase. Administration of the extract to the streptozotocin induced diabetic rats decreased the blood glucose levels. The study showed that Puerariae radix was effective in the amelioration of diabetes, which may be a consequence of its antioxidant potential.

Compounds isolated from kudzu roots, puerariafuran and coumestrol exhibited superior inhibitory activity against advanced glycation end products (AGEs) formation in-vitro with IC_{50} values of 0.53 and 0.19 μM respectively while daidzein and genistein showed significant inhibitory activity with IC_{50} of 47.2 and 260 μM (Jang et al. 2006). Puerariafuran and coumestrol appeared to be potential therapeutic agents for diabetic complications. Two isoflavone C-glucosides, puerarin (1) and PG-3 (2), a but-2-enolide, (+/−)-puerol B (3), two isoflavone O-glucosides, daidzin (4) and genistin (5), and three pterocarpans, (−)-medicarpin (6), (−)-glycinol (7) and (−)-tuberosin (8), were isolated from a methanol extract of *Pueraria lobata* roots using an in vitro bioassay based on the inhibition of the formation of advanced glycation end products (AGEs) (Kim et al. 2006a). Of these, puerarin (1), PG-3 (2), and (+/−)-puerol B (3) exhibited more potent AGEs inhibitory activity than the positive control aminoguanidine. High dose and medium dose of Radix Puerariae could decrease the level of blood glucose and the activity of aldose reductase in red blood cells, inhibit the formation of glycation products significantly in model rats induced by D-galactose increase insulin sensitivity and activity of superoxide dismutase, and decrease the amount of maleic dialdehyde (Zhang et al. 2006c).

Puerarin dose-dependently enhanced the uptake of radioactive glucose into cultured myoblast C2C12 in an insulin deficient state, which was abolished by prazosin pre-treatment (Hsu et al. 2002). The obtained data suggested that an activation of alpha 1A -adrenoceptor (alpha 1A -AR) by puerarin in C2C12 cells may increase the glucose uptake via phospholipase C – protein kinase C pathway. Bolus intravenous injection of puerarin decreased the plasma glucose concentrations in a dose-dependent manner in streptozotocin-induced diabetic rats (STZ-diabetic rats) (Hsu et al. 2003). Similar treatment with puerarin also decreased the plasma glucose in normal rats, although the effect was not as great as that in STZ-diabetic rats. In the isolated soleus muscle of STZ-diabetic rats, puerarin enhanced the uptake of radioactive glucose in a concentration-dependent manner and increased mRNA and protein levels of the subtype 4 form of glucose transporter (GLUT4) in soleus muscle The results suggested that puerarin could increase the glucose utilization to lower plasma glucose in diabetic rats lacking insulin. Chen et al. (2004) found that may activate alpha (1)-adrenoceptors on the adrenal gland to enhance the secretion of beta-endorphin to result in a decrease of plasma glucose in streptozotocin-diabetic rats. In contrast, puerarin failed to lower the plasma glucose in opioid micro-receptor knockout diabetic mice. Song and Bi (2004) found that puerarin injection could ameliorate insulin resistance in rats fed a high-fat diet. The mechanism was postulated to involved increase in cell-surface level of glucose transporter GLUT4 through decreasing fasting blood glucose and fasting serum insulin levels thus, improving GLUT4 transportation and intracellular insulin signaling. Puerarin exhibited anti-metabolic syndrome effect in-vitro (Xu et al. 2005a). They reported that abundant evidence indicated a complicated interplay among insulin resistance, adipocytes and endothelial dysfunction associated with abnormalities of metabolic syndrome such as obesity, Type II diabetes and cardiovascular diseases. They found that puerarin could potentiate insulin-induced preadipocyte differentiation, promote glucose-uptake of adipocytes that had been induced insulin resistance by high glucose, and prevent TNF-a-induced apoptosis and viability loss of endothelial cells. These effects were postulated to be due to promotion of PPARgamma expression and partly through inhibiting abnormal TNF-a-induced intracellular-free Ca(2+) accumulation of endothelial cells.

Pre-treatment of rat pancreatic islets with puerarin for 48 h resulted in a reduction in viability loss and apoptosis of islets induced by H_2O_2

(Xiong et al. 2006). In addition, pre-incubation with puerarin could restore the H_2O_2-induced decrease in basal and glucose-stimulated insulin secretion in pancreatic islets. Puerarin was also found to inhibit the free radicals production induced by H_2O_2 and to increase catalase and superoxide dismutase (SOD) activities in the isolated pancreatic islets. The results suggested that puerarin may be effective in preventing islet cells from the toxic action of reactive oxygen species in diabetes. Lee et al. (2010) found that puerarin treatment significantly enhanced differentiation of 3 T3-L1 pre-adipocytes accompanying increased lipid accumulation and glucose-6-phosphate dehydrogenase (G6PDH) activity. The results suggested that the hypoglycaemic effects of puerarin could be attributed to the upregulation of PPARγ and its downstream target genes, GLUT4 and adiponectin expression, leading to increased glucose utilization. Puerarin may also be effective in preventing the rise of oxidative stress during adipocyte differentiation by increasing endogenous antioxidant responses.

Studies showed that glycaemia in streptozotocin (STZ)-diabetogenic mice were significantly reduced following puerarin administration, while serum insulin concentration was increased (Wu et al. 2013). Puerarin also improved dyslipidaemia conditions and intrapancreatic protein levels of insulin receptor substrate-1 (IRS-1) and insulin-like growth factor-1 (IGF-1) were up-regulated, respectively. STZ-lesioned pancreas tissue in puerarin -administrated mice was effectively alleviated. In another study, puerarin treatment of streptozotocin-induced diabetic rats dose-dependently and significantly decreased kidney index, blood glucose, triglyceride, total cholesterol, malondialdehyde, fasting blood insulin, interferon IL-4, superoxide dismutase, catalase, glutathione peroxidase and nitric oxide levels and improved the renal function lowered blood urea nitrogen, serum creatinine, urine protein levels and glomerular extracellular matrix (relative area) (She et al. 2014). In addition, the mRNA and protein expression of TGF-β1, Smad2, CTGF and FN was downregulated. It was concluded that puerarin exerted its anti-diabetic effect on the STZ-treated rats through the inhibition of the TGF-β1/Smad2 pathway.

KIOM-79, a mixture of ethanol extracts from four herbs (parched Puerariae radix, gingered Magnoliae cortex, Glycyrrhizae radix and Euphorbiae radix), had been developed for the potential therapeutic application to diabetic symptoms (Park et al. 2009). The study found that the influence of KIOM-79 on cardiac ion channels were minor at concentrations effective for the diabetic models (0.1–10 μg/mL). The results suggested safety in terms of the risk of cardiac arrhythmia. Studies demonstrated that *Pueraria* root flavonoids suspended and delayed the absorption of 1-deoxynojirimycin (DNJ), the main active ingredient of Cortex Mori but did not affect the total amount of DNJ in the rat (Xiao et al. 2014). The resulting higher concentration of DNJ in the small intestine produced a relatively stronger effect of depressing the elevation of the postprandial blood glucose level. The data supported the important role of *Pueraria* root flavonoids in the application of Cortex Mori on blood glucose control.

Anti-atherosclerotic/Anti-hypercholesterolaemic Activity

Administration of total isoflavones from *Pueraria lobata* for 7 months significantly decreased concentrations of serum total cholesterol and ratios of TC/HDL-C, and liver triglycerides in ovariectomized (oestrogen deficient) rats (Zheng et al. 2002a). An inhibitor of β-hydroxy-β-methylglutaryl (HMG) coenzyme A (CoA) reductase was isolated from *Pueraria thunbergiana* with a yield of 1.3 % and an IC_{50} of 0.9 μg (4.25 μM). It was identified to be daidzein (Kim et al. 2005). The results of in-vivo studies showed that the ovariectomized hamsters orally given soybean and kudzu phytoestrogen extracts had significantly decreased serum total cholesterol (TC) and non-high-density lipoprotein cholesterol (non-HDL-C) with HDL cholesterol (HDL-C) being unaffected (Guan et al. 2006). It was demonstrated that administration of soybean but not kudzu phytoestrogen extracts decreased sig-

nificantly serum TC. However, administration of kudzu phytoestrogens caused redistribution of cholesterol among lipoproteins, leading to a significant decrease in the ratio of non-HDL-C to HDL-C. It was concluded that both soybean and kudzu phytoestrogens could modify favourably lipoprotein profiles in ovariectomized and castrated hamsters.

Supplementation of puerarin markedly attenuated the increased total cholesterol induced by hypercholesterolaemic diet in both serum and liver in Sprague-Dawley rats (Yan et al. 2006). It caused a significant reduction in the atherogenic index. Expression of mRNA for hepatic 7alpha-hydroxylase (CYP7A1) was significantly enhanced but not for those of 3-hydroxy-3-methylglutaryl-CoA reductase (HMGR) and lanosterol 14alpha-demethylase (CYP51). Further, rats administered with puerarin suppressed the hypercholesterolaemic diet induced impairment of eNOS expression, whereas there was no significant difference in the endothelium-dependent vasorelaxation among various groups of animals. Its hypocholesterolaemic function may be due to the promotion of cholesterol and bile acids excretion in liver. Both the oral and intraperitoneal administrations of kakkalide and irisolidone, with the exception of intraperitoneally treated kakkalide, potently lowered the serum levels of total cholesterol (TC) and triglyceride (TG) in Triton WR1339-induced hyperlipidaemic mice (Min and Kim 2007). The oral administrations of kakkalide and irisolidone in hyperlipidaemic mice induced, by the long-term feeding of a high fat diet, also potently reduced the serum levels of TC and TG and epididymal fat pad weight. These findings suggested that *P. thunbergiana* could improve hyperlipidaemia, and the hypolipidaemic effect may be due to inhibition of 3-hydroxy-3-methylglutaryl (HMG)-CoA reductase.

Another study showed that puerarin inhibited C-reactive protein expression via suppression of nuclear factor kappaB activation in lipopolysaccharide-induced peripheral blood mononuclear cells of patients with stable angina pectoris (Yang et al. 2010). It was concluded that puerarin acted as an anti-inflammatory agent by blocking NF-kappaB signalling, and may possibly be developed as a useful agent for the chemoprevention of atherosclerosis. Palmitate stimulation impaired insulin-mediated vasodilation in the rat aorta and puerarin treatment effectively restored the impaired vasodilation in a concentration-dependent manner (1, 10 and 50 μM) (Huang et al. 2012). Puerarin suppressed palmitate-induced inflammatory response in endothelial cells by inhibiting IKKβ/NF-κB activation and decreased TNF-α and IL-6 production with the downregulation of relative gene overexpression. Puerarin attenuated palmitate-induced phosphorylation of insulin receptor substrate-1 (IRS-1) at S307 and effectively ameliorated insulin-mediated tyrosine phosphorylation of IRS-1. The results suggested that puerarin inhibited inflammation and attenuated endothelial insulin resistance in an IKKβ/IRS-1-dependent manner.

Herbal formula DGW [Salviae Miltiorrhizae Radix (Danshen) and Puerariae Lobatae Radix (Gegen)]significantly inhibited A7r5 proliferation and exhibited G1/S cell cycle arrest by suppressing both p-ERK and cyclin D expression (Koon et al. 2011). Additionally, DGW showed anti-migratory effect against platelet-derived growth factor (PDGF)-induced A7r5 migration. Also, DGW inhibited the cell adhesion as well as the expression of ICAM-1 and VCAM-1, the production of MCP-1 but not IL-6 in TNF-α stimulated HUVECs. The study provided strong scientific evidence on carotid intimal-media thickening reduction in patients by modulating the key atherogenic events in both vascular smooth muscle cell and endothelial cells. The herbal combination of roots of *Salvia miltiorrhiza* (Danshen) and *Pueraria lobata* (Gegen) exhibited anti-atherogenic effect including anti-inflammation, anti-foam cell formation and anti-vascular smooth muscle cell (vSMC) proliferation using threes assays lipopolysaccharide (LPS)-induced nitric oxide production model, macrophage foam cell formation model and platelet-derived growth factor (PDGF)-induced vSMC proliferation model, respectively (Cheung et al. 2012). Their combination effects in anti-inflammation, anti-foam cell formation and anti-

vSMC proliferation were found to be synergistic, additive and antagonistic, respectively. The results supported the combination use of Danshen and Gegen on atherosclerosis. Gu et al. (2013) showed that oral administration of ethyl acetate or aqueous root extracts of *P. lobata* plus *Salvia miltiorrhiza* (PSRM) to cholesterol/high fat diet-induced atherosclerotic quails reduced serum cholesterol, triglycerides, low density lipoprotein cholesterol and very low density lipoprotein cholesterol levels as well as the weight of liver and liver index, and increase the serum level of high density lipoprotein cholesterol. Further, reduced levels of apolipoprotein B and elevated levels of apolipoprotein A1 were observed in ethyl acetate extract and aqueous extract of PRSM treated atherosclerotic quails. In-vivo studies showed that puerarin lowered the levels of total cholesterol, triglyceride, low density lipoprotein-cholesterol and increased high density lipoprotein-cholesterol in fat diet-induced atherosclerotic rabbits (Bao et al. 2014). Puerarin inhibited the formation and development of atherosclerotic plaques and suppressed the migration and reproduction of vascular smooth muscle cells by decreasing proliferating cell nuclear antigen (PCNA) and platelet-derived growth factor (PDGF) -A expressions in the rabbit. This is encouraging in terms of cardiovascular disease prevention/treatment.

Anti-obesity Activity

Pueraria flower extract (PFE) treatment to Western diet-loaded, spontaneously obese type 2 diabetic mice suppressed body weight gain and visceral fat accumulation, alleviated the abnormal glucose tolerance and hyperinsulinaemia, as well as causing an increase in blood adiponectin (Kubo et al. 2012). Furthermore, the suppression of liver enlargement was observed in PFE-treated mice, with suppression of fatty degeneration and anti-inflammatory effect. The results suggested that PFE acted to alleviate the effects of various metabolic diseases based on the accumulation of visceral adipose tissue, including obesity, diabetes, and hyperlipidaemia, with the promotion of catabolization/excretion of cholesterol in the liver being a key mechanism of action.

Eight weeks administration of Pueraria flower extract (PFE) was found to have anti-obesity effect in a double blind controlled study of 8 mildly obese subjects (Kamiya et al. 2011). Body Mass Index value, body weights, total fat area, and sub-cutaneous fat area of male subjects in 300 mg/day (PFE) group were significantly low compared with male subjects of placebo group. No adverse effects were found in haematological and biochemical markers of blood, urinary markers, and physical markers.

Cardiovascular /Cerebrovascular Protective Activities

In cardioprotective effect i.e. anti-myocardiac ischemia induced by pituitrin, *P. omeiensis* appeared to be the most potent, and *P. thomsonii* the least potent, while *P. lobata* and puerarin were intermediate (Zhou et al. 1995). Administration of puerarin for 21 days improved coronary collateral circulation by augmenting capillaries and distribution vessel density in the ischemic and infarctive zone in dogs with acute myocardial infarction and protecting ischemic myocardium (Liu et al 1999).

Puerarin exerted inhibitory effect only after NO production in neonatal rat cardiomyocytes was enhanced sharply during hypoxia/reperfusion injury in a dose-dependent manner (Zhu et al. 2001). Guo et al. (2004) found that puerarin could inhibit L-type calcium current of isolated rat ventricular myocytes in a time-dependent manner. The results suggested that puerarin participated in anti-myocardial ischemia and anti-arrhythmics partly due to the inhibition of L-type calcium channel. Puerarin from dried *P. lobata* root (100 mg/kg) markedly decreased the infarct volume by 34 % in cerebral cortex and improved the neurological functions after middle cerebral artery occlusion in rats (Xu et al. 2005b). neuroprotection of puerarin against cerebral ischemia was associated with a reduction in apoptosis and necrosis in the dorsolateral cortex. Studies using porcine coronary artery rings showed that expo-

sure to puerarin enhanced vasorelaxation to endothelium-independent relaxing agents, sodium nitroprusside and cromakalim possibly via the cyclic AMP-dependent pathway. (Yeung et al. 2006). However, puerarin had no effect on vasorelaxation induced by endothelium-dependent relaxing agents, bradykinin and calcium ionophore A23187. Puerarin, from *Pueraria* roots, had been suggested to be useful in the management of various cardiovascular disorders. In in-vitro studies, puerarin suppressed the proliferation and DNA synthesis of vascular smooth muscle cells (VSMC) induced by thrombin (Xu et al. 2006). The inhibitory effects of puerarin were associated with the suppression of c-fos and bcl-2 protein. Puerarin and daidzein were found to inhibit the proliferation of vascular smooth muscle cells (Han et al. 2004). Studies demonstrated that puerarin effectively prevented and reversed aconitine-induced arrhythmias in perfused heart in-vitro and in rats in-vivo (Zhang et al. 2011b). Puerarin (1.2 mM) significantly inhibited the I(K1) current in rat ventricular cells. Puerarin was found to be a novel antagonist to inward rectifier potassium channel (IK1) which underpinned its anti-arrhythmic action. Studies found that puerarin could prevent isoprenaline-induced myocardial fibrosis in mice, and its mechanisms might be related to reduction of transforming growth factor-β1 expression via activation of peroxisome proliferator-activated receptor α/γ and subsequent inhibition of nuclear factor-κB in myocardial tissue (Chen et al. 2012a).

Hydroxyethylpuerarin protected cultured bovine cerebral microvascular endothelial cells against hydrogen peroxide-induced injury and apoptosis in a concentration-dependent manner via its antioxidant potential (Guang et al. 2005). Zhang et al. (2006a) found that hydroxyethylpuerarin could reduce the occurrence of hydrogen peroxide-induced apoptosis and improve neurotrophic function of astrocytes, which may be related with its antioxidant effects during oxidative stress. Another study showed that pretreatment with hydroxyethylpuerarin (HEP) at doses of 15, 30, 60 mg/kg exhibited significant neuroprotective effects on rats against focal cerebral ischemia-reperfusion injury in adult male Wistar rats by markedly decreasing neurological deficit scores and the percentage of infarct area, reducing necrosis and apoptosis of neurons (Wang et al. 2007b). The findings suggested that HEP might provide neuroprotection against focal cerebral ischemia/reperfusion injury probably through its antioxidant and anti-inflammatory property.

Daidzein was found to be markedly effective in preventing ventricular fibrillation induced by chloroform in mice and arrhythmia induced by aconitine in rats (Ye et al. 2003). The arrhythmia induced by adrenalin in rabbits was antagonized by Daidzein and it also inhibit the action potential amplitude of isolated sciatic nerves in toads. It also prevented ventricular fibrillation induced by calcium chloride in rats, and thus reduced mortality rate of rats. Its anti-arrhythmic effect was dose-dependent and may be related to its inhibition of Na+or Ca2+ influx and its blocking beta-adrenergic receptor. Han et al. (2004) found daidzein exerted stronger inhibition than puerarin and daidzein treated groups had higher ratio of Bax/ glyceraldehyde-3-phosphate dehydrogenase (GAPDH) and Bcl-xl/GAPDH. An unidentified *Pueraria* compound markedly reduced the brain water content and the infarct size in middle cerebral artery occlusion in rats, and improved motor abilities in the cerebral ischemia-reinfusion model of gerbils (Zhao et al. 2005). It also decreased the contents of lactic acid (LA) and lipid peroxide and increased the activities of lactic dehydrogenase, glutathione peroxidase (GPx) and Na^+ -K^+ -ATPase in cerebral ischemia-reinfusion model of mice. The *Pueraria* compound protective ischemic brain tissue via its anti-oxidation effects.

Treatment with puerarin before and after cerebral ischemia-reperfusion in rats, attenuated ischemia-reperfusion injury through inhibition of NF-kappaB activation (Ding et al. 2007). Puerarin was found to inhibit intercellular adhesion molecule 1 (ICAM-1), vascular cell adhesion molecule 1 (VCAM-1) and endothelial leukocyte adhesion molecule 1 (E-selectin) expressions, by suppression of nuclear factor kappaB (NF-kappaB) activation in human umbilical vein

endothelial cells (HUVECs) (Hu et al. 2010). The results of studies suggested that Puerariae flos extract directly stimulated angiogenesis through the activation of mitogen activated protein kinase/ extracellular signal-regulated kinase (MEK/ERK)-, phosphatidylinositol 3-kinase/ protein kinase B/ endothelial nitric oxide synthase (PI3K/Akt/eNOS)-, and Src/FAK (focal adhesion kinase)-dependent pathways, without altering vascular endothelial growth factor (VEGF) expression, vascular inflammation, and permeability in-vitro and in-vivo and may be used as a therapeutic agent for ischemic disease and tissue regeneration (Chung et al. 2010).

In a 6-months, randomized, double-blind parallel, open-label, placebo-controlled study involving 100 coronary patients (mean age 58 ± 8 years), treatment with the herbal combination medicine of danshen (*Salvia miltiorrhiza*) and gegen (*Pueraria lobata*) resulted in significant improvement in brachial flow-mediated dilation and carotid intima-media thickness (Tam et al. 2009). The herbal combination treatment in coronary patients was well tolerated and effective in improving vascular function and structure and may become a novel agent for secondary prevention.

Antihypertensive Activity

Acetone (PA), methanol (PM) and water fractions (PW) of *Pueraria* root extract PA3, PA5, PM1, PM3, PM4, PM5 and PW2 decreased blood pressure in anesthetized dogs, while PM2 elevated it (Harada and Ueno 1975). PA3, PA5, PM1, PM3, PM4, PM5 and PW2 increased femoral arterial blood flow in anesthetized dogs. In the cat femoral veins and renal arteries isoproterenol caused relaxation of methoxamine-induced contraction (0.01 mmol/L) in a concentration-related manner (Wang et al. 1994). Puerarin (0.01–0.1 mmol/L) inhibited the relaxation response to isoproterenol in a concentration-response fashion. Puerarin (0.1 mmol/L) did not alter the relaxation response to nitroglycerin (1 μmol/L). The results indicated that puerarin acted as adrenergic beta-antagonist in isolated arteries and veins. Puerarin induced an endothelium-independent relaxation in rat aortic rings in a concentration-dependent manner (Dong et al. 2004). The mechanisms may involve the reduction in Ca^{2+} influx through the calcium channels operated by alpha-adrenergic receptor and the activation of the potassium channels [Kv (voltage sensitive K channels) and BKCa (large K channels), but not KATP (potassium adenosine 5′-triphosphate sensitive) channels]. In-vitro studies showed that puerarin activated Ca2+–activated potassium BK(Ca) channels, especially BK-alpha+beta1 channels (sun et al. 2007). Puerarin (0.1–1000 μM) caused concentration-dependent relaxations of rat thoracic aortic rings contracted with 1 μM noradrenaline bitartrate (EC_{50} = 1.1 μM). These were significantly inhibited by 50 nM iberiotoxin, a specific blocker of BK(Ca) channels, It was found that activation of the BK(Ca) channel probably contributed to the puerarin-mediated vasodilation action.

Long term (2 months) of dietary kudzu root extract supplementation could improve glucose, lipid, and blood pressure control in intact and ovariectomized stroke-prone spontaneously hypertensive rats (SP-SHR) (Peng et al. 2009). Results of studies by Yan et al. (2009) suggested that the anti-vasoconstriction elicited by puerarin in rat aortic rings was endothelium-dependent. NO/NO-cGMP pathway, PGI(2) and the opening of K^{+} channels sensitive to glibenclamide, tetraethylammonium, and Ba^{2+}, which might be triggered by the extracellular Ca^{2+} influx in the endothelium, appeared to contribute to the anti-vasoconstriction of puerarin. Kudzu root extract supplementation (compared to control diet) significantly lowered arterial pressure (11–15 mmHg), plasma cholesterol, fasting blood glucose (20–30 %), and fasting plasma insulin in both the ovariectomized and intact SP-SHR rats. Administration of an aqueous extract comprising roots of Danshen (*Salvia miltiorrhiza*) and Gegen (*Pueraria lobata*) in the ratio of 7:3 (DG) significantly reduced systolic blood pressure in both pre- and post-spontaneously hypertensive rat (SHR) which could be explained by its endothelium-independent vasodilation via the opening of K_{ATP}, Kir and K_v channels (Ng et al.

2011). Total flavone extracts from *Pueraria* roots (100, 200 and 400 mg/kg, i.v.) notably reduced the blood pressure of spontaneous hypertensive rats (SHRs) in a short period (Cai et al. 2011b). A 2-week administration of the extract (45, 90 and 180 mg/kg, p.o.) decreased the blood pressure of both reno-hypertensive 2K1C (two kidneys, one clip) rats and SHRs. The extract significantly and dose dependently inhibited the angiotensin converting enzyme activities in-vitro, and inhibited the plasma renin activity in-vivo. Intravenous injection of puerarin exerted effects of antihypertension and stroke prevention by improved microcirculation in 12-week old spontaneously hypertensive rats as a consequence of increase in cerebral blood perfusion both by arteriole relaxation and p42/44 MAPKs-mediated angiogenesis (Wu et al. 2014).

Antithrombotic Activity

Oral administration of *Pueraria* isoflavone (500, 1000 mg/kg) for 7 days significantly lowered blood viscosity and platelet adhesion rate, inhibited thrombosis and ADP-induced platelet aggregation in rats, and showed obvious antagonism for platelet thrombosis in ADP-treated mice (Yu et al. 1997). puerarin and especially daidzin from *Pueraria lobata* root strongly inhibited ADP- and collagen-induced platelet aggregation (Choo et al. 2002). when puerarin and daidzin were intraperitoneally administered, their anti-aggregation activities were weaker than when these compounds were administered orally. Both compounds exerted significant protection from death due to pulmonary thrombosis in mice. Puerarin injection ameliorated the haemorheology in acute blood-stasis model rats in a dose-response (Pan et al. 2003). Both the high dose and the low dose of Puerarin injection could reduce the viscosity of whole blood and plasma, blood yield stress and the maximum rate of platelet aggregation in the acute blood-stasis model rats. The high dose could also reduce the erythrocyte aggregation and the deformed Index of red blood cell. Kaikasaponin III, from kudzu flower, prolonged the bleeding time and plasma clotting time in streptozotocin (STZ)-treated rats and increased the tissue factor activity, suggesting that this compound had anti-thrombosis activity in STZ-induced rats (Choi et al. 2004).

Immunomodulating Activity

Triterpenoid saponins from *Pueraria lobata* and their hydrolytic analogues exhibited anti-complementary activities in vitro (Oh et al. 2000). Diglycosidic saponins [kaikasaponin I, soyasaponin III] showed most potent anti-complementary activities, followed by monoglycosidic saponins [soyasapogenol B monoglucuronide, sophoradiol monoglucuronide] and triglycosidic saponins [soyasaponin I, kaikasaponin III], whereas sophoradiol and soyasapogenol B showed enhancement of haemolysis under the presence of serum on the classical pathway of complement system. But all of them showed very weak or no anti-complementary activities on the alternative pathway of complement system. Administration of *Pueraria* root decoction to mice promoted the formation of anti- sheep red blood cell (SRBC), anti-ovalbumin antibodies (Ma et al. 2002)

Studies by Kim et al. (2013) found that a polysaccharide (PLP) isolated from *Pueraria lobata* enhanced the maturation of murine dendritic cells through TLR4 (toll-like receptor 4) signalling pathways. PLP induced functional maturation of dendritic cells, as shown by increased production of interleukin (IL)-12, IL-1β, and tumour necrosis factor-α, decreased antigen capture capacity, and enhanced allogenic T cell stimulation.

Vasculogenesis Activity

Incubation of isolated human mononuclear cells with puerarin dose increased the number of endothelial progenitor cells (EPC), maximum at 3 mmol/L, 24 h (Zhang et al. 2004). In addition, puerarin also promoted EPC proliferative, migratory, adhesive and in-vitro vasculogenesis capacity.

Estrogenic Activities

Administration kudzu root extract enhanced the weight of mammary gland and uterus growth in rats (Xue et al. 2003). Also it increase serum follicle stimulating hormone, luteinizing hormone, oestradiol, and decreased prolactin. Administration of *Pueraria*-isoflavone for 30 days was found to increase oestrogen level to normal in ovariectomized rats by way of increasing the level of gonadotropin-releasing hormone (Qi and Qi 2002). Conversely in normal rats, it exhibited anti-estrogenic effect. Studies showed that puerarin and total isoflavones of *P. lobata* (TIP) exerted weak oestrogen-like effect on oestrogen-deficient (ovariectomized) and infant rats, exerted no effect on normal-oestrogen level rats, but exerted anti-oestrogen-like effect in rats with high-oestrogen level (Zheng et al. 2002c). *Pueraria* root flavones (PRF) exhibited oestrogen-like effect on lipid metabolism in liver and adipose tissue of ovariectomized female rats (Wang et al. 2004). PRF inhibited the increase in the rat abdominal fat, the increase in liver total cholesterol (TC) and triglyceride (TG) and uterine atrophy caused by ovariectomy. The acid hydrolysate of *Pueraria* sprouts (HPS) increased rat growth hormone release concentration-dependently, and its EC_{50} was approximately 10.4 μg/ml (Jung et al. 2004). T_{max} for the HPS was 60 min, while C_{max} was increased approximately to 5.8 fold compared to control (244.1 pM). C_{max} for puerarin was 1028.6 7 pM, equivalent to approximately 5.2 times as high as the control level. However, tectorigenin (20 μg/kg) was of no statistical significance. The results suggested that the HPS and puerarin acted either on growth hormone secretagogue receptors or on growth hormone-releasing hormone receptor of somatotropin as possible agonists or an inhibitor on somatostatin receptor to release rat growth hormone, respectively.

The main isoflavones of *Pueraria thunbergiana* (PT) root, puerarin and daidzin were hydrolysed by human faecal and intestinal bacteria to daidzein (Park et al. 2006). With regard to estrogenic effects, the metabolites more potently increased proliferation of MCF-7 cells than PT, puerarin and daidzin. The metabolite daidzein also potently increased oestrogen-response c-fos mRNA and PR protein expressions. Studies found that kakkalide and tectoridin, main isoflavones in *P. thunbergiana* flowers, may be metabolized mainly to irisolidone and tectorigenin, respectively, by intestinal microflora in the intestines, and which may be subsequently absorbed into the blood where they can express their estrogenic effect (Shin et al. 2006). When the estrogenic effects of kakkalide and tectoridin were compared with those of their metabolites irisolidone and tectorigenin, the metabolites more potently increased proliferation of MCF-7 cells than kakkalide and tectoridin. Administration of puerarin exerted weak estrogenic activity in female rats (Malaivijitnond et al. 2010), Injection of puerarin (0.7 mg/kg BW/day) for 14 days into immature ovariectomized rats did not increase uterus weights, endometrium and myometrium areas, and the percent of cornified cells (%Co), but it increased the number of uterine glands. In long-term treatment, injection of mature rats with 7.0 mg/kg BW/day of puerarin for 140 days did not increase uterus weights, endometrium and myometrium areas, and the number of uterine glands, but a significant increase in the cornified cells was observed from day 98 onwards.

The clinical study of Yu et al. (2008) showed that puerarin could be used in the treatment of endometriosis by alleviating pain and improving infertility. Endometriosis is a gynaecological condition in which aromatase P450 (P450(arom)) is overexpressed; P450(arom) is also overexpressed in endometrial cancers and uterine fibroids. They demonstrated a significant decrease of P450(arom) expression at both mRNA and protein levels by low dose puerarin treatment in both human endometrial cancer Ishikawa and RL95-2 cells. The suppression of P450(arom) expression and activity by puerarin treatment may be associated with the downregulation of transcription factor AP-1 or c-jun.

Of the isoflavones isolated from kudzu roots, genistein exhibited the highest estrogenic activity at 10−6 M followed by daidzein (10−5 M), biochanin A (10−5 M), daidzin (10−5 M), 3′,4′,7-trihydroxyisoflavone (10−3 M), and for-

Kakkalide a major constituent of the flower of *Pueraria thunbergiana* and it metabolite irisolidone down-regulated the gene expression of cytokines [tumour necrosis factor alpha (TNF-α) and interleukin-1 beta (IL-1β)] and cyclooxygenase-2 (COX-2) and the production of pro-inflammatory cytokines, TNF-α and IL-1β, and inflammatory mediators, NO and prostaglandin E_2 (PGE_2), in LPS-stimulated peritoneal macrophages (Min et al. 2011). These agents also inhibited the phosphorylation of IκB-α and the nuclear translocation of nuclear factor-kappa B (NF-κB). Orally administered kakkalide and irisolidone significantly reduced carrageenan-induced inflammatory markers, leukocyte number, and protein amount in the exudates of the air pouch. These constituents also inhibited PGE_2 production and COX-2 inducible nitric oxide synthase, IL-1β, and TNF-α expression. These agents also inhibited NF-κB activation. The anti-inflammatory effects of irisolidone were more potent than those of kakkalide. Kakkalide, a predominant isoflavone from *P. lobata* flowers, inhibited reactive oxygen species (ROS) overproduction and effectively restored mitochondrial membrane potential, demonstrating its chemoprotection of mitochondrial function (Zhang et al. 2013a). In addition, kakkalide inhibited ROS-associated inflammation in the endothelium by inhibiting tumour necrosis factor-α and interleukin-6 production and gene expression, as well as suppressing the phosphorylation of c-Jun N-terminal kinase and IκB kinase β/nuclear factor-κB. Kakkalide facilitated PI3K signalling by positively regulating serine/tyrosine phosphorylation of IRS-1.

Singh et al. (2013) found that a bifunctional gold nanoparticle (AuNP) loaded with puerarin and curcumin may effectively suppress the lipopolysaccharide -induced inflammation and cytotoxicity under the following conditions: (1) The AuNP dose is at or below the no-effect dose; (2) the nanoparticles release a therapeutic dose of puerarin and curcumin in-vivo; and (3) the active ingredients are released into the intracellular component of the brain.

Anti-allergic Activity

Daidzein exhibited potent inhibitory activity on the beta-hexosaminidase release induced by DNP-BSA (2,4-dinitrophenylated bovine serum albumin) and potently inhibited the passive cutaneous anaphylaxis (PCA) reaction in rats (Choo et al. 2002). Daidzein administered intraperitoneally showed the strongest inhibitory activity and significantly inhibited the PCA reaction at doses of 25 and 50 mg/kg with inhibitory activity of 37 and 73 %, respectively. The inhibitory activity of intraperitoneally administered daidzein was stronger than those of intraperitoneally and orally administered puerarin and daidzin. Tectoridin, isolated from *P. thunbergiana* flowers, was metabolized to tectorigenin by human intestinal microflora (Park et al. 2004). When tectoridin was orally administered to rats, tectorigenin, but not tectoridin, was detected in urine after beta-glucuronidase hydrolysis. The main metabolite tectorigenin potently inhibited the passive cutaneous anaphylaxis reaction and inhibited in-vitro the release of beta-hexosaminidase from RBL-2H3 cells induced by IgE. The results suggested tectoridin to be a prodrug, which could be transformed into the active agent tectorigenin by human intestinal bacteria and could be a candidate for anti-allergic agent. Two isoflavones form *Pueraria* flower extract, tectorigenin and genistein were found to be inhibitors for expression of IgE receptor (FcepsilonRI), the key molecule triggering the allergic reactions, on human mast cells (Tamura et al. 2010). The synthesized isoflavone, 7-O-methyl glycitein was disclosed as the more potent inhibitor than tectorigenin.

Antiviral Activity

The isoflavone, irisilidone significantly inhibited expression of human polyomavirus JC virus, the etiologic agent of the fatal disease demyelinating progressive multifocal leukoencephalopathy, in primary cultured human astrocytes and glial cell lines (Kim et al. 2006b). The inhibitory effect of irisolidone against the JC virus may be attributed at least in part to the suppression of Sp1 binding

Anti-melanogenesis Activity

The aerial part of *P. thunbergiana* decreased tyrosinase activity significantly in B16F10 melanoma cell cultures and also showed great efficacy on anti-pigmentation in-vivo (Han et al. 2014). The anti-melanogenesis activity was modulated by two mechanisms (1) downgrading microphthalmia-associated transcription factor by activating Akt/GSK-3β, consequently decreasing transcription of tyrosinase and tyrosinase-related protein 1 and (2) interrupting maturation of tyrosinase through inhibiting α-glucosidase.

Anti-graying Activity

P. thunbergiana extract and its active constituent puerarin stimulated melanogenesis by upregulation of cAMP/MITF-M signalling pathway in-vitro and prevented the follicular depigmentation and vitiligo by stimulating melanin synthesis (Park et al. 2014).

Antimicrobial Activity

Isoflavone glycosides from the flowers and rhizomes of *Pueraria thunbergiana* did not inhibit the growth of *Helicobacter pylori* in-vitro (Bae et al. 2001). However, their aglycones, irisolidone, tectorigenin and genistein, inhibited *Helicobacter pylori* growth. Among them, irisolidone exerted the most potent inhibitory activity and its MIC was 12.5–25 μg/ml. Genistein only weakly inhibited the urease of *Helicobacter pylori* and H^+/K^+-ATPase of the rat stomach: its IC_{50} was 0.43 and 0.89 mg/ml, respectively. Crude water-soluble *Pueraria* root tea extract strongly inhibited microbial growth of *Escherichia coli* O157:H7, *Salmonella enterica* serovar *enteritidis, Listeria monocytogenes* and *Staphylococcus aureus* in liquid medium (Kim and Fung 2004). *Pueraria* root tea extract (5 %) was the most inhibitory concentration with a 6–7 log reduction in counts for all the strains compared with the control on day 3. Also, 0·63 % *Pueraria* root tea extract also strongly inhibited microbial growth of all the strains to the same extent as seen with 1·25, 2·5 and 5·0 % on day 5. One percent of β-chitosan containing 60 % kudzu starch (w/w chitosan) composite films possessed better mechanical and water barrier properties than pure β-chitosan films, and showed strong antibacterial activity against both Gram-positive and Gram-negative bacteria (Zhong et al. 2012). Reduced antibacterial activity might attribute to the interaction of amino groups in β-chitosan with the hydroxyl groups in kudzu starch. The films may be used as wraps or coatings to prolong the shelf life of different foods or other similar applications.

Anti-Hair Loss Activity

The 50 % ethanolic extract of Puerariae Flos (PF-ext) showed inhibitory activity of 60.2 % at 500 μg/ml against testosterone 5α-reductase (Murata et al. 2012). Interestingly, it was more potent than that of Puerariae Radix (roots of *Pueraria lobat*a). PF-ext also showed in-vivo anti-androgenic activity using a hair growth assay in testosterone-sensitive male C57Black/6NCrSlc strain mice. Saponins, including soyasaponin I and kaikasaponin III, were found to be the active components in PF-ext. In addition, hair growth promotion activity in C3H/He mice at 2 mg/mouse/day of the topical administration of PF-ext was demonstrated. Thus, Puerariae Flos appeared to be a promising crude drug for treating androgenic alopecia.

Herb/Drug Interaction Activity

Pueraria lobata root (gegen) crude extract and its main isoflavone puerarin exerted antioxidant properties and impaired rat liver cytochrome P450 (CYP)-catalysed drug metabolism (Guerra et al. 2000). Both biological samples inhibited the steady-state chemiluminescent reaction in a dose-dependent fashion. However, different inhibition mechanisms were postulated, since only gegen behaved like conventional antioxidants. Intragastric administration of gegen and puerarin

to male Wistar rats, both CYP content and NADPH-(CYP)-c-reductase activity were significantly increased in all situations, a complex pattern of CYP modulation was observed, including both induction (puerarin: CYP2A1, 1A1/2, 3A1, 2C11; gegen: CYP1A2, 3A1, 2B1) and inactivation (puerarin and gegen : CYP3A, 2E1, 2B1). Zheng et al. (2010) conducted a randomized, two-phase-cross-over study of 19 healthy male volunteers of different CYP2D6 genotypes to determine the effect of puerarin on hepatic CYP-linked drug metabolising enzymes such as CYP2D6and CYP1A2 in-vivo. The logarithm value of metabolic rate was found to decrease from −0.0055 to −0.1754 implying that puerarin inhibited activity of CYP2D6. There was no significant relationship between the inhibition with the CYP2D6 genotypes. The paraxanthine/caffeine ratio in the plasma sample at 6th hour after puerarin ingestion was increased by 30 %, implying that puerarin induced the activity of CYP1A2. Results of in-vitro human and rat liver microsomal and in-vivo (rat) pharmacokinetic studies suggested the possible inhibition of hepatic CYP3A-mediated drug (buspirone) metabolism by puerarin administration, potentially leading to metabolism-mediated herb-drug interactions with clinical significance (Kim et al. 2014). Studies found that the absorption of puerarin significantly increased when combined with *Angelica dahurica* root extracts, as shown by the increase in concentration of puerarin in blood from the hepatic portal vein, supporting the concept of *Angelica dahurica* root and *Pueraria lobata* root as a compatible herb-pair (Liao et al. 2014).

Pharmacokinetic Studies

Phytoestrogen levels in acidified kudzu-root samples were 5- to 10-fold greater than those in nonacidified samples (Benlhabib et al. 2002). Puerarin accounted for 80 % of total phytoestrogens in kudzu-root. When serum was dialyzed with phytoestrogen standards in a buffer, the protein binding of phytoestrogens correlated negatively with their polarity. When serum was dialyzed with kudzu-root almost all of the phytoestrogens present in the extract were bound to serum protein. The study revealed differences in the bioavailability of phytoestrogens when they were ingested as purified compounds compared with crude plant extract. After oral administration of 14C-labelled isoflavone extract to male Sprague-Dawley rats, analysis of bone tissues revealed that radiolabel accumulated in the femur, tibia and vertebrae at 0.04, 0.03 and 0.01 % of the administered dose, respectively (Mun et al. 2009). The liver accumulated the greatest concentration of radiolabel among the tissues tested, at 1.99 % of the administered kudzu extract. Urine and faeces contained 8.53 and 9.06 % of the kudzu dose respectively. The predominant isoflavones in the kudzu extract were the glycosides puerarin, daidzin and malonyl daidzin. Serum pharmacokinetics suggested that extracts from kudzu may undergo enterohepatic circulation.

The urine of rats administered daidzin orally contained four major metabolites, daidzein 7,4′-di-*O*-sulfate (M-1), daidzein 7-*O*-β-D-glucuronide (M-2), daidzein 4′-*O*-sulfate (M-3), daidzein (M-4), (Yasuda et al. 1994). The urine of rats treated with daidzein contained M-2 to M-4 metabolites. Total cumulative amounts of the four metabolites excreted in the urine at 48 h following the oral administration of daidzin and daidzein were approximately 4.8 % and 4.6 % of the doses administered, respectively. The bile of rats administered daidzin orally contained M-1to M-4. Daidzein 7-*O*-β-D-glucuronide 4′-*O*-sulfate (M-5), a major biliary metabolite, was also identified. In another study, they found that the urine of rats administered puerarin orally contained puerarin and four major metabolites, daidzein 4′,7-di-*O*-sulfate, daidzein 7-*O*-β-D-glucuronide, daidzein 4′-*O*-sulfate, and daidzein (Yasuda et al. 1995). The bile of rats administered puerarin orally contained puerarin and two major metabolites, which were identified as puerarin 4′-*O*-sulfate and puerarin 7-*O*-β-D-glucuronide. The experimental data suggested that C-glycoside puerarin was partially hydrolyzed to aglycone in the body, but mainly excreted in the urine as unchanged puerarin. The urine and bile of rats

administered genistein orally contained eight metabolites; besides genistein, three of these metabolites were identified as, genistein 4′-O-sulfate; genistein 7-*O*-β-D-glucuronide and genistein 4′-*O*-sulfate 7-*O*-β-D-glucuronide (Yasuda et al. 1996). After oral administration of *Pueraria* isoflavone daidzin to rats, the aglycone daidzein and three other metabolites, 3′,4′,7-trihydroxyisoflavone; 4′,7-dihydroxyisoflavanone and 4′,7-dihydroxyisoflavan (M4) were isolated from the urine following treatment with enzymes beta-glucuronidase and arylsulphatase (Yasuda and Ohsawa 1998). After oral administration of *Pueraria* isoflavone genistein to rats, genistein and its metabolites, 4′,5,7-trihydroxyisoflavanone; 4′,7-dihydroxyisoflavan and *p*-ethylphenol were isolated from the urine following treatment with enzymes beta-glucuronidase and arylsulfatase (Yasuda et al. 2001). After oral administration of puerarin to rats, two new metabolites of puerarin, mono- and dihydroxylated derivatives, were detected in the urine and feces of rats (Prasain et al. 2004). The persistence of puerarin in blood and urine as the principal metabolic form for the period of 4–72 h after oral administration suggested that puerarin was rapidly absorbed from the intestine without metabolism. Puerarin was also detected in organs such as the brain. Besides puerarin, equol, diadzein and dihydrodaidzein were detected in the urine. Studies demonstrated that Puerarin-nano-particles could increase the absorption of the drug in intestinal epithelia (Liu et al. 2009b). The main mechanism of puerarin and its nanoparticles across Caco-2 monolayer model was passive transference. After intravenous administration of puerarin to rats two glucuronidated metabolites in the rat plasma were identified as puerarin-7-*O*-glucuronide and puerarin-4′-*O*-glucuronide (Luo et al. 2010).

After oral administration of a single dose (100 mg/kg) of tectoridin to healthy rats, faeces and urine samples were collected for 0–48 h and 0–24 h for identification of metabolites (Chen et al. 2008b). Tectoridin was also incubated with rat intestinal flora and rat liver microsomes. The results revealed six metabolites of tectoridin in urine (tectorigenin, hydrogenated tectorigenin, mono-hydroxylated tectorigenin, di-hydroxylated tectorigenin, glucuronide-conjugated tectorigenin and sulphate-conjugated tectorigenin); three metabolites in faeces (tectorigenin, di-hydroxylated tectorigenin and sulphate-conjugated tectorigenin); one metabolite in the intestinal flora incubation mixture (tectorigenin), and four in the liver microsomal incubation mixture (tectorigenin, hydrogenated tectorigenin, mono-hydroxylated tectorigenin and di-hydroxylated tectorigenin).

Puerarin was extracted from rat plasma after intravenous administration of puerariae radix isoflavone (Yan et al. 2005). It was found that the elimination rate of puerarin was significantly slower in the cerebral ischemia reperfusion rat than in the normal rat, judging by the pharmacokinetic parameters obtained. After acute and repeated administration of a novel kudzu extract to human volunteers, its principal constituent puerarin was found to be rapidly absorbed via the oral route, reached peak levels at 2 h, and presented a half-life of approximately 4.3 h (Penetar et al. 2006). The elimination half-life was not significantly altered after repeated administration. The self-microemulsifying ability and dissolution behaviour of *Pueraria lobata* isoflavone in vitro and the pharmacokinetic behaviour in rats was compared with commercial tablets (Cui et al. 2007). There was an 82 % increase in the relative bioavailability of the isoflavone was observed for the self microemulsifying drug delivery systems compared with commercial Yufengningxin tablets. The results suggested that the self-microemulsifying drug delivery systems could increase drug dissolution in-vitro and absorption in-vivo significantly. Pelletization of *Pueraria lobata* flavonoids enhanced the bioavailability of the flavonoids in the rat plasma by 4.4-fold compared to Yufengningxin tablets (Jia et al. 2007)

Prasain et al. (2009) found that after oral administration of puerarin to rats, puerarin was found to be widely distributed in rat tissues with highest concentrations in lungs (799 ng/g wet tissues). Puerarin was excreted into the bile predominantly in the form of unconjugated puerarin and was also found in several organs including

kidney and pancreas and may explain its beneficial effects in diabetes. Gestation especially during the early stages of pregnancy was found to influence the pharmacokinetics of puerarin at different levels, after oral ingestion of puerarin (Cao et al. 2013). Additionally, puerarin was found to penetrate the placental barrier and maintain high concentrations in foetal rat plasma. Therefore, puerarin administration should be carefully considered in pregnant women.

Oral administration of Kakkalide afforded 13 urinary metabolites in rats (Bai et al. 2010). Four new compounds were identified as irisolidone-7-*O*-glucuronide, tectorigenin-7-*O*-sulfate, tectorigenin-4′-*O*-sulfate, and biochanin A-6-*O*-sulfate together with nine known compounds identified as irisolidone, tectorigenin, tectoridin, 5,7-dihydroxy-8,4′-dimethoxyisoflavone, isotectorigenin, biochanin A, genistein, daidzein and equol. Rat plasma contained three glucuronide metabolites, irisolidone-7-*O*-glucuronide, tectorigenin-7-*O*-glucuronide and 6-OH biochanin A-glucuronide, as well as kakkalide and trace amount of irisolidone after oral administration of 200 mg/kg kakkalide (Bai et al. 2011). Seven metabolites, tectorigenin-7-*O*-glucuronide, tectorigenin-7-*O*-sulfate, tectorigenin-4′-*O*-sulfate, 6-OH biochanin A-glucuronide, irisolidone-7-*O*-glucuronide, tectorigenin, and irisolidone were identified in rat urine after oral administration of kakkalide, a major isoflavone of *P. lobata* flowers (Wang et al. 2013a). In addition, irisolidone-7-*O*-glucuronide was found in bile, and irisolidone and kakkalide were found in feces. Over a 72-h period, 13.2 % of the kakkalide was excreted as seven metabolites in urine. Over the same time period, irisolidone-7-*O*-glucuronide excretion in bile accounted for 3.8 1 % of the dose, while kakkalide and irisolidone excretion in feces accounted for 2.1 % and 0.7 % of the dose, respectively. The results indicated that urine was the primary route of kakkalide elimination in-vivo and that extensive metabolism may be one of the reasons for the low bioavailability of kakkalide.

The metabolic profiles of kakkalide and irisolidone after incubation with human and rat intestinal bacteria were reported by Zhang et al. (2014). A total of 17 metabolites, including parent compounds, were detected in human and rat intestinal bacteria incubated samples. The results obtained indicated that hydrolysis, dehydroxylation, demethoxylation, demethylation, hydroxylation, decarbonylation, and reduction were the detected metabolic pathways of kakkalide and irisolidone in-vitro. The conversion rate of irisolidone in human and rat bacteria was 8.57 % and 6.51 %, respectively. Biochanin A was the relatively main metabolite of irisolidone, and the content of biochanin A in human and rat bacteria was 3.68 % and 4.25 %, respectively. The conversion rate of kakkalide in human and rat bacteria was 99.92 % and 98.58 %, respectively. Irisolidone was the main metabolite of kakkalide, and the content of irisolidone in human and rat bacteria was 89.58 % and 89.38 %, respectively. A total of 46 metabolites of irisolidone, from *P. lobata* flowers, were detected in plasma, bile/urine, and feces after administration to rats (Zhang et al. 2013b). The metabolic pathways of irisolidone in rats included decarbonylation, reduction, demethylation, demethoxylation, dehydroxylation, hydroxylation, sulfation, and glucuronidation.

Studies in 10 healthy male volunteers showed that *Pueraria lobata* root isoflavones were exclusively metabolized to sulphates/glucuronides of daidzein following administration of traditional decoction (TD) and concentrated powders (CP) of *Pueraria* root (Hou et al. 2011). The parent forms of puerarin. daidzin and daidzein were not detected in urine, whereas daidzein sulphates/glucuronides were predominant, mainly sulphates. The half-lives of daidzein sulphates/glucuronides were 5–7 h.

Adverse Toxicity Issues

Studies showed that co-administration of *Pueraria lobata* root decoction significantly decreased the elimination and resulted in markedly increased exposure time of methotrexate (MTX), a bicarboxylate anti-metabolite in rats resulting in high rat mortality (Chiang et al. 2005). Studies by Wang and Xu (2011) that puerarin injection could induce acute intravascular haemolysis and could seriously impact on the use

of puerarin injection for the treatment of cardiac/cerebral blood vascular diseases.

A recent case of a middle-aged woman developing acute interstitial nephritis following the ingestion of kudzu root juice for health well being was reported by Jung et al. (2013). She developed symptoms of anorexia, epigastric discomfort and azotemia but improved after discontinuation of the suspected offending agent and conservative treatment.

Traditional Medicinal Uses

Pueraria lobata (Willd.) Ohwi, a plant in traditional oriental medicine, is widely distributed in China, Korea and Japan (Chen et al. 2012b). *Pueraria lobata* and *P. thomsonii* are widely used in Chinese traditional herbal medicines; both are listed in the Chinese Pharmacopoeia 2000 edition as Radix Puerariae (Zhang et al. 2011a). However, *Pueraria lobata* and *Puerariae thomsonii* are listed as two herbal medicines in the 2005 edition due to different content of puerarin. *Pueraria lobata* is a main plant resource for clinical applications, while *Pueraria thomsonii* is widely used as a soup material in south of China. The kudzu vine, known as Ge Gen in China, is commonly used in Chinese herbalism, where it is considered to be one of the 50 fundamental herbs (Duke and Ayensu 1985). The plant is often used in combination with *Chrysanthemum x morifolium* in treating alcohol abuse (Chevallier 1996).

In traditional Chinese medicine, the flower of *Pueraria lobat*a (Puerariae Flos) has been used in therapy to counteract the problems associated with alcohol drinking and liver injury (Xiong et al. 2010). Pueraria Flos, which enhances acetaldehyde removal, is the traditional hangover remedy (McGregor 2007). Puerariae Flos is a traditional herbal medicine that has long been used as a treatment for colds, diabetes, and hangovers (Kim et al. 2003). The herbal medicine contained a wide variety of isoflavones such as kakkalide, tectoridin, and tectorigenin. The quantitative composition of these isoflavones varied according to the storage period (hydrolysed fresh, fresh<5 years and old>5 years). Kudzu flower buds are used as a diaphoretic and febrifuge medicine (Burkill 1966). The flowers and tubers are antidote, antiemetic, antipyretic, antispasmodic, demulcent, diaphoretic, digestive, febrifuge, hypoglycaemic and hypotensive (Yeung 1985; Duke and Ayensu 1985; Foster and Duke 1998). A concoction of the flowers and tubers is used to treat alcoholism, fever, colds, diarrhoea, dysentery, acute intestinal obstruction etc. It is also useful in the treatment of angina pectoris and migraine (Duke and Ayensu 1985).

Yege (Gegen or Radix Puerariae lobatae), the dried root of *Pueraria lobata* (Wild.) Ohwi, has been widely used in traditional medicine in China and, to a lesser extent, in Japan, Korea, and the United States. Radix puerariae is one of the most important oriental crude drugs that increases coronary artery blood flow and is used as an antipyretic, anti-diarrhetic, diaphoretic, antiemetic, antispasmodic, and anti-microbic remedy (Tang and Eisenbrand 1992). It is also an active agent against angina pectoris, hypertension, deafness, optic nerve atrophy, retinitis, and alcohol abuse. Tuber starch is used in Japan to restore intestinal and digestives disorders taken in soups or teas. Groen et al. 1996). Tea from tubers is used against influenza, diarrhoea, dysentery and hangovers The root is frequently used as a remedy for measles, often in combination with *Cimicifuga foetida* (Chevallier 1996). The dried roots of *P. lobata* have been used as the main ingredient of a traditional prescription for the treatment of early symptoms of common cold and as antipyretic, anti-diarrhetic, diaphoretic and antiemetic agents (Keung and Vallee 1998, Rong et al. 1998b). *P. lobat*a root is also used for the treatment of hypertension and alcoholism or as an antioxidant. Daidzin, a major active principle of an ancient Chinese herbal treatment (Radix puerariae) for alcohol abuse as it possess anti-dipsotropic activity (Keung and Vallee 1998). Daidzin suppresses ethanol intake by inhibiting aldehyde dehydrogenase (ALDH2). *Pueraria lobata* is used as a traditional Chinese herbal remedy for menopausal symptoms, as well as an ingredient in preparations for conditions affecting menopausal women, such as osteoporosis, coronary heart disease, and some hormone-dependent cancers

(Woo et al. 2003). The combined extracts obtained from three Chinese herb medicine, *Kalopanax pictus, Pueraria thunbergiana* and *Rhus verniciflua*, have been used as therapeutics for diabetes mellitus in Korea (Kim et al. 2004). *Pueraria* root (*Pueraria mirifica* from Thailand and *Pueraria lobata* from Korea), is used as a rejuvenating folk medicine in Thailand and China (Jeon et al. 2005).

Puerariae radix, as an edible plant, has been used for centuries in China to treat alcohol-related problems, including alcoholic liver disease (Zhang et al. 2009c). The radix of *Pueraria thunbergiana* is traditionally prescribed to attenuate the clinical manifestation of inner ear dysfunction and various clinical situations including fevers, gastrointestinal disorders, skin problems, migraine headaches, lowering cholesterol, and treating chronic alcoholism in oriental medicine (Yu et al. 2010). For more than 2000 years, kudzu root has been used as a herbal medicine for the treatment of fever, acute dysentery, diarrhoea, diabetes and cardiovascular diseases (Wong et al. 2011). According to ancient Chinese medicine, kudzu root has been used as an ingredient to treat alcohol intoxication for centuries (Bracken et al. 2011 Liu et al. 2012). The roots of *Salvia miltiorrhiza* (Danshen) and *Pueraria lobata* (Gegen) are principle herbs of Chinese herbal formulae which have long been used to treat cardiovascular diseases (Cheung et al. 2012). *Pueraria lobata*, is a common ingredient of Ge-Gen-Tang (Kakkon-to) and Sheng-Ma-Ge-Gen-Tang (Shoma-kakkon-to), prescriptions of Chinese traditional medicine proven to have antiviral activity against human respiratory syncytial virus (HRSV) (Lin et al. 2013). Ge-Gen-Tang (GGT) has been used against adult viral respiratory tract infection for thousand years in ancient China (Chang et al. 2012). Radix Puerariae has been traditionally used for the treatment of diarrhoea, acute dysentery, deafness and cardiovascular diseases (Zhang et al. 2013b). Among various commercially available products of Radix Puerariae, injection of puerarin, the major isoflavone from Radix Puerariae, has been most widely used as a vasodilator for the treatment of angina and myocardial infarction.

The stems are galactogogue and are also applied as a poultice to incipient boils, swellings, sore mouths etc ; the seed is used in the treatment of hangover and dysentery (Duke and Ayensu 1985; Foster and duke 19980. The leaves are styptic (Duke and Ayensu 1985). Pueraria shoots are used as a lactagogue.

Other Uses

Kudzu is a useful erosion-controlling soil cover and shade plant. Kudzu was introduced to the southern United States in the 1930s to help restore the soil and reduce erosion (Tanner et al. 1979). The plant had been considered useful for several uses: the root starch as a source of carbohydrate and as a medium for yeast and ethanol production; the fibre for use in paper, in grass wallpaper, and in textiles and clothing; and the leaves for a high-protein animal fodder. Preliminary studies indicated that the root provided a vitamin-enriched source of starch for ethanol and yeast fermentations. Kudzu root starch can be fermented to produce a lysine-enriched bakers' yeast and ethanol. High tensile strength fibres can be recovered from kudzu vine by fermentation for use in textiles and apparels..

The stems, shoot and leaves are also used as silage or hay as the plant is highly palatable to livestock and nutritious. Kudzu is nearly equal to alfalfa in nutritive value. Kudzu can be used by grazing animals as it is high in quality as a forage. Kudzu has been used as a form of erosion control to stabilize banks, slopes and gullies and also to enhance soil fertility. As a legume, it increases the nitrogen in the soil via a symbiotic relationship with nitrogen-fixing bacteria in the soil (Mitich 2000). In Japan, young kudzu vines are harvested to provide supple waterproof fibres for weaving sturdy wicker baskets or trunks (Shurtleff and Aoyagi 1977). The cellulose fibre from large vines and roots is used as the basic raw material for making fine traditional paper. The fibre is also used to stuff cushions, beds and chairs. When burned, it acts as a mosquito repellent (Mitich 2000). The stem fibres are still used for durable textiles and ropes in Korea. In the

southern United States, kudzu is used to make soaps, lotions, jelly and compost. It may become a valuable asset for the production of cellulosic ethanol.

Studies by Sage et al. (2009) found kudzu (*P. montana* var. *montana*) to be a new source of carbohydrate for bioethanol production. Belowground biomass in Alabama exceeded 13 t/ha, and contained an average of 37 % fermentable carbohydrates (sucrose, glucose and starch). Roots from Georgia of all size classes contained over 60 % fermentable carbohydrates. Based on the yield and carbohydrate content, they estimated wild kudzu stands in Alabama and Georgia could produce 5–10 t/ha of carbohydrate, which would rival carbohydrate production from maize and sugar cane fields. If economical harvesting and processing techniques could be developed, the kudzu infesting North America has the potential to supplement existing bioethanol feedstocks. Studies found that in comparison with the traditional fermentation technology, the production of ethanol and isoflavones by simultaneous saccharification and solid state fermentation of steam-pre-treatment *Puerariae* roots was clean and energy-saving (Fu et al. 2008). It provided a new way for the production of ethanol from the non-food starch material.

Kudzu was found to be an effective adsorbent for removal of heavy metals (Brown et al. 2001). Though its capacity for metals removal was less than commercial grade ion exchange resins, it could be used at much lower cost, and may find application in the treatment of dilute mixed-matrix metal waste streams, such as urban runoff, where the application of resins would be expensive and subject to premature poisoning by interfering contaminants.

Qi et al. (2006) found sodium chloride to be a promising isotonicity agent for the thermosensitive poloxamer gel containing puerarin, hydroxypropyl-β-cyclodextrin (HPCD) and benzalkonium chloride (BC) for ophthalmic use. HPCD enhanced the gelation temperature and reduced the gel strength and the bioadhesive force, while puerarin and BC had a comparatively smaller influence.

Cis, trans-xanthoxin and *trans,trans*-xanthoxin from *P. thunbergiana* leaves inhibited the root growth of cress (*Lepidium sativum*) seedlings at concentrations greater than 0.3 and 3 μM, respectively (Kato-Noguchi 2003). The doses required for 50 % inhibition on the cress roots were 1.1 and 14 μM for *cis,trans*- and *trans,trans*-xanthoxin, respectively. The results suggested that xanthoxins may contribute to the growth inhibitory effect of *P. thunbergiana* and may play an important role in the allelopathy of *P. thunbergiana* after being released into the soil. Studies showed that the kudzu (*P. montana*) leaf and root extracts sufficiently lowered the total growth and the speed of germination of the dandelion, white clover seeds and ryegrass seeds, but had a less profound effect on Bermuda grass seeds. The results suggested that the kudzu leaves and roots could be considered to be allelopathic and may have potential as a weed control agent (Mathur and Mathur 2013).

Comments

Kudzu can be readily propagated from stem cuttings.

The IUCN (2013) has listed kudzu among the world's 100 worst invasive species (IUCN Global Invasive Species Database 2002).

Selected References

Arao T, Kinjo J, Nohara T, Isobe R (1995) Oleanene-type triterpene glycosides from Puerariae Radix. II. Isolation of saponins and the application of tandem mass spectrometry to their structure determination. Chem Pharm Bull 43(7):1176–1179

Arao T, Kinjo J, Nohara T, Isobe R (1997a) Oleanene-type triterpene glycosides from Puerariae Radix. IV. Six new saponins from *Pueraria lobata*. Chem Pharm Bull 45(2):362–366

Arao T, Udayama M, Kinjo J, Nohara T, Funakoshi T, Kojima S (1997b) Preventive effects of saponins from puerariae radix (the root of *Pueraria lobata* Ohwi) on in vitro immunological injury of rat primary hepatocyte cultures. Biol Pharm Bull 20(9):988–991

Arao T, Udayama M, Kinjo J, Nohara T (1998) Preventive effects of saponins from the *Pueraria lobata* root on

in vitro immunological liver injury of rat primary hepatocyte cultures. Planta Med 64(5):413–416

Bae EA, Han MJ, Lee KT, Choi JW, Park HJ, Kim DH (1999) Metabolism of 6″-O-xylosyltectoridin and tectoridin by human intestinal bacteria and their hypoglycemic and in vitro cytotoxic activities. Biol Pharm Bull 22(12):1314–1318

Bae EA, Han MJ, Kim DH (2001) In vitro anti-*Helicobacter pylori* activity of irisolidone isolated from the flowers and rhizomes of *Pueraria thunbergiana*. Planta Med 67(2):161–163

Bai X, Xie YY, Liu J, Qu JL, Kano Y, Yuan D (2010) Isolation and identification of urinary metabolites of kakkalide in rats. Drug Metab Dispos 38(2):281–286

Bai X, Qu JL, Lu J, Kano Y, Yuan D (2011) Pharmacokinetics of kakkalide and its main metabolites in rat plasma determined by HPLC-DAD and LC–MS. J Chromatogr B 879(5–6):395–402

Bao LD, Zhang Y, Wei GM, Wang Y, Ma RL, Cheng R, Ren XH (2014) The anti-atherosclerotic effects of puerarin on induced-atherosclerosis in rabbits. Biomed Pap Med Fac Univ Palacky Olomouc Czech Repub (in press)

Bebrevska L, Bravo L, Vandervoort J, Pieters L, Vlietinck A, Apers S (2007) Development and validation of an HPLC method for quality control of *Pueraria lobata* flower. Planta 73(15):1606–1613

Bebrevska L, Foubert K, Hermans N, Chatterjee S, Van Marck E, De Meyer G, Vlietinck A, Pieters L, Apers S (2010) In vivo antioxidative activity of a quantified *Pueraria lobata* root extract. J Ethnopharmacol 127(1):112–117

Benlhabib E, Baker JI, Keyler DE, Singh AK (2002) Composition, red blood cell uptake, and serum protein binding of phytoestrogens extracted from commercial kudzu-root and soy preparations. J Med Food 5(3):109–123

Benlhabib E, Baker JI, Keyler DE, Singh AK (2004) Kudzu root extract suppresses voluntary alcohol intake and alcohol withdrawal symptoms in P rats receiving free access to water and alcohol. J Med Food 7(2):168–179

Bo J, Ming BY, Gang LZ, Lei C, Jia AL (2005) Protection by puerarin against MPP+–induced neurotoxicity in PC12 cells mediated by inhibiting mitochondrial dysfunction and caspase-3-like activation. Neurosci Res 53(2):183–188

Bracken BK, Penetar DM, Maclean RR, Lukas SE (2011) Kudzu root extract does not perturb the sleep/wake cycle of moderate drinkers. J Altern Complement Med 17(10):961–966

Brown PA, Brown JM, Allen SJ (2001) The application of kudzu as a medium for the adsorption of heavy metals from dilute aqueous wastestreams. Bioresour Technol 78(2):195–201

Burkill IH (1966) A dictionary of the economic products of the Malay Peninsula. Revised reprint of 1st ed 1935, 2 vols. Ministry of Agriculture and Co-operatives, Kuala Lumpur, Malaysia. Vol. 1 (A-H) pp 1–1240, Vol. 2 (I-Z). pp 1241–2444

Cai H, Wu LL, Sun XC, Zhu W, Qian H, Hu JB, Xu WR (2011a) Effects of puerarin on proliferation and differentiation of umbilical cord mesenchymal stem cells into osteoblasts in vitro. Yao Xue Xue Bao 46(6):738–741 (In Chinese)

Cai RL, Li M, Xie SH, Song Y, Zou ZM, Zhu CY, Qi Y (2011b) Antihypertensive effect of total flavone extracts from Puerariae Radix. J Ethnopharmacol 133(1):177–183

Cao L, Pu J, Cao QR, Chen BW, Lee BJ, Cui JH (2013) Pharmacokinetics of puerarin in pregnant rats at different stages of gestation after oral administration. Fitoterapia 86:202–207

Chang Q, Sun L, Zhao RH, Chow MS, Zuo Z (2008) Simultaneous determination of ten active components in traditional Chinese medicinal products containing both Gegen (*Pueraria lobata*) and Danshen (*Salvia miltiorrhiza*) by high-performance liquid chromatography. Phytochem Anal 19(4):368–375

Chang JS, Wang KC, Shieh DE, Hsu FF, Chiang LC (2012) Ge-Gen-Tang has anti-viral activity against human respiratory syncytial virus in human respiratory tract cell lines. J Ethnopharmacol 139(1):305–310

Chen G, Li L (2007) Nutrient consumption and production of isoflavones in bioreactor cultures of *Pueraria lobata* (Willd). J Environ Biol 28(2):321–326

Chen G, Zhang JX, Ye JN (2001) Determination of puerarin, daidzein and rutin in *Pueraria lobata* (Wild.) Ohwi by capillary electrophoresis with electrochemical detection. J Chromatogr A 923(1–2):255–262

Chen WC, Hayakawa S, Yamamoto T, Su HC, Liu IM, Cheng JT (2004) Mediation of beta-endorphin by the isoflavone puerarin to lower plasma glucose in streptozotocin-induced diabetic rats. Planta Med 70(2):113–116

Chen SB, Yang DJ, Chen SL, Xu HX, Chan ASC (2007) Seasonal variations in the isoflavonoids of radix Puerariae. Phytochem Anal 18(3):245–250

Chen HT, Yao CH, Chao PD, Hou YC, Chiang HM, Hsieh CC, Ke CJ, Chen YS (2008a) Effect of serum metabolites of *Pueraria lobata* in rats on peripheral nerve regeneration: in vitro and in vivo studies. J Biomed Mater Res B Appl Biomater 84(1):256–262

Chen Y, Song W, Peng ZH, Ge BY, Han FM (2008b) Identification of metabolites of tectoridin in-vivo and in-vitro by liquid chromatography-tandem mass spectrometry. J Pharm Pharmacol 60:709–716

Chen G, Wu X, Zhou WL, Li L (2010a) Preparation and assay of C-glucosyltransferase from roots of *Pueraria lobata*. J Environ Biol 31(5):655–660

Chen TR, Chen LA, Wei QK (2010b) Evaluation of quality of Radix Puerariae herbal medicine by isoflavonoids. J Pharm Pharmacol 62(5):644–650

Chen R, Xue J, Xie ML (2012a) Puerarin prevents isoprenaline-induced myocardial fibrosis in mice by

reduction of myocardial TGF-β1 expression. J Nutr Biochem 23(9):1080–1085

Chen TR, Shih SC, Ping HP, Wei QK (2012b) Antioxidant activity and isoflavonoid components in different sections of *Pueraria lobata* root. J Food Drug Anal 20(3):681–685

Cherdshewasart W, Sutjit W (2008) Correlation of antioxidant activity and major isoflavonoid contents of the phytoestrogen-rich *Pueraria mirifica* and *Pueraria lobata* tubers. Phytomed 15(1–2):38–43

Cherdshewasart W, Subtang S, Dahlan W (2007) Major isoflavonoid contents of the phytoestrogen rich-herb *Pueraria mirifica* in comparison with *Pueraria lobata*. J Pharm Biomed Anal 43(2):428–434

Cherdshewasart W, Sutjit W, Pulcharoen K, Chulasiri M (2009) The mutagenic and antimutagenic effects of the traditional phytoestrogen-rich herbs, *Pueraria mirifica* and *Pueraria lobata*. Brazilian J Med Biol Res 42(9):816–823

Cheung WSD, Koon CM, Ng CF, Leung PC, Fung KP, Poon KSS, Lau BSC (2012) The roots of *Salvia miltiorrhiza* (Danshen) and *Pueraria lobat*a (Gegen) inhibit atherogenic events: a study of the combination effects of the 2-herb formula. J Ethnopharmacol 143(3):859–866

Chevallier A (1996) The Encyclopedia of Medicinal Plants. Dorling Kindersley, London, 336 pp

Chiang HM, Fang SH, Wen KC, Hsiu SL, Tsai SY, Hou YC, Chi YC, Chao PD (2005) Life-threatening interaction between the root extract of Pueraria lobata and methotrexate in rats. Toxicol Appl Pharmacol 209(3):263–268

Cho HJ, Jun HJ, Lee JH, Jia Y, Hoang MH, Shim JH, Park KH, Lee SJ (2012) Acute effect of high-dose isoflavones from *Pueraria lobata* (Willd.) Ohwi on lipid and bone metabolism in ovariectomized mice. Phytother Res 26(12):1864–1871

Choi J, Shin MH, Park KY, Lee KT, Jung HJ, Lee MS, Park HJ (2004) Effect of kaikasaponin III obtained from *Pueraria thunbergiana* flowers on serum and hepatic lipid peroxides and tissue factor activity in the streptozotocin-induced diabetic rat. J Med Food 7(1):31–37

Choi YH, Hong SS, Shin YS, Hwang BY, Park SY, Lee D (2010) Phenolic compounds from *Pueraria lobata* protect PC12 cells against Aβ-induced toxicity. Arch Pharm Res 33(10):1651–1654

Choo MK, Park EK, Yoon HK, Kim DH (2002) Antithrombotic and antiallergic activities of daidzein, a metabolite of puerarin and daidzin produced by human intestinal microflora. Biol Pharm Bull 25(10):1328–1332

Chueh FS, Chang CP, Chio CC, Lin MT (2004) Puerarin acts through brain serotonergic mechanisms to induce thermal effects. J Pharmacol Sci 96(4):420–427

Chung BH, Cho YL, Kim JD, Jo HS, Won MH, Lee H, Ha KS, Kwon YG, Kim YM (2010) Promotion of direct angiogenesis in vitro and in vivo by Puerariae flos extract via activation of MEK/ERK-, PI3K/Akt/eNOS-, and Src/FAK-dependent pathways. Phytother Res 24(6):934–940

Corley RN, Woldeghebriel A, Murphy MR (1997) Evaluation of the nutritive value of kudzu (*Pueraria lobat*a) as a feed for ruminants. Animal Feed Sci Technol 68:183–188

Cui SM, Zhao CS, He ZG (2007) Assessment of *Pueraria lobata* isoflavone with self-microemulsifying drug delivery systems in vitro and in vivo. Zhong Yao Cai 30(6):684–687 (In Chinese)

Cui HX, Liu Q, Tao YZ, Zhang HF, Zhang L, Ding K (2008) Structure and chain conformation of a (1→6)-α-d-glucan from the root of *Pueraria lobata* (Willd.) Ohwi and the antioxidant activity of its sulfated derivative. Carbohydr Polym 74(4):771–778

Cui YH, Wang ZP, Fan H, Yang K (2013) Adsorption and isolation of macroporous resin for three isoflavonoids from Puerariae Lobatae radix. Zhong Yao Cai 36(4):650–653 (In Chinese)

Deng Y, Ng ES, Yeung JH, Kwan YW, Lau CB, Koon JC, Zhou L, Zuo Z, Leung PC, Fung KP, Lam FF (2012) Mechanisms of the cerebral vasodilator actions of isoflavonoids of Gegen on rat isolated basilar artery. J Ethnopharmacol 139(1):294–304

Ding MP, Feng F, Hu HT (2007) Effects of puerarin on expression of nuclear factor kappaB after cerebral ischemia/reperfusion in rats. Zhongguo Zhong Yao Za Zhi 32(23):2515–2518 (In Chinese)

Dong K, Tao QM, Xia Q, Shan QX, Pan GB (2004) Endothelium-independent vasorelaxant effect of puerarin on rat thoracic aorta. Zhongguo Zhong Yao Za Zhi 29(10):981–984 (In Chinese)

Du XF, Xu SY, Wang Z (2002) The morphology and characterization of starch from *Pueraria lobata* (Willd.) Ohwi. Int J Food Sci Technol 37(6):697–701

Du G, Zhao HY, Zhang QW, Li GH, Yang FQ, Wang Y, Li YC, Wang YT (2010) A rapid method for simultaneous determination of 14 phenolic compounds in Radix Puerariae using microwave-assisted extraction and ultra high performance liquid chromatography coupled with diode array detection and time-of-flight mass spectrometry. J Chromatogr A 1217:705–714

Du G, Zhao HY, Song Y, Zhang QW, Wang Y (2011) Rapid simultaneous determination of isoflavones in *Radix puerariae* using high-performance liquid chromatography–triple quadrupole mass spectrometry with novel shell-type column. J Sep Sci 34:2576–2585

Duke JA (1983) *Pueraria lobata* (Willd.) Ohwi. Handbook of energy crops. unpublished. https://www.hort.purdue.edu/newcrop/duke_energy/Pueraria_lobata.html

Duke JA, Ayensu ES (1985) Medicinal Plants of China, vol 1 & 2. Reference Publications, Inc, Algonac, 705 pp

Duke JA, Bogenschutz-Godwin MJ, Du Cellier J, Duke PA (2002) CRC Handbook of Medicinal Plants, 2nd edn. CRC Press, Boca Raton, FL, 936 pp

Facciola S (1990) Cornucopia. A Source Book of Edible Plants. Kampong Publ, Vista. California, 677 pp

Fang CB, Wan XC, Tan HR, Jaing CJ (2006a) Identification of isoflavonoids in several kudzu samples by high performance liquid chromatography coupled with electrospray ionization tandem mass spectrometry. J Chromatogr Sci 44:57–63

Fang CB, Wan XC, Tan HR, Jiang CJ (2006b) Separation and determination of isoflavonoids in several kudzu samples by high-performance capillary electrophoresis (HPCE). Ann Chim 96(1–2):117–124

Fosberg FR, Sachet MH, Oliver RL (1979) A geographical checklist of the Micronesian dicotyledonae. Micronesica 15:11

Foster S, Duke JA (1998) A Field Guide to Medicinal Plants in Eastern and Central N. America, Houghton Mifflin Co, 366 pp

Fotsis T, Pepper M, Adlercreutz H, Fleischmann G, Hase T, Montesano R, Schweigerer L (1993) Genistein, a dietary derived inhibitor of in vitro angiogenesis. Proc Natl Acad Sci U S A 90(7):2690–2694

Fu X, Chen H, Wang W (2008) Production of ethanol and isoflavones from steam-pretreated Radix Puerariae by solid state fermentation. Sheng Wu Gong Cheng Xue Bao 24(6):957–961, In Chinese

Groen LE, Siemonsma JS, Jansen PCM (1996) *Pueraria lobata* (Willdenow) Ohwi. In: Flach M, Rumawas F (eds) Plant resources of south-east Asia No 9. Plants yielding non-seed carbohydrates. Prosea Foundation, bogor, Indonesia, pp 178–179

Gu L, Yang Y, Sun YG, Zheng XX (2010) Puerarin inhibits acid-sensing ion channels and protects against neuron death induced by acidosis. Planta Med 76(6):583–588

Gu J, Li JC, Tan R, Fan LN, Zhang X (2013) Effects of combination of Puerariae Lobatae radix and Salviae Miltiorrhizae radix et rhizoma on lipid metabolism in atherosclerotic quails. Zhongguo Zhong Yao Za Zhi 38(22):3939–3942 (In Chinese)

Guan L, Yeung SY, Huang Y, Chen ZY (2006) Both soybean and kudzu phytoestrogens modify favorably the blood lipoprotein profile in ovariectomized and castrated hamsters. J Agric Food Chem 54(13):4907–4912

Guang HM, Zhang XM, Li YQ, Wei XB, Wang ZY, Liu HQ (2005) Protective effects of hydroxyethylpuerarin on cultured bovine cerebral microvascular endothelial cells damaged by hydrogen peroxide. Yao Xue Xue Bao 40(3):220–224

Guerra MC, Speroni E, Broccoli M, Cangini M, Pasini P, Minghett A, Crespi-Perellino N, Mirasoli M, Cantelli-Forti G, Paolini M (2000) Comparison between Chinese medical herb *Pueraria lobata* crude extract and its main isoflavone puerarin antioxidant properties and effects on rat liver CYP-catalysed drug metabolism. Life Sci 67(24):2997–3006

Guo XG, Chen JZ, Zhang X, Xia Q (2004) Effect of puerarin on L-type calcium channel in isolated rat ventricular myocytes. Zhongguo Zhong Yao Za Zhi 29(3):248–251 (In Chinese)

Guo C, Xu LY, He QL, Liang T, Duan XQ, Li R (2013) Anti-fibrotic effects of puerarin on CCl4-induced hepatic fibrosis in rats possibly through the regulation of PPAR-γ expression and inhibition of PI3K/Akt pathway. Food Chem Toxicol 56:436–442

Hakamatsuka T, Noguchi H, Ebizuka Y, Sankawa U (1988) Deoxychalcone synthase from cell suspension cultures of *Pueraria lobata*. Chem Pharm Bull 36(10):4225–4228

Hakamatsuka T, Noguchi H, Ebizuka Y, Sankawa U (1990) Isoflavone synthase from cell suspension cultures of *Pueraria lobata*. Chem Pharm Bull 38(7):1942–1945

Hakamatsuka T, Ebizuka Y, Sankawa U (1994) *Pueraria lobata* (kudzu vine): in vitro culture and the production of isoflavonoids. In: Medicinal and Aromatic Plants VII. Springer, Berlin, Heidelberg, pp 386–400

Han YO, Han MJ, Park SH, Kim DH (2003) Protective effects of kakkalide from Flos puerariae on ethanol-induced lethality and hepatic injury are dependent on its biotransformation by human intestinal microflora. J Pharmacol Sci 93(3):331–336

Han J, Wang W, Wang LY, Liu S, Kang TD (2004) Effect of puerarin and daidzein on proliferating vascular smooth muscle cells. Zhongguo Zhong Yao Za Zhi 29(5):437–440 (In Chinese)

Han P, Li J, Li WJ, Yu ZL, Wang Q, Wu DS (2005a) Potential antiviral drug *Pueraria* crude extract and puerarin protect against ethanol-induced cytotoxicity in embryonic mouse hippocampal cultures. Zhonghua Shi Yan He Lin Chuang Bing Du Xue Za Zhi 19(3):244–247 (In Chinese)

Han P, Liu LE, Liu J, Zhang HQ, Wu YM, Wu YJ, Yu F, Wu DS (2005b) Determination of copper, zinc, iron, calcium and magnesium in *Pueraria lobata* ohwi by FAAS. Guang Pu Xue Yu Guang Pu Fen Xi 25(9):1507–1509 (In Chinese)

Han P, Wu DS, Li WJ, Yu ZL, Wang Q (2005c) Effect of *Pueraria* crude extract and puerarin on ethanol-induced expression of heat shock protein 70 in embryonic mouse hippocampal cultures. Zhonghua Yi Xue Za Zhi 85(41):2930–2933 (In Chinese)

Han T, Cheng G, Liu Y, Yang H, Hu YT, Huang W (2012) In vitro evaluation of tectoridin, tectorigenin and tectorigenin sodium sulfonate on antioxidant properties. Food Chem Toxicol 50(2):409–414

Han EB, Chang BY, Kim DS, Cho HK, Kim SY (2014) Melanogenesis inhibitory effect of aerial part of *Pueraria thunbergiana* in vitro and in vivo. Arch Dermatol Res 307(1):57–72

Harada M, Ueno K (1975) Pharmacological studies on Pueraria root. I. Fractional extraction of *Pueraria* root and identification of its pharmacological effects. Chem Pharm Bull 23(8):1798–1805

Hayakawa J, Noda N, Yamada S, Uno K (1984) Studies on physical and chemical quality evaluation of crude drug preparations. I. Analysis of *Puerariae radix* and species *Puerariae* Yakugaku Zasshi 104(1):50–56 (In Japanese)

Heo HJ, Suh YM, Kim MJ, Choi SJ, Mun NS, Kim HK, Kim E, Kim CJ, Cho HY, Kim YJ, Shin DH (2006) Daidzein activates choline acetyltransferase from MC-IXC cells and improves drug-induced amnesia. Biosci Biotechnol Biochem 70(1):107–111

Hien TT, Kim HG, Han EH, Kang KW, Jeong HG (2010) Molecular mechanism of suppression of MDR1 by puerarin from *Pueraria lobata* via NF-kappaB pathway and cAMP-responsive element transcriptional activity-dependent up-regulation of AMP-activated protein kinase in breast cancer MCF-7/adr cells. Mol Nutr Food Res 54(7):918–928

Hou YC, Chao PDL, Yeh YR, Yang SY, Tsai SY (2011) Comparison of urinary kinetics between traditional decoction and concentrated powders of Puerariae Radix in healthy men. J Food Drug Anal 19(4):517–522

Hsiang SW, Lee HC, Tsai FJ, Tsai CC, Yao CH, Chen YS (2011) Puerarin accelerates peripheral nerve regeneration. Am J Chin Med 39(6):1207–1217

Hsieh MT, Kuo LH, Tsai FH, Wang WH, Wu CR (2002) Effects of puerarin on scopolamine-, mecamylamine-, p-chloroamphetamine- and dizocilpine-induced inhibitory avoidance performance impairment in rats. Planta Med 68(10):901–905

Hsu HH, Chang CK, Su HC, Liu IM, Cheng JT (2002) Stimulatory effect of puerarin on alpha1A-adrenoceptor to increase glucose uptake into cultured C2C12 cells of mice. Planta Med 68(11):999–1003

Hsu FL, Liu IM, Kuo DH, Chen WC, Su HC, Cheng JT (2003) Antihyperglycemic effect of puerarin in streptozotocin-induced diabetic rats. J Nat Prod 66(6):788–792

Hu SY (2005) Food Plants of China. The Chinese University Press, Hong Kong, 844pp

Hu WZ, Zhang Q, Yang XJ, Wang Y, Sun L (2010) Puerarin inhibits adhesion molecule expression in TNF-alpha-stimulated human endothelial cells via modulation of the nuclear factor kappaB pathway. Pharmacology 85(1):27–35

Hu WZ, Yang XJ, Zhe C, Zhang Q, Sun L, Cao KJ (2011) Puerarin inhibits iNOS, COX-2 and CRP expression via suppression of NF-κB activation in LPS-induced RAW264.7 macrophage cells. Pharmacol Rep 63(3):781–789

Huang F, Liu K, Du H, Kou JP, Liu BL (2012) Puerarin attenuates endothelial insulin resistance through inhibition of inflammatory response in an IKKβ/IRS-1-dependent manner. Biochimie 94(5):1143–1150

Hwang YP, Jeong HG (2008) Mechanism of phytoestrogen puerarin-mediated cytoprotection following oxidative injury: estrogen receptor-dependent up-regulation of PI3K/Akt and HO-1. Toxicol Appl Pharmacol 233(3):371–381

Hwang YP, Choi CY, Chung YC, Jeon SS, Jeong HG (2007) Protective effects of puerarin on carbon tetrachloride-induced hepatotoxicity. Arch Pharm Res 30(10):1309–1317

Inoue T, Fujita M (1977) Biosynthesis of puerarin in *Pueraria* root. Chem Pharm Bull 25(12):3226–3231

IUCN (2013) Global invasive Species database. http://www.issg.org/database/species/search.asp?st=100ss

Jang MH, Shin MC, Kim YJ, Chung JH, Yim SV, Kim EH, Kim Y, Kim CJ (2001) Protective effects of puerariae flos against ethanol-induced apoptosis on human neuroblastoma cell line SK-N-MC. Jpn J Pharmacol 87(4):338–342

Jang DS, Kim JM, Lee YM, Kim YS, Kim JH, Kim JS (2006) Puerariafuran, a new inhibitor of advanced glycation end products (AGEs) isolated from the roots of *Pueraria lobata*. Chem Pharm Bull 54(9):1315–1317

Jeon GC, Park MS, Yoon DY, Shin CH, Sin HS, Um SJ (2005) Antitumor activity of spinasterol isolated from Pueraria roots. Exp Mol Med 37(2):111–120

Ji ZN, Liu GQ (2002) Daidzin protects PC12 cells from serum deprivation-induced apoptosis. J Asian Nat Prod Res 494:265–269

Jia XB, Cai Y, Chen Y, Cheng XD, Fan CY, Shi YF (2007) Pharmacokinetics of pueraria flavonoids from *Pueraria* lobata by anti-oxidant pharmacodynamics effect method. Zhongguo Zhong Yao Za Zhi 32(22):2370–2373 (In Chinese)

Jiang RW, Lau KM, Lam HM, Yam WS, Leung LK, Choi KL, Waye MM, Mak TC, Woo KS, Fung KP (2005) A comparative study on aqueous root extracts of *Pueraria thomsonii* and *Pueraria lobata* by antioxidant assay and HPLC fingerprint analysis. J Ethnopharmacol 96(1):133–138

Jin SE, Son YK, Min BS, Jung HA, Choi JS (2012) Anti-inflammatory and antioxidant activities of constituents isolated from *Pueraria lobata* roots. Arch Pharm Res 35(5):823–837

Jun M, Fu HY, Hong J, Wan X, Yang CS, Ho CT (2003) Comparison of antioxidant activities of isoflavones from kudzu root (*Pueraria lobata* Ohwi). J Food Sci 68(6):2117–2122

Jun M, Hong J, Jeong WS, Ho CT (2005) Suppression of arachidonic acid metabolism and nitric oxide formation by kudzu isoflavones in murine macrophages. Mol Nutr Food Res 49(12):1154–1159

Jung DY, Ha H, Kim C (2004) Induction of growth hormone release by *Pueraria thunbergiana* Benth. Horm Metab Res 36(2):86–91

Jung JM, Kwon SH, Noh H, Han DC, Jeon JS, Jin SY (2013) Acute interstitial nephritis following kudzu root juice ingestion. Clin Nephrol 80(4):298–300

Kamiya T, Matsuzuka Y, Kusaba N, Ikegichi M, Takagaki T, Kondo K (2011) Preliminary research for the anti-obesity effect of Puerariae Flos extract in humans. J Health Sci 57(6):521–531

Kang KA, Chae S, Koh YS, Kim JS, Lee JH, You HJ, Hyun JW (2005a) Protective effect of puerariae radix on oxidative stress induced by hydrogen peroxide and streptozotocin. Biol Pharm Bull 28(7):1154–1160

Kang KA, Lee KH, Chae S, Zhang R, Jung MS, Kim SY, Kim HS, Kim DH, Hyun JW (2005b) Cytoprotective effect of tectorigenin, a metabolite formed by transfor-

mation of tectoridin by intestinal microflora, on oxidative stress induced by hydrogen peroxide. Eur J Pharmacol 519(1–2):16–23

Kato-Noguchi H (2003) Allelopathic substances in *Pueraria thunbergiana*. Phytochemistry 63(5):577–580

Kaufman PB, Duke JA, Brielmann H, Boik J, Hoyt JE (1997) A comparative survey of leguminous plants as sources of the isoflavones, genistein and daidzein: implications for human nutrition and health. J Altern Complement Med 3(1):7–12

Kayano S, Matsumura Y, Kitagawa Y, Kobayashi M, Nagayama A, Kawabata N, Kikuzaki H, Kitada Y (2012) Isoflavone C-glycosides isolated from the root of kudzu (*Pueraria lobata*) and their estrogenic activities. Food Chem 134(1):282–287

Keung WM, Vallee BL (1998) Kudzu root: an ancient Chinese source of modern antidipsotropic agents. Phytochemistry 47(4):499–506

Kim S, Fung DY (2004) Antibacterial effect of crude water-soluble arrowroot (Puerariae radix) tea extracts on food-borne pathogens in liquid medium. Lett Appl Microbiol 39(4):319–325

Kim CS, Shin SM, Ha HK, Kim JM (2003) Study of substance changes in flowers of *Pueraria thunbergiana* Benth. during storage. Arch Pharm Res 26(3):210–213

Kim IT, Park YM, Shin KM, Ha J, Choi J, Jung HJ, Park HJ, Lee KT (2004) Anti-inflammatory and antinociceptive effects of the extract from *Kalopanax pictus, Pueraria thunbergiana* and *Rhus verniciflua*. J Ethnopharmacol 94(1):165–173

Kim HJ, Lee DH, Hwang YY, Lee KS, Lee JS (2005) Characterization of beta-hydroxy-beta-methylglutaryl coenzyme A reductase inhibitor from *Pueraria thunbergiana*. J Agric Food Chem 53(15):5882–5888

Kim JM, Lee YM, Lee GY, Jang DS, Bae KH, Kim JS (2006a) Constituents of the roots of *Pueraria lobata* inhibit formation of advanced glycation end products (AGEs). Arch Pharm Res 29(10):821–825

Kim SY, Kim DH, Hyun JW, Henson JW, Kim HS (2006b) Irisolidone, an isoflavone metabolite, represses JC virus gene expression via inhibition of Sp1 binding in human glial cells. Biochem Biophys Res Commun 344(1):3–8

Kim YS, Kim NH, Jung DH, Kim JM, Kim JS (2007) Suppressive effects of PO41-14-42-K1, a component of *Pueraria lobata*, on expression of TGF-beta2 in Human lens epithelial cells cultured under high glucose and on lens opacity in organ culture. Diabetes 56(Sup1):A154–A154

Kim YS, Kim NH, Jung DH, Jang DS, Lee YM, Kim JM, Kim JS (2008) Genistein inhibits aldose reductase activity and high glucose-induced TGF-beta2 expression in human lens epithelial cells. Eur J Pharmacol 594(1–3):18–25

Kim SH, Heo JH, Kim YS, Kang SS, Choi JS, Lee SM (2009) Protective effect of daidzin against D-galactosamine and lipopolysaccharide-induced hepatic failure in mice. Phytother Res 23(5):701–706

Kim DH, Jung HA, Park SJ, Kim JM, Lee S, Choi JS, Cheong JH, Ko KH, Ryu JH (2010a) The effects of daidzin and its aglycon, daidzein, on the scopolamine-induced memory impairment in male mice. Arch Pharm Res 33(10):1685–1690

Kim NH, Kim YS, Lee YM, Jang DS, Kim JS (2010b) Inhibition of aldose reductase and xylose-induced lens opacity by puerariafuran from the roots of *Pueraria lobata*. Biol Pharm Bull 33(9):1605–1609

Kim HS, Shin BR, Lee HK, Kim YJ, Park MJ, Kim SY, Lee MK, Hong JT, Kim Y, Han SB (2013) A polysaccharide isolated from *Pueraria lobata* enhances maturation of murine dendritic cells. Int J Biol Macromol 52:184–191

Kim SB, Yoon IS, Kim KS, Cho SJ, Kim YS, Cho HJ, Chung SJ, Chong S, Kim DD (2014) In vitro and in vivo evaluation of the effect of puerarin on hepatic cytochrome p450-mediated drug metabolism. Planta Med 80(7):561–567

Kinjo J, Miyamoto I, Murakami K, Kida K, Tomimatsu T, Yamasaki M, Nohara T (1985) Oleanene-sapogenols from Puerariae Radix. Chem Pharm Bull 33(3):1293–1296

Kinjo J, Furusawa J, Baba J, Takeshita T, Yamasaki M, Nohara T (1987) Studies on the constituents of *Pueraria lobata*. III. Isoflavonoids and related compounds in the roots and the voluble stems. Chem Pharm Bull 35(12):4846–4850

Kinjo J, Takeshita T, Abe Y, Terada N, Yamashita H, Yamasaki M, Takeuchi K, Murakami K, Tomimatsu T, Nohara T (1988a) Studies on the constituents of *Pueraria lobata*. IV. : Chemical constituents in the flowers and the leaves. Chem Pharm Bull 36(3):1174–1179

Kinjo J, Takeshita T, Nohara T (1988b) Studies on the constituents of *Pueraria lobata*. V. A Tryptophan derivative from Puerariae Flos. Chem Pharm Bull 36(10):4171–4173

Kinjo J, Aoki K, Okawa M, Shii Y, Hirakawa T, Nohara T, Nakajima Y, Yamazaki T, Hosono T, Someya M, Niiho Y, Kurashige T (1999) HPLC profile analysis of hepatoprotective oleanene-glucuronides in Puerariae Flos. Chem Pharm Bull 47(5):708–710

Kirakosyan A, Kaufman PB, Warber S, Bolling S, Chang SC, Duke JA (2003) Quantification of major isoflavonoids and-canavanine in several organs of kudzu vine (*Pueraria montana*) and in starch samples derived from kudzu roots. Plant Sci 164:883–888

Koon CM, Woo KS, Leung PC, Fung KP (2011) Salviae Miltiorrhizae Radix and Puerariae Lobatae Radix herbal formula mediates anti-atherosclerosis by modulating key atherogenic events both in vascular smooth muscle cells and endothelial cells. J Ethnopharmacol 138(1):175–183

Kubo M, Fujita K, Nishimura H, Naruto S, Namba K (1973) A new irisolidone-7-0-glucoside and tectoridin

from *Pueraria* species. Phytochemistry 12(10):2547–2548

Kubo M, Sasaki M, Namba K, Naruto S, Nishimura H (1975) Isolation of a new isoflavone from Chinese Pueraria flowers. Chem Pharm Bull 23(10):2449–2451

Kubo K, Shimada T, Onishi R, Tsubata M, Kamiya T, Nagamine R, Iizuka S, Sai Y, Amagaya S, Aburada M, Miyamoto K (2012) Puerariae flos alleviates metabolic diseases in Western diet-loaded, spontaneously obese type 2 diabetic model mice. J Nat Med 66(4):622–630

Kueng WM, Lazo O, Kunze L, Valee BL (1996) Potentiation of the bioavailability of daidzin by an extract of Radix puerariae. Proc Natl Acad Sci U S A 93(9):4284–4288

Kurihara T, Kikuchi M (1975) Studies on the constituents of flowers. V. On the components of flower of *Pueraria thunbergiana* Benth. (2). Isolation of a new isoflavone glycoside. Yakugaku Zasshi 95(11):1283–1285 (In Japanese)

Kurihara T, Kikuchi M (1976) Studies on the constituents of flowers. VI. On the components of the flower of *Pueraria thunbergiana* Benth. Yakugaku Zasshi 96(12):1486–1488 (In Japanese)

Kurihara T, Kiruchi M (1973) Studies on the constituents of flowers. I. On the components of flower of *Pueraria thunbergiana* Benth. Yakugaku Zasshi 93(9):1201–1205 (In Japanese)

Kwon IB, Park HH (1995) Isoflavonoids of kudzu (*Pueraria lobata*) and bioconversion of exogenous compounds into their malonylglucosides by its cell cultures. Foods Food Ingred J Jpn 163:86–93

Lam FF, Deng SY, Ng ES, Yeung JH, Kwan YW, Lau CB, Koon JC, Zhou L, Zuo Z, Leung PC, Fung KP (2010) Mechanisms of the relaxant effect of a danshen and gegen formulation on rat isolated cerebral basilar artery. J Ethnopharmacol 132(1):186–192

Lau CS, Carrier DJ, Beitle RR, Howard LR, Lay JO, Liyanage R, Clausen EC (2005) A glycoside flavonoid in kudzu (*Pueraria lobata*): identification, quantification, and determination of antioxidant activity. Appl Biochem Biotechnol 121–124:783–794

Lau CC, Chan CO, Chau FT, Mok DK (2009) Rapid analysis of Radix puerariae by near-infrared spectroscopy. J Chromatogr A 1216(11):2130–2135

Lee JS (2004) Supplementation of Pueraria radix water extract on changes of antioxidant enzymes and lipid profile in ethanol-treated rats. Clin Chim Acta 347(1–2):121–128

Lee KT, Sohn IC, Kong EA, Kim DH, Choi SK, Choi JW, Park HJ (1999) Antioxidative and cytoprotective effects of isoflavones isolated from *Pueraria thunbergiana* flowers. Yakhak Hoeji 43(6):736–742

Lee KT, Sohn IC, Kim DH, Choi JW, Kwon SH, Park HJ (2000) Hypoglycemic and hypolipidemic effects of tectorigenin and kaikasaponin III in the streptozotocin-induced diabetic rat and their antioxidant activity in vitro. Arch Pharm Res 23(5):461–466

Lee KT, Sohn IC, Kim YK, Choi JH, Choi JW, Park HJ, Itoh Y, Miyamoto K (2001a) Tectorigenin, an isoflavone of *Pueraria thunbergiana* Benth., induces differentiation and apoptosis in human promyelocytic leukemia HL-60 cells. Biol Pharm Bull 24(10):1117–1121

Lee MK, Cho SY, Jang JY, Cho MS, Jeon SM, Jang MK, Kim MJ, Park YB (2001b) Effects of Puerariae Flos and Puerariae Radix extracts on antioxidant enzymes in ethanol-treated rats. Am J Chin Med 29(2):343–354

Lee HW, Choo MK, Bae EA, Kim DH (2003) Beta-glucuronidase inhibitor tectorigenin isolated from the flower of *Pueraria thunbergiana* protects carbon tetrachloride-induced liver injury. Liver Int 23(4):221–226

Lee HU, Bae EA, Kim DH (2005a) Hepatoprotective effects of irisolidone on tert-butyl hyperoxide-induced liver injury. Biol Pharm Bull 28(3):531–533

Lee HU, Bae EA, Kim DH (2005b) Hepatoprotective effect of tectoridin and tectorigenin on tert-butyl hyperoxide-induced liver injury. J Pharmacol Sci 97(4):541–544

Lee OH, Seo DH, Park CS, Kim YC (2010) Puerarin enhances adipocyte differentiation, adiponectin expression, and antioxidant response in 3T3-L1 cells. Biofactors 36(6):459–467

Leung WTW, Butrum RR, Huang Chang F, Narayana Rao M, Polacchi W (1972) Food composition table for use in East Asia. FAO, Rome, Italy, 347 pp

Li BB, Yu SF (2003) Effect of puerarin on the bone metabolism in vitro. Beijing Da Xue Xue Bao Yi Xue Ban 35(1):74–77 (In Chinese)

Li L, Zhang CR (2006) Production of puerarin and isoflavones in cell suspension cultures of *Pueraria lobata* (Willd.): effects of medium supplementation with casein hydrolysate and coconut milk. J Environ Biol 27(1):21–26

Li XX, Zou HF, Yan RL (1998) Analysis and evaluation of nutritional composition in wild kudzu roots and kudzu powder from the market. Acta Hort (ISHS) 467:65–70

Li GH, Zhang QW, Wang YT (2010a) Chemical constituents from roots of *Pueraria lobata*. Zhongguo Zhong Yao Za Zhi 35(23):3156–3160 (In Chinese)

Li JM, Wang G, Liu JC, Zhou L, Dong MX, Wang R, Li XY, Li XM, Lin CR, Niu YC (2010b) Puerarin attenuates amyloid-beta-induced cognitive impairment through suppression of apoptosis in rat hippocampus in vivo. Eur J Pharmacol 649(1–3):195–201

Li R, Liang T, He QL, Guo C, Xu LY, Zhang KF, Duan XQ (2013a) Puerarin, isolated from Kudzu root (Willd.), attenuates hepatocellular cytotoxicity and regulates the GSK-3β/NF-κB pathway for exerting the hepatoprotection against chronic alcohol-induced liver injury in rats. Int Immunopharmacol 17(1):71–78

Li R, Xu LY, Liang T, Li YW, Zhang SJ, Duan XQ (2013b) Puerarin mediates hepatoprotection against CCl_4-induced hepatic fibrosis rats through attenuation of

inflammation response and amelioration of metabolic function. Food Chem Toxicol 52:69–75

Li J, Li ZB, Li CF, Gou JB, Zhang YS (2014) Molecular cloning and characterization of an isoflavone 7-O-glucosyltransferase from *Pueraria lobata*. Plant Cell Rep 33(7):1173–1185

Liao ZG, Liang XL, Zhu JY, Zhao GW, Guan YM, Cao YC, Zhao LJ (2014) Transport properties of puerarin and effect of extract of Radix Angelicae dahuricae on puerarin intestinal absorption using in situ and in vitro models. Phytother Res 28(9):1288–1294

Lim DW, Kim JG, Kim YT (2013a) Effects of dietary isoflavones from Puerariae radix on lipid and bone metabolism in ovariectomized rats. Nutrients 5(7):2734–2746

Lim DW, Lee C, Kim IH, Kim YT (2013b) Anti-inflammatory effects of total isoflavones from *Pueraria lobata* on cerebral ischemia in rats. Molecules 18(9):10404–10412

Lin RC, Guthrie S, Xie CY, Mai K, Lee DY, Lumeng L, Li TK (1996) Isoflavonoid compounds extracted from *Pueraria lobata* suppress alcohol preference in a pharmacogenetic rat model of alcoholism. Alcohol Clin Exp Res 20(4):659–663

Lin CC, Wu CI, Shue SJ (2005) Determination of 12 *Pueraria* components by high- performance liquid chromatography-mass spectrometry. J Sep Sci 28(14):1785–1795

Lin YJ, Hou YC, Lin CH, Hsu YA, Sheu JJ, Lai CH, Chen BH, Lee Chao PD, Wan L, Tsai FJ (2009) Puerariae radix isoflavones and their metabolites inhibit growth and induce apoptosis in breast cancer cells. Biochem Biophys Res Commun 378(4):683–688

Lin TJ, Yeh CF, Wang KC, Chiang LC, Tsai JJ, Chang JS (2013) Water extract of *Pueraria lobata* Ohwi has anti-viral activity against human respiratory syncytial virus in human respiratory tract cell lines. Kaohsiung J Med Sci 29(12):651–657

Liu Q, Wang L, Lu Z, Li S, Xiong Y (1999) Effect of puerarin on coronary collateral circulation in dogs with experimental acute myocardial infarction. Zhongguo Zhong Yao Za Zhi 24(5):304–306, 320. (In Chinese)

Liu H, Li ZL, Guo J, Hua HM (2009a) Chemical constituents of the root of *Pueraria lobata*. J Shenyang Pharm Univ 35(11):882–899 (In Chinese)

Liu XJ, Jia Q, Wang CY, Wang NL (2009b) Study on absorption mechanism of puerarin and its nanoparticles across Caco-2 cell model. Zhong Yao Cai 32(8):1252–1255 (In Chinese)

Liu LH, Li WM, Gao Y (2010) Simultaneous determination of tectorigenin-7-O-xylosylglucosid, tectoridin and tectorigenin in flowers of *Pueraria lobata* by HPLC. Zhongguo Zhong Yao Za Zhi 35(17):2308–2311 (In Chinese)

Liu LL, Ma YJ, Chen XQ, Xiong X, ShuyunShi SY (2012) Screening and identification of BSA bound ligands from *Puerariae lobata* flower by BSA functionalized Fe_3O_4 magnetic nanoparticles coupled with HPLC-MS/MS. J Chromatogr B 887–888:55–60

Lu J, Xie Y, Tan Y, Qu J, Matsuda H, Yoshikawa M, Yuan D (2013) Simultaneous determination of isoflavones, saponins and flavones in Flos Puerariae by ultra performance liquid chromatography coupled with quadrupole time-of-flight mass spectrometry. Chem Pharm Bull (Tokyo) 61(9):941–951

Lukas SE, Penetar D, Berko J, Vicens L, Palmer C, Mallya G, Macklin EA, Lee DY (2005) An extract of the Chinese herbal root kudzu reduces alcohol drinking by heavy drinkers in a naturalistic setting. Alcohol Clin Exp Res 29(5):756–762

Lukas SE, Penetar D, Su Z, Geaghan T, Maywalt M, Tracy M, Rodolico J, Palmer C, Ma Z, Lee DY (2013) A standardized kudzu extract (NPI-031) reduces alcohol consumption in nontreatment-seeking male heavy drinkers. Psychopharmacology (Berl) 226(1):65–73

Luo CF, Yuan M, Chen MS, Liu SM, Ji H (2010) Metabolites of puerarin identified by liquid chromatography tandem mass spectrometry: Similar metabolic profiles in liver and intestine of rats. J Chromatogr B 878(3–4):363–370

Lv JH, Zhang SP, Shen FH, Pan JQ, Tan HR (2006) Improvement effects of puerarin on glycated brain damages in rats induced by D-galactose. Zhongguo Zhong Yao Za Zhi 31(14):1184–1187 (In Chinese)

Ma J, Cheng GZ, Li D, Liu H (2002) The effects of Radix Puerariae decoction on levels of antibody and interleukin-2 in mice. Zhong Yao Cai 25(10):731–732 (In Chinese)

Mahdy HM, Mohamed MR, Emam MA, Karim AM, Abdel-Naim AB, Khalifa AE (2014) The anti-apoptotic and anti-inflammatory properties of puerarin attenuate 3-nitropropionic-acid induced neurotoxicity in rats. Can J Physiol Pharmacol 92(3):252–258

Malaivijitnond S, Tungmunnithum D, Gittarasanee S, Kawin K, Limjunyawong N (2010) Puerarin exhibits weak estrogenic activity in female rats. Fitoterapia 81(6):569–576

Mathur S, Mathur S (2013) Allelopathic effects of kudzu (*Pueraria montana*) on seed germination and their potential use as a natural herbicide. J Emerg Investig Dec 2013:1–4

Matkowski A (2004) In vitro isoflavonoid production in callus from different organs of *Pueraria lobata* (Wild.) Ohwi. J Plant Physiol 161(3):343–346

McGregor NR (2007) Pueraria lobata (Kudzu root) hangover remedies and acetaldehyde-associated neoplasm risk. Alcohol 41(7):469–478

Michihara S, Tanaka T, Uzawa Y, Moriyama T, Kawamura Y (2012) Puerarin exerted anti-osteoporotic action independent of estrogen receptor-mediated pathway. J Nutr Sci Vitaminol (Tokyo) 58(3):202–209

Min SW, Kim DH (2007) Kakkalide and irisolidone: HMG-CoA reductase inhibitors isolated from the flower of *Pueraria thunbergiana*. Biol Pharm Bull 30(10):1965–1968

Min SW, Park YJ, Kim DH (2011) Kakkalide and its metabolite irisolidone ameliorate carrageenan-induced inflammation in mice by inhibiting NF-κB pathway. Inflammation 34(5):344–351

Mitich LW (2000) Intriguing world of weeds: Kudzu [*Pueraria lobata* (Wild.) Ohwi]. Weed Technol 14:231–235

Miyazawa M, Kameoka H (1988) Volatile flavor components of Puerariae radix (*Pueraria lobata* Ohwi). Agric Biol Chem 52(4):1053–1055

Miyazawa M, Sakano K, Nakamura S, Kosaka H (2001) Antimutagenic Activity of Isoflavone from *Pueraria lobata*. J Agric Food Chem 49(1):336–341

Mun JG, Grannan MD, Lachcik PJ, Reppert A, Yousef GG, Rogers RB, Janle EM, Weaver CM, Lila MA (2009) In vivo metabolic tracking of 14C-radiolabelled isoflavones in kudzu (*Pueraria lobata*) and red clover (*Trifolium pratense*) extracts. Br J Nutr 102(10):1523–1530

Mun JG, Grannan MD, Lachcik PJ, Rogers RB, Yousef GG, Grace MH, Janle EM, Wu QL, Simon JE, Weaver CM, Lila MA (2010) Tracking deposition of a 14C-radiolabeled kudzu hairy root-derived isoflavone-rich fraction into bone. Exp Biol Med (Maywood) 235(10):1224–1235

Murakami T, Nishikawa Y, Ando T (1960) Studies on the constituents of Japanese and Chinese crude drugs. IV. On the Constituents of *Pueraria* root. (2). Chem Pharm Bull 8(8):688–691

Murata K, Noguchi K, Kondo M, Onishi M, Watanabe N, Okamura K, Matsuda H (2012) Inhibitory activities of Puerariae Flos against testosterone 5α-reductase and its hair growth promotion activities. J Nat Med 66(1):158–165

Nakamoto H, Iwasaki Y, Kizu H (1977) The study of aqueous extract of Puerariae Radix. IV. The Isolation of the antispasmodic action and its body temperature-lowering effect and daidzin isolation from (MTF-101) activity of the extract (Part 4) Puerariae study of the water-soluble extract of kakkonto daidzin from the active extract (MTF-101) and its antifebrile and spasmolytic effect. Yakugaku Zasshi 97(1):103–105 (In Japanese)

Nakamura K, Nishihata T, Jin JS, Ma CM, Komatsu K, Iwashima M, Hattori M (2011) The C-glucosyl bond of puerarin was cleaved hydrolytically by a human intestinal bacterium strain PUE to yield its aglycone daidzein and an intact glucose. Chem Pharm Bull 59(1):23–27

National Institute of Materia Medica (NIMM) (1999) Selected Medicinal Plants in Vietnam, vol 2. Science and Technology Publishing House, Hanoi, Vietnam, 460 pp

Ng CF, Koon CM, Cheung DWS, Lam M, Leung PC, Lau CBS, Fung KP (2011) The anti-hypertensive effect of Danshen (*Salvia miltiorrhiza*) and Gegen (*Pueraria lobata*) formula in rats and its underlying mechanisms of vasorelaxation. J Ethnopharmacol 137(3):1366–1372

Nguyen VD, Min BC, Kyung MO, Park JT, Lee BH, Choi CH, Kim JW, Park KH (2009) Identification of a naturally-occurring 8-[α-D-glucopyranosyl-(1→6)-β-D-glucopyranosyl] daidzein from cultivated kudzu root. Phytochem Anal 20(6):450–455

Niiho Y, Nakajima Y, Yamazaki T, Okamoto M, Tsuchihashi R, Kodera M, Kinjo J, Nohara T (2010) Simultaneous analysis of isoflavones and saponins in *Pueraria* flowers using HPLC coupled to an evaporative light scattering detector and isolation of a new isoflavone diglucoside. J Nat Med 64(3):313–320

Nohara T (2004) Search for functions of natural oligoglycosides—Solanaceae and Leguminosae origin glycosides. Yakugaku Zasshi 124:183–205

Oh SR, Kinjo J, Shii Y, Ikeda T, Nohara T, Ahn KS, Kim JH, Lee HK (2000) Effects of triterpenoids from *Pueraria lobata* on immunohemolysis: β-d-glucuronic acid plays an active role in anticomplementary activity in vitro. Planta Med 66(6):506–510

Ohshima Y, Okuyama T, Takahashi K, Takizawa T, Shibata S (1988) Isolation and high performance liquid chromatography (HPLC) of isoflavonoids from the *Pueraria* root. Planta Med 54(3):250–254

Pan HP, Yang JZ, Li LL, Yi F, Huang ZQ, Huang KW (2003) Experimental study of puerarin injection on the hemorheology in acute blood-stasis model rats. Zhongguo Zhong Yao Za Zhi 12:1178–1180 (In Chinese)

Park HH, Hakamatsuka T, Noguchi H, Sankawa U, Ebizuka Y (1992) Isoflavone glucosides exist as their 6″-*O*-malonyl esters in *Pueraria lobata* and its cell suspension cultures. Chem Pharm Bull 40(7):1978–1980

Park HJ, Park JH, Moon JO, Lee KT, Jung WT, Oh SR, Lee HK (1999) Isoflavone glycosides from the flowers of *Pueraria thunbergiana*. Phytochemistry 51(1):147–151

Park KY, Jung GO, Choi J, Lee KT, Park HJ (2002) Potent antimutagenic and their anti-lipid peroxidative effect of kaikasaponin III and tectorigenin from the flower of *Pueraria thunbergiana*. Arch Pharm Res 25(3):320–324

Park EK, Shin YW, Lee HU, Lee CS, Kim DH (2004) Passive cutaneous anaphylaxis-inhibitory action of tectorigenin, a metabolite of tectoridin by intestinal microflora. Biol Pharm Bull 27(7):1099–1102

Park EK, Shin J, Bae EA, Lee YC, Kim DH (2006) Intestinal bacteria activate estrogenic effect of main constituents puerarin and daidzin of *Pueraria thunbergiana*. Biol Pharm Bull 29(12):2432–2435

Park SJ, Choi KS, Shin DH, Kim JS, Jang DS, Youm JB, Choe H, Earm YE, Kim SJ (2009) Effects of mixed herbal extracts from parched Puerariae radix, gingered Magnoliae cortex, Glycyrrhizae radix and Euphorbiae radix (KIOM-79) on cardiac ion channels and action potentials. J Korean Med Sci 24(3):403–412

Park WS, Kwon O, Yoon TJ, Chung JH (2014) Anti-graying effect of the extract of *Pueraria thunbergiana* via upregulation of cAMP/MITF-M signaling pathway. J Dermatol Sci 75(2):153–155

Penetar DM, Teter CJ, Ma Z, Tracy M, Lee DY, Lukas SE (2006) Pharmacokinetic profile of the isoflavone puerarin after acute and repeated administration of a novel kudzu extract to human volunteers. J Altern Complement Med 12(6):543–548

Penetar DM, Toto LH, Farmer SL, Lee DY, Ma Z, Liu Y, Lukas SE (2012) The isoflavone puerarin reduces alcohol intake in heavy drinkers: a pilot study. Drug Alcohol Depend 126(1–2):251–256

Peng N, Prasain JK, Dai Y, Moore R, Arabshahi A, Barnes S, Carlson S, Wyss JM (2009) Chronic dietary kudzu isoflavones improve components of metabolic syndrome in stroke-prone spontaneously hypertensive rats. J Agric Food Chem 57(16):7268–7273

Peng JH, Cui T, Huang F, Chen L, Zhao Y, Xu L, Xu LL, Feng Q, Hu YY (2013) Puerarin ameliorates experimental alcoholic liver injury by inhibition of endotoxin gut leakage, Kupffer cell activation, and endotoxin receptors expression. J Pharmacol Exp Ther 344(3):646–654

Peterson G, Barnes S (1991) Genistein inhibition of the growth of human breast cancer cells: independence from estrogen receptors and the multi-drug resistance gene. Biochem Biophys Res Commun 179(1):661–667

Pham VH, Morita N (2007) Chemical compositions, fine structure and physicochemical properties of kudzu (*Pueraria lobata*) starches from different regions. Food Chem 105(2):749–755

Powell JM (1976) Ethnobotany. In: Paijmans K (ed) New Guinea vegetation. Elsevier, Amsterdam, pp 106–183, 212 pp

Praptiwi (1999) *Pueraria montana* (Lour.) Merr. In: de Padua LS, Bunyapraphatsara N, Lemmens RHMJ (eds) Plant Resources of South East Asia No 12(1). Medicinal and poisonous plants 1. Prosea Foundation, Bogor, Indonesia. pp 417–420.

Prasain JK, Jones K, Kirk M, Wilson L, Smith-Johnson M, Weaver C, Barnes S (2003) Profiling and quantification of isoflavonoids in kudzu dietary supplements by high-performance liquid chromatography and electrospray ionization tandem mass spectrometry. J Agric Food Chem 51(15):4213–4218

Prasain JK, Jones K, Brissie N, Moore R, Wyss JM, Barnes S (2004) Identification of puerarin and its metabolites in rats by liquid chromatography-tandem mass spectrometry. J Agric Food Chem 52(12):3708–3712

Prasain JK, Reppert A, Jones K, Moore DR 2nd, Barnes S, Lila MA (2007) Identification of isoflavone glycosides in *Pueraria lobata* cultures by tandem mass spectrometry. Phytochem Anal 18(1):50–59

Prasain JK, Peng N, Moore R, Arabshahi A, Barnes S, Wyss JM (2009) Tissue distribution of puerarin and its conjugated metabolites in rats assessed by liquid chromatography–tandem mass spectrometry. Phytomedicine 16(1):65–71

Qi BL, Qi BM (2002) Effect of the purariae-isoflavones on estrogen level in normal and ovariectomized rats. Zhongguo Zhong Yao Za Zhi 27(11):850–852 (In Chinese)

Qi HY, Li L, Huang CY, Li WM, Wu CJ (2006) Optimization and physicochemical characterization of thermosensitive poloxamer gel containing puerarin for ophthalmic use. Chem Pharm Bull 54(1):1500–1507

Reppert A, Yousef GG, Rogers RB, Lila MA (2008) Isolation of radiolabeled isoflavones from kudzu (*Pueraria lobata*) root cultures. J Agric Food Chem 56(17):7860–7865

Rong HJ, De Keukeleire D, De Cooman L, Baeyens WR, Van der Weken G (1998a) Narrow-bore HPLC analysis of isoflavonoid aglycones and their O- and C-glycosides from *Pueraria lobata*. Biomed Chromatogr 12(3):170–171

Rong HJ, Stevens JF, Deinzer ML, De Cooman L, De Keukeleire D (1998b) Identification of isoflavones in the roots of *Pueraria lobata*. Planta Med 64(7):620–627

Sage RF, Coiner HA, Way DA, Runion GB, Prior SA, Torbert AH S, Sicher R, Ziska L (2009) Kudzu [*Pueraria montana* (Lour.) Merr. variety *lobata*]: A new source of carbohydrate for bioethanol production. Biomass Bioenergy 33(1):57–61

Saĭiad SA, Borysov MI, Koval'ov VM (1979) Isolation, identification and quantitative determination of robinin in the leaves of Pueraria lobata. Farm Zh 4:52–55 (In Ukrainian)

Sato T, Kawamoto A, Tamura A, Tatsumi Y, Fujii T (1992) Mechanism of antioxidant action of pueraria glycoside (PG)-1 (an isoflavonoid) and mangiferin (a xanthonoid). Chem Pharm Bull 40(3):721–724

Shao HM, Tang YH, Shen Q, Zhu HQ, Ji O, Zhang YC, Ji JM, Jiang PJ, Si YJ, Li ZR (2012) Apoptosis of Kasumi-1 cells induced by puerariae radix flavones and its molecular mechanism. Zhonghua Xue Ye Xue Za Zhi 33(1):43–46 (In Chinese)

She SY, Liu WJ, Li T, Hong YK (2014) Effects of puerarin in STZ-induced diabetic rats by oxidative stress and the TGF-β1/Smad2 pathway. Food Funct 5(5):944–950

Shibata S, Harada M, Murakami T (1959a) Studies on the constituents of Japanese and Chinese crude drugs. II. Antispasmodic action of the constituent of Pueraria root. Yakugaku Zasshi 79(7):863–866

Shibata S, Murakami T, Nishikawa Y, Harada M (1959b) The constituents of *Pueraria* root. Chem Pharm Bull 7(1):134–136

Shin JE, Bae EA, Lee YC (2006) Estrogenic effect of main components kakkalide and tectoridin of Puerariae Flos and their metabolites. Biol Pharm Bull 29(6):1202–1206

Shurtleff W, Aoyagi A (1977) The Book of Kudzu: A Culinary and Healing Guide. Autumn Press, Brookline, MA

Singh AK, Jiang Y, Gupta S, Younus M, Ramzan M (2013) Anti-inflammatory potency of nano-formulated puerarin and curcumin in rats subjected to the

lipopolysaccharide-induced inflammation. J Med Food 16(10):899–911

Song CY, Bi HM (2004) Effects of puerarin on plasma membrane GLUT4 content in skeletal muscle from insulin-resistant Sprague-Dawley rats under insulin stimulation. Zhongguo Zhong Yao Za Zhi 29(2):172–175 (In Chinese)

Su FM, Chang JS, Wang KC, Tsai JJ, Chiang LC (2008) A water extract of *Pueraria lobata* inhibited cytotoxicity of enterovirus 71 in a human foreskin fibroblast cell line. Kaohsiung J Med Sci 24(10):523–530

Sun XH, Ding JP, Li H, Pan N, Gan L, Yang XL, Xu HB (2007) Activation of large-conductance calcium-activated potassium channels by puerarin: the underlying mechanism of puerarin-mediated vasodilation. J Pharmacol Exp Ther 323(1):391–397

Sun YG, Wang SS, Feng JT, Xue XY, Liang XM (2008) Two new isoflavone glycosides from *Pueraria lobata*. J Asian Nat Prod Res 10(7–8):729–733

Takeya K, Itokawa H (1982) Isoflavonoids and the other constituents in callus tissues of *Pueraria Lobata*. Chem Pharm Bull 30(4):249–252

Tam WY, Chook P, Qiao M, Chan LT, Chan TY, Poon YK, Fung KP, Leung PC, Woo KS (2009) The efficacy and tolerability of adjunctive alternative herbal medicine (*Salvia miltiorrhiza* and *Pueraria lobata*) on vascular function and structure in coronary patients. J Altern Complement Med 15(4):415–421

Tamura S, Yoshihira K, Tokumaru M, Zisheng X, Murakami N (2010) Inhibitors for expression of IgE receptor on human mast cell from Puerariae Flos. Bioorg Med Chem Lett 20(13):3872–3875

Tanaka T, Tang HJ, Yu FN, Michihara S, Uzawa Y, Zaima N, Moriyama T, Kawamura Y (2011) Kudzu (*Pueraria lobata*) vine ethanol extracts improve ovariectomy-induced bone loss in female mice. J Agric Food Chem 59(24):13230–13237

Tang WC, Eisenbrand G (1992) *Pueraria lobata* (Willd.) Ohwi. Chinese Drugs of Plant Origin. Springer, Berlin, Heidelberg, pp 797–803

Tanner RD, Hussain SS, Hamilton LA, Wolf FT (1979) Kudzu (*Pueraria lobata*): Potential agricultural and industrial resource. Econ Bot 33(4):400–412

The Plant List (2013) *Pueraria montana* (Lour.) Merr. var. *lobata* (Willd.) Sanjappa & Pradeep. http://www.theplantlist.org/

Tian HZ, Wang H, Guan YF (2005) Separation and identification of isoflavonoids in *Pueraria lobata* extracts and its preparations by reversed-phase capillary liquid chromatography coupled with electrospray ionization quadrupole time of flight mass spectrometry. Se Pu 23(5):477–481 (In Chinese)

Tsuchihashi R, Kodera M, Sakamoto S, Nakajima Y, Yamazaki T, Niiho Y, Nohara T, Kinjo J (2009) Microbial transformation and bioactivation of isoflavones from Pueraria flowers by human intestinal bacterial strains. J Nat Med 63(3):254–260

USDA-ARS (2015) National genetic resources program. Germplasm resources information network – (GRIN) [Online Database]. National Germplasm Resources Laboratory, Beltsville, Maryland. https://npgsweb.ars-grin.gov/gringlobal/taxonomydetail.aspx?23373

Wang Q, Xu XY (2011) Progresses in research of hemolysis induced by puerarin injection. Zhongguo Zhong Yao Za Zhi 36(10):1402–1405 (In Chinese)

Wang LY, Zhao AP, Chai XS (1994) Effects of puerarin on cat vascular smooth muscle in vitro. Zhongguo Yao Li Xue Bao 15(2):180–182

Wang X, Wu J, Chiba H, Umegaki K, Yamada K, Ishimi Y (2003) Puerariae radix prevents bone loss in ovariectomized mice. J Bone Miner Metab 21(5):268–275

Wang JF, Guo YX, Niu JZ, Liu J, Wang LQ, Li PH (2004) Effects of Radix Puerariae flavones on liver lipid metabolism in ovariectomized rats. World J Gastroenterol 10(13):1967–1970

Wang X, Wu J, Chiba H, Yamada K, Ishimi Y (2005) Puerariae radix prevents bone loss in castrated male mice. Metabolism 54(11):1536–1541

Wang FW, Li J, Hu ZL, Xie YY (2006) Protective effect of puerarin on stress-induced gastric mucosal injury in rats. Zhongguo Zhong Yao Za Zhi 31(6):504–506 (In Chinese)

Wang YZ, Feng WS, Shi RB, Liu B (2007a) A new chemical component of *Pueraria lobata* (Willd.). Ohwi Yao Xue Bao 42(9):964–967 (In Chinese)

Wang ZY, Wei XB, Chen L, Liu P, Wang XL, Zhang B, Sun X, Zhang XM (2007b) Neuroprotective effects of hydroxyethylpuerarin against focal cerebral ischemia-reperfusion in rats. Chin J Physiol 50(5):211–216

Wang XM, Lei L, Zhang JH (2011) Antioxidant and anti-fatigue activities of flavonoids from Puerariae radix. Afr J Tradit Complement Altern Med 9(2):221–227

Wang H, Bai X, Sun JH, Kano Y, Makino T, Yuan D (2013a) Metabolism and excretion of kakkalide and its metabolites in rat urine, bile, and feces as determined by HPLC/UV and LC/MS/MS. Planta Med 79(16):1552–1557

Wang Y, Wang WL, Xie WL, Li LZ, Sun J, Sun WJ, Gong HY (2013b) Puerarin stimulates proliferation and differentiation and protects against cell death in human osteoblastic MG-63 cells via ER-dependent MEK/ERK and PI3K/Akt activation. Phytomedicine 20(10):787–796

Wei SY, Chen Y, Xu XY (2014) Progress on the pharmacological research of puerarin: a review. Chin J Nat Med 12(6):407–414

Wen KC, Lin SP, Yu CP, Chiang HM (2010) Comparison of Puerariae Radix and its hydrolysate on stimulation of hyaluronic acid production in NHEK cells. Am J Chin Med 38(1):143–155

Wong R, Rabie B (2007) Effect of puerarin on bone formation. Osteoarthr Cart 15(8):894–899

Wong KH, Li GQ, Li KM, Razmovski-Naumovski V, Chan K (2011) Kudzu root: traditional uses and potential medicinal benefits in diabetes and cardiovascular diseases. J Ethnopharmacol 134(3):584–607

Woo J, Lau E, Ho SC, Cheng F, Chan C, Chan AS, Haines CJ, Chan TY, Li M, Sham A (2003) Comparison of

*Pueraria lobat*a with hormone replacement therapy in treating the adverse health consequences of menopause. Menopause 10(4):352–361

Wu LM, Qiao HX, Li YK, Li LD (2007) Protective roles of puerarin and Danshensu on acute ischemic myocardial injury in rats. Phytomedicine 14(10):652–658

Wu YF, Wang XS, Fan EG (2012) Optimisation of ultrasound-assisted extraction of puerarin and total isoflavones from Puerariae Lobatae Radix (*Pueraria lobata* (Wild.) Ohwi) with response surface methodology. Phytochem Anal 23(5):513–519

Wu K, Liang T, Duan XQ, Xu LY, Zhang KF, Li R (2013) Anti-diabetic effects of puerarin, isolated from *Pueraria lobata* (Willd.), on streptozotocin-diabetogenic mice through promoting insulin expression and ameliorating metabolic function. Food Chem Toxicol 60:341–347

Wu XD, Wang C, Zhang ZY, Fu Y, Liu FY, Liu XH (2014) Puerarin attenuates cerebral damage by improving cerebral microcirculation in spontaneously hypertensive rats. Evid Based Complement Alternat Med 2014:408501

Xiao BX, Wang Q, Fan LQ, Kong LT, Guo SR, Chang Q (2014) Pharmacokinetic mechanism of enhancement by Radix Pueraria flavonoids on the hyperglycemic effects of Cortex Mori extract in rats. J Ethnopharmacol 151(2):846–851

Xie CI, Lin RC, Antony V, Lumeng L, Li TK, Mai K, Liu C, Wang QD, Zhao ZH, Wang GF (1994) Daidzin, an antioxidant isoflavonoid, decreases blood alcohol levels and shortens sleep time induced by ethanol intoxication. Alcohol Clin Exp Res 18(6):1443–1447

Xing GH, Dong MX, Li XM, Zou Y, Fan L, Wang XL, Cai DF, Li CC, Zhou L, Liu JC, Niu YC (2011) Neuroprotective effects of puerarin against beta-amyloid-induced neurotoxicity in PC12 cells via a PI3K-dependent signaling pathway. Brain Res Bull 85(3–4):212–218

Xiong FL, Sun XH, Gan L, Yang XL, Xu HB (2006) Puerarin protects rat pancreatic islets from damage by hydrogen peroxide. Eur J Pharmacol 529(1–3):1–7

Xiong YY, Yang YQ, Yang J, Chai HY, Li Y, Yang J, Jia ZM, Wang ZR (2010) Tectoridin, an isoflavone glycoside from the flower of *Pueraria lobata*, prevents acute ethanol-induced liver steatosis in mice. Toxicology 276(1):64–72

Xu XH (2003) Effects of puerarin on fatty superoxide in aged mice induced by D-galactose. Zhongguo Zhong Yao Za Zhi 28(1):66–69 (In Chinese)

Xu XH, Zhang ZG (2007) Effects of puerarin on synaptic structural modification in hippocampus of ovariectomized mice. Planta Med 73(10):1047–1053

Xu XH, Hu YY, Ruan Q (2004) Effects of puerarin on learning-memory and amino acid transmitters of brain in ovariectomized mice. Planta Med 70(7):627–631

Xu ME, Xiao SZ, Sun YH, Zheng XX, Ou-Yang Y, Guan C (2005a) The study of anti-metabolic syndrome effect of puerarin in vitro. Life Sci 77(25):3183–3196

Xu XH, Zhang SM, Zhang L, Yan WM, Zheng XX (2005b) The neuroprotection of puerarin against cerebral ischemia is associated with the prevention of apoptosis in rats. Planta Med 71(7):585–591

Xu YZ, Gao Y, Li PZ, Wang NF, Xu HY, Tong GX (2006) Puerarin suppresses the proliferation of vascular smooth muscle cells and c-fos and bcl-2 protein expression. Zhongguo Zhong Yao Za Zhi 31(6):490–493 (In Chinese)

Xue XO, Jin H, Niu JZ, Wang JF (2003) Effects of extracts of root of kudzu vine on mammary gland and uterus development in rats. Zhongguo Zhong Yao Za Zhi 28(6):560–562 (In Chinese)

Yamaki K, Kim DH, Ryu N, Kim YP, Shin KH, Ohuchi K (2002) Effects of naturally occurring isoflavones on prostaglandin E2 production. Planta Med 68(2):97–100

Yamazaki T, Nakajima Y, Niho Y, Hosono T, Kurashige T, Kinjo J, Nohara T (1997) Pharmacological studies on Puerariae flos III: protective effects of kakkalide on ethanol-induced lethality and acute hepatic injury in mice. J Pharm Pharmacol 49(8):831–833

Yan B, Wang DY, Xing DM, Ding Y, Wang RF, Lei F, Du LJ (2004) The antidepressant effect of ethanol extract of radix puerariae in mice exposed to cerebral ischemia reperfusion. Pharmacol Biochem Behav 78(2):319–325

Yan B, Xing DM, Ding Y, Tao JL, Du LJ (2005) HPLC method for the determination and pharmacokinetic studies on puerarin in cerebral ischemia reperfusion rat plasma after intravenous administration of puerariae radix isoflavone. J Pharm Biomed Anal 37(2):297–301

Yan LP, Chan SW, Chan AS, Chen SL, Ma XJ, Xu HX (2006) Puerarin decreases serum total cholesterol and enhances thoracic aorta endothelial nitric oxide synthase expression in diet-induced hypercholesterolemic rats. Life Sci 79(4):324–330

Yan LP, Zhuang YL, Chan SW, Chen SL, Shi GG (2009) Analysis of the mechanisms underlying the endothelium-dependent antivasoconstriction of puerarin in rat aorta. Naunyn Schmiedebergs Arch Pharmacol 379(6):587–597

Yang XJ, Hu WZ, Zhang Q, Wang YY, Sun L (2010) Puerarin inhibits C-reactive protein expression via suppression of nuclear factor kappaB activation in lipopolysaccharide-induced peripheral blood mononuclear cells of patients with stable angina pectoris. Basic Clin Pharmacol Toxicol 107(2):637–642

Yao MC, Liao YT, Li GQ, Law FCP, Tang Y (2010) Quantitative analysis of two isoflavones in *Pueraria lobata* flowers from eleven Chinese provinces using high performance liquid chromatography. Chinese Med 5:14

Yasuda T, Ohsawa K (1998) Urinary metabolites of daidzin orally administered in rats. Biol Pharm Bull 21(9):953–957

Yasuda T, Kano Y, Saito K, Ohsawa K (1994) Urinary and biliary metabolites of daidzin and daidzein in rats. Biol Pharm Bull 17(10):1369–1374

Yasuda T, Kano Y, Saito K, Ohsawa K (1995) Urinary and biliary metabolites of puerarin in rats. Biol Pharm Bull 18(2):300–303

Yasuda T, Mizunuma S, Kano Y, Saito K, Ohsawa K (1996) Urinary and biliary metabolites of genistein in rats. Biol Pharm Bull 19(3):413–417

Yasuda T, Ueda J, Ohsawa K (2001) Urinary metabolites of genistein administered orally to rats. Chem Pharm Bull 49(11):1495–1497

Yasuda T, Endo M, Konno T, Kato T, Mitsuzuka M, Ohsawa K (2005) Antipyretic, analgesic and muscle relaxant activities of pueraria isoflavonoids and their metabolites from *Pueraria lobata* Ohwi – a traditional Chinese drug. Biol Pharm Bull 28(7):1224–1228

Ye HY, Qiu F, Zeng J, Yiao XS, Lai F (2003) Effect of daidzein on antiarrhythmia. Zhongguo Zhong Yao Za Zhi 28(9):853–856 (In Chinese)

Yeung HC (1985) Handbook of Chinese Herbs and Formulas. Institute of Chinese Medicine, Los Angeles

Yeung DK, Leung SW, Xu YC, Vanhoutte PM, Man RY (2006) Puerarin, an isoflavonoid derived from Radix puerariae, potentiates endothelium-independent relaxation via the cyclic AMP pathway in porcine coronary artery. Eur J Pharmacol 552(1–3):105–111

Yu ZL, Li WJ (2006) Induction of apoptosis by puerarin in colon cancer HT-29 cells. Cancer Lett 238(1):53–60

Yu Z, Zhang G, Zhao H (1997) Effects of Puerariae isoflavone on blood viscosity, thrombosis and platelet function. Zhong Yao Cai 20(9):468–469 (In Chinese)

Yu CQ, Li YW, Chen H, Yang SS, Xie GR (2008) Decreased expression of aromatase in the Ishikawa and RL95-2 cells by the isoflavone, puerarin, is associated with inhibition of c-jun expression and AP-1 activity. Food Chem Toxicol 46(12):3671–3676

Yu HH, Jung SY, Shin MK, Park R, So HS, You YO (2010) *Pueraria thunbergiana* inhibits cisplatin-induced damage of HEI-OC1 auditory cells through scavenging free radicals. Phytother Res 24(6):834–839

Yu YL, Liao YT, Li X, Ye Y, Ke CQ, Li XQ, Yang XZ, Yao MC (2011) Isoflavonoid glycosides from the flowers of *Pueraria lobata*. J Asian Nat Prod Res 13(4):284–289

Yuan D, Xie YY, Bai X, Wu X, Yang JY, Wu CF (2009) Inhibitory activity of isoflavones of Pueraria flowers on nitric oxide production from lipopolysaccharide-activated primary rat microglia. J Asian Nat Prod Res 11(6):471–481

Zeng AG, Xing JF, Wang CH, Song J, Li C, Yang X, Yang GD (2012) Simultaneous analysis and retention behavior of major isoflavonoids in *Radix Puerariae lobatae* and *Radix Puerariae thomsonii* by high performance liquid chromatography with cyclodextrins as a mobile phase modifier. Anal Chim Acta 712:145–151

Zhang GC, Fang SM (1997) Antioxidation of *Pueraria lobata* isoflavones (PLIs). Zhong Yao Cai 20(7):358–360 (In Chinese)

Zhang FR, Chen JZ, Zhu JH, Wang XX, Zhu JH, Shang YP, Guo XG, Dai HM, Sun J (2004) Effects of puerarin on number and activity of endothelial progenitor cells from peripheral blood. Zhongguo Zhong Yao Za Zhi 29(8):777–781 (In Chinese)

Zhang Y, Xu Q, Zhang XZ, Chen JP, Liang XM, Kettrup A (2005) High-performance liquid chromatography–tandem mass spectrometry for identification of isoflavones and description of the biotransformation of kudzu root. Anal Bioanal Chem 383(5):787–796

Zhang B, Wei XB, Liu HQ, Wang LX, Sun R, Zhang XM (2006a) Protective effects of hydroxyethylpuerarin against brain astrocytes injury induced by hydrogen peroxide. Yao Xue Xue Bao 41(2):171–174 (In Chinese)

Zhang SH, Ji G, Liu JW (2006b) Reversal of chemical-induced liver fibrosis in Wistar rats by puerarin. J Nutr Biochem 17(7):485–491

Zhang SP, Fang WJ, Lü JH, Tan HR, Pan JQ (2006c) The experimental study of Radix Puerariae inhibiting glycation in rats induced by D-galactose. Zhong Yao Cai 29(3):266–269 (In Chinese)

Zhang Y, Zeng XH, Zhang L, Zheng XX (2007) Stimulatory effect of puerarin on bone formation through activation of PI3K/Akt pathway in rat calvaria osteoblasts. Planta Med 73(4):341–347

Zhang HY, Liu YH, Wang HQ, Xu JH, Hu HT (2008) Puerarin protects PC12 cells against beta-amyloid-induced cell injury. Cell Biol Int 32(10):1230–1237

Zhang DW, Ren Y, Dai SJ, Liu WH, Li GH (2009a) Isoflavones from vines of *Pueraria lobata*. Zhongguo Zhong Yao Za Zhi 34(24):3217–3220 (In Chinese)

Zhang GF, Wu X, Bai X, Kano Y, Yuan D (2009b) Simultaneous determination of four isoflavone compounds in flowers of *Pueraria lobata* by HPLC. J Shenyang Pharm Univ 26(1):40–44 (In Chinese)

Zhang RR, Hu YYY, Yuan J, Wu DZ (2009c) Effects of Puerariae radix extract on the increasing intestinal permeability in rat with alcohol-induced liver injury. J Ethnopharmacol 126(2):207–214

Zhang HJ, Yang XP, Wang KW (2010) Isolation of two new C-glucofuranosyl isoflavones from *Pueraria lobata* (Wild.) Ohwi with HPLC-MS guiding analysis. J Asian Nat Prod Res 12(4):293–299

Zhang CL, Ding XP, Hu ZF, Wang XT, Chen LL, Qi J, Yu BY (2011a) Comparative study of *Puerariae lobatae* and *Puerariae thomsonii* by HPLC-diode array detection-flow injection-chemiluminescence coupled with HPLC-electrospray ionization-MS. Chem Pharm Bull (Tokyo) 59(5):541–545

Zhang H, Zhang L, Zhang Q, Yang XC, Yu JY, Shun SH, Wu Y, Zeng QH, Wang TZ (2011b) Puerarin: a novel antagonist to inward rectifier potassium channel (IK1). Mol Cell Biochem 352(1–2):117–123

Zhang HY, Liu YH, Lao ML, Ma ZJ, Yi XN (2011c) Puerarin protects Alzheimer's disease neuronal cybrids from oxidant-stress induced apoptosis by inhibiting pro-death signaling pathways. Exp Gerontol 46(1):30–37

Zhang MY, Qiang H, Yang HQ, Dang XQ, Wang KZ (2012a) In vitro and in vivo effects of puerarin on pro-

motion of osteoblast bone formation. Chin J Integr Med 18(4):276–282

Zhang YP, Shi SY, Xiong X, Chen XQ, Peng MJ (2012b) Comparative evaluation of three methods based on high-performance liquid chromatography analysis combined with a 2,2′-diphenyl-1-picrylhydrazyl assay for the rapid screening of antioxidants from *Pueraria lobata* flowers. Anal Bioanal Chem 402(9):2965–2976

Zhang DY, Gao XJ, Wang Q, Qin MJ, Liu K, Huang F, Liu BL (2013a) Kakkalide ameliorates endothelial insulin resistance by suppressing reactive oxygen species-associated inflammation. J Diabetes 5(1):13–24

Zhang GF, Sun J, Kano Y, Yuan D (2013b) Metabolic profile of irisolidone in rats obtained by ultra-high performance liquid chromatography/quadrupole time-of-flight mass spectrometry. J Chromatogr B Analyt Technol Biomed Life Sci 941:1–9

Zhang Z, Lam TN, Zuo Z (2013c) Radix Puerariae: an overview of its chemistry, pharmacology, pharmacokinetics, and clinical use. J Clin Pharmacol 53(8):787–811

Zhang GZ, Gong TX, Kano Y, Yuan D (2014) Screening for *in vitro* metabolites of kakkalide and irisolidone in human and rat intestinal bacteria by ultra-high performance liquid chromatography/quadrupole time-of-flight mass spectrometry. J Chromatogr B 947:117–124

Zhao SP, Zhang YZ (1985) Quantitative TLC-densitometry of isoflavones in *Pueraria lobata* (Willd.) Ohwi. Yao Xue Xue Bao 20(3):203–208

Zhao Y, Du GY, Cui HF, Cao CY, Wang XR, Zhang CY (2005) Experimental study of protective effect of Pueraria compound on the cerebral ischemic injury. Zhongguo Zhong Yao Za Zhi 30(7):548–551 (In Chinese)

Zheng G, Zhang X, Meng Q, Gong W, Wen X, Xie H (2002a) Protective effect of total isoflavones from *Pueraria lobata* on secondary osteoporosis induced by dexamethasone in rats. Zhong Yao Cai 25(9):643–646 (In Chinese)

Zheng GL, Zhang XY, Zheng JW, Gong WG, Zheng XX, Chen AL (2002b) Hypocholesterolemic effect of total isoflavones from *Pueraria lobata* in ovariectomized rats. Zhong Yao Cai 25(4):273–275 (In Chinese)

Zheng GL, Zhang XY, Zheng JW, Meng QC, Zheng DS (2002c) Estrogen-like effects of puerarin and total isoflavones from *Pueraria lobata*. Zhong Yao Cai 25(8):566–568 (In Chinese)

Zheng PY, Ji G, Ma ZS, Liu T, Xin L, Wu HZ, Liang X, Liu JW (2009) Therapeutic effect of puerarin on non-alcoholic rat fatty liver by improving leptin signal transduction through JAK2/STAT3 pathways. Am J Chin Med 37(1):69–83

Zheng J, Chen B, Jiang B, Zeng L, Tang ZR, Fan L, Zhou HH (2010) The effects of puerarin on CYP2D6 and CYP1A2 activities in vivo. Arch Pharm Res 33(2):243–246

Zheng GM, Yu C, Yang Z (2012) Puerarin suppresses production of nitric oxide and inducible nitric oxide synthase in lipopolysaccharide-induced N9 microglial cells through regulating MAPK phosphorylation, O-GlcNAcylation and NF-κB translocation. Int J Oncol 40(5):1610–1618

Zhong Y, Li YF (2011) Effects of storage conditions and acid solvent types on structural, mechanical and physical properties of kudzu starch (*Pueraria lobata*)-chitosan composite films. Starch-Starke 63:579–586

Zhong G, Chen ZD, We YM (2007) Physicochemical properties of lotus (*Nelumbo nucifera* Gaertn.) and kudzu (*Pueraria hirsuta* Matsum.) starches. Int J Food Sci Technol 42(12):1449–1455

Zhong Y, Li YF, Zhao YY (2012) Physicochemical, microstructural, and antibacterial properties of β-chitosan and kudzu starch composite films. J Food Sci 77(10):E280–E286

Zhou YP, Su XL, Cheng B, Jiang J, Chen H (1995) Comparative study on pharmacological effects of various species of *Pueraria*. Zhongguo Zhong Yao Za Zhi 20(10):619–621, 640. (In Chinese)

Zhou J, Wang HX, Xiong YF, Li ZX, Feng YM, Chen J (2010) Puerarin attenuates glutamate-induced neurofilament axonal transport impairment. J Ethnopharmacol 132(1):150–156

Zhou YX, Zhang H, Peng C (2014) Puerarin: a review of pharmacological effects. Phytother Res 28(7):961–975

Zhu ZT, Li HQ, Lu Y, Lou JS, Yao Z (2001) Inhibitive effect of puerarin on increased NO production by neonatal cardiomyocytes during hypoxia/reoxygenation injury. Zhongguo Zhong Yao Za Zhi 26(12):856–859 (In Chinese)

Zhu L, Luo CF, Yuan M, Chen MS, Ji H (2010a) Molecular mechanism of protective effect of puerarin solid lipid nanoparticle on cerebral ischemia-reperfusion injury in gerbils. Zhong Yao Cai 33(12):1900–1904

Zhu GQ, Wang XC, Chen YF, Yang S, Cheng H, Wang N, Li QL (2010b) Puerarin protects dopaminergic neurons against 6-hydroxydopamine neurotoxicity via inhibiting apoptosis and upregulating glial cell line-derived neurotrophic factor in a rat model of Parkinson's disease. Planta Med 76(16):1820–1826

Zhu GH, Zhang Q, Dai HP, JI O, Shen Q (2013) Molecular mechanism of SHI-1 cell apoptosis induced by Puerariae Radix flavones in vitro. Zhongguo Shi Yan Xue Ye Xue Za Zhi 21(6):1423–1428 (In Chinese)

Zou Y, Zhao T, Mao GH, Zhang M, Zheng DH, Feng WW, Wang W, Wu XG, Yang LQ (2014) Isolation, purification and characterisation of selenium-containing polysaccharides and proteins in selenium-enriched Radix puerariae. J Sci Food Agric 94(2):349–358

Medical Glossary

AAD Allergic airway disease, an inflammatory disorder of the airways caused by allergens.

AAPH 2,2′-Azobis(2-amidinopropane) dihydrochloride, a water-soluble azo compound used extensively as a free radical generator, often in the study of lipid peroxidation and the characterisation of antioxidants.

Abeta aggregation Amyloid beta protein (Abeta) aggregation is associated with Alzheimer's disease (AD); it is a major component of the extracellular plaque found in AD brains.

Abdominal distension Referring to generalised distension of most or all of the abdomen. Also referred to as stomach bloating often caused by a sudden increase in fibre from consumption of vegetables, fruits and beans.

Ablation therapy The destruction of small areas of myocardial tissue, usually by application of electrical or chemical energy, in the treatment of some tachyarrhythmias.

Abortifacient A substance that causes or induces abortion.

Abortivum A substance inducing abortion.

Abscess A swollen infected, inflamed area filled with pus in body tissues.

ABTS 2.2-Azinobis-3-ethylhenthiazoline-6--sulphonic acid, a type of mediator in chemical reaction kinetics of specific enzymes.

ACAT Acyl CoA: cholesterol acyltransferase.

ACE See Angiotensin-converting enzyme.

ACTH (adrenocorticotropic hormone) Also known as 'corticotropin', is a polypeptide tropic hormone produced and secreted by the anterior pituitary gland.

Acetogenins Natural products from the plants of the family Annonaceae, are very potent inhibitors of the NADH-ubiquinone reductase (complex I) activity of mammalian mitochondria.

Acetyl-CoA carboxylase (ACC) Enzyme that-catalyses the biotin-dependent carboxylation of acetyl-CoA to produce malonyl-CoA.

Acetylcholinesterase (AChE) Is an enzyme that degrades (through its hydrolytic activity) the neurotransmitter acetylcholine, producing choline.

Acne vulga´ris Also known as chronic acne, usually occurring in adolescence, with comedones (blackheads), papules (red pimples), nodules (inflamed acne spots) and pustules (small inflamed pus-filled lesions) on the face, neck and upper part of the trunk.

Acidosis Increased acidity, an excessively acid condition of the body fluids.

Acquired immunodeficiency syndrome (AIDS) An epidemic disease caused by an infection by human immunodeficiency virus (HIV-1, HIV-2), retrovirus that causes immune system failure and debilitation and is often accompanied by infections such as tuberculosis.

Acridone An organic compound based on the acridine skeleton, with a carbonyl group at the 9 position.

ACTH Adrenocorticotropic hormone (or corticotropin), a polypeptide tropic hormone produced and secreted by the anterior pituitary gland. It palys a role in the synthesis and secretion of gluco- and mineralo-corticosteroids and androgenic steroids.

T.K. Lim, *Edible Medicinal and Non-Medicinal Plants: Volume 10, Modified Stems, Roots, Bulbs*,
DOI 10.1007/978-94-017-7276-1

Activating transcription factor (ATF) A protein (gene) that binds to specific DNA sequences regulating the transfer or transcription of information from DNA to mRNA.

Activator protein-1 (AP-1) A heterodimeric protein transcription factor that regulates gene expression in response to a variety of stimuli, including cytokines, growth factors, stress and bacterial and viral infections. AP-1 in turn regulates a number of cellular processes including differentiation, proliferation and apoptosis.

Actoprotective Increasing the body's physical performance.

Actoprotectors Preparations that increase the mental performance and enhance body stability against physical loads without increasing oxygen consumption. Actoprotectors are regarded as a subclass of adaptogens that hold a significant capacity to increase physical performance.

Acute otitis media (AOM) see Otitis media.

Acyl-CoA dehydrogenases A group of enzymes that catalyses the initial step in each cycle of fatty acid **β**-oxidation in the mitochondria of cells.

Adaptogen A term used by herbalists to refer to a natural herb product that increases the body's resistance to stresses such as trauma, stress and fatigue.

Adaptogenic Increasing the resistance of the body to stress.

Addison's disease Is a rare endocrine disorder. It occurs when the adrenal glands cannot produce sufficient hormones (corticosteroids). It is also known as chronic adrenal insufficiency, hypocortisolism or hypocorticism.

Adenocarcinoma A cancer originating in glandular tissue.

Adenoma A benign tumour from a glandular origin.

Adenoidectomy Surgical removal of the adenoids.

Adenopathy Abnormal enlargement or swelling of the lymph node.

Adenosine receptors A class of purinergic, G protein-coupled receptors with adenosine as endogenous ligand. In humans, there are four adenosine receptors. A_1 receptors and A_{2A} play roles in the heart, regulating myocardial oxygen consumption and coronary blood flow, while the A_{2A} receptor also has broader anti-inflammatory effects throughout the body. These two receptors also have important roles in the brain, regulating the release of other neurotransmitters such as dopamine and glutamate, while the A_{2B} and A_3 receptors are located mainly peripherally and are involved in inflammation and immune responses.

ADH See Alcohol dehydrogenase.

Adipocyte A fat cell involved in the synthesis and storage of fats.

Adipocytokine Bioactive cytokines produced by adipose tissues

Adiponectin A protein in humans that modulates several physiological processes, such as metabolism of glucose and fatty acids and immune responses.

Adipose tissues Body fat, loose connective tissue composed of adipocytes (fat cells).

Adaptogen Containing smooth pro-stressors which reduce reactivity of host defence systems and decrease damaging effects of various stressors due to increased basal level of mediators involved in the stress response.

Adrenal glands Star-shaped endocrine glands that sit on top of the kidneys.

Adrenalectomised Having had the adrenal glands surgically removed.

Adrenergic Having to do with adrenaline (epinephrine) and/or noradrenaline (norepinephrine).

Adrenergic receptors A class of G protein-coupled receptors that are targets of the noradrenaline (norepinephrine) and adrenaline (epinephrine).

Adulterant An impure ingredient added into a preparation.

Advanced glycation end products (AGEs) Resultant products of a chain of chemical reactions after an initial glycation

reaction. AGEs may play an important adverse role in process of atherosclerosis, diabetes, aging and chronic renal failure.

Aegilops An ulcer or fistula in the inner corner of the eye.

Aerophagia Excessive air swallowing.

Afferent Something that so conducts or carries towards, such as a blood vessel, fibre, or nerve.

Agammaglobulinaemia An inherited disorder in which there are very low levels of protective immune proteins called immunoglobulins. *cf.* x-linked agammaglobulinaemia.

Agalactia Lack of milk after parturition (birth).

Age-related macular degeneration (AMD) a medical condition of elderly adults that results in a loss of vision in the centre of the visual field (the macula) because of damage to the retina.

Agglutinin A protein substance, such as an antibody, that is capable of causing agglutination (clumping) of a particular antigen.

Agglutination Clumping of particles.

Agonist A drug that binds to a receptor of a cell and triggers a response by the cell.

Agoraphobia An anxiety disorder characterised by anxiety in situations where the sufferer perceives certain environments (openness or crowdedness) as dangerous or uncomfortable.

Ague A fever (such as from malaria) that is marked by paroxysms of chills, fever and sweating that recurs with regular intervals.

AHR AhR, aryl hydrocarbon receptor, a cytosolic protein transcription factor.

AIDS See Acquired immunodeficiency syndrome.

Akathisia A movement disorder in which there is an urge or need to move the legs to stop unpleasant sensations. Also called restless leg syndrome, the disorder is often caused by long-term use of antipsychotic medications.

AKT Serine/threonine kinase (also known as protein kinase B or PKB) plays a critical regulatory role in diverse cellular processes, including cancer progression and insulin metabolism.

Akt signalling pathway Akt are protein kinases involved in mammalian cellular signalling and inhibit apoptotic processes.

Akt/FoxO pathway Cellular processes involving Akt and FoxO transcription factors that play a role in angiogenesis and vasculogenesis.

Akt/GSK-3β/eNOS phosphorylation Amplifies serotonin 5-HT2B receptor blockade mediated anti-hypertrophic effects.

Alanine transaminase (ALT) Also called serum glutamic pyruvate transaminase (SGPT) or alanine aminotransferase (ALAT), an enzyme present in hepatocytes (liver cells). When a cell is damaged, it leaks this enzyme into the blood.

ALAT (alanine aminotransferase) See Alanine transaminase.

Albumin Water-soluble proteins found in egg white, blood serum, milk, various animal tissues and plant juices and tissues.

Albuminuria Excessive amount of albumin in the urine, a symptom of severe kidney disease.

Alcohol dehydrogenase (ADH) An enzyme involved in the breakdown of alcohol.

Aldose reductase, aldehyde reductase An enzyme in carbohydrate metabolism that converts glucose to sorbitol.

Aldosterone Is a steroid hormone. Its main role is to regulate salt and water in the body, thus having an effect on blood pressure.

Aldosteronism A condition in which there is excessive secretion of aldosterone, which disturbs the balance of sodium, potassium and water in the blood and so leads to high blood pressure.

Aldosteronopenia Deficiency of aldosterone production with normal secretion of cortisol.

Alexipharmic An antidote, remedy for poison.

Alexiteric A preservative against contagious and infectious diseases and the effects of poisons.

Algesic Endogenous substances involved in the production of pain that is associated with inflammation, e.g. serotonin, bradykinin and prostaglandins.

Alkaline phosphatase (ALP) An enzyme in the cells lining the biliary ducts of the liver. ALP levels in plasma will rise with large bile duct obstruction, intrahepatic cholestasis or infiltrative diseases of the liver. ALP is also present in bone and placental tissues.

Alkalosis Is a condition in which the body fluids have excess base (alkali).

Allergenic Having the properties of an antigen (allergen), immunogenic.

Allergic Pertaining to, caused, affected with, or the nature of the allergy.

Allergic conjunctivitis Inflammation of the tissue lining the eyelids (conjunctiva) due to allergy.

Allergy A hypersensitivity state induced by exposure to a particular antigen (allergen) resulting in harmful immunologic reactions on subsequent exposures. The term is usually used to refer to hypersensitivity to an environmental antigen (atopic allergy or contact dermatitis) or to drug allergy.

Allodynia A painful response to a normally innocuous stimulus.

Allogeneic Cells or tissues which are genetically different because they are derived from separate individuals of the same species. Also refers to a type of immunological reaction that occurs when cells are transplanted into a genetically different recipient.

Allografts Or homografts, a graft between individuals of the same species, but of different genotypes.

Alloknesis Itch produced by innocuous mechanical stimulation.

Allostasis The process of achieving stability, or homeostasis, through physiological or behavioural change.

Alopecia Is the loss of hair on the body.

Alopecia areata Type of hair loss that typically causes patches of baldness.

ALP See Alkaline phosphatase.

Alpha-adrenoceptor Receptors postulated to exist on nerve cell membranes of the sympathetic nervous system in order to explain the specificity of certain agents that affect only some sympathetic activities (such as vasoconstriction and relaxation of intestinal muscles and contraction of smooth muscles).

Alpha amylase (α-amylase) A major form of amylase found in humans and other mammals that cleaves alpha-bonds of large, alpha-linked polysaccharides, such as starch and glycogen, yielding glucose and maltose.

ALT See Alanine transaminase.

Alterative A medication or treatment which gradually induces a change and restores healthy functions without sensible evacuations.

Alveolar macrophage A vigorously phagocytic macrophage on the epithelial surface of lung alveoli that ingests carbon and other inhaled particulate matters. Also called coniophage or dust cell.

Alzheimer's disease A degenerative, organic, mental disease characterised by progressive brain deterioration and dementia, usually occurring after the age of 50.

Amastigote Refers to a cell that does not have any flagella, used mainly to describe a certain phase in the life cycle of trypanosome protozoans.

Amenorrhea The condition when a woman fails to have menstrual periods.

Amidolytic Cleavage of the amide structure.

Amoebiasis State of being infected by amoeba such as *Entamoeba histolytica.*

Amoebicidal Lethal to amoeba.

AMPK (5′ AMP-activated protein kinase) Or 5' adenosine monophosphate-activated protein kinase, enzyme that plays a role in cellular energy homeostasis.

Amygdalitis Inflammation of one or both tonsils, tonsillitis.

Amyloid beta (Aβ or Abeta) A peptide of 39–43 amino acids that appear to be the main constituent of amyloid plaques in the brains of Alzheimer's disease patients.

Amyloidosis A disorder that results from abnormal deposition of the protein, amyloid, in various tissues of the body.

Amyotrophic lateral sclerosis Or ALS, is a disease of the motor neurons in the brain and spinal cord that control voluntary muscle movement.

Amyotrophy Progressive wasting of muscle tissues. *adj.* amyotrophic.

Anaemia A blood disorder in which the blood is deficient in red blood cells and in haemoglobin.

Anaesthesia Condition of having sansation temporarily suppressed.

Anaesthetic A substance that decreases partially or totally nerve the sense of pain.

Analeptic A central nervous system (CNS) stimulant medication.

Analgesia Term describing relief, reduction or suppression of pain. *adj.* analgetic.

Analgesic A substance that relieves or reduces pain.

Anamnesis Patient's account of their medical history.

Anaphoretic An antiperspirant.

Anaphrodisiac Or antiaphrodisiac is something that reduces or blunts the libido.

Anaphylaxis A severe, life-threatening allergic response that may be characterised by symptoms such as reduced blood pressure, wheezing, vomiting or diarrhoea.

Anaphylactic *adj.* see Anaphylaxis.

Anaphylatoxins Are fragments (C3a, C4a or C5a) that are produced during the pathways of the complement system. They can trigger release of substances of endothelial cells, mast cells or phagocytes, which produce a local inflammatory response.

Anaplasia A reversion of differentiation in cells and is characteristic of malignant neoplasms (tumours).

Anaplastic *adj.* see Anaplasia.

Anasarca Accumulation of great quantity of fluid in body tissues.

Anencephaly A cephalic disorder that results from a neural tube defect that occurs when the cephalic (head) end of the neural tube fails to close, resulting in the absence of a major portion of the brain, skull and scalp.

Androgen Male sex hormone in vertebrates. Androgens may be used in patients with breast cancer to treat recurrence of the disease.

Androgenic alopecia Hair loss in men and women.

Android adiposity Centric fat distribution patterns with increased disposition towards the abdominal area, visceral fat—apple shaped. *cf.* gynoid adiposity.

Andrology Branch of medicine concerned with the reproductive diseases in men.

Aneugen An agent that affects cell division and the mitotic spindle apparatus, causing the loss or gain of whole chromosomes, thereby inducing aneuploidy. *adj.* aneugenic.

Angina pectoris, angina Chest pain or chest discomfort that occurs when the heart muscle does not get enough blood.

Angioedema Rapid swelling of the dermis, subcutaneous tissues, mucosa and submucosal tissues caused by small blood vessels leaking fluid into these tissues.

Angiogenic *adj.* see Angiogenesis.

Angiogenesis A physiological process involving the growth of new blood vessels from preexisting vessels.

Angiotensin An oligopeptide hormone in the blood that causes blood vessels to constrict and drives blood pressure up. It is part of the renin–angiotensin system.

Angiotensin-converting enzyme (ACE) An exopeptidase, a circulating enzyme that participates in the body's renin–angiotensin system (RAS) which mediates extracellular volume (i.e. that of the blood plasma, lymph and interstitial fluid) and arterial vasoconstriction.

Angioplasty Medical procedure used to open obstructed or narrowed blood vessel resulting usually from atherosclerosis.

Anguillulosis A parasitosis caused by the intestinal nematode *Strongyloides stercoralis* (round worm).

Anisakiasis A human parasitic infection of the gastrointestinal tract caused by the consumption of raw or undercooked seafood containing larvae of the nematode *Anisakis simplex*.

Anisonucleosis A morphological manifestation of nuclear injury characterised by variation in the size of the cell nuclei.

Ankylosing spondylitis (AS) Is a type of inflammatory arthritis that targets the joints of the spine.

Annexin V or annexin A5 Is a member of the annexin family of intracellular proteins that binds to phosphatidylserine (PS) in a calcium-dependent manner.

Annexitis Also called adnexitis, a pelvic inflammatory disease involving the inflammation of the ovaries or fallopian tubes.

Anodyne A substance that relieves or soothes pain by lessening the sensitivity of the brain or nervous system. Also called an analgesic.

Anoikis Apoptosis that is induced by inadequate or inappropriate cell-matrix interactions.

Anophthalmia Medical term for the absence of one or both eyes.

Anorectal Relating to the rectum and anus.

Anorectics Appetite suppressants, substances which reduce the desire to eat. Used on a

short-term basis clinically to treat obesity. Also called anorexigenics.

Anorexia Lack or loss of desire to eat.

Anorexia nervosa Is a psychiatric disorder characterised by an unrealistic fear of weight gain, self-starvation and conspicuous distortion of body image.

Anorexic Having no appetite to eat.

Anorexigenics See Anorectics.

Anosmia Inabity to perceive odour, reduced sense of smell.

Anoxia Absence of oxygen supply.

Antagonist A substance that acts against and blocks an action.

Antalgic A substance used to relive a painful condition.

Antecubital vein This vein is located in the antecubital fossa—the area of the arm in front of the elbow.

Anterior uveitis Is the most common form of ocular inflammation that often causes a painful red eye.

Anthelmintic An agent or substance that is destructive to worms and used for expulsion of internal parasitic worms in animals and humans.

Anthocyanins A subgroup of antioxidant flavonoids, are glucosides of anthocyanidins, which are beneficial to health. They occur as water-soluble vacuolar pigments that may appear red, purple or blue according to pH in plants.

Anthrax A bacterial disease of cattle and ship that can be transmitted to man though unprocessed wool.

Anthropometric Pertaining to the study of human body measurements.

Antiamoebic A substance that destroys or suppresses parasitic amoebae.

Antiamyloidogenic Compounds that inhibit the formation of Alzheimer's **β**-amyloid fibrils (fA**β**) from amyloid **β**-peptide (A**β**) and destabilise fA**β**.

Antianaphylactic Agent that can prevent the occurrence of anaphylaxis (life-threatening allergic response).

Antiangiogenic A drug or substance used to stop the growth of tumours and progression of cancers by limiting the pathologic formation of new blood vessels (angiogenesis).

Antiarrhythmic A substance to correct irregular heartbeats and restore the normal rhythm.

Antiasthmatic Drug that treats or ameliorates asthma.

Antiatherogenic That protects against atherogenesis, the formation of atheromas (plaques) in arteries.

Antibacterial Substance that kills or inhibits bacteria.

Antibilious An agent or substance which helps remove excess bile from the body.

Antibiotic A chemical substance produced by a microorganism which has the capacity to inhibit the growth of or to kill other microorganisms.

Antiblennorrhagic A substance that treats blennorrhagia a conjunctival inflammation resulting in mucus discharge.

Antibody A gamma globulin protein produced by a kind of white blood cell called the plasma cell in the blood used by the immune system to identify and neutralise foreign objects (antigen).

Anticarcinomic A substance that kills or inhibits carcinomas (any cancer that arises in epithelium/tissue cells).

Anticephalalgic Headache relieving or preventing.

Anticestodal A chemical destructive to tapeworms.

Anticholesterolemic A substance that can prevent the build-up of cholesterol.

Anticlastogenic Having a suppressing effect of chromosomal aberrations.

Anticoagulant A substance that thins the blood and acts to inhibit blood platelets from sticking together.

Antidepressant A substance that suppresses depression or sadness.

Antidiabetic A substance that prevents or alleviates diabetes. Also called antidiabetogenic.

Antidiarrhoeal Having the property of stopping or correcting diarrhoea, an agent having such action.

Antidipsotropic Antialcohol abuse; medication to reduce alcohol consumption.

Antidote A remedy for counteracting a poison.

Antidopaminergic A term for a chemical that prevents or counteracts the effects of dopamine.

Antidrepanocytary Anti-sickle cell anaemia.

Antidysenteric An agent used to reduce or treat dysentery and diarrhoea.

Antidyslipidemic Agent that will reduce the abnormal amount of lipids and lipoproteins in the blood.

Anti-oedematous Reduces or suppresses oedema.

Antiemetic An agent that stops vomiting and nausea.

Anti-epileptic A drug used to treat or prevent convulsions, anticonvulsant.

Antifebrile A substance that reduces fever, also called antipyretic.

Antifeedant Preventing something from being eaten.

Antifertility Agent that inhibits formation of ova and sperm and disrupts the process of fertilisation (antizygotic).

Anti-fibrosis Preventing/retarding the development of fibrosis, i.e. excessive growth and activity of fibroblasts.

Antifilarial Effective against human filarial worms.

Antifungal An agent that kills or inhibits the growth of fungi.

Antigen A substance that prompts the production of antibodies and can cause an immune response. *adj.* antigenic.

Antigenotoxic An agent that inhibits DNA adduct formation, stimulates DNA repair mechanisms and possesses antioxidant functions.

Antiganacratia Anti-menstruation.

Antigastralgic Preventing or alleviating gastric colic.

Antiemaetic Agent that stops vomiting.

Antihaemorrhagic An agent which stops or prevents bleeding.

Antihepatotoxic Counteracting injuries to the liver.

Antiherpetic Having activity against herpes simplex virus (HSV).

Antihistamine An agent used to counteract the effects of histamine production in allergic reactions.

Antihyperalgesia The ability to block enhanced sensitivity to pain, usually produced by nerve injury or inflammation, to nociceptive stimuli. *adj.* antihyperalgesic.

Antihypercholesterolemia Term to describe lowering of cholesterol level in the blood or blood serum.

Antihypercholesterolemic Agent that lowers cholesterol level in the blood or blood serum.

Antihyperlidemic Promoting a reduction of lipid levels in the blood, or an agent that has this action.

Antihypersensitive A substance used to treat excessive reactivity to any stimuli.

Antihypertensive A drug used in medicine and pharmacology to treat hypertension (high blood pressure).

Anti-inflammatory A substance used to reduce or prevent inflammation.

Antileishmanial Inhibiting the growth and proliferation of *Leishmania*, a genus of flagellate protozoans that are parasitic in the tissues of vertebrates.

Antileprotic Therapeutically effective against leprosy.

Antilithiatic An agent that reduces or suppresses urinary calculi (stones) and acts to dissolve those already present.

Antileukaemic Anticancer drugs that are used to treat leukaemia.

Antilithogenic Inhibiting the formation of calculi (stones).

Antimalarial An agent used to treat malaria and/or kill the malaria-causing organism, *Plasmodium* spp.

Antimelanogenesis Obstruct production of melanin.

Antimicrobial A substance that destroys or inhibits growth of disease-causing bacteria, viruses, fungi and other microorganisms.

Antimitotic Inhibiting or preventing mitosis.

Antimutagenic An agent that inhibits mutations.

Antimycotic Antifungal.

Antineoplastic Said of a drug intended to inhibit or prevent the maturation and proliferation of neoplasms that may become malignant, by targeting the DNA.

Antineuralgic A substance that stops intense intermittent pain, usually of the head or face, caused by neuralgia.

Antinociception Reduction in pain: a reduction in pain sensitivity produced within neurons when an endorphin or similar opium-containing substance opioid combines with a receptor.

Antinociceptive Having an analgesic effect.

Antioxytocic Inhibiting premature labour. *cf.* tocolytic.

Antinutrient Are natural or synthetic compounds that interfere with the absorption of nutrients and are commonly found in food sources and beverages.

Antioestrogen A substance that inhibits the biological effects of female sex hormones.

Antiophidian Antivenoms of snake.

Antiosteoporotic Substance that can prevent osteoporosis.

Antiovulatory Substance suppressing ovulation.

Antioxidant A chemical compound or substance that inhibits oxidation and protects against free radical activity and lipid oxidation such as vitamin E, vitamin C or beta-carotene (converted to vitamin B) and carotenoids and flavonoids which are thought to protect body cells from the damaging effects of oxidation. Many foods including fruit and vegetables contain compounds with antioxidant properties. Antioxidants may also reduce the risks of cancer and age-related macular degeneration (AMD).

Antipaludic Antimalarial.

Antiperiodic Substance that prevents the recurrence of symptoms of a disease, e.g. malaria.

Antiperspirant A substance that inhibits sweating. Also called antisudorific, anaphoretic.

Antiphlogistic A traditional term for a substance used against inflammation, an anti-inflammatory.

Antiplatelet agent Drug that decreases platelet aggregation and inhibits thrombus formation.

Antiplasmodial Suppressing or destroying plasmodia.

Antiproliferative Preventing or inhibiting the reproduction of similar cells.

Antiprostatic Drug to treat the prostate.

Antiprotozoal Suppressing the growth or reproduction of protozoa.

Antipruritic Alleviating or preventing itching.

Antipyretic A substance that reduces fever or quells it. Also known as antithermic.

Antirheumatic Relieving or preventing rheumatism.

Antiscorbutic A substance or plant rich in vitamin C that is used to counteract scurvy.

Antisecretory Inhibiting or diminishing secretion.

Antisense Refers to antisense RNA strand because its sequence of nucleotides is the complement of message sense. When mRNA forms a duplex with a complementary antisense RNA sequence, translation of the mRNA into the protein is blocked. This may slow or halt the growth of cancer cells.

Antiseptic Preventing decay or putrefaction, a substance inhibiting the growth and development of microorganisms.

Anti-sickling agent An agent used to prevent or reverse the pathological events leading to sickling of erythrocytes in sickle cell conditions.

Antispasmodic A substance that relieves spasms or inhibits the contraction of smooth muscles; smooth muscle relaxant, muscle relaxer.

Antispermatogenic Preventing or suppressing the production of semen or spermatozoa.

Antisudorific See Antiperspirant.

Antisyphilitic A drug (or other chemical agents) that is effective against syphilis.

Antithermic A substance that reduces fever and temperature. Also known as antipyretic.

Antithrombotic Preventing or interfering with the formation of thrombi.

Antitoxin An antibody with the ability to neutralise a specific toxin.

Antitumoral Substance that acts against the growth, development or spread of a tumour.

Antitussive A substance that depresses coughing.

Antiulcerogenic An agent used to protect against the formation of ulcers or is used for the treatment of ulcers.

Antivenin An agent used against the venom of a snake, spider or other venomous animals or insects.

Antivinous An agent or substance that treats addiction to alcohol.

Antiviral Substance that destroys or inhibits the growth and viability of infectious viruses.

Antivomitive A substance that reduces or suppresses vomiting.

Antizygotic See Antifertility.

Anuria Absence of urine production and excretion. *adj.* anuric.

Anxiogenic Substance that causes anxiety.

Anxiolytic A drug prescribed for the treatment of symptoms of anxiety.

APAF-1 Apoptotic protease-activating factor 1.

Apelin Also known as APLN, a peptide which in humans is encoded by the APLN gene.

Aperient A substance that acts as a mild laxative by increasing fluids in the bowel.

Aperitif An appetite stimulant.

Aphonia Loss of the voice resulting from disease, injury to the vocal cords, or various psychological causes, such as hysteria.

Aphrodisiac An agent that increases sexual activity and libido and/or improves sexual performance.

Aphthae White, painful oral ulcer of unknown cause.

Aphthous ulcer Also known as a canker sore, is a type of oral ulcer, which presents as a painful open sore inside the mouth or upper throat.

Aphthous stomatitis A canker sore, a type of painful oral ulcer or sore inside the mouth or upper throat, caused by a break in the mucous membrane. Also called aphthous ulcer.

Aplastic anaemia (AA) is a disease in which the bone marrow and the blood stem cells that reside there are damaged and do not make enough new blood cells.

Apnoea Suspension of external breathing.

Apoprotein The protein moiety of a molecule or complex, as of a lipoprotein.

Apolipoprotein A-I (APOA1) A major protein component of high-density lipoprotein (HDL) in plasma. The protein promotes cholesterol efflux from tissues to the liver for excretion.

Apolipoprotein B (APOB) Is the primary apolipoprotein of low-density lipoproteins (LDL or 'bad cholesterol'), which is responsible for carrying cholesterol to tissues.

Apolipoprotein E (APOE) The apolipoprotein found on intemediate density lipoprotein and chylomicron that binds to a specific receptor on liver and peripheral cells.

Apoplexy Unconsciousness or incapacity of the brain to function resulting from a cerebral haemorrhage or stroke.

Apoptogenic Ability to cause death of cells.

Apoptosis Death of cells.

Appendicitis Is a condition characterised by inflammation of the appendix. Also called epityphilitis.

Appetite stimulant A substance to increase or stimulate the appetite. Also called aperitif.

aPPT (activated partial thromboplastin time) A blood test, a measure of the part of the blood clotting pathway.

Apurinic lyase A DNA enzyme that catalyses a chemical reaction.

Arachidonate cascade Includes the cyclooxygenase (COX) pathway to form prostanoids and the lipoxygenase (LOX) pathway to generate several oxygenated fatty acids, collectively called eicosanoids.

ARE Antioxidant response element, is a transcriptional control element that mediates expression of a set of antioxidant proteins.

Ariboflavinosis A condition caused by the dietary deficiency of riboflavin that is characterised by mouth lesions, seborrhea and vascularisation.

Aromatase An enzyme involved in the production of oestrogen that acts by catalysing the conversion of testosterone (an androgen) to estradiol (an oestrogen). Aromatase is located in oestrogen-producing cells in the adrenal glands, ovaries, placenta, testicles, adipose (fat) tissue and brain.

Aromatic Having a pleasant, fragrant odour.

Aromatherapy A form of alternative medicine that uses volatile liquid plant materials, such as essential oils and other scented compounds from plants for the purpose of affecting a person's mood or health.

ARPE-19 cells A human retinal pigment epithelial cell line with differentiated properties.

Arrhythmias Abnormal heart rhythms that can cause the heart to pump less effectively. Also called dysrhythmias.

Arsenicosis See Arsenism.

Arsenism An incommunicable disease resulting from the ingestion of groundwater containing unsafe levels of arsenic, also known as arsenicosis.

Arteriogenic erectile dysfunction A penis dysfunction caused by the narrowing of the arteries in the penis, decreasing blood inflow to it, thus making erection impossible.

Arteriosclerosis Imprecise term for various disorders of arteries, particularly hardening due to fibrosis or calcium deposition, often used as a synonym for atherosclerosis.

Arthralgia Is pain in the joints from many possible causes.

Arthritis Inflammation of the joints of the body.

Arthrodynia An affection characterised by pain in or about a joint.

Arthus reaction An allergic reaction of the immediate hypersensitive type that results from the union of antigen and antibody, with complement present in blood vessel walls.

Aryl hydrocarbon receptor (AhR) A ligand-activated transcription factor best known for mediating the toxicity of dioxin and other exogenous contaminants and is responsible for their toxic effects, including immunosuppression.

ASATor AST Aspartate aminotransferase; see Aspartate transaminase.

ASBT Apical sodium-dependent bile acid transporter, belongs to the solute carrier family (SLC) of transporters and is an important carrier protein expressed in the small intestine.

Ascaris A genus of parasitic intestinal round worms.

Ascites Abnormal accumulation of fluid within the abdominal or peritoneal cavity.

Ascorbic acid See Vitamin C.

Aspartate transaminase (AST) Also called serum glutamic oxaloacetic transaminase (SGOT) or aspartate aminotransferase (ASAT) is similar to ALT in that it is another enzyme associated with liver parenchymal cells. It is increased in acute liver damage but is also present in red blood cells and cardiac and skeletal muscle and is therefore not specific to the liver.

Asphyxia Failure or suppression of the respiratory process due to obstruction of airflow to the lungs or due to the lack of oxygen in inspired air.

Asphyxiation The process of undergoing asphyxia.

Asthenia A nonspecific symptom characterised by loss of energy, strength and feeling of weakness.

Asthenopia Weakness or fatigue of the eyes, usually accompanied by headache and dimming of vision. *adj.* asthenopic.

Asthenozoospermia (asthenospermia) Reduced sperm motility.

Asthma A chronic illness involving the respiratory system in which the airway occasionally constricts, becomes inflamed and is lined with excessive amounts of mucus, often in response to one or more triggers.

Astringent A substance that contracts blood vessels and certain body tissues (such as mucous membranes) with the effect of reducing secretion and excretion of fluids and/or has a drying effect.

Astrocytes Collectively called astroglia, are characteristic star-shaped glial cells in the brain and spinal cord.

Ataxia (Loss of coordination) results from the degeneration of nerve tissue in the spinal cord and of nerves that control muscle movement in the arms and legs.

Ataxic polyneuropathy Is a syndrome characterised by problems with coordination and balance (sensory ataxia) and disturbances in nerve function (sensory neuropathy), bilateral optic atrophy and bilateral sensorineural deafness.

Ataxia telangiectasia and Rad3-related protein (ATR) Also known as serine/threonine-protein kinase ATR, FRAP-related protein 1 (FRP1), is an enzyme encoded by the ATR gene. It is involved in sensing DNA damage and activating the DNA damage checkpoint, leading to cell cycle arrest

Atelectasis The collapse or closure of the lung resulting in reduced or absent gas exchange.

ATF-2 Activating transcription factor 2.

Athlete's foot A contagious skin disease caused by parasitic fungi affecting the foot and hands, causing itching, blisters and cracking. Also called dermatophytosis.

Atherogenic Having the capacity to start or accelerate the process of atherogenesis.

Atherogenesis The formation of lipid deposits in the arteries.

Atheroma A deposit or degenerative accumulation of lipid-containing plaques on the innermost layer of the wall of an artery.

Atherosclerosis The condition in which an artery wall thickens as the result of a build-up of fatty materials such as cholesterol.

Atherothrombosis Medical condition characterised by an unpredictable, sudden disruption (rupture or erosion/fissure) of an atherosclerotic plaque, which leads to platelet activation and thrombus formation.

Athymic mice Laboratory mice lacking a thymus gland.

Atonic Lacking normal tone or strength.

Atony Insufficient muscular tone.

Atopic dermatitis An inflammatory, non-contagious, pruritic skin disorder of unknown aetiology; often called eczema.

Atresia A congenital medical condition in which a body orifice or passage in the body is abnormally closed or absent.

Atretic follicle Follicular atresia is the breakdown of the ovarian follicles.

Atretic ovarian follicles An involuted or closed ovarian follicle.

Atrial fibrillation Is the most common cardiac arrhythmia (abnormal heart rhythm) and involves the two upper chambers (atria) of the heart; the most serious consequence of atrial fribillation is ischemic stroke.

Atrioventricular node A node of specialised heart muscle located in the septal wall of the right atrium; receives impulses from the sinoatrial node and directs them to the walls of the ventricles.

Attention-deficit hyperactivity disorder (ADHD, ADD or AD/HD) Is a neurobehavioural developmental disorder, primarily characterised by the coexistence of attentional problems and hyperactivity.

Auditory brainstem response (ABR) Also called brainstem evoked response (BSER), is an electrical signal evoked from the brainstem of a human by the presentation of a sound such as a click.

Augmerosen A drug that may kill cancer cells by blocking the production of a protein that makes cancer cells live longer. Also called bcl-2 antisense oligonucleotide.

Auricular Of or relating to the auricle or the ear in general.

Aurones [2-Benzylidenebenzofuran-3(2H)-ones] are the secondary plant metabolites and are a subgroup of flavonoids. See Flavonoids.

Autoantibodies Antibodies manufactured by the immune system that mistakenly target and damage specific tissues and organs of the body.

Autolysin An enzyme that hydrolyses and destroys the components of a biological cell or a tissue in which it is produced.

Autonomic disorder A neurological disease in which the autonomic nervous system ceases to function properly.

Autonomic neuropathy Is a nerve disorder due to damage to the autonomic nerves that affects involuntary body functions, including heart rate, blood pressure, perspiration and digestion.

Autophagy Digestion of the cell contents by enzymes in the same cell.

Autopsy Examination of a cadaver to determine or confirm the cause of death.

Avenanthramides Low molecular weight, soluble phenolic compounds found in oats.

Avidity index Describes the collective interactions between antibodies and a multivalent antigen.

Avulsed teeth Is tooth that has been knocked out.

Ayurvedic Traditional Hindu system of medicine based largely on homeopathy and naturopathy.

Azoospermia Is the medical condition of a male not having any measurable level of sperm in his semen.

Azotaemia A higher than normal blood level of urea or other nitrogen-containing compounds in the blood.

B Cell-activating factor (BAFF) Also called tumour necrosis factor ligand superfamily member 13B. It plays an important role in the proliferation and differentiation of B cells.

Babesia A protozoan parasite (malaria-like) of the blood that causes a haemolytic disease known as babesiosis.

Babesiosis Malaria-like parasitic disease caused by *Babesia*, a genus of protozoal piroplasms.

Back tonus Normal state of balanced tension in the tissues of the back.

Bactericidal Lethal to bacteria.

BAFF A cytokine that belongs to the tumour necrosis factor (TNF) ligand family.

Balanitis Is an inflammation of the glans (head) of the penis.

BALB/c mice Balb/c mouse was developed in 1923 by McDowell. It is a popular strain and is used in many different research disciplines, but most often in the production of monoclonal antibodies.

Balm Aromatic oily resin from certain trees and shrubs used in medicine.

Barbiturates Are drugs that act as central nervous system depressants and can therefore produce a wide spectrum of effects, from mild sedation to total anaesthesia. They are also effective anxiolytics, hypnotics and anticonvulsants.

Baroreceptor A type of interoceptor that is stimulated by pressure changes, as those in blood vessel wall.

Barrett's oesophagus (Barrett oesophagitis) A disorder in which the lining of the oesophagus is damaged by stomach acid.

Basophil A type of white blood cell with coarse granules within the cytoplasm and a bilobate (two-lobed) nucleus.

Bax/Bad Proapoptotic proteins.

BCL-2 A family of apoptosis regulator proteins in humans encoded by the B cell lymphoma 2 (BCL-2) gene.

BCL-2 antisense oligonucleotide See Augmereson.

BCR/ABL A chimeric oncogene, from fusion of BCR and ABL cancer genes associated with chronic myelogenous leukaemia.

Bechic A remedy or treatment of cough.

Bed nucleus of the stria terminalis (BNST) Act as a relay site within the hypothalamic-pituitary-adrenal axis and regulate its activity in response to acute stress.

Belching, or burping Refers to the noisy release of air or gas from the stomach through the mouth.

Beriberi Is a disease caused by a deficiency of thiamine (vitamin B_1) that affects many systems of the body, including the muscles, heart, nerves and digestive system.

Beta-carotene Naturally occurring retinol (vitamin A) precursor obtained from certain fruits and vegetables with potential antineoplastic and chemopreventive activities. As an antioxidant, beta-carotene inhibits free radical damage to DNA. This agent also induces cell differentiation and apoptosis of some tumour cell types, particularly in early stages of tumorigenesis, and enhances immune system activity by stimulating the release of natural killer cells, lymphocytes and monocytes.

Beta-catenin Is a multifunctional oncogenic protein that contributes fundamentally to cell development and biology. It has been implicated as an integral component in the Wnt signalling pathway.

Beta cells A type of cell in the pancreas in areas called the islets of Langerhans.

Beta glucans Polysaccharides of D-glucose monomers linked by β-glycosidic bonds, (1→3), (1→4)-β-D-glucan, soluble, viscous component of fibres found in cereals like oats.

Beta-thalassemia An inherited blood disorder that reduces the production of haemoglobin.

Beta-lactamase Enzymes produced by some bacteria that are responsible for their resistance to beta-lactam antibiotics like penicillins.

BHT Butylated hydroxytoluene (phenolic compound), an antioxidant used in foods, cosmetics, pharmaceuticals and petroleum products.

BID The only known Bcl-2 family member that can function as an agonist of proapoptotic Bcl-2-related proteins such as Bax and Bak.

Bifidobacterium Is a genus of Gram-positive, nonmotile, often branched anaerobic bacteria. Bifidobacteria are one of the major genera of bacteria that make up the gut flora. Bifidobacteria aid in digestion are associated with a lower incidence of allergies and also prevent some forms of tumour growth. Some bifidobacteria are being used as probiotics.

Bifidogenic Promoting the growth of (beneficial) bifidobacteria in the intestinal tract.

Bile Fluid secreted by the liver and discharged into the duodenum where it is integral in the digestion and absorption of fats.

Bilharzia, bilharziasis See Schistosomiasis.

Biliary Relating to the bile or the organs in which the bile is contained or transported.

Biliary infections Infection of organ(s) associated with bile, comprise:
(a) acute cholecystitis, an acute inflammation of the gallbladder wall
(b) cholangitis, inflammation of the bile ducts

Biliousness Old term used in the eighteenth and nineteenth centuries pertaining to bad digestion, stomach pains, constipation and excessive flatulence.

Bilirubin A breakdown product of haem (a part of haemoglobin in red blood cells) produced by the liver that is excreted in bile which causes a yellow discoloration of the skin and eyes when it accumulates in those organs.

Biotin Also known as vitamin B7. See Vitamin B7.

Bitter A medicinal agent with a bitter taste and used as a tonic, alterative or appetiser.

Blackhead See Comedone.

Blackwater fever Dangerous complication of malarial whereby the red blood cells burst in the blood stream (haemolysis) releasing haemoglobin directly into the blood.

Blain See Chilblain.

Blastocyst Blastocyst is an embryonic structure formed in the early embryogenesis of mammals, after the formation of the morula, but before implantation.

Blastocystotoxic Agent that suppresses further development of the blastocyst through to the ovum stage.

Blebbing Bulging, e.g. membrane blebbing also called membrane bulging or ballooning.

Bleeding diathesis Is an unusual susceptibility to bleeding (haemorrhage) due to a defect in the system of coagulation.

Blennorrhagia Gonorrhoea.

Blennorrhoea Inordinate discharge of mucus, especially a gonorrhoeal discharge from the urethra or vagina.

Blepharitis Inflammation of the eyelids.

Blepharospasm Involuntary twitching, blinking closure or squeezing of the eyelids.

Blister Thin vesicle on the skin containing serum and caused by rubbing, friction or burn.

Blood–brain barrier (BBB) Is a separation of circulating blood and cerebrospinal fluid (CSF) in the central nervous system (CNS). It allows essential metabolites, such as oxygen and glucose, to pass from the blood to the brain and central nervous system (CNS) but blocks most molecules that are more massive than about 500 Da.

Blood stasis syndrome Blood stagnation or slowing of blood, an important underlying pathology of many disease processes according to traditional Chinese medicine.

BMPs (bone morphogenetic proteins) A family of secreted signalling molecules that can induce ectopic bone growth.

BNIP3 A pro-apoptotic BH3-only protein which is associated with mitochondrial dysfunction and cell death.

Boil Localised pyrogenic, painful infection, originating in a hair follicle.

Borborygmus Rumbling noise caused by the muscular contractions of peristalsis, the process that moves the contents of the stomach and intestines downwards.

Bowman–Birk inhibitors Type of serine proteinase inhibitor.

Bouillon A broth in French cuisine.

Bradicardia As applied to adult medicine, is defined as a resting heart rate of under 60 beats per minute.

Bradyphrenia Referring to the slowness of thought common to many disorders of the brain.

Brain-derived neurotrophic factor (BDNF) A protein member of the neurotrophin family that plays an important role in the growth, maintenance, function and survival of neurons. The protein molecule is involved in the modulation of cognitive and emotional functions and in the treatment of a variety of mental disorders.

Bright's disease Chronic nephritis.

Bronchial inflammation See Bronchitis.

Bronchiectasis A condition in which the airways within the lungs (bronchial tubes) become damaged and widened.

Bronchitis Is an inflammation of the main air passages (bronchi) to the lungs.

Bronchoalveolar lavage (BAL) A medical procedure in which a bronchoscope is passed through the mouth or nose into the lungs and fluid is squirted into a small part of the lung and then recollected for examination.

Bronchopneumonia Or bronchial pneumonia; inflammation of the lungs beginning in the terminal bronchioles.

Broncho-pulmonary Relating to the bronchi and lungs.

Bronchospasm Is a difficulty in breathing caused by a sudden constriction of the muscles in the walls of the bronchioles as occurs in asthma.

Brown fat Brown adipose tissue (BAT) in mammals; its primary function is to generate body heat in animals or newborns that do not shiver.

Bubo Inflamed, swollen lymph node in the neck or groin.

Buccal Of or relating to the cheeks or the mouth cavity.

Bulbectomy Removal of the olfactory bulb.

Bulimia An emotional disorder characterised by a distorted body image and an obsessive desire to lose weight, in which bouts of extreme overeating are followed by fasting or self-induced vomiting or purging.

Bullae Blisters; circumscribed, fluid-containing, elevated lesions of the skin, usually more than 5 mm in diameter.

Bursa A fluid-filled sac or saclike cavity situated in areas subjected to friction.

Bursitis Condition characterised by inflammation of one or more bursae (small sacs) of synovial fluid in the body.

C fibres Afferent fibres found in the nerve of the somatic sensory system.

c-FOS A cellular proto-oncogene belonging to the immediate early gene family of transcription factors.

C-jun NH(2)-terminal kinase Enzymes that belong to the family of the MAPK superfamily of protein kinases. These kinases mediate a plethora of cellular responses to such stressful stimuli, including apoptosis and production of inflammatory and immunoregulatory cytokines in diverse cell systems. *cf* MAPK.

c-Jun-I (Ser 73) Substrate of JNK-1 activated by phosphorylation at Ser73.

c-Jun-I (Ser 63) Substrate of JNK-1 activated by phosphorylation at Ser63.

C-reactive protein A protein found in the blood the levels of which rise in response to inflammation.

c-Src A cellular non-receptor tyrosine kinase.

CAAT element-binding proteins-alpha (c/EBP-alpha) Regulates gene expression in adipocytes in the liver.

Cachexia Physical wasting with loss of weight, muscle atrophy, fatigue and weakness caused by disease.

Caco-2 cell line A continuous line of heterogeneous human epithelial colorectal adenocarcinoma cells.

Cadaver A dead body, corpse.

Ca^{2+} ATPase (PMCA) is a transport protein in the plasma membrane of cells that serves to remove calcium (Ca^{2+}) from the cell.

Calcitonin gene-related peptide (CGRP) is a 37-amino acid neuropeptide that is abundant in the sensory neurons which innervate bone.

Calcium (Ca) Is the most abundant mineral in the body found mainly in bones and teeth. It is required for muscle contraction, blood vessel expansion and contraction, secretion of hormones and enzymes and transmitting impulses throughout the nervous system. Dietary sources include milk, yoghurt, cheese, Chinese cabbage, kale, broccoli, some green leafy vegetables, fortified cereals, beverages and soybean products.

Calcium ATPase Is a form of P-ATPase which transfers calcium after a muscle has contracted.

Calcium channel blockers (CCBs) A class of drugs and natural substances that disrupt the calcium (Ca2+) conduction of calcium channels.

Calciuria Abnormal presence of calcium in the urine.

Calculosis The tendency or deposition to form calculi or stones.

Calculus (calculi) Hardened, mineral deposits that can form a blockage in the urinary system.

Calculi infection Most calculi arise in the kidney when urine becomes supersaturated with a salt that is capable of forming solid crystals. Symptoms arise as these calculi become impacted within the ureter as they pass towards the urinary bladder.

Calefacient Substance that gives a sensation of warmth.

Caligo Dimness or obscurity of sight, dependent upon a speck on the cornea.

Calmodulin Is a calcium-modulated protein that can bind to and regulate a multitude of different protein targets, thereby affecting many different cellular functions.

cAMP-dependent pathway Cyclic adenosine monophosphate is a G protein-coupled receptor triggered signalling cascade used in cell communication in living organisms.

CAMP factor Diffusible, heat-stable, extracellular protein produced by Group B *Streptococcus* that enhances the haemolysis of sheep erythrocytes by *Staphylococcus aureus.* It is named after Christie, Atkins and Munch-Petersen, who described it in 1944.

Campylobacteriosis Is a gastrointestinal disease (gastroenteritis) caused by bacteria called *Campylobacter* which is frequently associated with the consumption of contaminated poultry.

Cancer a malignant neoplasm or tumour in any part of the body.

Candidiasis Infections caused by members of the fungus genus *Candida* that range from superficial, such as oral thrush and vaginitis, to systemic and potentially life-threatening diseases.

Canker See Chancre.

Cannabinoid receptor family Includes CB1 cannabinoid receptors found predominantly in the brain and nervous sytem and CB2 cannabinoid receptors mainly associated with immune tissues and expressed at low levels in the brain.

Cannabinoid receptor type 2 (CB 2 receptor) A G protein-coupled receptor from the cannabinoid receptor family that is mainly expressed on T cells of the immune system, on macrophages and B cells and in haematopoietic cells.

Carboxypeptidase An enzyme that hydrolyses the carboxy-terminal (C-terminal) end of a peptide bond. It is synthesised in the pancreas and secreted into the small intestine.

Carbuncle Is an abscess larger than a boil, usually with one or more openings draining pus onto the skin.

Carcinogenesis Production of carcinomas. *adj.* carcinogenic.

Carcinoma Any malignant cancer that arises from epithelial cells.

Carcinosarcoma A rare tumour containing carcinomatous and sarcomatous components.

Cardiac Relating to, situated near or affecting the heart.

Cardiac asthma Acute attack of dyspnoea with wheezing resulting from a cardiac disorder.

Cardiac hypertrophy Is a thickening of the heart muscle (myocardium) resulting in a decrease chamber size, including the left and right ventricles. Common causes of cardiac hypertrophy include high blood pressure (hypertension) and heart valve stenosis.

Cardialgia Heartburn.

Cardinolides Cardiac glycosides with a five-membered lactone ring in the side chain of the steroid aglycone.

Cardinolide glycoside Cardenolides that contain structural groups derived from sugars.

Cardioactive Having an effect on the heart.

Cardiogenic shock Is characterised by a decreased pumping ability of the heart that causes a shock-like state associated with an inadequate circulation of blood due to primary failure of the ventricles of the heart to function effectively.

Cardiomyocytes Cardiac muscle cells.

Cardiomyopathy Heart muscle disease.

Cardiopathy Disease or disorder of the heart.

Cardioplegia Stopping the heart so that surgical procedures can proceed in a still and bloodless field.

Cardiotonic Something which strengthens, tones or regulates heart functions without overt stimulation or depression.

Cardiovascular Pertaining to the heart and blood vessels.

Caries Tooth decay, commonly called cavities.

Cariogenic Leading to the production of caries.

Carminative Substance that stops the formation of intestinal gas and helps expel gas that has already formed, relieving flatulence or colic by expelling gas.

Carnitine palmitoyltransferase I (CPT1) Also known as carnitine acyltransferase I or CAT1, is a mitochondrial enzyme, involved in converting long-chain fatty acid into energy.

Carotenes Are a large group of intense red and yellow pigments found in all plants; these are hydrocarbon carotenoids (subclass of tetraterpenes) and the principal carotene is beta-carotene which is a precursor of vitamin A.

Carotenoids A class of natural fat-soluble pigments found principally in plants, belonging

to a subgroup of terpenoids containing eight isoprene units forming a C40 polyene chain. Carotenoids play an important potential role in human health by acting as biological antioxidants. See also Carotenes.

Carotenodermia Yellow skin discoloration caused by excess blood carotene.

Carpopedal spasm Spasm of the hand or foot, or of the thumbs and great toes.

Capases Cysteine–aspartic acid proteases, are a family of cysteine proteases, which play essential roles in apoptosis (programmed cell death), necrosis and inflammation.

Catalase (CAT) Enzyme in living organism that catalyses the decomposition of hydrogen peroxide to water and oxygen.

Catalepsy Indefinitely prolonged maintenance of a fixed body posture; seen in severe cases of catatonic schizophrenia.

Catamenia Menstruation.

Cataplasia Degenerative reversion of cells or tissue to a less differentiated form.

Cataplasm A medicated poultice or plaster. A soft moist mass, often warm and medicated, that is spread over the skin to treat an inflamed, aching or painful area, to improve the circulation.

Cataractogenesis Formation of cataracts.

Catarrh, catarrhal Inflammation of the mucous membranes especially of the nose and throat.

Catechins Are polyphenolic antioxidant plant metabolites. They belong to the family of flavonoids; tea is a rich source of catechins. See Flavonoids.

Catecholamines Hormones that are released by the adrenal glands in response to stress.

Cathartic Is a substance which accelerates defaecation.

Caustic Having a corrosive or burning effect.

Cauterisation A medical term describing the burning of the body to remove or close a part of it.

Caveolae Tiny (50–100nm) invaginations of the plasma membrane of the cell.

CB-1 receptor Cannabinoid receptor type 1 held to be one of the most widely expressed G protein-coupled receptors in the brain.

CCAAT/enhancer-binding proteins (C/EBP) Family of of transcription factors that interact with CCAAT (cytidine–cytidine–adenosine–adenosine–thymidine) box motif.

CCAAT/enhancer-binding protein (C/EBP)-α A key adipogenic transcription factor.

cdc2 kinase A member of the cyclin-dependent protein kinases (CDKs).

CDKs Cyclin-dependent protein kinases, a family of serine/threonine kinases that mediate many stages in mitosis.

CD4T cell Helper T cell with CD4 receptor that recognises antigens on the surface of a virus-infected cell and secretes lymphokines that stimulate B cells and killer T cells.

CD 28 Is one of the molecules expressed on T cells that provide co-stimulatory signals, which are required for T cell (lymphocytes) activation.

CD31 Also known as PECAM-1 (platelet endothelial cell adhesion molecule-1), a member of the immunoglobulin superfamily, that mediates cell-to-cell adhesion.

CD36 An integral membrane protein found on the surface of many cell types in vertebrate animals.

CD40 An integral membrane protein found on the surface of B lymphocytes, dendritic cells, follicular dendritic cells, haematopoietic progenitor cells, epithelial cells and carcinomas.

CD68 A glycoprotein expressed on monocytes/macrophages which binds to low-density lipoprotein.

Cecal ligation Tying up the caecum.

Celiac disease An autoimmune disorder of the small intestine, triggered in genetically susceptible individuals by ingested gluten from wheat, rye, barley and other closely related cereal grains.

Peptides resulting from partially digested gluten of wheat, barley or rye cause inflammation of the small intestinal mucosa.

Cell adhesion molecules (CAM) Glycoproteins located on the surface of cell membranes involved with binding of other cells or with the extracellular matrix.

Cellular respiration Is the set of the metabolic reactions and processes that take place in organisms' cells to convert biochemical energy from nutrients into adenosine triphosphate (ATP) and then release waste products. The reactions involved in respiration are catabolic reactions that involve the oxidation of one molecule and the reduction of another.

Cellulitis A bacterial infection of the skin that tends to occur in areas that have been damaged or inflamed.

Central nervous system Part of the vertebrate nervous system comprising the brain and spinal cord.

Central serous chorioretinopathy (CSCR) is a disease in which a serous detachment of the neurosensory retina occurs over an area of leakage from the choriocapillaris through the retinal pigment epithelium.

Central venous catheter A catheter placed into the large vein in the neck, chest or groin.

Cephalgia Pain in the head, a headache.

Cephalic Relating to the head.

Ceramide oligosides Oligosides with an *N*-acetyl-sphingosine moiety.

Cercariae A free swimming larva of the parasitic schistosome worm that has a tail and suckers on its head for penetration into a host.

Cerebral embolism A blockage of blood flow through a vessel in the brain by a blood clot that formed elsewhere in the body and travelled to the brain.

Cerebral ischemia Is the localised reduction of blood flow to the brain or parts of the brain due to arterial obstruction or systematic hyperfusion.

Cerebral infarction Is the ischemic kind of stroke due to a disturbance in the blood vessels supplying blood to the brain.

Cerebral tonic Substance that can alleviate poor concentration and memory, restlessness, uneasiness and insomnia.

Cerebrosides Are glycosphingolipids which are important components in animal muscle and nerve cell membranes.

Cerebrovascular disease Is a group of brain dysfunctions related to disease of the blood vessels supplying the brain.

Cervical spondylotic myelopathy A common cause of spinal cord dysfunction in older persons.

Cerumen Ear wax, a yellowish waxy substance secreted in the ear canal of humans and other mammals.

cFLIP Cellular FLICE-inhibitory protein, an inhibitor of death ligand-induced apoptosis.

cGMP Cyclic guanosine monophosphate is a cyclic nucleotide derived from guanosine triphosphate (GTP). cGMP is a common regulator of ion channel conductance, glycogenolysis and cellular apoptosis. It also relaxes smooth muscle tissues.

CGRP calcitonin gene-related peptide A vasodilator neuropeptide that is expressed in a subgroup of small neurons in the dorsal root, trigeminal and vagal ganglia. This neuropeptide has been postulated to play a role in the pathophysiology of migraine.

Chalcones A subgroup of flavonoids.

Chancre A painless lesion formed during the primary stage of syphilis.

Chaperones Are proteins that assist the noncovalent folding or unfolding and the assembly or disassembly of other macromolecular structures.

Chemoembolisation A procedure in which the blood supply to the tumour is blocked surgically or mechanically and anticancer drugs are administered directly into the tumour.

Chemokines Are chemotactic cytokines, which stimulate migration of inflammatory cells towards tissue sites of inflammation.

Chemonociceptors Nociceptors or sensory pheripheral neurons that are sensitive to chemical stimuli.

Chemosensitiser A drug that makes tumour cells more sensitive to the effects of chemotherapy.

Chemosis Oedema of the conjunctiva of the eye.

Chickenpox Is also known as varicella and is a highly contagious illness caused by primary infection with varicella zoster virus (VZV). The virus causes red, itchy bumps on the body.

Chilblains Small, itchy, painful lumps that develop on the skin. They develop as an abnormal response to cold. Also called perniosis or blain.

Chlorosis Iron deficiency anaemia characterised by greenish yellow colour.

Cholagogue Is a medicinal agent which promotes the discharge of bile from the system.

Cholecalcifereol A form of vitamin D, also called vitamin D3. See Vitamin D.

Cholecyst Gallbladder.

Cholecystitis Inflammation of the gallbladder.

Cholecystokinin A peptide hormone that plays a key role in facilitating digestion in the small intestine.

Cholera An infectious gastroenteritis caused by enterotoxin-producing strains of the bacterium *Vibrio cholera* and characterised by severe, watery diarrhoea.

Choleretic Stimulation of the production of bile by the liver.

Cholestasis A condition caused by rapidly developing (acute) or long-term (chronic) interruption in the excretion of bile from the liver to the duodenum.

Cholesterol A soft, waxy, steroid substance found among the lipids (fats) in the bloodstream and in all our body's cells.

Cholethiasis Presence of gall stones (calculi) in the gallbladder.

Choline A water-soluble, organic compound, usually grouped within the vitamin B complex. It is an essential nutrient and is needed for physiological functions such as structural integrity and signalling roles for cell membranes, cholinergic neurotransmission (acetylcholine synthesis).

Cholinergic Activated by or capable of liberating acetylcholine, especially in the parasympathetic nervous system.

Cholinergic system A system of nerve cells that uses acetylcholine in transmitting nerve impulses.

Cholinomimetic Having an action similar to that of acetylcholine; called also parasympathomimetic.

Chronotropic Affecting the time or rate, as the rate of contraction of the heart.

Choriocarcinoma A quick-growing malignant, trophoblastic, aggressive cancer that occurs in a woman's uterus (womb).

Choroidal neovascularisation (CNV) is the creation of new blood vessels in the choroid layer of the eye.

Chromium (Cr) Is required in trace amounts in humans for sugar and lipid metabolism. Its deficiency may cause a disease called chromium deficiency. It is found in cereals, legumes, nuts and animal sources.

Chromoblastomycosis A chronic fungal infection of the skin and the subcutaneous tissue caused by traumatic inoculation of a specific group of dematiaceous fungi (such as *Fonsecaea pedrosoi, Phialophora verrucosa, Fonsecaea compacta*) through the skin.

Chromosome Long pieces of DNA found in the centre (nucleus) of cells.

Chronic Persisting over extended periods.

Chronic anterior uveitis Inflammation of the iris and middle coat of the eyeball.

Chronic obstructive pulmonary disease (COPD) A progressive disease that makes it hard to breathe.

Chronic venous insufficiency (CVI) A medical condition where the veins cannot pump enough oxygen-poor blood back to the heart.

Chronotropic Affecting the rate of rhythmic movements (e.g. heartbeat).

Chyle A milky bodily fluid consisting of lymph and emulsified fats, or free fatty acids.

Chylomicrons Are large lipoprotein particles that transport dietary lipids from the intestines to other locations in the body. Chylomicrons are one of the five major groups of lipoproteins (chylomicrons, VLDL, IDL, LDL, HDL) that enable fats and cholesterol to move within the water-based solution of the bloodstream.

Chylorus Milky (having fat emulsion).

Chyluria Also called chylous urine, is a medical condition involving the presence of chyle (emulsified fat) in the urine stream, which results in urine appearing milky.

Chymase Member of the family of serine proteases found primarily in mast cell.

Chymopapain An enzyme derived from papaya, used in medicine and to tenderise meat.

Cicatrizant The term used to describe a product that promotes healing through the formation of scar tissue.

C-Kit receptor A protein tyrosine kinase receptor that is specific for stem cell factor. This interaction is crucial for the development of haematopoietic, gonadal and pigment stem cells.

Cirrhosis Chronic liver disease characterised by replacement of liver tissue by fibrous scar tissue and regenerative nodules/lumps leading progressively to loss of liver function.

Clastogen Is an agent that can cause one of two types of structural changes and breaks in chromosomes that results in the gain, loss, or rear-

rangements of chromosomal segments. *adj.* clastogenic.

Claudication Limping, impairment in walking.

Climacterium Refers to menopause and the bodily and mental changes associated with it.

Clonic seizures Consist of rhythmic jerking movements of the arms and legs, sometimes on both sides of the body.

Clonus A series of involuntary muscular contractions and relaxations.

Clyster Enema.

C-myc Codes for a protein that binds to the DNA of other genes and is therefore a transcription factor.

CNS depressant Anything that depresses, or slows, the sympathetic impulses of the central nervous system (i.e. respiratory rate, heart rate).

Coagulopathy A defect in the body's mechanism for blood clotting, causing susceptibility to bleeding.

Cobalamin Vitamin B12. See Vitamin B12.

Co-carcinogen A chemical that promotes the effects of a carcinogen in the production of cancer.

Cold An acute inflammation of the mucous membrane of the respiratory tract, especially of the nose and throat caused by a virus and accompanied by sneezing and coughing.

Collagen Protein that is the major constituent of cartilage and other connective tissues; comprises the amino acids, hydroxyproline, proline, glycine and hydroxylysine.

Collagenases Enzymes that break the peptide bonds in collagen.

Colibacillosis Infection with *Escherichia coli.*

Colic A broad term which refers to episodes of uncontrollable, extended crying in a baby who is otherwise healthy and well fed.

Colitis Inflammatory bowel disease affecting the tissue that lines the gastrointestinal system.

Collyrium A lotion or liquid wash used as a cleanser for the eyes, particularly in diseases of the eye.

Colorectal Relating to the colon or rectum.

Coma A state of unconsciousness from which a patient cannot be aroused.

Comedone A blocked, open sebaceous gland where the secretions oxidise, turning black. Also called blackhead.

Comitogen Agent that is considered not to induce cell growth alone but to promote the effect of the mitogen.

Concoction A combination of crude ingredients that is prepared or cooked together.

Condyloma, condylomata acuminata Genital wart, venereal wart, anal wart or anogenital wart, a highly contagious sexually transmitted infection caused by epidermotropic human papillomavirus (HPV).

Congestive heart failure Heart failure in which the heart is unable to maintain adequate circulation of blood in the tissues of the body or to pump out the venous blood returned to it by the venous circulation.

Conglutination Becoming stuck together.

Conjunctival hyperemia Enlarged blood vessels in the eyes.

Conjunctivitis Sore, red and sticky eyes caused by eye infection.

Conn's syndrome Extremely rare condition characterised by adenoma, carcinoma or hyperplasia of the zona glomerulosa of the adrenal cortex, resulting in excessive production of aldosterone and leading to sodium retention and hydrogen loss.

Constipation A very common gastrointestinal disorder characterised by the passing of hard, dry bowel motions (stools) and difficulty of bowel motion.

Constitutive androstane receptor (CAR, NR113) Is a nuclear receptor transcription factor that regulates drug metabolism and homoeostasis.

Consumption Term used to describe wasting of tissues including but not limited to tuberculosis.

Consumptive Afflicted with or associated with pulmonary tuberculosis.

Contraceptive An agent that reduces the likelihood of or prevents conception.

Contraindication A condition which makes a particular treatment or procedure inadvisable.

Contralateral muscle Muscle of opposite limb (leg or arm).

Contralateral rotation Rotation occurring or originating in a corresponding part on an opposite side.

Contusion Another term for a bruise. A bruise, or contusion, is caused when blood vessels are

damaged or broken as the result of a blow to the skin.

Convulsant A drug or physical disturbance that induces convulsion.

Convulsion Rapid and uncontrollable shaking of the body.

Coolant That which reduces body temperature.

Copper (Cu) Is essential in all plants and animals. It is found in a variety of enzymes, including the copper centres of cytochrome C oxidase and the enzyme superoxide dismutase (containing copper and zinc). In addition to its enzymatic roles, copper is used for biological electron transport. Because of its role in facilitating iron uptake, copper deficiency can often produce anaemia-like symptoms. Dietary sources include curry powder, mushroom, nuts, seeds, wheat germ, whole grains and animal meat.

Copulation To engage in coitus or sexual intercourse. *adj.* copulatory.

Cor pulmonale Or pulmonary heart disease is enlargement of the right ventricle of the heart as a response to high blood pressure or increased resistance in the lungs.

Cordial A preparation that is stimulating to the heart.

Corn Or callus, is a patch of hard, thickened skin on the foot that is formed in response to pressure or friction.

Corpora lutea A yellow, progesterone-secreting body that forms from an ovarian follicle after the release of a mature egg.

Corticosteroids A class of steroid hormones that are produced in the adrenal cortex, used clinically for hormone replacement therapy, for suppressing ACTH secretion, for suppression of immune response and as antineoplastic, anti-allergic and anti-inflammatory agents.

Corticosterone A 21-carbon steroid hormone of the corticosteroid type produced in the cortex of the adrenal glands.

Cortisol Is a corticosteroid hormone made by the adrenal glands and plays an essential role in homeostasis .

Cornification Is the process of forming an epidermal barrier in stratified squamous epithelial tissue.

Coryza A word describing the symptoms of a head cold. It describes the inflammation of the mucus membranes lining the nasal cavity which usually gives rise to the symptoms of nasal congestion and loss of smell, among other symptoms.

COX-1 See Cyclooxygenase-1.

COX-2 See Cyclooxygenase-2.

CpG islands Genomic regions that contain a high frequency of CpG sites.

CpG sites The cytosine–phosphate–guanine nucleotide that links two nucleosides together in DNA.

cPLA(2) Cytosolic phospholipases A2, these phospholipases are involved in cell signalling processes, such as inflammatory response.

CPY1B1, CPY1A1 A member of the cytochrome P450 superfamily of haem-thiolate monooxygenase enzymes.

Creatin A nitrogenous organic acid that occurs naturally in vertebrates and helps to supply energy to muscle.

Creatine phosphokinase (CPK, CK) Enzyme that catalyses the conversion of creatine and consumes adenosine triphosphate (ATP) to create phosphocreatine and adenosine diphosphate (ADP).

CREB cAMP response element binding, a protein that is a transcription factor that binds to certain DNA sequences called cAMP response elements.

Crohn disease An inflammatory disease of the intestines that affect any part of the gastrointestinal tract.

CRP (C-reactive protein) A substance produced by the liver that increases in the presence of inflammation in the body.

Crossover study A longitudinal, balance study in which participants receive a sequence of different treatments or exposures.

Croup Is an infection of the throat (larynx) and windpipe (trachea) that is caused by a virus (also called laryngotracheobronchitis).

Cryptococcal meningitis A fungal infection of the membranes covering the brain and spinal cord (meninges).

Cryptorchidism (cryptorchism) A developmental defect characterised by the failure of

one or both testes to move into the scrotum as the male fetus develops.

Curettage Surgical procedure in which a body cavity or tissue is scraped with a sharp instrument or aspirated with a cannula.

Cutaneous Pertaining to the skin.

CXC8 Also known as interleukin 8, IL-8.

Cyanogenesis Generation of cyanide. *adj.* cyanogenetic.

Cyclooxygenase (COX) An enzyme that is responsible for the formation of prostanoids—prostaglandins, prostacyclins and thromboxanes that are each involved in the inflammatory response. Two different COX enzymes existed, now known as COX-1 and COX-2.

Cyclooxygenase-1 (COX-1) Is known to be present in most tissues. In the gastrointestinal tract, COX-1 maintains the normal lining of the stomach. The enzyme is also involved in kidney and platelet function.

Cyclooxygenase-2 (COX-2) Is primarily present at sites of inflammation.

Cysteine proteases Are enzymes that degrade polypeptides possessing a common catalytic mechanism that involves a nucleophilic cysteine thiol in a catalytic triad. They are found in fruits like papaya, pineapple and kiwifruit.

Cystitis A common urinary tract infection that occurs when bacteria travel up the urethra, infect the urine and inflame the bladder lining.

Cystorrhea Discharge of mucus from the bladder.

Cytochrome bc-1 complex Ubihydroquinone: Cytochrome c oxidoreductase.

Cytochrome P450 3A CYP3A A very large and diverse superfamily of haem-thiolate proteins found in all domains of life. This group of enzymes catalyses many reactions involved in drug metabolism and synthesis of cholesterol, steroids and other lipids.

Cytokine Non-antibody proteins secreted by certain cells of the immune system which carry signals locally between cells. They are a category of signalling molecules that are used extensively in cellular communication.

Cytopathic Any detectable, degenerative changes in the host cell due to infection.

Cytoprotective Protecting cells from noxious chemicals or other stimuli.

Cytosolic Relates to the fluid of the cytoplasm in cells.

Cytostatic Preventing the growth and proliferation of cells.

Cytotoxic Of or relating to substances that are toxic to cells; cell killing.

d-Galactosamine An amino sugar with unique hepatotoxic properties in animals.

Dandruff Scurf, dead, scaly skin among the hair.

Dartre Condition of dry, scaly skin

Debility Weakness, relaxation of muscular fibre.

Debridement Is the process of removing non-living tissue from pressure ulcers, burns and other wounds.

Debriding agent Substance that cleans and treats certain types of wounds, burns and ulcers.

Deciduogenic Relating to the uterus lining that is shed off at childbirth.

Deciduoma Decidual tissue induced in the uterus (as by trauma) in the absence of pregnancy.

Deciduomata Plural of deciduoma.

Decidual stromal cells Like endometrial glands and endothelium, express integrins that bind basement components.

Decoction A medical preparation made by boiling the ingredients.

Decongestant A substance that relieves or reduces nasal or bronchial congestion.

Deep venous thrombosis Is a blood clot that forms in a vein deep inside a part of the body.

Defibrinated plasma Blood whose plasma component has had fibrinogen and fibrin removed.

Degranulation Cellular process that releases antimicrobial cytotoxic molecules from secretory vesicles called granules found inside some cells.

Delayed after depolarisations (DADs) Abnormal depolarisation that begins during phase 4—after depolarisation is completed, but before another action potential would normally occur.

Delirium Is common, sudden severe confusion and rapid changes in brain function that occur with physical or mental illness; it is reversible and temporary.

Demulcent An agent that soothes internal membranes. Also called emollient.

Dendritic cells Are immune cells and form part of the mammalian immune system, functioning as antigen-presenting cells.

Dentition A term that describes all of the upper and lower teeth collectively.

Deobstruent A medicine which removes obstructions; also called an aperient.

Deoxypyridinoline (Dpd) A crosslink product of collagen molecules found in bone and excreted in urine during bone degradation.

Depilatory An agent for removing or destroying hair.

Depressant A substance that diminishes functional activity, usually by depressing the nervous system.

Depurative An agent used to cleanse or purify the blood; it eliminates toxins and purifies the system.

Dermatitis Inflammation of the skin causing discomfort such as eczema.

Dermatitis herpetiformis An autoimmune chronic blistering skin disorder characterised by blisters filled with a watery fluid.

Dermatophyte A fungus parasitic on the skin.

Dermatosis Is a broad term that refers to any disease of the skin, especially one that is not accompanied by inflammation.

Dermonecrotic Pertaining to or causing necrosis of the skin.

Dermopathy A skin disorder characterised by discoloured patches and small papules that often become pigmented and ulcerated and result in scars, most commonly occurring on the shins of people with diabetes mellitus.

Desmutagen Substances that inactivate mutagens (cancer-causing agents).

Desquamation The shedding of the outer layers of the skin.

Desquamative gingivitis Red, painful, glazed and friable gingivae which may be a manifestation of some mucocutaneous conditions such as lichen planus or the vesiculobullous disorders.

Detoxifier A substance that promotes the removal of toxins from a system or organ.

Diabetes A metabolic disorder associated with inadequate secretion or utilisation of insulin and characterised by frequent urination and persistent thirst. See Diabetes mellitus.

Diabetes mellitus (DM) (Sometimes called 'sugar diabetes') is a set of chronic, metabolic disease conditions characterised by high blood sugar (glucose) levels that result from defects in insulin secretion, or action, or both. Diabetes mellitus appears in two forms.

Diabetes mellitus type I (Formerly known as juvenile onset diabetes) caused by deficiency of the pancreatic hormone insulin as a result of destruction of insulin-producing beta cells of the pancreas. Lack of insulin causes an increase of fasting blood glucose that begins to appear in the urine above the renal threshold.

Diabetes mellitus type II (Formerly called non-insulin-dependent diabetes mellitus or adult-onset diabetes) the disorder is characterised by high blood glucose in the context of insulin resistance and relative insulin deficiency in which insulin is available but cannot be properly utilised.

Diabetic autonomic neuropathy (DAN) is a serious and common complication of diabetes involving damage of the autonomic nerves. Major clinical manifestations of DAN include resting tachycardia, exercise intolerance, orthostatic hypotension, constipation, gastroparesis, erectile dysfunction, sudomotor dysfunction, impaired neurovascular function, 'brittle diabetes' and hypoglycaemic autonomic failure.

Diabetic foot Any pathology that results directly from diabetes mellitus or any long-term or chronic complication of diabetes mellitus.

Diabetic neuropathy A neuropathic disorder that is associated with diabetes mellitus. It affects all peripheral nerves including pain fibres, motor neurons and the autonomic nervous system.

Diabetic retinopathy Damage to the retina caused by complications of diabetes mellitus, which can eventually lead to blindness.

Diads Two adjacent structural units in a polymer molecule.

Dialysis Is a method of removing toxic substances (impurities or wastes) from the blood when the kidneys are unable to do so.

Diaphoresis Is profuse sweating commonly associated with shock and other medical emergency conditions.

Diaphoretic A substance that induces perspiration. Also called sudorific.

Diaphyseal Pertaining to or affecting the shaft of a long bone (diaphysis).

Diaphysis The main or midsection (shaft) of a long bone.

Diarrhoea A profuse, frequent and loose discharge from the bowels.

Diastolic Referring to the time when the heart is in a period of relaxation and dilatation (expansion). *cf.* systolic.

Dieresis Surgical separation of parts.

Dietary fibre Is a term that refers to a group of food components that pass through the stomach and small intestine undigested and reach the large intestine virtually unchanged. Scientific evidence suggest that a diet high in dietary fibre can be of value for treating or preventing such disorders as constipation, irritable bowel syndrome, diverticular disease, hiatus hernia and haemorrhoids. Some components of dietary fibre may also be of value in reducing the level of cholesterol in blood and thereby decreasing a risk factor for coronary heart disease and the development of gallstones. Dietary fibre is beneficial in the treatment of some diabetics.

Digalactosyl diglycerides Are the major lipid components of chloroplasts.

Diosgenin A steroid-like substance that is involved in the production of the hormone progesterone, extracted from roots of *Dioscorea* yam.

Dipsia Sensation of dryness in the mouth and throat related to a desire to drink.

Dipsomania Pathological use of alcohol.

Discutient An agent (as a medicinal application) which serves to disperse morbid matter.

Disinfectant An agent that prevents the spread of infection, bacteria or communicable disease.

Distal sensory polyneuropathy (DSPN) Or peripheral neuropathy, is the most common neurological problem in HIV disease. DSPN also represents a complex symptom that occurs because of peripheral nerve damage related to advanced HIV disease.

Diuresis Increased urination.

Diuretic A substance that increases urination (diuresis).

Diverticular disease Is a condition affecting the large bowel or colon and is thought to be caused by eating too little fibre.

Diverticulitis Common, sometimes painful digestive disease which involves the formation of pouches (diverticula) within the bowel wall.

DMBA 7,12-Dimethylbenzanthracene. A polycyclic aromatic hydrocarbon found in tobacco smoke that is a potent carcinogen.

DNA Deoxyribonucleic acid, a nucleic acid that contains the genetic instructions used in the development and functioning of all known living organisms.

DOCA Desoxycorticosterone acetate—a steroid chemical used as replacement therapy in Addison's disease.

Dopamine A catecholamine neurotransmitter that occurs in a wide variety of animals, including both vertebrates and invertebrates.

Dopaminergic Relating to, or activated by the neurotransmitter, dopamine.

Double blind Refers to a clinical trial or experiment in which neither the subject nor the researcher knows which treatment any particular subject is receiving.

Douche A localised spray of liquid directed into a body cavity or onto a part.

DPPH 2,2-Diphenyl-1-picryl-hydrazyl—a crystalline, stable free radical used as an inhibitor of free radical reactions.

Dracunculiasis Also called guinea worm disease (GWD), is a parasitic infection caused by the nematode, *Dracunculus medinensis*.

Dropsy An old term for the swelling of soft tissues due to the accumulation of excess water. *adj.* dropsical.

Drug metabolising enzymes Play central roles in the biotransformation, metabolism and/or detoxification of xenobiotics or foreign compounds that are introduced to the human body.

Drusen Tiny yellow or white deposits of extracellular materials in the retina of the eye or on the optic nerve head.

Dry eye syndrome Also called keratoconjunctivitis sicca, occurs when there are not enough tears on the front of the eyes.

DT diaphorase Also called DTD or NAD(P)H Quinone oxidoreductase, is an obligate two-electron reductase which bioactivates chemotherapeutic quinones.

Dysarthria Is a motor speech disorder.

Dysbiosis Also called dysbacteriosis, refers to a condition with microbial imbalances on or inside the body.

Dysentery (Formerly known as flux or the bloody flux) is a disorder of the digestive system that results in severe diarrhoea containing

mucus and blood in the faeces. It is caused usually by a bacterium called *Shigella*.

Dysesthesia An unpleasant abnormal sensation produced by normal stimuli.

Dysgeusia Distortion of the sense of taste.

Dyshomeostasis An imbalance or other breakdown of a homeostasis system.

Dyskinesia The impairment of the power of voluntary movement, resulting in fragmentary or incomplete movements. *adj.* dyskinetic.

Dyslipidemia Abnormality in or abnormal amount of lipids and lipoproteins in the blood.

Dysmenorrhoea Is a menstrual condition characterised by severe and frequent menstrual cramps and pain associated with menstruation.

Dysmotility syndrome A vague, descriptive term used to describe diseases of the muscles of the gastrointestinal tract (oesophagus, stomach, small and large intestines).

Dysosmia Qualitative alteration or distortion of the perception of smell.

Dyspareunia Painful sexual intercourse.

Dyspedia Indigestion followed by nausea.

Dyspepsia Refers to a symptom complex of epigastric pain or discomfort. It is often defined as chronic or recurrent discomfort centred in the upper abdomen and can be caused by a variety of conditions. *cf.* functional dyspepsia.

Dysphagia Difficulty in swallowing.

Dysphonia A voice disorder, an impairment in the ability to produce voice sounds using the vocal organs.

Dysplasia Refers to abnormality in development.

Dyspnoea Shortness of breath, difficulty in breathing.

Dysrhythmias See Arrhythmias.

Dystocia Abnormal or difficult childbirth or labour.

Dystonia A neurological movement disorder characterised by prolonged, repetitive muscle contractions that may cause twisting or jerking movements of muscles.

Dysuria Refers to difficult and painful urination.

E-Cadherin Has traditionally been categorised as a tumour suppressor.

E-Selectin Also known as endothelial leucocyte adhesion molecule-1 (ELAM-1), CD62E, a member of the selectin family. It is transiently expressed on vascular endothelial cells in response to IL-1 beta and TNF-alpha.

EC 50 Median effective concentration that produces desired effects in 50 % of the test population.

Ecbolic A drug (as an ergot alkaloid) that tends to increase uterine contractions and that is used especially to facilitate delivery.

Ecchymosis Skin discolouration caused by the escape of blood into the tissues from ruptured blood vessels.

ECG See electrocardiography.

EC–SOD Extracellular superoxide dismutase, a tissue enzyme mainly found in the extracellular matrix of tissues. It participates in the detoxification of reactive oxygen species by catalysing the dismutation of superoxide radicals.

Ectopic heartbeats Small changes in an otherwise normal heartbeat that lead to extra or skipped heartbeats.

Ectrodactyly Involves the absence of one or more central digits of the hand or foot.

Eczema Is broadly applied to a range of persistent skin conditions. These include dryness and recurring skin rashes which are characterised by one or more of these symptoms: redness, skin oedema, itching and dryness, crusting, flaking, blistering, cracking, oozing or bleeding.

Eczematous rash Dry, scaly, itchy rash.

ED 50 Is defined as the dose producing a response that is 50 % of the maximum obtainable.

Edema Formerly known as dropsy or hydropsy, is characterised as swelling caused by abnormal accumulation of fluid beneath the skin, or in one or more cavities of the body. It usually occurs in the feet, ankles and legs, but it can involve the entire body.

Edematogenic Producing or causing edema.

EGFR proteins Epidermal growth factor receptor (EGFR) proteins. Protein kinases are enzymes that transfer a phosphate group from a phosphate donor onto an acceptor amino acid in a substrate protein.

EGR-1 Early growth response 1, a human gene.

Eicosanoids Are signalling molecules made by oxygenation of arachidonic acid, a twenty-carbon essential fatty acid, and include prostaglandins and related compounds.

Elastase A serine protease that also hydrolyses amides and esters.

Electrocardiography Or ECG, is a transthoracic interpretation of the electrical activity of the heart over time captured and externally recorded by skin electrodes.

Electromyogram (EMG) A test used to record the electrical activity of muscles. An electromyogram (EMG) is also called a myogram.

Electuary A medicinal paste composed of powders, or other medical ingredients, incorporated with sweeteners to hide the taste, suitable for oral administration.

Elephantiasis A disorder characterised by chronic thickened and oedematous tissue on the genitals and legs due to various causes.

11β-Hydroxysteroid dehydrogenase (HSD-11β or 11β-HSD) is the name of a family of enzymes that catalyse the conversion of inert 11 keto-products (cortisone) to active cortisol, or vice versa.

Embrocation Lotion or liniment that relieves muscle or joint pains.

Embryonation Formation of embryo in the egg.

Embryotoxic Term that describes any chemical which is harmful to an embryo.

Emesis Vomiting, throwing up.

Emetic An agent that induces vomiting. *cf.* antiemetic

Emetocathartic Causing vomiting and purging.

Emmenagogue A substance that stimulates, initiates and/or promotes menstrual flow. Emmenagogues are used in herbal medicine to balance and restore the normal function of the female reproductive system.

Emollient An agent that has a protective and soothing action on the surfaces of the skin and membranes.

Emphysema A long-term, progressive disease of the lungs that primarily causes shortness of breath.

Emulsion A preparation formed by the suspension of very finely divided oily or resinous liquid in another liquid.

Encephalitis Inflammation of the brain caused by a virus.

Encephalocele A neural tube defect characterised by saclike protrusions of the brain tissue through a congenital fissure in the skull.

Encephalomalacia Cerebral softening, a localised softening of the brain substance, due to haemorrhage or inflammation.

Encephalopathy A disorder or disease of the brain.

Endocrine *adj.* of or relating to endocrine glands or the hormones secreted by them.

Endocytosis Is the process by which cells absorb material (molecules such as proteins) from outside the cell by engulfing it with their cell membrane.

Endometrial cancer Cancer that arises in the endometrium, the lining of the uterus (womb).

Endometriosis Is a common and often painful disorder of the female reproductive system in which the endometrium, the tissue that normally lines the womb (uterus), grows outside the uterus. The two most common symptoms of endometriosis are pain and infertility.

Endometritis Refers to inflammation of the endometrium, the inner lining of the uterus.

Endometrium The inner lining of the uterus.

Endoplasmic reticulum Is a network of tubules, vesicles and sacs around the nucleus that are interconnected.

Endostatin A naturally occurring 20-kDa C-terminal protein fragment derived from type XVIII collagen. It is reported to serve as an antiangiogenic agent that inhibits the formation of the blood vessels that feed cancer tumours.

Endosteum The thin layer of cells lining the medullary cavity of a bone.

Endosteal Pertaining to the endosteum.

Endothelial progenitor cells Population of rare cells that circulate in the blood with the ability to differentiate into endothelial cells, the cells that make up the lining of blood vessels.

Endothelin Any of a group of vasoconstrictive peptides produced by endothelial cells that constrict blood vessels and raise blood pressure.

Endotoxaemia The presence of endotoxins in the blood, which may result in shock. *adj.* endotoxemic.

Endotoxin Toxins associated with certain bacteria, unlike an 'exotoxin' that is not secreted in soluble form by live bacteria, but is a structural component in the bacteria which is released mainly when bacteria are lysed.

Enema Liquid injected into the rectum either as a purgative or medicine. Also called clyster.

Enophthalmos A condition in which the eye falls back into the socket and inhibits proper eyelid function.

eNOS (Endothelial nitric oxide synthase) the enzyme responsible for most of the vascular nitric oxide produced.

Enteral Term used to describe the intestines or other parts of the digestive tract.

Enteralgia Pain in the intestines; intestinal colic.

Enteral administration Involves the oesophagus, stomach and small and large intestines (i.e. the gastrointestinal tract).

Enteritis Refers to inflammation of the small intestine.

Enterocolic disorder Inflamed bowel disease.

Enterocytes Tall columnar cells in the small intestinal mucosa that are responsible for the final digestion and absorption of nutrients.

Enterohaemorrhagic Causing bloody diarrhoea and colitis, said of pathogenic microorganisms.

Enterohepatonephropathy Hepatorenal lesions accompanied by renal failure.

Enterolactone A lignin formed by the action of intestinal bacteria on lignan precursors found in plants; acts as a phytoestrogen.

Enteropooling Increased fluids and electrolytes within the lumen of the intestines due to increased levels of prostaglandins.

Enterotoxigenic Of or being an organism containing or producing an enterotoxin.

Enterotoxin Is a protein toxin released by a microorganism in the intestine.

Entheogen A substance taken to induce a spiritual experience.

Enuresis Bed-wetting, a disorder of elimination that involves the voluntary or involuntary release of urine into bedding, clothing or other inappropriate places.

Envenomation Is the entry of venom into a person's body, and it may cause localised or systemic poisoning.

Eosinophilia The state of having a high concentration of eosinophils (eosinophil granulocytes) in the blood.

Eosinophils (Or, less commonly, acidophils) are white blood cells that are one of the immune system components.

Epidermal growth factor receptor (EGFR) Belongs to the ErbB family of receptor tyrosine kinases (RTK). EGFR are involved in the pathogenesis and progression of different carcinoma types.

Epididymis A structure within the scrotum attached to the backside of the testis and whose coiled duct provides storage, transit and maturation of spermatozoa.

Epididymitis A medical condition in which there is inflammation of the epididymis.

Epidural haematoma Accumulation of blood in the potential space between dura and bone and may be intracranial or spinal .

Epigastralgia Pain in the epigastric region.

Epigastric discomfort Bloated abdomen, swelling of abdomen, abdominal ditension.

Epilepsy A common chronic neurological disorder that is characterised by recurrent unprovoked seizures.

Epileptiform Resembling epilepsy or its manifestations. *adj.* epileptiformic.

Epileptogenesis A process by which a normal brain develops epilepsy, a chronic condition in which seizures occur. *adj.* epileptogenic.

Episiotomy A surgical incision through the perineum made to enlarge the vagina and assist childbirth.

Epithelial–mesenchymal transition or transformation (EMT) A process by which epithelial cells lose their cell polarity and cell–cell adhesion and gain migratory and invasive properties to become mesenchymal cells.

Epithelioma A usually benign skin disease most commonly occurring on the face, around the eyelids and on the scalp.

Epitope A single antigenic site on a protein against which an antibody reacts.

Epitrochlearis The superficial-most muscle of the arm anterior surface.

Epistaxis Acute haemorrhage from the nostril, nasal cavity or nasopharynx (nosebleed).

Epstein–Barr virus Herpesvirus that is the causative agent of infectious mononucleosis. It is also associated with various types of human cancers.

ERbeta Oestrogen receptor beta, a nuclear receptor which is activated by the sex hormone, oestrogen.

Ergocalciferol A form of vitamin D, also called vitamin D2. See Vitamin D.

Ergogenic Increasing capacity for bodily or mental labour, especially by eliminating fatigue symptoms.

Ergonic Increasing capacity for bodily or mental labour, especially by eliminating fatigue symptoms.

ERK (extracellular signal regulated kinases) Widely expressed protein kinase intracellular signalling molecules which are involved in functions including the regulation of meiosis, mitosis and postmitotic functions in differentiated cells.

Eructation The act of belching or of casting up wind from the stomach through the mouth.

Eruption A visible rash or cutaneous disruption.

Eryptosis Suicidal death of erythrocytes, characterised by cell shrinkage, membrane blebbing, activation of proteases and phosphatidylserine exposure at the outer membrane leaflet.

Erysipelas Is an intensely red *Streptococcus* bacterial infection that occurs on the face and lower extremities.

Erythema Abnormal redness and inflammation of the skin, due to vasodilation.

Erythema multiforme Is a skin disorder due to an allergic reaction or infection; characterised by fever, general ill feeling, skin itching, joint aches and multiple skin lesions.

Erythematous Characterised by erythema.

Erythroderma Exfoliative dermatitis.

Erythroleukoplakia An abnormal patch of red and white tissue that forms on mucous membranes in the mouth and may become cancer. Tobacco (smoking and chewing) and alcohol may increase the risk of erythroleukoplakia.

Erythropoiesis Is the process whereby erythroid precursor cells proliferate and differentiate into red blood cells.

Erythropoietin (EPO) A hormone produced by the kidney that promotes the formation of red blood cells (erythrocytes) in the bone marrow.

Eschar A slough or piece of dead tissue that is cast off from the surface of the skin.

Escharotic Capable of producing an eschar; a caustic or corrosive agent.

Estradiol Is the predominant sex hormone present in females, also called oestradiol.

Estrogenic Relating to oestrogen or producing oestrus.

Euglycaemia Normal blood glucose concentration.

Eupeptic Conducive to digestion.

Exanthema Sudden widespread rash.

Exanthematous Characterised by or of the nature of an eruption or rash.

Excitotoxicity Is the pathological process by which neurons are damaged and killed by glutamate and similar substances.

Excipient A pharmacologically inert substance used as a diluent or vehicle for the active ingredients of a medication.

Exencephaly A type of cephalic disorder wherein the brain is located outside of the skull.

Exfoliative cheilitis Is a reactive process, in which upper, lower or both lips become chronically inflamed, crusted and sometimes fissured.

Exocytosis The cellular process by which cells excrete waste products or chemical transmitters.

Exophthalmos or exophthalmia or proptosis Is a bulging of the eye anteriorly out of the orbit. *adj.* exophthalmic.

Exotoxin A toxin secreted by a microorganism and released into the medium in which it grows.

Expectorant An agent that increases bronchial mucous secretion by promoting liquefaction of the sticky mucous and expelling it from the body.

Experimental allergic encephalomyelitis (EAE) Is an animal model of brain inflammation.

Exteroceptive Responsiveness to stimuli that are external to an organism.

Extrapyramidal side effects Are a group of symptoms (tremor, slurred speech, akathisia, dystonia, anxiety, paranoia and bradyphrenia) that can occur in persons taking antipsychotic medications.

Extravasation Discharge or escape, as of blood from the vein into the surrounding tissues; discharge or escape from a vessel or channel.

Eyelid oedema Swollen eyelid caused by inflammation or excess fluid.

Fabry disease Is a rare X-linked (inherited) lysosomal storage disease caused by alpha-

galactosidase A deficiency, which can cause a wide range of systemic symptoms such as pain in the extremities, papules on the lower body parts, cornea clouding, fatigue, neuropathy, renal and cardiac complications.

FAC chemotherapy Fluorouracil, doxorubicin (adriamycin) and cyclophosphamide chemotherapy.

FADD Fas-associated protein with death domain; the protein encoded by this gene is an adaptor molecule which interacts with other death cell surface receptors and mediates apoptotic signals.

Familial amyloid polyneuropathy (FAP) Also called Corino de Andrade's disease, a neurodegenerative autosomal dominant genetically transmitted, fatal, incurable disease.

Familial adenomatous polyposis (FAP) Is an inherited condition in which numerous polyps form mainly in the epithelium of the large intestine.

Familial dysautonomia A genetic disorder that affects the development and survival of autonomic and sensory nerve cells.

Fanconi syndrome Is a disease of the proximal renal tubes in which certain substances normally absorbed into the bloodstream by the kidneys are released into the urine instead.

FasL or CD95L Fas ligand is a type II transmembrane protein that belongs to the tumour necrosis factor (TNF) family.

FAS, fatty acid synthase (FAS) A multienzyme that plays a key role in fatty acid synthesis.

Fas molecule A member of the tumour necrosis factor receptors, which mediates apoptotic signal in many cell types.

Fauces The passage leading from the back of the mouth into the pharynx.

Favus A chronic skin infection, usually of the scalp, caused by the fungus *Trichophyton schoenleinii* and characterised by the development of thick, yellow crusts over the hair follicles. Also termed tinea favosa.

Febrifuge An agent that reduces fever. Also called an antipyretic.

Febrile Pertaining to or characterised by fever.

Febrile neutropenia The development of fever, often with other signs of infection, in an individual with neutropenia, an abnormally low number of neutrophil granulocytes in the blood.

Felon A purulent infection in the bulbous distal end of a finger.

Fetotoxic Toxic to the fetus.

Fibrates Hypolipidemic agents primarily used for decreasing serum triglycerides, while increasing high-density lipoprotein (HDL).

Fibril A small slender fibre or filament.

Fibrin Insoluble protein that forms the essential portion of the blood clot.

Fibrinolysis A normal ongoing process that dissolves fibrin and results in the removal of small blood clots.

Fibrinolytic Causing the dissolution of fibrin by enzymatic action.

Fibroblast Type of cell that synthesises the extracellular matrix and collagen, the structural framework (stroma) for animal tissues, and plays a critical role in wound healing.

Fibrogenic Promoting the development of fibres.

Fibromyalgia A common and complex chronic, body-wide pain disorder that affects people physically, mentally and socially. Symptoms include debilitating fatigue, sleep disturbance and joint stiffness. Also referred to as FM or FMS.

Fibronectin A high molecular weight (~440kDa) glycoprotein of the extracellular matrix (ECM) that adheres to membrane-spanning receptor proteins called integrins.

Fibrosarcoma A malignant tumour derived from fibrous connective tissue and characterised by immature proliferating fibroblasts or undifferentiated anaplastic spindle cells.

Fibrosis The formation of fibrous tissue as a reparative or reactive process.

Filarial Pertaining to a thread-like nematode worm.

Filariasis A parasitic and infectious tropical disease that is caused by thread-like filarial nematode worms in the superfamily Filarioidea.

Fistula An abnormal connection between two organs inside of the body.

Fistula-in-ano A track connecting the internal anal canal to the skin surrounding the anal orifice.

5′-Nucleotidase (5′-Ribonucleotide phosphohydrolase) an intrinsic membrane glycoprotein present as an ectoenzyme in a wide variety of mammalian cells, hydrolyses 5′-nucleotides to their corresponding nucleosides.

5-HT1A receptor A serotonin protein that binds to 5-5-hydroxytryptamine (5-HT) with high affinity to exert subtle control over emotion and behaviour.

Flash electroretinogram or flash ERG (fERG) Is a test which measures the electrical response of the eye's light-sensitive cells (rods and cones). It also checks other cell layers in the retina.

Flatulence Is the presence of a mixture of gases known as flatus in the digestive tract of mammals expelled from the rectum. Excessive flatulence can be caused by lactose intolerance, certain foods or a sudden switch to a high fibre.

Flavans A subgroup of flavonoids. See Flavonoids.

Flavanols A subgroup of flavonoids, are a class of flavonoids that use the 2-phenyl-3,4-dihydro-2H-chromen-3-ol skeleton. These compounds include the catechins and the catechin gallates. They are found in chocolate, fruits and vegetables. See Flavonoids.

Flavanones A subgroup of flavonoids, constitute >90% of total flavonoids in citrus. The major dietary flavanones are hesperetin, naringenin and eriodictyol.

Flavivirus A family of viruses transmitted by mosquitoes and ticks that cause some important diseases, including dengue, yellow fever, tick-borne encephalitis and West Nile fever.

Flavones A subgroup of flavonoids based on the backbone of 2-phenylchromen-4-one (2-phenyl-1-benzopyran-4-one). Flavones are mainly found in cereals and herbs.

Flavonoids (Or bioflavonoids) are a group of polyphenolic antioxidant compounds that occur in plant as secondary metabolites. They are responsible for the colour of fruit and vegetables. Twelve basic classes (chemical types) of flavonoids have been recognised: flavones, isoflavones, flavans, flavanones, flavanols, flavanolols, anthocyanidins, catechins (including proanthocyanidins), leucoanthocyanidins, chalcones, dihydrochalcones and aurones. Apart from their antioxidant activity, flavonoids are known for their ability to strengthen capillary walls, thus assisting circulation and helping to prevent and treat bruising, varicose veins, bleeding gums and nosebleeds and heavy menstrual bleeding, and are also anti-inflammatory.

Fluorine (F) is an essential chemical element that is required for the maintenance of healthy bones and teeth and to reduce tooth decay. It is found in seaweeds, tea, water, seafood and dairy products.

Fluorosis A dental health condition caused by a child receiving too much fluoride during tooth development.

Flux An excessive discharge of fluid.

FMD (flow-mediated dilation) A measure of endothelial dysfunction which is used to evaluate cardiovascular risk. Also called FMVD (flow-mediated vasodilation).

Focal adhesion kinase (FAK) Is a protein tyrosine kinase which is recruited at an early stage to focal adhesions and which mediates many of the downstream regulatory responses.

Follicle-stimulating hormone (FSH) A hormone produced by the pituitary gland. In women, it helps control the menstrual cycle and the production of eggs by the ovaries.

Follicular atresia The breakdown of the ovarian follicles.

Fomentation Treatment by the application of war, moist substance.

Fontanelle Soft spot on an infant's skull.

Forkhead box-O transcription factors (FOXOs) Are a family of transcription factors that play important roles in regulating the expression of genes involved in cell growth, proliferation, differentiation and longevity. It also plays an important role in tumour suppression by regulating the expression of genes involved in stress resistance, DNA damage repair, cell cycle arrest and apoptosis.

Framboesia See Yaws.

FRAP Ferric reducing ability of plasma, an assay used to assess antioxidant property.

Fibrillation Is the rapid, irregular and unsynchronised contraction of muscle fibres, especially with regard to the heart.

5-Dihydroaldosterone A hormone secreted by the adrenal cortex that regulates electrolyte and water balance by increasing the renal retention.

Friedreich's ataxia Is a genetic inherited disorder that causes progressive damage to the nervous system resulting in symptoms ranging

from muscle weakness and speech problems to heart disease. *cf.* ataxia.

Fulminant hepatitis Acute liver failure.

Functional dyspepsia A non-ulcer condition that causes an upset stomach or pain or discomfort in the upper belly, near the ribs.

Functional food Is any fresh or processed food claimed to have a health-promoting or disease-preventing property beyond the basic function of supplying nutrients. Also called medicinal food.

Furuncle Is a skin disease caused by the infection of hair follicles usually caused by *Staphylococcus aureus*, resulting in the localised accumulation of pus and dead tissue.

Furunculosis Skin condition characterised by persistent, recurring boils.

G protein-coupled receptor kinases (GRKs, GPCRKs) A family of protein kinases which regulate the activity of G protein-coupled receptors (GPCRs) by phosphorylating their intracellular domains after their associated G proteins have been released and activated.

GABA Gamma aminobutyric acid, required as an inhibitory neurotransmitter to block the transmission of an impulse from one cell to another in the central nervous system, which prevents over-firing of the nerve cells. It is used to treat both epilepsy and hypertension.

GADD 152 A proapoptotic gene.

Galctifuge Or lactifuge, causing the arrest of milk secretion.

Galactagogue A substance that promotes the flow of milk.

Galactophoritis Inflammation of the milk ducts.

Galactopoietic Increasing the flow of milk; milk producing.

Gallbladder A small, pear-shaped muscular sac, located under the right lobe of the liver, in which bile secreted by the liver is stored until needed by the body for digestion. Also called cholecyst, cholecystitis.

Gallic acid equivalent (GAE) Measures the total phenol content in terms of the standard gallic acid by the Folin–Ciocalteu assay.

Galphai proteins or G alpha I proteins Are heterotrimeric guanine nucleotide-regulatory (G) proteins associated with a variety of intracellular membranes and specific plasma membrane domains.

Gamma GT (GGT) Gamma-glutamyl transpeptidase, a liver enzyme.

Gap junction intercellular communication Is considered to be the sole means by which low molecular weight factors inside a cell can pass directly into the interior of neighbouring cells. Gap junctions are considered to play an essential role in the maintenance of homeostasis.

Gastralgia (Heartburn) pain in the stomach or abdominal region. It is caused by excess of acid, or an accumulation of gas, in the stomach.

Gastric Pertaining to or affecting the stomach.

Gastric emptying Refers to the speed at which food and drink leave the stomach.

Gastritis Inflammation of the stomach.

Gastrocnemius muscle The big calf muscle at the rear of the lower leg.

Gastrodynia Pain in the stomach.

Gastroparesis Also called **delayed gastric emptying**, a medical condition consisting of a paresis (partial paralysis) of the stomach, resulting in food remaining in the stomach for an abnormally long time.

Gastroprokinetic See Prokinetic.

Gastrotonic (gastroprotective) Substance that strengthens, tones or regulates gastric functions (or protects from injury) without overt stimulation or depression.

Gavage Forced feeding.

Gene silencing Suppression of the expression of a gene.

Genotoxic Describes a poisonous substance which harms an organism by damaging its DNA, thereby capable of causing mutations or cancer.

Genotoxin A chemical or another agent that damages cellular DNA, resulting in mutations or cancer.

Geriatrics Is a subspecialty of internal medicine that focuses on healthcare of elderly people.

Gestational hypertension Development of arterial hypertension in a pregnant woman after 20 weeks of gestation.

Ghrelin A gastrointestinal peptide hormone secreted by epithelial cells in the stomach lining; it stimulates appetite and gastric emptying and increases cardiac output.

Gingival index An index describing the clinical severity of gingival inflammation as well as its location.

Gingivitis Refers to gingival inflammation induced by bacterial biofilms (also called plaque) adherent to tooth surfaces.

Gin-nan sitotoxism Toxicity caused by ingestion of ginkgotoxin and characterised mainly by epileptic convulsions, paralysis of the legs and loss of consciousness.

GIP Gastric inhibitory polypeptide also known as the glucose-dependent insulinotropic peptide, a member of the secretin family of hormones.

Glaucoma A group of eye diseases in which the optic nerve at the back of the eye is slowly destroyed, leading to impaired vision and blindness.

Gleet A chronic inflammation (as gonorrhoea) of a bodily orifice usually accompanied by an abnormal discharge.

Glial cells Support non-neuronal cells in the central nervous system that maintain homeostasis, form myelin and provide protection for the brain's neurons.

Glioma Is a type of tumour that starts in the brain or spine. It is called a glioma because it arises from glial cells.

Glioblastoma Common and most lethal form of brain tumour.

Glioblastoma multiforme Most common and most aggressive type of primary brain tumour in humans, involving glial cells.

Glomerulonephritis (GN) A renal disease characterised by inflammation of the glomeruli, or small blood vessels in the kidneys. Also known as glomerular nephritis. *adj.* glomerulonephritic.

Glomerulopathy Any disease of the renal glomeruli.

Glomerulosclerosis A hardening (fibrosis) of the glomerulus in the kidney.

Glossal Pertaining to the tongue.

GLP-1 Glucagon-like peptide-1.

Glucagon-like peptide-1 (GLP-1) is derived from the transcription product of the proglucagon gene, reduces insulin requirement in diabetes mellitus and promotes satiety.

Gluconeogenesis A metabolic pathway that results in the generation of glucose from non-carbohydrate carbon substrates such as lactate. *adj.* gluconeogenic.

Glucose-6-phosphate dehydrogenase (G6PD or G6PDH) Is a cytosolic enzyme in the pentose phosphate metabolic pathway.

Glucose transporter type 4 (GLUT 4) Insulin-regulated glucose transporter found in adipose tissues and striated muscles that modulates insulin-related translocation into the cell.

Glucose transporters (GLUT or SLC2A family) are a family of membrane proteins found in most mammalian cells.

Glucosuria or glycosuria Is the excretion of glucose into the urine.

Glucosyltranferase An enzyme that enables the transfer of glucose.

Glucuronidation A phase II detoxification pathway occurring in the liver in which glucuronic acid is conjugated with toxins.

Glutamic oxaloacetic transaminase An enzyme that catalyzes the reversible transfer of an amino group from aspartate to α-ketoglutarate to form glutamate and oxaloacetate.

Glutamic pyruvate transaminase (GPT) See Alanine aminotransferase.

Glutathione (GSH) A tripeptide produced in the human liver and plays a key role in intermediary metabolism, immune response and health. It plays an important role in scavenging free radicals and protects cells against several toxic oxygen-derived chemical species.

Glutathione peroxidase (GPX) The general name of an enzyme family with peroxidase activity whose main biological role is to protect the organism from oxidative damage.

Glutathione S-transferase (GST) A major group of detoxification enzymes that participate in the detoxification of reactive electrophilic compounds by catalysing their conjugation to glutathione.

Glycaemic index (GI) Measures carbohydrates according to how quickly they are absorbed and raise the glucose level of the blood.

Glycaemic load (GL) Is a ranking system for carbohydrate content in food portions based on their glycaemic index and the amount of available carbohydrate, i.e. GI x available carbohydrate divided by 100. Glycaemic load combines both the quality and quantity of carbohydrate in one 'number'. It's the best way to

predict blood glucose values of different types and amounts of food.

Glycation or glycosylation A chemical reaction in which glycosyl groups are added to a protein to produce a glycoprotein.

Glycogenolysis Is the catabolism of glycogen by removal of a glucose monomer through cleavage with inorganic phosphate to produce glucose-1-phosphate.

Glycometabolism Metabolism (oxidation) of glucose to produce energy.

Glycosuria Or glucosuria is an abnormal condition of osmotic diuresis due to excretion of glucose by the kidneys into the urine.

Glycosylases A family of enzymes involved in base excision repair.

Goitre An enlargement of the thyroid gland leading to swelling of the neck or larynx.

Goitrogen Substance that suppresses the function of the thyroid gland by interfering with iodine uptake, causing enlargement of the thyroid, i.e. goitre.

Goitrogenic *adj.* causing goitre.

Gonadotroph A basophilic cell of the anterior pituitary specialised to secrete follicle-stimulating hormone or luteinising hormone.

Gonadotropins Protein hormones secreted by gonadotrope cells of the pituitary gland of vertebrates.

Gonorrhoea A common sexually transmitted bacterial infection caused by the bacterium *Neisseria gonorrhoeae*.

Gout A disorder caused by a build-up of a waste product, uric acid, in the bloodstream. Excess uric acid settles in joints causing inflammation, pain and swelling.

G protein-coupled receptors (GPCRs) Constitute the largest family of cell surface molecules involved in signal transmission. These receptors play key physiological roles and their dysfunction results in several diseases.

Granulation The condition or appearance of being granulated (becoming grain-like).

Gravel Sand-like concretions of uric acid, calcium oxalate and mineral salts formed in the passages of the biliary and urinary tracts.

Gripe water Is a home remedy for babies with colic, gas, teething pain or other stomach ailments. Its ingredients vary and may include alcohol, bicarbonate, ginger, dill, fennel and chamomile.

Grippe An epidemic catarrh; older term for influenza.

GSH See Glutathione.

GSH-Px Glutathione peroxidase, general name of an enzyme family with peroxidase activity whose main biological role is to protect the organism from oxidative damage.

GSSG Glutathione disulfides are biologically important intracellular thiols, and alterations in the GSH/GSSG ratio are often used to assess exposure of cells to oxidative stress.

GSTM Glutathione S transferase M1, a major group of detoxification enzymes.

GSTM 2 Glutathione S transferase M2, a major group of detoxification enzymes.

G2-M cell cycle The phase where the cell prepares for mitosis and where chromatids and daughter cells separate.

Guillain–Barre syndrome Is a serious disorder that occurs when the body's defence (immune) system mistakenly attacks part of the nervous system, leading to nerve inflammation, muscle weakness and other symptoms.

Gynecomastia Enlargement of the gland tissue of the male breast, resulting from an imbalance of hormones.

Gynecopathy Any or various diseases specific to women.

Gynoid adiposity Fat distribution mainly to the hips and thighs, pear shaped.

Haemagogic Promoting a flow of blood.

Haematemesis, hematemesis Is the vomiting of blood.

Haematinic Improving the quality of the blood, its haemoglobin level and the number of erythrocytes.

Haematochezia Passage of stools containing blood.

Haematochyluria, hematochyluria the discharge of blood and chyle (emulsified fat) in the urine; see also Chyluria.

Haematoma, hematoma A localised accumulation of blood in a tissue or space composed of clotted blood.

Haematometra, hematometra A medical condition involving bleeding of or near the uterus.

Haematopoiesis, hematopoiesis Formation of blood cellular components from the haematopoietic stem cells.

Haematopoietic *adj.* relating to the formation and development of blood cells.

Haematuria, Hematuria Is the presence of blood in the urine. Haematuria is a sign that something is causing abnormal bleeding in a person's genitourinary tract.

Haeme oxygenase (HO-1, encoded by Hmox1) is an inducible protein activated in systemic inflammatory conditions by oxidant stress, an enzyme that catalyses degradation of haem.

Haemochromatosis iron overload in the body with a hereditary or primary cause.

Haemodialysis, hemodialysis A method for removing waste products such as potassium and urea, as well as free water from the blood when the kidneys are in renal failure.

Haemolyis Lysis of red blood cells and the release of haemoglobin into the surrounding fluid (plasma). *adj.* haemolytic.

Haemoptysis, hemoptysis Is the coughing up of blood from the respiratory tract. The blood can come from the nose, mouth, throat and the airway passages leading to the lungs.

Haemorrhage, hemaorrhage bleeding, discharge of blood from blood vessels.

Haemorrhoids, hemorrhoids A painful condition in which the veins around the anus or lower rectum are enlarged, swollen and inflamed. Also called piles.

Haemostasis, hemostasis A complex process which causes the bleeding process to stop.

Haemostatic, hemostatic Something that stops bleeding.

Halitosis (Bad breath) a common condition caused by sulphur-producing bacteria that live within the surface of the tongue and in the throat.

Hallucinogen Drug that produces hallucinogen.

Hallucinogenic Inducing hallucinations.

Hallux abducto valgus Commonly called bunion is an abnormal bending of the big toe towards the other toes of the foot.

Haplotype A set of alleles of closely linked loci on a chromosome that tend to be inherited together.

Hapten A small molecule that can elicit an immune response only when attached to a large carrier such as a protein.

HATs Histone acetyl transferases, enzymes that regulate the acetylation of histones and transcription factors, playing a major role in the growth and differentiation of cells.

HbA1c Glycosylated haemoglobin.

HBeAg Hepatitis B e antigen.

HBsAg Hepatitis B s antigen.

HBD-2 (human β-defensin 2) A member of the defensin family of antimicrobial peptides that plays important roles in the innate and adaptive immune system of both vertebrates and invertebrates.

Heartburn Burning sensation in the stomach and oesophagus caused by excessive acidity of the stomach fluids.

Heat rash Any condition aggravated by heat or hot weather such as intertrigo.

Heat shock chaperones (HSC) Ubiquitous molecules involved in the modulation of protein conformational and complexation states, associated with heat stress or other cellular stress responses.

Heat shock proteins (HSP) A group of functionally related proteins, the expression of which is increased when the cells are exposed to elevated temperatures or other cellular stresses.

Helminthiasis A disease in which a part of the body is infested with worms such as pinworm, roundworm or tapeworm.

Haemagglutination A specific form of agglutination that involves red blood cells.

Haemagglutination–inhibition test Measures the ability of soluble antigen to inhibit the agglutination of antigen-coated red blood cells by antibodies.

Haemagglutinin Refers to a substance that causes red blood cells to agglutinate.

Haemangioma Blood vessel.

Haematocrit Is a blood test that measures the percentage of the volume of the whole blood that is made up of red blood cells.

Haematopoietic Pertaining to the formation of blood or blood cells.

Haematopoietic stem cell Is a cell isolated from the blood or bone marrow that can renew itself and can differentiate to a variety of specialised cells.

Haem oxygenase-1 (HO-1) An enzyme that catalyses the degradation of haem; an inducible stress protein, confers cytoprotection against oxidative stress in-vitro and in-vivo.

Hemiplegia Paralysis of the arm, leg and trunk on the same side of the body.

Haemoglobinopathies Genetic defects that produce abnormal haemoglobins and anaemia.

Haemolytic anaemia Anaemia due to haemolysis, the breakdown of red blood cells in the blood vessels or elsewhere in the body.

Hemorheology Study of blood flow and its elements in the circulatory system. *adj.* hemorheological.

Haemorrhagic colitis An acute gastroenteritis characterised by overtly bloody diarrhoea that is caused by *Escherichia coli* infection.

Haemolysin Certain proteins and lipids that cause lysis of red blood cells by damaging their cell membranes.

Haemolytic uremic syndrome Is a disease characterised by haemolytic anaemia, acute renal failure (uraemia) and a low platelet count.

Hepa-1c1c7 A type of hepatoma cells.

Hepatalgia Pain or discomfort in the liver area.

Hepatectomy The surgical removal of part or all of the liver.

Hepatic Relating to the liver.

Hepatic cirrhosis Affecting the liver, characterised by hepatic fibrosis and regenerative nodules.

Hepatic encephalopathy Is the loss of brain function that occurs when the liver is unable to remove toxins from the blood.

Hepatic fibrosis Is overly profuse wound healing in which excessive connective tissue builds up in the liver.

Hepatitis Inflammation of the liver.

Hepatitis A (Formerly known as infectious hepatitis) is an acute infectious disease of the liver caused by the hepatovirus hepatitis A virus.

Hepatocarcinogenesis Represents a linear and progressive cancerous process in the liver in which successively more aberrant monoclonal populations of hepatocytes evolve.

Hepatocellular carcinoma (HCC) Also called malignant hepatoma, is a primary malignancy (cancer) of the liver.

Hepatocytolysis Cytotoxicity (dissolution) of liver cells.

Hepatoma Cancer of the liver.

Hepatomegaly Condition of enlarged liver.

Hepatopathy A disease or disorder of the liver.

Hepatoprotective (Liver protector) a substance that helps protect the liver from damage by toxins, chemicals or other disease processes.

Hepatoregenerative A compound that promotes hepatocellular regeneration, repairs and restores liver function to optimum performance.

Hepatotonic (Liver tonic) a substance that is tonic to the liver—usually employed to normalise liver enzymes and function.

Hernia Occurs when part of an internal organ bulges through a weak area of muscle.

HER-2 Human epidermal growth factor receptor 2, a protein giving higher aggressiveness in breast cancer, also known as ErbB-2, ERBB2.

Herpes A chronic inflammation of the skin or mucous membrane characterised by the development of vesicles on an inflammatory base.

Herpes circinatus Dermatitis herpetiformis (resembling herpes).

Herpes simplex virus 1 and 2 (HSV-1 and HSV-2) Are two species of the herpesvirus family which cause a variety of illnesses/infections in humans such as cold sores, chickenpox or varicella, shingles or herpes zoster (VZV), cytomegalovirus (CMV) and various cancers and can cause brain inflammation (encephalitis). HSV-1 is commonly associated with herpes outbreaks of the face known as cold sores or fever blisters, whereas HSV-2 is more often associated with genital herpes. They are also called human herpesvirus 1 and 2 (HHV-1 and HHV-2) and are neurotropic and neuroinvasive viruses; they enter and hide in the human nervous system, accounting for their durability in the human body.

Herpes zoster Or simply zoster, commonly known as shingles and also known as zona, is a viral disease characterised by a painful skin rash with blisters.

Herpes zoster ophthalmicus (HZO) Is a viral ocular disease characterised by a painful skin rash in one or more dermatome distributions of the fifth cranial nerve, shared by the eye and orbit.

Heterophobia Term used to describe irrational fear of, aversion to, or discrimination against heterosexuals.

HDL-C (HDL cholesterol) High-density lipoprotein cholesterol, also called 'good cholesterol'. See also High-density lipoprotein.

Hiatus hernia Occurs when the upper part of the stomach pushes its way through a tear in the diaphragm.

High-density lipoprotein (HDL) Is one of the five major groups of lipoproteins which enable cholesterol and triglycerides to be transported within the water-based blood stream. HDL can remove cholesterol from atheroma within arteries and transport it back to the liver for excretion or reutilisation—which is the main reason why HDL-bound cholesterol is sometimes called 'good cholesterol', or HDL-C. A high level of HDL-C seems to protect against cardiovascular diseases. *cf.* LDL.

HGPRT, HPRT (hypoxanthine-guanine phosphoribosyl transferase) An enzyme that catalyses the conversion of 5-phospho ribosyl-1-pyrophosphate and hypoxanthine, guanine or 6-mercaptopurine to the corresponding 5′-mononucleotides and pyrophosphate. The enzyme is important in purine biosynthesis as well as central nervous system functions.

Hippocampus A ridge in the floor of each lateral ventricle of the brain that consists mainly of grey matter.

Hippocampal Pertaining to the hippocampus.

Hirsutism A condition where women have excess facial and body hair that is dark and coarse.

Histaminergic Liberated or activated by histamine, relating to the effects of histamine at histamine receptors of target tissues.

Histaminergic receptors Are types of G protein-coupled receptors with histamine as their endogenous ligand.

Histone acetyltransferases (HAT) Are enzymes that acetylate conserved lysine amino acids on histone proteins by transferring an acetyl group from acetyl CoA to form e-*N*-acetyl lysine. HATs act as transcriptional coactivators.

Histone lysine demethylases (KDMs) Enzymes that play a key role in the amplification of hypoxia-inducible-factor signalling and expression of proangiogenic genes in cancer and neurological disorders.

HIV See Human immunodeficiency virus.

Hives (Urticaria) is a skin rash characterised by circular wheals of reddened and itching skin.

HLA Human leucocyte antigen system, name of the major histocompatibility complex (MHC) in humans.

HLA-DQB1 Human leucocyte antigen beta chain.

HLA-DR A major histocompatibility complex (MHC) class II cell surface receptor encoded by the human leucocyte antigen complex on chromosome 6 region 6p21.31.

HMG-CoAr 3-Hydroxy-3-methyl-glutaryl--CoA reductase or HMGCR is the rate-controlling enzyme (EC 1.1.1.88) of the mevalonate pathway.

HMG-CoA 3-Hydroxy-3-methylglutaryl-coenzyme A, an intermediate in the mevalonate pathway.

Hodgkin's disease Disease characterised by enlargement of the lymph glands, spleen and anaemia.

Homeodomain transcription factor A protein domain encoded by a homeobox. Homeobox genes encode transcription factors which typically switch on cascades of other genes.

Homeostasis The maintenance of a constant internal environment of a cell or an organism, despite fluctuations in the external.

Homeotherapy Treatment or prevention of disease with a substance similar but not identical to the causative agent of the disease.

Homocysteine An amino acid in the blood.

Homograft See Allograft.

Hormesis a Term used by toxicologists to refer to a biphasic dose response to an environmental agent characterised by a low-dose stimulation or beneficial effect and a high-dose inhibitory or toxic effect.

Hormonal (female) Substance that has a hormone-like effect similar to that of oestrogen and/or a substance used to normalise female hormone levels.

Hormonal (male) Substance that has a hormone-like effect similar to that of testosterone and/or a substance used to normalise male hormone levels.

HRT Hormone replacement therapy, the administration of the female hormones, oestrogen and progesterone and sometimes testosterone.

HSF-1 factor Major regulator of heat shock protein transcription in eukaryotes.

HSP27 Is an ATP-independent, 27kDa heat shock protein chaperone that confers protection against apoptosis.

HSP70 Heat shock protein chaperone that confers protection against heat-induced apoptosis.

HSP90 A 90kDa heat shock protein chaperone that has the ability to regulate a specific subset of cellular signalling proteins that have been implicated in disease processes.

HSPD 1 Heat shock 60kDa protein 1

hTERT (TERT) Telomerase reverse transcriptase is a catalytic subunit of the enzyme telomerase in humans. It exerts a novel protective function by binding to mitochondrial DNA, increasing respiratory chain activity and protecting against oxidative stress-induced damage.

HT29 cells Are human intestinal epithelial cells which produce the secretory component of immunoglobulin A (IgA) and carcinoembryonic antigen (CEA).

Human cytomegalovirus (HCMV) A DNA herpesvirus which is the leading cause of congenital viral infection and mental retardation.

Human factor X A coagulation factor also known by the eponym Stuart–Prower factor or as thrombokinase, is an enzyme involved in blood coagulation. It synthesised in the liver and requires vitamin K for its synthesis.

Human immunodeficiency virus (HIV) A retrovirus that can lead to acquired immunodeficiency syndrome (AIDS), a condition in humans in which the immune system begins to fail, leading to life-threatening opportunistic infections.

Humoral immune response (HIR) Is the aspect of immunity that is mediated by secreted antibodies (as opposed to cell-mediated immunity, which involves T lymphocytes) produced in the cells of the B lymphocyte lineage (B cell).

HUVEC Human umbilical vein endothelial cells.

Hyaluronidase Enzymes that catalyse the hydrolysis of certain complex carbohydrates like hyaluronic acid and chondroitin sulphates.

Hydatidiform A rare mass or growth that forms inside the uterus at the beginning of a pregnancy.

Hydrocele Abnormal accumulation of fluid inside the scrotum.

Hydrocholeretic An agent that stimulates an increased output of bile of low specific gravity.

Hydragogue A purgative that causes an abundant watery discharge from the bowel.

Hydronephrosis Is distension and dilation of the renal pelvis and calyces, usually caused by obstruction of the free flow of urine from the kidney.

Hydrophobia A viral neuroinvasive disease that causes acute encephalitis (inflammation of the brain) in warm-blooded animals. Also called rabies.

Hydropsy See Dropsy.

Hydrothorax Accumulation of serous fluid in the pleural cavity.

Hyperaemia The increase of blood flow to different tissues in the body.

Hyperalgesia An increased sensitivity to pain (enhanced pricking pain), which may be caused by damage to nociceptors or peripheral nerves.

Hyperammonemia, hyperammonaemia A metabolic disturbance characterised by an excess of ammonia in the blood.

Hypercalciuria (*Idiopathic*) presence of excess calcium in the urine without obvious cause.

Hypercholesterolemia High levels of cholesterol in the blood that increase a person's risk for cardiovascular disease leading to stroke or heart attack.

Hyperdipsia Intense thirst that is relatively temporary.

Hyperemia Is the increased blood flow that occurs when tissue is active.

Hyperemesis Severe and persistent nausea and vomiting (morning sickness) during pregnancy.

Hyperemesis gravidarum Is a pregnancy complication characterised by severe nausea, vomiting, weight loss and electrolyte disturbance.

Hyperfibrinogenemia Excessive fibrinogen in the blood.

Hyperglycaemia hyperglycaemia High blood sugar; is a condition in which an excessive amount of glucose circulates in the blood plasma.

Hyperglycaemic A substance that raises blood sugar levels.

Hyperhomocysteinemia Is a medical condition characterised by an abnormally large level of homocysteine in the blood.

Hyperinsulinemia A condition in which there are excess levels of circulating insulin in the blood; also known as prediabetes.

Hyperkalemia Is an elevated blood level of the electrolyte potassium.

Hyperkeratosis Abnormal thickening of the outer layer of the skin. *adj.* hyperkeratotic.

Hyperknesis Enhanced itch to pricking.

Hyperleptinemia Increased serum leptin level.

Hyperlipoproteinemia A metabolic disorder characterised by abnormally elevated concentrations of lipid/lipoprotein in the plasma; also known as hyperlipidaemia and hyperlipemia.

Hypermenorrhea Abnormally heavy or prolonged menstruation.

Hypermethylation An increase in the inherited methylation of cytosine and adenosine residues in DNA.

Hypermineralocorticoidism Excessive mineralocorticoid activity.

Hyperoxaluria An excessive urinary excretion of oxalate.

Hyperphagia Or polyphagia abnormally large ingestion of food beyond that needed for basic energy requirements.

Hyperpiesia Persistent and pathological high blood pressure for which no specific cause can be found.

Hyperplasia Increased cell production in a normal tissue or organ.

Hyper-pre-beta-lipoproteinaemia Increased concentrations of pre-beta-lipoproteins in the blood.

Hyperpropulsion Using water pressure as a force to move objects; used to dislodge calculi in the urethra.

Hyperpyrexia Is an abnormally high fever.

Hypertension Commonly referred to as 'high blood pressure' or HTN, is a medical condition in which the arterial blood pressure is chronically elevated.

Hypertensive Characterised or caused by increased tension or pressure as abnormally high blood pressure.

Hypertonia Abnormal increase in muscle tension and a reduced ability of the muscle to stretch.

Hypertriglyceridaemia or hypertriglycaemia A disorder that causes high triglycerides in the blood.

Hypertrophy Enlargement or overgrowth of an organ.

Hyperuricaemia Is a condition characterised by abnormally high level of uric acid in the blood.

Hypoadiponectinemia The state of having too low level of adiponectin, a major metabolic endocrine, responsible for regulating things like glucose uptake and lipolysis (the breakdown of fat deposits); low adiponectin is a risk factor for both type II diabetes and metabolic syndrome.

Hypoalbuminemia A medical condition where levels of albumin in blood serum are abnormally low.

Hypocalcaemic tetany A disease caused by an abnormally low level of calcium in the blood and characterised by hyperexcitability of the neuromuscular system and results in carpopedal spasms.

Hypochlorhydria Refer to states where the production of gastric acid in the stomach is absent or low.

Hypocholesterolemic (Cholesterol reducer) a substance that lowers blood cholesterol levels.

Hypocitraturia Low amount of citrate in the urine, an important risk factor for kidney stone formation.

Hypocorticism See Addison's disease.

Hypocortisolism See Addison's disease.

Hypoesthesia (Or hypesthesia) refers to a reduced sense of touch or sensation, or a partial loss of sensitivity to sensory stimuli.

Hypoglycaemic An agent that lowers the concentration of glucose (sugar) in the blood.

Hypogonadism syndrome Characterised by defects of the gonads, a diminished functional activity of the gonads—the testes and ovaries in males and females, respectively.

Hypokalemia Medical condition in which the concentration of potassium (K^+) in the blood is low.

Hypoparathyroidism An uncommon condition in which your body secretes abnormally low levels of parathyroid hormone (PTH). PTH plays a key role in modulating the balance of calcium and phosphorus levels in the body.

Hypoperfusion Decreased blood flow through an organ, characterised by an imbalance

of oxygen demand and oxygen delivery to tissues.

Hypophagic Undereating.

Hypophysectomy The surgical removal of the hypophysis (pituitary gland).

Hypospadias An abnormal birth defect in males in which the urethra opens on the under surface of the penis.

Hypotensive Characterised by or causing diminished tension or pressure, as abnormally low blood pressure.

Hypothermia A condition in which an organism's temperature drops below that required for normal metabolism and body functions.

Hypothermic Relating to hypothermia, with subnormal body temperature.

Hypoxaemia Is the reduction of oxygen specifically in the blood.

Hypoxia A shortage of oxygen in the body. *adj.* hypoxic.

Hypoxia-inducible factors (HIFs) Transcription factors that respond to changes in available oxygen in the cellular environment, specifically, to deficiency in oxygen.

ICAM-1 (inter-cellular adhesion molecule 1) Also known as CD54 (cluster of differentiation 54), is a protein that in humans is encoded by the ICAM1 gene.

IC_{50} The median maximal inhibitory concentration; a measure of the effectiveness of a compound in inhibiting biological or biochemical function.

I.C.V. (Intra-cerebroventricular) injection of chemical into the right lateral ventricle of the brain.

Icterus Jaundice, yellowish pigmentation of the skin.

Ichthyosis Dry, rectangular, fishlike scales on the skin.

Ichthyotoxic A substance which is poisonous to fish.

Icteric hepatitis An infectious syndrome of hepatitis characterised by jaundice, nausea, fever, right-upper quadrant pain, enlarged liver and transaminitis (increase in alanine aminotransferase (ALT) and/or aspartate aminotransferase (AST)).

Icterus neonatorum Jaundice in newborn infants.

Idiopathic Of no apparent physical cause.

Idiopathic mesenteric phlebosclerosis (IMP) a rare disease, characterised by thickening of the wall of the right hemicolon with calcification of mesenteric veins.

Idiopathic sudden sensorineural hearing loss (ISSHL) Is a sudden hearing loss where clinical assessment fails to reveal a cause.

I.g. Gastric intubation, insertion of Levin tube through the nasal passage to the stomach.

IgE Immunoglobulin E, a class of antibody that plays a role in allergy.

IGFs Insulin-like growth factors, polypeptides with high sequence similarity to insulin.

IgG Immunoglobulin G—the most abundant immunoglobulin (antibody) and is one of the major activators of the complement pathway.

IgM Immunoglobulin M, primary antibody against A and B antigens on red blood cells.

IKAP Is a scaffold protein of the IvarKappaBeta kinase complex and a regulator for kinases involved in pro-inflammatory cytokine signalling.

IKappa B Or IkB-beta, a protein of the NF-Kappa-B inhibitor family.

Ileus A temporary disruption of intestinal peristalsis due to nonmechanical causes.

Immune modulator A substance that affects or modulates the functioning of the immune system.

Immunodeficiency A state in which the immune system's ability to fight infectious disease is compromised or entirely absent.

Immunogenicity The property enabling a substance to provoke an immune response, *adj.* immunogenic.

Immunoglobulin class switching Ig class switching A biological mechanism that changes a B cell's production of antibody from one class to another.

Immunomodulatory Capable of modifying or regulating one or more immune functions.

Immunoreactive Reacting to particular antigens or haptens.

Immunostimulant Agent that stimulates an immune response.

Immunosuppression Involves a process that reduces the activation or efficacy of the immune system.

Immunotoxin A man-made protein that consists of a targeting portion linked to a toxin.

Impaired glucose tolerance (IGT) a prediabetic state of dysglycaemia associated with insulin resistance, increased risk of cardiovascular pathology and also a risk factor for mortality.

Impetigo A contagious, bacterial skin infection characterised by blisters that may itch, caused by a *Streptococcus* bacterium or *Staphylococcus aureus* and mostly seen in children.

Impotence A sexual dysfunction characterised by the inability to develop or maintain an erection of the penis.

Incontinence (faecal) The inability to control bowel's movement.

Incontinence (urine) The inability to control urine excretion.

Incretin A group of gastrointestinal hormones that cause an increase in the amount of insulin released from the beta cells of the islets of Langerhans after a meal; members include GIP and GLP-1.

Index of structural atypia (ISA) Index of structural abnormality.

Induration Hardened, as a soft tissue that becomes extremely firm, sclerosis.

Infarct An area of living tissue that undergoes necrosis as a result of obstruction of local blood supply.

Infarction Is the process of tissue death (necrosis) caused by blockage of the tissue's blood supply.

Inflammasomes Are large intracellular caspase-1-activating multiprotein complexes that play a central role in innate immunity.

Inflammation A protective response of the body to infection, irritation or other injuries, aimed at destroying or isolating the injuries and characterised by redness, pain, warmth and swelling.

Influenza A viral infection that affects mainly the nose, throat, bronchi and, occasionally, lungs.

Infusion A liquid extract obtained by steeping something (e.g. herbs) that are more volatile or dissolve readily in water, to release their active ingredients without boiling.

Inguinal hernia A hernia into the inguinal canal of the groin.

Inhalant A medicinal substance that is administered as a vapour into the upper respiratory passages.

iNOS, inducible nitric oxide synthases Through its product, nitric oxide (NO), may contribute to the induction of germ cell apoptosis. It plays a crucial role in early sepsis-related microcirculatory dysfunction.

Inotropic Affecting the force of muscle contraction.

Insecticide An agent that destroys insects. *adj.* insecticidal.

Insomnia A sleeping disorder characterised by the inability to fall asleep and/or the inability to remain asleep for a reasonable amount of time.

Insulin A peptide hormone composed of 51 amino acids produced in the islets of Langerhans in the pancreas causes cells in the liver, muscle and fat tissue to take up glucose from the blood, storing it as glycogen in the liver and muscle. Insulin deficiency is often the cause of diabetes and exogenous insulin is used to control diabetes.

Insulin homeostasis Blood sugar regulation.

Insulin-likegrowthfactors(IGFs) Polypeptides with high sequence similarity to insulin. They are part of a complex system that cells employ to communicate with their physiologic environment.

Insulin mimetic To act like insulin.

Insulin resistance A condition where the natural hormone insulin becomes less effective at reducing blood sugars.

Insulinogenic Associated with or stimulating the production of insulin.

Insulinotropic Stimulating or affecting the production and activity of insulin.

Integrase An enzyme produced by a retrovirus (such as HIV) that enables its genetic material to be integrated into the DNA of the infected cell.

Intercellular adhesion molecule (ICAM) A part of the immunoglobulin superfamily. They are important in inflammation, in immune responses and in intracellular signalling events.

Interferons (IFNs) Are natural cell-signalling glycoproteins known as cytokines produced by the cells of the immune system of most vertebrates in response to challenges such as viruses, parasites and tumour cells.

Interleukins A group of naturally occurring proteins and is a subset of a larger group of cellular messenger molecules called cytokines, which are modulators of cellular behaviour.

Interleukin-1 (IL-1) A cytokine that could induce fever, control lymphocytes, increase the number of bone marrow cells and cause degeneration of bone joints. Also called endogenous pyrogen, lymphocyte-activating factor, haemopoetin-1 and mononuclear cell factor, among others that IL-1 is composed of two distinct proteins, now called IL-1α and IL-1β.

Interleukin 1 beta (IL-1β) A cytokine protein produced by activated macrophages. Cytokine is an important mediator of the inflammatory response and is involved in a variety of cellular activities, including cell proliferation, differentiation and apoptosis.

Interleukin 2 (IL-2) A type of cytokine immune system signalling molecule that is instrumental in the body's natural response to microbial infection.

Interleukin-2 receptor (IL-2R) A heterotrimeric protein expressed on the surface of certain immune cells, such as lymphocytes, that binds and responds to a cytokine called IL-2.

Interleukin-6 (IL-6) An interleukin that acts as both a pro-inflammatory and anti-inflammatory cytokine.

Interleukin 8 (I-8) A cytokine produced by macrophages and other cell types such as epithelial cells and is one of the major mediators of the inflammatory response.

Intermediate-density lipoproteins (IDL) Is one of the five major groups of lipoproteins (chylomicrons, VLDL, IDL, LDL and HDL) that enable fats and cholesterol to move within the water-based solution of the bloodstream. IDL is further degraded to form LDL particles and, like LDL, can also promote the growth of atheroma and increase cardiovascular diseases.

Intermittent claudication An aching, crampy, tired and sometimes burning pain in the legs that comes and goes, caused by peripheral vascular disease. It usually occurs when walking and disappears after rest.

Interoceptive Relating to stimuli arising from within the body.

Interstitium The space between cells in a tissue.

Interstitial Pertaining to the interstitium.

Intertrigo An inflammation (rash) caused by microbial infection in skin folds.

Intima Innermost layer of an artery or vein.

Intimal hyperplasia The thickening of the tunica intima of a blood vessel as a complication of a reconstruction procedure.

Intoxicant Substance that produces drunkenness or intoxication.

Intracavernosal Within the corpus cavernosum, columns of erectile tissues forming the body of the penis.

Intraperitoneal (i.p.) The term used when a chemical is contained within or administered through the peritoneum (the thin, transparent membrane that lines the walls of the abdomen).

Intrathecal (i.t.) Through the theca of the spinal cord into the subarachnoid space.

Intromission The act of putting one thing into another.

Intubation Refers to the placement of a tube into an external or internal orifice of the body.

Iodine (I) Is an essential chemical element that is important for hormone development in the human body. Lack of iodine can lead to an enlarged thyroid gland (goitre) or other iodine deficiency disorders including mental retardation and stunted growth in babies and children. Iodine is found in dairy products, seafood, kelp, seaweeds, eggs, some vegetables and iodised salt.

IP See Intraperitoneal.

IP3R3 (Inositol 1,4,5-triphosphate receptor type 3) is an intracellular calcium release channel that mediates calcium release from the endoplasmic reticulum.

Iron (Fe) Is essential to most life forms and to normal human physiology. In humans, iron is

an essential component of proteins involved in oxygen transport and for haemoglobin. It is also essential for the regulation of cell growth and differentiation. A deficiency of iron limits oxygen delivery to cells, resulting in fatigue, poor work performance and decreased immunity. Conversely, excess amounts of iron can result in toxicity and even death. Dietary sources include certain cereals, dark green leafy vegetables, dried fruit, legumes, seafood, poultry and meat.

Ischemia An insufficient supply of blood to an organ, usually due to a blocked artery.

Ischuria Retention or suppression of urine.

Isoflavones A subgroup of flavonoids in which the basic structure is a 3-phenyl chromane skeleton. They act as phytoestrogens in mammals. See Flavonoids.

Isomers Substances that are composed of the same elements in the same proportions and hence have the same molecular formula but differ in properties because of differences in the arrangement of atoms.

Isoprostanes Unique prostaglandin-like compounds generated in-vivo from the free radical-catalysed peroxidation of essential fatty acids.

Jamu Traditional Indonesian herbal medicine.

Janus kinase (JAK)-signal transducer and activator of transcription (STAT) signalling Are essential molecules in cytokine signal transduction pathways involved in cancer development and progression.

Jaundice Refers to the yellow colour of the skin and whites of the eyes caused by excess bilirubin in the blood.

JNK (Jun N-terminal kinase), also known as stress-activated protein kinase (SAPK), belongs to the family of MAP kinases.

Jurkat cells A line of T lymphocyte cells that are used to study acute T cell leukaemia.

KB cell A cell line derived from a human carcinoma of the nasopharynx, used as an assay for antineoplastic (antitumour) agents.

Kainate receptors Or KARs, are non-NMDA (N-methyl-D-aspartate) ionotropic receptors which respond to the neurotransmitter glutamate.

Kaliuresis The presence of excess potassium in the urine.

Kallikreins Peptidases (enzymes that cleave peptide bonds in proteins), a subgroup of the serine protease family; they liberate kinins from kininogens. Kallikreins are targets of active investigation by drug researchers as possible biomarkers for cancer.

Kaposi sarcoma A cancerous tumour of the connective tissues caused by the human herpesvirus 8 and is often associated with AIDS.

Kaposi sarcoma herpesvirus (KSHV) Also known as human herpesvirus 8, is a gamma 2 herpesvirus or rhadinovirus. It plays an important role in the pathogenesis of Kaposi sarcoma (KS), multicentric Castleman disease (MCD) of the plasma cell type and primary effusion lymphoma and occurs in HIV patients.

Karyolysis Dissolution and disintegration of the nucleus when a cell dies.

Karyorrhexis Destructive fragmentation of the nucleus of a dying cell whereby its chromatin disintegrates into formless granules.

Keloids Benign dermal tumours characterised by fibroblastic proliferation and excessive accumulation of collagen.

Keratin A sulphur-containing protein which is a major component in skin, hair, nails, hooves, horns and teeth.

Keratinocyte Is the major constituent of the epidermis, constituting 95% of the cells found there.

Keratinophilic Having an affinity for keratin.

Keratitis Inflammation of the cornea.

Keratoconjunctivitis sicca Also called keratitis sicca, xerophthalmia or dry eye syndrome (DES), is an eye disease characterised by a deficiency of aqueous tear film over the surface of the eye and in the lining of the lids.

Keratolysis Softening and separation of the horny layer of the epidermis.

Keratolytic Pertaining to keratolysis.

Keratomalacia An eye disorder that leads to a dry cornea.

Kidney stones (Calculi) are hardened mineral deposits that form in the kidney.

Kinin Is any of various structurally related polypeptides, such as bradykinin, that act locally to induce vasodilation and contraction of smooth muscle.

Kininogen Either of two plasma **α**2-globulins that are kinin precursors.

Ki-67 Human protein associated with cell proliferation.

Knockout Gene knockout is a genetic technique in which an organism is engineered to carry genes that have been made inoperative.

Konzo Is an epidemic paralytic disease occurring in outbreaks in remote rural areas of low-income African countries.

Kunitz protease inhibitors A type of protein contained in legume seeds which functions as a protease inhibitor.

Kupffer cells Are resident macrophages of the liver and play an important role in its normal physiology and homeostasis as well as participating in the acute and chronic responses of the liver to toxic compounds.

l-Dopa (L-*3,4*-Dihydroxyphenylalanine) is an amino acid that is formed in the liver and converted into dopamine in the brain.

Labour Process of childbirth involving muscular contractions.

Lacrimation Secretion and discharge of tears.

Lactagogue An agent that increases or stimulates milk flow or production. Also called a galactagogue.

Lactate dehydrogenase (LDH) Enzyme that catalyses the conversion of lactate to pyruvate.

Lactation Secretion and Production of milk.

Lactic acidosis Is a condition caused by the build-up of lactic acid in the body. It leads to acidification of the blood (acidosis) and is considered a distinct form of metabolic acidosis.

LAK cell A lymphokine-activated killer cell, i.e. a white blood cell that has been stimulated to kill tumour cells.

Lamella In cell biology, it refers to numerous plate or disc-like structures at both a tissue and cellular level.

Laminin A glycoprotein component of connective tissue basement membrane that promotes cell adhesion.

Laparoscopic cholecystectomy Is a procedure in which the gallbladder is removed by laparoscopic techniques.

Laparotomy A surgical procedure involving an incision through the abdominal wall to gain access into the abdominal cavity. *adj.* laparotomised.

Larvacidal An agent which kills insect or parasite larva.

Laryngitis Is an inflammation of the larynx.

Laxation Bowel movement.

Laxatives Substances that are used to promote bowel movement.

LC 50 Median lethal concentration; see LD 50.

LD 50 Median lethal dose—the dose required to kill half the members of a tested population. Also called LC 50 (median lethal concentration).

LDL See Low-density lipoprotein.

LDL cholesterol See Low-density lipoprotein.

LDL receptor (LDLr) A low-density lipoprotein receptor gene.

Lectins Are sugar-binding proteins that are highly specific for their sugar moieties, which agglutinate cells and/or precipitate glycoconjugates. They play a role in biological recognition phenomena involving cells and proteins.

Leiomyoma Benign smooth muscle neoplasm that is very rarely (0.1%) premalignant.

Leishmaniasis A disease caused by protozoan parasites that belong to the genus *Leishmania* and is transmitted by the bite of certain species of sand fly.

Lenitive Palliative; easing pain or discomfort.

Lenticular opacity Also known as or related to cataract.

Leprosy A chronic bacterial disease of the skin and nerves in the hands and feet and, in some cases, the lining of the nose. It is caused by the *Mycobacterium leprae.* Also called Hansen's disease.

Leptin Is a 16-kDa protein hormone with important effects in regulating body weight, metabolism and reproductive function.

Lequesne algofunctional index Is a widespread international instrument (ten-question survey) and recommended by the World Health Organization (WHO) for outcome measurement in hip and knee diseases such as osteoarthritis.

Leucocyte White blood corpuscles, colourless, without haemoglobin that helps to combat infection.

Leucocytopenia Abnormal decrease in the number of leucocytes (white blood cells) in the blood.

Leucocytosis Increase in white blood cell count above its normal range.

Leucoderma A skin abnormality characterised by white spots, bands and patches on the skin; they can also be caused by fungus and tinea. Also see Vitiligo.

Leucomyelopathy Any diseases involving the white matter of the spinal cord.

Leucopenia A decrease in the number of circulating white blood cells.

Leucorrhoea Commonly known as whites, refers to a whitish discharge from the female genitals.

Leukemia, leukaemia A cancer of the blood or bone marrow and is characterised by an abnormal proliferation (production by multiplication) of blood cells, usually white blood cells (leucocytes).

Leukemogenic Relating to leukaemia, causing leukaemia.

Leukoplakia Condition characterised by white spots or patches on mucous membranes, especially of the mouth and vulva.

Leukotriene A group of hormones that cause the inflammatory symptoms of hay-fever and asthma.

Leydig cells Also called interstitial cells of Leydig, are found adjacent to the seminiferous tubules in the testicle. They produce testosterone in response to luteinising hormone.

Levarterenol See Norepinephrine.

LexA repressor Or repressor LexA, is repressor enzyme that represses SOS response genes coding for DNA polymerases required for repairing DNA damage

Libido Sexual urge.

Lichen planus A chronic mucocutaneous disease that affects the skin, tongue and oral mucosa.

Ligroin A volatile, inflammable fraction of petroleum, obtained by distillation and used as a solvent.

Limbic system Complex set of brain structures, including the hypothalamus, amygdala, hippocampus, anterior thalamic nuclei, septum, limbic cortex and fornix that control various functions such as emotion, behaviour, motivation, memory and olfaction.

Liniment Liquid preparation rubbed on skin, used to relieve muscular aches and pains.

Linterised starch Starch that has undergone prolonged acid treatment.

Lipodiatic Having lipid and lipoprotein lowering property.

Lipodystrophy A medical condition characterised by abnormal or degenerative conditions of the body's adipose tissue.

Lipofuscin Finely granular yellow-brown pigment granules composed of lipid-containing residues of lysosomal digestion.

Lipogenesis Is the process by which acetyl-CoA is converted to fats; *adj.* lipogenic.

Lipolysis Is the breakdown of fat stored in fat cells in the body.

Liposomes Artificially prepared vesicles made of lipid bilayer.

Lipotoxicity Refers to tissue diseases that may occur when fatty acids spill over in excess of the oxidative needs of those tissues and enhances metabolic flux into harmful pathways of nonoxidative metabolism.

Lipotropic Refers to compounds that help catalyse the breakdown of fat during metabolism in the body, e.g. chlorine and lecithin.

Lipoxygenase A family of iron-containing enzymes that catalyse the dioxygenation of polyunsaturated fatty acids in lipids containing a *cis,cis*-1,4-pentadiene structure.

Lithiasis Formation of urinary calculi (stones) in the renal system (kidneys, ureters, the urinary bladder, urethra) can be of any one of several compositions.

Lithogenic Promoting the formation of calculi (stones).

Lithontriptic Removes stones from the kidney, gallbladder.

Liver X receptors Nuclear hormones that function as central transcriptional regulators for lipid homeostasis.

Lochia Vaginal discharge containing blood, mucus and uterine tissues, during the postpartum period

Lotion A liquid suspension or dispersion of chemicals for external application to the body.

Lovo cells Colon cancer cells.

Low-density lipoprotein (LDL) Is a type of lipoprotein that transports cholesterol and triglycerides from the liver to peripheral tissues. High levels of LDL cholesterol can signal medical problems like cardiovascular disease, and it is sometimes called 'bad cholesterol'.

LRP1 Low-density lipoprotein receptor-related protein-1, plays a role in intracellular signalling functions as well as in lipid metabolism.

LTB4 A type of leukotriene, a major metabolite in neutrophil polymorphonuclear leucocytes. It stimulates polymorphonuclear cell function (degranulation, formation of oxygen-centred free radicals, arachidonic acid release and metabolism). It induces skin inflammation.

Luciferase Is a generic name for enzymes commonly used in nature for bioluminescence.

Lumbago Is the term used to describe general lower back pain.

Lung abscess Necrosis of the pulmonary tissue and formation of cavities containing necrotic debris or fluid caused by microbial infections.

Lusitropic An agent that affects diastolic relaxation.

Lutein A carotenoid, occurs naturally as yellow or orange pigment in some fruits and leafy vegetables. It is one of the two carotenoids contained within the retina of the eye. Within the central macula, zeaxanthin predominates, whereas in the peripheral retina, lutein predominates. Lutein is necessary for good vision and may also help prevent or slow down atherosclerosis, the thickening of arteries, which is a major risk for cardiovascular disease.

Luteinising hormone (LH) A hormone produced by the anterior pituitary gland. In females, it triggers ovulation. In males, it stimulates the production of testosterone to aid sperm maturation.

Luteolysis Is the structural and functional degradation of the corpus luteum (CL) that occurs at the end of the luteal phase of both the estrous and menstrual cycles in the absence of pregnancy. *adj*. luteolytic.

Luteotropic Stimulating the formation of the corpus luteum.

Lymphadenitis The inflammation or enlargement of a lymph node casued by microbial infection.

Lymphadenitis, cervical Inflammation of the lymph nodes in the neck, usually caused by an infection.

Lymphatitis Inflammation of lymph vessels and nodes.

Lymphadenopathy A term meaning 'disease of the lymph nodes'—lymph node enlargement.

Lymphadenomegaly Is the enlargement of the lymph node/nodes.

Lymphangitis An inflammation or bacterial infection of the lymphatic channels, mostly commonly caused by the bacterium *Streptococcus pyogenes* in humans.

Lymphoblastic Pertaining to the production of lymphocytes.

Lymphocyte A small white blood cell (leucocyte) that plays a large role in defending the body against disease. Lymphocytes are responsible for immune responses. There are two main types of lymphocytes: B cells and T cells. Lymphocytes secrete products (lymphokines) that modulate the functional activities of many other types of cells and are often present at sites of chronic inflammation.

Lymphocyte B cells The B cells make antibodies that attack bacteria and toxins.

Lymphocyte T cells T cells attack body cells themselves when they have been taken over by viruses or have become cancerous.

Lymphoma A type of cancer involving cells of the immune system, called lymphocytes.

Lymphopenia Abnormally low in the number of lymphocytes in the blood.

Lysosomes Are small, spherical organelles containing digestive enzymes (acid hydrolases) and other proteases (cathepsins).

mTOR, the mammalian (or mechanistic) target of rapamycin Regulates a wide range of cellular and developmental processes by coordinating signalling responses to mitogens, nutrients and various stresses.

Maceration Softening or separation of parts by soaking in a liquid.

Macrophage A type of large leucocyte that travels in the blood but can leave the bloodstream and enter tissue; like other leucocytes it protects the body by digesting debris and foreign cells.

Macular degeneration A disease that gradually destroys the macula, the central portion of the retina, reducing central vision.

Macules Small circumscribed changes in the colour of skin that are neither raised (elevated) nor depressed.

Maculopapular Describes a rash characterised by raised, spotted lesions.

Magnesium (M g) Is the fourth most abundant mineral in the body and is essential to good health. It is important for normal muscle and nerve function, steady heart rhythm, immune system and strong bones. Magnesium also helps regulate blood sugar levels, promotes normal blood pressure and is known to be involved in energy metabolism and protein synthesis and plays a role in preventing and managing disorders such as hypertension, cardiovascular disease and diabetes. Dietary sources include legumes (e.g. soya bean and by-products), nuts, whole unrefined grains, fruit (e.g. banana, apricots), okra and green leafy vegetables.

MAK cell Macrophage-activated killer cell, activated nacrophage that is much more phagocytic than monocytes.

Malaise A feeling of weakness, lethargy or discomfort as of impending illness.

Malaria Is an infection of the blood by *Plasmodium* parasite that is carried from person to person by mosquitoes. There are four species of malaria parasites that infect man.

Plasmodium falciparum So-called malignant tertian fever, is the most serious disease, *Plasmodium vivax*, causing a relapsing form of the disease, *Plasmodium malariae* and *Plasmodium ovale*.

Malassezia A fungal genus (previously known as *Pityrosporum*) classified as yeasts, naturally found on the skin surfaces of many animals including humans. It can cause hypopigmentation on the chest or back if it becomes an opportunistic infection.

Mammalian target of rapamycin (mTOR) Pathway that regulates mitochondrial oxygen consumption and oxidative capacity.

Mammogram An X-ray of the breast to detect tumours.

Mandibular Relating to the mandible, the human jaw bone.

Manganese Is an essential element for health. It is an important constituent of some enzymes and an activator of other enzymes in physiologic processes. Manganese superoxide dismutase (MnSOD) is the principal antioxidant enzyme in the mitochondria. Manganese-activated enzymes play important roles in the metabolism of carbohydrates, amino acids and cholesterol. Manganese is the preferred cofactor of enzymes called glycosyltransferases which are required for the synthesis of proteoglycans that are needed for the formation of healthy cartilage and bone. Dietary source includes whole grains, fruit, legumes (soybean and by-products), green leafy vegetables, beetroot and tea.

MAO activity Monoamine oxidase activity.

MAPK (mitogen-activated protein kinase) These kinases are strongly activated in cells subjected to osmotic stress, UV radiation, dysregulated K^+ currents, RNA-damaging agents and a multitude of other stresses, as well as inflammatory cytokines, endotoxin and withdrawal of a trophic factor. The stress-responsive MAPKs mediate a plethora of cellular responses to such stressful stimuli, including apoptosis and production of inflammatory and immunoregulatory cytokines in diverse cell systems.

Marasmus Is one of the three forms of serious protein–energy malnutrition.

Mastalgia Breast pain.

Mastectomy Surgery to remove a breast.

Masticatory A substance chewed to increase salivation. Also called sialogogue.

Mastitis A bacterial infection of the breast which usually occurs in breastfeeding mothers.

Matrix metalloproteinases (MMP) A member of a group of enzymes that can break down proteins, such as collagen, that are normally found in the spaces between cells in tissues (i.e. extracellular matrix proteins). Matrix metalloproteinases are involved in wound healing, angiogenesis and tumour cell metastasis. See also Metalloproteinase.

MBC Minimum bacterial concentration—the lowest concentration of antibiotic required to kill an organism.

MCP-1 Monocyte chemotactic protein-1, plays a role in the recruitment of monocytes to sites

of infection and injury. It is a member of small inducible gene (SIG) family.

MDA Malondialdehyde is one of the most frequently used indicators of lipid peroxidation.

Measles An acute, highly communicable rash illness due to a virus transmitted by direct contact with infectious droplets or, less commonly, by airborne spread.

Mechanonociceptors Sensory neurons that are mechanically sensitive found in all of the paraspinal connective tissues including ligament, joint capsule, annulus fibrosus of the intervertebral disc, muscle, tendon and skin. They respond to a noxius (damaging) mechanical load.

Medial preoptic area Is located at the rostral end of the hypothalamus; it is important for the regulation of male sexual behaviour.

Megaloblastic anaemia An anaemia that results from inhibition of DNA synthesis in red blood cell production, often due to a deficiency of vitamin B12 or folate, and is characterised by many large immature and dysfunctional red blood cells (megaloblasts) in the bone marrow.

Melaena (melena) Refers to the black, 'tarry' faeces that are associated with gastrointestinal haemorrhage.

Melanogenesis Production of melanin by living cells.

Melanoma Malignant tumour of melanocytes which are found predominantly in skin but also in the bowel and the eye and appear as pigmented lesions.

Melatonin A hormone produced in the brain by the pineal gland; it is important in the regulation of the circadian rhythms of several biological functions.

Menarche The first menstrual cycle, or first menstrual bleeding, in female human beings.

Menorrhagia Heavy or prolonged menstruation, too frequent menstrual periods.

Menopausal Refers to permanent cessation of menstruation.

Menses See Menstruation.

Menstruation The approximately monthly discharge of blood from the womb in women of childbearing age who are not pregnant. Also called menses. *adj.* menstrual.

Mesangial cells Are specialised cells around blood vessels in the kidneys, at the mesangium.

Mesencephalon Midbrain.

Mesothelioma Is an aggressive cancer affecting the membrane lining of the lungs and abdomen.

Metabolic syndrome (MetS) represents a combination of cardiometabolic risk factors, including visceral obesity, glucose intolerance or type 2 diabetes, elevated triglycerides, reduced HDL cholesterol and hypertension.

Metabolome Complete set of metabologically regulated elements in cells.

Metabolomics Is the scientific study of chemical processes involving metabolites.

Metalloproteinase Enzymes that break down proteins and require zinc or calcium atoms for proper function.

Metallothionein (MT) a family of cysteine-rich, low molecular weight (500 to 14000 Da) proteins.

Meta-analysis A statistical procedure that combines the results of several studies that address a set of related research hypotheses.

Metaphysis Is the portion of a long bone between the epiphyses and the diaphysis of the femur.

Metaphyseal Pertaining to the metaphysis.

Metaplasia Transformation of one type of one mature differentiated cell type into another mature differentiated cell type.

Metastasis Is the movement or spreading of cancer cells from one organ or tissue to another.

Metestrus The quiescent period of sexual inactivity between oestrus cycles.

Methaemoglobinemia Is a disorder characterised by the presence of a higher than normal level of methaemoglobin (ferric [Fe^{3+}] rather than ferrous [Fe^{2+}] haemoglobin) in red blood cells. This results in a decreased availability of oxygen to the tissues.

Metropathy Any disease of the uterus especially of the myometrium.

Metroptosis The slipping or falling out of place of an organ (as the uterus)

Metrorrhagia Uterine bleeding at irregular intervals, particularly between the expected menstrual periods.

Mevinolin A potent inhibitor of 3-hydroxy-3-methylglutaryl coenzyme A reductase (HMG-CoA reductase).

MHC Acronym for major histocompatibility complex, a large cluster of genes found on the short arm of chromosome 6 in most vertebrates that encodes MHC molecules. MHC molecules play an important role in the immune system and autoimmunity.

MHC 11 molecules Class II MHC molecules belong to a group of molecules known as the immunoglobulin supergene family, which includes immunoglobulins, T cell receptors, CD4, CD8 and others.

MIC Minimum inhibitory concentration, lowest concentration of an antimicrobial that will inhibit the visible growth of a microorganism.

Micelle A submicroscopic aggregation of molecules.

Micellisation Formation process of micelles.

Michael acceptors See Michael reaction.

Micahel donors See Michael reaction.

Michael reaction Conjugate addition of a carbon nucleophile to an **α,β**-unsaturated acceptor; a thermodynamically controlled reaction between unusually acidic donors (**β**-ketoesters or **β**-diketones) and unhindered **α,β**-unsaturated acceptors. Stable enolates, active methylenes such as malonates and nitroalkanes are Michael donors, and activated olefins such as **α,β**-unsaturated carbonyl compounds are Michael acceptors.

Microangiopathy (Or microvascular disease) is an angiopathy affecting small blood vessels in the body.

Microfilaria A pre-larval parasitic worm of the family Onchocercidae, found in the vector and in the blood or tissue fluid of human host.

Micronuclei Small particles consisting of acentric fragments of chromosomes or entire chromosomes, which lag behind at anaphase of cell division.

Microphthalmia-associated transcription factor (MITF) A basic helix-loop-helix leucine zipper transcription factor protein that plays a role in the development, survival and function of melanocytes and osteoclast.

Microsomal PGE2 synthase Is the enzyme that catalyses the final step in prostaglandin E2 (PGE2) biosynthesis.

Microvasculature The finer vessels of the body, as the arterioles, capillaries and venules.

Micturition Urination, act of urinating.

Migraine A neurological syndrome characterised by altered bodily perceptions; severe, painful headaches; and nausea.

Mimosine Is an alkaloid, **β**-3-hydroxy-4 pyridone amino acid; it is a toxic nonprotein free amino acid and is an antinutrient.

Mineral apposition rate MAR, rate of addition of new layers of mineral on the trabecular surfaces of bones.

Mineralcorticoids A group of steroid hormones that are secreted by the adrenal cortex and regulate the balance of water and electrolytes (sodium, potassium) in the body.

Miscarriage Spontaneous abortion.

Mitochondrial complex I The largest enzyme in the mitochondrial respiratory oxidative phosphorylation system.

Mitochondrial permeability transition (MPT) Is an increase in the permeability of the mitochondrial membranes to molecules of less than 1500 Da in molecular weight. MPT is one of the major causes of cell death in a variety of conditions.

Mitogen An agent that triggers mitosis and elicits all the signals necessary to induce cell proliferation.

Mitogenic Able to induce mitosis or transformation.

Mitogenicity Process of induction of mitosis.

Mitomycin A chemotherapy drug that is given as a treatment for several different types of cancer, including breastl, stomach, oesophagus and bladder cancers.

Mitosis Cell division in which the nucleus divides into nuclei containing the same number of chromosomes.

Mitral valve prolapse The most common heart valve abnormality. Symptoms could include palpitations, shortness of breath, cough, fatigue, dizziness or anxiety, migraine headaches and chest discomfort.

MMP Matrix metalloproteinases, a group of peptidases involved in the degradation of the extracellular matrix (ECM).

Mnestic Pertaining to memory.

Molecular docking Is a key tool in structural molecular biology and computer-assisted drug design.

Molluscidal Destroying molluscs like snails.

Molt 4 cells MOLT4 cells are lymphoblast-like in morphology and are used for studies of apoptosis, tumour cytotoxicity and tumorigenicity, as well as for antitumour testing.

Molybdenum (Mo) Is an essential element that forms part of several enzymes such as xanthine oxidase involved in the oxidation of xanthine to uric acid and use of iron. Molybdenum concentrations also affect protein synthesis, metabolism and growth. Dietary sources include meat, green beans, eggs, sunflower seeds, wheat flour, lentils and cereal grain.

Monoamine oxidase A (MAOA) Is an isozyme of monoamine oxidase. It preferentially deaminates norepinephrine (noradrenaline), epinephrine (adrenaline), serotonin and dopamine.

Monoaminergic Of or pertaining to neurons that secrete monoamine neurotransmitters (e.g. dopamine, serotonin).

Monoclonal antibodies Are produced by fusing single antibody-forming cells to tumour cells grown in culture.

Monocyte Large white blood cell that ingests microbes, other cells and foreign matter.

Monogalactosyl diglyceride Are the major lipid components of chloroplasts.

Menorrhagia Is heavy bleeding and that is usually defined as periods lasting longer than 7 days or excessive bleeding.

Morbidity A diseased state or symptom or can refer either to the incidence rate or to the prevalence rate of a disease.

Morelloflavone A biflavonoid extracted from *Garcinia dulcis*, has shown antioxidative, antiviral and anti-inflammatory properties.

Morphine The major alkaloid of opium and a potent narcotic analgesic.

MTTP Microsomal triglyceride transfer protein that is required for the assembly and secretion of triglyceride-rich lipoproteins from both enterocytes and hepatocytes.

MUC 5AC Mucin 5AC, a secreted gel-forming protein mucin with a high molecular weight of about 641kDa.

Mucositis Painful inflammation and ulceration of the mucous membranes lining the digestive tract.

Mucous Relating to mucus.

Mucolytic Capable of reducing the viscosity of mucus, or an agent that so acts.

Mucus Viscid secretion of the mucous membrane.

Multidrug resistance (MDR) Ability of a living cell to show resistance to a wide variety of structurally and functionally unrelated compounds.

Muscarinic receptors Are G protein-coupled acetylcholine receptors found in the plasma membranes of certain neurons and other cells.

Musculotropic Affecting or acting upon muscular tissue.

Mutagen An agent that induces genetic mutation by causing changes in the DNA.

Mutagenic Capable of inducing mutation (used mainly for extracellular factors such as X-rays or chemical pollution).

Myalgia Muscle pain.

Myc Codes for a protein that binds to the DNA of other genes and is therefore a transcription facor, found on chromosome 8 in human.

Mycosis An infection or disease caused by a fungus.

Mydriasis Abnormal, excexsive dilation of the pupil caused by disease or drug.

Myelocyte Is a young cell of the granulocytic series, occurring normally in bone marrow, but not in circulating blood.

Myeloid leukaemia (chronic) A type of cancer that affects the blood and bone marrow, characterised by excessive number of white blood cells.

Myeloma Cancer that arises in the plasma cells, a type of white blood cells.

Myelopathy Refers to pathology of the spinal cord.

Myeloperoxidase (MPO) Is a peroxidase enzyme most abundantly present in neutrophil granulocytes (a subtype of white blood cells). It is an inflammatory enzyme produced by activated leucocytes that predicts risk of coronary heart disease.

Myeloproliferative disorder Disease of the bone marrow in which excess cells are produced.

Myelosuppressive Causing bone marrow suppression.

Myelotoxicity State of being toxic to myeloid tissues, the bone marrow.

Myiasis Parasitic infestation of the body of a live mammal by fly larvae.

Myocardial Relating to heart muscle tissues.

Myocardial infarction (MI) Is the rapid development of myocardial necrosis caused by a critical imbalance between oxygen supply and demand of the myocardium.

Myocardial ischemia An intermediate condition in coronary artery disease during which the heart tissue is slowly or suddenly starved of oxygen and other nutrients.

Myocardial lipidosis Is the accumulation of fat droplets in myocardial fibres.

Myoclonus Brief, involuntary twitching of a muscle or a group of muscles.

Myogenesis The formation of muscular tissue, especially during embryonic development.

Myoglobin A red, iron- and oxygen-binding protein which carries and stores oxygen in muscle tissues; this haemoprotein resembles a single subunit of haemoglobin.

Myoglobinuria Is the presence of myoglobin in the urine, usually associated with rhabdomyolysis or muscle destruction.

Myopathy A muscular disease wherein the muscle fibres do not function for any one of many reasons, resulting in muscular weakness.

Myopia Near- or short-sightedness.

Myosarcoma A malignant muscle tumour.

Myotonia dystrophica An inherited disorder of the muscles and other body systems characterised by progressive muscle weakness, prolonged muscle contractions (myotonia), clouding of the lens of the eye (cataracts), cardiac abnormalities, balding and infertility.

Myotube A developing skeletal muscle fibre or cell with a tubular appearance and a centrally located nucleus.

Myringosclerosis Also known as tympanosclerosis or intratympanic tympanosclerosis, is a condition caused by calcification of collagen tissues in the tympanic membrane of the middle ear.

Myotonia A symptom of certain neuromuscular disorders characterised by the slow relaxation of the muscles after voluntary contraction or electrical stimulation.

***N*-Nitrosomorpholine** A human carcinogen.

***N*-Nitrosoproline** An indicator for *N*-nitrosation of amines.

Nicotinamide adenine dinucleotide phosphate (NADP) A coenzyme comprising nicotinamide mononucleotide coupled by pyrophosphate linkage to adenosine 2′,5′-bisphosphate; it acts as an electron carrier in numerous reactions, being alternately oxidised (NADP+) and reduced (NADPH).

NADPH The reduced form of nicotinamide adenine dinucleotide phosphate that serves as an electron carrier.

NAFLD Non-alcoholic fatty liver disease.

Narcosis A state of stupor, drowsiness or unconsciousness produced by drugs.

Narcotic An agent that produces narcosis; in moderate doses it dulls the senses, relieves pain and induces sleep; in excessive dose it causes stupor, coma, convulsions and death.

Nasopharynx Upper part of the alimentary continuous with the nasal passages.

Natriorexia Excessive intake of sodium evoked by sodium depletion. *adj.* natriorexic, natriorexigenic.

Natriuresis The discharge of excessive large amount of sodium through urine. *adj.* natriuretic.

Natural killer cells (NK cells) A type of cytotoxic lymphocyte that constitutes a major component of the innate immune system.

Natural killer T (NKT) cells A heterogeneous group of T cells that share properties of both T cells and natural killer (NK) cells.

Nausea Sensation of unease and discomfort in the stomach with an urge to vomit.

Necropsy See Autopsy.

Necrosis Morphological changes that follow cell death, usually involving nuclear and cytoplasmic changes.

Neointima A new or thickened layer of arterial intima formed especially on a prosthesis or in atherosclerosis by migration and proliferation of cells from the media.

Neonatal *adj.* of or relating to newborn infants or an infant.

Neoplasia Abnormal growth of cells, which may lead to a neoplasm, or tumour.

Neoplasm Tumour; any new and abnormal growth, specifically one in which cell multiplication is uncontrolled and progressive. Neoplasms may be benign or malignant.

Neoplastic transformation Conversion of a tissue with a normal growth pattern into a malignant tumour.

Neovascularisation Is the development of tiny, abnormal, leaky blood vessels inside the eye.

Neovasculature Formation of new blood vessels.

Nephrectomised Kidneys surgically removed.

Nephrectomy Surgical removal of the kidney.

Nephric Relating to or connected with a kidney.

Nephrin Is a protein necessary for the proper functioning of the renal filtration barrier.

Nephritic syndrome Is a collection of signs (known as a syndrome) associated with disorders affecting the kidneys, more specifically glomerular disorders.

Nephritis Is inflammation of the kidney.

Nephrolithiasis Process of forming a kidney stone in the kidney or lower urinary tract.

Nephropathy A disorder of the kidney.

Nephrotic syndrome Nonspecific disorder in which the kidneys are damaged, causing them to leak large amounts of protein from the blood into the urine.

Nephrotoxicity Poisonous effect of some substances, both toxic chemicals and medication, on the kidney.

Nerve growth factor (NGF) A small protein that induces the differentiation and survival of particular target neurons (nerve cells).

Nervine A nerve tonic that acts therapeutically upon the nerves, particularly in the sense of a sedative that serves to calm ruffled nerves.

Neural tube defects (NTDs) Are common birth defects of the brain and spinal cord.

NEU 4 sialidase This protein belongs to a family of glycohydrolytic enzymes, which remove terminal sialic acid residues from various sialo derivatives, such as glycoproteins, glycolipids, oligosaccharides and gangliosides.

Neuralgia Is a sudden, severe painful disorder of the nerves.

Neuraminidase Glycoside hydrolase enzymes that cleaves the glycosidic linkages of neuraminic acids.

Neuraminidase inhibitors A class of antiviral drugs targeted at the influenza viruses whose mode of action consists of blocking the function of the viral neuraminidase protein, thus preventing the virus from reproducing.

Neurasthenia A condition with symptoms of fatigue, anxiety, headache, impotence, neuralgia and impotence.

Neurasthenic A substance used to treat nerve pain and/or weakness (i.e. neuralgia, sciatica, etc).

Neurectomy Surgical cutting through or removal of a nerve or a section of a nerve.

Neurite Refers to any projection from the cell body of a neuron.

Neuritis An inflammation of the nerve characterised by pain, sensory disturbances and impairment of reflexes. *adj.* neuritic.

Neuritogenesis The formation of neuritis. a*dj.* neuritogenic.

Neuroblastoma A common extracranial cancer that forms in nerve tissues, common in infancy.

Neuroendocrine *adj.* of, relating to, or involving the interaction between the nervous system and the hormones of the endocrine glands.

Neurogenesis Process by which neurons are generated from neural stem and progenitor cells.

Neurogenic Originating from the nerves of the nervous system.

Neurolathyrism Is a neurodegenerative disease that is caused by heavy consumption of *Lathyrus* legumes, resulting in weakness and paralysis of the legs.

Neuroleptic Refers to the effects on cognition and behaviour of antipsychotic drugs that reduce confusion, delusions, hallucinations and psychomotor agitation in patients with psychoses.

Neuroma Is a growth or tumour of nerve tissue.

Neuropathy A collection of disorders that occurs when the peripheral nervous sytems are damaged causing pain and numbness in the hands and feet.

Neuropharmacological Relating the effects of drugs on the neurosystem**.**

Neuroradiology Is a subspecialty of radiology focusing on the diagnosis and characterisation of abnormalities of the central and peripheral nervous system. *adj.* neuroradiologic.

Neurotrophic Relating to the nutrition and maintenance of nervous tissue (neurons).

Neutropenia a disorder of the blood, characterised by abnormally low levels of neutrophils.

Neutrophil Type of white blood cell, specifically a form of granulocyte.

Neurutrophin Protein that induces the survival, development and function of neurons.

***NF*-Kappa B (*NF*-kB)** Nuclear factor kappa-B, is an ubiquitous rapid response transcription factor in cells involved in immune and inflammatory reactions.

Niacin Vitamin B3. See Vitamin B3.

Niacinamide An amide of niacin, also known as nicotinamide. See Vitamin B3.

NIH3T3 cells A mouse embryonic fibroblast cell line used in the cultivation of keratinocytes.

Nidation Implantation.

Niosomes Are novel, vesicular, drug delivery systems composed of nonionic surfactants instead of phospholipids; they are capable of entrapping hydrophilic and hydrophobic drugs.

Nitrogen (N) Is an essential building block of amino and nucleic acids and proteins and is essential to all living organisms. Protein-rich vegetables like legumes are rich food sources of nitrogen.

NK cells Natural killer cells, a type of cytotoxic lymphocyte that constitutes a major component of the innate immune system.

NK1.1+ T (NKT) cells A type of natural killer T (NKT) cells. See Natural killer T cells.

NMDA receptor *N*-Methyl-D-aspartate receptor, the predominant molecular device for controlling synaptic plasticity and memory function. A brain receptor activated by the amino acid glutamate, which when excessively stimulated may cause cognitive defects in Alzheimer's disease.

Nociceptive Causing pain, responding to a painful stimulus.

Nociceptors Specialised peripheral sensory neurons that respond to potentially damaging stimuli by sending nerve signals to the spinal cord and brain.

Non-osteogenic Fibromata of bone, a benign tumour of bone which shows no evidence of ossification.

Non-alcoholic fatty liver disease One cause of a fatty liver, occurring when fat is deposited (steatosis) in the liver not due to excessive alcohol use

Nootropics Are substances which are claimed to boost human cognitive abilities (the functions and capacities of the brain). Also popularly referred to as 'smart drugs', 'smart nutrients', 'cognitive enhancers' and 'brain enhancers'.

Noradrenalin See Norepinephrine.

Norepinephrine A substance, both a hormone and neurotransmitter, secreted by the adrenal medulla and the nerve endings of the sympathetic nervous system to cause vasoconstriction and increases in heart rate, blood pressure and the sugar level of the blood. Also called levarterenol, noradrenalin.

Normoglycaemic Having the normal amount of glucose in the blood.

Normotensive Having normal blood pressure.

Nocebo A harmless substance that when taken by a patient is associated with unpleasant or harmful effects due to negative expectations or the psychological state of the person.

Nosocomial infections Infections which are a result of treatment in a hospital or a healthcare service unit, but secondary to the patient's original condition.

NPC1L1 Niemann–Pick C1-like 1 gene that plays a major role in cholesterol homeostasis. It is critical for the uptake of cholesterol across the plasma membrane of the intestinal enterocyte.

Nrf2 Nuclear erythroid 2-related factor 2, a transcription factor that activates ARE-containing genes.

Nrf2/ARE pathway Plays an important role in inducing phase II detoxifying enzymes and antioxidant proteins and has been considered a potential target for cancer chemoprevention because it eliminates harmful reactive oxygen species or reactive intermediates generated from carcinogens.

Nuclear factor erythroid 2-related factor 2 (Nrf2) A transcription factor that plays a major role in response to oxidative stress by binding to antioxidant-responsive elements that regulate many hepatic phase I and II enzymes as well as hepatic efflux transporters.

Nucleosomes Fundamental repeating subunits of all eukaryotic chromatin, consisting of a DNA chain coiled around a core of histones.

Nulliparous Term used to describe a woman who has never given birth.

Nyctalopia Night blindness, impaired vision in dim light and in the dark, due to impaired function of certain specialised vision cells.

Nystagmus Fast, involuntary movements of the eyes.

Nocturia Excessive urination at night; especially common in older men.

Obsessive–compulsive disorder (OCD) A common psychiatric disorder defined by the presence of obsessive thoughts and repetitive compulsive actions; self-grooming.

Occludin A novel integral membrane protein localising at tight junctions. *cf.* tight junction.

Occlusion Closure or blockage (as of a blood vessel).

Occlusive peripheral arterial disease (PAOD) Also known as peripheral vascular disease (PVD), or peripheral arterial disease (PAD), refers to the obstruction of large arteries not within the coronary, aortic arch vasculature, or the brain. PVD can result from atherosclerosis, inflammatory processes leading to stenosis, an embolism, or thrombus formation.

Oculomotor nerve The third of twelve paired cranial nerves.

Odds ratio A statistical measure of effect size, describing the strength of association or non-independence between two binary data values.

Odontalgia Toothache. *adj.* odontalgic.

Odontopathy Any disease of the teeth.

Oedema See Edema.

Oestrogen Female hormone produced by the ovaries that play an important role in the estrous cycle in women.

Oestrogen receptor (ER) Is a protein found in high concentrations in the cytoplasm of breast, uterus, hypothalamus and anterior hypophysis cells; ER levels are measured to determine a breast CA's potential for response to hormonal manipulation.

Oestrogen receptor positive (ER+) Means that oestrogen is causing the tumour to grow and that the breast cancer should respond well to hormone suppression treatments.

Oestrogen receptor negative (ER-) Tumour is not driven by oestrogen and needs another test to determine the most effective treatment.

Oestrus Sexual excitement or heat of female; or period of this characterised by changes in the sex organs.

Oligoarthritis An inflammation of two, three or four joints.

Oligoasthenoteratozoospermia A combination of asthenozoospermia (reduced sperm motility) and oligozoospermia (low spermatozoon count).

Oligonucleosome A series of nucleosomes.

Oligospermia or oligozoospermia Refers to semen with a low concentration of sperm, commonly associated with male infertility.

Oliguria Decreased production of urine.

Oligoanuria Insufficient urine volume to allow for administration of necessary fluids, etc.

Omega-3 fatty acids Are essential polyunsaturated fatty acids that have in common a final carbon–carbon double bond in the n–3 position. Dietary sources of omega-3 fatty acids include fish oil and certain plant/nut oils. The three most nutritionally important omega-3 fatty acids are alpha-linolenic acid, eicosapentaenoic acid (EPA) and docosahexaenoic acid (DHA). Research indicates that omega-3 fatty acids are important in health promotion and disease and can help prevent a wide range of medical problems, including cardiovascular disease, depression, asthma and rheumatoid arthritis.

Omega-6 fatty acids Are essential polyunsaturated fatty acids that have in common a final carbon–carbon double bond in the n–6 position. Omega-6 fatty acids are considered essential fatty acids (EFAs) found in vegetable oils, nuts and seeds. They are essential to human health but cannot be made in the body. Omega-6 fatty acids—found in vegetable oils, nuts and seeds—are a beneficial part of a heart-healthy eating. Omega-6 and omega-3 PUFA play a crucial role in heart and brain function and in normal growth and development. Linoleic acid (LA) is the main omega-6 fatty acid in foods, accounting for 85% to 90% of the dietary omega-6 PUFA. Other omega-6 acids include gamma-linolenic acid or GLA, sometimes called gamoleic acid, eicosadienoic acid, arachidonic acid and docosadienoic acid.

Omega-9 fatty acids Are not essential polyunsaturated fatty acids that have in common a final carbon–carbon double bond in the n–9 position. Some n–9s are common components of animal fat and vegetable oil. Two n–9 fatty acids important in industry are:

Oleic acid (18, 1, n–9), which is a main component of olive oil

Erucic acid (22, 1, n–9), which is found in rapeseed, wallflower seed and mustard seed

Oncogenes Genes carried by tumour viruses that are directly and solely responsible for the neoplastic (tumorous) transformation of host cells.

Oncosis Accidental cell death, also referred to swelling necrosis.

Ophthalmia Severe inflammation of the eye, or the conjunctiva or deeper structures of the eye. Also called ophthalmitis.

Ophthalmia (sympathetic) Inflammation of both eyes following trauma to one eye.

Ophthalmopathy An autoimmune disease where the thyroid gland is overactive leading to ocular manifestations.

Opiate Drug derived from the opium plant.

Opioid receptors A group of G protein-coupled receptors located in the brain and various organs that bind opiates or opioid substances.

Oppilation Obstruction particularly of the lower intestines.

Optic placode An ectodermal placode from which the lens of the embryonic eye develops; also called lens placode.

ORAC (oxygen radical absorbance capacity) A method of measuring antioxidant capacities in biological samples.

Oral submucous fibrosis A chronic debilitating disease of the oral cavity characterised by inflammation and progressive fibrosis of the submucosa tissues.

Oral thrush An infection of yeast fungus, *Candida albicans*, in the mucous membranes of the mouth.

Orchidectomy Surgery to remove one or both testicles.

Orchidectomised With the testis removed.

Orchitis An acute painful inflammatory reaction of the testis secondary to infection by different bacteria and viruses.

Orexigenic Increasing or stimulating the appetite.

Orofacial dyskinesia Abnormal involuntary movements involving muscles of the face, mouth, tongue, eyes and, occasionally, the neck—may be unilateral or bilateral and constant or intermittent.

Oropharyngeal Relating to the oropharynx.

Oropharynx Part of the pharynx between the soft palate and the epiglottis.

Osmophobia A fear, aversion or psychological hypersensitivity to odours.

Ostalgia, ostealgia Pain in the bones. Also called osteodynia.

Osteoarthritis Is the deterioration of the joints that becomes more common with age.

Osteoarthrosis Chronic noninflammatory bone disease.

Osteoblast A mononucleate cell that is responsible for bone formation.

Osteoblastic Relating to osteoblasts.

Osteocalcin A noncollagenous protein found in bone and dentin, also refers to as bone gamma-carboxyglutamic acid-containing protein.

Osteoclasts A kind of bone cell that removes bone tissue by removing its mineralised matrix.

Osteoclastogenesis The production of osteoclasts.

Osteodynia Pain in the bone.

Osteogenic Derived from or composed of any tissue concerned in bone growth or repair.

Osteomalacia Refers to the softening of the bones due to defective bone mineralisation.

Osteomyelofibrosis A myeloproliferative disorder in which fibrosis and sclerosis finally lead to bone marrow obliteration.

Osteopenia Reduction in bone mass, usually caused by a lowered rate of formation of new bone that is insufficient to keep up with the rate of bone destruction.

Osteoporosis A disease of bone that leads to an increased risk of fracture.

Osteoprotegerin Also called osteoclastogenesis inhibitory factor (OCIF), a cytokine, which can inhibit the production of osteoclasts.

Osteosacrcoma A malignant bone tumour. Also called osteogenic sarcoma.

Otalgia Earache, pain in the ear.

Otic placode A thickening of the ectoderm on the outer surface of a developing embryo from which the ear develops.

Otitis Inflammation of the inner or outer parts of the ear.

Otitis media Inflammation of the middle ear.

Otorrhea Running drainage (discharge) exiting the ear.

Otopathy Disease of the ear.

Ovariectomised With one or two ovaries removed.

Ovariectomy Surgical removal of one or both ovaries.

Oxidation The process of adding oxygen to a compound, dehydrogenation or increasing the electro-negative charge.

Oxidoreductase activity Catalysis of an oxidation–reduction (redox) reaction, a reversible chemical reaction. One substrate acts as a hydrogen or electron donor and becomes oxidised, while the other acts as hydrogen or electron acceptor and becomes reduced.

Oxygen radical absorbance capacity (ORAC) A method of measuring antioxidant capacities in biological samples.

Oxytocic *adj.* hastening or facilitating childbirth, especially by stimulating contractions of the uterus.

Oxytocin Is a mammalian hormone that also acts as a neurotransmitter in the brain. It is best known for its roles in female reproduction. It is released in large amounts after distension of the cervix and vagina during labour and after stimulation of the nipples, facilitating birth and breastfeeding, respectively*****.

Oxyuriasis Infestation by pinworms.

Ozoena Discharge of the nostrils caused by chronic inflammation of the nostrils.

p.o. Per os, oral administration.

***P*-Glycoprotein (P-gp, ABCB1, MDR1)** A cell membrane-associated drug-exporting protein that transports a variety of drug substrates from cancer cells.

P-Selectin Also known as CD62P, GMP-140, LLECAM-3 and PADGEM, a member of the selectin family. It is expressed by activated platelets and endothelial cells.

P65 transcription factor Is a protein that in humans is encoded by the RELA gene. Its alternative name is nuclear factor NF-kappa-B p65 subunit.

P300/CBP Are transcriptional coactivators that play critical roles in integrating multiple signal-dependent transcription events and may have specific roles in tumour suppression pathways.

p21waf1/cip1 Encodes a cyclin-dependent kinase inhibitor that is transcriptionally activated by the p53 tumour suppressor gene, transforming growth factor beta 1 (TGF-beta 1), AP2 and other pathways, all regulating apoptosis and the cell cycle.

Palliative Relieving pain without alleviating the underlying problem.

Palinosmia Olfactory perservation.

Palpebral ptosis The abnormal drooping of the upper lid, caused by partial or total reduction in levator muscle function.

Palpitation Rapid pulsation or throbbing of the heart.

Paludism State of having symptoms of malaria characterised by high fever and chills.

Pancreatectomised Having undergone a pancreatectomy.

Pancreatectomy Surgical removal of all or part of the pancreas.

Pancreatitis Inflammation of the pancreas.

Pancytopenia A haematological condition in which there is a reduction in the number of red and white blood cells, as well as platelets.

Pantothenic acid Vitamin B5. See Vitamin B5.

Papain A protein-degrading enzyme used medicinally and to tenderise meat.

Papilloma A benign epithelial tumour growing outwardly like in fingerlike fronds.

Papule A small, solid, usually inflammatory elevation of the skin that does not contain pus.

Paradontosis Is the inflammation of gums and other deeper structures, including the bone.

Parageusia Abnormal sense of taste.

Paralytic Person affected with paralysis, pertaining to paralysis.

Paraoxonase An enzyme that protects against oxidation of low-density lipoprotein and affects the risk of coronary artery disease.

Paraplegia An impairment in motor or sensory function of the lower extremities.

Parasitemia Presence of parasites in blood. *adj.* parasitemic.

Parasympathetic nervous system Subsystem of the nervous systems that slows the heart rate and increases intestinal and gland activity and relaxes the sphincter muscles.

Parasympathomimetic Having an action resembling that caused by stimulation of the parasympathetic nervous system.

Parenteral administration Administration by intravenous, subcutaneous or intramuscular routes.

Paresis A condition characterised by partial loss of movement, or impaired movement.

Paresthesia A sensation of tingling, burning, pricking or numbness of a person's skin with no apparent long-term physical effect. Also known as 'pains and needles'.

Parotitis Inflammation of salivary glands.

Paroxysm A sudden outburst of emotion or action and a sudden attack, recurrence or intensification of a disease.

Paroxystic Relating to an abnormal event of the body with an abrupt onset and an equally sudden return to normal.

PARP See Poly-(ADP-ribose) polymerase.

Pars compacta Is a portion of the substantia nigra (a brain structure located in the midbrain).

Parturition Act of childbirth.

Pathognomonic Distinctively characteristic of a particular disease.

PCAF P300/CBP-associated factor, a histone acetyl transferase (HAT) that plays an important role in the remodelling of chromatin and the regulation of gene expression, transcription, cell cycle progression and differentiation.

PCE/PCN ratio Polychromatic erythrocyte/normochromatic erythrocyte ratio used as a measure of cytotoxic effects.

PCNA Proliferating cell nuclear antigen, an auxiliary protein of DNA polymerase delta involved in modulating eukaryotic DNA replication.

pCREB Phosphorylated cAMP (adenosine 3′5′ cyclic monophosphate)-response element binding protein.

PDEF Acronym for prostate-derived ETS factor, an ETS (epithelial-specific E26 transforming sequence) family member that has been identified as a potential tumour suppressor.

PDGR receptor (platelet-derived growth factor receptor) Are cell surface tyrosine kinase receptors for members of the platelet-derived growth factor (PDGF) family.

PDGFs Platelet-derived growth factors constitute a group of growth factors that play a significant role in blood vessel formation and the growth of blood vessels.

Pectoral Pertaining to or used for the chest and respiratory tract.

Pectoralgia Pain experienced in the thorax or chest.

pERK Phosphorylated extracellular signal-regulated kinase, protein kinases involved in many cell functions.

P53 Also known as protein 53 or tumour protein 53, is a tumour suppressor protein that in humans is encoded by the TP53 gene.

Peliosis See Purpura.

Pellagra Is a systemic nutritional wasting disease caused by a deficiency of vitamin B3 (niacin).

Pemphigus Describes a group of autoimmune disorders in which there is blistering of the skin and/or mucosal surfaces.

Pemphigus neonatorum Staphylococcal scalded skin syndrome, a bacterial disease of infants, characterised by elevated vesicles or blebs on a normal or reddened skin .

Peptic ulcer A sore in the lining of the stomach or duodenum, the first part of the small intestine.

Peptide YY A short (36-amino acid) pancreatic protein released by cells in the ileum and colon in response to feeding.

Percutaneous Pertains to a medical procedure where access to inner organs or tissues is done via needle puncture of the skin.

Perfusion To force fluid through the lymphatic system or blood vessels to an organ or tissue.

Periapical periodontitis Is the inflammation of the tissue adjacent to the tip of the tooth's root.

Perifuse To flush a fresh supply of bathing fluid around all of the outside surfaces of a small piece of tissue immersed in it.

Perilipins Highly phosphorylated adipocyte proteins that are localised at the surface of the lipid droplet.

Perimenopause Is the phase before menopause actually takes place, when ovarian hormone production is declining and fluctuating. *adj.* perimenopausal.

Perineum The region between the thighs inferior to the pelvic diaphragm.

Perineal Pertaining to the perineum.

Periodontal ligament (PDL) Is a group of specialised connective tissue fibres that essentially attach a tooth to the bony socket.

Periodontitis Is a severe form of gingivitis in which the inflammation of the gums extends to the supporting structures of the tooth. Also called pyorrhea.

Perioral paresthesias Are sensations of numbness and tingling around the mouth.

Peripheral arterial disease (PAD) Is a disease in which plaque builds up in the arteries that carry blood to your head, organs and limbs.

Peripheral neuropathy Refers to damage to nerves of the peripheral nervous system.

Peripheral neuropathic pain (PNP) Refers to situations where nerve roots or peripheral nerve trunks have been damaged by mechanical and/or chemical stimuli that exceeded the physical capabilities of the nervous system. Symptoms may include pain, parestesia, dysesthesia, spasm, weakness, hypoesthesia or anaesthesia.

Peripheral vascular disease (PVD) See Peripheral artery occlusive disease .

Peristalsis A series of organised, wave-like muscle contractions that occur throughout the digestive tract.

PERK A transmembrane protein kinase of the PEK family resident in the endoplasmic reticulum (ER) membrane and is linked to insulin processing.

Perlingual Through or by way of the tongue.

Perniosis An abnormal reaction to cold that occurs most frequently in women, children and the elderly. Also called chilblains.

Per os (P.O.) Oral administration.

Peroxisome proliferator-activated receptors (PPARs) A family of nuclear receptors that are involved in lipid metabolism, differentiation, proliferation, cell death and inflammation.

Peroxisome proliferator-activated receptor alpha (PPAR-alpha) A nuclear receptor protein, transcription factor and a major regulator of lipid metabolism in the liver.

Peroxisome proliferator-activated receptor gamma (PPAR-γ) A type II nuclear receptor protein that regulates fatty acid storage and glucose metabolism.

Pertussis Whooping cough, sever cough.

Peyers patches Patches of lymphoid tissue or lymphoid nodules on the walls of the ileal-small intestine.

PGE-2 Prostaglandin E2, a hormone-like substance that is released by blood vessel walls in response to infection or inflammation that acts on the brain to induce fever.

Phagocytes Are the white blood cells that protect the body by ingesting (phagocytosing) harmful foreign particles, bacteria and dead or dying cells. *adj.* phagocytic.

Phagocytosis Is a process the human body uses to destroy dead or foreign cells.

Phantosmia A form of olfactory hallucination.

Pharmacognosis The branch of pharmacology that studies the composition, use and history of drugs.

Pharmacodynamics Branch of pharmacology dealing with the effects of drugs and the mechanism of their action.

Pharmacokinetics Branch of pharmacology concerned with the movement of drugs within the body including processes of absorption, distribution, metabolism and excretion in the body.

Pharmacopoeia Authoritative treatise containing directions for the identification of drug samples and the preparation of compound medicines, published by the authority of a government or a medical or pharmaceutical society and in a broader sense is a general reference work for pharmaceutical drug specifications.

Pharyngitis, pharyngolaryngitis Inflammation of the pharynx and the larynx.

Pharyngolaryngeal Pertaining to the pharynx and larynx.

Pharyngopathy Disease of the pharynx.

Phase II drug metabolising enzymes Play an important role in biotransformation of endogenous compounds and xenobiotics to more easily excretable forms as well as in the metabolic inactivation of pharmacologically active compounds. Phase II drug metabolising enzymes are mainly transferases.

Phenolics Class of chemical compounds consisting of a hydroxyl group (–OH) bonded directly to an aromatic hydrocarbon group.

Pheochromocytoma Is a rare neuroendocrine tumour that usually originates from

the adrenal glands' chromaffin cells, causing overproduction of catecholamines, powerful hormones that induce high blood pressure and other symptoms.

Phlebitis Is an inflammation of a vein, usually in the legs.

Phlegm Abnormally viscid mucus secreted by the mucosa of the respiratory passages during certain infectious processes.

Phlegmon A spreading, diffuse inflammation of the soft or connective tissue due to infection by streptococci bacteria.

Phonophobia Fear of loud sound.

Phloroglucinol A white, crystalline compound used as an antispasmodic, analytical reagent and decalcifier of bone specimens for microscopic examination.

Phosphatidylglycerol Is a glycerophospholipid found in pulmonary active surface lipoprotein and consists of an L-glycerol 3-phosphate backbone ester bonded to either saturated or unsaturated fatty acids on carbons 1 and 2.

Phosphatidylinositol 3-kinases (PI 3-kinases or PI3Ks) A group of enzymes involved in cellular functions such as cell growth, proliferation, differentiation, motility, survival and intracellular trafficking, which in turn are involved in cancer.

Phosphatidylserine A phosphoglyceride phospholipid that is one of the key building blocks of cellular membranes, particularly in the nervous system. It is derived from soy lecithin.

Phosphaturia A urinary tract condition of excessive urine phosphorus, causing urine to appear cloudy or murky colour; also called hypophosphatemia.

Phosphodiesterases A diverse family of enzymes that hydrolyse cyclic nucleotides and thus play a key role in regulating intracellular levels of the second messengers, cAMP and cGMP, and hence cell function.

Phosphoenolpyruvate C kinase (PEPCK) An enzyme in the lyase family used in the metabolic pathway of gluconeogenesis.

Phospholipase An enzyme that hydrolyses phospholipids into fatty acids and other lipophilic substances.

Phospholipase A2 (PLA2) A small lipolytic enzyme that releases fatty acids from the second carbon group of glycerol. Plays an essential role in the synthesis of prostaglandins and leukotrienes.

Phospholipase C Enzymes that cleaves phospholipase.

Phospholipase C gamma (PLC gamma) Enzymes that cleaves phospholipase in cellular proliferation and differentiation, and its enzymatic activity is upregulated by a variety of growth factors and hormones.

Phosphorus (P) Is an essential mineral that makes up 1% of a person's total body weight and is found in the bones and teeth. It plays an important role in the body's utilisation of carbohydrates and fats; in the synthesis of protein for the growth, maintenance, and repair of cells and tissues. It is also crucial for the production of ATP, a molecule the body uses to store energy. Main sources are meat and milk; fruits and vegetables provide small amounts.

Photoaging Is the term that describes damage to the skin caused by intense and chronic exposure to sunlight resulting in premature aging of the skin.

Photocarcinogenesis Represents the sum of a complex of simultaneous and sequential biochemical events that ultimately lead to the occurrence of skin cancer caused by exposure to the sun.

Photodermatoses Skin disorders caused by exposure to sunlight.

Photophobia Abnormal visual intolerance to light.

Photopsia An affection of the eye, in which the patient perceives luminous rays, flashes, coruscations, etc.

Photosensitivity Sensitivity towards light.

Phthisis An archaic name for tuberculosis.

Phytohemagglutinin A lectin found in plant that is involved in the stimulation of lymphocyte proliferation.

Phytonutrients Certain organic components of plants that are thought to promote human health. Fruits, vegetables, grains, legumes, nuts and teas are rich sources of phytonutrients. Phytonutrients are not 'essential' for life. Also called phytochemicals.

Phytosterols A group of steroid alcohols, cholesterol-like phytochemicals naturally occurring in plants like vegetable oils, nuts and legumes.

Pica The persistent eating of substances with no nutrition, such as dirt, chalk, sand, ice, clay or paint.

Piebaldism Rare autosomal dominant disorder of melanocyte development characterised by distinct patches of skin and hair that contain no pigment.

Piles See Haemorrhoids.

PI3K Phosphoinositide 3-kinase.

PI13K/AKT signalling pathways Are involved in the modulation of cell survival, cell cycle progression and cellular growth in cancer.

Pityriasis lichenoides Is a rare skin disorder of unknown aetiology characterised by multiple papules and plaques.

Pityriasis versicolor Common fungal infection of the skin; the fungus interferes with the normal pigmentation of the skin, resulting in small, discoloured patches.

PKC Protein kinase C, a membrane-bound enzyme that phosphorylates different intracellular proteins and raised intracellular Ca levels.

PKC delta inhibitors Protein kinase C delta inhibitors that induce apoptosis of haematopoietic cell lines.

Placebo A sham or simulated medical intervention.

Placode A platelike epithelial thickening in the embryo where some organ or structure later develops.

Plantar verruca Wart occurring on the sole of the foot.

Plasma The yellow-coloured liquid component of blood, in which blood cells are suspended.

Plasma kallikrein A serine protease, synthesised in the liver and circulated in the plasma.

Plasmalemma Plasma membrane.

Plasmin A proteinase enzyme that is responsible for digesting fibrin in blood clots.

Plasminogen The proenzyme of plasmin, whose primary role is the degradation of fibrin in the vasculature.

Plasminogen activator inhibitor-1 (PAI-1) Also known as endothelial plasminogen activator inhibitor or serpin E1, is a serine protease inhibitor (serpin) that functions as the principal inhibitor of tissue plasminogen activator (tPA) and urokinase (uPA), the activators of plasminogen and hence fibrinolysis (the physiological breakdown of blood clots).

Plaster Poultice.

Platelet-activating factor (PAF) Is an acetylated derivative of glycerophosphorylcholine, released by basophils and mast cells in immediate hypersensitive reactions and macrophages and neutrophils in other inflammatory reactions. One of its main effects is to induce platelet aggregation.

Platelet-derived growth factor (PDGF) Is one of the numerous growth factors or proteins that regulate cell growth and division.

PLC gamma Phospholipase C gamma plays a central role in signal transduction.

Pleurisy Is an inflammation of the pleura, the lining of the pleural cavity surrounding the lungs, which can cause painful respiration and other symptoms. Also known as pleuritis.

Pneumonia An inflammatory illness of the lung caused by bacteria or viruses.

Pneumotoxicity Damage to lung tissues.

Poliomyelitis Is a highly infectious viral disease that may attack the central nervous system and is characterised by symptoms that range from a mild non-paralytic infection to total paralysis in a matter of hours; also called polio or infantile paralysis.

Pollakiuria Extraordinary daytime urinary frequency.

Poly-(ADP-ribose) polymerase (PARP) A protein involved in a number of cellular processes, especially DNA repair and programmed cell death.

Polyarthritis Is any type of arthritis which involves five or more joints.

Polychromatic erythrocyte (PCE) An immature red blood cell containing RNA that can be differentiated by appropriate staining techniques from a normochromatic erythrocyte (NCE), which lacks RNA.

Polycystic kidney disease Is a kidney disorder passed down through families in which multiple cysts form on the kidneys, causing them to become enlarged.

Polycystic ovary syndrome Imbalance of woman's sex hormone; this imbalance may cause changes in menstrual cycle, excessive hair growth, acne, obesity, reduced fertility and an increased risk of diabetes. The ovaries are larger and have many cysts or follicles that rarely grow to maturity or produce eggs capable of being fertilised.

Polycythaemia A type of blood disorder characterised by the production of too many red blood cells.

Polymorphnuclear Having a lobed nucleus. Used especially of neutrophilic white blood cells.

Polyneuritis Widespread inflammation of the nerves.

Polyneuritis gallinarum A nervous disorder in birds and poultry.

Polyneuropathy Simultaneous malfunction of many peripheral nerves throughout the body.

Polyp A growth that protrudes from a mucous membrane.

Polyphagia Medical term for excessive hunger or eating.

Polyposis Describes a condition where there are a lot of polyps.

PolyQ disease Polyglutamine repeat diseases are neurodegenerative ailments elicited by glutamine-encoding CAG nucleotide expansions within endogenous human genes.

Polyuria A condition characterised by the passage of large volumes of urine with an increase in urinary frequency.

Pomade A thick oily dressing.

Porphyria A disorder wherein the body cannot convert naturally occurring compounds (porphyrins) into haem which contains iron.

Porphyrin Any of a class of water-soluble, nitrogenous biological pigments, derivatives of which include the haemoproteins.

Postherpetic neuralgia (PHN) is neuralgia (pain in the nerves) caused by the varicella herpes zoster virus. The pain may last for more than a month or more after a shingles infection occurred.

Postpartum depression Depression after pregnancy; also called postnatal depression.

Postprandial After mealtime.

Postural hypotension Also called orthostatic hypotension—a condition of low blood pressure that can cause dizziness.

Potassium (K) Is an element that is essential for the body's growth and maintenance. It is necessary to keep a normal water balance between the cells and body fluids, for cellular enzyme activities, and plays an essential role in the response of nerves to stimulation and in the contraction of muscles. Potassium is found in many plant foods and fish (tuna, halibut): chard, mushrooms, spinach, fennel, kale, mustard greens, Brussels sprouts, broccoli, cauliflower, cabbage winter squash, eggplant, cantaloupe, tomatoes, parsley, cucumber, bell pepper, turmeric, ginger root, apricots, strawberries, avocado and banana.

Poultice Is a soft moist mass, often heated and medicated, that is spread on cloth over the skin to treat an aching, inflamed or painful part of the body. Also called cataplasm.

PPARs Peroxisome proliferator-activated receptors—a group of nuclear receptor proteins that function as transcription factors regulating the expression of genes.

PR interval Is the time (in seconds) from the beginning of the P wave (onset of atrial depolarisation) to the beginning of the QRS complex.

Prebiotics A category of functional food, defined as non-digestible food ingredients that beneficially affect the host by selectively stimulating the growth and/or activity of one or a limited number of bacteria in the colon, and thus improve host health. *cf.* probiotics.

Pre-eclampsia Toxic condition of pregnancy characterised by high blood pressure, abnormal weight gain, proteinuria and oedema.

Prenidatory phase Preimplantation phase.

Prepubertal Before puberty; pertaining to the period of accelerated growth preceding gonadal maturity.

Pregnane X receptor (PXR, NR1I2) is a ligand-activated transcription factor that plays a role not only in drug metabolism and transport but also in various other biological processes.

Pregnenolone A steroid hormone produced by the adrenal glands, involved in the steroidogenesis of other steroid hormones like progesterone, mineralocorticoids, glucocorticoids, androgens and oestrogens.

Prenidatory Referring to the time period between fertilisation and implantation.

Prenylated flavones Flavones with an isoprenyl group in the 8 position, has been reported to have good anti-inflammatory properties.

Presyncopal sensation State consisting of light-headedness, muscular weakness, blurred vision and feeling faint.

Primiparous Relating to a woman who has given birth once.

Proangiogenic Promote angiogenesis (formation and development of new blood vessels).

Probiotication Enhancement with beneficial probiotic bacteria such as *Lactobacillus* species that can prevent the growth of intestinal pathogenic microflora.

Probiotics Are dietary supplements and live microorganisms containing potentially beneficial bacteria or yeasts that are taken into the alimentary system for healthy intestinal functions. *cf.* prebiotics.

Proctitis An inflammation of the rectum that causes discomfort, bleeding and, occasionally, a discharge of mucus or pus.

Procyanidin Also known as proanthocyanidin, oligomeric proanthocyanidin, leukocyanidin and leucoanthocyanin, is a class of flavanols found in many plants. It has antioxidant activity and plays a role in the stabilisation of collagen and maintenance of elastin.

Progestational Of or relating to the phase of the menstrual cycle immediately following ovulation, characterised by secretion of progesterone.

Proglottid One of the segments of a tapeworm.

Prognosis Medical term to describe the likely outcome of an illness.

Prokinetic Or gastroprokinetic, substance that enhances gastrointestinal motility by increasing the frequency of contractions in the small intestine or making them stronger.

Prolactin A hormone produced by the pituitary gland; it stimulates the breasts to produce milk in pregnant women. It is also present in males, but its role is not well understood.

Prolapse A common condition where the bladder, uterus and/or bowel protrudes into the vagina.

Prolapsus To fall or slip out of place.

Prolapsus ani Eversion of the lower portion of the rectum and protruding through the anus, common in infancy and old age.

Proliferating cell nuclear antigen (PCNA) A new marker to study human colonic cell proliferation.

Proliferative vitreoretinopathy (PVR) Is a blinding retinal condition. It involves the formation of pathological membranes, which dislodges the retina and thereby compromises an individual's ability to see.

Prolyl-4-hydroxylase (P4H) Key enzyme in collagen synthesis.

Promastigote The flagellate stage in the development of trypanosomatid protozoa, characterised by a free anterior flagellum.

Promyelocytic leukaemia A subtype of acute myelogenous leukaemia (AML), a cancer of the blood and bone marrow.

Pro-oxidants Chemicals that induce oxidative stress, either through creating reactive oxygen species or inhibiting antioxidant systems.

Prophylaxis Prevention or protection against disease.

Proptosis See Exophthalmos.

Prostacyclin A prostaglandin that is a metabolite of arachidonic acid, inhibits platelet aggregation and dilates blood vessels.

Prostaglandins A family of C 20 lipid compounds found in various tissues, associated with muscular contraction and the inflammation response such as swelling, pain, stiffness, redness and warmth.

Prostaglandin E2 (PEG-2) One of the prostaglandins, a group of hormone-like substances that participate in a wide range of body functions such as the contraction and relaxation of smooth muscle, the dilation and constriction of blood vessels, control of blood pressure and modulation of inflammation.

Prostaglandin E synthase An enzyme that in humans is encoded by the glutathione-dependent PTGES gene.

Prostanoids Term used to describe a subclass of eicosanoids (products of COX pathway) consisting of the prostaglandins (mediators of inflammatory and anaphylactic reactions), the thromboxanes (mediators of vasoconstriction) and the prostacyclins (active in the resolution phase of inflammation.)

Prostanoid EP 4 A prostaglandin receptor that may be involved in the neonatal adaptation of circulatory system, osteoporosis as well as initiation of skin immune responses.

Prostate A gland that surrounds the urethra at the bladder in the male.

Prostate cancer A disease in which cancer develops in the prostate, a gland in the male reproductive system. Symptoms include pain,

difficulty in urinating, erectile dysfunction and other symptoms.

Prostate-specific antigen (PSA) A protein produced by the cells of the prostate gland.

Protein kinase C (PKC) A family of enzymes involved in controlling the function of other proteins through the phosphorylation of hydroxyl groups of serine and threonine amino acid residues on these proteins. PKC enzymes play important roles in several signal transduction cascades.

Protein tyrosine phosphatase (PTP) A group of enzymes that remove phosphate groups from phosphorylated tyrosine residues on proteins.

Proteinase A protease (enzyme) involved in the hydrolytic breakdown of proteins, usually by splitting them into polypeptide chains.

Proteinuria Means the presence of an excess of serum proteins in the urine.

Proteolysis Cleavage of the peptide bonds in protein forming smaller polypeptides. *adj.* proteolytic.

Proteomics The large-scale study of proteins, particularly their structures and functions.

Prothrombin Blood-clotting protein that is converted to the active form, factor IIa, or thrombin, by cleavage.

Prothyroid Good for thyroid function.

Protheolithic Proteolytic. See Proteolysis.

Proto-oncogene A normal gene which, when altered by mutation, becomes an oncogene that can contribute to cancer.

Prurigo A general term used to describe itchy eruptions of the skin.

Pruritus Defined as an unpleasant sensation on the skin that provokes the desire to rub or scratch the area to obtain relief; itch, itching. *adj.* pruritic.

PSA Prostate-specific antigen, a protein which is secreted into ejaculate fluid by the healthy prostate. One of its functions is to aid sperm movement.

Pseudoaldosteronism Is a medical condition characterised by hypertension, reduced aldosterone secretion, hypokalemia and metabolic acidosis and associated with low plasma renin activity.

Pseudohyperaldosteronism (also pseudoaldosteronism) Is a medical condition that mimics hyperaldosteronism. Like hyperaldosteronism, it produces hypertension associated with low plasma renin activity and metabolic alkalosis associated with hypokalemia.

Pseudohypoaldosteronism A hereditary disorder of infancy characterised by severe salt and water depletion and other signs of aldosterone deficiency, although aldosterone secretion is normal or increased; causes include aldosterone receptor defects and renal dysfunction.

Psoriasis A common chronic, non-contagious autoimmune dermatosis that affects the skin and joints.

Psychoactive Having effects on the mind or behaviour.

Psychonautics Exploration of the psyche by means of approaches such as meditation, prayer, lucid dreaming, brain wave entrainment, etc.

Psychotomimetic Hallucinogenic.

Psychotropic Capable of affecting the mind, emotions and behaviour.

PTEN Phosphatase and tensin homolog, a tumour suppressor gene.

Ptosis Also known as drooping eyelid; caused by weakness of the eyelid muscle and damage to the nerves that control the muscles or looseness of the skin of the upper eyelid.

P13-K Is a lipid kinase enzyme involved in the regulation of a number of cellular functions such as cell growth, proliferation, differentiation, motility, survival and intracellular trafficking, which in turn are involved in cancer.

P13-K/AKT signalling pathway Shown to be important for an extremely diverse array of cellular activities—most notably cellular proliferation and survival.

Phthisis Silicosis with tuberculosis.

Ptosis Drooping of the upper eye lid.

PTP Protein tyrosine phosphatase.

PTPIB Protein tyrosine phosphatase 1B.

P21 Also known as cyclin-dependent kinase inhibitor 1 or CDK-interacting protein 1, is a potent cyclin-dependent kinase inhibitor.

Puerperal Pertaining to childbirth.

Puerperium Postpartum period.

Pulmonary embolism A blockage (blood clot) of the main artery of the lung.

Purgative A substance used to cleanse or purge, especially causing the immediate evacuation of the bowel.

Purpura Is the appearance of red or purple discolourations on the skin that do not blanch on applying pressure. Also called peliosis.

Purulent Containing pus discharge.

Purulent sputum Sputum containing, or consisting of, pus.

Pustule Small, inflamed, pus-filled lesions.

Pyelitis Acute inflammation of the pelvis of the kidney caused by bacterial infection.

Pyelonephritis An ascending urinary tract infection that has reached the pyelum (pelvis) of the kidney.

Pyoderma Bacterial skin infection.

Pyodermatitis Refers to inflammation of the skin.

Pyorrhea See Periodontitis.

Pyretic Referring to fever.

Pyrexia Fever of unknown origin.

Pyridoxal A chemical form of vitamin B6. See Vitamin B6.

Pyridoxamine A chemical form of vitamin B6. See Vitamin B6.

Pyridoxine A chemical form of vitamin B6. See Vitamin B6.

Pyrolysis Decomposition or transformation of a compound caused by heat. *adj.* pyrolytic.

PYY peptide A 36-amino acid peptide secreted by L cells of the distal small intestine and colon that inhibits gastric and pancreatic secretion.

QSR complex Series of deflections in an electrocardiogram that represent electrical activity generated by ventricular depolarisation prior to contraction of the ventricle.

QT interval Is a measure of the time between the start of the Q wave and the end of the T wave in the heart's electrical cycle. A prolonged QT interval is a biomarker for ventricular tachyarrhythmias and a risk factor for sudden death.

Quorum sensing (QS) The control of gene expression in response to cell density, is used by both Gram-negative and Gram-positive bacteria to regulate a variety of physiological functions.

Radiculitis Inflammation of the radicle of a nerve.

Radiodermatitis A skin disease associated with prolonged exposure to ionising radiation.

Radiolysis The dissociation of molecules by radiation.

Radioprotective Serving to protect or aiding in protecting against the injurious effect of radiations.

RAD23B UV excision repair protein RAD23 homolog B

RAGE Is the receptor for advanced glycation end products, a multiligand receptor that propagates cellular dysfunction in several inflammatory disorders, in tumours and in diabetes.

RANKL Receptor activator of nuclear factor kappa-B ligand, a type II membrane protein and a member of the tumour necrosis factor (TNF) superfamily.

RAS See Renin–angiotensin system or recurrent aphthous stomatitis.

Rash A temporary eruption on the skin; see Uticaria.

Reactive oxygen species Species such as superoxide, hydrogen peroxide and hydroxyl radical. At low levels, these species may function in cell signalling processes. At higher levels, these species may damage cellular macromolecules (such as DNA and RNA) and participate in apoptosis (programmed cell death).

Rec A Is a 38-kDa *Escherichia coli* protein essential for the repair and maintenance of DNA.

Receptor for advanced glycation end products (RAGE) Is a member of the immunoglobulin superfamily of cell surface molecules; mediates neurite outgrowth and cell migration upon stimulation with its ligand, amphoterin.

Reticulocyte Non-nucleated stage in the development of the red blood cell.

Reticulocyte lysate Cell lysate produced from reticulocytes, used as an in-vitro translation system.

Reticuloendothelial system Part of the immune system, consists of the phagocytic cells located in reticular connective tissue, primarily monocytes and macrophages.

Recurrent aphthous stomatitis, or RAS Is a common, painful condition in which recurring ovoid or round ulcers affect the oral mucosa.

Redox homeostasis Is considered as the cumulative action of all free radical reactions and antioxidant defences in different tissues.

Refrigerant A medicine or an application for allaying heat, fever or its symptoms.

Renal calculi Kidney stones.

Renal interstitial fibrosis Damage sustained by the kidneys' renal tubules and interstitial capillaries due to accumulation of extracellular waste in the wall of the small arteries and arterioles.

Renal resistive index (RRI) Measures the resistance of renal arterial flow to the kidney.

Renin Also known as an angiotensinogenase, is an enzyme that participates in the body's renin–angiotensin system (RAS). It regulates the body's mean arterial blood pressure.

Renin–angiotensin system (RAS) Also called the renin–angiotensin–aldosterone system (RAAS), is a hormone system that regulates blood pressure and water (fluid) balance.

Reperfusion The restoration of blood flow to an organ or tissue that has had its blood supply cut off, after a heart attack.

Reporter gene A transfected gene that produces a signal, such as green fluorescence, when it is expressed.

Resistin A cysteine-rich protein secreted by adipose tissue of mice and rats.

Resolutive A substance that induces subsidence of inflammation.

Resolvent Reduce inflammation or swelling.

Respiratory burst Is the rapid release of reactive oxygen species (superoxide radical and hydrogen peroxide) from different cells.

Resorb To absorb or assimilate a product of the body such as an exudate or cellular growth.

Restenosis Is the reoccurrence of stenosis, a narrowing of a blood vessel, leading to restricted blood flow.

Resveratrol Is a phytoalexin produced naturally by several plants when under attack by pathogens such as bacteria or fungi. It is a potent antioxidant found in red grapes and other plants.

Reticuloendothelial system Part of the immune system that consists of the phagocytic cells located in reticular connective tissue. Also called macrophage system or mononuclear phagocyte system.

Retinal ischemia Is a common cause of visual impairment and blindness.

Retinitis pigmentosa (RP) An inherited, degenerative eye disease that causes severe vision impairment and may lead to blindness.

Retinol A form of vitamin A; see Vitamin A.

Retinoblastoma protein A tumour suppressor protein that is dysfunctional in several major cancers.

Retinopathy A general term that refers to some form of noninflammatory damage to the retina of the eye.

Revulsive Counterirritant, used for swellings.

Reye's syndrome A potentially fatal disease that has numerous detrimental effects to many organs, especially the brain and liver, and occurs commonly in children after a viral infection.

Rhabdomyolysis Breakdown of muscle fibres leading to the release of muscle fibre content (myoglobin) into the bloodstream.

Rheumatic Pertaining to rheumatism or to abnormalities of the musculoskeletal system.

Rheumatism, rheumatic disorder, rheumatic diseases Refers to various painful medical conditions which affect bones, joints, muscles and tendons. Rheumatic diseases are characterised by the signs of inflammation—redness, heat, swelling and pain.

Rheumatoid arthritis (RA) Is a chronic, systemic autoimmune disorder that most commonly causes inflammation and tissue damage in joints (arthritis) and tendon sheaths, together with anaemia.

Rhinitis Irritation and inflammation of some internal areas of the nose and the primary symptom of rhinitis is a runny nose.

Rhinopathy Disease or malformation of the nose.

Rhinoplasty Is surgery to repair or reshape the nose.

Rhinorrhea Commonly known as a runny nose, characterised by an unusually significant amount of nasal discharge.

Rhinosinusitis Inflammation of the nasal cavity and sinuses.

Rho GTPases Rho-guanosine triphosphate hydrolase enzymes are molecular switches that regulate many essential cellular processes, including actin dynamics, gene transcription, cell cycle progression and cell adhesion.

Ribosome-inactivating proteins Proteins that are capable of inactivating ribosomes.

Rickets Is a softening of the bones in children potentially leading to fractures and deformity.

Ringworm Dermatophytosis, a skin infection caused by fungus.

Roborant Restoring strength or vigour, a tonic.

Rotavirus The most common cause of infectious diarrhoea (gastroenteritis) in young children and infants, one of several viruses that causes infections called stomach flu.

Rubefacient A substance for external application that produces redness of the skin, e.g. by causing dilation of the capillaries and an increase in blood.

Ryanodine receptor Intracellular Ca ++ channels in animal tissues like muscles and neurons.

S.C. Abbreviation for subcutaneous, beneath the layer of skin.

S-T segment The portion of an electrocardiogram between the end of the QRS complex and the beginning of the T wave. Elevation or depression of the S-T segment is the characteristics of myocardial ischemia or injury and coronary artery disease.

Salve Medical ointment used to soothe the head or body surface.

Sapraemia See Septicaemia.

Sarcoma Cancer of the connective or supportive tissue (bone, cartilage, fat, muscle, blood vessels) and soft tissues.

Sarcopenia Degenerative loss of skeletal muscle mass and strength associated with aging.

Sarcoplasmic reticulum A special type of smooth endoplasmic reticulum found in smooth and striated muscle.

SARS Severe acute respiratory syndrome, the name of a potentially fatal new respiratory disease in humans which is caused by the SARS coronavirus (SARS-CoV)

Satiety State of feeling satiated, fully satisfied (appetite or desire).

Scabies A transmissible ectoparasite skin infection characterised by superficial burrows, intense pruritus (itching) and secondary infection.

Scarlatina Scarlet fever, an acute, contagious disease caused by infection with group A streptococcal bacteria.

Schwann cells Or neurolemmocytes, are the principal supporting cells of the peripheral nervous system; they form the myelin sheath of a nerve fibre.

Schistosomiasis Is a parasitic disease caused by several species of fluke of the genus *Schistosoma*. Also known as bilharzia, bilharziosis or snail fever.

Schizophrenia A psychotic disorder (or a group of disorders) marked by severely impaired thinking, emotions and behaviours.

Sciatica A condition characterised by pain deep in the buttock often radiating down the back of the leg along the sciatic nerve.

Scleroderma A disease of the body's connective tissue. The most common symptom is a thickening and hardening of the skin, particularly of the hands and face.

Scrofula A tuberculous infection of the skin on the neck caused by the bacterium *Mycobacterium tuberculosis*.

Scrofulosis See Scrofula.

Scurf Abnormal skin condition in which small flakes or sales become detached.

Scurvy A state of dietary deficiency of vitamin C (ascorbic acid) which is required for the synthesis of collagen in humans.

Sebotropic Having an affinity for or a stimulating effect on sebaceous glands; promoting the excretion of sebum.

Sebum Oily secretion of the sebaceous glands.

Secretagogue A substance that causes another substance to be secreted.

Sedative Having a soothing, calming or tranquilising effect; reducing or relieving stress, irritability or excitement.

Seizure The physical findings or changes in behaviour that occur after an episode of abnormal electrical activity in the brain.

Selectins Are a family of cell adhesion molecules, e.g. selectin E, selectin L and selectin P.

Selenium (Se) A trace mineral that is essential to good health but required only in tiny amounts; it is incorporated into proteins to make selenoproteins, which are important antioxidant enzymes. It is found in avocado, brazil nut, lentils, sunflower seeds, tomato, whole grain cereals, seaweed, seafood and meat.

Sensorineural bradyacusia Hearing impairment of the inner ear resulting from damage

to the sensory hair cells or to the nerves that supply the inner ear.

Sepsis Potentially fatal whole-body inflammation caused by severe infection.

Sequela An abnormal pathological condition resulting from a disease, injury or trauma.

Serine proteinase Peptide hydrolases which have an active centre histidine and serine involved in the catalytic process.

Serotonergic Liberating, activated by, or involving serotonin in the transmission of nerve impulses.

Serotonin A monoamine neurotransmitter synthesised in serotonergic neurons in the central nervous system.

Sepsis Is a potentially fatal medical condition characterised by a whole-body inflammatory response (called a systemic inflammatory response syndrome or SIRS) that is triggered by an infection.

Septicaemia A systemic disease associated with the presence and persistence of pathogenic microorganisms or their toxins in the blood.

Sequelae A pathological condition resulting from a prior disease, injury or attack.

Sexual potentiator Increases sexual activity and potency and enhances sexual performance due to increased blood flow and efficient metabolism.

Sexually transmitted diseases (STD) Infections that are transmitted through sexual activity.

SGOT, serum glutamic oxaloacetic transaminase An enzyme that is normally present in liver and heart cells. SGOT is released into blood when the liver or heart is damaged. Also called aspartate transaminase (AST).

SGPT, serum glutamic pyruvic transaminase An enzyme normally present in serum and body tissues, especially in the liver; it is released into the serum as a result of tissue injury, also called alanine transaminase (ALT).

Shiga-like toxin A toxin produced by the bacterium *Escherichia coli* which disrupts the function of ribosomes, also known as verotoxin.

Shiga toxigenic *Escherichia coli* (STEC) Comprises a diverse group of organisms capable of causing severe gastrointestinal disease in humans.

Shiga toxin A toxin produced by the bacterium *Shigella dysenteriae*, which disrupts the function of ribosomes.

Shingles Skin rash caused by the zoster virus (same virus that causes chicken pox) and is medically termed herpes zoster.

Sialogogue Salivation promoter, a substance used to increase or promote the excretion of saliva.

Sialoproteins Glycoproteins that contain sialic acid as one of their carbohydrates.

Sialorrhea Excessive production of saliva.

Sialylation Reaction with sialic acid or its derivatives; used especially with oligosaccharides.

Sialyltransferases Enzymes that transfer sialic acid to nascent oligosaccharide.

Sickle cell disease Is an inherited blood disorder that affects red blood cells. People with sickle cell disease have red blood cells that contain mostly haemoglobin S, an abnormal type of haemoglobin. Sometimes these red blood cells become sickle shaped (crescent shaped) and have difficulty passing through small blood vessels.

Side stitch Is an intense stabbing pain under the lower edge of the ribcage that occurs while exercising.

Signal transduction cascade Refers to a series of sequential events that transfer a signal through a series of intermediate molecules until final regulatory molecules, such as transcription factors, and is modified in response to the signal.

Silicon (Si) Is required in minute amounts by the body and is important for the development of healthy hair and the prevention of nervous disorders. Lettuce is the best natural source of silicon.

Sinapism Signifies an external application, in the form of a soft plaster, or poultice.

Sinusitis Inflammation of the nasal sinuses.

SIRC cells Statens Seruminstitut rabbit cornea (SIRC) cell line.

SIRT 1 Stands for sirtuin (silent mating type information regulation 2 homolog) 1. It is an enzyme that deacetylates proteins that contribute to cellular regulation.

Sirtuin Also called Sir2 proteins, a class of proteins that possess either histone deacetylase or mono-ribosyltransferase activity.

6-Keto-PGF1 alpha A physiologically active and stable hydrolysis product of epoprostenol, found in nearly all mammalian tissues.

Sjögren's syndrome An autoimmune disease that mainly affects the eyes and salivary glands, but can affect different parts of the body. Symptoms include dry and itchy eyes, a dry mouth, thirst and swallowing difficulties.

Skp1 (S-Phase kinase-associated protein 1) is a core component of SCF ubiquitin ligases and mediates protein degradation.

Smads A family of intracellular proteins that mediate signalling by members of the TGF-beta (transforming growth factor beta) superfamily.

Smad2/3 A key signalling molecule for TGF-beta.

Smad7 A TGF**β** type 1 receptor antagonist.

Smallpox Is an acute, contagious and devastating disease in humans caused by *Variola* virus and has resulted in high mortality over the centuries.

Snuff Powder inhaled through the nose.

SOCE (store-operated Ca^{2+} Is a receptor-regulated Ca^{2+} entry pathway.

SOD Superoxide dismutase, is an enzyme that repairs cells and reduces the damage done to them by superoxide, the most common free radical in the body.

Sodium (Na) Is an essential nutrient required for health. Sodium cations are important in neuron (brain and nerve) function and in influencing osmotic balance between cells and the interstitial fluid and in maintenance of total body fluid homeostasis. Extra intake may cause a harmful effect on health. Sodium is naturally supplied by salt intake with food.

Soleus muscle Smaller calf muscle lower down the leg and under the gastrocnemius muscle.

Somites Mesodermal structures formed during embryonic development that give rise to segmented body parts such as the muscles of the body wall.

Soporific A sleep-inducing drug.

SOS response A global response to DNA damage in which the cell cycle is arrested and DNA repair and mutagenesis are induced.

Soyasapogenins Triterpenoid products obtained from the acid hydrolysis of soyasaponins, designated soyasapogenols A,B, C, D and E.

Soyasaponins Bioactive saponin compounds found in many legumes.

Spasmogenic Inducing spasm.

Spasmolytic Checking spasms; see Antispasmodic.

Spastic paraparesis A disorder that causes gradual weakness with muscle spasms (spastic weakness) in the legs.

Spermatogenic Giving rise to sperms.

Spermatorrhoea Medically an involuntary ejaculation/drooling of semen usually nocturnal emissions.

Spermidine An important polyamine in DNA synthesis and gene expression.

Spina bifida A congenital birth defect caused by the incomplete closing of the embryonic neural tube.

Sphingolipid A member of a class of lipids derived from the aliphatic amino alcohol, sphingosine.

Spinocerebellar ataxia (SCA) is a progressive, degenerative, genetic disease with multiple types.

Spleen Organ that filters blood and prevents infection.

Spleen tyrosine kinase (SYK) Is an enigmatic protein tyrosine kinase functional in a number of diverse cellular processes such as the regulation of immune and inflammatory responses.

Splenitis Inflammation of the spleen.

Splenocyte Is a monocyte, one of the five major types of white blood cell, and is characteristically found in the splenic tissue.

Splenomegaly Is an enlargement of the spleen.

Spongiosis Abnormal accumulation of fluid in the epidermis.

Sprain To twist a ligament or muscle of a joint without dislocating the bone.

Sprue Is a chronic disorder of the small intestine caused by sensitivity to gluten, a protein found in wheat and rye and to a lesser extent oats and barley. It causes poor absorption by the intestine of fat, protein, carbohydrates, iron, water and vitamins A, D, E and K.

Sputum Matter coughed up and usually ejected from the mouth, including saliva, foreign material and substances such as mucus or phlegm, from the respiratory tract.

SREBP-1 See Sterol regulatory element-binding protein-1.

Stanch To stop or check the flow of a bodily fluid like blood from a wound.

Statin A type of lipid-lowering drug.

STAT3 Signal transducer and activator of transcription 3, a transcription factor, plays a key role in many cellular processes such as cell growth and apoptosis.

Status epilepticus Refers to a life-threatening condition in which the brain is in a state of persistent seizure.

STD Sexually transmitted disease.

Steatorrhea Is the presence of excess fat in faeces which appear frothy, foul smelling and floats because of the high fat content.

Steatohepatitis Liver disease, characterised by inflammation of the liver with fat accumulation in the liver.

Steatosis Refer to the deposition of fat in the interstitial spaces of an organ like the liver, fatty liver disease.

Stereotypy Excessive repetitive or ritualistic movement, posture or utterance.

Sterility Inability to produce offspring, also called asepsis.

Sternutatory Causing or tending to cause sneezing.

Steroidogenic Relating to steroidogenesis.

Steroidogenesis The production of steroids, as by the adrenal glands.

Sterol regulatory element-binding protein-1 (SREBP1) Is a key regulator of the transcription of numerous genes that function in the metabolism of cholesterol and fatty acids.

Stimulant A substance that promotes the activity of a body system or function.

Stomachic (Digestive stimulant) an agent that stimulates or strengthens the activity of the stomach; used as a tonic to improve the appetite and digestive processes.

Stomatitis Oral inflammation and ulcers, may be mild and localised or severe, widespread and painful.

Stomatology Medical study of the mouth and its diseases.

Stool Faeces.

Strangury Is the painful passage of small quantities of urine which are expelled slowly by straining with severe urgency; it is usually accompanied with the unsatisfying feeling of a remaining volume inside and a desire to pass something that will not pass.

Straub tail Condition in which an animal carries its tail in an erect (vertical or nearly vertical) position.

STREPs Sterol regulatory element-binding proteins, a family of transcription factors that regulate lipid homeostasis by controlling the expression of a range of enzymes required for endogenous cholesterol, fatty acid, triacylglycerol and phospholipid synthesis.

Stria terminalis A structure in the brain consisting of a band of fibres running along the lateral margin of the ventricular surface of the thalamus.

Striae gravidarum A cutaneous condition characterised by stretch marks on the abdomen during and following pregnancy.

Stricture An abnormal constriction of the internal passageway within a tubular structure such as a vessel or duct

Strongyloidiasis An intestinal parasitic infection in humans caused by two species of the parasitic nematode *Strongyloides*. The nematode or round worms are also called thread worms.

Styptic A short stick of medication, usually anhydrous aluminium sulphate (a type of alum) or titanium dioxide, which is used for stanching blood by causing blood vessels to contract at the site of the wound. Also called haemostatic pencil. See Antihaemorrhagic.

Subarachnoid haemorrhage Is bleeding in the area between the brain and the thin tissues that cover the brain.

Substance P A neuropeptide that functions as a neurotransmitter and neuromodulator and is associated with the sensation of pain.

Substantia nigra Is a dark coloured brain structure located in the midbrain that plays an important role in reward, addiction and movement.

Sudatory Medicine that causes or increases sweating. Also see Sudorific.

Sudorific A substance that causes sweating.

Sulphur Sulphur is an essential component of all living cells. Sulphur is important for the synthesis of sulphur-containing amino acids, all polypeptides, proteins and enzymes such as glutathione, an important sulphur-containing tripeptide which plays a role in cells as a source of chemical reduction potential. Sulphur is also important for hair formation.

Good plant sources are garlic, onion, leeks and other alliaceous vegetables and brassicaceous vegetables like cauliflower, cabbages, Brussels sprout, kale; legumes, beans, green and red gram and soybeans; horse radish; water cress; and wheat germ.

Superior mesenteric artery (SMA) Arises from the anterior surface of the abdominal aorta, just inferior to the origin of the celiac trunk, and supplies the intestine from the lower part of the duodenum to the left colic flexure and the pancreas.

Superoxidae dismutase (SOD) Antioxidant enzyme.

Suppuration The formation of pus, the act of becoming converted into and discharging pus.

Supraorbital Located above the orbit of the eye.

Sural nerve Sensory nerve comprising collateral branches off of the common tibial and common fibular nerve.

SYK, spleen tyrosine kinase Is a human protein and gene. Syk plays a similar role in transmitting signals from a variety of cell surface receptors including CD74, Fc receptor and integrins.

Sympathetic nervous system The part of the autonomic nervous system originating in the thoracic and lumbar regions of the spinal cord that in general inhibits or opposes the physiological effects of the parasympathetic nervous system, as in tending to reduce digestive secretions or speed up the heart.

Sympathomimetic Mimicking the effects of impulses conveyed by adrenergic postganglionic fibres of the sympathetic nervous system.

Synaptic plasticity The ability of neurons to change the number and strength of their synapses.

Synaptogenesis The formation of synapses.

Synaptoneurosomes Purified synapses containing the pre- and postsynaptic termini.

Synaptosomes Isolated terminal of a neuron.

Syncope Fainting, sudden loss of consciousness followed by the return of wakefulness.

Syndactyly Webbed toes, a condition where two or more digits are fused together.

Syneresis Expulsion of liquid from a gel, as contraction of a blood clot and expulsion of liquid.

Syngeneic Genetically identical or closely related, so as to allow tissue transplant; immunologically compatible.

Synovial Lubricating fluid secreted by synovial membranes, as those of the joints.

Synoviocyte Located in the synovial membrane; there are two types. Type A cells are more numerous, have phagocytic characteristics and produce degradative enzymes. Type B cells produce synovial fluid, which lubricates the joint and nourishes the articular cartilage.

Syphilis Is perhaps the best known of all the STDs. Syphilis is transmitted by direct contact with infection sores, called chancres, syphilitic skin rashes or mucous patches on the tongue and mouth during kissing, necking, petting or sexual intercourse. It can also be transmitted from a pregnant woman to a fetus after the fourth month of pregnancy.

System lupus erythematosus A long-term autoimmune disorder that may affect the skin, joints, kidneys, brain and other organs. Symptoms may include chest pain, fatigue, fever, hair loss, malaise, mouth sores, sensitivity to sunlight, skin rash (butterfly rash).

Systolic The blood pressure when the heart is contracting. It is specifically the maximum arterial pressure during contraction of the left ventricle of the heart.

T cells Or T lymphocytes, a type of white blood cell that plays a key role in the immune system and attacks virus-infected cells, foreign cells and cancer cells.

Tachyarrhythmia Any disturbance of the heart rhythm in which the heart rate is abnormally increased.

Tachycardia A false heart rate applied to adults to rates over 100 beats per minute.

Tachykinins Neuropeptide transmitters that are widely distributed and active in the central nervous system and periphery, rapidly acting secretagogues, and cause smooth muscle contraction and vasodilation (hypotension).

Tachyphylaxia A decreased response to a medicine given over a period of time so that larger doses are required to produce the same response.

Tachypnea Abnormally fast breathing.

Taenia A parasitic tapeworm or flatworm of the genus *Taenia.*

Taeniacide An agent that kills tapeworms.

Tardive dyskinesia A disorder characterised by repetitive, involuntary, purposeless movements in the body such as grimacing, tongue protrusion, lip smacking, puckering and pursing of the lips and rapid eye blinking. Rapid, involuntary movements of the limbs, torso and fingers may also occur.

Tau Is a class of microtubule-associated protein (MAP) in neuronal and glial cells.

Tau-1 (Ser198/199/202), pS396 (Ser396) and pS214 (Ser214) epitopes Serine phosphorylation sites of tau-1.

Tau phosphorylation Plays an important role in neurodegenerative diseases and regulated by protein kinases and phosphatases.

TBARS See thiobarbituric acid reactive substances.

TCA cycle See Tricarboxylic acid cycle.

TCID50 Median tissue culture infective dose; that amount of a pathogenic agent that will produce pathological change in 50% of cell cultures.

Telencephalon The cerebral hemispheres, the largest divisions of the human brain.

Teletherapy A noninvasive procedure using external beam radiotherapy treatments.

Telomerase Enzyme that acts on parts of chromosomes known as telomeres.

Temporomandibular joint disorder (TMJD or TMD syndrome) A disorder characterised by acute or chronic inflammation of the temporomandibular joint that connects the mandible to the skull.

Tendonitis Is inflammation of a tendon.

Tenesmus A strong desire to defecate.

Teras (Medicine) a grossly malformed and usually nonviable fetus. *plural* terata.

Teratogen Is an agent that can cause malformations of an embryo or fetus. *adj.* teratogenic.

Testicular torsion Twisting of the spermatic cord, which cuts off the blood supply to the testicle and surrounding structures within the scrotum.

Tetanus An acute, potentially fatal disease caused by tetanus bacilli multiplying at the site of an injury and producing an exotoxin that reaches the central nervous system producing prolonged contraction of skeletal muscle fibres. Also called lockjaw.

Tete Acute dermatitis caused by both bacterial and fungal infections.

Tetraparesis Weakness of muscles of all four limbs.

Tetter Any of a number of skin diseases.

TGF-beta Transforming growth factor beta is a protein that controls proliferation, cellular differentiation and other functions in most cells.

Th cells or T helper cells A subgroup of lymphocytes that helps other white blood cells in immunologic processes.

Th 1 cells Helper cells that play an important role in the immune system.

Th 17 cells A subset of T helper cells producing interleukin 17.

Thalassemia major Is a genetic blood disorder that causes the body to manufacture an abnormal form of haemoglobin.

Thelarche The beginning of secondary (postnatal) breast development, usually occurring at the beginning of puberty in girls.

Thermogenic Tending to produce heat, applied to drugs or food (fat burning food).

Thermogenesis Is the process of heat production in organisms.

Thermonociceptors Or thermal nociceptors, sensory receptors that are stimulated by noxius heat or cold at various temperatures.

Thiobarbituric acid reactive substances (TBARS) A well-established method for screening and monitoring lipid peroxidation.

Thixotropy The property exhibited by certain gels of becoming fluid when stirred or shaken and returning to the semisolid state upon standing.

Thoracodynia Pain in the chest.

3-β-HSD (Or 3-β-hydroxysteroid dehydrogenase/δ-5-4 isomerase) is an enzyme that catalyses the synthesis of progesterone from pregnenolone.

3-Nitrotyrosine (3-NT) protein Used as a marker for oxidative damage or nitrosative stress.

Thrombocythaemia A blood condition characterised by a high number of platelets in the blood.

Thrombocytopenia A condition when the bone marrow does not produce enough platelets (thrombocytes) like in leukaemia.

Thromboembolism Formation in a blood vessel of a clot (thrombus) that breaks loose and is carried by the blood stream to plug another vessel. *cf.* deep vein thrombosis.

Thrombogenesis Formation of a thrombus or blood clot.

Thrombophlebitis Occurs when there is inflammation and clot in a surface vein.

Thromboplastin An enzyme liberated from blood platelets that converts prothrombin into thrombin as blood starts to clot, also called thrombokinase.

Thrombosis The formation or presence of a thrombus (clot).

Thromboxanes Any of several compounds, originally derived from prostaglandin precursors in platelets that stimulate aggregation of platelets and constriction of blood vessels.

Thromboxane B2 The inactive product of thromboxane.

Thrombus A fibrinous clot formed in a blood vessel or in a chamber of the heart.

Thrush A common mycotic infection caused by yeast, *Candida albicans*, in the digestive tract or vagina. In children it is characterised by white spots on the tongue.

Thymocytes Are T cell precursors which develop in the thymus.

Thyrotoxicosis Or hyperthyroidism—an overactive thyroid gland, producing excessive circulating free thyroxine and free triiodothyronine, or both.

Tight junction Associated areas of two cells whose membranes join together forming a virtually impermeable barrier to fluid.

TIMP-3 A human gene belongs to the tissue inhibitor of matrix metalloproteinases (MMP) gene family. See MMP.

Tincture Solution of a drug in alcohol.

Tinea Ringworm, fungal infection on the skin.

Tinea favosa See Favus.

Tinea cruris Ringworm of the groin.

Tinea imbricata (Also called Tokelau) an eruption characterised by concentric rings of overlapping scales forming papulosquamous patches scattered over the body; it occurs in tropical climates especially prevalent in southwest Polynesia and is caused by the fungus *Trichophyton concentricum*.

Tinea pedis Fungal infection of the foot, also called athletes' foot.

Tinnitus A noise in the ears, as ringing, buzzing, roaring, clicking, etc.

Tisane An herbal infusion used as tea or for medicinal purposes.

Tissue plasminogen activator (t-PA) A serine protease involved in the breakdown of blood clots.

TNF alpha Cachexin or cachectin and formally known as tumour necrosis factor-alpha, a cytokine involved in systemic inflammation. Primary role of TNF is in the regulation of immune cells. TNF is also able to induce apoptotic cell death, to induce inflammation and to inhibit tumorigenesis and viral replication.

Tocolytics Medications used to suppress premature labour.

Tocopherol Fat-soluble organic compounds belonging to vitamin E group. See Vitamin E.

Tocotrienol Fat-soluble organic compounds belonging to vitamin E group. See Vitamin E.

Tolerogenic Producing immunological tolerance.

Toll-like receptors (TLRs) A class of proteins that play a key role in the innate immune system.

Tonic Substance that acts to restore, balance, tone, strengthen or invigorate a body system without overt stimulation or depression

Tonic–clonic seizure A type of generalised seizure that affects the entire brain.

Tonsillitis An inflammatory condition of the tonsils due to bacteria, allergies or respiratory problems.

TOP2A Topoisomerase II alpha enzyme.

Topoisomerases A class of enzymes involved in the regulation of DNA supercoiling.

Topoiosmerase inhibitors A new class of anticancer agents with a mechanism of action aimed at interrupting DNA replication in cancer cells.

Torsade de Pointes An uncommon condition of the heart. It is a polymorphic ventricular tachycardia occurring in the context of QT prolongation.

Total parenteral nutrition (TPN) Is a method of feeding that bypasses the gastrointestinal tract.

Toxaemia Is the presence of abnormal substances in the blood, but the term is also used for a serious condition in pregnancy that involves hypertension and proteinuria. Also called pre-eclampsia.

Tracheitis Is a bacterial infection of the trachea; also known as bacterial tracheitis or acute bacterial tracheitis.

Trachoma A contagious disease of the conjunctiva and cornea of the eye, producing painful sensitivity to strong light and excessive tearing.

TRAIL Acronym for tumour necrosis factor-related apoptosis-inducing ligand, is a cytokine that preferentially induces apoptosis in tumour cells.

Tranquiliser A substance drug used in calming person suffering from nervous tension or anxiety.

Transaminase Also called aminotransferase, is an enzyme that catalyses a type of reaction between an amino acid and an α-keto acid.

Transaminitis Increase in alanine aminotransferase (ALT) and/or aspartate aminotransferase (AST) to > 5 times the upper limit of normal.

Transcatheter arterial chemoembolisation (TACE) Is an interventional radiology procedure involving percutaneous access to the hepatic artery and passing a catheter through the abdominal artery aorta followed by radiology. It is used extensively in the palliative treatment of unresectable hepatocellular carcinoma (HCC).

Transcriptional activators Are proteins that bind to DNA and stimulate transcription of nearby genes.

Transcriptional coactivator PGC-1 A potent transcriptional coactivator that regulates oxidative metabolism in a variety of tissues.

Transcriptome Is a set of all RNA molecules, including mRNA, rRNA, tRNA and other non-coding RNA transcribed in one cell or a population of cells.

Transcriptome profiling To identify genes involved in peroxisome assembly and function.

Transoesophageal echocardiogram Uses sound wave (ultrasound) technology to examine heart function.

Transforming growth factor beta (TGF-β) A protein that controls proliferation, cellular differentiation and other functions in most cells.

Transient receptor potential ankyrin 1 (TRPA1) Is a $Ca^{(2+)}$-permanent, non-selective cationic channel that may play a role in nociception.

Transient receptor potential vanilloid 1 (TRPV1) Receptor also known as capsaicin receptor and vanilloid receptor, is a Ca 2+-permeable non-selective cation channel localised on a subset of primary sensory neurons and can be activated by physical and chemical stimuli.

TRAP 6 Thrombin receptor-activating peptide with 6 amino acids.

Tremorine A chemical that produces a tremor resembling Parkinsonian tremor.

Tremulous Marked by trembling, quivering or shaking.

Triacylglycerols Or triacylglyceride, is a glyceride in which the glycerol is esterified with three fatty acids.

Tricarboxylic acid cycle (TCA cycle) A series of enzymatic reactions in aerobic organisms involving oxidative metabolism of acetyl units and producing high-energy phosphate compounds, which serve as the main source of cellular energy. Also called citric acid cycle, Krebs cycle.

Trichophytosis Infection by fungi of the genus *Trichophyto*n.

Trigeminal neuralgia (TN) Is a neuropathic disorder of one or both of the facial trigeminal nerves, also known as prosopalgia.

Triglycerides A type of fat (lipids) found in the blood stream.

Trismus Continuous contraction of the muscles of the jaw, specifically as a symptom of tetanus, or lockjaw; inability to open mouth fully.

TrKB receptor Also known as TrKB tyrosine kinase, a protein in humans that acts as a catalytic receptor for several neurotrophils.

Trolox equivalent Measures the antioxidant capacity of a given substance, as compared to the standard, Trolox also referred to as TEAC (Trolox equivalent antioxidant capacity).

Trypanocidal Destructive to trypanosomes.

Trypanosomes Protozoan of the genus *Trypanosoma.*

Trypanosomiasis Human disease or an infection caused by a trypanosome.

Trypsin An enzyme of pancreatic juice that hydrolyses proteins into smaller polypeptide units.

Trypsin inhibitor Small protein synthesised in the exocrine pancreas which prevents conversion of trypsinogen to trypsin, so protecting itself against trypsin digestion.

TRPV1 See Transient receptor potential vanilloid 1.

Tuberculosis (TB) Is a bacterial infection of the lungs caused by a bacterium called *Mycobacterium tuberculosis*, characterised by the formation of lesions (tubercles) and necrosis in the lung tissues and other organs.

Tubulopathy Any disease of the renal tubules of the nephron.

Tumorigenesis Formation or production of tumours.

Tumour An abnormal swelling of the body other than those caused by direct injury.

Tussis A cough.

Tympanic membrane Ear drum.

Tympanitis Infection or inflammation of the inner ear.

Tympanophonia Increased resonance of one's own voice, breath sounds, arterial murmurs, etc., noted especially in disease of the middle ear.

Tympanosclerosis See Myringosclerosis.

Tyrosinase A copper containing enzyme found in animals and plants that catalyses the oxidation of phenols (such as tyrosine) and the production of melanin and other pigments from tyrosine by oxidation.

Ubiquitin ligase Also called an E3 ubiquitin ligase, is a protein that targets other proteins to be broken down (degraded) within cells.

UCP1 An uncoupling protein found in the mitochondria of brown adipose tissue used to generate heat by non-shivering thermogenesis.

UCP: 2 enzyme Uncoupling protein 2 enzyme, a mitochondrial protein expressed in adipocytes.

Ulcer An open sore on an external or internal body surface usually accompanied by disintegration of tissue and pus.

Ulcerative colitis Is one of two types of inflammatory bowel disease—a condition that causes the bowel to become inflamed and red.

Ulemorrhagia Bleeding of the gums.

Ulitis Inflammation of the gums.

Unguent Ointment.

Unilateral ureteral obstruction Unilateral blockage of urine flow through the ureter of one kidney, resulting in a backup of urine, distension of the renal pelvis and calyces and hydronephrosis.

Uraemia An excess in the blood of urea, creatinine and other nitrogenous end products of protein and amino acid metabolism, more correctly referred to as azotaemia.

Urethra Tube conveying urine from the bladder to the external urethral orifice.

Urethritis Is an inflammation of the urethra caused by infection.

Uricaemia An excess of uric acid or urates in the blood.

Uricosuric Promoting the excretion of uric acid in the urine.

Urinary Pertaining to the passage of urine.

Urinary incontinence Sudden and strong need to urinate because of poor bladder control.

Urinogenital Relating to the genital and urinary organs or functions.

Urodynia Pain on urination.

Urokinase Also called urokinase-type plasminogen (u-PA), is a serine protease enzyme in human urine that catalyses the conversion of plasminogen to plasmin. It is used clinically as a thrombolytic agent.

Urokinase-type plasminogen (u-PA) Plays a key role in tumour invasion and metastasis; also see Urokinase.

Urolithiasis Formation of stone in the urinary tract (kidney bladder or urethra).

Urticant A substance that causes wheals to form.

Urticaria (Or hives) is a skin condition, commonly caused by an allergic reaction, that is characterised by raised red skin welts.

Uterine Relating to the uterus.

Uterine myomas Also called fibroids, tumours that grown from the uterine wall.

Uterine prolapse Occurs when weakened or damaged muscles and ligaments allow the uterus to slip into the vagina.

Uterine relaxant An agent that relaxes the muscles in the uterus.

Uterine stimulant An agent that stimulates the uterus (and often employed during active childbirth).

Uterotonic Giving muscular tone to the uterus.

Uterotrophic Causing an effect on the uterus.

Uterus Womb.

Vaginal dystrophy A condition in which the outer part of the vagina becomes dry and the skin thickens or thins.

Vaginitis Infectious or non-infectious inflammation of the vaginal mucosa.

Vaginopathy Any disease of the vagina.

Vagotomy The surgical cutting of the vagus nerve to reduce acid secretion in the stomach.

Vagus nerve A cranial nerve, that is, a nerve connected to the brain. The vagus nerve has branches to most of the major organs in the body, including the larynx, throat, windpipe, lungs, heart and most of the digestive system.

Variola Or smallpox, a contagious disease unique to humans, caused by either of two virus variants, *Variola major* and *Variola minor.* The disease is characterised by fever, weakness and skin eruption with pustules that form scabs that leave scars.

Varicose veins Are veins that have become enlarged and twisted.

Vasa vasorum Is a network of small blood vessels that supply large blood vessels. *pl.* vasa vasori.

Vascular cell adhesion molecule (VCAM) A part of the immunoglobulin superfamily. They are important in inflammation, immune responses and intracellular signalling events.

Vascular endothelial growth factor (VEGF) A polypeptide chemical produced by cells that stimulates the growth of new blood vessels.

Vasculitis Group of disorders that destroy blood vessels by inflammation.

Vasculogenesis The process of blood vessel formation occurring by a de novo production of endothelial cells.

Vasoconstrictor Drug that causes constriction of blood vessels.

Vasodilator Drug that causes dilation or relaxation of blood vessels.

Vasodilatory Causing the widening of the lumen of blood vessels.

Vasomotor symptoms Menopausal symptoms characterised by hot flushes and night sweats.

Vasospasm Refers to a condition in which blood vessels spasm, leading to vasoconstriction and subsequently to tissue ischemia and death (necrosis).

VCAM-1 (vascular cell adhesion molecule-1) Also known as CD106, contains six or seven immunoglobulin domains and is expressed on both large and small vessels only after the endothelial cells are stimulated by cytokines.

VEGF Vascular endothelial growth factor.

Venereal disease (VD) Term given to the diseases syphilis and gonorrhoea.

Venule A small vein, especially one joining capillaries to larger veins.

Vermifuge A substance used to expel worms from the intestines.

Verotoxin S Shiga-like toxin produced by *Escherichia coli*, which disrupts the function of ribosomes, causing acute renal failure.

Verruca A contagious and painful wart on the sole of the foot.

Verruca plana Is a reddish-brown or flesh-coloured, slightly raised, flat-surfaced, well-demarcated papule on the hand and face, also called flat wart.

Verruca vulgaris Small painless warts on the skin caused by the human papillomavirus.

Vertigo An illusory, sensory perception that the surroundings or one's own body is revolving; dizziness.

Very low-density lipoprotein (VLDL) A type of lipoprotein made by the liver. VLDL is one of the five major groups of lipoproteins (chylomicrons, VLDL, intermediate-density lipoprotein, low-density lipoprotein, high-density lipoprotein (HDL)) that enable fats and cholesterol to move within the water-based solution of the bloodstream. VLDL is converted in the bloodstream to low-density lipoprotein (LDL).

Vesical calculus Calculi (stones) in the urinary bladder

Vesicant A substance that causes tissue blistering.

Vestibular Relating to the sense of balance.

Vestibular disorders Includes symptoms of dizziness, vertigo and imbalance; it can be

result from or worsened by genetic or environmental conditions.

Vestibular schwannoma Also called acoustic neuroma is a benign tumour that may develop from an overproduction of Schwann cells that press on the hearing and balance nerves in the inner ear.

Vestibular system Includes parts of the inner ear and brain that process sensory information involved with controlling balance and eye movement.

Vibrissa Stiff hairs that are located especially about the nostrils.

Vimentin A type III intermediate filament protein that is expressed in mesenchymal cells.

Viremia A medical condition where viruses enter the bloodstream and hence have access to the rest of the body.

Visceral fat Intra-abdominal fat, is located inside the peritoneal cavity, packed in between internal organs and torso.

Visual entopia Visual disturbances

Vitamin Any complex, organic compound, found in various food or sometimes synthesised in the body, required in tiny amounts and are essential for the regulation of metabolism, normal growth and function of the body.

Vitamin A Retinol, fat-soluble vitamins that play an important role in vision, bone growth, reproduction, cell division and cell differentiation, and help regulate the immune system in preventing or fighting off infections. Vitamin A that is found in colourful fruits and vegetables is called provitamin A carotenoid. They can be made into retinol in the body. Deficiency of vitamin A results in night blindness and keratomalacia.

Vitamin B1 Also called thiamine, water-soluble vitamins, dissolve easily in water and, in general, are readily excreted from the body they are not readily stored; consistent daily intake is important. It functions as coenzyme in the metabolism of carbohydrates and branched chain amino acids and other cellular processes. Deficiency results in beriberi disease.

Vitamin B2 Also called riboflavin, an essential water-soluble vitamin that functions as coenzyme in redox reactions. Deficiency causes ariboflavinosis.

Vitamin B3 Comprises niacin and niacinamide, water-soluble vitamin that functions as coenzyme or co-substrate for many redox reactions and is required for energy metabolism. Deficiency causes pellagra.

Vitamin B5 Also called pantothenic acid, a water-soluble vitamin that functions as coenzyme in fatty acid metabolism. Deficiency causes paresthesia.

Vitamin B6 water-soluble vitamin, exists in three major chemical forms: pyridoxine, pyridoxal and pyridoxamine. Vitamin B6 is needed in enzymes involved in protein metabolism, red blood cell metabolism, efficient functioning of nervous and immune systems and haemoglobin formation. Deficiency causes anaemia and peripheral neuropathy.

Vitamin B 7 Also called biotin or vitamin H, an essential water-soluble vitamin, is involved in the synthesis of fatty acids, amino acids and glucose, in energy metabolism. Biotin promotes normal health of sweat glands, bone marrow, male gonads, blood cells, nerve tissue, skin and hair. Deficiency causes dermatitis and enteritis.

Vitamin B9 Also called folic acid, an essential water-soluble vitamin. Folate is especially important during periods of rapid cell division and growth such as infancy and pregnancy. Deficiency during pregnancy is associated with birth defects such as neural tube defects. Folate is also important for production of red blood cells and prevents anaemia. Folate is needed to make DNA and RNA, the building blocks of cells. It also helps prevent changes to DNA that may lead to cancer.

Vitamin B12 A water-soluble vitamin, also called cobalamin as it contains the metal cobalt. It helps maintain healthy nerve cells and red blood cells and DNA production. Vitamin B12 is bound to the protein in food. Deficiency causes megaloblastic anaemia.

Vitamin C Also known as ascorbic acid is an essential water-soluble vitamin. It functions as cofactor for reactions requiring reduced copper or iron metalloenzyme and as a protective antioxidant. Deficiency of vitamin C causes scurvy.

Vitamin D A group of fat-soluble, prohormone vitamin, the two major forms of which are vitamin D2 (or ergocalciferol) and vitamin D3 (or cholecalciferol). Vitamin D obtained from sun exposure, food and supplements is biolog-

ically inert and must undergo two hydroxylations in the body for activation. Vitamin D is essential for promoting calcium absorption in the gut and maintaining adequate serum calcium and phosphate concentrations to enable normal growth and mineralisation of bone and prevent hypocalcaemic tetany. Deficiency causes rickets and osteomalacia. Vitamin D has other roles in human health, including modulation of neuromuscular and immune function, reduction of inflammation and modulation of many genes encoding proteins that regulate cell proliferation, differentiation and apoptosis.

Vitamin E Is the collective name for a group of fat-soluble compounds and exists in eight chemical forms (alpha-, beta-, gamma- and delta-tocopherol and alpha-, beta-, gamma- and delta-tocotrienol). It has pronounced antioxidant activities stopping the formation of reactive oxygen species when fat undergoes oxidation and helps prevent or delay the chronic diseases associated with free radicals. Besides its antioxidant activities, vitamin E is involved in immune function, cell signalling, regulation of gene expression and other metabolic processes. Deficiency is very rare but can cause mild haemolytic anaemia in newborn infants.

Vitamin K A group of fat-soluble vitamin and consists of vitamin K_1 which is also known as phylloquinone or phytomenadione (also called phytonadione) and vitamin K_2 (menaquinone, menatetrenone). Vitamin K plays an important role in blood clotting. Deficiency is very rare but can cause bleeding diathesis.

Vitamin P A substance or mixture of substances obtained from various plant sources, identified as citrin or a mixture of bioflavonoids, thought to but not proven to be useful in reducing the extent of haemorrhage.

Vitiligo A chronic skin disease that causes loss of pigment, resulting in irregular pale patches of skin. It occurs when the melanocytes, cells responsible for skin pigmentation, die or are unable to function. Also called leucoderma.

Vitreoretinopathy See Proliferative vitreoretinopathy.

VLA-4 Very late antigen-4, expressed by most leucocytes, but it is observed on neutrophils under special conditions.

VLDL See Very low-density lipoproteins.

Vomitive Substance that causes vomiting.

Vulnerary (Wound healer) a substance used to heal wounds and promote tissue formation.

Vulva-vaginal erythema Abnormal redness and inflammation of the skin in the vagina.

Wart An infectious skin tumour caused by a viral infection.

Welt See Wheal.

Wheal A firm, elevated swelling of the skin. Also called a weal or welt.

White fat White adipose tissue (WAT) in mammals, store of energy. *cf.* brown fat.

Whitlow Painful infection of the hand involving one or more fingers that typically affects the terminal phalanx.

Whooping cough Acute infectious disease usually in children caused by a *Bacillus* bacterium and accompanied by catarrh of the respiratory passages and repeated bouts of coughing.

Wnt proteins Are a diverse family of secreted lipid-modified signalling glycoproteins that are 350–400 amino acids in length

Wnt signalling pathway Is a network of proteins involved in embryogenesis and cancer and also in normal physiological processes.

X-Linked agammaglobulinemia Also known as X-linked hypogammaglobulinemia, XLA, Bruton-type agammaglobulinemia, Bruton syndrome or sex-linked agammaglobulinemia; a rare x-linked genetic disorder that affects the body's ability to fight infection.

Xanthine oxidase A flavoprotein enzyme containing a molybdenum cofactor (Moco) and (Fe_2S_2) clusters, involved in purine metabolism. In humans, inhibition of xanthine oxidase reduces the production of uric acid and prevents hyperuricaemia and gout.

Xanthones Unique class of biologically active phenol compounds with the molecular formula C13H8O2 possessing antioxidant properties, discovered in the mangosteen fruit.

Xenobiotics A chemical (as a drug, pesticide or carcinogen) that is foreign to a living organism.

Xenograft A surgical graft of tissue from one species to an unlike species.

Xerophthalmia A medical condition in which the eye fails to produce tears.

Xerostomia Dryness in the mouth due to lack of saliva production.

Yaws An infectious tropical infection of the skin, bones and joints caused by the spirochete bacterium *Treponema pertenue*, characterised by papules and papilloma with subsequent deformation of the skins, bone and joints; also called framboesia.

yGCN5 A histone acetyl transferase (HAT) that plays a role in regulation of transition, cell cycle progression and differentiation.

Yellow fever Is a viral disease that is transmitted to humans through the bite of infected mosquitoes. Illness ranges in severity from an influenza-like syndrome to severe hepatitis and haemorrhagic fever. Yellow fever virus (YFV) is maintained in nature by mosquito-borne transmission between nonhuman primates.

Zeaxanthin A common carotenoid, found naturally as coloured pigments in many fruit vegetables and leafy vegetables. It is important for good vision and is one of the two carotenoids contained within the retina of the eye. Within the central macula, zeaxanthin predominates, whereas in the peripheral retina, lutein predominates.

Zinc (Zn) Is an essential mineral for health. It is involved in numerous aspects of cellular metabolism: catalytic activity of enzymes, immune function, protein synthesis, wound healing, DNA synthesis and cell division. It also supports normal growth and development during pregnancy, childhood and adolescence and is required for proper sense of taste and smell. Dietary sources include beans, nuts, pumpkin seeds, sunflower seeds, whole wheat bread and animal sources.

ZK1 Kruppel-type zinc finger protein—binds DNA and, through this binding, regulates gene transcription.

ZO1 protein A high molecular weight tight junction-associated protein.

Scientific Glossary

Abaxial Facing away from the axis, as of the surface of an organ.

Abortive Imperfectly formed.

Abscission Shedding of leaves, flowers or fruits following the formation of the abscission zone.

Acaulescent Lacking a stem, or stem very much reduced.

Accrescent Increasing in size after flowering or with age.

Achene A dry, small, one-seeded, indehiscent fruit formed from a superior ovary of one carpel as in sunflower.

Acid soil Soil that maintains a pH of less than 7.0.

Acidulous Acid or sour in taste.

Actinomorphic Having radial symmetry, capable of being divided into symmetrical halves by any plane, referring to a flower, calyx or corolla.

Aculeate Having sharp prickles.

Acuminate Tapering gradually to a sharp point.

Acute (Botany) tapering at an angle of less than 90° before terminating in a point as of leaf apex and base.

Adaxial Side closest to the stem axis.

Adelphous Having stamens united together by their filaments.

Adherent Touching without organic fusion as of floral parts of different whorls.

Adnate United with another unlike part as of stamens attached to petals.

Adpressed Lying close to another organ but not fused to it.

Adventitious Arising in abnormal positions, e.g. roots arising from the stem, branches or leaves, buds arising elsewhere than in the axils of leaves.

Adventive Not native to and not fully established in a new habitat or environment; locally or temporarily naturalised, e.g. an adventive weed.

Aestivation Refers to positional arrangement of the floral parts in the bud before it opens.

Akinete A thick-walled dormant cell derived from the enlargement of a vegetative cell. It serves as a survival structure.

Alfisols Soil with a clay-enriched subsoil and relatively high native fertility, having undergone only moderate leaching, containing aluminium, iron and with at least 35% base saturation, meaning that calcium, magnesium and potassium are relatively abundant.

Alkaline soil Soil that maintains a pH above 7.0, usually containing large amounts of calcium, sodium and magnesium, and is less soluble than acidic soils.

Alkaloids Naturally occurring bitter, complex organic-chemical compounds containing basic nitrogen and oxygen atoms and having various pharmacological effects on humans and other animals.

Alternate Leaves or buds that are spaced along opposite sides of stem at different levels.

Allomorphic With a shape or form different from the typical.

Alluvial soil A fine-grained fertile soil deposited by water flowing over flood plains or in river beds.

T.K. Lim, *Edible Medicinal and Non-Medicinal Plants: Volume 10, Modified Stems, Roots, Bulbs*,
DOI 10.1007/978-94-017-7276-1

Alluvium Soil or sediments deposited by a river or other running water.

Amplexicaul Clasping the stem as base of certain leaves.

Anatomising Interconnecting network as applied to leaf veins.

Andisols Are soils formed in volcanic ash and containing high proportions of glass and amorphous colloidal materials.

Androdioecious With male flowers and bisexual flowers on separate plants.

Androecium Male parts of a flower; comprising the stamens of one flower.

Androgynophore A stalk bearing both the androecium and gynoecium above the perianth of the flower.

Androgynous With male and female flowers in distinct parts of the same inflorescence.

Andromonoecious Having male flowers and bisexual flowers on the same plant.

Angiosperm A division of seed plants with the ovules borne in an ovary.

Annual A plant which completes its life cycle within a year.

Annular Shaped like or forming a ring.

Annulus Circle or ringlike structure or marking; the portion of the corolla which forms a fleshy, raised ring.

Anthelate An open, paniculate cyme.

Anther The part of the stamen containing pollen sac which produces the pollen.

Antheriferous Containing anthers.

Anthesis The period between the opening of the bud and the onset of flower withering.

Anthocarp A false fruit consisting of the true fruit and the base of the perianth.

Anthocyanidins Are common plant pigments. They are the sugar-free counterparts of anthocyanins.

Anthocyanins A subgroup of antioxidant flavonoids, are glucosides of anthocyanidins. They occur as water-soluble vacuolar pigments that may appear red, purple or blue according to pH in plants.

Antipetala Situated opposite petals.

Antisepala Situated opposite sepals.

Antrorse Directed forward and upward.

Apetalous Lacking petals as of flowers with no corolla.

Apical meristem Active growing point. A zone of cell division at the tip of the stem or the root.

Apically Towards the apex or tip of a structure.

Apiculate Ending abruptly in a short, sharp, small point.

Apiculum A short, pointed, flexible tip.

Apocarpous Carpels separate in single individual pistils.

Apopetalous With separate petals, not united to other petals.

Aposepalous With separate sepals, not united to other sepals.

Appendiculate Having small appendages

Appressed Pressed closely to another structure but not fused or united.

Aquatic A plant living in or on water for all or a considerable part of its life span.

Arachnoid (Botany) formed of or covered with long, delicate hairs or fibres.

Arborescent Resembling a tree; applied to nonwoody plants attaining tree height and to shrubs tending to become tree-like in size.

Arbuscular mycorrhiza (AM) A type of mycorrhiza in which the fungus (of the phylum Glomeromycota) penetrates the cortical cells of the roots of a vascular plant and forms unique structures such as arbuscules and vesicles. These fungi help plants to capture nutrients such as phosphorus and micronutrients from the soil.

Archegonium A flask-shaped female reproductive organ in mosses, ferns and other related plants.

Areolate With areolae.

Areole (Botany) a small, specialised, cushion-like area on a cactus from which hairs, glochids, spines, branches or flowers may arise; an irregular angular specs marked out on a surface, e.g. fruit surface. *pl.* areolae.

Aril Specialised outgrowth from the funiculus (attachment point of the seed) (or hilum) that encloses or is attached to the seed. *adj.* arillate.

Arillode A false aril; an aril originating from the micropyle instead of from the funicle or chalaza of the ovule, e.g. mace of nutmeg.

Aristate Bristlelike part or appendage, e.g. awns of grains and grasses.

Aristulate Having a small, stiff, bristlelike part or appendage; a diminutive of aristate

Articulate Jointed; usually breaking easily at the nodes or point of articulation into segments.

Ascending Arched upwards in the lower part and becoming erect in the upper part.

Ascospore Spore produced in the ascus in ascomycete fungi.

Ascus Is the sexual spore-bearing cell produced in ascomycete fungi. *pl.* asci.

Asperulous Refers to a rough surface with short, hard projections.

Attenuate Tapered or tapering gradually to a point.

Auricle An ear-like appendage that occurs at the base of some leaves or corolla.

Auriculate Having auricles.

Awn A hair-like or bristlelike appendage on a larger structure.

Axil Upper angle between a lateral organ, such as a leaf petiole, and the stem that bears it.

Axile Situated along the central axis of an ovary having two or more locules, as in axile placentation.

Axillary Arising or growing in an axil.

Baccate Beery-like, pulpy or fleshy.

Barbate Bearded, having tufts of hairs.

Barbellae Short, stiff, hair-like bristles. *adj.* barbellate.

Bark Is the outermost layers of stems and roots of woody plants.

Basal Relating to, situated at, arising from or forming the base.

Basaltic soil Soil derived from basalt, a common extrusive volcanic rock.

Basidiospore A reproductive spore produced by basidiomycete fungi.

Basidium A microscopic, spore-producing structure found on the hymenophore of fruiting bodies of basidiomycete fungi.

Basifixed Attached by the base, as certain anthers are to their filaments.

Basionym The synonym of a scientific name that supplies the epithet for the correct name.

Beak A prominent apical projection, especially of a carpel or fruit. *adj.* beaked.

Bearded Having a tuft of hairs.

Berry A fleshy or pulpy indehiscent fruit from a single ovary with the seed(s) embedded in the fleshy tissue of the pericarp.

Biconvex Convex on both sides.

Biennial Completing the full cycle from germination to fruiting in more than one, but not more than two years.

Bifid **Forked,** divided into two parts.

Bifoliolate Having two leaflets.

Bilabiate Having two lips as of a corolla or calyx with segments fused into an upper and lower lip.

Bipinnate Twice pinnate; the primary leaflets being again divided into secondary leaflets.

Bipinnatisect Refers to a pinnately compound leaf, in which each leaflet is again divided into pinnae.

Biserrate Doubly serrate; with smaller, regular, asymmetric teeth on the margins of larger teeth.

Bisexual Having both sexes, as in a flower bearing both stamens and pistil, hermaphrodite or perfect.

Biternate Twice ternate; with three pinnae each divided into three pinnules.

Blade Lamina; part of the leaf above the sheath or petiole.

Blotched See Variegated.

Bole Main trunk of tree from the base to the first branch.

Brachyblast A short, axillary, densely crowded branchlet or shoot of limited growth, in which the internodes elongate little or not at all.

Bracket fungus Shelf fungus.

Bract A leaflike structure, different in form from the foliage leaves, associated with an inflorescence or flower. *adj.* bracteate.

Bracteate Possessing bracts.

Bracteolate Having bracteoles.

Bracteole A small, secondary, bract-like structure borne singly or in a pair on the pedicel or calyx of a flower. *adj.* bracteolate.

Bran Hard outer layer of grain and comprises the aleurone and pericarp. It contains important antioxidant, vitamins and fibre.

Bristle A stiff hair.

Bulb A modified underground axis that is short and crowned by a mass of usually fleshy, imbricate scales. *adj.* bulbous.

Bulbil A small bulb or bulb-shaped body, especially one borne in the leaf axil or an inflorescence and usually produced for asexual reproduction.

Bullate Puckered, blistered.

Burr Type of seed or fruit with short, stiff bristles or hooks or may refer to a deformed type of wood in which the grain has been misformed.

Bush Low, dense shrub without a pronounced trunk.

Buttress Supporting, projecting outgrowth from base of a tree trunk as in some Rhizophoraceae and Moraceae.

Caducous Shedding or falling early before maturity refers to sepals and petals.

Caespitose Growing densely in tufts or clumps; having short, closely packed stems.

Calcareous Composed of or containing lime or limestone.

Calcrete A hardpan consisting gravel and sand cemented by calcium.

Callus A condition of thickened raised mass of hardened tissue on leaves or other plant parts often formed after an injury but sometimes a normal feature. A callus also can refer to an undifferentiated plant cell mass grown on a culture medium. *n.* callosity**.** *pl.* calli, callosities. *adj.* callose.

Calyptra The protective cap or hood covering the spore case of a moss or related plant.

Calyptrate Operculate, having a calyptra.

Calyx Outer floral whorl usually consisting of free sepals or fused sepals (calyx tube) and calyx lobes. It encloses the flower while it is still a bud. *adj.* calycine.

Calyx lobe One of the free upper parts of the calyx which may be present when the lower part is united into a tube.

Calyx tube The tubular fused part of the calyx, often cup shaped or bell shaped, when it is free from the corolla.

Campanulate Shaped like a bell refers to calyx or corolla.

Canaliculate Having groove or grooves.

Candelabriform Having the shape of a tall branched candle stick.

Canescent Covered with short, fine whitish or greyish hairs or down.

Canopy Uppermost leafy stratum of a tree.

Cap See Pileus.

Capitate Growing together in a head. Also means enlarged and globular at the tip.

Capitulum A flower head or inflorescence having a dense cluster of sessile, or almost sessile, flowers or florets.

Capsule A dry, dehiscent fruit formed from two or more united carpels and dehiscing at maturity by sections called valves to release the seeds. *adj.* capsular.

Carinate Keeled.

Carpel A simple pistil consisting of ovary, ovules, style and stigma. *adj.* carpellary.

Carpogonium Female reproductive organ in red algae. *pl.* carpogonia.

Carpophore Part of the receptacle which is lengthened between the carpels as a central axis; any fruiting body or fruiting structure of a fungus.

Cartilaginous Sinewy, having a firm, tough, flexible texture (in respect of leaf margins).

Caruncle (Bot) fleshy structure attached to the seed of certain plants.

Caryopsis A simple dry, indehiscent fruit formed from a single ovary with the seed coat united with the ovary wall as in grasses and cereals.

Cataphyll A reduced or scarcely developed leaf at the start of a plant's life (i.e. cotyledons) or in the early stages of leaf development.

Catkin A slim, cylindrical, pendulous flower spike usually with unisexual flowers.

Caudate Having a narrow, tail-like appendage.

Caudex Thickened, usually underground base of the stem.

Caulescent Having a well-developed aerial stem.

Cauliflory Botanical term referring to plants which flower and fruit from their main stems or woody trunks. *adj.* cauliflorous.

Cauline Borne on the aerial part of a stem.

Chaffy Having thin, membranous scales in the inflorescence as in the flower heads of the sunflower family.

Chalaza The basal region of the ovule where the stalk is attached.

Chamaephyte A low-growing perennial plant whose dormant overwintering buds are borne at or just above the surface of the ground.

Chartaceous Papery, of paper-like texture.

Chasmogamous Describing flowers in which pollination takes place while the flower is open.

Chatoyant Having a velvety sheen or lustre.

Chloroplast A chlorophyll-containing organelle (plastid) that gives the green colour to leaves and stems. Plastids harness light energy that is used to fix carbon dioxide in the process called photosynthesis.

Chromoplast Plastid containing coloured pigments apart from chlorophyll.

Chromosomes Thread-shaped structures that occur in pairs in the nucleus of a cell, containing the genetic information of living organisms.

Cilia Hairs along the margin of a leaf or corolla lobe.

Ciliate With a fringe of hairs on the margin as of the corolla lobes or leaf.

Ciliolate Minutely ciliate.

Cilium A straight, usually erect hair on a margin or ridge. *pl.* cilia.

Cincinnus A monochasial cyme in which the lateral branches arise alternately on opposite sides of the false axis.

Circinate Spirally coiled, with the tip innermost.

Circumscissile Opening by a transverse line around the circumference as of a fruit.

Cladode The modified photosynthetic stem of a plant whose foliage leaves are much reduced or absent. *cf.* cladophyll, phyllode.

Cladophyll A photosynthetic branch or portion of a stem that resembles and functions as a leaf, like in asparagus. *cf.* cladode, phyllode.

Clamp connection In the basidiomycetes fungi, a lateral connection or outgrowth formed between two adjoining cells of a hypha and arching over the septum between them.

Clavate Club shaped thickened at one end referring to fruit or other organs.

Claw The conspicuously narrowed basal part of a flat structure.

Clay A naturally occurring material composed primarily of fine-grained minerals like kaolinite, montmorillonite-smectite or illite which exhibit plasticity through a variable range of water content and which can be hardened when dried and/or fired.

Clayey Resembling or containing a large proportion of clay.

Cleft Incised halfway down.

Cleistogamous Refers to a flower in which fertilisation occurs within the bud, i.e. without the flower opening. *cf.* chasmogamous.

Climber Growing more or less upwards by leaning or twining around another structure.

Clone All the plants reproduced, vegetatively, from a single parent, thus having the same genetic make-up as the parent.

Coccus One of the sections of a distinctly lobed fruit which becomes separate at maturity; sometimes called a mericarp. *pl.* cocci.

Coenocarpium A fleshy, multiple pseudocarp formed from an inflorescence rather than a single flower.

Coherent Touching without organic fusion, referring to parts normally together, e.g. floral parts of the same whorl. *cf.* adherent, adnate, connate.

Collar Boundary between the above- and below-ground parts of the plant axis.

Colliculate Having small elevations.

Column A structure formed by the united style, stigma and stamen(s) as in Asclepiadaceae and Orchidaceae.

Comose Tufted with hairs at the ends as of seeds.

Composite Having two types of florets as of the flowers in the sunflower family, Asteraceae.

Compost Organic matter (like leaves, mulch, manure, etc.) that breaks down in soil releasing its nutrients.

Compound Describe a leaf that is further divided into leaflets or pinnae or flower with more than a single floret.

Compressed Flattened in one plane.

Conceptacles Specialised cavities of marine algae that contain the reproductive organs.

Concolorous Uniformly coloured, as in upper and lower surfaces. *cf.* discolorous

Conduplicate Folded together lengthwise.

Cone A reproductive structure composed of an axis (branch) bearing sterile bract-like organs and seed- or pollen-bearing structures. Applied to Gymnospermae, Lycopodiaceae, Casuarinaceae and also in some members of Proteaceae.

Conic Cone shaped, attached at the broader end.

Conic–capitate A cone-shaped head of flowers.

Connate Fused to another structure of the same kind. *cf.* adherent, adnate, coherent.

Connective The tissue separating two lobes of an anther.

Connivent Converging.

Conspecific Within or belonging to the same species.

Contorted Twisted.

Convolute Refers to an arrangement of petals in a bud where each has one side overlapping the adjacent petal.

Cordate Heart shaped as of leaves.

Core Central part.

Coriaceous Leathery texture as of leaves.

Corm A short, swollen, fleshy, underground plant stem that serves as a food storage organ used by some plants to survive winter or other adverse conditions

Cormel A miniature, new corm produced on a mature corm.

Corn silk The long, filamentous styles that grow as a silky tuft or tassel at the tip of an ear of corn.

Corolla The inner floral whorl of a flower, usually consisting of free petals or petals fused forming a corolla tube and corolla lobes. *adj.* corolline.

Corona A crown-like section of the staminal column, usually with the inner and outer lobes as in the **Stapelieae**.

Coroniform Crown shaped, as in the pappus of Asteraceae.

Cortex The outer of the stem or root of a plant, bounded on the outside by the epidermis and on the inside by the endodermis containing undifferentiated cells.

Corymb A flat-topped, short, broad inflorescence, in which the flowers, through unequal pedicels, are in one horizontal plane and the youngest in the centre. *adj.* corymbose

Costa A thickened, linear ridge or the midrib of the pinna in ferns. *adj.* costate.

Costapalmate Having definite costa (midrib) unlike the typical palmate leaf, but the leaflets are arranged radially like in a palmate leaf.

Cotyledon The primary seed leaf within the embryo of a seed.

Cover crop Crop grown in between trees or in fields primarily to protect the soil from erosion, to improve soil fertility and to keep off weeds.

Crenate Round toothed or scalloped as of leaf margins.

Crenulate Minutely crenate, very strongly scalloped.

Crested Frilled and ruffled edge.

Crispate Weakly undulating edge.

Crisped With a curled or twisted edge.

Cristate Having or forming a crest or crista.

Crozier Shaped like a shepherd's crook.

Crustaceous Like a crust; having a hard crust or shell.

Cucullate Having the shape of a cowl or hood, hooded.

Culm The main aerial stem of the Gramineae (grasses, sedges, rushes and other monocots).

Culm sheath The plant casing (similar to a leaf) that protects the young bamboo shoot during growth, attached at each node of culm.

Cultigen Plant species or race known only in cultivation.

Cultivar Cultivated variety; an assemblage of cultivated individuals distinguished by any characters significant for the purposes of agriculture, forestry or horticulture and which, when reproduced, retains its distinguishing features.

Cuneate Wedge-shaped, obtriangular.

Cupular Cup shaped, having a cupule.

Cupule A small cup-shaped structure or organ, like the cup at the base of an acorn.

Cusp An elongated, usually rigid, acute point. *cf.* mucro.

Cuspidate Terminating in or tipped with a sharp firm point or cusp. *cf.* mucronate.

Cuspidulate Constricted into a minute cusp. *cf.* cuspidate.

Cyathiform In the form of a cup, a little widened at the top.

Cyathium A specialised type of inflorescence of plants in the genera *Euphorbia* and *Chamaesyce* in which the unisexual flowers

are clustered together within a bract-like envelope. *pl.* cyathia.

Cylindric Tubular or rod shaped.

Cylindric–acuminate Elongated and tapering to a point.

Cymbiform Boat shaped and elongated and having the upper surface decidedly concave.

Cyme An inflorescence in which the lateral axis grows more strongly than the main axis with the oldest flower in the centre or at the ends. *adj.* cymose

Cymule A small cyme or one or a few flowers.

Cystidium A relatively large cell found on the hymenium of a basidiomycete, for example, on the surface of a mushroom.

Cystocarp Fruitlike structure (sporocarp) developed after fertilisation in the red algae.

Deciduous Falling off or shedding at maturity or a specific season or stage of growth.

Decorticate To remove the bark, rind or husk from an organ; to strip of its bark; to come off as a skin.

Decompound As of a compound leaf; consisting of divisions that are themselves compound.

Decumbent Prostrate, laying or growing on the ground but with ascending tips. *cf.* ascending, procumbent.

Decurrent Having the leaf base tapering down to a narrow wing that extends to the stem.

Decussate Having paired organs with successive pairs at right angles to give four rows as of leaves.

Deflexed Bent downwards.

Degumming Removal of gum deposits (phosphatides, entrained oil and meal particles) from crude edible oils traditionally done with water. Water degumming process also removes hydrophilic substances such as sugars from the oil.

Dehisce To split open at maturity, as in a capsule.

Dehiscent Splitting open at maturity to release the contents. *cf.* indehiscent.

Deltate Triangular shape.

Deltoid Shaped like an equilateral triangle.

Dendritic Branching from a main stem or axis like the branches of a tree.

Dentate With sharp, rather coarse teeth perpendicular to the margin.

Denticulate Finely toothed.

Diageotropic The tendency of growing parts, such as roots, to grow at right angle to the line of gravity.

Diadelphous Having stamens in two bundles as in Papilionaceae flowers.

Dichasium A cymose inflorescence in which the branches are opposite and approximately equal. *pl.* dichasia. *adj.* dichasial.

Dichotomous Divided into two parts.

Dicotyledon Angiosperm with two cotyledons.

Didymous Arranged or occurring in pairs as of anthers, having two lobes.

Digitate Having digits or fingerlike projections.

Dikaryophases Or dendrophydia, irregularly, strongly branched terminal hyphae in the hymenomycetes (class of basidiomycetes) fungi.

Dimorphic Having or occurring in two forms, as of stamens of two different lengths or a plant, having two kinds of leaves.

Dioecious With male and female unisexual flowers on separate plants. *cf.* monoecious.

Diploid a condition in which the chromosomes in the nucleus of a cell exist as pairs, one set being derived from the female parent and the other from the male.

Diplobiontic life cycle Life cycle that exhibits alternation of generations, which features spore-producing multicellular sporophytes and gamete-producing multicellular gametophytes. Mitoses occur in both the diploid and haploid phases.

Diplontic life cycle Or gametic meiosis, wherein instead of immediately dividing meiotically to produce haploid cells, the zygote divides mitotically to produce a multicellular diploid individual or a group of more diploid cells.

Diplochory Seed dispersal involving two or more modes.

Dipterocarpus Trees of the family Dipterocarpaceae, with two-winged fruit found mainly in tropical lowland rainforest.

Disc (Botany) refers to the usually disc-shaped receptacle of the flower head in Asteraceae; also the fleshy nectariferous organ usually between the stamens and ovary; also used for the enlarged style end in Proteaceae.

Disc floret The central, tubular four- or five-toothed or lobed floret on the disc of an inflorescence, as of flower head of Asteraceae.

Disciform Flat and rounded in shaped. *cf.* discoid, radiate.

Discoid Resembling a disc; having a flat, circular form; disc shaped. *cf.* disciform, radiate.

Discolorous Having two colours, as of a leaf which has different colours on the two surfaces. *cf.* concolorous.

Disomic Having one or more chromosomes present twice but without the entire genome doubled.

Dispersal Dissemination of seeds.

Distal Site of any structure farthest from the point of attachment. *cf.* proximal.

Distichous Referring to two rows of upright leaves in the same plane.

Dithecous Having two thecae.

Divaricate Diverging at a wide angle.

Domatium A part of a plant (e.g. a leaf) that has been modified to provide protection for other organisms. *pl.* domatia.

Dormancy A resting period in the life of a plant during which growth slows or appears to stop.

Dorsal Referring to the back surface.

Dorsifixed Attached to the back as of anthers.

Drupaceous Resembling a drupe.

Drupe A fleshy fruit with a single seed enclosed in a hard shell (endocarp) which is tissue embedded in succulent tissue (mesocarp) surrounded by a thin outer skin (epicarp). *adj.* drupaceous.

Drupelet A small drupe.

Ebracteate Without bracts.

Echinate Bearing stiff, stout, bristly, prickly hairs.

Edaphic Refers to plant communities that are distinguished by soil conditions rather than by the climate.

Eglandular Without glands. *cf.* glandular.

Elaioplasts A type of leucoplast that is specialised for the storage of lipids in plants.

Elaiosome Fleshy lipid-rich structures that are attached to the seeds of many plant species.

Ellipsoid A three-dimensional shape; elliptic in outline.

Elliptic Having a two-dimensional shape of an ellipse or flattened circle.

Elongate Extended, stretched out.

Emarginate Refers to leaf with a broad, shallow notch at the apex. *cf.* retuse.

Embryo (Botany) a minute rudimentary plant contained within a seed or an archegonium, composed of the embryonic axis (shoot end and root end).

Endemic Prevalent in or peculiar to a particular geographical locality or region.

Endocarp The hard innermost layer of the pericarp of many fruits.

Endosperm Tissue that surrounds and nourishes the embryo in the angiosperm seed. It contains starchy carbohydrates, proteins and small amounts of vitamins and minerals.

Endospermous Refers to seeds having an endosperm.

Ensiform Shaped like the blade of a sword, long and narrow with sharp edges and a pointed tip.

Endotrophic As of mycorrhiza obtaining nutrients from inside.

Ensilage The process of preserving green food for livestock in an undried condition in airtight conditions. Also called silaging.

Entire Having a smooth, continuous margin without any incisions or teeth as of a leaf.

Entisols Soils that do not show any profile development other than an A horizon.

Ephemeral Transitory, short lived.

Epicalyx A whorl of bracts, subtending and resembling a calyx.

Epicarp Outermost layer of the pericarp of a fruit.

Epicormic Attached to the corm.

Epicotyl The upper portion of the embryonic axis, above the cotyledons and below the first true leaves.

Epigeal Above grounds with cotyledons raised above ground.

Epiparasite An organism parasitic on another that parasitizes a third.

Epipetalous Borne on the petals, as of stamens.

Epiphyte A plant growing on, but not parasitic on, another plant, deriving its moisture and nutrients from the air and rain, e.g. some Orchidaceae. *adj.* epiphytic.

Epithet Name.

Equitant In a loose fan pattern.

Erect Upright, vertical.

Essential oils Volatile products obtained from a natural source; refers to volatile products

obtained by steam or water distillation in a strict sense.

Etiolation To cause (a plant) to develop without chlorophyll by preventing exposure to sunlight.

Eutrophic Having waters rich in mineral and organic nutrients that promote a proliferation of plant life, especially algae, which reduces the dissolved oxygen content and often causes the extinction of other organisms.

Excentric Off the true centre.

Excrescence Abnormal outgrowth.

Excurrent Projecting beyond the tip, as the midrib of a leaf or bract.

Exserted Sticking out, protruding beyond some enclosing organ, as of stamens which project beyond the corolla or perianth.

Exstipulate Without stipules. *cf.* stipulate.

Extra-floral Outside the flower.

Extrorse Turned outwards or away from the axis as of anthers. *cf.* introrse, latrorse.

Falcate Sickle shaped, crescent shaped.

Fascicle A cluster or bundle of stems, flowers and stamens. *adj.* fasciculate.

Fasciclode Staminode bundles.

Fastigiate A tree in which the branches grow almost vertically.

Ferrosols Soils with an iron oxide content of greater than 5%.

Ferruginous Rust coloured, reddish-brown.

Fertile Having functional sexual parts which are capable of fertilisation and seed production. *cf.* sterile.

Filament The stalk of a stamen supporting and subtending the anther.

Filiform Having the form of or resembling a thread or filament.

Fimbriate Fringed.

Fixed oils Non-volatile oils, triglycerides of fatty acids.

Flaccid Limp and weak.

Flag leaf The uppermost leaf on the stem.

Flaky In the shape of flakes or scales.

Flexuous Zigzagging, sinuous, bending, as of a stem.

Floccose Covered with tufts of soft woolly hairs.

Floral tube A flower tube usually formed by the basal fusion of the perianth and stamens.

Floret One of the small individual flowers of sunflower family or the reduced flower of the grasses, including the lemma and palea.

Flower The sexual reproductive organ of flowering plants, typically consisting of gynoecium, androecium and perianth or calyx and/ or corolla and the axis bearing these parts.

Fluted As of a trunk with grooves and folds.

Fodder Plant material, fresh or dried fed to animals.

Foliaceous Leaflike.

Foliage Leaves of the plant.

Foliar Pertaining to a leaf.

Foliolate Pertaining to leaflets, used with a number prefix to denote the number of leaflets.

Foliose Leaflike.

Follicle (Botany) a dry fruit, derived from a single carpel and dehiscing along one suture.

Forb Any herb that is not grass or grass-like.

Foveolate Surface pitted with shallow depressions.

Free central placentation The arrangement of ovules on a central column that is not connected to the ovary wall by partitions, as in the ovaries of the carnation and primrose.

Frond The leaf of a fern or cycad.

Fruit Ripened ovary with adnate parts.

Frutescent Shrubby.

Fugacious Shedding off early.

Fulvous Yellow, tawny.

Funiculus (Botany) short stalk which attaches the ovule to the ovary wall.

Fuscescent Dusky.

Fusiform A three-dimensional shape; spindle shaped, i.e. broad in the centre, but tapering at both thick ends.

Galea A part of the calyx or corolla having the form of a helmet.

Gall flower Short-styled flower that does not develop into a fruit but is adapted for the development of a specific wasp within the fruit, e.g. in the fig.

Gamete A reproductive cell that fuses with another gamete to form a zygote. Gametes are haploid (they contain half the normal (diploid) number of chromosomes); thus when two fuse, the diploid number is restored.

Gametophyte The gamete-producing phase in a plant characterised by alternation of generations.

Gamosepalous With sepals united or partially united.

Genome Complete set of genetic material of an organism.

Geniculate Bent like a knee, refer to awns and filaments.

Geocarpic Where the fruit is pushed into the soil by the gynophore and matures.

Geophyte A plant that stores food in an underground storage organ, e.g. a tuber, bulb or rhizome, and has subterranean buds which form aerial growth.

Geotextile Are permeable fabrics which, when used in association with soil, have the ability to separate, filter, reinforce, protect or drain.

Germ Of cereal is the embryo of the seed or kernel. It contains vitamins B and E, folic acid, some protein, minerals and polyunsaturated fats.

Glabrescent Becoming glabrous.

Glabrous Smooth, hairless without pubescence.

Gland A secretory organ, e.g. a nectary, extrafloral nectary or a gland tipped, hair-like or wartlike organ. *adj.* glandular. *cf.* eglandular.

Glaucous Pale blue-green in colour, covered with a whitish bloom that rubs off readily.

Gley soils A hydric soil which exhibits a greenish-blue-grey soil colour due to wetland conditions.

Globose Spherical in shape.

Globular A three-dimensional shape; spherical or orbicular; circular in outline.

Glochids Tiny, finely barbed hair-like spines found on the areoles of some cacti and other plants.

Glochidiate Having glochids.

Glochidote Plant having glochids.

Glume One of the two small, sterile bracts at the base of the grass spikelet, called the lower and upper glumes, due to their position on the rachilla. Also used in Apiaceae, Cyperaceae for the very small bracts on the spikelet in which each flower is subtended by one floral glume. *adj.* glumaceous.

Grits Consist of coarsely ground corn, or sometimes alkali-treated corn.

Groats Hulled, whole grains of various cereals, such as oats, wheat, barley or buckwheat; it includes the cereal germ, fibre-rich bran portion and endosperm of the grain.

Guttation The appearance of drops of xylem sap on the tips or edges of leaves of some vascular plants, such as grasses and bamboos.

Guttule Small droplet.

Gymnosperm A group of spermatophyte seed-bearing plants with ovules on scales, which are usually arranged in cone-like structures and not borne in an ovary. *cf.* angiosperm.

Gynoecium The female organ of a flower; a collective term for the pistil, carpel or carpels.

Gynomonoecious Having female flowers and bisexual flowers on the same plant. *cf.* andromonoecious.

Gynophore Stalk that bears the pistil/carpel.

Habit The general growth form of a plant, comprising its size, shape, texture and stem orientation, the locality in which the plant grows..

Halophyte A plant adapted to living in highly saline habitats. Also a plant that accumulates high concentrations of salt in its tissues. *adj.* halophytic.

Hapaxanthic Refer to palms which flower only once and then die. c.f. pleonanthic.

Haploid Condition where nucleus or cell has a single set of unpaired chromosomes; the haploid number is designated as n.

Haplontic life cycle Or zygotic meiosis wherein meiosis of a zygote, immediately after karyogamy, produces haploid cells which produces more or larger haploid cells ending its diploid phase.

Hastate Having the shape of an arrowhead but with the basal lobes pointing outwards at right angles as of a leaf.

Hastula A piece of plant material at the junction of the petiole and the leaf blade; the hastula can be found on the top of the leaf, adaxial or the bottom, abaxial or both sides.

Heartwood Wood from the inner portion of a tree.

Heliophilous Sun loving, tolerates high level of sunlight.

Heliotropic Growing towards sunlight.

Herb A plant which is nonwoody or woody at the base only, the above-ground stems usually being ephemeral. *adj.* herbaceous.

Herbaceous Resembling a herb, having a habit of a herb.

Hermaphrodite Bisexual, bearing flowers with both androecium and gynoecium in the same flower. *adj.* hermaphroditic.

Heterocyst A differentiated cyanobacterial cell that carries out nitrogen fixation.

Heterogamous Bearing separate male and female flowers, or bisexual and female flowers, or florets in an inflorescence or flower head, e.g. some Asteraceae in which the ray florets may be neuter or unisexual and the disc florets may be bisexual. *cf.* homogamous.

Heteromorphous Having two or more distinct forms. *cf.* homomorphous.

Heterophyllous Having leaves of different forms.

Heterosporous Producing spores of two sizes, the larger giving rise to megagametophytes (female) and the smaller giving rise to microgametophytes (male). Refer to the ferns and fern allies. *cf.* homosporous.

Heterostylous Having styles of two different lengths or forms.

Heterostyly The condition in which flowers on polymorphous plants have styles of different lengths, thereby facilitating cross-pollination.

Hilar Of or relating to a hilum.

Hilum The scar on a seed, indicating the point of attachment to the funiculus.

Hirsute Bearing long coarse hairs.

Hispid Bearing stiff, short, rough hairs or bristles.

Hispidulous Minutely hispid.

Histosol Soil comprising primarily of organic materials, having 40 cm or more of organic soil material in the upper 80 cm.

Hoary Covered with a greyish layer of very short, closely interwoven hairs.

Holdfast An organ or structure of attachment, especially the basal, root-like formation by which certain seaweeds or other algae are attached to a substrate.

Holocarpic Having the entire thallus developed into a fruiting body or sporangium.

Homochromous Having all the florets of the same colour in the same flower head *cf.* heterochromous.

Homogamous Bearing flowers or florets that do not differ sexually. *cf.* heterogamous.

Homogeneous endosperm Endosperm with even surface that lacks invaginations or infoldings of the surrounding tissue.

Homomorphous Uniform, with only one form. *cf.* heteromorphous.

Homosporous Producing one kind of spores. Refer to the ferns and fern allies. *cf.* heterosporous.

Hormogonium A part of a filament of a cyanobacterium that detaches and grows by cell division into a new filament. *pl.* hormogonia.

Hurd fibre Long pith fibre of the stem.

Hyaline Colourless, almost transparent.

Hybrid The first-generation progeny of the sexual union of plants belonging to different taxa.

Hybridisation The crossing of individuals from different species or taxa.

Hydathode A type of secretory tissue in leaves, usually of Angiosperms, that secretes water through pores in the epidermis or margin of the leaf.

Hydrophilous Water loving; requiring water in order to be fertilised, referring to many aquatic plants.

Hygrochastic Applied to plants in which the opening of the fruits is caused by the absorption of water.

Hygrophilous Living in water or moist places.

Hymenial cystidia The cells of the hymenium develop into basidia or asci, while in others some cells develop into sterile cells called cystidia.

Hymenium Spore-bearing layer of cells in certain fungi containing asci (ascomycetes) or basidia (basidiomycetes).

Hypanthium Cup-like receptacles of some dicotyledonous flowers formed by the fusion of the calyx, corolla and androecium that surrounds the ovary which bears the sepals, petals and stamens. *adj.* relating to or of the nature of a hypanthium.

Hypha Is a long, branching filamentous cell of a fungus and also of unrelated Actinobacteria. *pl.* hyphae.

Hypocotyl The portion of the stem below the cotyledons.

Hypodermis The cell layer beneath the epidermis of the pericarp.

Hypogeal Below ground as of germination of seed.

Hysteresis Refers to systems that may exhibit path dependence.

Imbricate Closely packed and overlapping. *cf.* valvate.

Imparipinnate Pinnately compound with a single terminal leaflet and hence with an odd number of leaflets. *cf.* paripinnate.

Inceptisols Old soils that have no accumulation of clays, iron, aluminium or organic matter.

Incised Cut jaggedly with very deep teeth.

Included Referring to stamens which do not project beyond the corolla or to valves which do not extend beyond the rim of a capsular fruit. *cf.* exserted.

Incurved Curved inwards; curved towards the base or apex.

Indefinite Numerous and variable in number.

Indehiscent Not opening or splitting to release the contents at maturity as of fruit. *cf.* dehiscent.

Indumentum covering of fine hairs or bristles commonly found on external parts of plants.

Indurate To become hard, often the hardening developed only at maturity.

Indusium An enclosing membrane, covering the sorus of a fern. Also used for the modified style end or pollen-cup of some Goodeniaceae (including Brunoniaceae). *adj.* indusiate.

Inferior Said of an ovary or fruit that has sepals, petals and stamens above the ovary. *cf.* superior.

Inflated Enlarged and hollow except in the case of a fruit which may contain a seed. *cf.* swollen.

Inflexed Bent or curved inwards or downwards, as petals or sepals.

Inflorescence A flower cluster or the arrangement of flowers in relation to the axis and to each other on a plant.

Infrafoliar Located below the leaves.

Infraspecific Referring to any taxon below the species rank.

Infructescence The fruiting stage of an inflorescence.

Infundibulum Funnel-shaped cavity or structure.

Inrolled Curved inwards.

Integuments Two distinct tissue layers that surround the nucleolus of the ovule, forming the testa or seed coat when mature.

Intercalary Of growth, between the apex and the base; of cells, spores, etc., between two cells.

Interfoliar Inter-leaf.

Internode Portion of the stem, culm, branch or rhizome between two nodes or points of attachment of the leaves.

Interpetiolar As of stipules positioned between petioles of opposite leaves.

Intrastaminal Within the stamens.

Intricate Entangled, complex.

Introduced Not indigenous; not native to the area in which it now occurs.

Introrse Turned inwards or towards the axis or pistil as of anthers. *cf.* extrorse, latrorse.

Involucre A whorl of bracts or leaves that surround one to many flowers or an entire inflorescence.

Involute Having the margins rolled inwards, referring to a leaf or other flat organ.

Jugate Of a pinnate leaf; having leaflets in pairs.

Juvenile Young or immature, used here for leaves formed on a young plant which are different in morphology from those formed on an older plant.

Keel A longitudinal ridge, at the back of the leaf. Also the two lower fused petals of a 'pea' flower in the Papilionaceae, which form a boat-like structure around the stamens and styles, also called carina. *adj.* keeled. *cf.* standard, wing.

Labellum The modified lowest of the three petals forming the corolla of an orchid, usually larger than the other two petals, and often spurred.

Lacerate Irregularly cleft.

Laciniate Fringed; having a fringe of slender, narrow, pointed lobes cut into narrow lobes.

Lamella A gill-shaped structure: fine sheets of material held adjacent to one another.

Lamina The blade of the leaf or frond.

Lanate Woolly, covered with long hairs which are loosely curled together like wool.

Lanceolate Lance shaped in outline, tapering from a broad base to the apex.

Landrace: landrace Plants adapted to the natural environment in which they grow, developing naturally with minimal assistance or guidance from humans and usually possess more diverse phenotypes and genotypes. They have not been improved by formal breeding programmes.

Laterite Reddish-coloured soils rich in iron oxide, formed by weathering of rocks under oxidising and leaching conditions, commonly

found in tropical and subtropical regions. *adj.* lateritic.

Latex A milky, clear or sometimes coloured sap of diverse composition exuded by some plants.

Latrorse Turned sideways, i.e. not towards or away from the axis as of anthers dehiscing longitudinally on the side. *cf.* extrorse, introrse.

Lax Loose or limp, not densely arranged or crowded.

Leaflet One of the ultimate segments of a compound leaf.

Lectotype A specimen chosen after the original description to be the type.

Lemma The lower of two bracts (scales) of a grass floret, usually enclosing the palea, lodicules, stamens and ovary.

Lenticel Is a lens shaped opening that allows gases to be exchanged between air and the inner tissues of a plant, commonly found on young bark, or the surface of the fruit.

Lenticellate Dotted with lenticels.

Lenticular Shaped like a biconvex lens. *cf.* lentiform.

Lentiform Shaped like a biconvex lens, *cf.* lenticular.

Leptomorphic Temperate, running bamboo rhizome; usually thinner than the culms they support and the internodes are long and hollow.

Liane A woody climbing or twining plant.

Ligneous Woody.

Lignotuber A woody, usually underground, tuberous rootstock often giving rise to numerous aerial stems.

Ligulate Small and tongue shaped or with a little tongue-shaped appendage or ligule, star shaped as of florets of Asteraceae.

Ligule A strap-shaped corolla in the flowers of Asteraceae; also a thin membranous outgrowth from the inner junction of the grass leaf sheath and blade. *cf.* ligulate.

Limb The expanded portion of the calyx tube or the corolla tube, or the large branch of a tree.

Linear A two-dimensional shape, narrow with nearly parallel sides.

Linguiform Tongue shaped. *cf.* ligulate.

Lipotubuloids Are cytoplasmic domains containing aggregates of lipid bodies surrounded by a network of microtubules, which join one lipid body with the others.

Lithosol A kind of shallow soils lacking well-defined horizons and composed of imperfectly weathered fragments of rock.

Littoral Of or on a shore, especially seashore.

Loam A type of soil made up of sand, silt and clay in relative concentration of 40–40–20% respectively.

Lobed Divided but not to the base.

Loculicidal Opening into the cells, when a ripe capsule splits along the back.

Loculus Cavity or chamber of an ovary. *pl.* loculi.

Lodicules Two small structures below the ovary which, at flowering, swell up and force open the enclosing bracts, exposing the stamens and carpel.

Lorate Strap shaped with obtuse tip.

Lyrate Pinnately lobed, with a large terminal lobe and smaller lateral ones which become progressively smaller towards the base.

Macronutrients Chemical elements which are needed in large quantities for growth and development by plants and include nitrogen, phosphorus, potassium and magnesium.

Maculate Spotted.

Mallee A growth habit in which several to many woody stems arise separately from a lignotuber; usually applied to certain low-growing species of *Eucalyptus.*

Mangrove A distinctive vegetation type of trees and shrubs with modified roots, often viviparous, occupying the saline coastal habitats that are subject to periodic tidal inundation.

Marcescent Withering or to decay without falling off.

Margin The edge of the leaf blade.

Medulla The pith in the stems or roots of certain plants; or the central portion of a thallus in certain lichens.

Megasporangium The sporangium containing megaspores in fern and fern allies. *cf.* microsporangium.

Megaspore The large spore which may develop into the female gametophyte in heterosporous ferns and fern allies. *cf.* microspore.

Megasporophyll A leaflike structure that bears megasporangia.

Megastrobilus Female cone, seed cone or ovulate cone and contains ovules within which, when fertilised by pollen, become seeds. The

female cone structure varies more markedly between the different conifer families.

Meiosis The process of cell division that results in the formation of haploid cells from diploid cells to produce gametes.

Mericarp A one-seeded portion of an initially syncarpous fruit (schizocarp) which splits apart at maturity. *cf.* coccus.

Meristem The region of active cell division in plants, from which permanent tissue is derived. *adj.* meristematic

-merous Used with a number prefix to denote the basic number of the three outer floral whorls, e.g. a five-merous flower may have 5 sepals, 10 petals and 15 stamens.

Mesic Moderately wet.

Mesocarp The middle layer of the fruit wall derived from the middle layer of the carpel wall. *cf.* endocarp, exocarp and pericarp.

Mesophytes Terrestrial plants which are adapted to neither a particularly dry nor particularly wet environment.

Micropyle The small opening in a plant ovule through which the pollen tube passes in order to effect fertilisation.

Microsporangium The sporangium containing microspores in pteridophytes. *cf.* megasporangium.

Microspore A small spore which gives rise to the male gametophyte in heterosporous pteridophytes. Also for a pollen grain. *cf.* megaspore.

Midvein The main vascular supply of a simple leaf blade or lamina. Also called midrib.

Mitosis Is a process of cell division which results in the production of two daughter cells from a single parent cell.

Mollisols Soils with deep, high organic matter, nutrient-enriched surface soil (a horizon), typically between 60 and 80 cm thick.

Monadelphous Applied to stamens united by their filaments into a single bundle.

Monocarpic Refer to plants that flower, set seeds and then die.

Monochasial A cyme having a single flower on each axis.

Monocotyledon Angiosperm having one cotyledon.

Monoecious Having both male and female unisexual flowers on the same individual plant. *cf.* dioecious.

Monoembryonic seed The seed contains only one embryo, a true sexual (zygotic) embryo. Polyembryonic seed.

Monolete A spore that has a simple linear scar.

Monopodial With a main terminal growing point producing many lateral branches progressively. *cf.* sympodial.

Monostichous Forming one row.

Monotypic Of a genus with one species or a family with one genus; in general, applied to any taxon with only one immediately subordinate taxon.

Montane Refers to highland areas located below the subalpine zone.

Mucilage A soft, moist, viscous, sticky secretion. *adj.* mucilaginous.

Mucous (Botany) slimy.

Mucro A sharp, pointed part or organ, especially a sharp terminal point, as of a leaf.

Mucronate Ending with a short, sharp tip or mucro, resembling a spine. cf. cuspidate, muticous.

Mucronulate With a very small mucro; a diminutive of mucronate.

Mulch Protective cover of plant (organic) or non-plant material placed over the soil, primarily to modify and improve the effects of the local microclimate and to control weeds.

Multiple fruit A fruit that is formed from a cluster of flowers.

Muricate Covered with numerous short hard outgrowths. *cf.* papillose.

Muriculate With numerous minute hard outgrowths; a diminutive of muricate.

Muticous Blunt, lacking a sharp point. *cf.* mucronate.

MYB proteins Are a superfamily of transcription factors that play regulatory roles in developmental processes and defence responses in plants.

Mycorrhiza The mutualistic symbiosis (non-pathogenic association) between soilborne fungi with the roots of higher plants.

Mycorrhiza (vesicular arbuscular) Endomycorrhiza living in the roots of higher plants producing inter- and intracellular fungal growth in root cortex and forming specific fungal structures, referred to as vesicles and arbuscules. *abbrev.* VAM.

Myrmecochory Seed dispersal by ants.

Native A plant indigenous to the locality or region.

Naviculate Boat shaped.

Necrotic Applied to dead tissue.

Nectariferous Having one or more nectaries.

Nectary A nectar secretory gland; commonly in a flower, sometimes on leaves, fronds or stems.

Nervation Venation, a pattern of veins or nerves as of leaf.

Nixtamalisation Refers to a process for the preparation of maize (corn), or other grains, in which the grains are soaked and cooked in an alkaline solution, usually limewater, and hulled.

Node The joint between segments of a culm, stem, branch or rhizome; the point of the stem that gives rise to the leaf and bud.

Nodule A small knoblike outgrowth, as those found on the roots of many leguminous, containing *Rhizobium* bacteria which fixes nitrogen in the soil.

Nom. ambig. Nomen ambiguum (Latin) ambiguous name used in different senses which has become a long-persistent source of error.

Nom. cons. Nomen nonservandum (Latin) name conserved in International Code of Botanical Nomenclature.

Nom. dub. Nomen dubium (Latin) an invalid proposed taxonomic name because it is not accompanied by a definition or description of the taxon to which it applies.

Nom. illeg. Nomen illegitimum (Latin) illegitimate taxon deemed as superfluous at its time of publication either because the taxon to which it was applied already has a name, or because the name has already been applied to another plant.

Nom. invalid. Nomen invalidum (Latin) invalid name according to International Code of Botanical Nomenclature.

Nom. nud. Nomen nudum (Latin) the name of a taxon which has never been validated by a description.

Nom. rej. Nomen rejiciendum (Latin) name rejected in International Code of Botanical Nomenclature.

Notho- (Subsp. or var.) prefix to the rank of a hybrid taxon below the rank of species.

Nucellus Central portion of an ovule in which the embryo sac develops.

Nucellar embryony A form of seed reproduction in which the nucellar tissue which surrounds the embryo sac can produce additional embryos (polyembryony) which are genetically identical to the parent plant. This is found in many citrus species and in mango.

Nut A dry indehiscent one-celled fruit with a hard pericarp.

Nutlet A small, one-seeded, indehiscent lobe of a divided fruit.

Ob- Prefix meaning inversely or opposite to.

Obconic A three-dimensional shape; inversely conic; cone shaped, conic with the vertex pointing downwards.

Obcordate Inversely cordate, broad and notched at the tip; heart shaped but attached at the pointed end.

Obdeltate Inversely deltate; deltate with the broadest part at the apex.

Oblanceolate Inversely lanceolate, lance shaped but broadest above the middle and tapering towards the base as of leaf.

Oblate Having the shape of a spheroid with the equatorial diameter greater than the polar diameter; being flattened at the poles.

Oblong Longer than broad with sides nearly parallel to each other.

Obovate Inversely ovate, broadest above the middle.

Obpyramidal Resembling a four-sided pyramid attached at the apex with the square base facing away from the attachment.

Obpyriform Inversely pyriform, resembling a pear which is attached at the narrower end. *cf.* pyriform.

Obspatulate Inversely spatulate; resembling a spoon but attached at the broadest end. *cf.* spatulate.

Obtriangular Inversely triangular; triangular but attached at the apex. *cf.* triangular.

Obtrullate Inversely trullate; resembling a trowel blade with the broadest axis above the middle. *cf.* trullate.

Obtuse With a blunt or rounded tip, the converging edges separated by an angle greater than 90°.

-oid Suffix denoting a three-dimensional shape, e.g. spheroid.

Ochraceous A dull yellow colour.

Ocreate Having a tube-like covering around some stems, formed of the united stipules; sheathed.

Oleaginous Oily.

Oligotrophic Lacking in plant nutrients and having a large amount of dissolved oxygen throughout.

Operculum A lid or cover that becomes detached at maturity by abscission, e.g. in *Eucalyptus*, also a cap or lid covering the bud and formed by fusion or cohesion of sepals and/or petals. *adj.* operculate.

Opposite Describing leaves or other organs which are borne at the same level but on opposite sides of the stem. *cf.* alternate.

Orbicular Of circular outline, disc-like.

Order A taxonomic rank between class and family used in the classification of organisms, i.e. a group of families believed to be closely related.

Orifice An opening or aperture.

Organosols Soils not regularly inundated by marine waters and containing a specific thickness of organic materials within the upper part of the profile.

Orth. var. Orthographic variant, i.e. an incorrect alternate spelling of a name.

Ovary The female part of the pistil of a flower which contains the ovules (immature seeds).

Ovate Egg shaped, usually with reference to two dimensions.

Ovoid Egg shaped, usually with reference to three dimensions.

Ovule The young, immature seed in the ovary which becomes a seed after fertilisation. *adj.* ovular.

Ovulode A sterile reduced ovule borne on the placenta, commonly occurring in Myrtaceae.

Oxisols Refer to ferralsols.

Pachymorphic Describes the short, thick rhizomes of clumping bamboos with short, thick and solid internode (except the bud-bearing internodes, which are more elongated). *cf.* sympodial.

Palate (Botany) a raised appendage on the lower lip of a corolla which partially or completely closes the throat.

Palea The upper of the two membraneous bracts of a grass floret, usually enclosing the lodicules, stamens and ovary. *pl.* paleae. *adj.* paleal. *cf.* lemma.

Paleate Having glumes.

Palm heart Refers to soft, tender inner core and growing bud of certain palm trees which are eaten as vegetables. Also called heart of palm, palmito, burglar's thigh, chonta or swamp cabbage.

Palmate Describing a leaf which is divided into several lobes or leaflets which arise from the same point. *adv.* palmately.

Palmito See Palm heart.

Palustrial Paludal, swampy, marshy.

Palustrine Marshy, swampy.

Palustrine herb Vegetation that is rooted below water but grows above the surface in wetland system.

Panduriform Fiddle shaped, usually with reference to two dimensions.

Panicle A compound, indeterminate, racemose inflorescence in which the main axis bears lateral racemes or spikes. *adj.* paniculate.

Pantropical Distributed throughout the tropics.

Papilionaceous Butterfly-like, said of the pea flower or flowers of Papilionaceae, flowers which are zygomorphic with imbricate petals, one broad upper one, two narrower lateral ones and two narrower lower ones.

Papilla A small, superficial protuberance on the surface of an organ being an outgrowth of one epidermal cell. *pl.* papillae. *adj.* papillose.

Papillate Having papillae.

Papillose Covered with papillae.

Pappus A tuft (or ring) of hairs, bristles or scales borne above the ovary and outside the corolla as in Asteraceae often persisting as a tuft of hairs on a fruit. *adj.* pappose, pappous.

Papyraceous Resembling parchment of paper.

Parenchyma Undifferentiated plant tissue composed of more or less uniform cells.

Parietal Describes the attachment of ovules to the outer walls of the ovaries.

Paripinnate Pinnate with an even number of leaflets and without a terminal leaflet. *cf.* imparipinnate.

-partite Divided almost to the base into segments, the number of segments written as a prefix.

Patelliform Shaped like a limpet shell; cap shaped and without whorls.

Patent Diverging from the axis almost at right angles.

Peat Is an accumulation of partially decayed vegetation matter.

Pectin A group of water-soluble colloidal carbohydrates of high molecular weight found in certain ripe fruits.

Pectinate Pinnatifid with narrow segments resembling the teeth of a comb.

Pedicel The stalk of the flower or stalk of a spikelet in Poaceae. *adj.* pedicellate.

Pedicellate Having pedicel.

Peduncle A stalk supporting an inflorescence. *adj.* pedunculate

Pellucid Allowing the passage of light; transparent or translucent.

Pellucid dotted Copiously dotted with immersed, pellucid, resinous glands.

Peltate With the petiole attached to the lower surface of the leaf blade.

Pendant Hanging down.

Pendulous Drooping, as of ovules.

Penniveined or penninerved Pinnately veined.

Pentamerous In five parts.

Perennial A plant that completes its life cycle or lives for more than two years. *cf.* annual, biennial.

Perfoliate A leaf with the basal lobes united around—and apparently pierced by—the stem.

Pergamentaceous Parchment-like.

Perianth The two outer floral whorls of the angiosperm flower; commonly used when the calyx and the corolla are not readily distinguishable (as in monocotyledons).

Pericarp (Botany). The wall of a ripened ovary; fruit wall composed of the exocarp, mesocarp and endocarp.

Persistent Remaining attached; not falling off. *cf.* caduceus.

Petal Free segment of the corolla. *adj.* petaline. *cf.* lobe.

Petiolar Relating to the petiole.

Petiolate Having petiole.

Petiole Leaf stalk. *adj.* petiolate.

Petiolulate Supported by its own petiolule.

Petiolule The stalk of a leaflet in a compound leaf. *adj.* petiolulate.

pH Is a measure of the acidity or basicity of a solution. It is defined as the cologarithm of the activity of dissolved hydrogen ions (H+).

Phenology The study of periodic plant life cycle events as influenced by seasonal and interannual variations in climate.

Phyllary A bract of the involucre of a composite plant, term for one of the scalelike bracts beneath the flower head in Asteraceae.

Phylloclade A flattened, photosynthetic branch or stem that resembles or performs the function of a leaf, with the true leaves represented by scales.

Phyllode A petiole that functions as a leaf. *adj.* phyllodineous. *cf.* cladode.

Phyllopodia Refer to the reduced, scalelike leaves found on the outermost portion of the corm where they seem to persist longer than typical sporophylls as in the fern *Isoetes*.

Phytoremediation Describes the treatment of environmental problems (bioremediation) through the use of plants which mitigate the environmental problem without the need to excavate the contaminant material and dispose of it elsewhere.

Pileus (Botany) cap of mushroom.

Piliferous (Botany) bearing or producing hairs, as of an organ with the apex having long, hairlike extensions.

Pilose Covered with fine soft hairs.

Pinna A primary division of the blade of a compound leaf or frond. *pl.* pinnae.

Pinnate Bearing leaflets on each side of a central axis of a compound leaf; divided into pinnae.

Pinnatifid, pinnatilobed A pinnate leaf parted approximately halfway to midrib; when divided to almost to the midrib described as deeply pinnatifid or pinnatisect.

Pinnatisect Lobed or divided almost to the midrib.

Pinnule A leaflet of a bipinnate compound leaf.

Pistil Female part of the flower comprising the ovary, style and stigma.

Pistillate Having one or more pistils; having pistils but no stamens.

Placenta The region within the ovary to which ovules are attached. *pl.* placentae.

Placentation The arrangement of the placentae and ovules in the ovary.

Plano- A prefix meaning level or flat.

Pleonanthic Refer to palms in which the stem does not die after flowering.

Plicate Folded like a fan.

Plumose Feather-like, with fine hairs arising laterally from a central axis; feathery.

Pneumatophore Modified root which allows gaseous exchange in mud-dwelling shrubs, e.g. mangroves.

Pod A dry one- to many-seeded dehiscent fruit, as applied to the fruit of Fabaceae, i.e. Caesalpiniaceae, Mimosaceae and Papilionaceae.

Podzol, podzolic soil Any of a group of acidic, zonal soils having a leached, light-coloured, grey and ashy appearance. Also called spodosol.

Pollen cone Male cone or microstrobilus or pollen cone is structurally similar across all conifers; extending out from a central axis are microsporophylls (modified leaves). Under each microsporophyll is one or several microsporangia (pollen sacs).

Pollinia The paired, waxy pollen masses of flowers of orchids and milkweeds.

Polyandrous (Botany) having an indefinite number of stamens.

Polyembryonic seed Seeds contain many embryos, most of which are asexual (nucellar) in origin and genetically identical to the maternal parent.

Polygamous With unisexual and bisexual flowers on the same or on different individuals of the same species.

Polymorphic With different morphological variants.

Polypetalous (Botany) having a corolla composed of distinct, separable petals.

Pome A fleshy fruit where the succulent tissues are developed from the receptacle.

Pore A tiny opening.

Premorse Abruptly truncated, as though bitten or broken off as of a leaf.

Procumbent Trailing or spreading along the ground but not rooting at the nodes, referring to stems. *cf.* ascending, decumbent, erect.

Pro hyb. (Latin) as a hybrid.

Pro parte (Latin) in part

Pro parte majore (Latin) for the greater part.

Pro parte minore (Latin) for a small part.

Pro sp. (Latin) as a species.

Pro subsp. (Latin) as a subspecies.

Pro syn. (Latin) as a synonym.

Prophyll A plant structure that resembles a leaf.

Prostrate Lying flat on the ground.

Protandrous Relating to a flower in which the anthers release their pollen before the stigma of the same flower becomes receptive.

Proximal End of any structure closest to the point of attachment. *cf.* distal.

Pruinose Having a thick, waxy, powdery coating or bloom.

Pseudocarp A false fruit, largely made up of tissue that is not derived from the ovary but from floral parts such as the receptacle and calyx.

Pseudostem The false, herbaceous stem of a banana plant composed of overlapping leaf bases.

Pteridophyte A vascular plant which reproduces by spores; the ferns and fern allies.

Puberulent Covered with minute hairs or very fine down; finely pubescent.

Puberulous Covered with a minute down.

Pubescent Covered with short, soft hairs.

Pulvinate Having a swell, pulvinus at the base as a leaf stalk.

Pulvinus Swelling at the base of leaf stalk.

Pulviniform Swelling or bulging.

Punctate Marked with translucent dots or glands.

Punctiform Marked by or composed of points or dots.

Punctulate Marked with minute dots; a diminutive of punctate.

Purpurascens Purple or becoming purple.

Pusticulate Characterised by small pustules.

Pyrene The stone or pit of a drupe, consisting of the hardened endocarp and seed.

Pyriform Pear shaped, a three-dimensional shape; attached at the broader end. *cf.* obpyriform.

Pyxidium Seed capsule having a circular lid (operculum) which falls off to release the seed.

Raceme An indeterminate inflorescence with a simple, elongated axis and pedicellate flowers, youngest at the top. *adj.* racemose.

Rachilla The main axis of a grass spikelet.

Rachis The main axis of the spike or other inflorescence of grasses or a compound leaf.

Radiate Arranged around a common centre; as of an inflorescence of Asteraceae with marginal, female or neuter, ligulate ray florets and central, perfect or functionally male, tubular, disc florets. *cf.* disciform, discoid.

Radical Arising from the root or its crown, or the part of a plant embryo that develops into a root.

Ray The marginal portion of the inflorescence of Asteraceae and Apiaceae when distinct from the disc. Also, the spreading branches of a compound umbel.

Receptacle The region at the end of a pedicel or on an axis which bears one or more flowers. *adj.* receptacular.

Recurved Curved downwards or backwards.

Reflexed Bent or turned downward.

Regosol Soil that is young and undeveloped, characterised by medium to fine-textured unconsolidated parent material that maybe alluvial in origin and lacks a significant horizon layer formation.

Reniform Kidney shaped in outline.

Repand With slightly undulate margin.

Replicate Folded back, as in some corolla lobes.

Resinous Producing sticky resin.

Resupinate Twisted through 180°.

Reticulate Having the appearance of a network.

Retrorse Bent or directed downwards or backwards. *cf.* antrorse.

Retuse With a very blunt and slightly notched apex. *cf.* emarginated.

Revolute With the margins inrolled on the lower (abaxial) surface.

Rhizine A root-like filament or hair growing from the stems of mosses or on lichens.

Rhizoid Root-like filaments in a moss, fern, fungus, etc., that attach the plant to the substratum.

Rhizome A prostrate or underground stem consisting of a series of nodes and internodes with adventitious roots and which generally grows horizontally.

Rhizophore A stilt-like outgrowth of the stem which branches into roots on contact with the substrate.

Rhombic Shaped like a rhombus.

Rhomboid Shaped like a rhombus.

Rib A distinct vein or linear marking, often raised as a linear ridge.

Riparian Along the river margins, interface between land and a stream.

Rosette A tuft of leaves or other organs arranged spirally like petals in a rose, ranging in form from a hemispherical tuft to a flat whorl. *adj.* rosetted, rosulate.

Rostrate Beaked; the apex tapered into a slender, usually obtuse point.

Rostrum A beak-like extension.

Rosulate Having a rosette.

Rotate Wheel shaped; refers to a corolla with a very short tube and a broad upper part which is flared at right angles to the tube. *cf.* salverform.

Rotundate Rounded; especially at the end or ends.

Rugae Refers to a series of ridges produced by folding of the wall of an organ.

Rugose Deeply wrinkled.

Rugulose Finely wrinkled.

Ruminate (Animal) chew repeatedly over an extended period.

Ruminate endosperm Uneven endosperm surface that is often highly enlarged by ingrowths or infoldings of the surrounding tissue. *cf.* homogeneous endosperm.

Rz value Is a numerical reference to the mesh/emulsion equalisation on the screen.

Saccate Pouched.

Sagittate Shaped like an arrow head.

Saline soils Soils that contain excessive levels of salts that reduce plant growth and vigour by altering water uptake and causing ion-specific toxicities or imbalances.

Salinity Is characterised by high electrical conductivities and low sodium ion concentrations compared to calcium and magnesium

Salverform Applies to a gamopetalous corolla having a slender tube and an abruptly expanded limb.

Samara An indehiscent, winged, dry fruit.

Sand A naturally occurring granular material composed of finely divided rock and mineral particles range in diameter from 0.0625μm to 2 mm. *adj.* sandy

Saponins Are plant glycosides with a distinctive foaming characteristic. They are found in

many plants, but get their name from the soapwort plant (*Saponaria)*.

Saprophytic Living on and deriving nourishment from dead organic matter.

Sapwood Outer woody layer of the tree just adjacent to and below the bark.

Sarcotesta Outermost fleshy covering of Cycad seeds below which is the sclerotesta.

Scabrid Scurfy, covered with surface abrasions, irregular projections or delicate scales.

Scabrous Rough to the touch because of scattered rough hairs.

Scale Dry bract or leaf.

Scandent Refer to plants, climbing.

Scape Erect flowering stem, usually leafless, rising from the crown or roots of a plant. *adj.* scapose.

Scapigerous With a scape.

Scarious Dry, thin and membranous.

Schizocarp A dry fruit which splits into longitudinally multiple parts called mericarps or cocci. *adj.* schizocarpous.

Sclerotesta The innermost fleshy coating of cycad seeds, usually located directly below the sarcotesta.

Scorpoid Refers to a cymose inflorescence in which the main axis appears to coil.

Scutellum (Botany) any of various parts shaped like a shield.

Scutiform Shaped like a shield.

Secondary venation Arrangement of the lateral veins arising from the midrib in the leaf lamina.

Secund With the flowers all turned in the same direction.

Sedge A plant of the family Apiaceae, Cyperaceae.

Segmented Constricted into divisions.

Seminal root Or seed root originates from the scutellar node located within the seed embryo and is composed of the radicle and lateral seminal roots.

Senescence Refers to the biological changes which take place in plants as they age.

Sepal Free segment of the calyx. *adj.* sepaline.

Septum A partition or cross wall. *pl.* septa. *adj.* septate.

Seriate Arranged in rows.

Sericeous Silky; covered with close-pressed, fine, straight silky hairs.

Serrate Tooth like a saw; with regular, asymmetric teeth pointing forward.

Serrated Toothed margin.

Serratures Serrated margin.

Serrulate With minute teeth on the margin.

Sessile Without a stalk.

Seta A bristle or stiff hair. *pl.* setae. *adj.* setose, setaceous.

Setaceous Bristlelike.

Setate With bristles.

Setiform Bristle shaped.

Setulose With minute bristles.

Sheathing Clasping or enveloping the stem.

Shrub A woody plant usually less than 5 m high and with many branches without a distinct main stem except at ground level.

Silicula A broad, dry, usually dehiscent fruit derived from two or more carpels which usually dehisce along two sutures. *cf.* siliqua.

Siliqua A silicula which is at least twice as long as broad.

Silt Is soil- or rock-derived granular material of a grain size between sand and clay, grain particles ranging from 0.004 to 0.06 mm in diameter. *adj.* silty.

Simple Refer to a leaf or other structure that is not divided into parts. *cf.* compound.

Sinuate With deep wavy margin.

Sinuous Wavy.

Sinus An opening or groove, as occurs between the bases of two petals.

Sodicity Is characterised by low electrical conductivities and high sodium ion concentrations compared to calcium and magnesium.

Sodic soils Contains high levels of sodium salts that affects soil structure, inhibits water movement and causes poor germination and crop establishment and plant toxicity.

Soil pH Is a measure of the acidity or basicity of the soil. See pH.

Solitary Usually refer to flowers which are borne singly and not grouped into an inflorescence or clustered.

Sorocarp Fruiting body formed by some cellular slime moulds and has both stalk and spore mass.

Sorophore Stalk bearing the sorocarp.

Sorosis Fleshy multiple fruit formed from flowers that are crowded together on a fleshy stem, e.g. pineapple and mulberry.

Sorus A discrete aggregate of sporangia in ferns. *pl.* sori

Spadix Fleshy spikelike inflorescence with an unbranched, usually thickened axis and small embedded flowers often surrounded by a spathe. *pl.* spadices.

Spathe A large bract ensheathing an inflorescence or its peduncle. *adj.* spathaceous.

Spatheate Like or with a spathe.

Spatulate Spatula or spoon shaped; broad at the tip and narrowed towards the base.

Spicate Borne in or forming a spike.

Spiculate Spikelet bearing.

Spike An unbranched, indeterminate inflorescence with sessile flowers or spiklets. *adj.* spicate, spiciform.

Spikelet A small or secondary spike characteristics of the grasses and sedges and, generally, composed of two glumes and one or more florets. Also applied to the small spikelike inflorescence or inflorescence units commonly found in Apiaceae.

Spine A stiff, sharp, pointed structure, formed by modification of a plant organ. *adj.* spinose.

Spinescent Ending in a spine; modified to form a spine

Spinulate Covered with small spines.

Spinulose With small spines over the surface.

Spodosol See Podzol.

Sporidia Asexual spores of smut fungi.

Sporangium A spore bearing structure found in ferns, fern allies and gymnosperms. *pl.* sporangia. *adj.* sporangial.

Sporocarp A stalked specialised fruiting structure formed from modified sporophylls, containing sporangia or spores as found in ferns and fern allies.

Sporophore A spore-bearing structure, especially in fungi.

Sporophyll A leaf or bract which bears or subtends sporangia in the fern allies, ferns and gymnosperms.

Sporophyte The spore-producing phase in the life cycle of a plant that exhibits alternation of generations.

Spreading Bending or spreading outwards and horizontally.

Spur A tubular or saclike extension of the corolla or calyx of a flower.

Squama Structure shaped like a fish scale. *pl.* squamae.

Squamous Covered in scales.

Squarrose Having rough or spreading scalelike processes.

Stamen The male part of a flower, consisting typically of a stalk (filament) and a pollen-bearing portion (anther). *adj.* staminal, staminate .

Staminate Unisexual flower bearing stamens but no functional pistils.

Staminode A sterile or abortive stamen, often reduced in size and lacked anther. *adj.* staminodial.

Standard Refers to the adaxial petal in the flower of Papilionaceae. cf. keel, wing.

Starch A polysaccharide carbohydrate consisting of a large number of glucose units joined together by glycosidic bonds **α**-1-4 linkages.

Stellate Star shaped, applies to hairs.

Stem The main axis of a plant, developed from the plumule of the embryo and typically bearing leaves.

Sterile Lacking any functional sexual parts which are capable of fertilisation and seed production.

Stigma The sticky receptive tip of an ovary with or without a style which is receptive to pollen.

Stilt root A supporting root arising from the stem some distance above the ground as in some mangroves, sometimes also known as a prop root.

Stipe A stalk that supports some other structures like the frond, ovary or fruit.

Stipel Secondary stipule at the base of a leaflet. *pl.* stipellae. *adj.* stipellate.

Stipitate Having a stalk or stipe, usually of an ovary or fruit.

Stipulated Having stipules.

Stipule Small leaflike, scalelike or bristlelike appendages at the base of the leaf or on the petiole. *adj.* stipulate.

Stolon A horizontal, creeping stem rooting at the nodes and giving rise to another plant at its tip.

Stoloniferous Bearing stolon or stolons.

Stoma A pore in the epidermis of the leaf or stem for gaseous exchange. *pl.* stomata.

Stone The hard endocarp of a drupe, containing the seed or seeds.

Stramineous Chaffy; straw-like.

Striae Parallel longitudinal lines or ridges. *adj.* striate.

Striate Marked with fine longitudinal parallel lines or ridges.

Strigose Bearing stiff, straight, closely appressed hair; often the hairs have swollen bases.

Strobilus A cone-like structure formed from sporophylls or sporangiophores. *pl.* strobili.

Strophile An appendage at the hilum of certain plant seeds.

Strophiolate Furnished with a strophile or caruncle.

Style The part of the pistil between the stigma and ovary.

Sub- A prefix meaning nearly or almost, as in subglobose or subequal.

Subcarnose Nearly fleshy.

Subfamily Taxonomic rank between the family and tribe.

Subglobose Nearly spherical in shape.

Subretuse Faintly notched at the apex.

Subsessile Nearly stalkless or sessile.

Subshrub Intermediate between a herb and shrub.

Subspecies A taxonomic rank subordinate to species.

Substrate Surface on which a plant or organism grows or attached to.

Subtend Attached below of something.

Subulate Narrow and tapering gradually to a fine-point, awl shaped.

Succulent Fleshy, juicy, soft in texture and usually thickened.

Suckers Young plants sprouting from the underground roots of a parent plant and appearing around the base of the parent plant.

Suffrutescent stem Stem woody at the base.

Sulcate Grooved longitudinally with deep furrows.

Sulcus A groove or depression running along the internodes of culms or branches.

Superior Refers to the ovary that is free and mostly above the level of insertion of the sepals and petals. *cf.* inferior.

Suture Line of dehiscence.

Swidden Slash-and-burn or shifting cultivation.

Syconium A type of pseudocarp formed from a hollow receptacle with small flowers attached to the inner wall. After fertilisation the ovaries of the female flowers develop into one-seeded achenes, e.g. fig.

Symbiosis Describes close and often long-term mutualistic and beneficial interactions between different organisms.

Sympetalous Having petals united.

Sympodial Refers to a specialised lateral growth pattern in the apical meristem. *cf* monopodial.

Synangium An organ composed of united sporangia, divided internally into cells, each containing spores. *pl.* synangia.

Syncarp An aggregate or multiple fruit formed from two or more united carpels with a single style. *adj.* syncarpous.

Syncarpous Carpels fused forming a compound pistil.

Synteny Presence of two or more genetic loci on the same chromosome.

Tannins Group of plant-derived phenolic compounds.

Taxon The taxonomic group of plants of any rank, e.g. a family, genus, species or any infraspecific category. *pl.* taxa.

Tendril A slender, thread-like organ formed from a modified stem, leaf or leaflet which, by coiling around objects, supports a climbing plant.

Tepal A segment of the perianth in a flower in which all the perianth segments are similar in appearance and are not differentiated into calyx and corolla; a sepal or petal.

Tetrasporangium A sporangium containing four haploid spores as found in some algae.

Terete Having a circular shape when cross-sectioned or a cylindrical shape that tapers at each end.

Terminal At the apex or distal end.

Ternate In threes as of leaf with three leaflets.

Testa A seed coat, outer integument of a seed.

Thallus Plant body of algae, fungi and other lower organisms.

Thyrse A dense, panicle-like inflorescence, as of the lilac, in which the lateral branches terminate in cymes.

Tomentose Refers to plant hairs that are bent and matted forming a woolly coating.

Tomentellose Mildly tomentose.

Torus Receptacle of a flower.

Transpiration Evaporation of water from the plant through leaf and stem pores.

Tree That has many secondary branches supported clear of the ground on a single main stem or trunk.

Triangular Shaped like a triangle, three angled and three sided.

Tribe A category intermediate in rank between subfamily and genus.

Trichome A hair-like outgrowth of the epidermis.

Trichotomous Divided almost equally into three parts or elements.

Tridentate Three toothed or three pronged.

Trifid Divided or cleft into three parts or lobes.

Trifoliate Having three leaves.

Trifoliolate A leaf having three leaflets.

Trifurcate Having three forks or branches.

Trigonous Obtusely three angled; triangular in cross section with plane faces.

Tripartite Consisting of three parts.

Tripinnate Relating to leaves, pinnately divided three times with pinnate pinnules.

Tripliveined Main laterals arising above base of lamina.

Triploid Describing a nucleus or cell that has three times (3n) the haploid number (n) of chromosomes.

Triveined Main laterals arising at the base of lamina.

Triquetrous Three edged; acutely three angled.

Trullate With the widest axis below the middle and with straight margins; ovate but margins straight and angled below middle, trowel, angular, ovate shaped.

Truncate With an abruptly transverse end as if cut off.

Tuber A stem, usually underground, enlarged as a storage organ and with minute scalelike leaves and buds. *adj.* tuberous.

Tubercle A wartlike protuberance. *adj.* tuberculate.

Tuberculate Bearing tubercles; covered with warty lumps.

Tuberisation Formation of tubers in the soil.

Tuft A densely packed cluster arising from an axis. *adj.* tufted.

Turbinate Having the shape of a top; cone shaped, with the apex downwards, inversely conic.

Turgid Distended by water or other liquid.

Turion The tender young, scaly shoot such as asparagus, developed from an underground bud without branches or leaves.

Turnery Articles made by the process of turning.

Twining Winding spirally.

Ultisols Mineral soils with no calcareous material, having less than 10% weatherable minerals in the extreme top layer of soil and with less the 35% base saturation throughout the soil.

Umbel An inflorescence of pedicellate flowers of almost equal length arising from one point on top of the peduncle. *adj.* umbellate.

Umbellet A secondary umbel of a compound umbel. *cf.* umbellule.

Umbellule A secondary umbel of a compound umbel.*cf.* umbellet.

Uncinate Bent at the end like a hook; unciform.

Undershrub Subshrub; a small, usually sparsely branched woody shrub less than 1 m high. *cf.* shrub.

Undulate With an edge/margin or edges wavy in a vertical plane; may vary from weakly to strongly undulate or crisped. *cf.* crisped.

Unifoliolate A compound leaf which has been reduced to a single, usually terminal leaflet.

Uniform With one form, e.g. having stamens of a similar length or having one kind of leaf. *cf.* dimorphic.

Uniseriate Arranged in one row or at one level.

Unisexual With one sex only, either bearing the anthers with pollen, or an ovary with ovules, referring to a flower, inflorescence or individual plant. *cf.* bisexual.

Urceolate Shaped like a jug, urn or pitcher.

Utricle A small bladdery pericarp.

Vaginate Forming or enclosed in a sheath.

Valvate Meeting without overlapping, as of sepals or petals in bud. *cf.* imbricate.

Valve One of the sections or portions into which a capsule separates when ripe.

Variant Any definable individual or group of individuals which may or may not be regarded as representing a formal taxon after examination.

Variegate, variegated Diverse in colour or marked with irregular patches of different colours, blotched.

Variety A taxonomic rank below that of subspecies.

Vein (Botany) a strand of vascular bundle tissue.

Veinlets Small veins.

Velum A flap of tissue covering the sporangium in the fern, *Isoetes*.

Velutinous Having the surface covered with a fine and dense silky pubescence of short fine hairs; velvety. *cf.* sericeous

Venation Distribution or arrangement of veins in a leaf.

Veneer Thin sheet of wood.

Ventral (Botany) facing the central axis, opposed to dorsal.

Vernation The arrangement of young leaves or fronds in a bud or at a stem apex. *cf.* circinate.

Verrucose Warty.

Verticil A circular arrangement, as of flowers, leaves or hairs, growing about a central point; a whorl.

Verticillaster False whorl composed of a pair of opposite cymes as in Lamiaceae.

Verticillate Whorled, arranged in one or more whorls.

Vertisol A soil with a high content of expansive montmorillonite clay that forms deep cracks in drier seasons or years.

Vertosols Soils that both contain more than 35% clay and possess deep cracks wider than 5mm during most years.

Vesicle A small bladdery sac or cavity filled with air or fluid. *adj.* vesicular.

Vestigial The remaining trace or remnant of an organ which seemingly lost all or most of its original function in a species through evolution.

Vestiture Covering; the type of hairiness, scaliness or other covering commonly found on the external parts of plants. *cf.* indumentums.

Vibratile Capable of to and fro motion.

Villose Covered with long, fine, soft hairs, finer than in pilose.

Villous Covered with soft, shaggy unmatted hairs.

Vine A climbing or trailing plant.

Violaxanthin Is a natural xanthophyll pigment with an orange colour found in a variety of plants like pansies.

Viscid Sticky, being of a consistency that resists flow.

Viviparous Describes seeds or fruit which sprout before they fall from the parent plant.

Whorl A ringlike arrangement of leaves, sepals, stamens or other organs around an axis.

Winged Having a flat, often membranous expansion or flange, e.g. on a seed, stem or one of the two lateral petals of a papilionaceous flower or one of the petal-like sepals of Polygalaceae. *cf.* keel, standard.

Xanthophylls Are yellow, carotenoid pigments found in plants. They are oxidised derivatives of carotenes.

Xeromorphic Plant with special modified structure to help the plant to adapt to dry conditions.

Xerophyte A plant which naturally grows in dry regions and is often structurally modified to withstand dry conditions.

Zygomorphic Having only one plane of symmetry, usually the vertical plane, referring to a flower, calyx or corolla. *cf.* actinomorphic.

Zygote The fist cell formed by the union of two gametes in sexual reproduction. *adj.* zygotic.

Common Name Index

A
A375 and A2058 skin cancer cells, 378
Abrus precatorius, 359
Achira, 4, 70, 72, 76
Acute myeloid leukemia cell line Kasumi-1 cells, 498
Adzuki bean, 19
Aerial yam, 10, 235
African arrowfoot, 70
African arrowroot, 70
African turmeric, 70
African winged bean, 20
Africa yam bean, 22, 291
Aha bayere (Twi, Gold Coast), 12
Ahipa, 2, 19, 458–463
A2780 human ovarian cancer cell line, 332
Aibika, 220
Air potato, 10, 235, 275
Air yam, 10, 235
Akar kemnyan paya, 13
Alaska cotton, 7
Alkali bulrush, 8
Alliums, 2
Almor-ira, 5
Aloti (Gabon), 15
Alshi, 12
Altissima group, 60
Amazonian yam bean, 19
American licorice, 17
American potato bean, 15
Andean yam bean, 458
Ants, 199
Añu, 2
Arctic lupine, 18
Aroi huwi curuk (Sumatra), 11
Arrowhead, 1
Arrowroot, 1, 301, 303
Asiatic bitter yam, 11, 266
Asiatic yam, 218, 253
Asparagus bean root, 20
Astragalus, 16
Astragal Yasenelistnyi (Russian), 16
Australian bindweed, 4

B
Badak, 13
Bafra (Arabic, Sudan), 15
Baihuasheshecao, 286
Baklaiakang (Bisaya), 10
Balisier, 3
Balun (Thai Sakai), 11
Banana, 220, 227
Bangkwaun, 19
Barley, 95
Bat's wing coral tree, 16
Bayae (Thai Sakai), 10
Beach moonflower, 6
Beach morning glory, 5
Beach pea, 22
Bean aphid, 477
Bearing yam, 235
Bee, 200
Beetroot, 2–3, 23, 27, 28, 30–31, 34–36, 40, 42–46, 48–54, 56–58
Bell peppers, 51
Bengkoang, 470, 476
Bengkuang, 467
Betel yam, 275
B16-F10 melanoma, 378
Bilai-kand, 22
Birali panwa, 22
Biralu, 22
Birch rind yam, 253
Bitter Brazilian arrowroot, 309
Bitter cassava, 15, 309
Bitter tapioca meal, 309
Bitter yam, 9, 10, 298
Black pepper stem borer, 477
Black salsify, 2
Black sugar, 354
B16 murine melanoma cells, 412
Boll weevil, 215
Bonvista bean, 17
Brazilian arrowroot, 309
Breadfruit, 227, 303
Breadroot, 2

T.K. Lim, *Edible Medicinal and Non-Medicinal Plants: Volume 10, Modified Stems, Roots, Bulbs*,
DOI 10.1007/978-94-017-7276-1

Bread-root, 20
Breast (MCF-7) cancer cell, 48, 51, 52
Breast cancer HS578T, 497
Breast cancer MDA-MB-231 cell, 380
Brown nutsedge, 179
Brown salwood, 15
B Tiny morning glory, 5
Bubil bitter yam, 235
Buckroot, 20
Bulrush, 8, 9, 172
Burdock, 2
Bush nut, 209
Bush potato, 2, 4, 5
Buta (Hindi), 5

C
Cairo morning glory, 5
Calenda, 70
California bulrush, 8
California rose, 4
Canna lily, 3, 70, 72, 75–79
Carrot, 2, 95
Cassava, 2, 15, 227, 309–313, 315–320, 322–343
Cassava mealy bug, 343
Celeriac, 2
Celery cabbage, 210, 476
Cerea rubber, 15
Chago, 2
Chai, 12
Chain fern rhizome, 82
Chaoli, 22
Chard, 23, 27, 44, 46, 48, 50, 54, 58
Chatong (Thai Sakai), 9
Chavi, 12
Chechur (Assamese), 8
Cheeky yam, 10, 235
Chenopodium, 54, 58
Chik, 21
Chillies, 220
Chinese artichoke, 2
Chinese bindweed, 4
Chinese eriosoma, 16
Chinese potato, 275
Chinese water chestnut, 1, 209, 212–215, 471
Chinese yam, 11–13, 253, 275, 277–278, 280–287, 298
Chioggia beet, 23
Chop suey bean, 465
Chufa, 2, 6
Cibota, 82
Cibot rhizome, 82
Cibotum, 82
Cicla group, 60
Cinnamon vine, 275, 312, 341
Cinnamon yam, 275
Citrus lime, 96
Ciu mai (Vietnamese), 12
Clangporn (Thai Sakai), 12
Cloves, 312
Cluster yam, 10
Coast morning glory, 5
Coco grass, 179
Coconut, 95, 220, 303, 312
Cocoyam, 1
Colon cancer (DLD-1), 136
Colon cancer cells (SW480, HT-29, HCT-15), 378
Common liquorice, 354
Common yam, 275
Conditiva group, 60
Connie yam, 291
Corkwood wattle, 15
Cowpea, 201
Coxsackievirus A16 (CVA16), 388
Cranberry, 51
Crassa group, 56, 60
Crosne, 2
Cucumber, 273
Cudzú-tropical (Portuguese, Brazil), 21
Cush-cush, 14
Cyperi rhizoma, 195, 197

D
Daikon, 2
Dalrymple vigna, 22
Daly river club rush, 8
Dangs (Bombay), 16
Danshen, 506, 507, 509, 515, 526
Dasheen, 227
Deba, 224
Deer pea, 22
Desert yam, 2, 5
Dhak, 16
Diamondback moth, 200, 272
DNA and RNA viruses, 388
Dolichos bean, 17
Downs nutgrass, 6
DU145 and LNCaP prostate cancer cell lines, 379
Dudung (Java), 11
Dutch runner bean, 20

E
Earthnut pea, 2, 17
Earthworm, 77, 247
East Indian arrowroot, 14, 301
Eboe yam, 291
Eddoes, 227
Edible
    arrowroot, 70
    canna, 4, 70
    kudzu, 21
    spike rush, 209
E-Dyeny (Bedik, Gold coast), 10
Egyptian paper reed plant, 6, 172
Egyptian papyrus, 172
Ehrlich carcinoma (EAC) cells, 48
Eight months guinea yam, 13, 291

Electric eel, 201
English shot (South America), 70
Enterovirus 71 (EV71), 388
Epstein-Barr virus, 51

F
Fancy yam, 9
Field beet, 3, 23
Fiji arrowroot, 301
Finger root, 1
Fish poison yam, 13
Five-leaved yam, 12–13
Flavescens group, 60
Florida hammock milkpea, 17
Fodder
  beet group, 23, 60
  sugar beet, 23
Foliage beet, 3, 23
Foot-A-night vine, 482
Forage beet, 23
Four-lobed-root yam bean, 465
French beans, 467
Fu-Fu-Miao, 4

G
Gadog, 11
Gadong, 11
Gadong lilin, 11
Gadong mabuk, 11
Gadung (Malay), 11
Gang E Teng (Chinese), 21
Garden bee, 3
Garden beet group, 3, 37, 41, 56, 60
Garden beetroot, 23
Garlic, 220
Gegen, 507, 509, 515, 526
Ge Gen Fen (Chinese), 21
Ge Ma Mu (Chinese), 21
Gemsbok bean, 22
Genthi (Kumaon region, Western Himalayas), 10
German cockroach, 199
G. glabra, 394
Giant bulrush, 8
Giant potato, 5
Ginger, 1, 96, 270
Glutinous yam, 11
Goa bean root, 20
Goa potato, 253
Goat's foot convolvulus, 5
Golden beet, 23
Golden chicken fern, 4, 82
Golden lamb, 82
Golden moss, 4, 82
Goolabura, 5
Gosh alu (Assamese), 10
Gourd, 79
Great bulrush, 8
Greater club rush, 8
Greater galangal, 1
Greater yam, 9, 218, 219
Grey sedge, 7
Ground almond, 179
Ground chestnut, 7, 209
Groundnut, 2, 15
Gruya, 4
Guinea yam, 292, 298, 299
Gun (Kumaon region, Western Himalayas), 10
Guppies, 200
Guyana arrowroot, 218

H
Hai Nan Shu (Chinese), 5
Hairypod cowpea, 22
Half-A-yam, 291
Halgia (Bombay), 22
Hausa potato, 253
Hazara (Hausa, Nigeria), 12
HCT 116, 136
Hedge bindweed, 4
Hela and HepG2 tumour cells, 137
Hepatitis A virus (HAV), 386
Hepatitis B virus (HBV), 387, 388
Hepatitis C virus (HCV), 394, 395
Hepatocellular carcinoma (HCC), 394, 395
Herpes simplex-1 virus, 197
Herpes simplex virus type 1 (HSV-1), 387, 388
HHV-6 and HHV-7, 387
H4IIE murine hepatoma cells, 415
HIV, 77, 137
HIV-1, 77
HIV (NSI-HIV), 387
Hngo (Tahi Sakai), 13
Ho (Hawaiian), 10
Hog potato, 2
Horse gram, 16, 18
HT-29 colon cancer cells, 136, 378
Huaishan, 286
Huangchoy (Chinese), 16
Huang Hua Tu Gua, 6
Huangqi, 286
Hua shan, 286
Hubei Yang Ti Jia (Chinese), 16
Human acute myeloid leukaemia SHI-1 cells, 498
Human breast carcinoma cell lines, 297
Human breast MDR cancer cell line, 498
Human cervical carcinoma (HeLa) cell lines, 388
Human colon cancer (RKO) cells, 51
Human colon carcinoma cell lines, 136
Human colonic SW480 cancer cells, 137
Human colorectal cancer HCT-8 cells, 137
Human colorectal cancer SW480 cells, 137
Human cytomegalovirus, 388
Human cytomegalovirus (HCMV), 388
Human gastric cancer cell line SGC7901, 475
Human glioblastoma (AMGM), 190

Human hepatocellular carcinoma cell line Bel-7402, 244
Human hepatoma cell line (HepG2), 392
Human herpes virus, 387
Human immunodeficiency virus (HIV), 387
  type 1, 388
Human LNCaP prostate cancer cells, 380
Human lymphoma NB4 cells, 136
Human pancreatic (PaCa), 51
Human polyomavirus JC virus, 518
Human promyelocytic leukaemia (HL-60) cell, 379
Human prostate cancer PC-3 cells, 52, 137
Human respiratory syncytial virus (HRSV), 519
Human stomach cancer (Kato III), 136
Human tongue cancer Tca8113 cells, 136
Huwi Upas (Sundanese), 13
Hyacinth, 2
Hyacinth bean, 17

I

Ichoimo, 276
Igname, 9
Ilus (Javanese), 13
Indian arrowroot, 301
Indian breadroot, 19, 20
Indian canna arrowroot, 70
Indian kudzu, 22
Indian potato, 15, 465
Indian shot, 3, 4, 70
Indian-shot, 70
Indian spinach, 23, 51
Indian turnip, 19
Indian yam, 14
Influenza virus, 387, 388
Ingwese, 15
Intoxicating yam, 11, 266
Irish potato, 103, 227
Italian beetroot, 23
Izi Ntondo (Zulu), 15

J

Jactupe, 19
Jaglia Che Kand, 12
Japanese arrowroot, 21, 482
Japanese encephalitis virus (JEV), 388
Japanese mountain yam, 12
Japanese yam, 11, 275
Java grass, 179
Jerusalem artichoke, 2
Jicama, 2, 19, 465–468, 471–473
Jurkat cell line (human T-cell leukemia cells), 190

K

Kacang botol, 332
Kadawa kand, 14
Kadia kand, 14
Kadu kand (Bombay Presidency), 12
Kakalu (Bengali), 9
Kala kand, 12
Kangkong, 4
Kanta-alu (Western Rajasthan), 12
Kapuang (Thai Sakai), 10
Karalasana, 20
Karalsona (Tamil), 20
Karen potato, 253
Karu alachandra (Telugu), 20
Kath alu (Assamese), 9, 14
Katil (Indonesia), 16
KB oral cancer cells, 380
K562 (human erythromyeloblastoid leukaemia) cell lines, 332
Kedut (Malaysia), 13
Kedut (Sumatra), 13
Kedvo kand (Bombay Presidency), 14
Kelana (Malaysia), 13
Kelunoh, 13
Kencur, 1
Khitro (Bodo), 8
Khoai Mai, 12
K562 human chronic myelogenous leukemia cells, 190
K562 human leukaemia cells, 475
Kiroi (Tagalog), 10
Kiyak (Thai Sakai), 13
Ko-Hemp, 482
Kohlrabi, 2
Koji, 115
Korean yam, 12, 275
Kuanga, 22
Kudsu, 482
Kudzu, 21–22, 482–487, 489–490, 494–496, 499–500, 502–505, 509–513, 517, 522, 525–527
Kudzu bean, 482
Kudzu hemp, 482
Kudzu vine, 482
Kungkwad (Thai Sakai), 9
Kuro Gawai (Japanese), 7
K'Uuwaap (Teenek), 4

L

Lablab bean, 17
Lamb of tartary, 82
Large bindweed, 4
Largebract Indian breadroot, 19, 20
L1210 cells, 190
Leaf beet group, 3, 23, 56, 60
Leeks, 2
Lesser asiatic yam, 9, 11
Lesser galangal, 1
Lesser glory, 5
Lesser yam, 9, 253, 256, 271
Lettuce, 55, 59, 201
Licorice, 17, 354, 356, 358, 359, 361, 362, 365–367, 369, 371, 372, 374–377, 380–384, 386, 388–391, 393, 394, 396, 397, 405, 408–418, 421–433
Licorice-root, 17, 354

Lima-Lima (Tagalog, Kasi (Igorot), Pari (Bagobo), 10
L 'Indo (Sandawe), 22
Lingzhi, 286
Liquorice, 17, 354, 364, 404, 406, 413, 419, 421, 422, 424, 425, 430, 431
Liquorice root, 354
Little Indian breadroot, 19
LNCaP and PC-3 prostate cancer cells, 87
Long Chinese yam, 275
Long yam, 9
Lotus root, 1
Luntak (Thai sakai), 11
L5178y mouse lymphoma cells, 190

M
Maca, 2
Madras gram, 18
Magiya, 13
Magoraza, 12
Maize, 76, 273, 341–343
Malacca yam, 235
Man-alu (Kumaon Region, Western Himalayas), 9
Mandioca brava, 15
Mandong (Thai), 11
Mangel-Wurzel, 23
Mangold, 23
Manicoba, 15
Manicot, 309
Manidoca, 309
Manihot, 309
Manila yam, 218
Manioc bean, 2, 15, 309, 320, 340, 465
Man-of-the-Earth, 5
Mano kand, 14
Ma-Nyeny, 10
Marama bean, 22
Marsh club rush, 6
Mashua, 2
Mauka, 2
Mayabang (Tagalog) Kamengeg (Ilooko), 12
MCF-7/ adr cells, 498
MCF-7 and MCF-7-D-40 (oestrogen receptor positive), 297
MCF-7 and MDA-MB-231 breast cells, 378
MCF-7 breast cancer cells, 380
MCF7 cancerous cell line, 377
MCF-7 cell lines, 410, 497
MDA-468 (oestrogen receptor negative), 297
MDA-MB-231, 497
Mediterranean sedge, 172
Messina creeper, 5
Mexican bean beetle, 477
Mexican jicama, 465
Mexican potato, 465
Mexican turnip, 465
Mexican yam bean, 465
Mice mammary adenocarcenoma (AMN3) cancer cells, 190
Mile-A-minute vine, 5
Ming-Chien yam, 224
Mint weed, 476
MLL prostate cancer cells, 378
Mng'oko (Tanzania), 11
Moa alu (Assamese), 11
Morning glory, 5
Mosquito, 78
Motha (Rajasthan), 7
Moth bean, 19
Mothee, 7
Mountain yam, 261
Mouse melanoma B16F1 cells, 48, 52
Mousenuts, 7
Mouse skin JB6 cell line, 136
Mulli balli (Kannada), 5
Mungel wurzel, 3
Munia (Kumaon region, Western Himalayas India), 13
Murine lymphoblastic leukaemia, 190
Mustard, 273

N
Nagaimo, 12, 276, 279, 283, 286
Nakayama, 388
Nalgoo (Australia), 6
NCI-H460 and A549 non-small lung cancer cells, 379
NCI-H460 non-small lung cancer cells, 379
Nepalese kudzu, 22
Nepal yam, 10
Newcastle disease, 388
New Zealand yam, 2
Nian Shan Yao (Chinese), 11
Niluva pendalum (Telugu), 9
Nootka lupine, 18
Northern wattle, 15
Notched cowpea, 22
Nung Gum (Assamese), 4
Nutgrass, 7, 8, 179, 184–187, 190–192, 194, 196, 197, 199–201
Nutsedge, 179, 201
Nüzhenzi, 286

O
Oats, 201
Oca, 2
Oestrogen receptor (ER)-positive human breast cancer cell, MCF-70, 513
Onions, 2, 220
Oral squamous cell carcinoma, 244, 380
Ordinary beet, 23
Ordinary garden beet, 23
Otaheite potato, 235
Otaheite yam, 235

P

Pachpotia Alu, 12
Pakit, 10, 12
Pala (Rajasthan, Western India), 16
Panc-1, 136
Pancreatic cancer line, 136
P-20778 and 821564XY48, 388
Pani alu (Assamese), 21, 22
Panil book, 12
Papaya, 332
Paper reed, 6, 172
Papua New Guinea red wattle, 15
Papyrus sedge, 172–176
Par Baul Fauk, 4
Parsley, 2
Parsnips, 2
Pawpaw, 220
PC-3 prostate cancer cells, 87
Peanut, 95
Pegaga, 332
Pencil yam maloga bean, 22
Perpetual beet, 23
Persian and Turkish licorice, 356
Persian licorice, 354
Phan Karo (Meghalaya), 5
Pi'A (Hawaiian), 12
Pignut, 2
Pineapple, 313
P388 leukaemia cells, 78
Polynesian arrowroot, 14, 301
Pomegranate, 105
Pond onion, 7
Potato, 2, 76, 304
Potato bean, 15
Potato yam, 9–11, 235
Prairie potato, 20
Prairie turnip, 2, 19, 20
Preuss' Dioscorea, 13
Prickly yam, 12
Promyelocytic leukaemia cell (HL-60), 136
Prostate cancer cell line DU-145 cells, 378
Prostate (PC-3) tumour cells, 47, 51
Pueraria, 21
Pumpkin, 95
Purple arrowroot, 70
Purple nutgrass, 7
Purple nutsedge, 179, 180
Purple prairie clover, 16
Purpletassels, 16
Purple yam, 9, 218
Purslane, 59

Q

Qian Jin Ba (Chinese), 18
Qi Zhao Long (Chinese), 5
Queensland arrowroot, 4, 70
Quinoa, 72

R

Radish, 2, 273
Rainbow chard, 23
Red beet, 23, 28, 29, 31, 33, 35–37, 39–45, 47–49, 54–55, 59
Red beetroot, 23, 29, 33, 38, 40–42, 44, 46, 48, 51, 53, 57
Red cabbage, 137
Red-fleshed beetroot, 23
Red nutsedge, 179
Red onion, 47, 51
Rhizoma Cyperi, 196
Rhizoma glycyrrhizae, 354
Rice, 201
Rice moth, 477
Rice weevil, 477
River club rush, 9
River spinach, 4
Rogon biri (Hausa Nigeria), 10
Roman kale, 23
Root beet, 23
Root-knot nematode, 111
Rotavirus, 389
Round white yam, 13
Roundworm, 247
Rui-Chilong, 13
Ruidok (Assam), 5
Rui Nihang (Assamese), 14
Ruipheng (Assamese), 12
Ruipheng selu (Assamese), 11
Runner bean, 20
Russian licorice, 356, 374, 384
Russian liquorice, 354
Rutabaga, 2

S

Salep, 14
Salsify, 2
Sand ginger, 1
S180 and H22 tumour cells, 244
Sarsaparilla vine, 17
SARS-associated coronavirus (FFM-1 and FFM-2), 389
Savida dumpa (Telugu), 9
Savoy beet, 23
Scarlet runner bean, 20
Scurfpea, 19
Sea beet, 23, 27
Sea club rush, 6
Seakale beet, 3, 23
Sea moonflower, 6
Sedge, 7
Sehod (Thai Sakai), 13
Seim bean, 17
Semi-sweet sugarbeet, 23
Sengkuang, 19
Ser (Thai Sakai), 12
SGC-7901 cancer line, 244

Shahada, 12
Shan Lian Ou (Chinese), 18
Shan Tu Gua (Chinese), 6
Shen Zi Ye, 16
Shoebutton air potato, 235
Short-podded yam bean, 465
Showy milletia, 18
Sichuan pepper, 210
Sicilian broad-rib beet, 23
Sierra Leone arrowroot, 70
Silk worm, 477
Silky cow vine, 5
Silver beet, 23
Silver beet spinach beet, 23
Silver leaf Indian breadroot, 20
Silver leaf scurf pea, 20
Sinkiang licorice, 354
Si-pei licorice, 354
SK-BR-3 and Hs578T, 497
Skirret, 2
Slender scurfy pea, 20
Slimflower scurfpea, 19, 20
Small Indian breadroot, 20
Snail, 78
Soh Lah, 4
Soh Phlang, 17
Sohplong (India), 17
Sorghum, 201
Southern astragalus, 18
Soy, 312
Soybean, 201, 505
Spanish and Italian licorice, 356
Spanish and Russian licorice, 363
Spanish juice, 354
Spanish licorice, 354, 355
Spanish liquorice sweet root, 354
Spanish woodbine, 5
Spear-leaved yam, 9
Spike rush, 209
Spinach, 273
Spinach beet, 3, 27, 29, 60
Spinach chard, 23
Spiny Chinese yam, 253
Spirit yam, 266
Spring onions, 95
Subterranean Indian breadroot, 19
Sugar beet group, 3, 23, 27, 29–39, 41–43, 45, 46, 51–60
Sugarcane, 341
Sulian (Iloko), 10
Suna (Thai Sakai), 9
Sunchoke, 2
Sural (Kumaon Region, Western Himalayas), 22
SW480, 136
Swamp cotton, 7
SW837 cell lines, 136
Sweet Brazilian arrowroot, 309
Sweet broom, 17
Sweet cassava, 309
Sweet clover, 18
Sweet kudzu, 21
Sweet potato, 2, 3, 5, 92, 94–97, 99–117, 119–154, 220
Sweet-potato tree, 309
Sweet potato weevil, 121, 125
Sweet-root, 17
Sweet sorghum, 341, 342
Sweet tapioca meal, 309
Sweet vetch, 17
Sweet wood liquorice, 354
Sweet yam, 11, 253
Swiss chard, 3, 23, 27–30, 32, 34, 37, 40, 41, 43–44, 48, 51, 56, 60
Sycthian lamb, 82

T

Table beetroot, 42, 46–47, 51–52, 55
Tacca, 301, 303–305
Tacca arrowroot, 304
Tahiti arrowroot, 301
Taigun, 12
Taiwanese yam, 11
Taiwan kudzu, 21, 482
Takob (Thai Sakai), 12
Takuli (Kumaon Region, Western Himalayas), 12
Tall sedge, 8
Tannia, 1, 227
Tapewrom, 247
Tapioca, 15, 309, 341
Tapioca manihot, 309
Target beetroot, 23
Taro, 1
Tartarian lamb, 82
Tarur (Kumaon region, Western Himalayas, India), 13
Ten months yam, 218
Thick-podded salwood, 15
Thomson kudzu, 21
Three-leaved yam, 297
Three-lobed-leaved yam bean, 465
Tian Xuan Hua (Chinese), 4
Tigernut, 2, 6
Timpsula, 19
Tipsin, 2, 19
Tobacco, 356, 432
Tobacco caterpillar, 477
Tomato, 95, 201
Tongkat ali, 2
Tororo, 282, 285
Tororo Imo, 276
Totora, 8
Totoro, 279
Tropical kudzu, 482
True licorice, 354
True liquorice, 354
Tsukune Imo, 276
Tsukuneimo, 279
Tuba Gunjo (Indonesia), 13
Tuba Ubi (Malaysia), 13
Tuberous bitter vetch, 19

Tuberous pea, 17
Tule, 8, 9
Tulips, 2
Tupe, 18
Turmeric, 1
Turnip, 2
Tuskuneimo, 280

U
Ubi Babi, 13
Ubi Garam (Malaysia) Man Tayong (Thai), 12
Ubi Kelonak, 13
Ulluco, 2
Urahi alu, 22

V
Vaccinia, 388
Vaj kand, 14
Varicella-zoster virus, 388
Vegetable lamb of tartary, 82
Velvety morning glory, 5
Vesicular stomatitis, 388
Vine-that-ate-the-south, 482
Voandzeia subterranea, 339

W
Wapato, 1
Ware sedge, 6
Warrine, 11
Water chestnut, 7, 209–211, 215
Water morning glory, 4
Water nut, 209
Water spinach, 4
Water yam, 9, 218, 222, 297, 298
Wauh (Malaysia), 11
West Indian sweet potato weevil, 125
Wheat, 95, 201, 342, 477
White beetroot, 23, 43, 47, 49, 53
White clover, 22, 201
White-fleshed beetroot, 23
White guinea yam, 13, 291, 292, 294, 298
White prairie clover, 16, 19
Whiterim scurfpea, 19
White yam, 9, 10, 13, 291, 296–298
Wier vine, 5
Wild bean, 15
Wild cowpea, 22
Wild kudzu, 482
Wild pea adzuki bean, 19
Wild potato vine, 5
Wild tapioca, 70
Wild yam, 10, 11, 266, 271, 275
Williams arrowroot, 301
Winged bean root, 20
Winged yam, 9, 218
Woolly fern, 4, 82
Wu Zhao Jin Long (Chinese), 5

X
Xiakucao, 286
Xinjiang licorice, 374

Y
Yacon, 72
Yam, 220, 228, 261, 270
Yamaimo, 12
Yama-No-Imo, 11
Yamatoimo, 276, 279
Yam bean, 19, 265, 458, 467–469, 474–477
Yampee, 14
Yams, 2
Yams kalung (Tamil), 9
Yam tsukuneimo, 282, 285
Yangmingshan yam, 224
Yarex (Thai Sakai), 14
Yeast, 264
Yellow beetroot, 23, 37–38, 41, 42
Yellow-fleshed beetroot, 23
Yellow guinea yam, 10, 298
Yellow nutgrass, 6
Yellow sweet clover, 18
Yellow yam, 10, 296
You Sha Cao (Chinese), 6
Yuca, 2, 15, 309
Yucca, 340

Z
Zanzibar yam, 12
ZR-75-1, 497

Scientific Name Index

A

Abelmochus spp., 2
Abelmoschus, 332
  A. esculentus, 331
  A. manihot, 220
Abrus precatorius, 474
Absidia pseudocylinderospora, 392
Acacia
  A. bidwillii, 15
  A. crassicarpa, 15
  A. holosericea, 15
Acinetobacter baumannii, 245
Actinomyces viscosus, 385
Actinoscirpus grossus, 8
Aedes
  A. aegypti, 198, 476
  A. albopictus, 199
Aegle marmelos, 191
Aerobacter aerogenes, 215
Agrobacterium tumefaciens, 57, 384
Alisma orientale, 284
Allium
  A. cepa, 383
  A. fistulosum, 383
  A. sativum, 2
Alpinia galanga, 1
Alternaria
  A. alternata, 200
  A. brassicola, 200
  A. chearanthi, 200
  A. solani, 200
Amaranthus sp., 273
Amorphallus spp., 1
Angelica dahurica, 384, 385, 522
Anopheles
  A. culicifacies, 198
  A. stephensi, 198
Anthonomus grandis, 215
Antiaris toxicaria, 273
Apios
  A. americana, 2, 15
  A. fortunei, 15
  A. tuberosa, 15
Apis florea, 200
Arabidopsis thaliana, 325, 372
Argyrolobium marginatum, 15
Arracacia xanthorriza, 76
Artocarpus altilis, 227
Ascardia galli, 247
Aspergillus
  A. awamori, 115, 341
  A. flavus, 193
  A. fumigatus, 193, 384
  A. niger, 153, 193, 264, 298, 316, 341, 384
Aspidium barometz, 82
Asplenium
  A. aureum, 86
  A. nidus, 86, 87
Astragalus, 286
  A. fraxinifolius, 16
  A. membranaceus, 16
  A. propinquus, 16
Atriplex canescens, 36
Atylosia reticulata, 16

B

Bacillus
  B. cereus, 56, 87, 298, 384
  B. subtilis, 77, 193, 244, 245, 264, 384–386
Bacopa monniera, 413
Bacteroides, 419, 420
Batatas edulis, 92
Bauhinia
  B. glauca subsp. *hupehana*, 16
  B. hupehana, 16
Beta
  B. alba, 26
  B. altissima, 26
  B. atriplicifolia, 26
  B. bengalensis, 26
  B. brasiliensis, 26
  B. carnulosa, 26
  B. cicla, 26
  B. cicla var. *argentea*, 26
  B. cicla var. *viridis*, 26

T.K. Lim, *Edible Medicinal and Non-Medicinal Plants: Volume 10, Modified Stems, Roots, Bulbs*,
DOI 10.1007/978-94-017-7276-1

Beta (*cont.*)
B. crispa, 26
B. decumbens, 26
B. esculenta, 26
B. foliosa, 26
B. hortensis, 26
B. hybrida, 26
B. incarnata, 26
B. lutea, 26
B. marina, 26
B. maritima, 26
B. maritima subsp. *atriplicifolia*, 26
B. maritima subsp. *danica*, 26
B. maritima subsp. *marcosii*, 26
B. maritima subsp. *orientalis*, 26
B. maritima var. *atriplicifolia*, 26
B. maritima var. *erecta*, 26
B. maritima var. *glabra*, 26
B. maritima var. *pilosa*, 26
B. maritima var. *prostrata*, 26
B. noeana, 26
B. orientalis, 26
B. purpurea, 26
B. rapa, 26
B. rapacea, 26
B. rosea, 26
B. sativa, 26
B. stricta, 26
B. sulcata, 26
B. triflora, 26
B. vulgaris, 2–3, 26, 28, 35–38, 40–51, 53–60
B. vulgaris cicla, 51
B. vulgaris cv Crassa Group, 3
B. vulgaris cv Group Garden Beet, 3
B. vulgaris cv. Group Spinach Beet, 3
B. vulgaris rubra, 51
B. vulgaris ssp. *esculenta* var. *rubra*, 51–52
B. vulgaris subsp. *asiatica*, 26
B. vulgaris subsp. *cicla*, 26
B. vulgaris subsp. *cycla*, 44, 48
B. vulgaris subsp. *foliosa*, 26
B. vulgaris subsp. *lomatogonoides*, 26
B. vulgaris subsp. *maritima*, 26, 28
B. vulgaris subsp. *orientalis*, 26
B. vulgaris subsp. *perennis*, 44, 51
B. vulgaris subsp. *provulgaris*, 26
B. vulgaris subsp. *vulgaris*, 27, 28, 41, 60
B. vulgaris var. *altissima*, 26
B. vulgaris var. *annua*, 26
B. vulgaris var. *aritima*, 26
B. vulgaris var. *asiatica*, 26
B. vulgaris var. *atriplicifolia*, 26
B. vulgaris var. *aurantia*, 26
B. vulgaris var. *bengalensis*, 51
B. vulgaris var. *cicla*, 26, 35, 46, 50, 53, 55
B. vulgaris var. *conditiva*, 3
B. vulgaris var. *coniciformis*, 26
B. vulgaris var. *crassa*, 26
B. vulgaris var. *debeauxii*, 26
B. vulgaris var *esculenta*, 3
B. vulgaris var. *foliosa*, 26
B. vulgaris var. *glabra*, 26
B. vulgaris var. *grisea*, 26
B. vulgaris var. *ilosa*, 26
B. vulgaris var. *lutea*, 26, 40
B. vulgaris var. *marcosii*, 26
B. vulgaris var. *mediasiatica*, 26
B. vulgaris var. *orientalis*, 26
B. vulgaris var. *ovaliformis*, 26
B. vulgaris var. *perennis*, 26
B. vulgaris var. *rapa*, 3
B. vulgaris var. *rapacea*, 26
B. vulgaris var. *rosea*, 26
B. vulgaris var. *rubidus*, 26
B. vulgaris var. *rubra*, 26, 40, 58
B. vulgaris var. *rubrifolia*, 26
B. vulgaris var. *saccharifera*, 26
B. vulgaris var. *virescens*, 26
B. vulgaris var. *viridifolia*, 26
B. vulgaris var. *vulgaris*, 3, 27
Bifidobacterium
B. bifidum, 259
B. breve, 259, 503, 519
B. longum, 259
Biomophalairia alexandrina, 78
Blattella germanica, 199
Blechnum orientale, 86, 87
Boesenbergia rotunda, 1
Bolboschoenus
B. caldwellii, 6
B. fluviatis, 6
B. maritimus, 6
B. maritimus subsp. *paludosus*, 8
B. maritimus subsp. *tuberosus*, 209
Botryodiplodia theobromae, 248, 297, 298
Brassica sp., 273
Bridelia ferruginea, 152
Brugua malayi, 417
Bunium persicum, 2
Butea
B. frondosa, 16
B. monosperma, 16

C
Cacara
C. bulbosa, 465
C. bulbous, 465
C. erosa, 465
C. palmatiloba, 465
Callosobruchus analis, 477
Calystegia
C. japonica, 4
C. pubescens, 4
C. sepium, 4
Candida, 199
C. albicans, 193, 305, 333, 384, 385
C. parapsilosis, 193
C. utilis, 384

Canna
C. achiras, 3, 69
C. altensteinii, 69
C. amabilis, 69
C. ascendens, 69
C. aurantiaca, 69
C. aureovittata, 69
C. barbadica, 69
C. bidentata, 3, 69
C. bifida, 69
C. brasiliensis, 69
C. carnea, 69
C. cearensis, 69
C. chinensis, 69
C. cinnabarina, 69
C. coccinea, 69
C. coccinea f. *flaviflora*, 69
C. coccinea var. *bicolor*, 69
C. coccinea var. *concolor*, 69
C. coccinea var. *floribunda*, 69
C. coccinea var. *limbata*, 69
C. commutata, 69
C. compacta, 69
C. concinna, 69
C. crocea, 69
C. densifolia, 69
C. discolor, 69, 79
C. discolor var. *rubripunctata*, 69
C. discolor var. *viridifolia*, 69
C. edulis, 4, 69, 76, 77
C. ehrenbergii, 69
C. elegans, 69
C. ellipticifolia, 69
C. ellipticifolia var. *coccinea*, 69
C. ellipticifolia var. *lutea*, 69
C. ellipticifolia var. *patens*, 69
C. ellipticifolia var. *rubra*, 69
C. esculenta, 69
C. exigua, 69
C. eximia, 69
C. floribunda, 69
C. formosa, 69
C. fulgida, 69
C. heliconiifolia, 69
C. heliconiifolia var. *xalapensis*, 69
C. humilis, 69
C. indica, 3–4, 69, 72, 74, 75, 77–79
C. indica f. *rubroaurantiaca*, 69
C. indica subsp. *orientalis*, 69
C. indica var. *coccinea*, 69
C. indica var. *edwardsii*, 69
C. indica var. *flava*, 69
C. indica var. *indica*, 79
C. indica var. *karsteniana*, 69
C. indica var. *limbata*, 69
C. indica var. *lutea*, 69
C. indica var. *maculata*, 69
C. indica var. *nepalensis*, 69
C. indica var. *patens*, 69
C. indica var. *rubra*, 69
C. indica var. *sanctaerosae*, 69
C. indica var. *variegata*, 69
C. indica var. *warszewiczii*, 69
C. juncea, 69
C. laeta, 69
C. lagunensis, 69
C. lambertii, 69
C. lanuginosa, 69
C. lavescens, 69
C. leptochila, 69
C. limbata, 69
C. lutea, 4, 69
C. lutea var. *aurantiaca*, 69
C. lutea var. *maculata*, 69
C. lutea var. *pallida*, 69
C. macrophylla, 69–70
C. maculata, 70
C. maxima, 70
C. montana, 70
C. moritziana, 70
C. nepalensis, 70
C. occidentalis, 70
C. orientalis, 4, 70
C. orientalis var. *flava*, 70
C. pallida, 70
C. pallida var. *maculata*, 70
C. patens, 70
C. patens var. *limbata*, 70
C. pentaphylla, 70
C. platyphylla, 70
C. plurituberosa, 70, 79
C. poeppigii, 70
C. polyclada, 70
C. polymorpha, 70
C. portoricensis, 70
C. pruinosa, 70
C. pulchra, 70
C. recurvata, 70
C. roscoeana, 70
C. rotundifolia, 70
C. rubra, 70
C. rubricaulis, 70
C. sanctae-rosae, 70
C. sanguinea, 70
C. saturate-rubra, 70
C. schubertii, 70
C. seleriana, 70
C. sellowii, 70
C. speciosa, 70
C. spectabilis, 70
C. sulphurea, 70
C. surinamensis, 70
C. tenuiflora, 70
C. texensis, 70
C. textoria, 70
C. thyrsiflora, 70
C. tinei, 70
C. variabilis, 70
C. variegata, 70
C. variegatifolia, 70
C. ventricosa, 70
C. warszewiczii, 70
C. warszewiczii var. *flameus*, 70
C. xalapensis, 70

Cannacorus
C. indicus, 70
C. ovatus, 70
Cantharospermum reticulatum, 16
Capsella bursa-pastoris, 385
Carex
C. spp., 6
C. tuberosa, 209
Carica papaya, 332
Cassia occidentalis, 152
Centella asiatica, 332
Ceratocystis fimbriata, 109, 122, 124, 125
Ceratostomella fimbriata, 123
Cercospora beticola, 35, 43, 56
Chaitaea tacca, 301
Chlorocyperus
C. papyrus, 172
C. rotundus, 178
C. salaamensis, 178
Chrysanthemum x morifolium, 525
Cibotium barometz, 84, 89
Cimicifuga foetida, 525
Cinnamomum cassia, 141
Cissampelos owariensis, 305
Citrobacter freundii, 56
Citrus clementina, 36
Cladosporium
C. cladosporioides, 298
C. herebaceum, 193
Clostridium
C. beijerinckii, 342
C. innocuum, 419
C. thermocellum, 342
Coffea canephora, 152
Colletotrichum
C. gloeosporioides, 478
C. musae, 200
Colletotrichum sp., 200
Colocasia
C. esculenta, 1, 220, 227
C. esculenta var. *antiquorum*, 227
Conopodium majus, 2
Convolvulus
C. arvensis, 4
C. batatas, 92
C. candicans, 92
C. chinensis, 4
C. chrysorrhizus, 92
C. edulis, 92
C. erubescens, 4
Coptis chinensis, 385
Corcyra cephalonica, 477
Coriandrum sativum, 191
Cornus officinalis, 284
Corydalis yanhusuo, 198
Corynebacterium diphtheria, 333
Cosmos caudatus, 332
Cucumis sp., 273
Cucurma longa, 1
Culex quinquefasciatus, 78, 198
Cullen badocanum, 20
Cunninghamella echinulata, 392
Curtobacterium ammoniigene, 215
Curvularia
C. lunata, 248
C. maculans, 200
C. pallescens, 200
C. pennisetti, 200
Cylas
C. formicarius, 125
C. formicarius elegantulus, 118, 121, 125
Cyperus, 2, 172
C. agrestis, 178
C. antiquorum var. *palaestinae*, 172
C. arabicus, 178
C. badius var. *inconspicuus*, 178
C. bicolor, 178
C. bifax, 6, 178
C. bulbosostoloniferus, 178
C. bulbosus, 6
C. comosus, 178
C. cyperoides, 8
C. disruptus, 178
C. elapsus, 172
C. elongatus, 178
C. esculentus, 2, 6
C. esculentus var. sativus, 6
C. herbicavus, 178
C. hexastachyos, 178
C. hildra, 178
C. hydra, 178
C. inconspicuus, 178
C. jeminicus, 6
C. laevissimus, 178
C. leptostachyus, 178
C. longus, 178
C. merkeri, 178
C. micreilema, 178
C. nubicus, 178
C. ochreoides, 178
C. officinalis, 178
C. oliganthus, 178
C. olivaris, 178
C. olivaris var. *brevibracteatus*, 178
C. pallescens, 178
C. panormitanus, 172
C. papyraceus, 172
C. papyrus, 6, 172
C. papyrus subsp. *antiquorum*, 172
C. papyrus subsp. *hadidii*, 172
C. papyrus subsp. *niliacus*, 172
C. papyrus subsp. *papyrus*, 172
C. papyrus subsp. *siculus*, 172
C. papyrus subsp. *ugandensis*, 172
C. papyrus var. *antiquorum*, 172
C. papyrus var. *niliacus*, 172
C. papyrus var. *palaestinae*, 172
C. patulus, 178

C. platystachys, 178
C. procerulus, 178
C. proteinolepis, 178
C. pseudovariegatus, 178
C. purpureovariegatus, 178
C. radicosus, 178
C. retzii, 178
C. rotundus, 6, 7, 178, 181, 183, 185–202
C. rotundus f. *comosus*, 178
C. rotundus f. *contractus*, 178
C. rotundus f. *depallescens*, 178
C. rotundus f. *inconspicuus*, 178
C. rotundus f. *latifolius*, 178
C. rotundus f. *latimarginatus*, 178
C. rotundus subsp. *brevibracteatus*, 178
C. rotundus subsp. *comosus*, 178
C. rotundus subsp. *divaricatus*, 178
C. rotundus subsp. *inconspicuus*, 178
C. rotundus subsp. *merkeri*, 178
C. rotundus subsp. *retzii*, 179
C. rotundus subsp. *rotundus*, 179
C. rotundus subsp. *tuberosus*, 179
C. rotundus var. *acutus*, 178
C. rotundus var. *amaliae*, 178
C. rotundus var. *brevibracteatus*, 178
C. rotundus var. *carinalis*, 178
C. rotundus var. *carinatus*, 178
C. rotundus var. *centiflorus*, 178
C. rotundus var. *comosus*, 178
C. rotundus var. *disruptus*, 178
C. rotundus var. *elongatus*, 178
C. rotundus var. *hydra*, 178
C. rotundus var. *inconspicuus*, 178
C. rotundus var. *macrostachyus*, 178
C. rotundus var. *major*, 178
C. rotundus var. *nubicus*, 178
C. rotundus var. *pallidus*, 178
C. rotundus var. *platystachys*, 178
C. rotundus var. *procerula*, 178
C. rotundus var. *quimoyensis*, 179
C. rotundus var. *rotundus*, 179
C. rotundus var. *salsolus*, 179
C. rotundus var. *spadiceus*, 179
C. rotundus var. *taylorii*, 179
C. rotundus var. *tetrastachyos*, 179
C. rubicundus, 179
C. rudioi, 179
C. rudioi var. *minor*, 179
C. siculus, 172
C. stoloniferus var. *pallidus*, 179
C. syriacus, 172
C. taylorii, 179
C. tenuifolius, 179
C. tetrastachyos, 179
C. ugandensis, 172
C. viridis, 179
C. weinlandii, 179
C. yoshinagae, 179
Cyprinus carpio, 272

D
Dalea
D. gattingeri, 16
D. purpurea, 16
Dalea candida, 16, 19
Derris elliptica, 199
Dianthus caryophyllus, 36
Diaporthe batatatis, 124
Dicksonia barometz, 82
Dictamnus dasycarpus, 244
Dioscorea
D. abyssinica, 298, 299
D. aculeata, 9, 253
D. aculeata var. *spinosa*, 253
D. alata, 9, 218–220, 222–231, 257, 293, 296–298
D. alata var. *globosa*, 218
D. alata var. *purpurea*, 218, 223
D. alata var. *purpurea*, 230
D. alata var. *tarri*, 218
D. alata var. *vera*, 218
D. anguina, 9
D. anthropophagorum, 235
D. atropurpurea, 218
D. belophylla, 9
D. bulbifera, 10, 235, 237, 239–248
D. bulbifera forma *spontanea*, 239
D. bulbifera var. *anthropophagorum*, 235
D. bulbifera var. *crispata*, 235
D. bulbifera var. *elongata*, 235
D. bulbifera var. *pulchella*, 235
D. bulbifera var. *sativa*, 235, 246, 248
D. bulbifera var. *suavia*, 235
D. bulbifera var. *vera*, 235, 237
D. burkilliana, 298, 299
D. calcicola, 10
D. cayenensis, 10, 292, 296, 298, 299
D. cayenensis-rotundata, 296–298
D. cayennensis, 9
D. cayennensis subsp. *rotundata*, 13, 291
D. cinnamomifolia, 14
D. colocasiifolia, 218
D. crispata, 235
D. cumingii, 10
D. daemona, 266
D. daemona var. *reticulata*, 266
D. daunea, 9
D. deltoidea, 10
D. divaricata, 10
D. dumetorum, 10, 298
D. eburina, 218
D. eburnea, 218
D. esculenta, 11, 224, 253, 256–259, 271
D. esculenta var. *fasciculata*, 11, 253
D. esculenta var. *fulvidotomentosa*, 253
D. esculenta var. *spinosa*, 253
D. esculenta var. *tiliifolia*, 253
D. fasciculata, 253
D. fasciculata var. *spinosa*, 253
D. filiformis, 11

Dioscorea (*cont.*)
D. gibbiflora, 11
D. glabra, 11
D. globosa, 218
D. hamiltonii, 12, 261, 264, 270
D. hastifolia, 11
D. hemsleyi, 11
D. heterophylla, 235
D. hirsuta, 266
D. hirtiflora, 11
D. hispida, 11, 264, 266, 268–273
D. hispida var. *daemona*, 266
D. hispida var. *reticulata*, 266
D. hoffa, 235
D. hofika, 235
D. japonica, 11, 228, 230, 279
D. javanica, 218
D. korrorensis, 235
D. latifolia, 235
D. laurifolia, 12
D. liebrechtsiana, 298, 299
D. longipetiolata, 235
D. lunata, 266
D. luzonensis, 12
D. macrostachya, 12
D. macroura, 12
D. membranacea, 9
D. mexicana, 12
D. minutiflora, 12, 298, 299
D. mollissima, 266
D. nummularia, 12
D. opposita, 12, 264, 275, 278–286
D. oppositifolia, 12, 275, 276, 279, 282, 284, 286, 298
D. oppositifolia var. *dukhunensis*, 275
D. oppositifolia var. *linnae*, 275
D. oppositifolia var. *oppositifolia*, 277
D. oppositifolia var. *thwaitesii*, 275
D. orbiculata, 12
D. owenii, 12
D. papillaris, 253
D. papuana, 253
D. pentaphylla, 12, 14
D. perrieri, 235
D. persimilis, 12, 261, 262, 264, 265, 281
D. persimilis var. *pubscens*, 261
D. persimilis var. *wukangensis*, 261
D. piscatorum, 13
D. polyclados, 13
D. polystachya, 13
D. praehensilis, 298, 299
D. prainiana, 13
D. prazeri, 13
D. preussii, 13
D. pseudojaponica, 224
D. pubera, 9, 13, 264, 270
D. pulchella, 235
D. purpurea, 218, 224
D. pyrifolia, 13
D. quinata, 13
D. raishaensis, 261
D. rogersii, 235
D. rotundata, 13, 224, 226, 291–293, 295–299
D. rubella, 218
D. sagitatta, 13
D. sansibarensis, 12
D. sapinii, 218
D. sativa, 14
D. sativa f. *domestica*, 235
D. sativa var. *elongata*, 235
D. sativa var. *rotunda*, 235
D. schimperiana, 14
D. smilacifolia, 298, 299
D. spicata, 264, 270
D. spinosa, 253
D. spp., 2
D. stemonoides, 9
D. sylvestris, 235
D. tamifolia, 235
D. tenuiflora, 235
D. tiliifolia, 253
D. togoensis, 299
D. tranversa, 9
D. trifida, 14
D. triphylla, 14
D. triphylla var. *daemona*, 266
D. triphylla var. *mollissima*, 266
D. triphylla var. *reticulata*, 266
D. tuberosa, 14
D. tugui, 253
D. villosa, 13, 14
D. vulgaris, 218
D. wallichii, 14
D. zingiberensis, 264, 281
Diplazium esculentum, 331
Diplodia tubericola, 124, 125
Distemon brasiliensis, 70
Distemon grandis, 70
Dolichos
D. ahipa, 458
D. articulatus, 465
D. biflorus, 16
D. bulbosus, 465
D. erosus, 465
D. hirsutus, 482
D. japonicus, 482
D. lobatus, 482
D. palmatilobus, 465
Drosophila melanogaster, 408
Drynaria linearis, 86, 87

E

Echinacea purpurea, 408
Eicinia fetida, 247
Eleocharis
E. austrocaledonica, 209
E. dulcis, 1, 7, 8, 209, 211, 214, 215
E. dulcis var. *tuberosa*, 209
E. equisetina, 209
E. esculenta, 209
E. indica, 209

E. kuroguwai, 7
E. plantaginea, 209
E. plantaginea var. *stolonifera*, 209
E. plantagineiformis, 209
E. plantaginoidea, 209
E. plantaginoides, 209
E. sphacelata, 209
E. tuberosa, 209, 213–215
E. tumida, 209
Elephantodon eburnea, 218
Eleuterococcus, 56
Emblica officinalis, 399, 413
Entamoeba histolytica, 334
Enterobacter aerogenes, 244
Enterococcus faecalis, 193, 245, 264, 384, 385
Ephedra sinica, 242
Epilachna varivestis, 477
Eriobotrya japonica, 242
Eriophorum gracile, 7
Eriosema chinense, 16
Erwinia carotovora, 124
Erythrina vespertilio, 16
Escherichia coli, 56, 137, 190, 193, 194, 215, 230, 244, 245, 259, 264, 298, 305, 329, 333, 334, 372, 383–386, 423
Eschericia coli, 77
Eubacterium sp., 420, 503
Eurycoma longifolia, 2
Euscepes postfasciatus, 125

F

Fimbristylis
F. kysoor, 7
F. subbispicata, 7
F. tristachya var. *subbispicata*, 7
Flemingia
F. procumbens, 17
F. prostrata, 18
F. vestita, 17
Fusarium sp., 213
F. javanicum, 151
F. moniliforme, 231, 248
F. oxysporum, 193, 324, 478
F. oxysporum f. sp. *batatas*, 124
F. solani, 124, 125, 151, 324

G

Galactia tenuiflora, 17
Ganoderma lucidum, 286
Geotrichum sp., 213
Ginkgo biloba, 376
Gliocladium viride, 392
Glycyrrhiza, 395, 418
G. brachycarpa, 354
G. echinata, 358, 365, 366, 423
G. glabra, 17, 354–356, 358, 359, 361–377, 379–392, 396–402, 404–415, 417, 421, 423, 429, 432, 433
G. glabra subsp. *glandulifera*, 354
G. glabra var. *caduca*, 354
G. glabra var. *glabra*, 354, 369, 371, 405
G. glabra var. *glandulifera*, 354–356, 358, 362, 369, 371, 372, 374, 384, 405, 407, 415
G. glabra var. *glandulosa*, 354
G. glabra var. *laxifoliolata*, 354
G. glabra var. *typica*, 354–356, 369, 384
G. glabra var. *typica*, 433
G. glabra var. *violacea*, 354, 356, 433
G. glandulifera, 354
G. glauca, 246
G. hirsuta, 354
G. inflata, 355, 365–367, 374, 433
G. labra var. *glandulifera*, 433
G. lepidota, 17
G. macedonica, 366
G. pallida, 354
G. pallidiflora, 366
G. uralensis, 355, 366, 367, 370, 374, 433
G. violacea, 354
Gnidia glauca, 245
Grateloupia filicina, 248
Griffonia simplicifolia, 412

H

Hardenbergia retusa, 17
Hedysarum
H. boreale, 17
H. boreale subsp. *mackenzii*, 17
H. mackenzii, 17
H. occidentale, 17
Helianthus tuberosus, 2
Helicobacter pylori, 405, 406, 521
Helmia
H. bulbifera, 235
H. daemona, 266
H. hirsuta, 266
Helminthosporium
H. carbonum, 369, 475
H. colocasiae, 200
H. echinoclova, 200
H. pennissetti, 200
H. spiciferum, 200
Hibiscus acetosella, 331, 332
Hippuris indica, 209
Hordeum vulgare, 36
Hyptis suaveolens, 476

I

Imperata cylindrica, 2
Ipomoea
I. aquatica, 4
I. batatas, 2, 5, 92, 117, 128–130, 136, 140, 149, 152, 220
I. batatas var. *edulis*, 92
I. batatas var. *lobata*, 92
I. cairica, 5
I. calobra, 5
I. cheirophylla, 5
I. chrysorrhiza, 92

Ipomoea (*cont.*)
I. costata, 2, 5
I. digitata, 5
I. edulis, 92
I. eriocarpa, 5
I. fastigiata, 92
I. gracilis, 5
I. graminea, 4
I. mauritiana, 5
I. pandurata, 5
I. pes-caprae subsp. *brasiliensis*, 5
I. polpha, 5
I. polymorpha, 5
I. racemosa, 4
I. staphylina, 5
I. trifida, 94
I. velutina, 5
I. violacea, 6

J

Janipha
J. aipi, 308
J. manihot, 308
Jatropha, 308
J. aipi, 308
J. curcas, 36
J. diffusa, 308
J. digitiformis, 308
J. dulcis, 308
J. flabellifolia, 308
J. glauca, 308
J. janipha, 308
J. lobata var. *richardiana*, 308
J. loureiroi, 308
J. manihot L., 308
J. mitis, 308
J. paniculata, 308
J. silvestris Vell., 308
J. stipulata Vell., 308

K

Kalopanax pictus, 517, 526
Kampferia galanga, 1
Klebsiella pneumoniae, 193, 244

L

Labichea buettneriana, 17
Lablab purpureus, 17
Lactobacillus, 153, 334
L. acidophilus, 279, 283, 385
L. bulgaricus, 279
L. casei, 259
L. coryniformis, 118
L. delbrueckii subsp lactis, 264
L. murinus, 283
L. paracasei, 118
L. plantarum, 118
L. reuteri, 283
Lactococcus lactis spp. *lactis*, 341
Lathyrus
L. linifolius, 19
L. tuberosus, 2, 17
Leishmania donovani, 401, 417
Leontice leontopetaloides L., 301
Leonurus heterophyllus, 197
Lepidium meyenni, 2
Lepironia articulata, 7
Lilium, 2
Limnochloa
L. plantaginea, 209
L. tumida, 209
Listeria monocytogenes, 333
Lophobaris piperis, 478
Lotus siloquosus, 18
Ludwigia adscendens, 215
Lupinus
L. arcticus, 18
L. nootkatensis, 18
Lymnaea acuminata, 78

M

Macrophomina
M. phaseoli, 124
M. phaseolina, 248
Macrotyloma uniflorum, 18
Magnolia officinalis, 402
Mandioca
M. aipi, 308
M. dulcis, 308
M. utilissima, 308
Mangifera indica, 413
Manihot
M. aesculifolia, 311
M. aipi, 308
M. aipi var. *lanceolata*, 308
M. aipi var. *latifolia*, 308
M. aipi var. *lutescens*, 308
M. brachyloba, 311
M. cannabina, 308
M. carthaginensis, 311
M. carthaginensis subsp. *glaziovii*, 15
M. cassava, 308
M. diffusa, 308
M. digitiformis, 308
M. dulcis, 308
M. dulcis var. *aipi*, 308
M. dulcis var. *diffusa*, 308
M. dulcis var. *flabellifolia*, 308
M. edule, 308
M. edulis, 308
M. esculenta, 2, 15, 227, 308, 310, 311, 315, 322, 331, 332
M. esculenta subsp. *flabellifolia*, 308, 310, 311
M. esculenta subsp. *grandifolia*, 308
M. esculenta subsp. *peruviana*, 310, 311
M. esculenta var. *argentea*, 308
M. esculenta var. *coalescens*, 308
M. esculenta var. *debilis*, 308

M. esculenta var. *digitifolia*, 308
M. esculenta var. *flavicaulis*, 308
M. esculenta var. *fuscescens*, 308
M. esculenta var. *grandifolia*, 308
M. esculenta var. *nodosa*, 308
M. esculenta var. *sprucei*, 308
M. flabellifolia, 308
M. flexuosa, 308
M. glaziovii, 15
M. guyanensis, 308
M. loureiroi, 308
M. manihot, 308
M. melanobasis, 308
M. oligantha, 315
M. palmata var. *aipi*, 308
M. palmata var. *diffusa*, 308
M. palmata var. *digitiformis*, 308
M. palmata var. *flabellifolia*, 308
M. pruinosa, 310
M. sprucei Pax, 308
M. utilissima, 15, 308, 331
M. utilissima var. *castellana*, 308
M. utilissima var. *sutinga*, 308
Maranta arundinacea, 1
Mariscus sieberianus, 8
Melilotus
M. albus, 18
M. officinalis, 18
Meloidogyne incognita, 111
Mentha longifolia, 194, 200
Merremia hungaiensis, 6
Micrococcus
M. luteus, 193, 384
M. pyogenes var. *aureus*, 215
Microdesmis puberula, 340
Milletia speciosa, 18
Mirabilis extensa, 2
Miscanthidium violaceum, 176
Moghania philippinensis, 18
Monilochaetes infuscan, 124
Mucor sp., 272, 384
M. pusillus, 384
Mucuna glabra, 18
Musa spp., 227
Mycobacterium
M. smegmatis, 244
M. tuberculosis, 244, 385

N

Nelumbo nucifera, 1
Nephrodium barometz, 82
Neustanthus chinensis, 482
Nicotiana benthamiana, 372

O

Oesophagostomum dentatum, 57
Oncorhiza esculentus, 253
Oncus esculentus, 253
Orobus tuberosus, 19
Orthosiphon aristatus, 198
Oxalis tuberosa, 2, 76

P

Pachyrhizus
P. ahipa, 19, 458, 459, 461, 462, 466
P. ahipa var. *albiflora*, 458
P. ahipa var. *violacea*, 458
P. angulatus, 19, 465
P. articulatus, 465
P. bulbosus, 465
P. erosus, 19, 462, 465–467, 470, 473–478
P. erosus var. *erosus*, 465
P. erosus var. *palmatilobus*, 465
P. erosus var. *typicus*, 465
P. jicamas, 465
P. palmatilobus, 465
P. spp., 2
P. strigosus, 465
P. thunbergianus, 482
P. tuberosus, 19
Paoenia ostii, 284
Papyrus
P. antiquorum, 172
P. domesticus, 172
P. mossambicensis, 172
P. siculus, 172
Paramecium caudatum, 56
Passiflora edulis, 152
Pediomelum
P. cuspidatum, 19
P. esculentum, 19
P. hypogaeum var. *hypogaeum*, 19
P. subacaule, 19
P. tenuiflorum, 19
Penicillium
P. chrysogenum, 264
P. marneffei, 384
P. schlerotgenum, 298
Petalostemon candidum, 19
Phaseolus, 473
P. adenanthus, 19
P. coccineus, 20
P. rostratus, 20
P. trilobus, 482
Pheritima posthuma, 77
Pichia nakazawae, 384
Piper nigrum, 399
Pisum sativum, 372
Pityrosporium ovale, 333
Plasmodium falciparum, 78, 417
Plectranthus esculenta, 2
Plenodomus destruens, 124
Plumbago zeylanica, 152
Plutella xylostella, 200, 272
Poecillia reticulata, 200
Polygonum bistorta, 244

Polynome
P. alata, 218
P. bulbifera, 235
Polypodium barometz, 82
Pomacea canaliculata, 199
Poria cocos, 284
Porphyromonas gingivalis, 385
Portulaca oleracea, 59
Propionibacterium acnes, 333, 384, 402
Proteus vulgaris, 193, 264, 305
Prunella vulgaris, 244, 286
Pseudeminia comosa, 22
Pseudococcus jackbeardsleyi, 343
Pseudomonas
P. aeroginosa, 385
P. aeruginosa, 56, 193, 244, 245, 264, 298, 329
P. syringae pv. *tomato*, 384
Psophocarpus
P. palustris, 20
P. tetragonolobus, 20, 332
Psoralea
P. argophylla, 20
P. badocana, 20
P. canescens, 20
P. cuspidata, 19, 20
P. esculenta, 2, 19, 20
P. hypogaea, 19, 20
P. tenuiflora, 20
Psuedomonas aeruginosa, 333
Pteridium, 332
P. aquilinum, 331
Pueraria
P. argyi, 482
P. bodinieri, 482
P. caerulea, 482
P. edulis, 21
P. harmsii, 482
P. hirsuta, 21, 482
P. koten, 482
P. lobata, 21, 482, 484–499, 501, 502, 504–507, 509–520, 522–526
P. lobata var. *montana*, 21
P. lobata var. *thomsonii*, 21
P. mirifica, 489, 497, 498, 526
P. montana, 21, 527
P. montana var. *chinensis*, 21
P. montana var. *lobata*, 21, 22, 482, 483
P. montana var. *montana*, 21, 527
P. neo-caledonica, 482
P. omeiensis, 507, 520
P. phaseoloides, 21
P. phaseoloides var. *phaseoloides*, 21
P. pseudohirsuta, 482
P. pseudo-hirsuta, 482, 486
P. stricta, 21
P. thomsonii, 21, 486, 489, 491–493, 501, 502, 507
P. thunbergiana, 22, 482, 486, 490, 491, 495–497, 503, 506, 511, 514, 517, 518, 521, 526, 527
P. triloba, 482
P. tuberosa, 22
P. volkensii, 482
Puerariae thomsonii, 525
Pycreus rotundus, 179
Pythium ultimum, 384

R

Raillietina spiralis, 247
Raphanus sp., 273
Rehmannia glutinosa, 284
Rhizoctonia solani, 43
Rhizopus stolonifer, 124, 149, 478
Rhus verniciflua, 517
Rhynchosia comosa, 22
Robynsia
R. lobata, 465
R. macrophylla, 465
Ruminococcus sp., 419, 420

S

Saccharomyces cerevisiae, 59, 264, 384
Sagittaria spp., 1
Sahnonella typhimurium, 381
Salmonella, 190, 193
S. choleraesuis, 341
S. enteritidis, 137, 193
S. paratyphi A, 244
S. paratyphi B, 244
S. typhi, 244, 245, 305, 333, 385
S. typhimurium, 56, 136, 138, 286, 383, 423, 498
Salvia miltiorrhiza, 501, 506, 507, 509, 515, 526
Sauropus androgynus, 331
Schisandra chinensis, 392
Schoenoplectus
S. americanus, 8
S. californicus, 8
S. lacustris, 8
S. litoralis, 8
S. subterminalis, 8
S. tabernaemontani, 9
Schoenus tuberosus, 179
Scirpus
S. americanus, 8
S. californicus, 8
S. dubius, 209
S. grossus, 8
S. lacustris, 8
S. microcarpus, 8
S. paludosus, 8
S. plantagineus, 209
S. plantaginoides, 209
S. subterminalis, 8
S. tuberosus, 8, 209
S. tumidus, 209
S. validus, 9
Sclerotium rolfsii, 124, 248
Shigella
S. dysenteriae, 305
S. flexneri, 194, 333, 385
S. sonnei, 245, 385
Silybum marianum, 141

Sium sisarum, 2
Smallanthus sonchifolius, 2
Solanum tuberosum, 227
Sonchus arvensis, 244
Sophora tonkinensis, 244
Sphenostylis stenocarpa, 22
Spodoptera
  S. littoralis, 198, 200
  S. litura, 477
Stachys affinis, 2
Staphylococcus
  S. aureus, 56, 77, 87, 193, 298, 305, 329, 333, 341, 384–386, 411, 521
  S. aureus (MRSA), 385
  S. epidermidis, 385
  S. flexneri, 385
  S. paratyphi B, 385
  S. sanguis, 385
  S. sciuri, 56
Stizolobium
  S. bulbosum, 465
  S. domingense, 465
Streptococcus, 329, 401, 420
  S. aureus, 244
  S. faecalis, 419
  S. mutans, 193, 385, 386
  S. pyogene, 305
  S. pyogenes, 193, 264
  S. thermophilus, 264
Streptomyces ipomoea, 124
Suctellaria baicalensis, 402
Syzygium aromaticum, 413

T
Tabernaemontana elegans, 153
Tacca, 306
  T. abyssinica, 301
  T. artocarpifolia, 301
  T. brownii, 301
  T. browni var. *paeoniifolia*, 301
  T. dubia, 301
  T. gaogao, 301
  T. guineensis G., 301
  T. hawaiiensis, 301
  T. involucrata, 14, 301
  T. involucrata var. *acutifolia*, 301
  T. leontopetaloides, 14, 301, 302, 305
  T. maculata, 301
  T. madagascariensis, 301
  T. oceanica, 301
  T. phallifera, 301
  T. pinnatifida, 14, 301
  T. pinnatifida subsp. *eupinnatifida*, 301
  T. pinnatifida subsp. *interrupta*, 301
  T. pinnatifida subsp. *involucrata*, 301
  T. pinnatifida subsp. *madagascariensis*, 301
  T. pinnatifida subsp. *minor*, 301
  T. pinnatifida var. *acutifolia*, 301
  T. pinnatifida var. *brownii*, 301
  T. pinnatifida var. *maculata*, 301
  T. pinnatifida var. *obtusata*, 301
  T. pinnatifida var. *paeoniifolia*, 301
  T. pinnatifida var. *permagna*, 301
  T. pinnatifolia, 301
  T. quanzensis, 301
  T. umbrarum, 301
  T. viridis, 301
Taeniocarpum articulatum, 465
Thermoanaerobacterium aotearoense, 342
Tinospora cordifolia, 408
Tragopogon porrifolius, 2
Trichoderma
  T. harzianum, 324
  T. koningii, 477
Trichophyton mentagrophytes, 384
Trichostronyglus, 334
Trifolium repens, 22
Trigonella foenum-graecum, 141
Tropaeolum tuberosum, 2
Turbina racemosa, 4
Tussilago farfara, 242
Tylosema esculentum, 22
Typha spp., 1

U
Ullucus tuberosus, 2

V
Vandasina retusa, 17
Vernonia amygdalina, 152
Vetiveria zinzanioids, 191
Vibrio
  V. cholerae, 333
  V. parahaemolyticus, 264
Vicia faba, 383
Vigna
  V. adenantha, 19, 20
  V. lanceolata, 2, 22
  V. luteola, 22
  V. marina, 22
  V. radiata, 22
  V. unguilata, 16
  V. vexillata, 22
Vitis vinifera, 36

X
Xanthosoma spp., 1
  X. sagittifolium, 227
Xyphostylis lutea, 70

Z
Zea mays, 273
Zingiber
  Z. officinale, 194, 200
  Z. spp., 1
Zymomonas mobilis, 154

Printed by Printforce, the Netherlands